U.S. Customary and SI Unit Systems.

	System of Units	
Base Dimension	U.S. Customary	SI
force	pound (lb)	newton[a](N) $\equiv$ kg·m/s^2
mass	slug[a] $\equiv$ lb·s^2/ft	kilogram (kg)
length	foot (ft)	meter (m)
time	second (s)	second (s)

[a] Derived unit.

Conversion factors between U.S. Customary and SI unit systems.

	U.S. Customary		SI
length	1 in.	=	0.0254 m (2.54 cm, 25.4 mm)[a]
	1 ft (12 in.)	=	0.3048 m[a]
	1 mi (5280 ft)	=	1.609 km
force	1 lb	=	4.448 N
	1 kip (1000 lb)	=	4.448 kN
mass	1 slug (1 lb·s^2/ft)	=	14.59 kg

[a] Exact.

Common prefixes used in the SI unit systems.

Multiplication Factor		Prefix	Symbol
1 000 000 000 000 000 000 000 000	10^{24}	yotta	Y
1 000 000 000 000 000 000 000	10^{21}	zetta	Z
1 000 000 000 000 000 000	10^{18}	exa	E
1 000 000 000 000 000	10^{15}	peta	P
1 000 000 000 000	10^{12}	tera	T
1 000 000 000	10^{9}	giga	G
1 000 000	10^{6}	mega	M
1 000	10^{3}	kilo	k
100	10^{2}	hecto	h
10	10^{1}	deka	da
0.1	10^{-1}	deci	d
0.01	10^{-2}	centi	c
0.001	10^{-3}	milli	m
0.000 001	10^{-6}	micro	μ
0.000 000 001	10^{-9}	nano	n
0.000 000 000 001	10^{-12}	pico	p
0.000 000 000 000 001	10^{-15}	femto	f
0.000 000 000 000 000 001	10^{-18}	atto	a
0.000 000 000 000 000 000 001	10^{-21}	zepto	z
0.000 000 000 000 000 000 000 001	10^{-24}	yocto	y

Engineering Mechanics:
Statics

Michael E. Plesha
Department of Engineering Physics
University of Wisconsin–Madison

Gary L. Gray
Department of Engineering Science and Mechanics
Penn State

Francesco Costanzo
Department of Engineering Science and Mechanics
Penn State

McGraw Hill Higher Education

Boston Burr Ridge, IL Dubuque, IA New York San Francisco St. Louis
Bangkok Bogotá Caracas Kuala Lumpur Lisbon London Madrid Mexico City
Milan Montreal New Delhi Santiago Seoul Singapore Sydney Taipei Toronto

The McGraw-Hill Companies

Higher Education

ENGINEERING MECHANICS: STATICS

Published by McGraw-Hill, a business unit of The McGraw-Hill Companies, Inc., 1221 Avenue of the Americas, New York, NY 10020.

This book is printed on acid-free paper.

1 2 3 4 5 6 7 8 9 0 DOW/DOW 0 9

ISBN 978–0–07–282865–8
MHID 0–07–282865–X

Global Publisher: *Raghothaman Srinivasan*
Senior Sponsoring Editor: *Bill Stenquist*
Vice-President New Product Launches: *Michael Lange*
Developmental Editor: *Darlene M. Schueller*
Senior Marketing Manager: *Curt Reynolds*
Senior Project Manager: *Sheila M. Frank*
Senior Production Supervisor: *Sherry L. Kane*
Lead Media Project Manager: *Stacy A. Patch*
Digital Product Manager: *Daniel Wallace*
Senior Designer: *David W. Hash*
Cover/Interior Designer: *Greg Nettles/Squarecrow Design*
(USE) Cover Image: *Moonlight on Very Large Array Telescopes ©Roger Ressmeyer/CORBIS*
Lead Photo Research Coordinator: *Carrie K. Burger*
Photo Research: *Sabina Dowell*
Compositor: *Aptara®, Inc.*
Typeface: *10/12 Times Roman*
Printer: *R. R. Donnelley Willard, OH*

Library of Congress has Catalogued the main title as follows:

Costanzo, Francesco, 1964-
Engineering mechanics : statics & dynamics / Francesco Costanzo, Michael E. Plesha, Gary L. Gray. -- 1st ed.
p. cm.
Includes index.
Summary: This is a full version; do not confuse with 2 vol. set version (Statics 9780072828658 and Dynamics 9780072828719) which LC will not retain.
ISBN 978–0–07–313412–3 — ISBN 0–07–313412–0 (hard copy : alk. paper) 1. Mechanics, Applied—Textbooks. I. Plesha, Michael E. II. Gray, Gary L. III. Title.
TA350.C79 2010
620.1--dc22

2008054286

www.mhhe.com

ABOUT THE AUTHORS

Michael E. Plesha is a Professor of Engineering Mechanics in the Department of Engineering Physics at the University of Wisconsin-Madison. Professor Plesha received his B.S. from the University of Illinois-Chicago in structural engineering and materials, and his M.S. and Ph.D. from Northwestern University in structural engineering and applied mechanics. His primary research areas are computational mechanics, focusing on the development of finite element and discrete element methods for solving static and dynamic nonlinear problems, and the development of constitutive models for characterizing behavior of materials. Much of his work focuses on problems featuring contact, friction, and material interfaces. Applications include nanotribology, high temperature rheology of ceramic composite materials, modeling geomaterials including rock and soil, penetration mechanics, and modeling crack growth in structures. He is co-author of the book *Concepts and Applications of Finite Element Analysis* (with R. D. Cook, D. S. Malkus, and R. J. Witt). He teaches courses in statics, basic and advanced mechanics of materials, mechanical vibrations, and finite element methods.

Gary L. Gray is an Associate Professor of Engineering Science and Mechanics in the Department of Engineering Science and Mechanics at Penn State in University Park, PA. He received a B.S. in Mechanical Engineering (cum laude) from Washington University in St. Louis, MO, an S.M. in Engineering Science from Harvard University, and M.S. and Ph.D. degrees in Engineering Mechanics from the University of Wisconsin-Madison. His primary research interests are in dynamical systems, dynamics of mechanical systems, mechanics education, and multi-scale methods for predicting continuum-level properties of materials from molecular calculations. For his contributions to mechanics education, he has been awarded the Outstanding and Premier Teaching Awards from the Penn State Engineering Society, the Outstanding New Mechanics Educator Award from the American Society for Engineering Education, the Learning Excellence Award from General Electric, and the Collaborative and Curricular Innovations Special Recognition Award from the Provost of Penn State. In addition to dynamics, he also teaches mechanics of materials, mechanical vibrations, numerical methods, advanced dynamics, and engineering mathematics.

Francesco Costanzo is an Associate Professor of Engineering Science and Mechanics in the Engineering Science and Mechanics Department at Penn State. He received the Laurea in Ingegneria Aeronautica from the Politecnico di Milano, Milan, Italy. After coming to the U.S. as a Fulbright scholar he received his Ph.D. in aerospace engineering from Texas A&M University. His primary research interest is the mathematical and numerical modeling of material behavior. He has focused on the theoretical and numerical characterization of dynamic fracture in materials subject to thermo-mechanical loading via the use of cohesive zone models and various finite element methods, including space-time formulations. His research has also focused on the development of multi-scale methods for predicting continuum-level material properties from molecular calculations, including the development of molecular dynamics methods for the determination of the stress-strain response of nonlinear elastic systems. In addition to scientific research, he has contributed to various projects for the advancement of mechanics education under the sponsorship of several organizations, including the National Science Foundation. For his contributions, he has received various awards, including the 1998 and the 2003 GE Learning Excellence Awards, and the 1999 ASEE Outstanding New Mechanics Educator Award. In addition to teaching dynamics, he also teaches statics, mechanics of materials, continuum mechanics, and mathematical theory of elasticity.

The authors thank their families for their patience, understanding, and, most importantly, encouragement during the long years it took to bring these books to completion. Without their support, none of this would have been possible.

Brief descriptions of the chapter-opening photos and how they relate to statics are given below.

Chapter 1: The Millau Viaduct is a cable-stayed bridge that spans the River Tarn near Millau in southern France. Concepts from statics formed the basis for the analysis and design of this bridge.

Chapter 2: A skyward view of the cables supporting the Leonard P. Zakim Bunker Hill Bridge, in Boston, Massachusetts. Each of the cables supports a tensile force whose size and direction can be described using a vector.

Chapter 3: Tower cranes lifting beams and other items at a construction site in Houston, Texas. Many of the components in this photo can be idealized as particles in equilibrium.

Chapter 4: A hovering rescue helicopter lifting a man to safety. Concepts of the moment of a force and equivalent force systems are needed to determine the forces developed by the main rotor and tail rotor of the helicopter.

Chapter 5: The Space Shuttle Columbia sits on a launch pad at Kennedy Space Center, Florida, before its maiden flight in 1981. For many purposes, the Space Shuttle with its attached external fuel tank and two solid fuel rocket boosters may be modeled as a single body.

Chapter 6: The Prince Felipe Science Museum at the City of Arts and Sciences, in Valencia, Spain. The building is supported by an arrangement of trusses that surround its exterior.

Chapter 7: Hoover Dam, in Boulder City, Nevada. Concepts of centroid, center of mass, center of gravity, and fluid pressure loading were needed to design this structure.

Chapter 8: The glass roof at Canada Place, Vancouver, British Columbia, Canada. The design of the beams that support the weight of the roof and the forces from wind, rain, and snow requires determination of the internal forces throughout all of the beams in the roof system.

Chapter 9: The driver of a car spins its tires before a drag race, generating intense heat due to frictional slip between the tires and pavement. The increase in temperature of the tires increases the frictional resistance between the tires and pavement, thus improving performance during the race.

Chapter 10: A precast section of the Bay Bridge Skyway, near Oakland, California, is raised into position after being lifted nearly 200 ft from a barge. The area moment of inertia for the precast section is one of the factors controlling the strength of the bridge and its deflection due to the weight of vehicles and other forces.

BRIEF CONTENTS

Statics

TABLE OF CONTENTS

Welcome to statics! We assume you are embarking on the study of statics because you are interested in engineering and science. The major objectives of this book are to help you

1. learn the fundamental principles of statics; and
2. gain the skills needed to apply these principles in the modeling of real-life problems and for carrying out engineering design.

The need for thorough coverage of the fundamental principles is paramount, and, as such, a substantial portion of this book is devoted to these principles. Because the development of problem-solving skills is equally important, we focus a great deal of attention to these skills, especially in the context of real-life problems. Indeed, the emphasis on problem-solving skills is a major difference in the treatment of statics between engineering and physics. It is only through the mastering of these skills that a true, deep understanding of fundamentals can be achieved. You must be flawless in your ability to apply the concepts of statics to real-life problems. When mistakes are made, structures and machines will fail, money and time will be lost, and worst of all, people may be hurt or killed.

What should you take away from this book?

First and foremost, you should gain a thorough understanding of the fundamental principles, and, at a minimum, key points should remain in your memory for the rest of your life. We say this with a full appreciation that some of you will have careers with new and unexpected directions. Regardless of your eventual professional responsibilities, knowledge of the fundamentals of statics will help you to be technically literate. By contrast, if you *are* actively engaged in the practice of engineering and/or the sciences, then your needs go well beyond mere technical literacy. In addition to understanding the fundamentals, you must also be accomplished at applying these fundamentals. This ability is needed so that you can study more advanced subjects that build on statics, and because you will apply concepts of statics on a daily basis in your career.

Why Another Statics and Dynamics Series?

These books provide thorough coverage of all the pertinent topics traditionally associated with statics and dynamics. Indeed, many of the currently available texts also provide this. However, the new books by Plesha/Gray/Costanzo offer several major innovations that enhance the learning objectives and outcomes in these subjects.

What Then Are the Major Differences between Plesha/Gray/Costanzo and Other Engineering Mechanics Texts?

• A Consistent Approach to Problem Solving

The example problems in Plesha/Gray/Costanzo follow a structured five-step problem-solving methodology that will help you develop your problem-solving skills not only in statics and dynamics, but also in almost all other mechanics subjects that follow. This structured problem-solving approach consists of the following steps: Road Map, Modeling, Governing Equations, Computation, and Discussion & Verification. The Road Map provides some of the general objectives of the problem and develops a strategy for how the solution will be developed. Modeling is next, where a real-life problem is idealized by a model. This step results in the creation of a free body diagram and the selection of the balance laws needed to solve the problem. The Governing Equations step is devoted to writing all the equations needed to solve the problem. These equations typically include the Equilibrium Equations, and, depending upon the particular problem, Force Laws (e.g. spring law, failure criteria, frictional sliding criteria) and Kinematic Equations. In the Computation step, the governing equations are solved. In the final step, Discussion & Verification, the solution is interrogated to ensure that it is meaningful and accurate. This five-step problem-solving methodology is followed for all examples that involve equilibrium concepts. Some problems (e.g., determination of the center of mass for an object) do not involve equilibrium concepts, and for these the Modeling step is not needed.

• Contemporary Examples, Problems, and Applications

The examples, problem sets, and design problems were carefully constructed to help show you how the various topics of statics and dynamics are used in engineering practice. Statics and dynamics are immensely important subjects in modern engineering and science, and one of our goals is to excite you about these subjects and the career that lies ahead of you.

• A Focus on Design

A major difference between Plesha/Gray/Costanzo and other books is the systematic incorporation of design and modeling of real-life problems throughout. Topics include important discussions on design, ethics, and professional responsibility. These books show you that meaningful engineering design is possible using the concepts of statics and dynamics. Not only is the ability to develop a design very satisfying, but it also helps you develop a greater understanding of basic concepts and helps sharpen your ability to apply these concepts. Because the main focus of statics and dynamics textbooks should be the establishment of a firm understanding of basic concepts and correct problem-solving techniques, design topics do not have an overbearing presence in the books. Rather, design topics are included where they are most appropriate. While some of the discussions on design could be described as "common sense," such a characterization trivializes the importance and necessity for discussing pertinent issues such as safety, uncertainty in determining loads, the designer's responsibility to anticipate uses, even unintended uses, communications, ethics, and uncertainty in workmanship. Perhaps the most important feature of our inclusion of design and modeling topics is that you get a glimpse of what engineering is about and where your career in engineering is headed. The book is structured so that design topics and design problems are offered in a variety of places, and it is possible to pick when and where the coverage of design is most effective.

• Problem-Based Introductions

Many topics are presented using a problem-based introduction. By this approach, we hope to pique your interest and curiosity with a problem that has real-life significance and/or offers physical insight into the phenomena to be discussed. Using an interesting problem as a springboard, the necessary theory and/or tools needed to address the problem are developed. Problem-based introductions are used where they are especially effective, namely, topics that are challenging to visualize or understand.

• Computational Tools

Some examples and problems are appropriate for solution using computer software. The use of computers extends the types of problems that can be solved while alleviating the burden of solving equations. Such examples and problems give you insight into the power of computer tools and further insight into how statics and dynamics are used in engineering practice.

• Modern Pedagogy

Numerous modern pedagogical elements have been included. These elements help reinforce concepts and they provide you with additional information to help you understand concepts. Marginal notes (including Helpful Information, Common Pitfalls, Interesting Facts, and Concept Alerts) help place topics, ideas, and examples in a larger context. These notes will help you study (e.g., Helpful Information and Common Pitfalls), will provide real-world examples of how different aspects of statics and dynamics are used (e.g., Interesting Facts), and will drive home important concepts or dispel misconceptions (e.g., Concept Alerts and Common Pitfalls). Mini Examples are used throughout the text to immediately and quickly illustrate a point or concept without having to wait for the worked-out examples at the end of the section.

• Answers to Problems

Answers to most even-numbered problems are posted as a freely downloadable PDF file at www.mhhe.com/pgc. Providing answers in this manner allows for the inclusion of more complex information than would otherwise be possible. In addition to final numerical or symbolic answers, selected problems have more extensive information such as free body diagrams and/or shear and moment diagrams. This feature not only provides more complete answers in selected circumstances, but also provides a kick start that might help you on some homework problems. Furthermore, the multitude of free body diagram answers give you ample opportunity to practice constructing these on your own for extra problems. Appendix B gives an example of the additional information provided for a particular problem.

A Note To The Instructor

Statics is the first engineering course taken by most students en route to an undergraduate degree in engineering. As such, you are presented with numerous challenges when choosing the text you use. Because statics is so fundamental to subsequent engineering coursework and professional practice, a text must be accurate, thorough, and comprehensive. Statics also presents an opportunity to excite students and show them what engineering is about early in their education. Further, if this opportunity is missed and students do not receive an accurate picture of where their career is heading, they may make a poorly-informed decision to change their major away from engineering. These statements are recognized in the current Accreditation Board for Engineering and Technology (ABET) criteria for accreditation of engineering programs, which requires design to be integrated throughout an engineering curriculum. This book provides thorough coverage of the principles of statics. It also includes discussions on the theory and the more subtle points of statics. Such discussions usually follow an introductory treatment of a topic so that students have a grasp of concepts and their application before covering more subtle topics. For example, the concepts of particle equilibrium are presented in Section 3.1, with common assumptions such as cables being inextensible and pulleys being frictionless. In that section, the emphasis is on drawing free body diagrams, writing equilibrium equations, solution of equations, application of failure criteria, and interrogation of solutions. In Section 3.2, the reasons for the typical assumptions are thoroughly discussed, including the limitations of these assumptions for modeling real-life problems. This will help students develop an appreciation for the fact that, despite these assumptions, statics is an immensely useful and widely applicable subject. Further, the discussion in Section 3.2 is used to present more advanced topics such as springs and static indeterminacy.

Design topics include ethics, professional responsibility, pertinent codes and standards, and much more. Design problems are open ended and allow students to show creativity in developing solutions that solve important and realistic engineering problems. The design problems in this book may take students several hours to complete. It is recommended that students write a short report, suitable for reading by an engineer. A brief discussion of technical writing is included in Appendix A since many students have not yet studied technical writing. Perhaps the most important feature of our inclusion of design and modeling topics is that students get a glimpse of what engineering is about and where their career in engineering is headed. The book is structured so that you may pick when and where design is most appropriately covered.

McGraw-Hill's 360° Development

McGraw-Hill's 360° Development Process is a continuous, market-oriented approach to building accurate and innovative print and digital products. It is dedicated to continual improvement driven by multiple customer feedback loops and checkpoints. This is initiated during the early planning stages of our new products and intensifies during the development and production stages, then begins again upon publication, in anticipation of the next edition.

This process is designed to provide a broad, comprehensive spectrum of feedback for refinement and innovation of our learning tools for both student and instructor. The 360° Development Process includes market research, content reviews, faculty and student focus groups, course- and product-specific symposia, accuracy checks, art reviews, and a Board of Advisors.

Here is a brief overview of the initiatives included in the 360° Development Process of the new statics and dynamics books.

Board of Advisors A hand-picked group of trusted instructors active in teaching engineering mechanics courses served as chief advisors and consultants to the authors and editorial team during manuscript development. The Board of Advisors reviewed parts of the manuscript; served as a sounding board for pedagogical, media, and design concerns; and consulted on organizational changes.

Manuscript Review Panels Numerous instructors reviewed the various drafts of the manuscript to give feedback on content, design, pedagogy, and organization. This feedback was summarized by the book team and used to guide the direction of the text.

Symposia McGraw-Hill conducted several engineering mechanics symposia attended by instructors from across the country. These events are an opportunity for McGraw-Hill editors and authors to gather information about the needs and challenges of instructors teaching these courses. They also offered a forum for the attendees to exchange ideas and experiences with colleagues they might not have otherwise met.

Focus Group In addition to the symposia, McGraw-Hill held a focus group with the authors and selected engineering mechanics professors. These engineering mechanics professors provided ideas on improvements and suggestions for fine tuning the content, pedagogy, and art.

Accuracy Check A select group of engineering mechanics instructors reviewed the entire final manuscript for accuracy and clarity of the text and solutions.

Class Tests Over a number of years, both books have been class tested by thousands of students at schools such as Texas A&M University, Penn State, and the University of Wisconsin-Madison. In addition to the class testing done by the authors, the books have been class tested by David M. Hoerr at the University of Wisconsin and Arun R. Srinivasa at Texas A&M University. We are especially grateful to Professor Srinivasa for the excellent suggestions he has made as a result of his extensive class testing.

Student Focus Groups Student focus groups provided the editorial team with an understanding of how content and the design of a textbook impacts students' homework and study habits in the engineering mechanics course area.

Manuscript Preparation The authors developed the manuscripts on Apple Macintosh laptop computers using LaTeX and Adobe Illustrator. This approach is novel to the publishing industry. The code generated by the authors was used to typeset the final manuscript. This approach eliminates the usual source of errors where the original authors' manuscript is rekeyed by the publisher to obtain the final manuscript.

The following individuals have been instrumental in ensuring the highest standard of content and accuracy. We are deeply indebted to them for their tireless efforts. We are especially grateful to Professor Robert J. Witt for his frequent advice, deep insights, and the amazing precision of his work.

Board of Advisors

Janet Brelin-Fornari
Kettering University

Manoj Chopra
University of Central Florida

Pasquale Cinnella
Mississippi State University

Ralph E. Flori
Missouri University of Science and Technology

Christine B. Masters
Penn State

Mark Nagurka
Marquette University

David W. Parish
North Carolina State University

Gordon R. Pennock
Purdue University

Michael T. Shelton
California State Polytechnic University-Pomona

Joseph C. Slater
Wright State University

Arun R. Srinivasa
Texas A&M University

Carl R. Vilmann
Michigan Technological University

Ronald W. Welch
The University of Texas at Tyler

Robert J. Witt
University of Wisconsin-Madison

Reviewers

Shaaban Abdallah
University of Cincinnati

Makola M. Abdullah
Florida Agricultural and Mechanical University

Mohamed Aboul-Seoud
Rensselaer Polytechnic Institute

George G. Adams
Northeastern University

Shahid Ahmad
SUNY-Buffalo

Riad Alakkad
University of Dayton

Manohar L. Arora
Colorado School of Mines

Kenneth Belanus
Oklahoma State University

Glenn Beltz
University of California-Santa Barbara

Sherrill Biggers
Clemson University

Neeraj Buch
Michigan State University

Manoj Chopra
University of Central Florida

Pasquale Cinnella
Mississippi State University

K. L. (Larry) DeVries
University of Utah

Roger M. Di Julio, Jr.
California State University-Northridge

Frank J. Fronczak
University of Wisconsin-Madison

Barry Goodno
Georgia Institute of Technology

Walter Haisler
Texas A&M University

Roy J. Hartfield, Jr.
Auburn University

G.A. Hartley
Carleton University

Scott L. Hendricks
Virginia Tech University

Paul R. Heyliger
Colorado State University

Lih-Min Hsia
California State University, Los Angeles

Nancy Hubing
University of Missouri-Rolla

David A. Jenkins
University of Florida

Erik A. Johnson
University of Southern California

James D. Jones
Purdue University

Jeff Kuo
California State University, Fullerton

Shaofan Li
University of California-Berkeley

Mohammad Mahinfalah
Milwaukee School of Engineering

Francisco Manzo-Robledo
Washington State University

Christine B. Masters
Penn State

Karim Nohra
University of South Florida

David W. Parish
North Carolina State University

Gordon R. Pennock
Purdue University

Gary A. Pertmer
University of Maryland

W. Tad Pfeffer
University of Colorado at Boulder

John J. Phillips
Oklahoma State University

Scott D. Schiff
Clemson University

Michael T. Shelton
California State Polytechnic University-Pomona

Joseph C. Slater
Wright State University

Ahmad Sleiti
University of Central Florida

Arun R. Srinivasa
Texas A&M University

Ganesh Thiagarajan
Louisiana State University

Jeffrey Thomas
Northwestern University

David G. Ullman
Professor Emeritus, Oregon State University

Steven A. Velinsky
University of California, Davis

Carl R. Vilmann
Michigan Technological University

Ronald W. Welch
The University of Texas at Tyler

Claudia M. D. Wilson
Florida State University

Robert J. Witt
University of Wisconsin-Madison

T. W. Wu
University of Kentucky

Joseph R. Zaworski
Oregon State University

Xiangwu (David) Zeng
Case Western Reserve University

Symposium Attendees

Farid Amirouche
University of Illinois at Chicago

Subhash C. Anand
Clemson University

Manohar L. Arora
Colorado School of Mines

Stephen Bechtel
Ohio State University

Sherrill Biggers
Clemson University

J. A. M. Boulet
University of Tennessee

Janet Brelin-Fornari
Kettering University

Louis M. Brock
University of Kentucky

Amir Chaghajerdi
Colorado School of Mines

Manoj Chopra
University of Central Florida

Pasquale Cinnella
Mississippi State University

Adel ElSafty
University of North Florida

Ralph E. Flori
Missouri University of Science and Technology

Walter Haisler
Texas A&M University

Kimberly Hill
University of Minnesota

James D. Jones
Purdue University

Yohannes Ketema
University of Minnesota

Charles Krousgrill
Purdue University

Jia Lu
The University of Iowa

Mohammad Mahinfalah
Milwaukee School of Engineering

Tom Mase
California Polytechnic State University, San Luis Obispo

Christine B. Masters
Penn State

Daniel A. Mendelsohn
The Ohio State University

Faissal A. Moslehy
University of Central Florida

LTC Mark Orwat
United States Military Academy at West Point

David W. Parish
North Carolina State University

Arthur E. Peterson
Professor Emeritus, University of Alberta

W. Tad Pfeffer
University of Colorado at Boulder

David G. Pollock
Washington State University

Robert L. Rankin
Professor Emeritus, Arizona State University

Mario Rivera-Borrero
University of Puerto Rico at Mayaguez

Hani Salim
University of Missouri

Brian P. Self
California Polytechnic State University, San Luis Obispo

Michael T. Shelton
California State Polytechnic University-Pomona

Lorenz Sigurdson
University of Alberta

Larry Silverberg
North Carolina State University

Joseph C. Slater
Wright State University

Arun R. Srinivasa
Texas A&M University

David G. Ullman
Professor Emeritus, Oregon State University

Carl R. Vilmann
Michigan Technological University

Anthony J. Vizzini
Mississippi State University

Andrew J. Walters
Mississippi State University

Ronald W. Welch
The University of Texas at Tyler

Robert J. Witt
University of Wisconsin-Madison

T. W. Wu
University of Kentucky

Musharraf Zaman
University of Oklahoma-Norman

Joseph R. Zaworski
Oregon State University

Focus Group Attendees

Shaaban Abdallah
University of Cincinnati

Manohar L. Arora
Colorado School of Mines

Manoj Chopra
University of Central Florida

Christine B. Masters
Penn State

Michael T. Shelton
California State Polytechnic University-Pomona

Accuracy Checkers

Richard McNitt
Penn State

Carl R. Vilmann
Michigan Technological University

Robert J. Witt
University of Wisconsin-Madison

ACKNOWLEDGMENTS

The authors would like to thank Jonathan Plant, former editor at McGraw-Hill, for his guidance in the early years of this project.

We are grateful to a number of students for thoroughly proofreading the manuscript and for making suggestions to improve it. Especially among these are Robert Shebesta and Holly Powell. In addition, we would like to thank Chris Punshon, Andrew Miller, Chris O'Brien, Chandan Kumar, and Joseph Wyne for their substantial contributions to the solutions manual.

Most of all, we would like to thank Andrew Miller for infrastructure he created to keep the authors, manuscript, and solutions manual in sync. His knowledge of programming, scripting, subversion, and many other computer technologies made a gargantuan task feel just a little more manageable.

4 Moment of a Force and Equivalent Force Systems

Additional concepts of forces and systems of forces are discussed in this chapter. These concepts are used extensively in the analysis of equilibrium and motion of bodies and throughout more advanced mechanics subjects.

4.1 Moment of a Force

To help demonstrate some of the features of the *moment of a force*, we will consider an example of a steering wheel in a car. Figure 4.1 shows a classic Ferrari sports car, and Fig. 4.2 shows the steering wheel in this car. The wheel lies in a plane that is perpendicular to the steering column AB (it would not be very comfortable to use if this were not the case), and the steering wheel offers "resistance" to being turned (for most vehicles, this resistance increases for slower speeds and as a turn becomes sharper). Imagine you are driving this car, and you wish to execute a right-hand turn. Figure 4.2 shows two possible locations where you could position your left hand to turn the steering wheel, and the directions of forces F_1 and F_2 that you would probably apply, where both of these forces lie in the plane of the steering wheel. For a given speed and sharpness of turn, clearly position C will require a lower force to turn the wheel than position D (i.e., $F_1 < F_2$). Both forces F_1 and F_2 produce a *moment* (i.e., twisting action) about line AB of the steering column, and the size of this moment increases as the force becomes larger *and/or* as the distance from the force's line of action to line AB increases. If the line of action of F_1 is perpendicular to line AB (as we have assumed here), and if we let d be the

Chapter Introduction

Each chapter begins with an introductory section setting the purpose and goals of the chapter.

Computation This system of equations is not as easy to solve as that in Part (a). Systems of equation that are difficult to solve are a fact of life in engineering, and you must be proficient in solving them. The basic strategy for hand computation (where one of the equations is solved for one of the unknowns in terms of the others and then this result is substituted into the remaining equations, and so on) is workable for systems of three equations, but it rapidly becomes very tedious for larger systems of equations. You should take this opportunity to use a programmable calculator or one of the software packages mentioned earlier to find the solutions to Eqs. (14)–(16), which are

$$\boxed{F_{AB} = 1027\,\text{lb}, \quad F_{AC} = 930\,\text{lb}, \quad \text{and} \quad F_{AO} = -1434\,\text{lb}.} \tag{17}$$

In addition to the results in Eq. (17), the force supported by cable AD is equal to W, therefore $\boxed{F_{AD} = 1000\,\text{lb}.}$

Computer Solutions

We make use of computer solutions in some problems and it is important that you be able to easily identify when this is the case. Therefore, anywhere a computer is used for a solution, you will see the symbol. If this occurs within part of an example problem or within the discussion, then the part requiring the use of a computer will be enclosed in the following symbols. If one of the exercises requires a computer for its solution, then the computer symbol and its mirror image will appear on either side of the problem heading.

Mini-Examples

Mini-examples are used throughout the text to immediately and quickly illustrate a point or concept without having to wait for the worked-out examples at the end of the section. Mini-examples begin with the text **■ Mini-Example. and end with the symbol ——■.**

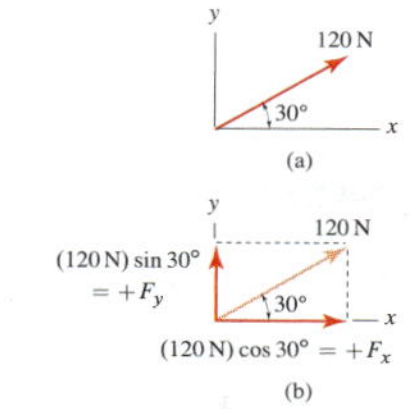

Figure 2.13
Determination of the Cartesian components of a 120 N force vector.

■ Mini-Example. Express the vector shown in Fig. 2.13(a), using Cartesian representation.

Solution. Let $\vec{F}$ denote the 120 N magnitude force vector shown in Fig. 2.13(a) and let F_x and F_y denote the components of $\vec{F}$ in the x and y directions, respectively. Components F_x and F_y are obtained by constructing projections of $\vec{F}$ that are parallel to the x and y axes, respectively, with the resulting magnitudes (120 N) cos 30° and (120 N) sin 30°, as shown in Fig. 2.13(b). We next use these projections to assign vector components in the x and y directions, paying careful notice that F_x and F_y are positive when acting in the positive x and y directions, respectively, and are negative when acting in the negative x and y directions. Thus,

$$F_x = (120\,\text{N})\cos 30° = 104\,\text{N}, \quad F_y = (120\,\text{N})\sin 30° = 60\,\text{N}. \tag{2.15}$$

Thus, the Cartesian representation for $\vec{F}$ is

$$\vec{F} = (104\,\hat{\imath} + 60\,\hat{\jmath})\,\text{N}. \tag{2.16}$$

Some useful checks. There are a few useful checks that may help you avoid errors when resolving a vector into Cartesian components. Foremost is to verify that the components have reasonable size and are in proper directions. The next check is to verify that the vector's components give the correct magnitude. Thus, for Eq. (2.16), we evaluate $\sqrt{(104\,\text{N})^2 + (60\,\text{N})^2} = 120\,\text{N}$ to find that indeed it has the correct magnitude. While these checks are reassuring, they do not guarantee the components are correct. Nonetheless, if we find an incorrect magnitude, then certainly an error has been made. ——■

168 *Equilibrium of Particles* Chapter 3

EXAMPLE 3.10 *Engineering Design*

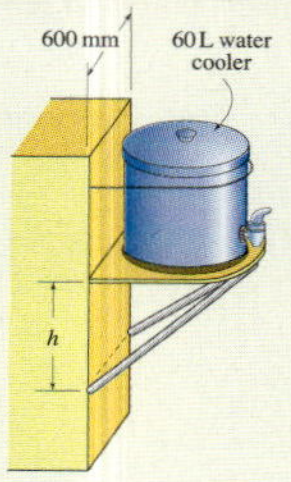

Figure 1

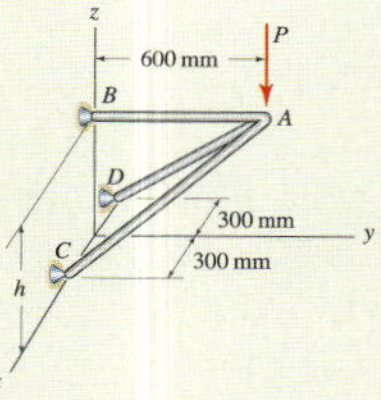

Figure 2
Model for the water cooler support shown in Fig. 1.

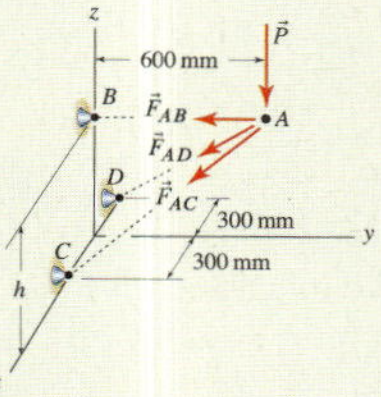

Figure 3
Free body diagram for point A.

A large construction company plans to add plastic 60 L water coolers to the back of most of its trucks, as shown in Fig. 1. Your supervisor asks you to use the model shown in Fig. 2 to specify the dimension h and the diameter of round steel bars to be used for members AB, AC, and AD. Although a more refined model is possible using concepts discussed in Chapter 5, the model shown in Fig. 2 is useful for this design problem. In this model, load P is vertical and member AB is parallel to the y axis. Your supervisor also tells you to allow for loads up to twice the weight of the cooler in a crude attempt to allow for dynamic forces, and to use the allowable loads given in Table 3.2.

SOLUTION

Road Map The most poorly defined part of this problem is the loading the water cooler support will be subjected to. Thus, assumptions on the size of loads will need to be made. Furthermore, different loading conditions may need to be considered, and our design should be safe for all of these. Finally, this design problem does not have a unique solution. That is, there is not a single value of h that will work. Rather, we may need to assume reasonable values for some parameters and then calculate the values of others based on this.

Modeling When full, the weight of water in the cooler is*

$$W = \underbrace{1000\,\frac{\text{kg}}{\text{m}^3}}_{\rho}\;\underbrace{60\,\text{L}\,\frac{(0.1\,\text{m})^3}{\text{L}}}_{V}\;\underbrace{9.81\,\frac{\text{m}}{\text{s}^2}}_{g} = 589\,\text{N}. \tag{1}$$

We next use our judgment to add to the above value a nominal amount of 60 N to account for the weight of the container itself. Thus, the total weight to be supported is 589 N + 60 N = 649 N.

Doubling the weights (or use of other multiplicative factors) is an imprecise but common approach to account for inertial forces produced when the truck hits bumps in the road. Thus, we obtain the approximate load P = 1300 N. Before continuing with this load, we should consider other possible load scenarios. For example, if the cooler is removed, the support might make a convenient step for a worker to use while climbing on or off the truck. Consider a person weighing 890 N (200 lb). Since it is extremely unlikely a person would be standing on the support when the truck is moving and hitting bumps, it is not necessary to further increase this value as was done earlier. Since the weight of this person is lower than the force determined earlier, we will proceed with the design, using P = 1300 N.

Small values of dimension h will lead to large forces in all members, while larger values of h will decrease these forces. However, the allowable load data in Table 3.2 shows that allowable compressive load decreases rapidly as the length of members increases. Thus, we will select for an initial design the value h = 250 mm, in which case the length of members AC and AD is 716 mm.

The FBD for point A is shown in Fig. 3.

Governing Equations Vector expressions for forces are

$$\vec{P} = 1300\,\text{N}\,(-\hat{k}), \tag{2}$$

* One liter (1 L) is defined to be the volume of 1 kg of pure water at 4°C and pressure of 76 cm of mercury. For practical purposes, however, the transformation 1 L = 1000 cm³ = 0.001 m³ may be used with less than 0.003% error.

Examples

Consistent Problem-Solving Methodology

Every problem in the text employs a carefully defined problem-solving methodology to encourage systematic problem formulation while reinforcing the steps needed to arrive at correct and realistic solutions.

Each example problem contains these five steps:

- **Road Map**
- **Modeling**
- **Governing Equations**
- **Computation**
- **Discussion & Verification**

Some examples include a Closer Look (noted with a magnifying glass icon 🔍) that offers additional information about the example.

Concept Alerts and Concept Problems

Two additional features are the Concept Alert and the Concept Problems. These have been included because research has shown (and it has been our experience) that even though you may do quite well in a science or engineering course, your conceptual understanding may be lacking. **Concept Alerts** are marginal notes and are used to drive home important concepts (or dispel misconceptions) that are related to the material being developed at that point in the text. **Concept Problems** are mixed in with the problems that appear at the end of each section. These are questions designed to get you thinking about the application of a concept or idea presented within that section. They should never require calculation and should require answers of no more than a few sentences.

Concept Alert

Applications of the cross product. The cross product between two vectors produces a result that is a vector. The cross product is frequently used to determine the normal direction to a surface, the area of a parallelogram formed by two vectors, and (as discussed in Chapter 4) the *moment* produced by a force. The last application is especially important in statics and mechanics.

Problem 2.94

Describe how the cross product operation can be used to determine (or "test") whether two vectors $\vec{A}$ and $\vec{B}$ are orthogonal. Is this test as easy to use as the test based on the dot product? Explain, perhaps using an example to support your remarks.
Note: Concept problems are about *explanations*, not computations.

Problem 2.95

Imagine a left-hand coordinate system has inadvertently been used for a problem. That is, if the x and y directions have been selected first, the z direction has been taken in the wrong direction for a right-hand coordinate system. What consequences will this have for dot products and cross products? Perhaps use an example to support your remarks.
Note: Concept problems are about *explanations*, not computations.

Common Pitfall

Failure loads. A common error in solving problems with failure criteria, such as Part (b) of this example, is to assume that *all* members are at their failure loads at the same time. With reference to the FBD of Fig. 3, you will be making this error if you take $F_{AB} = 1200\,\mathrm{N}$ and $F_{AC} = -1600\,\mathrm{N}$; in fact, if you do this, you will find that $\sum F_x = 0$ (Eq. (5)) cannot be satisfied! Another way to describe this problem is to consider slowly increasing force P from zero. Eventually, one of the members will reach its failure load first, while the other will be below its failure

Interesting Fact

Springs. Springs are important structural members in their own right, but they are also important for laying the groundwork for characterizing more general engineering materials and members, which you will study in subjects that follow statics. Simply stated, almost all materials are idealized ... more complex than that ... 5, over at least some

Helpful Information

Free body diagram (FBD). An FBD is an essential aid, or tool, for applying Newton's law of motion $\sum \vec{F} = m\vec{a}$. Among the many skills you will need to be successful in statics, and in the subjects that follow, and as a practicing engineer, the ability to draw accurate FBDs is essential. An incorrect FBD is the most common source of errors in an analysis.

Marginal Notes

Marginal notes have been implemented that will help place topics, ideas, and examples in a larger context. This feature will help students study (using **Helpful Information** and **Common Pitfalls**) and will provide real-world examples of how different aspects of statics are used (using **Interesting Facts**).

Sections and End of Section Summary

Each chapter is organized into several sections. There is a wealth of information and features within each section, including examples, problems, marginal notes, and other pedagogical aids. Each section concludes with an end of section summary that succinctly summarizes the section. In many cases, cross-referenced important equations are presented again for review and reinforcement before the student proceeds to the examples and homework problems.

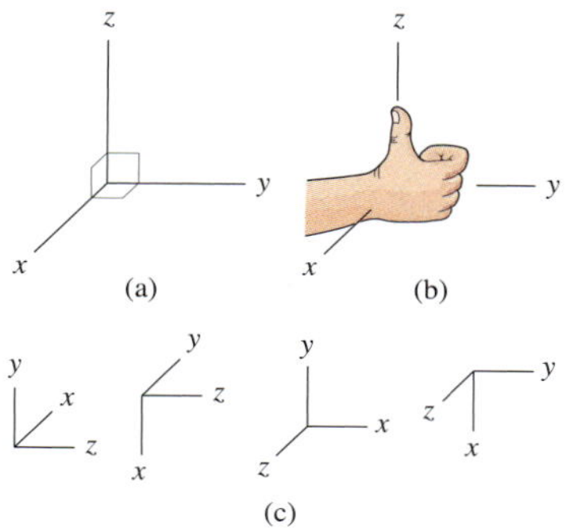

Figure 2.17
(a) Cartesian coordinate system in three dimensions. (b) A scheme for constructing a right hand coordinate system. Position your right hand so the positive x direction passes into your palm and the positive y direction passes through your finger tips. Your thumb then indicates the positive z direction. (c) More examples of right-hand coordinate systems.

2.3 Cartesian Representation of Vectors in Three Dimensions

For problems in three dimensions, vectors are especially powerful and without them many problems would be intractable. Concepts of Section 2.2 apply, with some additional enhancements needed for three dimensions. These include definitions of a right-handed coordinate system, direction angles, and direction cosines.

Right-hand Cartesian coordinate system

In three dimensions a Cartesian coordinate system uses three orthogonal reference directions. These will consist of x, y, and z directions as shown in Fig. 2.17(a). Proper interpretation of many vector operations, such as the cross product to be discussed in Section 2.5, requires the x, y, and z directions be arranged in a consistent manner. For example, when you are constructing the coordinate system shown in Fig. 2.17(a), imagine the x and y directions are chosen first. Then, should z be taken in the direction shown, or can it be in the opposite direction? The universal convention in mechanics and vector mathematics in general is z must be taken in the direction shown, and the result is called a *right-hand coordinate system*. Figure 2.17(b) describes a scheme for constructing a right-hand coordinate system. You should study this scheme and become comfortable with its use.

Cartesian vector representation

We define vectors $\hat{\imath}$, $\hat{\jmath}$, and $\hat{k}$ to be unit vectors that point in the positive x, y, and z directions, respectively. A vector $\vec{v}$ can then be written as

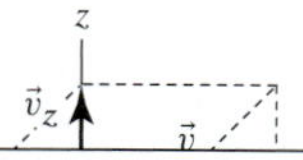

$$\vec{v} = \vec{v}_x + \vec{v}_y + \vec{v}_z$$
$$= v_x\,\hat{\imath} + v_y\,\hat{\jmath} + v_z\,\hat{k}. \qquad (2.23)$$

y, and z components is shown in Fig. 2.18. The *mag-*

$$|\vec{v}| = \sqrt{v_x^2 + v_y^2 + v_z^2}. \qquad (2.24)$$

d using the construction shown in Fig. 2.19 as follows. ies in the xy plane is defined. Because v_x, v_y, and , the Pythagorean theorem provides $v_a^2 = v_x^2 + v_y^2$. form a right triangle, and the Pythagorean theorem . Substituting for v_a^2 in this latter expression yields d thus Eq. (2.24) follows.

nd direction cosines

acterizing a vector's orientation is by use of *direction* θ_x, θ_y, and θ_z are shown in Fig. 2.20 and are defined d from the positive x, y, and z directions, respectively, ctor.

End of Section Summary

In this section, Cartesian coordinate systems and Cartesian representation for vectors in three dimensions have been defined. Some of the key points are:

- The xyz coordinate system you use must be a *right-hand coordinate system*. Proper interpretation of some vector operations requires this.
- *Direction angles* provide a useful way for specifying a vector's orientation in three dimensions. A vector has three direction angles θ_x, θ_y, and θ_z, but only two of these are independent. Direction angles satisfy the equation $\cos^2\theta_x + \cos^2\theta_y + \cos^2\theta_z = 1$, so that if two direction angles are known, the third may be determined.
- Structural members such as cables, ropes, and bars support forces whose lines of action have the same orientation as the member's geometry. Thus, if $\vec{r}$ describes a member's geometry, a vector expression for the force supported by the member may be written as $\vec{F} = F(\vec{r}/|\vec{r}|)$.

PROBLEMS

Problem 2.1

For each vector, write two expressions using polar vector representations, one using a positive value of θ and the other a negative value, where θ is measured counterclockwise from the right-hand horizontal direction.

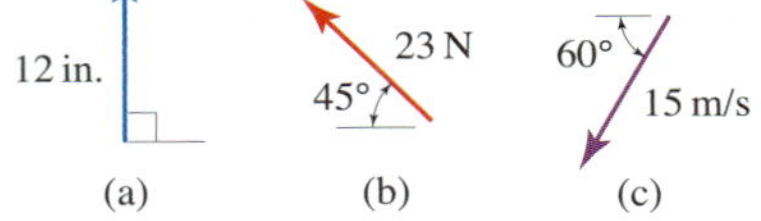

Figure P2.1

Problems 2.2 and 2.3

Add the two vectors shown to form a resultant vector $\vec{R}$, and report your result using polar vector representation.

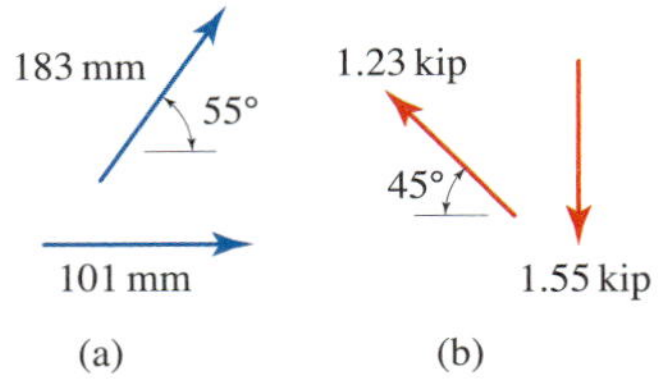

Figure P2.2

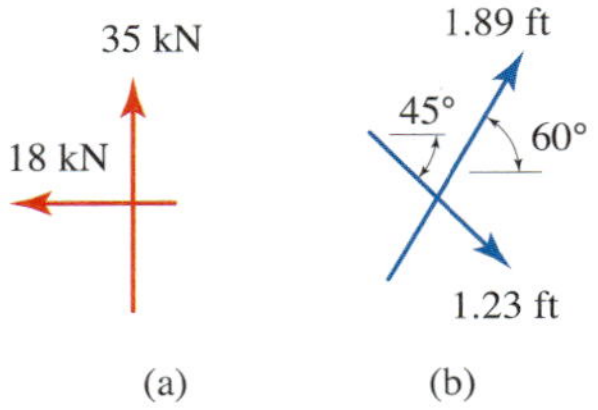

Figure P2.3

Problem 2.4

Let $\vec{A} = 2\,\text{m} @ 0°\measuredangle$ and $\vec{B} = 6\,\text{m} @ 90°\measuredangle$. Sketch the vector polygons and evaluate $\vec{R}$ for the following, reporting your answer using polar vector representation.

(a) $\vec{R} = \vec{A} + \vec{B}$,

(b) $\vec{R} = 2\vec{A} - \vec{B}$,

(c) $\vec{R} = |\vec{A}|\,\vec{B} + |\vec{B}|\,\vec{A}$,

(d) $\vec{R} = \dfrac{\vec{A}}{|\vec{A}|} + \dfrac{\vec{B}}{|\vec{B}|}$.

Problem 2.5

A tow truck applies forces $\vec{F}_1$ and $\vec{F}_2$ to the bumper of an automobile where $\vec{F}_1$ is horizontal. Determine the magnitude of $\vec{F}_2$ that will provide a vertical resultant force, and determine the magnitude of this resultant.

Problem 2.6

Arm OA of a robot is positioned as shown. Determine the value for angle α of arm AB so that the distance from point O to the actuator at B is 650 mm.

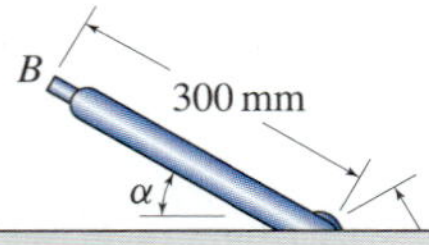

Modern Problems

Problems of varying difficulty follow each section. These problems allow students to develop their ability to apply concepts of statics on their own. Statics is not an easy subject, and the most common question asked by students is "How do I set this problem up?" What is really meant by this question is "How do I develop a good mathematical model for this problem?" The only way to develop this ability is by practicing numerous problems. Answers to most even-numbered problems are posted as a freely downloadable PDF file at www.mhhe.com/pgc. Providing answers in this manner allows for more complex information than would otherwise be possible. In addition to final numerical or symbolic answers, selected problems have more extensive information such as free body diagrams and/or shear and moment diagrams. Not only does this feature provide more complete answers in selected circumstances, but it also provides the kick start needed to get students started on some homework problems. Furthermore, the multitude of free body diagram answers give students ample opportunity to practice constructing FBDs on their own for extra problems. Appendix B gives an example of the extensive information provided for a particular problem. Each problem in the book is accompanied by a thermometer icon that indicates the approximate level of difficulty. Those considered to be "introductory" are indicated with the symbol. Problems considered to be "representative" are indicated with the symbol, and problems that are considered to be "challenging" are indicated with the symbol.

Engineering Design and Design Problems

Throughout the book, in appropriate places, engineering design is discussed including topics such as methods of design, issues of professional responsibility, and ethics. Design Problems are also presented. These problems are open ended and allow students to show creativity in developing a solution that solves an important and realistic real-life engineering problem.

DESIGN PROBLEMS

General Instructions. In problems requiring the specification of sizes for steel cable, bar, or pipe, selections should be made from Tables 3.1–3.3. In all problems, write a brief technical report following the guidelines of Appendix A, where you summarize all pertinent information in a well-organized fashion. It should be written using proper, simple English that is easy to read by another engineer. Where appropriate, sketches, along with critical dimensions, should be included. Discuss the objectives and constraints considered in your design, the process used to arrive at your final design, safety issues if appropriate, and so on. The main discussion should be typed, and figures, if needed, can be computer-drawn or neatly hand-drawn. Include a neat copy of all supporting calculations in an appendix that you can refer to in the main discussion of your report. A length of a few pages, plus appendix, should be sufficient.

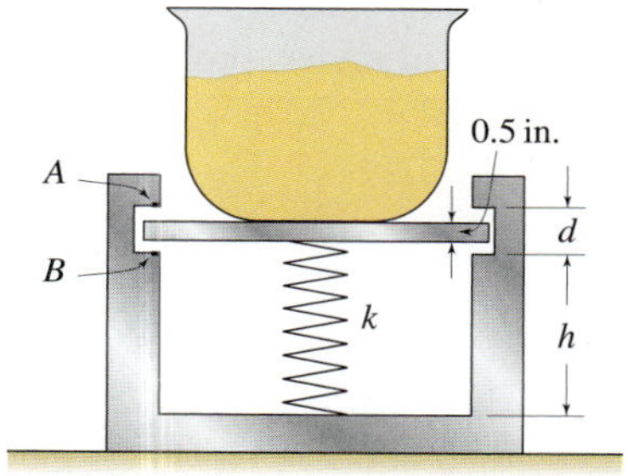

Figure DP3.1

Design Problem 3.1

A scale for rapidly weighing ingredients in a commercial bakery operation is shown. An empty bowl is first placed on the scale. Electrical contact is made at point A, which illuminates a light indicating the bowl's contents are underweight. A bakery ingredient, such as flour, is slowly poured into the bowl. When a sufficient amount is added, the contact at A is broken. If too much is added, contact is made at B, thus indicating an overweight condition. If the contents of the bowl are to weigh 18 lb $\pm$ 0.25 lb, specify dimensions h and d, spring stiffness k, and the unstretched length of the spring L_0. The bowl and the platform on which it rests have a combined weight of 5 lb. Assume the scale has guides or other mechanisms so that the platform on which the bowl rests is always horizontal.

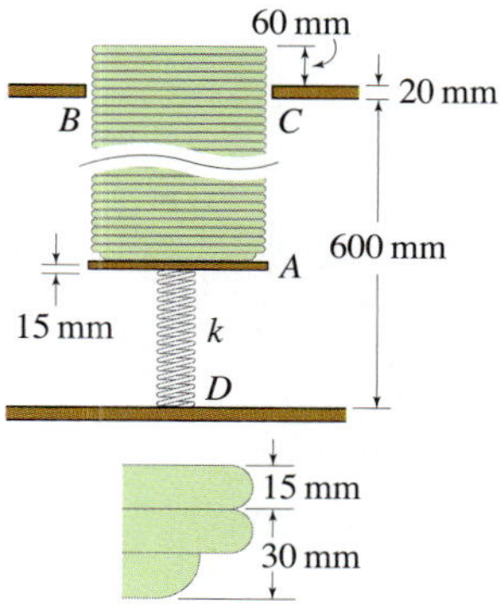

Figure DP3.2 and DP3.3

Design Problem 3.2

A plate storage system for a self-serve salad bar in a restaurant is shown. As plates are added to or withdrawn from the stack, the spring force and stiffness are such that the plates always protrude above the tabletop by about 60 mm. If each plate has 0.509 kg mass, and if the support A also has 0.509 kg mass, determine the stiffness k and unstretched length L_0 of the spring. Assume the spring can be compressed by a maximum of 40% of its initial unstretched length before its coils begin to touch. Also specify the number of plates that can be stored. Assume the system has guides or other mechanisms so the support A is always horizontal.

Design Problem 3.3

In Design Problem 3.2, the spring occupies valuable space that could be used to store additional plates. Repeat Design Problem 3.2, employing cable(s) and pulley(s) in conjunction with one or more springs to design a different system that will allow more plates to be stored. Pulleys, cables, and springs can be attached to surfaces A, B, C, and D. For springs in compression, assume they may not contract by more than 40% of their initial unstretched length before their coils begin to touch.

3.5 Chapter Review

Important definitions, concepts, and equations of this chapter are summarized. For equations and/or concepts that are not clear, you should refer to the original equation numbers cited for additional details.

Equilibrium of a particle. A particle is in *static equilibrium* if

Eq. (3.20), p. 148

$$\sum \vec{F} = \vec{0},$$

$$\text{or} \quad \left(\sum F_x\right)\hat{\imath} + \left(\sum F_y\right)\hat{\jmath} + \left(\sum F_z\right)\hat{k} = \vec{0},$$

$$\text{or} \quad \sum F_x = 0 \text{ and } \sum F_y = 0 \text{ and } \sum F_z = 0.$$

The above summations must include all forces applied to the particle. For equilibrium of a system of particles, the above equilibrium equations are written for each particle, and then the resulting system of simultaneous equations is solved.

Free body diagram. The *free body diagram* (FBD) is an essential aid for helping ensure that all forces applied to a particle are accounted for when you write equations of equilibrium. When you draw an FBD, it is helpful to imagine enclosing the particle by a closed line in two dimensions, or a closed surface in three dimensions. Wherever the cut passes through a structural member, the forces supported by that member must be introduced in the FBD. Wherever the cut passes through a support, the reaction forces that the support applies to the particle must be introduced in the FBD.

Cables and bars. Cables and straight bars are structural members that support forces that are collinear with their axis. We assume cables may support tensile forces only and may be freely bent, such as when wrapped around a pulley. We usually assume cables have negligible weight. Bars may support both tensile and compressive forces.

Pulleys. A pulley is a device that changes the direction of a cable, and hence changes the direction of the force that is supported by the cable. If the pulley is frictionless and the cable is weightless, then the magnitude of the force throughout the cable is uniform.

Springs. Behavior of a *linear elastic spring* is shown in Fig. 3.22 and is described by the *spring law*

Eq. (3.18), p. 137

$$F_s = k\delta$$
$$= k(L - L_0)$$

where k is the spring stiffness (units: force/length), δ is the elongation of the spring from its unstretched length, L_0 is the initial (unstretched) spring length,

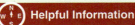

Helpful Information

Spring law sign conventions. The sign conventions for the spring law given in Eq. (3.18) are as follows:

$F_s > 0$ tension,
$F_s < 0$ compression,
$\delta = 0$ unstretched position,
$\delta > 0$ extension,
$\delta < 0$ contraction.

End-of-Chapter Review and Problems

Every chapter concludes with a succinct, yet comprehensive chapter review and a wealth of review problems.

Section 3.5 *Chapter Review* **177**

REVIEW PROBLEMS

Problem 3.71

Consider a problem involving cables and bars only. For the conditions listed below, is the solution obtained from $\sum \vec{F} = \vec{0}$ using geometry of the structure before loads are applied approximate or exact? Explain.

(a) Cables are modeled as inextensible, and bars are modeled as rigid.

(b) Cables and bars are modeled as linear elastic springs.

Note: Concept problems are about *explanations*, not computations.

Problem 3.72

The frictionless pulley A weighs 20 N and supports a box B weighing 60 N. When you solve for the force in cable CD, a "problem" arises. Describe this problem and its physical significance.

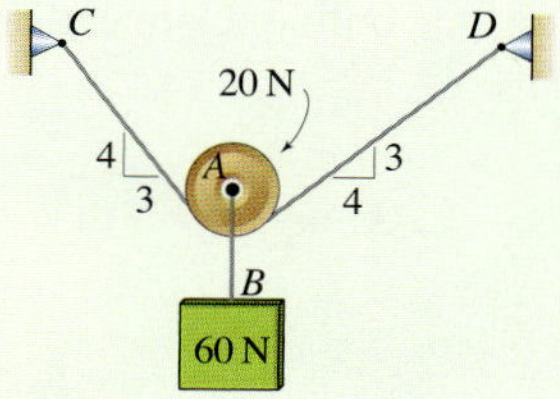

Figure P3.72

Problem 3.73

To produce a force $P = 40$ N in horizontal member CD of a machine, a worker applies a force F to the handle B. Determine the smallest value of F that can be used and the angle α it should be applied at.

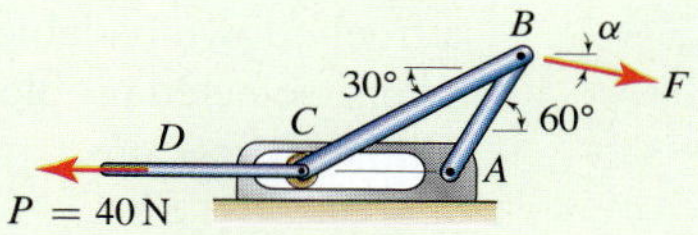

Figure P3.73

Problem 3.74

The structure shown consists of five cables. Cable $ABCD$ supports a drum having weight $W = 200$ lb. Cable DF is horizontal, and cable segments AB and CD are vertical. If contact between the drum and cable $ABCD$ is frictionless, determine the force in each cable.

Problem 3.75

In Prob. 3.74, if cable $ABCD$ has 600 lb breaking strength and all other cables have 200 lb breaking strength, determine the largest value W may have.

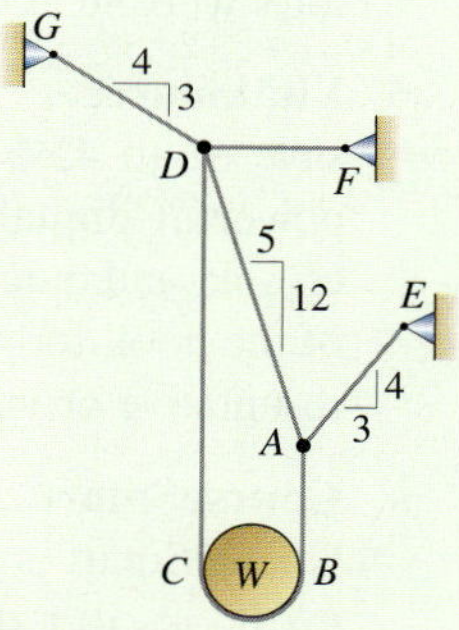

Figure P3.74 and P3.75

Problem 3.76

Two frictionless pulleys connected by a weightless bar AB support the 200 and 300 N forces shown. The pulleys rest on a wedge that is fixed in space. Determine the angle θ when the system is in equilibrium and the force in bar AB.

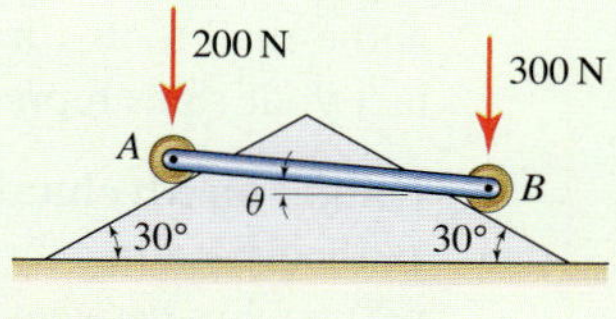

Figure P3.76

What Resources Support This Textbook?

McGraw-Hill offers various tools and technology products to support ***Engineering Mechanics: Statics*** and ***Engineering Mechanics: Dynamics.*** Instructors can obtain teaching aids by calling the McGraw-Hill Customer Service Department at 1-800-338-3987, visiting our online catalog at www.mhhe.com, or contacting their local McGraw-Hill sales representative.

McGraw-Hill Connect Engineering. McGraw-Hill Connect Engineering is a web-based assignment and assessment platform that gives students the means to better connect with their coursework, with their instructors, and with the important concepts that they will need to know for success now and in the future. With Connect Engineering, instructors can deliver assignments, quizzes and tests easily online. Connect Engineering is available at www.mhhe.com/pgc.

Problem Answers. For the use of students, answers to most even-numbered problems are posted as a freely downloadable PDF file at www.mhhe.com/pgc. In addition to final numerical or symbolic answers, selected problems have more extensive information such as free body diagrams and/or shear and moment diagrams.

Solutions Manual. The Solutions Manual that accompanies the first edition features typeset solutions to the homework problems. The Solutions Manual is available on the password-protected instructor's website at www.mhhe.com/pgc.

VitalSource. VitalSource is a downloadable eBook. Students that choose the VitalSource eBook can save up to 45% off the cost of the print book, reduce their impact on the environment, and access powerful digital learning tools. Students can share notes with others, customize the layout of the eBook, and quickly search their entire eBook library for key concepts. Students can also print sections of the book for maximum portability. To learn more about VitalSource options, contact your sales representative or visit www.vitalsource.com.

CourseSmart. This text is offered through CourseSmart for both instructors and students. CourseSmart is an online browser where students can purchase access to this and other McGraw-Hill textbooks in a digital format. Through their browser, students can access the complete text online at almost half the cost of a traditional text. Purchasing the eTextbook also allows students to take advantage of CourseSmart's web tools for learning, which include full text search, notes and highlighting, and e-mail tools for sharing notes among classmates. To learn more about CourseSmart options, contact your sales representative or visit www.coursesmart.com.

Hands-on Mechanics. Hands-on Mechanics is a website designed for instructors who are interested in incorporating three dimensional, hands-on teaching aids into their lectures. Developed through a partnership between the McGraw-Hill Engineering Team and the Department of Civil and Mechanical Engineering at the United States Military Academy at West Point, this website not only provides detailed instructions on how to build 3-D teaching tools using materials found in any lab or local hardware store but also provides a community where educators can share ideas, trade best practices, and submit their own demonstrations for posting on the site. Visit www.handsonmechanics.com.

Custom Publishing. Did you know that you can design your own text using any McGraw-Hill text and your personal materials to create a custom product that correlates specifically to your syllabus and course goals? Contact your McGraw-Hill sales representative to learn more about this option.

Engineering Mechanics:

Statics

1 Introduction

In *statics* we study the equilibrium of bodies under the action of forces that are applied to them. Our goal is to provide an introduction to the science, skill, and art involved in modeling and designing real life mechanical systems. We begin the study of statics with an overview of the relevant history of the subject. In subsequent sections and chapters, we cover those elements of physics and mathematics (especially vectors) needed to analyze the equilibrium of particles and rigid bodies. Throughout the book are discussions and applications of engineering design.

1.1 Engineering and Statics

Engineers design structures, machines, processes, and much more for the benefit of humankind. In the process of doing this, an engineer must answer questions such as "Is it strong enough?" "Will it last long enough?" and "Is it safe enough?" To answer these questions requires the ability to quantify important phenomena present in the design or system at hand, and to compare these measures with known criteria for what is acceptable and what is not. To do this requires an engineer to have thorough knowledge of science, mathematics, and computational tools, and the creativity to exploit the laws of nature to develop new designs. Central to all of this is the ability to idealize real life problems with mathematical models that capture the essential science of the problem, yet are tractable enough to be analyzed. Proficiency in doing this is a characteristic that sets engineering apart from the pure sciences.

In most engineering disciplines, understanding the response of materials or objects subjected to forces is important, and the fundamental science concepts governing such response are known as *Newtonian physics*.* This book examines applications of this topic to engineering problems under the special

* When the velocity of an object is close to the speed of light, relativistic physics is required.

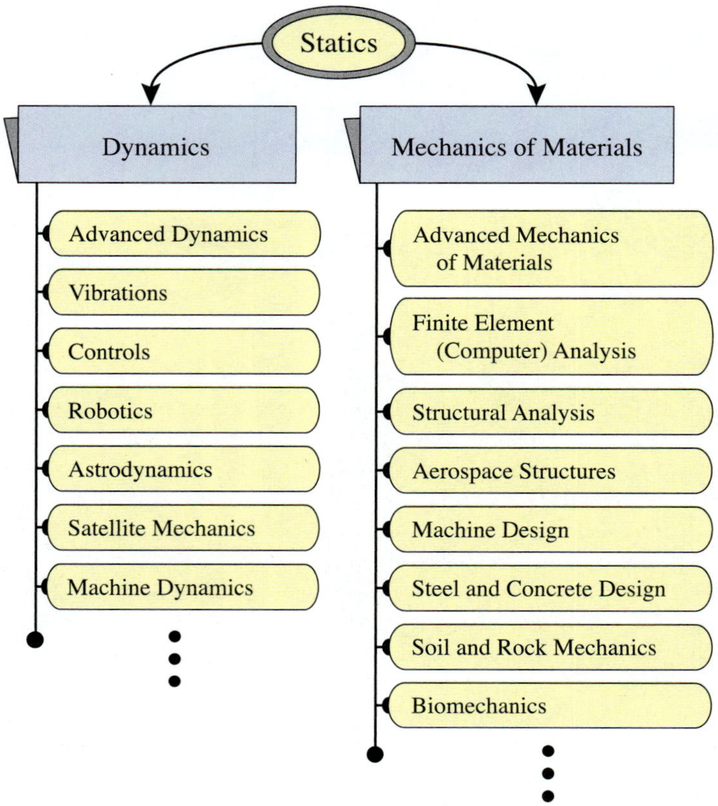

Figure 1.1. Hierarchy of subject matter and courses studied by many engineering students. Courses in statics, dynamics, and mechanics of materials provide fundamental concepts and a basis for more advanced study. Many subjects, such as vibrations and finite element analysis, draw heavily on concepts from both dynamics and mechanics of materials.

circumstances in which a system is in force equilibrium, and this body of material is called *statics*. As such, statics is usually the first engineering course that students take. Statics is an important subject in its own right, and develops essential groundwork for more advanced study.

If you have read this far, then we presume you are embarking on a study of statics, using this book as an aid. Figure 1.1 shows a hierarchy of subjects, many of which you are likely to study en route to an education in engineering. Following a course in statics are introductory courses in *dynamics* and *mechanics of materials*. Dynamics studies the motion of particles and bodies subjected to forces that are not in equilibrium. Mechanics of materials introduces models for material behavior and methods for determining stresses and deformations in structures. The concepts learned in these three basic courses are used on a daily basis by almost all engineers who are concerned with the mechanical response of structures and materials! Following these basic courses are a wide variety of advanced subjects such as vibrations, stress analysis, robotics, finite element analysis, machine design, design of steel and concrete structures, and so on.

The purpose of this book is to provide you with a solid and comprehensive education in statics. Very often, when engineering problems are boiled down to their essential elements, they are remarkably simple to analyze. In fact, throughout most of this book, the mathematics needed to analyze problems is straightforward. The bigger challenge usually lies in the idealization of a real life problem by a model, and we hope this book helps you cultivate your ability to do this.

1.2 A Brief History of Statics*

The history of statics is not a distinct subject, as it is closely intertwined with the development of dynamics and mechanics of materials. Early scientists and engineers were commonly called *philosophers*, and their noble undertaking was to use thoughtful reasoning to provide explanations for natural phenomena. Much of their focus was on understanding and describing equilibrium of objects and motion of celestial bodies. With few exceptions, their studies had to yield results that were intrinsically beautiful and/or compatible with the dominant religion of the time and location. What follows is a short historical survey of the major figures who profoundly influenced the development of key aspects of mechanics that are especially significant to statics.

For centuries, philosophers studied the equilibrium and motion of bodies with less than full understanding, and sometimes incorrect understanding. Notable early contributors include:

- Aristotle (384–322 B.C.) wrote about science, politics, economics, and biology, and he proposed what is often called a "physics of common sense." He studied levers and although he attributed their efficiency to the "magical" properties of the circle, he understood some basic concepts of the moment of a force and its effect on equilibrium. He classified objects as being either light or heavy, and he believed that light objects fall slower than heavy objects. He recognized that objects can move in directions other than up or down and said that such motion is contrary to the natural motion of the body and that some force must continuously act on the body for it to move this way. Most importantly, he said that the natural state of objects is for them to be at rest.

- Archimedes (287–212 B.C.) postulated several axioms based on experimental observations of the equilibrium of levers, and using these, he proved several propositions. His work shows further understanding of the effects of the moment of a force on equilibrium. Archimedes is perhaps best known for his pioneering work on hydrostatic fluid mechanics where one of his discoveries is that a body that floats in fluid will displace a volume of fluid whose weight is equal to that of the body. Recently, evidence has been found that he discovered some elementary concepts of calculus.

* This history is culled from the excellent works of C. Truesdell, *Essays in the History of Mechanics*, Springer-Verlag, Berlin, 1968; I. Bernard Cohen, *The Birth of a New Physics*, revised and updated edition, W. W. Norton & Company, New York, 1985; R. Dugas, *A History of Mechanics*, Dover, Mineola, NY, 1988; and James H. Williams, Jr., *Fundamentals of Applied Dynamics*, John Wiley & Sons, New York, 1996.

Interesting Fact

Early structural design codes. While most of our discussion focuses on accomplishments of philosophers, there were also significant accomplishments in the development of structural design codes over a period of thousands of years. Some of these include the ancient books of Ezekiel and Vitruvius and the secret books of the medieval masonic lodges. Additional history is given in J. Heyman, "Truesdell and the History of the Theory of Structures," a chapter in *Essays on the History of Mechanics*, edited by A. Becchi, M. Corradi, F. Foce, and O. Pedemonte, Birkhauser, Boston, 2003. These codes were largely empirical rules of proportion that provided for efficient design and construction of masonry structures. The great Greek temples, Roman aqueducts, and Gothic cathedrals are testament to their effectiveness. While the writers of these codes were not philosophers, their engineering accomplishments were impressive.

The Parthenon in Athens, Greece, was completed in 438 B.C. and is an example of early column and beam masonry construction.

- Leonardo da Vinci (1452–1519) had bold imagination and tackled a wide variety of problems. He correctly understood the moment of a force and used the terminology *arm of the potential lever* to describe what we today call the *moment arm*. While his conclusions were wrong, he studied the equilibrium of a body supported by two strings. He also conducted experiments on the strength of structural materials.

Following the progress of these and many other early philosophers came the work of Galileo and Newton. With their work came rapid progress in achieving the essential elements of a theory for the motion of bodies, and their accomplishments represent the most important milestone in the history of mechanics until the work of Einstein. The contributions of Galileo and Newton are discussed in some detail in the remainder of this section.

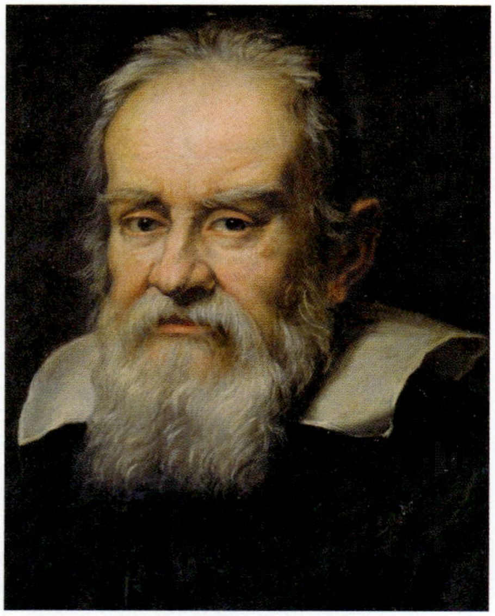

Figure 1.2
A portrait of Galileo painted in 1636 by Justus Sustermans.

Galileo Galilei (1564–1642)

Galileo Galilei (1564–1642) had a strong interest in mathematics, mechanics, astronomy, heat, and magnetism. He made important contributions throughout his life, despite persecution from the church for his support of the Copernican theory that the Earth was not the center of the universe. One of his most important contributions was his thought experiment in which he concluded that a body in its natural state of motion has *constant velocity*. Galileo discovered the correct law for freely falling bodies; that is, the distance traveled by a body is proportional to the square of time. He also concluded that two bodies of different weight would fall at the same rate and that any differences are due to air resistance. Galileo developed a theory (with some minor errors) for the strength of beams, such as that shown in Fig. 1.3. He was the first to use the concept of stress as a fundamental measure of the loading a material supports, and he is viewed as the father of mechanics of materials. He also discovered that the strength of structures does not scale linearly; that is, if the dimensions of a beam are doubled, the load the beam can support does not double. He speculated that it is for this reason that trees, animals, and so on have natural limits to the size they may have before they would fail under their own weight. More importantly, his work showed that newer, larger structures could not necessarily be built by simply scaling the dimensions of smaller structures that were successfully built.

Figure 1.3
A sketch from Galileo's last book *Discourses on Two New Sciences*, published in 1638, where he studies the strength of beams, among several other topics.

Isaac Newton (1643*–1727)

Newton was one of the greatest scientists of all time. He made important contributions to optics, astronomy, mathematics, and mechanics, and his collection of three books entitled *Philosophiæ Naturalis Principia Mathematica*, or *Principia* as they are generally known, which were published in 1687, is considered by many to be the greatest collection of scientific books ever written.

In the *Principia*, Newton analyzed the motion of bodies in "resisting" and "non-resisting media." He applied his results to orbiting bodies, projectiles, pendula, and free fall near the Earth. By comparing his "law of centrifugal force" with Kepler's third law of planetary motion, Newton further demonstrated that the planets were attracted to the Sun by a force varying as the

* This birth date is according to the Gregorian, or "modern," calendar. According to the older Julian calendar, which was used in England at that time, Newton's birth was in 1642.

inverse square of the distance, and he generalized that all heavenly bodies mutually attract one another in the same way. In the first book of the *Principia* Newton develops his three laws of motion; in the second book he develops some concepts in fluid mechanics, waves, and other areas of physics; and in the third book he presents his law of universal gravitation. His contributions in the first and third books are especially significant to statics and dynamics.

Newton's *Principia* was the final brick in the *foundation* of the laws that govern the motion of bodies. We say *foundation* because it took the work of Daniel Bernoulli (1700–1782), Johann Bernoulli (1667–1748), Jean le Rond d'Alembert (1717–1783) Joseph-Louis Lagrange (1736–1813), and Leonhard Euler (1707–1783) to clarify, refine, and advance the theory of dynamics into the form used today. Euler's contributions are especially notable since he used Newton's work to develop the theory for rigid body dynamics.* Newton's work, along with Galileo's, also provided the foundation for the theory of mechanical behavior of deformable bodies, which is more commonly called mechanics of materials. However, it took the work of Charles-Augustin Coulomb (1736–1806), Claude Louis Marie Henri Navier (1785–1857), and Augustin Cauchy (1789–1857) to further refine the concept of stress into the form used today; the work of Robert Hooke (1635–1703) and Thomas Young (1773–1829) to develop a theory for elastic deformation of materials; and the work of Leonhard Euler (1707–1783) to consider the deformations of a structure (an elastic strip in particular).†

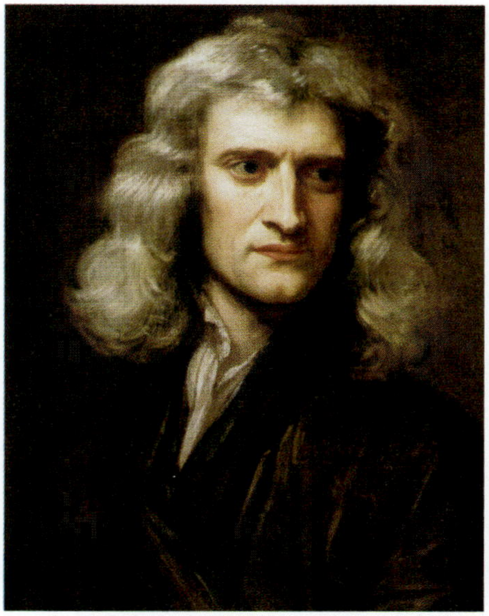

Figure 1.4
A portrait of Newton painted in 1689 by Sir Godfrey Kneller, which is owned by the 10th Earl of Portsmouth. It shows Newton before he went to London to take charge of the Royal Mint and when he was at his scientific peak.

1.3 Fundamental Principles

Space and time. Most likely you already have a good intuitive understanding of the concepts of space and time. In fact, to refine concepts of space and time is not easy and may not provide the clarification we would like. *Space* is the collection of all positions in our universe that a point may occupy. The location of a point is usually described using a coordinate system where measurements are made from some reference position using the coordinate system's reference directions. While selection of a reference position and directions is arbitrary, it is usually based on convenience. Because space is three-dimensional, three pieces of information, called *coordinates*, are required to locate a point in space. Most often we will use a rectangular Cartesian coordinate system where the distances to a point are measured in three orthogonal directions from a reference location. Other coordinate systems, such as spherical and cylindrical coordinates (and polar coordinates in two dimensions), are sometimes more convenient. All engineering problems are three-dimensional, but often we will be able to idealize a problem as being two-dimensional, or one-dimensional. *Time* provides a measure of when an event, or sequence of events, occurs.

* Additional comments on the history of mechanics as it pertains to dynamics are given in G. L. Gray, F. Costanzo, and M. E. Plesha, *Engineering Mechanics: Dynamics*, McGraw-Hill, New York, 2009.

† Additional comments on the history of mechanics as it pertains to mechanics of materials are given in M. Vable, *Mechanics of Materials*, Oxford University Press, New York, 2002.

Mass and force. *Mass* is the amount of matter, or material, in an object. *Force* is an agency that is capable of producing motion of an object. Forces can arise from contact or interaction between objects, from gravitational attraction, from magnetic attraction, and so on. As discussed in Section 1.5, interpretation and quantification of mass and force should be viewed as being related by Newton's second law of motion. Force is discussed further in Section 1.4.

Particle. A *particle* is an object whose mass is concentrated at a point. For this reason, a particle is also called a *point mass*, and it is said to have zero volume. An important consequence of this definition is that the notion of rotational motion of a particle is meaningless. Clearly there are no true particles in nature, but under the proper circumstances it is possible to idealize real life objects as particles. Objects that are small compared to other objects and/or dimensions in a problem can often be idealized as particles. For example, to determine the orbit of a satellite around the Earth, it is probably reasonable to idealize the satellite as a particle. Objects do not necessarily need to be small to be accurately idealized as particles. For example, for the satellite orbiting Earth, the Earth is not small, but for many purposes the Earth can also be idealized as a particle.

Body and rigid body. A *body* has mass and occupies a volume of space. In nature, all bodies are deformable. That is, when a body is subjected to forces, the distances between points in the body may change. A *rigid body* is a body that is not deformable, and hence the distance between any two points in the body never changes. There are no true rigid bodies in nature, but very often we may idealize an object to be a rigid body, and this provides considerable simplification because the intricate details of how the body deforms do not need to be accounted for in an analysis. Furthermore, in statics we will be able to make precise statements about the behavior of rigid bodies, and will establish methods of analysis that are exact.

Scalars and vectors. A *scalar* is a quantity that is completely characterized by a single number. For example, temperature, length, and density are scalars. In this book, scalars are denoted by italic symbols, such as s. A *vector* is an entity that has both size (or magnitude) and direction. Much will be said about vectors in Chapter 2, but basic notions of vectors will be useful immediately. Statements such as "my apartment is 1 mile northeast of Engineering Hall" and "I'm walking north at 3 km/h" are statements of vector quantities. In the first example, position of one location relative to another is stated, while in the second example velocity is stated. In both examples, commonly used reference directions of north and east are employed. Vectors are immensely useful for describing many entities in mechanics. Vectors offer compact representation and easy manipulation, and they can be transformed. That is, if a vector is known referred to one set of coordinate directions, then using established rules for transformation, the vector is known in any other set of coordinate directions. In this book, vectors are denoted by placing an arrow above the symbol for the vector, such as $\vec{v}$.

> **Concept Alert**
>
> **Vectors.** A *vector* is an entity that has both size and direction. Vectors are immensely useful in mechanics, and the ability to use vectors to represent force, position, and other entities is essential.

Position, velocity, and acceleration. Position, velocity, and acceleration are all examples of vectors. If we consider a particle that has *position* $\vec{r}$ relative to some location, then the *velocity* of the particle is the time rate of change of its position

$$\vec{v} = d\vec{r}/dt, \tag{1.1}$$

where d/dt denotes the derivative with respect to time.* Similarly, the *acceleration* is the time rate of change of velocity

$$\vec{a} = d\vec{v}/dt. \tag{1.2}$$

Since statics is concerned with situations where $\vec{a} = \vec{0}$, our discussion of Eqs. (1.1) and (1.2) will be brief. If a particle's acceleration is zero, then integration of Eq. (1.2) shows the particle has constant velocity, which may be zero or nonzero. If the velocity is zero, then Eq. (1.1) shows the particle's position does not change, while if the velocity is not zero, integration of Eq. (1.2) shows the particle's position changes as a linear function of time. If the acceleration is not zero, then the particle will move with velocity and position that change with time.

Newton's laws of motion

Inspired by the work of Galileo and others before him, Newton postulated his three laws of motion in 1687:

First Law. A particle remains at rest, or continues to move in a straight line with uniform velocity, if there is no unbalanced force acting on it.

Second Law. The acceleration of a particle is proportional to the resultant force acting on the particle and is in the direction of this force. The mathematical statement of this law† is

$$\boxed{\vec{F} = m\vec{a},} \tag{1.3}$$

where $\vec{F}$ is the resultant force acting on the particle, $\vec{a}$ is the acceleration of the particle, and the constant of proportionality is the mass of the particle m. In Eq. (1.3), $\vec{F}$ and $\vec{a}$ are vectors, meaning they have both size (or magnitude) and direction. Vectors are discussed in detail in Chapter 2.

Third Law The forces of action and reaction between interacting bodies are equal in magnitude, opposite in direction, and collinear.

Concept Alert

Newton's second law. Newton's second law, $\vec{F} = m\vec{a}$, is the most important fundamental principle upon which statics, dynamics, and mechanics in general are based.

Newton's laws of motion, especially Eq. (1.3), are the basis of mechanics. They are postulates whose validity and accuracy have been borne out by countless experiments and applications for more than three centuries. Unfortunately, there is not a fundamental proof of their validity, and we must accept these as rules that nature follows. The first law was originally stated by Galileo. Of the three laws, only the second two are independent. In Eq. (1.3), we see that

* Equations (1.1) and (1.2) are valid regardless of how a vector might be represented. However, the details of how the time derivative is evaluated depend on the particular vector representation (e.g., Cartesian, spherical, etc.) that is used. Dynamics explores these details further.

† Actually, Newton stated his second law in a more general form as $\vec{F} = d(m\vec{v})/dt$, where $\vec{v}$ is the velocity of the particle and $d(m\vec{v})/dt$ denotes the time rate of change of the product $m\vec{v}$, which is called the *momentum* of the particle. When mass is constant, this equation specializes to Eq. (1.3). For problems in which mass is not constant, such as in the motion of a rocket that burns a substantial mass of fuel, the more general form of Newton's second law is required.

if the resultant force $\vec{F}$ acting on a particle is zero, then the acceleration of the particle is zero, and hence the particle may move with uniform velocity, which may be zero or nonzero in value. Hence, when there is no acceleration (i.e., $\vec{a} = \vec{0}$), the particle is said to be in *static equilibrium*. The third law will play an important role when drawing free body diagrams, which we will see are an essential aid for applying $\vec{F} = m\vec{a}$.

1.4 Force

Forces are of obvious importance to us. In statics, we are usually interested in how structures support the forces that are applied to them, and how to design structures so they can accomplish the goal of supporting forces. In dynamics, we are usually interested in the motions of objects that are caused by forces that are applied to them. In this section, we discuss force in some detail, examine some different types of forces, and discuss how forces are produced.

Simply stated, a *force* is any agency that is capable of producing an acceleration of an unsupported body.* While this definition may seem vague, it is comprehensive. All forces are produced from the interaction of two or more bodies (or collections of matter), and the interaction between the bodies can take several forms which gives rise to different ways that forces can be produced.

For many purposes, a force can be categorized as being either a *contact force* or a *field force*:

- **Contact force.** When two bodies touch, *contact forces* develop between them. In general, the contact forces are distributed over a finite area of contact, and hence, they are *distributed forces* with dimensions of force/area. If the bodies touch over only a small region, or if we replace the distributed force by an equivalent concentrated force as discussed in Chapter 7, then the contact forces are concentrated at a point. Contact forces are made up of two parts: a normal-direction force and a tangential-direction force, which is also called the *friction force*. Examples of contact forces include the forces between your feet and ground when you are standing, and the force applied by air to a building during a blowing wind.

- **Field force.** A force between bodies that acts through space is called a *field force*. Field forces act throughout the volume of an object and thus have dimensions of force/volume. Field forces are often called *body forces*. For many applications, we can represent a field force by a concentrated force that acts at a point. Examples of field forces include the weight of an object, the attractive force between the Earth and Moon, and the force of attraction between a magnet and an iron object.

Some examples of contact and field forces are given in Fig. 1.5.

Although the definition of contact forces given above is useful, more careful consideration of contact at an atomic length scale shows that contact forces are in fact a special case of a field force. As an atom from one surface comes

* Whether or not a particular body does accelerate depends upon the combined action of *all* forces that are applied to the body.

Interesting Fact

Measuring force. In addition to the capability of producing an acceleration of an unsupported body, a force causes a body to deform, or change shape. This suggests two ways to measure force. First, for an accelerating body with known mass m, by measuring the acceleration $\vec{a}$, we may then determine the force $\vec{F}$ applied to the body, using Newton's law $\vec{F} = m\vec{a}$. This approach is common in celestial mechanics and projectile motion, but it cannot be used for objects that are in static equilibrium.
A second approach that is more common for both static and dynamic applications is by measuring the *deformation* (i.e., shape change) that a force produces in an object whose behavior is known. An example is the handheld spring scale shown which is being used to weigh bananas.

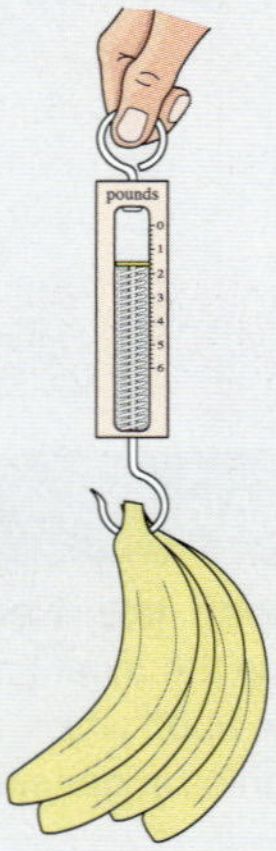

The weight of the bananas causes the spring's length to change, and because the spring's stiffness is known, the force the bananas apply to the scale can be determined. A brief historical discussion of mass and force measurements is given in J. C. Maxwell's notes on dynamics entitled *Matter and Motion*, Dover Publications, Inc., New York, 1991, the preface of which is dated 1877. A more contemporary discussion of force measurements (and measurements in general) is available from the *National Institute of Standards and Technology* (NIST) (see `http://www.nist.gov/`).

very close to an atom on the opposite surface, the atoms never touch one another, but rather they develop a repulsive field force that increases rapidly as the two atoms come closer. However, the range of distances over which these forces act is very small (on the order of atomic dimensions), and for macroscopic applications, our definition of contact forces is useful.

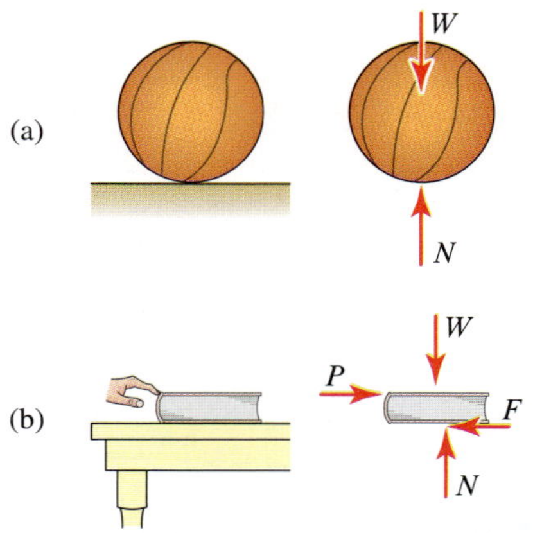

Figure 1.5
Examples of contact forces and field forces. (a) A basketball rests on a hard level surface. (b) A book is pushed across a table with your finger. In both examples, the field force is the weight W of the object, and the contact forces are the normal force N, the friction force F, and the force P applied by your finger to the book. For the basketball, contact occurs over a very small region, and it is reasonable to idealize this as a point. For the book, contact occurs over the entire surface of the book cover, but it is nonetheless possible to model the contact forces by concentrated forces acting at a point.

1.5 Units and Unit Conversions

Units are an essential part of any quantifiable measure. Newton's law $F = ma$, written here in scalar form, provides for the formulation of a consistent and unambiguous system of units. We will employ both U.S. Customary units, and SI units (International System*) as shown in Table 1.1. Each system has three *base units* and a fourth *derived unit*. In the U.S. Customary system, the base units measure force, length, and time, using lb, ft, and s, respectively, and the derived unit is obtained from the equation $m = F/a$ which gives the mass unit as $\text{lb}\cdot\text{s}^2/\text{ft}$, which is defined as 1 *slug*. In the SI system, the base units measure mass, length, and time, using kg, m, and s, respectively, and the derived unit is obtained from the equation $F = ma$ which gives the force unit as $\text{kg}\cdot\text{m/s}^2$, which is defined as 1 *newton*, N. For both systems, we may occasionally use different, but consistent, measures for some units. For example, we may use minutes rather than seconds, inches instead of feet, grams instead of kilograms, and so on. Nonetheless, the definitions of 1 newton and 1 slug are always as shown in Table 1.1.

Table 1.1. U.S. Customary and SI unit systems.

	System of Units	
Base Dimension	U.S. Customary	SI
force	pound (lb)	newton[a](N) $\equiv \text{kg}\cdot\text{m/s}^2$
mass	slug[a] $\equiv \text{lb}\cdot\text{s}^2/\text{ft}$	kilogram (kg)
length	foot (ft)	meter (m)
time	second (s)	second (s)

[a] Derived unit.

Common Pitfall

Weight and mass are different. It is unfortunately common for people, especially laypeople, to refer to weight using mass units. For example, when a person says, "I weigh 70 kg," the person really means "My mass is 70 kg." In this book, as well as throughout engineering, we must be precise with our nomenclature. Weights and forces will always be reported using appropriate force units, and masses will always be reported using appropriate mass units.

Dimensional homogeneity and unit conversions

Of course, the symbol "=" means that what is on the left-hand side of the symbol is the same as what is on the right-hand side. Hence, for an expression to be correct, it must be numerically correct and dimensionally correct. Normally this means that the left- and right-hand sides have the same numerical value and the same units.† All too often units are not carried along during a calculation, only to be incorrectly assumed at the end. Our strong recommendation is that you always use appropriate units in all equations. Such practice

Helpful Information

Dimensions versus units. *Dimensions* and *units* are different. Dimensions are a measurable extent of some kind, while units are used to measure a dimension. For example, length and time are both dimensions, and meter and second, respectively, are units used to measure these dimensions.

* SI has been adopted as the abbreviation for the French *Le Système International d'Unités*.

† A simple example of an exception to this is the equation 12 in. = 1 ft. Such equations play a key role in performing unit conversions.

helps avoid catastrophic blunders and provides a useful check on a solution, for if an equation is found to be dimensionally inconsistent, then an error has certainly been made.

Unit conversions are frequently needed, and are easily accomplished using conversion factors such as those found in Table 1.2 and rules of algebra. The basic idea is to multiply either or both sides of an equation by dimensionless factors of unity, where each factor of unity embodies an appropriate unit conversion. This description perhaps sounds vague, and the procedure is better illustrated by the examples that follow.

Interesting Fact

Abbreviation for inch. Notice in Table 1.2 that the abbreviation for inch is "in.", which contains a period. This is unusual, but is done because without the period, the abbreviation would also be the same as a word in the English language, and this might lead to confusion.

Table 1.2. Conversion factors between U.S. Customary and SI unit systems.

	U.S. Customary		SI
length	1 in.	=	0.0254 m (2.54 cm, 25.4 mm)[a]
	1 ft (12 in.)	=	0.3048 m[a]
	1 mi (5280 ft)	=	1.609 km
force	1 lb	=	4.448 N
	1 kip (1000 lb)	=	4.448 kN
mass	1 slug ($1\ \text{lb}\cdot\text{s}^2/\text{ft}$)	=	14.59 kg

[a] Exact.

Prefixes

Prefixes are a useful alternative to scientific notation for representing numbers that are very large or very small. Common prefixes and a summary of rules for use are given in Table 1.3.

Rules for Prefix Use

1. With few exceptions, use prefixes only in the numerator of unit combinations. One common exception is kg, which may appear in numerator or denominator.

2. Use a dot or dash to denote multiplication of units. For example, use N·m or N-m.

3. Exponentiation applies to both the unit and prefix. For example, $\text{mm}^2 = (\text{mm})^2$.

4. When the number of digits on either side of a decimal point exceeds 4, it is common to group the digits into groups of 3, with the groups separated by commas or thin spaces. Since many countries use a comma to represent a decimal point, the thin space is sometimes preferable. For example, 1234.0 could be written as is, and 12345.0 should be written as 12,345.0 or as 12 345.0.

While prefixes can often be incorporated in an expression by inspection, the rules for accomplishing this are identical to those for performing unit transformations, as shown in the examples of this section.

Table 1.3. Common prefixes used in the SI unit systems.

Multiplication Factor		Prefix	Symbol
1 000 000 000 000 000 000 000 000	10^{24}	yotta	Y
1 000 000 000 000 000 000 000	10^{21}	zetta	Z
1 000 000 000 000 000 000	10^{18}	exa	E
1 000 000 000 000 000	10^{15}	peta	P
1 000 000 000 000	10^{12}	tera	T
1 000 000 000	10^{9}	giga	G
1 000 000	10^{6}	mega	M
1 000	10^{3}	kilo	k
100	10^{2}	hecto	h
10	10^{1}	deka	da
0.1	10^{-1}	deci	d
0.01	10^{-2}	centi	c
0.001	10^{-3}	milli	m
0.000 001	10^{-6}	micro	μ
0.000 000 001	10^{-9}	nano	n
0.000 000 000 001	10^{-12}	pico	p
0.000 000 000 000 001	10^{-15}	femto	f
0.000 000 000 000 000 001	10^{-18}	atto	a
0.000 000 000 000 000 000 001	10^{-21}	zepto	z
0.000 000 000 000 000 000 000 001	10^{-24}	yocto	y

Angular measure

Angles are usually measured using either radians (rad) or degrees (°). The radian measure of the angle θ shown in Fig. 1.6 is defined to be the ratio of the circumference c of a circular arc to the radius r of the arc. Thus, as seen in the examples of Fig. 1.7, the angle for one-quarter of a circular arc is $\theta = \pi/2$ rad (or 1.571 rad), and for a full circular arc the angle is $\theta = 2\pi$ rad (or 6.283 rad). Degree measure arbitrarily chooses the angle for a full circular arc to be 360°, in which case 1° is the angle of an arc that is 1/360 parts of a full circle. Thus, the transformation between radian and degree measure is

$$2\pi \text{ rad} = 360°. \tag{1.4}$$

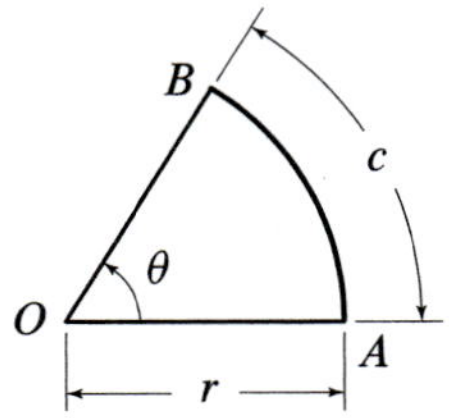

$\theta = \dfrac{c}{r}$

definition of radian measure

Figure 1.6
Definition of radian measure for angles.

Transformations are carried out using the procedures described in this section. For example, to convert the angle $\theta = 12°$ to radian measure, we use Eq. (1.4) to write

$$\theta = (12°)\,\frac{2\pi \text{ rad}}{360°} = 0.209 \text{ rad}. \tag{1.5}$$

Radians are a measure of angle that naturally arises throughout mathematics and science, and most equations derived from fundamental principles use radian measure. Nonetheless, degree measure has intuitive appeal and is used widely.

When writing angles, we will always label these as radians or degrees. However, radians and degrees are not units in the same way as those discussed

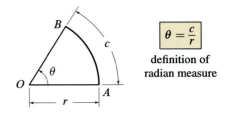

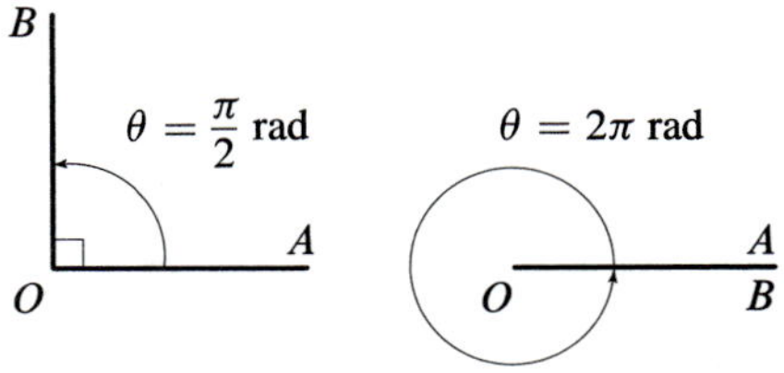

Figure 1.7
Examples of angles measured in radians.

earlier and, while perhaps puzzling, both of these measures are dimensionless. This can be seen by examining the definition of radian measure shown in Fig. 1.6, namely $\theta = c/r$. With c and r having the same units of length, angle θ is clearly dimensionless. Thus, radians and degrees are not really units, but rather are statements of the convention used for measuring an angle. Nonetheless, for practical purposes we may consider these to be units, and we will transform them using our usual procedures. Further, if we derive an expression that we expect to be dimensionless and we discover it has units of radians or degrees, then we should not necessarily be alarmed.

Accuracy of calculations

The accuracy of answers obtained for a particular problem is only as precise as the coarsest, or least accurate, information used in the analysis. For example, consider the numbers 1.23 and 45.67. By writing these numbers using three and four digits, respectively, the implication is that they are known to three and four significant digits of accuracy. The exact product of these numbers is 56.1741. But it is wrong to imply that the product is known to six-digit accuracy. Rather, it is appropriate to report the product to the same number of significant digits as the least accurate piece of information used. Hence, we would round the exact product to three significant digits and report the answer as 56.2.

The use of number of digits to imply precision, as suggested above, is ambiguous. Consider the number 6000; it is not clear if this number is known to one, two, three, or four significant digits. To embody accuracy information in numbers, it is probably best to use scientific notation. Thus, for example, if the number 6000 were known to three significant digits, we could write 6.00×10^3 with the convention that the number of digits used indicates the accuracy of the number. In this book, we will use a more pragmatic approach and will generally assume that data is known to three significant digits. When you are performing computations, it is good practice to carry a few extra digits of accuracy for intermediate computations; and if an electronic device such as a calculator or computer is used, then you certainly want to use the full precision that is available. Nonetheless, final answers should be interpreted as having precision that is commensurate with the precision of data used. The margin note on this page describes the convention for accuracy of numbers that is used for the calculations carried out in this book.

Helpful Information

Accuracy of numbers in calculations. Throughout this book, we will generally assume that the data given for problems is known to three significant digits of accuracy. When calculations are performed, such as in example problems, all intermediate results are stored in the memory of a calculator or computer, using the full precision these machines offer. However, when these intermediate results are reported in this book, they are usually rounded to four significant digits. Final answers are usually reported with three significant digits. If you verify the calculations described in this book using the rounded numbers that are reported, you may occasionally calculate results that are slightly different from those shown.

EXAMPLE 1.1 *Unit Conversion*

Convert the speed $s = 5.12\,\text{ft/s}$ to the SI units m/s and km/h.

SOLUTION

Road Map Starting with $s = 5.12\,\text{ft/s}$, we will multiply the right-hand side of this expression by appropriate conversion factors to achieve the desired unit conversion.

Governing Equations & Computation Referring to Table 1.2, we find

$$1\,\text{ft} = 0.3048\,\text{m}. \tag{1}$$

Dividing both sides of Eq. (1) by 1 ft provides the middle term of the following equation

$$1 = \frac{0.3048\,\text{m}}{1\,\text{ft}} = \frac{1\,\text{ft}}{0.3048\,\text{m}}, \tag{2}$$

whereas dividing both sides of Eq. (1) by 0.3048 m provides the last term of the above equation. Regardless of which form of Eq. (2) is used, the left-hand side is the number 1, with no units. The form of Eq. (2) that is used in a particular unit transformation will depend on what units need to be replaced, or canceled. To accomplish our unit conversion for $s = 5.12\,\text{ft/s}$, we write

$$s = 5.12\,\frac{\text{ft}}{\text{s}}\,(1) = 5.12\,\frac{\cancel{\text{ft}}}{\text{s}}\,\underbrace{\frac{0.3048\,\text{m}}{1\,\cancel{\text{ft}}}}_{=1} = \boxed{1.56\,\frac{\text{m}}{\text{s}}}. \tag{3}$$

In writing Eq. (3), we first multiply 5.12 ft/s by the dimensionless number 1; this changes neither the value nor the units of s. Since we want to eliminate the foot unit, we elect to substitute for the dimensionless number 1 using the first form of transformation in Eq. (2), namely $1 = 0.3048\,\text{m}/1\,\text{ft}$. Finally, we cancel the foot unit in the numerator and denominator to obtain the speed $s = 1.56\,\text{m/s}$ in the desired SI units.

To obtain s in units of km/h, we continue with Eq. (3) and perform the following transformations:

$$s = 1.56\,\frac{\cancel{\text{m}}}{\cancel{\text{s}}}\,\underbrace{\frac{\text{km}}{10^3\,\cancel{\text{m}}}}_{=1}\,\underbrace{\frac{60\,\cancel{\text{s}}}{\cancel{\text{min}}}}_{=1}\,\underbrace{\frac{60\,\cancel{\text{min}}}{\text{h}}}_{=1} = \boxed{5.62\,\frac{\text{km}}{\text{h}}}. \tag{4}$$

Discussion & Verification When possible, answers should be checked to verify that they are reasonable. For example, starting with $s = 5.12\,\text{ft/s}$, the result in Eq. (3) is reasonable since a meter is about 3 feet.

Common Pitfall

Omitting units in equations. The most serious mistake made when performing unit conversions (as well as when writing equations in general) is to omit units in equations. Although writing units in equations takes a few moments longer, doing so will help avoid the errors that are sure to result if you do not make this a practice.

EXAMPLE 1.2 *Unit Conversion*

The universal gravitational constant, whose physical significance we discuss later in this chapter, is $G = 66.74 \times 10^{-12}\ \mathrm{m^3/(kg{\cdot}s^2)}$. Express G in base U.S. Customary units.

SOLUTION

Road Map Perhaps the most straightforward solution strategy is to first convert mass in kilograms to mass in slugs and then replace the unit of slug with its fundamental definition.

Governing Equations & Computation Beginning our calculation with $G = 66.74\times 10^{-12}\ \mathrm{m^3/(kg{\cdot}s^2)}$, we multiply the right-hand side by appropriate conversion factors to achieve the desired unit conversion. Thus,

$$G = 66.74 \times 10^{-12}\,\frac{\mathrm{m^3}}{\mathrm{kg{\cdot}s^2}}\underbrace{\left(\frac{14.59\ \mathrm{kg}}{\mathrm{slug}}\right)}_{=1}\underbrace{\left(\frac{\mathrm{ft}}{0.3048\ \mathrm{m}}\right)^3}_{=(1)^3}\underbrace{\left(\frac{\mathrm{slug}}{\mathrm{lb{\cdot}s^2/ft}}\right)}_{=1}$$

$$= \boxed{34.39 \times 10^{-9}\,\frac{\mathrm{ft^4}}{\mathrm{lb{\cdot}s^4}}}. \tag{1}$$

Alternatively, we could also perform the unit transformation by first introducing the SI force measure newton, followed by conversion to force measure in pounds, followed by conversion of length. Thus,

$$G = 66.74 \times 10^{-12}\,\frac{\mathrm{m^3}}{\mathrm{kg{\cdot}s^2}}\underbrace{\left(\frac{\mathrm{kg{\cdot}m/s^2}}{\mathrm{N}}\right)}_{=1}\underbrace{\left(\frac{4.448\ \mathrm{N}}{\mathrm{lb}}\right)}_{=1}\underbrace{\left(\frac{\mathrm{ft}}{0.3048\ \mathrm{m}}\right)^4}_{=(1)^4}$$

$$= \boxed{34.39 \times 10^{-9}\,\frac{\mathrm{ft^4}}{\mathrm{lb{\cdot}s^4}}}. \tag{2}$$

Discussion & Verification Because of the complexity of the unit combinations for G, it is not possible to use inspection to verify that Eqs. (1) and (2) are reasonable. Rather, the accuracy of our results relies solely on the use of appropriate conversion factors and accurate cancellation of units in Eqs. (1) and (2). For this reason, it is essential that you carry units throughout all equations.

1.6 Newton's Law of Gravitation

Because weight produced by gravity is so omnipresent, it is worthwhile to examine the source of such forces closely, and to understand the limitations of common expressions such as $W = mg$ where m is an object's mass, g is acceleration due to gravity, and W is the object's weight. In 1666, Newton developed his *law of universal gravitational attraction* as

$$F = G\frac{m_1 m_2}{r^2} \tag{1.6}$$

where

m_1, m_2 = masses of particles 1 and 2;
r = distance between the particles;
G = universal gravitational constant, found to be approximately $66.74\times10^{-12}\ \mathrm{m^3/(kg{\cdot}s^2)}$;
F = force of attraction between two particles.

Concept Alert

Force due to gravity. Force due to gravitational attraction between two objects is a vector, hence it has both magnitude and direction. Equation (1.6) gives the magnitude, and the direction is along a line connecting the centers of gravity of the two objects.

It has been widely reported that Newton's inspiration for this law was the motion of an apple falling from a tree, but he also recognized that the same law should apply to the attraction of celestial bodies to one another. Although Newton postulated the law in 1666, it was not until 1687 that he published his ideas in the *Principia*. This delay was due in part to the need to prove that objects such as the Earth (if assumed to be spherical and uniform) could be treated as a point mass for gravitational effects on neighboring particles, and in the course of proving this he developed calculus.* The first accurate measurement of G was by Lord Cavendish in 1798, and this value has been refined by more careful experiments over the last two centuries, leading to the value reported here. The law of universal gravitational attraction is a postulate, and as with Newton's three laws of motion, we must accept this as a rule that nature follows without a fundamental proof of its validity.

For the vast majority of applications on Earth, Eq. (1.6) takes the simple and convenient form $W = mg$, as follows. Let m_1 in Eq. (1.6) denote the mass m of an object, and let m_2 denote the mass of the Earth (with an approximate value $m_{\text{Earth}} = 5.9736 \times 10^{24}$ kg). If the object is on or near the surface of the Earth, then its position r is about the same as the mean radius of the Earth (with an approximate value 6.371×10^6 m). The force F in Eq. (1.6) is then called the *weight* W of the object, and Eq. (1.6) can be rewritten as

$$W = mg \quad \text{where} \quad g \equiv Gm_{\text{Earth}}/r^2. \tag{1.7}$$

From Eq. (1.7), we see that g is not a constant because it depends on the value of r. However, for the vast majority of applications where objects are near

* Calculus was also developed independently by Gottfried Wilhelm Leibniz (1646–1716), and he and Newton had a long-standing dispute over who was the true originator. The historical records show that while Newton was the first to discover calculus (about 10 years before Leibniz), Leibniz was the first to publish his discovery (about 15 years before Newton). In some respects, Leibniz won since it is his superior notation that we use in calculus today.

the surface of the Earth, effects of small changes in r are negligible, and the commonly used values for *acceleration due to gravity* are

$$g = 9.81\ \text{m/s}^2 = 32.2\ \text{ft/s}^2. \tag{1.8}$$

Note that if the values reported above for G, Earth's mass, and Earth's mean radius are used in Eq. (1.7), the value of g produced is slightly different than $9.81\ \text{m/s}^2$. The difference between the accepted value of g and the theoretically computed value provided by Eq. (1.7) has several sources, including that the Earth is not perfectly spherical and does not have uniform mass distribution, and the effects of centripetal acceleration due to the Earth's rotation are not accounted for. Because of these sources, the actual acceleration due to gravity is about 0.3% lower at the equator, and 0.3% higher at the poles, relative to the numbers given in Eq. (1.8) which are for a north or south latitude of 45° at sea level. In addition, there may be small local variations in acceleration due to gravity due to the effects of geology. Nonetheless, throughout this book we will use the standard values of g given in Eq. (1.8).

Helpful Information

Center of gravity. The *center of gravity* is the point through which the weight of a body, or a collection of bodies, may be considered to act. In figures, we will often denote the center of gravity by using the symbol. To illustrate, imagine a waiter at a restaurant brings you wine and pasta on a tray. Obviously, the waiter must position his hand so that the combined weight of the tray and everything on it is located over his hand.

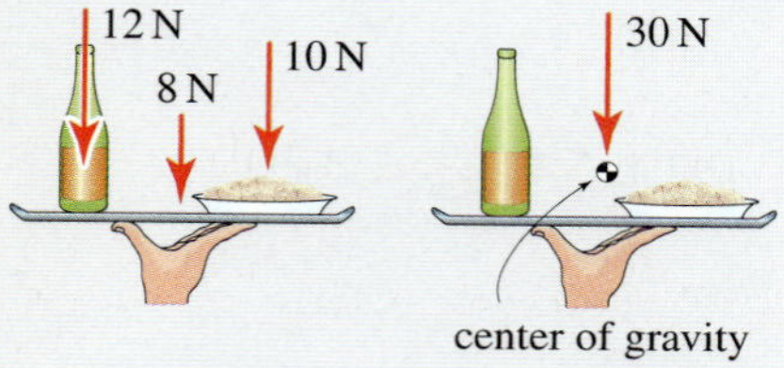

The weight of the wine (12 N), pasta (10 N), and tray (8 N) can be thought of as a single 30 N force acting through the center of gravity for the collection of objects. Center of gravity and how it is determined are discussed thoroughly in Chapter 7, where it is seen that the two force systems shown above are *equivalent force systems*. In the meantime, a working knowledge of this definition will be useful.

Relationship between specific weight and density

The specific weights and densities of some common materials are given in Table 1.4. When using U.S. Customary units, it is common to characterize the

Table 1.4. Specific weight and density for some common materials. Except for water and ice, numbers reported are generally at 20 °C. Data may vary depending on composition, alloying, temperature, moisture content for wood, etc.

Material	Specific Weight γ (lb/ft^3)	Density ρ (kg/m^3)
iron (pure)	491	7860
iron (cast)	450 ± 15	7210 ± 240
aluminum (pure)	169	2710
aluminum (alloy)	170 ± 10	2710 ± 160
steel	490	7850
stainless steel	500	8010
brass	537 ± 8	8610 ± 130
titanium	280	4480
rubber	70 ± 10	1120 ± 160
nylon	70	1120
concrete	150	2400
rock (dry granite)	165	2640
cortical bone (adult)	119	1900
wood (dry Douglas fir)	32 ± 2	510 ± 30
water (fresh, 4 °C, 1 atm)	62.4	1000
ice	57	920
JP–4 jet fuel	48	770

density of materials using *specific weight* (sometimes also called *weight density*, or *unit weight*) which is defined to be the weight on Earth of a unit volume of material. For example, the specific weight of steel is $\gamma = 490\,\text{lb/ft}^3$ ($= 0.284\,\text{lb/in.}^3$). However, specific weight *is not* the same as density, although they are related. *Density* is defined to be the mass of a unit volume of material, and when SI units are used, it is most common to directly report a material's density. Thus, for steel, the density is $\rho = 7850\ \text{kg/m}^3$. These measures are related by Eq. (1.7) as follows. Imagine a certain volume V of material has weight (on Earth) W and mass m. Dividing Eq. (1.7) by volume V provides

$$\frac{W}{V} = \frac{m}{V} g. \tag{1.9}$$

In this expression, W/V is the definition of specific weight γ, and m/V is the definition of density ρ. Thus, Eq. (1.9) becomes

$$\boxed{\gamma = \rho g \quad \text{or} \quad \rho = \frac{\gamma}{g}.} \tag{1.10}$$

EXAMPLE 1.3 *Weight and Force of Mutual Attraction*

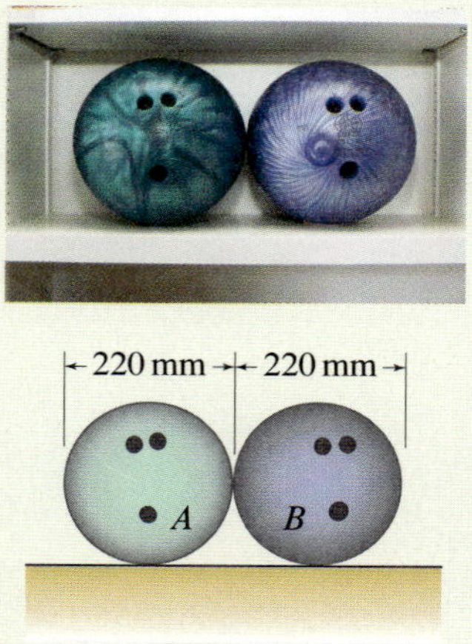

Figure 1

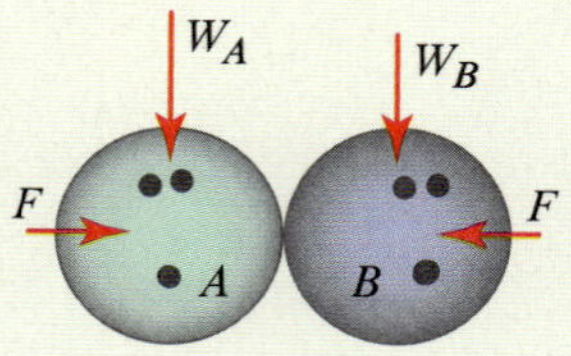

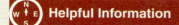

Figure 2
The weight of each ball and the force of mutual attraction are vectors with the directions shown. **Important Note:** The bowling balls are also subjected to other forces that are not shown (see the Helpful Information margin note below).

Two bowling balls resting on a shelf touch one another. The balls have 220 mm diameter and are made of plastic with density $\rho_A = 1170\ \text{kg/m}^3$ for ball A and $\rho_B = 980\ \text{kg/m}^3$ for ball B. Determine the weight of each ball and the force of mutual attraction, expressing both in SI units and U.S. Customary units.

SOLUTION

Road Map The forces to be determined are shown in Fig. 2. The weights of balls A and B are forces (vectors) with magnitudes W_A and W_B, respectively, and these forces act in the downward vertical direction. The force of mutual attraction between the two balls has magnitude F, with directions as shown in Fig. 2. Note that Newton's third law requires the force of mutual attraction between the two balls to have equal magnitude and opposite direction. We will assume both balls are uniform (i.e., the density is the same throughout each ball), and we will neglect the presence of the finger holes. We will first determine the mass of each ball. We will then determine the weight of each ball, using $W_A = m_A g$ and $W_B = m_B g$, and then the force of mutual attraction, using Newton's law of gravitational attraction.

Governing Equations & Computation The mass m_A of ball A is the product of the material's density ρ_A and the ball's volume V_A, and similarly for ball B. Thus

$$m_A = \rho_A V_A = \left(1170\,\frac{\text{kg}}{\text{m}^3}\right)\frac{4}{3}\pi\left(\frac{0.220\ \text{m}}{2}\right)^3 = 6.523\ \text{kg}, \tag{1}$$

$$m_B = \rho_B V_B = \left(980\,\frac{\text{kg}}{\text{m}^3}\right)\frac{4}{3}\pi\left(\frac{0.220\ \text{m}}{2}\right)^3 = 5.464\ \text{kg}. \tag{2}$$

The weight of each ball is

$$W_A = m_A g = (6.523\ \text{kg})\left(9.81\,\frac{\text{m}}{\text{s}^2}\right) = 64.0\,\frac{\text{kg}\cdot\text{m}}{\text{s}^2} = \boxed{64.0\ \text{N},} \tag{3}$$

$$W_B = m_B g = (5.464\ \text{kg})\left(9.81\,\frac{\text{m}}{\text{s}^2}\right) = 53.6\,\frac{\text{kg}\cdot\text{m}}{\text{s}^2} = \boxed{53.6\ \text{N}.} \tag{4}$$

In U.S. Customary units, $W_A = (64.0\ \text{N})(1\ \text{lb}/4.448\ \text{N}) = \boxed{14.4\ \text{lb}}$ and $W_B = (53.6\ \text{N})(1\ \text{lb}/4.448\ \text{N}) = \boxed{12.1\ \text{lb}.}$

The force of mutual attraction is given by Eq. (1.6) (with subscripts 1 and 2 replaced by A and B) as

$$F = G\,\frac{m_A m_B}{r^2} = \left(66.74\times10^{-12}\,\overbrace{\frac{\text{m}^3}{\text{kg}\cdot\text{s}^2}}^{\text{m}^2\cdot\text{m}}\right)\frac{(6.523\ \text{kg})(5.464\ \text{kg})}{(0.220\ \text{m})^2}$$

$$= 4.91\times10^{-8}\,\frac{\text{kg}\cdot\text{m}}{\text{s}^2} = \boxed{4.91\times10^{-8}\ \text{N}.} \tag{5}$$

In Eq. (5), $r = 0.220\ \text{m}$ is the distance between the center of each ball. In U.S. Customary units, $F = (4.91\times10^{-8}\ \text{N})(1\ \text{lb}/4.448\ \text{N}) = \boxed{1.10\times10^{-8}\ \text{lb}.}$

Discussion & Verification As you might have expected, the force of mutual attraction between the two balls is very small compared to the weight of the balls (9 orders of magnitude smaller). In developing models for engineering problems, the force of mutual attraction will usually be small compared to other forces, and when this is the case, it will be neglected.

> **Helpful Information**
>
> **Additional forces.** The balls shown in Fig. 2 are subjected to additional forces that are not shown. For example, the shelf applies a force to each ball, and there are probably contact forces between the two balls where they touch. Clearly, without these additional forces, the bowling balls could not be in static equilibrium. Chapter 3 will thoroughly discuss these additional forces and how they may be determined.

EXAMPLE 1.4 *Specific Weight and Density*

The specific weight of a particular aluminum alloy is $\gamma = 0.099\,\text{lb/in.}^3$. Determine the density of this alloy, and report this in U.S. Customary units.

SOLUTION

Road Map Beginning with weight per unit volume for an aluminum alloy, we will determine its mass per unit volume.

Governing Equations & Computation We use Eq. (1.10), with appropriate unit transformations

$$\rho = \frac{\gamma}{g} = \frac{0.099\,\text{lb/in.}^3}{32.2\,\text{ft/s}^2}\left(\frac{\text{ft}}{12\,\text{in.}}\right) = 2.562\times10^{-4}\,\frac{\text{lb}\cdot\text{s}^2}{\text{in.}^4}$$

$$= 2.562\times10^{-4}\,\underbrace{\frac{\text{lb}\cdot\text{s}^2}{\text{in.}^4}}_{\text{in.}\cdot\,\text{in.}^3}\left(\frac{\text{slug}}{\text{lb}\cdot\text{s}^2/\text{ft}}\right)\left(\frac{12\,\text{in.}}{\text{ft}}\right) = \boxed{3.07\times10^{-3}\,\frac{\text{slug}}{\text{in.}^3}}. \quad (1)$$

Discussion & Verification The first expression in Eq. (1) does not use the conventional U.S. Customary unit for mass, but is otherwise a perfectly acceptable and useful answer for the density of this aluminum alloy. The second expression in Eq. (1) incorporates the mass unit slug and provides the density in the expected form of mass per unit volume.

1.7 Failure

Among all of the goals confronting engineers when they design structures and machines, the most crucial goal is to develop designs that are as safe as possible. Unfortunately, despite all human efforts to meet this goal, sometimes we do not, and for reasons that are almost always unexpected, failure occurs. When failure occurs, we must learn from it so that our mistakes and/or lack of foresight is not repeated in the future.* In this section, some examples of engineering failures are highlighted.

Figure 1.8
Failure of the Tacoma Narrows bridge in Tacoma, Washington, in 1940, due to severe vibrations produced by a 42 mph wind.

Figure 1.9
Failure of the Escambia Bay bridge in Pensacola, Florida, during Hurricane Ivan in September 2004.

- **Tacoma Narrows bridge.** Only four months after its opening in 1940, the Tacoma Narrows suspension bridge in Washington collapsed violently due to severe vibrations produced by aerodynamic forces that were not fully anticipated and accounted for in its design (see Fig. 1.8). Interestingly, the Deer Isle bridge along the coast of Maine, while smaller, was of similar construction. It opened one year earlier and also experienced severe wind-induced vibrations. However, the designer of this bridge had the foresight and perhaps sufficient time to add wind fairings along the bridge's length to give it better aerodynamic properties, and additional diagonal cable bracing to provide greater stiffness. This bridge is still in service today.†

- **Escambia Bay bridge.** Fifty-six sections of the Interstate 10 bridge crossing Escambia Bay in Pensacola, Florida, were dislodged by Hurricane Ivan in September 2004, including numerous sections that were completely washed into the bay (see Fig. 1.9). Each of these sections weighed about 220 tons. The National Weather Service categorizes the intensity of hurricanes using a scale of 1 to 5. When Ivan struck the Escambia Bay bridge, it was a category 3 hurricane with sustained winds of 111 to 130 mph. While Ivan was not an extreme hurricane according to this scale, the damage caused to the Escambia Bay bridge was extreme.

- **Airbus A300 failure.** On November 12, 2001, only minutes after take-off, American Airlines flight 587, an Airbus A300, crashed into a residential area of Belle Harbor, New York, because the airplane's vertical stabilizer separated in flight due to failure of the attachment lugs between the stabilizer and fuselage (see Fig. 1.10). All 260 people on board and 5 people on the ground were killed. The National Transportation Safety Board‡ (NTSB) investigated the accident and attributed the cause to high aerodynamic loads resulting from unnecessary and excessive rudder pedal inputs as the first officer reacted to turbulence

* Interesting case studies of failures and how we can learn from these are given in H. Petroski, *Design Paradigms: Case Histories of Error and Judgment in Engineering*, Cambridge University Press, New York, 1994.

† For additional reading, see B. Moran (1999), "A Bridge That Didn't Collapse," *Invention and Technology*, **15**(2), pp. 10–18.

‡ The National Transportation Safety Board (NTSB) is an independent federal agency charged by Congress with investigating every civil aviation accident in the United States and significant accidents in other modes of transportation including railroad, highway, marine and pipeline, and issuing safety recommendations aimed at preventing future accidents. Although implementation of the NTSB's recommendations is not mandatory, over 80% of their recommendations have been adopted.

caused by another aircraft. The airline's pilot training program and the airplane's rudder design were also cited as contributing factors. Among the recommendations made by the NTSB were to modify the rudder control systems to increase protection from high forces due to hazardous rudder pedal inputs at high speeds.

Figure 1.10
The vertical stabilizer of an Airbus A300 airplane separated in flight and was recovered from Jamaica Bay, about 1 mile from the crash site.

- **Kansas City Hyatt Regency Hotel.** On July 17, 1981, two suspended walkways at the Kansas City Hyatt Regency Hotel collapsed during a dance party, killing 114 people and seriously injuring many more. The collapse was caused by connections that failed, as shown in Fig. 1.11(a). The original connection design, shown in Fig. 1.11(b), was changed during construction to the design shown in Fig. 1.11(c), with the agreement of all parties involved. While the original design had satisfactory strength, the revised design was easier to fabricate, featured shorter bars that were more readily available, and was more straightforward than the potentially confusing original design. However, the revised design was never analyzed to determine its adequacy.*

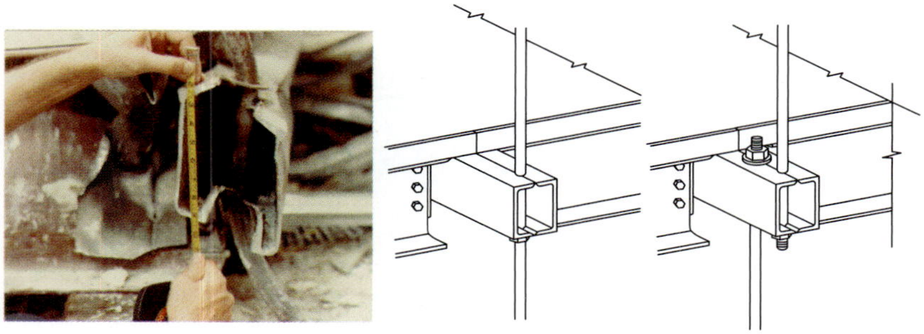

(a) (b) original design (c) as constructed

Figure 1.11. (a) Failure of a connection supporting a walkway at the Kansas City Hyatt Regency Hotel, where a support rod has pulled through a box beam, allowing the walkways to collapse. (b) The original design, which had satisfactory strength. (c) The revised design, which was easier to fabricate.

- **Tropicana Casino parking garage.** On October 30, 2003, a 10-story parking garage under construction at the Tropicana Casino and Resort in Atlantic City, New Jersey, collapsed, killing 4 workers and injuring 21 others (see Fig. 1.12). The failure occurred as concrete was being poured for one of the upper floor decks. The Occupational Safety and Health Administration† (OSHA) investigated the failure and fined the concrete contractor for intentional disregard of safety standards for failing to erect, support, brace, and maintain framework that would be capable of supporting all vertical and lateral loads that may reasonably be anticipated during construction. The design of the building itself was adequate, but the design of structures needed for fabrication was not. Note that concrete requires time after pouring (28 days is common) to reach its full design strength.

Figure 1.12
Inspectors survey a five-story collapsed section of a parking garage under construction at the Tropicana Casino and Resort in Atlantic City, New Jersey, October 30, 2003.

* Additional aspects of this failure are discussed in H. Petroski, *Design Paradigms: Case Histories of Error and Judgment in Engineering*, Cambridge University Press, New York, 1994.

† The mission and regulatory power of the Occupational Safety and Health Administration is described on p. 315.

1.8 Chapter Review

Important definitions, concepts, and equations of this chapter are summarized. For equations and/or concepts that are not clear, you should refer to the original equation numbers cited for additional details.

Scalars and Vectors. A *scalar* is a quantity that is completely characterized by a single number. A *vector* has both size (or magnitude) and direction. In this book, scalars are denoted by italic symbols such as s, and vectors are denoted by placing an arrow above the symbol for the vector, such as $\vec{v}$.

Position, Velocity, and Acceleration. Position, velocity, and acceleration are all vector quantities. If $\vec{r}$ denotes the *position* of a particle relative to some location, then the *velocity* and *acceleration* of the particle are defined by

Eq. (1.1), p. 7

$$\vec{v} = d\vec{r}/dt,$$

Eq. (1.2), p. 7

$$\vec{a} = d\vec{v}/dt.$$

When $\vec{a} = \vec{0}$, the particle is said to be in *static equilibrium*, and it either moves with constant velocity or remains stationary in space. If $\vec{a} \neq \vec{0}$, then the particle will move with velocity and position that change with time.

Laws of Motion. Newton's three laws of motion are as follows:

First Law. A particle remains at rest, or continues to move in a straight line with uniform velocity, if there is no unbalanced force acting on it.

Second Law. The acceleration of a particle is proportional to the resultant force acting on the particle, and is in the direction of this force.

Eq. (1.3), p. 7

$$\vec{F} = m\vec{a}.$$

Third Law. The forces of action and reaction between interacting bodies are equal in magnitude, opposite in direction, and collinear.

Static Equilibrium. In Eq. (1.3), if the resultant force $\vec{F}$ acting on a particle is zero, then the acceleration of the particle is zero, and hence the particle may move with uniform velocity in a straight line, which may be zero or nonzero in value. Hence, when there is no acceleration (i.e., $\vec{a} = \vec{0}$), the particle is said to be in *static equilibrium*.

Newton's Law of Gravitation. Newton's law of universal gravitational attraction is

Eq. (1.6), p. 15

$$F = G\frac{m_1 m_2}{r^2}$$

where

m_1, m_2 = masses of particles 1 and 2;
r = distance between the particles;
G = universal gravitational constant, found to be approximately $66.74\times 10^{-12}\,\mathrm{m^3/(kg{\cdot}s^2)}$;
F = force of attraction between two particles.

When written for objects resting on or near the surface of Earth, this law takes the simple and useful form

Eq. (1.7), p. 15

$$W = mg$$

where m is an object's mass, g is *acceleration due to gravity* ($g = 9.81\,\mathrm{m/s^2} = 32.2\,\mathrm{ft/s^2}$), and W is the object's *weight*.

Relationship Between Specific Weight and Density. The *density* ρ of a material is defined to be the material's mass per unit volume. The *specific weight* γ of a material (sometimes also called *weight density*, or *unit weight*) is defined to be the material's weight on Earth per unit volume. The relation between these is

Eq. (1.10), p. 17

$$\gamma = \rho g \quad \text{or} \quad \rho = \frac{\gamma}{g}.$$

Attention to Units. It is strongly recommended that you always use appropriate units in all equations. Such practice helps avoid catastrophic blunders and provides a useful check on a solution, for if an equation is found to be dimensionally inconsistent, then an error has certainly been made.

REVIEW PROBLEMS

Problem 1.1

(a) Consider a situation in which the force F applied to a particle of mass m is zero. Multiply the scalar form of Eq. (1.2) on page 7 (i.e., $a = dv/dt$) by dt, and integrate both sides to show that the velocity v (also a scalar) is constant. Then use the scalar form of Eq. (1.1) to show that the (scalar) position r is a linear function of time.

(b) Repeat Part (a) when the force applied to the particle is a nonzero constant, to show that the velocity and position are linear and quadratic functions of time, respectively.

Problem 1.2

Using the length and force conversion factors in Table 1.2 on p. 10, verify that 1 slug = 14.59 kg.

Problems 1.3 through 1.5

Convert the numbers given in U.S. Customary units to the corresponding SI units indicated.

Problem 1.3

(a) *Length*: Convert $l = 2.35$ in. to m.

(b) *Mass*: Convert $m = 0.156$ slug to kg.

(c) *Force* (weight): Convert $F = 100$ lb to N.

(d) *Moment* (torque): Convert $M = 32.9$ ft·lb to N·m.

Problem 1.4

(a) *Length*: Convert $l = 0.001$ in. to μm.

(b) *Mass*: Convert $m = 0.305$ lb·s^2/in. to kg.

(c) *Force* (weight): Convert $F = 2.56$ kip to kN. (Recall: 1 kip = 1000 lb.)

(d) *Mass moment of inertia*: Convert $I_{\text{mass}} = 23.0$ in.·lb·s^2 to N·m·s^2.

Problem 1.5

(a) *Pressure*: Convert $p = 25$ lb/ft^2 to N/m^2.

(b) *Elastic modulus*: Convert $E = 30 \times 10^6$ lb/in.^{2} to GN/m^2.

(c) *Area moment of inertia*: Convert $I_{\text{area}} = 63.2$ in.^{4} to mm^4.

(d) *Mass moment of inertia*: Convert $I_{\text{mass}} = 15.4$ in.·lb·s^2 to kg·m^2.

Problems 1.6 through 1.8

Convert the numbers given in SI units to the corresponding U.S. Customary units indicated.

Problem 1.6

(a) *Length*: Convert $l = 1.53$ m to in.

(b) *Mass*: Convert $m = 65$ kg to slug.

(c) *Force* (weight): Convert $F = 89.2$ N to lb.

(d) *Moment* (torque): Convert $M = 32.9$ N·m to in.·lb.

Problem 1.7

(a) *Length*: Convert $l = 122$ nm to in.

(b) *Mass*: Convert $m = 3.21$ kg to lb·s^2/in.

(c) *Force* (weight): Convert $F = 13.2$ kN to lb.

(d) *Mass moment of inertia*: Convert $I_{\text{mass}} = 93.2\ \text{kg}\cdot\text{m}^2$ to slug·in.^{2}.

Problem 1.8

(a) *Pressure*: Convert $p = 25\ \text{kN/m}^2$ to lb/in.^{2}.

(b) *Elastic modulus*: Convert $E = 200\ \text{GN/m}^2$ to lb/in.^{2}.

(c) *Area moment of inertia*: Convert $I_{\text{area}} = 23.5 \times 10^5\ \text{mm}^4$ to in.^{4}.

(d) *Mass moment of inertia*: Convert $I_{\text{mass}} = 12.3\ \text{kg}\cdot\text{m}^2$ to in.·lb·s^2.

Problem 1.9

(a) Convert the kinetic energy $T = 0.379\ \text{kg}\cdot\text{m}^2/\text{s}^2$ to slug·in.^{2}/s^2.

(b) Convert the kinetic energy $T = 10.1$ slug·in.^{2}/s^2 to kg·m^2/s^2.

Problem 1.10

If the weight of a certain object on the surface of the Earth is 0.254 lb, determine its mass in kilograms.

Problem 1.11

If the mass of a certain object is 69.1 kg, determine its weight on the surface of the Earth in pounds.

Problem 1.12

Use Eq. (1.7) on p. 15 to compute a theoretical value of acceleration due to gravity g, and compare this value with the actual acceleration due to gravity at the Earth's poles, which is about 0.3% higher than the value reported in Eq. (1.8). Comment on the agreement.

Problem 1.13

Two identical asteroids travel side by side while touching one another. If the asteroids are composed of homogeneous pure iron and are spherical, what diameter in feet must they have for their mutual gravitational attraction to be 1 lb?

Problem 1.14

The mass of the Moon is approximately 7.35×10^{22} kg, and its mean distance from the Earth is about 3.80×10^8 km. Determine the force of mutual gravitational attraction in newtons between the Earth and Moon. In view of your answer, discuss why the Moon does not crash into the Earth.

Problem 1.15

If a person standing at the first-floor entrance to the Sears Tower in Chicago, weighs exactly 150 lb, determine the weight while he or she is standing on top of the building, which is 1450 ft above the first-floor entrance. How high would the top of the building need to be for the person's weight to be 99% of its value at the first-floor entrance?

Problem 1.16

The specific weights of several materials are given in U.S. Customary units. Convert these to specific weights in SI units (kN/m^3), and also compute the densities of these materials in SI units (kg/m^3).

(a) Zinc die casting alloy, $\gamma = 0.242$ lb/in.^{3}.

(b) Oil shale (30 gal/ton rock), $\gamma = 133$ lb/ft^3.

(c) Styrofoam (medium density), $\gamma = 2.0$ lb/ft^3.

(d) Silica glass, $\gamma = 0.079$ lb/in.^{3}.

Problem 1.17

The densities of several materials are given in SI units. Convert these to densities in U.S. Customary units (slug/ft^3), and also compute the specific weights of these materials in U.S. Customary units (lb/ft^3).

(a) Lead (pure), $\rho = 11.34$ g/cm^3.

(b) Ceramic (alumina Al_2O_3), $\rho = 3.90$ Mg/m^3.

(c) Polyethylene (high density), $\rho = 960$ kg/m^3.

(d) Balsa wood, $\rho = 0.2$ Mg/m^3.

Problem 1.18

A Super Ball is a toy ball made of hard synthetic rubber called Zectron. This material has a high coefficient of restitution so that if it is dropped from a certain height onto a hard fixed surface, it rebounds to a substantial portion of its original height. If the Super Ball has 5 cm diameter and the density of Zectron is about 1.5 Mg/m^3, determine the weight of the Super Ball on the surface of the Earth in U.S. Customary units.

Problem 1.19

An ice hockey puck is a short circular cylinder, or disk, of vulcanized rubber with 3.00 in. diameter and 1.00 in. thickness, with weight between 5.5 and 6.0 oz (16 oz = 1 lb). Compute the range of densities for the rubber, in conventional SI units, that will provide for a puck that meets these specifications.

Problem 1.20

Convert the angles given to the units indicated.

(a) Convert $\theta = 35.6°$ to rad.

(b) Convert $\theta = (1.08 \times 10^{-3})°$ to mrad.

(c) Convert $\theta = 4.65$ rad to degrees.

(d) Convert $\theta = 0.254$ mrad to degrees.

Problem 1.21

Many of the examples of failure discussed in Section 1.7 have common causes, such as loads that were not anticipated, overestimation of the strength of materials, unanticipated use, etc. Using several paragraphs, identify those examples that have common causes of failure and discuss what these causes were.

Note: Concept problems are about *explanations*, not computations.

2 Vectors: Force and Position

Vectors are immensely useful for describing many entities in mechanics. Vectors can be compactly represented, are easy to manipulate, and can be transformed from one component description to another using established rules. In this chapter vectors are used to describe force and position. In later chapters, vectors will be used to represent other entities. Early sections of this chapter focus on vectors in two dimensions, and later sections treat vectors in three dimensions. Vector dot product and cross product operations are also presented.

2.1 Basic Concepts

Introduction – force, position, vectors, and tides

One phenomenon in which vectors play an important role is ocean tides. High and low tides each occur approximately twice per day at most locations on the Earth due to the rotation of the Earth and the Moon's orbit around the Earth—let's see how vectors help explain this.

Figure 2.1 shows the Earth and Moon in a coordinate system that has its origin at point O (for measuring the motion of objects in our solar system, the origin is usually located at the center of the Sun). Your intuitive notion of *position* is probably close to its mathematical definition, but an important point needs to be made. A description of *position*, or *location*, is always relative to the coordinate system that is used. For example, if the coordinates of the Moon (x_M, y_M) in Fig. 2.1 are known, along with a coordinate system whose origin is at a known point (perhaps the center of the Sun), then the absolute location of the Moon is known. However, if the location of the origin is unknown, then obviously the coordinates of the Moon have little meaning.

Rather than using coordinates, we can describe the position of a point P by specifying the distance from a reference point to P, and by specifying the direction from the reference point to P. The reference point referred to here

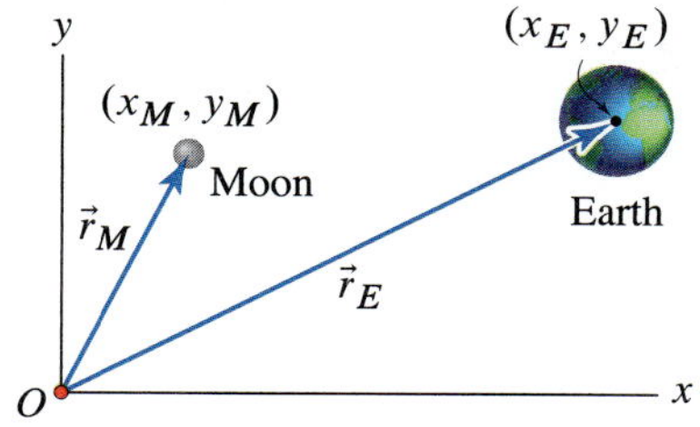

Figure 2.1
A coordinate system showing the positions of the Earth and Moon as well as their position vectors. The relative sizes of the Earth and Moon are correct, but the distance between them is not drawn to scale.

could be the origin of the coordinate system, or any other point we choose. This description combines statements of both *size* and *direction*, and has the same characteristics as an invention of mathematics called a *vector*. A vector has both size and direction, and to emphasize this content, vectors are represented in this book* by placing an arrow above the symbol used for the vector, such as $\vec{r}$. Figure 2.1 uses arrows to represent the position vectors of the Earth and Moon, $\vec{r}_E$ and $\vec{r}_M$, respectively, where these vectors are measured from the origin of our coordinate system. Figure 2.2 defines some useful nomenclature. A vector has a *head* and a *tail*, as well as a *line of action*, which is the line of infinite extent along which the vector is positioned. For force vectors especially, the line of action is important, and it is most often within this context that we refer to it. The *magnitude* of a vector $\vec{r}$ is the measure of its size, or length, including appropriate units, and is denoted by $|\vec{r}|$. The magnitude of a vector is a positive scalar for any vector that has nonzero size, and is zero only for a vector of zero size. The magnitude can never be negative.

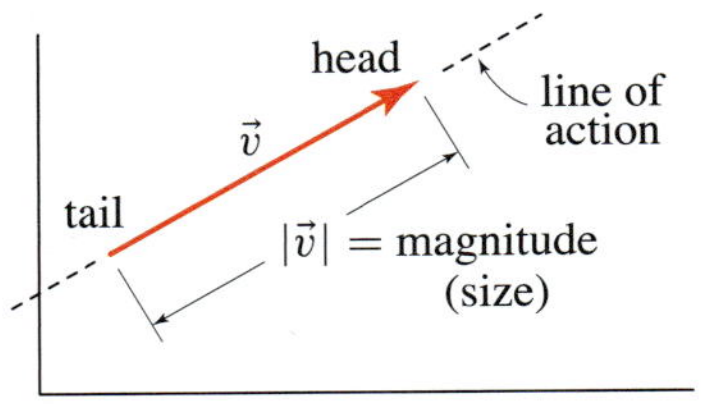

Figure 2.2
Vector nomenclature.

Now, let's look at how the Earth and Moon attract one another as a result of Newton's law of universal gravitation [Eq. (1.6)], which states that the gravitational attraction between two bodies is proportional to the product of their masses and is inversely proportional to the square of the distance between them. Unfortunately, this equation doesn't tell the whole story, the remainder of which is found in Fig. 2.3(a). While Eq. (1.6) tells us *how much* two bodies attract one another, it doesn't tell us the *direction* of the attraction. Figure 2.3(a) shows that these bodies attract one another toward their centers (actually, toward their centers of gravity). Thus, the gravitational attraction between two bodies is a force that has magnitude [given by Eq. (1.6)] and direction (the center of gravity of one body is pulled toward the center of gravity of the other body), and therefore it possesses the characteristics of a vector.

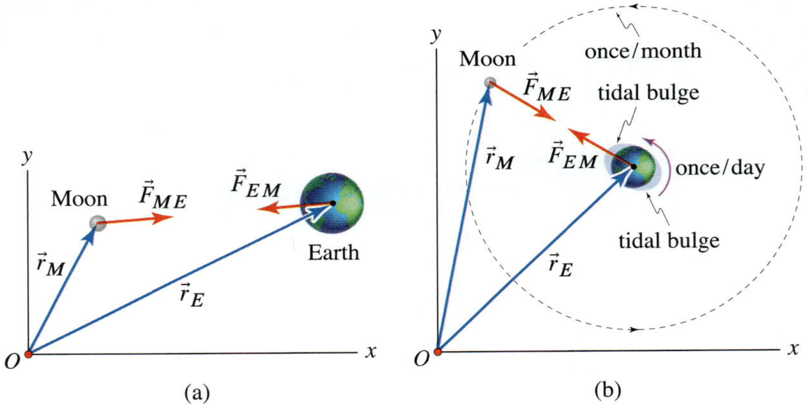

Figure 2.3. (a) The forces of gravitational attraction between the Earth and Moon. The force $\vec{F}_{EM}$ is the force on E due to M, and similarly $\vec{F}_{ME}$ is the force on M due to E. (b) Orbit of the Moon about the Earth and the rotation of the Earth as seen from above their orbital planes. We also show the positions of the Earth and Moon, the gravitational force between the Earth and the Moon, and the tidal bulge (greatly exaggerated) of the Earth due to the Moon.

Interesting Fact

Why don't the Earth and Moon "crash" into one another? If the Earth and Moon were not moving, then the forces $\vec{F}_{ME}$ and $\vec{F}_{EM}$ shown in Fig. 2.3 would cause them to accelerate toward one another, and they would eventually collide. However, the Earth and Moon are not stationary, and as you well know, they have not collided. When you study dynamics, you will learn that if objects have appropriate motion, the acceleration $\vec{a}$ of each of them is such that, according to Newton's second law $\vec{F} = m\vec{a}$, the objects will orbit their common center of mass (such as the Earth and Moon do) even though there is a force attracting them.

* The notation used throughout the engineering literature to denote vectors is very diverse. Hence, in other books you may see vectors denoted by the use of a bar placed above or below a given symbol, such as $\bar{r}$ or $\underline{r}$ or by bold type such as **r**.

In addition, Newton's third law (see p. 7) tells us that the forces of attraction between these bodies must be equal, opposite, and collinear. With all of this as background, let's see how the positions of the Earth and Moon work in concert with the forces between them to cause the ocean tides on Earth.

The Moon orbits the Earth about once per month, and the Earth rotates on its axis once per day, as shown in Fig. 2.3(b). The position of the Moon as seen by the Earth* determines the direction and magnitude of the gravitational force exerted on the Earth by the Moon $\vec{F}_{EM}$. As you might expect, the side of the Earth closest to the Moon experiences a tidal bulge, but you might not expect that the side of the Earth farthest from the Moon does too! Why is this? The reason that tides occur is the fact that the Earth's shape elongates in the direction of $\vec{F}_{EM}$ due to the gravitational effect of the Moon. This elongation is the result of *differential* forces exerted by the Moon on the Earth. These differential forces arise because the materials of the Earth (i.e., soil, rock, magma, etc.) that are closer to the Moon are pulled by the Moon more strongly than the materials of the Earth that are farther from the Moon. The net effect of these differential forces is that the Earth is stretched *in both directions* along the line of action of the force $\vec{F}_{EM}$ and is compressed in the directions perpendicular to $\vec{F}_{EM}$. This causes a tidal bulge on the side of the Earth facing the Moon as well as a tidal bulge on the side of the Earth opposite the Moon. Now, since the Earth rotates about its axis once per day, there are approximately two tides per day in most locations on the surface of the Earth.† We say approximately because the Moon orbits the Earth as the Earth simultaneously rotates [see Fig. 2.3(b)], and this results in tides occurring about 50 minutes later each day.‡

We hope this helps to explain why there are two high tides and two low tides at most locations on the Earth every day and why force and position vectors play such an important role in determining these tides.

> **Interesting Fact**
>
> **More tidal tidbits.** Tides on Earth are caused mostly by the Moon, but the Sun also has an effect. Although the Sun's gravitational force on the Earth is about 180 times that of the Moon, because the Moon is much closer to the Earth, the *differential* force of the Moon on the Earth is about twice that of the Sun.
>
> At full Moon (when the Earth is between the Sun and Moon) and new Moon (when the Moon is between the Earth and Sun), the Sun, Earth, and Moon line up, producing higher than normal tides (called *spring tides*). When the Moon is at first or last quarter, smaller *neap tides* form. Since the Moon's 29.5 day orbit around Earth is not quite circular, when the Moon is closest to Earth (called *perigee*), spring tides are even higher (called *perigean spring tides*).
>
> Interestingly, because the Moon is continually deforming the Earth (changing its shape), the Earth's rotational energy is continually being lost, causing the Earth's rate of rotation to decrease so that the length of a day increases by approximately 1.5 milliseconds per century.
>
> Finally, some books and websites claim that the "centrifugal force" on the water due to the rotation of the Earth and/or the rotation of the Earth-Moon system are factors in the creation of tides. This is simply not true. When you study dynamics, you will learn that these effects are negligible compared with gravitational effects, and that there is no such thing as a "centrifugal force."

Denoting vectors in figures

In figures, we represent vectors by using arrows. Furthermore, the arrows will follow a consistent color scheme to indicate what the vector physically represents. For example, force vectors will be shown in red ■, position vectors will be shown in blue ■, and when vectors with other physical significance are introduced, they will use other colors. Vectors with no particular physical significance will be black ■, magenta ■, or gray ■. This practice makes comprehension of a figure quicker, but otherwise is not needed. Next to each arrow shown in a figure, we will provide a symbol to identify the vector; or if the numerical value of a vector's magnitude is known, we will often simply show this value with the appropriate units.

Consider the example shown in Fig. 2.4(a) where a crate is held in equilibrium on a ramp by two people who apply forces to it. We will assume the person on the left pulls on the rope with a force having 25 lb magnitude

* Later in this section we discuss subtraction of vectors, after which you may show that the position of the Moon as seen by the Earth is given by $\vec{r}_M - \vec{r}_E$.

† Since the orbital plane of the Moon is tilted with respect to the spin axis of the Earth, some locations (at mainly higher latitudes) only experience one significant tide per day.

‡ The Moon orbits around the Earth every 29.5 days, so it takes $24/29.5$ hours $= 48.8$ minutes longer each day for the Moon to reach the same position above the surface of the Earth.

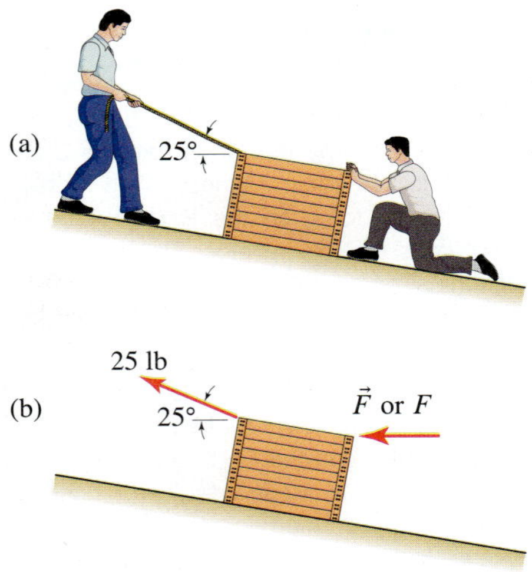

Figure 2.4
Vector representation used in figures. (a) Two people apply forces to a crate to keep it in equilibrium on a ramp, where we assume the person pulling the rope applies a 25 lb force oriented at 25° from the horizontal and the person pushing the crate applies a horizontal force with unknown magnitude. (b) Forces applied by the two people to the crate using various forms of representation. For further examples contrasting these forms of representation and the implications of Newton's third law, you should compare Fig. 2 on p. 38 with Fig. 3 on p. 65.

oriented at 25° from the horizontal, and the person on the right pushes the crate with a horizontal force whose magnitude is unknown. In Fig. 2.4(b), the crate is sketched showing the forces applied by the two people, using the following forms for labeling vectors.*

- In the case of the force applied by the rope to the crate, the magnitude is known to be 25 lb, and in such cases we will simply list this magnitude next to its arrow in the figure, and the direction of the arrow gives the direction of the force.
- When a vector symbol is used in a figure (e.g., $\vec{F}$), the symbol represents an expression that fully describes the magnitude *and* direction for the vector. This form of labeling will generally be used when we want to emphasize the arbitrary directional nature of a particular vector.
- When a scalar symbol is used (e.g., F), it refers to the *component* of the vector in the direction of the arrow shown in the figure. We will provide a full definition for the word *component* later, but for the present, this is the amount of a vector that acts in the direction of the arrow. With this definition, a symbol such as F can be positive, zero, or negative, as follows. If F is positive, then indeed the direction of the vector is the same as the direction of the arrow shown in the figure. If F is negative, then the direction of the vector is actually opposite to the direction of the arrow shown in the figure. A vector of zero size will have $F = 0$.

Observe that the notations F and $|\vec{F}|$ are different, but they are closely related. As described above, F shown in Fig. 2.4 can be positive, zero, or negative, but in any event, the absolute value of F is the same as $|\vec{F}|$. Labeling vectors by using a symbol such as F will be done often, and it is especially useful when the line of action for a vector is known but we are unsure of its direction along this line of action. To illustrate, consider again the crate in Fig. 2.4(b). We have assigned the direction of $\vec{F}$ to correspond to the person *pushing* the crate. However, it is possible that this person may actually need to *pull* the crate to keep it in equilibrium. Only after the equilibrium equations are solved, taking into account the weight of the crate, friction between the crate and ramp, steepness of the ramp, and so on, will we know the value (or range of values) F must have for equilibrium. For Fig. 2.4(b), if $F > 0$, the person is pushing the crate while if $F < 0$, the person is pulling the crate.

> **Helpful Information**
>
> **Jargon.** Consider the following statement: "The widget is subjected to a force with 123 N magnitude." This statement is precise in its nomenclature, but a practicing engineer would probably have said, "The widget is subjected to a 123 N force." The second statement, if taken literally, is imprecise because it states that a force (which is a vector) is equal to 123 N (a scalar), and furthermore it does not explicitly say that the force has 123 N magnitude. Nonetheless, the second expression is widely used among engineers, and it is understood that the statement means the magnitude of the force is 123 N.

Basic vector operations

The following remarks on equivalent vectors and vector addition, subtraction, and multiplication by a scalar are true regardless of the type of vector representation that is used. However, the details of how these operations are carried out will depend on the vector representation that is employed, and we will see that some forms of vector representation allow for easier manipulation than others.

* Note that in addition to the forces shown in Fig. 2.4(b), the crate is subjected to other forces that are not shown, such as the weight of the crate and the contact forces applied by the ramp.

Equivalent vectors

Two vectors are said to be *equivalent*, or equal, if they have the same magnitude and orientation. Note that two equivalent vectors may have different lines of action, provided the lines of action are parallel.

Vector addition

Addition of two vectors $\vec{A}$ and $\vec{B}$ produces a new vector $\vec{R}$, and this operation is denoted by $\vec{R} = \vec{A} + \vec{B}$. Figure 2.5 illustrates two methods for performing vector addition.

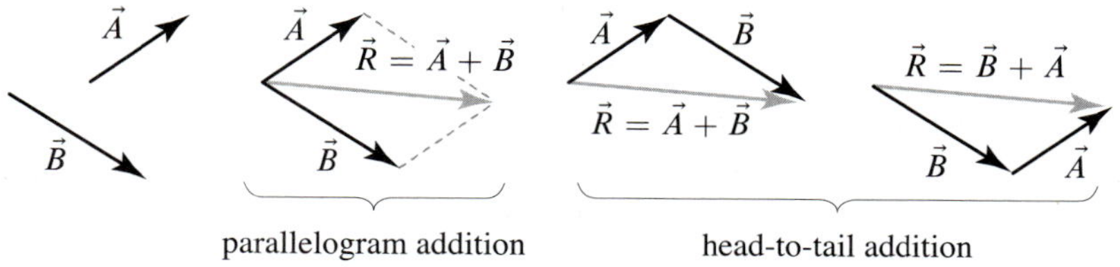

Figure 2.5. Addition of two vectors using the parallelogram method and the head-to-tail method.

With the *parallelogram method* of addition, the outcome of $\vec{A} + \vec{B}$ is determined by arranging the vectors $\vec{A}$ and $\vec{B}$ tail to tail to form a parallelogram. The vector $\vec{R} = \vec{A} + \vec{B}$ is then the vector whose tail coincides with the tails of $\vec{A}$ and $\vec{B}$ and whose head coincides with the parallelogram's opposite vertex. Alternatively, $\vec{R}$ can also be determined using the *head-to-tail method* by which we slide the tail of $\vec{B}$ to the head of $\vec{A}$, and the resulting triangle provides $\vec{R}$. Alternatively, the tail of $\vec{A}$ can be slid to the head of $\vec{B}$, and the resulting triangle provides the same $\vec{R}$.

Vector addition has the following properties:

$$\vec{A} + \vec{B} = \vec{B} + \vec{A} \qquad \text{commutative property,} \tag{2.1}$$

$$(\vec{A} + \vec{B}) + \vec{C} = \vec{A} + (\vec{B} + \vec{C}) \qquad \text{associative property,} \tag{2.2}$$

where $\vec{A}$, $\vec{B}$, and $\vec{C}$ are three arbitrary vectors. Equations (2.1) and (2.2) imply that the result of adding an arbitrary number of vectors is independent of the order in which the addition is carried out. Also, the head-to-tail method of addition generalizes so that an arbitrary number of vectors can be added simultaneously simply by arranging them head to tail, one after another. Doing this gives rise to a *vector polygon*, which is very useful for visualizing the arrangement of vectors. These comments are explored in greater detail in Example 2.1.

Multiplication of a vector by a scalar

Multiplication of a vector $\vec{A}$ and a scalar s produces a new vector $\vec{R}$ where $\vec{R} = s\vec{A} = \vec{A}s$. The magnitude of $\vec{R}$ is equal to the magnitude of $\vec{A}$ multiplied by $|s|$. If s is positive, then $\vec{R}$ and $\vec{A}$ have the same direction; if s is negative, then $\vec{R}$ has direction opposite to $\vec{A}$. Multiplication of a vector by a scalar is common, and this operation allows us to change a vector's size. Multiplication

Helpful Information

Polar vector representation. A simple representation we will sometimes use for writing vectors in two dimensions is called *polar vector representation*. This representation consists of a statement of the vector's magnitude and direction referred to a right-hand horizontal reference direction, with a positive angle being measured counterclockwise. For example:

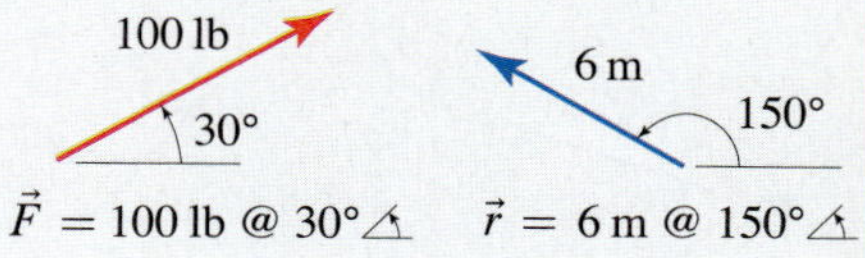

Common Pitfall

Addition of vectors versus addition of scalars. Addition of vectors is an operation that is very different, and more complex, than addition of scalars. A common error is to attempt to add vectors as if they were scalars.

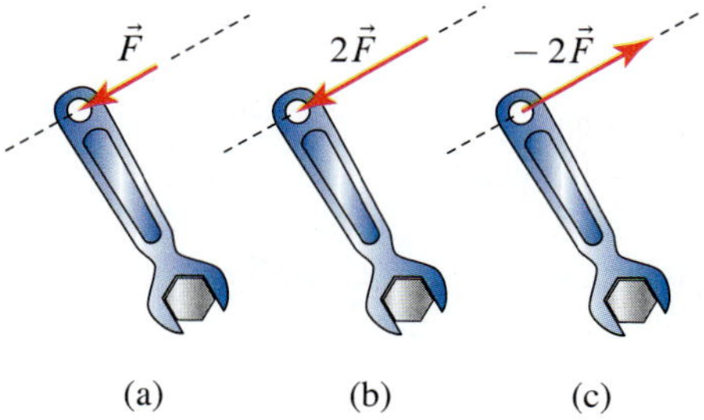

Figure 2.6
Examples of multiplication of a vector by a scalar. (a) A force $\vec{F}$ is applied to the handle of a wrench. (b) $\vec{F}$ is multiplied by 2, with the resulting vector drawn to scale. (c) $\vec{F}$ is multiplied by −2, with the resulting vector drawn to scale and with reversed direction.

of a vector by a scalar does not change the vector's line of action; but if the scalar is negative, then the vector's direction along its line of action is reversed. Figure 2.6 shows some examples.

Let $\vec{A}$ and $\vec{B}$ be vectors and s and t be scalars. Multiplication of a vector by a scalar has the following properties:

$$s(\vec{A}+\vec{B}) = s\vec{A}+s\vec{B} \quad \text{distributive property with respect to vector addition,} \tag{2.3}$$

$$(s+t)\vec{A} = s\vec{A}+t\vec{A} \quad \text{distributive property with respect to addition of scalars,} \tag{2.4}$$

$$(st)\vec{A} = s(t\vec{A}) \quad \text{associative property with respect to multiplication by a scalar.} \tag{2.5}$$

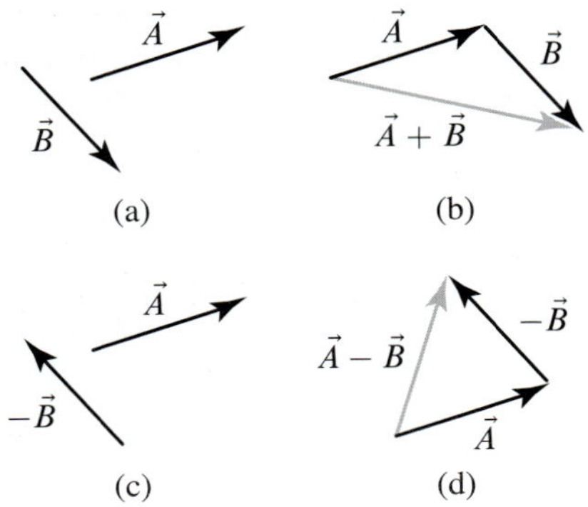

Figure 2.7
Comparison of addition and subtraction of two vectors. (a) Vectors $\vec{A}$ and $\vec{B}$ are defined. (b) $\vec{A}+\vec{B}$ is evaluated using the head-to-tail method. (c) The direction of $\vec{B}$ is reversed by multiplying it by -1. (d) $\vec{A}-\vec{B}$ is evaluated using the head-to-tail method.

Equation (2.3) states $\vec{A}$ and $\vec{B}$ may be first added and then multiplied by s; or each vector may be multiplied by s first, and then the vector addition is performed. Similar comments apply to Eqs. (2.4) and (2.5).

Vector subtraction

Subtracting a vector $\vec{B}$ from a vector $\vec{A}$ is denoted by $\vec{A}-\vec{B}$ and is defined as

$$\vec{A}-\vec{B} = \vec{A}+(-1)\,\vec{B}. \tag{2.6}$$

In this definition, first the direction of $\vec{B}$ is reversed by multiplying it by -1, and then the result is added to $\vec{A}$. In Fig. 2.7, addition and subtraction of vectors are contrasted.

Performing vector operations

To use parallelogram addition or head-to-tail addition, we could add vectors graphically using a ruler and protractor with very careful drawings, but it is more precise and appropriate to use analytical methods. Since addition of two vectors will generally involve a triangle or parallelogram of complex geometry, the laws of sines and cosines will be useful. With reference to Fig. 2.8, for a general triangle the *law of sines* and the *law of cosines* are

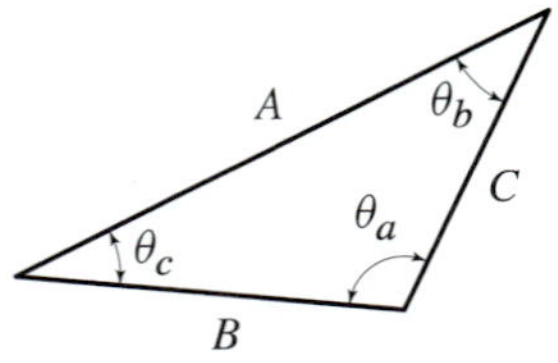

Figure 2.8
A general triangle.

$$\frac{\sin\theta_a}{A} = \frac{\sin\theta_b}{B} = \frac{\sin\theta_c}{C} \quad \text{law of sines,} \tag{2.7}$$

$$\begin{aligned} A &= \sqrt{B^2+C^2-2BC\cos\theta_a} \\ B &= \sqrt{A^2+C^2-2AC\cos\theta_b} \\ C &= \sqrt{A^2+B^2-2AB\cos\theta_c} \end{aligned} \quad \text{law of cosines.} \tag{2.8}$$

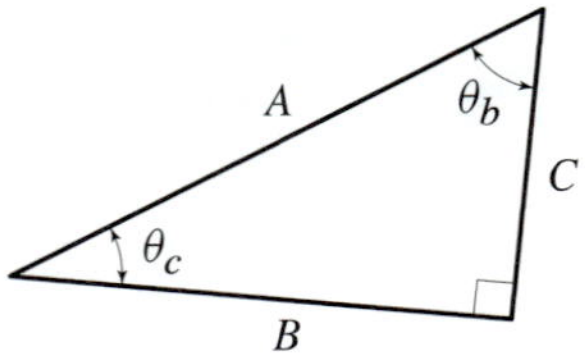

Figure 2.9
A right triangle.

In the special case that the triangle has a right angle, as in Fig. 2.9 where $\theta_a = 90°$, the laws of sines and cosines simplify to the familiar expressions

$$\begin{aligned} B &= A\cos\theta_c = A\sin\theta_b \\ C &= A\sin\theta_c = A\cos\theta_b \end{aligned} \tag{2.9}$$

$$A = \sqrt{B^2 + C^2} \qquad \text{Pythagorean theorem.} \tag{2.10}$$

If it has been some time since you have used trigonometry, you should thoroughly refamiliarize yourself with Eqs. (2.7)–(2.10).

Resolution of a vector into vector components

Consider a vector $\vec{F}$ that might represent a force, position, or some other entity. When $\vec{F}$ is expressed as a sum of a set of vectors, then each vector of this set is called a *vector component* of $\vec{F}$, and the process of representing $\vec{F}$ as a sum of other vectors is often called *resolution of $\vec{F}$ into vector components*. For example, we will often ask, "What two vectors $\vec{A}$ and $\vec{B}$, when added together, will yield $\vec{F}$?" In answering this question, the vectors $\vec{A}$ and $\vec{B}$ that we find are called the *vector components* of $\vec{F}$. Of course, as shown in Fig. 2.10, there is not a unique answer to this problem, as there are an infinite number of vectors $\vec{A}$ and $\vec{B}$ such that $\vec{A} + \vec{B} = \vec{F}$. However, if we place certain constraints or restrictions on $\vec{A}$ and/or $\vec{B}$, such as they must have certain prescribed directions and/or magnitudes, then the vectors $\vec{A}$ and $\vec{B}$ whose addition yields $\vec{F}$ may be unique. In statics, we will usually want to find vector components that have directions we specify, such as the directions of structural members or the directions of coordinate axes.

Referring to Fig. 2.10, we see that the vector triangles that arise when resolving a vector into vector components may have general triangular shape or right triangular shape. General triangular shapes occur for cases $\vec{A}_1$ and $\vec{B}_1$, and $\vec{A}_2$ and $\vec{B}_2$, and for such vector triangles the laws of sines and cosines are usually needed to determine the vector components. If the vector components are taken to be orthogonal (i.e., perpendicular), then the vector triangles are right triangles as shown for cases $\vec{A}_3$ and $\vec{B}_3$, and $\vec{A}_4$ and $\vec{B}_4$. For right triangles, elementary trigonometry is sufficient, and resolution of the vector into vector components is usually straightforward. In fact, it is for this reason that Cartesian vector representation, to be discussed in Section 2.2, is so convenient and effective.

Common Pitfall

Law of sines and obtuse angles. You will avoid errors if you avoid using the law of sines to determine obtuse angles! The pitfall is that the inverse sine function provided by an electronic calculator yields angles between $-90°$ and $+90°$ only (right or acute angles only). For example, in Fig. 2.8, let $A = 9\,\text{mm}$, $B = 6\,\text{mm}$, and $\theta_b = 30°$ (with these values, Fig. 2.8 is drawn roughly to scale). Using the law of sines, Eq. (2.7), we obtain $\sin\theta_a = (A/B)\sin\theta_b = 0.7500$. Using an electronic calculator to evaluate $\theta_a = \sin^{-1}(0.7500)$ provides $48.59°$, which is not the correct value of θ_a for this triangle! Of course, the equation $\theta_a = \sin^{-1}(0.7500)$ has an infinite number of solutions, and the correct solution for this triangle is $\theta_a = 131.4°$, which is an obtuse angle.

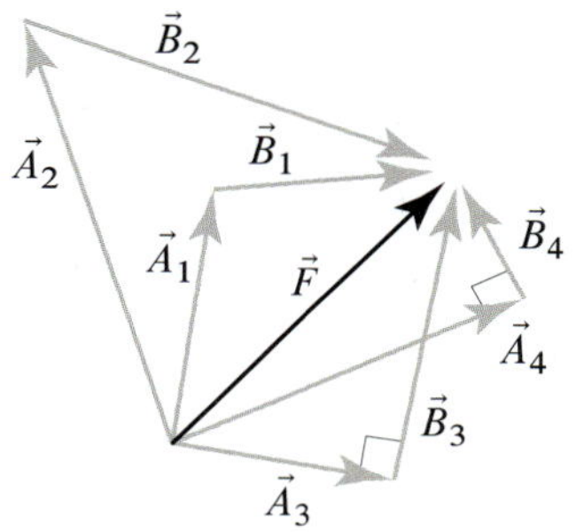

Figure 2.10
Four examples of the resolution of a vector $\vec{F}$ into two vectors $\vec{A}$ and $\vec{B}$ such that $\vec{A} + \vec{B} = \vec{F}$.

End of Section Summary

In this section, basic properties of vectors and some operations using vectors have been described, including

- Vector addition, subtraction, and multiplication by a scalar.
- Resolution of a vector into *vector components*.
- As an aid to help manipulate vectors, the laws of sines and cosines were reviewed.

EXAMPLE 2.1 *Addition of Vectors*

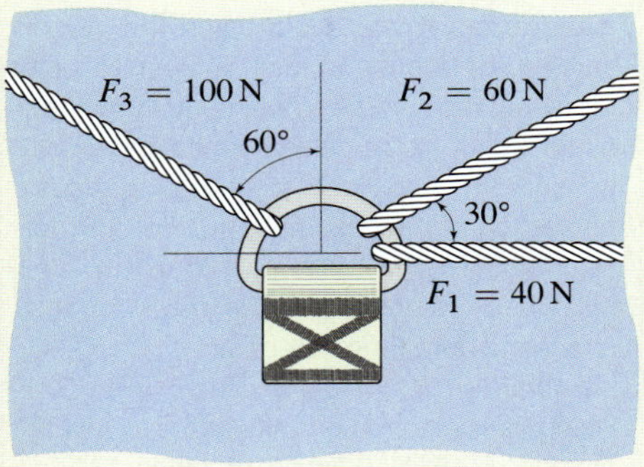

Figure 1

A D ring is sewn on a backpack for use in securing miscellaneous items to the outside of the backpack. If the D ring has three cords tied to it and the cords support the forces shown, determine the resultant force applied to the D ring by the cords, expressing the result as a vector.

SOLUTION

Road Map We first note that the force supported by a cord has the same direction as the cord, and thus, in Fig. 2 we redraw the D ring, showing the forces that are applied to it by the cords. The resultant force vector is the sum of the three force vectors applied to the D ring. Thus, we will add the vectors, choosing the head-to-tail method, and apply the laws of sines and cosines to determine the results of the vector addition.

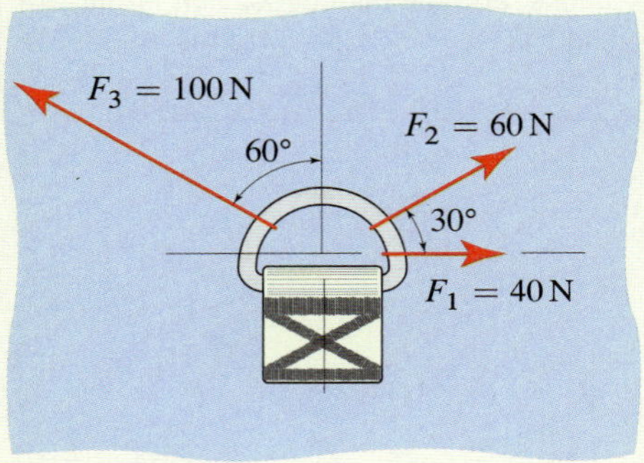

Figure 2
D ring showing the forces applied by the cords to the D ring.

Governing Equations & Computation The resultant force vector is

$$\vec{R} = \vec{F}_1 + \vec{F}_2 + \vec{F}_3. \tag{1}$$

The addition is illustrated graphically in Fig. 3, where the first illustration shows forces added in the order $\vec{F}_1 + \vec{F}_2 + \vec{F}_3$, while the second illustration shows addition in the order $\vec{F}_3 + \vec{F}_2 + \vec{F}_1$, and the third illustration shows addition in the order $\vec{F}_1 + \vec{F}_3 + \vec{F}_2$; according to Eq. (2.2), all three sums provide the same resultant force $\vec{R}$. The polygons shown in Fig. 3 are often called *force polygons* (or more generally *vector polygons*), and they are useful in statics as well as many areas of mechanics, not just for computation, but also for visualizing the spatial relationship among forces.

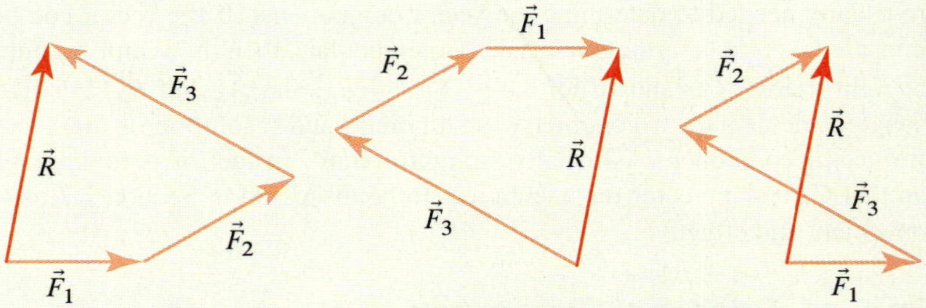

Figure 3. Vector polygons illustrating head-to-tail addition of $\vec{F}_1$, $\vec{F}_2$, and $\vec{F}_3$ in different orders. In all cases, the same resultant vector $\vec{R}$ is obtained.

To compute $\vec{R}$, we will use the first force polygon shown in Fig. 3. Unfortunately, the geometry of a general polygon is usually too complex to allow a direct evaluation of $\vec{R}$. Thus, we usually must break the polygon into smaller elements, namely, triangles, each of which may be analyzed using the laws of sines and cosines as given in Eqs. (2.7) and (2.8), or in the case of right triangles as given in Eqs. (2.9) and (2.10). Thus, we will first define an intermediate vector sum to be $\vec{P} = \vec{F}_1 + \vec{F}_2$, as shown in Fig. 4. Using the law of cosines, we write

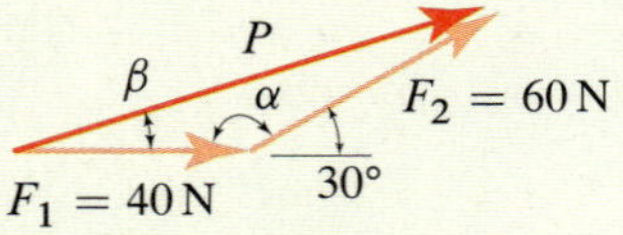

Figure 4
Addition of $\vec{F}_1$ and $\vec{F}_2$ to obtain an intermediate result $\vec{P}$ (i.e., $\vec{P} = \vec{F}_1 + \vec{F}_2$).

$$\begin{aligned} P &= \sqrt{F_1^2 + F_2^2 - 2F_1F_2\cos\alpha} \\ &= \sqrt{(40\,\mathrm{N})^2 + (60\,\mathrm{N})^2 - 2(40\,\mathrm{N})(60\,\mathrm{N})\cos(150°)} = 96.73\,\mathrm{N}, \end{aligned} \tag{2}$$

where angle α was easily found from Fig. 4 as $\alpha = 180° - 30° = 150°$. Now that magnitude P is known, we use the law of sines, Eq. (2.7), to find angle β shown in Fig. 4 as

$$\frac{P}{\sin\alpha} = \frac{F_2}{\sin\beta}, \tag{3}$$

hence

$$\sin\beta = \frac{F_2}{P}\sin(150°) = \frac{60\,\text{N}}{96.73\,\text{N}}\sin(150°) = 0.3101. \tag{4}$$

Solving Eq. (4) for β provides

$$\beta = \sin^{-1}(0.3101) = 18.1°. \tag{5}$$

Next, as shown in Fig. 5, we evaluate $\vec{P} + \vec{F}_3$ to obtain $\vec{R}$. Defining the angle γ to be the angle between force vectors $\vec{P}$ and $\vec{F}_3$, we use the geometry shown in Fig. 2 and the value of β previously found to write $\gamma = 90° - 60° + 18.1° = 48.1°$. The law of cosines provides

$$\begin{aligned} R &= \sqrt{P^2 + F_3^2 - 2PF_3\cos\gamma} \\ &= \sqrt{(96.73\,\text{N})^2 + (100\,\text{N})^2 - 2(96.73\,\text{N})(100\,\text{N})\cos(48.1°)} \\ &= 80.23\,\text{N}. \end{aligned} \tag{6}$$

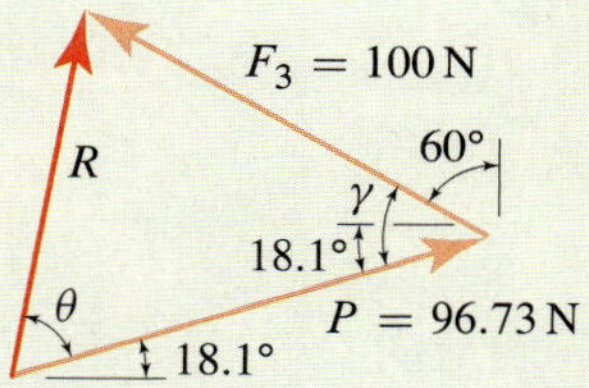

Figure 5
Addition of $\vec{P}$ and $\vec{F}_3$ to obtain the final result $\vec{R}$ (i.e., $\vec{R} = \vec{P} + \vec{F}_3$).

The law of sines is then used to write

$$\frac{F_3}{\sin\theta} = \frac{R}{\sin\gamma}. \tag{7}$$

Hence

$$\sin\theta = \frac{F_3}{R}\sin\gamma = \frac{100\,\text{N}}{80.2\,\text{N}}\sin 48.1° = 0.928, \tag{8}$$

which provides

$$\theta = \sin^{-1}(0.928) = 68.1°. \tag{9}$$

To complete our solution, we report $\vec{R}$ in vector form using polar vector representation. The magnitude of $\vec{R}$ is $R = 80.2$ N, and its orientation is $\theta + \beta = 68.1° + 18.1° = 86.2°$ counterclockwise from the right-hand horizontal direction. Thus,

$$\boxed{\vec{R} = 80.2\,\text{N} @ 86.2° \measuredangle.} \tag{10}$$

Discussion & Verification

- Addition of vectors by the method illustrated here is tedious. In fact, as the number of vectors to be added increases, the number of sides of the vector polygon also increases. The Cartesian vector representation, introduced in Section 2.2, will provide for considerably more straightforward addition of vectors. Nonetheless, it is often very helpful to be able to sketch vector polygons, as this provides a visual understanding of the spatial relationship among vectors which in turn may offer significant simplifications in analysis.
- This problem focused on the resultant force applied by the three cords to the D ring. However, the fabric that attaches the D ring to the backpack also applies forces to the D ring. In Chapter 3, we consider the combined effect of all forces applied to objects such as the D ring in this example.

EXAMPLE 2.2 Resolution of a Vector into Components

Two people apply forces to push a child's play structure resting on a patio. The woman at A applies a force in the negative a direction and the man at B applies a force in the b direction, with the goal of producing a resultant force of 250 N in the c direction. Determine the forces the two people must apply, expressing the results as vectors.

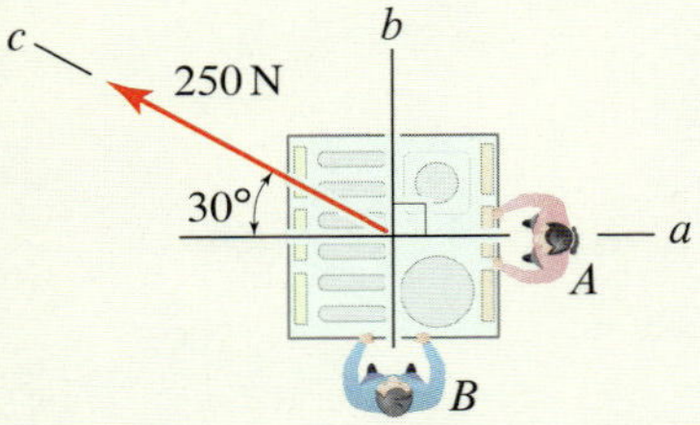

Figure 1

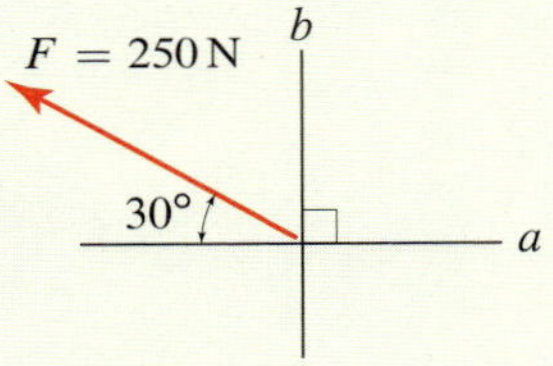

Figure 2
The resultant force applied to the play structure by the two people has 250 N magnitude.

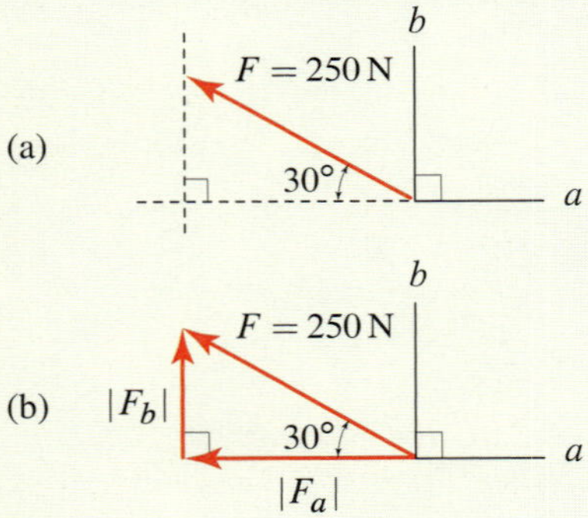

Figure 3
Resolution of the 250 N force vector into components in the a and b directions.

SOLUTION

Road Map The force applied by the woman at A will be called $\vec{F}_a$, and the force applied by the man at B will be called $\vec{F}_b$, such that the sum of these is the 250 N force vector $\vec{F}$ shown in Fig. 2

$$\vec{F} = \vec{F}_a + \vec{F}_b. \tag{1}$$

Thus, our goal is to resolve $\vec{F}$ into the component vectors $\vec{F}_a$ and $\vec{F}_b$. Since the a and b directions are orthogonal, determining the component vectors will be straightforward using basic trigonometry.

Governing Equations & Computation We first note that both vectors $\vec{F}_a$ and $\vec{F}_b$ have known direction and unknown magnitude. To determine the components, you may find the process shown in Fig. 3 to be helpful. In Fig. 3(a), we begin by sketching a dashed line parallel to the a direction that passes through the tail of the vector, followed by sketching a dashed line parallel to the b direction that passes through the head of the vector. These dashed lines form a triangle, which leads to the sketch shown in Fig. 3(b). Using elementary trigonometry, the absolute values of the components are

$$|F_a| = (250\,\text{N})\cos 30^\circ = 216.5\,\text{N}, \tag{2}$$

$$|F_b| = (250\,\text{N})\sin 30^\circ = 125.0\,\text{N}. \tag{3}$$

The absolute value signs are used in Eqs. (2) and (3) and in Fig. 3 because, by definition, the components of the vector are positive in the positive a and b directions. The vector components may now be written as

$$\boxed{\begin{aligned} \vec{F}_a &= -217\ \text{N} \quad \text{in the } a \text{ direction}, && (4)\\ \vec{F}_b &= 125\ \text{N} \quad \text{in the } b \text{ direction}. && (5)\end{aligned}}$$

where the negative sign is needed in Eq. (4) because this component, as seen in Fig. 3(b), acts in the negative a direction. Further, we state that the scalar components (or simply components) of the 250 N force in the a and b directions are -217 and 125 N, respectively.

Equations (4) and (5) state both magnitude and direction and hence are vector expressions. These results may alternatively be stated using polar vector representation as

$$\boxed{\begin{aligned} \vec{F}_a &= 217\,\text{N} @ 180^\circ \measuredangle, && (6)\\ \vec{F}_b &= 125\,\text{N} @ 90^\circ \measuredangle. && (7)\end{aligned}}$$

Discussion & Verification By examining Fig. 3(b), you should verify that F_a and F_b have reasonable values. That is, we see in Fig. 3(b) that the "length" of $\vec{F}_a$ is somewhat smaller than the "length" of $\vec{F}$; thus we expect F_a (found above to be 217 N) to be somewhat smaller than 250 N. Similarly, we see in Fig. 3(b) that the "length" of $\vec{F}_b$ is somewhat smaller than the "length" of $\vec{F}_a$, thus we expect F_b (found above to be 125 N) to be somewhat smaller than 217 N. Furthermore, since the forces in Fig. 3(b) form a right triangle, we can use the Pythagorean theorem, Eq. (2.10), to verify that $\sqrt{F_a^2 + F_b^2} = 250\,\text{N}$.

EXAMPLE 2.3 *Resolution of a Vector into Components*

Repeat Example 2.2 to determine the forces the two people must apply in the new a and b directions shown such that a resultant force of 250 N in the c direction is produced.

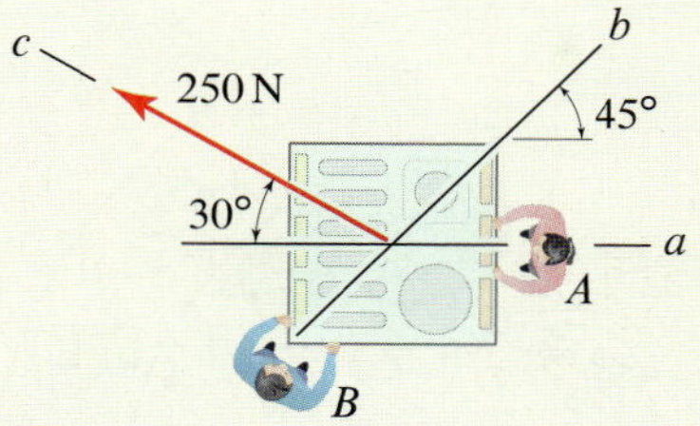

Figure 1

SOLUTION

Road Map The people at A and B apply forces $\vec{F}_a$ and $\vec{F}_b$, respectively, to the child's play structure such that the sum of these is the 250 N force vector $\vec{F}$ shown in Fig. 2

$$\vec{F} = \vec{F}_a + \vec{F}_b. \tag{1}$$

Thus, our goal is to resolve $\vec{F}$ into the component vectors $\vec{F}_a$ and $\vec{F}_b$. Since the a and b directions are not orthogonal, determining the component vectors will be more work than in Example 2.2, and the law of sines and/or cosines will be needed.

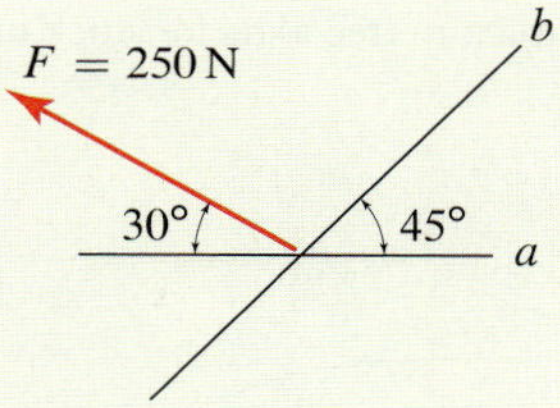

Figure 2
The resultant force applied to the play structure by the two people has 250 N magnitude.

Governing Equations & Computation As in Example 2.2, both vectors $\vec{F}_a$ and $\vec{F}_b$ have known direction and unknown magnitude. To determine the components, you may find the process shown in Fig. 3 to be helpful. In Fig. 3(a), we begin by sketching a dashed line parallel to the a direction that passes through the tail of the vector, followed by sketching a dashed line parallel to the b direction that passes through the head of the vector. These dashed lines form a triangle, which leads to the sketch shown in Fig. 3(b). Angle α shown in Fig. 3(b) is easily found to be $\alpha = 180° - 30° - 45° = 105°$. Using the law of sines, the absolute values of the components $|F_a|$ and $|F_b|$ are related by

$$\frac{250\text{ N}}{\sin 45°} = \frac{|F_a|}{\sin\alpha} = \frac{|F_b|}{\sin 30°}, \tag{2}$$

hence

$$|F_a| = 341.5\text{ N} \quad \text{and} \quad |F_b| = 176.8\text{ N}. \tag{3}$$

The absolute value signs are used in Eqs. (2) and (3) and in Fig. 3 because, by definition, the components of a vector are positive in the positive a and b directions. The vector components may now be written as

$$\vec{F}_a = -342\text{ N} \quad \text{in the } a \text{ direction}, \tag{4}$$

$$\vec{F}_b = 177\text{ N} \quad \text{in the } b \text{ direction}. \tag{5}$$

where the negative sign is needed in Eq. (4) because this component, as seen in Fig. 3(b), acts in the negative a direction. Further, the scalar components (or simply components) of the 250 N force in the a and b directions are −342 and 177 N, respectively. Equations (4) and (5) can alternatively be rewritten using polar vector representation as

$$\vec{F}_a = 342\text{ N} \ @\ 180°\measuredangle, \tag{6}$$

$$\vec{F}_b = 177\text{ N} \ @\ 45°\measuredangle. \tag{7}$$

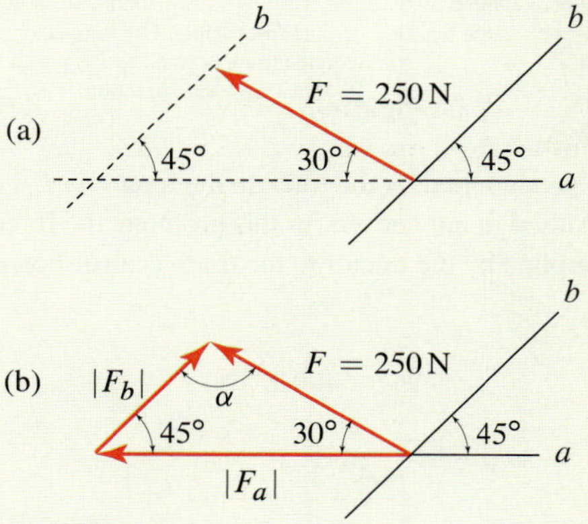

Figure 3
Resolution of the 250 N force vector into components in the a and b directions.

Discussion & Verification

- By examining Fig. 3(b), you should verify that F_a and F_b have reasonable values. That is, we expect F_a to be somewhat larger than 250 N and F_b to somewhat smaller than 250 N.
- To achieve the objective of producing a 250 N resultant force in the c direction shown in Fig. 1, the woman at A in this example problem must apply a substantially higher force than in the previous example problem (342 N versus 217 N). Obviously, the man at B in this example is applying force in a direction that is not as effective as in Example 2.2.

EXAMPLE 2.4 *Parallel and Perpendicular Vector Components*

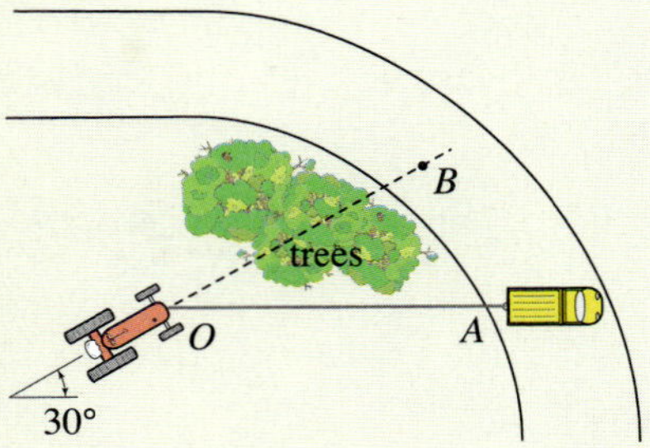

Figure 1
Use of a truck to free a tractor stuck in mud.

A tractor becomes stuck in mud, and the operator plans to use a truck positioned at point A to pull it free.

(a) If a 400 lb force in the direction of the tractor's chassis (i.e., in the direction from O to B) is sufficient to free it, determine the force the truck must produce in the cable. Also, determine the portion of the cable force that is perpendicular to the tractor's direction, since in the tractor's precarious condition this force will tend to tip it.

(b) If the trees were not present, determine the ideal location on the road where the truck should be positioned.

SOLUTION

Road Map The force applied by the truck to the tractor will be called F_{OA}, and this force is shown in Fig. 2. Note that according to Newton's third law, the force applied by the tractor to the truck has equal magnitude and opposite direction, and this force is also shown in Fig. 2. We will assume that points O, A, and B lie in a horizontal plane so that this is a two-dimensional problem; if this is not the case, then vectors in three dimensions are needed.

With the truck positioned at A, we ask for the cable force F_{OA} needed so that the component of this force in the tractor's direction (line OB) is 400 lb. We denote by $\vec{F}_\parallel$ and $\vec{F}_\perp$ the vector components of $\vec{F}_{OA}$; i.e., $\vec{F}_\parallel + \vec{F}_\perp = \vec{F}_{OA}$, where $\vec{F}_\parallel$ is the force in the direction OB, and $\vec{F}_\perp$ is the force in the direction perpendicular to OB. In this problem, $\vec{F}_\parallel$ is known and has 400 lb magnitude in the direction of OB. Vector $\vec{F}_\perp$ has unknown magnitude but known direction that is perpendicular to OB, and $\vec{F}_{OA}$ has unknown magnitude but known direction OA. For Part (b), we will determine the location of the truck so that the magnitude of $\vec{F}_\perp$ will be zero, and hence there will be no tendency for the tractor to tip while it is being pulled free.

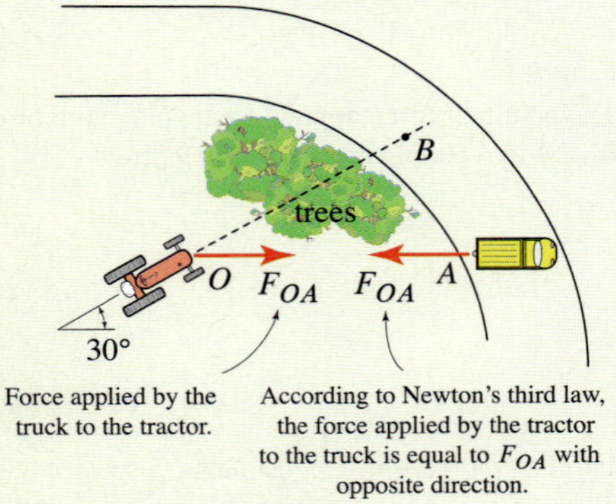

Figure 2
Force applied by the truck to the tractor is F_{OA}. Although not needed for this problem, the force applied by the tractor to the truck is also shown.

Part (a)

Governing Equations & Computation The force polygon for the addition $\vec{F}_\parallel + \vec{F}_\perp = \vec{F}_{OA}$ is shown in Fig. 3 (alternatively, the force polygon for the addition $\vec{F}_\perp + \vec{F}_\parallel = \vec{F}_{OA}$ could be used). Because this force polygon is a right triangle, the basic trigonometric relations in Eq. (2.9) can be used to provide

$$F_{OA} \cos 30° = 400\,\text{lb} \quad \Rightarrow \quad \boxed{F_{OA} = 462\,\text{lb},} \tag{1}$$

and

$$F_\perp = F_{OA} \sin 30° = \boxed{231\,\text{lb}.} \tag{2}$$

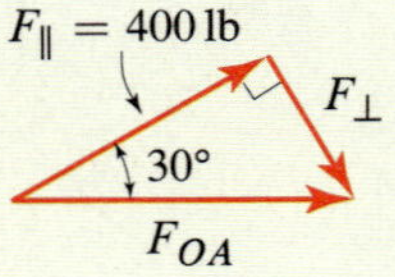

Figure 3
Force polygon for the force applied by the truck to the tractor.

Discussion & Verification By examining the force polygon in Fig. 3, you should verify that F_{OA} and $F_\perp$ have reasonable values. That is, we expect F_{OA} to be somewhat larger than 400 lb, and $F_\perp$ to be somewhat smaller than 400 lb. Furthermore, since the forces in Fig. 3 form a right triangle, we can use the Pythagorean theorem to verify that $\sqrt{(400\,\text{lb})^2 + (F_\perp)^2} = F_{OA}$.

Part (b)

Discussion & Verification If the trees were not present, then the optimal location for the truck would be $\boxed{\text{point } B.}$ To see that this is true, we consult Fig. 3 and observe that if the truck is positioned at point B, then $F_\perp = 0$ and the entire cable force F_{OA} is in the direction of the tractor. Thus, to free the tractor requires $F_{OA} = F_\parallel = 400\,\text{lb}$, which is smaller than the value found in Eq. (1). Furthermore, $F_\perp = 0$ so there is no force tending to tip the tractor.

EXAMPLE 2.5 *Vector Components and Optimization*

Repeat Example 2.4, now using *two* trucks to free a tractor that is stuck in mud. By using two trucks, it is possible to avoid the tendency to tip the tractor. To free the tractor, the two trucks must produce a combined force of 400 lb in the direction of the tractor's chassis (i.e., in the direction from O to B).

(a) One truck is positioned at point A and a second truck is positioned at point C. Determine the cable forces so that there is no force perpendicular to the tractor's direction (i.e., there is no tipping force).

(b) Determine the point C' where the truck originally at point C should be relocated so that there is no force perpendicular to the tractor's direction and the force in this truck's cable is as small as possible.

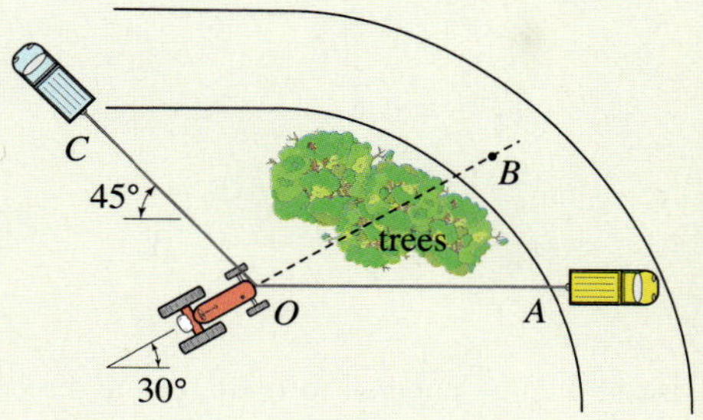

Figure 1
Use of two trucks to free a tractor stuck in mud.

SOLUTION

Road Map The main benefit of using two trucks to free the tractor is that it is possible to simultaneously produce a 400 lb force in the direction of the tractor *and* zero force perpendicular to the tractor. The forces applied by the trucks to the tractor are called F_{OA} and F_{OC}, as shown in Fig. 2. Although not needed for this problem, Newton's third law has been used to also show the forces applied by the tractor to the trucks in Fig. 2. We will assume that points O, A, B, and C lie in a horizontal plane so that this is a two-dimensional problem; if this is not the case, then vectors in three dimensions are needed.

In both parts of this problem, we ask for two force vectors, one in direction OA and the other in direction OC, such that $\vec{F}_{OA} + \vec{F}_{OC} = \vec{F}_{\parallel}$. In Part (a), both $\vec{F}_{OA}$ and $\vec{F}_{OC}$ have known direction but unknown magnitude, and $\vec{F}_{\parallel}$ is fully known. In Part (b), the only known information is the direction of $\vec{F}_{OA}$ and the direction and magnitude of $\vec{F}_{\parallel}$.

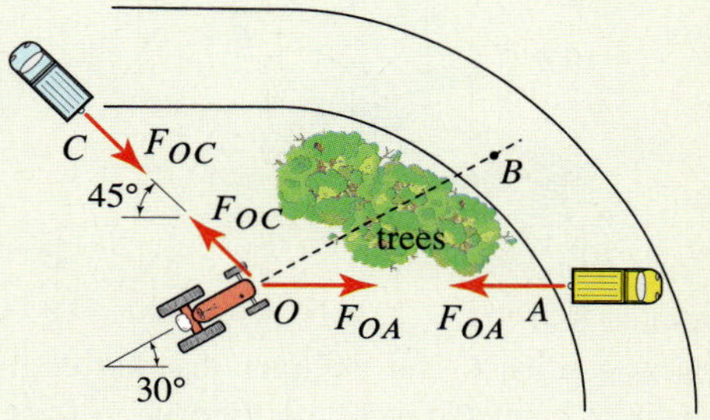

Figure 2
Forces applied by the trucks to the tractor are F_{OA} and F_{OC}. Although not needed for this problem, the forces applied by the tractor to the trucks are also shown.

Part (a)

Governing Equations & Computation The force polygon is shown in Fig. 3, and it has general triangular shape (i.e., it does not have a right angle). First, angle α is easily found as $\alpha = 180° - 30° - 45° = 105°$. Then the law of sines provides

$$\frac{400\,\text{lb}}{\sin 45°} = \frac{F_{OA}}{\sin\alpha} = \frac{F_{OC}}{\sin 30°}, \tag{1}$$

hence

$$\boxed{F_{OA} = 546\,\text{lb} \quad \text{and} \quad F_{OC} = 283\,\text{lb}.} \tag{2}$$

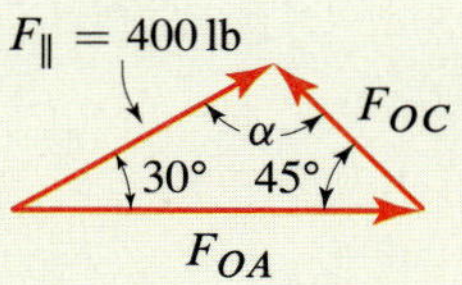

Figure 3
Force polygon for the forces applied by the trucks to the tractor.

Discussion & Verification By examining the force polygon in Fig. 3, you should verify that F_{OA} and F_{OC} have reasonable values. That is, we expect F_{OA} to be somewhat larger than 400 lb, and F_{OC} to be somewhat smaller than 400 lb.

Part (b)

Governing Equations & Computation In this part, we ask where the truck originally at C should be relocated so that the force in its cable is as small as possible, while still producing the necessary 400 lb in the direction of the tractor (from points O to B) and zero force perpendicular to the tractor's direction; the new location for this truck is denoted by C'. As in Part (a), we seek two vectors $\vec{F}_{OA}$ and $\vec{F}_{OC'}$ such that $\vec{F}_{OA} + \vec{F}_{OC'} = \vec{F}_{\parallel}$ and as before, $\vec{F}_{\parallel}$ is known and $\vec{F}_{OA}$ has known direction but unknown magnitude. However, this problem is more complicated because $\vec{F}_{OC'}$ has unknown magnitude and direction. It is possible to use calculus to determine the optimal position for the truck at C' (Problem 2.16 explores this solution). However, consideration of force polygons corresponding to possible choices for $\vec{F}_{OA}$ and $\vec{F}_{OC'}$ will offer a more straightforward solution as follows.

In Fig. 4 we first sketch vector $\vec{F}_{\parallel}$ since it has known magnitude and direction. Since $\vec{F}_{OA}$ has known direction, we sketch some possible choices for it that have different magnitude, and three such choices are shown in Fig. 4. For each $\vec{F}_{OA}$ we sketch the required vector $\vec{F}_{OC'}$ to complete the force polygon, and these three vectors are also shown in Fig. 4. Examination of the vectors $\vec{F}_{OC'}$ in Fig. 4 clearly shows that the smallest magnitude of $\vec{F}_{OC'}$ is achieved when $\vec{F}_{OC'}$ is perpendicular to $\vec{F}_{OA}$. Hence, to minimize the magnitude of the force in cable OC', the truck originally at C should be repositioned to point C' shown in Fig. 5. Now that the direction of $\vec{F}_{OC'}$ is known, the magnitudes of $\vec{F}_{OA}$ and $\vec{F}_{OC'}$ can be found, as was done in Part (a), except now Eqs. (2.9) and (2.10) for a right triangle may be used. Doing so, we find $F_{OA} = 346$ lb and $F_{OC'} = 200$ lb.

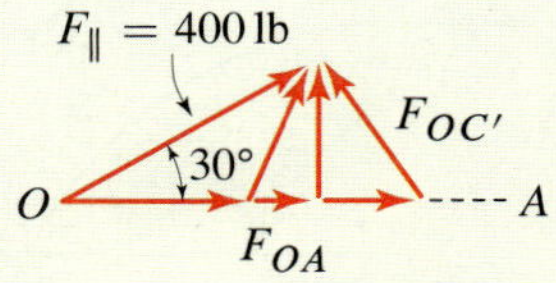

Figure 4
Use of a force polygon to optimize the direction of a force. Three possible choices for $\vec{F}_{OA}$ are shown, along with the corresponding vectors $\vec{F}_{OC'}$ that complete the force polygon.

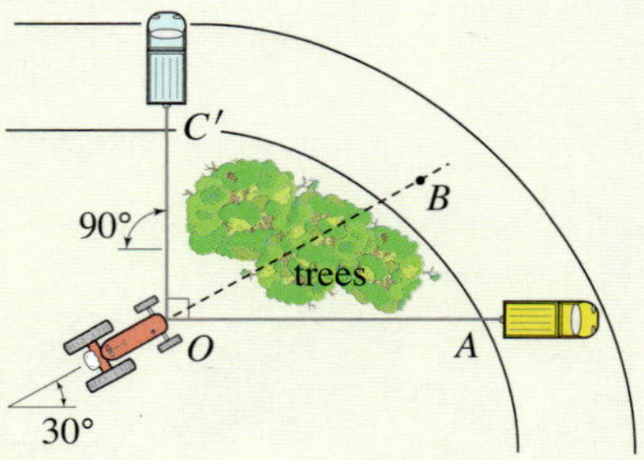

Figure 5
New position C' for the truck that was originally at point C, such that there is no force perpendicular to the tractor's direction and the force in cable OC' is as small as possible.

Discussion & Verification

- **Usefulness of force polygons.** The solution to Part (b) requires a vector with unknown magnitude and direction be determined (i.e., $\vec{F}_{OC'}$). This may be determined using calculus, as described in Problem 2.16, but nonlinear equations must be solved. By drawing a force polygon, the solution for the optimal orientation of $\vec{F}_{OC'}$ was found by inspection, with the result that it was easier to solve for the remaining unknown, namely, the magnitude of $\vec{F}_{OC'}$.
- **Tip.** Although in principle vectors can be added in any order, from a practical standpoint, when constructing vector polygons, you should begin with those vectors whose magnitude and direction are known. Vectors that have unknown magnitude and/or direction should be added last. For example, in Part (b), the force polygon resulting from addition of forces in the order $\vec{F}_{OC'}$, and then $\vec{F}_{OA}$, followed by $\vec{F}_{\parallel}$ would be considerably less straightforward to visualize than Fig. 4.

PROBLEMS

Problem 2.1

For each vector, write two expressions using polar vector representations, one using a positive value of θ and the other a negative value, where θ is measured counterclockwise from the right-hand horizontal direction.

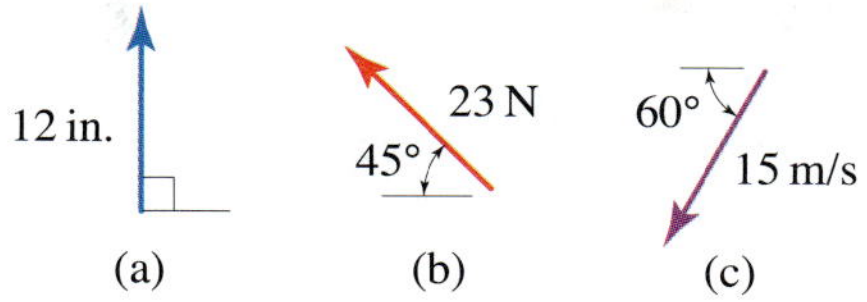

Figure P2.1

Problems 2.2 and 2.3

Add the two vectors shown to form a resultant vector $\vec{R}$, and report your result using polar vector representation.

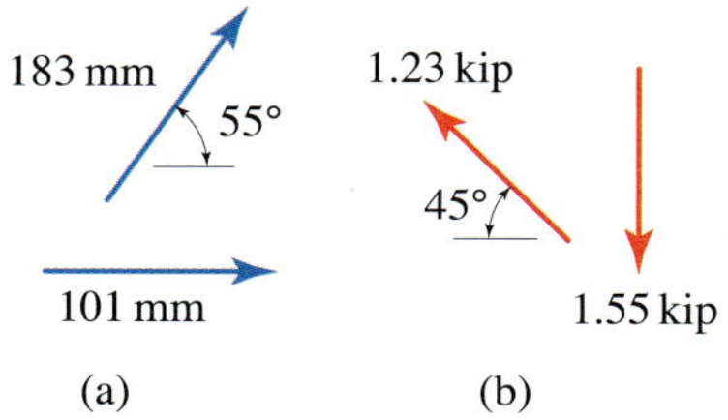

Figure P2.2

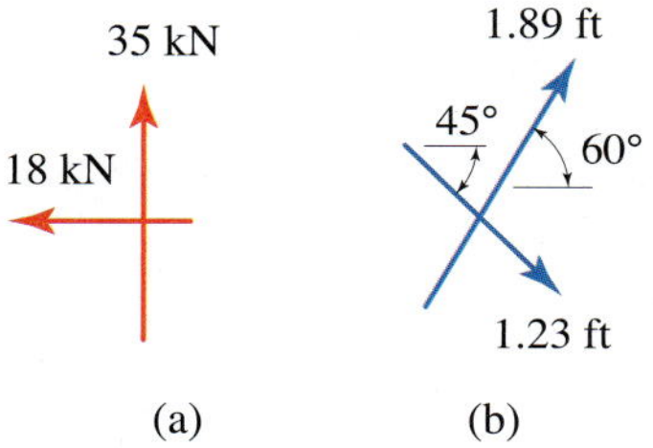

Figure P2.3

Problem 2.4

Let $\vec{A} = 2\,\text{m} @ 0°\measuredangle$ and $\vec{B} = 6\,\text{m} @ 90°\measuredangle$. Sketch the vector polygons and evaluate $\vec{R}$ for the following, reporting your answer using polar vector representation.

(a) $\vec{R} = \vec{A} + \vec{B}$,

(b) $\vec{R} = 2\vec{A} - \vec{B}$,

(c) $\vec{R} = |\vec{A}|\,\vec{B} + |\vec{B}|\,\vec{A}$,

(d) $\vec{R} = \dfrac{\vec{A}}{|\vec{A}|} + \dfrac{\vec{B}}{|\vec{B}|}$.

Problem 2.5

A tow truck applies forces $\vec{F}_1$ and $\vec{F}_2$ to the bumper of an automobile where $\vec{F}_1$ is horizontal. Determine the magnitude of $\vec{F}_2$ that will provide a vertical resultant force, and determine the magnitude of this resultant.

Figure P2.5

Problem 2.6

Arm OA of a robot is positioned as shown. Determine the value for angle α of arm AB so that the distance from point O to the actuator at B is 650 mm.

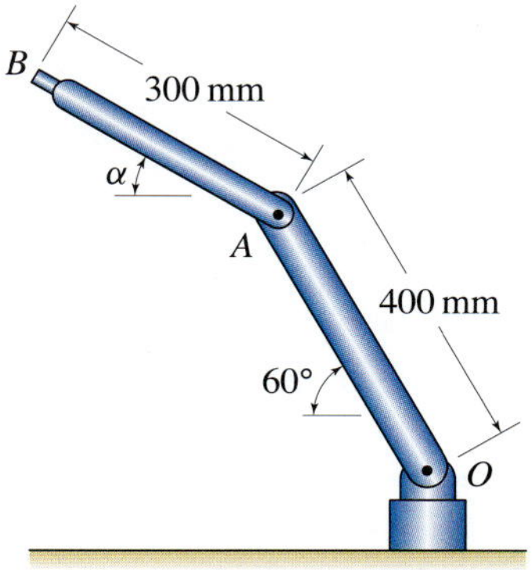

Figure P2.6

Problem 2.7

Add the three vectors shown to form a resultant vector $\vec{R}$, and report your result using polar vector representation.

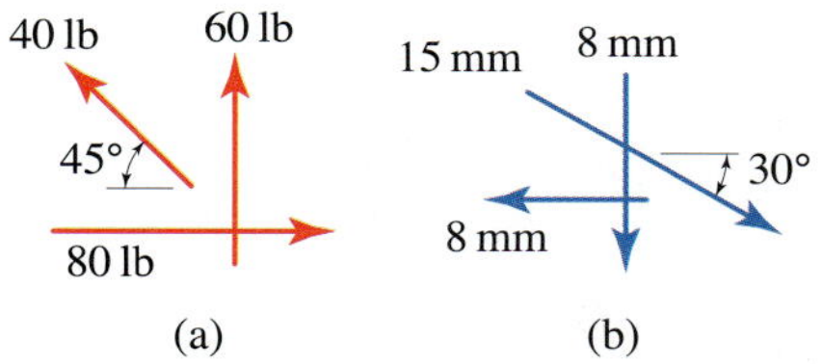

Figure P2.7

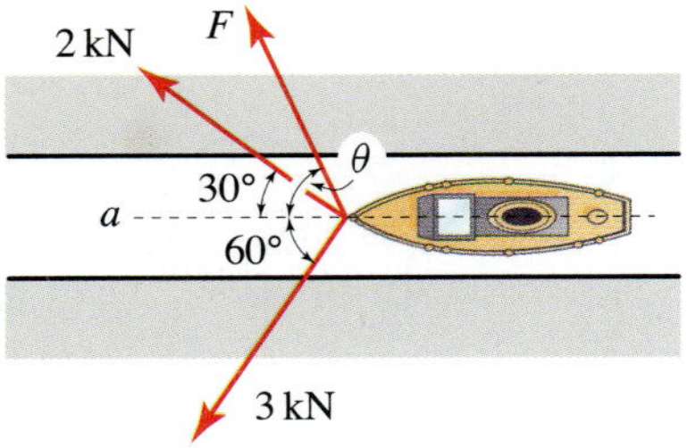

Figure P2.8

Problem 2.8

A ship is towed through a narrow channel by applying forces to three ropes attached to its bow. Determine the magnitude and orientation θ of the force $\vec{F}$ so that the resultant force is in the direction of line a and the magnitude of $\vec{F}$ is as small as possible.

Problem 2.9

A surveyor needs to plant a marker directly northeast from where she is standing. Because of obstacles, she walks a route in the horizontal plane consisting of 200 m east, followed by 400 m north, followed by 300 m northeast. From this position, she would like to take the shortest-distance route back to the line that is directly northeast of her starting position. What direction should she travel and how far, and what will be her final distance from her starting point?

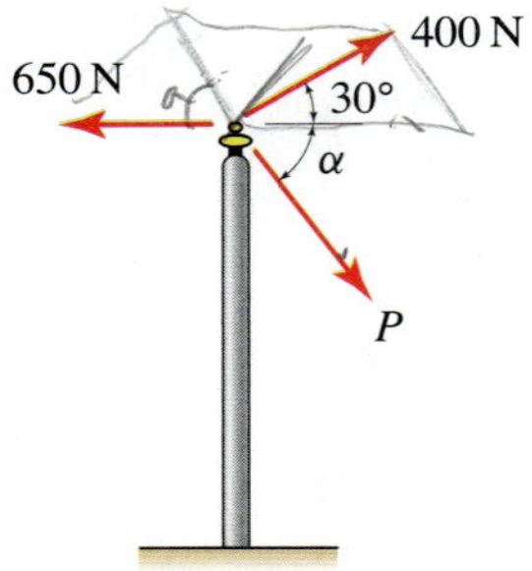

Figure P2.10

Problem 2.10

A utility pole supports a bundle of wires that apply the 400 and 650 N forces shown, and a guy wire applies the force $\vec{P}$.

(a) If $\vec{P} = \vec{0}$, determine the resultant force applied by the wires to the pole and report your result using polar vector representation.

(b) Repeat Part (a) if $P = 500$ N and $\alpha = 60°$.

(c) With $\alpha = 60°$, what value of P will produce a resultant force that is vertical?

(d) If the resultant force is to be vertical and P is to be as small as possible, determine the value α should have and the corresponding value of P.

Problem 2.11

Determine the smallest force F_1 such that the resultant of the three forces F_1, F_2, and F_3 is vertical, and the angle α at which F_1 should be applied.

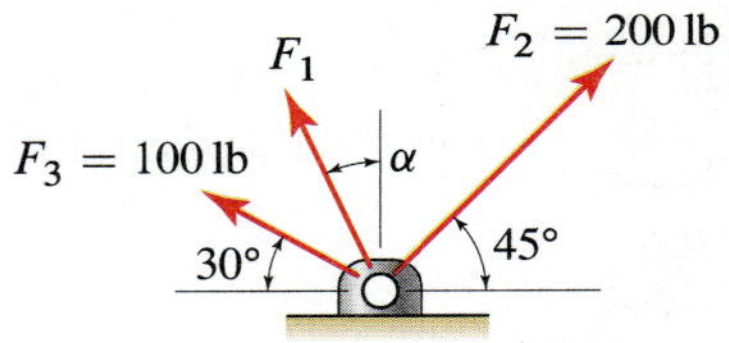

Figure P2.11

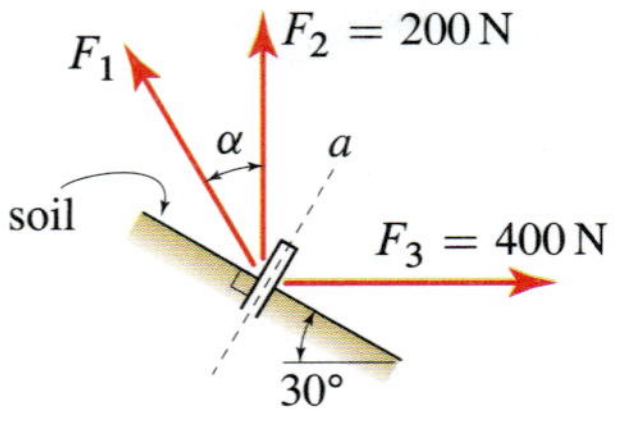

Figure P2.12

Problem 2.12

Forces F_1, F_2, and F_3 are applied to a soil nail to pull it out of a slope. If F_2 and F_3 are vertical and horizontal, respectively, with the magnitudes shown, determine the magnitude of the smallest force F_1 that can be applied and the angle α so that the resultant force applied to the nail is directed along the axis of the nail (direction a).

Problem 2.13

Determine the magnitudes of vectors $\vec{r}_a$ and $\vec{r}_b$ in the a and b directions, respectively, such that their sum is the 2 km position vector shown.

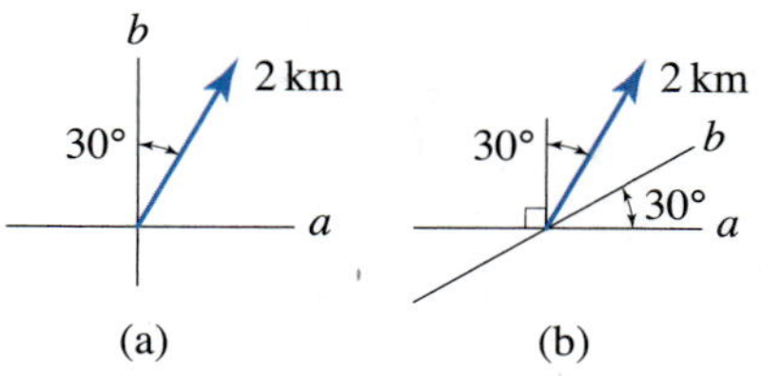

Figure P2.13

Problem 2.14

Determine the magnitudes of vectors $\vec{F}_a$ and $\vec{F}_b$ in the a and b directions, respectively, such that their sum is the 100 lb force vector shown.

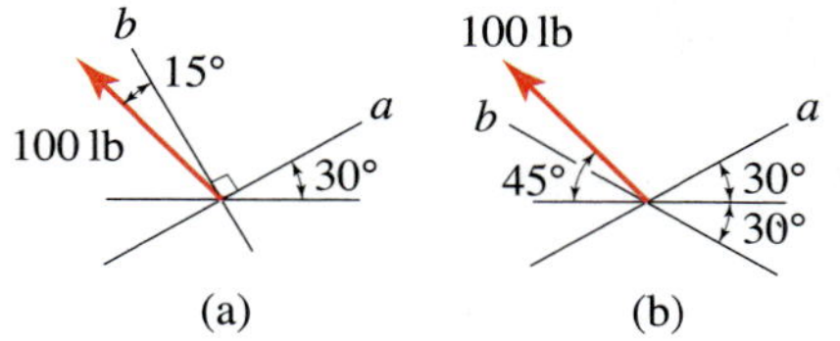

Figure P2.14

Problem 2.15

While canoes are normally propelled by paddle, if there is a favorable wind from the stern, adventurous users will sometimes employ a small sail. If a canoe is sailing northwest and the wind applies a 40 lb force perpendicular to the sail in the direction shown, determine the components of the wind force parallel and perpendicular to the keel of the canoe (direction a).

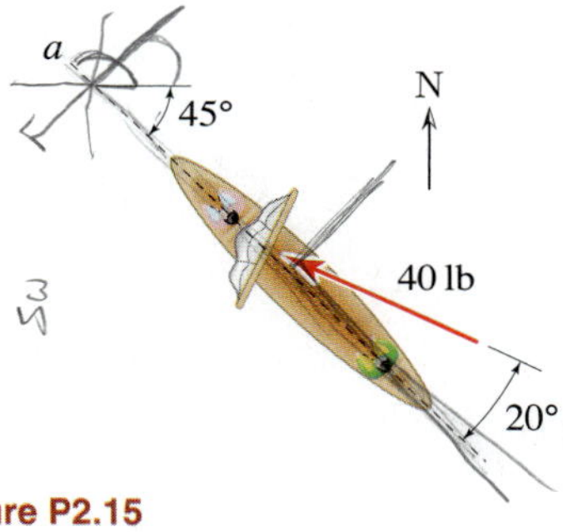

Figure P2.15

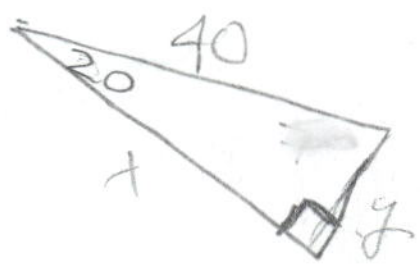

Problem 2.16

Repeat Part (b) of Example 2.5, using the optimization methods of calculus. *Hint:* Redraw the force polygon of Fig. 3 and rewrite Eq. (1) on p. 39 with the 45° angle shown there replaced by β, where β defined in Fig. P2.16. Rearrange this equation to obtain an expression for $F_{OC'}$ as a function of β, and then determine the value of β that makes $dF_{OC'}/d\beta = 0$. While the approach described here is straightforward to carry out "by hand," you might also consider using symbolic algebra software such a Mathematica or Maple.

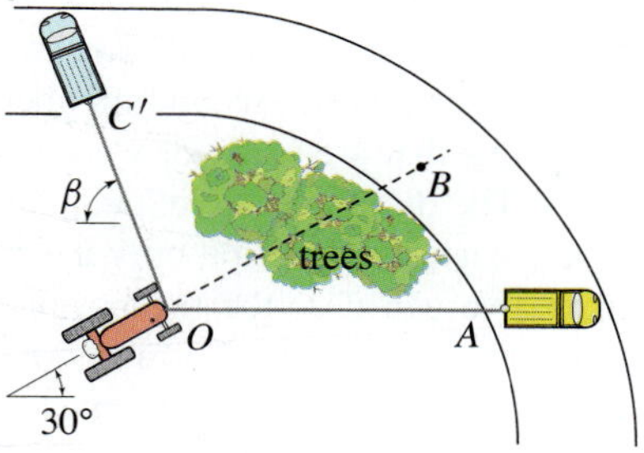

Figure P2.16

2.2 Cartesian Representation of Vectors in Two Dimensions

There are a variety of ways, or representations, that may be used to embed magnitude and direction information in a vector expression. Cartesian vector representation is straightforward and is the most widely used form for expressing vectors. For special classes of problems, such as when geometry is circular or spherical, other forms of vector representation may be more convenient. Some of these representations are discussed in dynamics.

Introduction – Cartesian representation and a walk to work

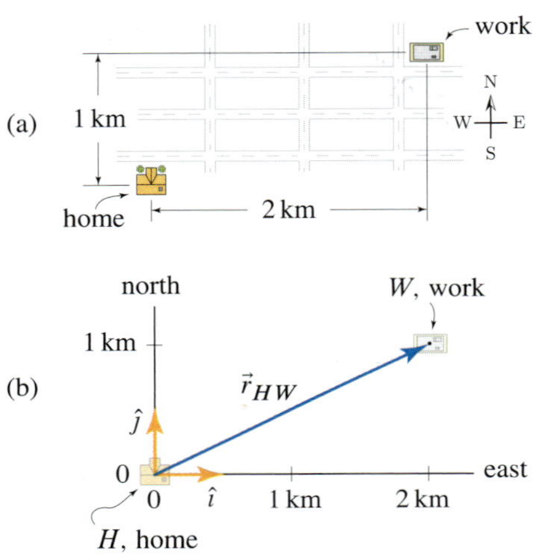

Figure 2.11
(a) A map shows the locations of home and work. (b) A Cartesian coordinate system is used to describe positions and vectors.

Imagine that to get from your home to where you work, you walk 2 km east and 1 km north, as shown on the map of Fig. 2.11(a). This simple statement implies a Cartesian coordinate system, unit vectors, and a position vector that describes the location of where you work relative to your home. Figure 2.11(b) shows these ingredients, with the following explanations. We will use east and north to describe directions (we could just as well use west and/or south, respectively), and since these directions are orthogonal (that is, their intersection forms a right angle), they constitute a *Cartesian coordinate system*. We may select the *origin* of this coordinate system to be any point we choose, and once this is done, the location of other points can be quantified using *coordinates*. For example, if we take the origin of our coordinate system to be our home, as shown in Fig. 2.11(b), then the coordinates of where we work are (2 km, 1 km), where this pair of numbers is ordered such that the first number is always the east coordinate and the second number is always the north coordinate. Note that positions that are south or west of our origin will have some negative coordinates. The directions "east" and "north" define *unit vectors*. That is, when we say east, we describe only a direction, and no statement of distance is made. We can describe the position vector from home to work by $\vec{r}_{HW} = 2\,\text{km east} + 1\,\text{km north}$, or in more compact notation by $\vec{r}_{HW} = 2\,\text{km}\,\hat{\imath} + 1\,\text{km}\,\hat{\jmath}$, where $\hat{\imath}$ means east and $\hat{\jmath}$ means north. The concepts introduced in this example are more thoroughly described in the remainder of this section.

Unit vectors

The concept of a unit vector is useful. A *unit vector* is defined to be a dimensionless vector that has unit magnitude. Given any arbitrary vector $\vec{v}$ having nonzero magnitude, we may construct a unit vector $\hat{u}$ that has the same direction, using

$$\hat{u} = \frac{\vec{v}}{|\vec{v}|}. \tag{2.11}$$

We will use a "hat" symbol ($\hat{}$) over unit vectors, whereas all other vectors will use an arrow symbol ($\vec{}$). Note that vector $\vec{v}$ and its magnitude $|\vec{v}|$ have the same units, thus unit vector $\hat{u}$ is dimensionless. In this book, we will reserve the symbol $\hat{u}$ to represent unit vectors, although we will also use the more primitive form $\vec{v}/|\vec{v}|$. In figures, unit vectors will be orange ■.

Cartesian coordinate system

Most likely you already have considerable experience using Cartesian coordinate systems, but to be complete, we will state exactly what ingredients are needed. In two dimensions a *Cartesian coordinate system* uses two orthogonal reference directions, which we will usually call the x and y directions, as shown in Fig. 2.12(a). The intersection of the reference directions is the *origin* of the coordinate system, and any point in the xy plane is identified by its x and y coordinates. We denote the *coordinates* of a point P using an ordered pair of numbers (x_P, y_P) where x_P and y_P are the x and y coordinates of the point, respectively, measured from the origin. Although we will often take the x and y directions to be horizontal and vertical, respectively, other orientations, provided they are orthogonal, are acceptable and may be more convenient.

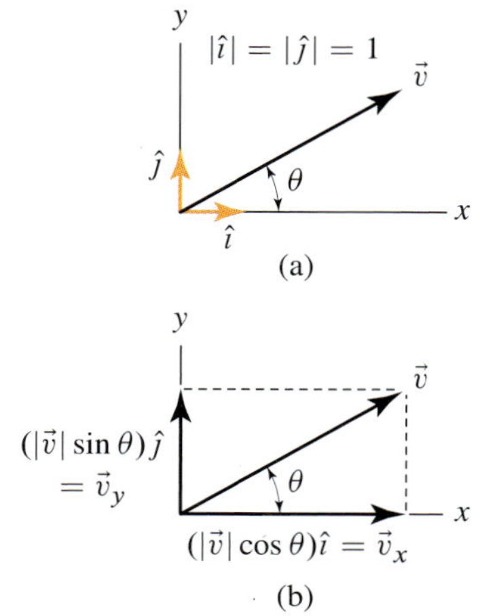

Figure 2.12
(a) Cartesian coordinate system with unit vectors $\hat{\imath}$ and $\hat{\jmath}$ in the x and y directions, respectively. (b) Resolution of a vector $\vec{v}$ into vector components $\vec{v}_x$ and $\vec{v}_y$ in the x and y directions, respectively.

Cartesian vector representation

We begin by defining unit vectors $\hat{\imath}$ and $\hat{\jmath}$ that point in the positive x and y directions, respectively, as shown in Fig. 2.12(a). Both $\hat{\imath}$ and $\hat{\jmath}$ are dimensionless and are used to describe direction in exactly the same way you would use east and north. Now consider a vector $\vec{v}$ in the xy plane. Following the ideas used in Section 2.1, we ask for two vectors, one parallel to the x axis and one parallel to the y axis, such that their vector sum yields $\vec{v}$. This can be stated in equation form as

$$\vec{v} = \vec{v}_x + \vec{v}_y \tag{2.12}$$

where $\vec{v}_x$ and $\vec{v}_y$ are vectors parallel to the x and y directions, respectively, as shown in Fig. 2.12(b). Vectors $\vec{v}_x$ and $\vec{v}_y$ are called the *vector components* of $\vec{v}$. Noting that vectors $\vec{v}_x$ and $\hat{\imath}$ are parallel (although possibly with opposite direction), we may write $\vec{v}_x = v_x\,\hat{\imath}$, and similarly, we may write $\vec{v}_y = v_y\,\hat{\jmath}$. Thus, Eq. (2.12) becomes

$$\vec{v} = v_x\,\hat{\imath} + v_y\,\hat{\jmath}. \tag{2.13}$$

An expression in the form of Eq. (2.13) is called a *Cartesian representation* of the vector $\vec{v}$. The scalars v_x and v_y are called the *scalar components* (or simply the *components*) of $\vec{v}$, and these may be positive, zero, or negative. Because the vector components $\vec{v}_x$ and $\vec{v}_y$ are always orthogonal, basic trigonometric relations and the Pythagorean theorem, Eqs. (2.9) and (2.10), are all that is needed to resolve a vector into its Cartesian components.

Magnitude and orientation of a vector. For a vector in two dimensions, the *magnitude* $|\vec{v}|$ and *orientation* θ from the horizontal ($\pm x$ direction) are related to its Cartesian components by

$$|\vec{v}| = \sqrt{v_x^2 + v_y^2} \quad \text{and} \quad \theta = \tan^{-1}\left(\frac{v_y}{v_x}\right). \tag{2.14}$$

The magnitude of a vector is a scalar that is positive, or is zero in the case of a *zero vector* (i.e., a vector having zero values for all its components). As discussed in Fig. 2.4, rather than using the magnitude, we will often represent a vector's "size" by using a scalar symbol such as v, where v is *the component of the vector in an assumed direction for the vector*. Because v is a component, it can be positive, zero, or negative, as follows. If v is positive or zero, then v equals the magnitude of the vector and the direction of the vector is the same as the direction we assumed it had. If v is negative, then $-v$ is equal to the magnitude of the vector and the direction of the vector is opposite the direction we assumed it had. Thus, under all circumstances, the absolute value of v gives the magnitude of the vector. The main benefit of this notation is that in many problems, we will know the line of action for a vector, but at the outset of a problem we may not know its direction along that line of action. Thus, for a vector $\vec{v}$ we will assume a direction along its line of action and will often find after our analysis is complete that $v < 0$, which means that in fact the vector's direction is opposite to the direction we assumed it had. Note that v and $|\vec{v}|$ are defined differently, although they both measure size and, under the circumstances described here, yield the same value. Throughout this book, we will be careful to call v the magnitude of a vector *only* when we are sure it is nonnegative.

Remark. The angle θ defined in Eq. (2.14) gives the orientation of a vector measured from the positive x direction if the vector's x component is positive (i.e., $v_x > 0$), and gives the orientation measured from the negative x direction if the vector's x component is negative (i.e., $v_x < 0$). If $v_x = 0$, then $\theta = \pm 90°$.

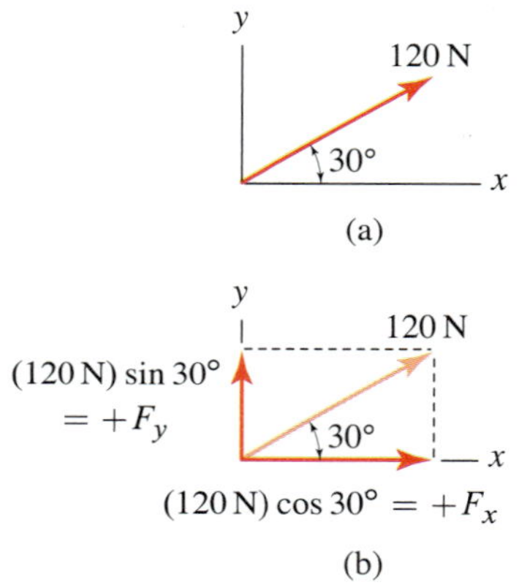

Figure 2.13
Determination of the Cartesian components of a 120 N force vector.

■ Mini-Example. Express the vector shown in Fig. 2.13(a), using Cartesian representation.

Solution. Let $\vec{F}$ denote the 120 N magnitude force vector shown in Fig. 2.13(a) and let F_x and F_y denote the components of $\vec{F}$ in the x and y directions, respectively. Components F_x and F_y are obtained by constructing projections of $\vec{F}$ that are parallel to the x and y axes, respectively, with the resulting magnitudes (120 N) cos 30° and (120 N) sin 30°, as shown in Fig. 2.13(b). We next use these projections to assign vector components in the x and y directions, paying careful notice that F_x and F_y are positive when acting in the positive x and y directions, respectively, and are negative when acting in the negative x and y directions. Thus,

$$F_x = (120\,\mathrm{N})\cos 30° = 104\,\mathrm{N}, \quad F_y = (120\,\mathrm{N})\sin 30° = 60\,\mathrm{N}. \tag{2.15}$$

Thus, the Cartesian representation for $\vec{F}$ is

$$\vec{F} = (104\,\hat{\imath} + 60\,\hat{\jmath})\,\mathrm{N}. \tag{2.16}$$

Some useful checks. There are a few useful checks that may help you avoid errors when resolving a vector into Cartesian components. Foremost is to verify that the components have reasonable size and are in proper directions. The next check is to verify that the vector's components give the correct magnitude. Thus, for Eq. (2.16), we evaluate $\sqrt{(104\,\mathrm{N})^2 + (60\,\mathrm{N})^2} = 120\,\mathrm{N}$ to find that indeed it has the correct magnitude. While these checks are reassuring, they do not guarantee the components are correct. Nonetheless, if we find an incorrect magnitude, then certainly an error has been made. ■

Mini-Example. Express the vector shown in Fig. 2.14(a), using Cartesian representation.

Solution. Geometric information for the 12 m position vector shown in Fig. 2.14(a), is in the form of a 3–4–5 triangle. While it is possible to evaluate the angle θ and then proceed as in the previous mini-example, this is not necessary and the geometry of the 3–4–5 triangle (or any other right triangle where the edge lengths are known) can be exploited to directly obtain the vector's components as follows. From Fig. 2.14(a), the vector's orientation is such that a "run" of 3 units is always accompanied by a "rise" of 4 units. If we imagine traversing 3 units of horizontal motion followed by 4 units of vertical motion, the total motion is $\sqrt{3^2+4^2}=5$ units. Of this 5 units total, 3 parts have been horizontal in the negative x direction, and 4 parts have been vertical in the positive y direction, and this constitutes a statement of the vector's components, as shown in Fig. 2.14(b). Hence we may write

$$\vec{r} = (12\,\mathrm{m})\left(\frac{-3}{5}\,\hat{\imath} + \frac{4}{5}\,\hat{\jmath}\right). \tag{2.17}$$

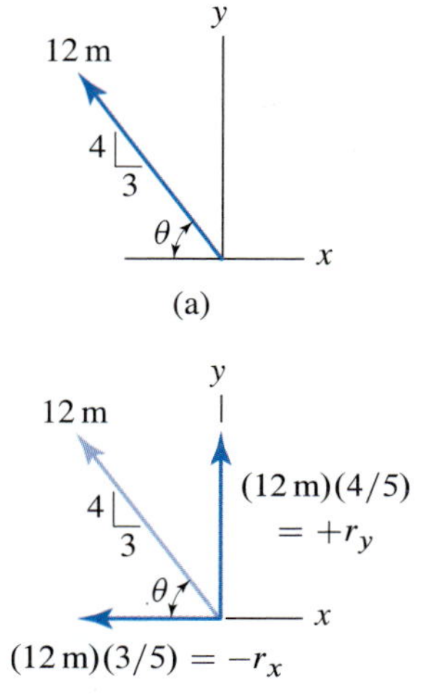

Figure 2.14
Determination of the Cartesian components of a 20 m position vector.

The negative sign in Eq. (2.17) and Fig. 2.14(b) has been included because the horizontal component of the vector is in the negative x direction. Note that the vector within the parentheses on the right-hand side of Eq. (2.17) is a unit vector, and hence $\vec{r}$ has 12 m magnitude, as expected.

Remarks. In this example, the geometry was especially nice (i.e., 3–4–5 triangle). Nonetheless, even if the rise, run, and/or hypotenuse is noninteger, this method of determining the Cartesian components of a vector is still very straightforward. If you find the heuristic argument used to construct Eq. (2.17) unconvincing, then we can use a more formal construction to show its validity. With θ as defined in Fig. 2.14(a), we write

$$\begin{aligned}\vec{r} &= -(12\,\mathrm{m})\,(\cos\theta)\,\hat{\imath} + (12\,\mathrm{m})\,(\sin\theta)\,\hat{\jmath}\\ &= (12\,\mathrm{m})\,(-\cos\theta\,\hat{\imath} + \sin\theta\,\hat{\jmath}).\end{aligned} \tag{2.18}$$

Next using the geometry for the vector's orientation (3–4–5 triangle), we note $\cos\theta = 3/5$ and $\sin\theta = 4/5$, and combining these with Eq. (2.18) provides the same result as Eq. (2.17).

Addition of vectors using Cartesian components

Consider the addition of the two vectors shown in Fig. 2.15(a). We first write the vectors using Cartesian representation. Thus

$$\begin{aligned}\vec{v}_1 &= v_{1x}\,\hat{\imath} + v_{1y}\,\hat{\jmath},\\ \vec{v}_2 &= v_{2x}\,\hat{\imath} + v_{2y}\,\hat{\jmath}.\end{aligned} \tag{2.19}$$

The resultant vector $\vec{R}$ is the sum of $\vec{v}_1$ and $\vec{v}_2$, hence

$$\begin{aligned}\vec{R} &= \vec{v}_1 + \vec{v}_2\\ &= (v_{1x}\,\hat{\imath} + v_{1y}\,\hat{\jmath}) + (v_{2x}\,\hat{\imath} + v_{2y}\,\hat{\jmath})\\ &= (v_{1x} + v_{2x})\,\hat{\imath} + (v_{1y} + v_{2y})\,\hat{\jmath}.\end{aligned} \tag{2.20}$$

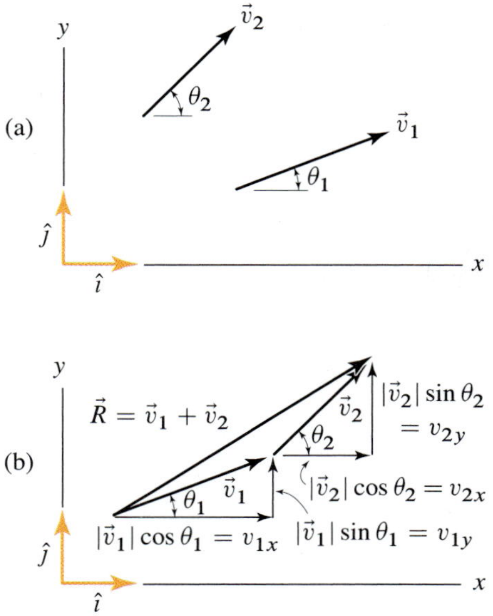

Figure 2.15
(a) Two vectors $\vec{v}_1$ and $\vec{v}_2$ are to be added. (b) The result of the vector addition is $\vec{R}$.

The addition is illustrated graphically in Fig. 2.15(b) where we slide the vectors $\vec{v}_1$ and $\vec{v}_2$ head to tail to yield a resultant vector $\vec{R}$. Notice the x and y

components of $\vec{R}$ are, respectively, the sums of the x and y components of the vectors being added. To complete the addition, we use the magnitudes $|\vec{v}_1|$ and $|\vec{v}_2|$ and orientations θ_1 and θ_2 provided in Fig. 2.15(b) to determine the components v_{1x}, v_{1y}, v_{2x}, and v_{2y}. Addition of an arbitrary number n of vectors is similarly straightforward, and the last of Eqs. (2.20) generalizes to

$$\vec{R} = \vec{v}_1 + \vec{v}_2 + \cdots + \vec{v}_n = \left(\sum_{i=1}^{n} v_{ix}\right)\hat{\imath} + \left(\sum_{i=1}^{n} v_{iy}\right)\hat{\jmath}. \tag{2.21}$$

In words, Eq. (2.21) states that the x component of the resultant is the sum of the x components of all vectors being added, and the y component of the resultant is the sum of the y components of all vectors being added.

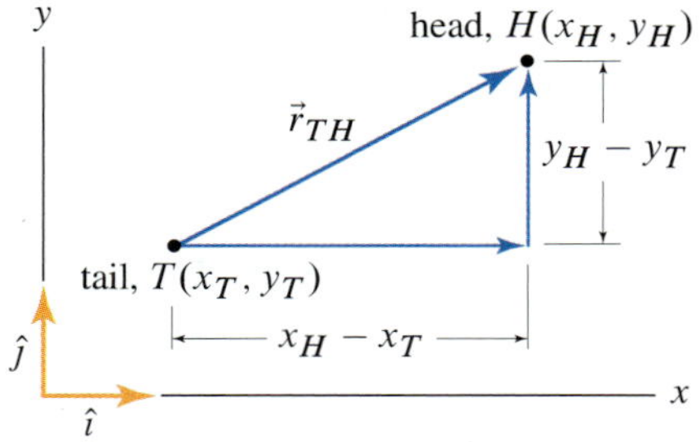

Figure 2.16
Construction of a position vector using Cartesian coordinates of the vector's head and tail.

Position vectors

The spatial position of one point (head of vector) relative to another point (tail of vector) is provided by a *position vector*. For the example of Fig. 2.16, the two points T and H denote the tail and head of a position vector and have coordinates (x_T, y_T) and (x_H, y_H), respectively. The position vector from T to H is denoted by $\vec{r}_{TH}$, and it has the x and y components shown in Fig. 2.16; hence we may write

$$\vec{r}_{TH} = (x_H - x_T)\,\hat{\imath} + (y_H - y_T)\,\hat{\jmath}. \tag{2.22}$$

In words, Eq. (2.22) states that the components of the position vector are given by "*coordinates of the head* minus *coordinates of the tail*," and this phrase is worth committing to memory. Another useful method for constructing a position vector is to imagine being positioned at point T and then asking what distances in the x and y directions must be traversed to arrive at point H; this also leads to Eq. (2.22).

What does a position vector really tell us? A position vector provides information on *relative* position only. That is, it tells us where the head of a vector is located relative to its tail. It does not tell us the *absolute* position of the head and tail. That is, it does not tell us where the head and tail are located.

For example, walking 1 mile west and 2 miles north is a statement of a position vector. With this position vector, we know only where our traveling ends relative to where it started, and exactly where our initial or final positions are located is information that is not contained in the position vector. On the other hand, if we state that we begin our traveling at Sears Tower in Chicago, and we walk 1 mile west and 2 miles north, then the position vector *together with* the coordinates of the starting point provides the absolute position of our final location.

End of Section Summary

In this section, Cartesian coordinate systems and Cartesian representation for vectors in two dimensions have been defined. Some of the key points are:

- The x and y directions you choose for a Cartesian coordinate system are arbitrary, but these directions must be orthogonal.

- The Cartesian representation for a vector $\vec{v}$ is $\vec{v} = \vec{v}_x + \vec{v}_y = v_x\,\hat{\imath} + v_y\,\hat{\jmath}$. In these expressions, $\vec{v}_x$ and $\vec{v}_y$ are called the *vector components* of $\vec{v}$, and v_x and v_y are called the *scalar components* (or simply the *components*) of $\vec{v}$.

- Addition of vectors with Cartesian representation is accomplished by simply summing the components in the x direction of all vectors being added to obtain the x component of the resultant, and then summing the components in the y direction of all vectors being added to obtain the y component of the resultant. Addition of even a large number of vectors is straightforward.

- A *position vector* provides the location of one point relative to another point. Given the coordinates of the two endpoints of the vector, Eq. (2.22) can be used to obtain the position vector. An easy way to remember Eq. (2.22) is that the components of the position vector are given by "*coordinates of the head* minus *coordinates of the tail*." Another useful method for constructing a position vector is to imagine being positioned at the tail of a vector and then asking what distances in the x and y directions must be traversed to arrive at the head of the vector.

EXAMPLE 2.6 *Addition of Vectors and Working Loads*

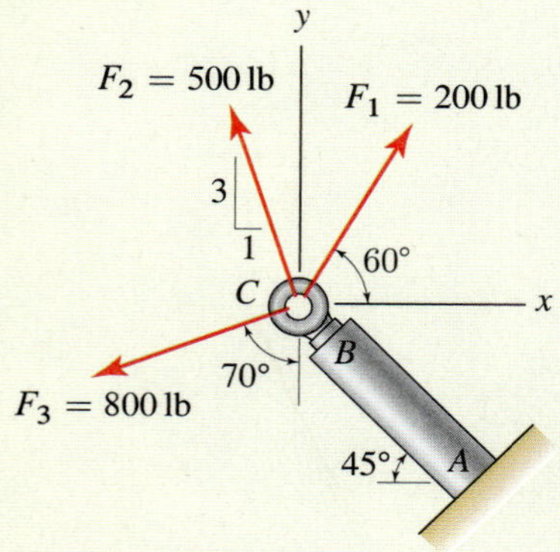

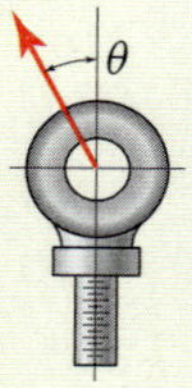

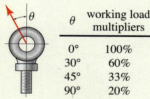

θ	working load multipliers
0°	100%
30°	60%
45°	33%
90°	20%

Figure 1

A short post AB has a commercially manufactured eyebolt screwed into its end. Three cables attached to the eyebolt apply the forces shown.

(a) Determine the resultant force applied to the eyebolt by the three cables, using a Cartesian vector representation.

(b) The manufacturer of the eyebolt specifies a maximum working load* of 2100 lb in the direction of the eyebolt's axis. When loads are not in the direction of the eyebolt's axis, the manufacturer specifies reduction of the working load using the multipliers given in Fig. 1. Determine if this size eyebolt is satisfactory.

SOLUTION

Road Map The resultant force vector is the sum of the three force vectors applied to the eyebolt. First we will express each of the three force vectors using Cartesian representation, and then we will add these to obtain the resultant force. For Part (b), we will compare this resultant force to the working load criteria for the eyebolt to determine if it is acceptable.

Part (a)

Governing Equations & Computation We first express the three forces using Cartesian representations as

$$\vec{F}_1 = (200\,\text{lb})(\cos 60°\,\hat{\imath} + \sin 60°\,\hat{\jmath}), \tag{1}$$

$$\vec{F}_2 = (500\,\text{lb})(-1/\sqrt{10}\,\hat{\imath} + 3/\sqrt{10}\,\hat{\jmath}), \tag{2}$$

$$\vec{F}_3 = (800\,\text{lb})(-\sin 70°\,\hat{\imath} - \cos 70°\,\hat{\jmath}). \tag{3}$$

The construction of $\vec{F}_2$ and $\vec{F}_3$ in Eqs. (2) and (3) is accomplished as described in earlier examples, with the only point of notice being that the 70° orientation of $\vec{F}_3$ is measured from the vertical which leads to sine and cosine functions providing the x and y components, respectively. While $\vec{F}_1$ can be constructed in an identical manner to $\vec{F}_3$, the following slightly different construction is useful. We first write $\vec{F}_1$ in the form $\vec{F}_1 = (200\,\text{lb})\hat{u}_1$ where $\hat{u}_1$ is a unit vector in the direction of $\vec{F}_1$. We next note that $\hat{u}_1$ is described by taking a step of $\cos 60°$ in the positive x direction followed by a step of $\sin 60°$ in the positive y direction. Hence $\hat{u}_1 = \cos 60°\,\hat{\imath} + \sin 60°\,\hat{\jmath}$ and, because $(\cos\theta)^2 + (\sin\theta)^2 = 1$ for any angle θ, $\hat{u}_1$ is clearly a unit vector and the expression for $\vec{F}_1$ in Eq. (1) follows. Adding the force vectors yields the resultant force $\vec{R}$ as

$$\begin{aligned}
\vec{R} &= \vec{F}_1 + \vec{F}_2 + \vec{F}_3 \\
&= \left[200\cos 60° + 500(-1/\sqrt{10}) + 800(-\sin 70°)\right]\,\text{lb}\,\hat{\imath} \\
&\quad + \left[200\sin 60° + 500(3/\sqrt{10}) + 800(-\cos 70°)\right]\,\text{lb}\,\hat{\jmath} \\
&= \boxed{(-810\,\hat{\imath} + 374\,\hat{\jmath})\,\text{lb}.}
\end{aligned} \tag{4}$$

* *Working loads* are loads a certain piece of hardware is designed to be subjected to on a repeated basis. These loads are obtained by reducing the actual failure strength by an appropriate factor, called the *factor of safety*. For the particular model of eyebolt cited here, the actual breaking strength is about 9800 lb for loads in the direction of the bolt's axis. While a working load of 2100 lb provides a reasonable margin of safety, the margin is not quite as generous as you might believe because of *fatigue*: materials and structures that must withstand many load cycles can be subjected to only a fraction of their one-time load strength. While working load specifications for eyebolts always have a margin of safety, it is dangerous and irresponsible to use any structural component in an application where the manufacturer's working loads are exceeded.

As an alternative to reporting our answer using Cartesian vector representation, we may use Eq. (2.14) to report the magnitude and orientation of the resultant force as

$$R = \sqrt{(-810\,\text{lb})^2 + (374\,\text{lb})^2} = \boxed{892\,\text{lb},} \quad \text{and} \tag{5}$$

$$\theta = \tan^{-1}\left(\frac{374\,\text{lb}}{-810\,\text{lb}}\right) = \boxed{-24.8°,} \tag{6}$$

and this force is shown in Fig. 2.

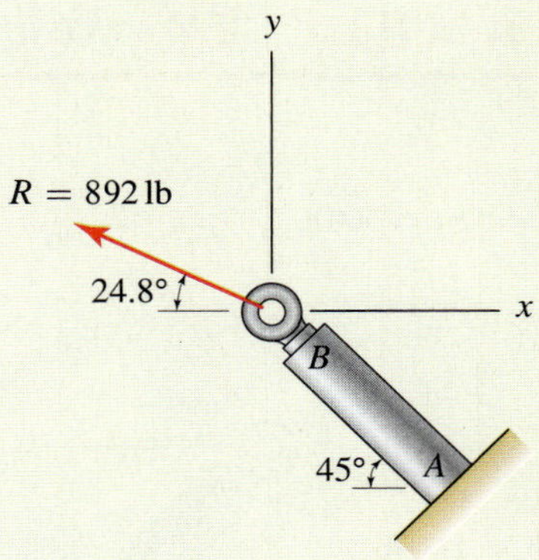

Figure 2
Resultant $\vec{R}$ of forces $\vec{F}_1$, $\vec{F}_2$, and $\vec{F}_3$.

Part (b)

Governing Equations & Computation To determine if the eyebolt is adequate to support the 892 lb resultant force, the orientation of $\vec{R}$ with respect to the bolt's axis is needed. Defining a new Cartesian coordinate system tn as shown in Fig. 3, where n is in the direction AB and t is perpendicular to direction AB, we find angle $\theta = 45° - 24.8° = 20.2°$. Consulting Fig. 1, we find the eyebolt's load multiplier for loading angles up to 30° is 60%, which yields a net working load of $(2100\,\text{lb})(60\%) = 1260\,\text{lb}$. This working load is larger than our resultant force $R = 892\,\text{lb}$, and thus we conclude that **this size eyebolt is satisfactory** for supporting the system of three forces considered here.

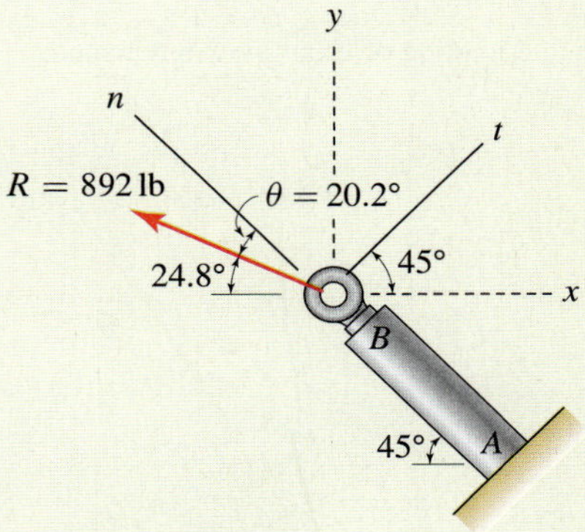

Figure 3
Orientation of resultant force $\vec{R}$ with respect to t and n directions.

Discussion & Verification Our analysis thus far indicates the eyebolt is acceptable when all three forces $\vec{F}_1$, $\vec{F}_2$, and $\vec{F}_3$ are present simultaneously. However, we should also consider other possible loading scenarios in which forces are applied sequentially, and in various orders, to make sure that at no time are the manufacturer's working load specifications exceeded. In addition, the manufacturer's specifications refer specifically to eyebolts subjected to just a single force. We have applied these criteria to the resultant of three forces, which is reasonable, provided good judgment is used. To clarify further, consider an eyebolt subjected to two equal forces with opposite direction: $\vec{F}_1 = P\,\hat{\imath}$ and $\vec{F}_2 = -P\,\hat{\imath}$. The resultant force is zero, regardless of how large P is, and thus the manufacturer's working load specifications are always met. However, if P is too large, the eyebolt will clearly fail. Thus, we should consult the manufacturer to see if additional criteria are available for multiple loads; and if these are not, then we must perform more detailed analysis of the strength properties of the eyebolt to ensure that it is acceptable for our application.

Figure 4. An example of a commercially manufactured eyebolt, made of forged steel.

EXAMPLE 2.7 *Resolution of a Vector into Components*

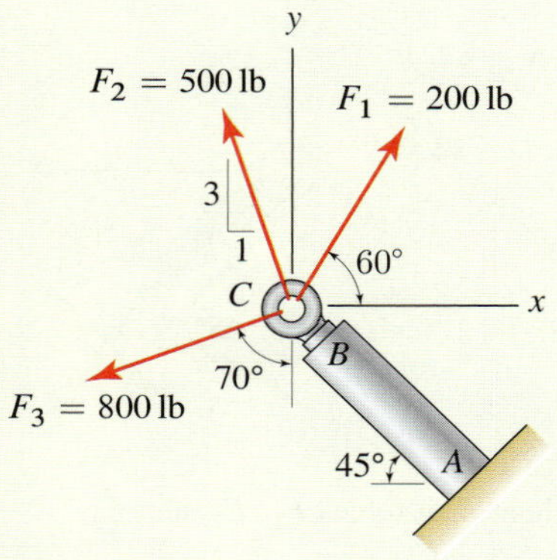

Figure 1
Post and loading of Example 2.6 repeated.

In Example 2.6 we eventually need to determine if post AB is strong enough to support the forces applied to it. Such problems are discussed in later chapters of this book where it will be necessary to resolve the forces applied to the post into components parallel and perpendicular to the post's axis. Thus, resolve the resultant of forces $\vec{F}_1$, $\vec{F}_2$, and $\vec{F}_3$ into components parallel and perpendicular to the post's axis AB.

SOLUTION

Road Map Several strategies may be used, and we will consider two solutions here. Our first solution will simply repeat the procedure of Example 2.6, except with vectors written in terms of a new tn coordinate system where t and n are perpendicular and parallel to the post's axis, respectively. The second solution will use transformation of the results of Example 2.6.

Solution 1

Governing Equations & Computation Using the geometry shown in Fig. 1, we compute angles with respect to the t and n directions with the results shown in Fig. 2. Computation of the 15° and 65° angles is straightforward. For the orientation of $\vec{F}_2$,

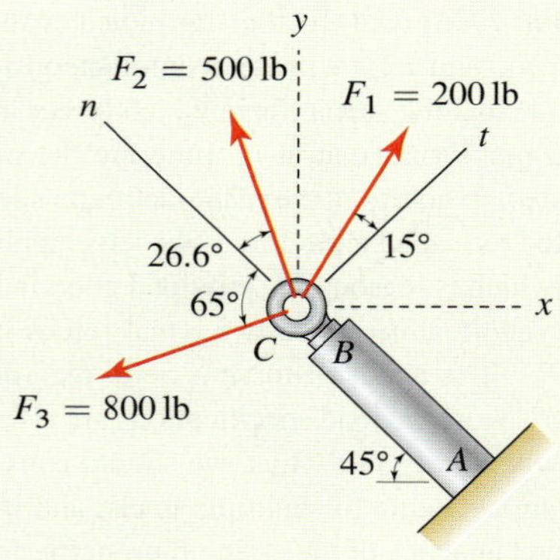

Figure 2. Forces of Fig. 1 redrawn with orientations with respect to t and n directions.

we first evaluate $\tan^{-1}(3/1) = 71.6°$ to obtain the orientation of $\vec{F}_2$ with respect to the negative x direction, followed by evaluation of $71.6° - 45° = 26.6°$ to obtain the orientation of $\vec{F}_2$ with respect to the n direction. We then write the force vectors using Cartesian representation as

$$\vec{F}_1 = (200\,\text{lb})(\cos 15°\,\hat{t} + \sin 15°\,\hat{n}), \tag{1}$$

$$\vec{F}_2 = (500\,\text{lb})(\sin 26.6°\,\hat{t} + \cos 26.6°\,\hat{n}), \tag{2}$$

$$\vec{F}_3 = (800\,\text{lb})(-\sin 65°\,\hat{t} + \cos 65°\,\hat{n}), \tag{3}$$

where $\hat{t}$ and $\hat{n}$ are defined to be unit vectors in the t and n directions, respectively. The resultant force is then

$$\begin{aligned}\vec{R} &= \vec{F}_1 + \vec{F}_2 + \vec{F}_3\\ &= \left[200\cos 15° + 500\sin 26.6° + 800(-\sin 65°)\right]\hat{t}\,\text{lb}\\ &\quad + \left[200\sin 15° + 500\cos 26.6° + 800(\cos 65°)\right]\hat{n}\,\text{lb}\\ &= \boxed{(-308\,\hat{t} + 837\,\hat{n})\,\text{lb}.}\end{aligned} \tag{4}$$

Thus, the resultant force has the t and n components $R_t = -308\,\mathrm{lb}$ and $R_n = 837\,\mathrm{lb}$, respectively, which are shown in Fig. 3. As a partial check on our solution, we evaluate the magnitude $R = \sqrt{(-308\,\mathrm{lb})^2 + (837\,\mathrm{lb})^2} = 892\,\mathrm{lb}$ which agrees with the results found in Example 2.6.

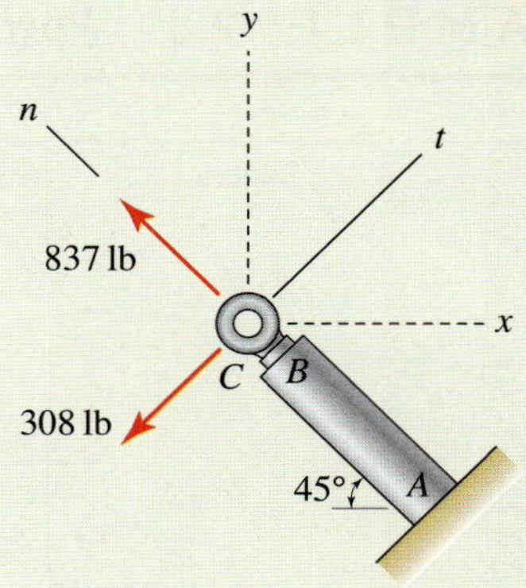

Figure 3
Components of the resultant force in t and n directions.

Solution 2

Governing Equations & Computation This solution takes advantage of the results of Example 2.6 by resolving $\vec{R}$ in Eq. (4) on p. 50 into t and n components. Referring to Fig. 3 on p. 51, angle θ is easily found to be $\theta = 45° - 24.8° = 20.2°$. Thus, the components of $\vec{R}$ in the t and n directions are

$$R_t = -(892\,\mathrm{lb})(\sin 20.2°) = \boxed{-308\,\mathrm{lb},} \tag{5}$$

$$R_n = (892\,\mathrm{lb})(\cos 20.2°) = \boxed{837\,\mathrm{lb},} \tag{6}$$

which agrees with the results reported in Eq. (4) and shown in Fig. 3.

Discussion & Verification

- When you are resolving a vector into orthogonal components, such as R_t and R_n in this example, a useful and quick check of accuracy is to evaluate $\sqrt{R_t^2 + R_n^2}$ and verify that it agrees with the magnitude of the original vector.
- A disadvantage of Solution 1 is that if the resultant force must be determined for several different coordinate systems, the work for each of these is fully repeated. The advantage of Solution 2 is that once the resultant force is known for one coordinate system, it is straightforward to transform the result to other coordinate systems.
- A third solution that is not presented here is discussed in Problem 2.35. This approach is similar to Solution 2, but is more elegant and effective. It uses *vector transformation* to take the x and y components of $\vec{R}$ found in Example 2.6 and transform these into t and n components.

EXAMPLE 2.8 Position Vectors

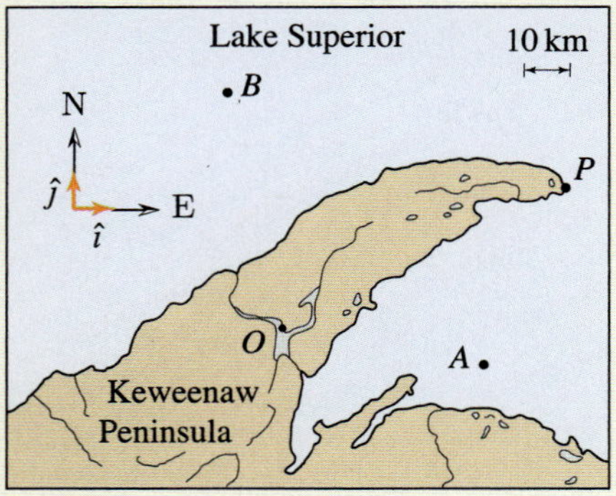

Figure 1

The Keweenaw Peninsula on the south shore of Lake Superior is shown. Point O is the location of a U.S. Coast Guard base, point A is the location of a Coast Guard rescue craft (either a helicopter or a ship), point B is the location of a boat in need of help, and point P is the tip of the Keweenaw Peninsula. Letting $\hat{\imath}$ and $\hat{\jmath}$ be unit vectors in the east and north directions, respectively, the position vectors are

$$\vec{r}_{OA} = (44\,\hat{\imath} - 9\,\hat{\jmath})\,\text{km}, \quad \vec{r}_{OP} = (61\,\hat{\imath} + 31\,\hat{\jmath})\,\text{km}, \quad \vec{r}_{OB} = (-12\,\hat{\imath} + 53\,\hat{\jmath})\,\text{km}. \tag{1}$$

(a) If the rescue craft is a helicopter, determine the shortest distance it must fly from A to reach the boat at B.

(b) If the rescue craft is a ship, determine the shortest distance it must cruise from A to reach the boat at B.

(c) In Part (b), if the ship cruises 20 km from A toward P, determine the ship's coordinates and position vector relative to the base at O.

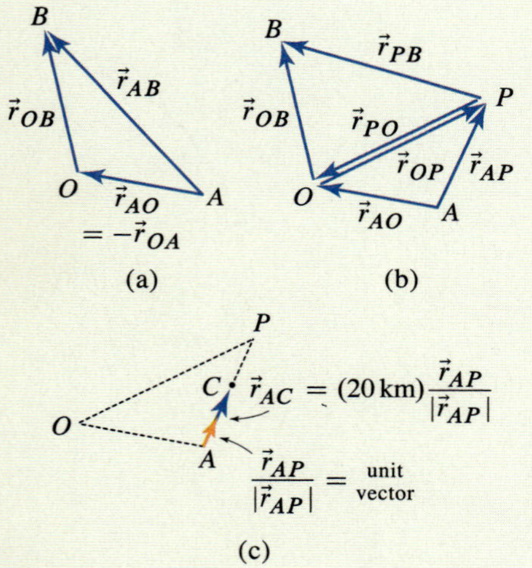

Figure 2

Addition of vectors. Figure (a) is used to obtain the answer to Part (a) of this example, wherein vectors $\vec{r}_{AO}$ and $\vec{r}_{OB}$ are added to obtain $\vec{r}_{AB}$. Figure (b) shows various vectors that are added to obtain the answer to Part (b). Figure (c) is used to obtain the answer to Part (c), wherein a position vector $\vec{r}_{AC}$ is constructed using the unit vector $\vec{r}_{AP}/|\vec{r}_{AP}|$.

SOLUTION

Road Map A strategy that could be used for all three parts of this example is to first determine the coordinates of points A, B, and P and then use Eq. (2.22) on p. 48 to write position vectors $\vec{r}_{AB}$ and so on, where the magnitudes of these vectors give the distances asked for in each question. A better strategy is to use vector addition to obtain the desired position vectors, and this is the approach we will follow.

Part (a)

Governing Equations & Computation The shortest route for the helicopter is the straight line from A to B. Thus we seek vector $\vec{r}_{AB}$, which may be written as

$$\vec{r}_{AB} = \vec{r}_{AO} + \vec{r}_{OB}. \tag{2}$$

A useful way to think of Eq. (2) is shown in Fig. 2(a). Rather than travel directly from A to B, we may first travel from A to O, followed by travel from O to B. Of course, this is not the path to be taken by the helicopter; but if this path were followed, the same final position would result. The merit in writing Eq. (2) is that both vectors on the right-hand side are known: $\vec{r}_{OB}$ is specified in Eq. (1), and $\vec{r}_{AO} = -\vec{r}_{OA}$ where $\vec{r}_{OA}$ is also specified in Eq. (1). Thus, Eq. (2) becomes

$$\begin{aligned}\vec{r}_{AB} &= -\vec{r}_{OA} + \vec{r}_{OB}\\ &= -(44\,\hat{\imath} - 9\,\hat{\jmath})\,\text{km} + (-12\,\hat{\imath} + 53\,\hat{\jmath})\,\text{km}\\ &= (-56\,\hat{\imath} + 62\,\hat{\jmath})\,\text{km},\end{aligned} \tag{3}$$

and the distance traveled is the magnitude

$$r_{AB} = \sqrt{(-56\,\text{km})^2 + (62\,\text{km})^2} = \boxed{83.5\,\text{km}.} \tag{4}$$

Part (b)

Governing Equations & Computation The shortest route for a rescue ship consists of the straight-line distances from A to P and from P to B. Thus we seek vectors $\vec{r}_{AP}$ and $\vec{r}_{PB}$. Using the addition shown in Fig. 2(b), we write

$$\vec{r}_{AP} = \vec{r}_{AO} + \vec{r}_{OP}, \tag{5}$$

$$\vec{r}_{PB} = \vec{r}_{PO} + \vec{r}_{OB}. \tag{6}$$

Vectors $\vec{r}_{OP}$ and $\vec{r}_{OB}$ are specified in the problem statement, $\vec{r}_{AO} = -\vec{r}_{OA}$, and $\vec{r}_{PO} = -\vec{r}_{OP}$. Thus

$$\begin{aligned} \vec{r}_{AP} &= -\vec{r}_{OA} + \vec{r}_{OP} \\ &= -(44\,\hat{\imath} - 9\,\hat{\jmath})\,\text{km} + (61\,\hat{\imath} + 31\,\hat{\jmath})\,\text{km} \\ &= (17\,\hat{\imath} + 40\,\hat{\jmath})\,\text{km}, \end{aligned} \tag{7}$$

$$\begin{aligned} \vec{r}_{PB} &= -\vec{r}_{OP} + \vec{r}_{OB} \\ &= -(61\,\hat{\imath} + 31\,\hat{\jmath})\,\text{km} + (-12\,\hat{\imath} + 53\,\hat{\jmath})\,\text{km} \\ &= (-73\,\hat{\imath} + 22\,\hat{\jmath})\,\text{km}. \end{aligned} \tag{8}$$

The magnitudes of the vectors in Eqs. (7) and (8) are

$$r_{AP} = \sqrt{(17\,\text{km})^2 + (40\,\text{km})^2} = 43.5\,\text{km}, \tag{9}$$

$$r_{PB} = \sqrt{(-73\,\text{km})^2 + (22\,\text{km})^2} = 76.2\,\text{km}. \tag{10}$$

Thus, the total distance cruised by the rescue ship is $r_{AP} + r_{PB} = \boxed{119.7\,\text{km.}}$ As a check on the accuracy of our solution, you may want to verify that $\vec{r}_{AP} + \vec{r}_{PB} = \vec{r}_{AB}$.

Part (c)

Governing Equations & Computation Let point C shown in Fig. 2(c) denote the position of the rescue ship after it cruises 20 km from A toward P. The position vector from A to C is $\vec{r}_{AC}$, and it is collinear with $\vec{r}_{AP}$ and can be constructed as shown in Fig. 2(c) as follows:

$$\vec{r}_{AC} = (20\,\text{km})\,\frac{\vec{r}_{AP}}{r_{AP}} = (20\,\text{km})\,\frac{(17\,\hat{\imath} + 40\,\hat{\jmath})\,\text{km}}{\sqrt{(17\,\text{km})^2 + (40\,\text{km})^2}} \tag{11}$$

$$= (7.8\,\hat{\imath} + 18.4\,\hat{\jmath})\,\text{km}. \tag{12}$$

Note that $\vec{r}_{AP}/r_{AP}$ in Eq. (11) is a unit vector that provides the proper direction for the 20 km distance the ship cruises. To find the position vector $\vec{r}_{OC}$, we use vector addition to write

$$\vec{r}_{OC} = \vec{r}_{OA} + \vec{r}_{AC}. \tag{13}$$

Vectors on the right-hand side are known, and hence Eq. (13) provides

$$\begin{aligned} \vec{r}_{OC} &= (44\,\hat{\imath} - 9\,\hat{\jmath})\,\text{km} + (7.8\,\hat{\imath} + 18.4\,\hat{\jmath})\,\text{km} \\ &= \boxed{(51.8\,\hat{\imath} + 9.4\,\hat{\jmath})\,\text{km.}} \end{aligned} \tag{14}$$

Finally, to obtain the coordinates of C, we use Eq. (2.22) to write

$$\vec{r}_{OC} = (x_C - x_O)\,\hat{\imath} + (y_C - y_O)\,\hat{\jmath}. \tag{15}$$

Selecting the coordinates of O to be $(0, 0)$ and equating Eqs. (14) and (15) provides

$$(51.8\,\hat{\imath} + 9.4\,\hat{\jmath})\,\text{km} = x_C\,\hat{\imath} + y_C\,\hat{\jmath}. \tag{16}$$

Equation (16) is a vector equation. For it to be satisfied requires terms on both sides of the equation that multiply $\hat{\imath}$ to be equal, and similarly terms on both sides of the equation that multiply $\hat{\jmath}$ must be equal. Hence

$$\boxed{x_C = 51.8\,\text{km} \quad \text{and} \quad y_C = 9.4\,\text{km.}} \tag{17}$$

Discussion & Verification Vector addition, as illustrated in this example, is a powerful application of vectors for determining positions, and these techniques will be used throughout this book and subjects that follow statics.

PROBLEMS

Problems 2.17 through 2.24

For the following problems, use an xy Cartesian coordinate system where x is horizontal, positive to the right, and y is vertical, positive upward. For problems where the answers require vector expressions, report the vectors using Cartesian representations.

Problem 2.17 Repeat Prob. 2.2 on p. 41.

Problem 2.18 Repeat Prob. 2.3 on p. 41.

Problem 2.19 Repeat Prob. 2.7 on p. 41.

Problem 2.20 Repeat Prob. 2.8 on p. 42.

Problem 2.21 Repeat Prob. 2.9 on p. 42.

Problem 2.22 Repeat Prob. 2.10 on p. 42.

Problem 2.23 Repeat Prob. 2.11 on p. 42.

Problem 2.24 Repeat Prob. 2.12 on p. 42.

Problem 2.25

Let $\vec{A} = (150\,\hat{\imath} - 200\,\hat{\jmath})$ lb and $\vec{B} = (200\,\hat{\imath} + 480\,\hat{\jmath})$ lb. Evaluate the following, and for Parts (a), (b), and (c) state the magnitude of $\vec{R}$.

(a) $\vec{R} = \vec{A} + \vec{B}$.

(b) $\vec{R} = 2\vec{A} - (1/2)\vec{B}$.

(c) Find a scalar s such that $\vec{R} = s\vec{A} + \vec{B}$ has an x component only.

(d) Determine a dimensionless unit vector in the direction $\vec{B} - \vec{A}$.

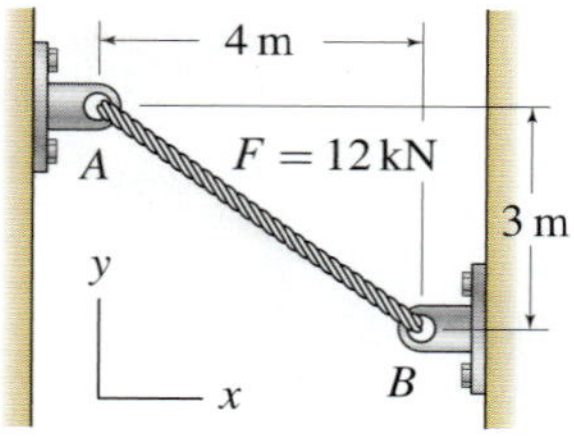

Figure P2.26

Problems 2.26 and 2.27

A rope connecting points A and B supports the force F shown in the figure. Write expressions using Cartesian vector representation for the following:

(a) $\vec{r}_{AB}$: the position vector from A to B.

(b) $\vec{r}_{BA}$: the position vector from B to A.

(c) $\hat{u}_{AB}$: the unit vector in the direction from A to B.

(d) $\hat{u}_{BA}$: the unit vector in the direction from B to A.

(e) $\vec{F}_{AB}$: the force vector the rope applies to A.

(f) $\vec{F}_{BA}$: the force vector the rope applies to B.

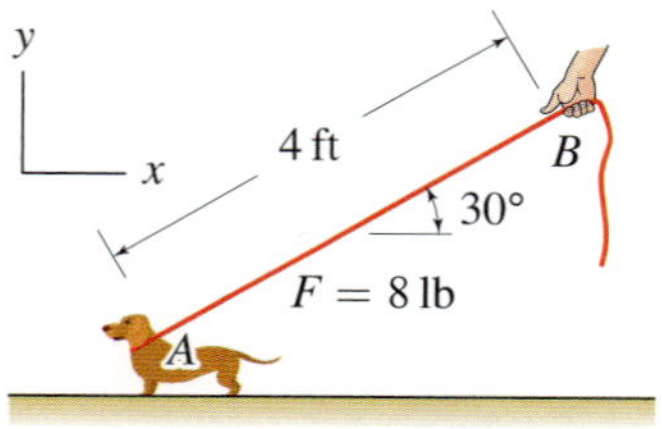

Figure P2.27

Problem 2.28

A cleat on a boat is used to support forces from three ropes as shown. Determine the resultant force vector $\vec{R}$, using Cartesian representation, and determine the magnitude R. Also express $\vec{R}$ in polar vector representation.

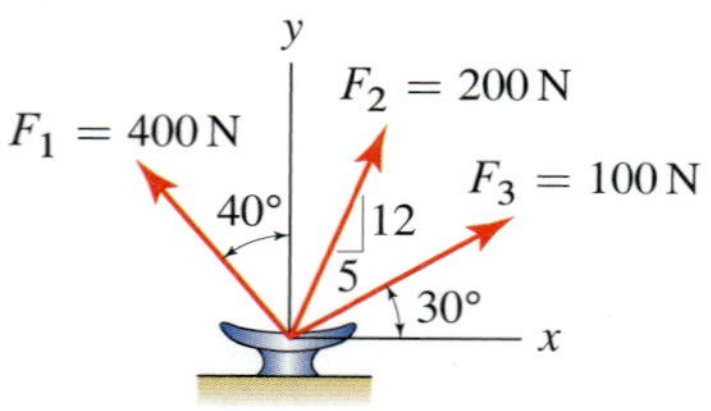

Figure P2.28

Problem 2.29

A model of a person's arm is used for ergonomics studies. Distances are $r_{AB} = 35$ cm, $r_{BC} = 28$ cm, and $r_{CD} = 19$ cm.

(a) Determine the position vector $\vec{r}_{AD}$ and its magnitude r_{AD}.

(b) Express $\vec{r}_{AD}$ using polar vector representation, measuring orientation positive counterclockwise from the right-hand horizontal direction.

(c) Determine a unit vector $\hat{u}_{AD}$ in the direction from A to D.

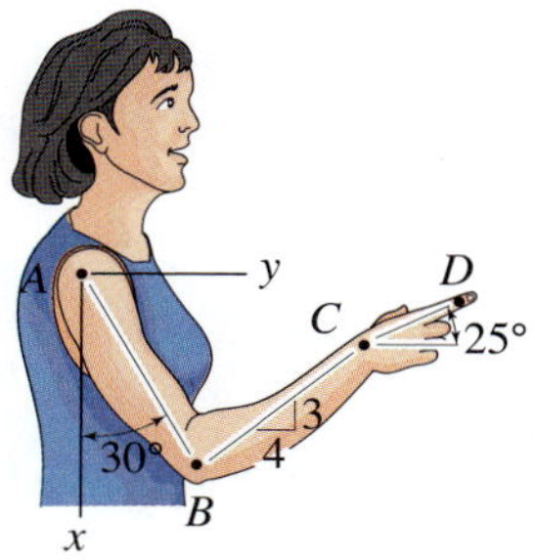

Figure P2.29

Problem 2.30

Two ropes are used to lift a pipe in a congested region. Determine the ratio F_2/F_1 so that the resultant of $\vec{F}_1$ and $\vec{F}_2$ is vertical.

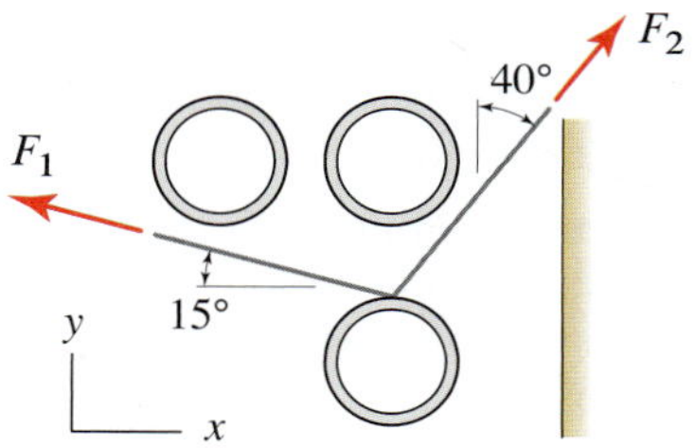

Figure P2.30

Problem 2.31

A welded steel tab is subjected to forces F and P. Determine the largest value P may have if $F = 1000$ lb and the magnitude of the resultant force cannot exceed 1500 lb.

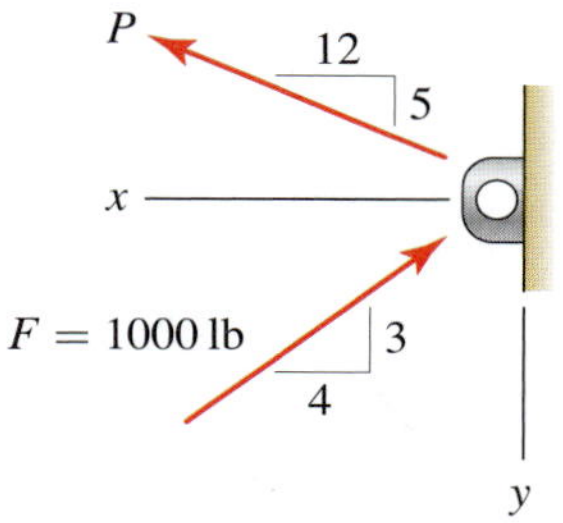

Figure P2.31

Problem 2.32

A short cantilever beam is subjected to three forces. If $F = 8$ kN, determine the value of α that minimizes the magnitude of the resultant of the three forces. Also, determine the magnitude of that resultant.

Problem 2.33

A short cantilever beam is subjected to three forces. If $\alpha = 45°$, determine the value of F that will make the magnitude of the resultant of the three forces smallest.

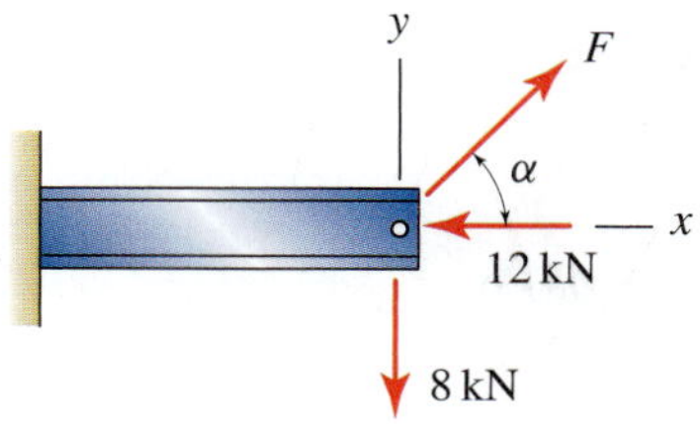

Figure P2.32 and P2.33

Problem 2.34

An eyebolt is loaded by forces F_1 and F_2. If the eyebolt has a maximum working load of 1200 lb, determine if the working load specifications given in Fig. 1 of Example 2.6 are met for the following loading scenarios.

(a) Only F_1 is applied.

(b) Only F_2 is applied.

(c) Both F_1 and F_2 are applied simultaneously.

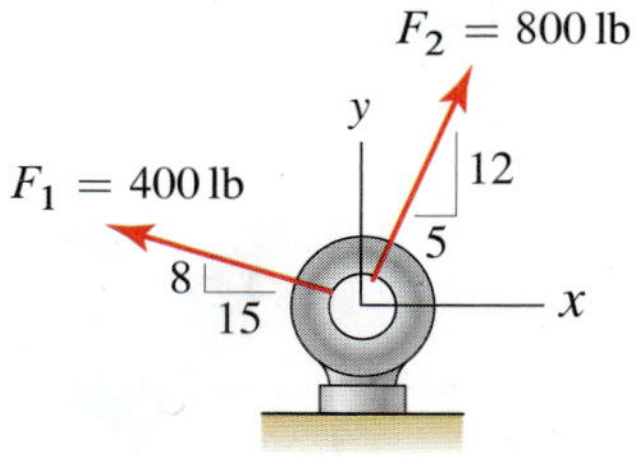

Figure P2.34

Problem 2.35

An important and useful property of vectors is they may be easily *transformed* from one Cartesian coordinate system to another. That is, if the x and y components of a vector are known, the t and n components can be found (or vice versa) by applying the formulas

$$\vec{v} = v_x\,\hat{\imath} + v_y\,\hat{\jmath} = v_t\,\hat{t} + v_n\,\hat{n}, \tag{1}$$

$$\text{where} \quad v_t = v_x \cos\phi + v_y \sin\phi, \tag{2}$$

$$v_n = -v_x \sin\phi + v_y \cos\phi, \tag{3}$$

$$\text{or} \quad v_x = v_t \cos\phi - v_n \sin\phi, \tag{4}$$

$$v_y = v_t \sin\phi + v_n \cos\phi. \tag{5}$$

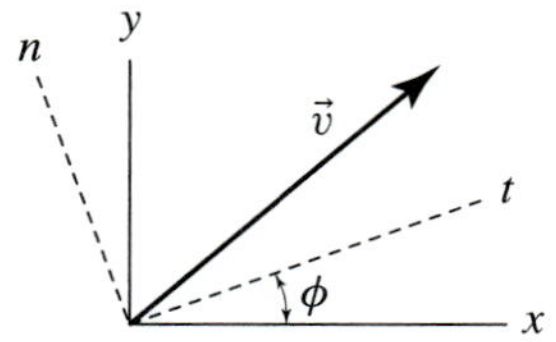

Figure P2.35

In these equations, $\hat{t}$ and $\hat{n}$ are unit vectors in the t and n directions, respectively; ϕ is measured positive counterclockwise from the positive x direction to the positive t direction; and the y and n directions must be oriented 90° counterclockwise from the positive x and t directions, respectively.

(a) Derive the above transformation that gives v_t and v_n in terms of v_x and v_y. *Hint:* First consider a vector $\vec{v}_x$ that acts in the x direction, and resolve this into components in t and n directions. Then consider a vector $\vec{v}_y$ that acts in the y direction, and resolve this into components in t and n directions. Vectorially adding these results yields the transformation.

(b) For the eyebolt and post of Example 2.7, the x and y components of the resultant force are given by Eq. (4) of Example 2.6. Use these x and y components with the above transformation equations to obtain the t and n components of the resultant force, and verify these are the same as those in Eq. (4) of Example 2.7.

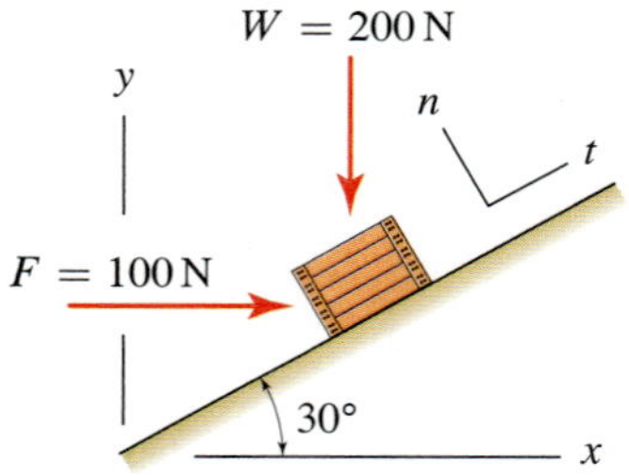

Figure P2.36

Problem 2.36

A box weighing 200 N rests on an inclined surface. A worker applies a horizontal force F to help position the box. Determine the x and y components of the resultant of forces W and F. Also determine the t and n components of the resultant force vector. Comment on why the t and n components might be useful to know.

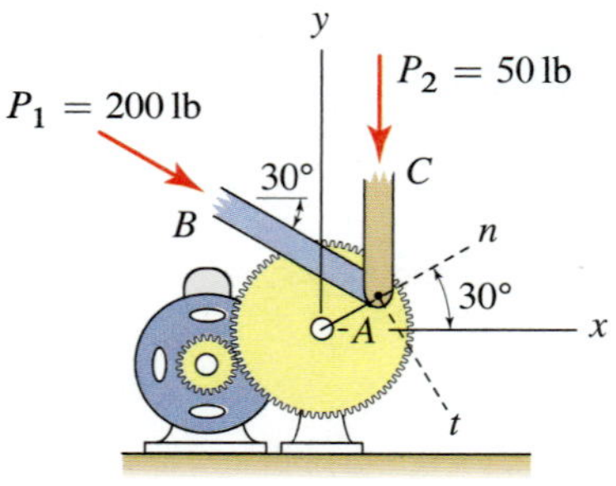

Figure P2.37

Problem 2.37

A motor-driven gear is used to produce forces P_1 and P_2 in members AB and AC of a machine. Member AC is parallel to the y axis.

(a) Determine the x and y components of the resultant force vector at A due to forces P_1 and P_2.

(b) Determine the t and n components of the resultant force vector. Comment on why the t and n components might be useful to know.

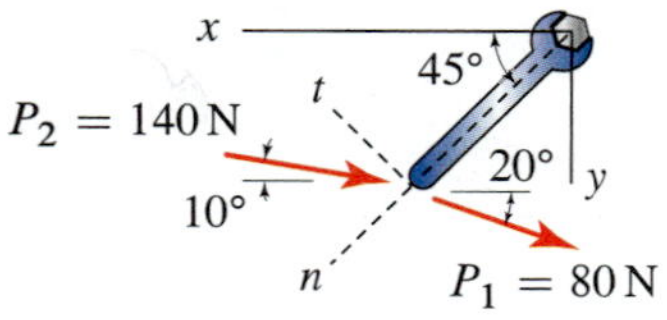

Figure P2.38

Problem 2.38

Two people apply forces P_1 and P_2 to the handle of a wrench as shown.

(a) Determine the x and y components of the resultant force vector applied to the handle of the wrench.

(b) Determine the t and n components of the resultant force vector. Comment on why the t and n components might be useful to know.

Problem 2.39

Bar AC is straight and has 106 in. length, B is a pulley that supports forces W and F, W is vertical, and the t direction is parallel to bar AC.

(a) If $F = 150\,\text{lb}$ and $\alpha = 30°$, determine the coordinates of point C so that the t component of the resultant of F and W is zero.

(b) If $F = 150\,\text{lb}$ and C is located at $(56, 90)$ in., determine angle α so that the t component of the resultant of F and W is zero.

(c) If C is located at $(56, 90)$ in. and $\alpha = 30°$, determine F so that the t component of the resultant of F and W is zero.

Hint: For each of these questions, first find the x and y components of the resultant force and then use the transformation given in Prob. 2.35 to obtain the t component.

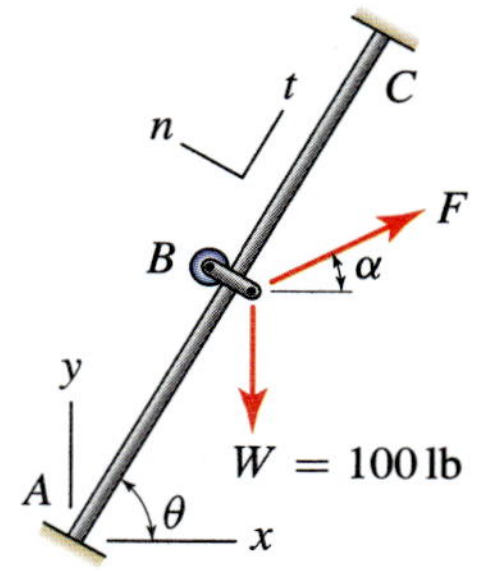

Figure P2.39

Problem 2.40

Screw AC is used to position point D of a machine. Points A and C have coordinates $(185, 0)$ mm and $(125, 144)$ mm, respectively, and are fixed in space by the bearings that support the screw. If point B is 52 mm from point A, determine the position vector $\vec{r}_{AB}$ and the coordinates of point B.

Problem 2.41

Repeat Prob. 2.40 if point B is 39 mm from point C.

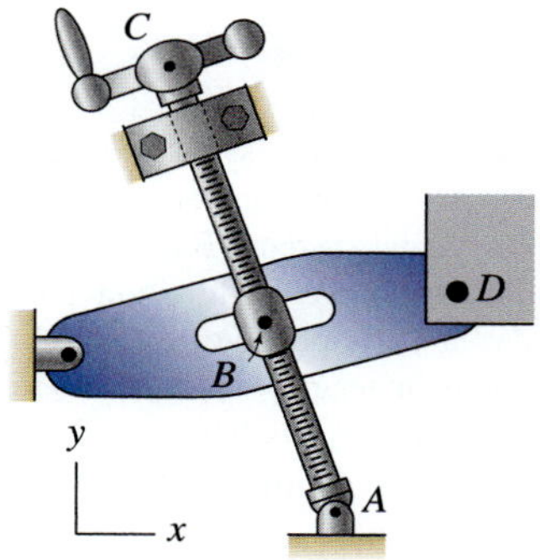

Figure P2.40 and P2.41

Problem 2.42

Screw AC is used to position point D of a machine. Point A has coordinates $(185, 0)$ mm and is fixed in space by a bearing that supports the screw. The screw nut at point B is supported by lever ED and at the instant shown has coordinates $(160, 60)$ mm. If the length of the screw from A to C is 130 mm, determine the position vector $\vec{r}_{AC}$ and the coordinates of point C.

Problem 2.43

In Prob. 2.42 if point B is equidistant from points A and C and if $\vec{r}_{EA} = (185\,\hat{\imath} - 50\,\hat{\jmath})$ mm and $\vec{r}_{EB} = (155\,\hat{\imath} + 22\,\hat{\jmath})$ mm, determine the position vector $\vec{r}_{AC}$.

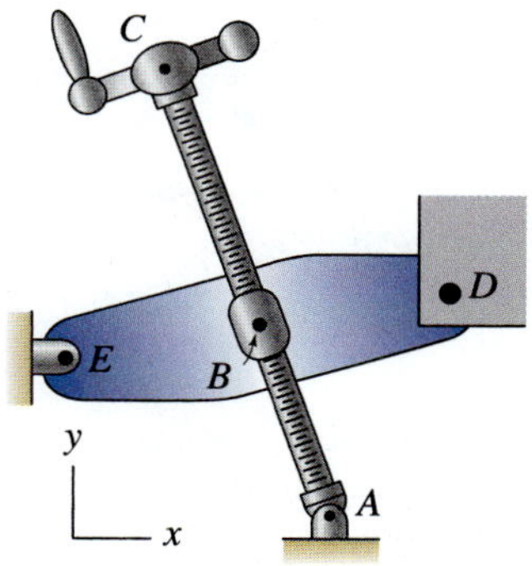

Figure P2.42 and P2.43

Problem 2.44

A computer numerical control (CNC) milling machine is used to cut a slot in the component shown. The cutting tool starts at point A and advances to point B where it pauses while the depth of cut is increased, and then the tool continues on to point C. If the coordinates of points A and C are $(95, 56)$ mm and $(17, 160)$ mm, respectively, and B is located 45 mm from A, determine the position vectors $\vec{r}_{OA}$, $\vec{r}_{OB}$, and $\vec{r}_{OC}$ and the coordinates of B.

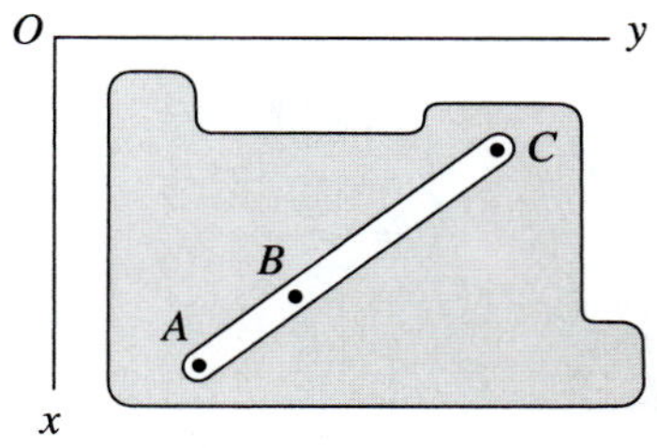

Figure P2.44

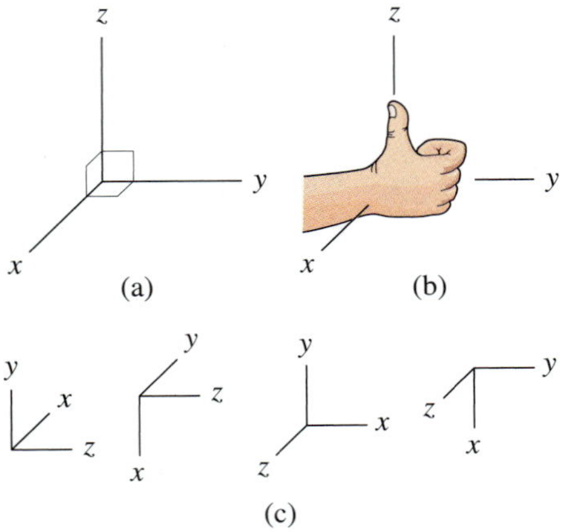

Figure 2.17
(a) Cartesian coordinate system in three dimensions. (b) A scheme for constructing a right hand coordinate system. Position your right hand so the positive x direction passes into your palm and the positive y direction passes through your finger tips. Your thumb then indicates the positive z direction. (c) More examples of right-hand coordinate systems.

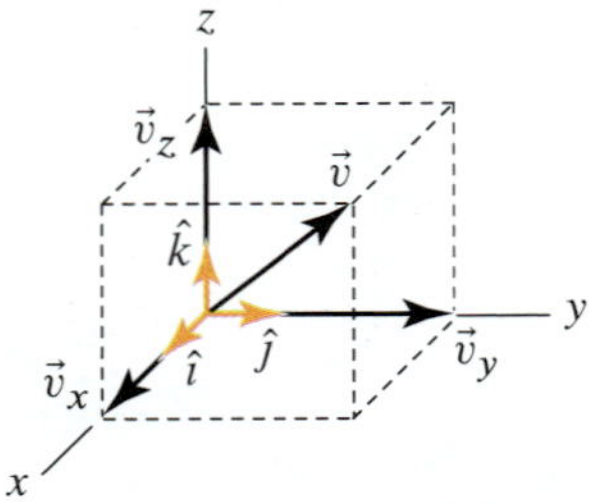

Figure 2.18
Right-hand Cartesian coordinate system with unit vectors $\hat{\imath}$, $\hat{\jmath}$, and $\hat{k}$ in the x, y, and z directions, respectively, and resolution of a vector $\vec{v}$ into vector components $\vec{v}_x$, $\vec{v}_y$, and $\vec{v}_z$.

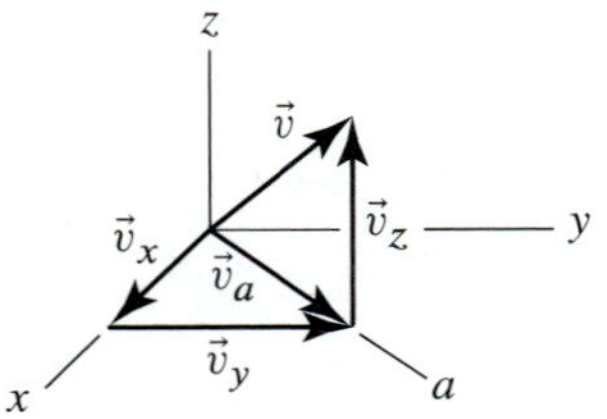

Figure 2.19
Evaluation of a vector's magnitude in terms of its components.

2.3 Cartesian Representation of Vectors in Three Dimensions

For problems in three dimensions, vectors are especially powerful and without them many problems would be intractable. Concepts of Section 2.2 apply, with some additional enhancements needed for three dimensions. These include definitions of a right-handed coordinate system, direction angles, and direction cosines.

Right-hand Cartesian coordinate system

In three dimensions a Cartesian coordinate system uses three orthogonal reference directions. These will consist of x, y, and z directions as shown in Fig. 2.17(a). Proper interpretation of many vector operations, such as the cross product to be discussed in Section 2.5, requires the x, y, and z directions be arranged in a consistent manner. For example, when you are constructing the coordinate system shown in Fig. 2.17(a), imagine the x and y directions are chosen first. Then, should z be taken in the direction shown, or can it be in the opposite direction? The universal convention in mechanics and vector mathematics in general is z must be taken in the direction shown, and the result is called a *right-hand coordinate system*. Figure 2.17(b) describes a scheme for constructing a right-hand coordinate system. You should study this scheme and become comfortable with its use.

Cartesian vector representation

We define vectors $\hat{\imath}$, $\hat{\jmath}$, and $\hat{k}$ to be unit vectors that point in the positive x, y, and z directions, respectively. A vector $\vec{v}$ can then be written as

$$\begin{aligned}\vec{v} &= \vec{v}_x + \vec{v}_y + \vec{v}_z \\ &= v_x\,\hat{\imath} + v_y\,\hat{\jmath} + v_z\,\hat{k}.\end{aligned} \tag{2.23}$$

Resolution of $\vec{v}$ into x, y, and z components is shown in Fig. 2.18. The *magnitude* of $\vec{v}$ is given by

$$|\vec{v}| = \sqrt{v_x^2 + v_y^2 + v_z^2}. \tag{2.24}$$

This equation is obtained using the construction shown in Fig. 2.19 as follows. First, a vector $\vec{v}_a$ that lies in the xy plane is defined. Because v_x, v_y, and v_a form a right triangle, the Pythagorean theorem provides $v_a^2 = v_x^2 + v_y^2$. Then v_a, v_z, and v also form a right triangle, and the Pythagorean theorem provides $v^2 = v_a^2 + v_z^2$. Substituting for v_a^2 in this latter expression yields $v^2 = v_x^2 + v_y^2 + v_z^2$, and thus Eq. (2.24) follows.

Direction angles and direction cosines

An effective way of characterizing a vector's orientation is by use of *direction angles*. Direction angles θ_x, θ_y, and θ_z are shown in Fig. 2.20 and are defined to be the angles measured from the positive x, y, and z directions, respectively, to the direction of the vector.

Direction angles can be used to obtain a vector's components, and vice versa, as follows. Consider the vector polygon shown in Fig. 2.21. This polygon is a right triangle that consists of the vector's magnitude $|\vec{v}|$, the y component v_y, and another vector component that we are not especially interested in, but which is nonetheless present. Although this triangle has rather complicated orientation in space, since it is a right triangle, the relationship between $|\vec{v}|$, v_y, and θ_y is given by elementary trigonometry as $v_y = |\vec{v}|\cos\theta_y$. Now imagine sketching a new vector triangle that contains $|\vec{v}|$, v_x, and angle θ_x; again, elementary trigonometry provides $v_x = |\vec{v}|\cos\theta_x$. Finally, consideration of the vector triangle containing $|\vec{v}|$, v_z, and angle θ_z provides $v_z = |\vec{v}|\cos\theta_z$. Collectively, these may be written as

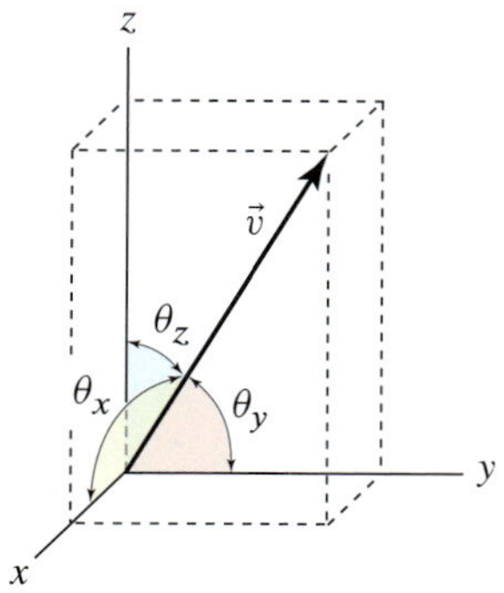

Figure 2.20
Definition of direction angles θ_x, θ_y, and θ_z.

$$\begin{aligned}\vec{v} &= v_x\,\hat{\imath} + v_y\,\hat{\jmath} + v_z\,\hat{k}\\ &= |\vec{v}|\,\cos\theta_x\,\hat{\imath} + |\vec{v}|\,\cos\theta_y\,\hat{\jmath} + |\vec{v}|\,\cos\theta_z\,\hat{k}\\ &= |\vec{v}|\,(\cos\theta_x\,\hat{\imath} + \cos\theta_y\,\hat{\jmath} + \cos\theta_z\,\hat{k}).\end{aligned} \tag{2.25}$$

In Eq. (2.25), note that $(\cos\theta_x\,\hat{\imath} + \cos\theta_y\,\hat{\jmath} + \cos\theta_z\,\hat{k})$ is a unit vector that points in the direction of $\vec{v}$. Because cosines of the direction angles play such an important role in writing Eq. (2.25), the quantities $\cos\theta_x$, $\cos\theta_y$, and $\cos\theta_z$ are called *direction cosines*.

Examination of Eq. (2.25) shows that the direction cosines constitute components of a unit vector. That is, substituting the expression for v_x, v_y, and v_z from Eq. (2.25) into Eq. (2.24) provides

$$\cos^2\theta_x + \cos^2\theta_y + \cos^2\theta_z = 1 \tag{2.26}$$

where $\cos^2\theta \equiv (\cos\theta)^2$. A subtle point about direction angles is that although there are three of them, only two are independent. For example, if θ_x and θ_y are known, then θ_z may be determined from Eq. (2.26), although care is usually needed to ensure the desired solution is obtained among the multiple solutions this equation has. These subtleties are illustrated in some of the example problems of this section. To summarize, important facts regarding direction angles and direction cosines for a vector $\vec{v}$ are as follows:

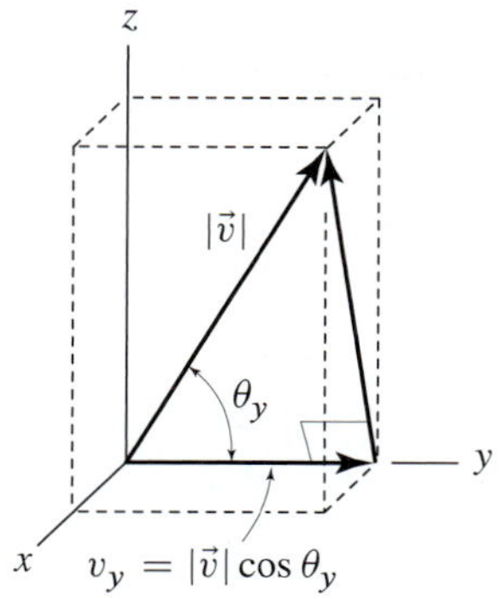

Figure 2.21
Relation between direction angle θ_y and vector component v_y. Similar sketches of triangles using direction angles θ_x and θ_z provide $v_x = |\vec{v}|\cos\theta_x$ and $v_z = |\vec{v}|\cos\theta_z$.

Summary Box

$$\begin{aligned}&\theta_x = \text{angle between positive } x \text{ direction and vector},\\ &\theta_y = \text{angle between positive } y \text{ direction and vector},\\ &\theta_z = \text{angle between positive } z \text{ direction and vector},\\ &\vec{v} = |\vec{v}|\,(\cos\theta_x\hat{\imath} + \cos\theta_y\hat{\jmath} + \cos\theta_z\hat{k}),\\ &\cos^2\theta_x + \cos^2\theta_y + \cos^2\theta_z = 1,\\ &\cos\theta_x = v_x/|\vec{v}|, \quad \cos\theta_y = v_y/|\vec{v}|, \quad \cos\theta_z = v_z/|\vec{v}|.\end{aligned} \tag{2.27}$$

Position vectors

Construction of a position vector in three dimensions is analogous to the procedure described in the previous section. Given points T and H having coordinates (x_T, y_T, z_T) and (x_H, y_H, z_H), respectively, as shown in Fig. 2.22,

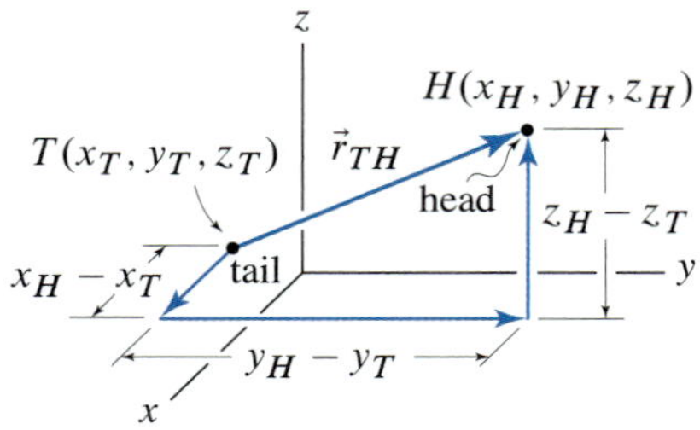

Figure 2.22
Construction of a position vector using Cartesian coordinates of the vector's head and tail.

Figure 2.23
The cross section of a steel cable is shown where the many individual strands that make up the cable can be seen.

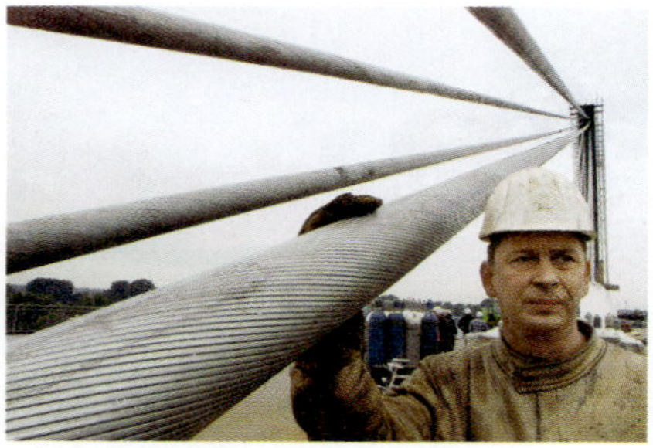

Figure 2.24
A worker stands next to one of the steel cables that support a bridge.

the *position vector* from tail T to head H is denoted by $\vec{r}_{TH}$ and is written as

$$\vec{r}_{TH} = (x_H - x_T)\,\hat{\imath} + (y_H - y_T)\,\hat{\jmath} + (z_H - z_T)\,\hat{k}. \tag{2.28}$$

The components of the position vector are given by "*coordinates of the head* minus *coordinates of the tail*," and this is a useful phrase to remember.

Use of position vectors to write expressions for force vectors

A position vector can often be a useful aid for writing an expression for a force vector. The basic idea is if a force vector $\vec{F}$ lies along the line of action of a position vector $\vec{r}$, or is known to be parallel to the position vector, then a vector expression for the force may be written as $\vec{F} = F\vec{r}/|\vec{r}|$. In this expression, $\vec{r}/|\vec{r}|$ is a unit vector that gives $\vec{F}$ the proper direction, and F is the component of the force $\vec{F}$ along the direction $\vec{r}$. Often, the location of two points on the line of action of a force will be known, and the components of $\vec{r}$ can be obtained by taking the difference between coordinates of the position vector's head and tail, as given by Eq. (2.28). Note that in an expression such as $\vec{F} = F\vec{r}/|\vec{r}|$, if $F > 0$, then the direction of the force is the same as the direction of $\vec{r}$, whereas if $F < 0$, then the direction of the force is opposite the direction of $\vec{r}$.

In statics, most of our applications of this idea will be for writing force vectors and, starting in Chapter 4, moment vectors also. However, this idea easily generalizes so that if any two vectors have the same line of action, or are parallel, then one of them may be used to construct an expression for the other, regardless of the physical interpretations each vector might have. For example, imagine a particle initially at rest is subjected to a known force $\vec{F}$ that has constant orientation, but perhaps magnitude that varies with time. Due to this force, the particle moves with acceleration $\vec{a}$ and velocity $\vec{v}$ where all three of these vectors share the same line of action. Since $\vec{F}$ is known, vector expressions for $\vec{a}$ and $\vec{v}$ may be written as $\vec{a} = a\vec{F}/|\vec{F}|$ and $\vec{v} = v\vec{F}/|\vec{F}|$, where a and v, which are functions of time, need to be determined.

Some simple structural members

One category of forces we deal with in mechanics is forces that develop within structural members. These is called *internal forces*, and much will be said about these in later chapters of this book. Simply stated, it is because structural members have the capability to develop internal forces that they are used to make structures, and these structures are in turn called upon to support *external forces* that are applied to them. The question to be briefly examined here is for what types of structural members do we know that the forces they support (internal forces) are collinear with their orientation.

Shown in Fig. 2.25 are structural members consisting of a cable (or rope, string, or cord), a wire, and a straight bar.* Consider a simple experiment with a piece of light string. When tensile forces are applied to the two ends of the string, the string obviously assumes the shape of a straight-line segment. Thus

* Bars can also be curved or have other geometry, but the remarks made here apply to straight bars only.

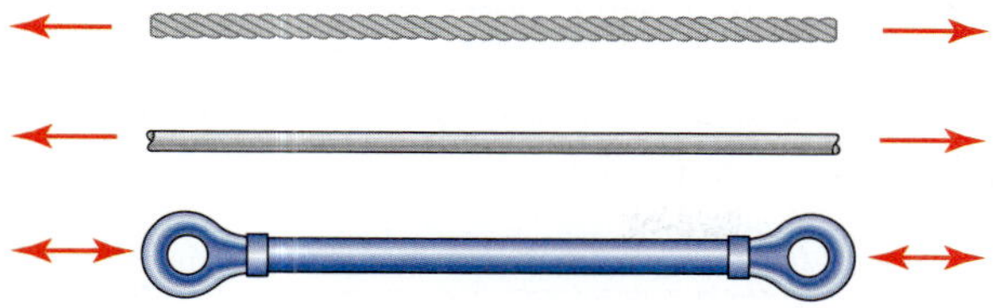

Figure 2.25. Some common structural members whose internal forces are collinear with their geometry.

we may infer that the forces it supports are directed from one endpoint of the string to the other. Another experiment would be to attach an object to one end of a string and hold the other end of the string with your hand. Again, the string takes the shape of a straight-line segment that is vertical. Newton's law of gravity tells us that the weight of the object, which is a vector quantity, is in the vertical direction, pointing from the object toward the center of the Earth. The vertical direction of the weight is identical to the orientation of the string, so once again we may infer that the internal force in the string is collinear with the line connecting the string's endpoints. Cables, ropes, wires, and the like are idealized to behave the same way. Further, all of these members are assumed to be incapable of supporting compressive forces as this would cause them to immediately bend or buckle. A straight bar is a member that is loaded by forces at its two ends as shown in Fig. 2.25. Its behavior is similar to that for cables and ropes, but a bar can also support compressive forces.

To summarize, for the simple, but important and common structural members shown in Fig. 2.25, it is possible to characterize the forces they support by expressions such as $\vec{F} = F\vec{r}/|\vec{r}|$, where $\vec{r}$ is a position vector that describes the orientation of the member. These structural members belong to a category of members called *two-force members*, which are examined in detail in Chapter 5.

End of Section Summary

In this section, Cartesian coordinate systems and Cartesian representation for vectors in three dimensions have been defined. Some of the key points are:

- The xyz coordinate system you use must be a *right-hand coordinate system*. Proper interpretation of some vector operations requires this.
- *Direction angles* provide a useful way for specifying a vector's orientation in three dimensions. A vector has three direction angles θ_x, θ_y, and θ_z, but only two of these are independent. Direction angles satisfy the equation $\cos^2\theta_x + \cos^2\theta_y + \cos^2\theta_z = 1$, so that if two direction angles are known, the third may be determined.
- Structural members such as cables, ropes, and bars support forces whose lines of action have the same orientation as the member's geometry. Thus, if $\vec{r}$ describes a member's geometry, a vector expression for the force supported by the member may be written as $\vec{F} = F(\vec{r}/|\vec{r}|)$.

EXAMPLE 2.9 *Direction Angles*

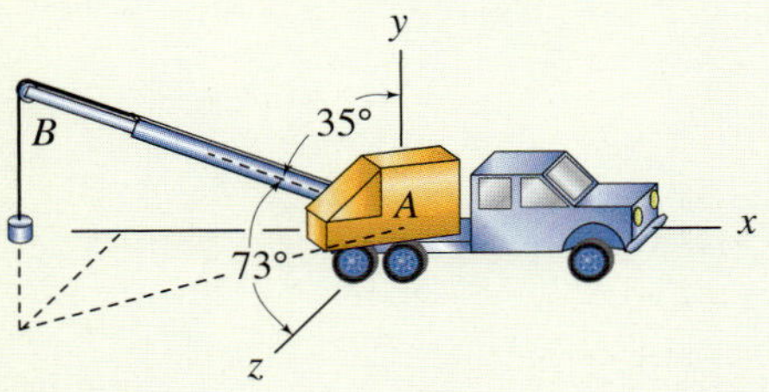

Figure 1

Determine the direction angles for boom AB of the crane.

SOLUTION

Road Map We denote the position vector of the crane's boom from point A to point B by $\vec{r}_{AB}$. The 35° and 73° angles reported in the figure are indeed direction angles: the 35° angle is measured from the positive y direction to the direction of $\vec{r}_{AB}$, and the 73° angle is measured from the positive z direction to the direction of $\vec{r}_{AB}$. Thus, $\theta_y = 35°$ and $\theta_z = 73°$. The remaining direction angle θ_x must satisfy Eq. (2.26).

Governing Equations & Computation Knowing that $\theta_y = 35°$ and $\theta_z = 73°$, we solve Eq. (2.26) to obtain

$$\cos^2\theta_x = 1 - \cos^2\theta_y - \cos^2\theta_z, \tag{1}$$

$$\cos\theta_x = \pm\sqrt{1 - \cos^2\theta_y - \cos^2\theta_z}, \tag{2}$$

$$\begin{aligned}\theta_x &= \cos^{-1}\left(\pm\sqrt{1 - \cos^2\theta_y - \cos^2\theta_z}\right)\\ &= \cos^{-1}\left(\pm\sqrt{1 - \cos^2 35° - \cos^2 73°}\right)\\ &= \cos^{-1}(\pm 0.493).\end{aligned} \tag{3}$$

Using the inverse cosine function on an electronic calculator provides the solutions $\theta_x = \cos^{-1}(+0.493) = 60.4°$ and $\theta_x = \cos^{-1}(-0.493) = 119.6°$. In fact, there are an infinite number of additional solutions, but the only two solutions in the interval $0° \le \theta_x \le 180°$ are 60.4° and 119.6°. Inspection will usually allow us to easily determine which of the solutions for the direction angle is correct for our physical problem, and you should examine Fig. 1 now to see if you can make this determination. If there is uncertainty, then an unequivocal answer can be obtained by computing the unit vectors corresponding to the two sets of direction angles to see which of the two vectors points in the proper direction. To show this, we calculate these unit vectors as

$$\text{for } \theta_x = 60.4°: \quad \begin{aligned}\hat{u}_1 &= \cos(60.4°)\,\hat{\imath} + \cos(35.0°)\,\hat{\jmath} + \cos(73.0°)\,\hat{k}\\ &= 0.493\,\hat{\imath} + 0.819\,\hat{\jmath} + 0.292\,\hat{k},\end{aligned} \tag{4}$$

$$\text{for } \theta_x = 119.6°: \quad \begin{aligned}\hat{u}_2 &= \cos(119.6°)\,\hat{\imath} + \cos(35.0°)\,\hat{\jmath} + \cos(73.0°)\,\hat{k}\\ &= -0.493\,\hat{\imath} + 0.819\,\hat{\jmath} + 0.292\,\hat{k}.\end{aligned} \tag{5}$$

Notice that the only difference between $\hat{u}_1$ and $\hat{u}_2$ in Eqs. (4) and (5) is the sign of the x component. Figure 1 clearly shows $\vec{r}_{AB}$ has a negative x component. So the solution for θ_x and $\hat{u}_1$ in Eq. (4) is physically meaningless and must be discarded. Thus, the direction angles for boom AB are

$$\boxed{\theta_x = 119.6°, \quad \theta_y = 35.0°, \quad \text{and} \quad \theta_z = 73.0°.} \tag{6}$$

Discussion & Verification As a quick partial check of solution accuracy, you should verify that the values for θ_x, θ_y, and θ_z in Eq. (6) satisfy the equation $\cos^2\theta_x + \cos^2\theta_y + \cos^2\theta_z = 1$; if this equation is not satisfied, then at least one of the angles is not a direction angle and an error has been made.

> **Common Pitfall**
>
> **Square roots.** When we take a square root, it is common to overlook the possibility of a negative solution. For example, if we desire the solution for s where $s^2 = 9$, the solution for s is always $s = \pm\sqrt{9} = \pm 3$. Depending on the physics of the problem, both solutions for s, or perhaps only one solution, may be meaningful. You will avoid serious blunders if you always check to be sure your direction angles and/or direction cosines are reasonable.

EXAMPLE 2.10 *Position Vectors and Force Vectors*

Cable AB supports a tensile force of magnitude $P = 2\,\mathrm{kN}$, and cable CD supports a tensile force of magnitude $F = 1\,\mathrm{kN}$. Write expressions for the forces $\vec{W}$, $\vec{F}$, and $\vec{P}$ applied to the cantilever I beam.

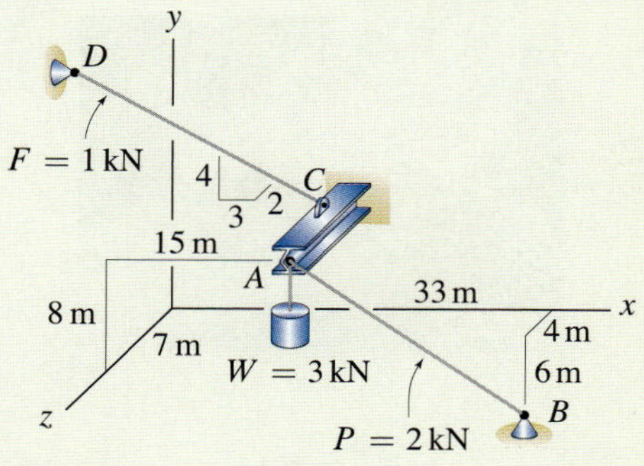

Figure 1

SOLUTION

Road Map The directions of the forces applied to the I beam by the weight and two cables are shown in Fig. 2. Because the y axis is vertical and weight due to gravity always acts in the downward vertical direction, writing an expression for the weight $\vec{W}$ will be straightforward. For the cables at points A and C, first we will write position vectors that describe the orientation of each cable, and from these we will construct expressions for the forces each cable applies to the I beam.

Governing Equations & Computation The weight is a force with 3 kN magnitude that acts in the downward vertical direction, which is the $-y$ direction in this example. Thus,

$$\boxed{\vec{W} = -3\,\hat{\jmath}\,\mathrm{kN}.} \tag{1}$$

The 2 kN tensile force cable AB applies to the I beam is directed from point A to point B. In other words, the force in cable AB tends to draw point A closer to point B. This vector will be called $\vec{P}$, and to construct a vector expression for it, we note that its direction is the same as that of the position vector from points A to B, $\vec{r}_{AB}$. Figure 1 shows the coordinates of these points to be $A(15, 8, 7)$ m and $B(33, -6, 4)$ m, and Eq. (2.28) can be applied to write

$$\begin{aligned}\vec{r}_{AB} &= [(33-15)\,\hat{\imath} + (-6-8)\,\hat{\jmath} + (4-7)\,\hat{k}]\,\mathrm{m} \\ &= (18\,\hat{\imath} - 14\,\hat{\jmath} - 3\,\hat{k})\,\mathrm{m}.\end{aligned} \tag{2}$$

An alternative way to construct $\vec{r}_{AB}$ is to imagine being positioned at point A, the tail of the vector, and then ask what distances in the x, y, and z directions must be traversed to arrive at point B, the head of the vector; Eq. (2) is again obtained. Finally, we use $\vec{r}_{AB}$ to write the vector expression for $\vec{P}$ as

$$\begin{aligned}\vec{P} &= (2\,\mathrm{kN})\frac{\vec{r}_{AB}}{|\vec{r}_{AB}|} = (2\,\mathrm{kN})\frac{(18\,\hat{\imath} - 14\,\hat{\jmath} - 3\,\hat{k})\,\mathrm{m}}{23\,\mathrm{m}} \\ &= \boxed{(1.57\,\hat{\imath} - 1.22\,\hat{\jmath} - 0.261\,\hat{k})\,\mathrm{kN},}\end{aligned} \tag{3}$$

where the magnitude of $\vec{r}_{AB}$ is $|\vec{r}_{AB}| = \sqrt{(18)^2 + (-14)^2 + (-3)^2}\,\mathrm{m} = 23\,\mathrm{m}$.

The 1 kN tensile force that cable CD applies to the I beam is directed from point C to D. In other words, the force in cable CD tends to draw point C closer to point D. This vector will be called $\vec{F}$, and Fig. 2 shows it has orientation that is the same as a vector consisting of 2 units in the positive z direction, plus 3 units in the negative x direction, plus 4 units in the positive y direction. Thus, we may immediately write

$$\vec{F} = (1\,\mathrm{kN})\frac{(-3\,\hat{\imath} + 4\,\hat{\jmath} + 2\,\hat{k})}{\sqrt{29}} = \boxed{(-557\,\hat{\imath} + 743\,\hat{\jmath} + 371\,\hat{k})\,\mathrm{N}.} \tag{4}$$

Equation (4) uses the unit vector $(-3\,\hat{\imath} + 4\,\hat{\jmath} + 2\,\hat{k})/\sqrt{29}$ to give $\vec{F}$ the proper direction, and the 1 kN term gives $\vec{F}$ the proper magnitude and units.

Discussion & Verification As a quick check, you should verify that each force vector points in the proper direction, and that the expressions for $\vec{P}$ and $\vec{F}$ have 2 kN and 1 kN magnitudes, respectively.

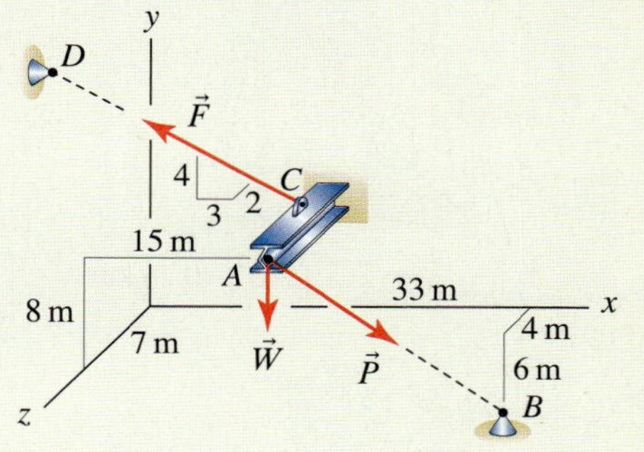

Figure 2

Directions of forces applied by the weight and cables to the I beam.

Helpful Information

Comments on Newton's third law. In our solution, $\vec{P}$ is the force vector that cable AB applies to the I beam. This cable also applies force to the support at B, and Newton's third law tells us this force vector is $-\vec{P}$, as illustrated below. Similar comments apply to $\vec{F}$.

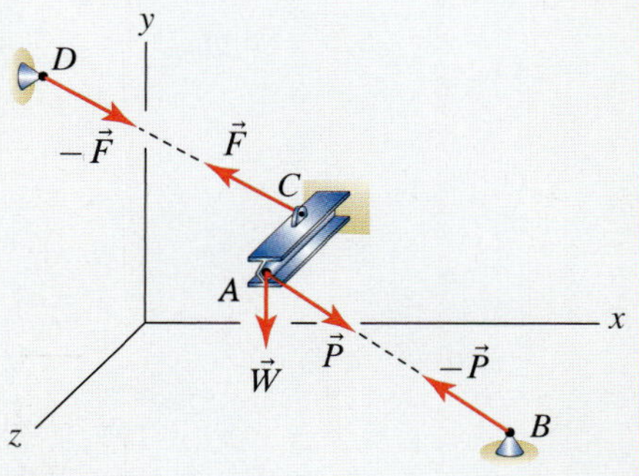

Figure 3

EXAMPLE 2.11 Direction Angles and Position Vectors

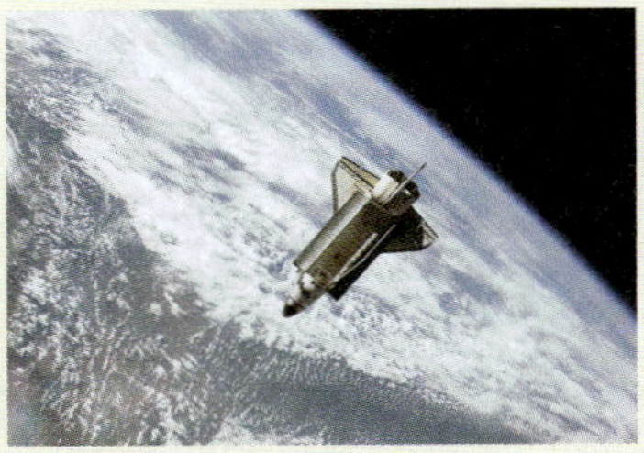

Observers on Earth at points A and B measure direction cosines of position vectors to the Space Shuttle C as

$$\text{for } \vec{r}_{AC}: \quad \cos\theta_x = 0.360, \quad \cos\theta_y = 0.480, \quad \cos\theta_z = 0.800, \tag{1}$$

$$\text{for } \vec{r}_{BC}: \quad \cos\theta_x = -0.515, \quad \cos\theta_y = -0.606, \quad \cos\theta_z = 0.606. \tag{2}$$

Determine the xyz coordinates of the Space Shuttle.

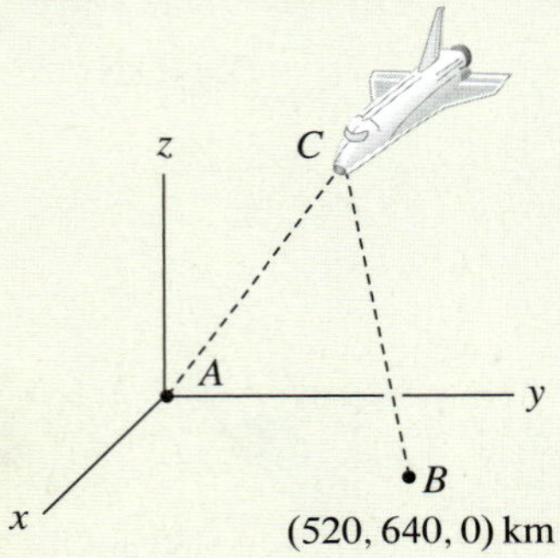

Figure 1

SOLUTION

Road Map Some careful thought before putting pencil to paper will help you create a successful strategy for this problem. The orientations of position vectors $\vec{r}_{AC}$ and $\vec{r}_{BC}$ are known, but their magnitudes are not. Given the direction cosines cited above, if the distance between the observers at points A and B (this will be called r_{AB}) were to increase, then the magnitudes of $\vec{r}_{AC}$ and $\vec{r}_{BC}$ would clearly also increase. Thus, our solution must incorporate the relative position of points A and B. A straightforward way of doing this is to write the vector equation $\vec{r}_{AC} = \vec{r}_{AB} + \vec{r}_{BC}$.

Governing Equations & Computation The addition $\vec{r}_{AC} = \vec{r}_{AB} + \vec{r}_{BC}$ is shown in Fig. 2. Note that this vector equation contains three scalar equations; that is, equality must be achieved independently for each of the x, y and z components, and hopefully this will provide enough equations so the unknown coordinates of point C may be determined.

Using the specified direction cosines, and letting r_{AC} and r_{BC} denote the magnitudes of vectors $\vec{r}_{AC}$ and $\vec{r}_{BC}$, respectively, we find the vector expressions for $\vec{r}_{AC}$ and $\vec{r}_{BC}$ are

$$\vec{r}_{AC} = r_{AC}\,(0.360\,\hat{\imath} + 0.480\,\hat{\jmath} + 0.800\,\hat{k}), \tag{3}$$

$$\vec{r}_{BC} = r_{BC}\,(-0.515\,\hat{\imath} - 0.606\,\hat{\jmath} + 0.606\,\hat{k}). \tag{4}$$

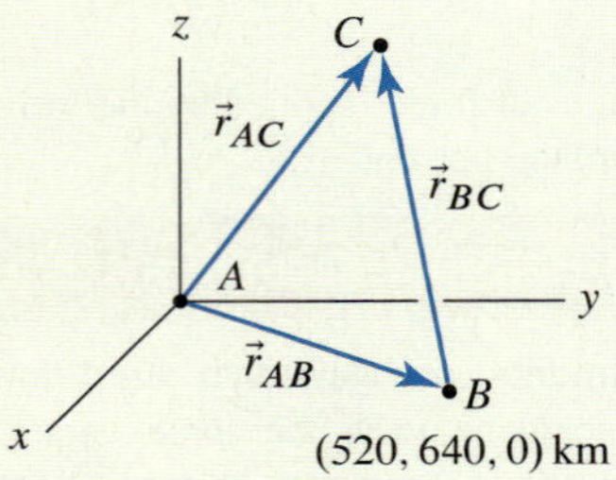

Figure 2
Vector addition that provides the position of C.

The coordinates of points A and B are given in Fig. 1, so we may write

$$\vec{r}_{AB} = (520\,\hat{\imath} + 640\,\hat{\jmath})\,\text{km}. \tag{5}$$

Then

$$\vec{r}_{AC} = \vec{r}_{AB} + \vec{r}_{BC},$$

$$r_{AC}(0.360\,\hat{\imath} + 0.480\,\hat{\jmath} + 0.800\,\hat{k}) = (520\,\hat{\imath} + 640\,\hat{\jmath})\,\text{km} + r_{BC}(-0.515\,\hat{\imath} - 0.606\,\hat{\jmath} + 0.606\,\hat{k}). \tag{6}$$

As described earlier, Eq. (6) contains three scalar equations as follows:

$$r_{AC}(0.360) = 520\,\text{km} + r_{BC}(-0.515), \tag{7}$$

$$r_{AC}(0.480) = 640\,\text{km} + r_{BC}(-0.606), \tag{8}$$

$$r_{AC}(0.800) = r_{BC}(0.606). \tag{9}$$

Solving Eq. (9) for r_{AC} provides $r_{AC} = r_{BC}(0.606/0.800)$, and substituting this into Eq. (7) [alternatively, Eq. (8) could be used] and rearranging provides the solution for r_{BC} as

$$r_{BC} = \frac{520\,\text{km}}{(0.606/0.800)(0.360) + 0.515} = 660\,\text{km}. \tag{10}$$

Then, using Eq. (9) [alternatively, Eq. (7) or (8) could be used] provides the solution for r_{AC} as

$$r_{AC} = r_{BC}(0.606/0.800) = 500\,\text{km}. \tag{11}$$

To determine the coordinates of the spacecraft at point C, we use Eq. (2.28) to write an expression for $\vec{r}_{AC}$ (alternatively, an expression for $\vec{r}_{BC}$ could be used)

$$\vec{r}_{AC} = x_C\,\hat{\imath} + y_C\,\hat{\jmath} + z_C\,\hat{k}. \tag{12}$$

Equating our two expressions for $\vec{r}_{AC}$, Eq. (3) with $r_{AC} = 500\,\text{km}$ and Eq. (11), provides

$$500\,\text{km}\,(0.360\,\hat{\imath} + 0.480\,\hat{\jmath} + 0.800\,\hat{k}) = x_C\,\hat{\imath} + y_C\,\hat{\jmath} + z_C\,\hat{k}. \tag{13}$$

Equating x, y, and z terms provides the solution for the coordinates of point C as

$$x_C = (500\,\text{km})(0.360) = 180\,\text{km}, \tag{14}$$

$$y_C = (500\,\text{km})(0.480) = 240\,\text{km}, \tag{15}$$

$$z_C = (500\,\text{km})(0.800) = 400\,\text{km}. \tag{16}$$

Discussion & Verification In view of the direction cosines given in Eqs. (1) and (2), and the coordinates of points A and B given in Fig. 2, the coordinates found in Eqs. (14)–(16) are reasonable. Furthermore, z_C is the altitude of the Space Shuttle (neglecting the effects of the Earth's curvature, which are minor for the dimensions of this problem), and the result $z_C = 400\,\text{km}$ is also reasonable for real life missions that space shuttles fly.

A Closer Look You might wonder why this example has more equations than unknowns. That is, there are three equations available, Eqs. (7)–(9), but they contain only two unknowns, r_{AC} and r_{BC}. Normally, such a situation indicates an *overdetermined* problem, where more information is available than is needed to determine the unknowns. Generally, overdetermined problems do not have a single, unique answer for each unknown. Such situations will rarely arise in statics, except when we are working with direction angles (recall that, as discussed on p. 61, of the three direction angles for a vector, only two are independent).

In this example, Eqs. (7)–(9) are not overdetermined. Rather, only two of Eqs. (7)–(9) are independent. To see this, if Eqs. (7) and (8) are multiplied by −640 and 520, respectively, and added, the result is an equation that has exactly the same proportions as Eq. (9). In mathematical nomenclature, we say that only two of the three equations are *linearly independent*.

The message of this short discussion is that you should not be surprised or alarmed when subtleties such as those pointed out here arise when working with direction angles and direction cosines.

EXAMPLE 2.12 *Force Vectors and Vector Addition*

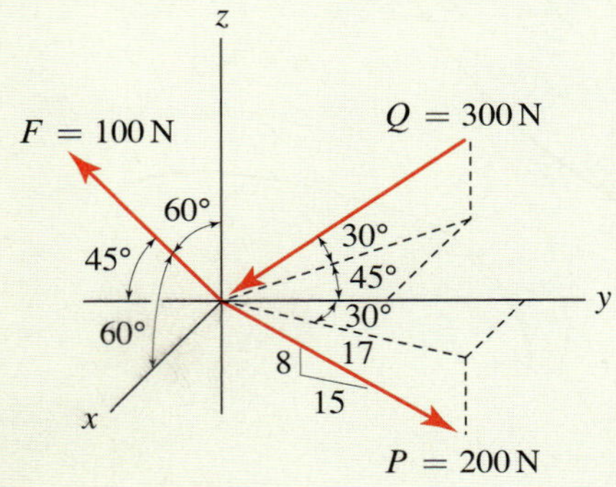

Figure 1

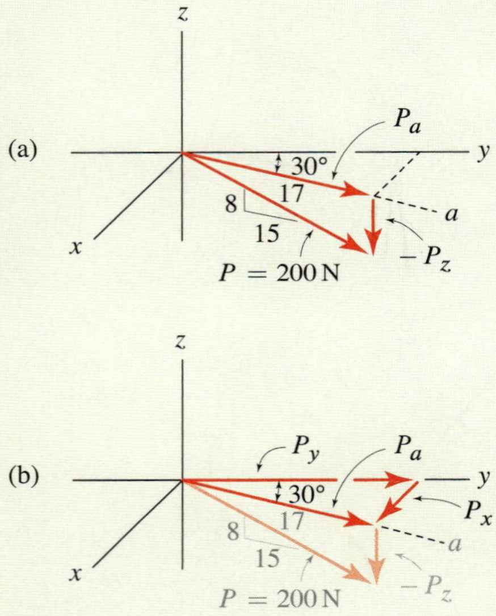

Figure 2
(a) Resolution of force P into z and a direction components. (b) Resolution of the a component into x and y components.

Forces $\vec{F}$, $\vec{P}$, and $\vec{Q}$ are given using a variety of different forms of orientation information that are commonly encountered.

(a) Write a vector expression for each force.

(b) Determine the resultant of the three forces.

(c) Determine the direction angles for the resultant.

SOLUTION

Road Map The strategy to be used to write a vector expression depends on the details of the geometry information provided. Once expressions for the three force vectors are obtained, we will add these to obtain the resultant force vector. Once the resultant force vector is known, it will be straightforward to determine the direction angles for the resultant.

Part (a)

Governing Equations & Computation

Force $\vec{F}$: We carefully examine Fig. 1 to see if the angles reported for $\vec{F}$ are direction angles. Indeed, the two 60° angles are measured from the positive x and z directions to the direction of $\vec{F}$, so both of these are direction angles and thus $\theta_x = 60°$ and $\theta_z = 60°$. The remaining 45° angle is measured from the negative y direction to the vector, so it is not a direction angle. However, the appropriate direction angle, as measured from the positive y direction, is easily found as $\theta_y = 180° - 45° = 135°$. As a partial check, we evaluate Eq. (2.26) to find $\cos^2 60° + \cos^2 135° + \cos^2 60° = 1$, and thus, these angles are indeed direction angles. Note, however, that if we incorrectly took θ_y to be 45°, Eq. (2.26) would still have been satisfied, but the resulting vector $\vec{F}$ would have a positive y component, rather than the correct value which is negative. Given the direction angles, a vector expression for $\vec{F}$ may be written as

$$\vec{F} = (100\,\text{N})\,(\cos 60°\,\hat{\imath} + \cos 135°\,\hat{\jmath} + \cos 60°\,\hat{k})$$
$$= \boxed{(50.00\,\hat{\imath} - 70.71\,\hat{\jmath} + 50.00\,\hat{k})\,\text{N}.} \tag{1}$$

Force $\vec{P}$: For this force, we observe the 30° angle shown in Fig. 1 is not a direction angle; while it is measured from the positive y direction, it is not measured to the vector's direction. When geometry is given in this fashion, we use a two-step process to determine the vector's components. We first use the 8-15 rise-run geometry to compute the components P_z and P_a shown in Fig. 2(a):

$$P_z = -(200\,\text{N})\frac{8}{17} = -94.12\,\text{N} \quad \text{and} \quad P_a = (200\,\text{N})\frac{15}{17} = 176.5\,\text{N}. \tag{2}$$

The negative sign is included in the expression for P_z because this force component acts in the negative z direction. For the second step of this procedure, we use the 30° angle to resolve P_a into x and y components, as shown in Fig. 2(b). Thus

$$P_x = P_a \sin 30° = 88.24\,\text{N} \quad \text{and} \quad P_y = P_a \cos 30° = 152.8\,\text{N}. \tag{3}$$

Collecting results for P_x, P_y, and P_z allows us to write

$$\boxed{\vec{P} = (88.24\,\hat{\imath} + 152.8\,\hat{\jmath} - 94.12\,\hat{k})\,\text{N}.} \tag{4}$$

As a partial check on our work, we compute the magnitude of the above expression to find $P = \sqrt{(88.24\,\text{N})^2 + (152.8\,\text{N})^2 + (-94.12\,\text{N})^2} = 200.0\,\text{N}$, which is the expected result.

Force $\vec{Q}$: For this force, we observe the 30° and 45° angles are not direction angles; the 45° angle is not measured to the vector's direction, and the 30° angle is not measured from a coordinate direction. As with force $\vec{P}$, we again use a two-step process to obtain the vector's components. As shown in Fig. 3(a), we first use the 30° angle to resolve $\vec{Q}$ into z and b direction components

$$Q_z = -(300\,\mathrm{N})\sin 30° = -150.0\,\mathrm{N} \quad \text{and} \tag{5}$$

$$Q_b = (300\,\mathrm{N})\cos 30° = 259.8\,\mathrm{N}. \tag{6}$$

The negative sign is included in the expressions for Q_z because this component acts in the negative z direction. We could similarly include a negative sign for Q_b because it acts in the negative b direction, but subsequent resolution of Q_b will take into account the appropriate directions in a straightforward manner. As shown in Fig. 3(b), we next use the 45° angle to resolve Q_b into the x and y components

$$Q_x = Q_b \sin 45° = 183.7\,\mathrm{N} \quad \text{and} \quad Q_y = -Q_b \cos 45° = -183.7\,\mathrm{N}. \tag{7}$$

Collecting results for Q_x, Q_y, and Q_z allows us to write

$$\boxed{\vec{Q} = (183.7\,\hat{\imath} - 183.7\,\hat{\jmath} - 150.0\,\hat{k})\,\mathrm{N}.} \tag{8}$$

As a partial check of correctness, you should evaluate the magnitude of the above expression to verify it is 300 N.

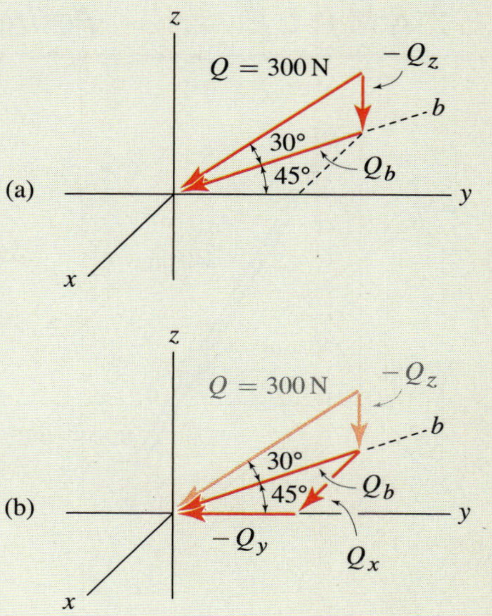

Figure 3
(a) Resolution of force Q into z and b direction components. (b) Resolution of the b component into x and y components.

Part (b)

Governing Equations & Computation The resultant of the three forces is

$$\begin{aligned}\vec{R} = \vec{F} + \vec{P} + \vec{Q} &= (50.00\,\hat{\imath} - 70.71\,\hat{\jmath} + 50.00\,\hat{k})\,\mathrm{N} \\ &\quad + (88.24\,\hat{\imath} + 152.8\,\hat{\jmath} - 94.12\,\hat{k})\,\mathrm{N} \\ &\quad + (183.7\,\hat{\imath} - 183.7\,\hat{\jmath} - 150.0\,\hat{k})\,\mathrm{N} \\ &= \boxed{(321.9\,\hat{\imath} - 101.6\,\hat{\jmath} - 194.1\,\hat{k})\,\mathrm{N},}\end{aligned} \tag{9}$$

and the magnitude of the resultant is

$$|\vec{R}| = \sqrt{(321.9\,\mathrm{N})^2 + (-101.6\,\mathrm{N})^2 + (-194.1\,\mathrm{N})^2} = 389.4\,\mathrm{N}. \tag{10}$$

Part (c)

Governing Equations & Computation Once $\vec{R}$ is known, obtaining its direction angles is straightforward. Applying the equations $\cos\theta_x = R_x/|\vec{R}|$, $\cos\theta_y = R_y/|\vec{R}|$, and $\cos\theta_z = R_z/|\vec{R}|$ gives

$$\cos\theta_x = 321.9\,\mathrm{N}/389.4\,\mathrm{N} \Rightarrow \theta_x = \cos^{-1}(321.9/389.4) = \boxed{34.2°,} \tag{11}$$

$$\cos\theta_y = -101.6\,\mathrm{N}/389.4\,\mathrm{N} \Rightarrow \theta_y = \cos^{-1}(-101.6/389.4) = \boxed{105.1°,} \tag{12}$$

$$\cos\theta_z = -194.1\,\mathrm{N}/389.4\,\mathrm{N} \Rightarrow \theta_z = \cos^{-1}(-194.1/389.4) = \boxed{119.9°.} \tag{13}$$

Discussion & Verification As simple checks of accuracy, you should verify that $\vec{F}$, $\vec{P}$, and $\vec{Q}$ each point in the proper direction and have the correct magnitude. Depending on the number of forces being added and their relative sizes and magnitudes, it may also be possible to verify that some (or all) of the components of $\vec{R}$ have reasonable direction and size.

EXAMPLE 2.13 Position Vectors and Force Vectors

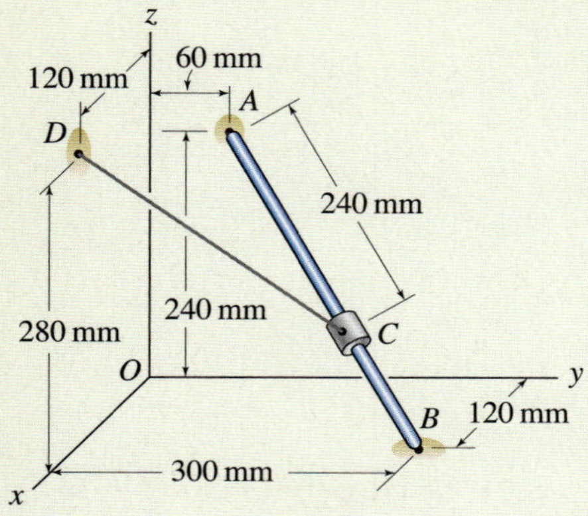

Figure 1

The collar C slides on a straight bar AB and is supported by a cable attached to point D.

(a) Determine the coordinates of point C.

(b) If the tensile force in the cable is known to be 150 N, write vector expressions using Cartesian representations for the force the cable exerts on point C and for the force the cable exerts on point D.

SOLUTION

Road Map To determine the coordinates of point C, our strategy will be to write the position vector from points A to B, and then using this, we will write the position vector from A to C. Then the coordinates of point C may be determined. With the coordinates of points C and D known, we will write the position vector $\vec{r}_{CD}$, and from this the desired force vectors $\vec{F}_{CD}$ and $\vec{F}_{DC}$ can be written.

Part (a)

Governing Equations & Computation To construct the position vector from A to B, we imagine being positioned at point A and asking what distances in the x, y, and z directions must be traversed to arrive at point B. Referring to Fig. 1, we find these distances to be 120 mm, 300 mm − 60 mm = 240 mm, and −240 mm, respectively. Thus, the position vector and its magnitude are

$$\vec{r}_{AB} = (120\,\hat{\imath} + 240\,\hat{\jmath} - 240\,\hat{k})\text{ mm}, \tag{1}$$

$$|\vec{r}_{AB}| = \sqrt{(120)^2 + (240)^2 + (-240)^2}\text{ mm} = 360\text{ mm}. \tag{2}$$

Alternatively, we could tabulate the coordinates of points B and A and apply Eq. (2.28) with B and A being the head and tail of the position vector, respectively, to arrive at the same result.

Noting that position vectors $\vec{r}_{AC}$ and $\vec{r}_{AB}$ have the same direction, we may now write

$$\begin{aligned}\vec{r}_{AC} &= (240\text{ mm})\frac{\vec{r}_{AB}}{|\vec{r}_{AB}|} = (240\text{ mm})\frac{120\,\hat{\imath} + 240\,\hat{\jmath} - 160\,\hat{k}}{360} \\ &= (80\,\hat{\imath} + 160\,\hat{\jmath} - 160\,\hat{k})\text{ mm}.\end{aligned} \tag{3}$$

Since $\vec{r}_{AC}$ is known, we may use its components to evaluate the difference between the coordinates of the head and tail of $\vec{r}_{AC}$, where C is the head and A is the tail. The coordinates of point C are (x_C, y_C, z_C), and from Fig. 1 the coordinates of point A are (0, 60, 240) mm, so we may write

$$\begin{aligned}\vec{r}_{AC} &= (80\,\hat{\imath} + 160\,\hat{\jmath} - 160\,\hat{k})\text{ mm} \\ &= (x_C - 0\text{ mm})\,\hat{\imath} + (y_C - 60\text{ mm})\,\hat{\jmath} + (z_C - 240\text{ mm})\,\hat{k}.\end{aligned} \tag{4}$$

Equality must be achieved independently for each of the x, y, and z components; thus our solutions are

$$x_C = \boxed{80\text{ mm},} \tag{5}$$

$$y_C = 160\text{ mm} + 60\text{ mm} = \boxed{220\text{ mm},} \tag{6}$$

$$z_C = -160\text{ mm} + 240\text{ mm} = \boxed{80\text{ mm}.} \tag{7}$$

Part (b)

Governing Equations & Computation Cable CD exerts a force $\vec{F}_{CD}$ of magnitude 150 N on point C, in the direction from C to D, as shown in Fig. 2. Using D and C as the head and tail, respectively, of a position vector where the coordinates of these are $D(120, 0, 280)$ mm and $C(80, 220, 80)$ mm, we write

$$\vec{r}_{CD} = [(120 - 80)\,\hat{\imath} + (0 - 220)\,\hat{\jmath} + (280 - 80)\,\hat{k}]\,\text{mm},$$

$$= (40\,\hat{\imath} - 220\,\hat{\jmath} + 200\,\hat{k})\,\text{mm} \tag{8}$$

$$|\vec{r}_{CD}| = \sqrt{(40)^2 + (-220)^2 + (200)^2}\,\text{mm} = 300\,\text{mm}. \tag{9}$$

The cable force exerted on C is

$$\vec{F}_{CD} = (150\,\text{N})\,\frac{\vec{r}_{CD}}{|\vec{r}_{CD}|} = (150\,\text{N})\,\frac{40\,\hat{\imath} - 220\,\hat{\jmath} + 200\,\hat{k}}{300}$$

$$= \boxed{(20\,\hat{\imath} - 110\,\hat{\jmath} + 100\,\hat{k})\,\text{N}.} \tag{10}$$

The force that cable CD exerts on the support at D is $\vec{F}_{DC}$ and, according to Newton's third law, it is of equal magnitude and opposite direction to the force the cable exerts on C. Thus, $\vec{F}_{DC} = -\vec{F}_{CD}$ and we obtain

$$\boxed{\vec{F}_{DC} = (-20\,\hat{\imath} + 110\,\hat{\jmath} - 100\,\hat{k})\,\text{N}.} \tag{11}$$

Discussion & Verification As quick, partial checks of solution accuracy, you should verify that $\vec{r}_{AC}$ and $\vec{F}_{CD}$ point in proper directions and have correct magnitudes.

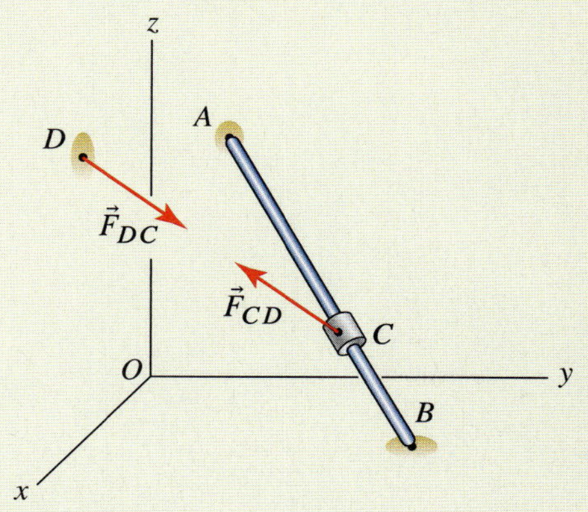

Figure 2
The force applied by the cable to the collar C is $\vec{F}_{CD}$, and the force applied by the cable to the support at D is $\vec{F}_{DC}$. Newton's third law requires $\vec{F}_{DC} = -\vec{F}_{CD}$.

PROBLEMS

Problem 2.45

A position vector $\vec{r}$ has direction angles $\theta_x = 36°$ and $\theta_y = 72°$. Determine the possible values for direction angle θ_z, and describe the differences in the orientation of $\vec{r}$. Provide a sketch showing these different vectors.

Problems 2.46 through 2.49

Write an expression for each force using Cartesian representation, and evaluate the resultant force vector. Sketch the resultant force in the xyz coordinate system.

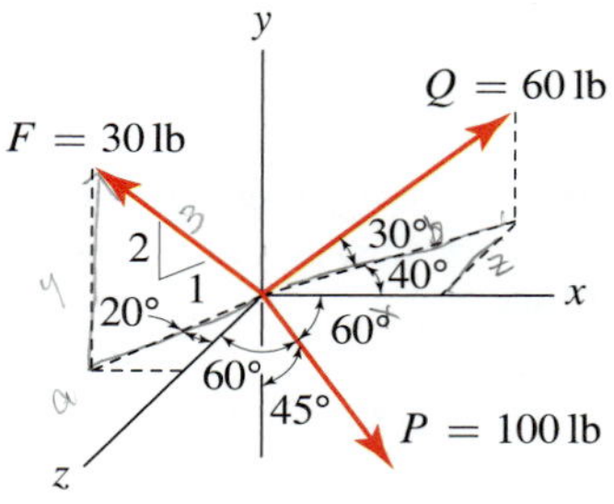

Figure P2.46

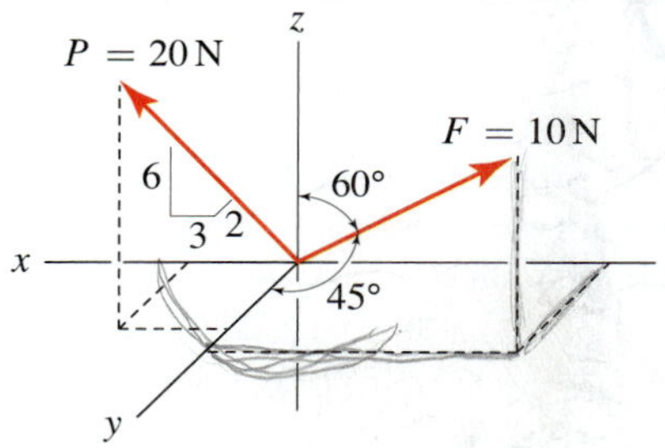

Figure P2.47

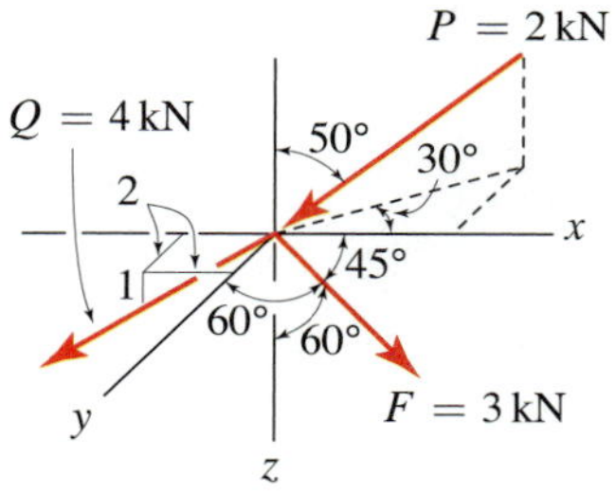

Figure P2.48

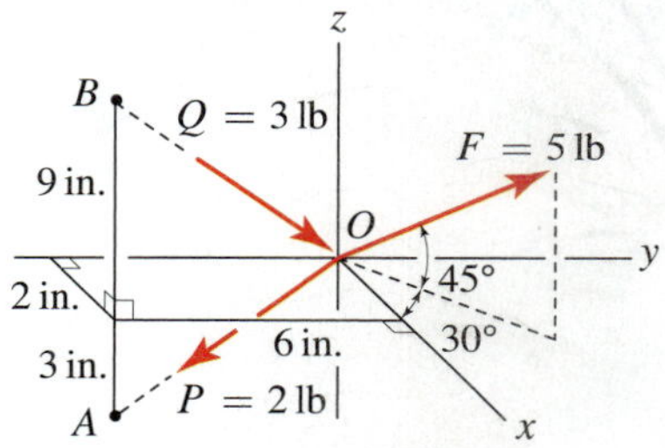

Figure P2.49

Problems 2.50 and 2.51

For each vector, find the direction angles and write an expression for the vector using Cartesian representation. Evaluate the sum of the two vectors, and report the direction angles for the resultant. Also, sketch the resultant in the xyz coordinate system.

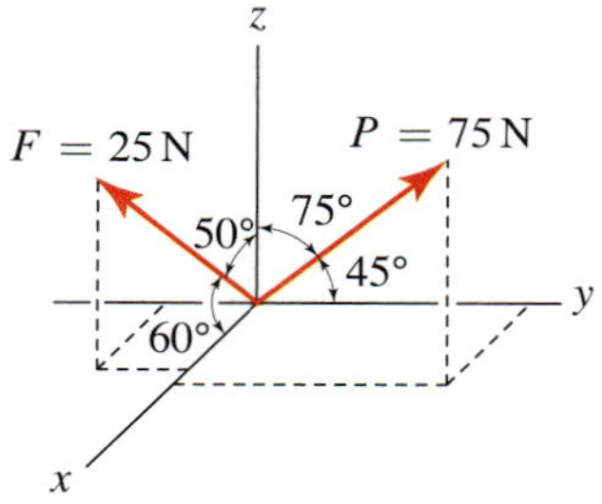

Figure P2.50

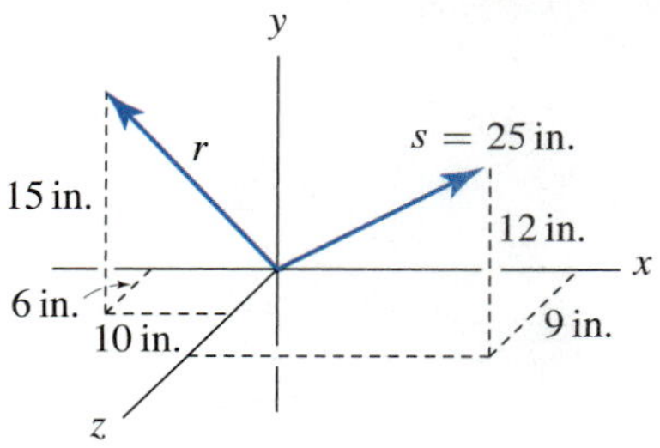

Figure P2.51

Problem 2.52

The Space Shuttle uses radar to determine the magnitudes and direction cosines of position vectors to satellites A and B as

$$\text{for } \vec{r}_{OA}\text{: } |\vec{r}_{OA}| = 2\,\text{km},\ \cos\theta_x = 0.768,\ \cos\theta_y = -0.384,\ \cos\theta_z = 0.512$$

$$\text{for } \vec{r}_{OB}\text{: } |\vec{r}_{OB}| = 4\,\text{km},\ \cos\theta_x = 0.743,\ \cos\theta_y = 0.557,\ \cos\theta_z = -0.371.$$

Determine the distance between the satellites.

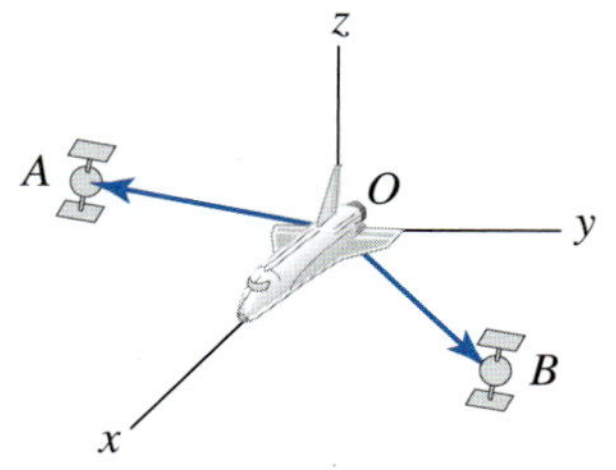

Figure P2.52

Problem 2.53

A cube of material with 1 mm edge lengths is examined by a scanning electron microscope, and a small inclusion (i.e., a cavity) is found at point P. It is determined that the direction cosines for a vector from points A to P are $\cos\theta_x = -0.485$, and $\cos\theta_y = 0.485$, and $\cos\theta_z = -0.728$; and the direction cosines for a vector from points B to P are $\cos\theta_x = -0.667$, $\cos\theta_y = -0.667$, and $\cos\theta_z = 0.333$. Determine the coordinates of point P.

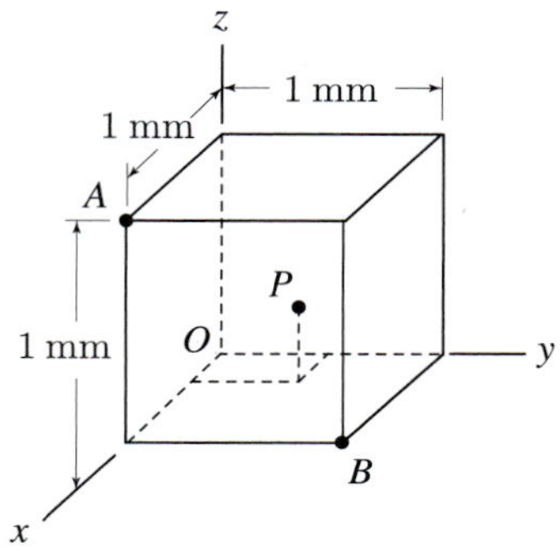

Figure P2.53

Problem 2.54

A theodolite is an instrument that measures horizontal and vertical angular orientations of a line of sight. The line of sight may be established optically, or by laser, and many forms of theodolite are used in surveying, construction, astronomy, and manufacturing. A simple optical theodolite is shown. After the instrument is aligned so that its base is in a horizontal plane and a desired reference direction is selected (such as perhaps north in the case of a surveying instrument), the telescopic sight is used to establish a line of sight, and then horizontal and vertical angles θ_h and θ_v are measured.

(a) If θ_h and θ_v are known, derive an expression that gives the direction angles θ_x, θ_y, and θ_z for the line of sight.

(b) If $\theta_h = 30°$ and $\theta_v = 60°$, use your answer to Part (a) to determine θ_x, θ_y, and θ_z.

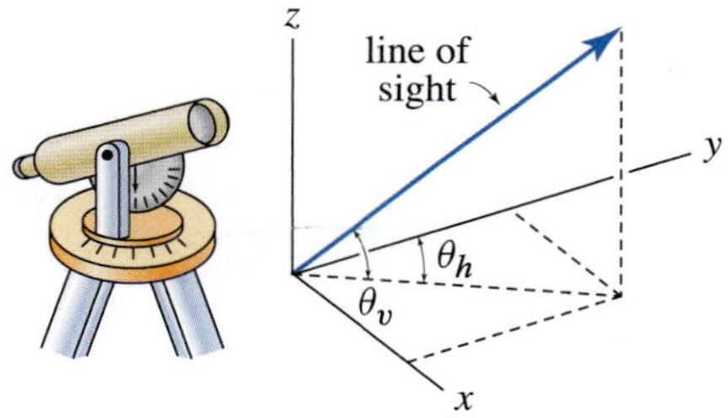

Figure P2.54

Problem 2.55

Using a theodolite with a laser range finder, a researcher at point A locates a pair of rare birds nesting at point B and determines the direction angles $\theta_x = 53°$, $\theta_y = 81°$, and $\theta_z = 38°$ and distance 224 m for a position vector from point A to B. For subsequent observations, the researcher would like to use position C, and thus while at point A, she also measures the direction angles $\theta_x = 135°$, $\theta_y = 47°$, and $\theta_z = 79°$ and distance 507 m for a position vector from point A to C. For observation from point C, determine the direction angles and distance to the nest at B.

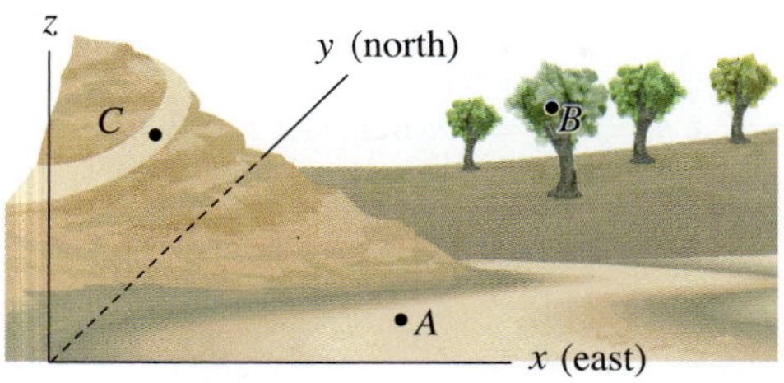

Figure P2.55

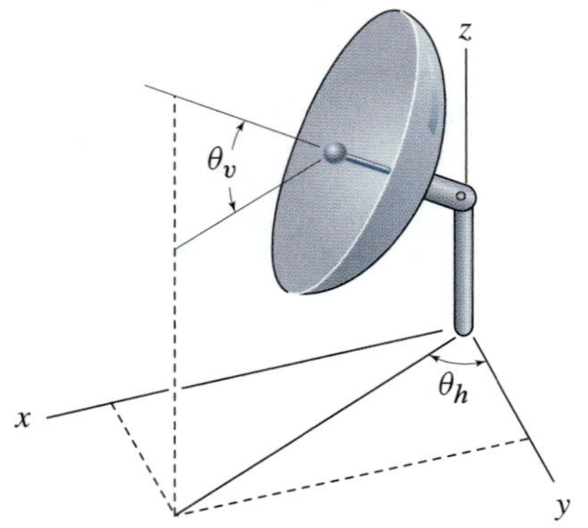

Figure P2.56

Problem 2.56

If the direction cosines for a satellite antenna are to be $\cos\theta_x = 0.286$, $\cos\theta_y = 0.429$, and $\cos\theta_z = 0.857$, determine values of angles θ_h and θ_v.

Problem 2.57

With direction angles, general practice is to use values of θ_x, θ_y, and θ_z between 0° and 180°, and this is sufficient to characterize the orientation of any vector. When direction angles are measured in this way, the sum of any two direction angles is always 90° or greater. Offer an argument that shows this statement is true.

Problems 2.58 and 2.59

Bars AC and DG are straight and parallel to the x and y axes, respectively. BE is a cable whose tensile force is 100 lb. For the dimensions given, determine expressions for the force the cable exerts on B and the force the cable exerts on E using Cartesian vector representation.

Problem 2.58 $x = 4$ in. and $y = 7$ in.

Problem 2.59 $x = 6$ in. and $y = 12$ in.

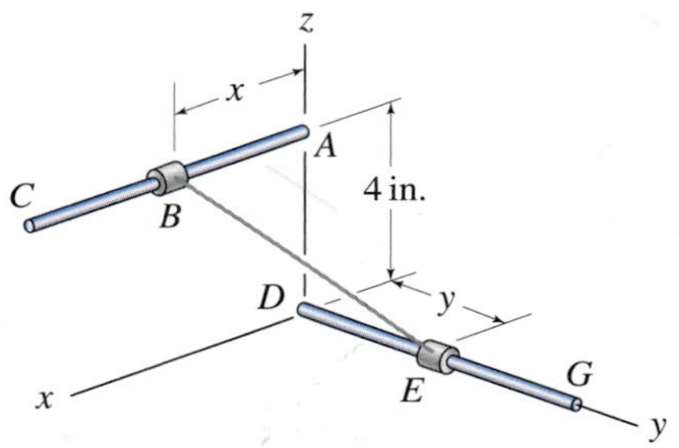

Figure P2.58 and P2.59

Problems 2.60 and 2.61

The structure consists of a quarter-circular rod AB with radius 150 mm that is fixed in the xy plane. CD is a straight rod where D may be positioned at different locations on the circular rod. GE is an elastic cord whose support at G lies in the yz plane, and the bead at E may have different positions d. For the values of b, h, d, and z provided, determine the coordinates of E, and if the elastic cord supports a tensile force of 100 N, write a vector expression using Cartesian vector representation for the force $\vec{F}_{EG}$ the cord exerts on bead E.

Problem 2.60 $b = 4$, $h = 3$, $d = 260$ mm, and $z = 240$ mm.

Problem 2.61 $b = 3$, $h = 4$, $d = 195$ mm, and $z = 270$ mm.

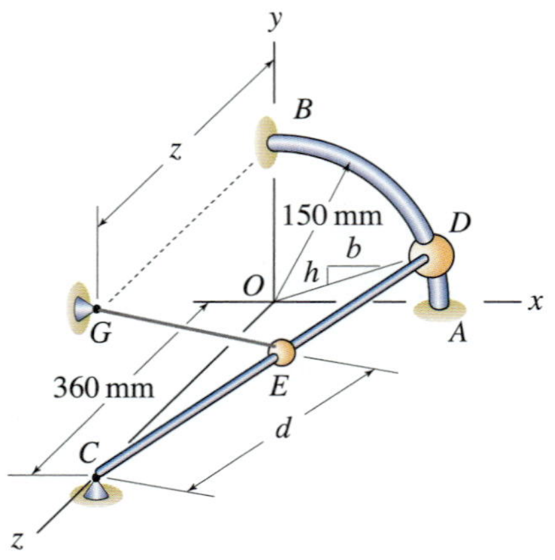

Figure P2.60 and P2.61

Problem 2.62

The rear wheel of a multispeed bicycle is shown. The wheel has 32 spokes, with one-half being on either side (spokes on sides A and B are shown in the figure in black and red, respectively). For the tire to be properly centered on the frame of the bicycle, points A and B of the hub must be positioned at the same distance d from the centerline of the tire. To make room for the sprocket cluster, bicycle manufacturers give spokes on side B of the wheel different orientation than spokes on side A. For the following questions, assume the tire is in the process of being manufactured, so that all spokes on side A have the same force F_A and all spokes on side B have the same force F_B.

(a) Determine the ratio of spoke forces F_B/F_A so that the resultant force in the x direction applied to the hub by all 32 spokes is zero. *Hint:* Although each spoke has different orientation, all spokes on side A have the same length, and similarly all spokes on side B have the same length. Furthermore, all spokes on side A have the same x component of force, and all spokes on side B have the same x component of force.

(b) On which side of the wheel are the spokes most severely loaded?

(c) Briefly describe a new design in which spokes on both sides of the wheel are equally loaded and points A and B are at the same distance d from the centerline of the tire.

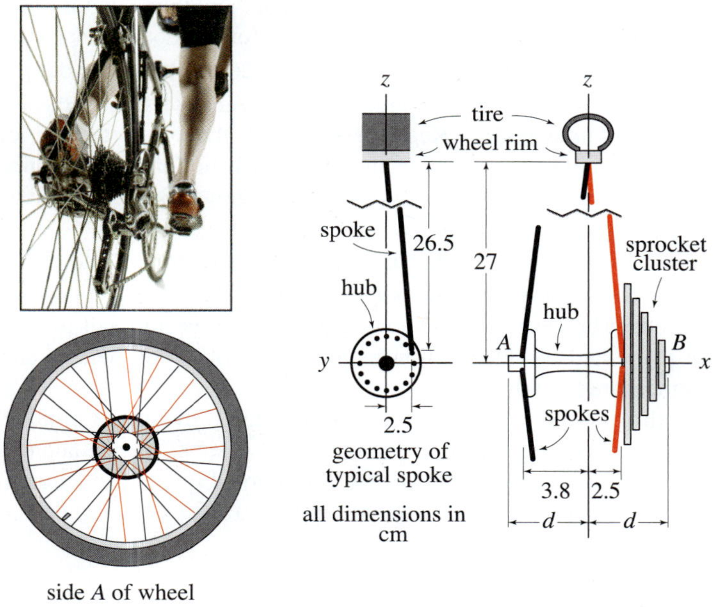

Figure P2.62

2.4 Vector Dot Product

The dot product between two vectors may be used to (1) determine the angle between the lines of action for the two vectors and (2) determine the component of one vector that acts in the direction of the other. Throughout statics and subjects that follow we must frequently find such information, and the dot product provides a straightforward way to do this, especially for vectors in three dimensions.

The *dot product* between two vectors $\vec{A}$ and $\vec{B}$ is an operation defined as

$$\vec{A} \cdot \vec{B} = |\vec{A}||\vec{B}| \cos\theta \tag{2.29}$$

where:

θ is the angle between the lines of action of $\vec{A}$ and $\vec{B}$ where* $0 \le \theta \le 180°$.

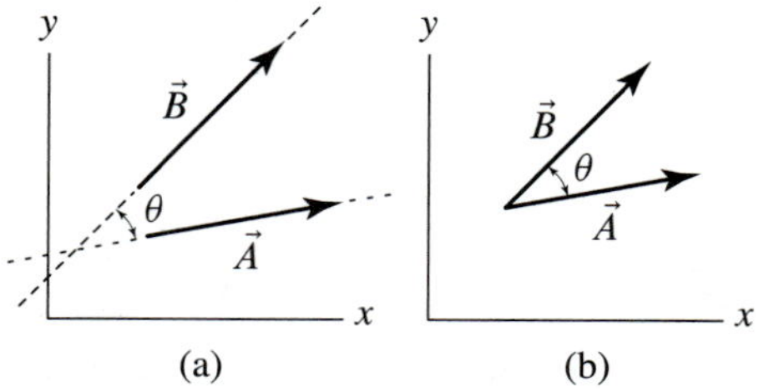

Figure 2.26
(a) Vectors $\vec{A}$ and $\vec{B}$ in two dimensions. The intersection of the lines of action of the two vectors defines angle θ for evaluation of the dot product. (b) Often we imagine sliding the vectors tail to tail for evaluation of the dot product.

In words, Eq. (2.29) states "$\vec{A}$ dot $\vec{B}$" equals the product of the magnitude of $\vec{A}$, magnitude of $\vec{B}$, and cosine of angle θ between the lines of action of $\vec{A}$ and $\vec{B}$. The dot product yields a result that is a scalar, and thus the dot product is sometimes called the *scalar product*. The units for the dot product are equal to the product of the units for $\vec{A}$ and $\vec{B}$. Figure 2.26 illustrates the dot product between vectors $\vec{A}$ and $\vec{B}$ in two dimensions. The intersection of the vectors' lines of action defines angle θ. Regardless of where the two vectors are located in space, and regardless of whether their lines of action intersect, to perform (or interpret) the dot product it is helpful if we imagine sliding the vectors tail to tail, which again identifies angle θ.

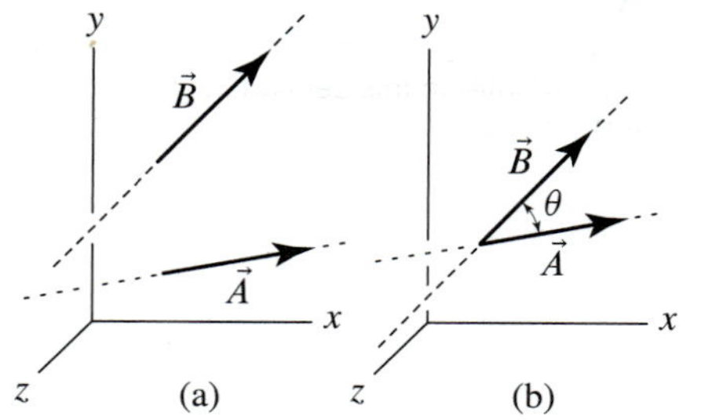

Figure 2.27
(a) Vectors $\vec{A}$ and $\vec{B}$ in three dimensions. (b) To evaluate the dot product the vectors are arranged tail to tail to define angle θ, which is measured in the plane containing the two vectors.

Figure 2.27 illustrates the dot product between vectors $\vec{A}$ and $\vec{B}$ in three dimensions. Note that $\vec{A}$ and $\vec{B}$ do not need to lie in the same plane to compute the dot product between them. After we arrange the vectors tail to tail, they define a plane and the intersection of their lines of action defines angle θ, which is measured in this plane.

Depending on the value of θ, the scalar produced by the dot product may be positive, zero, or negative. For example, $0 \le \theta < 90°$ (acute angle) gives a dot product that is positive, $\theta = 90°$ (orthogonal or perpendicular angle) gives a dot product that is zero, and $90° < \theta \le 180°$ (obtuse angle) gives a dot product that is negative.

The dot product has the following properties:

$\vec{A} \cdot \vec{B} = \vec{B} \cdot \vec{A}$	commutative property,	(2.30)
$s(\vec{A} \cdot \vec{B}) = (s\vec{A}) \cdot \vec{B} = \vec{A} \cdot (s\vec{B})$	associative property with respect to multiplication by a scalar,	(2.31)
$(\vec{A} + \vec{B}) \cdot \vec{C} = (\vec{A} \cdot \vec{C}) + (\vec{B} \cdot \vec{C})$	distributive property with respect to vector addition.	(2.32)

* The dot product, and cross product described in Section 2.5, can be defined to allow values of θ outside this range. However, the description of direction for the cross product becomes more intricate. Thus, with little loss of generality it is common to use $0 \le \theta \le 180°$ for both definitions.

The foregoing remarks are true regardless of the type of vector representation that is used. For example, consider the two position vectors shown in Fig. 2.28(a). We first slide the vectors tail to tail as shown in Fig. 2.28(b) and then identify the angle between $\vec{A}$ and $\vec{B}$ as 60°. Applying Eq. (2.29) provides the dot product between $\vec{A}$ and $\vec{B}$ as

$$\vec{A} \cdot \vec{B} = (2\,\mathrm{m})(3\,\mathrm{m}) \cos 60° = 3\,\mathrm{m}^2. \tag{2.33}$$

Observe the result is a scalar, with units of m^2. Although these are appropriate units for area, that does not necessarily mean that this result has the interpretation of being an area. The physical significance of the result of the dot product will be discussed in detail soon.

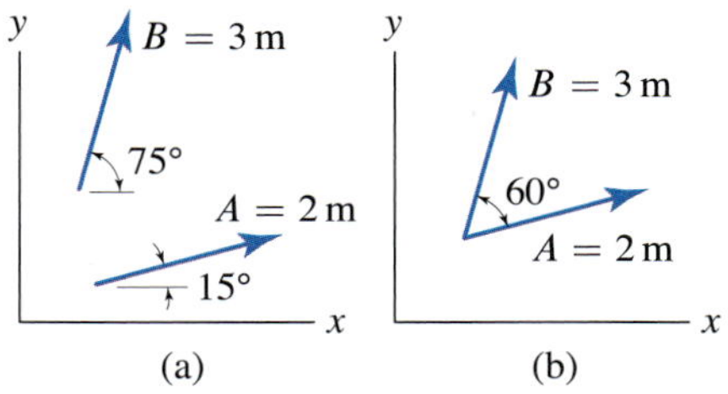

Figure 2.28
Dot product between two vectors.

Dot product using Cartesian components

When vectors $\vec{A}$ and $\vec{B}$ have Cartesian representations, the dot product between them may be evaluated using the following convenient computation

$$\vec{A} \cdot \vec{B} = A_x B_x + A_y B_y + A_z B_z. \tag{2.34}$$

Simply stated, for two vectors with Cartesian representation, the dot product between them is the sum of the products of their components. Equation (2.34) can be obtained using the following derivation. We first write $\vec{A}$ and $\vec{B}$, using Cartesian representation

$$\vec{A} = A_x \hat{\imath} + A_y \hat{\jmath} + A_z \hat{k}, \tag{2.35}$$

$$\vec{B} = B_x \hat{\imath} + B_y \hat{\jmath} + B_z \hat{k}. \tag{2.36}$$

We then take the dot product between $\vec{A}$ and $\vec{B}$, using the distributive law of Eq. (2.32) to expand the product term by term, as follows:

$$\begin{aligned} \vec{A} \cdot \vec{B} &= (A_x \hat{\imath} + A_y \hat{\jmath} + A_z \hat{k}) \cdot (B_x \hat{\imath} + B_y \hat{\jmath} + B_z \hat{k}) \\ &= (A_x \hat{\imath} \cdot B_x \hat{\imath}) + (A_x \hat{\imath} \cdot B_y \hat{\jmath}) + (A_x \hat{\imath} \cdot B_z \hat{k}) \\ &\quad + (A_y \hat{\jmath} \cdot B_x \hat{\imath}) + (A_y \hat{\jmath} \cdot B_y \hat{\jmath}) + (A_y \hat{\jmath} \cdot B_z \hat{k}) \\ &\quad + (A_z \hat{k} \cdot B_x \hat{\imath}) + (A_z \hat{k} \cdot B_y \hat{\jmath}) + (A_z \hat{k} \cdot B_z \hat{k}). \end{aligned} \tag{2.37}$$

To proceed further we must address dot products between combinations of unit vectors. Equation (2.29) shows $\hat{\imath} \cdot \hat{\imath} = (1)(1)\cos 0° = 1$, and similarly $\hat{\jmath} \cdot \hat{\jmath} = 1$ and $\hat{k} \cdot \hat{k} = 1$. Further, the dot product between combinations of different unit vectors is always zero because the vectors are orthogonal. For example, $\hat{\imath} \cdot \hat{\jmath} = (1)(1)\cos 90° = 0$. In summary, dot products between combinations of unit vectors are

$$\begin{array}{lll} \hat{\imath} \cdot \hat{\imath} = 1 & \hat{\imath} \cdot \hat{\jmath} = 0 & \hat{\imath} \cdot \hat{k} = 0 \\ \hat{\jmath} \cdot \hat{\imath} = 0 & \hat{\jmath} \cdot \hat{\jmath} = 1 & \hat{\jmath} \cdot \hat{k} = 0 \\ \hat{k} \cdot \hat{\imath} = 0 & \hat{k} \cdot \hat{\jmath} = 0 & \hat{k} \cdot \hat{k} = 1. \end{array} \tag{2.38}$$

Substituting Eqs. (2.38) into Eq. (2.37) then provides Eq. (2.34).

Helpful Information

Alternatives for evaluation of the dot product. Regardless of the representation used to express vectors, Eq. (2.29) can always be used to evaluate the dot product. For vectors with Cartesian representation, either Eq. (2.29) or Eq. (2.34) can be used, with the latter usually being more convenient. In fact, sometimes *both* methods are used to help determine useful information, such as the angle between two vectors, as discussed later in this section in connection with Eq. (2.39).

Determination of the angle between two vectors

Given two vectors $\vec{A}$ and $\vec{B}$ with Cartesian representation, we use the dot product given by Eq. (2.34) to provide the left-hand side of Eq. (2.29), from which the angle θ between the two vectors is then solved for. This provides

$$\cos\theta = \frac{\vec{A}\cdot\vec{B}}{|\vec{A}||\vec{B}|} = \frac{A_x B_x + A_y B_y + A_z B_z}{|\vec{A}||\vec{B}|} \quad \text{and}$$
$$\theta = \cos^{-1}\frac{\vec{A}\cdot\vec{B}}{|\vec{A}||\vec{B}|} = \cos^{-1}\frac{A_x B_x + A_y B_y + A_z B_z}{|\vec{A}||\vec{B}|}. \tag{2.39}$$

Determination of the component of a vector in a particular direction

> **Concept Alert**
>
> **Dot product.** In statics, the most important and frequent use of the dot product is for determining the component, or amount, of a vector that acts in a particular direction. It is essential that the direction of interest be described by a unit vector [i.e., in Eq. (2.40), the dot product of $\vec{F}$ must be taken with a unit vector].

Often, we must determine the component of a vector $\vec{A}$ that acts in the direction of another vector. In other words, we must determine how much, or what portion, of $\vec{A}$ acts in a particular direction. If that direction happens to be the same as the positive x, y, or z direction, then the answer is easy: the portion of $\vec{A}$ in the direction we are interested in is simply the x, y, or z component of $\vec{A}$, respectively. However, we are often not so fortunate and the direction in which we must find the component of a vector has complicated orientation, and this leads to perhaps the most useful application of the dot product.

A common situation is the determination of the component, or the amount, of a force vector that acts in a particular direction. Thus, consider a force vector $\vec{F}$ and a direction $\vec{r}$. The *parallel component of* $\vec{F}$, or in other words, the amount of $\vec{F}$ that acts in direction $\vec{r}$, is denoted by $F_{\parallel}$, where the subscript $\parallel$ means "parallel to $\vec{r}$," and is given by*

$$F_{\parallel} = \vec{F}\cdot\frac{\vec{r}}{|\vec{r}|}. \tag{2.40}$$

In the above expression, $\vec{F}$ has the usual physical significance: it is a force and has appropriate force units, such as lb or N. The physical significance of $\vec{r}$ is not important, other than it defines a direction of interest. Often $\vec{r}$ will be a position vector, but it could have other physical meaning. Notice in Eq. (2.40) that regardless of the physical significance of $\vec{r}$, it is converted into a dimensionless unit vector. Thus, the only purpose it serves is to specify direction. $F_{\parallel}$ is a scalar and may be positive, zero, or negative: a positive value indicates $F_{\parallel}$ is in the same direction as $\vec{r}$, a zero value indicates $\vec{F}$ is orthogonal (or perpendicular) to $\vec{r}$, and a negative value indicates $F_{\parallel}$ is in the opposite direction to $\vec{r}$. Although most of our applications of Eq. (2.40) will be for finding the component of a force in a particular direction, the idea is directly applicable to vectors with other physical interpretation.

To show why Eq. (2.40) accomplishes the task of finding how much of a vector acts in a particular direction, we use the following argument, the

* Rather than use subscript $\parallel$ to denote "parallel to $\vec{r}$," we will occasionally use subscript t, where this letter means "tangent to $\vec{r}$." Similarly, rather than using the subscript $\perp$ to denote "perpendicular to $\vec{r}$" (discussed later in this section), we will occasionally use subscript n, where this letter means "normal to $\vec{r}$."

beginning of which has nothing to do with the dot product. Consider the two-dimensional example shown in Fig. 2.29(a), where we wish to determine the component of $\vec{F}$ that acts in the direction $\vec{r}$. We arrange $\vec{F}$ and $\vec{r}$ tail to tail which leads to Fig. 2.29(b). Given angle θ between the two vectors, basic trigonometry gives us the component of $\vec{F}$ in the direction of $\vec{r}$ as

$$F_{\|} = |\vec{F}| \cos\theta. \tag{2.41}$$

Now consider the scalar s produced by taking the dot product between $\vec{F}$ and $\vec{r}$ according to Eq. (2.29)

$$s = \vec{F} \cdot \vec{r} = |\vec{F}||\vec{r}| \cos\theta. \tag{2.42}$$

For s in Eq. (2.42) to be the same as $F_{\|}$ in Eq. (2.41), and hence for the dot product to be capable of telling us the portion of $\vec{F}$ that acts in the direction $\vec{r}$, it is necessary that $\vec{r}$ be a unit vector so that $|\vec{r}| = 1$ in Eq. (2.42). To elaborate further, if a unit vector is not used in Eq. (2.40), the value s produced by the dot product has inappropriate units and uncertain physical interpretation. To summarize, if we want the dot product to tell us how much of a vector acts in a particular direction, then the direction must be described by a unit vector.

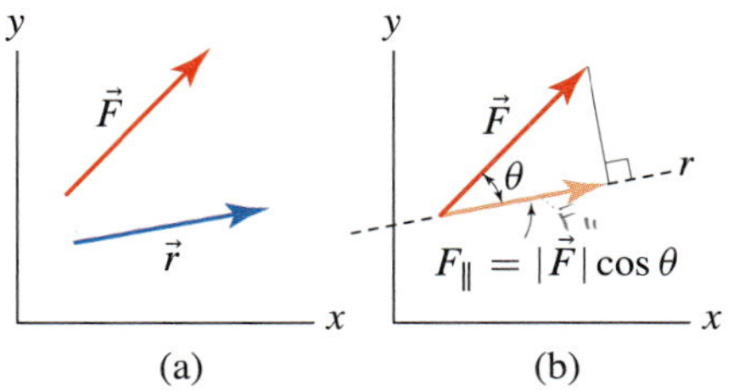

Figure 2.29
Use of basic trigonometry to determine the component of $\vec{F}$ acting in the direction of $\vec{r}$.

Common Pitfall

Not using a unit vector. When we use the dot product to determine the component, or amount, of a vector that acts in a particular direction, the most common error is to not use a unit vector. For example, in Eq. (2.40), an incorrect result is produced if $\vec{F} \cdot \vec{r}$ is evaluated.

Determination of the component of a vector perpendicular to a direction

Once we have determined the component of a vector parallel to a particular direction $\vec{r}$, it is straightforward to then determine the component of the vector that is perpendicular, or orthogonal, to $\vec{r}$, and we call this the *perpendicular component of the vector*. Consider the resolution of $\vec{F}$ into the parallel and perpendicular components shown in Fig. 2.30(a) and (b). The Pythagorean theorem may be used to write

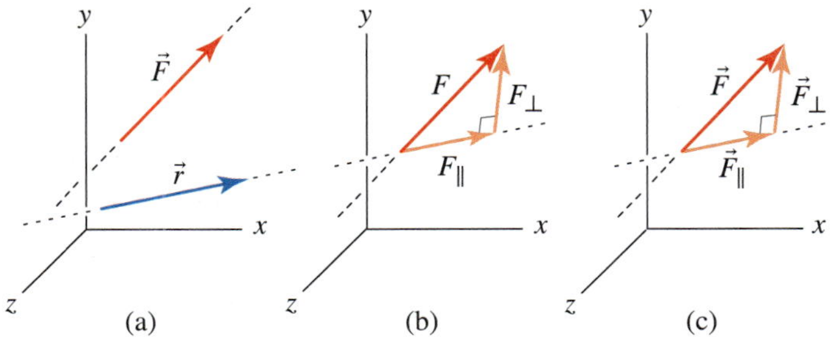

Figure 2.30. (a) Vectors $\vec{F}$ and $\vec{r}$ in three dimensions. (b) Resolution of $\vec{F}$ into *scalar components* in directions parallel and perpendicular to $\vec{r}$. (c) Resolution of $\vec{F}$ into *vector components* in directions parallel and perpendicular to $\vec{r}$.

$$F^2 = F_{\|}^2 + F_{\perp}^2. \tag{2.43}$$

Once $F_{\|}$ is known from Eq. (2.40), the perpendicular component may be obtained from the above equation as

$$F_{\perp} = \sqrt{F^2 - F_{\|}^2}. \tag{2.44}$$

Helpful Information

Line of action for a force vector. When we analyze equilibrium of objects, the position of the line of action for each force vector is important. While we may think of repositioning the lines of action of vectors for purposes of evaluating the dot product, equilibrium of objects depends on the actual line of action that each force has.

Thus, the magnitude of the perpendicular component is easily determined, but its direction is not yet known. For many problems, however, this direction might not be needed. If the direction for the perpendicular component is needed, then consider the resolution of $\vec{F}$ into the parallel and perpendicular vector components shown in Fig. 2.30(c). Vector addition provides

$$\vec{F} = \vec{F}_{\parallel} + \vec{F}_{\perp}. \tag{2.45}$$

If $F_{\parallel}$ is known, then $\vec{F}_{\parallel} = F_{\parallel}\,(\vec{r}/|\vec{r}|)$, and Eq. (2.45) may be rearranged to obtain

$$\vec{F}_{\perp} = \vec{F} - \vec{F}_{\parallel}. \tag{2.46}$$

Important equations for using the dot product to resolve a vector $\vec{F}$ into components parallel and perpendicular to a direction $\vec{r}$ are collected in the following summary box:

Summary Box (See Fig. 2.30.)

$F_{\parallel}$ = component of $\vec{F}$ acting in direction $\vec{r}$,

$\vec{F}_{\parallel}$ = vector component of $\vec{F}$ acting in direction $\vec{r}$,

$F_{\perp}$ = component of $\vec{F}$ acting perpendicular to $\vec{r}$,

$\vec{F}_{\perp}$ = vector component of $\vec{F}$ acting perpendicular to $\vec{r}$, (2.47)

$$F_{\parallel} = \vec{F} \cdot \frac{\vec{r}}{|\vec{r}|}, \qquad \vec{F}_{\parallel} = F_{\parallel}\frac{\vec{r}}{|\vec{r}|},$$

$$F_{\perp} = \sqrt{F^2 - F_{\parallel}^2}, \qquad \vec{F}_{\perp} = \vec{F} - \vec{F}_{\parallel}.$$

End of Section Summary

In this section, the dot product between two vectors has been defined. Some of the key points are as follows:

- The *dot product* between two vectors $\vec{A}$ and $\vec{B}$ is a scalar s. The units of s are the product of the units of $\vec{A}$ and $\vec{B}$, and s may be positive, zero, or negative.
- The dot product can be used to find the angle between the lines of action of two vectors.
- The dot product can be used to determine the component (or amount) of one vector that acts in the direction of another vector $\vec{r}$. This component is often called the *parallel component*. In addition, the component of a vector perpendicular to the direction $\vec{r}$, which we call the *perpendicular component*, can be determined.

EXAMPLE 2.14 *Angle Between Two Vectors*

Bar AB supports two cables at its end. The position vector from A to B is $\vec{r}_{AB}$, the position vector from B to some point along the length of cable 1 is $\vec{r}_1$, and the position vector from B to some point along the length of cable 2 is $\vec{r}_2$. Determine the angles between bar AB and the cables.

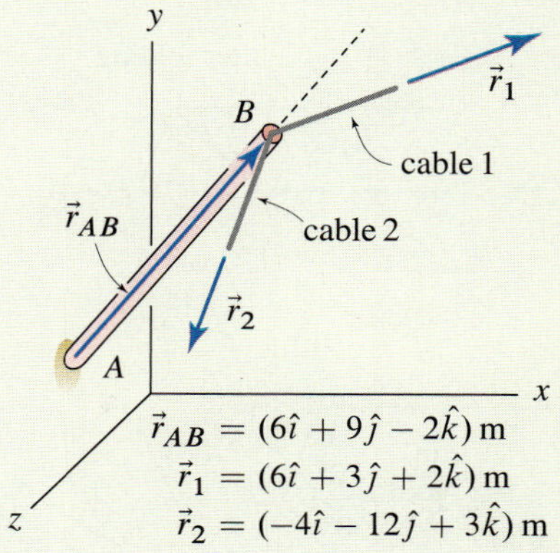

Figure 1

SOLUTION

Road Map We will use the dot product, as given by Eq. (2.39), to evaluate the angles between the bar and cables.

Governing Equations & Computation Evaluating the magnitudes of the vectors shown in Fig. 1 provides $r_{AB} = 11$ m, $r_1 = 7$ m, and $r_2 = 13$ m. The desired angles are then obtained from Eq. (2.39) as

$$\begin{aligned}\theta_1 &= \cos^{-1} \frac{\vec{r}_{AB} \cdot \vec{r}_1}{|\vec{r}_{AB}|\,|\vec{r}_1|} \\ &= \cos^{-1} \frac{(6\,\text{m})(6\,\text{m}) + (9\,\text{m})(3\,\text{m}) + (-2\,\text{m})(2\,\text{m})}{(11\,\text{m})(7\,\text{m})} \\ &= \cos^{-1} \frac{59}{77} = \boxed{40°,} \end{aligned} \tag{1}$$

$$\begin{aligned}\theta_2 &= \cos^{-1} \frac{\vec{r}_{AB} \cdot \vec{r}_2}{|\vec{r}_{AB}|\,|\vec{r}_2|} \\ &= \cos^{-1} \frac{(6\,\text{m})(-4\,\text{m}) + (9\,\text{m})(-12\,\text{m}) + (-2\,\text{m})(3\,\text{m})}{(11\,\text{m})(13\,\text{m})} \\ &= \cos^{-1} \frac{-138}{143} = \boxed{164.8°.} \end{aligned} \tag{2}$$

Discussion & Verification

- As shown in Fig. 2, and as expected, θ_1 is an acute angle and θ_2 is an obtuse angle.
- This problem asks for the angles between the cables and the direction from A to B, and these angles are given in Eqs. (1) and (2). The angles between the cables and the direction from B to A are different than these; and to see how they differ, and how they can be evaluated, we consider the angle between the direction from B to A and cable 2, using one of the following approaches:
  - Use the value of θ_2 found in Eq. (2) to write $180° - 164.8° = 15.2°$.
  - Write an expression for $\vec{r}_{BA}$ and evaluate $\vec{r}_{BA} \cdot \vec{r}_2$ when solving for the angle.
  - Because $\vec{r}_{BA} = -\vec{r}_{AB}$, we could evaluate $-\vec{r}_{AB} \cdot \vec{r}_2$ when solving for the angle.

You may wish to verify one of the last two computations for yourself.

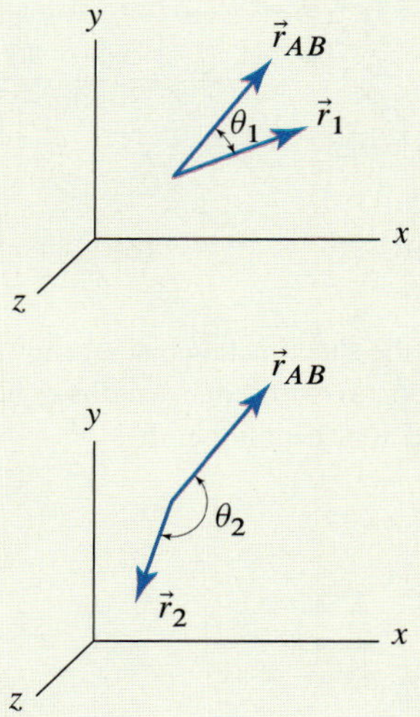

Figure 2
The angle θ_1 between $\vec{r}_{AB}$ and cable 1 ($\vec{r}_1$) is expected to be acute, and the angle θ_2 between $\vec{r}_{AB}$ and cable 2 ($\vec{r}_2$) is expected to be obtuse.

EXAMPLE 2.15 *Angle Between Two Vectors*

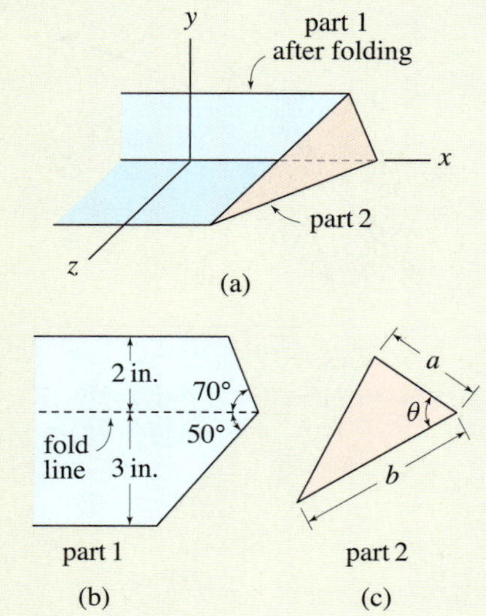

Figure 1

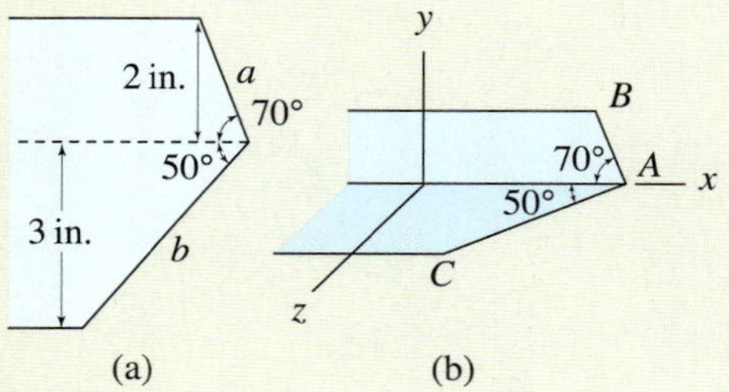

Figure 2
Part 1 of the sheet metal channel shown before bending in (a) with edge lengths a and b defined, and after bending in (b).

A sheet metal channel is to be fabricated as follows. Part 1, after being cut to the dimensions shown in Fig. 1(b), is bent to a 90° angle along the fold line. Then part 2 is welded to the end of part 1. Determine the dimensions a, b, and θ of part 2 so that it fits the end of the channel.

SOLUTION

Road Map Using basic trigonometry, determination of edge lengths a and b for part 2 will be straightforward. The angle θ will be determined using the dot product once the position vectors for the two edges of part 1 (after folding) have been obtained.

Governing Equations & Computation Edge lengths a and b for part 2 can be determined using the geometry shown in Fig. 2(a):

$$a \sin 70^\circ = 2\text{ in.} \quad \Rightarrow \quad a = 2\text{ in.}/\sin 70^\circ = \boxed{2.128\text{ in.}} \tag{1}$$

$$b \sin 50^\circ = 3\text{ in.} \quad \Rightarrow \quad b = 3\text{ in.}/\sin 50^\circ = \boxed{3.916\text{ in.}} \tag{2}$$

Determination of θ is more challenging. To aid our discussion, points A, B, and C on the end of the channel are labeled in Fig. 2(b). If we can determine vectors in the direction A to B, and A to C, then the dot product between these vectors will provide the angle θ needed for part 2. We may construct a position vector for edge AB as follows: beginning at point A, taking a step of $\cos 70^\circ$ in the negative x direction, followed by a step of $\sin 70^\circ$ in the positive y direction, gives a point that lies on the line connecting points A and B. Calling this vector $\hat{u}_{AB}$, we write

$$\hat{u}_{AB} = -\cos 70^\circ\,\hat{\imath} + \sin 70^\circ\,\hat{\jmath}. \tag{3}$$

Similarly, beginning at point A, a step of $\cos 50^\circ$ in the negative x direction, followed by a step of $\sin 50^\circ$ in the positive z direction, gives a point that lies on the line connecting points A and C. Calling this vector $\hat{u}_{AC}$, we write

$$\hat{u}_{AC} = -\cos 50^\circ\,\hat{\imath} + \sin 50^\circ\,\hat{k}. \tag{4}$$

Note that $\hat{u}_{AB}$ and $\hat{u}_{AC}$ are unit vectors. Instead of these unit vectors, you could use the position vectors from point A to B, $\vec{r}_{AB} = (2.128\text{ in.})\,\hat{u}_{AB}$, and from point A to C, $\vec{r}_{AC} = (3.916\text{ in.})\,\hat{u}_{AC}$. Finally, we determine the angle θ needed for part 2 using Eq. (2.39)

$$\begin{aligned}
\theta &= \cos^{-1}\frac{\hat{u}_{AB}\cdot\hat{u}_{AC}}{|\hat{u}_{AB}||\hat{u}_{AC}|} \\
&= \cos^{-1}\frac{(-\cos 70^\circ)(-\cos 50^\circ) + (\sin 70^\circ)(0) + (0)(\sin 50^\circ)}{(1)(1)} \\
&= \cos^{-1} 0.220 = \boxed{77.3^\circ.}
\end{aligned} \tag{5}$$

Discussion & Verification The values for a, b, and θ in Eqs. (1), (2), and (5) appear to be reasonable. That is, we expect a to be somewhat larger than 2 in., b to be somewhat larger than 3 in., and θ to be somewhat smaller than 90°.

EXAMPLE 2.16 *Component of a Force in a Particular Direction*

A tractor is stuck in mud, and to free it, a cable applying a force with magnitude F and direction $\vec{r}_1$ is attached to the front of the tractor as shown. The operator estimates that a 400 lb force applied in the direction of the tractor's chassis, which is $\vec{r}_2$, will be sufficient to free it. Determine the cable force F that should be applied and the component of this force perpendicular to the direction of the tractor's chassis. **Note:** This problem is similar to Example 2.4 except here the force and position vectors are not in a horizontal plane.

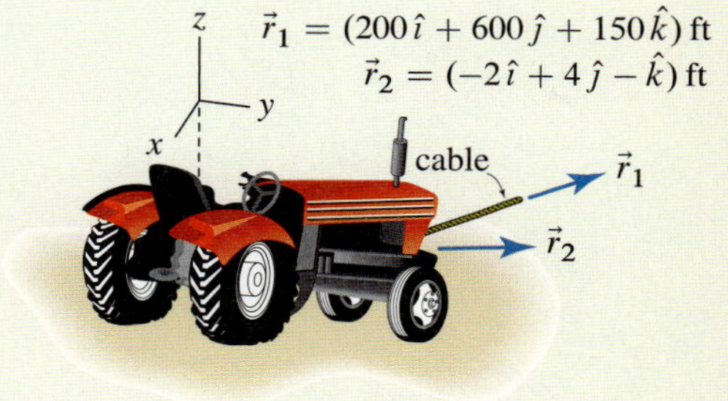

Figure 1

SOLUTION

Road Map The direction of the cable force is known, but its magnitude is unknown. Thus, $\vec{F} = F(\vec{r}_1/|\vec{r}_1|)$ where F is to be determined. We will use the dot product to determine the component of the cable force that acts in the direction of the tractor, and according to the problem statement, this must equal 400 lb.

Governing Equations & Computation The dot product between the cable force $\vec{F}$ and a unit vector in the direction of the tractor's chassis must equal 400 lb:

$$F_{\|} = \vec{F} \cdot \frac{\vec{r}_2}{|\vec{r}_2|} = F\frac{\vec{r}_1}{|\vec{r}_1|} \cdot \frac{\vec{r}_2}{|\vec{r}_2|}$$

$$400\text{ lb} = F\frac{(200\,\hat{\imath} + 600\,\hat{\jmath} + 150\,\hat{k})\text{ ft}}{650\text{ ft}} \cdot \frac{(-2\,\hat{\imath} + 4\,\hat{\jmath} - \hat{k})\text{ ft}}{\sqrt{21}\text{ ft}}$$

$$= F\frac{1850}{650\sqrt{21}}$$

$$= F(0.6211). \tag{1}$$

Solving the above equation provides the cable force

$$\boxed{F = 644.0\text{ lb.}} \tag{2}$$

Determination of only the magnitude of the perpendicular component of force is straightforward using the Pythagorean theorem, Eq. (2.44):

$$F_{\perp} = \sqrt{F^2 - F_{\|}^2} = \sqrt{(644.0\text{ lb})^2 - (400\text{ lb})^2} = \boxed{504.8\text{ lb.}} \tag{3}$$

If the direction of the perpendicular component is also desired, then we may use vector addition to obtain it, as expressed by Eq. (2.46):

$$\vec{F}_{\perp} = \vec{F} - \vec{F}_{\|} = F\frac{\vec{r}_1}{|\vec{r}_1|} - F_{\|}\frac{\vec{r}_2}{|\vec{r}_2|}$$

$$= (644\text{ lb})\frac{(200\,\hat{\imath} + 600\,\hat{\jmath} + 150\,\hat{k})\text{ ft}}{650\text{ ft}} - (400\text{ lb})\frac{(-2\,\hat{\imath} + 4\,\hat{\jmath} - \hat{k})\text{ ft}}{\sqrt{21}\text{ ft}}$$

$$= (373\,\hat{\imath} + 245\,\hat{\jmath} + 236\,\hat{k})\text{ lb}. \tag{4}$$

Discussion & Verification As a partial check of our solution, the magnitude of Eq. (4) is $|\vec{F}_{\perp}| = \sqrt{(373)^2 + (245)^2 + (236)^2}$ lb $= 505$ lb, which agrees with the value found in Eq. (3).

EXAMPLE 2.17 Component of a Force in a Particular Direction

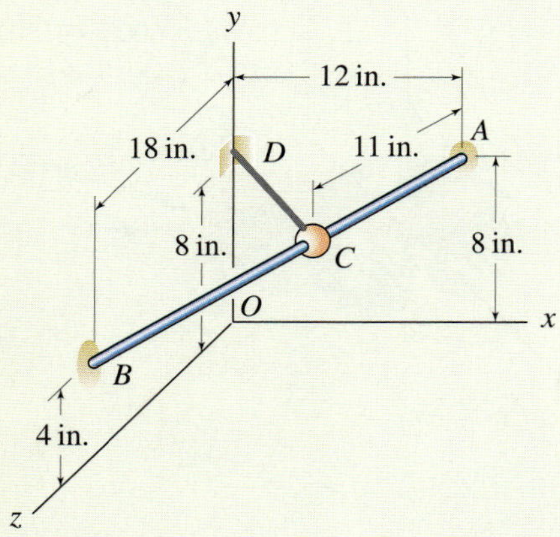

Figure 1

Rod AB is straight and has a bead at C. An elastic cord having a 3 lb tensile force is attached between the bead and a support at D.

(a) Determine the components of the cord force in directions parallel and perpendicular to rod AB.

(b) Determine the vector components of the cord force in directions parallel and perpendicular to rod AB.

(c) If the bead at C is free to slide on rod AB and is released from rest, will the cord force tend to make the bead slide toward A or B?

SOLUTION

Road Map After writing expressions for position and force vectors, we will use the dot product to determine the component of the cord force that acts in the direction of the rod. It will then be straightforward to use the Pythagorean theorem to determine the component of the cord force perpendicular to the rod. Both of these results are scalars. In Part (b), we will write vector expressions for the forces parallel and perpendicular to the rod. Consideration of the sign of the parallel component of the force, as provided by the dot product, will determine the direction along the bar in which the bead will tend to slide.

Part (a)

Governing Equations & Computation By imagining being positioned at A, the distances to be traversed in the x, y, and z directions to arrive at B may be determined by inspection of Fig. 1, allowing us to write the position vector from A to B, and its magnitude, as

$$\vec{r}_{AB} = (-12\,\hat{\imath} - 4\,\hat{\jmath} + 18\,\hat{k})\text{ in.}, \tag{1}$$

$$|\vec{r}_{AB}| = \sqrt{(-12)^2 + (-4)^2 + (18)^2}\text{ in.} = 22\text{ in.} \tag{2}$$

The position vector from A to C has 11 in. magnitude and has the same direction as $\vec{r}_{AB}$, thus

$$\vec{r}_{AC} = (11\text{ in.})\frac{\vec{r}_{AB}}{|\vec{r}_{AB}|} = (11\text{ in.})\frac{-12\,\hat{\imath} - 4\,\hat{\jmath} + 18\,\hat{k}}{22}$$
$$= (-6\,\hat{\imath} - 2\,\hat{\jmath} + 9\,\hat{k})\text{ in.} \tag{3}$$

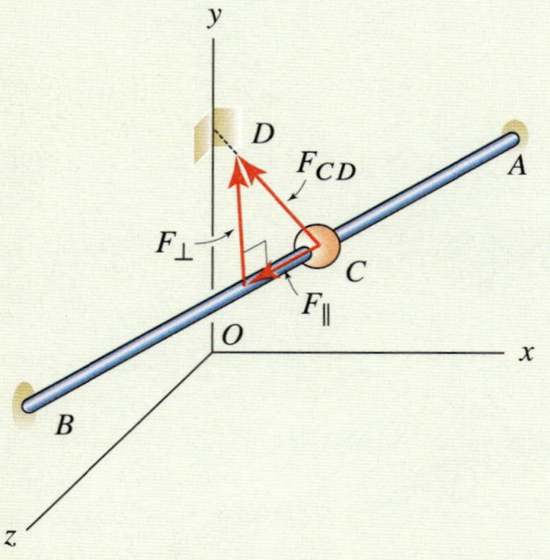

Figure 2
The force supported by the cord has component F_{CD} in the direction shown. The component of the cord force in the direction from A to B is denoted by $F_\parallel$, and the component of the cord force perpendicular to the direction from A to B is denoted by $F_\perp$.

To write a vector expression for the cord force from C to D, we will first obtain a position vector from C to D, $\vec{r}_{CD}$. This can be easily accomplished by computing the coordinates of C and then taking the difference between the coordinates of D and C as the head and tail, respectively, of $\vec{r}_{CD}$. Alternatively, we will evaluate $\vec{r}_{CD}$ using the vector addition

$$\vec{r}_{CD} = \vec{r}_{CA} + \vec{r}_{AD} = -\vec{r}_{AC} + \vec{r}_{AD}$$
$$= -(-6\,\hat{\imath} - 2\,\hat{\jmath} + 9\,\hat{k})\text{ in.} + (-12\,\hat{\imath})\text{ in.}$$
$$= (-6\,\hat{\imath} + 2\,\hat{\jmath} - 9\,\hat{k})\text{ in.}, \tag{4}$$

$$|\vec{r}_{CD}| = \sqrt{(-6)^2 + (2)^2 + (-9)^2}\text{ in.} = 11\text{ in.} \tag{5}$$

Thus, the cord force, which is shown in Fig. 2, is

$$\vec{F}_{CD} = (3\text{ lb})\frac{\vec{r}_{CD}}{|\vec{r}_{CD}|} = (3\text{ lb})\frac{-6\,\hat{\imath} + 2\,\hat{\jmath} - 9\,\hat{k}}{11}$$

$$= (-1.636\,\hat{\imath} + 0.545\,\hat{\jmath} - 2.455\,\hat{k})\text{ lb} \tag{6}$$

We denote the component of the cord force $\vec{F}_{CD}$ in direction A to B as $F_{\|}$, and this is given by the dot product of $\vec{F}_{CD}$ with a unit vector in direction A to B as follows:

$$F_{\|} = \vec{F}_{CD} \cdot \frac{\vec{r}_{AB}}{|\vec{r}_{AB}|}$$
$$= (3\text{ lb})\frac{-6\,\hat{\imath} + 2\,\hat{\jmath} - 9\,\hat{k}}{11} \cdot \frac{-12\,\hat{\imath} - 4\,\hat{\jmath} + 18\,\hat{k}}{22} = \boxed{-1.215\text{ lb.}} \tag{7}$$

Note that instead of $\vec{r}_{AB}/|\vec{r}_{AB}|$ in Eq. (7), we could have used $\vec{r}_{AC}/|\vec{r}_{AC}|$. To determine the magnitude of the perpendicular component of the cord force, the Pythagorean theorem, Eq. (2.44), may be applied to write

$$F_{\perp} = \sqrt{F_{CD}^2 - F_{\|}^2} = \sqrt{(3\text{ lb})^2 - (-1.215\text{ lb})^2} = \boxed{2.743\text{ lb.}} \tag{8}$$

Part (b)

Governing Equations & Computation Once the value of the parallel component of the force $F_{\|}$ has been determined from Eq. (7), we may write the vector quantity $\vec{F}_{\|}$ and then proceed to determine $\vec{F}_{\perp}$ as follows:

$$\vec{F}_{\|} = F_{\|}\frac{\vec{r}_{AB}}{|\vec{r}_{AB}|} = (-1.215\text{ lb})\frac{-12\,\hat{\imath} - 4\,\hat{\jmath} + 18\,\hat{k}}{22}$$
$$= \boxed{(0.663\,\hat{\imath} + 0.221\,\hat{\jmath} - 0.994\,\hat{k})\text{ lb,}} \tag{9}$$
$$\vec{F}_{\perp} = \vec{F}_{CD} - \vec{F}_{\|}$$
$$= (3\text{ lb})\frac{-6\,\hat{\imath} + 2\,\hat{\jmath} - 9\,\hat{k}}{11} - (0.663\,\hat{\imath} + 0.221\,\hat{\jmath} - 0.994\,\hat{k})\text{ lb}$$
$$= \boxed{(-2.299\,\hat{\imath} + 0.325\,\hat{\jmath} - 1.461\,\hat{k})\text{ lb.}} \tag{10}$$

Part (c)

Governing Equations & Computation The sign of $F_{\|}$ determines the direction in which the bead will tend to slide due to the cord force. A positive value of $F_{\|}$ indicates the parallel component of the force is in the same direction as the position vector used in the dot product, namely, from A to B, as shown in Fig. 2. A negative value of $F_{\|}$ indicates the parallel component of the force is in the opposite direction, while if $F_{\|}$ is zero, then the cord force is perpendicular to AB. According to Eq. (7), $F_{\|} = -1.215$ lb, and because this value is negative, the parallel component of the cord force is in the direction from B to A, and the bead will tend to slide toward A.

Discussion & Verification As a partial check on our solutions in Parts (a) and (b), we should compute the magnitudes of $\vec{F}_{\|}$ and $\vec{F}_{\perp}$ in Eqs. (9) and (10) to verify they are the same as those found in Eqs. (7) and (8). We may also examine the components of $\vec{F}_{\|}$ and $\vec{F}_{\perp}$ to see if they have reasonable directions and proportions. For $\vec{F}_{\|}$ this check is uncertain, but for $\vec{F}_{\perp}$ the components appear to be in proper directions. In view of our answers in this example, Fig. 2 may be redrawn to more accurately show the direction of $\vec{F}_{\|}$; that is, the actual direction of $\vec{F}_{\|}$ is from point C toward point A.

Helpful Information

Actual sliding direction of the bead. In Part (c), we considered the direction that the bead would slide due to the force provided by the cord *only*. Note that other forces are also applied to the bead, such as the bead's weight, reaction forces, and possible friction between the bead and rod AB – the motion of the bead is also influenced by these. The treatment of these other forces is considered in subsequent chapters of this book. Nonetheless, if the weight of the bead is negligible, the bead can slide without friction on the rod, and the bead has zero initial velocity, then as determined in Part (c), the bead will slide toward point A.

EXAMPLE 2.18 Smallest Distance Between a Point and Line

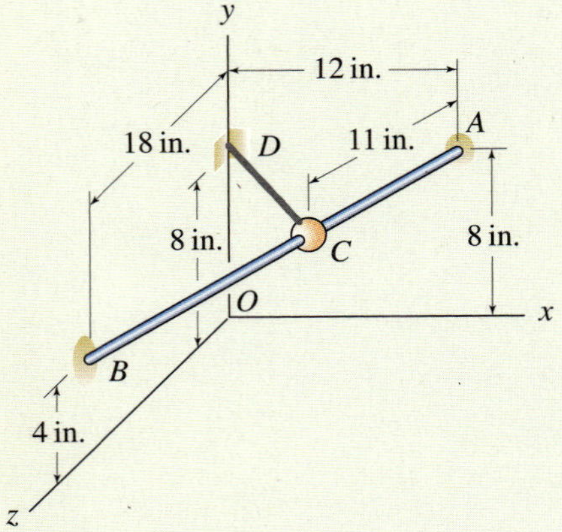

Figure 1

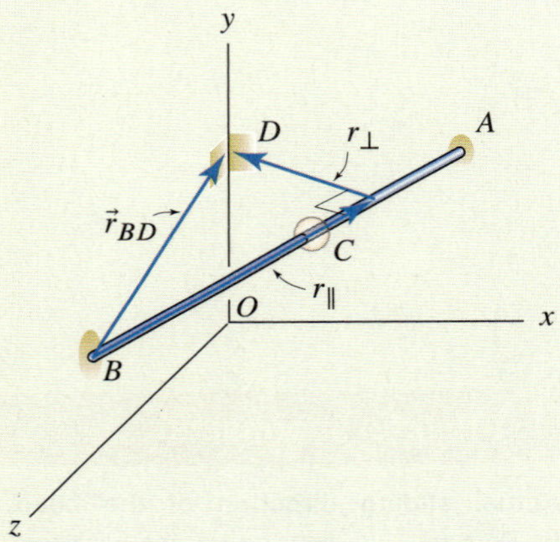

Figure 2

Resolution of $\vec{r}_{BD}$ into parallel and perpendicular components to determine the smallest distance between point D and rod AB.

In the rod and bead of Example 2.17, determine the smallest distance between D and rod AB.

SOLUTION

Road Map There are several strategies that can be used to determine the smallest distance between point D and rod AB. The technique we will use is identical in concept to the solution used in Part (b) of Example 2.17, except only position vectors are used. The idea is to write a position vector from any convenient point on rod AB to point D. Then, using the dot product, we resolve this position vector into components parallel and perpendicular to AB, as shown in Fig. 2 (where point B has been selected as a convenient location on rod AB). While the magnitude of the parallel component will depend on the particular point on AB we select, the perpendicular component is the same and is the shortest distance between D and rod AB.

Governing Equations & Computation We select point B as a convenient point on rod AB and write

$$\vec{r}_{BD} = (4\,\hat{\jmath} - 18\,\hat{k})\text{ in.}, \tag{1}$$

$$|\vec{r}_{BD}| = \sqrt{(4)^2 + (-18)^2}\text{ in.} = 18.44\text{ in.} \tag{2}$$

We now take the dot product of the above with $\vec{r}_{BA}$, where $\vec{r}_{BA}$ is easily written by examining Fig. 1 [or we may use Eq. (1) of Example 2.17 to write $\vec{r}_{BA} = -\vec{r}_{AB}$], to obtain the parallel component of $\vec{r}_{BD}$ as

$$r_\parallel = \vec{r}_{BD} \cdot \frac{\vec{r}_{BA}}{|\vec{r}_{BA}|} = [(4\,\hat{\jmath} - 18\,\hat{k})\text{ in.}] \cdot \frac{12\,\hat{\imath} + 4\,\hat{\jmath} - 18\,\hat{k}}{22} = 15.45\text{ in.} \tag{3}$$

Then using the Pythagorean theorem, we find

$$r_\perp = \sqrt{r_{BD}^2 - r_\parallel^2} = \sqrt{(18.44)^2 - (15.45)^2}\text{ in.} = 10.06\text{ in.} \tag{4}$$

Thus, **the smallest distance between point D and rod AB is 10.06 in.** You should repeat this problem starting with a different position vector, say $\vec{r}_{AD}$, to verify that the same 10.06 in. distance is obtained.

Discussion & Verification The solution presented here actually determines the smallest distance from point D to the infinite line passing through points A and B, which is not necessarily the same as the smallest distance to line segment AB. To determine if the smallest distance found here is to a point that is actually on rod AB, we compare the 15.45 in. magnitude $r_\parallel$ found in Eq. (3) to the length of member AB, which is 22 in. Hence, the distance found is indeed to a point that lies on rod AB.

PROBLEMS

Problems 2.63 and 2.64

(a) Determine the angle between vectors $\vec{A}$ and $\vec{B}$.

(b) Determine the components of $\vec{A}$ parallel and perpendicular to $\vec{B}$.

(c) Determine the vector components of $\vec{A}$ parallel and perpendicular to $\vec{B}$.

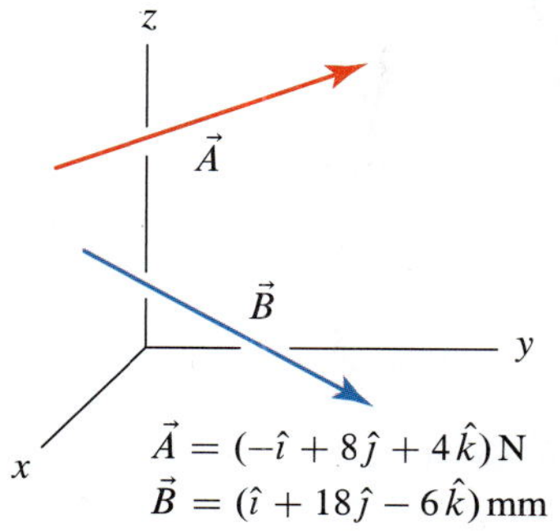

Figure P2.63

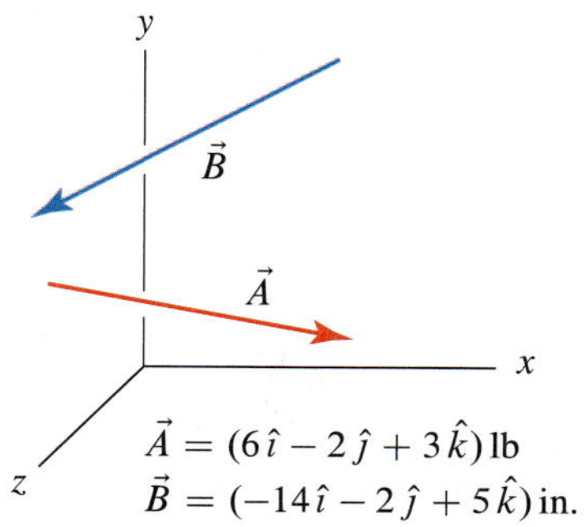

Figure P2.64

Problem 2.65

A slide on a child's play structure is to be supported in part by strut CD (railings are omitted from the sketch for clarity). End C of the strut is to be positioned along the outside edge of the slide, halfway between ends A and B. End D of the strut is to be positioned on the y axis so that the angle $\angle ACD$ between the slide and the strut is a right angle. Determine the distance h that point D should be positioned.

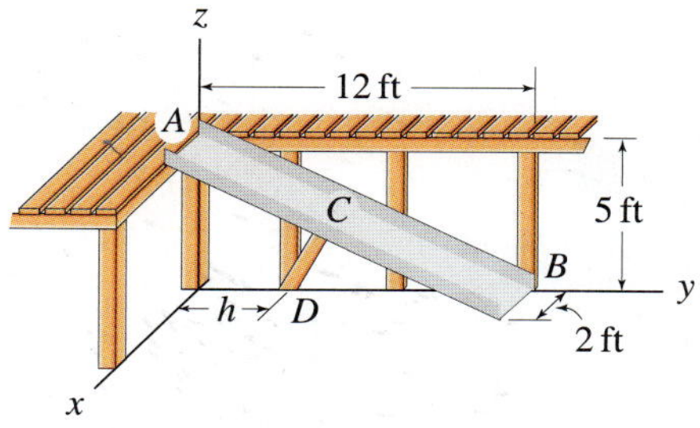

Figure P2.65

Problem 2.66

A whistle is made of square tube with a notch cut in its edge, into which a baffle is brazed. Determine the dimensions d and θ for the baffle.

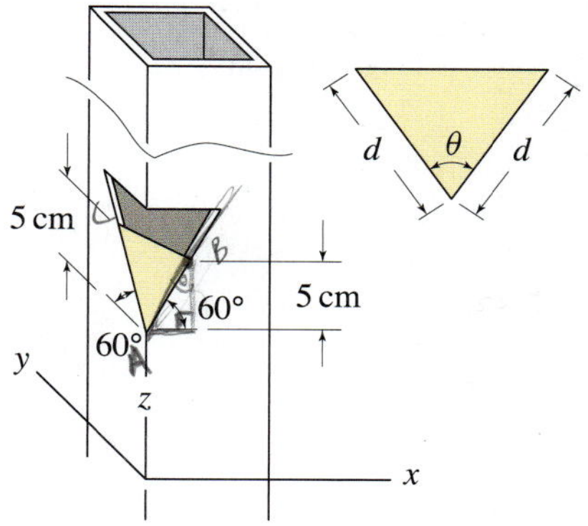

Figure P2.66

Problem 2.67

A flat triangular shape window for the cockpit of an airplane is to have the corner coordinates shown. Specify the angles θ_A, θ_B, and θ_C and dimensions d_A, d_B, and d_C for the window.

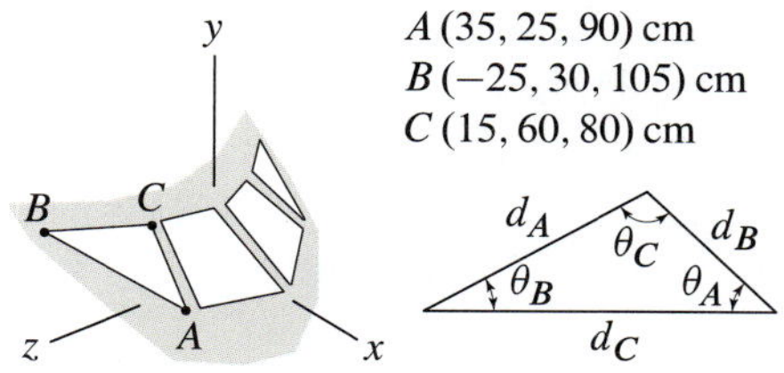

Figure P2.67

Problem 2.68

The corner of an infant's bassinet is shown. Determine angles α and β and dimensions a and b of the side and end pieces so the corners of the bassinet will properly meet when assembled.

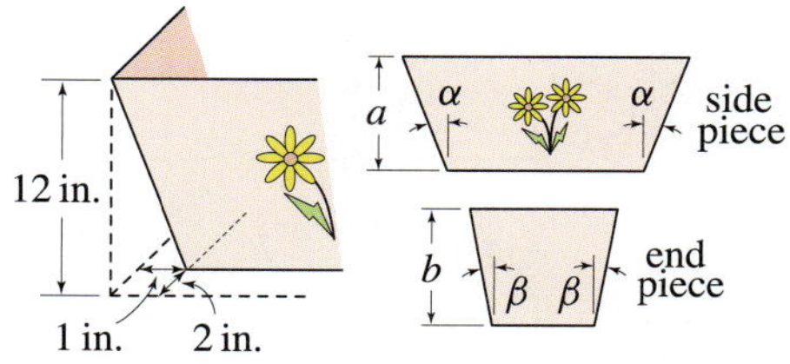

Figure P2.68

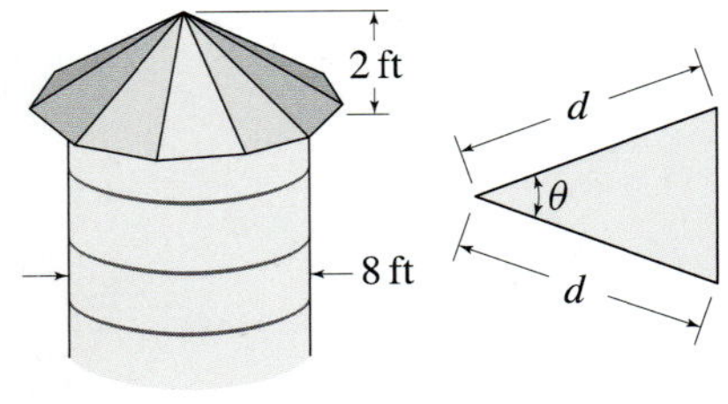

Figure P2.69

Problem 2.69

The roof of a grain silo is to be made using 12 identical triangular panels. Determine the value of angle θ needed and the smallest value of d that can be used.

Problems 2.70 and 2.71

For the description and figure indicated below, determine the components of the cord force in directions parallel and perpendicular to rod CD. If released from rest, will the cord force tend to make bead E slide toward C or D?

Problem 2.70 Use the description and figure for Prob. 2.60 on p 74.

Problem 2.71 Use the description and figure for Prob. 2.61 on p 74.

Problems 2.72 through 2.75

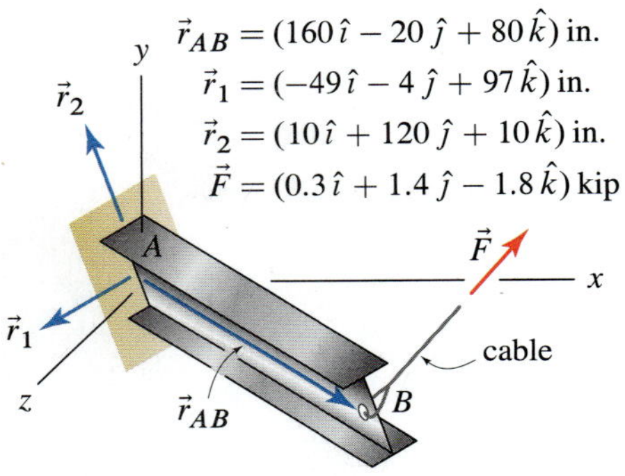

Figure P2.72–P2.75

A cantilever I beam has a cable at end B that supports a force $\vec{F}$, and $\vec{r}_{AB}$ is the position vector from end A of the beam to end B. Position vectors $\vec{r}_1$ and $\vec{r}_2$ are parallel to the flanges and web of the I beam, respectively. For determination of the internal forces in the beam (discussed in Chapter 8), and for mechanics of materials analysis, it is necessary to know the components of the force in the axial direction of the beam (AB) and in directions parallel to the web and flanges.

Problem 2.72 Using the dot product, show that $\vec{r}_1$, $\vec{r}_2$, and $\vec{r}_{AB}$ are orthogonal to one another.

Problem 2.73 Determine the scalar and vector components of $\vec{F}$ in direction $\vec{r}_{AB}$.

Problem 2.74 Determine the scalar and vector components of $\vec{F}$ in direction $\vec{r}_1$.

Problem 2.75 Determine the scalar and vector components of $\vec{F}$ in direction $\vec{r}_2$.

Problem 2.76

The gearshift lever AB for the transmission of a sports car has position vector $\vec{r}$ whose line of action passes through points A and B. The driver applies a force $\vec{F}$ to the knob of the lever to shift gears. If the component of $\vec{F}$ perpendicular to the gearshift lever must be 5 N to shift gears, determine the magnitude F of the force.

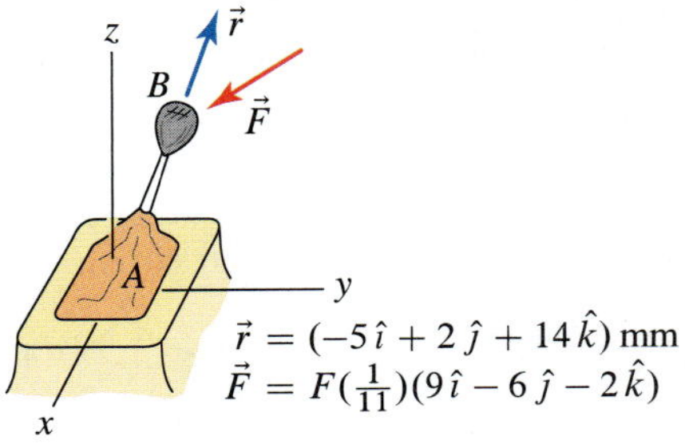

Figure P2.76

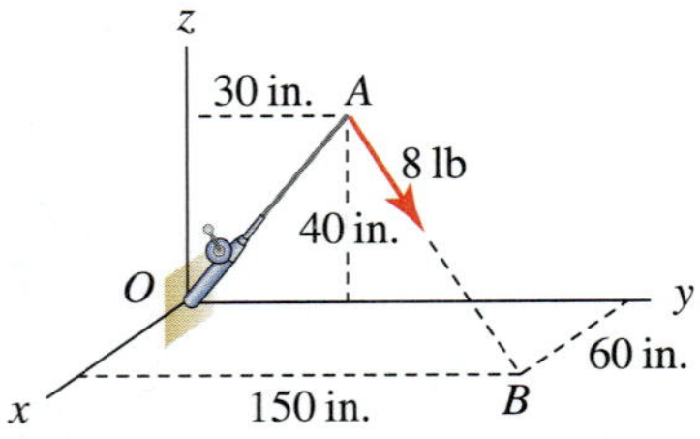

Figure P2.77

Problem 2.77

A force of magnitude 8 lb is applied to line AB of a fishing rod OA. Determine the vector components of the 8 lb force in directions parallel and perpendicular to the rod.

Problem 2.78

Rod AB is the firing pin for the right-hand barrel of a side-by-side shotgun. During firing, end A is impacted by the hammer, which imparts a force F to the firing pin. The force F lies in a plane parallel to the xy plane but the firing pin has a three-dimensional orientation as shown. Determine the components of F in directions parallel and perpendicular to the firing pin.

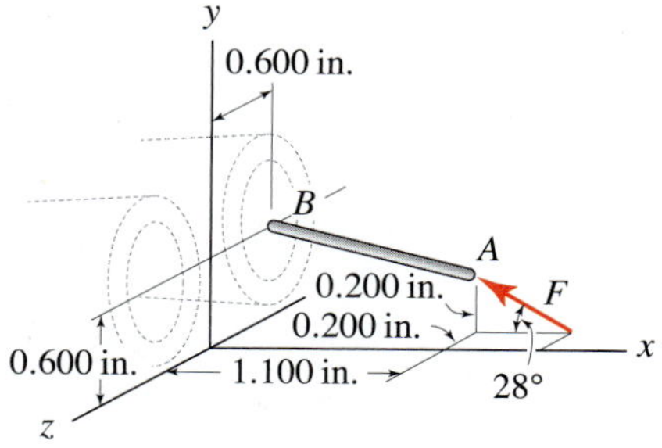

Figure P2.78

Problem 2.79

The collar C is fixed to rod AB and supports a weight $W = 15\,\text{N}$, acting in the negative z direction. Determine the vector components of the weight that are parallel and perpendicular to rod AB.

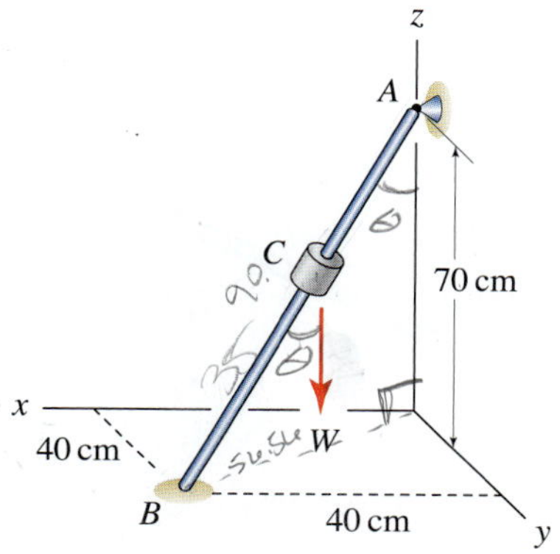

Figure P2.79 and P2.80

Problem 2.80

The collar C is fixed to rod AB using a glued bond that allows a maximum force of 35 N parallel to the axis of the rod. The collar has weight W acting in the negative z direction. Determine the weight W of the collar that will cause the glued bond to break.

Problem 2.81

The structure consists of a quarter-circular rod AB with radius 150 mm that is fixed in the xy plane. An elastic cord supporting a force of 100 N is attached to a support at D and a bead at C. Determine the components of the cord force in directions tangent and normal to the curved rod AB at point C.

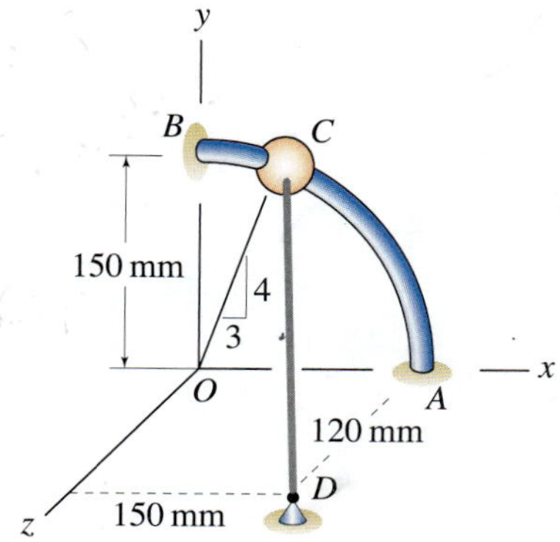

Figure P2.81

Problem 2.82

Bead B has negligible weight and slides without friction on rigid fixed bar AC. An elastic cord BD which supports a 60 N tensile force is attached to the bead. At the instant shown, the bead has zero velocity and is positioned halfway between points A and C. Determine the components of the cord tension that act parallel and perpendicular to direction AC of the bar. Due to the cord tension, will the bead slide toward point A or C?

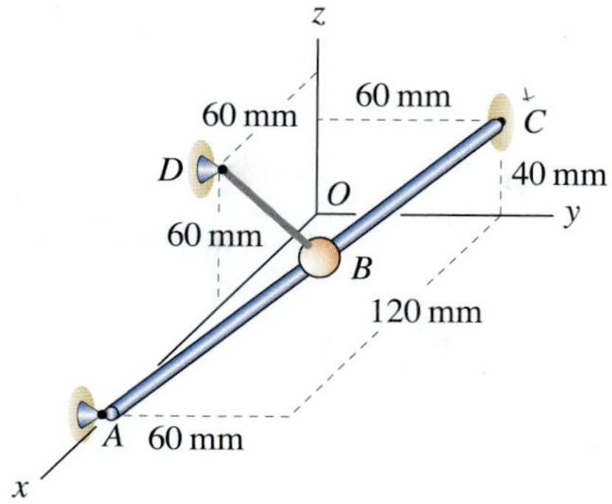

Figure P2.82

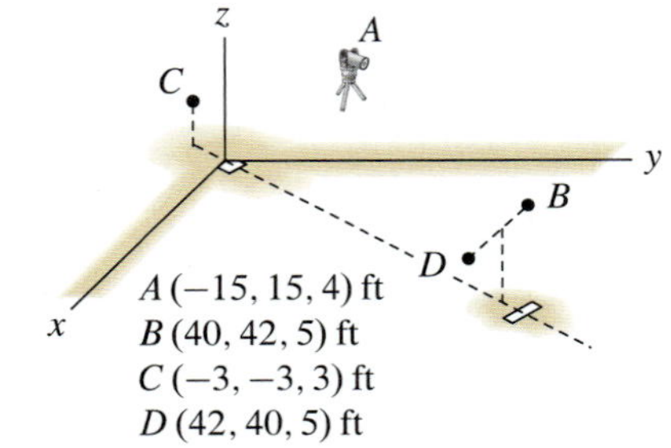

Figure P2.83 and P2.84

Problem 2.83

The manager of a baseball team plans to use a radar gun positioned at point A to measure the speed of pitches for a right-handed pitcher. If the person operating the radar gun measures a speed s when the baseball is one-third the distance from the release point at B to the catcher's glove at C, what is the actual speed of the pitch? Assume the pitch follows a straight-line path, and express your answer in terms of s. Note that the value s measured by the radar gun is the rate of change of the distance between point A and the ball.

Problem 2.84

Repeat Prob. 2.83 for a left-handed pitcher whose release point is D.

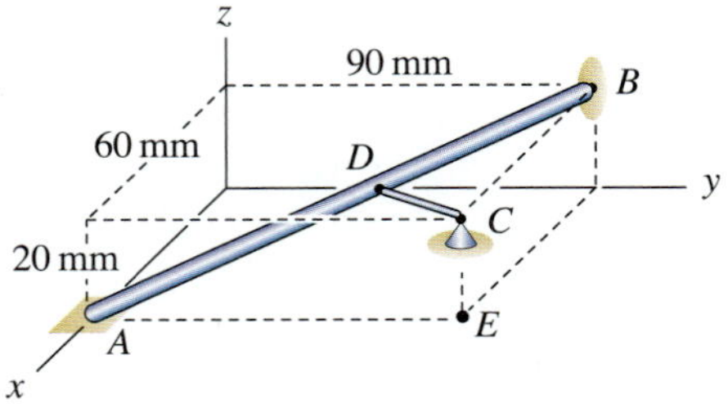

Figure P2.85 and P2.86

Problem 2.85

Structural member AB is to be supported by a strut CD. Determine the smallest length CD may have, and specify where D must be located for a strut of this length to be used.

Problem 2.86

Determine the smallest distance between member AB and point E.

Problem 2.87

In Example 2.13 on p. 70, determine the smallest distance between point D and the infinite line passing through points A and B. Is this distance the same as the smallest distance to rod AB? Explain.

Problem 2.88

In Example 2.18 on p. 86, determine the smallest distance between point O and the infinite line passing through points A and B. Is this distance the same as the smallest distance to rod AB? Explain.

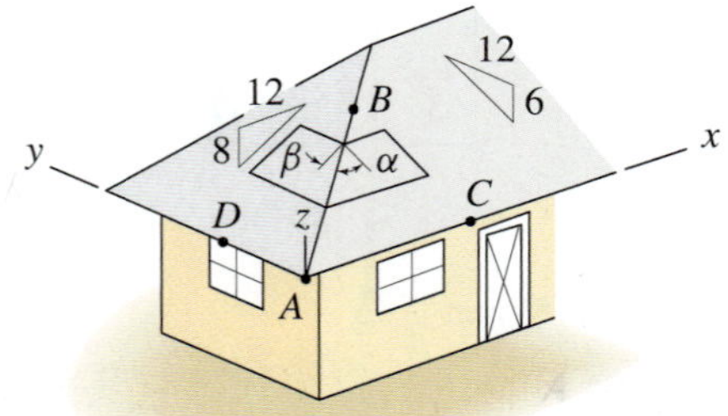

Figure P2.89

Problem 2.89

A building's roof has "6 in 12" slope in the front and back and "8 in 12" slope on the sides. Determine the angles α and β that should be used for cutting sheets of plywood so they properly meet along edge AB of the roof. *Hint:* Write the position vector $\vec{r}_{AB}$ (where B is some point along the edge of the roof) two ways: $\vec{r}_{AB} = \vec{r}_{AC} + \vec{r}_{CB}$ and $\vec{r}_{AB} = \vec{r}_{AD} + \vec{r}_{DB}$. Then use the roof slopes to help write $\vec{r}_{CB}$ and $\vec{r}_{DB}$ such that the magnitudes of the two expressions for $\vec{r}_{AB}$ are the same.

2.5 Vector Cross Product

The cross product between two vectors may be used to determine (1) the direction normal to the plane containing two vectors, (2) the area of a parallelogram formed by two vectors, and (3) the *moment* produced by a force. The last use is especially important in statics and mechanics and will be discussed extensively in Chapter 4. Compared to the dot product, the cross product is more intricate to evaluate. This section studies techniques for evaluating the cross product and interpreting its results.

The *cross product* between two vectors $\vec{A}$ and $\vec{B}$ is an operation defined as

$$\vec{A} \times \vec{B} = (|\vec{A}||\vec{B}| \sin\theta)\, \hat{u} \qquad (2.48)$$

where

θ = angle between the lines of action of $\vec{A}$ and $\vec{B}$ where $0 \le \theta \le 180°$,
$\hat{u}$ = unit vector normal to the plane containing $\vec{A}$ and $\vec{B}$ according to right-hand rule.

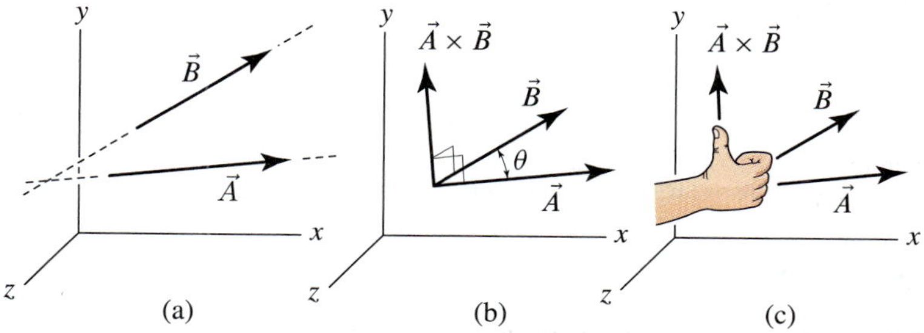

Figure 2.31. (a) Vectors $\vec{A}$ and $\vec{B}$ in three dimensions. (b) To evaluate the cross product between vectors $\vec{A}$ and $\vec{B}$, the vectors can be arranged tail to tail to define angle θ, which is measured in the plane containing the two vectors. (c) The result of $\vec{A} \times \vec{B}$ is a vector whose direction is governed by the right-hand rule.

In words, Eq. (2.48) states "$\vec{A}$ cross $\vec{B}$" equals a vector whose magnitude is the product of the magnitude of $\vec{A}$, magnitude of $\vec{B}$, and sine of angle θ between the lines of action of $\vec{A}$ and $\vec{B}$, and whose direction is perpendicular to the plane containing $\vec{A}$ and $\vec{B}$ as governed by the right-hand rule. The definition of θ in Eq. (2.48) is the same as that used for the dot product. Since the cross product yields a result that is a vector, the cross product is sometimes called the *vector product*. The units for the cross product are equal to the product of the units for $\vec{A}$ and $\vec{B}$.

Figure 2.31 illustrates the cross product between vectors $\vec{A}$ and $\vec{B}$. Although most often they will, $\vec{A}$ and $\vec{B}$ do not need to lie in the same plane to compute the cross product between them. After the vectors are arranged tail to tail, they define a plane, and the intersection of their lines of action defines angle θ, which is measured in this plane. The direction of the cross product is determined by applying the right-hand rule, where $\vec{A}$ is the first vector and $\vec{B}$

Concept Alert

Applications of the cross product. The cross product between two vectors produces a result that is a vector. The cross product is frequently used to determine the normal direction to a surface, the area of a parallelogram formed by two vectors, and (as discussed in Chapter 4) the *moment* produced by a force. The last application is especially important in statics and mechanics.

is the second vector. That is, if you let $\vec{A}$ pass in to the palm of your right-hand and $\vec{B}$ pass through your fingertips as shown in Fig. 2.31, then your thumb will point in the proper direction for $\vec{A} \times \vec{B}$. Immediately we see that if $\vec{B}$ is taken as the first vector and $\vec{A}$ the second, then the right-hand rule gives a direction for $\vec{B} \times \vec{A}$ that is opposite that for $\vec{A} \times \vec{B}$. Hence, the order in which vectors are taken in the cross product is important.

The cross product has the following properties:

$$\vec{A} \times \vec{B} = -\vec{B} \times \vec{A} \quad \text{anticommutative property,} \tag{2.49}$$

$$s(\vec{A} \times \vec{B}) = (s\vec{A}) \times \vec{B} = \vec{A} \times (s\vec{B}) \quad \text{associative property with respect to multiplication by a scalar,} \tag{2.50}$$

$$(\vec{A} + \vec{B}) \times \vec{C} = (\vec{A} \times \vec{C}) + (\vec{B} \times \vec{C}) \quad \text{distributive property with respect to vector addition.} \tag{2.51}$$

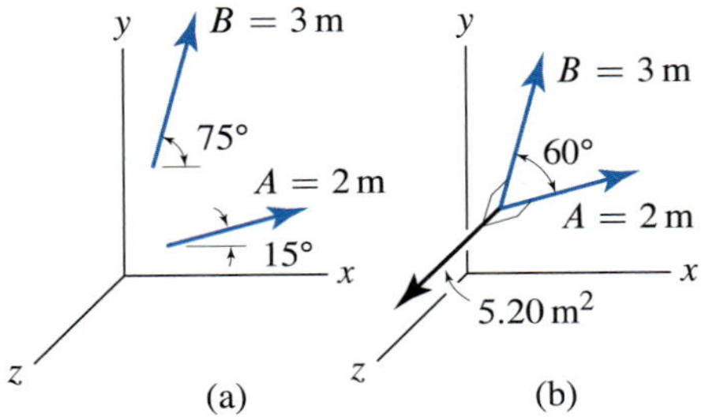

Figure 2.32
Cross product between two vectors $\vec{A}$ and $\vec{B}$ that lie in the xy plane. The result of $\vec{A} \times \vec{B}$ is a vector having magnitude $5.20\,\text{m}^2$ in the z direction.

The foregoing remarks are true regardless of the type of vector representation that is used. For example, consider the two position vectors $\vec{A}$ and $\vec{B}$ shown in Fig. 2.32(a), where both vectors lie in the xy plane. Sliding the vectors tail to tail as shown in Fig. 2.32(b) provides the angle between $\vec{A}$ and $\vec{B}$ as 60°, and applying Eq. (2.48) provides the cross product between $\vec{A}$ and $\vec{B}$ as

$$\vec{A} \times \vec{B} = (2\,\text{m})(3\,\text{m}) \sin 60°\, \hat{u} = 5.20\,\text{m}^2\, \hat{u} \tag{2.52}$$

where $\hat{u}$ is a unit vector normal to the plane containing $\vec{A}$ and $\vec{B}$ and whose direction is given by the right-hand rule, namely, the z direction shown in Fig. 2.32(b). Observe the cross product has units of m^2. Although not obvious, the magnitude of this result has the physical interpretation of being the area of the parallelogram formed by $\vec{A}$ and $\vec{B}$. This feature of the cross product is discussed in greater detail soon.

Helpful Information

Alternatives for evaluation of the cross product. Regardless of the representaion used to express vectors, Eq. (2.48) can always be used to evaluate the cross product. For vectors with Cartesian representation, either Eq. (2.48) or Eq. (2.53) can be used, with the latter usually being more convenient. In fact, sometimes *both* methods are used to help determine useful information.

Cross product using Cartesian components

For two vectors $\vec{A}$ and $\vec{B}$ with Cartesian representations, the cross product between them is given by

$$\vec{A} \times \vec{B} = (A_y B_z - A_z B_y)\,\hat{\imath} + (A_z B_x - A_x B_z)\,\hat{\jmath} + (A_x B_y - A_y B_x)\,\hat{k}. \tag{2.53}$$

Equation (2.53) can be obtained using the following derivation. We begin by writing $\vec{A}$ and $\vec{B}$ using Cartesian vector representation

$$\vec{A} = A_x\,\hat{\imath} + A_y\,\hat{\jmath} + A_z\,\hat{k} \tag{2.54}$$

$$\vec{B} = B_x\,\hat{\imath} + B_y\,\hat{\jmath} + B_z\,\hat{k} \tag{2.55}$$

We then take the cross product between $\vec{A}$ and $\vec{B}$, using the distributive law of Eq. (2.51) to expand the product term by term:

$$\begin{aligned}\vec{A} \times \vec{B} &= (A_x\,\hat{\imath} + A_y\,\hat{\jmath} + A_z\,\hat{k}) \times (B_x\,\hat{\imath} + B_y\,\hat{\jmath} + B_z\,\hat{k}) \\ &= (A_x\,\hat{\imath} \times B_x\,\hat{\imath}) + (A_x\,\hat{\imath} \times B_y\,\hat{\jmath}) + (A_x\,\hat{\imath} \times B_z\,\hat{k})\end{aligned}$$

$$+ (A_y\,\hat{\jmath} \times B_x\,\hat{\imath}) + (A_y\,\hat{\jmath} \times B_y\,\hat{\jmath}) + (A_y\,\hat{\jmath} \times B_z\,\hat{k})$$
$$+ (A_z\,\hat{k} \times B_x\,\hat{\imath}) + (A_z\,\hat{k} \times B_y\,\hat{\jmath}) + (A_z\,\hat{k} \times B_z\,\hat{k}) \qquad (2.56)$$

To proceed further we must address cross products between combinations of unit vectors. Equation (2.48) shows $\hat{\imath} \times \hat{\imath}$ has magnitude $(1)(1)\sin 0° = 0$, and similarly $\hat{\jmath} \times \hat{\jmath}$ and $\hat{k} \times \hat{k}$ have zero magnitude. The cross product between combinations of different unit vectors always provides the magnitude $(1)(1)\sin 90° = 1$ and direction that depends on the order in which the vectors are taken. For example, $\hat{\imath} \times \hat{\jmath} = \hat{k}$, while $\hat{\jmath} \times \hat{\imath} = -\hat{k}$. In summary, cross products between combinations of unit vectors are

$$\begin{array}{lll} \hat{\imath} \times \hat{\imath} = \vec{0} & \hat{\imath} \times \hat{\jmath} = \hat{k} & \hat{\imath} \times \hat{k} = -\hat{\jmath} \\ \hat{\jmath} \times \hat{\imath} = -\hat{k} & \hat{\jmath} \times \hat{\jmath} = \vec{0} & \hat{\jmath} \times \hat{k} = \hat{\imath} \\ \hat{k} \times \hat{\imath} = \hat{\jmath} & \hat{k} \times \hat{\jmath} = -\hat{\imath} & \hat{k} \times \hat{k} = \vec{0}, \end{array} \qquad (2.57)$$

where $\vec{0} = 0\,\hat{\imath} + 0\,\hat{\jmath} + 0\,\hat{k}$ is called the *zero vector*. Substituting Eqs. (2.57) into Eq. (2.56) then provides Eq. (2.53). While Eq. (2.53) can always be employed to evaluate the cross product between vectors with Cartesian representation, it is awkward to remember and use. However, if you want to program a digital machine such as your pocket calculator or a computer to evaluate cross products, then Eq. (2.53) is ideal. But for manual evaluation the following procedure is easier.

Helpful Information

A little trick. The following diagram is a helpful tool for quickly evaluating cross products between combinations of unit vectors $\hat{\imath}$, $\hat{\jmath}$, and $\hat{k}$.

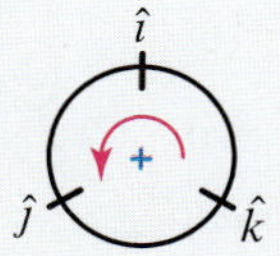

Cross products that move counterclockwise (positive direction) give "positive" results, while cross products that move clockwise (negative direction) give "negative" results. For example, when we evaluate $\hat{\imath} \times \hat{\jmath}$, the diagram shows that the result is counterclockwise, hence $\hat{\imath} \times \hat{\jmath} = +\hat{k}$. When we evaluate $\hat{\jmath} \times \hat{\imath}$, the diagram shows that the result is clockwise, hence $\hat{\jmath} \times \hat{\imath} = -\hat{k}$.

Evaluation of cross product using determinants

By arranging vectors $\vec{A}$ and $\vec{B}$ in a matrix, the determinant of the matrix can be evaluated to yield $\vec{A} \times \vec{B}$. A matrix is an arrangement of elements into a rectangular array of rows and columns. There is a precise and extensive science underlying matrices, and because matrices are of fundamental importance in engineering, you are sure to study these as you advance in your education. Here we present only those details that are needed for evaluation of the cross product.

To evaluate the cross product between vectors $\vec{A}$ and $\vec{B}$, we evaluate the determinant of the matrix

$$\vec{A} \times \vec{B} = \det \begin{bmatrix} \hat{\imath} & \hat{\jmath} & \hat{k} \\ A_x & A_y & A_z \\ B_x & B_y & B_z \end{bmatrix} = \begin{vmatrix} \hat{\imath} & \hat{\jmath} & \hat{k} \\ A_x & A_y & A_z \\ B_x & B_y & B_z \end{vmatrix} \qquad (2.58)$$

where det is an abbreviation indicating the determinant of the matrix that follows is to be evaluated. In the rightmost form of Eq. (2.58), a more abbreviated notation is used where the vertical bars indicate the determinant of the enclosed matrix. In this book we will primarily use vertical bars to denote determinant. The first row of the matrix in Eq. (2.58) consists of the basis vectors for the coordinate system, namely, $\hat{\imath}$, $\hat{\jmath}$, and $\hat{k}$. The second row consists of the components of $\vec{A}$, and the third row consists of the components of $\vec{B}$. Note that if $\vec{B} \times \vec{A}$ were to be evaluated, then the second row of the matrix would contain the components of $\vec{B}$ and the third row would contain the components of $\vec{A}$. While there are many ways to evaluate the determinant of a matrix, we discuss two methods that are effective for Eq. (2.58).

Determinant by Method 1. To help in the computation of the determinant, we start by repeating the first and second columns after writing the original matrix. We then take the sums and differences of products of the diagonal elements as follows: take the product of the three elements along diagonal 1 [i.e., $\hat{\imath}(A_y)(B_z)$], add the product of the three elements along diagonal 2, add the product of the three elements along diagonal 3, subtract the product of the three elements along diagonal 4, subtract the product of the three elements along diagonal 5, and finally subtract the product of the three elements along diagonal 6:

$$\vec{A}\times\vec{B} = \begin{vmatrix} \hat{\imath} & \hat{\jmath} & \hat{k} \\ A_x & A_y & A_z \\ B_x & B_y & B_z \end{vmatrix} = \begin{matrix} \hat{\imath} & \hat{\jmath} & \hat{k} & \hat{\imath} & \hat{\jmath} \\ A_x & A_y & A_z & A_x & A_y \\ B_x & B_y & B_z & B_x & B_y \end{matrix}$$

$$= \underbrace{\hat{\imath}\,A_y B_z}_{\text{diagonal 1}} + \underbrace{\hat{\jmath}\,A_z B_x}_{\text{diagonal 2}} + \underbrace{\hat{k}\,A_x B_y}_{\text{diagonal 3}} - \underbrace{\hat{k}\,A_y B_x}_{\text{diagonal 4}} - \underbrace{\hat{\imath}\,A_z B_y}_{\text{diagonal 5}} - \underbrace{\hat{\jmath}\,A_x B_z}_{\text{diagonal 6}}$$

$$= (A_y B_z - A_z B_y)\,\hat{\imath} + (A_z B_x - A_x B_z)\,\hat{\jmath} + (A_x B_y - A_y B_x)\,\hat{k} \tag{2.59}$$

Note that Eq. (2.59) is identical to Eq. (2.53).

Determinant by Method 2. Here we expand the determinant of the matrix by using minors. A *minor* is the determinant of a subset of the original matrix that is obtained as shown in Eq. (2.60) by eliminating the row and column corresponding to a particular element; the determinant of the elements remaining in the matrix constitutes the minor for that element. Each of the minors is found using the procedure described in method 1 for computing a determinant. That is, the determinant of the matrix multiplying $\hat{\imath}$ in Eq. (2.60) is the product of the two elements along diagonal 1, minus the product of the two elements along diagonal 2. The determinant of the matrix multiplying $\hat{\jmath}$ is the product of the two elements along diagonal 3, minus the product of the two elements along diagonal 4. Similarly, the determinant of the matrix multiplying $\hat{k}$ is the product of the two elements along diagonal 5, minus the product of the two elements along diagonal 6.

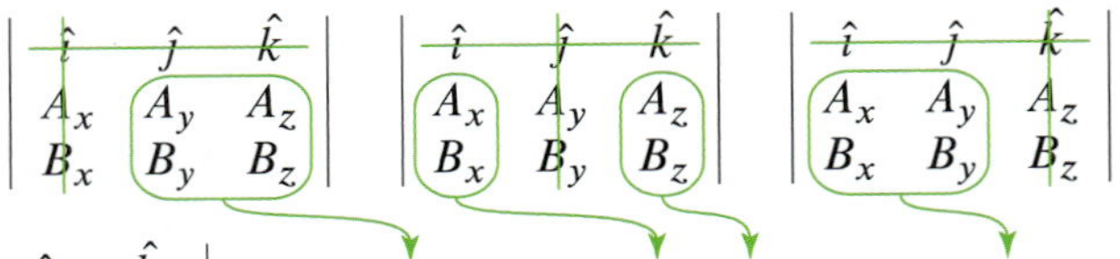

$$\vec{A}\times\vec{B} = \begin{vmatrix} \hat{\imath} & \hat{\jmath} & \hat{k} \\ A_x & A_y & A_z \\ B_x & B_y & B_z \end{vmatrix} = \hat{\imath}\begin{vmatrix} A_y & A_z \\ B_y & B_z \end{vmatrix} - \hat{\jmath}\begin{vmatrix} A_x & A_z \\ B_x & B_z \end{vmatrix} + \hat{k}\begin{vmatrix} A_x & A_y \\ B_x & B_y \end{vmatrix}$$

$$= \hat{\imath}\,(\underbrace{A_y B_z}_{\text{diagonal 1}} - \underbrace{A_z B_y}_{\text{diagonal 2}}) - \hat{\jmath}\,(\underbrace{A_x B_z}_{\text{diagonal 3}} - \underbrace{A_z B_x}_{\text{diagonal 4}}) + \hat{k}\,(\underbrace{A_x B_y}_{\text{diagonal 5}} - \underbrace{A_y B_x}_{\text{diagonal 6}}) \tag{2.60}$$

We have not discussed all of the details of why this procedure works, and some of these details explain why the negative sign associated with $\hat{\jmath}$ appears in Eq. (2.60). A common error is to forget this negative sign.

Common Pitfall

Don't forget the negative sign on $\hat{\jmath}$. When you use Method 2 to evaluate the determinant, a common error is to forget the negative sign for $\hat{\jmath}$.

Remarks. You should try both methods discussed here and adopt the method you prefer. An advantage of the second method is that terms are automatically grouped with vectors $\hat{\imath}$, $\hat{\jmath}$, and $\hat{k}$. In more advanced mathematics and mechanics subjects, you may have need to evaluate the determinant of larger matrices than those treated here. Method 1 is limited to matrices that have at most three rows and columns. With some additional details, Method 2 can be applied to larger matrices, but this may involve use of many minors and other methods (not discussed here) are usually more effective.

Determination of the normal direction to a plane

The cross product can be used to determine the normal direction $\vec{C}$ to the plane containing two vectors $\vec{A}$ and $\vec{B}$ by evaluating $\vec{C} = \vec{A} \times \vec{B}$. If we desire the normal direction to be a unit vector, then we normalize $\vec{C}$ by evaluating $\hat{u} = \vec{C}/|\vec{C}|$. Note that even if $\vec{A}$ and $\vec{B}$ are unit vectors, the cross product between these is not a unit vector unless $\vec{A}$ and $\vec{B}$ are orthogonal. A normal direction can also be obtained by evaluating $\vec{B} \times \vec{A}$, and of course the vector that is produced has opposite direction to $\vec{A} \times \vec{B}$. In many problems that call for a normal direction to be computed, either direction can be used, while in some problems it may be desirable or necessary to distinguish between these.

Determination of the area of a parallelogram

The magnitude of the cross product gives the area of the parallelogram formed by vectors $\vec{A}$ and $\vec{B}$ arranged tail to tail. We illustrate this using the example of Fig. 2.32, where the magnitude of the cross product was found in Eq. (2.52) to be $5.20\,\mathrm{m}^2$, and this area is shown in Fig. 2.33(a). To show that this statement is true, consider the shaded triangle shown in Fig. 2.33(b): it has base A and height $B \sin\theta$. Hence its area is (base)(height)/2 $= AB(\sin\theta)/2$. The remaining triangle shown in Fig. 2.33(b) has the same base and height and thus the same area. Hence, the area of the parallelogram is the same as the magnitude of the cross product given in Eq. (2.48), namely, $AB\sin\theta = 5.20\,\mathrm{m}^2$.

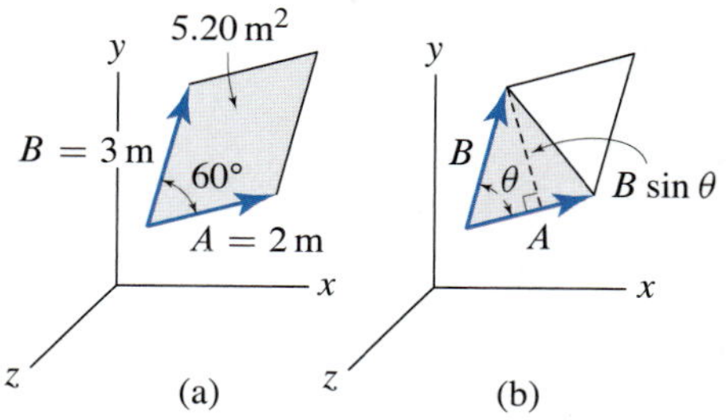

Figure 2.33
The magnitude $5.20\,\mathrm{m}^2$ of $\vec{A} \times \vec{B}$ corresponds to the area of the parallelogram formed by $\vec{A}$ and $\vec{B}$ arranged tail to tail.

Scalar triple product

Occasionally the cross product between two vectors $\vec{A}$ and $\vec{B}$ is to be immediately followed by a dot product with a third vector $\vec{C}$, and we call the expression $(\vec{A} \times \vec{B}) \cdot \vec{C}$ a *scalar triple product*, or a *mixed triple product*.* Of course this product can be evaluated by first computing the cross product, which produces a vector, and then taking the dot product of this with $\vec{C}$ to provide the final result, which is a scalar. Alternatively, the scalar triple product may be computed by evaluating

$$(\vec{A} \times \vec{B}) \cdot \vec{C} = \begin{vmatrix} C_x & C_y & C_z \\ A_x & A_y & A_z \\ B_x & B_y & B_z \end{vmatrix} \tag{2.61}$$

* The nomenclature *scalar triple product* is preferred to distinguish this from the triple product $\vec{A} \times \vec{B} \times \vec{C}$, which is more accurately called a *vector triple product*.

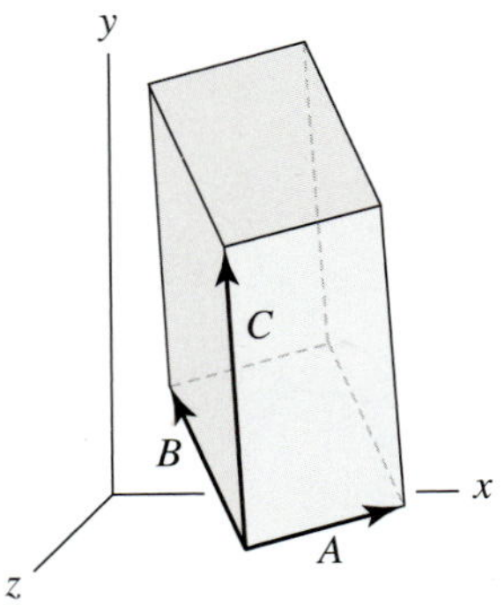

Figure 2.34
Volume of the parallelepiped formed by $\vec{A}$, $\vec{B}$, and $\vec{C}$ arranged tail to tail is given by the scalar triple product $(\vec{A} \times \vec{B}) \cdot \vec{C}$.

Although we do not prove this, the value produced by $(\vec{A}\times\vec{B})\cdot\vec{C}$ is the volume of the parallelepiped defined by $\vec{A}$, $\vec{B}$, and $\vec{C}$ arranged tail to tail as shown in Fig. 2.34. In Prob. 2.105, the implications of changing the ordering of vectors in the scalar triple product are explored.

End of Section Summary

In this section, the cross product between two vectors has been defined. Some of the key points are as follows:

- The *cross product* between two vectors $\vec{A}$ and $\vec{B}$ is a vector $\vec{C}$, where $\vec{C} = \vec{A} \times \vec{B}$. The unit of $\vec{C}$ is the product of the units of $\vec{A}$ and $\vec{B}$, and the direction of $\vec{C}$ is perpendicular to the plane containing $\vec{A}$ and $\vec{B}$ as governed by the right-hand rule.
- The cross product can be used to find the normal direction to a plane.
- The cross product can be used to determine the area of a parallelogram and the volume of a parallelepiped.
- One of the most important uses for the cross product is for computing the *moment of a force*, and this application is discussed extensively in Chapter 4.

EXAMPLE 2.19 *Evaluation of the Cross Product*

Evaluate the cross product between vectors $\vec{A}$ and $\vec{B}$ where

$$\vec{A} = (3\,\hat{\imath} + 5\,\hat{\jmath} + \hat{k})\,\text{mm},$$
$$\vec{B} = (-4\,\hat{\imath} + 6\,\hat{\jmath} + 2\,\hat{k})\,\text{mm}.$$

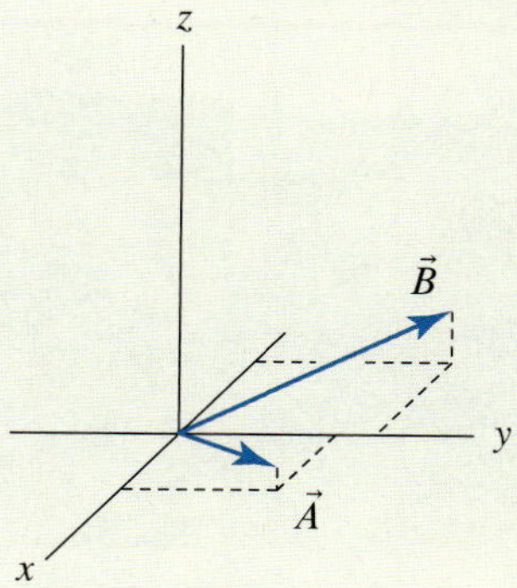

Figure 1

SOLUTION

Road Map Although we could use Eq. (2.53) to evaluate the cross product, it will generally be more effective to use your choice of Method 1 or 2 as shown here.

Method 1

Governing Equations & Computation Using Method 1 described in Eq. (2.59), the cross product between $\vec{A}$ and $\vec{B}$ is

$$\vec{C} = \vec{A} \times \vec{B} = \begin{vmatrix} \hat{\imath} & \hat{\jmath} & \hat{k} \\ 3 & 5 & 1 \\ -4 & 6 & 2 \end{vmatrix} \text{mm}^2 = \begin{matrix} \hat{\imath} & \hat{\jmath} & \hat{k} & \hat{\imath} & \hat{\jmath} \\ 3 & 5 & 1 & 3 & 5 \\ -4 & 6 & 2 & -4 & 6 \end{matrix}\ \text{mm}^2 \tag{1}$$

$$= [\hat{\imath}\,(5)(2) + \hat{\jmath}\,(1)(-4) + \hat{k}\,(3)(6) - \hat{k}\,(5)(-4) - \hat{\imath}\,(1)(6) - \hat{\jmath}\,(3)(2)]\,\text{mm}^2$$
$$= \boxed{(4\,\hat{\imath} - 10\,\hat{\jmath} + 38\,\hat{k})\,\text{mm}^2.} \tag{2}$$

The mm^2 term in Eq. (1) comes from factoring the mm dimensions out of both $\vec{A}$ and $\vec{B}$. The result of the cross product, vector $\vec{C}$, is shown in Fig. 2.

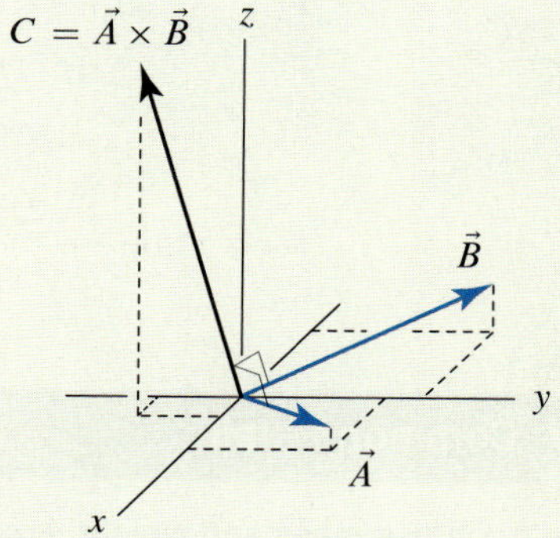

Figure 2
Evaluation of the cross product between vectors $\vec{A}$ and $\vec{B}$.

Method 2

Governing Equations & Computation Using Method 2 described in Eq. (2.60), the cross product between $\vec{A}$ and $\vec{B}$ is

$$\vec{C} = \vec{A} \times \vec{B} = \begin{vmatrix} \hat{\imath} & \hat{\jmath} & \hat{k} \\ 3 & 5 & 1 \\ -4 & 6 & 2 \end{vmatrix} \text{mm}^2$$

$$= \hat{\imath} \begin{vmatrix} 5 & 1 \\ 6 & 2 \end{vmatrix} \text{mm}^2 - \hat{\jmath} \begin{vmatrix} 3 & 1 \\ -4 & 2 \end{vmatrix} \text{mm}^2 + \hat{k} \begin{vmatrix} 3 & 5 \\ -4 & 6 \end{vmatrix} \text{mm}^2 \tag{3}$$

$$= \hat{\imath}\,[(5)(2) - (1)(6)]\,\text{mm}^2 - \hat{\jmath}\,[(3)(2) - (1)(-4)]\,\text{mm}^2$$
$$+ \hat{k}\,[(3)(6) - (5)(-4)]\,\text{mm}^2$$
$$= \boxed{(4\,\hat{\imath} - 10\,\hat{\jmath} + 38\,\hat{k})\,\text{mm}^2,} \tag{4}$$

which is the same result as obtained by Method 1.

Discussion & Verification To verify that $\vec{C}$ is indeed perpendicular to the plane containing $\vec{A}$ and $\vec{B}$, you should evaluate the dot products $\vec{C} \cdot \vec{A}$ and $\vec{C} \cdot \vec{B}$ to find they are both zero. You may also wish to evaluate $\vec{B} \times \vec{A}$ to show that $-\vec{C}$ is produced.

EXAMPLE 2.20 *Components of Force in Directions Normal and Tangent to a Plane*

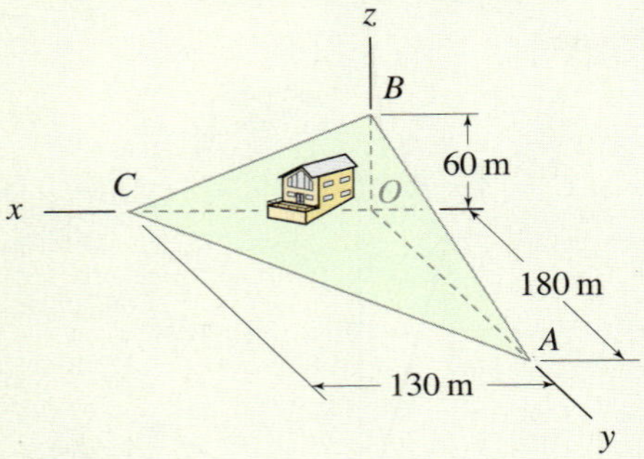

Figure 1

Common Pitfall

Don't confuse mass and weight. A common error is to mistake mass for weight. Mass must be multiplied by acceleration due to gravity to obtain weight. In Eq. (1) and Fig. 2, if you incorrectly gave the weight of the house a value of 95, then all of your answers are about an order of magnitude too small, and they have the wrong dimensions!

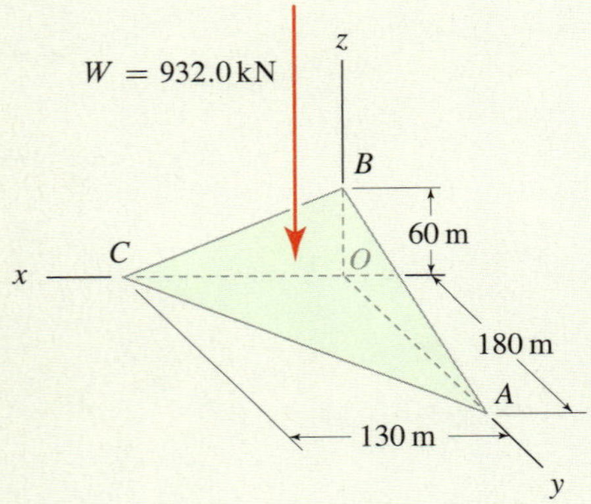

Figure 2
The weight of the house is a vertical force that is applied to the slope.

A house with 95 Mg mass is built on a steep slope defined by points A, B, and C. To help assess the possibility of slope failure (mud slide), it is necessary to

(a) Determine the components of the weight in directions normal and parallel (tangent) to the slope.

(b) Determine the component vectors of the weight in directions normal and parallel (tangent) to the slope.

(c) Determine the smallest distance from point O to the slope.

SOLUTION

Road Map The weight of the house is $(95\,\text{Mg})(9.81\,\text{m/s}^2) = 932.0\,\text{kN}$, as shown in Fig. 2, and the vector expression for this is

$$\vec{W} = (-932.0\,\text{kN})\,\hat{k}. \tag{1}$$

Our strategy for part (a) will be to use the cross product to determine the normal vector to the slope. Then the dot product between the weight vector and the normal direction vector will yield the component of weight in the normal direction. Once this is obtained, the component of the weight that is parallel, or tangent, to the slope can be determined using the Pythagorean theorem. For part (b), we will express the weights normal and parallel to the slope as vectors. For part (c), we write a position vector from point O to any convenient point on the slope, and then we use the dot product to determine the component of this vector that is normal to the slope. This result is the shortest distance from point O to the slope.

Part (a)

Governing Equations & Computation The normal direction can be computed using the cross product between a variety of different position vectors. We will use position vectors from A to B and from A to C

$$\vec{r}_{AB} = (-180\,\hat{\jmath} + 60\,\hat{k})\,\text{m}, \tag{2}$$

$$\vec{r}_{AC} = (130\,\hat{\imath} - 180\,\hat{\jmath})\,\text{m}. \tag{3}$$

The normal direction $\vec{n}$ to the surface and its magnitude $|\vec{n}|$ are then

$$\vec{n} = \vec{r}_{AB} \times \vec{r}_{AC} = \begin{vmatrix} \hat{\imath} & \hat{\jmath} & \hat{k} \\ 0 & -180 & 60 \\ 130 & -180 & 0 \end{vmatrix} \text{m}^2$$

$$= (10,800\,\hat{\imath} + 7800\,\hat{\jmath} + 23,400\,\hat{k})\,\text{m}^2, \tag{4}$$

$$|\vec{n}| = \sqrt{(10,800)^2 + (7800)^2 + (23,400)^2}\,\text{m}^2 = 26,930\,\text{m}^2. \tag{5}$$

Observe the units for both $\vec{r}_{AB}$ and $\vec{r}_{AC}$ have been factored out of the matrix expression in Eq. (4), and this is helpful to reduce the repetitious writing of units when expanding the determinant. Also, we selected $\vec{r}_{AB}$ as the first vector and $\vec{r}_{AC}$ as the second vector in the cross product so that the normal vector produced would be in the "outward" direction to the slope (i.e., toward the sky). The same result would have been obtained using $\vec{r}_{BC} \times \vec{r}_{BA}$, or $\vec{r}_{CA} \times \vec{r}_{CB}$. Alternatively, we could use the "inward" normal direction for this problem, as given by $\vec{r}_{AC} \times \vec{r}_{AB}$, or $\vec{r}_{BA} \times \vec{r}_{BC}$, or $\vec{r}_{CB} \times \vec{r}_{CA}$; this vector's direction is into the Earth.

Next, the component of the weight vector $\vec{W}$ acting in the direction $\vec{n}$ will be called W_n, and it is given by

$$W_n = \vec{W} \cdot \frac{\vec{n}}{|\vec{n}|} = \frac{(0)(10,800\,\text{m}^2) + (0)(7800\,\text{m}^2) + (-932.0\,\text{kN})(23,400\,\text{m}^2)}{26,930\,\text{m}^2}$$
$$= \boxed{-809.9\,\text{kN.}} \tag{6}$$

Since $\vec{n}$ is the outward normal direction, the negative sign for W_n indicates that the normal component of $\vec{W}$ acts into the surface, which is the expected result. Note that if an inward normal vector were used instead, then the sign of W_n would be positive.

The Pythagorean theorem is now used to obtain the component of the weight that is parallel, or tangent to the surface, W_t, as

$$W_t = \sqrt{W^2 - W_n^2} = \sqrt{(-932.0\,\text{kN})^2 - (-810.7\,\text{kN})^2} = \boxed{461.1\,\text{kN.}} \tag{7}$$

While the magnitude of the tangential component of the force W_t is known, we do not know its direction other than it lies in the plane defined by points A, B, and C.

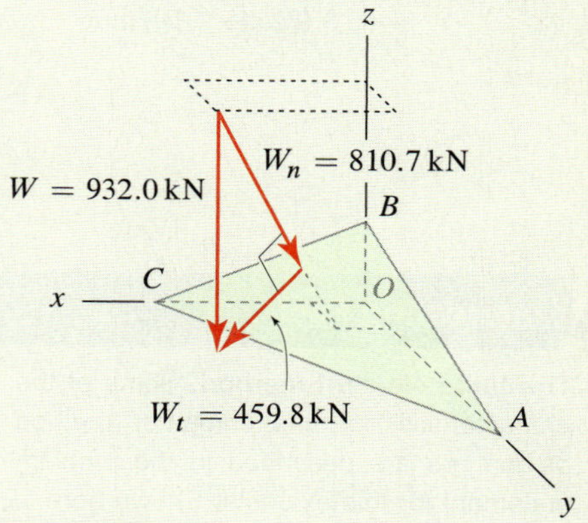

Figure 3
The weight of the house is resolved into components W_n and W_t that are normal and tangential to the slope, respectively.

Part (b)

Governing Equations & Computation Once W_n is known from Eq. (6), we may then find the vector components $\vec{W}_n$ and $\vec{W}_t$ as follows:

$$\vec{W}_n = W_n \frac{\vec{n}}{|\vec{n}|} = \boxed{(-325\,\hat{\imath} - 235\,\hat{\jmath} - 704\,\hat{k})\,\text{kN.}} \tag{8}$$

Since $\vec{W} = \vec{W}_n + \vec{W}_t$, we evaluate

$$\vec{W}_t = \vec{W} - \vec{W}_n = \boxed{(325\,\hat{\imath} + 235\,\hat{\jmath} - 228\,\hat{k})\,\text{kN.}} \tag{9}$$

The components W_n and W_t are shown in their proper orientations in Fig. 3. As a partial check on our solution, you should evaluate the magnitudes of Eqs. (8) and (9) to verify that they agree with Eqs. (6) and (7), respectively.

Part (c)

Governing Equations & Computation The procedure for finding the smallest distance from point O to the surface defined by points A, B, and C is very similar to that outlined in Section 2.4. We first write a position vector from point O to any convenient point on the surface. We then take the dot product of this vector with a unit vector that is normal to the surface, and the scalar that is produced is the smallest distance between point O and the surface. The position vector from O to C is

$$\vec{r}_{OC} = (130\,\text{m})\,\hat{\imath}. \tag{10}$$

The portion of $\vec{r}_{OC}$ in the normal direction is

$$r_n = \vec{r}_{OC} \cdot \frac{\vec{n}}{|\vec{n}|} = \frac{(130\,\text{m})(10,800) + (0)(7800) + (0)(23,400)}{26,930} = 52.1\,\text{m} \tag{11}$$

and thus, $\boxed{\text{the smallest distance between point } O \text{ and the surface is } 52.1\text{ m.}}$

Discussion & Verification The solution in part (c) finds the smallest distance from point O to the plane of infinite extent that contains points A, B, and C. If, however, the surface in question were finite in size, then the next step would be to determine the coordinates of the head of the vector $\vec{r}_n = r_n\,\vec{n}/|\vec{n}|$ to determine if this point lies on or off of the surface in question.

EXAMPLE 2.21 *Determination of the Normal Direction to a Plane*

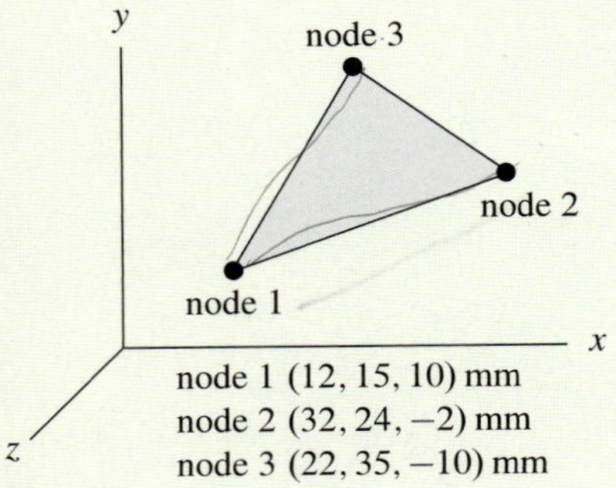

Figure 1

The *finite element method* is a computer technique that has revolutionized the way structural engineering is performed. With this method, a structure usually having complex geometry is subdivided into small regions called *finite elements*, each of which has simple geometry and thus whose behavior can be more easily characterized. Shown in Fig. 1 is a flat three-node plate finite element. Its geometry is fully described by the coordinates of the corners, which are called *nodes*. Determine the unit vector in the direction normal to the surface of the element, and the area of the element.

SOLUTION

Road Map We will use the cross product between two vectors oriented along the edges of the triangle to determine the normal direction $\vec{n}$. Because three points define a unique plane, the same normal direction $\vec{n}$ (or possibly $-\vec{n}$, depending on the order in which vectors are taken in the cross product) should be obtained regardless of which two vectors are used.

Governing Equations & Computation Position vectors from points (nodes) 1 to 2 and 1 to 3 will be used for computing the cross product. Thus, taking the differences between the coordinates of the head and tail, we write

$$\vec{r}_{12} = (20\,\hat{\imath} + 9\,\hat{\jmath} - 12\,\hat{k})\,\text{mm}, \qquad |\vec{r}_{12}| = 25\,\text{mm}, \tag{1}$$

$$\vec{r}_{13} = (10\,\hat{\imath} + 20\,\hat{\jmath} - 20\,\hat{k})\,\text{mm}, \qquad |\vec{r}_{13}| = 30\,\text{mm}. \tag{2}$$

The vector in the normal direction and its magnitude are

$$\vec{n} = \vec{r}_{12} \times \vec{r}_{13} = \begin{vmatrix} \hat{\imath} & \hat{\jmath} & \hat{k} \\ 20 & 9 & -12 \\ 10 & 20 & -20 \end{vmatrix}\,\text{mm}^2 = (60\,\hat{\imath} + 280\,\hat{\jmath} + 310\,\hat{k})\,\text{mm}^2, \tag{3}$$

$$|\vec{n}| = 422\,\text{mm}^2. \tag{4}$$

The unit vector $\hat{u}$ in the normal direction is obtained by normalizing $\vec{n}$ in Eq. (3) which provides

$$\hat{u} = \frac{\vec{n}}{|\vec{n}|} = \boxed{0.142\,\hat{\imath} + 0.664\,\hat{\jmath} + 0.735\,\hat{k}.} \tag{5}$$

To obtain the area of the element, we note the magnitude of the cross product in Eq. (3) is the area of the parallelogram formed by $\vec{r}_{12}$ and $\vec{r}_{13}$, which is twice the area of the triangle formed by these same vectors. Hence, the area A of the triangular region is

$$\boxed{A = \frac{422\,\text{mm}^2}{2} = 211\,\text{mm}^2.} \tag{6}$$

Discussion & Verification As a casual check, you should use inspection to verify that the direction of the normal vector is reasonable. However, in this example the geometry is complex enough that this is probably difficult or inconclusive. As a rigorous check, you could verify that $\hat{u}$ is perpendicular to both $\vec{r}_{12}$ and $\vec{r}_{13}$ by showing that $\hat{u} \cdot \vec{r}_{12} = 0$ and $\hat{u} \cdot \vec{r}_{13} = 0$.

In closing, we note that finite element computer programs typically perform these very same calculations for every element.

Interesting Fact

The finite element method. Some of the characteristics of this computer method of analysis are described in the problem statement for this example. Shown here is a finite element model with thousands of elements of a crash test dummy used to assess trauma in crash simulations.

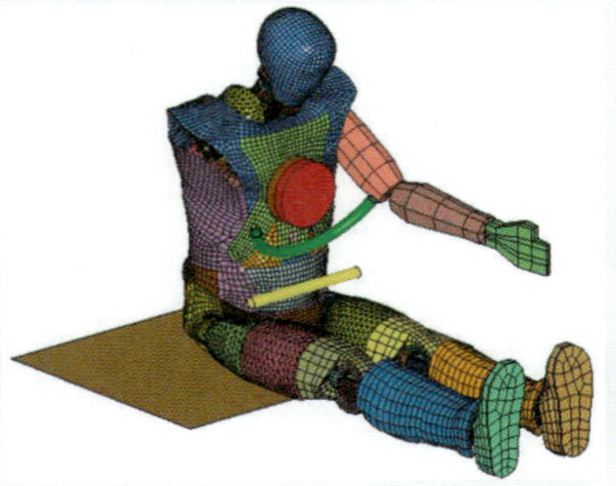

PROBLEMS

Problems 2.90 and 2.91

Vectors $\vec{A}$ and $\vec{B}$ lie in the xy plane.

(a) Use Eq. (2.48) on p. 91 to evaluate $\vec{A} \times \vec{B}$, expressing the resulting vector using Cartesian representation.

(b) Evaluate $\vec{A} \times \vec{B}$ by computing the determinant of a matrix, using either Method 1 or Method 2.

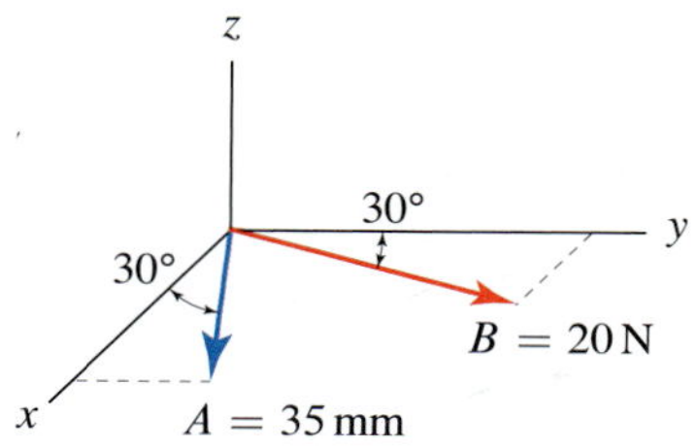

Figure P2.90

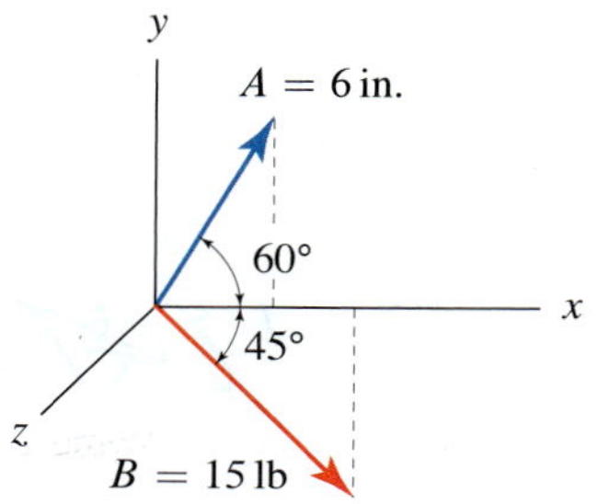

Figure P2.91

Problems 2.92 and 2.93

(a) Evaluate $\vec{A} \times \vec{B}$.

(b) Evaluate $\vec{B} \times \vec{A}$.

(c) Comment on any differences between the results of parts (a) and (b).

(d) Use the dot product to show the result of part (a) is orthogonal to vectors $\vec{A}$ and $\vec{B}$.

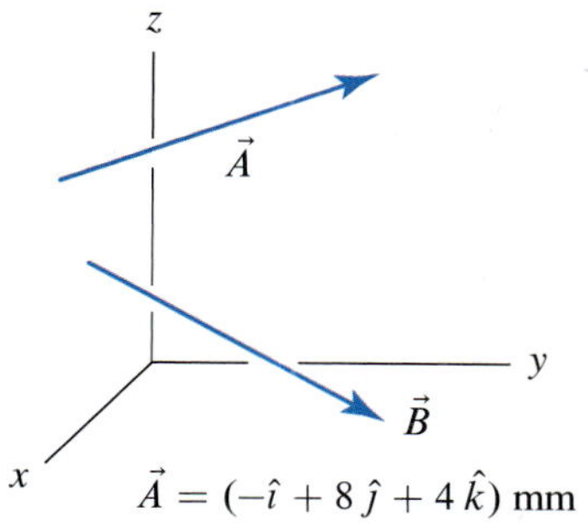

$\vec{A} = (-\hat{\imath} + 8\,\hat{\jmath} + 4\,\hat{k})$ mm
$\vec{B} = (\hat{\imath} + 18\,\hat{\jmath} - 6\,\hat{k})$ mm

Figure P2.92

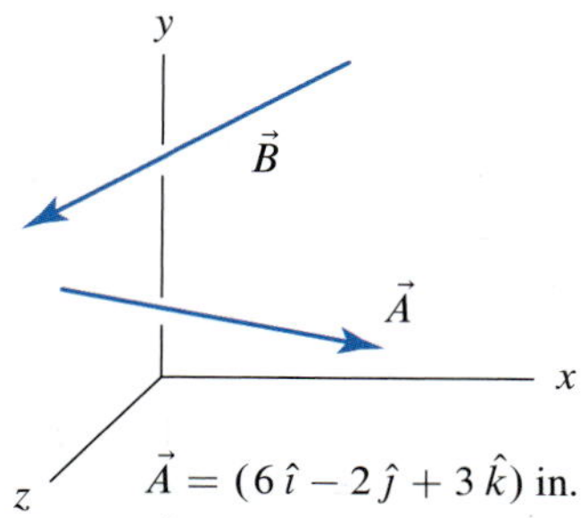

$\vec{A} = (6\,\hat{\imath} - 2\,\hat{\jmath} + 3\,\hat{k})$ in.
$\vec{B} = (-14\,\hat{\imath} - 2\,\hat{\jmath} + 5\,\hat{k})$ in.

Figure P2.93

Problem 2.94

Describe how the cross product operation can be used to determine (or "test") whether two vectors $\vec{A}$ and $\vec{B}$ are orthogonal. Is this test as easy to use as the test based on the dot product? Explain, perhaps using an example to support your remarks.
Note: Concept problems are about *explanations*, not computations.

Problem 2.95

Imagine a left-hand coordinate system has inadvertently been used for a problem. That is, if the x and y directions have been selected first, the z direction has been taken in the wrong direction for a right-hand coordinate system. What consequences will this have for dot products and cross products? Perhaps use an example to support your remarks.
Note: Concept problems are about *explanations*, not computations.

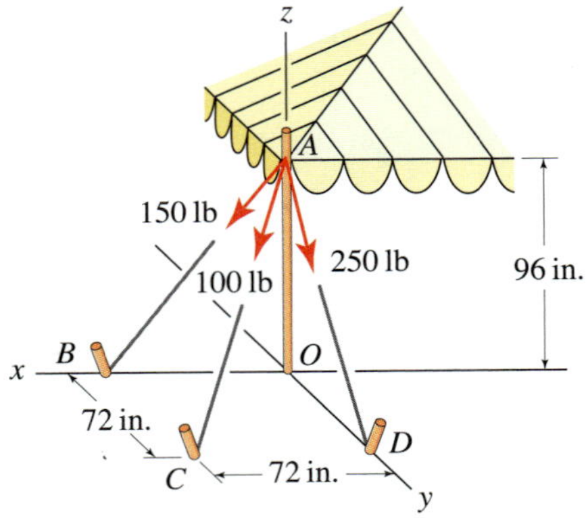

Figure P2.96

Problem 2.96

The corner of a tent is supported using three ropes having the forces shown. It is desired to compute the sum of the cross products $\vec{M}_O = \vec{r}_{OA}\times\vec{F}_{AB}+\vec{r}_{OA}\times\vec{F}_{AC}+\vec{r}_{OA}\times\vec{F}_{AD}$ where $\vec{r}_{OA}$ is the position vector from points O to A, $\vec{F}_{AB}$ is the force directed from points A to B, and so on.

(a) Rather than compute three separate cross products to find $\vec{M}_O$, do the properties of the cross product permit $\vec{M}_O$ to be found using just one cross product? Explain.

(b) Determine $\vec{M}_O$.

Problem 2.97

For the triangular shape window of Prob. 2.67 on p. 87, use the cross product to determine the outward normal unit vector (i.e., pointing away from the origin) and the area of the window.

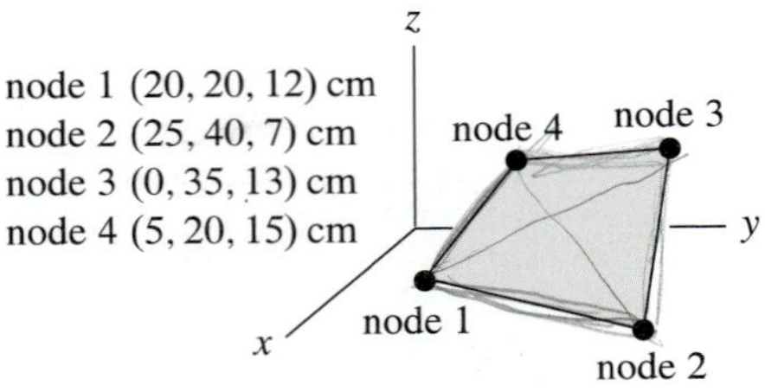

Figure P2.98

Problem 2.98

A flat quadrilateral plate finite element is shown.

(a) Describe and perform a test, using two cross products that will verify if all four nodes (corners) lie in the same plane.

(b) Determine the unit outward normal direction to the surface (i.e., pointing away from the origin).

(c) Determine the surface area of the element.

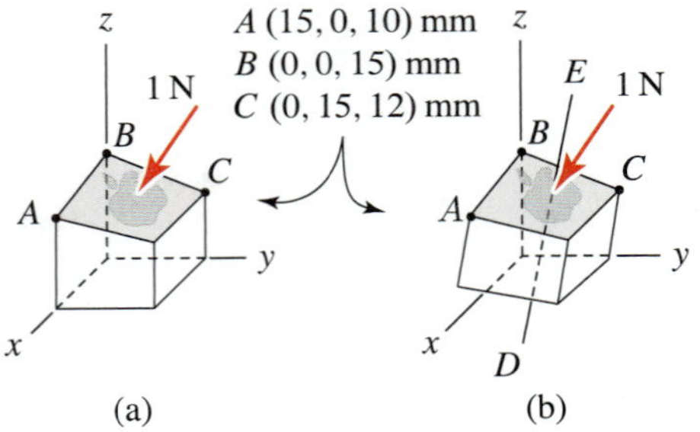

Figure P2.99

Problem 2.99

An ergonomically designed key for a computer keyboard has an approximately flat surface defined by points A, B, and C and is subjected to a 1 N force in the direction normal to the key's surface.

(a) In Fig. P2.99(a), motion of the key is in the z direction. Determine the components of the force in directions normal and tangent to the key's motion. Comment on why it might be important to know these components.

(b) In Fig. P2.99(b), the switch mechanism is repositioned so that motion of the key is in the direction of the line connecting points D and E, where this line has x, y, and z direction cosines of 0.123, 0.123, and 0.985, respectively. Determine the components of the force in directions normal and tangent to the key's motion. Comment on why this design might be more effective than that in Part (a).

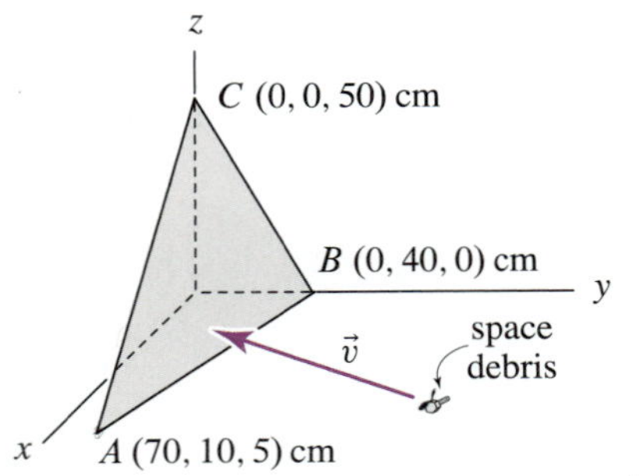

$\vec{v} = (12\,\text{km/s})(-0.545\,\hat{\imath} - 0.818\,\hat{\jmath} - 0.182\,\hat{k})$

Figure P2.100

Problem 2.100

Impact of debris, both natural and artificial, is a significant hazard for spacecraft. Space Shuttle windows are routinely replaced because of damage due to impact with small objects, and recent flights have employed evasive maneuvers to avoid impact with larger objects, whose orbits NASA constantly monitors. For the triangular shape window and the relative velocity of approach $\vec{v}$ shown, determine the components and vector components of the velocity in directions normal and tangent to the window. Note that this information is needed before an analysis of damage due to impact can be performed.

Problem 2.101

The velocity of air approaching the rudder of an aircraft has magnitude 900 ft/s in the y direction. The rudder rotates about line OA.

(a) The position vector from B to C has the x, y, and z direction cosines $\sin\alpha$, $(\cos\alpha)(\cos 20°)$, and $(-\cos\alpha)(\sin 20°)$, respectively. If the rudder is rotated so that $\alpha = 10°$, determine the components, and vector components, of the air velocity in directions normal and tangent to the surface of the rudder.

(b) Using the geometry shown in Fig. P2.101, verify the direction cosines stated in Part (a).

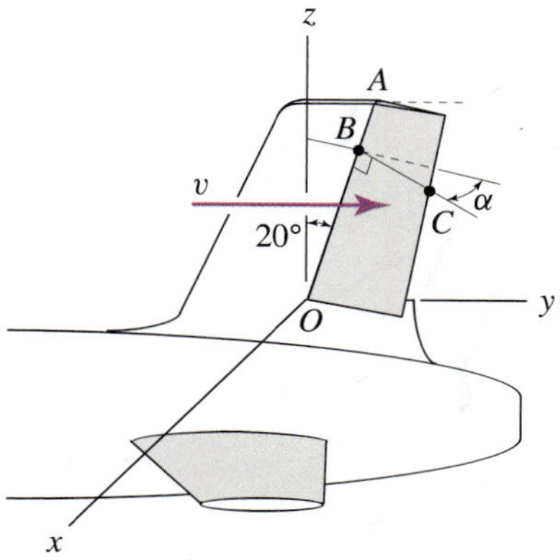

Figure P2.101

Problem 2.102

Determine the smallest distance between point O and the infinite plane containing points A, B and C.

Problem 2.103

The vector from points O to P has magnitude 40 mm and has equal direction angles with the x, y, and z axes. Determine the smallest distance from point P to the infinite plane containing points A, B, and C.

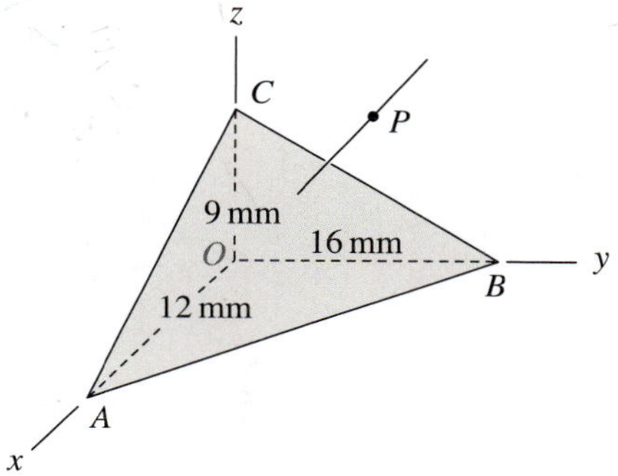

Figure P2.102 and P2.103

Problem 2.104

The product $\vec{r}_1 \times \vec{F}$ produces a vector $\vec{M}$. The product $\vec{M} \cdot \vec{r}_2/|\vec{r}_2|$ produces a scalar $M_{\|}$ which is the component of $\vec{M}$ in the direction of $\vec{r}_2$.

(a) Evaluate $M_{\|}$ by finding $\vec{M}$ first, followed by the dot product.

(b) Evaluate $M_{\|}$ using the scalar triple product.

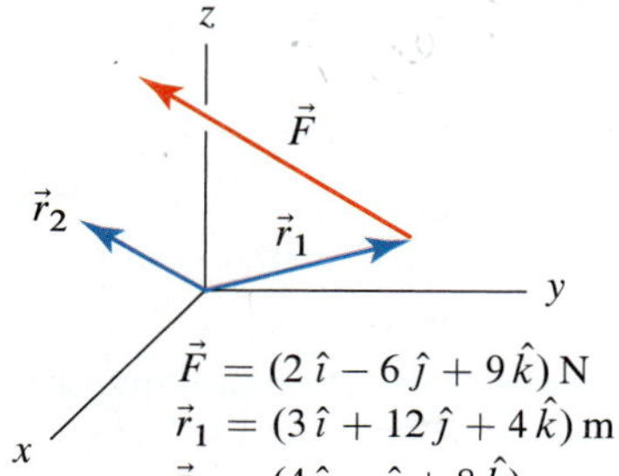

$$\vec{F} = (2\,\hat{\imath} - 6\,\hat{\jmath} + 9\,\hat{k})\,\text{N}$$
$$\vec{r}_1 = (3\,\hat{\imath} + 12\,\hat{\jmath} + 4\,\hat{k})\,\text{m}$$
$$\vec{r}_2 = (4\,\hat{\imath} - \hat{\jmath} + 8\,\hat{k})\,\text{m}$$

Figure P2.104

Problem 2.105

As described in connection with Fig. 2.34 on p. 96, the scalar triple product $(\vec{A} \times \vec{B}) \cdot \vec{C}$ provides the volume of the parallelepiped formed by $\vec{A}$, $\vec{B}$ and $\vec{C}$. Comment on how the results of the following triple products compare to the value provided by $(\vec{A} \times \vec{B}) \cdot \vec{C}$:

(a) $(\vec{A} \times \vec{C}) \cdot \vec{B}$.

(b) $(\vec{B} \times \vec{C}) \cdot \vec{A}$.

(c) $(\vec{C} \times \vec{B}) \cdot \vec{A}$.

(d) $(\vec{C} \times \vec{A}) \cdot \vec{B}$.

Note: Concept problems are about *explanations*, not computations.

2.6 Chapter Review

Important definitions, concepts, and equations of this chapter are summarized. For equations and/or concepts that are not clear, you should refer to the original equation and page numbers cited for additional details.

Laws of sines and cosines. See Figs. 2.35 and 2.36.

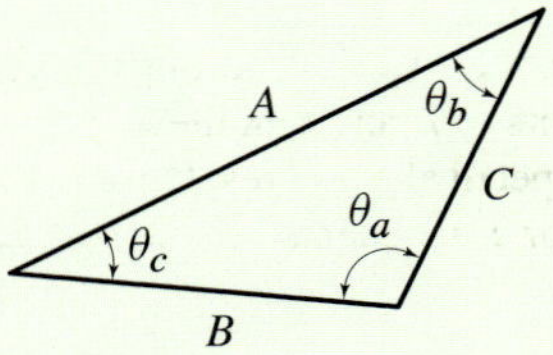

Figure 2.35
A general triangle.

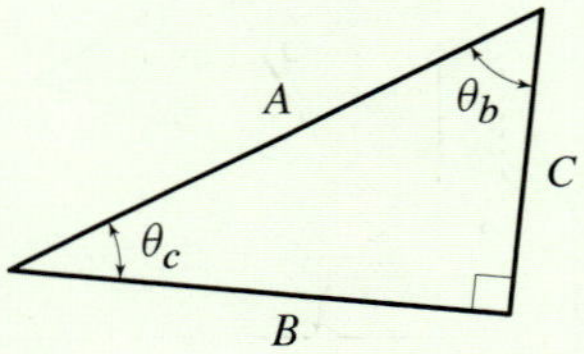

Figure 2.36
A right triangle.

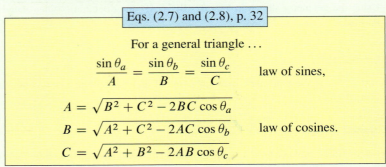

Eqs. (2.7) and (2.8), p. 32

For a general triangle . . .

$$\frac{\sin\theta_a}{A} = \frac{\sin\theta_b}{B} = \frac{\sin\theta_c}{C} \qquad \text{law of sines,}$$

$$A = \sqrt{B^2 + C^2 - 2BC\cos\theta_a}$$
$$B = \sqrt{A^2 + C^2 - 2AC\cos\theta_b} \qquad \text{law of cosines.}$$
$$C = \sqrt{A^2 + B^2 - 2AB\cos\theta_c}$$

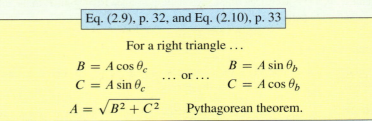

Eq. (2.9), p. 32, and Eq. (2.10), p. 33

For a right triangle . . .

$$B = A\cos\theta_c \qquad \ldots \text{ or } \ldots \qquad B = A\sin\theta_b$$
$$C = A\sin\theta_c \qquad\qquad\qquad C = A\cos\theta_b$$

$$A = \sqrt{B^2 + C^2} \qquad \text{Pythagorean theorem.}$$

Unit vectors. A *unit vector* has unit magnitude and is dimensionless. Given any vector $\vec{v}$ having nonzero magnitude, a unit vector $\hat{u}$ in the direction of $\vec{v}$ can be written as

Eq. (2.11), p. 44

$$\hat{u} = \frac{\vec{v}}{|\vec{v}|} \qquad \text{unit vector.}$$

Cartesian vector representation in two dimensions. A *Cartesian coordinate system* in two dimensions uses x and y axes that are orthogonal. A vector $\vec{v}$ can be written in terms of its Cartesian components as shown in Fig. 2.37 as

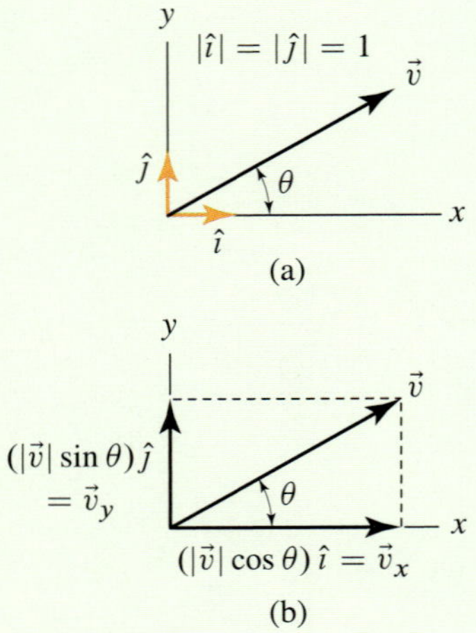

Figure 2.37
(a) Cartesian coordinate system with unit vectors $\hat{\imath}$ and $\hat{\jmath}$ in the x and y directions, respectively. (b) Resolution of a vector $\vec{v}$ into vector components in x and y directions.

Eqs. (2.12) and (2.13), p. 45

$$\vec{v} = \vec{v}_x + \vec{v}_y$$
$$= v_x\,\hat{\imath} + v_y\,\hat{\jmath}.$$

$\vec{v}_x$ and $\vec{v}_y$ are called the *vector components* of $\vec{v}$, and v_x and v_y are called the *scalar components* (or simply the *components*) of $\vec{v}$. The *magnitude* and orientation from the $\pm x$ direction are

Eq. (2.14), p. 45

$$|\vec{v}| = \sqrt{v_x^2 + v_y^2}, \quad \theta = \tan^{-1}\left(\frac{v_y}{v_x}\right) \qquad \text{magnitude and orientation.}$$

Cartesian vector representation in three dimensions and direction angles. Expressions for vectors in three dimensions are reported here. Vectors in two dimensions are a special case of the following equations with $v_z = 0$ and $\theta_z = 90°$.

Eq. (2.24), p. 60

$$|\vec{v}| = \sqrt{v_x^2 + v_y^2 + v_z^2} \qquad \text{magnitude.}$$

Eq. (2.25), p. 61

$$\begin{aligned}\vec{v} &= \vec{v}_x + \vec{v}_y + \vec{v}_z \\ &= v_x\,\hat{\imath} + v_y\,\hat{\jmath} + v_z\,\hat{k} \\ &= |\vec{v}|\cos\theta_x\,\hat{\imath} + |\vec{v}|\cos\theta_y\,\hat{\jmath} + |\vec{v}|\cos\theta_z\,\hat{k} \\ &= |\vec{v}|\,(\cos\theta_x\,\hat{\imath} + \cos\theta_y\,\hat{\jmath} + \cos\theta_z\,\hat{k}).\end{aligned}$$

Eq. (2.27), p. 61

Summary Box

θ_x = angle between positive x direction and vector,
θ_y = angle between positive y direction and vector,
θ_z = angle between positive z direction and vector,

$$\vec{v} = |\vec{v}|\,(\cos\theta_x\hat{\imath} + \cos\theta_y\hat{\jmath} + \cos\theta_z\hat{k}),$$

$$\cos^2\theta_x + \cos^2\theta_y + \cos^2\theta_z = 1,$$

$$\cos\theta_x = v_x/|\vec{v}|, \quad \cos\theta_y = v_y/|\vec{v}|, \quad \cos\theta_z = v_z/|\vec{v}|.$$

If two direction angles (or direction cosines) are known, the third may be determined using $\cos^2\theta_x + \cos^2\theta_y + \cos^2\theta_z = 1$.

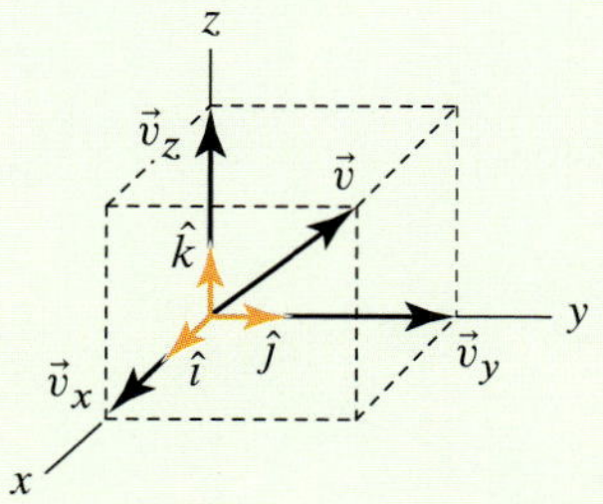

Figure 2.38
Right-hand Cartesian coordinate system with unit vectors $\hat{\imath}$, $\hat{\jmath}$, and $\hat{k}$ in the x, y, and z directions, respectively, and resolution of a vector $\vec{v}$ into vector components $\vec{v}_x$, $\vec{v}_y$, and $\vec{v}_z$.

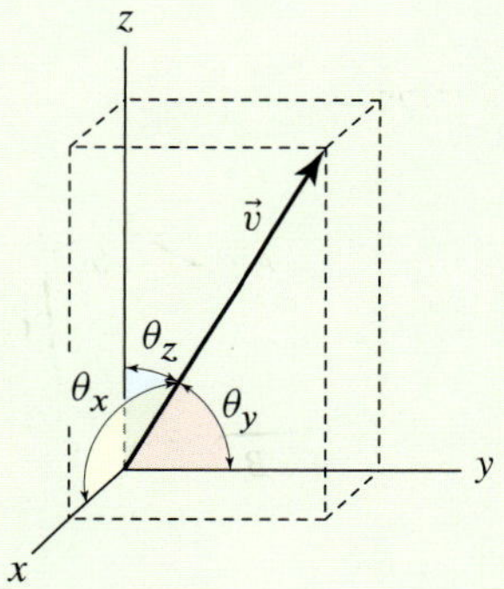

Figure 2.39
Definition of direction angles θ_x, θ_y, and θ_z.

Position vectors. The *position vector* from point T (tail) to point H (head) is shown in Fig. 2.40 and is written as

Eq. (2.28), p. 62

$$\vec{r}_{TH} = (x_H - x_T)\,\hat{\imath} + (y_H - y_T)\,\hat{\jmath} + (z_H - z_T)\,\hat{k}$$

where x_T, y_T, and z_T are coordinates of the tail and x_H, y_H, and z_H are coordinates of the head. For applications in two dimensions, Eq. (2.28) also applies with $z_H - z_T = 0$ [this expression is given explicitly by Eq. (2.22)].

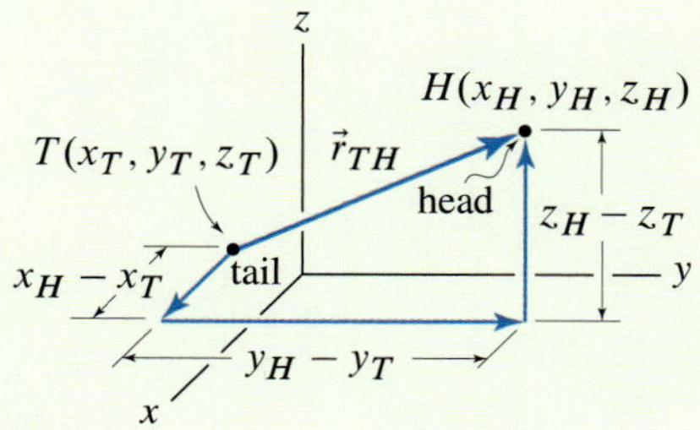

Figure 2.40
Construction of a position vector using Cartesian coordinates of the vector's head and tail.

Dot product. The *dot product* between two vectors $\vec{A}$ and $\vec{B}$ produces a scalar s and is defined as

Eq. (2.29), p. 76, and Eq. (2.34), p. 77

$$\begin{aligned} s = \vec{A}\cdot\vec{B} &= |\vec{A}||\vec{B}|\cos\theta \\ &= A_xB_x + A_yB_y + A_zB_z.\end{aligned}$$

The dot product can be used to determine the angle θ between two vectors using

Eq. (2.39), p. 78

$$\cos\theta = \frac{\vec{A}\cdot\vec{B}}{|\vec{A}||\vec{B}|} = \frac{A_x B_x + A_y B_y + A_z B_z}{|\vec{A}||\vec{B}|},$$

$$\theta = \cos^{-1}\frac{\vec{A}\cdot\vec{B}}{|\vec{A}||\vec{B}|} = \cos^{-1}\frac{A_x B_x + A_y B_y + A_z B_z}{|\vec{A}||\vec{B}|}.$$

Resolution of a vector into components. Important equations for using the dot product to resolve a vector $\vec{F}$ into *parallel and perpendicular components* to a direction $\vec{r}$ are collected in the following summary box:

Eq. (2.47), p. 80

Summary Box (See Fig. 2.41.)

$F_\parallel$ = component of $\vec{F}$ acting in direction $\vec{r}$,

$\vec{F}_\parallel$ = vector component of $\vec{F}$ acting in direction $\vec{r}$,

$F_\perp$ = component of $\vec{F}$ acting perpendicular to $\vec{r}$,

$\vec{F}_\perp$ = vector component of $\vec{F}$ acting perpendicular to $\vec{r}$,

$$F_\parallel = \vec{F}\cdot\frac{\vec{r}}{|\vec{r}|}, \qquad \vec{F}_\parallel = F_\parallel\frac{\vec{r}}{|\vec{r}|},$$

$$F_\perp = \sqrt{F^2 - F_\parallel^2}, \qquad \vec{F}_\perp = \vec{F} - \vec{F}_\parallel.$$

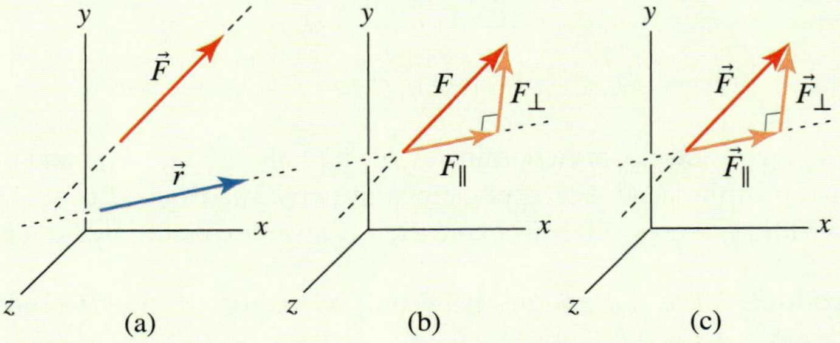

Figure 2.41. (a) Vectors $\vec{F}$ and $\vec{r}$ in three dimensions. (b) Resolution of $\vec{F}$ into *scalar components* in directions parallel and perpendicular to $\vec{r}$. (c) Resolution of $\vec{F}$ into *vector components* in directions parallel and perpendicular to $\vec{r}$.

Cross product. The *cross product* between two vectors $\vec{A}$ and $\vec{B}$ produces a vector $\vec{C}$ as shown in Fig. 2.42 and is defined as

Eq. (2.48), p. 91, and Eq. (2.53), p. 92

$$\begin{aligned}\vec{C} &= \vec{A} \times \vec{B} \\ &= (|\vec{A}||\vec{B}|\sin\theta)\,\hat{u} \\ &= (A_y B_z - A_z B_y)\,\hat{\imath} + (A_z B_x - A_x B_z)\,\hat{\jmath} + (A_x B_y - A_y B_x)\,\hat{k}\end{aligned}$$

For vectors with Cartesian representations, the cross product may conveniently be evaluated by expanding the determinant

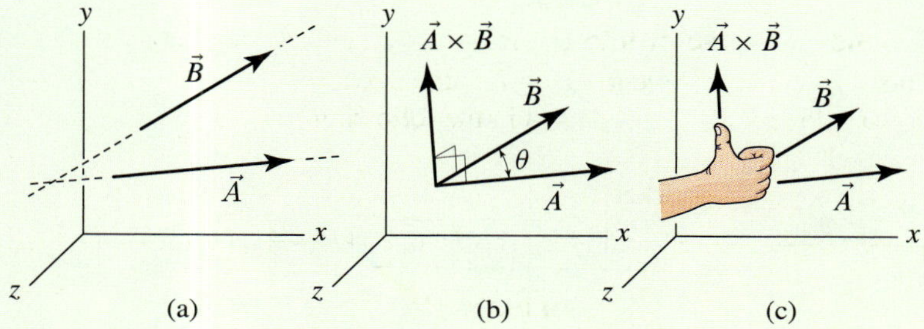

Figure 2.42. (a) Vectors $\vec{A}$ and $\vec{B}$ in three dimensions. (b) To evaluate the cross product between vectors $\vec{A}$ and $\vec{B}$, the vectors can be arranged tail to tail to define angle θ, which is measured in the plane containing the two vectors. (c) The result of $\vec{A} \times \vec{B}$ is a vector whose direction is governed by the right-hand rule.

Eq. (2.58), p. 93

$$\vec{A} \times \vec{B} = \det\begin{bmatrix} \hat{\imath} & \hat{\jmath} & \hat{k} \\ A_x & A_y & A_z \\ B_x & B_y & B_z \end{bmatrix} = \begin{vmatrix} \hat{\imath} & \hat{\jmath} & \hat{k} \\ A_x & A_y & A_z \\ B_x & B_y & B_z \end{vmatrix}$$

using Method 1 or 2 described in Section 2.5. Also, the magnitude of $\vec{A} \times \vec{B}$ is the area of the parallelogram formed by $\vec{A}$ and $\vec{B}$ arranged tail to tail (see Fig. 2.33).

Scalar triple product. A cross product that is followed by a dot product can be simultaneously evaluated using the *scalar triple product*

Eq. (2.61), p. 95

$$(\vec{A} \times \vec{B}) \cdot \vec{C} = \begin{vmatrix} C_x & C_y & C_z \\ A_x & A_y & A_z \\ B_x & B_y & B_z \end{vmatrix}.$$

The scalar triple product produces a scalar. Also, the value produced by $(\vec{A} \times \vec{B}) \cdot \vec{C}$ is the volume of the parallelepiped defined by $\vec{A}$, $\vec{B}$, and $\vec{C}$ arranged tail to tail (see Fig. 2.34).

REVIEW PROBLEMS

Problem 2.106

The manufacturer of a welded steel bracket specifies the working loads depicted in the figure of R_n versus R_t. Values of R_n and R_t that lie within the shaded region are allowable, while values that lie outside of the region are unsafe. Such a diagram is often called an *interaction diagram* because it characterizes the combined effect that multiple loads have on the strength of a component. For the loading and geometry shown, determine the range of values load P may have and still satisfy the manufacturer's combined loading criterion.

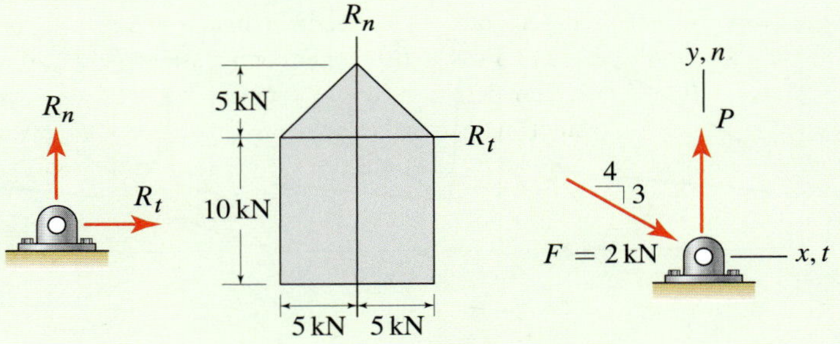

Figure P2.106

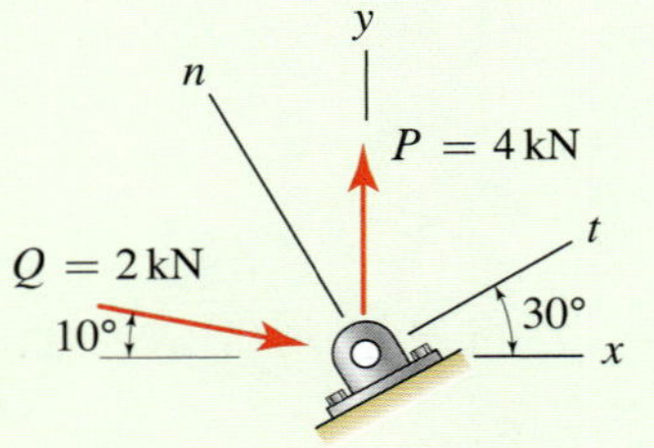

Figure P2.107

Problem 2.107

For the loading and geometry shown, use the interaction diagram of Prob. 2.106 to determine if the manufacturer's combined loading criterion is satisfied.

Problems 2.108 and 2.109

(a) Determine the resultant $\vec{R}$ of the three forces $\vec{F} + \vec{P} + \vec{Q}$.

(b) If an additional force $\vec{T}$ in the $\pm x$ direction is to be added, determine the magnitude it should have so that the magnitude of the resultant is as small as possible.

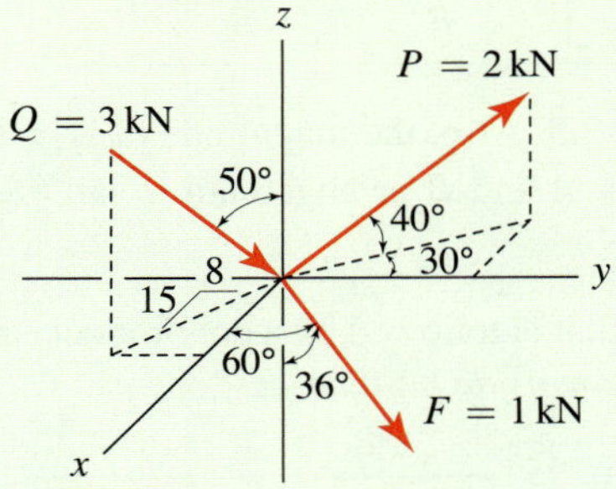

Figure P2.108

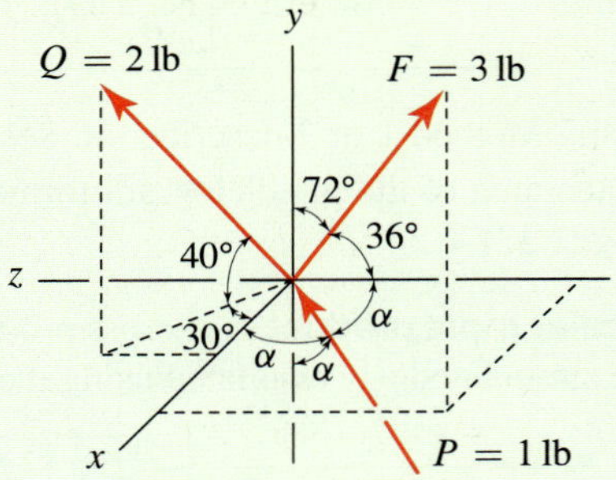

Figure P2.109

Problems 2.110 and 2.111

An architect specifies the roof geometry shown for a building. Each of lines EAB, BD, DCG, EG, and AC are straight. Two forces of magnitude P and Q acting in the $-y$ direction are applied at the positions shown.

Problem 2.110 Determine the components of the force P in directions normal and parallel to the roof at point A. Express your answer in terms of P.

Problem 2.111 Determine the components of the force Q in directions normal and parallel to the roof at point C. Express your answer in terms of Q.

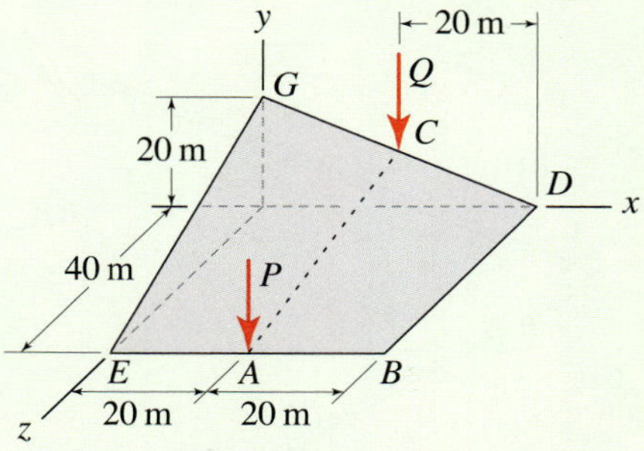

Figure P2.110 and P2.111

Problem 2.112

A specimen of composite material consisting of ceramic matrix and unidirectional ceramic fiber reinforcing is tested in a laboratory under compressive loading. If a 10 kN force is applied in the $-z$ direction, determine the components, and vector components, of this force in directions parallel and perpendicular to the fiber direction f, where this direction has direction angle $\theta_z = 40°$ and remaining direction angles that are equal (i.e., $\theta_x = \theta_y$).

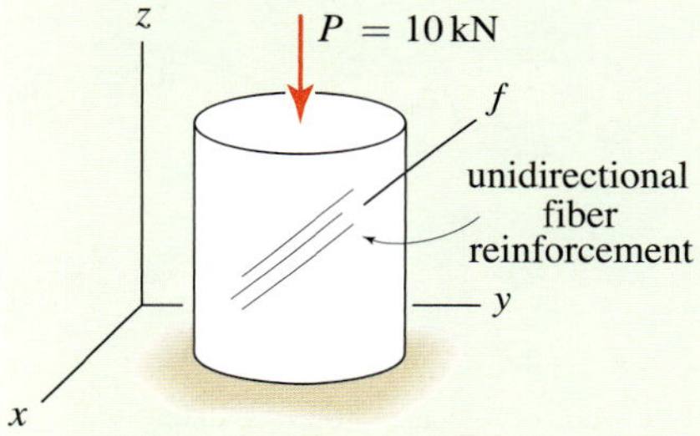

Figure P2.112

Problem 2.113

An automobile body panel is subjected to a force $\vec{F}$ from a stiffening strut. Assuming that region ABC of the panel is planar, determine the components, and vector components, of $\vec{F}$ in the directions normal and parallel to the panel.

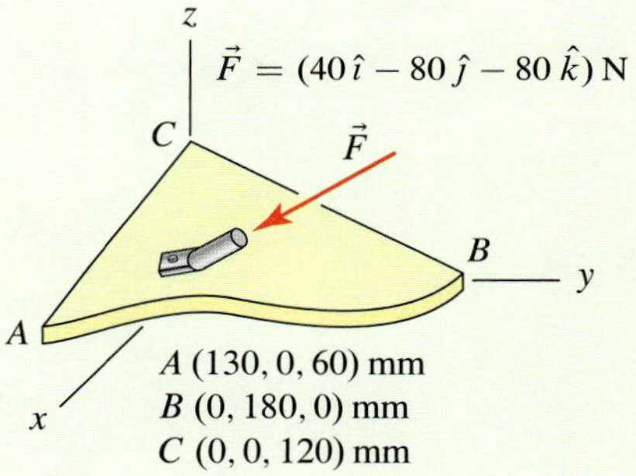

Figure P2.113

Problem 2.114

An I beam is positioned from points A to B. Because its strength and deformation properties for bending about an axis through the web of the cross section are different than those for bending about an axis parallel to the flanges, it is necessary to also characterize these directions. This can be accomplished by specifying just one of the direction angles for the direction of the web from A to C, which is perpendicular to line AB, plus the octant of the coordinate system in which line AC lies.

(a) If direction angle $\theta_z = 30°$ for line AC, determine the remaining direction angles for this line.

(b) Determine the unit vector in the direction perpendicular to the web of the beam (i.e., perpendicular to lines AB and AC).

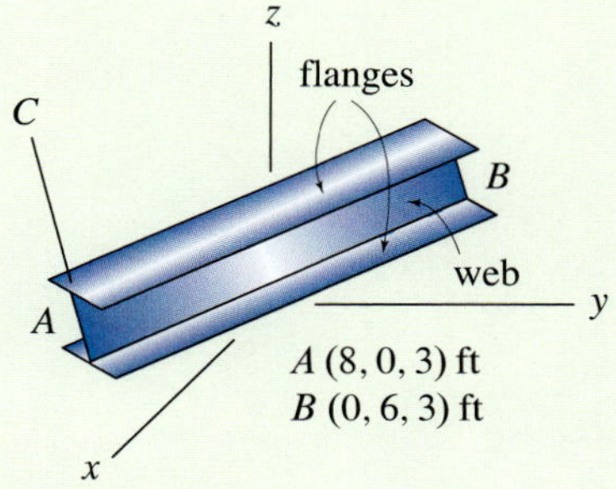

Figure P2.114

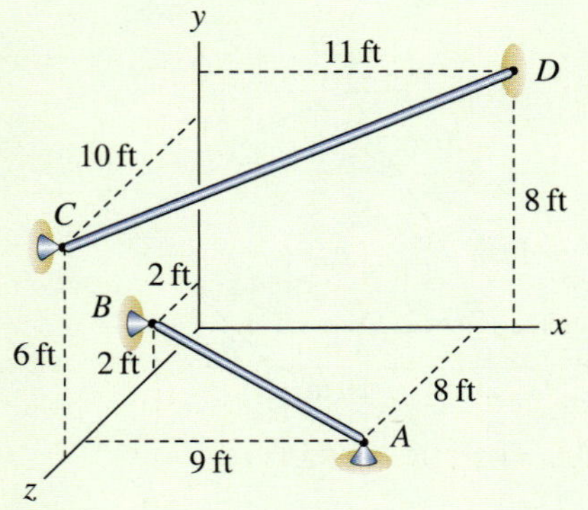

Figure P2.115

Problem 2.115

Determine the smallest distance between the infinite lines passing through bars AB and CD.

Problem 2.116

The tetrahedron shown arises in advanced mechanics, and it is necessary to relate the areas of the four surfaces. Show that the surface areas are related by $A_x = A\cos\theta_x$, $A_y = A\cos\theta_y$, and $A_z = A\cos\theta_z$ where A is the area of surface ABC, and $\cos\theta_x$, $\cos\theta_y$, and $\cos\theta_z$ are the direction cosines for normal direction $\vec{n}$. *Hint:* Find $\vec{n}$ by taking the cross product of vectors along edges AB, AC, and/or BC, and note the magnitude of this vector is $2A$. Then by inspection write expressions for A_x, A_y, and A_z (e.g., $A_x = yz/2$ and so on).

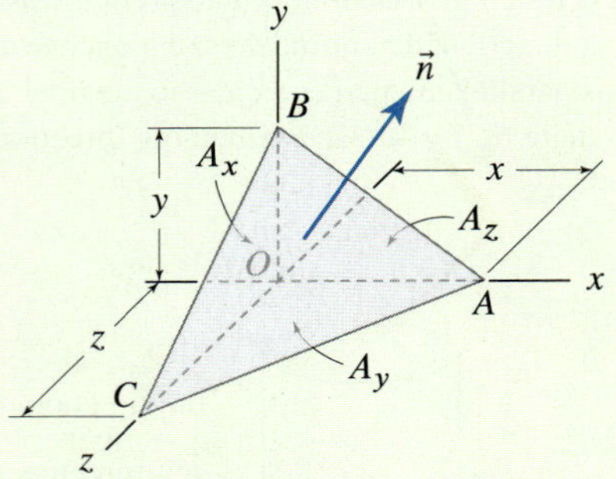

Figure P2.116

3 *Equilibrium of Particles*

Many engineering problems can be accurately idealized as a single particle or system of particles in equilibrium. This chapter discusses criteria for when a particle, or system of particles, is in equilibrium and presents a systematic analysis procedure that can be applied to general problems. Early sections of this chapter focus on two-dimensional problems, and later sections treat three-dimensional problems. Engineering design applications are also discussed.

3.1 Equilibrium of Particles in Two Dimensions

In mechanics, a particle is defined to have zero volume, but potentially may have mass. While there are no true particles in nature, under the proper circumstances it is possible to idealize a real life object as a particle. An object that is small compared to other objects and/or dimensions in a problem can often be idealized as a particle. A large object may often be idealized as a particle, and whether or not this is possible depends on the forces that are applied to the object. In the example shown in Fig. 3.1, the lines of action of all forces intersect at a common point, and this is called a *concurrent force system*. An object subjected to a concurrent force system may be idealized as a particle as shown in the figure. There are other circumstances in which an object, even if large, may be idealized as a particle. In this chapter you will begin to develop the ability to recognize when an object may be idealized as a particle, and as you progress through subsequent chapters of this book, this ability will be further sharpened.

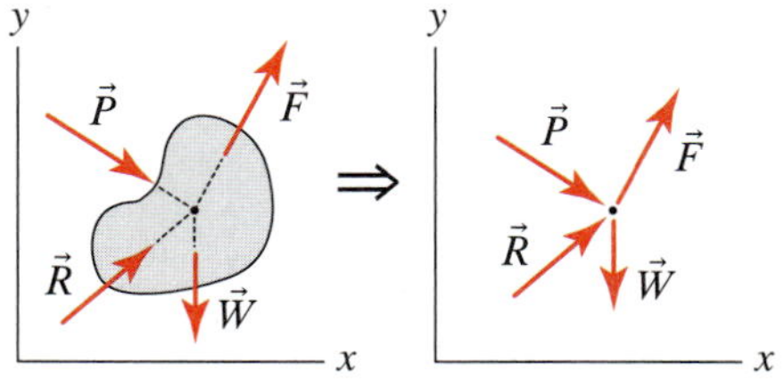

Figure 3.1
Example of a two-dimensional *concurrent force system* where the lines of action of all forces intersect at a common point. When subjected to a concurrent force system, an object – even if it is very large – can be idealized as a particle for many purposes.

Newton's laws, discussed in Chapter 1, provide the conditions under which a particle subjected to forces is in static equilibrium. In particular, Newton's second law states $\sum \vec{F} = m\vec{a}$, where we have included the summation sign to emphasize that *all* forces applied to the particle must be included. *Static equilibrium* means $\vec{a} = \vec{0}$. For brevity, throughout the rest of this book we will use the word *equilibrium* to mean *static equilibrium*. Hence the conditions for equilibrium of a particle are

$$\begin{aligned} &\sum \vec{F} = \vec{0}, \\ \text{or} \quad &\left(\sum F_x\right)\hat{\imath} + \left(\sum F_y\right)\hat{\jmath} = \vec{0}, \\ \text{or} \quad &\sum F_x = 0, \ \text{and} \ \sum F_y = 0. \end{aligned} \tag{3.1}$$

Helpful Information

More on static equilibrium. A particle is in *static equilibrium* if its acceleration is zero ($\vec{a} = \vec{0}$). Thus, a particle is in static equilibrium if it has no motion (i.e., is at rest) or if it moves with constant velocity (i.e., has uniform speed and uniform direction). A particle that has changing velocity (i.e., has nonuniform speed and/or nonuniform direction) is not in static equilibrium, and dynamics must be considered to determine its response.

In Eq. (3.1), the expression $\sum \vec{F} = \vec{0}$ is valid regardless of the type of vector representation used, while the remaining expressions result if Cartesian vector representation is used. The expressions $\sum \vec{F} = (\sum F_x)\hat{\imath} + (\sum F_y)\hat{\jmath} = \vec{0}$ states conditions for equilibrium in *vector form*, while the expressions $\sum F_x = 0$ and $\sum F_y = 0$ state conditions for equilibrium in *scalar form*. Both the vector and scalar forms are completely equivalent, and the choice of which of these to use for a particular problem is a matter of convenience. The vector form provides a compact and concise description of equilibrium. In complex problems, especially in three dimensions, and especially when rigid bodies are involved (these are addressed later in this book), it will often be advantageous to use this form. For many problems, especially in two dimensions where the geometry of forces is straightforward, the scalar form will be effective. When using the scalar form, we compute components of forces in x and y directions and sum forces in each of these directions.

Particles and forces. In particle equilibrium problems, the particle under consideration may represent an individual particle of a real body or structure, or the particle may represent a portion of the real body or structure, or the particle may represent the entire body or structure (as in the example of Fig. 3.1). When applying $\sum \vec{F} = \vec{0}$ to the particle, all forces that are applied to the particle must be included. These forces have a number of sources, as follows.

- Some of the forces may be due to interaction of the particle with its environment, such as weight due to gravity, force of wind blowing against a structure, forces from magnetic attraction of nearby objects, and so on.
- Some of the forces may be due to structural members that are attached to (or contain) the particle. For example, if a particular particle has a cable attached to it, the cable will usually apply a force to the particle.
- Some of the forces may be due to supports. For example, if a particle (or the body the particle represents) is glued to a surface, the glue will usually apply forces to the particle. We call forces such as these *reaction forces*, and more is said about these later in this section.

Free body diagram (FBD)

A *free body diagram* (FBD) is a sketch of a body or a portion of a body that is separated or made free from its environment and/or other parts of the structural system, and *all* forces that act on the body must be shown in the sketch. We often use the word *cut* to describe the path along which the free body is removed from its environment. In this chapter, the FBD will always result in a particle that is in equilibrium. In subsequent chapters, the FBD may result in an object of finite size in equilibrium.

An FBD is an essential aid for applying Newton's law of motion. It is a tool that helps ensure that all forces that are applied to the FBD are accounted for, including their proper directions. Once an FBD is drawn, application of $\sum \vec{F} = m\vec{a}$ may proceed. In statics we seek conditions so that $\vec{a} = \vec{0}$, whereas in subjects that follow statics, such as dynamics and vibrations, we seek to determine $\vec{a}$ and how the motion of a body or structure evolves with time. Among all of the concepts you will learn in mechanics, regardless of how advanced a level at which you eventually study, *the ability to draw accurate FBDs is one of the most important skills you will need.* For students and practicing engineers alike, FBDs are used on a daily basis.

Procedure for Drawing FBDs

1. Decide on the particle whose equilibrium is to be analyzed.
2. Imagine this particle is "cut" completely free (separated) from the body, structure, and/or its environment. That is:
   - In two dimensions, think of a closed line that completely encircles the particle.
   - In three dimensions, think of a closed surface that completely surrounds the particle.
3. Sketch the particle (i.e., draw a point).
4. Sketch the forces:
   (a) Sketch the forces that are applied to the particle by the environment (e.g., weight).
   (b) Wherever the cut passes through a structural member, sketch the forces that occur at that location.
   (c) Wherever the cut passes through a support, sketch the reaction forces that occur at that location.
5. Sketch the coordinate system to be used. Add pertinent dimensions and angles to the FBD to fully define the locations and orientations of all forces.

Helpful Information

Free body diagram (FBD). An FBD is an essential aid, or tool, for applying Newton's law of motion $\sum \vec{F} = m\vec{a}$. Among the many skills you will need to be successful in statics, and in the subjects that follow, and as a practicing engineer, the ability to draw accurate FBDs is essential. An incorrect FBD is the most common source of errors in an analysis.

Because this chapter deals with equilibrium of particles, Step 3 in the above procedure is trivial. But, with very small modification the same procedure will be used for rigid body equilibrium where this step will require slightly greater artistry. The order in which the forces are sketched in Step 4 is irrelevant. For complicated FBDs, it may be difficult to include all of the dimensions and/or angles in Step 5. When this is the case, some of this information may be obtained from a different sketch.

In many problems it will be relatively clear which particle should be used for drawing an FBD. In complex problems this selection may require some thought and perhaps some trial and error. For a single particle in two dimensions, Eq. (3.1) provides two scalar equations $\sum F_x = 0$ and $\sum F_y = 0$, and if the FBD involves two unknowns, then these equilibrium equations are sufficient to yield the solution.* In more complex problems, the FBD will often involve more than two unknowns, in which case FBDs must be drawn and equilibrium equations written for additional particles so that the final system of equations has as many equations as unknowns.

Mini-Example. A skier uses a tow rope as shown in Fig. 3.2(a) to reach the top of a ski hill. If the skier weighs 150 lb, if the snow-covered slope can be considered frictionless, and if the portion of the tow rope behind the skier is slack,† determine the force required in the tow rope to pull the skier at constant velocity.

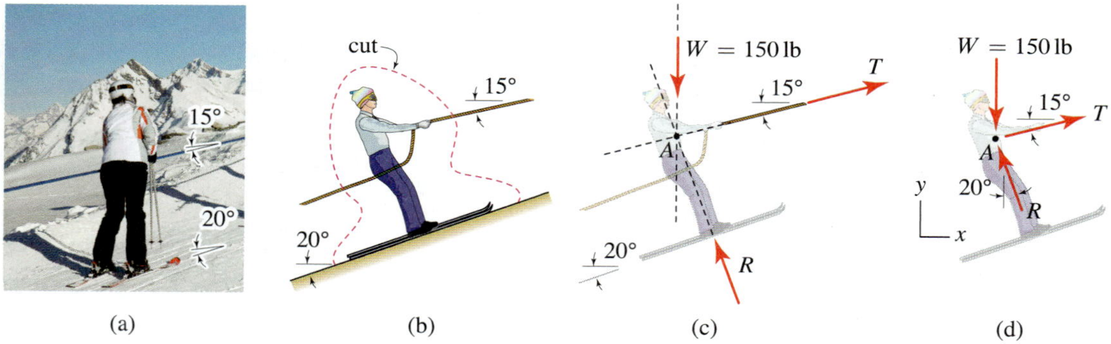

Figure 3.2. (a) Photograph of a skier being towed up a snow-covered slope at constant velocity. (b) Sketch of the skier showing the orientation of the slope and tow rope, and the cut to be used for drawing the FBD. (c) Free body diagram of the skier showing the lines of action of all forces intersecting at a common point. (d) Free body diagram where the skier is idealized as a particle. All FBDs assume the portion of the tow rope behind the skier is slack.

Solution.

Draw FBD. To draw the FBD and to determine if the skier can be idealized as a particle, we use the cut shown in Fig. 3.2(b) to separate the skier from the environment. The forces applied to the skier's body by the environment *outside* of the cut are shown in Fig. 3.2(c), as follows. The 150 lb force represents the skier's weight. Where the cut passes through the tow rope in front of the skier, the tow rope applies a force T to the skier's hands. Where the cut passes through the tow rope behind the skier, there is no force since we have assumed that portion of the rope to be slack. Where the cut passes between the skis and the slope, there is a reaction force R. Note that between the skis and the slope, the actual forces are *distributed* over the full contact area of the skis, and this distribution is probably complicated. In Fig. 3.2(c), we are modeling this distribution by a single force R.

As seen in Fig. 3.2(c), all of the forces applied to the skier by the environment *outside of the cut* intersect at a common point (point A). Thus, the skier's body may be idealized as a particle. That this is true will be more thoroughly explored when rigid bodies are discussed in Chapter 5. Furthermore, since the skier is to move with constant velocity, the skier's acceleration is zero and hence this is a problem of static equilibrium of a particle. To complete the

* There are occasional subtle exceptions to this statement, such as when a structure is simultaneously statically indeterminate and a mechanism. Such exceptions are discussed later in this book.

† In Problem 3.4, you will let the portion of the tow rope behind the skier be taut.

FBD, we select a coordinate system. Since the 15° and 20° orientations are given with respect to the horizontal direction, a coordinate system where x and y are horizontal and vertical is convenient.

Once the appropriate particle has been identified and its FBD has been drawn, we may then write the equations of equilibrium, followed by solving these to determine the unknowns (i.e., the force T in the tow rope and the reaction R between the skis and slope). In what follows, we demonstrate a vector solution, followed by a scalar solution, followed by comments on the merits of using an alternate coordinate system.

Vector solution of equilibrium equations. To carry out a vector solution, we begin by writing vector expressions for all forces as follows:

$$\vec{W} = -150\,\text{lb}\,\hat{\jmath}, \tag{3.2}$$

$$\vec{T} = T\left(\cos 15^\circ\,\hat{\imath} + \sin 15^\circ\,\hat{\jmath}\right), \tag{3.3}$$

$$\vec{R} = R\left(-\sin 20^\circ\,\hat{\imath} + \cos 20^\circ\,\hat{\jmath}\right). \tag{3.4}$$

Next we apply Newton's law with $\vec{a} = \vec{0}$ to write

$$\sum \vec{F} = \vec{0}: \quad \vec{W} + \vec{T} + \vec{R} = \vec{0}, \tag{3.5}$$

$$\left(T\cos 15^\circ - R\sin 20^\circ\right)\hat{\imath} + \left(-150\,\text{lb} + T\sin 15^\circ + R\cos 20^\circ\right)\hat{\jmath} = \vec{0}. \tag{3.6}$$

For the above vector equation to be satisfied, both the x and y components must be zero independently, which provides two scalar equations. Also, we observe that there are two unknowns, T and R. Thus the system of equations is *determinate*, meaning there are as many equations as unknowns, and a solution for the unknowns is obtainable. Writing the two scalar equations contained in Eq. (3.6) provides

$$T\cos 15^\circ - R\sin 20^\circ = 0, \tag{3.7}$$

$$T\sin 15^\circ + R\cos 20^\circ = 150\text{ lb}. \tag{3.8}$$

Basic algebra is used to solve these equations. For example, if Eq. (3.7) is multiplied by cos 20° and Eq. (3.8) is multiplied by sin 20° and the results are added, the terms containing R sum to zero and we obtain

$$T\left(\cos 15^\circ \cos 20^\circ + \sin 15^\circ \sin 20^\circ\right) = (150\text{ lb})\sin 20^\circ \tag{3.9}$$

$$\Rightarrow \quad T = 51.5\,\text{lb}. \tag{3.10}$$

Once one of the unknowns has been determined, T in this case, either of Eqs. (3.7) and (3.8) may be used to determine the other unknown. Using Eq. (3.7), we write

$$R = \frac{T\cos 15^\circ}{\sin 20^\circ} = 145\,\text{lb}, \tag{3.11}$$

which completes the solution.

Helpful Information

What forces should be included in the FBD? When drawing an FBD, only forces that are *external* to the FBD (i.e., external to the cut used to construct the FBD) are included on the FBD. A common source of error is to include forces that are *internal* to the FBD. For example, the skier's boots are attached to the skis, and there are forces between the boots and skis that keep them connected to one another. However, these forces are *internal* to the FBD, because the cut that was used to construct the FBD did not separate the boots from the skis. To take this idea a step further, every atom of material within the skier's body and clothing exerts forces on neighboring atoms. However, all such forces are *internal* to the FBD and hence do not appear on the FBD.

Helpful Information

Some useful checks of our solution. Intuitively, we know that the portion of the tow rope in front of the skier will be in tension. From Fig. 3.2(c), we observe the direction of T in the FBD has been assigned so that a positive value of T corresponds to tension in the rope. Thus, we expect the solution to this problem to display a positive value for T, and indeed this is the case. Further, the reaction R is positive in compression. Thus, we expect the solution to this problem to display a positive value for R, and indeed this is also the case. Had we obtained a negative value for either T or R, then we would suspect an error in our solution.

Scalar solution of equilibrium equations. We examine Fig. 3.2(d) and sum the x and y components of forces to write

$$\sum F_x = 0: \quad T\cos 15^\circ - R\sin 20^\circ = 0, \tag{3.12}$$

$$\sum F_y = 0: \quad T\sin 15^\circ + R\cos 20^\circ - 150\,\text{lb} = 0. \tag{3.13}$$

These equations are identical to Eqs. (3.7) and (3.8), and thus the solution is the same as that obtained using the vector approach.

Solution with alternate coordinate system. In the foregoing solutions we elected to use a horizontal-vertical Cartesian coordinate system, but this selection was arbitrary and in some problems a coordinate system with different orientation may be more convenient. For example, consider the tn coordinate system shown in Fig. 3.3. Using a scalar solution approach, we sum forces in t and n directions to obtain

$$\sum F_t = 0: \quad -(150\,\text{lb})\sin 20^\circ + T\cos 5^\circ = 0, \tag{3.14}$$

$$\sum F_n = 0: \quad -(150\,\text{lb})\cos 20^\circ - T\sin 5^\circ + R = 0. \tag{3.15}$$

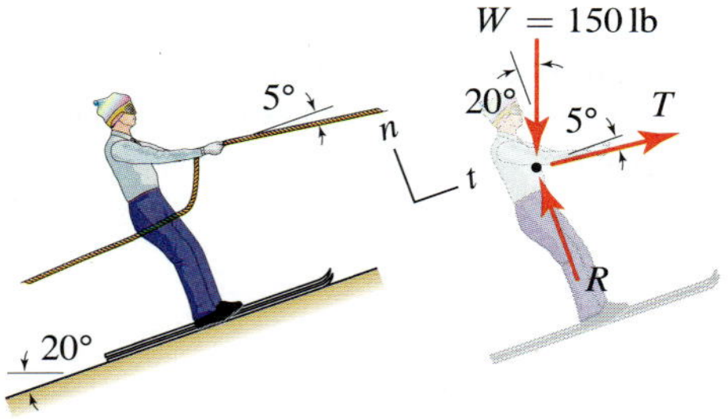

Figure 3.3
(a) Skier being towed up a slope at constant velocity. (b) Free body diagram with orientation of forces given with respect to a tn coordinate system where t and n are parallel and perpendicular to the slope, respectively. This FBD assumes the portion of the tow rope behind the skier is slack.

Notice that because of the choice of coordinate system, Eq. (3.14) has only one unknown, which is easily found to be

$$T = \frac{(150\,\text{lb})\sin 20^\circ}{\cos 5^\circ} = 51.5\,\text{lb}. \tag{3.16}$$

Then, using Eq. (3.15), the remaining unknown R is found to be

$$R = (150\,\text{lb})\cos 20^\circ + T\sin 5^\circ = 145\,\text{lb}. \tag{3.17}$$

As expected, both solutions are the same as those obtained earlier. ■

Modeling and problem solving

The process of drawing a FBD involves *modeling*, wherein a real life problem is replaced by an idealization. For example, in Fig. 3.2, the actual distribution of forces between the skis and slope was modeled by a single reaction force, friction between the skis and slope was neglected, the portion of the tow rope behind the skier was assumed to be slack, and so on. In modeling, reasonable assumptions are made about what is physically important in a system and what is not, with the goal of developing a model that contains the essential physics while hopefully being simple enough to allow for a tractable mathematical solution. Effective modeling is both an art and a science. Sometimes it may take several iterations to develop a good model for a problem. Once a model has been established, the next step of an analysis is largely an exercise in mathematics. In the example problems in this book that deal with equilibrium concepts,* the following structured problem solving approach will be used:

* In example problems that do not deal with equilibrium concepts, such as those in Chapters 1 and 2, the modeling step may not be required.

Structured Problem Solving Approach

- **Road Map:** An overall strategy for solving a problem is discussed.
- **Modeling:** Reasonable assumptions are made, leading to the development of an FBD for a problem.
- **Governing Equations:** Equations of equilibrium, plus perhaps equations for other physical phenomena pertinent to a problem, are written.
- **Computation:** The equations that govern the model for the problem are solved.
- **Discussion and Verification:** The results for the problem are critically examined and interpreted.

Occasionally, some of the above steps may be done in combination.

Helpful Information

More on Governing Equations. Some of the examples in this book will involve failure criteria, deformable members, friction, or other criteria related to the physical behavior of a structural system. When this is the case, the governing equations category in our structured problem solving approach will be subdivided to include, as appropriate,

- Equilibrium equations.
- Force laws.
- Kinematic equations.

The specific meanings of force laws and kinematic equations are discussed later in this and subsequent chapters.

Cables and bars

In the foregoing example of Fig. 3.2, the FBD was constructed so that a positive value of T corresponded to tension in the tow rope that pulls the skier up the slope. It may not be obvious that this statement is true, and it is important that you fully understand why, because construction of FBDs and interpretation of the results of many problems require this understanding. Consider the structure shown in Fig. 3.4(a), consisting of a bar and two cables that support a bucket weighing 100 N. As discussed in Section 2.3, the forces supported

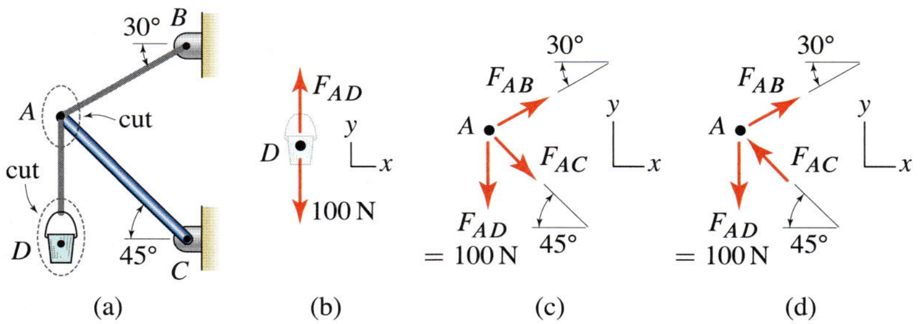

Figure 3.4. (a) A structure consisting of a bar and two cables supports a bucket weighing 100 N. Two cuts are taken to draw two FBDs. (b) FBD of the bucket. (c) FBD of point A where F_{AC} is defined to be positive in tension. (d) FBD of point A where F_{AC} is defined to be positive in compression.

by the bar and cables are collinear with their respective axes. To model the structure in Fig. 3.4(a), you might begin by considering the bucket, to arrive at the intuitively obvious conclusion that if the bucket weighs 100 N, then the force in cable AD must be 100 N. This conclusion is reached more formally by examining Fig. 3.4(a) to identify point D as the location of a concurrent force system. A cut is then taken to separate point D from its environment, and this cut passes through cable AD, leading to the FBD shown in Fig. 3.4(b). With this FBD, writing $\sum F_y = 0$ provides $F_{AD} - 100\,\text{N} = 0$, whose solution is $F_{AD} = 100\,\text{N}$. Thus, one of the cable forces has been determined.

Figure 3.6
Examples of pulleys in use.

To determine the forces supported by the bar and the remaining cable, we again examine Fig. 3.4(a) to identify point A as the location of another concurrent force system. A cut is then taken to separate point A from its environment, and this cut passes through one bar and two cables. Two possible FBDs resulting from this cut are shown in Fig. 3.4(c) and (d), where we have taken advantage of the result $F_{AD} = 100\,\text{N}$. In both Fig. 3.4(c) and Fig. 3.4(d), the direction for the cable force F_{AB} is such that a positive value corresponds to a tensile force in the cable. For the force in the bar, the FBD in Fig. 3.4(c) defines the direction of F_{AC} such that a positive value corresponds to a tensile force in the bar, while the FBD in Fig. 3.4(d) defines the direction such that a positive value corresponds to a compressive force in the bar.

If it is not clear that these statements are true, then drawing additional FBDs will fully clarify the situation. In fact if you are ever in doubt about whether a positive value for a particular cable or bar force means tension or compression, the construction of an additional FBD will always provide clarification. To illustrate, in Fig. 3.5(b) the FBD for point A is shown, along with FBDs for portions of the cable and bar. Note that in drawing these FBDs, we have invoked Newton's third law, which states that forces of action and reaction are equal in magnitude, opposite in direction, and collinear. Examining the FBD of cable AB in Fig. 3.5(b) unquestionably confirms that a positive value of F_{AB} corresponds to tension in the cable. Examining the FBD of bar AC in Fig. 3.5(b) also confirms that a positive value of F_{AC} corresponds to tension in the bar. Examining the FBD of bar AC in Fig. 3.5(c) shows that in this case a positive value of F_{AC} corresponds to compression in the bar.

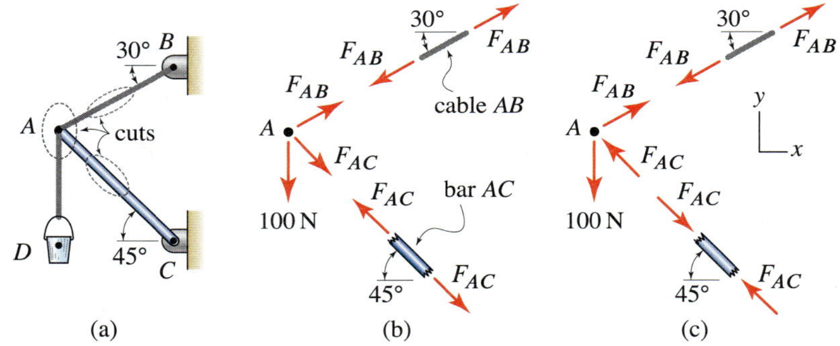

Figure 3.5. (a) The structure of Fig. 3.4 shown again with more extensive FBDs to clarify different sign conventions for the force supported by a bar. (b) A positive value of F_{AC} corresponds to tension in bar AC. (c) A positive value of F_{AC} corresponds to compression in bar AC.

Since cables buckle when subjected to very low compressive force, it is common to assume they can support only tensile forces. Thus we will always assign cable forces in FBDs so that positive values correspond to tension. If, after the analysis of a particular problem, you find that a cable supports a compressive load, then either you have made an error or the structure has a serious flaw; namely, equilibrium of the structure is relying on a cable to support a compressive force whereas in reality the cable cannot do this.

In contrast to cables, bars can support both tensile and compressive forces, and thus we always confront the question of what direction should be used to represent the forces they support when drawing FBDs. The answer is that it really does not matter what convention you follow, provided your FBDs are consistent within a particular problem, and that you understand the proper interpretation for the forces you compute. In a problem as simple as that of Fig. 3.4(a), where there are a small number of structural members and the loading is simple, it is probably obvious that the cables will be in tension and the bar in compression. Thus, many people would draw the FBD shown in Fig. 3.4(d) where the bar force is taken positive in compression. In more complicated problems, however, it is often not possible to identify beforehand which bars will be in tension and which will be in compression. Thus, a good practice when drawing FBDs is to assign the forces supported by all bars such that positive values correspond to tension. Mixed sign conventions where some bars have forces that are positive in tension and the others are positive in compression are manageable for simple problems, but are cumbersome for larger problems since you must always consult the original FBDs to discern which are in tension and compression because the sign of the member's force alone is not sufficient to make this determination.

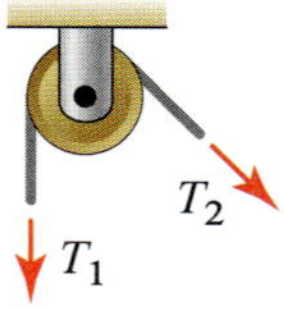

If the pulley has friction, then in general $T_1 \neq T_2$.
If the pulley is frictionless, then $T_1 = T_2$ always.

Figure 3.7
A cable with negligible weight wrapped around a pulley.

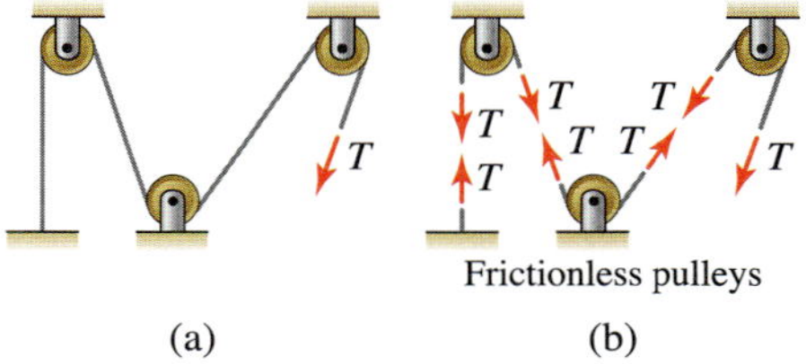

Figure 3.8
(a) A single cable wrapped around several pulleys and subjected to a force T at its end. (b) If the pulleys are frictionless and the cable has negligible weight, then the magnitude of the force throughout the entire cable is the same.

Pulleys

Cables are often used in conjunction with pulleys. A pulley, as shown in Figs. 3.6 and 3.7, is a simple device that changes the orientation of a cable and hence changes the direction of the force a cable supports. If a pulley is idealized as being frictionless (that is, the pulley can rotate on its bearing without friction) and the cable has negligible weight, then the magnitude of the force supported by the cable is unchanged as it wraps around the pulley. For the present, we must accept this statement without proof. In Chapter 5, when rigid bodies are discussed, we will show this statement is true. To carry this concept further, if a cable with negligible weight is wrapped around several frictionless pulleys, such as shown in Fig. 3.8(a), then the magnitude of the force is the same throughout the entire cable, as shown in Fig. 3.8(b). Note that the sketches of pulleys in Fig. 3.8(b) are not FBDs, because the pulleys have not been completely removed from their environment and thus there are additional forces acting on the pulleys that are not shown. If a cable has significant weight, then the force supported by the cable will vary throughout its length, even if frictionless pulleys are used. Throughout most of this book, cables are assumed to have negligible weight and pulleys are assumed to be frictionless.

To idealize a pulley as a particle, such as when drawing an FBD, you can look for the point of intersection of all forces applied to the pulley, or better yet, you can simply "shift" the cable forces to the bearing of the pulley, as follows. Figure 3.9(a) shows a pulley that is supported by a shackle. To see how the pulley and shackle may be idealized as a particle, first we remove the pulley from the shackle to begin drawing the FBDs in Fig. 3.9(b) and (c). In Fig. 3.9(b) we introduce two forces, equal in magnitude to the cable force T but in opposite directions, on the bearing A of the pulley. Then on the FBD of the shackle in Fig. 3.9(c) we use Newton's third law to include these same two forces on point A, but in opposite directions. The shackle shown in Fig. 3.9(c) is then easily modeled as a particle since it is a concurrent force system where all forces intersect at the bearing A of the pulley.

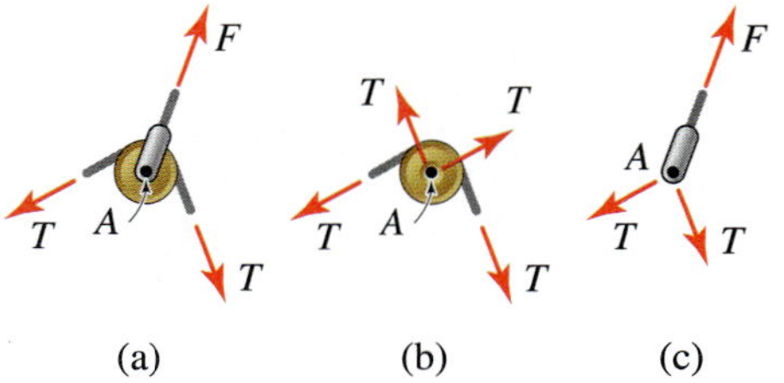

Figure 3.9
Demonstration of how a pulley may be idealized as a particle. (a) A frictionless pulley is supported by a shackle. (b) Free body diagram of the pulley removed from shackle. (c) Free body diagram of the shackle removed from the pulley, showing how the pulley's cable forces have been "shifted" to point A.

Reactions

As previously stated, a *reaction force*, or simply *reaction*, is a force exerted by a support on a body or a structure. We will study support reactions extensively later in this book, but some remarks in addition to those already made are needed now. We begin with an example that is easy to relate to. Consider yourself as you stand on a horizontal floor, as shown in Fig. 3.10(a). It is because the floor exerts a force on your body that you remain in equilibrium. Thus, the floor exerts on your body a vertical force of magnitude R, and your body in turn exerts on the floor a force of magnitude R in the opposite direction. The principal feature of the floor is that it prevents vertical motion of your body, and it will generate whatever force is needed to accomplish this. Thus, if you also wear a backpack while standing on the floor, the floor will produce an even larger reaction on your body so that you have no vertical motion.

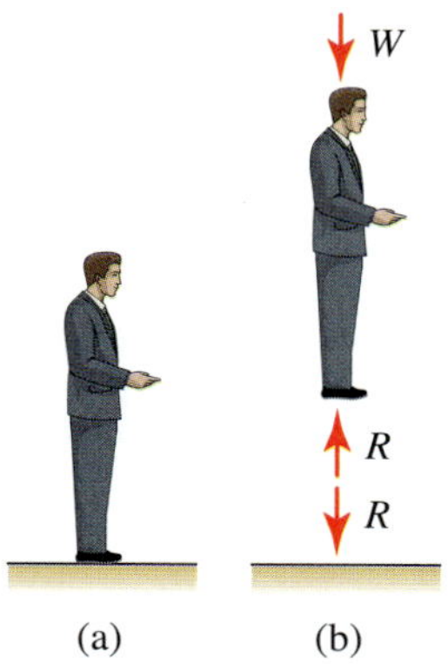

Figure 3.10
(a) Person standing on a horizontal surface. (b) Free body diagram of the person, showing the reaction the floor exerts on the person's body.

In the example shown in Fig. 3.11, imagine you are now standing on a slope, such as a sidewalk on a gentle hill. Assuming the sidewalk is sufficiently rough so that your feet do not slip, there are now two components of reaction forces in the directions shown. The reaction R arises because the sidewalk does not let your body move in the direction normal to the surface, and the tangential component F arises because the sidewalk does not let your feet slide. The sidewalk provides constraint of motion in two directions, and hence there are two reactions in those directions. If the sidewalk is covered with ice so that it is frictionless, then the sidewalk no longer prevents slip, $F = 0$, and an unpleasant experience results! When ice-covered, the sidewalk constrains motion in just one direction, and hence there is only one reaction, R.

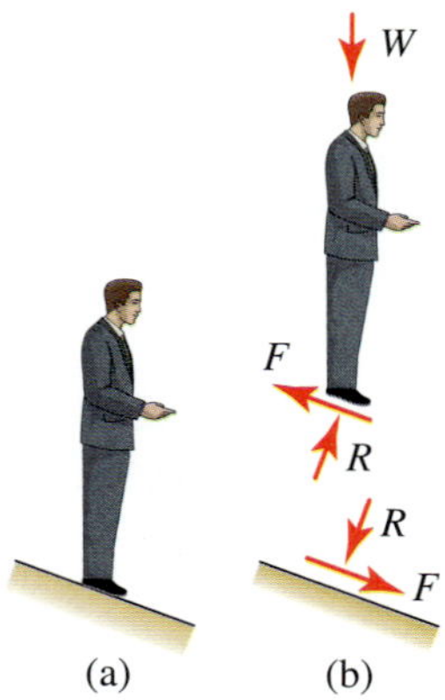

Figure 3.11
(a) Person standing on a rough slope. (b) Free body diagram of the person, showing the reactions the surface exerts on the person's body.

The foregoing examples illustrate a thought process that will always allow us to identify the number and direction of reaction forces associated with a particular support. Namely, if a support prevents motion in a certain direction, it can do so only by producing a reaction force in that direction. When we are solving particle equilibrium problems, the supports and associated reactions shown in Fig. 3.12 occur often. It is not necessary to memorize these reactions; rather, you should reconstruct these as needed. For example, consider the particle pinned to a surface. The particle may be subjected to external forces or may be a connection point between several cables and/or bars, which give rise to the forces F_1, F_2, and F_3 shown. The surface to which the particle is fixed prevents motion of the particle in the y direction, so there must be a reaction in this direction; and the surface also prevents motion of the particle in the x direction, so there must also be a reaction in this direction. For the slider on a frictionless bar shown in Fig. 3.12, the fixed bar prevents motion of the slider in the direction normal to the bar, so there must be a reaction in this direction. The slider is free to move along the frictionless bar, hence there is no reaction in that direction.

End of Section Summary

In this section, the equations governing static equilibrium of a particle were discussed, and analysis procedures were described. Some of the key points are as follows:

- For a particle in two dimensions, the equations of equilibrium written in vector form are $\sum \vec{F} = \vec{0}$ and in scalar form are $\sum F_x = 0$ and

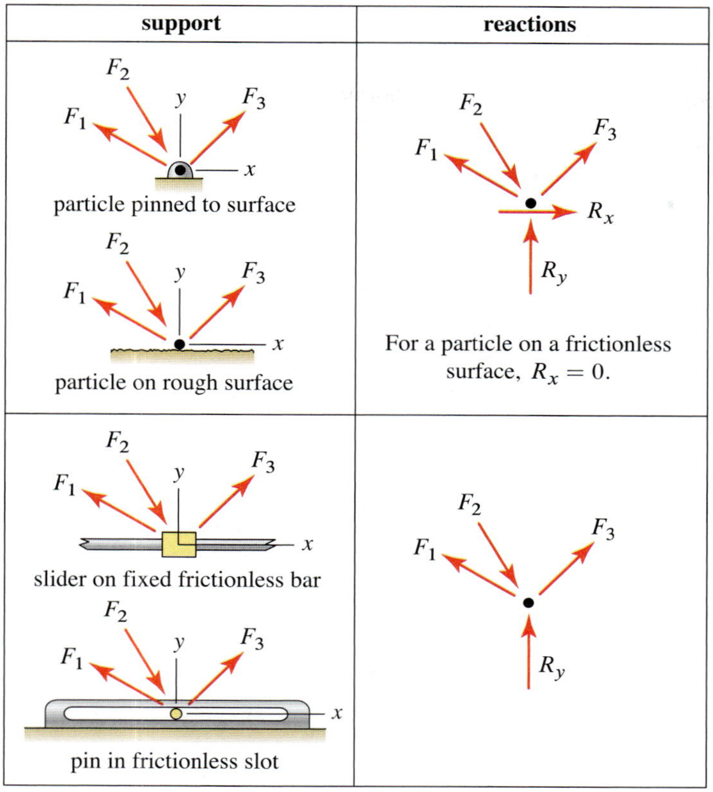

Figure 3.12. Some common supports and reactions in two dimensional particle equilibrium problems. Forces F_1, F_2, and F_3 are hypothetical forces applied to a particle by cables and/or bars that might be attached to the particle, and forces R_x and R_y are reactions.

$\sum F_y = 0$. In these summations, all forces applied to the particle must be included.

- A *free body diagram* (FBD) is a sketch of a particle and all forces applied to the particle. The FBD is an essential tool to help ensure that all forces are accounted for when you are writing the equilibrium equations.

- Complex problems may require more than one FBD. In two dimensions, each FBD allows two equilibrium equations to be written (i.e., $\sum \vec{F} = (\sum F_x)\,\hat{\imath} + (\sum F_y)\,\hat{\jmath} = \vec{0}$, or $\sum F_x = 0$ and $\sum F_y = 0$), thus allowing for an increased number of unknowns to be determined.

- Cables and straight bars are structural members supporting forces having the same orientation and line of action as the member's geometry.

- A pulley is used to change the direction of a cable and hence change the direction of the force supported by a cable. If a single cable with negligible weight is wrapped around any number of frictionless pulleys, then the magnitude of the force throughout the cable is the same everywhere.

EXAMPLE 3.1 *Cables, Bars, and Failure Criteria*

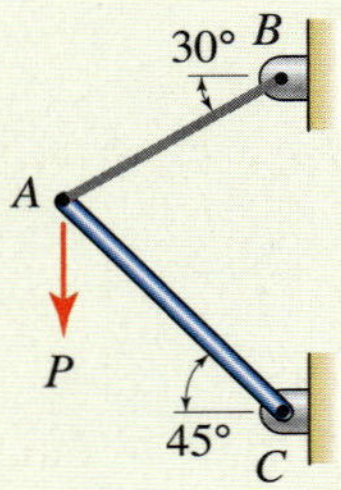

Figure 1

Consider the structure discussed earlier in this section (Figs. 3.4 and 3.5) shown again here in Fig. 1 where the loading now consists of a vertical force P applied to point A.

(a) If $P = 100\,\text{N}$, determine the forces supported by cable AB and bar AC.

(b) If cable AB can support a tensile force of 1200 N before breaking, and bar AC can support a compressive force of 1600 N before buckling, determine the largest force P that can be applied.

SOLUTION

Part (a)

Road Map An FBD of point A will involve three forces: the force P applied to the structure and the forces supported by the cable and bar. Because P is known to be 100 N, the FBD will contain two unknowns, and since there are two equilibrium equations available, we expect to be able to determine the forces supported by the cable and bar.

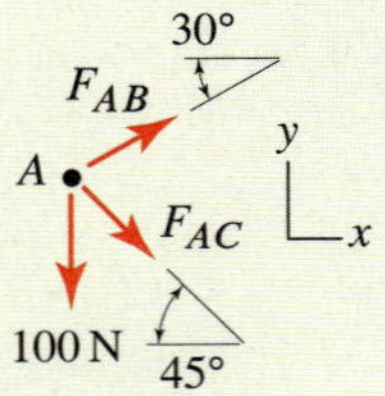

Figure 2
Free body diagram of point A where $P = 100\,\text{N}$.

Modeling We draw the FBD shown in Fig. 2, where positive values of F_{AB} and F_{AC} correspond to tensile forces in the cable and bar, respectively. We also select the xy coordinate system shown.

Governing Equations The equilibrium equations are written by summing forces in the x and y directions

$$\sum F_x = 0: \quad F_{AB}\cos 30^\circ + F_{AC}\cos 45^\circ = 0, \tag{1}$$

$$\sum F_y = 0: \quad F_{AB}\sin 30^\circ - F_{AC}\sin 45^\circ - 100\,\text{N} = 0. \tag{2}$$

Computation Equations (1) and (2) are easily solved to obtain

$$F_{AB} = 73.2\,\text{N}, \tag{3}$$

$$F_{AC} = -89.7\,\text{N}. \tag{4}$$

Part (b)

Road Map We are given the maximum loads the cable and bar can support. It is important to understand that it is unlikely that both the bar and cable will simultaneously be at their failure loads, and the margin note at the end of this example provides further discussion of this. One solution strategy is to use the results of Part (a) to determine the largest load the structure can support by exploiting linearity, as follows. In Part (a), the solutions for F_{AB} and F_{AC} were obtained using $P = 100\,\text{N}$. If P is doubled, then F_{AB} and F_{AC} are doubled, and so on. This assumes the angles shown in Fig. 2 remain the same as the load changes. Thus, we may scale the load P until $F_{AB} = 1200\,\text{N}$ or $F_{AC} = -1600\,\text{N}$.

A generally preferable way to solve problems with multiple failure criteria is to determine the forces supported by each member in terms of P, where P is yet to be determined, and this is the approach we will use here. A feature of this approach is that you do not need to guess which of several members will fail first.

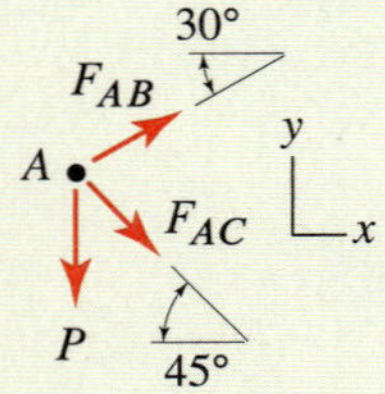

Figure 3
Free body diagram of point A where P is unknown.

Modeling The FBD shown in Fig. 3 is the same as that for Part (a) except that P is unknown.

Governing Equations & Computation

Equilibrium Equations The equilibrium equations are the same as Eqs. (1) and (2), except with the 100 N force replaced by P.

$$\sum F_x = 0: \qquad F_{AB}\cos 30^\circ + F_{AC}\cos 45^\circ = 0, \tag{5}$$

$$\sum F_y = 0: \qquad F_{AB}\sin 30^\circ - F_{AC}\sin 45^\circ - P = 0. \tag{6}$$

Solving Eqs. (5) and (6) provides

$$F_{AB} = (0.7321)\,P, \tag{7}$$

$$F_{AC} = (-0.8966)\,P. \tag{8}$$

Force Laws Now the various failure criteria can be applied using Eqs. (7) and (8):

$$\text{If } F_{AB} = 1200\,\text{N}, \quad \text{then } P = 1640\,\text{N}. \tag{9}$$

$$\text{If } F_{AC} = -1600\,\text{N}, \quad \text{then } P = 1790\,\text{N}. \tag{10}$$

Only the smaller value of P in Eqs. (9) and (10) will simultaneously satisfy the failure criteria for both the bar and cable. Thus, the maximum value P may have is

$$\boxed{P_{max} = 1640\,\text{N},} \tag{11}$$

and cable AB will fail first.

Discussion & Verification As expected from intuition, solutions to both parts of this example show F_{AB} is positive, meaning the cable is in tension, and F_{AC} is negative, meaning the bar is in compression. After this check, we should substitute the solutions for F_{AB} and F_{AC} into the original equilibrium equations; in Part (a) both of Eqs. (1) and (2) must be satisfied, and in Part (b) both of Eqs. (5) and (6) must be satisfied. This is a useful check to avoid algebra errors, but unfortunately it does not catch errors in drawing FBDs and writing equilibrium equations.

Common Pitfall

Failure loads. A common error in solving problems with failure criteria, such as Part (b) of this example, is to assume that *all* members are at their failure loads at the same time. With reference to the FBD of Fig. 3, you will be making this error if you take $F_{AB} = 1200\,\text{N}$ and $F_{AC} = -1600\,\text{N}$; in fact, if you do this, you will find that $\sum F_x = 0$ (Eq. (5)) cannot be satisfied! Another way to describe this problem is to consider slowly increasing force P from zero. Eventually, one of the members will reach its failure load first, while the other will be below its failure load.

EXAMPLE 3.2 *Cables, Pulleys, and Failure Criteria*

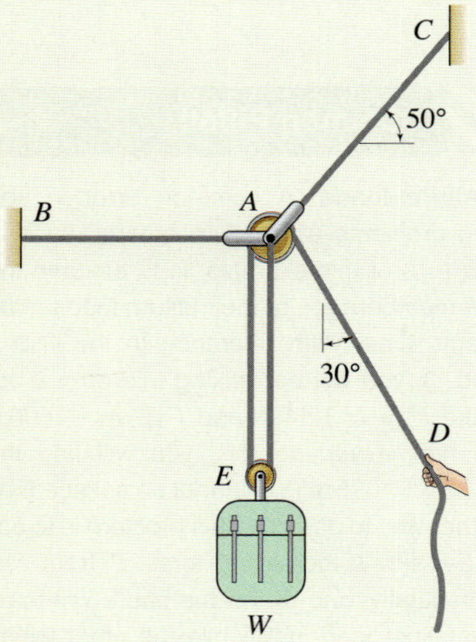

Figure 1

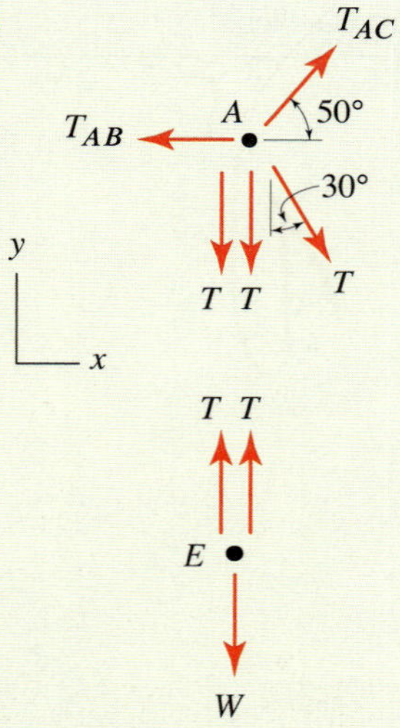

Figure 2
Free body diagrams of points A and E. The force systems for points A and E are both concurrent, although these figures show some of the forces to be slightly separated for clarity.

> **Helpful Information**
>
> **Notation for forces.** In problems where we take the force supported by a structural member to be positive in tension, such as T, T_{AB}, and T_{AC} in this example, we will often use the symbol T to emphasize that a positive value means *tension.*

The cable and pulley system shown is used by a camper to hoist a backpack into a tree to keep it out of the reach of bears. If cables AB and AC have breaking strengths of 200 lb and cable DAE has a breaking strength of 100 lb, determine the largest weight W that may be lifted.

SOLUTION

Road Map Our strategy is to find the forces in each cable in terms of weight W and then to apply the failure criteria to determine the largest value W may have.

Modeling The force the camper applies to the cable at point D will be called T, and assuming the cable's weight is negligible, the orientation of this force is the same as the 30° orientation of the cable segment. Cable DAE is a single continuous cable, and assuming the pulleys are frictionless, in addition to the assumption of negligible weight of the cable, the tensile force throughout this cable is the same with value T. The FBDs for points A and E are shown in Fig. 2,* where as discussed in Fig. 3.9 on p. 119, the cable forces applied to pulley A have been shifted to the bearing of that pulley, and similarly for pulley E.

Governing Equations & Computation

Equilibrium Equations The equilibrium equations for point A are

$$\sum F_x = 0: \qquad -T_{AB} + T_{AC}\cos 50^\circ + T\sin 30^\circ = 0, \tag{1}$$

$$\sum F_y = 0: \qquad T_{AC}\sin 50^\circ - T - T - T\cos 30^\circ = 0. \tag{2}$$

Equations (1) and (2) are a system of two equations with three unknowns (T, T_{AB}, and T_{AC}), and hence an additional equation is needed. Thus, we also need the FBD for point E in Fig. 2, and we write the equilibrium equation

$$\sum F_y = 0: \quad T + T - W = 0 \quad \Rightarrow \quad T = \frac{W}{2}. \tag{3}$$

Using the solution $T = W/2$ obtained in Eq. (3), Eqs. (1) and (2) may be solved for

$$T_{AB} = (1.452)W, \tag{4}$$

$$T_{AC} = (1.871)W. \tag{5}$$

Force Laws The various failure criteria can now be applied using Eqs. (3)–(5):

$$\text{If } T = 100\,\text{lb}, \quad \text{then } W = 200\,\text{lb}. \tag{6}$$

$$\text{If } T_{AB} = 200\,\text{lb}, \quad \text{then } W = 138\,\text{lb}. \tag{7}$$

$$\text{If } T_{AC} = 200\,\text{lb}, \quad \text{then } W = 107\,\text{lb}. \tag{8}$$

Only the smallest value of W in Eqs. (6)–(8) will simultaneously satisfy all three failure criteria. Thus the largest value W may have is

$$\boxed{W_{max} = 107\,\text{lb},} \tag{9}$$

and cable AC will be the first to fail.

Discussion & Verification As expected, Eqs. (3)–(5) show that all cable forces are tensile. After this check, we should substitute all our solutions into the equilibrium equations to verify that each of them is satisfied. A common error in problems such as this is to assume that all cables are at their failure loads simultaneously.

* The need for the FBD at E may not be apparent until the equilibrium equations for point A are written and we find there are too many unknowns.

EXAMPLE 3.3 *A Problem Requiring Multiple FBDs*

A small cable car is used to transport passengers across a river. If the cable car and its contents have a mass of 400 kg, determine the force in cables AC and ED and the force in bar AB. Point A is a pin.

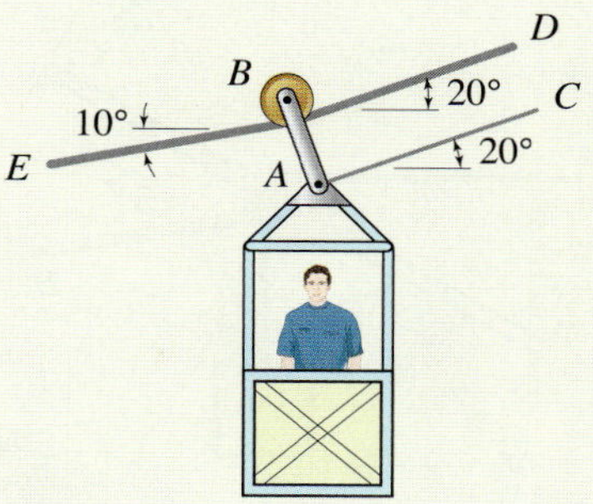

Figure 1

SOLUTION

Road Map We begin this problem by searching for an appropriate particle, or particles, whose equilibrium should be analyzed. Points A and B are likely choices, and we will see that both of these are needed to obtain enough equations for a solution.

Modeling The FBDs for points A and B are shown in Fig. 2 where for pulley B the cable forces have been shifted to the bearing of the pulley. The weight of the cable car and its contents is $(400\,\text{kg})(9.81\,\text{m/s}^2) = 3924\,\text{N} = 3.924\,\text{kN}$. We will assume pulley B is frictionless so that the tensile force throughout cable ED is the same. Note that the orientation of bar AB is unknown. An xy coordinate system is also selected.

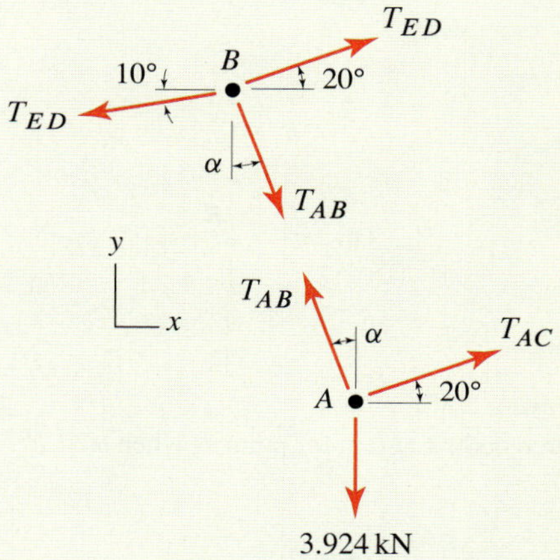

Figure 2
Free body diagrams of points A and B.

Governing Equations The equilibrium equations for point A are

$$\sum F_x = 0: \quad -T_{AB}\sin\alpha + T_{AC}\cos 20^\circ = 0, \tag{1}$$

$$\sum F_y = 0: \quad T_{AB}\cos\alpha + T_{AC}\sin 20^\circ - 3.924\,\text{kN} = 0. \tag{2}$$

While we would like to immediately solve these two equations, we observe that there are three unknowns, T_{AB}, T_{AC}, and α. Thus, additional equilibrium equations are needed, and to produce these we also consider the equilibrium of point B. Thus, from the FBD of point B shown in Fig. 2, we write

$$\sum F_x = 0: \quad T_{AB}\sin\alpha - T_{ED}\cos 10^\circ + T_{ED}\cos 20^\circ = 0, \tag{3}$$

$$\sum F_y = 0: \quad -T_{AB}\cos\alpha - T_{ED}\sin 10^\circ + T_{ED}\sin 20^\circ = 0. \tag{4}$$

Computation There are now four equations and four unknowns, T_{AB}, T_{AC}, T_{ED}, and α. Solving a system of four equations is rarely fun and is often tedious, especially when trigonometric functions of the unknowns are involved. While use of software such as Mathematica or Maple can considerably ease this burden, in many problems some careful study of the equation system will offer a simple solution strategy, and before reading further, you should examine these equations to see if you can identify such a strategy.

Solving for the term $T_{AB}\sin\alpha$ in Eq. (1) and substituting this into Eq. (3), and solving for the term $T_{AB}\cos\alpha$ from Eq. (2) and substituting this into Eq. (4), provide a system of two equations in two unknowns, T_{AC} and T_{ED}, which can then be solved to obtain

$$\boxed{T_{AC} = 1.019\,\text{kN} \quad \text{and} \quad T_{ED} = 21.23\,\text{kN}.} \tag{5}$$

Now, using Eqs. (1) and (2) to compute the ratio $T_{AB}\sin\alpha/(T_{AB}\cos\alpha)$, we obtain an expression for $\tan\alpha$ which then provides

$$\alpha = 15.00^\circ. \tag{6}$$

Finally, T_{AB} may be found from any of Eqs. (1)–(4) as

$$\boxed{T_{AB} = 3.70\,\text{kN}.} \tag{7}$$

Discussion & Verification As expected, the solution shows all cables are in tension. Furthermore, by intuition we also expect the force supported by bar AB to be tensile. After these checks, we should substitute all our solutions into the equilibrium equations to verify that each of them is satisfied.

Common Pitfall

Don't confuse mass and weight. A common error, especially when you are using SI units, is to mistake mass for weight when drawing FBDs, writing equilibrium equations, and so on. Mass must be multiplied by acceleration due to gravity to obtain weight. With reference to Fig. 2, if you incorrectly give the weight of the cable car and its contents a value of 400 , then the values you obtain for T_{AB}, T_{AC}, and T_{ED} are all about an order of magnitude too small!

EXAMPLE 3.4 *Reactions and Force Polygons*

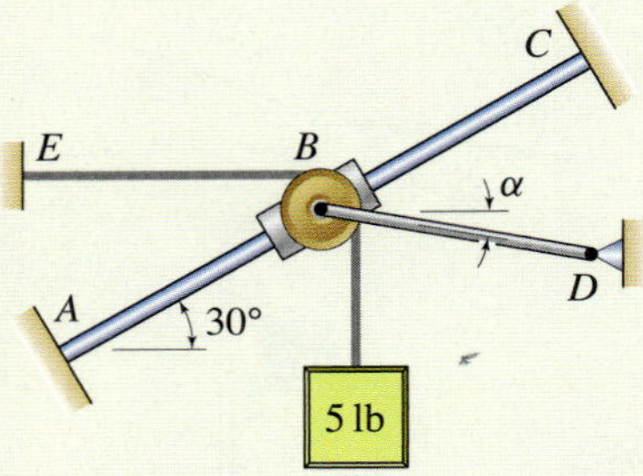

Figure 1

The structure consists of a collar at B that is free to slide along a straight fixed bar AC. Mounted on the collar is a frictionless pulley, around which a cable supporting a 5 lb weight is wrapped. The collar is further supported by a bar BD.

(a) If $\alpha = 0°$, determine the force in bar BD needed to keep the system in equilibrium.

(b) Determine the value of α that will provide for the smallest force in bar BD, and determine the value of this force.

SOLUTION

Road Map Part (a) will involve two unknowns, the force in bar BD and the reaction between the collar and the fixed bar AC. Because there are two equilibrium equations available, we expect the solution to be reasonably straightforward. Part (b) is more challenging: here there will be three unknowns, and in addition to the two equilibrium equations, a minimization criterion will be needed.

Part (a)

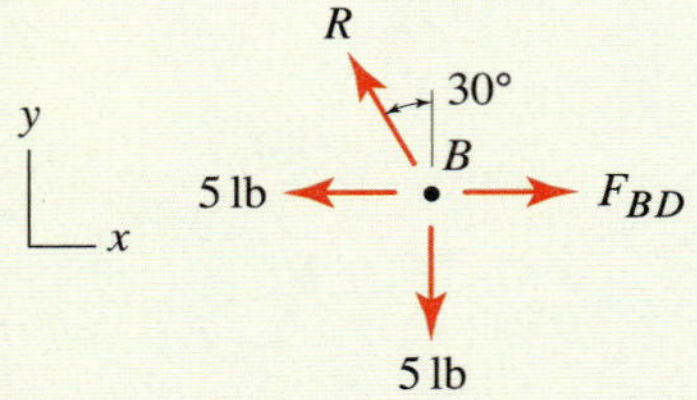

Figure 2
Free body diagram for point B when $\alpha = 0°$.

Modeling In drawing the FBD for the collar, we cut through the cable twice and bar BD once, and we separate the collar from bar AC. Assuming the pulley is frictionless and the cable is weightless, the tensile force is the same throughout the cable. Further, there is a reaction between the collar and bar AC. We could consult Fig. 3.12 to determine the direction for this reaction, but it is easy to construct this as follows. The collar may not move perpendicular to bar AC, thus there must be a reaction in this direction. The collar is free to slide along bar AC, so there is no reaction in this direction. Hence the FBD is as shown in Fig. 2.

Governing Equations The equilibrium equations for point B are

$$\sum F_x = 0: \qquad -R\sin 30° + F_{BD} - 5\text{ lb} = 0, \tag{1}$$

$$\sum F_y = 0: \qquad R\cos 30° - 5\text{ lb} = 0. \tag{2}$$

Computation Solving Eqs. (1) and (2) provides

$$\boxed{R = 5.77\text{ lb} \quad \text{and} \quad F_{BD} = 7.89\text{ lb}.} \tag{3}$$

Part (b)

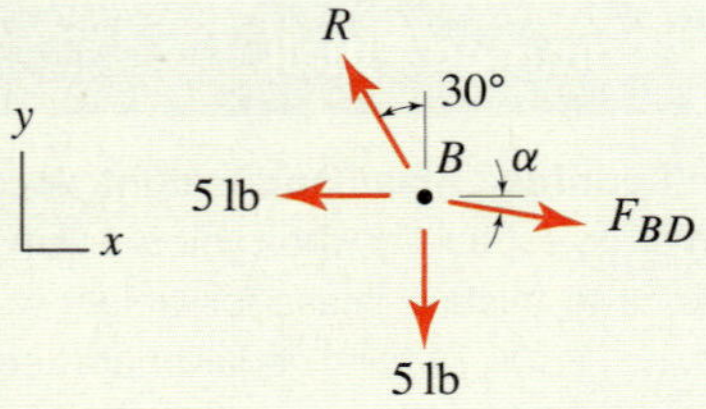

Figure 3
Free body diagram for point B when $\alpha \neq 0°$.

Modeling With the orientation of member BD unknown, the FBD for point B is shown in Fig. 3. The easiest and most insightful way to determine the orientation α for which the force in member BD is smallest is to use the force polygon concepts of Section 2.1.* Newton's law $\sum \vec{F} = \vec{0}$ can be evaluated by constructing a *closed* force polygon. Thus, in Fig. 4 we add all forces applied to point B in head-to-tail fashion to form a closed polygon. In principle, the order in which forces are added does not matter; but visualization and computation are more straightforward if forces having known magnitude and direction are added first, followed by forces having unknown magnitude and/or unknown direction. Thus, in Fig. 4(a) we assemble the two 5 lb cable forces first. Of the two remaining forces, the reaction R has known direction but unknown magnitude, whereas F_{BD} has unknown direction and magnitude. Thus, it is best to assemble R next and F_{BD} last. Note that if $\alpha = 0°$, then the polygon corresponds to the forces found in Part (a).

* Problem 3.32 guides you to use calculus to solve this problem another way.

Governing Equations & Computation We now study the force polygon in Fig. 4(a) to see what orientation α will give the smallest value of F_{BD}. Clearly that orientation is when F_{BD} is perpendicular to the direction of R, and the force polygon for this case is shown in Fig. 4(b). Hence, $\alpha = -30°$, where the negative sign* is used because the angle found here is opposite to the direction defined in Fig. 1.

To determine R and F_{BD}, we could use the geometry of the force polygon shown in Fig. 4(b), but it is easier to use the FBD in Fig. 3 to write the equilibrium equations

$$\sum F_x = 0: \qquad -R\sin 30° + F_{BD}\cos\alpha - 5\,\text{lb} = 0, \tag{4}$$

$$\sum F_y = 0: \qquad R\cos 30° - F_{BD}\sin\alpha - 5\,\text{lb} = 0. \tag{5}$$

Substituting $\alpha = -30°$ into the above expressions, we may solve them to obtain

$$\boxed{R = 1.83\,\text{lb} \quad \text{and} \quad F_{BD} = 6.83\,\text{lb}.} \tag{6}$$

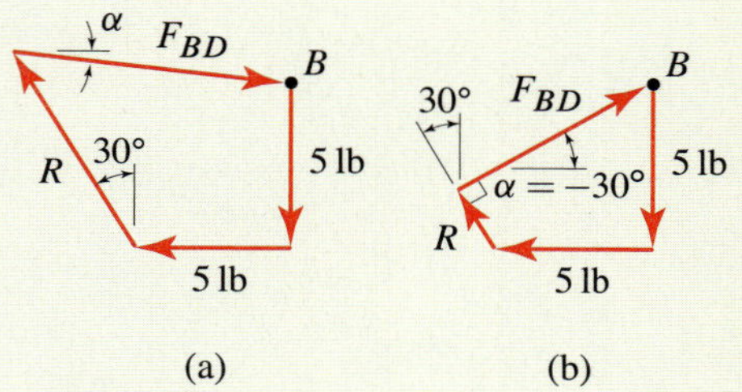

Figure 4
Force polygons for equilibrium of point B.

Alternate solution. Using a tn coordinate system where t and n are tangent and normal, respectively, to bar AC provides for an alternate solution that, for this problem, is effective. Using these tn directions, the FBD for point B is shown in Fig. 5, where for convenience an angle β is defined such that $\beta = 30° + \alpha$. The equilibrium equations are

$$\sum F_t = 0: \qquad -(5\,\text{lb})\cos 30° - (5\,\text{lb})\sin 30° + F_{BD}\cos\beta = 0, \tag{7}$$

$$\sum F_n = 0: \qquad R + (5\,\text{lb})\sin 30° - (5\,\text{lb})\cos 30° - F_{BD}\sin\beta = 0. \tag{8}$$

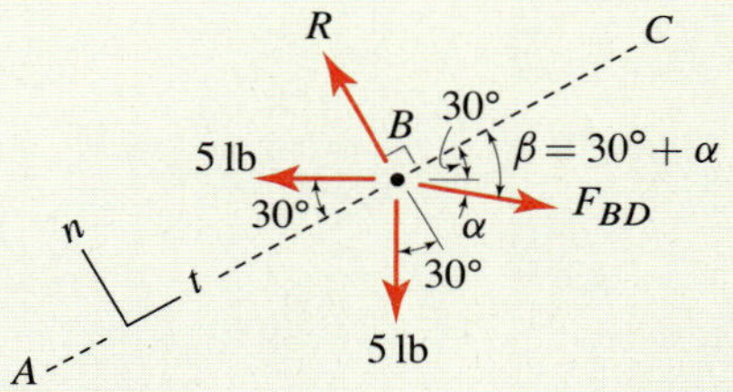

Figure 5
Free body diagram for point B using a tn coordinate system.

The merit of using this tn coordinate system is that Eq. (7) may be immediately solved to obtain

$$F_{BD} = \frac{1}{\cos\beta}(5\,\text{lb})(\cos 30° + \sin 30°). \tag{9}$$

Inspection of Eq. (9) shows that the smallest value of F_{BD} occurs when $\cos\beta = 1$. Hence, $\beta = 0$ and since $\beta = 30° + \alpha$, we determine that $\alpha = -30°$, which is the same conclusion obtained by using the force polygons in Fig. 4. To complete this solution, we use $\beta = 0$ in Eq. (9) to obtain $F_{BD} = 6.83$ lb, and then we solve Eq. (8) to obtain $R = 1.83$ lb, both of which agree with the solutions found in Eq. (6).

Discussion & Verification As discussed in the road map, Part (a) was reasonably straightforward while Part (b) was more difficult because we were asked to minimize a particular force. In Part (b), three equations were needed, and if you re-solve this problem using the calculus approach described in Prob. 3.32, you will see that the additional equation is the optimization condition $dF_{BD}/d\alpha = 0$. The feature of using the force polygon approach is that the solution for α can be found by inspection, which then leaves the two equilibrium equations that were easily solved.

* Rather than using $\alpha = -30°$, we could use $\alpha = 330°$ instead.

PROBLEMS

General Instructions. In all problems, draw FBDs and label all unknowns. It is recommended that you state forces in cables and bars using positive values for tension. In problems where forces due to gravity are present, unless otherwise stated, these forces are in the downward vertical direction in the illustration provided. **Tip:** To practice extra problems quickly, draw FBDs and write equilibrium equations, but do not solve for unknowns.

Problem 3.1

Consider an airplane whose motion is described below. For each case, state whether or not the airplane is in static equilibrium, with a brief explanation.

(a) The airplane flies in a straight line at a constant speed and at a constant altitude.

(b) The airplane flies in a straight line at a constant speed while climbing in altitude.

(c) The airplane flies at a constant speed and at a constant altitude while making a circular turn.

(d) After touching down on the runway during landing, the airplane rolls in a straight line at a constant speed.

(e) After touching down on the runway during landing, the airplane rolls in a straight line while its brakes are applied to reduce its speed.

Note: Concept problems are about *explanations*, not computations.

Problems 3.2 and 3.3

In a machining setup, workpiece B, which weighs 20 lb, is supported by a fixed V block E and a clamp at A. All contact surfaces are frictionless, and the clamp applies a vertical force of 35 lb to the workpiece. Determine the reactions at points C and D between the V block and the workpiece.

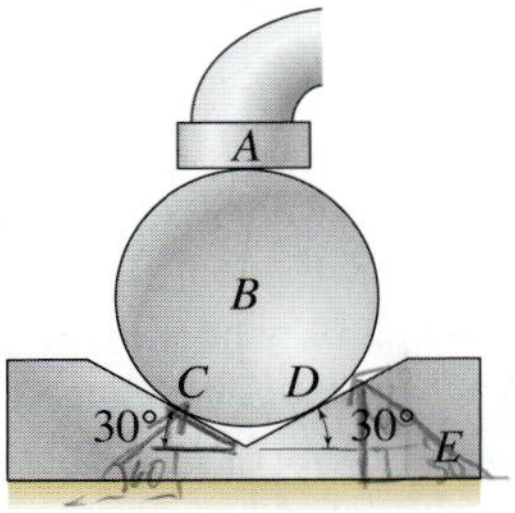

Figure P3.2

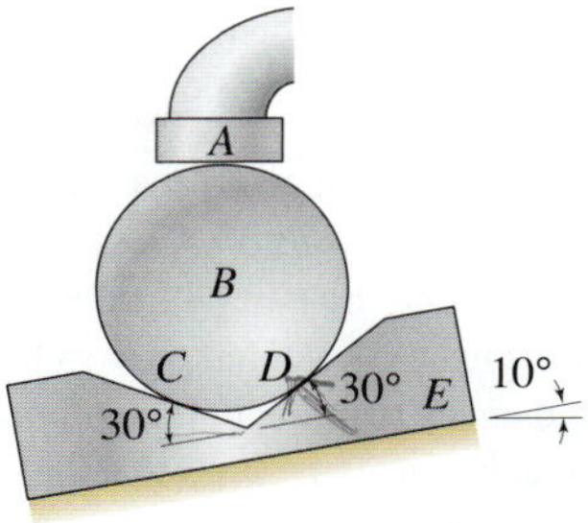

Figure P3.3

Problem 3.4

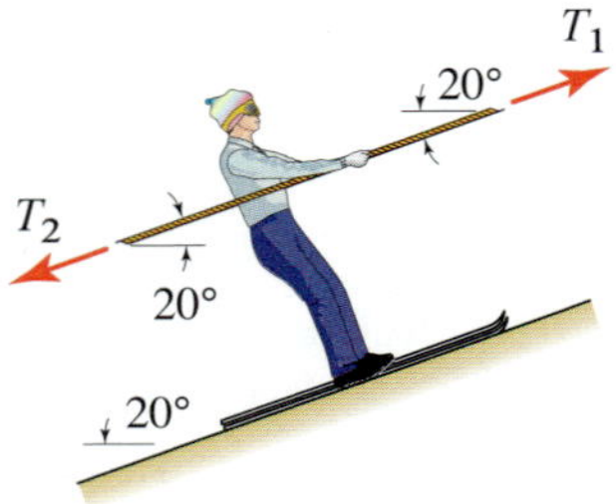

Figure P3.4

A skier uses a taut tow rope to reach the top of a ski hill. The skier weighs 150 lb, the snow-covered slope is frictionless, and the tow rope is parallel to the slope.

(a) If $T_2 = 200$ lb, determine the value of T_1 to move the skier up the slope at constant velocity and the reaction between the skier and the slope.

(b) If $T_1 = 200$ lb, determine the value of T_2 to move the skier up the slope at constant velocity and the reaction between the skier and the slope.

Problem 3.5

The dimension h is to be determined so that a worker can comfortably slide boxes weighing up to 100 lb up and down a frictionless incline. If the worker can apply a 50 lb horizontal force to the box, what is the largest value h should have?

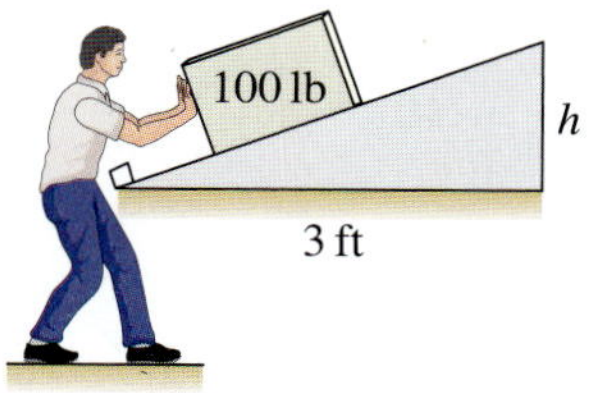

Figure P3.5

Problem 3.6

Blocks A and B each have 5 kg mass, and all contact surfaces are frictionless.

(a) Determine the force F needed to keep the blocks in static equilibrium and the forces on all contact surfaces.

(b) If the value of F determined in Part (a) is applied, will the blocks move? Explain.

(c) If F is smaller than the value determined in Part (a), describe what happens.

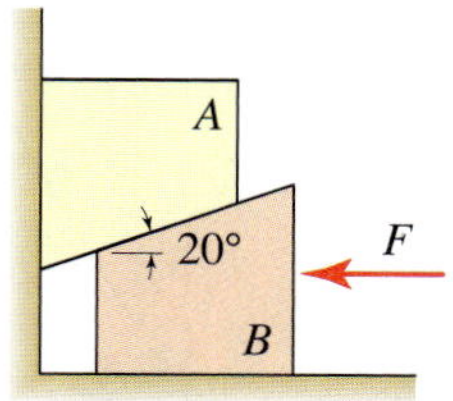

Figure P3.6

Problem 3.7

Bead A has 1 kg mass and slides without friction on bar BC.

(a) Determine the force F needed to keep the bead in static equilibrium and the reaction force between the bead and bar.

(b) If the value of F determined in Part (a) is applied, will the bead move? Explain.

(c) If F is larger than the value determined in Part (a), describe what happens.

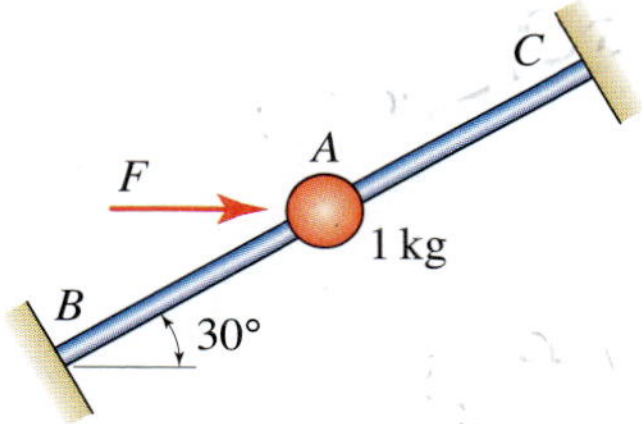

Figure P3.7

Problem 3.8

A crane is used to lift a concrete pipe weighing 5 kN into place. If $d = 0.25$ m, determine the tension in cables AB and AC.

Problem 3.9

A crane is used to lift a concrete pipe weighing 5 kN into place. For precision positioning, the worker at C can apply up to a 400 N force to cable AC. Determine the largest distance d the concrete pipe may be moved.

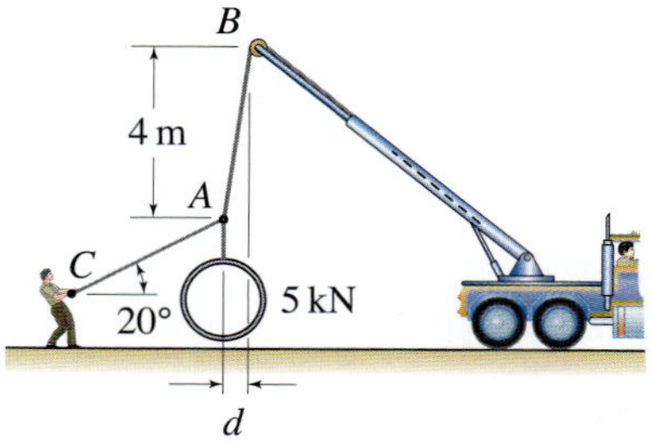

Figure P3.8 and P3.9

Problem 3.10

Guy wire AB is used to help support the utility pole AC. Assuming the force supported by the utility pole is directed along the line AC,* determine the force in wire AB and pole AC if $P = 800$ N.

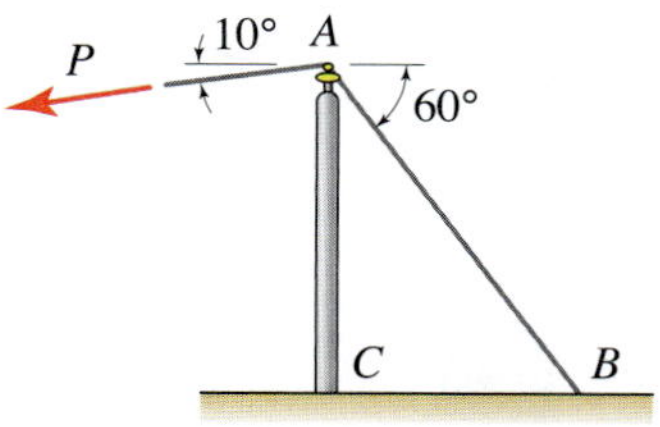

Figure P3.10 and P3.11

* Whether or not this statement is true depends on details of how end C of the utility pole is supported. Such issues are explored in Chapter 5.

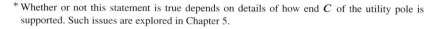

Problem 3.11

Guy wire AB is used to help support the utility pole AC. If the guy wire AB can support a maximum tensile force of 500 lb, and if the pole AC can support a maximum compressive force of 800 lb before buckling, determine the largest force P that can be supported.

Problem 3.12

Two schemes are shown for hanging a large number of flowerpots side by side on an outdoor porch. The flowerpots are to have 60 cm spacing. Each flowerpot weighs 175 N.

(a) Determine the force in wire AB.

(b) Determine the forces in wires CD and CE.

(c) Compared to the scheme using one wire, the scheme using two wires may be more resistant to adjacent flowerpots hitting one another in high winds. Do you believe this statement is valid? Explain. *Hint:* Consider the application of a horizontal wind force P to points A and C. Then compare the values of P needed to produce the same horizontal displacement of, say, 10 cm.

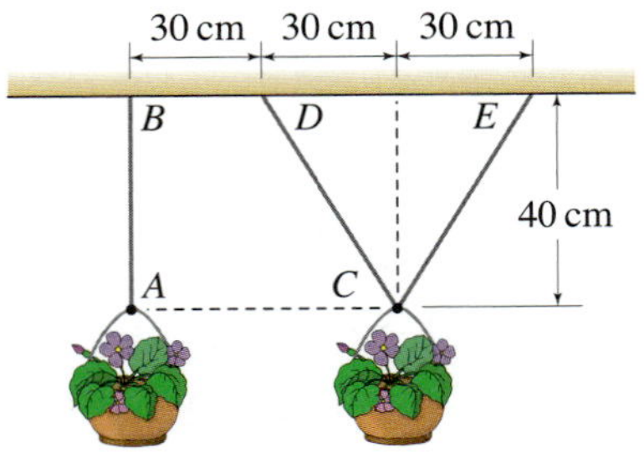

Figure P3.12

Problem 3.13

A hydraulic cylinder AB produces a 4500 lb compressive force. Determine the forces in members BD and BC.

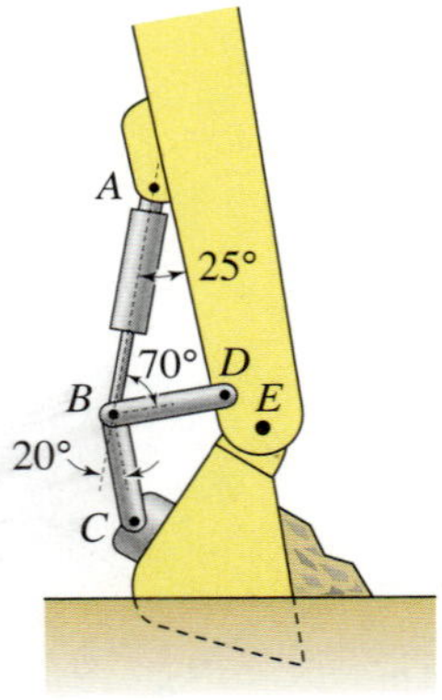

Figure P3.13

Problem 3.14

The pulley is frictionless and all weights are negligible.

(a) Show that $\theta = \alpha$.

(b) By drawing an FBD of the pulley and writing and solving equilibrium equations, determine the force F in terms of the force T and angle θ. Plot the ratio F/T versus θ for $0 \leq \theta \leq 90°$.

(c) Imagine a structure has the pulley and cable arrangement shown, and you carefully measure θ and α and find they are not equal. Explain possible circumstances that might cause this to occur.

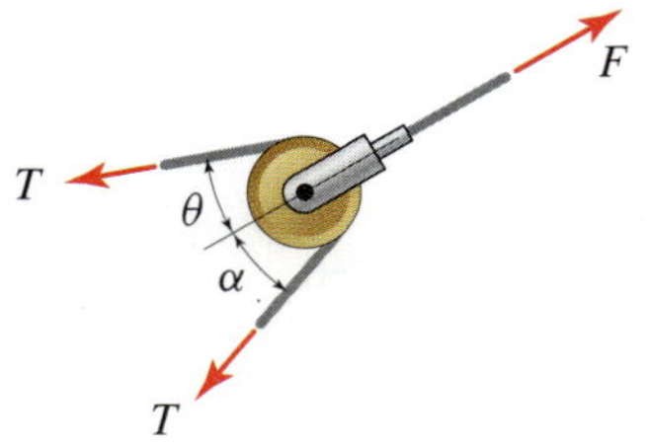

Figure P3.14

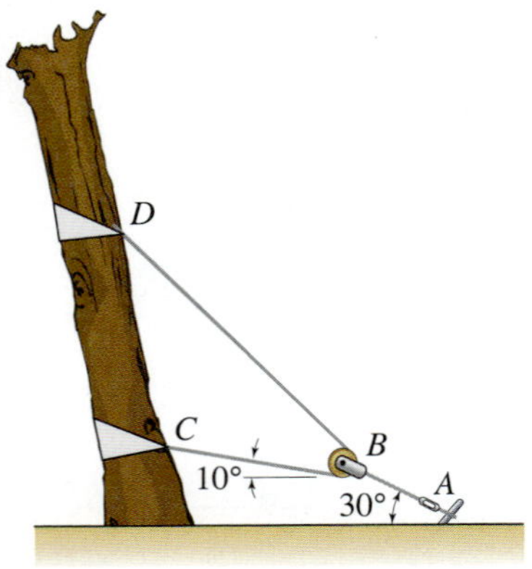

Figure P3.15

Problem 3.15

Due to settlement of soil, a recently planted tree has started to lean. To straighten it, the cable system shown is used, where a turnbuckle on cable AB is periodically tightened to keep the cable taut as the tree gradually straightens. If the force in cable AB is 450 N, determine the force in cable CBD.

Problems 3.16 and 3.17

The symmetric cable and pulley arrangement shown is used to lift a fragile architectural stone beam. If the only significant mass in the system is the 800 kg mass m of the beam, determine the forces in cables ACB and CDE.

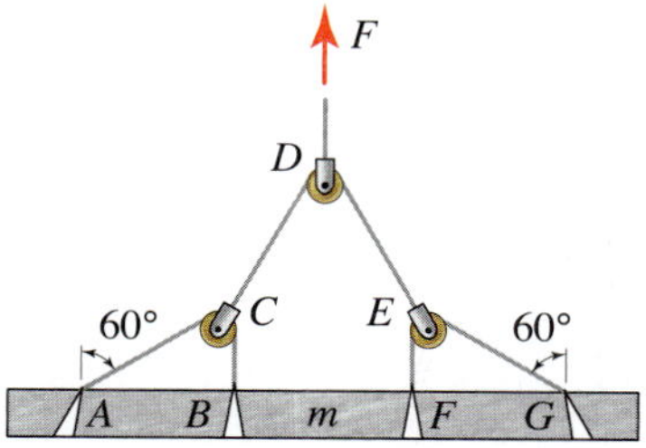

Figure P3.16

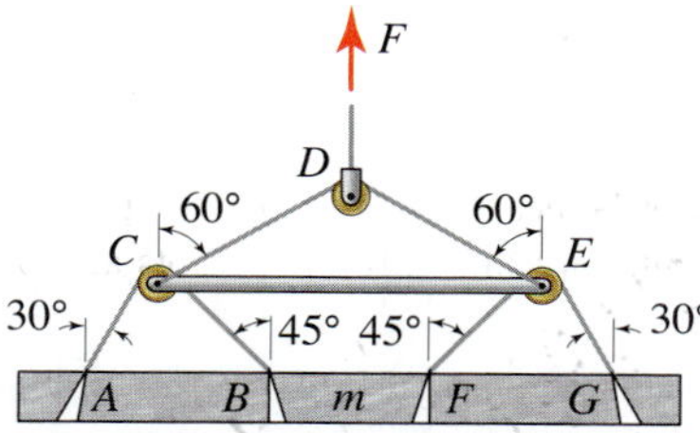

Figure P3.17 and P3.18

Problem 3.18

Cables ACB and FEG each can support a maximum force of 3 kN. Cable CDE can support a maximum force of 9 kN. The spreader bar CE can support a maximum compressive force of 5 kN. Determine the largest mass m of the stone beam that may be lifted, assuming this is the only significant mass in the system.

Problem 3.19

Blocks A and B each weigh 100 lb and rest on frictionless surfaces. They are connected to one another by cable AB. Determine the force P required to hold the blocks in the equilibrium position shown and the reactions between the blocks and surfaces.

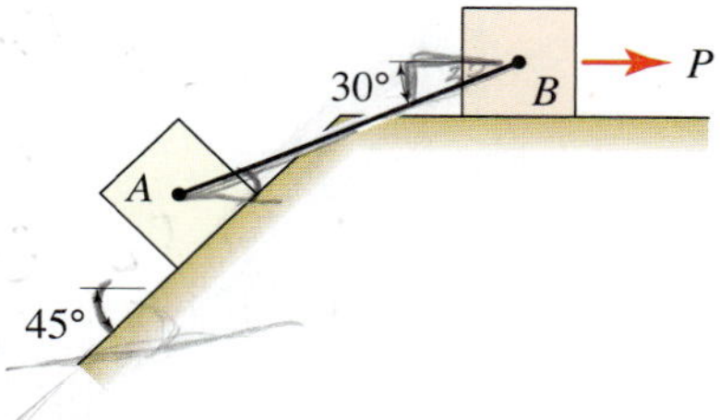

Figure P3.19

Problem 3.20

Two weights are suspended by cable $ABCD$. With the geometry shown, if one of the weights is 2000 N, determine the other weight W and the cable tensions.

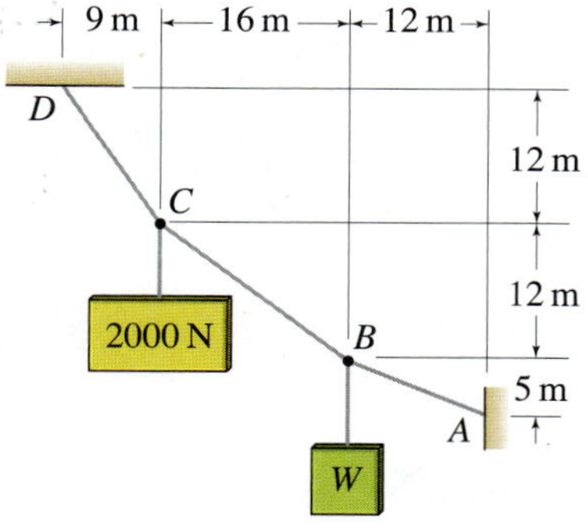

Figure P3.20

Problem 3.21

The system shown consists of cables AB and AC, horizontal cable CE, and vertical bar CD. If the cables and bar have the failure strengths shown, determine the largest load P that can be supported by the system.

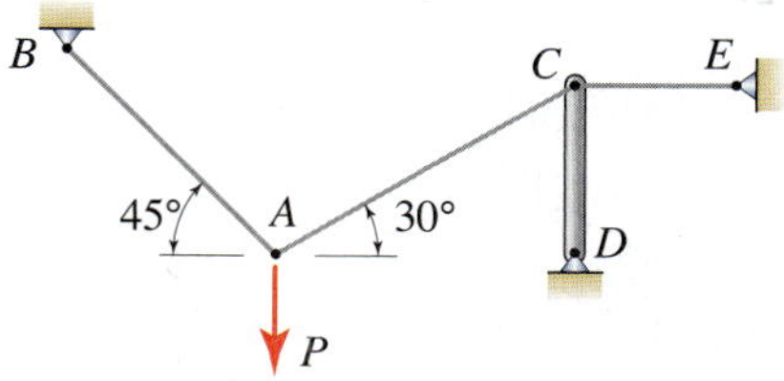

member	strength
AB	300 lb
AC	300 lb
CE	200 lb
CD	200 lb compression

Figure P3.21

Problem 3.22

The cable-pulley systems shown are used to support a weight W.

(a) Assuming the segments of cable between the pulleys are vertical, determine the cable tension T in terms of W.

(b) If the segments of cable between the pulleys are not precisely vertical, will your answer to Part (a) be affected? Explain.

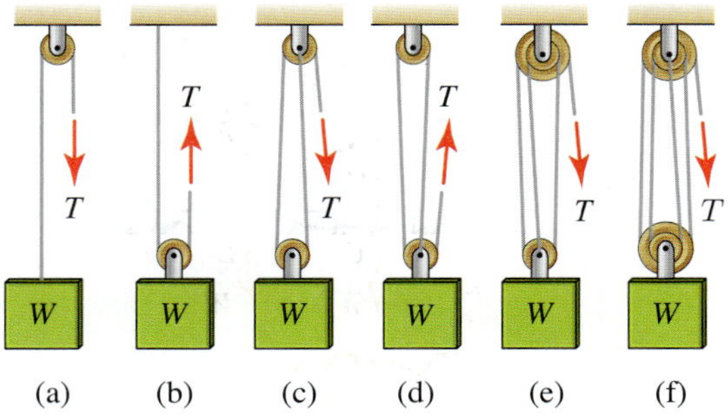

Figure P3.22

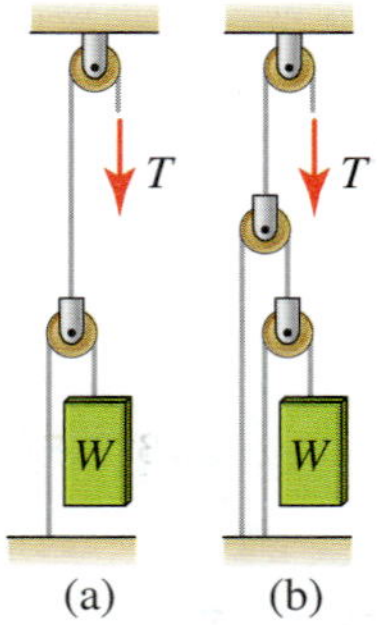

Figure P3.23

Problem 3.23

The cable-pulley systems shown are used to support a weight W. Determine the cable tension T in terms of W.

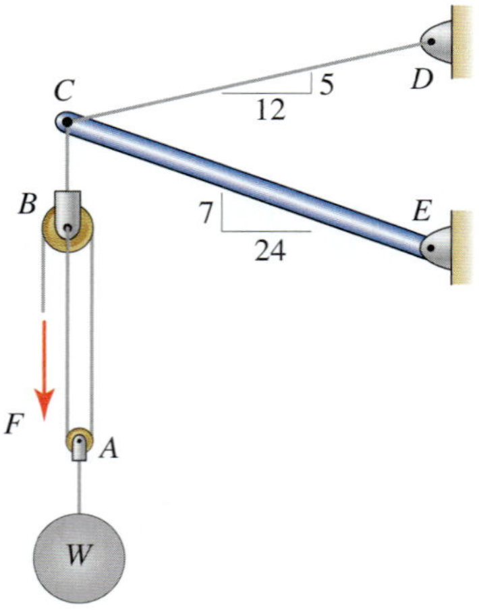

Figure P3.24

Problem 3.24

The hoist shown is used in a machine shop to position heavy workpieces in a lathe. If the cable between pulleys A and B can support a force of 300 lb, all other cables can support a force of 500 lb, and bar CE can support a compressive force of 600 lb, determine the largest weight W that may be lifted.

Problem 3.25

The cable system shown is used to help support a water pipe that crosses a creek. The water pipe applies forces W_A and W_D to the ends of cables AB and DE, respectively, where it is known that $W_A + W_D = 800$ lb. If cable BE is horizontal and cables AB and DE are vertical, determine the tension in each cable and the forces W_A and W_D.

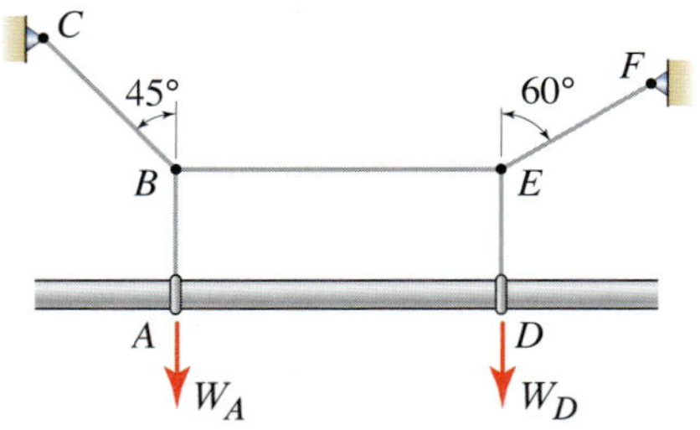

Figure P3.25

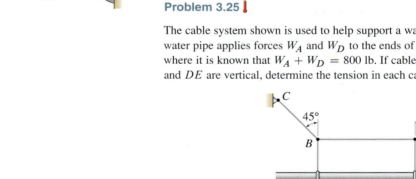

Problem 3.26

Compared to the cable system shown in Prob. 3.25, the system shown here may provide for more uniform support of a water pipe that crosses a creek. The water pipe applies forces W_A, W_D, and W_G to the ends of cables AB, DE, and GH, respectively, where it is known that $W_A + W_D + W_G = 800$ lb. If cables AB, DE, and GH are vertical, determine the tension in each cable and the forces W_A, W_D, and W_G.

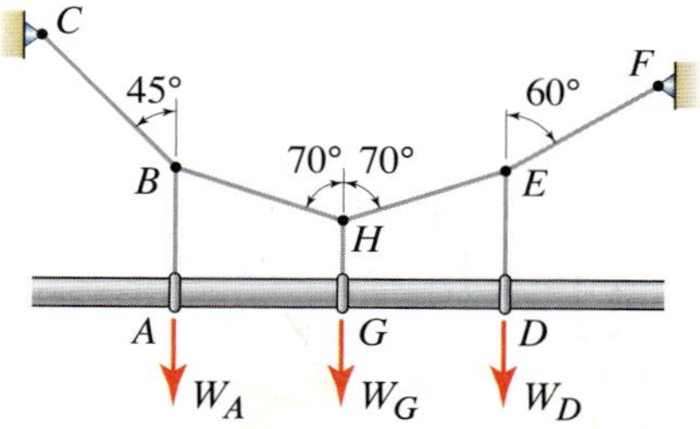

Figure P3.26

Problem 3.27

The frictionless slider B has a frictionless pulley mounted to it, around which is wrapped a cable that supports weight $W = 30$ lb as shown. The pulley at D is also frictionless, and member BE is a bar.

(a) If $\alpha = 20°$, determine the force in bar BE and the reaction between the slider and bar AC.

(b) Determine the value of α that will provide for the smallest force in bar BE, and determine the value of this force.

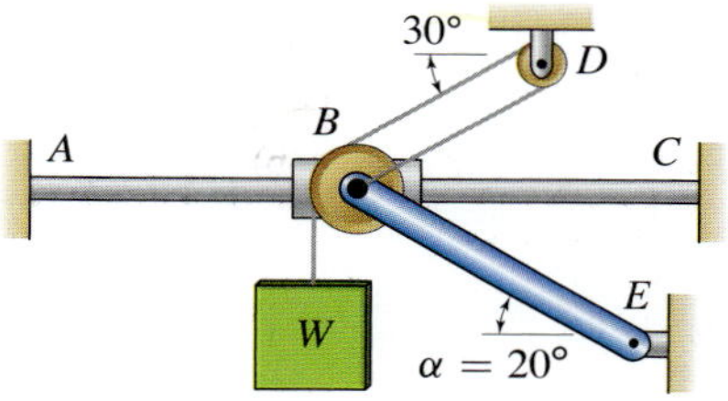

Figure P3.27

Problem 3.28

If $W = 225$ N, determine the tension in cable BCA and angle θ.

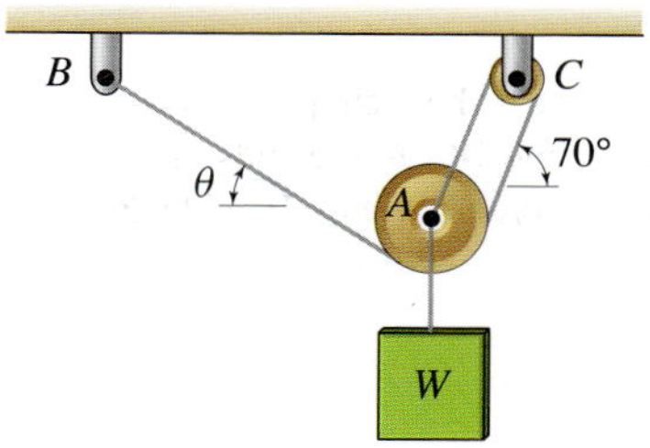

Figure P3.28 and P3.29

Problem 3.29

Consider a configuration where $\theta = 30°$. Is this configuration possible? Explain.

Problem 3.30

Determine the weights W_1 and W_2 needed for the pulley-cable structure to have the equilibrium configuration shown.

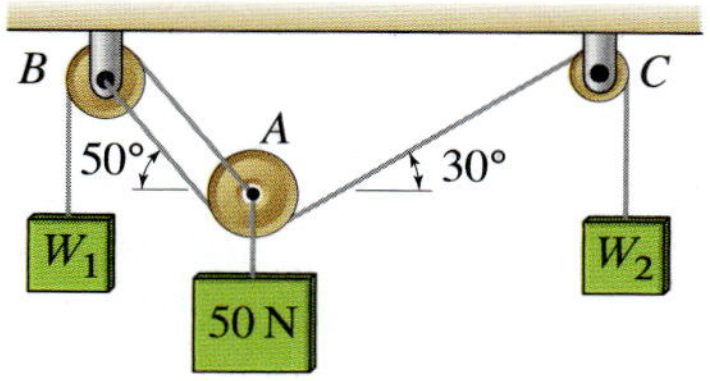

Figure P3.30

Problem 3.31

Load P is supported by cables AB, AC, and AD. Cable AC is vertical. When you solve for these cable tensions, a "problem" arises. Describe this problem and, if possible, its physical significance.

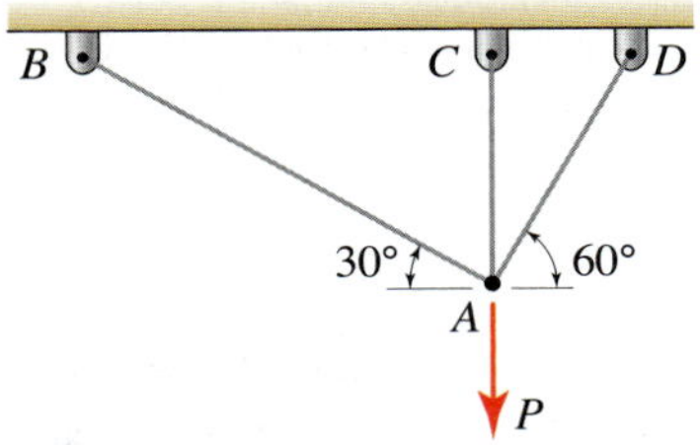

Figure P3.31

Problem 3.32

Repeat Part (b) of Example 3.4 on p. 126, using optimization methods of calculus. *Hint:* Solve Eqs. (4) and (5) in Example 3.4 for F_{BD} as a function of α. Then solve for the value of α that makes $dF_{BD}/d\alpha = 0$ (this equation is difficult to solve analytically, and you may need to solve it graphically or by other approximate means). This problem can also be effectively solved using computer mathematics programs such as Mathematica or Maple.

3.2 Behavior of Cables, Bars, and Springs

We begin this section by more thoroughly examining the theoretical underpinnings of the equilibrium analyses performed in the previous section. We then introduce springs and the inclusion of deformable members in equilibrium problems. Finally, we present a very brief introduction to statically indeterminate problems in an example problem.

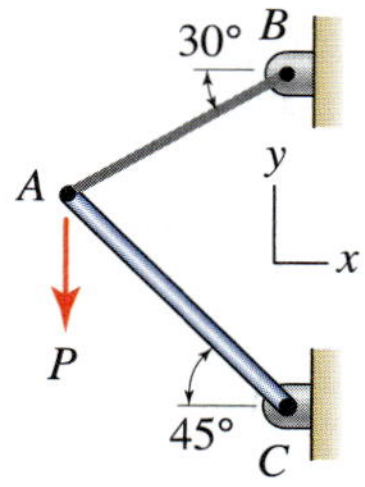

Figure 3.13
A structure consisting of a cable and bar subjected to a force P. This structure was analyzed in Example 3.1 on p. 122.

Equilibrium geometry of a structure

When analyzing equilibrium by summing forces in Newton's law $\sum \vec{F} = \vec{0}$, the orientation of forces *in the equilibrium configuration* must be used, and this generally means the orientation or shape of the structure after loads have been applied must be known, or must be determined. To explore the ramifications of this statement, reconsider the example shown in Fig. 3.13, which was previously analyzed in Example 3.1 on p. 122. The crucial question is, Do the 30° and 45° angles shown correspond to the geometry of the structure before force P is applied, after P is applied, or both?

To answer this question, we first note that all materials are *deformable*, meaning if a material is subjected to a force, the material will change shape. After force P is applied in Fig. 3.13, the cable AB will support a tensile force and hence will lengthen, or *elongate*, and the bar AC will support a compressive force and hence will shorten, or *contract*. These statements are true regardless of the material the cable and bar are made of, or the size of their cross sections.* Thus, if the 30° and 45° angles correspond to the geometry of the structure before P is applied, then the geometry of the structure after P is applied is different, or vice versa.

Typically, the geometry of a structure before loads are applied is known. For example, when constructing the structure shown in Fig. 3.13, we first cut the cable and bar to prescribed lengths and then assemble them, producing a structure with known initial geometry. After loads are applied, the structure's geometry changes and is unknown. So we immediately confront a problem when writing $\sum F_x = 0$ and $\sum F_y = 0$, namely, What angles should be used to resolve forces into x and y components?

Fortunately, for many problems, the members that constitute a structure are sufficiently stiff and appropriately arranged so that the structure's geometry changes very little after loads are applied; and accurate results may be obtained by taking the geometries of the structure before loads are applied and after loads are applied to be the same. Words such as *sufficiently stiff* and *adequately arranged* are imprecise and difficult to quantify, so in most of statics, we idealize materials to be *undeformable*, as described in the following subsections.

Cables

We generally assume cables to be *inextensible* and *perfectly flexible*. In other words, we assume a cable's length does not change, regardless of how large its tensile load is. Also, we assume a cable is perfectly flexible, so that it may

* Some materials are inherently more resistant to shape change than others, and increasing a cable's or bar's cross section size will also help increase its resistance to deformation.

be freely bent, such as when it is wrapped around a pulley. A perfectly flexible cable will immediately buckle if it is subjected to a compressive force. In addition, the cable may have a maximum tensile load limit, beyond which the cable will be unsafe or will fail.

Bars

We generally assume bars to be *inextensible*, in both tension and compression. In other words, we assume a bar's length does not change, regardless of how large its tensile or compressive load is. In Chapter 5, some additional features of the behavior of bars are discussed within the context of rigid bodies. In addition, a bar may have maximum tensile load and compressive load limits, beyond which the bar will be unsafe or will fail.

Modeling idealizations and solution of $\sum \vec{F} = \vec{0}$

The idealizations of behavior discussed above allow us to approximate a real life problem by a mathematical model where only the equilibrium equations $\sum \vec{F} = \vec{0}$ are needed to obtain the solution. Because of these assumptions, the geometry of a structure after loads are applied is assumed to be known, and the methods of analysis discussed in this book give exact results for the forces in structural members and the support reactions.

Springs

A *spring* is a mechanical device that produces a force when it undergoes a deformation. Springs come in myriad shapes, sizes, and materials. A *coil spring* is a particular type of spring that is constructed of wire (or other material) wound in the shape of a helix (usually). Figure 3.14 shows several examples of coil springs. A few examples of other types of springs include an elastic band, an elastic cord, and a *gas spring*, which consists of a column of gas in a sealed cylinder that is compressed by forces applied to it.

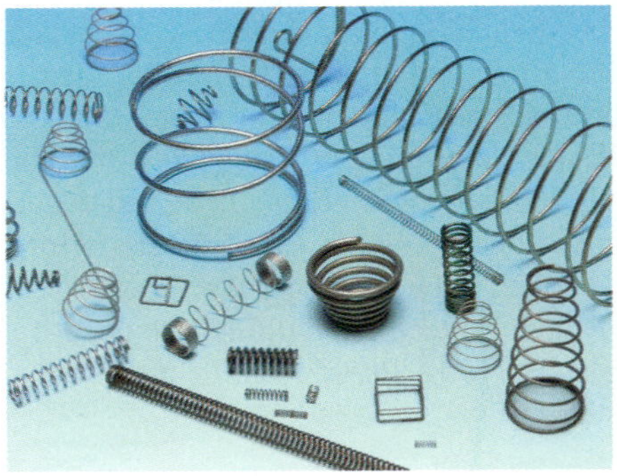

Figure 3.14. An assortment of small coil springs made of wire. While coil springs are usually wrapped in the shape of a helix, two of the springs shown here are wrapped in the shape of a square.

Springs such as those shown in Fig. 3.14 are important and common structural members. However, cables and bars are also springs. In problems where the deformation of a structure is large, it is usually necessary to model the members of the structure, especially those that are very flexible, using springs. Even if the deformations are not large, it may be necessary to determine the deformation of the structure, in which case its members may be modeled as springs.

The behavior of a linear elastic spring is characterized in Fig. 3.15. *Linear*

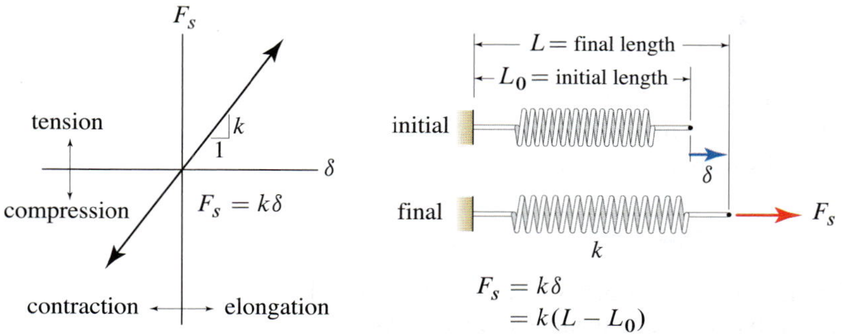

Figure 3.15. Spring law for a linear elastic spring.

means there is a linear relation between the force F_s supported by the spring and the change in its length δ. *Elastic* means the spring returns to its original length, or shape, after the force F_s is removed. The equation governing the spring's behavior is often called the *spring law* and is

$$\begin{aligned} F_s &= k\delta \\ &= k(L - L_0) \end{aligned} \tag{3.18}$$

where

- F_s is the force supported by the spring;
- δ is the elongation of the spring from its unstretched or undeformed length;
- k is the spring stiffness (units: force/length);
- L_0 is the initial (unstretched) spring length;
- L is the final spring length.

As stated in the Helpful Information margin note, in Eq. (3.19), the convention in writing the spring law is that positive values of force F_s correspond to tension and negative values to compression, and positive values of δ correspond to elongation and negative values to contraction. The constant k is called the *spring stiffness*: it is always positive, and a large value means the spring is stiff whereas a low value means the spring is flexible. It is possible to use other conventions for positive force and deformation. For example, if F_s and δ are positive in compression and contraction, respectively, the spring law is still given by Eq. (3.18). If F_s is positive in compression and δ is positive in elongation, or vice versa, then the spring law becomes $F_s = -k\delta$.

Springs are used to represent a variety of deformable members in engineering problems. When the deformations of cables and bars must be accounted for, they are usually represented by springs where their stiffness depends on

Helpful Information

Spring law sign conventions. The sign conventions for the spring law given in Eq. (3.18) are

$$\begin{aligned} F_s &> 0 && \text{tension,} \\ F_s &< 0 && \text{compression,} \\ \delta &= 0 && \text{unstretched position,} \\ \delta &> 0 && \text{extension,} \\ \delta &< 0 && \text{contraction.} \end{aligned} \tag{3.19}$$

Interesting Fact

Springs. Springs are important structural members in their own right, but they are also important for laying the groundwork for characterizing more general engineering materials and members, which you will study in subjects that follow statics. Simply stated, almost all materials are idealized as springs, albeit more complex than that shown in Fig. 3.15, over at least some range of forces.

the material, length, and cross-sectional area of the bar or cable.* Other deformable members or structures, even if very complex, may also be characterized by a simple spring, such as the examples shown in Fig. 3.16. The nail

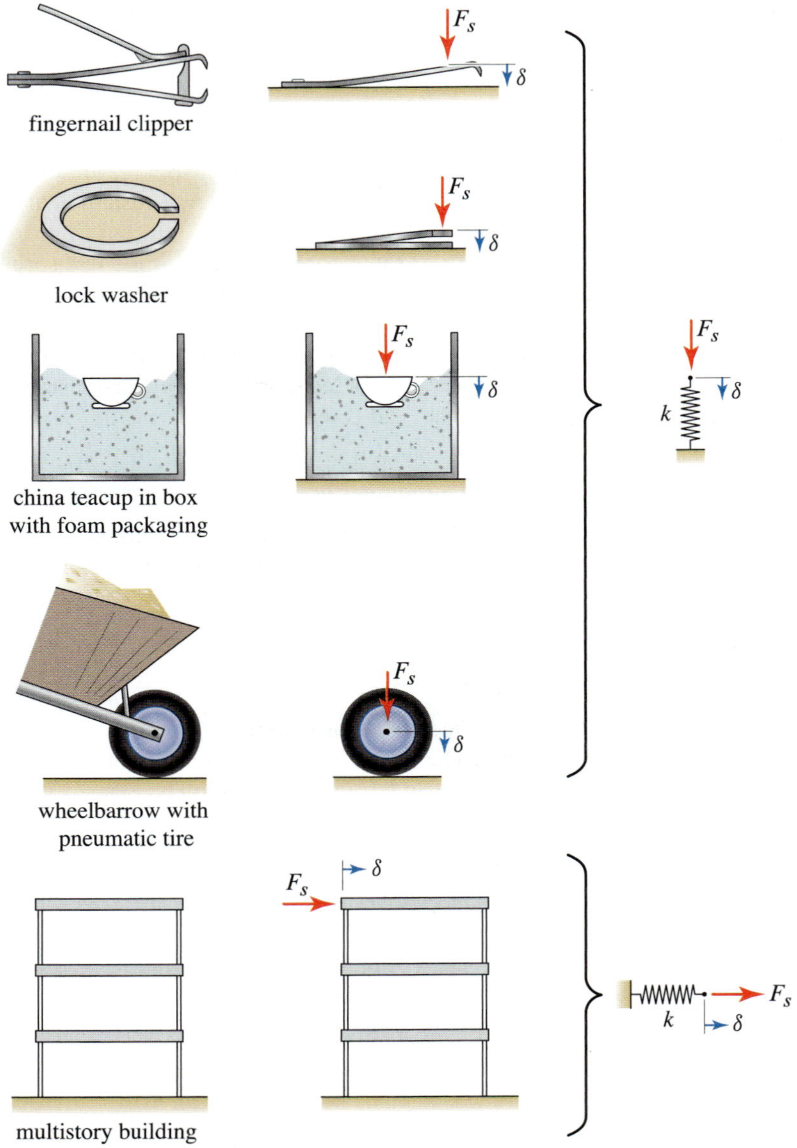

Figure 3.16. Examples of structures or devices that may be idealized as springs.

clipper and lock washer each have a metal part that deforms as loads are applied. The teacup rests on a cushion of foam packaging that deflects due to

* For both cables and bars, the spring stiffness is $k = AE/L$, where A and L are the cross-sectional area and length, respectively, of the cable or bar, and E, called the *elastic modulus* with units of force/area, is a property of the material that characterizes its inherent stiffness. For example, nylon has $E = 4\times10^5$ lb/in.^{2}, while steel, which is much stiffer, has $E = 30\times10^6$ lb/in.^{2}.

forces from vibrations during handling. The tire of the wheelbarrow supports loads by changing its shape and compressing the air within it. The multistory building deflects when a force, perhaps due to wind, is applied as shown. In all cases cited here, the linear elastic spring law $F_s = k\delta$ may be used over at least a portion of the full range of response that these devices display, although determining the value of k usually requires further analysis. Note that in all of these problems, as the deflection δ increases beyond some point, the response probably becomes *nonlinear*. For example, the nail clipper becomes very stiff once the cutting edges come into contact. The multistory building, on the other hand, will likely become less stiff as δ increases beyond some limit because the structural members within the building will sustain damage if they are loaded too severely.

Occasionally springs are designed to be nonlinear, such as the leaf spring suspension for a truck shown in Fig. 3.17. When the load supported by the axle is small, only a few of the leaves in the spring will engage and the stiffness is low. As the load increases, more leaves engage and the stiffness becomes progressively higher. Although nonlinear, a leaf spring is still elastic since it returns to its original shape when the load is removed.

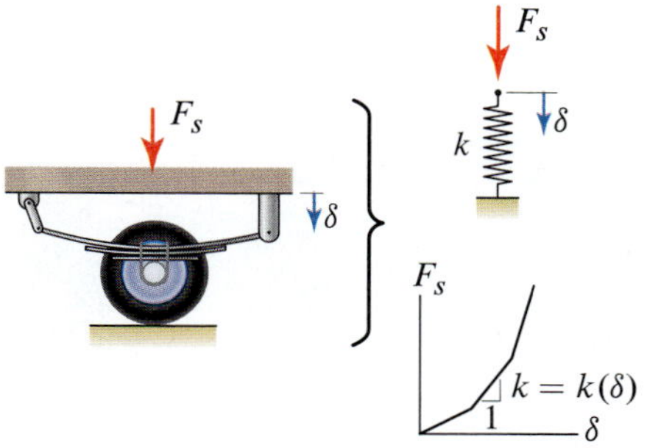

Figure 3.17
Truck axle with multileaf spring suspension. Such springs are elastic, but nonlinear.

End of Section Summary

In this section, some of the finer points regarding static equilibrium were reviewed, and springs were discussed in detail. Some of the key points are as follows:

- When you are writing equilibrium equations $\sum \vec{F} = \vec{0}$, the geometry of the structure in the equilibrium position must be used.

- A *spring* is a deformable member that undergoes a change of length when subjected to a force. The *spring law* is $F_s = k\delta$, where k is called the *spring stiffness* and $k \geq 0$. In writing this equation it is assumed that force F_s and displacement δ are positive in the same direction. The force F_s and displacement δ may be measured positive in opposite directions, but it may be necessary to introduce a negative sign in the spring law (i.e., $F_s = -k\delta$) as discussed on p. 137.

EXAMPLE 3.5 *Springs*

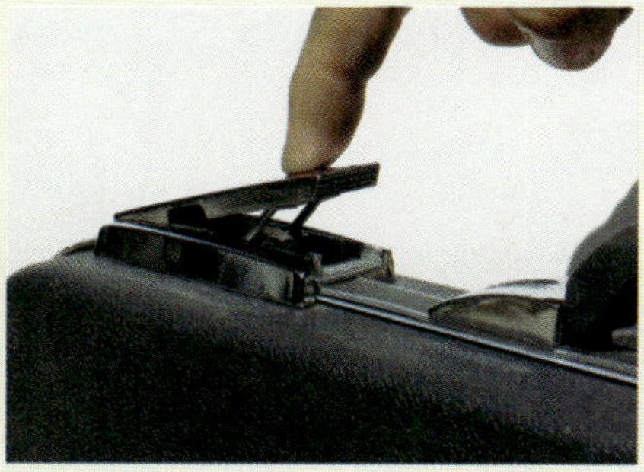

A model for the latch of a briefcase is shown. Spring AB has stiffness $k = 3\,\mathrm{N/cm}$ and 6 cm unstretched length.

(a) Determine the force F needed to begin closing the latch from the open position shown.

(b) Determine the force F needed to begin opening the latch from the closed position where member BC is horizontal.

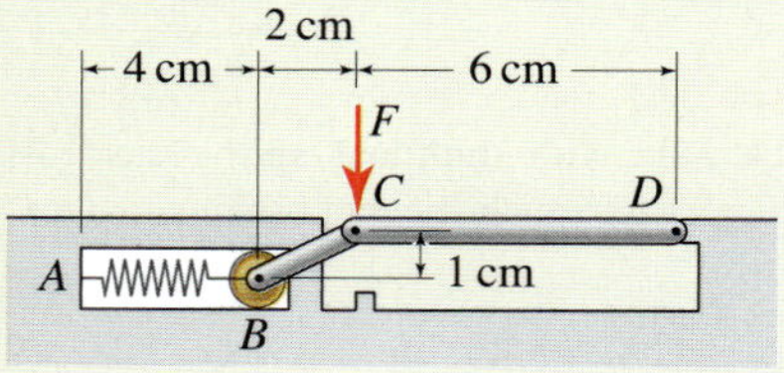

Figure 1
Briefcase latch in the open position.

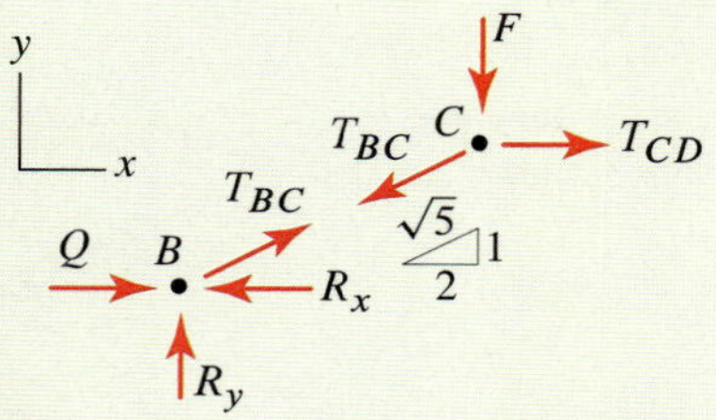

Figure 2
Free body diagrams of points B and C for closing the latch from the open position.

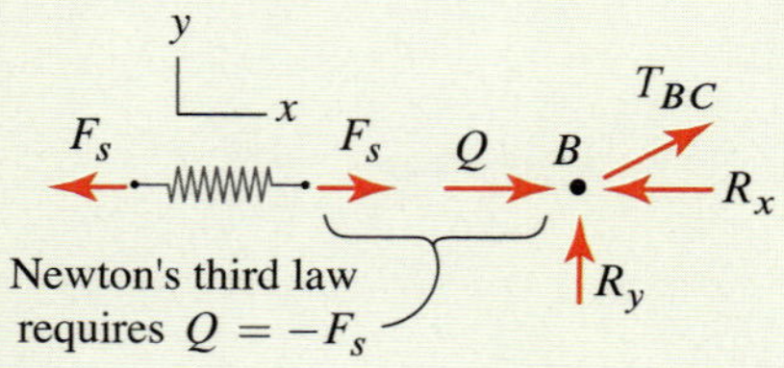

Figure 3
Free body diagrams of point B and the spring showing that force Q and spring force F_s are related by Newton's third law with the result $Q = -F_s$.

SOLUTION

Road Map This problem involves a spring, and to determine the force it supports requires that we use the spring law, Eq. (3.18) on p. 137. This equation is in addition to the equilibrium equations.

Part (a)

Modeling Free body diagrams for points B and C are shown in Fig. 2. Note in the FBD for B there are two reactions R_x and R_y between the roller at B and the track it rolls in. However, at the instant the latch begins to close, contact between the roller and vertical surface is broken and hence, $R_x = 0$.

Governing Equations & Computation

Force Laws To determine the spring force, one of the forms of Eq. (3.18) will always be applicable, with the choice depending on the form of data available in the problem. Here the initial length of the spring is $L_0 = 6\,\mathrm{cm}$, and the final length of the spring, from Fig. 1, is $L = 4\,\mathrm{cm}$. Hence, in this problem the second form of Eq. (3.18) is more useful. In the geometry shown, when the latch begins to close, the spring force F_s is

$$F_s = \left(3\,\frac{\mathrm{N}}{\mathrm{cm}}\right)(4\,\mathrm{cm} - 6\,\mathrm{cm}) = -6\,\mathrm{N}. \tag{1}$$

Recall that in developing Eq. (3.18), and hence in writing Eq. (1), the spring force F_s is positive in tension. Thus, the negative sign that appears in Eq. (1) indicates the spring is in compression. However, in the FBD of point B, the force Q applied by the spring to point B is taken to be positive in compression. Thus, $Q = -F_s = 6\,\mathrm{N}$. This mental accounting of sign is simple to perform; but if there is any confusion, then adding the FBD of the spring to that of point B, as shown in Fig. 3, provides full clarification, where it is seen that Newton's third law requires $Q = -F_s$.

Equilibrium Equations Writing equilibrium equations for point B provides

$$\sum F_x = 0: \qquad Q - R_x + T_{BC}(2/\sqrt{5}) = 0, \tag{2}$$

$$\sum F_y = 0: \qquad R_y + T_{BC}(1/\sqrt{5}) = 0. \tag{3}$$

With $R_x = 0$ and $Q = 6\,\mathrm{N}$, Eqs. (2) and (3) may be solved to obtain

$$T_{BC} = -6.708\,\mathrm{N} \quad \text{and} \quad R_y = 3.000\,\mathrm{N}. \tag{4}$$

Writing equilibrium equations for point C provides

$$\sum F_x = 0: \qquad -T_{BC}(2/\sqrt{5}) + T_{CD} = 0, \tag{5}$$

$$\sum F_y = 0: \qquad -T_{BC}(1/\sqrt{5}) - F = 0. \tag{6}$$

By using the result for T_{BC} in Eq. (4), Eqs. (5) and (6) may be solved for:

$$T_{CD} = -6.000\,\mathrm{N} \quad \text{and} \quad F = 3.000\,\mathrm{N}. \tag{7}$$

Hence, a downward force $\boxed{F = 3.00\,\mathrm{N}}$ will cause the latch to begin closing.

Part (b)

Modeling In the closed position, member BC is horizontal as shown in Fig. 4. The FBDs of points B and C are shown in Fig. 5 where, for convenience, we have reassigned F to be positive upward since clearly a force in this direction is needed to open the latch. Note in the FBD for C, there is a reaction R_C between point C and the surface it contacts. However, at the instant the latch begins to open, this contact is broken and hence, $R_C = 0$.

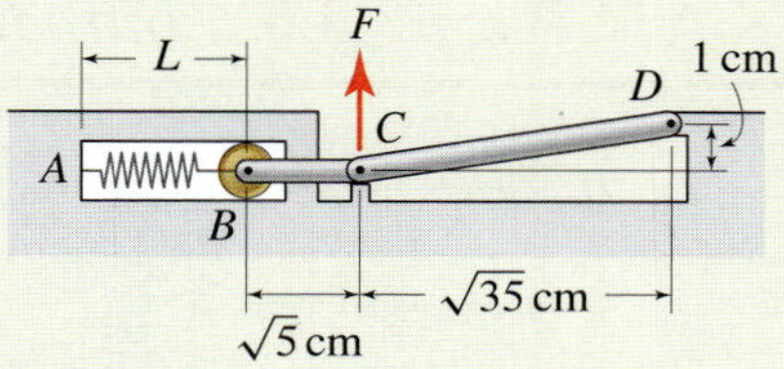

Figure 4
Briefcase latch in the closed position.

Governing Equations & Computation

Force Laws To obtain T_{BC}, the force Q exerted by the spring on point B is needed. The initial length of spring AB is known, and the final length L of the spring can be evaluated, using the geometry in Figs. 1 and 4, as

$$L = 4\,\text{cm} + 2\,\text{cm} + 6\,\text{cm} - (\sqrt{5}\,\text{cm} + \sqrt{35}\,\text{cm}) = 3.848\,\text{cm}. \tag{8}$$

In Eq. (8), $4\,\text{cm}+2\,\text{cm}+6\,\text{cm}$ is obtained from Fig. 1 as the horizontal distance between points A and D, and $\sqrt{5}\,\text{cm}+\sqrt{35}\,\text{cm}$ is obtained from Fig. 4 as the horizontal distance between points B and D. The second form of Eq. (3.18) then provides

$$F_s = \left(3\,\frac{\text{N}}{\text{cm}}\right)(3.848\,\text{cm} - 6\,\text{cm}) = -6.456\,\text{N}. \tag{9}$$

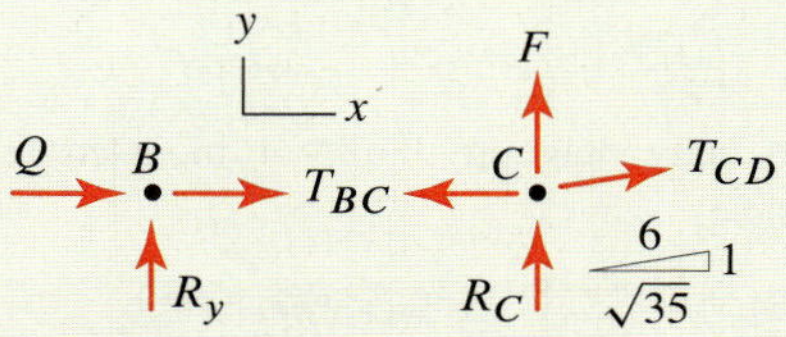

Figure 5
Free body diagrams for points B and C for opening the latch from the closed position.

Equilibrium Equations From the FBD of point B in Fig. 5, the equilibrium equations for point B are as follows:

$$\sum F_x = 0: \qquad Q + T_{BC} = 0, \tag{10}$$

$$\sum F_y = 0: \qquad R_y = 0. \tag{11}$$

As argued earlier in Fig. 3, $Q = -F_s = 6.456\,\text{N}$, and the solutions of Eqs. (10) and (11) are, respectively,

$$T_{BC} = -6.456\,\text{N} \quad \text{and} \quad R_y = 0. \tag{12}$$

Writing equilibrium equations for point C provides

$$\sum F_x = 0: \qquad -T_{BC} + T_{CD}\left(\frac{\sqrt{35}}{6}\right) = 0, \tag{13}$$

$$\sum F_y = 0: \qquad F + T_{CD}\left(\frac{1}{6}\right) + R_C = 0. \tag{14}$$

By using the results for T_{BC} from Eq. (12) and noting $R_C = 0$, Eqs. (13) and (14) may be solved for:

$$T_{CD} = -6.548\,\text{N} \quad \text{and} \quad F = 1.091\,\text{N}. \tag{15}$$

Hence, an upward force $\boxed{F = 1.09\,\text{N}}$ will cause the latch to begin opening.

Discussion & Verification In this example, it was possible to evaluate the material models (i.e., spring law) independently of solving the equilibrium equations. In some problems, such as Example 3.6, the material models and the equilibrium equations are coupled and must be solved simultaneously.

EXAMPLE 3.6 *Introduction to Statically Indeterminate Problems*

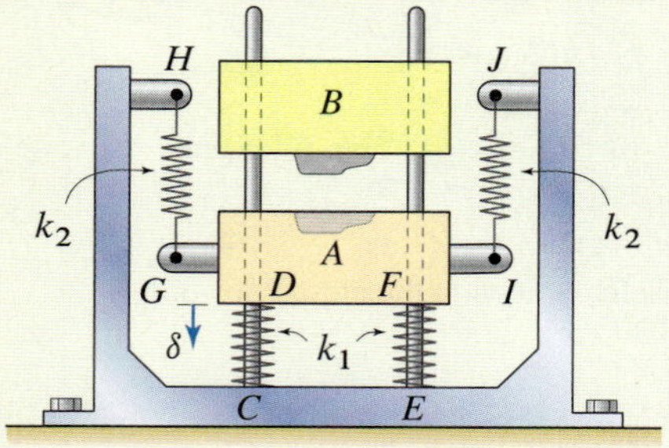

Figure 1

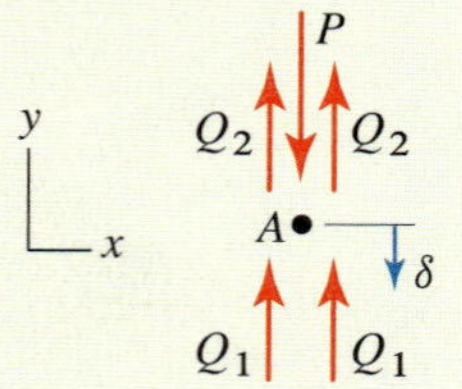

Figure 2

Free body diagram of die A modeled as a particle.

> **Helpful Information**
>
> **Statically indeterminate structures.** A characteristic of a statically indeterminate structure is that there are extra "load paths" available to support the applied loads. In this example, the springs at the base, by themselves, are adequate to support the applied loads, and equations of equilibrium will determine exactly what value of force they must support. Conversely, the springs on the sides, by themselves, are also adequate to support the applied loads, and equations of equilibrium will determine exactly what value of force they must support. With both sets of springs present, it is unclear from the equations of equilibrium alone what portions of the applied force are supported by each set of springs. Either set of springs could be viewed as providing an extra load path.
>
> Static indeterminacy makes analysis more difficult, but is often a desirable feature to have in structures. For example, if one load path in a statically indeterminate structure is lost because of an accident or failure, the remaining load paths may be sufficient for the structure to remain in static equilibrium. Static indeterminacy is studied in greater detail in later chapters of this book.

A drop hammer is used in some sheet metal forming and forging operations. With this machine, die B is dropped onto die A, and due to impact forces, the workpiece between A and B is hammered into the shape of the cavity or die between them. Both dies A and B slide on low-friction fixed vertical bars so they maintain proper alignment. Special coil springs CD and EF, called *die springs*, are used to help isolate the severe vibrations of the dies from being transmitted to the floor that supports the machine.

The user of the machine determines that additional stiffness is needed for vibration isolation. Although the die springs could be replaced with heavier springs, this requires disassembly of the machine. Rather, additional vertical springs such as GH and IJ are more easily added and removed when they are no longer needed. The springs have stiffnesses $k_1 = 5000$ lb/in. and $k_2 = 1500$ lb/in. If all springs are unstretched when $\delta = 0$, determine the deflection δ if die A is subjected to a downward vertical force $P = 1$ kip.

SOLUTION

Road Map A solution to this problem will require as many equations as there are unknowns. As described in the margin note, the equations of equilibrium alone are not sufficient in number, and additional equations must be added that describe the behavior of the springs.

Modeling Assuming the vertical guides allow die A to undergo only vertical motion with no rotation, both springs at the bottom of A are compressed by the same amount, and hence develop the same force Q_1. Similarly, both springs on the sides of A develop the same force Q_2. Idealizing die A as a particle allows the FBD shown in Fig. 2 to be drawn, where Q_1 and Q_2 are taken to be positive in compression and tension, respectively.

Equilibrium Equations The equilibrium equation for A is

$$\sum F_y = 0: \quad 2Q_1 + 2Q_2 - P = 0. \tag{1}$$

Since no forces act in the x direction, the equation $\sum F_x = 0$ is automatically satisfied regardless of the values for Q_1, Q_2, and P, and hence this equation provides no useful information.

Equation (1) is unlike any we have encountered thus far in this book. Even though $P = 1$ kip, there remain two unknowns and only one equation with which to determine them. Such a problem has no unique solution. Rather, there are an infinite number of solutions for Q_1 and Q_2 that will satisfy Eq. (1). We call such problems *statically indeterminate*, because the equations of static equilibrium alone are not sufficient to determine the unknowns. To obtain the appropriate solution among the infinite number of solutions that are possible, additional information is needed. In this problem, this information corresponds to descriptions of the deformability of the springs, which we address next.

Force Laws Since all the springs are uncompressed when $\delta = 0$, the first form of Eq. (3.18) on p. 137 allows us to write for each set of springs

$$Q_1 = k_1\delta = (5000\,\text{lb/in.})\,\delta, \tag{2}$$
$$Q_2 = k_2\delta = (1500\,\text{lb/in.})\,\delta. \tag{3}$$

In writing Eq. (2), Q_1 is positive in compression and positive δ gives contraction of the springs at the base. Hence, the negative signs associated with each of these quantities, compared to our normal convention of spring force being positive in tension and spring deformation being positive in extension, compensate one another.

Computation Combining Eqs. (1)–(3) provides

$$2\left(5000\,\frac{\text{lb}}{\text{in}} + 1500\,\frac{\text{lb}}{\text{in}}\right)\delta - P = 0. \tag{4}$$

With $P = 1000\,\text{lb}$, we now have one equation and one unknown, and Eq. (4) may be solved for

$$\boxed{\delta = 0.0769\,\text{in.}} \tag{5}$$

Once δ is known, if desired, we may return to Eqs. (2) and (3) to determine the portion of the force P that is supported by each spring to obtain

$$Q_1 = 385\,\text{lb} \quad \text{and} \quad Q_2 = 115\,\text{lb}. \tag{6}$$

Discussion & Verification We expect the die to move downward, and hence our solution should show $\delta > 0$, which it does. We expect springs CD and EF to be in compression, and since Q_1 was defined to be positive in compression, our solution should show $Q_1 > 0$, which it does. Similarly, we expect springs GH and IJ to be in tension, and since Q_2 was defined to be positive in tension, our solution should show $Q_2 > 0$, which it does. After these simple checks, we should verify that our solutions for δ, Q_1, and Q_2 satisfy all governing equations, which consist of the equilibrium equation, Eq. (1), and the force laws, Eqs. (2) and (3).

Figure 3. A drop hammer used for sheet metal forming. A finished part is shown in the lower left-hand corner.

PROBLEMS

Problem 3.33

A device for tensioning recording tape in a video cassette recorder is shown. The tape wraps around small pins at A, B, and C. The pins at A and C are fixed, and the pin at B is supported by a spring and can undergo vertical motion in the frictionless slot. Friction between the tape and pins is negligible. The spring has stiffness $k = 0.5\,\mathrm{N/mm}$ and is unstretched when $h = 25\,\mathrm{mm}$. Neglecting the size of the pins, determine the tension in the tape when

(a) $h = 18\,\mathrm{mm}$.

(b) $h = 10\,\mathrm{mm}$.

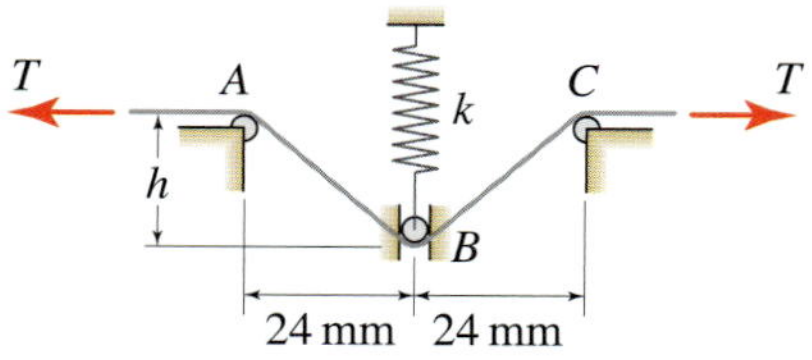

Figure P3.33

Problem 3.34

The brake linkage for a vehicle is actuated by pneumatic cylinder AB. Cylinder AB, springs CD, EF, and GH, and the slotted tracks at A, C, E, and G are all parallel. If the slotted tracks are frictionless, and cylinder AB produces a tensile force of 12 kip, determine the deflection δ and forces in members AC, AE, and AG. All springs are unstretched when $\delta = 0$.

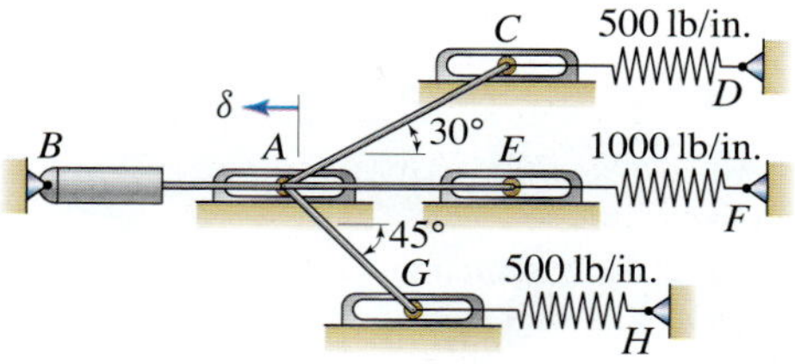

Figure P3.34

Problem 3.35

Collar A has negligible weight and slides without friction on the vertical bar CD. Determine the vertical force F that will produce $\theta = 30°$ if

(a) Spring AB is unstretched when $\theta = 0°$.

(b) Spring AB has an unstretched length of 2 ft.

(c) Spring AB has an unstretched length of 4 ft.

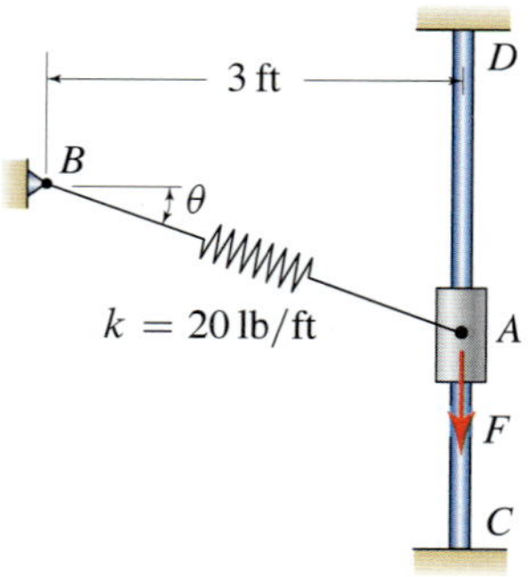

Figure P3.35

Problem 3.36

Point A is supported by springs BC and DE and cable segments AB and AD. The springs and cables have negligible weight, and the springs have identical stiffness $k = 10\,\mathrm{N/cm}$ and 20 cm unstretched length. The structure has the geometry shown when $F = 225\,\mathrm{N}$.

(a) Determine the lengths of cables AB and AD.

(b) Determine the coordinates of point A when $F = 0$.

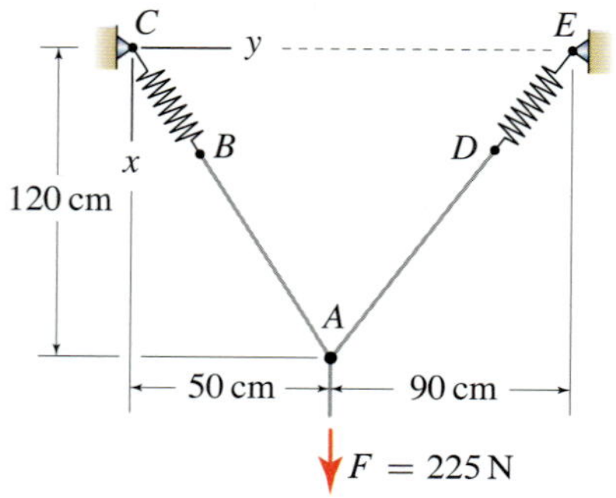

Figure P3.36

Problem 3.37

Spring AB is supported by a frictionless roller at B so that it is always vertical. If the spring is unstretched when $\theta = 0°$, determine θ and the forces in spring AB and bar AC when F has the value indicated. *Hint:* The force supported by bar AC is zero when $\theta < 90°$, and may be nonzero when $\theta = 90°$.

(a) $F = 25\,\text{lb}$.

(b) $F = 50\,\text{lb}$.

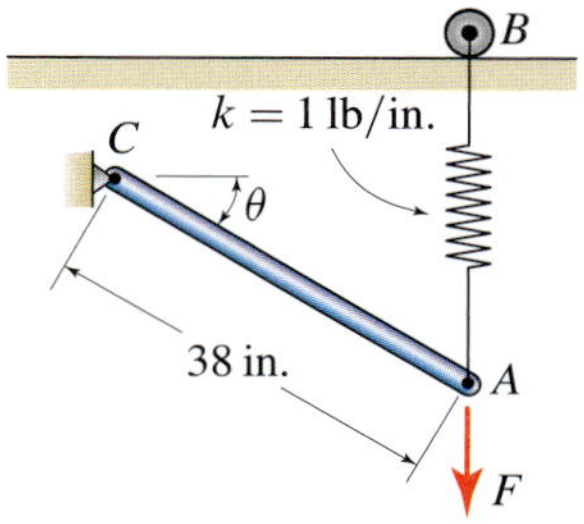

Figure P3.37

Problem 3.38

The machine shown is used for compacting powder. Collar C slides on plunger AB and is driven by a motor (not shown in the sketch) so that it has the oscillatory vertical motion $\delta = (10\,\text{mm})(\sin \pi t)$ where t is time in seconds. Plunger AB weighs 4 N and is pressed against the powder by the spring whose end is driven by the motion of collar C. The spring has stiffness $k = 0.1\,\text{N/mm}$ and 100 mm unstretched length. Assume there is no friction in the system (other than friction between individual grains of powder), and assume the motion of C is slow enough that there are no dynamic effects. Determine the largest and smallest forces the plunger applies to the powder over a full cycle of motion of C if

(a) The powder column is at its initial height $h = 110\,\text{mm}$.

(b) The powder column is at its compacted height $h = 80\,\text{mm}$.

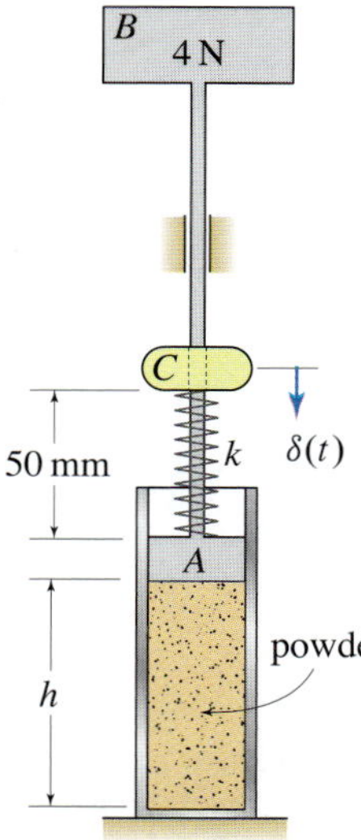

Figure P3.38

Problem 3.39

A fuel pump is driven by a motor-powered flywheel. The pump behaves as a spring with stiffness 2 N/mm that is unstretched when $\alpha = \pm 60°$. Neglect any possible dynamic effects.

(a) Determine the largest tensile and compressive forces spring AB experiences during one revolution of the flywheel, and state the positions α where these occur.

(b) Without further analysis, is it certain that the largest tensile and compressive forces in crank arm BC occur at the same positions α determined in Part (a)? Explain.

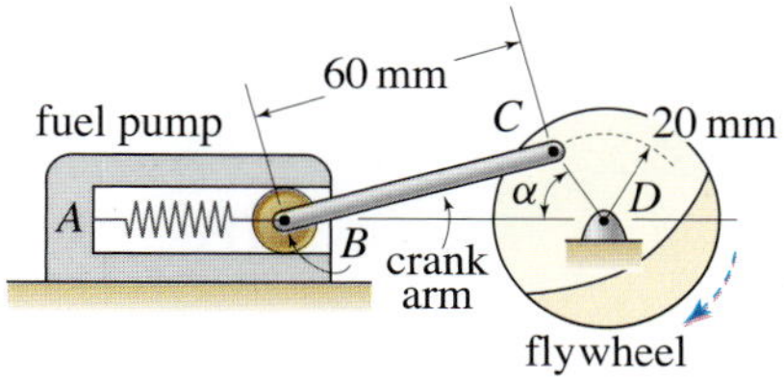

Figure P3.39

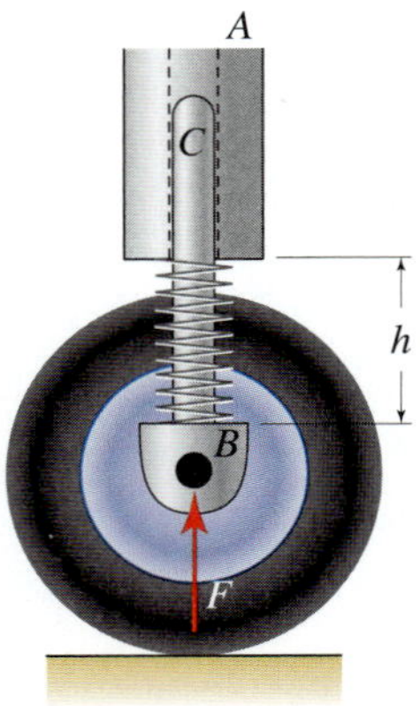

Figure P3.40

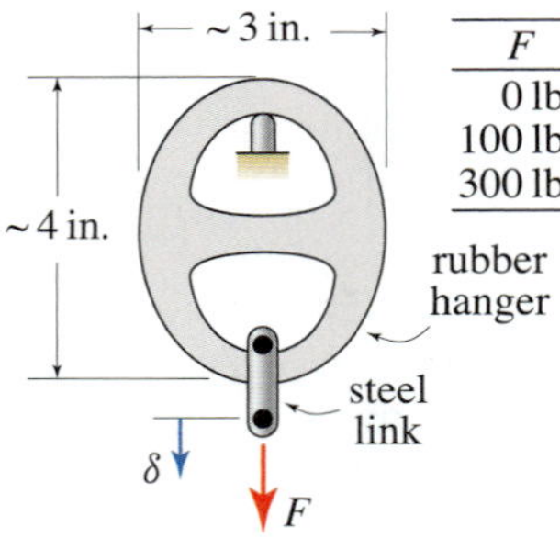

F	δ
0 lb	0 in.
100 lb	0.50 in.
300 lb	1.00 in.

Figure P3.41

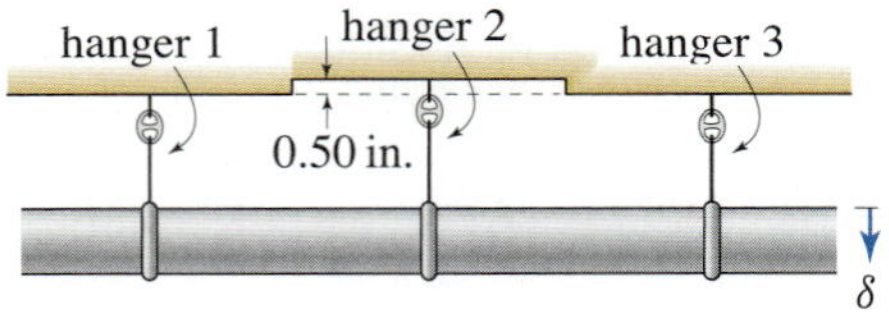

Figure P3.42

Problem 3.40

The suspension for the landing gear of an aircraft is shown. The wheel is attached to bar BC, which slides vertically without friction in housing A, which is fixed to the frame of the aircraft. The spring is precompressed so that it does not undergo additional deflection until the force supported by the landing gear is sufficiently large. Further, if the force supported by the landing gear exceeds a limit, the suspension "bottoms out" and deflects no more. Specify the spring stiffness and initial length if $h = 18$ in. for $F \leq 500$ lb and $h = 12$ in. for $F \geq 1500$ lb.

Problem 3.41

A hanger is made of cord-reinforced rubber. It is used as a spring support with limited travel for a wide variety of applications. If prototype samples are available, an effective means to characterize its nonlinear stiffness is by testing in a laboratory, where forces of known magnitude are applied and the deflections that result are measured. Imagine this produces the load-deflection data provided in the table of Fig. P3.41.

(a) Determine the constants a, b, and c that will fit the general quadratic equation $F = a + b\delta + c\delta^2$ to the load-deflection data for this hanger.

(b) Plot the load-deflection relation determined in Part (a).

(c) Speculate on the range of values for F for which the relation obtained in Part (a) will be reasonably accurate.

Problem 3.42

In Prob. 3.41, imagine that the load-deflection data for the hanger is such that $F = (100\,\text{lb/in.})\,\delta + (200\,\text{lb/in.}^2)\,\delta^2$ and that three hangers are used to support a straight rigid pipe from an uneven ceiling as shown. Assume the pipe may undergo vertical motion only. Hanger 2 displaces by 0.50 in. more than hangers 1 and 3. Because of symmetry, equal forces are supported by hangers 1 and 3. Further, the sum of the forces supported by all three hangers equals the weight of the pipe and its contents, which is 600 lb. Determine the deflection of the pipe and the forces supported by each of the three hangers.

Problems 3.43 and 3.44

If $k = 5\,\text{N/mm}$ and $W = 100\,\text{N}$, determine δ. Springs are unstretched when $\delta = 0$.

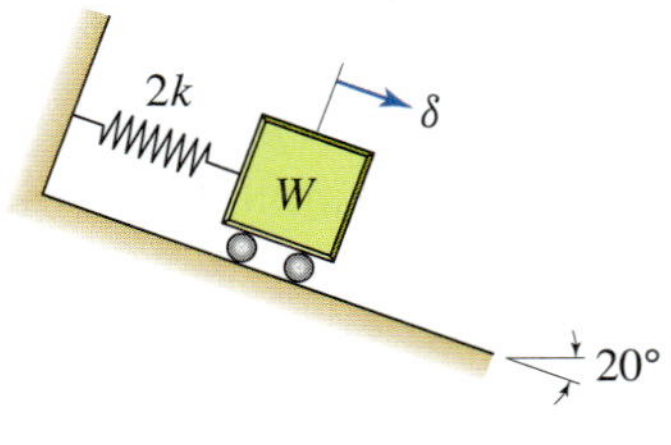

Figure P3.43

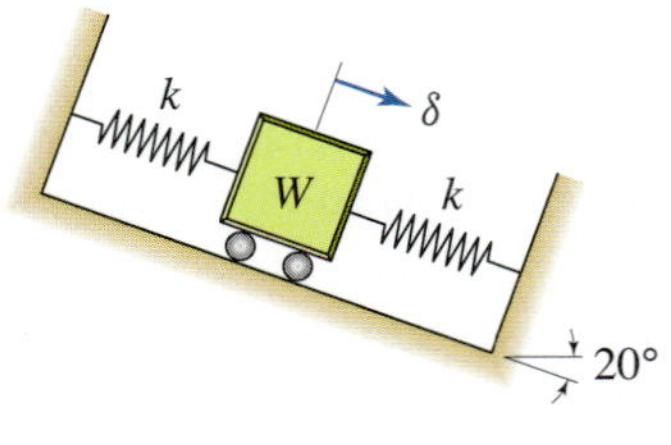

Figure P3.44

Problem 3.45

A fragile item A, with weight W, is to be shipped within a box B using the vertical spring suspension shown. Springs 1 and 2 have stiffnesses k_1 and k_2, respectively, and unstretched lengths L_1 and L_2, respectively. The springs are installed by stretching them to the same length h and then attaching them to A. The vertical deflection δ is measured such that $\delta = 0$ is the position where A is in the middle of box B.

(a) Show that W and δ are related by $\delta = \dfrac{W - k_1(h - L_1) + k_2(h - L_2)}{k_1 + k_2}$.

(b) Explain why $\delta \neq 0$ when $W = 0$.

(c) Suggest some mathematical tests you can perform to verify the accuracy of the expression in Part (a). For example, if $\delta = 0$ and $k_2 = 0$, show that W has the expected value.

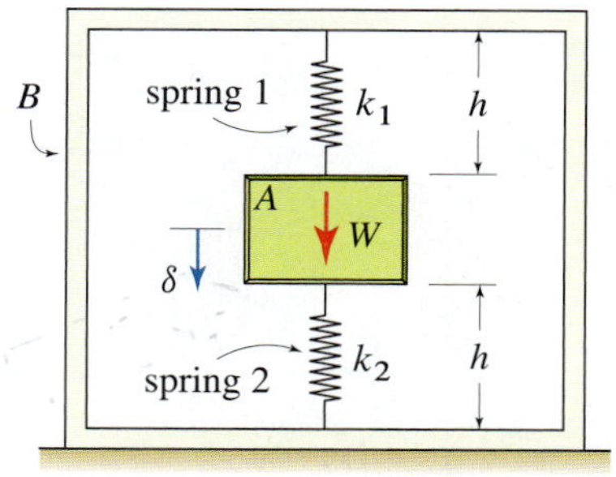

Figure P3.45

Problem 3.46

A model for the suspension of a truck is shown. Block A represents the chassis of the truck, and it may undergo vertical motion only with no rotation. Force F represents the portion of the truck's weight and payload that is supported by this suspension. Due to F, block A deflects by amount δ. When $F = 0$, both springs are undeformed and there is a gap h between the second spring and the chassis. When δ is less than h, only one spring supports F; and when δ exceeds h, the second spring engages and helps support F. If $k_1 = 1200$ lb/in., $k_2 = 600$ lb/in., and $h = 1.3$ in.,

(a) Determine δ when $F = 1400$ lb.

(b) Determine δ when $F = 2800$ lb.

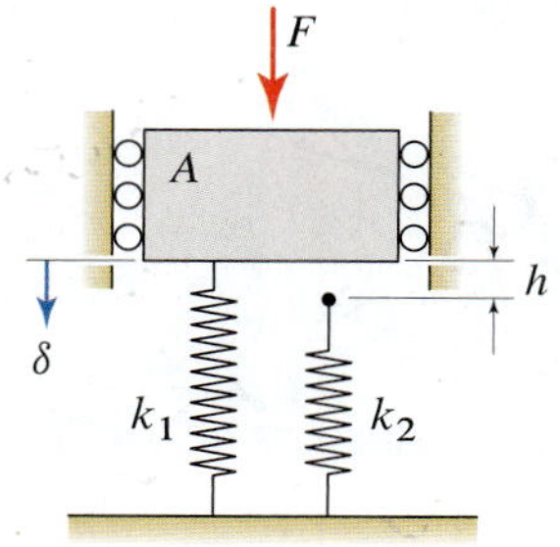

Figure P3.46

Problem 3.47

A model for a push button or key, such as on a calculator or computer keyboard, is shown. The model has the feature that if F is sufficiently large, point A snaps through to give the user positive tactile feedback that the keystroke was properly entered. Both springs have stiffness 2 N/mm, and when $F = 0$, the structure has the geometry shown and all springs are unstretched.

(a) Derive an expression that gives F as a function of δ.

(b) Plot F versus δ for $0 \le \delta \le 5$ mm.

(c) Determine the approximate largest value that F has for $0 \le \delta \le 2$ mm.

(d) What deficiency does this model display for representing a push button?

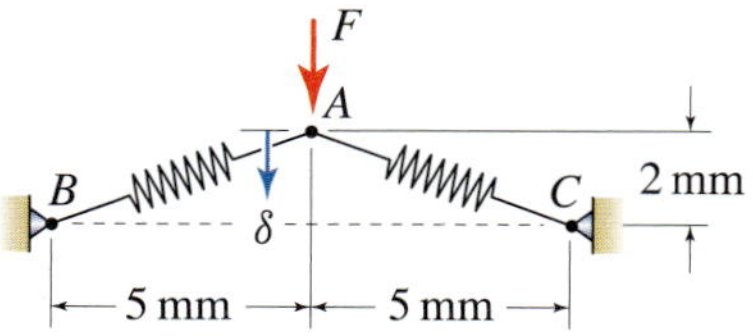

Figure P3.47

Problem 3.48

The model of Prob. 3.47 is revised to include a third spring. Springs AB and AC have stiffness 2 N/mm, and spring AD has stiffness 0.3 N/mm. When $F = 0$, the structure has the geometry shown and all springs are unstretched.

(a) Derive an expression that gives F as a function of δ.

(b) Plot F versus δ for $0 \le \delta \le 5$ mm.

(c) Determine the approximate largest value that F has for $0 \le \delta \le 2$ mm.

(d) Does spring AD serve a useful purpose in this model? Explain.

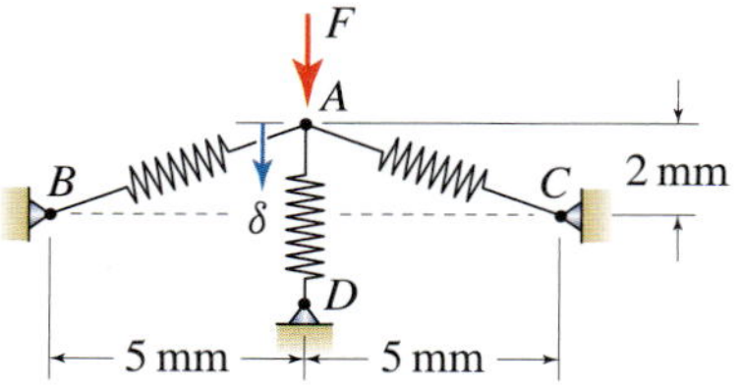

Figure P3.48

3.3 Equilibrium of Particles in Three Dimensions

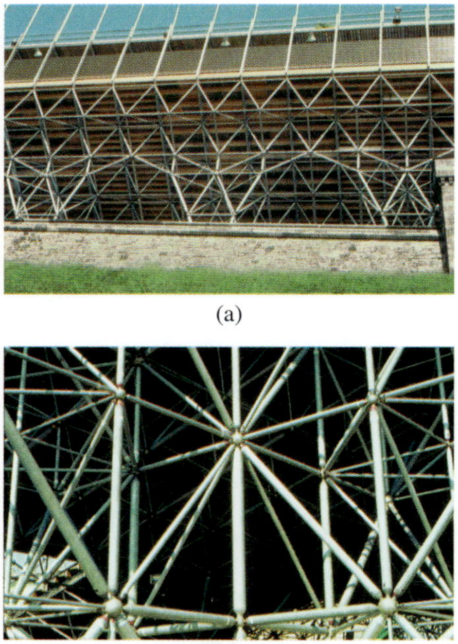

(a)

(b)

Figure 3.18
(a) Temporary grandstand erected at Edinburgh Castle, Scotland. (b) Close-up view showing numerous bar members that intersect at *joints*. Each of these joints may be modeled as a particle in equilibrium.

The conditions for static equilibrium of a particle in three dimensions are

$$\begin{aligned} &\sum \vec{F} = \vec{0}, \\ \text{or} \quad &\left(\sum F_x\right)\hat{\imath} + \left(\sum F_y\right)\hat{\jmath} + \left(\sum F_z\right)\hat{k} = \vec{0}, \\ \text{or} \quad &\sum F_x = 0 \text{ and } \sum F_y = 0 \text{ and } \sum F_z = 0. \end{aligned} \tag{3.20}$$

The first two expressions in Eq. (3.20) state conditions for equilibrium in *vector form*, while the last expression states conditions in *scalar form*. In three-dimensional problems, it will very often be advantageous to use the vector form.

Fundamentally, everything stated in earlier sections for equilibrium of particles in two dimensions also applies to equilibrium in three dimensions, except now all vectors have, in general, three components. The procedure for drawing an FBD of a particle is as described in Section 3.1, except now it is useful to think of the cut as being a closed surface that completely surrounds the particle. The procedure for analysis is also as described in Section 3.1, except now forces in the z direction must also sum to zero. All remarks made in Sections 3.1 and 3.2 on pulleys and the forces supported by bars, cables, and springs are still applicable. Some additional remarks on reactions and the solution of algebraic equations are helpful.

Reactions

We use the same thought process described in Section 3.1 to construct reactions for a particle in three dimensions. Namely, if a support prevents motion of a particle in a certain direction, it can do so only by producing a reaction force in that direction. When you are solving particle equilibrium problems, the supports and associated reactions shown in Fig. 3.19 occur often. It is not necessary to memorize these reactions; rather, you should reconstruct these as needed. For example, consider the slider on a fixed frictionless bar. The bar prevents motion of the slider in both the y and z directions, so there must be reactions in both of these directions. The slider is free to move in the x direction, hence there is no reaction in this direction.

Solution of algebraic equations

Analysis of equilibrium of one particle in three dimensions will typically produce three simultaneous linear algebraic equations (equilibrium equations) with three unknowns. More complicated problems may require several FBDs, and there will generally be three equations and three unknowns for each. If the problem has convenient geometry and/or symmetry, then solving these equations may be straightforward. In general, however, the solution of three or more equations is tedious.

Fortunately, efficient techniques for solving systems of simultaneous algebraic equations have been developed. Most scientific pocket calculators include programs for solving such equations, and you may wish to review the

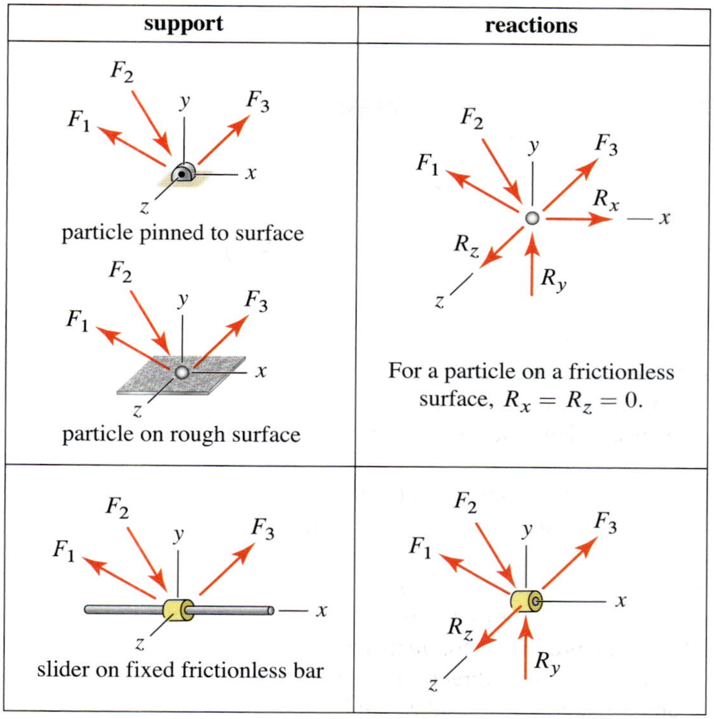

Figure 3.19. Some common supports and reactions in three-dimensional particle equilibrium problems. Forces F_1, F_2, and F_3 are hypothetical forces applied to a particle by cables and/or bars that might be attached to the particle; and forces R_x, R_y, and R_z are reactions.

owner's manual that came with your calculator to become familiar with this feature. In addition, programs such as Mathematica, Maple, Mathcad, EES (Engineering Equation Solver), MATLAB, and many others are available for use on personal computers and workstations. You are strongly encouraged to become familiar with some of these tools, as they will help you throughout statics and subjects that follow, and in your professional practice.

While computing tools are indispensable in engineering, it is your responsibility to be sure the answers produced by a calculator or computer are correct. Errors can arise from a variety of sources including poor modeling, errors in input data and/or program options that are selected, occasional software errors (these are rare, but do occur), outright blunders, and so on. Any solution you obtain, whether by computer or hand calculation, must be thoroughly interrogated so that you are certain it is correct. This is accomplished by substituting your solutions into *all* of the governing equations for the problem (i.e., the equilibrium equations plus perhaps other equations pertinent to the problem) and verifying that *all* of these equations are satisfied. This check does not verify that FBDs have been accurately drawn and equations have been accurately written. Thus, it is also necessary to scrutinize the solution to verify that it is physically reasonable. These comments hold for both two-dimensional and three-dimensional problems. However, it is generally more challenging to verify the validity of a solution to a three-dimensional problem.

Summing forces in directions other than x, y, or z

Most often, we choose coordinate directions x, y, and z for a particular problem because they provide a convenient description of geometry. As a result, writing vector expressions for forces and summing forces in these directions are usually straightforward. Occasionally, it will be useful to sum forces in a direction other than x, y, or z. While we could use a new coordinate system for this purpose where one of the new x, y, or z directions is coincident with the direction we are interested in, this is often not convenient because the necessary geometric information for writing vector expressions for forces may not be available or may be difficult to determine. An easier way is to use the dot product, as follows.

Newton's law in vector form is

$$\sum \vec{F} = \vec{0}: \quad \vec{F}_1 + \vec{F}_2 + \cdots + \vec{F}_n = \vec{0}. \tag{3.21}$$

Of course, because the force vectors in Eq. (3.21) are written in terms of x, y, and z components, Eq. (3.21) amounts to summation of forces in the x, y, and z directions. Let $\vec{r}$ be the direction in which we would like to sum forces. Often $\vec{r}$ will be a position vector. However, its physical significance is irrelevant, other than it describes a direction that is useful for summing forces. To sum forces in the direction of $\vec{r}$, we take the dot product of both sides of Eq. (3.21) with the unit vector* $\vec{r}/|\vec{r}|$ to obtain the scalar equilibrium equation

Helpful Information

Vector equation versus scalar equation. You should contrast Eqs. (3.21) and (3.22) to see that the first of these is a *vector equation* whereas the second is a *scalar equation.*

Summation of forces in the $\vec{r}$ direction:

$$\sum F_r = 0: \quad \vec{F}_1 \cdot \frac{\vec{r}}{|\vec{r}|} + \vec{F}_2 \cdot \frac{\vec{r}}{|\vec{r}|} + \cdots + \vec{F}_n \cdot \frac{\vec{r}}{|\vec{r}|} = 0. \tag{3.22}$$

In the above expression, $\vec{F}_1 \cdot \vec{r}/|\vec{r}|$ is the portion (or component) of force $\vec{F}_1$ that acts in the direction $\vec{r}$, $\vec{F}_2 \cdot \vec{r}/|\vec{r}|$ is the portion (or component) of $\vec{F}_2$ that acts in the direction $\vec{r}$, and so on. On the right-hand side is the scalar 0, which of course is the portion (or component) of the vector $\vec{0}$ that acts in the direction $\vec{r}$. Use of Eq. (3.22) is especially convenient in problems where we might know that some of the forces are perpendicular to $\vec{r}$, in which case the dot product of these forces with $\vec{r}$ is zero. This technique is illustrated in Example 3.8.

End of Section Summary

In this section, the equations governing static equilibrium of a particle in three dimensions were discussed. The main differences compared to equilibrium in two dimensions are that reactions are more complex, FBDs depict forces in three dimensions, and the equation systems that must be solved are larger. Some of the key points are as follows:

- Problems in three dimensions with simple geometry can often be effectively solved using a scalar approach. But very often a vector approach will be more tractable.

* Because the right-hand side of Eq. (3.22) is zero, a unit vector in the direction of $\vec{r}$ is not needed and we could just as well evaluate $\vec{F}_1 \cdot \vec{r} + \vec{F}_2 \cdot \vec{r} + \cdots + \vec{F}_n \cdot \vec{r} = 0$.

- The selection of an xyz coordinate system is dictated largely by the geometry that is provided for a particular problem. Often it will be effective to sum forces in directions other than the x, y, or z direction, and this is easily done using the dot product, as given by Eq. (3.22).
- Software for solving systems of simultaneous algebraic equations can be very helpful. Regardless of how you obtain your solution for a particular problem, whether it be analytically or by using software, it is your responsibility to verify the solution is correct, both mathematically and physically.

EXAMPLE 3.7 *Cables, Bars, and Failure Criteria*

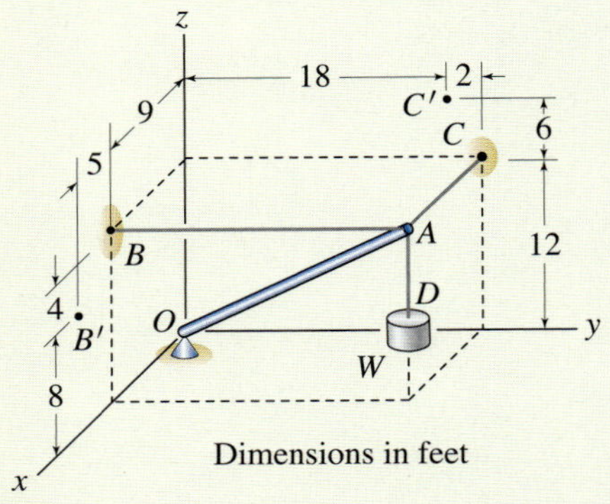

Figure 1

The weight W is supported by boom AO and cables AB, AC, and AD, which are parallel to the y, x, and z axes, respectively.

(a) If cables AB and AC can support maximum forces of 5000 lb each, and boom AO can support a maximum compressive force of 8000 lb before buckling, determine the largest weight W that can be supported. Assume that cable AD is sufficiently strong to support W.

(b) If the supports at points B and C are relocated to points B' and C', respectively, and $W = 1000$ lb, determine the forces supported by boom AO and cables AB, AC, and AD.

SOLUTION

Road Map In both parts of this problem, the solution approach is the same: make appropriate modeling decisions to draw an FBD, write vector expressions for forces, write equilibrium equations, and then solve these equations. In Part (a), we then apply failure criteria to each member to determine the largest load W that can be supported. In Part (b), we are given $W = 1000$ lb, and then we determine the force supported by each member. By inspection,* the force in cable AD is equal to the weight W.

Part (a)

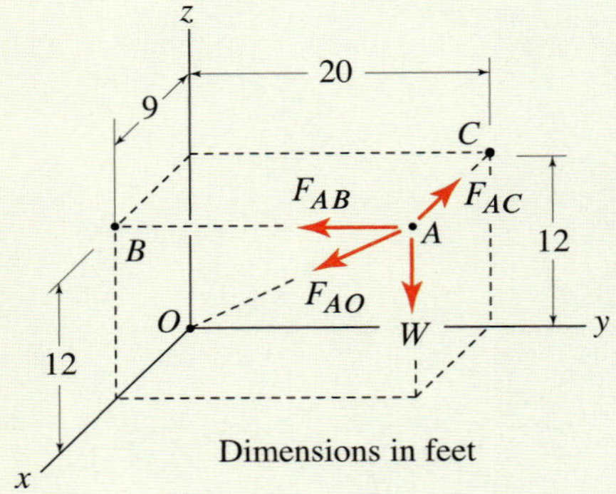

Figure 2
Free body diagram of point A where, by inspection, the force in cable AD is equal to the weight W.

Modeling The FBD for point A is shown in Fig. 2, where the cable forces and boom force are shown such that positive values correspond to tension. Vector expressions for the cable forces and weight may be written by inspection, while the force supported by the boom is written using position vector $\vec{r}_{AO}$, with the following results:

$$\vec{W} = W(-\hat{k}), \qquad \vec{F}_{AB} = F_{AB}(-\hat{\jmath}), \qquad \vec{F}_{AC} = F_{AC}(-\hat{\imath}),$$
$$\vec{F}_{AO} = F_{AO}\frac{\vec{r}_{AO}}{|\vec{r}_{AO}|} = F_{AO}\frac{-9\,\hat{\imath} - 20\,\hat{\jmath} - 12\,\hat{k}}{25}. \tag{1}$$

Governing Equations & Computation

Equilibrium Equations Applying Newton's law with terms grouped by direction provides

$$\sum \vec{F} = \vec{0}: \quad \left(F_{AO}\frac{-9}{25} - F_{AC}\right)\hat{\imath} + \left(F_{AO}\frac{-20}{25} - F_{AB}\right)\hat{\jmath} + \left(F_{AO}\frac{-12}{25} - W\right)\hat{k} = \vec{0}. \tag{2}$$

For Eq. (2) to be satisfied, each term must be zero independently. Thus, the following system of equations results:

$$F_{AO}\frac{-9}{25} - F_{AC} = 0, \tag{3}$$
$$F_{AO}\frac{-20}{25} - F_{AB} = 0, \tag{4}$$
$$F_{AO}\frac{-12}{25} - W = 0. \tag{5}$$

The solution to Eqs. (3)–(5) is easily obtained and is

$$F_{AO} = -\frac{25}{12}W, \quad F_{AB} = \frac{5}{3}W, \quad \text{and} \quad F_{AC} = \frac{3}{4}W. \tag{6}$$

* If this statement is not clear, you may formally arrive at this conclusion by drawing a FBD of point D (as in Fig. 3.4(b) on p. 117) and then use $\sum F_z = 0$.

With weight W being a positive number that still needs to be determined, Eq. (6) shows the forces supported by both cables are positive and hence they are in tension, and the force supported by the boom is negative and hence it is in compression.

Force Laws Using Eq. (6), the various failure criteria can be applied as follows:

$$\text{If } F_{AO} = -8000\text{ lb}, \quad \text{then } W = 3840\text{ lb}. \tag{7}$$
$$\text{If } F_{AB} = 5000\text{ lb}, \quad \text{then } W = 3000\text{ lb}. \tag{8}$$
$$\text{If } F_{AC} = 5000\text{ lb}, \quad \text{then } W = 6670\text{ lb}. \tag{9}$$

Only the smallest value of W in Eqs. (7)–(9) will simultaneously satisfy all three failure criteria, and thus the largest value W may have is $\boxed{3000\text{ lb.}}$

Part (b)

Modeling The FBD of point A is the same as that shown in Fig. 2, except the orientations of F_{AB} and F_{AC} are slightly different. Vector expressions for the weight, boom force, and cable force are

$$\vec{W} = 1000\text{ lb}(-\hat{k}), \tag{10}$$
$$\vec{F}_{AB} = F_{AB}\frac{\vec{r}_{AB}}{|\vec{r}_{AB}|} = F_{AB}\frac{5\,\hat{\imath} - 20\,\hat{\jmath} - 4\,\hat{k}}{21}, \tag{11}$$
$$\vec{F}_{AC} = F_{AC}\frac{\vec{r}_{AC}}{|\vec{r}_{AC}|} = F_{AC}\frac{-9\,\hat{\imath} - 2\,\hat{\jmath} + 6\,\hat{k}}{11}, \tag{12}$$
$$\vec{F}_{AO} = F_{AO}\frac{\vec{r}_{AO}}{|\vec{r}_{AO}|} = F_{AO}\frac{-9\,\hat{\imath} - 20\,\hat{\jmath} - 12\,\hat{k}}{25}. \tag{13}$$

Governing Equations Writing $\sum \vec{F} = \vec{0}$ provides the system of equations

$$F_{AB}\frac{5}{21} - F_{AC}\frac{9}{11} - F_{AO}\frac{9}{25} = 0, \tag{14}$$
$$-F_{AB}\frac{20}{21} - F_{AC}\frac{2}{11} - F_{AO}\frac{20}{25} = 0, \tag{15}$$
$$-F_{AB}\frac{4}{21} + F_{AC}\frac{6}{11} - F_{AO}\frac{12}{25} = 1000\text{ lb}. \tag{16}$$

Computation This system of equations is not as easy to solve as that in Part (a). Systems of equation that are difficult to solve are a fact of life in engineering, and you must be proficient in solving them. The basic strategy for hand computation (where one of the equations is solved for one of the unknowns in terms of the others and then this result is substituted into the remaining equations, and so on) is workable for systems of three equations, but it rapidly becomes very tedious for larger systems of equations. You should take this opportunity to use a programmable calculator or one of the software packages mentioned earlier to find the solutions to Eqs. (14)–(16), which are

$$\boxed{F_{AB} = 1027\text{ lb}, \quad F_{AC} = 930\text{ lb}, \quad \text{and} \quad F_{AO} = -1434\text{ lb}.} \tag{17}$$

In addition to the results in Eq. (17), the force supported by cable AD is equal to W, therefore $\boxed{F_{AD} = 1000\text{ lb.}}$

Discussion & Verification For both parts of this example, we expect the cables to be in tension, and the boom to be in compression, and indeed our solutions show this. A common error made in problems such as Part (a), where each member has its own failure criterion, is to assume that each member is at its failure load at the same time.

EXAMPLE 3.8 *Summing Forces in a Direction Other Than x, y, or z*

Bar AB is straight and is fixed in space. Spring CD has 3 N/mm stiffness and 200 mm unstretched length. If there is no friction between collar C and bar AB, determine

(a) The weight W of the collar that produces the equilibrium configuration shown.

(b) The reaction between the collar and bar AB.

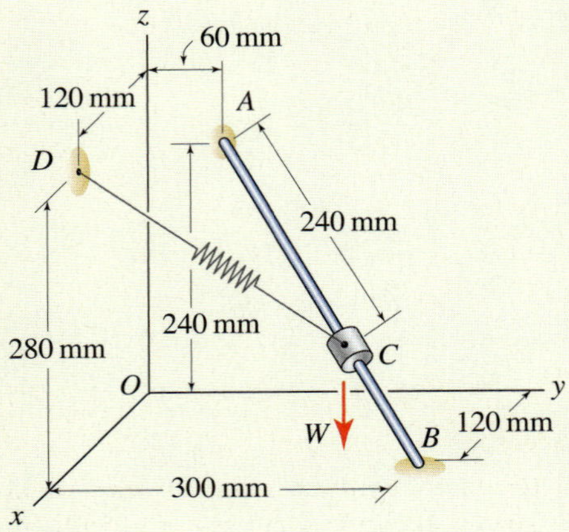

Figure 1

SOLUTION

Road Map The initial length of the spring is given, and sufficient information is provided to determine the final length of the spring. Thus, the spring force will be determined first. Then we will use the dot product to sum forces applied to the collar in the direction of bar AB. For Part (b), we will use the equilibrium equation to determine the reaction between the collar and bar.

Part (a)

Modeling The FBD for collar C is shown in Fig. 2. In accord with Fig. 3.19, there are two reactions components between the collar and bar AB. If we select the direction for $\vec{R}_1$ first, so that it is perpendicular to bar AB, whose direction is $\vec{r}_{AB}$, then the remaining reaction $\vec{R}_2$ must be perpendicular to both $\vec{R}_1$ and $\vec{r}_{AB}$. Observe that there are an infinite number of choices for the direction of $\vec{R}_1$ with the only requirement being that its direction must be perpendicular to $\vec{r}_{AB}$. Once $\vec{R}_1$ is selected, the requirement that $\vec{R}_2$ be perpendicular to both $\vec{R}_1$ and $\vec{r}_{AB}$ dictates the only direction it may have. With this approach, $\vec{R}_1$ and $\vec{R}_2$ have known directions but unknown magnitudes. Alternatively, we may call the reaction simply $\vec{R}$, where $\vec{R}$ is perpendicular to $\vec{r}_{AB}$, but otherwise has unknown direction and magnitude; this approach is employed later in this problem.

Governing Equations

Force Laws The spring's final length is the magnitude of the position vector from C to D (or vice versa). We could determine the coordinates of point C, and this along with the coordinates of point D would allow us write $\vec{r}_{CD}$. However, we will use the following approach:

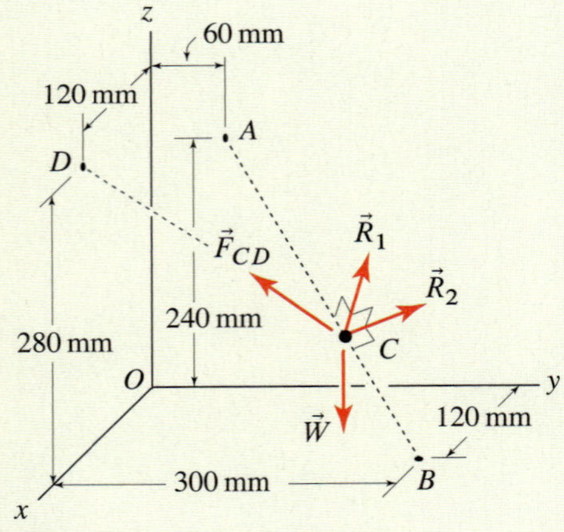

Figure 2
Free body diagram for the collar C.

$$\vec{r}_{CD} = \vec{r}_{CA} + \vec{r}_{AD} = (240\,\text{mm})\frac{\vec{r}_{BA}}{|\vec{r}_{BA}|} + \vec{r}_{AD}$$
$$= (240\,\text{mm})\frac{-120\,\hat{\imath} - 240\,\hat{\jmath} + 240\,\hat{k}}{360} + (120\,\hat{\imath} - 60\,\hat{\jmath} + 40\,\hat{k})\,\text{mm}$$
$$= (40\,\hat{\imath} - 220\,\hat{\jmath} + 200\,\hat{k})\,\text{mm}. \qquad (1)$$

Hence, the final length of the spring is $L = |\vec{r}_{CD}| = 300$ mm, and using Eq. (3.18) on p. 137, the force supported by the spring is

$$F_{CD} = k(L - L_0) = \left(3\,\frac{\text{N}}{\text{mm}}\right)(300\,\text{mm} - 200\,\text{mm}) = 300\,\text{N}. \qquad (2)$$

Equilibrium Equations Using the FBD in Fig. 2, Newton's law may be written for the collar as

$$\sum \vec{F} = \vec{0}: \quad \vec{F}_{CD} + \vec{R}_1 + \vec{R}_2 + \vec{W} = \vec{0}. \qquad (3)$$

Vector expressions for the weight and spring force are

$$\vec{W} = W(-\hat{k}), \qquad \vec{F}_{CD} = F_{CD}\frac{\vec{r}_{CD}}{|\vec{r}_{CD}|} = (300\,\text{N})\frac{40\,\hat{\imath} - 220\,\hat{\jmath} + 200\,\hat{k}}{300}. \qquad (4)$$

Vector expressions for $\vec{R}_1$ and $\vec{R}_2$ can be developed using a variety of strategies.* However, some careful thought will allow us to avoid the use of these vectors altogether for Part (a) of this example! Referring to the FBD shown in Fig. 2, observe that $\vec{R}_1$ and $\vec{R}_2$ are both perpendicular to direction BA. Thus, if we sum forces in this direction, we know $\vec{R}_1$ and $\vec{R}_2$ will not appear in this equation since they have no component in this direction. This can be accomplished by using the dot product on each force to determine how much of them, or what component, acts in the BA direction. Thus, as discussed in connection with Eq. (3.22) on p. 150, we take the dot product of both sides of Eq. (3) with the unit vector $\vec{r}_{BA}/|\vec{r}_{BA}|$ to write

$$\vec{F}_{CD}\cdot\frac{\vec{r}_{BA}}{|\vec{r}_{BA}|} + \underbrace{\vec{R}_1\cdot\frac{\vec{r}_{BA}}{|\vec{r}_{BA}|}}_{=0} + \underbrace{\vec{R}_2\cdot\frac{\vec{r}_{BA}}{|\vec{r}_{BA}|}}_{=0} + \vec{W}\cdot\frac{\vec{r}_{BA}}{|\vec{r}_{BA}|} = \underbrace{\vec{0}\cdot\frac{\vec{r}_{BA}}{|\vec{r}_{BA}|}}_{=0}. \tag{5}$$

In Eq. (5) the dot product with the zero vector on the right-hand side is of course zero. Because $\vec{R}_1$ and $\vec{R}_2$ are orthogonal to $\vec{r}_{BA}$, their dot products must also be zero and because of this knowledge we may altogether avoid evaluating those dot products in Eq. (5). Evaluating the remaining two dot products in Eq. (5) results in

$$\begin{aligned} 0 &= (300\,\text{N})\frac{40\,\hat{\imath} - 220\,\hat{\jmath} + 200\,\hat{k}}{300}\cdot\frac{-120\,\hat{\imath} - 240\,\hat{\jmath} + 240\,\hat{k}}{360} \\ &\quad + W(-\hat{k})\cdot\frac{-120\,\hat{\imath} - 240\,\hat{\jmath} + 240\,\hat{k}}{360} && (6) \\ &= (300\,\text{N})(0.8889) - W(0.6667). && (7) \end{aligned}$$

Computation Equation (7) is a scalar equation, and its solution is

$$\boxed{W = 400\,\text{N}.} \tag{8}$$

Note that as promised, this clever solution approach avoided having to compute the reactions between the collar and bar.

Part (b)

Computation To determine the reactions between the collar and bar, we return to Eq. (3) and solve for the reaction $\vec{R} = \vec{R}_1 + \vec{R}_2$ as

$$\begin{aligned} \vec{R} &= -\vec{F}_{CD} - \vec{W} \\ &= (-300\,\text{N})\frac{40\,\hat{\imath} - 220\,\hat{\jmath} + 200\,\hat{k}}{300} - (400\,\text{N})(-\hat{k}) \\ &= \boxed{(-40\,\hat{\imath} + 220\,\hat{\jmath} + 200\,\hat{k})\,\text{N},} && (9) \end{aligned}$$

and the magnitude of the above reaction is

$$\boxed{|\vec{R}| = 300\,\text{N}.} \tag{10}$$

Discussion & Verification As a partial check of our solution, we can verify that $\vec{R}$ given by Eq. (9) is perpendicular to bar BA by showing $\vec{R}\cdot\vec{r}_{BA} = 0$. While $\vec{R}$ must be given by the expression determined in Eq. (9) for equilibrium to prevail, there are an infinite number of vectors $\vec{R}_1$ and $\vec{R}_2$ that are perpendicular to bar AB and perpendicular to one another and will still sum to $\vec{R}$. If you construct $\vec{R}_1$ and $\vec{R}_2$ using the suggestions provided in the footnote on this page, then you will obtain one of these possible combinations. Problem 3.66 asks you to do this.

Helpful Information

Use of dot product to sum forces. Use of the dot product to sum forces in a direction other than a coordinate direction can be very helpful, as illustrated in this example. This example can also be solved by directly solving Eq. (3) (i.e., $\sum \vec{F} = \vec{0}$), but this requires writing expressions for $\vec{R}_1$ and $\vec{R}_2$ and some possibly tedious algebra.

---

* One strategy is to evaluate the cross product $\vec{r}_{BA}\times\vec{r}_{CD}$, then make this result a unit vector called $\hat{u}_1$, and then $\vec{R}_1 = R_1\hat{u}_1$. Next evaluate $\vec{r}_{BA}\times\vec{R}_1$, make this result a unit vector called $\hat{u}_2$, and then $\vec{R}_2 = R_2\hat{u}_2$.

PROBLEMS

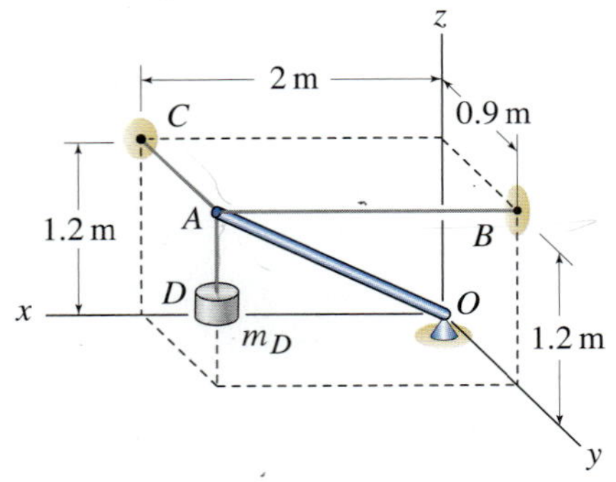

Figure P3.49 and P3.50

Problem 3.49

The object at D is supported by boom AO and by cables AB, AC, and AD, which are parallel to the x, y, and z axes, respectively. If the mass of object D is $m_D = 100$ kg, determine the forces supported by the boom and three cables.

Problem 3.50

In Prob. 3.49, if the cables and boom have the failure strengths given below, determine the largest mass m_D that can be supported.

Member	Strength
AO	2000 N compression
AB	1500 N
AC	1000 N
AD	2000 N

Problem 3.51

A weight W is supported at A by bars AB and AD and by cable AC. Bars AB and AD are parallel to the y and x axes, respectively. If $W = 1000$ N, determine the forces in each bar and the cable.

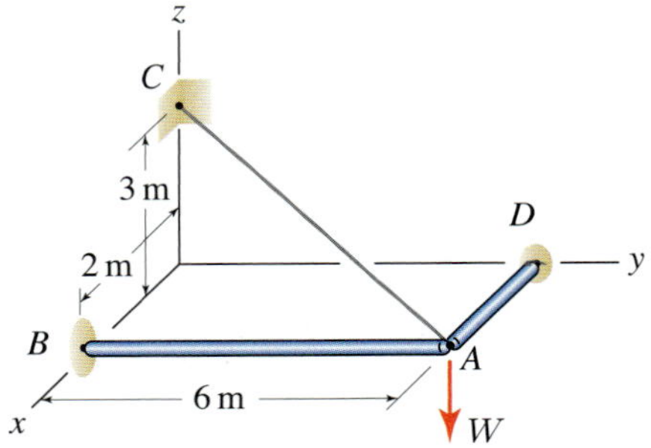

Figure P3.51

Problem 3.52

In Prob. 3.51, if the cable can support a maximum force of 2 kN, and bars AB and AD can support maximum compressive forces of 1 kN and 3 kN, respectively, determine the largest weight W that may be supported.

Problem 3.53

When in equilibrium, plate BCD is horizontal. If the plate weighs 1.6 kN, determine the forces in cables AB, AC, and AD.

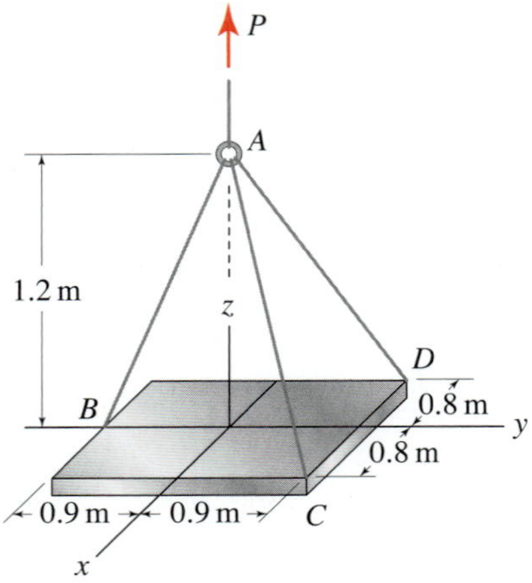

Figure P3.53

Problem 3.54

A circular ring with weight W and inside radius r is supported by three identical springs having stiffness k and that are unstretched when $d = 0$. When in equilibrium, the ring is horizontal.

(a) Derive an expression that relates the weight W to d and r.

(b) If $W = 300$ lb, $k = 25$ lb/in., and $r = 20$ in., determine d (an accurate approximate solution is acceptable).

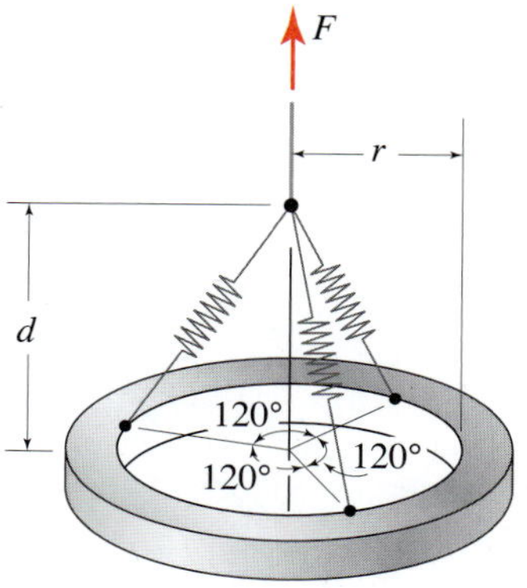

Figure P3.54

Problem 3.55

In the cable system shown, point D lies in the yz plane and force P is parallel to the z axis. If $P = 500$ lb, determine the force in cables AB, AC, AD, and AE.

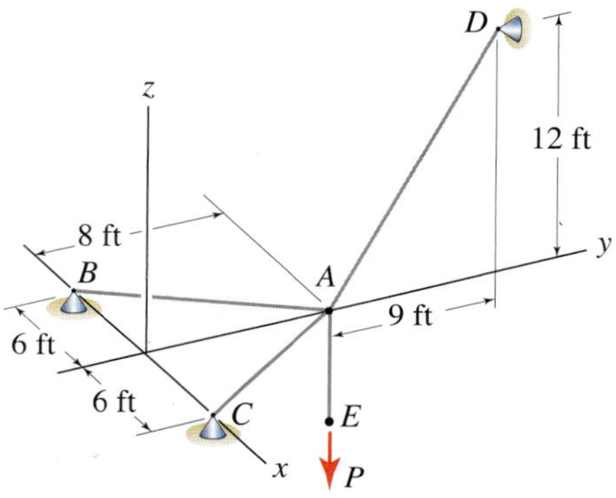

Figure P3.55 and P3.56

Problem 3.56

In the cable system shown, point D lies in the yz plane and force P is parallel to the z axis. If cables AB and AC have 500 lb breaking strength, and cables AD and AE have 1000 lb breaking strength, and if all cables are to have a factor of safety* of 1.5 against failure, determine the largest force P that may be supported.

Problem 3.57

Repeat Prob. 3.55 with point A having coordinates $A(1, 8, 0)$ ft.

Problem 3.58

Repeat Prob. 3.56 with point A having coordinates $A(1, 8, 0)$ ft.

* The *factor of safety* against failure is defined to be the failure load for a member divided by the allowable load for the member. Thus, the largest load the member may be subjected to is its failure load divided by the factor of safety.

Problem 3.59

Force P is supported by two cables and a bar. Point A lies in the yz plane, and points B and C lie in the xz plane. If $P = 3$ kip, determine the forces supported by the cables and bar.

Problem 3.60

Force P is supported by two cables and a bar. Point A lies in the yz plane, and points B and C lie in the xz plane. The compressive load that causes the bar to buckle and the breaking strength of each cable are specified below. If factors of safety against failure (see the footnote of Prob. 3.56) of 1.7 and 2.0 are to be used for cables and bars, respectively, determine the allowable force P that can be supported.

Member	Strength
AO	3000 lb compression
AB	6000 lb
AC	5000 lb

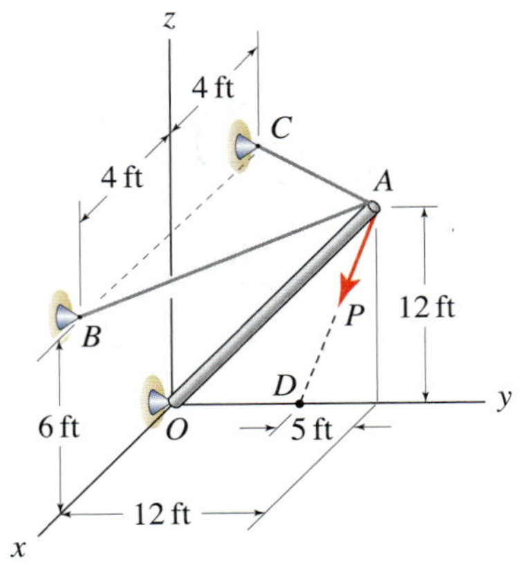

Figure P3.59 and P3.60

Problem 3.61

Member OA buckles when the compressive force it supports reaches 400 N. Cables AC and AD each have 300 N breaking strength. Assuming the cabling between A and B is sufficiently strong, determine the force T that will cause the structure to fail. Assume the pulleys are frictionless with diameters that are small enough so that all cables between A and B are parallel to line AB.

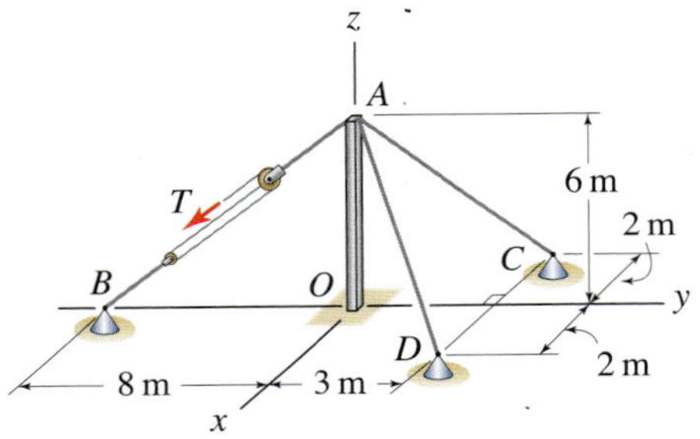

Figure P3.61

Problem 3.62

A portable tripod hoist for moving objects in and out of a manhole is shown. The hoist consists of identical-length bars AB, AC, and AD that are connected by a socket at A and are supported by equal 8 ft length cables BC, BD, and CD to prevent ends B, C, and D of the bars from slipping. Cable FAE passes around a frictionless pulley at A and terminates at winch E, which is fixed to bar AB. Idealize points A and E to be particles where all bar and cable forces pass through these points. If the tripod is erected on level ground and is 7 ft high, and the object being lifted in the manhole weighs 300 lb,

(a) Determine the forces in bars AC and AD and in portion AE of bar AB.

(b) Determine the force in portion EB of bar AB.

Hint: Define an xyz coordinate system where the x and y directions lie in the plane defined by points B, C, and D, and where the x or y direction coincides with one of cables BC, BD, or CD. Then determine the coordinates of points B, C, and D. The x and y coordinates of point A are then the averages of the coordinates of points B, C, and D.

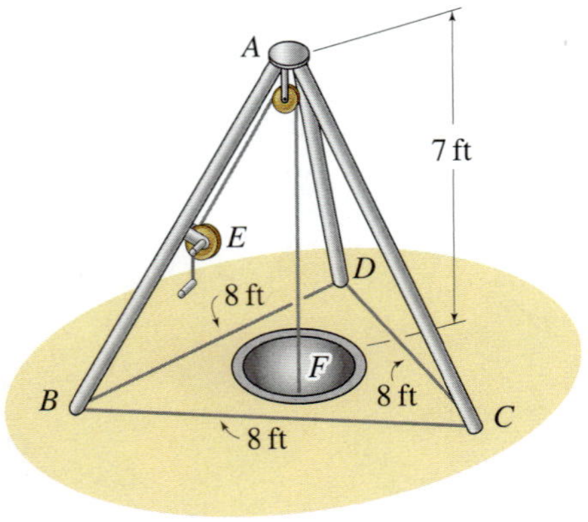

Figure P3.62

Problem 3.63

A worker standing on a truck uses rope AB to slowly lower object B down a chute. The object weighs 100 lb and fits loosely against the walls of the chute and slides with no friction. In the position shown, the center of object B is halfway down the chute. The person's hand A lies in the xy plane, and the chute lies in a plane parallel to the yz plane.

(a) Determine the tension in rope AB and the reactions between the object and the chute.

(b) If the worker wishes to slowly pull the object up the chute, explain how your answers to Part (a) change.

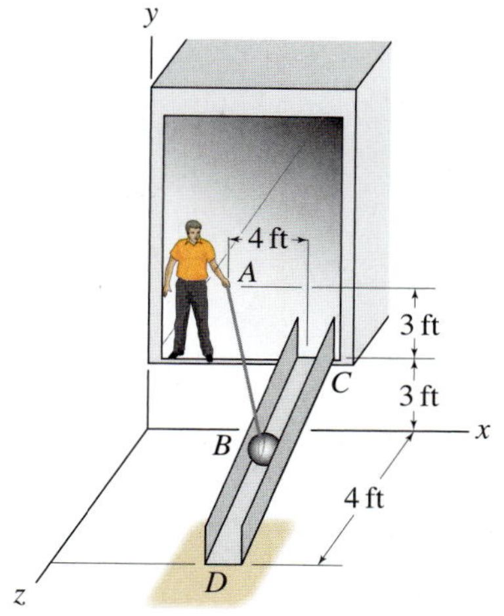

Figure P3.63

Problem 3.64

Due to a poorly designed foundation, the statue at point A slowly slides down a grass-covered slope. To prevent further slip, a cable is attached from the statue to point B, and another cable is attached from the statue to point C. The statue weighs 1000 lb, and idealize the surface on which the statue rests to be frictionless. Determine the minimum tensile strength each cable must have, and the magnitude of the reaction between the statue and the slope.

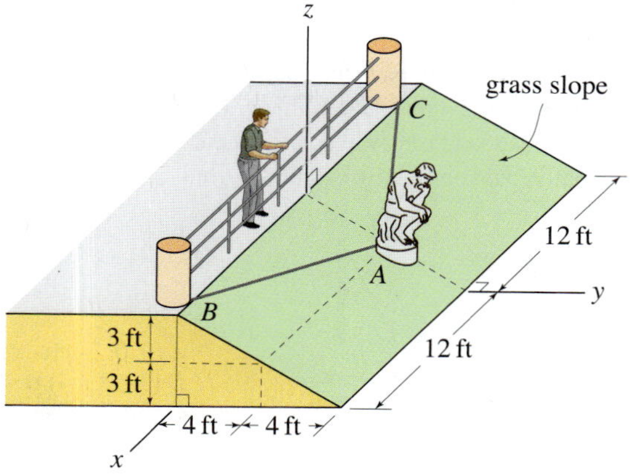

Figure P3.64

Problem 3.65

Channel AB is fixed in space, and its centerline lies in the xy plane. The plane containing edges AC and AD of the channel is parallel to the xz plane. If the surfaces of the channel are frictionless and the sphere E has 2 kg mass, determine the force supported by cord EF, and the reactions R_C and R_D between the sphere and sides C and D, respectively, of the channel.

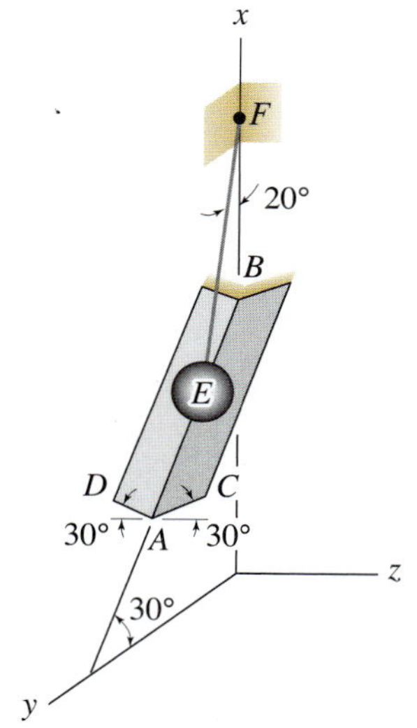

Figure P3.65

Problem 3.66

Follow the suggestions made in the footnote of Example 3.8 on p. 155 to write vector expressions for $\vec{R}_1$ and $\vec{R}_2$. Then determine the magnitude of these, and weight $\vec{W}$, by applying $\sum \vec{F} = \vec{0}$. Show that the magnitude of $\vec{W}$ and the vector sum $\vec{R}_1 + \vec{R}_2$ agree with the results reported in Example 3.8.

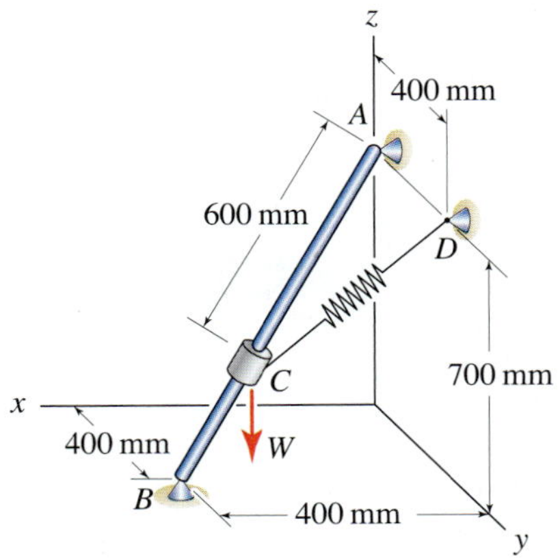

Figure P3.67

Problem 3.67

Rod AB is fixed in space. Spring CD has stiffness 1.5 N/mm and an unstretched length of 400 mm. If there is no friction between the collar and rod, determine the weight of the collar W that produces the equilibrium configuration shown, and the reaction between the collar and rod AB.

Problem 3.68

In Prob. 2.81 on p. 89, the elastic cord CD has stiffness $k = 2$ N/mm and 100 mm unstretched length. Bead C has negligible weight and is subjected to a force of magnitude P that lies in the xy plane and is tangent to the curved rod AB in the position shown. Determine the value of P needed for equilibrium and the reaction between the bead and curved rod AB.

Problem 3.69

In Prob. 2.82 on p. 89, elastic cord BD has spring constant $k = 3$ N/mm and 20 mm unstretched length, and bead B has a force of magnitude P in direction BC applied to it. Determine the value of P needed for equilibrium in the position shown, and the reaction between bead B and rod AC.

Problem 3.70

Two identical traffic lights each weighing 120 lb are to be suspended by cables over an intersection. For points E and F to have the coordinates shown, it is necessary to add an additional weight to one of the lights. Determine the additional weight needed, the light to which it should be added, and forces in each cable. *Remark:* The system of equations can be solved manually, but solution by calculator or computer is recommended.

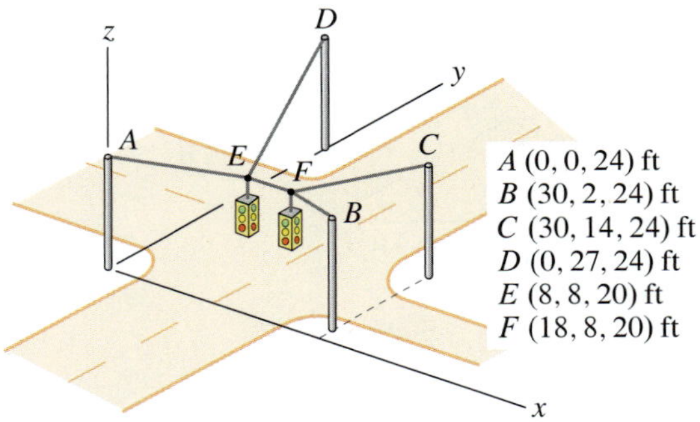

Figure P3.70

3.4 Engineering Design

While there are many different ways to define engineering design, the description that follows encompasses the key points. *Engineering design* is a process that culminates in the description and specifications of how a structure, machine, device, procedure, and so on is to be constructed so that needs and requirements that were identified in the design process are achieved. The word *engineering* implies that laws of nature, mathematics, and computing are at the core of the design. Design is a creative and exciting activity, and broadly speaking, proficiency in doing this is what sets engineering apart from the sciences. Indeed, the ability to design is a distinct point of pride among all engineers!

The word *process* in the above definition is important. Design should not be conducted solely by iterations of trial and error, but rather by iterations of identifying and prioritizing needs, making value decisions, and exploiting the laws of nature to develop a sound solution that optimizes all of the objectives while satisfying constraints that have been identified. As you progress in your education, you will learn about structured procedures and algorithms for design, including exposure to pertinent performance standards, safety standards, and design codes. In the meantime, your knowledge of particle equilibrium allows you to begin some meaningful engineering design.

Figure 3.20
An engineer examines the design of the Cosworth XF Champ car engine, and a photograph of this engine as manufactured is shown. This was the first Cosworth engine that was designed completely using computer aided design (CAD) tools. While CAD tools help alleviate the burden of performing complicated calculations, the engineers who use these tools must be experts in modeling, interpretation of results, ability to perform hand calculations on simpler models to help verify the results of more sophisticated models, and in the methodology of design.

Objectives of design

Our discussion of the objectives of design and the design process will be brief in this section. Chapter 5 contains additional discussion. At a minimum, structures, machines, mechanical devices, and so on must be designed to achieve the following criteria:

1. They will not fail during normal use.
2. Performance objectives are adequately met.
3. Hazards are adequately addressed.
4. All foreseeable uses must be anticipated and accounted for to the extent possible or reasonable.
5. Design work must be thoroughly documented and archived.

In addition, a good design will also consider

6. Cost.
7. Ease of manufacture and assembly.
8. Energy efficiency in both manufacture and use.
9. Impact on the environment in manufacture, use, and retirement.

The need for item 1 is obvious. Further, your design should not contain unnecessary hazards. Through good design, it may be possible to eliminate or reduce a hazard. Clearly some devices, due to their very nature, contain hazards that are unavoidable. For example, a machine to cut paper, whether

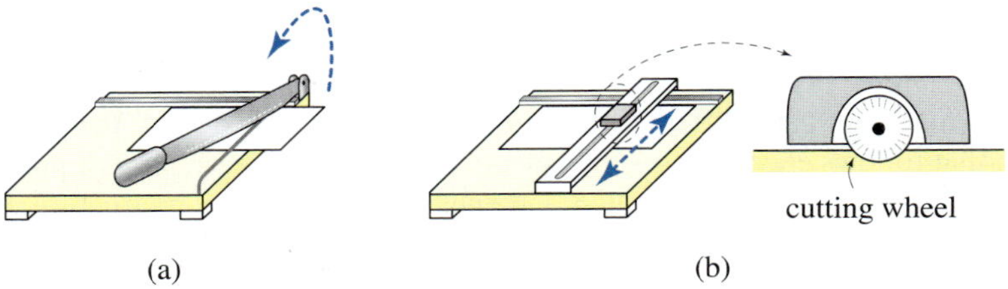

Figure 3.21. Two table-top paper cutters.

it be a simple scissors, a tabletop cutter, or a device for cutting large rolls of paper, must contain sharp cutting edges. Consider the two tabletop paper cutters shown in Fig. 3.21. The traditional cutter shown in Fig. 3.21(a) uses a long, heavy steel blade that is unavoidably exposed to a user's fingers and hands, whereas the more contemporary cutter shown in Fig. 3.21(b) employs a small circular cutting wheel that is less exposed and easier to guard and in fact performs better. When hazards are unavoidably present, they should be guarded to the extent feasible, and warnings should be included to pictorially show users what the hazards are. Thus, the hierarchy in managing hazards is as follows: eliminate or reduce the hazard through design, guard users from the hazard, and finally warn users of the hazard.

An extremely important aspect of design, and one that is unfortunately sometimes overlooked, is anticipation of and accounting for how a structure, machine, or device might be used, even if it was not intended for such use. While we may occasionally need to consider malicious misuse, here we are more concerned with a well-intentioned user who pushes the limits of a product in an effort to complete some task. For example, imagine you design a jack for use in lifting an automobile so a tire may be changed. If the jack is designed to have 2000 lb lifting capacity, most certainly someone will try to lift more. While your design does not need to be capable of lifting more than 2000 lb plus some reasonable margin for error, the main question is, What is the consequence of attempting to lift more? Will the jack collapse, allowing the vehicle to fall and potentially hurt someone, or can it be designed so the result is less catastrophic?

Regardless of how fertile and creative a mind you have, it will be difficult for your ideas to have an impact if you cannot communicate effectively, both orally and in writing, with other engineers, scientists, professionals, and laypeople. It is necessary that your design and analysis work be documented and archived. This information may be needed in the future by you or others, to show how your design was developed, to aid in modifications, to help support patent rights, and so on. Expectations of engineers are high, and it is necessary that you perform your work precisely and conduct your behavior to the highest standards of ethics and professionalism.

Particle equilibrium design problems

For the problems in this section, imagine you are employed as an intern working under the supervision of an engineer who asks you to conduct a design study, or to add details to a design that is started. You will be presented with a problem that is suitable to your level of knowledge along with some pertinent data. But, unlike most textbook homework problems, the information provided may not be complete, and you may not be instructed on everything that needs to be done. This is not an attempt to be artificially vague, but rather is a reflection of how design and modeling of real life problems are conducted. Typically, a design begins with an idea or the desire to accomplish something. What information is needed and the analysis tasks to be performed are often not fully known at the outset, but rather are discovered as the design and analysis process evolves. It is in this spirit that the problems of this section are presented, and you may need to make reasonable assumptions or seek out additional information on your own. Your work should culminate in a short written technical report that is appropriate for an engineer to read, where you present your design, state assumptions made, and so on. Appendix A of this book gives a brief discussion of technical writing that may be helpful.

Strength of steel cable, bar, and pipe

Many of the problems in this section involve steel cable, bar, and pipe. For these problems, use the allowable loads given in Tables 3.1–3.3, which are

Table 3.1 Allowable loads for steel cable. Includes suitable factor of safety for repeated loads.

Steel Cable[a] U.S. Customary and SI units			
Nominal diameter		Allowable load	
1/4 in.	6 mm	1,300 lb	6.0 kN
3/8 in.	10 mm	3,000 lb	13 kN
1/2 in.	13 mm	5,000 lb	20 kN
5/8 in.	16 mm	8,000 lb	35 kN
3/4 in.	19 mm	11,000 lb	50 kN
7/8 in.	22 mm	15,000 lb	70 kN
1 in.	25 mm	20,000 lb	90 kN

[a] 1/4 and 3/8 in. (6 and 10 mm) diameter cable is 6×19 construction; others are 6×26 construction. Allowable load is obtained using a factor of safety of 5 against breaking loads published by manufacturer: Performance Series 620 Rope, Wire Rope Industries, Ltd. If pulleys are used, pulley diameter is recommended to be at least 34 times the nominal cable diameter, and 51 times is suggested.

Table 3.2. Allowable loads for round steel bar. Includes suitable factors of safety for repeated loads.

Steel Bar[a] U.S. Customary units				
Nominal diameter	Tensile load not to exceed	Compressive load: not to exceed the smaller of[b]		
1/8 in.	200 lb	200 lb	or	$1,700\ \text{lb·in.}^2/L^2$
1/4 in.	850 lb	850 lb	or	$28(10)^3\ \text{lb·in.}^2/L^2$
1/2 in.	3,500 lb	3,500 lb	or	$450(10)^3\ \text{lb·in.}^2/L^2$
3/4 in.	7,500 lb	7,500 lb	or	$2.2(10)^6\ \text{lb·in.}^2/L^2$
1 in.	14,000 lb	14,000 lb	or	$7.2(10)^6\ \text{lb·in.}^2/L^2$

SI units				
Nominal diameter	Tensile load not to exceed	Compressive load: not to exceed the smaller of[c]		
3 mm	0.85 kN	0.85 kN	or	$3.9(10)^6\ \text{N·mm}^2/L^2$
6 mm	3.5 kN	3.5 kN	or	$62(10)^6\ \text{N·mm}^2/L^2$
12 mm	14 kN	14 kN	or	$1(10)^9\ \text{N·mm}^2/L^2$
18 mm	30 kN	30 kN	or	$5(10)^9\ \text{N·mm}^2/L^2$
24 mm	55 kN	55 kN	or	$16(10)^9\ \text{N·mm}^2/L^2$

[a] Allowable tension load and first column of allowable compressive load are obtained using a factor of safety of 2 against yielding for 36 ksi (250 MPa) steel. Second column of allowable compressive load is obtained using a factor of safety of 2 against the theoretical buckling load for a centrically loaded pin-supported bar.
[b] L represents the length of the member in inches.
[c] L represents the length of the member in mm.

Table 3.3. Allowable loads for standard weight steel pipe. Includes suitable factors of safety for repeated loads.

Steel Pipe[a]

U.S. Customary units

Nominal inside diameter	Tensile load not to exceed	Compressive load: not to exceed the smaller of[b]		
0.5 in.	4,500 lb	4,500 lb	or	$2.5(10)^6$ lb·in.$^2/L^2$
1 in.	8,500 lb	8,500 lb	or	$12(10)^6$ lb·in.$^2/L^2$
1.5 in.	14,000 lb	14,000 lb	or	$45(10)^6$ lb·in.$^2/L^2$
2 in.	19,000 lb	19,000 lb	or	$98(10)^6$ lb·in.$^2/L^2$

SI units

Nominal inside diameter	Tensile load not to exceed	Compressive load: not to exceed the smaller of[c]		
13 mm	20 kN	20 kN	or	$6.9(10)^9$ N·mm$^2/L^2$
25 mm	40 kN	40 kN	or	$36(10)^9$ N·mm$^2/L^2$
38 mm	60 kN	60 kN	or	$120(10)^9$ N·mm$^2/L^2$
51 mm	85 kN	85 kN	or	$270(10)^9$ N·mm$^2/L^2$

[a] Allowable tension load and first column of allowable compressive load are obtained using a factor of safety of 2 against yielding for 36 ksi (250 MPa) steel. Second column of allowable compressive load is obtained using a factor of safety of 2 against the theoretical buckling load for a centrically loaded pin-supported bar. Geometric data used to compute these forces is obtained from the *Manual of Steel Construction—Load and Resistance Factor Design*, vol. I, American Institute of Steel Construction, 1998, and the *Metric Conversion Volume* of the same title, 1999.

[b] L represents the length of the member in inches.

[c] L represents the length of the member in mm.

generally applicable for situations in which loads may be cyclically applied and removed many times. *Allowable loads*, or *working loads*, are the forces to which components can be safely subjected, and these are obtained by dividing the failure strength of a member by a *factor of safety*. Thus,

$$\text{Allowable load} = \frac{\text{failure load}}{\text{factor of safety}} \quad \text{or} \qquad \text{Factor of safety} = \frac{\text{failure load}}{\text{allowable load}}. \tag{3.23}$$

The failure load can be defined in different ways. In some applications it may be the load at which a member breaks or ruptures, while in other applications it may be the load at which a member starts to *yield* (i.e., take on permanent deformation) or otherwise fails to perform adequately. The factor of safety helps account for uncertainties in loads and strength, consequences of failure, and many other concerns. Determining a suitable factor of safety requires many value decisions to be made. Sometimes an appropriate factor of safety is recommended by the manufacturer of a component and is reflected in the allowable load data the manufacturer provides, or it may be specified by industry standards or design codes. Often you will need to make such decisions. Determining an appropriate factor of safety may be problem-dependent and if industry or government standards are available, you may be required to use these. Even if use of these is voluntary, it is obviously prudent to adhere to these as minimum standards. While the data of Tables 3.1–3.3 is safe and reasonable for many real life applications, more stringent factors of safety may be required for specific applications. However, you should not indiscriminately use excessive factors of safety, as this generally implies more massive components, which increases cost and may decrease performance.

Cables, bars, and pipes can support tensile forces, and the allowable tensile loads listed in Tables 3.1–3.3 should not be exceeded. Pipes and bars can also support compressive forces, as follows. If the member is short, then its allowable compressive load is similar in magnitude to, or even the same as, its allowable tensile load. However, if the member is long, then it may fail by buckling, and the susceptibility to buckling increases rapidly with increasing length. Thus, the allowable compressive load for bar and pipe is the smaller of the two values listed in Tables 3.2 and 3.3, where it is seen that the buckling load depends on the inverse of the member's length squared. For most applications of bars and pipes loaded in compression, buckling considerations usually govern allowable loads. Buckling is discussed in greater detail in Chapter 6.

A cable, sometimes called wire rope, is a complex composite structural member consisting of an arrangement of wires twisted into strands, which are then twisted into a cable, often having a core of different material that is intended to improve flexibility or performance. Cables are manufactured in an enormous variety of materials, sizes, and constructions for different purposes. Cables that are used in conjunction with pulleys are more severely loaded than cables used solely as straight, static structural tension members. When a cable is wrapped around a pulley, it is subjected to high bending loads and if the pulley's diameter is too small, the cable's life will be substantially reduced. The allowable loads reported in Table 3.1 are for a very common construction of steel cable and use a factor of safety of 5 against breakage of a straight segment of cable. These loads can also be employed for cables used with pulleys provided the pulleys have a diameter that is at least about 34 times the nominal diameter of the cable (51 is suggested). However, the factor of safety is then not as generous. For many applications, the factor of safety should be taken higher. For example, in passenger elevators a factor of safety of at least 10 is generally used.

EXAMPLE 3.9 *Engineering Design*

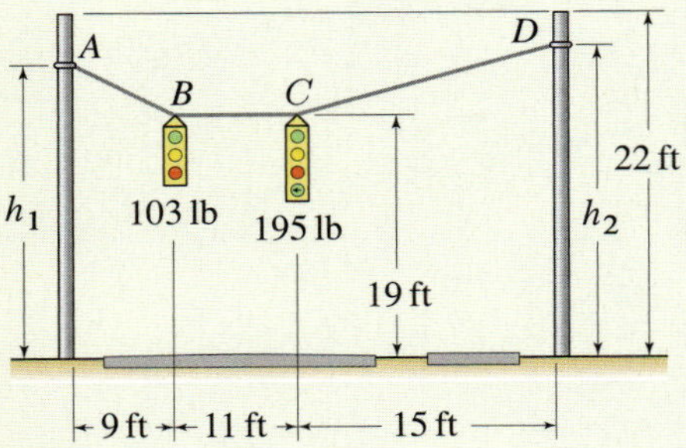

Figure 1

Cabling is to be designed to support two traffic lights such that both lights are at the same 19 ft height above the road, with the horizontal spacing shown in Fig. 1. Each of the supporting poles has 22 ft height. Specify dimensions h_1 and h_2 where ends of the steel cable should be attached to the support poles. Also specify the appropriate diameter of a single steel cable, from the selection in Table 3.1, that will safely support the weight of the two traffic lights.

SOLUTION

Road Map Possible values for h_1 and h_2 are between 19 and 22 ft. Further, these values are not independent. That is, if specific values for both h_1 and h_2 are assumed, it is unlikely the equilibrium configuration will have both traffic lights at the same 19 ft height above the roadway. Rather, if a value for h_1 is assumed, equations of equilibrium will determine the value h_2 must have, and vice versa. In this problem, the cost of the cabling is small and thus is not a major concern. Rather, our design priorities are driven primarily by performance objectives. The merit of having h_1 and h_2 as large as possible is that the forces in the cables will be smaller. Some thought about the geometry shown in Fig. 1 suggests that the lights may experience vibratory motion in the vertical direction, most likely due to aerodynamically produced forces from wind, where as one light moves up, the other moves down, in oscillatory fashion. Selecting small values of h_1 and h_2 will provide for shorter cabling which may help minimize the possible amplitude of such motion. Thus, for our initial design we will select $h_2 = 20$ ft.

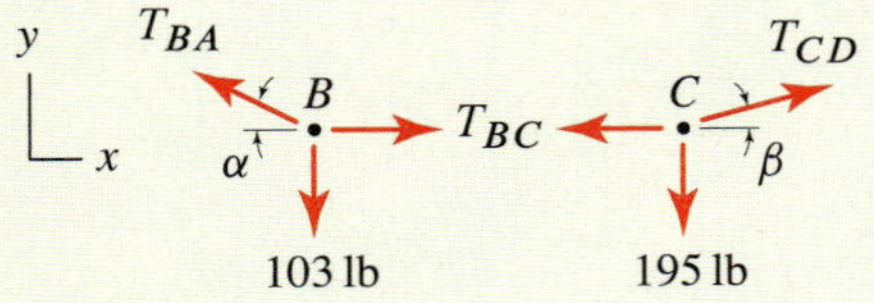

Figure 2
Free body diagrams of points B and C.

Modeling Neglecting the weight of the cables, FBDs are shown in Fig. 2.

Governing Equations & Computation The equilibrium equations for point C are

$$\sum F_y = 0: \qquad T_{CD} \sin\beta - 195\,\text{lb} = 0, \tag{1}$$

$$\sum F_x = 0: \qquad -T_{BC} + T_{CD} \cos\beta = 0, \tag{2}$$

where

$$\cos\beta = \frac{15\,\text{ft}}{\sqrt{(15\,\text{ft})^2 + (h_2 - 19\,\text{ft})^2}} = 0.9978, \tag{3}$$

$$\sin\beta = \frac{h_2 - 19\,\text{ft}}{\sqrt{(15\,\text{ft})^2 + (h_2 - 19\,\text{ft})^2}} = 0.0665. \tag{4}$$

Solving Eqs. (1) and (2) provides

$$T_{CD} = 2931\,\text{lb}, \tag{5}$$

$$T_{BC} = 2925\,\text{lb}. \tag{6}$$

The equilibrium equations for point B are

$$\sum F_y = 0: \qquad T_{BA} \sin\alpha - 103\,\text{lb} = 0, \tag{7}$$

$$\sum F_x = 0: \qquad -T_{BA} \cos\alpha + T_{BC} = 0. \tag{8}$$

Solving Eqs. (7) and (8) provides

$$\tan\alpha = \frac{103\,\text{lb}}{T_{BC}} \quad \Rightarrow \quad \alpha = 2.02°, \tag{9}$$

$$T_{BA} = 2927\,\text{lb}. \tag{10}$$

From the geometry of Fig. 1,

$$\tan\alpha = \frac{h_1 - 19\,\text{ft}}{9\,\text{ft}}, \tag{11}$$

which provides $h_1 = 19.32$ ft, which is less than the maximum permitted value of 22 ft.

To specify an appropriate cable, note the largest tensile force to be supported is 2931 lb in portion CD of the cable. Thus, consulting Table 3.1, we see that any cable with 3/8 in. nominal diameter or larger will be acceptable. However, because the 2931 lb cable force is very close to the 3000 lb allowable load for the 3/8 in. cable, we will select the 1/2 in. diameter size. To summarize, our initial design calls for

$$T_{BA} = 2927\,\text{lb}, \quad T_{BC} = 2925\,\text{lb}, \quad T_{CD} = 2931\,\text{lb}, \tag{12}$$

$$h_1 = 19.32\,\text{ft}, \quad h_2 = 20.00\,\text{ft}, \quad \text{and} \quad \text{cable diameter} = 1/2\,\text{in.} \tag{13}$$

Discussion & Verification The cable forces in this initial design are very large. While they may be acceptable, we might question the ability of the support poles to be subjected to these. Note that even if the support poles themselves were deemed to be strong enough, the soil that supports the bases of the poles may not be. These considerations go beyond information that is available and our analysis ability at this time. Nonetheless, we reanalyze this design, starting with $h_2 = 21$ ft, with the results

$$T_{BA} = 1466\,\text{lb}, \quad T_{BC} = 1463\,\text{lb}, \quad T_{CD} = 1475\,\text{lb}, \tag{14}$$

$$\boxed{h_1 = 19.63\,\text{ft}, \quad h_2 = 21.00\,\text{ft}, \quad \text{and} \quad \text{cable diameter} = 3/8\,\text{in.}} \tag{15}$$

Although a subjective decision, the design reported in Eq. (15) would seem preferable to that in Eq. (13), due to the substantially lower forces the support poles are subjected to.

To complete the design, we should write a brief technical report, following the guidelines of Appendix A, where we summarize all pertinent information. It should be written using proper, simple English that is easy to read. A sketch along with critical dimensions should be included. We should discuss the objectives and constraints considered in the design and the process used to arrive at our final design. Detailed calculations, such as Eqs.(1)–(11) should not be included in the main body of the report, but should be included in an appendix. In addition to the immediate use of the report in helping to erect the traffic lights, it will be archived for possible future reference.

EXAMPLE 3.10 Engineering Design

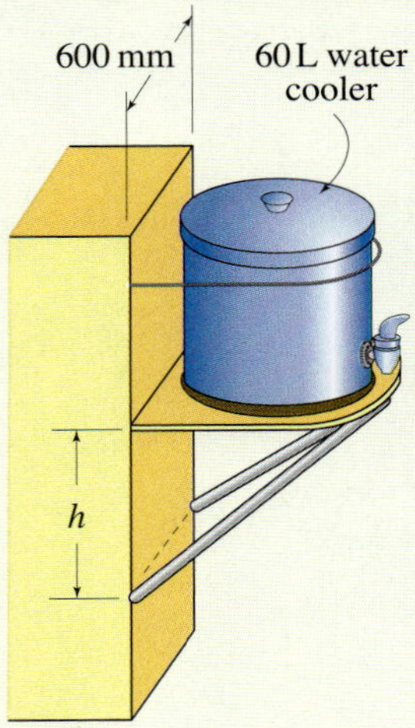

Figure 1

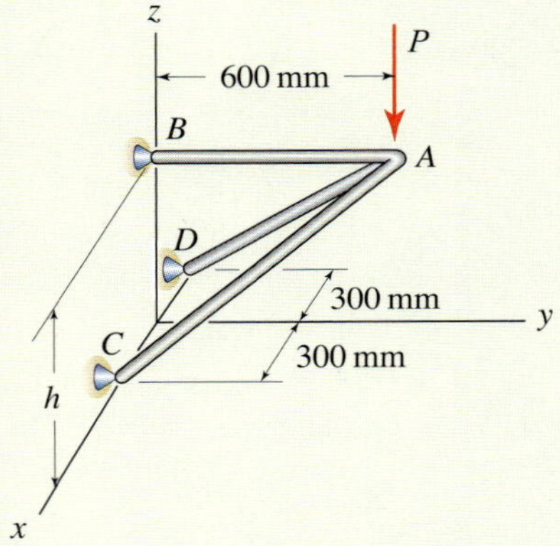

Figure 2
Model for the water cooler support shown in Fig. 1.

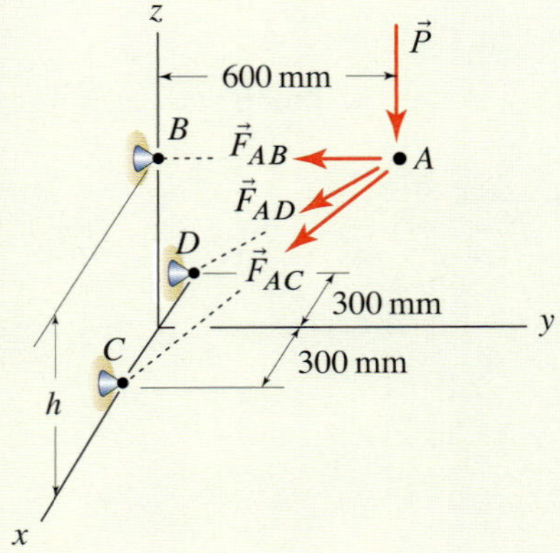

Figure 3
Free body diagram for point A.

A large construction company plans to add plastic 60 L water coolers to the back of most of its trucks, as shown in Fig. 1. Your supervisor asks you to use the model shown in Fig. 2 to specify the dimension h and the diameter of round steel bars to be used for members AB, AC, and AD. Although a more refined model is possible using concepts discussed in Chapter 5, the model shown in Fig. 2 is useful for this design problem. In this model, load P is vertical and member AB is parallel to the y axis. Your supervisor also tells you to allow for loads up to twice the weight of the cooler in a crude attempt to allow for dynamic forces, and to use the allowable loads given in Table 3.2.

SOLUTION

Road Map The most poorly defined part of this problem is the loading the water cooler support will be subjected to. Thus, assumptions on the size of loads will need to be made. Furthermore, different loading conditions may need to be considered, and our design should be safe for all of these. Finally, this design problem does not have a unique solution. That is, there is not a single value of h that will work. Rather, we may need to assume reasonable values for some parameters and then calculate the values of others based on this.

Modeling When full, the weight of water in the cooler is*

$$W = \underbrace{1000\,\frac{\text{kg}}{\text{m}^3}}_{\rho}\;\underbrace{60\,\text{L}\,\frac{(0.1\,\text{m})^3}{\text{L}}}_{V}\;\underbrace{9.81\,\frac{\text{m}}{\text{s}^2}}_{g} = 589\,\text{N}. \tag{1}$$

We next use our judgment to add to the above value a nominal amount of 60 N to account for the weight of the container itself. Thus, the total weight to be supported is $589\,\text{N} + 60\,\text{N} = 649\,\text{N}$.

Doubling the weights (or use of other multiplicative factors) is an imprecise but common approach to account for inertial forces produced when the truck hits bumps in the road. Thus, we obtain the approximate load $P = 1300\,\text{N}$. Before continuing with this load, we should consider other possible load scenarios. For example, if the cooler is removed, the support might make a convenient step for a worker to use while climbing on or off the truck. Consider a person weighing 890 N (200 lb). Since it is extremely unlikely a person would be standing on the support when the truck is moving and hitting bumps, it is not necessary to further increase this value as was done earlier. Since the weight of this person is lower than the force determined earlier, we will proceed with the design, using $P = 1300\,\text{N}$.

Small values of dimension h will lead to large forces in all members, while larger values of h will decrease these forces. However, the allowable load data in Table 3.2 shows that allowable compressive load decreases rapidly as the length of members increases. Thus, we will select for an initial design the value $h = 250\,\text{mm}$, in which case the length of members AC and AD is 716 mm.

The FBD for point A is shown in Fig. 3.

Governing Equations Vector expressions for forces are

$$\vec{P} = 1300\,\text{N}\,(-\hat{k}), \tag{2}$$

* One liter (1 L) is defined to be the volume of 1 kg of pure water at 4°C and pressure of 76 cm of mercury. For practical purposes, however, the transformation $1\,\text{L} = 1000\,\text{cm}^3 = 0.001\,\text{m}^3$ may be used with less than 0.003% error.

$$\vec{F}_{AB} = F_{AB}(-\hat{\jmath}), \tag{3}$$

$$\vec{F}_{AC} = F_{AC}\frac{300\,\hat{\imath} - 600\,\hat{\jmath} - 250\,\hat{k}}{716}, \tag{4}$$

$$\vec{F}_{AD} = F_{AD}\frac{-300\,\hat{\imath} - 600\,\hat{\jmath} - 250\,\hat{k}}{716}. \tag{5}$$

Writing $\sum \vec{F} = \vec{0}$ and grouping x, y, and z direction terms provide the system of equations

$$F_{AC}\frac{300}{716} + F_{AD}\frac{-300}{716} = 0, \tag{6}$$

$$F_{AB}(-1) + F_{AC}\frac{-600}{716} + F_{AD}\frac{-600}{716} = 0, \tag{7}$$

$$F_{AC}\frac{-250}{716} + F_{AD}\frac{-250}{716} = 1300\,\text{N}. \tag{8}$$

Computation The solution of Eqs. (6)–(8) is

$$F_{AB} = 3120\,\text{N}, \quad F_{AC} = -1860\,\text{N}, \quad \text{and} \quad F_{AD} = -1860\,\text{N}. \tag{9}$$

To determine appropriate members to support these loads, we consult Table 3.2 on p. 163. For the tensile load in member AB, any bar with diameter 6 mm or larger is acceptable. For the compressive load in members AC and AD, we first check the suitability of the 6 mm bar, which can support a maximum compressive force that is the smaller of 3500 N or $62(10)^6\,\text{N}\cdot\text{mm}^2/L^2 = 62(10)^6\,\text{N}\cdot\text{mm}^2/(716\,\text{mm})^2 = 120.9\,\text{N}$; hence this bar is not acceptable. The next larger size bar has 12 mm diameter, which allows a maximum compressive load that is the smaller of 14 and 1.95 kN; hence this bar is acceptable for members AC and AD.

Discussion & Verification Our design thus far requires

$$\begin{aligned} h &= 250\,\text{mm} \\ \text{member } AB&: \quad \text{use 6 mm diameter steel bar or larger.} \\ \text{members } AC, \text{ and } AD&: \quad \text{use 12 mm diameter steel bar or larger.} \end{aligned} \tag{10}$$

For convenience in fabrication, we will recommend that the same size bar be used for all three components, in which case 12 mm diameter bar or larger is to be used. Finally, we consider whether the use of a diameter greater than 12 mm is warranted. This is analogous to incorporating an even greater factor of safety in the design, which does not seem necessary, since the factors of safety incorporated in the allowable loads and forces applied to the structure are already generous. Further, we have applied the entire weight of the water cooler to point A whereas in reality, a portion of this weight will be supported by point B.

To summarize, our final design is

$$\boxed{\begin{aligned} h &= 250\,\text{mm} \\ \text{members } AB, AC, \text{ and } AD&: \quad \text{use 12 mm diameter steel bar.} \end{aligned}} \tag{11}$$

To complete our work, we should prepare a short report following the guidelines described in Appendix A.

DESIGN PROBLEMS

General Instructions. In problems requiring the specification of sizes for steel cable, bar, or pipe, selections should be made from Tables 3.1–3.3. In all problems, write a brief technical report following the guidelines of Appendix A, where you summarize all pertinent information in a well-organized fashion. It should be written using proper, simple English that is easy to read by another engineer. Where appropriate, sketches, along with critical dimensions, should be included. Discuss the objectives and constraints considered in your design, the process used to arrive at your final design, safety issues if appropriate, and so on. The main discussion should be typed, and figures, if needed, can be computer-drawn or neatly hand-drawn. Include a neat copy of all supporting calculations in an appendix that you can refer to in the main discussion of your report. A length of a few pages, plus appendix, should be sufficient.

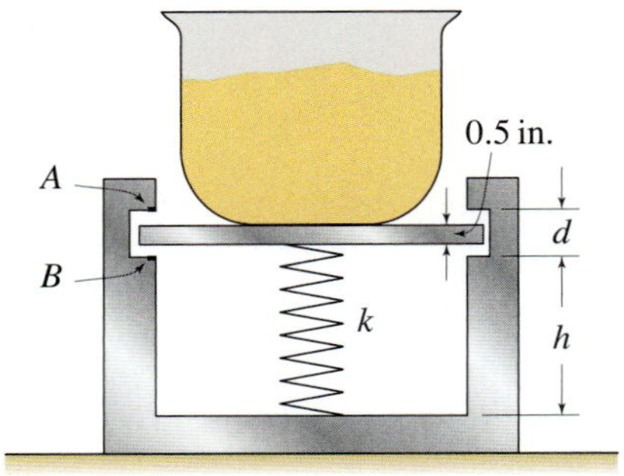

Figure DP3.1

Design Problem 3.1

A scale for rapidly weighing ingredients in a commercial bakery operation is shown. An empty bowl is first placed on the scale. Electrical contact is made at point A, which illuminates a light indicating the bowl's contents are underweight. A bakery ingredient, such as flour, is slowly poured into the bowl. When a sufficient amount is added, the contact at A is broken. If too much is added, contact is made at B, thus indicating an overweight condition. If the contents of the bowl are to weigh 18 lb $\pm$ 0.25 lb, specify dimensions h and d, spring stiffness k, and the unstretched length of the spring L_0. The bowl and the platform on which it rests have a combined weight of 5 lb. Assume the scale has guides or other mechanisms so that the platform on which the bowl rests is always horizontal.

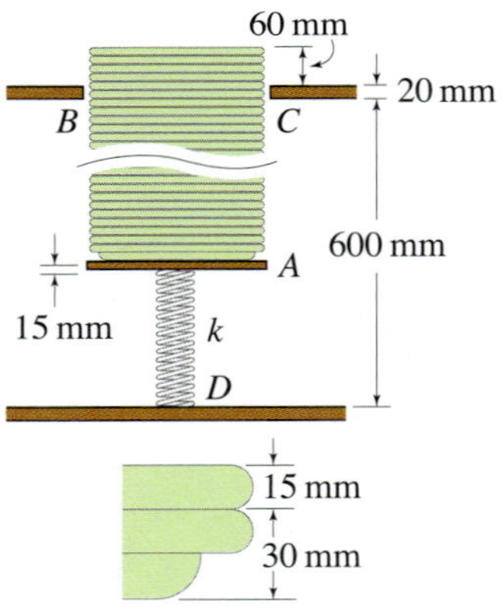

Figure DP3.2 and DP3.3

Design Problem 3.2

A plate storage system for a self-serve salad bar in a restaurant is shown. As plates are added to or withdrawn from the stack, the spring force and stiffness are such that the plates always protrude above the tabletop by about 60 mm. If each plate has 0.509 kg mass, and if the support A also has 0.509 kg mass, determine the stiffness k and unstretched length L_0 of the spring. Assume the spring can be compressed by a maximum of 40% of its initial unstretched length before its coils begin to touch. Also specify the number of plates that can be stored. Assume the system has guides or other mechanisms so the support A is always horizontal.

Design Problem 3.3

In Design Problem 3.2, the spring occupies valuable space that could be used to store additional plates. Repeat Design Problem 3.2, employing cable(s) and pulley(s) in conjunction with one or more springs to design a different system that will allow more plates to be stored. Pulleys, cables, and springs can be attached to surfaces A, B, C, and D. For springs in compression, assume they may not contract by more than 40% of their initial unstretched length before their coils begin to touch.

Design Problem 3.4

A push button for a pen is shown in cross section. Pressing the push button and releasing it advance the ink cartridge so that the pen may be used for writing. Pressing the push button and releasing it again retract the ink cartridge so that the pen may be stored. The mechanism that keeps the ink cartridge in the advanced or retracted position is not shown. Design the spring for the pen by specifying the spring's stiffness k (units: N/mm), the spring's unstretched length L_0, and the dimensions d and t (units for L_0, d, and t: mm). To prevent the coils of the spring from making contact with one another, the spring should not be compressed by more than 50% of its unstretched length.

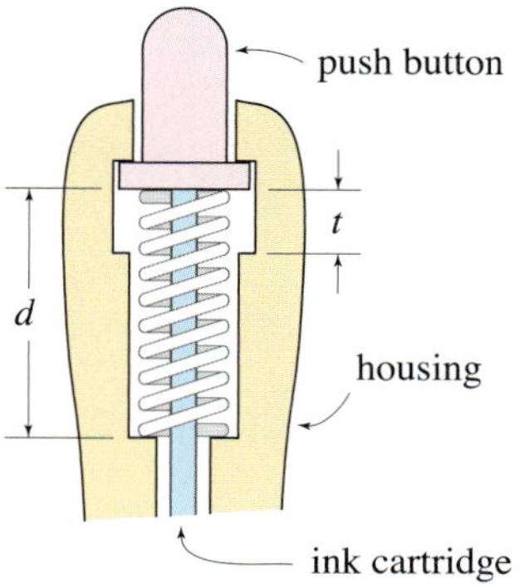

Figure DP3.4

Design Problem 3.5

You are working for the Peace Corps in an impoverished country that needs your help. Your job is to design a simple, but accurate scale for weighing bulk materials such as food and construction materials. A possible idea consists of a loading platform D, that is supported by cable AB, where end A is attached to a large tree limb. After loading platform D, weight W_3 is applied, which causes point B to move horizontally by a small distance δ that is to be measured with a yardstick.

Assume:

W_2 = weight of the loading platform = 100 lb,

W_1 = weight of materials to be weighed $\leq$ 500 lb,

Cable segment BC is horizontal,

Pulley C is frictionless.

You are to specify

L = length of cable AB, where 10 ft $\leq L \leq$ 15 ft,

W_3 = weight of the counterweight, where $W_3 \leq$ 40 lb.

For the specific values of L and W_3 you choose, produce a graph with W_1 and δ as the vertical and horizontal axes, respectively. Thus, by measuring the deflection δ, your graph will tell the user the weight W_1.

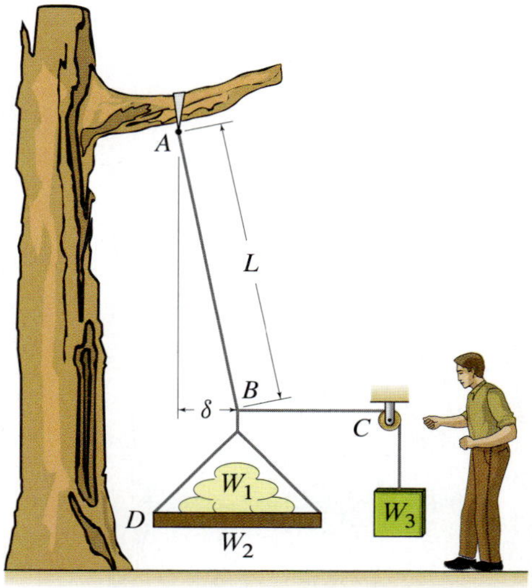

Figure DP3.5

Design Problem 3.6

An aviation museum has exhibits in a room that is 60 m wide by 35 m high. A cable system is to be designed to suspend a new display, namely, a Mercury space capsule. Using this cabling system, point A (where the two cables attach to the capsule) will be slowly maneuvered to a final position in the "permanent position region" indicated in Fig. DP3.6. The cabling system must be sufficiently strong that the factor of safety against breaking is at least 5 while the capsule is maneuvered, and is at least 10 when the capsule is anywhere in the permanent position region. Note that during maneuvering, the display will be closed to the public, so that the factor of safety during maneuvering can be lower than the final factor of safety. Your analysis may neglect the size of the pulleys at points B and C. You are to select the smallest cable diameter possible and to specify the dimensions d, l, h_1, and h_2.

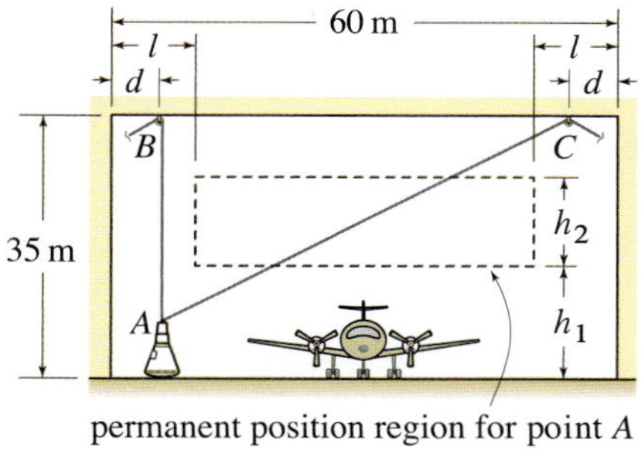

Figure DP3.6

Design Problem 3.7

A utility pole supports a wire that exerts a maximum horizontal force of 500 lb at the top of the pole as shown. Although the pole is to be buried in soil at end A, for in-plane behavior it is conservative to idealize this support to be a pin, and thus the net force supported by the pole is directed along the axis of the pole. Design a support system for the pole, considering that it is located in a congested region. Your design can utilize steel cable and/or steel pipe. Whatever support members you choose, only one support member may be attached to the utility pole, and this must be at the eyebolt C. Specify the size eyebolt to be used (if needed, see Section 2.2 for a description of these load multipliers). The support system you design cannot be attached to the adjacent building or to any point on the sidewalk or roadway. Safe pedestrian passage on the sidewalk should be considered.

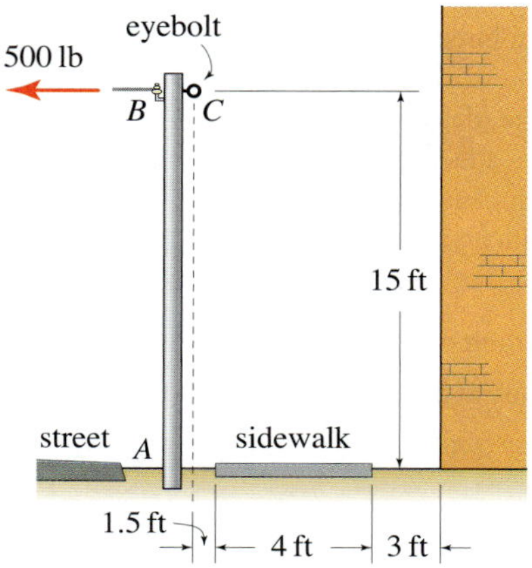

eyebolt working loads

nominal size	working load parallel to shank
3/4 in.	480 lb
1	1,200
1.5	4,800
2	8,500
2.5	17,800

θ	working load multipliers
0°	100%
30	60%
45	33%
90	20%

Figure DP3.7

Design Problem 3.8

A ferry is being designed to cross a small river having a strong current. The ferry has no engine and is supported by a steel cable *ABCA*. At B the cable is wrapped around a winch so that it does not slip, and at C the cable passes around a freely rotating pulley. Both ends of the cable are fixed to the ferry at A. The drag force on the ferry is highest at midstream and lowest at the banks, and it is given by $F(x) = (4\text{ kN})[1 + \sin(\pi x/98\text{ m})]$. The winch moves the ferry slowly, so there are no other in-plane forces applied to the ferry beyond F and the cable forces. Specify the necessary diameter for cable *ABCA*. The owner of the ferry also wants to know the position x (within ± a few meters) where the load on the cable is the greatest.

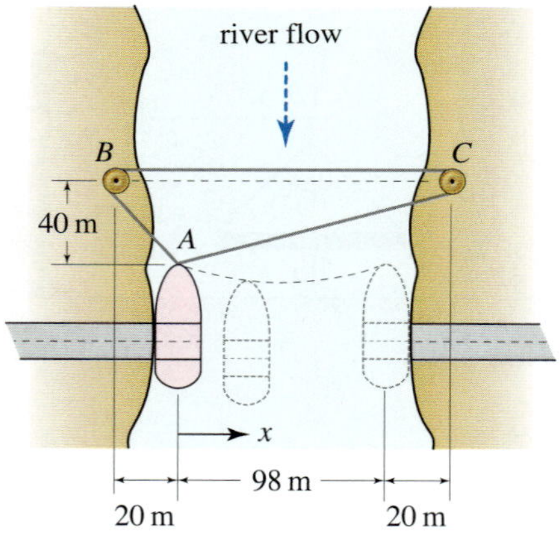

Figure DP3.8

Design Problem 3.9

A company is designing an after-market hoist to be installed on a pickup truck. Thus, the dimensions of the hoist have largely been determined by the available space inside the bed of the truck. The hoist consists of two pipes, AC and AD, and two cables, AE and $FABA$. The hoist is to be capable of lifting a maximum of 4000 lb. You are to specify

The diameter of the steel wire rope $FABA$,
The diameter of steel wire rope AE,
The diameter of steel pipes AC and AD,
The torque rating for the winch F (equal to the force cable $FABA$ supports multiplied by 6 in.).

In your calculations, you may assume cable segments AB are vertical, and the bed of the truck is horizontal. Although the hoist is to be designed for a maximum load of 4000 lb, surely some user will attempt to lift more. Discuss in your report the consequences of doing this and how your specifications account for this possibility.

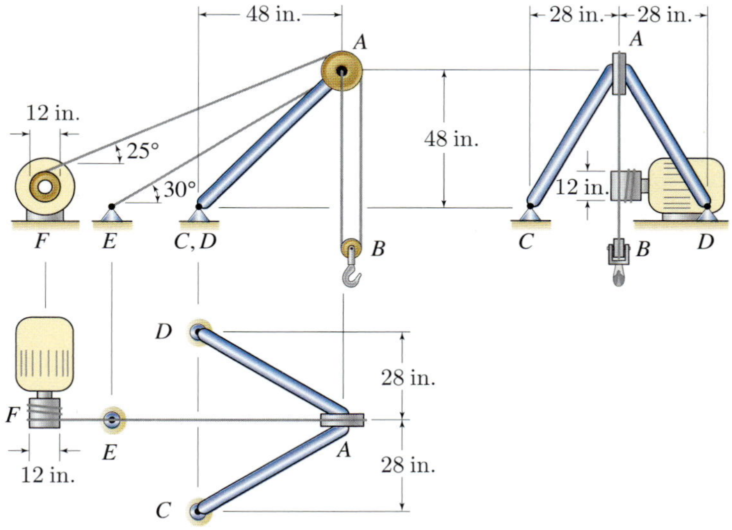

Figure DP3.9

Design Problem 3.10

A bar and cable system is to be designed to support a weight W suspended from point A. Due to handling and positioning of the weight, a small force equal to 10% of W laying in a horizontal plane is also applied at point D. If $W = 5$ kN, specify

- Dimensions d and h, where $d \leq 1.0$ m and $h \leq 1.4$ m.
- The diameter of the steel wire ropes AB and AC.
- The diameter of the steel pipe AO.

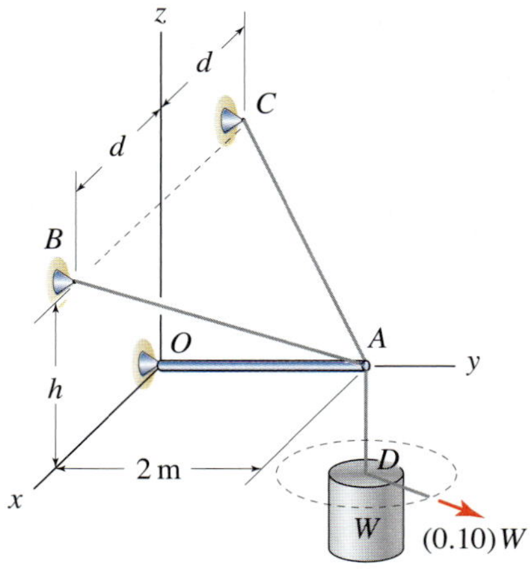

Figure DP3.10

Design Problem 3.11

Design the cable support system for a radio tower of height 25 m. This includes specifying the radius R of the supports at point B, C, and D (as measured from the base of the tower) and the diameter of the steel cable.

Remarks:

- All supports will be erected on a horizontal surface.
- Supports B, C, and D will each be located at the same radial distance R from the base of the tower (point O) such that $\angle BOC = \angle BOD = \angle COD = 120°$.
- The cables must be able to support a maximum horizontal storm force $F = 18\,\text{kN}$ from any direction in the horizontal plane.
- Neglect the weight of the cables in your calculations.
- Assume one of the cables is always slack and supports no load. For example, when force F lies in sector CAD (as shown), cable AB is slack.

In your report describe the merits of large versus small values of R and why you selected the value you did. State which orientation(s) of the storm force F lead to the most severe cable loads and how you determined these orientations.

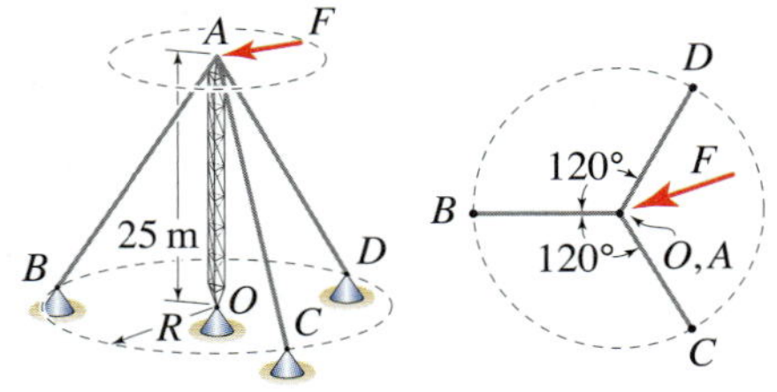

Figure DP3.11

3.5 Chapter Review

Important definitions, concepts, and equations of this chapter are summarized. For equations and/or concepts that are not clear, you should refer to the original equation numbers cited for additional details.

Equilibrium of a particle. A particle is in *static equilibrium* if

Eq. (3.20), p. 148

$$\sum \vec{F} = \vec{0},$$
$$\text{or} \quad \left(\sum F_x\right)\hat{\imath} + \left(\sum F_y\right)\hat{\jmath} + \left(\sum F_z\right)\hat{k} = \vec{0},$$
$$\text{or} \quad \sum F_x = 0 \text{ and } \sum F_y = 0 \text{ and } \sum F_z = 0.$$

The above summations must include all forces applied to the particle. For equilibrium of a system of particles, the above equilibrium equations are written for each particle, and then the resulting system of simultaneous equations is solved.

Free body diagram. The *free body diagram* (FBD) is an essential aid for helping ensure that all forces applied to a particle are accounted for when you write equations of equilibrium. When you draw an FBD, it is helpful to imagine enclosing the particle by a closed line in two dimensions, or a closed surface in three dimensions. Wherever the cut passes through a structural member, the forces supported by that member must be introduced in the FBD. Wherever the cut passes through a support, the reaction forces that the support applies to the particle must be introduced in the FBD.

Cables and bars. Cables and straight bars are structural members that support forces that are collinear with their axis. We assume cables may support tensile forces only and may be freely bent, such as when wrapped around a pulley. We usually assume cables have negligible weight. Bars may support both tensile and compressive forces.

Pulleys. A pulley is a device that changes the direction of a cable, and hence changes the direction of the force that is supported by the cable. If the pulley is frictionless and the cable is weightless, then the magnitude of the force throughout the cable is uniform.

Springs. Behavior of a *linear elastic spring* is shown in Fig. 3.22 and is described by the *spring law*

Eq. (3.18), p. 137

$$F_s = k\delta$$
$$= k(L - L_0)$$

where k is the spring stiffness (units: force/length), δ is the elongation of the spring from its unstretched length, L_0 is the initial (unstretched) spring length,

Helpful Information

Spring law sign conventions. The sign conventions for the spring law given in Eq. (3.18) are as follows:

$F_s > 0$ tension,
$F_s < 0$ compression,
$\delta = 0$ unstretched position,
$\delta > 0$ extension,
$\delta < 0$ contraction.

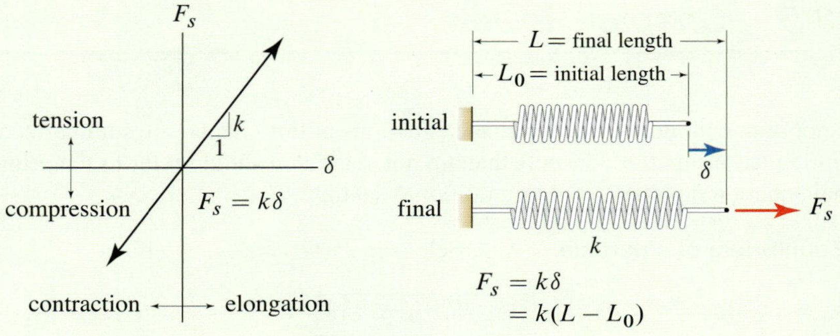

Figure 3.22. Spring law for a linear elastic spring.

and L is the final spring length. In solving problems with springs, one of the forms of Eq. (3.18) will usually be more convenient than the other, depending on what data is provided. These equations are written with the convention that positive values of force F_s correspond to tension and positive values of δ correspond to elongation. Other conventions are possible, such as force being positive in compression and/or deformation being positive in contraction, but it may be necessary to introduce a negative sign on the right-hand side of Eq. (3.18), as described in Section 3.2.

Summing forces in directions other than x, y, or z. To sum forces in a direction $\vec{r}$, we take the dot product of all vectors in the expression $\sum \vec{F} = \vec{0}$ with the unit vector* $\vec{r}/|\vec{r}|$ to obtain

Eq. (3.22), p. 150

Summation of forces in the $\vec{r}$ direction:

$$\sum F_r = 0: \quad \vec{F}_1 \cdot \frac{\vec{r}}{|\vec{r}|} + \vec{F}_2 \cdot \frac{\vec{r}}{|\vec{r}|} + \cdots + \vec{F}_n \cdot \frac{\vec{r}}{|\vec{r}|} = 0.$$

Use of $\sum F_r = 0$ is especially convenient when it is known that some forces are perpendicular to $\vec{r}$.

Allowable load. *Allowable loads*, sometimes called *working loads*, are forces structural components can safely be subjected to, and they are obtained by dividing the failure strength of a member by a *factor of safety*. Thus,

Eq. (3.23), p. 164

$$\text{Allowable load} = \frac{\text{failure load}}{\text{factor of safety}} \quad \text{or}$$

$$\text{Factor of safety} = \frac{\text{failure load}}{\text{allowable load}}.$$

* Because the right-hand side of Eq. (3.22) is zero, a unit vector in the direction of $\vec{r}$ is not needed and we could just as well evaluate $\vec{F}_1 \cdot \vec{r} + \vec{F}_2 \cdot \vec{r} + \cdots + \vec{F}_n \cdot \vec{r} = 0$.

REVIEW PROBLEMS

Problem 3.71

Consider a problem involving cables and bars only. For the conditions listed below, is the solution obtained from $\sum \vec{F} = \vec{0}$ using geometry of the structure before loads are applied approximate or exact? Explain.

(a) Cables are modeled as inextensible, and bars are modeled as rigid.

(b) Cables and bars are modeled as linear elastic springs.

Note: Concept problems are about *explanations*, not computations.

Problem 3.72

The frictionless pulley A weighs 20 N and supports a box B weighing 60 N. When you solve for the force in cable CD, a "problem" arises. Describe this problem and its physical significance.

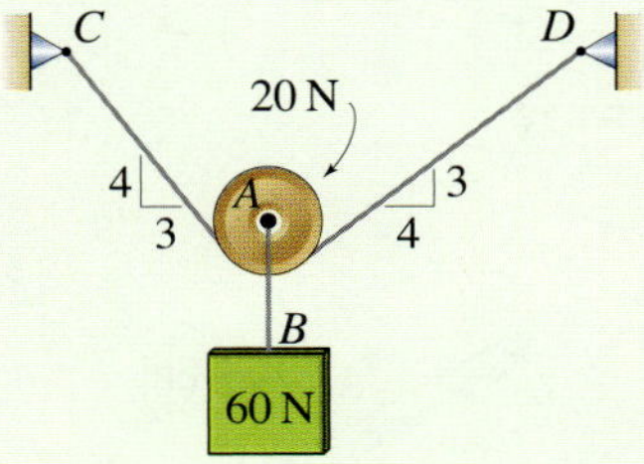

Figure P3.72

Problem 3.73

To produce a force $P = 40$ N in horizontal member CD of a machine, a worker applies a force F to the handle B. Determine the smallest value of F that can be used and the angle α it should be applied at.

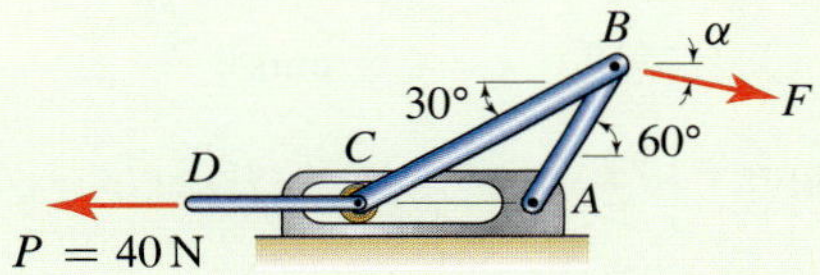

Figure P3.73

Problem 3.74

The structure shown consists of five cables. Cable $ABCD$ supports a drum having weight $W = 200$ lb. Cable DF is horizontal, and cable segments AB and CD are vertical. If contact between the drum and cable $ABCD$ is frictionless, determine the force in each cable.

Problem 3.75

In Prob. 3.74, if cable $ABCD$ has 600 lb breaking strength and all other cables have 200 lb breaking strength, determine the largest value W may have.

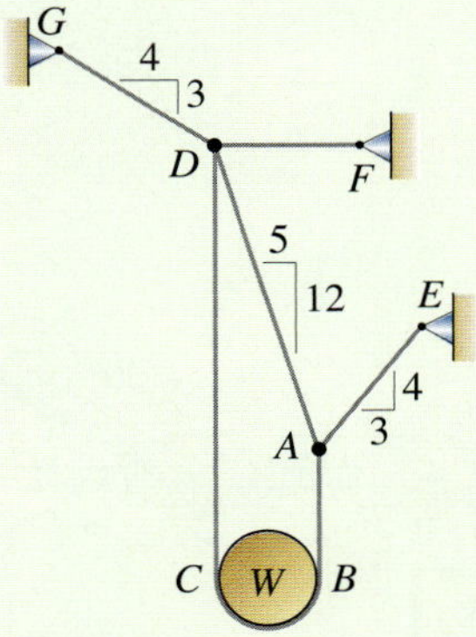

Figure P3.74 and P3.75

Problem 3.76

Two frictionless pulleys connected by a weightless bar AB support the 200 and 300 N forces shown. The pulleys rest on a wedge that is fixed in space. Determine the angle θ when the system is in equilibrium and the force in bar AB.

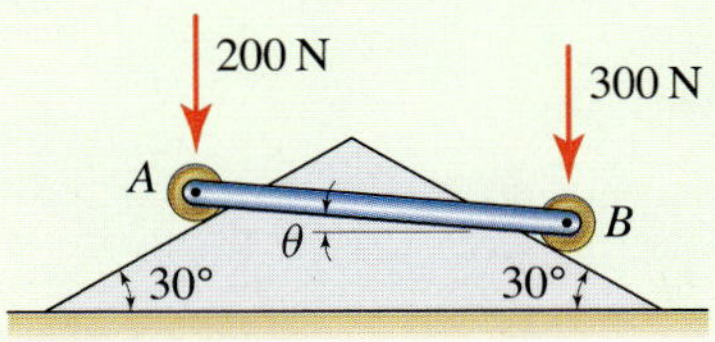

Figure P3.76

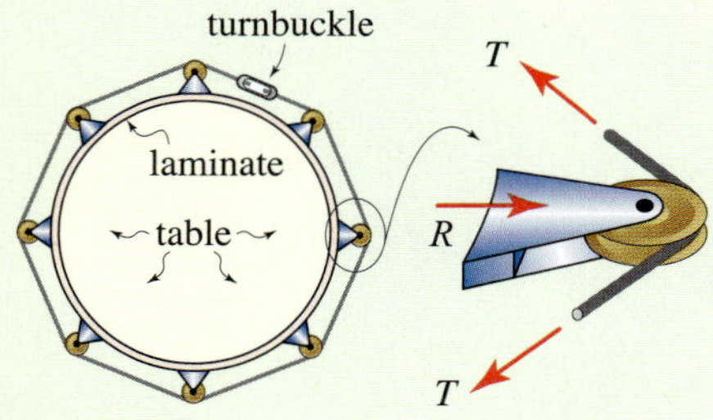

Figure P3.77

Problem 3.77

To glue a strip of laminate to the edge of a circular table, n clamps are evenly spaced around the perimeter of the table (the figure shows $n = 8$). Each clamp has a small pulley, and around all pulleys a cable is wrapped. A turnbuckle is used to tighten the cable, producing a force T. Assuming the diameter of each pulley is small, show that the force R each clamp applies to the edge of the table is given by $R = 2T\cos[(n-2)180°/(2n)]$. Does this expression give expected values when $n = 2$ and $n = \infty$? Explain. *Hint:* The sum of the interior angles of a polygon with m corners is $(m-2)180°$.

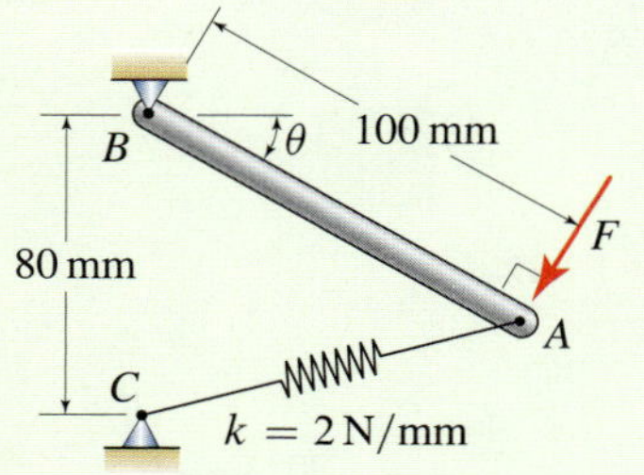

Figure P3.78

Problem 3.78

Spring AC is unstretched when $\theta = 0$. Force F is always perpendicular to bar AB. Determine the value of F needed for equilibrium when $\theta = 45°$.

Problems 3.79 and 3.80

If $W = 100\,\text{N}$ and $k = 5\,\text{N/mm}$, determine δ_1 and δ_2. Springs are unstretched when $\delta_1 = \delta_2 = 0$.

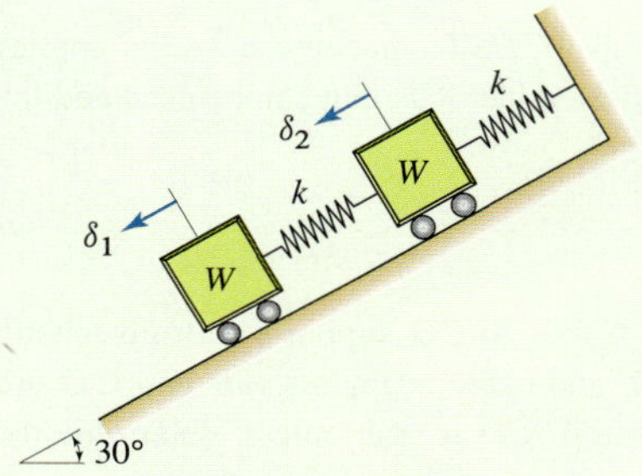

Figure P3.79

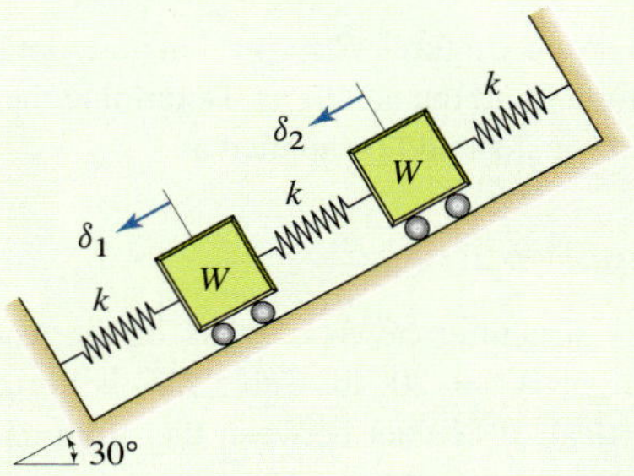

Figure P3.80

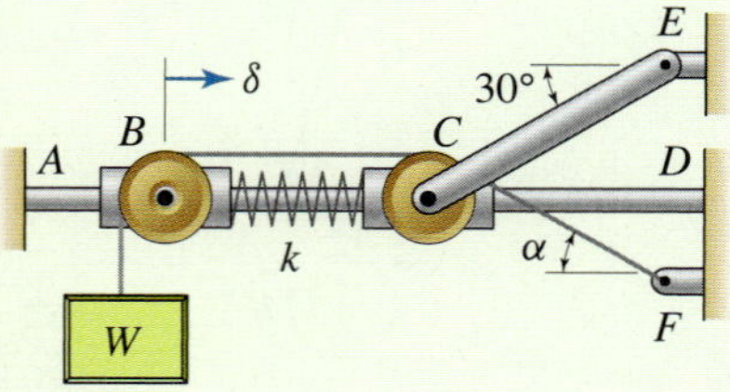

Figure P3.81

Problem 3.81

Frictionless sliders B and C have frictionless pulleys mounted to them, and are connected by a spring with stiffness $k = 12\,\text{N/mm}$. Around the pulleys is wrapped a cable that supports a weight $W = 100\,\text{N}$. Member CE is a bar.

(a) If $\alpha = 30°$, determine the force in bar CE and the motion δ of slider B.

(b) Determine the value of α that will provide for the smallest force in bar CE, and determine this force.

Problem 3.82

A weight W is supported by four cables. Points B and C lie in the yz plane. If $W = 8\,\text{kN}$, determine the force supported by each cable.

Problem 3.83

A weight W is suspended by four cables. Points B and C lie in the yz plane. If the allowable strength of each cable is as specified below, determine the largest allowable weight W that can be supported.

Cable	*Strength*
AB	12 kN
AC	10 kN
AD	5 kN
AE	15 kN

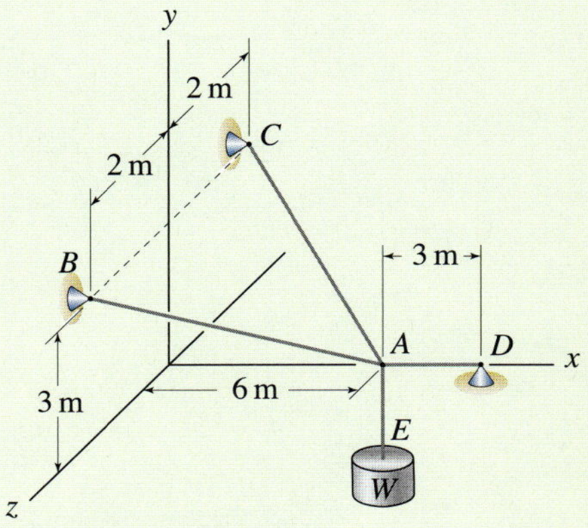

Figure P3.82 and P3.83

Problem 3.84

Repeat Prob. 3.82 with points B and C having coordinates $B(0, 6, 3)$ m and $C(0, 6, -7)$ m.

Problem 3.85

Repeat Prob. 3.83 with points B and C having coordinates $B(0, 6, 3)$ m and $C(0, 6, -7)$ m.

Problem 3.86

Bead C has 2 lb weight and slides without friction on straight bar AB. The tensile forces in elastic cords CD and CE are 0.5 and 1.5 lb, respectively. If the bead is released from the position shown with no initial velocity, will it slide toward point A or B or will it remain stationary?

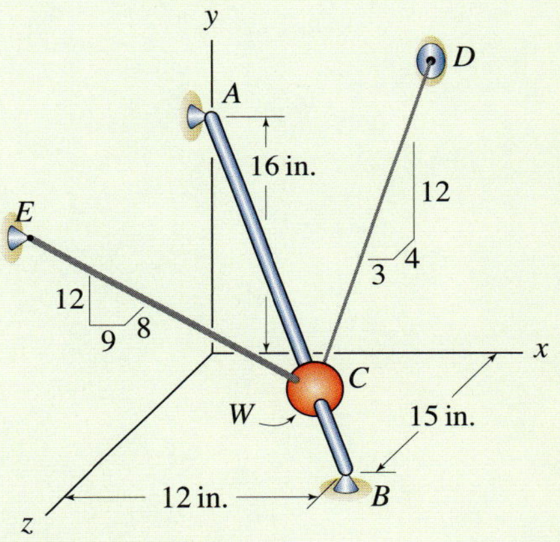

Figure P3.86 and P3.87

Problem 3.87

Bead C has 2 lb weight and slides without friction on straight bar AB, and the tensile force in elastic cord CE is 0.9 lb. Determine the force needed in cord CD for the bead to be in equilibrium, and the magnitude of the reaction between the bead and bar AB.

Problem 3.88

A hoist for lifting objects onto and off a truck is shown. All cables lie in the xy plane, cable segment AB is horizontal, and the plane formed by bars CE and CF is parallel to the yz plane. Pulleys are frictionless and force P is vertical. If the object is being slowly lifted and weighs $W = 600$ lb, determine the force P and forces supported by all cables and bars.

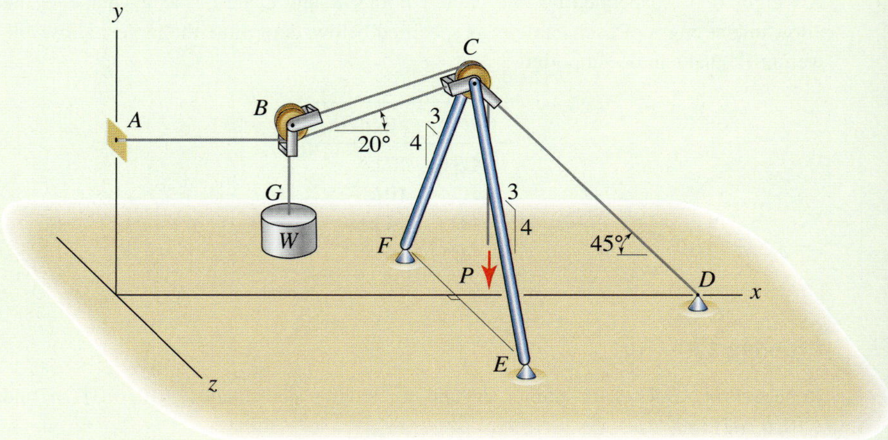

Figure P3.88 and P3.89

Problem 3.89

For Prob. 3.88, determine the largest weight W that may be lifted if forces supported by cables may not exceed 2000 lb and compressive forces supported by bars may not exceed 2800 lb.

4 Moment of a Force and Equivalent Force Systems

Additional concepts of forces and systems of forces are discussed in this chapter. These concepts are used extensively in the analysis of equilibrium and motion of bodies and throughout more advanced mechanics subjects.

4.1 Moment of a Force

To help demonstrate some of the features of the *moment of a force*, we will consider an example of a steering wheel in a car. Figure 4.1 shows a classic Ferrari sports car, and Fig. 4.2 shows the steering wheel in this car. The wheel lies in a plane that is perpendicular to the steering column AB (it would not be very comfortable to use if this were not the case), and the steering wheel offers "resistance" to being turned (for most vehicles, this resistance increases for slower speeds and as a turn becomes sharper). Imagine you are driving this car, and you wish to execute a right-hand turn. Figure 4.2 shows two possible locations where you could position your left hand to turn the steering wheel, and the directions of forces F_1 and F_2 that you would probably apply, where both of these forces lie in the plane of the steering wheel. For a given speed and sharpness of turn, clearly position C will require a lower force to turn the wheel than position D (i.e., $F_1 < F_2$). Both forces F_1 and F_2 produce a *moment* (i.e., twisting action) about line AB of the steering column, and the size of this moment increases as the force becomes larger *and/or* as the distance from the force's line of action to line AB increases. If the line of action of F_1 is perpendicular to line AB (as we have assumed here), and if we let d be the

Figure 4.1
Ferrari 250 GTO sports car, circa 1962–1964.

shortest distance between these two lines, then the moment produced by F_1 has size $F_1 d$ and the direction of the twisting action is about line AB. Thus,

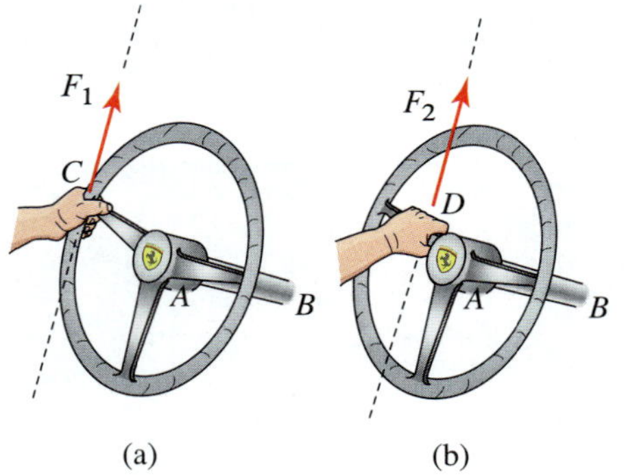

Figure 4.2. An accurate sketch of the Nardi Anni '60 steering wheel in the Ferrari 250 GTO, showing two possible locations where your hand could be positioned to turn the wheel.

we observe that the moment of a force has the properties of a vector, namely, magnitude (or size) and direction. In the remainder of this section, we provide a more precise definition and describe methods of evaluation.

The *moment of a force*, or simply *moment*, is a measure of a force's ability to produce twisting, or rotation about a point. Moment has both magnitude (or size) and direction and is a vector quantity. For example, consider using a wrench to twist a pipe, as shown in Fig. 4.3. By applying a force to the handle of the wrench, the force tends to make the pipe twist. Whether or not the pipe actually does twist depends on details of how the pipe is supported. To create a greater tendency to twist the pipe, either we could apply a larger force to the wrench or we could use a wrench with a longer handle. Moment can be evaluated using both scalar and vector approaches, as follows.

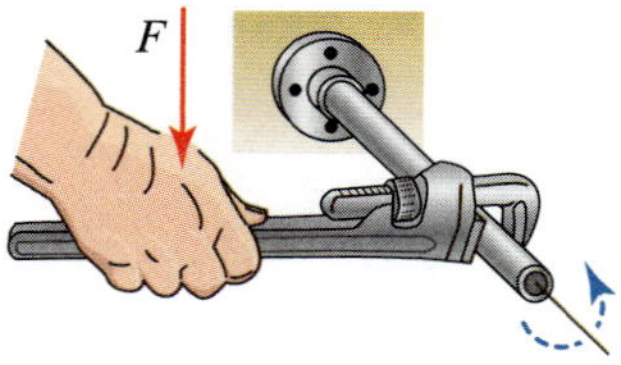

Figure 4.3
Force applied to a wrench to twist a pipe.

Scalar approach

The moment of a force is a vector and can be evaluated using the scalar approach described here. Consider a force $\vec{F}$ with magnitude F. As shown in Fig. 4.4, this force produces a moment vector about point O (the twisting action shown) where the magnitude of this moment is M_O, which is given by

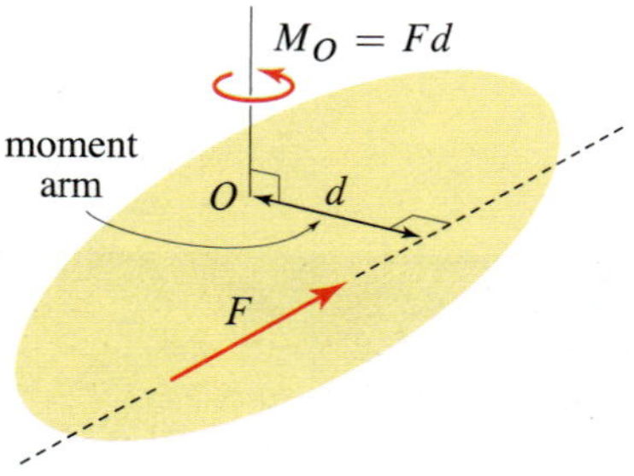

Figure 4.4
Scalar approach to evaluate the moment of a force.

$$M_O = Fd \tag{4.1}$$

where

F is the magnitude of the force;
d is the perpendicular distance from point O to the line of action of $\vec{F}$ and is called the *moment arm*; and
M_O has dimensions of *force* times *length*.

The moment of a force is a *vector*, and it has both magnitude and direction. In the scalar evaluation of the moment, Eq. (4.1) conveniently gives the magnitude. The direction of the moment is not provided by Eq. (4.1), but is understood to be as follows. The line of action of the moment is parallel to the axis through point O that is perpendicular to the plane containing $\vec{F}$ and the moment arm. The direction of the moment along the line of action is given by the direction of the thumb of your right hand when your fingers curl in the twisting direction of the moment. For summing multiple moments, Eq. (4.1) must be supplemented with the proper directions for each moment, as illustrated in the mini-example that follows and in Examples 4.1 and 4.2.

Vector approach

The magnitude and direction of the moment of a force can be obtained using the cross product as described here. As shown in Fig. 4.5, the moment of a force $\vec{F}$ about a point O is denoted by $\vec{M}_O$ and is given as

$$\vec{M}_O = \vec{r} \times \vec{F} \tag{4.2}$$

where

$\vec{F}$ is the force vector;
$\vec{r}$ is a position vector from point O to *any* point* on the line of action of $\vec{F}$.

In contrast to Eq. (4.1) for the scalar approach, Eq. (4.2) automatically provides both the magnitude and direction of the moment.

Remarks

- The order in which the vectors are taken when computing the cross product is important, and hence, vectors $\vec{r}$ and $\vec{F}$ in Eq. (4.2) may not be interchanged.
- In Fig. 4.5, we show the moment vector using a *double-headed arrow* to emphasize its physical differences compared to vectors that are represented using a single-headed arrow. That is, vectors describing force and position represent physical phenomena that are directed along a line, and we represent these using a single-headed arrow. Moment vectors represent physical phenomena that are directed *about* (or twisting around) a line, and we represent these using a double-headed arrow. The convention between the direction of the double-headed arrow and the twisting direction is given by the right-hand rule as described in Fig. 4.6.
- Observe that the moment of a force depends on the location of the point about which the moment is computed. Hence, the moments of a force $\vec{F}$ computed about two different points are usually different, in both magnitude and direction.

Mini-Example. For the balance mechanism shown in Fig. 4.7(a), determine the value of F so that the resultant moment of all forces about point O is zero.

Solution. Using a scalar approach first, we note that each force has a tendency to twist the structure about point O in either a clockwise or counterclockwise fashion. In other words, the moment of each force about point O is a vector

* Because the head of the position vector $\vec{r}$ can be *any* point on the line of action of $\vec{F}$, there is considerable flexibility in selecting $\vec{r}$. Problem 4.25 asks you to show this is true.

Helpful Information

Units for moment. The dimensions for moment are *force* times *length*, and the following table gives the units we will typically use.

Units for moment of a force	
U.S. Customary	**SI**
ft·lb *or* in.·lb	N·m

From a fundamental point of view, the order in which the units are written (e.g., ft·lb or lb·ft) is irrelevant. However, some people follow the convention that moments, whether in U.S. Customary units or SI units, have the force unit first, followed by the length unit.

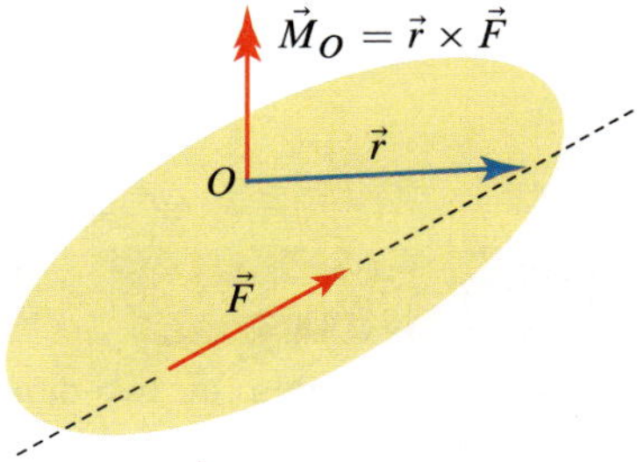

Figure 4.5
Vector approach to evaluate the moment of a force.

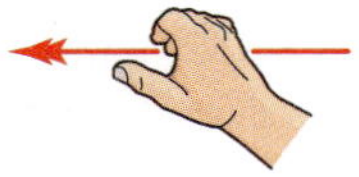

Figure 4.6
Use of the right-hand rule to determine the twisting direction for vectors with double-headed arrows. With your right hand, position your thumb in the direction of the vector's arrows; then your fingers define the direction of the twisting action. If the direction of the twisting action is known, wrap your fingers in this direction; then your thumb defines the direction of the vector.

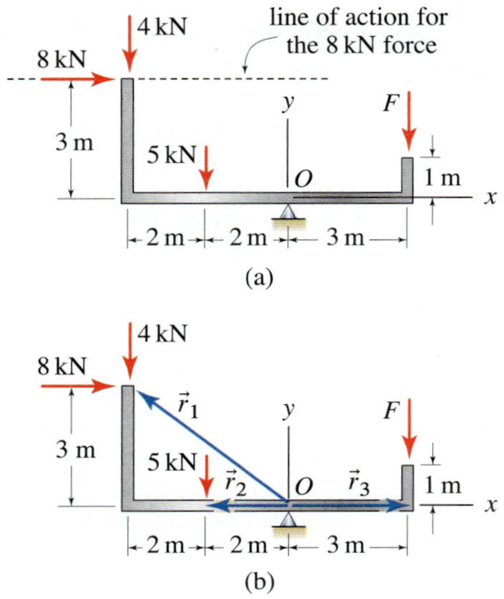

Figure 4.7
A balance mechanism subjected to several forces.

that points either into or out of the plane of Fig. 4.7. Furthermore, we may distinguish between these by taking the *counterclockwise* direction to be *positive* (see additional discussion in the margin note). Thus,

$$M_O = -(8\,\text{kN})(3\,\text{m}) + (4\,\text{kN})(4\,\text{m}) + (5\,\text{kN})(2\,\text{m}) - F(3\,\text{m}) = 0, \quad (4.3)$$

$$\Rightarrow \quad F = 0.667\,\text{kN}. \quad (4.4)$$

Each term of Eq. (4.3) consists of the product of the magnitude of a force with its corresponding moment arm, where the moment arm is the shortest (perpendicular) distance from point O to the line of action of the force. For example, in Fig. 4.7(a), the line of action for the 8 kN force is shown, and the shortest distance from point O to this line is 3 m. The sign of each term is positive for counterclockwise twisting about point O and negative for clockwise twisting.

A vector approach can also be used for this problem with the following force and position vectors (see Fig. 4.7(b)):

$$\vec{F}_1 = (8\,\hat{\imath} - 4\,\hat{\jmath})\,\text{kN}, \qquad \vec{r}_1 = (-4\,\hat{\imath} + 3\,\hat{\jmath})\,\text{m}, \quad (4.5)$$

$$\vec{F}_2 = -5\,\hat{\jmath}\,\text{kN}, \qquad \vec{r}_2 = -2\,\hat{\imath}\,\text{m}, \quad (4.6)$$

$$\vec{F}_3 = -F\,\hat{\jmath}\,\text{kN}, \qquad \vec{r}_3 = 3\,\hat{\imath}\,\text{m}, \quad (4.7)$$

$$\vec{M}_O = \vec{r}_1 \times \vec{F}_1 + \vec{r}_2 \times \vec{F}_2 + \vec{r}_3 \times \vec{F}_3,$$

$$= [2\,\text{kN}{\cdot}\text{m} - F(3\,\text{m})]\,\hat{k} = \vec{0}, \quad (4.8)$$

$$\Rightarrow F = 0.667\,\text{kN}. \quad (4.9)$$

Helpful Information

Direction for moments in two-dimensional problems. In two-dimensional problems, such as Fig. 4.7, the moment of a force about a given point will be either clockwise or counterclockwise. That is, the moment of a force about a given point is a vector in either the $-z$ direction or the $+z$ direction, respectively. In two-dimensional problems, when using a scalar approach, we will always take *counterclockwise* to be the *positive* direction for moments.

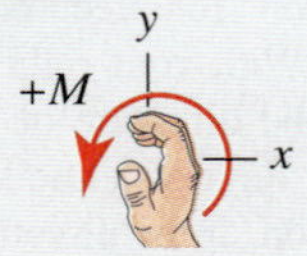

Figure 4.8

This choice is consistent with the right-hand rule for the xy coordinate system shown, where the z direction, and hence the direction for positive moment vectors, is out of the plane of the figure.

Remarks

- Since all forces and point O lie in the xy plane, the forces produce moment about the z axis only, and hence a scalar approach is very straightforward. In a fully three-dimensional problem the scalar approach may be difficult to use, and the vector approach is often more effective.

- In the vector approach, since the 8 kN and 4 kN forces share the same point of application, they are combined to yield a single force vector $\vec{F}_1$ in Eq. (4.5). Thus, their moment can be obtained by evaluating just one cross product.

- In the vector approach, the direction of positive moment is the positive z direction, which in this example is the direction out of the plane of the figure as provided by the right-hand rule. To see that this is true, use your right hand such that your thumb points in the positive z direction; then the direction in which your fingers curl provides the twisting sense of a positive moment.

- In the vector approach there is considerable flexibility in selecting the position vectors. In this example, $\vec{r}_1$ and $\vec{r}_2$ are taken from point O to the actual points of application of $\vec{F}_1$ and $\vec{F}_2$, while $\vec{r}_3$ is taken from point O to a convenient point on the line of action of $\vec{F}_3$. The same result for $\vec{M}_O$ is obtained using, for example, $\vec{r}_3 = (3\,\hat{\imath} + \hat{\jmath})$ m or $\vec{r}_3 = (3\,\hat{\imath} - \hat{\jmath})$ m.

- In general, there is no reason why the resultant moment of all forces about point O should be zero. In this example, if $F = 1.0\,\text{kN}$, then the scalar approach yields $M_O = -1\,\text{kN}\cdot\text{m}$ and the vector approach yields $\vec{M}_O = -1\,\text{kN}\cdot\text{m}\,\hat{k}$. The result of having a nonzero resultant moment about point O is that the balance mechanism will tend to rotate about point O. This topic is explored in detail in Chapter 5.

Varignon's theorem

Varignon's theorem, also known as the *principle of moments*, states that the moment of a force is equal to the sum of the moments of the vector components of the force. Thus, if $\vec{F}$ has vector components $\vec{F}_1$, $\vec{F}_2$, and so on, then the moment of $\vec{F}$ about a point A is given by

$$\begin{aligned}\vec{M}_A &= \vec{r} \times \vec{F}\\ &= \vec{r} \times (\vec{F}_1 + \vec{F}_2 + \cdots)\\ &= \vec{r} \times \vec{F}_1 + \vec{r} \times \vec{F}_2 + \cdots,\end{aligned} \tag{4.10}$$

where $\vec{r}$ is a position vector from point A to any point on the line of action of $\vec{F}$. This principle is simply a restatement of the distributive property of the cross product, but in fairness to Varignon (1654–1722), he discovered the concepts underlying Eq. (4.10) well before vector mathematics and the cross product were invented. Varignon's theorem remains very useful for evaluating moments, especially for scalar evaluations. While the component forces will often be Cartesian components, in which case $\vec{F}_1$, $\vec{F}_2$, and $\vec{F}_3$ would usually be called $\vec{F}_x$, $\vec{F}_y$, and $\vec{F}_z$, in general the components do not need to be orthogonal, and there may be an arbitrary number of them.

Mini-Example. With reference to Figs. 4.9 and 4.10, use scalar and vector approaches, respectively, to evaluate the moment of force $\vec{F}$ about point A.

Solution. Force $\vec{F}$ has magnitude F and components F_x and F_y. The magnitude of the moment of $\vec{F}$ about a point A can be evaluated by the two scalar approaches shown in Fig. 4.9. In the first of these, the magnitude of the moment is computed using the definition given in Eq. (4.1), where the moment arm d is the perpendicular distance from point A to the line of action of the force and positive values of M_A are taken to be counterclockwise. In the second evaluation in Fig. 4.9, the x and y components of the force are used, and the moment of each of these is evaluated and summed, again with counterclockwise being positive. In practice, you should use the first approach if it is easy to find the moment arm d, and the second approach if it is easier to obtain the moment arms for the components of the force.

The vector approaches shown in Fig. 4.10 can also be used. In the first of these, the position vector is taken from point A to the tail of $\vec{F}$, although a position vector from A to any point on the line of action of $\vec{F}$ is acceptable. In the second evaluation in Fig. 4.10, the force is resolved into vector components $\vec{F}_x$ and $\vec{F}_y$, and the moment of each of these is computed and summed. In practice, the first of these vector evaluations will almost always be more convenient.

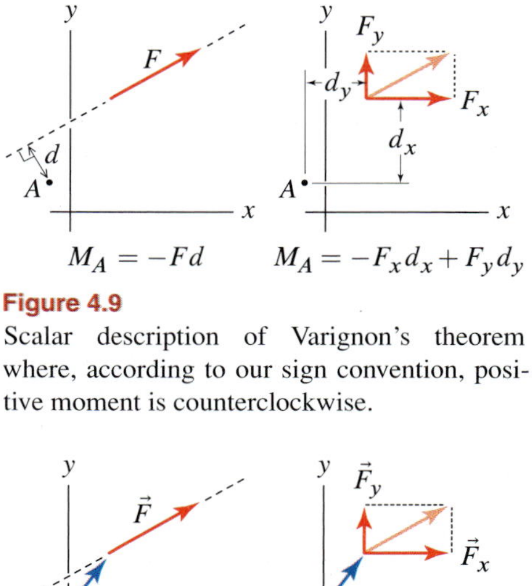

Figure 4.9
Scalar description of Varignon's theorem where, according to our sign convention, positive moment is counterclockwise.

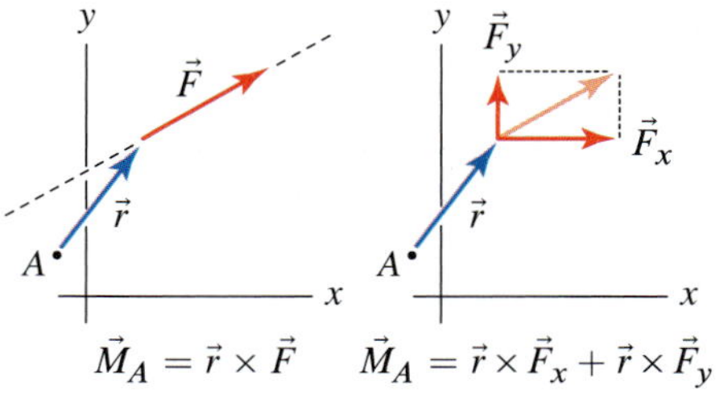

Figure 4.10
Vector description of Varignon's theorem.

Common Pitfall

What does the moment depend on? A common misconception is that the moment of a force is an inherent property of only the force. The moment of a force depends on both the force *and* the location of the point about which the moment is evaluated. For example, consider the moment of force $\vec{F}$ about point A and about point B.

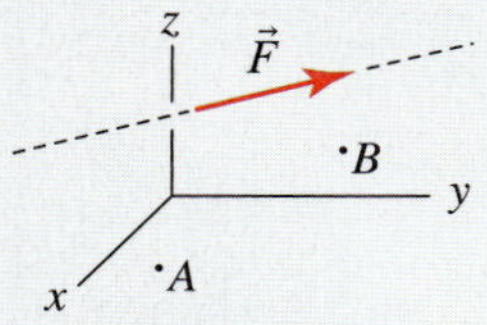

In general, the moment of $\vec{F}$ about these two points is different, and $\vec{M}_A \neq \vec{M}_B$.

Which approach should I use: scalar or vector?

In simpler problems, such as two-dimensional problems when all forces are *coplanar* (i.e., all forces lie in the same plane and hence produce moments about the same axis, which is perpendicular to that plane), the scalar approach for evaluating moments will usually be easier and faster, and use of *clockwise* and *counterclockwise* to distinguish moment direction is effective. For more complicated problems, such as in three dimensions, a scalar approach can sometimes be used effectively, but generally the vector approach is better. You should contrast the scalar and vector solutions for the example problems that follow to help refine your ability to select the easier approach for a particular problem.

End of Section Summary

In this section, the *moment of a force*, or simply *moment*, is defined and scalar and vector approaches for evaluation are discussed. Some of the key points are as follows:

- The moment is a vector quantity.
- When we evaluate the moment using the scalar approach, the magnitude of the moment of a force $\vec{F}$ about a point P is given by $M_P = Fd$ where d is the *moment arm*, which is the perpendicular distance from point P to the line of action of the force. The scalar approach does not automatically provide the direction of the moment; this must be assigned manually.
- When we evaluate the moment using the vector approach, the moment of a force $\vec{F}$ about a point P is given by $\vec{M}_P = \vec{r} \times \vec{F}$, where $\vec{r}$ is a position vector measured from point P to *anywhere* on the line of action of $\vec{F}$. The direction of $\vec{M}_P$ is automatically provided by this approach.
- The moment of a force depends on both the force *and* the location of the point about which the moment is evaluated. The important characteristics of the force are its magnitude, its orientation or direction, *and* the position in space of its line of action.
- *Varignon's theorem*, also known as the *principle of moments*, states that the moment of a force is equal to the sum of the moments of the components of the force.

EXAMPLE 4.1 *Moment: Vector and Scalar Evaluations for a Two-dimensional Problem*

A portion of a structure is acted upon by the 10 and 20 lb forces shown. Determine the resultant moment of these forces about point A.

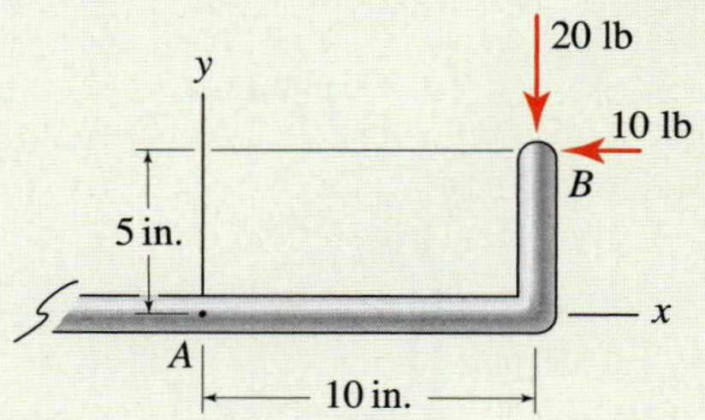

Figure 1

SOLUTION

Road Map Both scalar and vector approaches are effective for this problem. In the scalar approach, we must find the moment arm for each force, and after evaluating the moment of each force, we must be careful to sum the results according to a sign convention for positive moment. With the vector approach, we must develop vector expressions for forces and positions and then carry out cross products to determine the moment. For two-dimensional problems, the vector approach will usually require more computations, but does not require the careful visualization of the scalar approach.

Scalar solution

Governing Equations & Computation In Fig. 2, the moment arms for the 10 and 20 lb forces are determined. By applying Eq. (4.1) on p. 182 to each force and summing the results, using the convention that positive moment is counterclockwise, the resultant moment about point A is

$$M_A = (10\,\text{lb})(5\,\text{in.}) - (20\,\text{lb})(10\,\text{in.}) = \boxed{-150\,\text{in.}\cdot\text{lb.}} \tag{1}$$

When writing Eq. (1), we used Eq. (4.1) to evaluate the magnitude of the moment for the 10 lb force as (10 lb)(5 in.), followed by examination of Fig. 1 to determine that this force produces counterclockwise moment about point A and hence it is positive in accordance with our sign convention. We then repeated this process for the 20 lb force to determine the magnitude of its moment as $-(20\,\text{lb})(10\,\text{in.})$ where the negative sign is used because this force produces clockwise moment about point A. Carrying out the algebra in Eq. (1) yields the final answer: the moment of the two forces about point A is $\boxed{-150\,\text{in.}\cdot\text{lb counterclockwise.}}$

If desired, the results of Eq. (1) can be used to write a vector expression for the resultant moment about point A as

$$\boxed{\vec{M}_A = -150\,\text{in.}\cdot\text{lb} \circlearrowleft\!+} \tag{2}$$

where the $\circlearrowleft\!+$ symbol specifies the direction and is analogous to the $\measuredangle$ symbol used in polar vector representation (described in the margin note on p. 31).

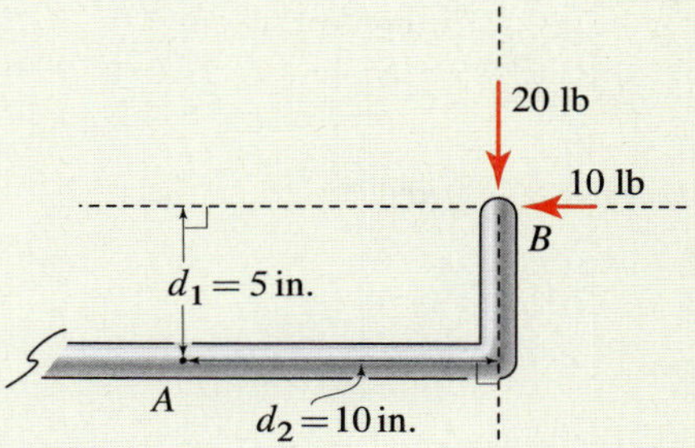

Figure 2
A sketch showing the lines of action of each force. The moment arm d_1 for the 10 lb force is the shortest distance from point A to the line of action of this force, hence $d_1 = 5$ in. The moment arm d_2 for the 20 lb force is the shortest distance from point A to the line of action of this force, hence $d_2 = 10$ in.

Vector solution

Governing Equations & Computation A vector approach can also be employed using the vectors $\vec{F}$ and $\vec{r}_{AB}$ shown in Fig. 3, which are

$$\vec{F} = (-10\,\hat{\imath} - 20\,\hat{\jmath})\,\text{lb}, \quad \vec{r}_{AB} = (10\,\hat{\imath} + 5\,\hat{\jmath})\,\text{in.} \tag{3}$$

Using Eq. (4.2) on p. 183, the moment vector of the two forces about point A is

$$\vec{M}_A = \vec{r}_{AB} \times \vec{F} = \boxed{-150\,\hat{k}\,\text{in.}\cdot\text{lb.}} \tag{4}$$

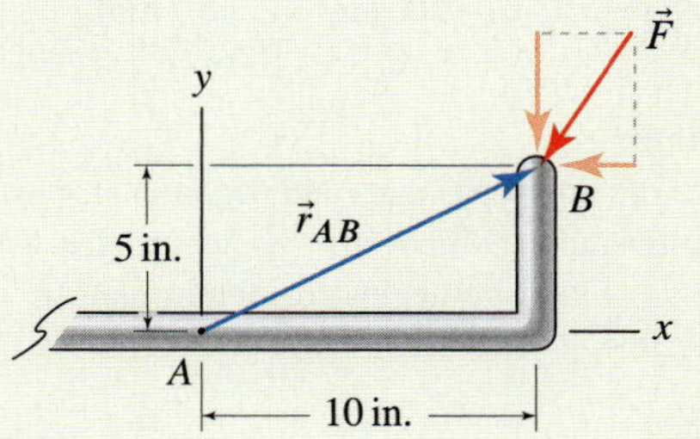

Figure 3
Force and position vectors used to determine the moment of $\vec{F}$ about point A.

Discussion & Verification While you should be proficient using both scalar and vector approaches, you should contrast these two solutions to help you recognize which of the two will be more efficient for a particular problem. You can also use both approaches for a particular problem to help verify the accuracy of your answers.

EXAMPLE 4.2 *Moment: Different Points Along the Line of Action of the Force May Be Used*

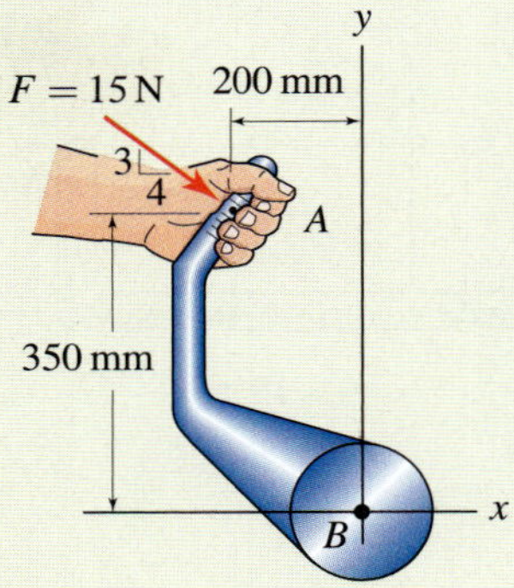

Figure 1

A machine handle is connected to a shaft at B. Determine the moment produced by the 15 N force about point B.

SOLUTION

Road Map Both scalar and vector approaches are effective for this problem, and several solutions will be carried out and contrasted. If a scalar solution is used, there are several options for treatment of the force and determination of moment arms. If a vector approach is used, there are options on selection of the position vector. Regardless of your approach, the results should be the same, but some solutions may be easier to carry out than others.

Solution 1

Governing Equations & Computation With the geometry provided, similar triangles and trigonometry can be used (detailed calculations are given in Fig. 2) to determine that the moment arm is $d = 160\,\text{mm}$, as shown in Fig. 3(a). With our convention that counterclockwise moments are positive, the moment about point B is

$$M_B = -(15\,\text{N})(160\,\text{mm}) = \boxed{-2400\,\text{N}{\cdot}\text{mm.}} \tag{1}$$

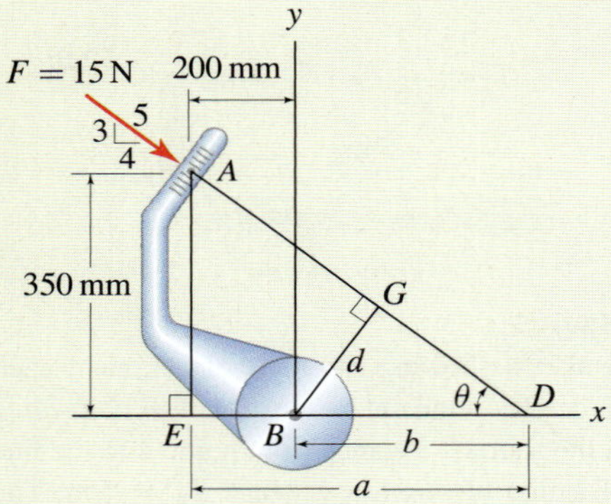

Triangle AED:

$$\frac{350\,\text{mm}}{a} = \frac{3}{4} \quad \Rightarrow \quad a = 466.7\,\text{mm}$$

$$\tan\theta = \frac{350\,\text{mm}}{a} \quad \Rightarrow \quad \theta = 36.9°$$

Triangle BDG:

$$b = a - 200\,\text{mm} = 266.7\,\text{mm}$$

$$d = b\sin\theta = \boxed{160\,\text{mm}}$$

Figure 2
Use of similar triangles and trigonometry to determine the moment arm d needed for Solution 1. Other strategies for determining d are possible.

Solution 2

Governing Equations & Computation We resolve the 15 N force into its x and y components as shown in Fig. 3(b) and use Varignon's theorem (the principle of moments) to sum the moments produced by each of these components. With our convention that positive moment is counterclockwise,

$$M_B = -(12\,\text{N})(350\,\text{mm}) + (9\,\text{N})(200\,\text{mm}) = \boxed{-2400\,\text{N}{\cdot}\text{mm.}} \tag{2}$$

Solution 3

Governing Equations & Computation The moment of a force does not depend on the specific location where the force is applied, but rather depends on the position and orientation of the force's line of action. Thus, we may consider the 15 N force as being positioned anywhere along its line of action, and we select point C shown in Fig. 3(c) as a convenient location (you should verify for yourself the coordinates of point C – Fig. 2 may be helpful). With our convention that positive moment is counterclockwise,

$$M_B = -(12\,\text{N})(200\,\text{mm}) + (9\,\text{N})(0\,\text{mm}) = \boxed{-2400\,\text{N}{\cdot}\text{mm.}} \tag{3}$$

Solution 4

Governing Equations & Computation We use the same approach employed in the previous solution, except here we "move" the force to point D shown in Fig. 3(d). With our convention that positive moment is counterclockwise,

$$M_B = (12\,\text{N})(0\,\text{mm}) - (9\,\text{N})(266.7\,\text{mm}) = \boxed{-2400\,\text{N}{\cdot}\text{mm.}} \tag{4}$$

Solution 5

Governing Equations & Computation A vector approach can also be employed using the force and position vectors shown in Fig. 3(e), which are

$$\vec{F} = (12\,\hat{\imath} - 9\,\hat{\jmath})\,\text{N}, \qquad \vec{r}_{BA} = (-200\,\hat{\imath} + 350\,\hat{\jmath})\,\text{mm}. \tag{5}$$

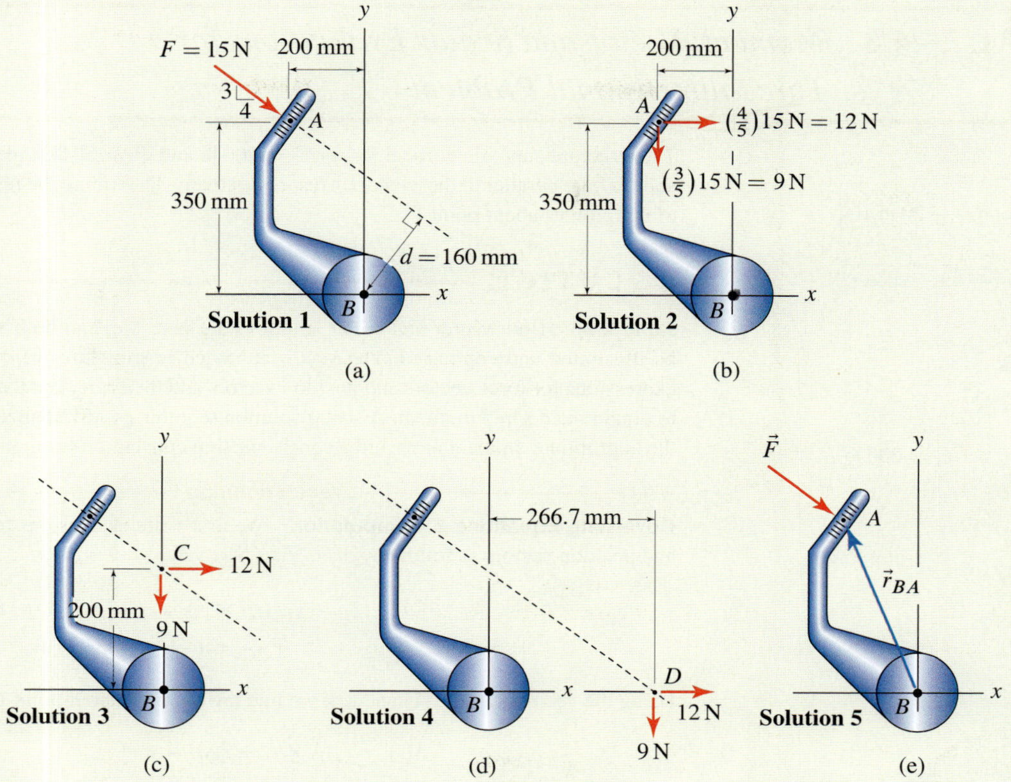

Figure 3. (a)–(d) Positioning of forces and/or resolution of forces into components so that a scalar approach can be used to determine the moment of the 15 N force about point B. (e) Force and position vectors that can be used in a vector approach.

Rather than using $\vec{r}_{BA}$, the position vectors $\vec{r}_{BC}$ and $\vec{r}_{BD}$ are convenient choices that could have been used instead. Evaluating the moment provides

$$\vec{M}_B = \vec{r}_{BA} \times \vec{F} = \boxed{-2400\,\hat{k}\ \text{N·mm}.} \tag{6}$$

Discussion & Verification As expected, all solutions produce the same result. For this example, Solution 1 is not especially convenient because the determination of the moment arm is tedious. Solutions 2–5 are all effective with Solutions 2 and 5 probably being the most straightforward.

Helpful Information

A useful tip. Sometimes, the distance between a point and the line of action of a force (or other vector) is desired, such as d in Fig. 2. For situations where a direct evaluation of d is tedious, such as in this example, you may use one of Solutions 2–5 to determine the moment of the force about point B (namely -2400 N·mm), then write Eq. (1) as $M_B = -(15\ \text{N})d = -2400\ \text{N·mm}$, and then solve for $d = 160$ mm.

EXAMPLE 4.3 *Moment: Vector and Scalar Evaluations for a Three-dimensional Problem*

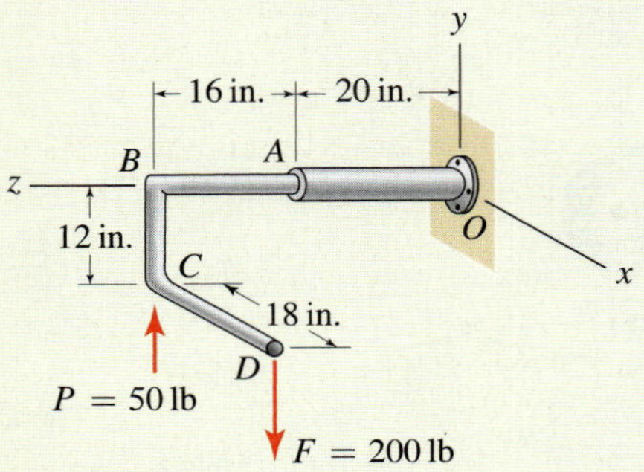

Figure 1

The structure supports vertical forces $F = 200$ lb and $P = 50$ lb. Pipe segments BC and CD are parallel to the y and x axes, respectively. Determine the resultant moment of both forces about point O.

SOLUTION

Road Map Both vector and scalar solutions are possible, and both approaches will be illustrated and contrasted. The vector approach is straightforward: we first write expressions for force vectors and position vectors and then carry out the cross products to obtain the desired moment. A scalar solution requires good visualization to identify the appropriate moment arms and a consistent sign convention for moments.

Vector solution

Governing Equations & Computation We first write expressions for force vectors and position vectors as follows:

$$\vec{F} = -200\,\hat{\jmath}\ \text{lb}, \qquad \vec{r}_{OD} = (18\,\hat{\imath} - 12\,\hat{\jmath} + 36\,\hat{k})\ \text{in.}, \tag{1}$$

$$\vec{P} = 50\,\hat{\jmath}\ \text{lb}, \qquad \vec{r}_{OC} = (-12\,\hat{\jmath} + 36\,\hat{k})\ \text{in.} \tag{2}$$

Using the vectors in Eqs. (1) and (2), we find the moment about point O is

$$\vec{M}_O = \vec{r}_{OD} \times \vec{F} + \vec{r}_{OC} \times \vec{P}$$
$$= \boxed{(5400\,\hat{\imath} - 3600\,\hat{k})\ \text{in.}\cdot\text{lb.}} \tag{3}$$

Remarks

- Rather than using $\vec{r}_{OD}$ and $\vec{r}_{OC}$ in Eq. (3), the position vectors $(18\,\hat{\imath} + 36\,\hat{k})$ in. and $\vec{r}_{OB}$, respectively, could have been used. These are slightly better choices since they each have fewer components, which will reduce the number of computations needed to evaluate the cross products.
- The result for $\vec{M}_O$ in Eq. (3) has x and z components, meaning that $\vec{F}$ and $\vec{P}$ combine to produce twisting action about both the x and z axes through point O as shown in Fig. 2(a). Alternatively, the x and z components can be added to give a single moment vector with magnitude M_O, as shown in Fig. 2(b).

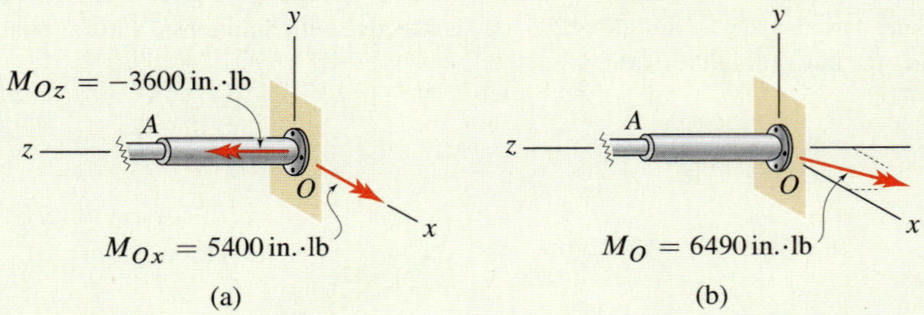

Figure 2. The resultant moment $\vec{M}_O$ of forces F and P about point O is shown. (a) The components of $\vec{M}_O$ are shown. (b) The vector $\vec{M}_O$ is shown in its proper orientation.

Scalar solution

Governing Equations & Computation A scalar solution is possible, but requires good visualization to identify the appropriate moment arms and a consistent sign convention for moments. In three-dimensional problems, it is very difficult to unambiguously categorize moments with words like *clockwise* and *counterclockwise*. Thus, we will usually take the positive coordinate directions to define the directions of positive moments (i.e., the right-hand rule governs the direction of positive moment).

As shown in Fig. 3, we extend the lines of action for each force to help identify the appropriate moment arms. Force F produces positive moment about the x axis where

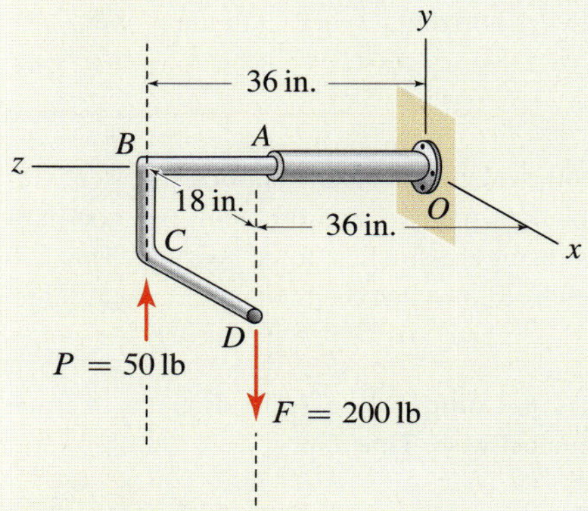

Figure 3. Moment arms to be used in a scalar approach for determining the moment of forces F and P about point O.

the moment arm is 36 in, and negative moment about the z axis where the moment arm is 18 in. Since F is parallel to the y axis, it produces no moment about this axis. Force P produces negative moment about the x axis where the moment arm is 36 in. and produces no moment about the other two axes. Thus,

$$M_{Ox} = (200\,\text{lb})(36\,\text{in.}) - (50\,\text{lb})(36\,\text{in.}) = \boxed{5400\,\text{in.}\cdot\text{lb},} \tag{4}$$

$$M_{Oy} = \boxed{0,} \tag{5}$$

$$M_{Oz} = -(200\,\text{lb})(18\,\text{in.}) = \boxed{-3600\,\text{in.}\cdot\text{lb}.} \tag{6}$$

Discussion & Verification As expected, the x, y, and z components of the moment about point O found by the scalar approach agree with those found using the vector approach. Note that in the scalar approach, if you do not take positive moments to be in the positive coordinate directions, then the signs of moment components may not agree with those in the vector approach. Our recommendation is that you always take moments to be positive in the positive coordinate directions.

Common Pitfall

Clockwise or counterclockwise? Earlier example problems in this section all have moments that are either into or out of the plane containing the forces for each problem, and when we used the scalar approach, use of *clockwise* and *counterclockwise* for describing moment directions was effective. For moments in three dimensions, *clockwise* and *counterclockwise* are ambiguous and have little meaning. When using the scalar approach to find moments in three dimensions, you should use the positive coordinate directions and the right-hand rule to define positive moments, as illustrated in this example.

EXAMPLE 4.4 *Maximizing the Moment of a Force*

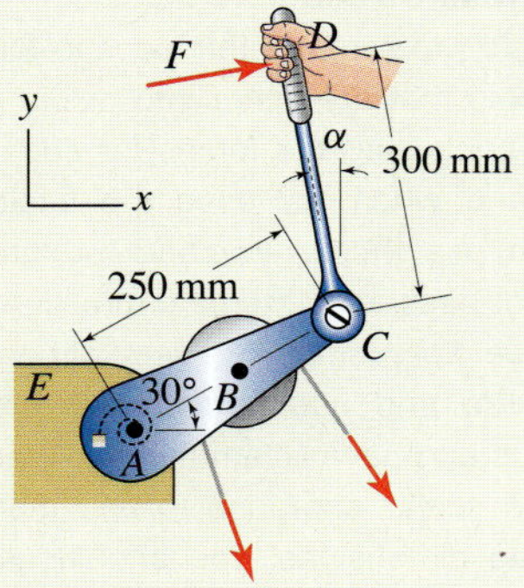

Figure 1

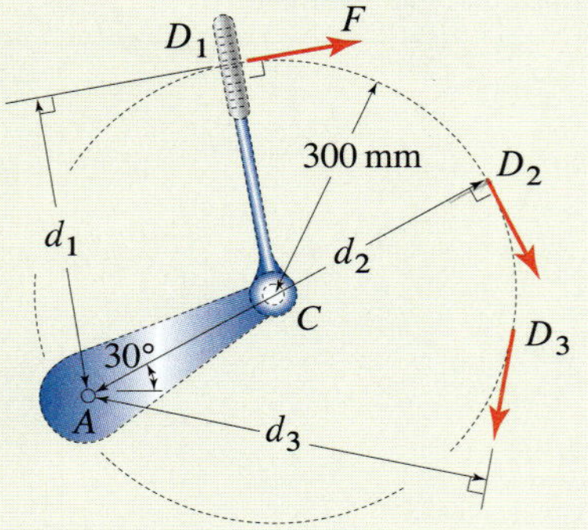

Figure 2
Depending on the value of α, there are an infinite number of possible positions for point D and force F, and three possible positions, denoted by D_1, D_2, and D_3, are shown with their corresponding moment arms d_1, d_2, and d_3, respectively.

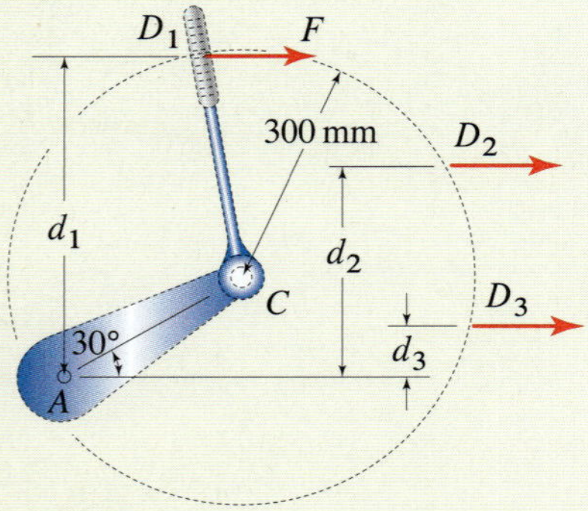

Figure 3
Among the infinite number of possible positions for point D, three possible positions, denoted by D_1, D_2, and D_3, are shown.

The belt tensioner ABC is attached to an engine using a bearing at A having a torsional spring. To release the belt tension, a ratchet wrench CD is applied to the tensioner at point C. If a moment about point A of $50\,\text{N}\cdot\text{m}$ is required to release the belt tension, determine the smallest force F required and the angle α at which the wrench should be positioned. Consider the following two cases:

(a) The force F is always perpendicular to the handle of the wrench.

(b) The force F is always horizontal (parallel to the x axis).

SOLUTION

Road Map While both scalar and vector solutions are effective, only a scalar solution will be used here. For both parts of this problem, our goal is to produce a $50\,\text{N}\cdot\text{m}$ moment about point A using the smallest force possible. This will be accomplished by positioning the wrench handle so that the moment arm is as large as possible.

Part (a)

Governing Equations & Computation The moment of force F about point A is required to be $50\,\text{N}\cdot\text{m}$ clockwise. Thus,

$$M_A = -Fd = -50\,\text{N}\cdot\text{m} \tag{1}$$

where the moment arm d is the distance from point A to the line of action of F, and because F produces clockwise moment about point A, the negative signs are included. To minimize F, we seek the position of the wrench handle D so that moment arm d is maximized. Shown in Fig. 2 is the circular locus of possible locations of point D, with three example locations shown (points D_1, D_2, and D_3) and their associated moment arms (d_1, d_2, and d_3). Examination of this figure shows that the largest moment arm is achieved when the handle of the wrench is at point D_2; thus $\boxed{\alpha = -60^\circ}$ and $d = d_2 = 550\,\text{mm}$. Equation (1) can now be solved to obtain

$$F = \frac{-M_A}{d} = \frac{50\,\text{N}\cdot\text{m}}{0.550\,\text{m}} = \boxed{90.9\,\text{N.}} \tag{2}$$

Part (b)

Governing Equations & Computation As with Part (a), we wish to maximize the moment arm d in Eq. (1) where now force F is always horizontal. The locus of possible locations for point D is shown in Fig. 3, with three example locations illustrated (points D_1, D_2, and D_3). Clearly, d is largest when the handle of the wrench is vertical. For this position $\boxed{\alpha = 0^\circ}$, the moment arm is $d = (0.250\,\text{m})\sin 30^\circ + 0.300\,\text{m} = 0.425\,\text{m}$. Solving Eq. (1) provides

$$F = \frac{-M_A}{d} = \frac{50\,\text{N}\cdot\text{m}}{0.425\,\text{m}} = \boxed{118\,\text{N.}} \tag{3}$$

Discussion & Verification Sketches such as Figs. 2 and 3 can be very helpful in problems where you must maximize or minimize the moment of a force about a particular point. Without the insight these figures offer, this problem would otherwise be more difficult.

PROBLEMS

General instructions. Unless otherwise stated, in the following problems you may use a scalar approach, a vector approach, or a combination of these.

Problems 4.1 and 4.2

Compute the moment of force F about point B, using the following procedures.

(a) Determine the moment arm d and then evaluate $M_B = Fd$.

(b) Resolve force F into x and y components at point A and use the principle of moments.

(c) Use the principle of moments with F positioned at point C.

(d) Use the principle of moments with F positioned at point D.

(e) Use a vector approach.

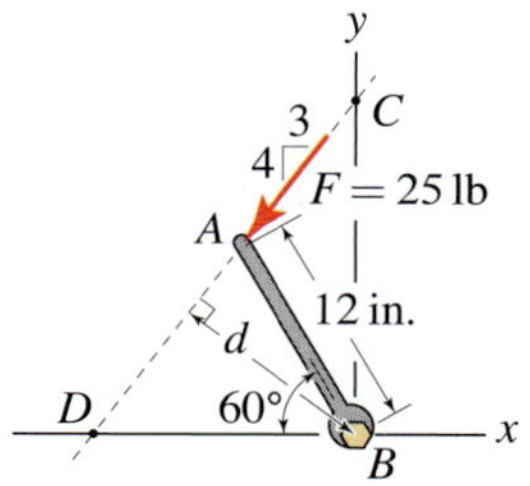

Figure P4.1

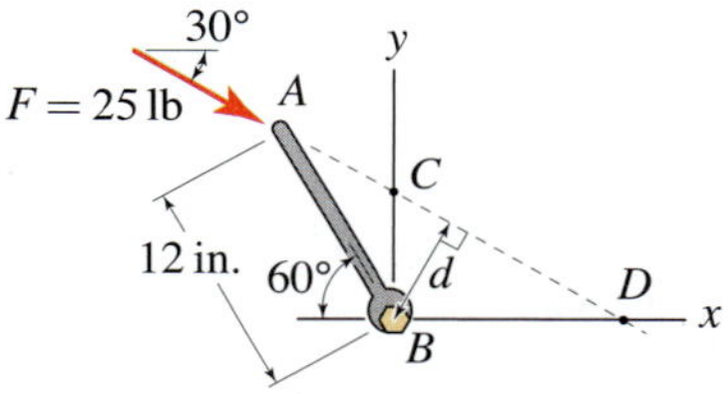

Figure P4.2

Problems 4.3 and 4.4

The cover of a computer mouse is hinged at point B so that it may be "clicked." Repeat Prob. 4.1 to determine the moment about point B.

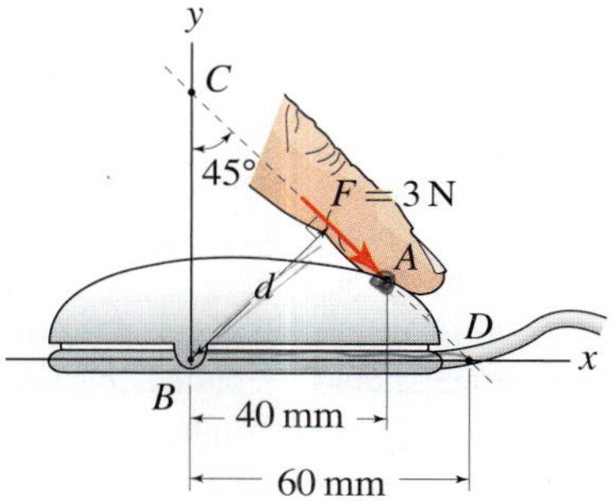

Figure P4.3

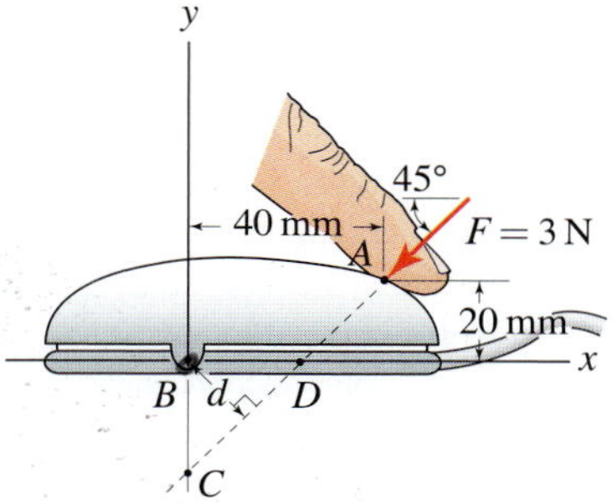

Figure P4.4

Problem 4.5

An atomic force microscope (AFM) is a state-of-the-art device used to study the mechanical and topological properties of surfaces on length scales as small as the size of individual atoms. The device uses a flexible cantilever beam AB with a very sharp, stiff tip BC that is brought into contact with the surface to be studied. Due to contact forces at C, the cantilever beam deflects. If the tip of the AFM is subjected to the forces shown, determine the resultant moment of both forces about point A. Use both scalar and vector approaches.

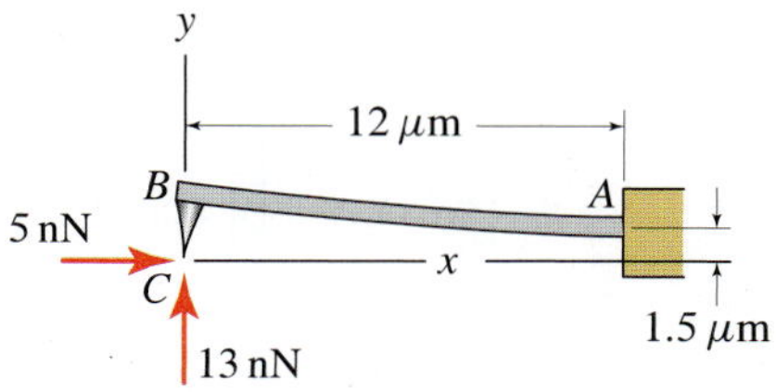

Figure P4.5

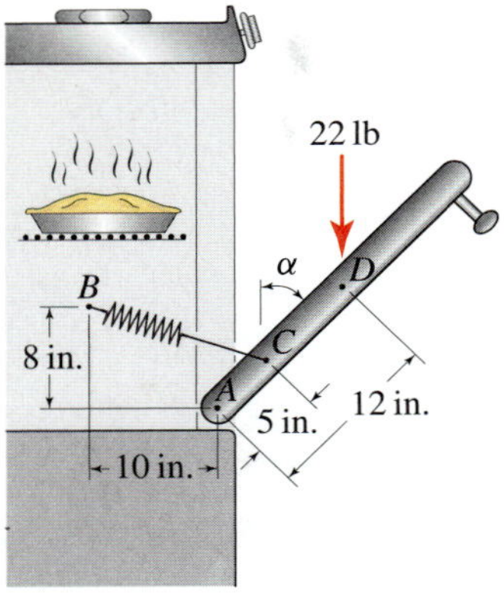

Figure P4.6

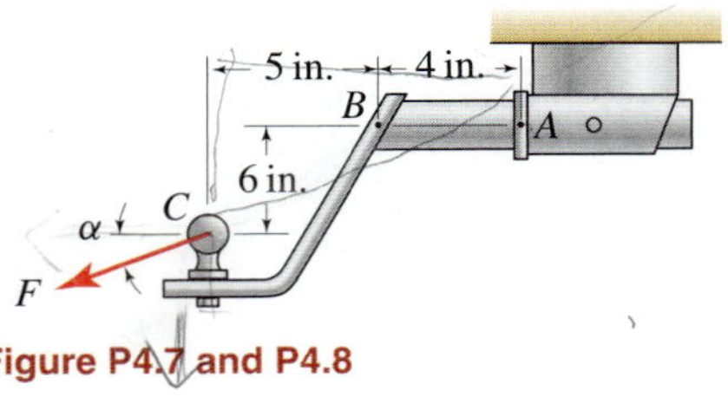

Figure P4.7 and P4.8

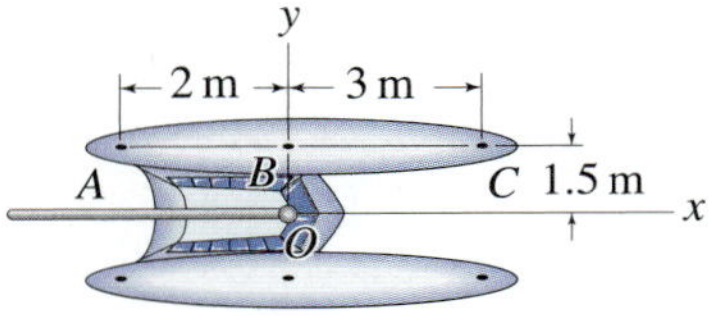

Figure P4.9

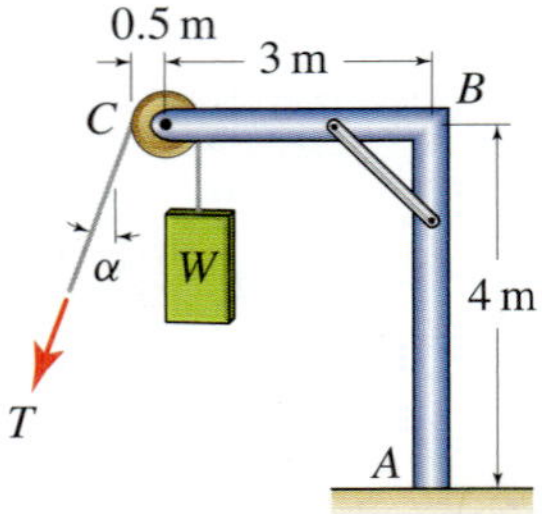

Figure P4.10 and P4.11

Problem 4.6

The door of an oven has 22 lb weight that acts vertically through point D. The door is supported by a hinge at point A and two springs that are symmetrically located on each side of the door. For the positions specified below, determine the force needed in each of the two springs if the resultant moment about point A of the weight and spring forces is to be zero.

(a) $\alpha = 45°$.

(b) $\alpha = 90°$.

(c) Based on your answers to Parts (a) and (b), determine an appropriate stiffness for the springs.

Problem 4.7

The ball of a trailer hitch is subjected to a force F. If failure occurs when the moment of F about point A reaches 10,000 in.·lb, determine the largest value F may have and specify the value of α for which the moment about A is the largest. Note that the value of F you determine must not produce a moment about A that exceeds 10,000 in.·lb for any possible value of α.

Problem 4.8

Repeat Prob. 4.7 if the moment at point B may not exceed 5000 in.·lb.

Problem 4.9

The port hull of a catamaran (top view shown) has cleats at points A, B, and C. A rope having 100 N force is to be attached to one of these cleats. If the force is to produce the largest possible counterclockwise moment about point O, determine the cleat to which the rope should be attached and the direction the rope should be pulled (measured positive counterclockwise from the positive x direction). Also, determine the value of M_O produced. Assume all cleats, point O, and the rope lie in the same plane.

Problem 4.10

Frame ABC has a frictionless pulley at C around which a cable is wrapped. Determine the resultant moment about point A produced by the cable forces if $W = 5$ kN and

(a) $\alpha = 0°$.

(b) $\alpha = 90°$.

(c) $\alpha = 30°$.

Problem 4.11

Frame ABC has a frictionless pulley at C around which a cable is wrapped. If the resultant moment about point A produced by the cable forces is not to exceed 20 kN·m, determine the largest weight W that may be supported and the corresponding value of α.

Problem 4.12

The load carrying capacity of the frame of Prob. 4.11 can be increased by placing a counterweight Q at point D as shown. The resultant moment about point A due to the cable forces and Q is not to exceed 20 kN·m.

(a) Determine the largest value of counterweight Q. *Hint:* Let $W = 0$ and determine Q so that $M_A = 20$ kN·m.

(b) With the value of Q determined in Part (a), determine the largest weight W that may be supported and the corresponding value of α.

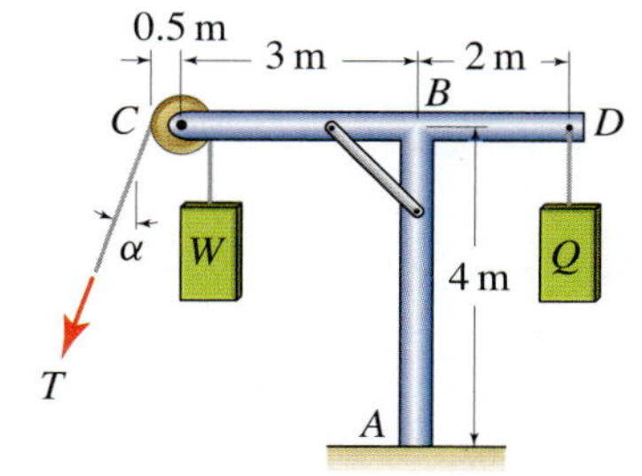

Figure P4.12

Problem 4.13

A flat rectangular plate is subjected to the forces shown, where all forces are parallel to the x or y axis. If $F = 200$ N and $P = 300$ N, determine the resultant moment of all forces about the

(a) z axis.

(b) a axis, which is parallel to the z axis.

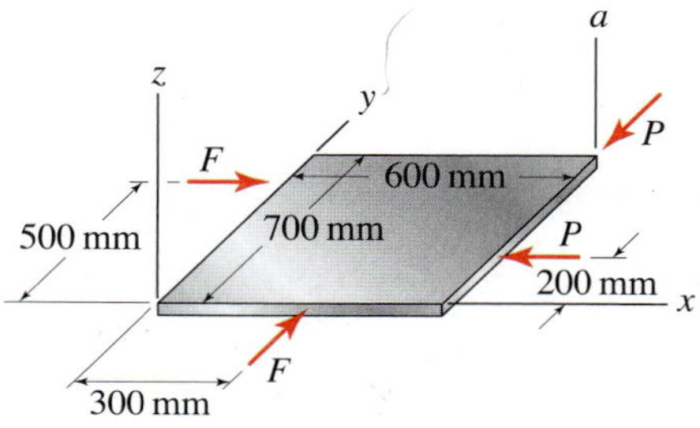

Figure P4.13–P4.15

Problem 4.14

A flat rectangular plate is subjected to the forces shown, where all forces are parallel to the x or y axis. If $P = 300$ N, determine F when the resultant moment of all forces about the z axis is $M_z = -100$ N·m.

Problem 4.15

Repeat Prob. 4.14 if the resultant moment of all forces about the a axis is $M_a = -100$ N·m.

Problem 4.16

Structure $OBCD$ is built in at point O and supports a 50 lb cable force at point C and 100 and 200 lb vertical forces at points B and D, respectively. Using a vector approach, determine the moment of these forces about

(a) point B.

(b) point O.

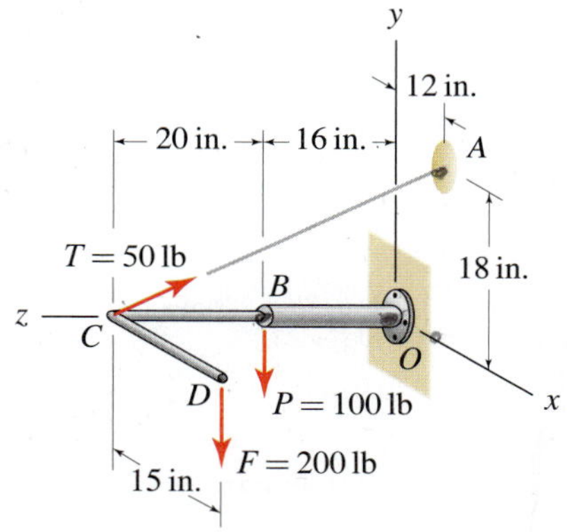

Figure P4.16 and P4.17

Problem 4.17

Repeat Prob. 4.16, using a scalar approach.

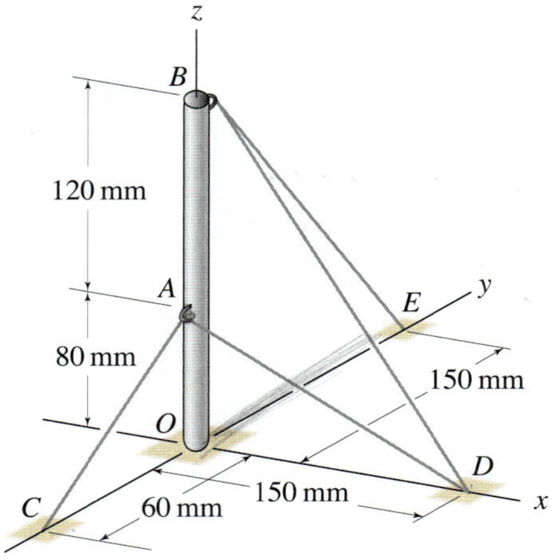

Figure P4.18 and P4.19

Problem 4.18

Structure OAB is built in at point O and supports forces from two cables. Cable CAD passes through a frictionless ring at point A, and cable DBE passes through a frictionless ring at point B. If the force in cable CAD is 250 N and the force in cable DBE is 100 N, use a vector approach to determine

(a) the moment of all cable forces about point A.

(b) the moment of all cable forces about point O.

Problem 4.19

Repeat Prob. 4.18, using a scalar approach.

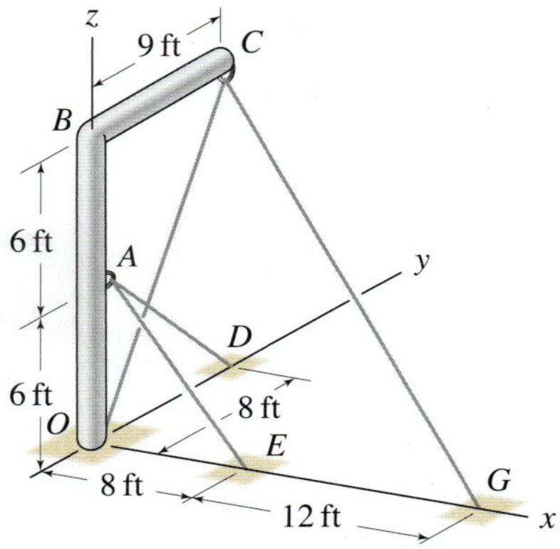

Figure P4.20 and P4.21

Problem 4.20

Structure $OABC$ is built in at point O and supports forces from two cables. Cable EAD passes through a frictionless ring at point A, and cable OCG passes through a frictionless ring at point C. If the force in cable EAD is 800 lb and the force in cable OCG is 400 lb, determine

(a) the moment of forces from cable OCG about point B.

(b) the moment of all cable forces about point A.

(c) the moment of all cable forces about point O.

Problem 4.21

Repeat Prob. 4.20, using a scalar approach.

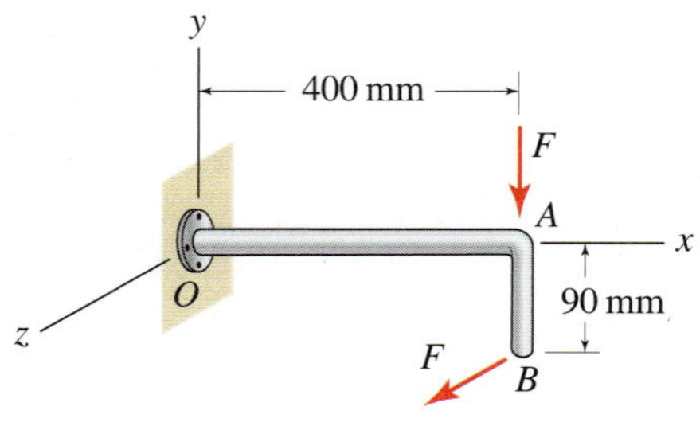

Figure P4.22 and P4.23

Problem 4.22

Structure OAB is built in at point O and supports two forces of magnitude F parallel to the y and z axes. If the magnitude of the moment about point O cannot exceed $1.0\,\text{kN·m}$, determine the largest value F may have.

Problem 4.23

Repeat Prob. 4.22 if the x component of the moment (torsional component) at point O may not exceed $0.5\,\text{kN·m}$ and the resultant of the y and z components (bending components) may not exceed $0.8\,\text{kN·m}$ (i.e., $\sqrt{M_y^2 + M_z^2} \le 0.8\,\text{kN·m}$).

Problem 4.24

Forces of 3 kN and 200 N are exerted at points B and C of the main rotor of a helicopter, and force F is exerted at point D on the tail rotor. The 3 kN forces are parallel to the z axis, the 200 N forces are perpendicular to the main rotor and are parallel to the xy plane, F is parallel to the y axis, and $\alpha = 45°$.

(a) Determine the value of F so that the z component of the moment about point O of all rotor forces is zero.

(b) Using the value of F found in Part (a), determine the resultant moment of all rotor forces about point O.

(c) If α is different than 45°, do your answers to Parts (a) and (b) change? Explain.

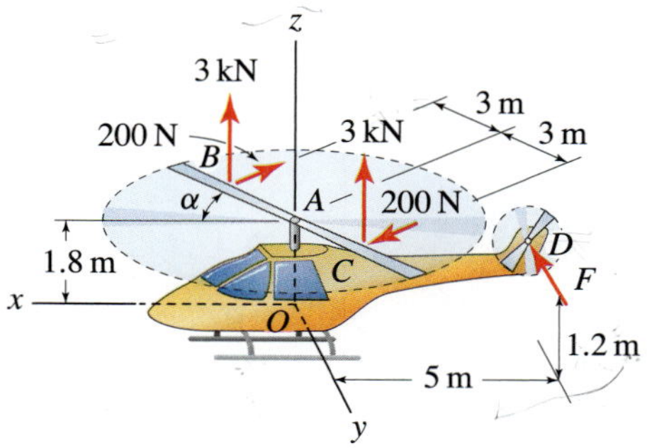

Figure P4.24

Problem 4.25

The moment of force $\vec{F}$ about point A can be computed using $\vec{M}_{A,1} = \vec{r}_{AB} \times \vec{F}$ or using $\vec{M}_{A,2} = \vec{r}_{AC} \times \vec{F}$, where B and C are points on the line of action of $\vec{F}$. Noting that $\vec{r}_{AC} = \vec{r}_{AB} + \vec{r}_{BC}$, show that $\vec{M}_{A,1} = \vec{M}_{A,2}$.

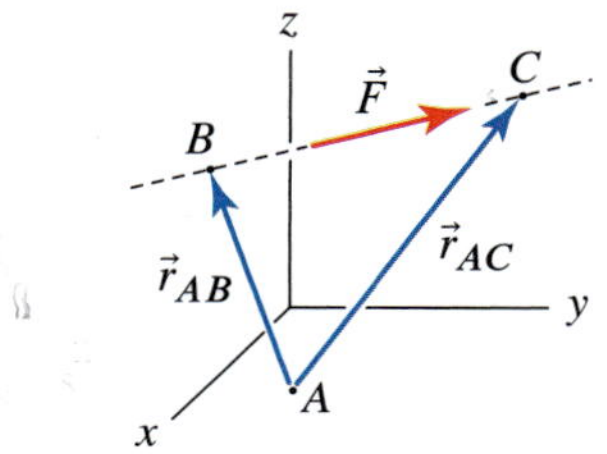

Figure P4.25

4.2 Moment of a Force About a Line

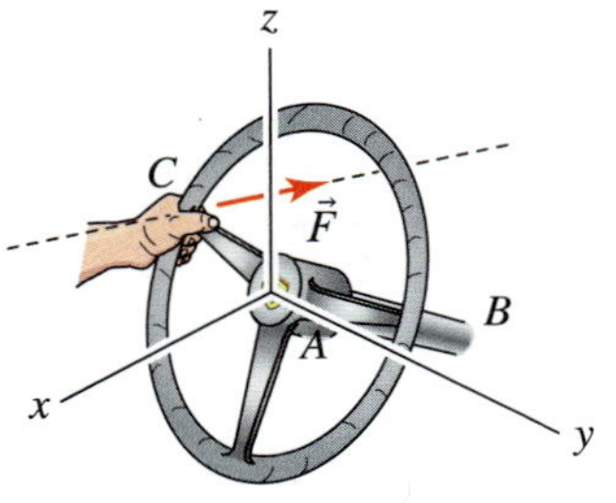

Figure 4.11
The steering wheel in the Ferrari 250 GTO (see the discussion beginning on p. 181), where the driver applies a force $\vec{F}$ at point C to turn the steering wheel.

We begin this section by summarizing the main result of the previous section: the moment of a force about a point (usually called simply the *moment*) is the tendency of the force to cause twisting of the point about which the moment is evaluated. As such, the moment of a force about a point is a vector quantity. In contrast to this, very often it is useful or necessary to determine the *moment of a force about a line* or about a *specific direction*. The moment of a force about a line is the tendency of the force to cause twisting about the line. To illustrate, we reconsider the steering wheel from the Ferrari sports car, shown again in Fig. 4.11, except that now the direction of the force $\vec{F}$ applied by the driver's hand is arbitrary. Clearly, the effectiveness of $\vec{F}$ to turn the steering wheel depends greatly on the orientation of $\vec{F}$. In fact, if the driver pushes on the steering wheel so that $\vec{F}$ is in the same direction as the steering column AB (i.e., $\vec{F} = F\vec{r}_{AB}/|\vec{r}_{AB}|$ where $\vec{r}_{AB}$ is the position vector from point A to B), the steering wheel will have no tendency to turn! Problem 4.26 explores this steering wheel further.

The *moment of a force about a line* is defined to be the *component of the moment that is in the direction of the line*. If the line happens to be parallel to a coordinate axis, then the answer is easily obtained, as described below. If the line has a more general direction, then obtaining the answer is more involved, and both vector and scalar approaches can be used as follows.

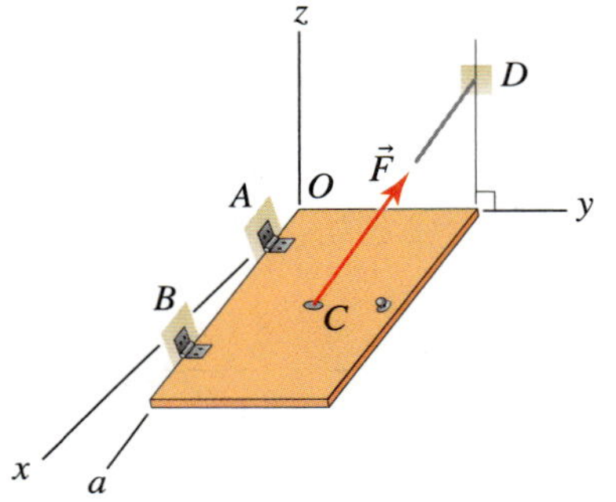

Figure 4.12
Hinged door supported by a rope.

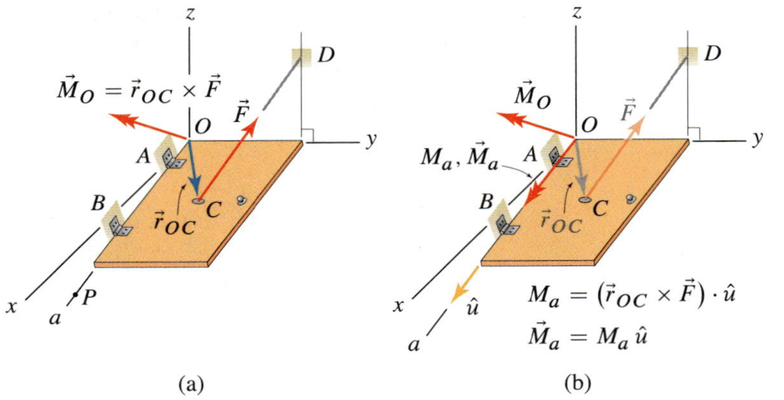

Figure 4.13. Vector approach to determine the moment of a force about a line a. (a) The moment of force $\vec{F}$ about some convenient point on line a (point O is selected here) is evaluated. (b) The dot product is then used to determine the component of the moment vector in the a direction M_a.

Vector approach

Consider the example shown in Fig. 4.12, where a door with rectangular shape is hinged along an axis with direction a lying in the xz plane. The door is supported by a cable between points C and D, where the tensile force in the cable has magnitude F. It will often be necessary to determine the moment of F about a direction such as the line passing through the hinges. To use a vector approach, first we find the moment at some convenient point along

EXAMPLE 4.5 *Moment of a Force About a Line – Vector & Scalar Solutions*

A door with rectangular shape is hinged along an axis having direction a lying in the xz plane. The door is supported by a cable that has a tensile force $F = 100\,\text{N}$. Determine the moment of F about line a.

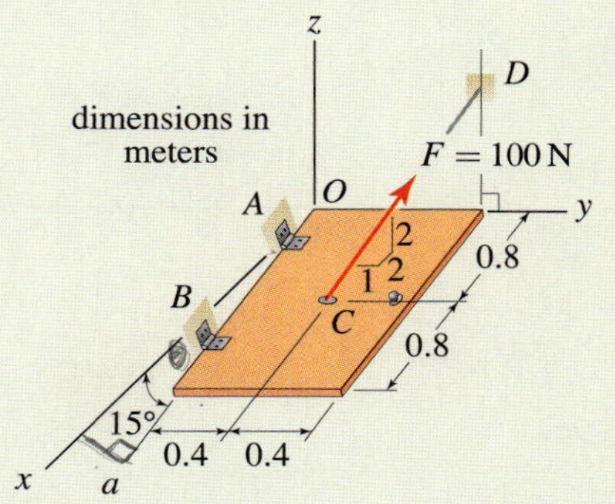

Figure 1

SOLUTION

Road Map Determining the moment of a force about a line is inherently a three-dimensional problem, and while both vector and scalar approaches can always be used, the vector approach will usually be more straightforward and methodical. Nonetheless, both solution approaches are carried out.

Vector solution

Governing Equations & Computation We first select point O as a convenient location on line a to compute the moment of F. Needed vectors are

$$\vec{F} = (100\,\text{N})\,\frac{-2\,\hat{\imath} + \hat{\jmath} + 2\,\hat{k}}{3}, \tag{1}$$

$$\begin{aligned}\vec{r}_{OC} &= (0.8\,\text{m})(\cos 15°)\,\hat{\imath} + (0.4\,\text{m})\,\hat{\jmath} - (0.8\,\text{m})(\sin 15°)\,\hat{k} \\ &= (0.7727\,\hat{\imath} + 0.4\,\hat{\jmath} - 0.2071\,\hat{k})\,\text{m}.\end{aligned} \tag{2}$$

The moment of $\vec{F}$ about point O is

$$\vec{M}_O = \vec{r}_{OC} \times \vec{F} = (33.57\,\hat{\imath} - 37.71\,\hat{\jmath} + 52.42\,\hat{k})\,\text{N·m}. \tag{3}$$

Finally, we take the dot product of $\vec{M}_O$ in Eq. (3) with a unit vector in the a direction, $\hat{u}_a = \cos 15°\,\hat{\imath} - \sin 15°\,\hat{k}$, to obtain

$$\boxed{M_a = \vec{M}_O \cdot \hat{u}_a = 18.9\,\text{N·m}.} \tag{4}$$

Thus, the portion of $\vec{M}_O$ that acts in the a direction, and hence the moment of F about the a axis, is 18.9 N·m. The fact that this result is positive means the twisting action of F about line a is in the positive a direction, as governed by the right-hand rule.

Scalar solution

Governing Equations & Computation The geometry of this problem is complex enough that a scalar evaluation of the perpendicular component of $\vec{F}$ using trigonometry is tedious. Rather we will start with a vector approach to obtain the perpendicular component of $\vec{F}$ (the magnitude of this vector is $F_\perp$ shown in Fig. 2) by taking the dot product of $\vec{F}$ in Eq. (1) with a unit vector $\hat{u}_n$ normal to the door, where $\hat{u}_n = \sin 15°\,\hat{\imath} + \cos 15°\,\hat{k}$. This provides

$$F_\perp = \vec{F} \cdot \hat{u}_n = 47.14\,\text{N}. \tag{5}$$

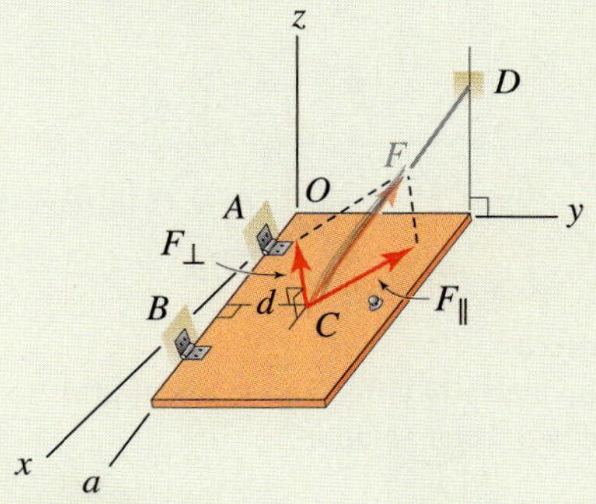

Figure 2
Resolution of $\vec{F}$ into perpendicular and parallel components $F_\perp$ and $F_\parallel$, respectively.

To calculate the moment of $F_\perp$ about line a, the moment arm needed is the perpendicular distance from line a to the line of action of $F_\perp$, namely, $d = 0.4\,\text{m}$. Thus

$$M_a = F_\perp d = (47.14\,\text{N})(0.4\,\text{m}) = \boxed{18.9\,\text{N·m}.} \tag{6}$$

In the scalar approach, the sign for M_a must be assigned manually. Examination of Fig. 2 shows that $F_\perp$ produces twisting action about line a, according to the right-hand rule, that is in the positive a direction. Thus, we conclude that $M_a = +18.9\,\text{N·m}$.

Discussion & Verification As expected, both solutions produce the same result.

EXAMPLE 4.6 *Moment of a Force About a Line – Vector & Scalar Solutions*

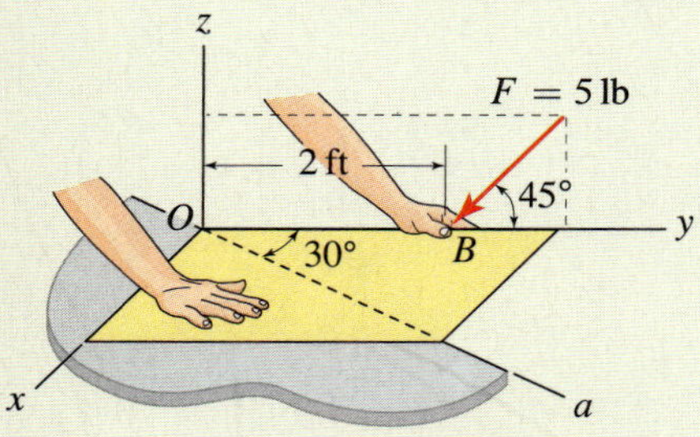

Figure 1

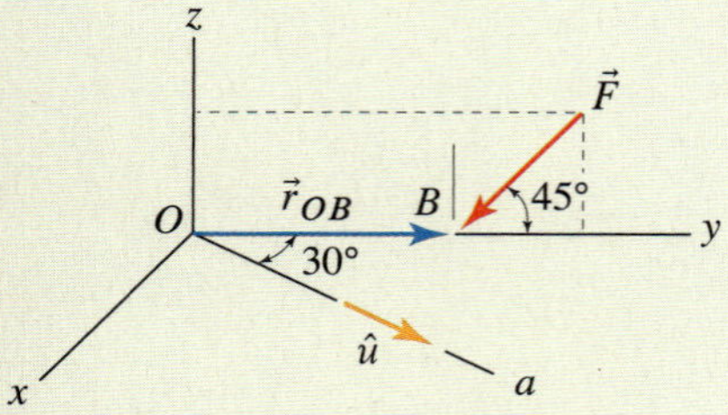

Figure 2
Force and position vectors to be used in a vector solution.

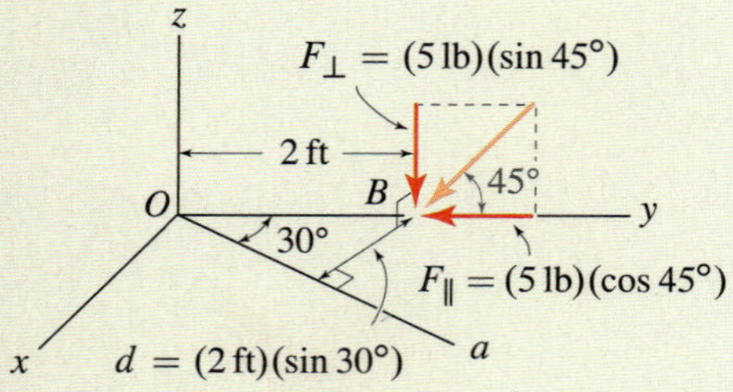

Figure 3
Components of force and moment arms to be used in a scalar solution.

A piece of cardboard is bent along the edge of a table. The a axis lies in the xy plane and force F lies in the yz plane. Determine the moment of F about the a axis.

SOLUTION

Road Map Both vector and scalar solutions are possible, and both are illustrated.

Vector solution

Governing Equations & Computation With point O selected as a convenient location, the necessary vectors, as shown in Fig. 2, are

$$\vec{r}_{OB} = 2\,\hat{\jmath}\text{ ft}, \tag{1}$$

$$\vec{F} = (5\,\text{lb})(-\cos 45°\,\hat{\jmath} - \sin 45°\,\hat{k}). \tag{2}$$

The moment of $\vec{F}$ about point O is

$$\vec{M}_O = \vec{r}_{OB} \times \vec{F} = (-7.071)\,\hat{\imath}\text{ ft·lb}. \tag{3}$$

Finally, using the unit vector $\hat{u} = \sin 30°\,\hat{\imath} + \cos 30°\,\hat{\jmath}$ to describe the a direction, the moment about line a is

$$M_a = \vec{M}_O \cdot \hat{u} = \boxed{-3.54\text{ ft·lb.}} \tag{4}$$

The negative sign for M_a means the twisting direction of F about line a, according to the right-hand rule, is the negative a direction.

If desired, the moment of F about line a can be stated as a vector by evaluating

$$\begin{aligned}\vec{M}_a &= M_a\,\hat{u}\\ &= (-3.54\text{ ft·lb})(\sin 30°\,\hat{\imath} + \cos 30°\,\hat{\jmath})\\ &= \boxed{(-1.77\,\hat{\imath} - 3.06\,\hat{\jmath})\text{ ft·lb.}}\end{aligned} \tag{5}$$

Scalar solution

Governing Equations & Computation We resolve F into perpendicular and parallel components, as shown in Fig. 3, and determine the moment arm d as the perpendicular distance from the a axis to the line of action of $F_\perp$. Then

$$\begin{aligned}M_a &= F_\perp d = (5\,\text{lb})(\sin 45°)(2\,\text{ft})(\sin 30°)\\ &= 3.54\text{ ft·lb.}\end{aligned} \tag{6}$$

In the scalar approach, we must manually provide the sign for our result. Examining Fig. 3, we observe the twisting action of $F_\perp$ about line a is in the negative a direction according to the right-hand rule. Hence, we conclude $\boxed{M_a = -3.54\text{ ft·lb.}}$

Discussion & Verification As expected, both solutions produce the same result.

EXAMPLE 4.7 *Moment of a Force About a Line – Vector Solution*

The axle and steering linkage for the front wheel of an off-road vehicle is shown. Force F causes the assembly to rotate about line a so that the vehicle can be steered. Point A is located at $(160, -20, 100)$ mm. Determine the force F needed to produce a moment about line a of $10\,\text{N}\cdot\text{m}$ if

(a) line a lies in the yz plane and has direction angle $\theta_z = 10°$.

(b) line a coincides with the z axis.

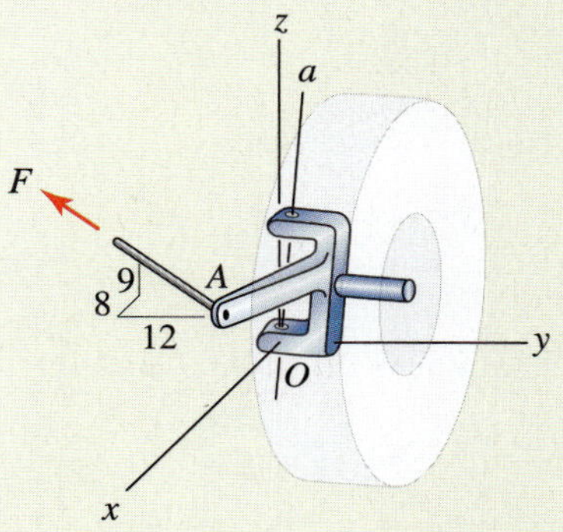

Figure 1

SOLUTION

Road Map A vector solution is straightforward and will be used here. To determine the moment of F about line a, we will first determine the moment of F about a convenient point on line a (we will select point O, and the result of this is a vector). We will then use the dot product to determine the portion of this moment vector that acts in the a direction, and according to the problem statement, this must be $10\,\text{N}\cdot\text{m}$. From this the value of F may be determined.

Part (a)

Governing Equations & Computation To find the moment of F about line a, a position vector from some point along a to some point along the line of action of F is needed, and $\vec{r}_{OA}$, as shown in Fig. 2, is an obvious choice. The necessary vectors are

$$\vec{r}_{OA} = (160\,\hat{\imath} - 20\,\hat{\jmath} + 100\,\hat{k})\,\text{mm}, \quad \vec{F} = F\frac{-8\,\hat{\imath} - 12\,\hat{\jmath} + 9\,\hat{k}}{17}. \tag{1}$$

The moment of $\vec{F}$ about point O is then

$$\vec{M}_O = \vec{r}_{OA} \times \vec{F} = F(60.00\,\hat{\imath} - 131.8\,\hat{\jmath} - 122.4\,\hat{k})\,\text{mm}. \tag{2}$$

The component of $\vec{M}_O$ in the direction a is given by the dot product between $\vec{M}_O$ and a unit vector in the direction a, $\hat{u}_a = \sin 10°\,\hat{\jmath} + \cos 10°\,\hat{k}$:

$$M_a = \vec{M}_O \cdot \hat{u}_a = -F(143.4\,\text{mm}). \tag{3}$$

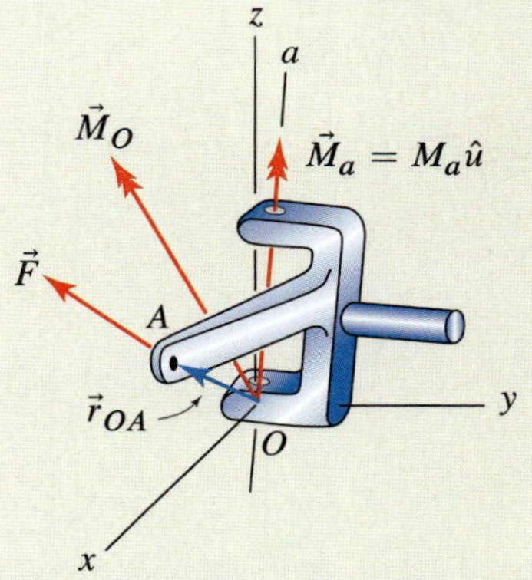

Figure 2

Force and position vectors $\vec{F}$ and $\vec{r}_{OA}$, respectively, used to determine the moment of $\vec{F}$ about point O and the moment of $\vec{F}$ about line a.

The negative sign in the above result indicates a positive force F produces a moment about line a that is in the negative a direction. To finish this problem, the moment about line a is required to have magnitude $10\,\text{N·m} = 10,000\,\text{N·mm}$. Thus, using the absolute value of Eq. (3), we obtain

$$F(143.4\,\text{mm}) = 10,000\,\text{N}\cdot\text{mm} \quad \Rightarrow \quad \boxed{F = 69.7\,\text{N}.} \tag{4}$$

Part (b)

Governing Equations & Computation Assuming the orientation of F and the location of point A are unchanged when line a coincides with the z axis, the moment of F about point O is unchanged and is given by Eq. (2). The unit vector $\hat{u}$ used in Eq. (3) becomes $\hat{u} = \hat{k}$, and reevaluation of Eq. (3) simply provides the z component of $\vec{M}_0$, which is

$$M_a = M_{Oz} = -F(122.4\,\text{mm}). \tag{5}$$

Requiring the absolute value of the above result to be equal to $10{,}000\,\text{N}\cdot\text{mm}$ provides

$$\boxed{F = 81.7\,\text{N}.} \tag{6}$$

Discussion & Verification The main difference between the two parts of this example is the orientation of line a. In Part (b), where line a is parallel to the z axis, the dot product is a trivial operation that produces the z component of the moment vector. The first remark on p. 199 gives additional comments.

PROBLEMS

General instructions. Unless otherwise stated, in the following problems you may use a scalar approach, a vector approach, or a combination of these.

Problem 4.26

The steering wheel of a Ferrari sports car has circular shape with 190 mm radius, and it lies in a plane that is perpendicular to the steering column AB. Point C, where the driver's hand applies force $\vec{F}$ to the steering wheel, lies on the y axis with y coordinate $y_C = -190$ mm. Point A is at the origin of the coordinate system, and point B has the coordinates $B\,(-120, 0, -50)$ mm. Determine the moment of $\vec{F}$ about line AB of the steering column if

(a) $\vec{F}$ has 10 N magnitude and lies in the plane of the steering wheel and has orientation such that its moment arm to line AB is 190 mm. Also determine the vector expression for this force.

(b) $\vec{F} = (10\,\text{N})\dfrac{18\,\hat{\imath} - 3\,\hat{\jmath} + 14\,\hat{k}}{23}$.

(c) $\vec{F} = (10\,\text{N})\dfrac{-12\,\hat{\imath} - 5\,\hat{k}}{13}$.

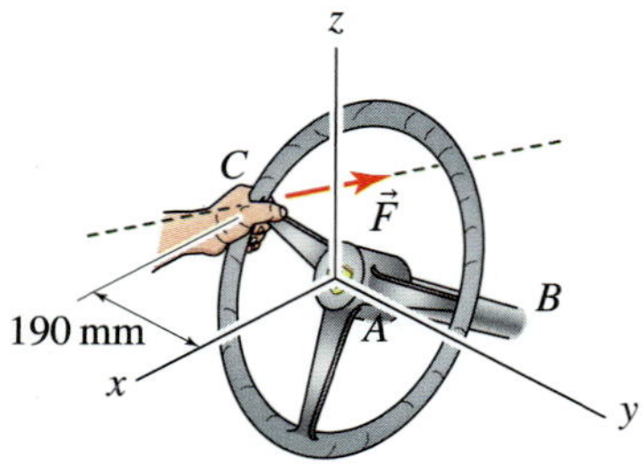

Figure P4.26

Problem 4.27

A rectangular piece of sheet metal is clamped along edge AB in a machine called a *brake*. The sheet is to be bent along line AB by applying a y direction force F. Determine the moment about line AB if $F = 200$ lb. Use both vector and scalar approaches.

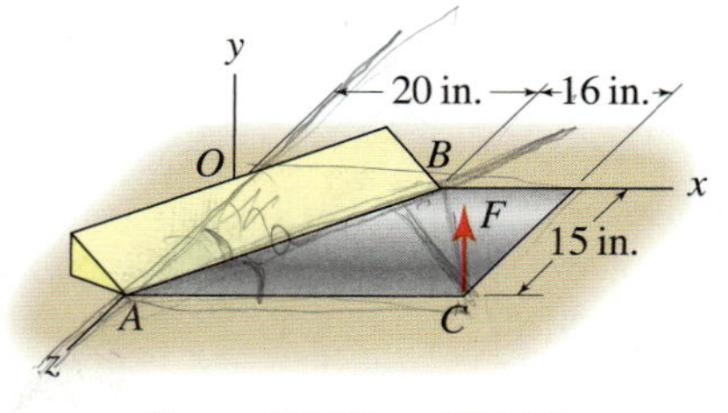

Figure P4.27 and P4.28

Problem 4.28

In Prob. 4.27 determine F if the moment about line AB is to be 2000 in.·lb.

Problem 4.29

In the pipe assembly shown, points B and C lie in the xy plane, and force F is parallel to the z axis. If $F = 150$ N, determine the moment of F about lines OA and AB. Use both vector and scalar approaches.

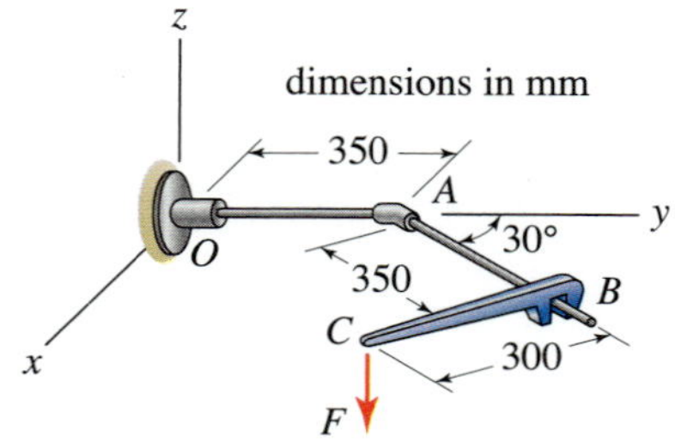

Figure P4.29 and P4.30

Problem 4.30

In the pipe assembly shown, points B and C lie in the xy plane and force F is parallel to the z axis. If a twisting moment (torque) of 50 N·m will cause a pipe to begin twisting in the flange fitting at O or at either end of the elbow fitting at A, determine the first fitting that twists and the value of F that causes it.

Problems 4.31 and 4.32

Determine the moment of force F about line AB as follows.

(a) Determine the moment of F about point A, $\vec{M}_A$, and then determine the component of this moment in the direction of line AB.

(b) Determine the moment of F about point B, $\vec{M}_B$, and then determine the component of this moment in the direction of line AB.

(c) Comment on differences and/or agreement between $\vec{M}_A$, $\vec{M}_B$, and the moment about line AB found in Parts (a) and (b). Also comment on the meaning of the sign (positive or negative) found for the moment about line AB.

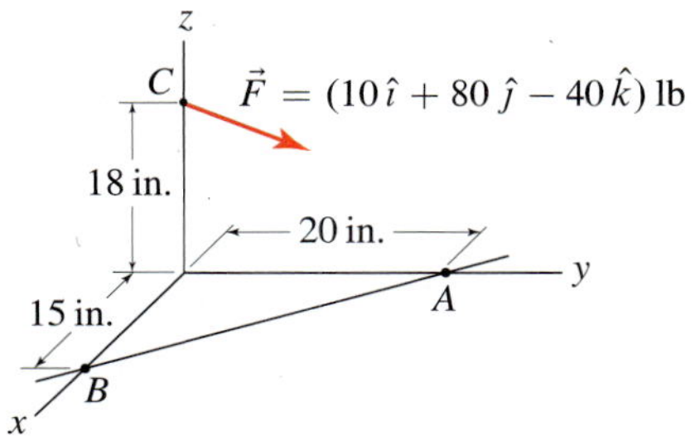

Figure P4.31

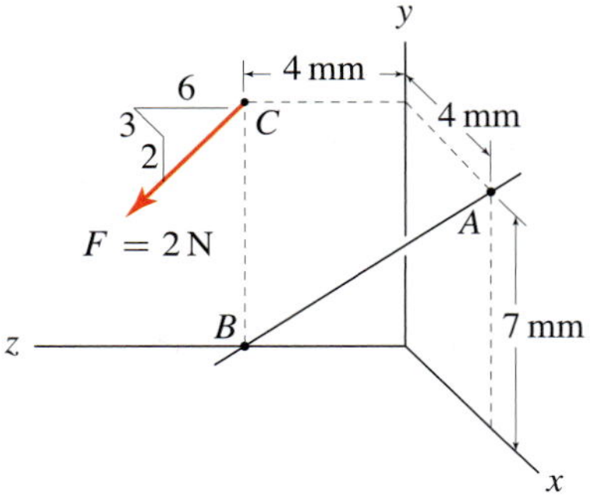

Figure P4.32

Problem 4.33

The moment of force $\vec{F}$ about line AB can be computed using $M_{AB,1}$ or $M_{AB,2}$ where

$$M_{AB,1} = (\vec{r}_{AC} \times \vec{F}) \cdot \frac{\vec{r}_{AB}}{|\vec{r}_{AB}|}, \qquad M_{AB,2} = (\vec{r}_{BC} \times \vec{F}) \cdot \frac{\vec{r}_{AB}}{|\vec{r}_{AB}|},$$

where C is a point on the line of action of $\vec{F}$. Noting that $\vec{r}_{AC} = \vec{r}_{AB} + \vec{r}_{BC}$, show that $M_{AB,1} = M_{AB,2}$.

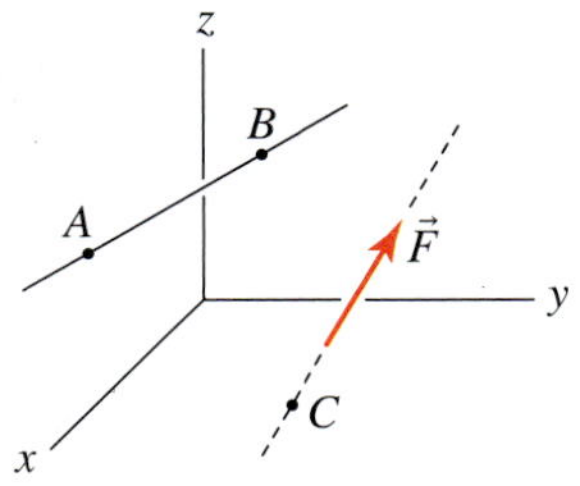

Figure P4.33

Problem 4.34

An automobile windshield wiper is actuated by a force $\vec{F} = (120\,\hat{\imath} - 50\,\hat{\jmath})$ N. Determine the moment of this force about shaft OB.

Problem 4.35

In the windshield wiper of Prob. 4.34, if $\vec{F}$ has 130 N magnitude, determine the direction in which it should be applied so that its moment about shaft OB is as large as possible, and determine the value of this moment.

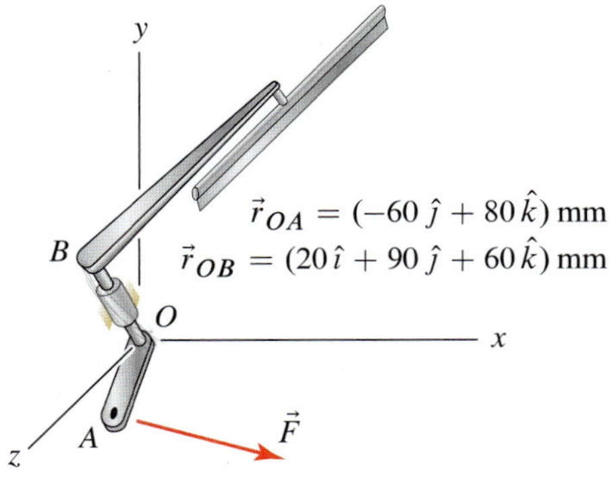

Figure P4.34 and P4.35

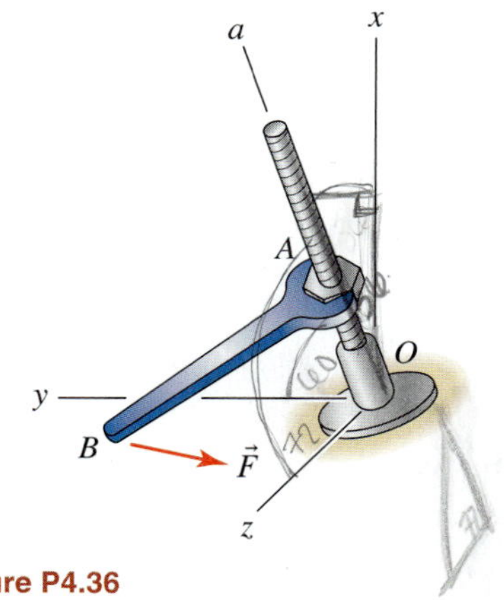

Figure P4.36

Problem 4.36

Wrench AB is used to twist a nut on a threaded shaft. Point A is located 120 mm from point O, and line a has x, y, and z direction angles of $36°$, $60°$, and $72°$, respectively. Point B is located at (0, 300, 100) mm, and the force is $\vec{F} = (-80\,\hat{\jmath} + 20\,\hat{k})$ N.

(a) Determine the moment of $\vec{F}$ about point A, $\vec{M}_A$, and then find the component of this moment about line a, M_a.

(b) Determine the moment of $\vec{F}$ about point O, $\vec{M}_O$, and then find the component of this moment about line a, M_a.

A (0, 28, 12) in.
B (24, 4, 12) in.
C (0, 4, 60) in.
D (−40, 24, 20) in.
E (−10, 24, 20) in.
G (−10, 8, 52) in.
H (−40, 8, 52) in.

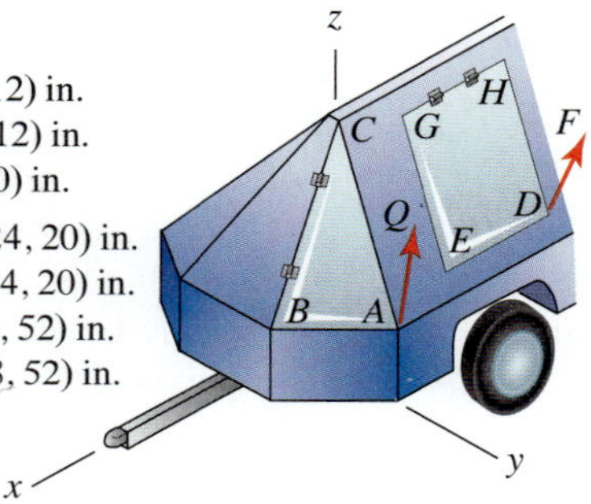

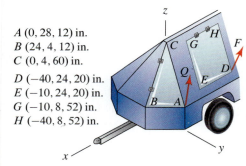

Figure P4.37–P4.40

Problem 4.37

A trailer has a rectangular door $DEGH$ hinged about edge GH. If $\vec{F} = (-2\,\hat{\imath} + 5\,\hat{\jmath} + 14\,\hat{k})$ lb, determine the moment of F about edge GH.

Problem 4.38

In the trailer of Prob. 4.37, if $F = 15$ lb, determine the direction in which it should be applied so that its moment about edge GH of the rectangular door is as large as possible, and determine the value of this moment.

Problem 4.39

A trailer has a triangular door ABC hinged about edge BC. If $\vec{Q} = (\hat{\imath} + 4\,\hat{\jmath} + 8\,\hat{k})$ lb, determine the moment of Q about edge BC.

Problem 4.40

In the trailer of Prob. 4.39, if $Q = 9$ lb, determine the direction in which it should be applied so that its moment about edge BC of the triangular door is as large as possible, and determine the value of this moment.

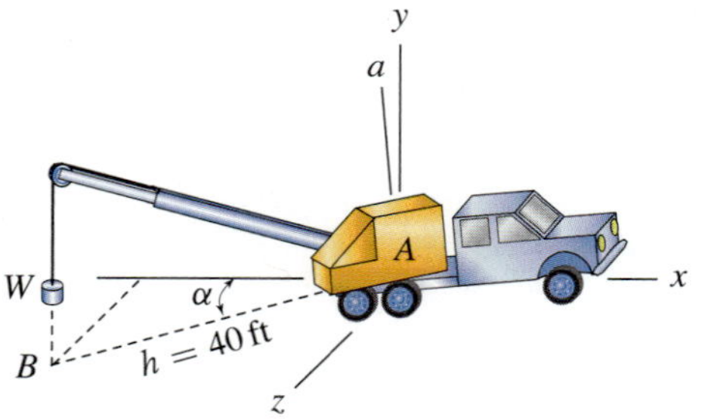

Figure P4.41 and P4.42

Problem 4.41

A poorly leveled crane rotates about line a, which lies in the xy plane and has a y direction angle of $5°$. Line AB lies in the xz plane. If the crane supports a weight $W = 5000$lb and $\alpha = 45°$, determine the moment of W about line a.

Problem 4.42

If the poorly leveled crane of Prob. 4.41 can have any position α, where $0 \le \alpha \le 135°$, determine the largest moment of the weight $W = 5000$ lb about line a and the position α that produces this moment. Assume $h = 40$ ft for any position α (this requires the operator to slightly change the boom's height and/or length as the crane rotates about line a).

Problem 4.43

Three-wheel and four-wheel all-terrain vehicles (ATVs) are shown. For both ATVs, the combined weight of the driver and vehicle is $W = 2600$ N. During a hard turn, the tendency for the ATV to tip is modeled* by the force F, which acts in the negative y direction. Determine the value of F such that the resultant moment of F and the weight about line AB is zero. Comment on which ATV is more prone to tip during hard turns. **Note:** One of these models of ATV is no longer manufactured because of its poor safety record.

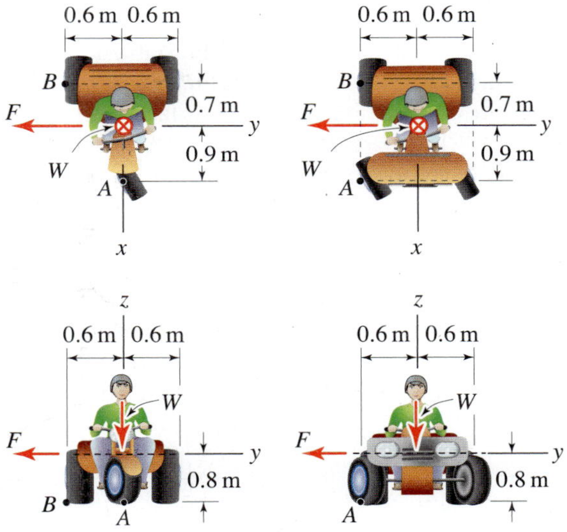

Figure P4.43

* When an ATV moves on a curved path, it is a dynamics problem. We can model this system as a statics problem by treating inertial effects as external forces, as we do in this problem. While this is a crude approximation, it is nevertheless a common modeling approach. You will see in dynamics how to correctly formulate these problems.

4.3 Moment of a Couple

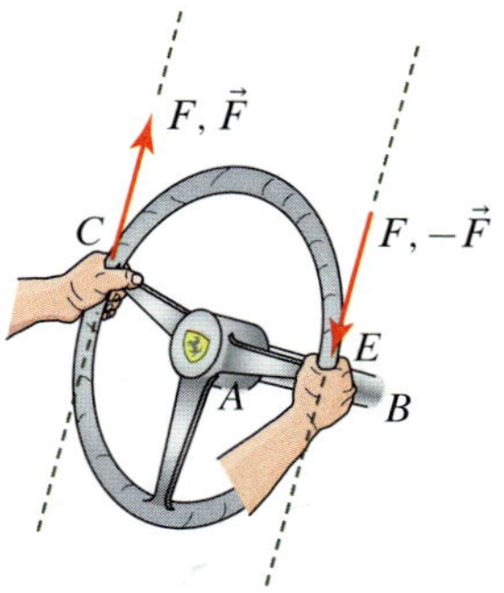

Figure 4.17
The steering wheel in the Ferrari 250 GTO (see the discussion beginning on p. 181), where the driver uses two hands to apply two forces of equal magnitude and opposite direction to the steering wheel. Such a force system is defined to be a *couple*.

Helpful Information

Vector notation. Notice in Fig. 4.17 that two types of vector notation are shown (these were described in Chapter 2 on p. 30). When forces are labeled using a scalar such as F, then the direction of the arrow in the figure provides the direction of the force. When forces are labeled using a vector such as $\vec{F}$, then the symbol $\vec{F}$ embodies both the magnitude and direction for the force. Hence, in this figure if the force at point C is labeled as $\vec{F}$ and if the force system is a couple, then the force at E must be labeled as $-\vec{F}$.

A *couple* is defined to be a system of two forces of equal magnitude and opposite direction and whose lines of action are separated by a distance. To understand better what a couple is, reconsider the steering wheel from the Ferrari sports car, shown again in Fig. 4.17, except that now the driver uses two hands to apply forces to turn the steering wheel. Of course, the driver could apply forces of different magnitudes and in any two different directions that he or she pleases. However, if the driver applies forces of equal magnitude and with opposite direction, as shown in Fig. 4.17, then the system of two forces is called a *couple*.

The *moment of a couple* (sometimes also called a *couple moment*) is the moment produced by the couple, that is, the moment produced by two forces of equal magnitude and opposite direction. The moment of a couple can always be evaluated using the procedures of Section 4.1. However, the moment of a couple has some special features, and in addition to the methods of Section 4.1, this section discusses other approaches that may be used for evaluation.

A couple that is applied to a body produces a moment, but does not apply any net force to the body. To help illustrate some of the implications of this, consider the air hockey table shown in Fig. 4.18(a), and imagine that a rectangular plate is resting motionless on the surface of the table.* If the two forces shown in Fig. 4.18(b) are applied to the plate, because the two forces have equal magnitude and opposite direction, there is no net force applied in the x direction and hence there is no net motion of the plate in the x direction. However, the two forces produce a moment, and due to this moment the plate will begin to spin. Thus, the essential feature of a couple is that it produces only a moment. Furthermore, when the moment has the proper magnitude and direction (as described below), then the two force systems shown in Fig. 4.18(b) and (c) are said to be *equivalent*, and if the plate can be idealized as being rigid, then the motion of the plate in Fig. 4.18(b) and (c) is identical. Section 4.4 discusses equivalent force systems in detail.

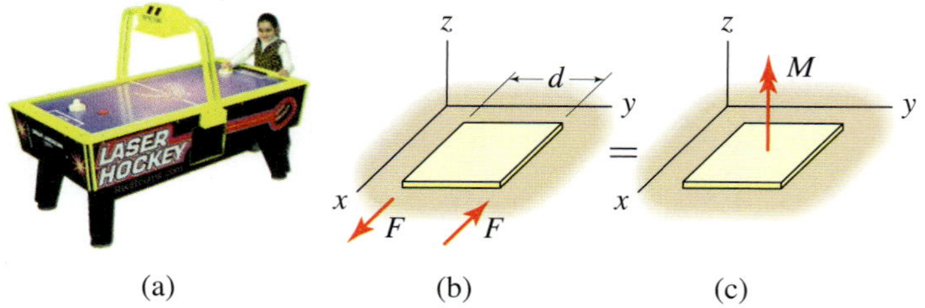

Figure 4.18. (a) An air hockey table. (b) A rectangular plate resting on the surface of the air hockey table is subjected to a *couple*, that is, two forces with equal magnitude and opposite direction, separated by a distance d. (c) The moment produced by the couple. The two force systems shown in (b) and (c) are *equivalent* when M has the proper value and direction, as described in this section.

The moment of a couple can be evaluated using both vector and scalar approaches as follows.

* If you are unfamiliar with an air hockey table, you can instead imagine the rectangular plate resting on any horizontal frictionless surface, such as a sheet of ice.

Vector approach

Referring to Fig. 4.19, consider a couple consisting of two parallel forces $\vec{F}$ and $-\vec{F}$. The moment $\vec{M}$ of this couple is

$$\begin{aligned}\vec{M} &= \vec{r}_{AB} \times \vec{F} \\ &= \vec{r}_{BA} \times (-\vec{F})\end{aligned} \tag{4.11}$$

where

$\vec{r}_{AB}$ and $\vec{r}_{BA}$ are position vectors,
A is *any* point on the line of action of $-\vec{F}$,
B is *any* point on the line of action of $\vec{F}$.

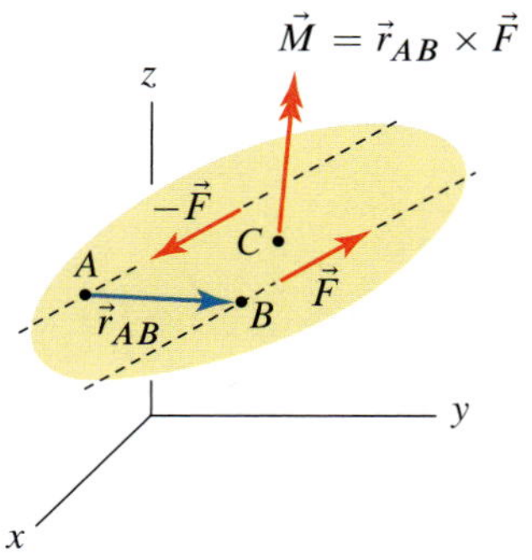

Figure 4.19
Moment of a couple: vector description.

To see that the two expressions in Eq. (4.11) are equal, note that $\vec{r}_{BA} = -\vec{r}_{AB}$ and thus, $\vec{M} = \vec{r}_{BA} \times (-\vec{F}) = -\vec{r}_{AB} \times (-\vec{F}) = \vec{r}_{AB} \times \vec{F}$.

When evaluating the moment of a couple using the vector approach, you will select one of the forces of the couple to be called $\vec{F}$. Then the position vector $\vec{r}$ must start somewhere on the line of action of the other force and terminate somewhere on the line of action of $\vec{F}$. It does not matter which of the two forces of the couple you choose to call $\vec{F}$; using the proper position vector as described here will result in the same moment $\vec{M}$.

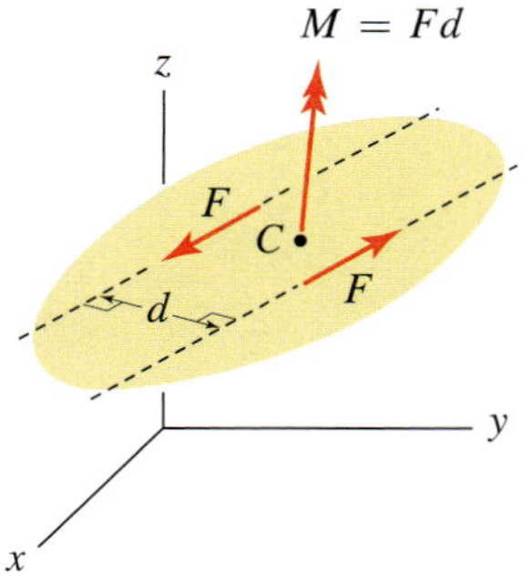

Figure 4.20
Moment of a couple: scalar description.

Scalar approach

Referring to Fig. 4.20, consider a couple consisting of two parallel forces having the same magnitude F. The magnitude of the moment of this couple is

$$M = Fd \tag{4.12}$$

where

d is the perpendicular (shortest) distance between the forces' lines of action,
the direction of the moment is perpendicular to the plane containing the forces.

Comments on the moment of a couple

To see that the vector and scalar descriptions of the moment of a couple are valid and to understand some subtle features of a couple and the moment it produces, consider the forces shown in Fig. 4.21. The moment produced by these forces about point C having arbitrary location is

$$\vec{M}_C = \vec{r}_{CA} \times (-\vec{F}) + \vec{r}_{CB} \times \vec{F}. \tag{4.13}$$

Noting that $\vec{r}_{CB} = \vec{r}_{CA} + \vec{r}_{AB}$, we see Eq. (4.13) becomes

$$\begin{aligned}\vec{M}_C &= \vec{r}_{CA} \times (-\vec{F}) + (\vec{r}_{CA} + \vec{r}_{AB}) \times \vec{F} \\ &= \vec{r}_{CA} \times (-\vec{F}) + \vec{r}_{CA} \times \vec{F} + \vec{r}_{AB} \times \vec{F} \\ &= \vec{r}_{AB} \times \vec{F}.\end{aligned} \tag{4.14}$$

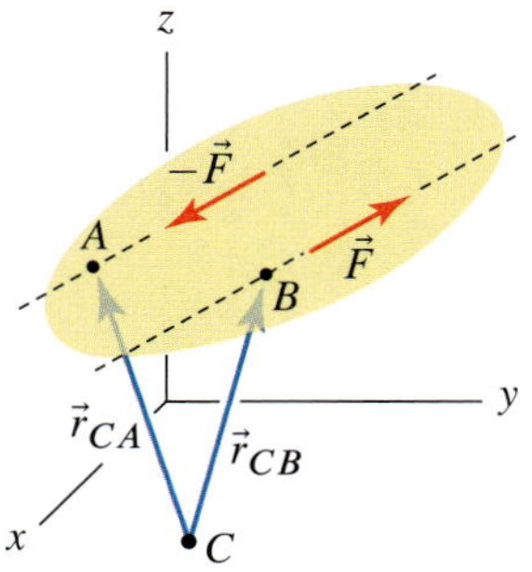

Figure 4.21
Vectors for computing the moment of forces $\vec{F}$ and $-\vec{F}$ about point C. Because $\vec{F}$ and $-\vec{F}$ are a *couple*, the moment they produce is seen to be the same regardless of where point C is located.

Notice from Eq. (4.14) that the moment about point C depends only on the position vector between the two forces' lines of action and does not depend on the location of C. Thus, regardless of where C is located, the moment produced by the couple is the same. In other words, a couple produces a moment vector that has a specific magnitude and orientation, but the vector's position in space (i.e., the location of its line of action) is arbitrary, because you can let point C be located anywhere. Thus we may consider the point at which a couple moment acts as being anywhere we choose. For the examples in Figs. 4.19 and 4.20 point C is shown as being positioned between the two forces, although any other location for point C is equally valid.

For the foregoing reasons, the moment of a couple is often called a *free vector*, and this is an important difference compared to all other vectors we have encountered in this book. A *free vector* is a vector that has a specific magnitude and direction, but its position in space is arbitrary. For example, a force vector has a specific magnitude and direction and a specific location in space for its line of action. The moment of a force about a point also has the same characteristics. In contrast, the moment of a couple has the first two characteristics, namely, a specific magnitude and direction, but its position, or where its line of action is located in space, is arbitrary.

Equivalent couples

Two couples are said to be *equivalent* if the moment vectors they produce are identical. Thus, the couples shown in the examples of Fig. 4.22(a)–(c) are all equivalent since they produce moments having the same magnitude and direction.

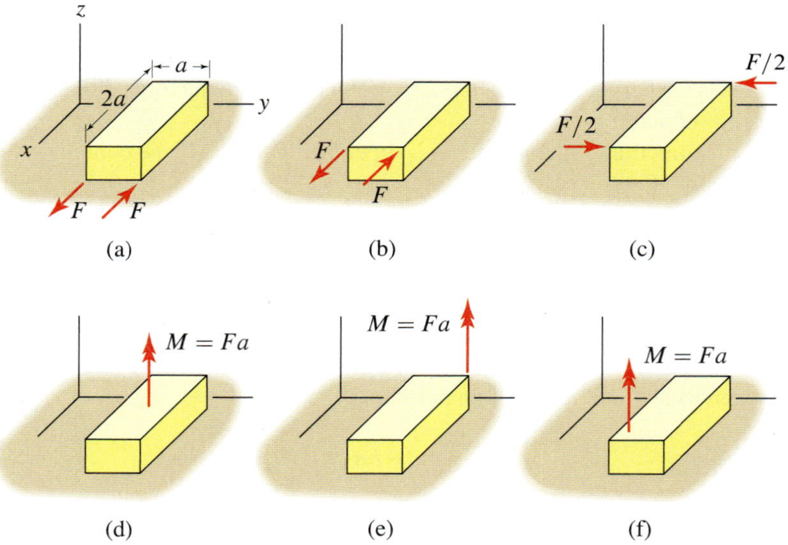

Figure 4.22. Equivalent couples and the moments they produce.

Equivalent force systems

A couple and the moment it produces are examples of *equivalent force systems*. This important concept is explored in greater detail in the next section, but for the present we will simply state that equivalent force systems applied to a body or structure produce the same effects for many purposes. For example, consider the identical prisms of material shown in Fig. 4.22. The couple shown in Fig. 4.22(a) produces a moment with magnitude $M = Fa$ that acts in the z direction. Since this moment is a free vector, it can be positioned anywhere on, or even off, of the prism of material, and three examples are shown in Fig. 4.22(d)–(f). Since the couples shown in Fig. 4.22(b) and (c) are equivalent to that in Fig. 4.22(a), the moments they produce can also be positioned as shown in Fig. 4.22(d)–(f). In summary, all six force systems shown in Fig. 4.22 are equivalent. If we imagine these prisms rest on a smooth frictionless horizontal surface such as the air hockey table discussed earlier, and if these prisms are initially motionless, then all six prisms of material shown in Fig. 4.22 will undergo the same motion, namely, spin about an axis parallel to the z direction and no net translation.

Resultant couple moment

Because the moment produced by a couple is a free vector, if an object has more than one couple applied to it, the moment vector produced by each couple can simply be added to yield a *resultant couple moment*, which is also a free vector. This process is illustrated in Fig. 4.23.

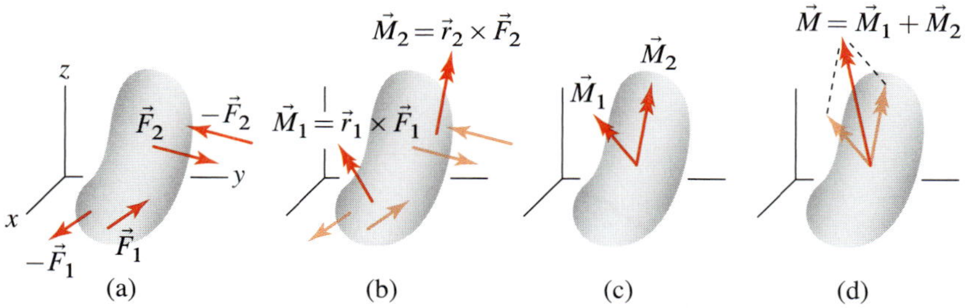

Figure 4.23. Addition of moments of couples to form a resultant couple moment. (a) Object with two couples. (b) Moments $\vec{M}_1$ and $\vec{M}_2$ are produced by the couples where $\vec{r}_1$ is a position vector from somewhere on the line of action of $-\vec{F}_1$ to somewhere on the line of action of $\vec{F}_1$, and similarly for $\vec{r}_2$. (c) Since $\vec{M}_1$ and $\vec{M}_2$ are free vectors, they may be positioned tail to tail at any location. (d) Addition of $\vec{M}_1$ and $\vec{M}_2$ to form the resultant couple moment $\vec{M}$.

Moments as free vectors

Consider the moment applied to a structure shown in Fig. 4.24(a). Because a moment can always be thought of as being produced by a couple, even if the actual agency that produced it is different, all moments are *free vectors*. For many purposes, moments can be thought of as being positioned anywhere on the structure, or even off of it. Thus the moment M in Fig. 4.24 can be

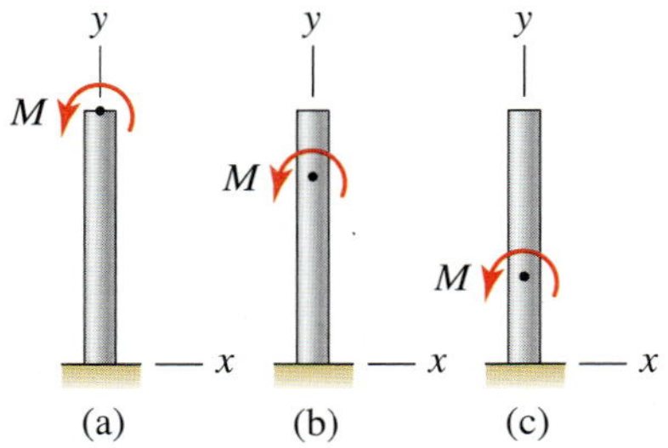

Figure 4.24
Moments are *free vectors* and can be positioned anywhere on a rigid structure with "equivalent" results, where the full meaning of this is discussed in Section 4.4.

positioned at any of the locations shown with "equivalent" effects, where the exact meaning of the word *equivalent* is discussed in Section 4.4.

End of Section Summary

In this section, the moment of a couple is described. Some of the key points are as follows:

- A *couple* is defined to be a system of two forces of equal magnitude and opposite direction and whose lines of action are separated by a distance.
- The *moment of a couple*, or *couple moment*, is a vector and can be evaluated using a vector or scalar approach.
- The moment of a couple is a *free vector*, meaning the line of action of the couple can be positioned anywhere on (or even off) an object.
- A *resultant couple moment* is the sum of all couple moments applied to an object.

EXAMPLE 4.8 *Moment of a Couple for a Two-dimensional Problem*

A cantilever beam is subjected to a couple. Determine the moment of the couple.

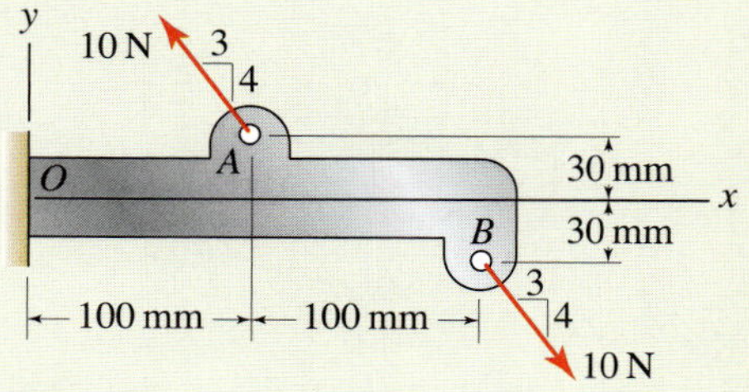

Figure 1

SOLUTION

Road Map The two forces have equal magnitude and opposite direction, so indeed they are a couple. Both scalar and vector approaches can be used to determine the moment of the couple, as follows.

Solution 1

Governing Equations & Computation The moment of the couple can be determined using the scalar approach in Eq. (4.12) on p. 209 as $M = -(10\,\text{N})\,d$, where d is the perpendicular distance between the lines of action of the forces as shown in Fig. 2, and the negative sign is included because the couple shown in Fig. 1 produces a clockwise moment. However, for this particular problem finding d is tedious, so we will select a different solution approach.

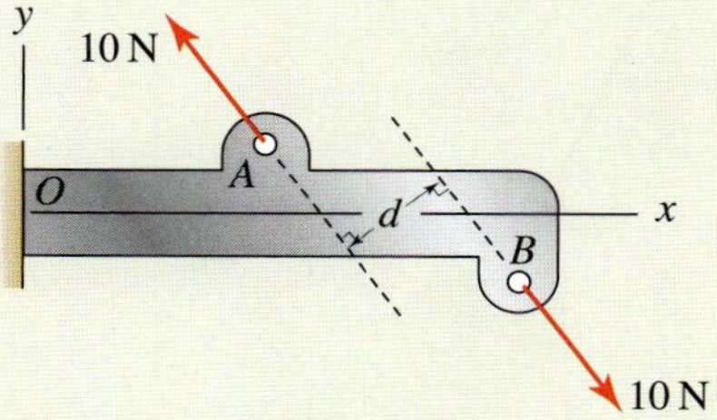

Figure 2
Forces and moment arm used to determine the moment of a couple.

Solution 2

Governing Equations & Computation The 10 N forces are resolved into the x and y components as shown in Fig. 3 to yield two couples. With $F_x = (10\,\text{N})(3/5) = 6\,\text{N}$ and $F_y = (10\,\text{N})(4/5) = 8\,\text{N}$, we then use Eq. (4.12) to sum the moment from each couple, taking counterclockwise to be positive as usual, to obtain

$$M = -(8\,\text{N})(100\,\text{mm}) + (6\,\text{N})(60\,\text{mm}) = \boxed{-440\,\text{N}\cdot\text{mm.}} \tag{1}$$

Solution 3

Governing Equations & Computation Although use of Eq. (4.12) as demonstrated in the preceding solutions will usually be convenient for two-dimensional problems, we may also evaluate the moment of the couple by summing moments about any convenient point we choose. Using, for example, point O shown in Fig. 3 with counterclockwise being positive, we obtain

$$\begin{aligned} M_O &= (6\,\text{N})(30\,\text{mm}) + (8\,\text{N})(100\,\text{mm}) + (6\,\text{N})(30\,\text{mm}) - (8\,\text{N})(200\,\text{mm}) \\ &= \boxed{-440\,\text{N}\cdot\text{mm.}} \end{aligned} \tag{2}$$

Since we know M_O is the resultant moment from two couples, we know that the same moment would be obtained if any other summation point had been used. On the other hand, if this approach is used for forces that are not couples, then the resultant moment will usually differ from point to point.

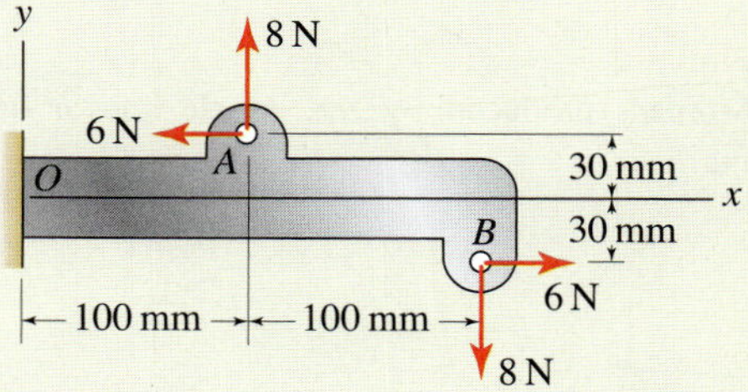

Figure 3
Resolution of the 10 N forces into components so that the resultant moment of two couples can be determined.

Scalar
M = Fd

vector
M = r x F

Solution 4

Governing Equations & Computation A vector solution can also be employed. Using Eq. (4.11) on p. 209 with $\vec{r}_{AB} = (100\,\hat{\imath} - 60\,\hat{\jmath})\,\text{mm}$ and $\vec{F} = (6\,\hat{\imath} - 8\,\hat{\jmath})\,\text{N}$, we obtain

$$\vec{M} = \vec{r}_{AB} \times \vec{F} = \boxed{-440\,\hat{k}\,\text{N}\cdot\text{mm.}} \tag{3}$$

The negative sign in the above result indicates the moment acts in the negative z direction according to the right-hand rule, which corresponds to clockwise in Figs. 1–3. You may wish to verify that the same result for $\vec{M}$ is obtained if expressions for the other force of the couple and the associated position vector are used.

Discussion & Verification As expected, all solutions agree. For this problem, solutions 2 and 3 were the quickest, and solution 4 was only slightly longer.

EXAMPLE 4.9 *Moment of a Couple for a Three-dimensional Problem*

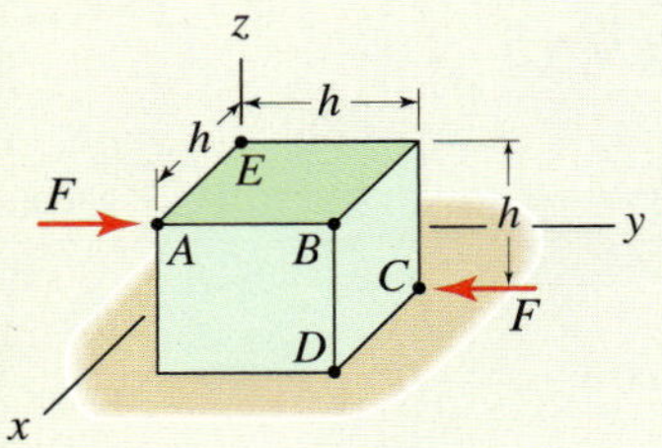

Figure 1

A cube with edge length h is subjected to a couple whose forces have magnitude F. Edges of the cube are parallel to their respective coordinate directions, and the forces are parallel to the y axis. Determine the moment of the couple. Express your answers in terms of parameters such as h and F.

SOLUTION

Road Map The two forces have equal magnitude and opposite direction, so indeed they are a couple. Although a vector solution will usually be the most straightforward for three-dimensional problems, the geometry here is simple enough that a scalar solution can be used.

Solution 1

Governing Equations & Computation The perpendicular distance between the lines of action of the forces is the distance between points B and C, which is $d = h\sqrt{2}$. Hence, Eq. (4.12) on p. 209 provides

$$\boxed{M = F(h\sqrt{2})} \tag{1}$$

where the direction of the moment is perpendicular to the plane containing the two forces, as shown in Fig. 2.

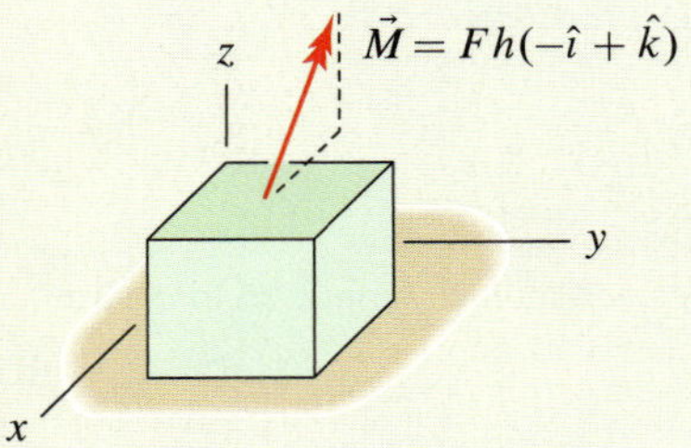

Figure 2
Moment produced by the couple shown in Fig. 1.

Solution 2

Governing Equations & Computation To use a vector solution we will use the force at point C and choose a position vector from points A to C. Equation (4.11) on p. 209 then provides

$$\vec{M} = \vec{r}_{AC} \times \vec{F}_C = h(-\hat{\imath} + \hat{\jmath} - \hat{k}) \times (-F\,\hat{\jmath}) = \boxed{Fh(-\hat{\imath} + \hat{k}).} \tag{2}$$

This moment vector is sketched in its proper orientation in Fig. 2.

Solution 3

Governing Equations & Computation In this alternative solution we use the force at point A, where $\vec{F}_A = -\vec{F}_C$, and choose a position vector from points C to B, where we note that B is on the line of action of the force at A. Thus

$$\vec{M} = \vec{r}_{CB} \times \vec{F}_A = h(\hat{\imath} + \hat{k}) \times (F\,\hat{\jmath}) = \boxed{Fh(-\hat{\imath} + \hat{k}).} \tag{3}$$

As expected, the result is the same as that obtained in Eq. (2).

Solution 4

Governing Equations & Computation In this solution we sum moments about a point that is not on either force's line of action. Using point D, we write

$$\begin{aligned} \vec{M} &= \vec{r}_{DC} \times \vec{F}_C + \vec{r}_{DA} \times \vec{F}_A \\ &= (-h\,\hat{\imath}) \times (-F\,\hat{\jmath}) + h(-\hat{\jmath} + \hat{k}) \times (F\,\hat{\jmath}) \\ &= \boxed{Fh(-\hat{\imath} + \hat{k}),} \end{aligned} \tag{4}$$

which again is the same result as those obtained earlier. You may wish to repeat this solution, summing moments about point E, or any other point, to find the same result.

Solution 5

Governing Equations & Computation In this solution we add two new forces at point D, as shown in Fig. 3(a). The argument for why we may do this is that the two

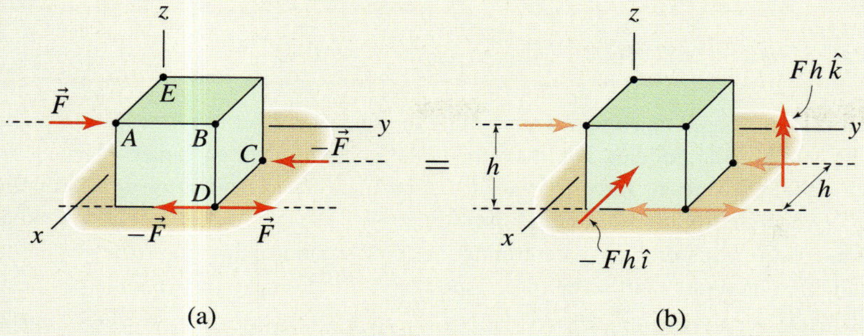

Figure 3. (a) Two forces of equal magnitude and opposite direction, and sharing the same line of action are introduced at point D, so that the object is now subjected to two couples. (b) The moment for each couple may be written by inspection.

forces added at D will have no net effect on the object if they have equal magnitude and opposite direction *and* they have the same line of action. Examining Fig. 3(a) shows that the object is now subjected to two couples, where the forces $\vec{F}$ and $-\vec{F}$ at points A and D, respectively, are one couple and the forces $-\vec{F}$ and $\vec{F}$ at points C and D, respectively, are the other couple. The merit of this solution strategy is that the moment for each of these couples may easily be evaluated by inspection, yielding the results shown in Fig. 3(b). Summing the two moments in Fig. 3(b) yields

$$\begin{aligned} \vec{M} &= -Fh\,\hat{\imath} + Fh\hat{k} \\ &= \boxed{Fh(-\hat{\imath} + \hat{k}).} \end{aligned} \tag{5}$$

Discussion & Verification As expected, all solutions agree, except that Eq. (1) gives only the magnitude of the moment, whereas the other solutions are vectors and hence also provide the direction of the moment. Solution 4 does not require that the two forces involved be a couple. If you did not recognize that the two forces were a couple, then you would expect Eq. (4) to be the moment at point D only. Because the two forces are a couple, the result in Eq. (4) is obtained for any other moment point that might be used.

EXAMPLE 4.10 *Resultant Couple Moment*

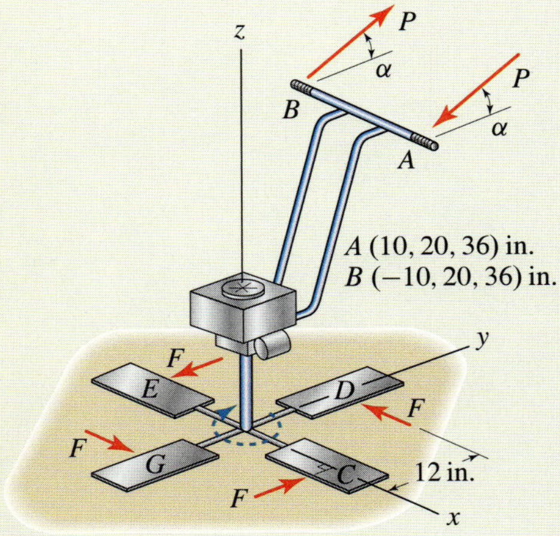

Figure 1

A gasoline-powered machine for smoothing the surface of wet concrete slabs is shown (the guard for the paddles is removed). If the concrete applies forces $F = 8\,\text{lb}$ to each paddle, where all forces F lie in the xy plane (the z direction forces applied by the concrete to the paddles, and the weight of the machine, are not shown and are not needed for this problem), determine the force P and orientation α needed so the resultant couple moment of the forces shown is zero. Forces P are parallel to the yz plane.

SOLUTION

Road Map Because this problem is three-dimensional, a vector solution will probably be the most straightforward. However, with some forethought, a scalar solution can also be used effectively.

Scalar solution

Governing Equations & Computation If the resultant couple moment is to be zero, then by inspection of Fig. 1 we see that we must have $\boxed{\alpha = 0}$ so that the moment of couple forces P has the same direction as the moments due to the paddle force couples. With this observation, a scalar solution can be used to sum the magnitudes of the couple moments in the z direction as

$$M = (8\,\text{lb})(24\,\text{in.}) + (8\,\text{lb})(24\,\text{in.}) - P(20\,\text{in.}) = 0 \quad \Rightarrow \quad \boxed{P = 19.2\,\text{lb},} \tag{1}$$

where moment is taken to be positive in the positive z direction and the distance between points A and B is 20 in. from the coordinate information provided. **Warning**: Equation (1) can be used to sum moment magnitudes only if we know that all moments share the same direction.

Vector solution

Governing Equations & Computation If you did not recognize that $\alpha = 0$, then a vector solution is better. The resultant couple moment is

$$\begin{aligned}\vec{M} &= 2\left(\vec{r}_{GD} \times \vec{F}_D\right) + \vec{r}_{AB} \times \vec{P}_B \\ &= 2\left[(24\,\text{in.}\,\hat{\jmath}) \times (-8\,\text{lb}\,\hat{\imath})\right] + (-20\,\text{in.}\,\hat{\imath}) \times P(\cos\alpha\,\hat{\jmath} + \sin\alpha\,\hat{k}) \\ &= P(20\,\text{in.})\sin\alpha\,\hat{\jmath} + \left[384\,\text{in.}\cdot\text{lb} - P(20\,\text{in.})\cos\alpha\right]\hat{k}.\end{aligned} \tag{2}$$

In the above expression, we have simply doubled the moment of the couple forces at points D and G, since by inspection, the moment from the couple forces at points C and E has the same magnitude and direction. To have $\vec{M} = \vec{0}$, each component of Eq. (2) must be zero. Thus

$$y \text{ component:} \quad P(20\,\text{in.})\sin\alpha = 0, \tag{3}$$

$$z \text{ component:} \quad 384\,\text{in.}\cdot\text{lb} - P(20\,\text{in.})\cos\alpha = 0. \tag{4}$$

Since $P \neq 0$, Eq. (3) is satisfied only if $\boxed{\alpha = 0.}$ Equation (4) can then be solved for $\boxed{P = 19.2\,\text{lb.}}$

Discussion & Verification Both solutions provide the same results. If we did not recognize by inspection that $\alpha = 0$, then solution 1 would have been a little more tedious, but could still be carried out.

PROBLEMS

Problem 4.44

An *open-end* wrench applies the forces F to the head of a bolt to produce the moment M, where each force F is normal to the surface on which it acts. Determine M if $F = 400$ lb.

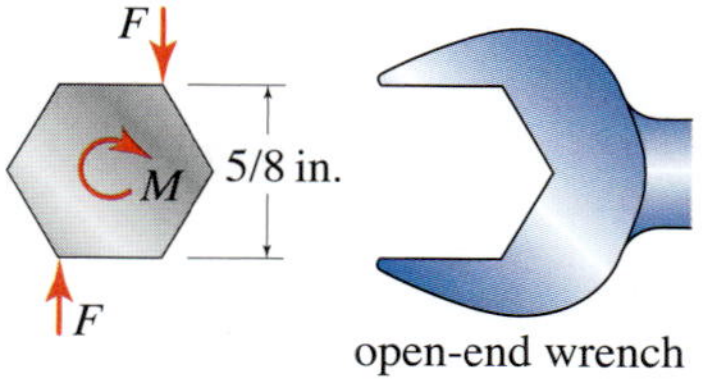

Figure P4.44

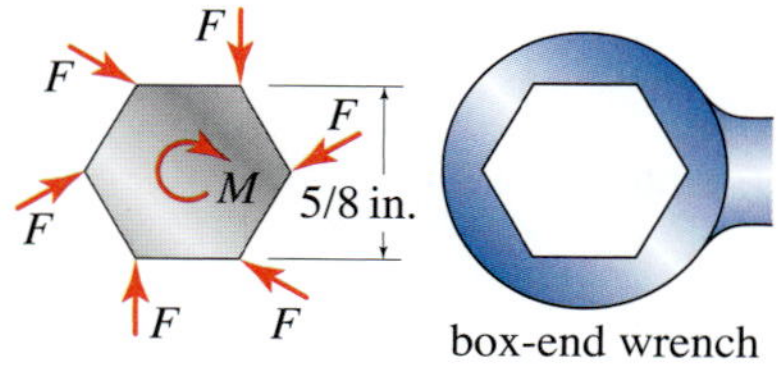

Figure P4.45

Problem 4.45

A *box-end* wrench applies the forces F to the head of a bolt to produce the moment M, where each force F is normal to the surface on which it acts. Determine F if $M = 20$ ft·lb.

Problem 4.46

The top view of a workpiece that fits loosely in a fixture for drilling is shown. The drill bit has two edges that apply in-plane cutting forces F to the workpiece.

(a) If $F = 600$ N, determine the forces Q between the workpiece and fixture so that the resultant couple moment is zero when $\alpha = 30°$.

(b) Does your answer for Q from Part (a) change if α has different value? If yes, then repeat Part (a) with $\alpha = 60°$.

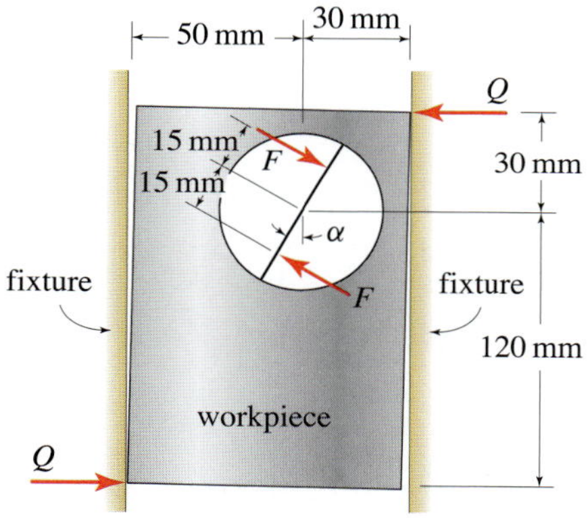

Figure P4.46

Problem 4.47

Three tugboats are used to turn a barge in a narrow channel. To avoid producing any net translation of the barge, the forces applied should be couples. The tugboat at point A applies a 400 lb force.

(a) Determine F_B and F_C so that only couples are applied.

(b) Using your answers to Part (a), determine the resultant couple moment that is produced.

(c) Resolve the forces at A and B into x and y components, and identify the pairs of forces that constitute couples.

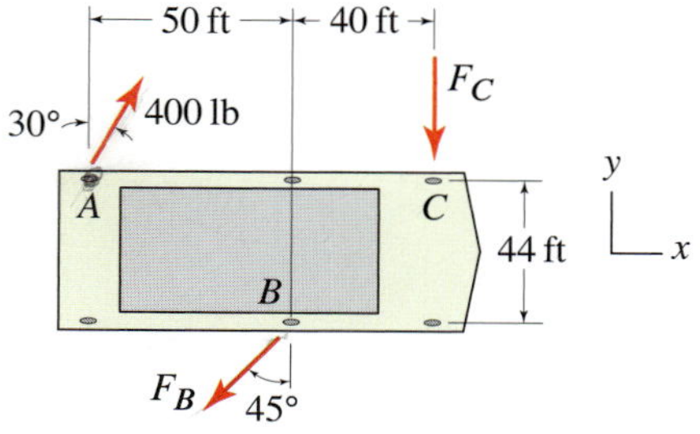

Figure P4.47

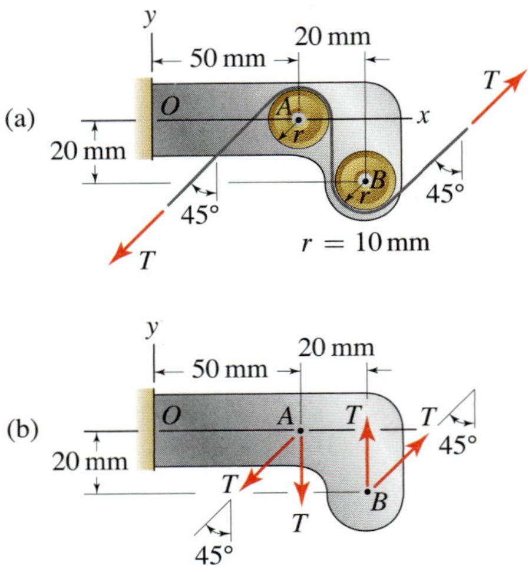

Figure P4.48

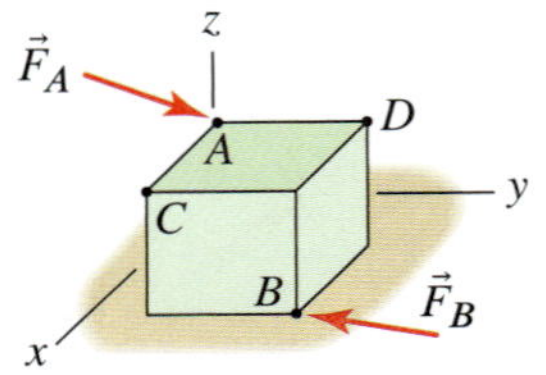

Figure P4.49

Problem 4.48

A bracket with pulleys for positioning magnetic tape in an electronic data storage device is shown. The pulleys are frictionless, the tape supports a force $T = 2\,\mathrm{N}$, and the thickness of the tape can be neglected.

(a) Use the two tape forces in Fig. P4.48(a) to compute the resultant couple moment.

(b) Replace the forces on the two pulleys with point forces on the pinions as shown in Fig. P4.48(b), and compute the resultant couple moment (replacement of pulley forces by bearing forces is discussed in Chapter 3 in connection with Fig. 3.9).

Problem 4.49

Consider an object with forces $\vec{F}_A$ and $\vec{F}_B$ applied. The first column of the following table lists resultant moments at various points due to both of these forces. For each statement select True or False.

	If $\vec{F}_A$ and $\vec{F}_B$ are not a couple	If $\vec{F}_A$ and $\vec{F}_B$ are a couple
$\vec{M}_C = \vec{r}_{CA} \times \vec{F}_A + \vec{r}_{CB} \times \vec{F}_B$	T or F?	T or F?
$\vec{M}_C = \vec{r}_{AB} \times \vec{F}_B$	T or F?	T or F?
$\vec{M}_C = \vec{r}_{BA} \times \vec{F}_A$	T or F?	T or F?
$\vec{M}_A = \vec{r}_{AB} \times \vec{F}_B$	T or F?	T or F?
$\vec{M}_D = \vec{r}_{CA} \times \vec{F}_A + \vec{r}_{CB} \times \vec{F}_B$	T or F?	T or F?

Note: Concept problems are about *explanations*, not computations.

Problem 4.50

A structure built in at point O supports 300 and 400 N couples. Determine the resultant couple moment vector, using both scalar and vector approaches.

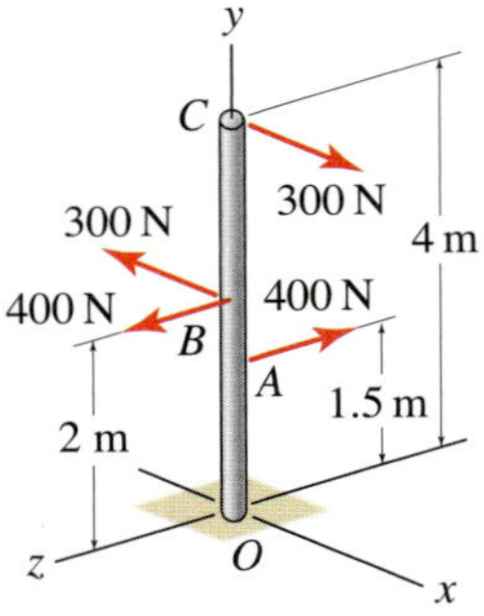

Figure P4.50

Problem 4.51

A structure built in at point O supports 70 and 85 N couples and a tip moment. Determine the resultant couple moment, using both scalar and vector approaches.

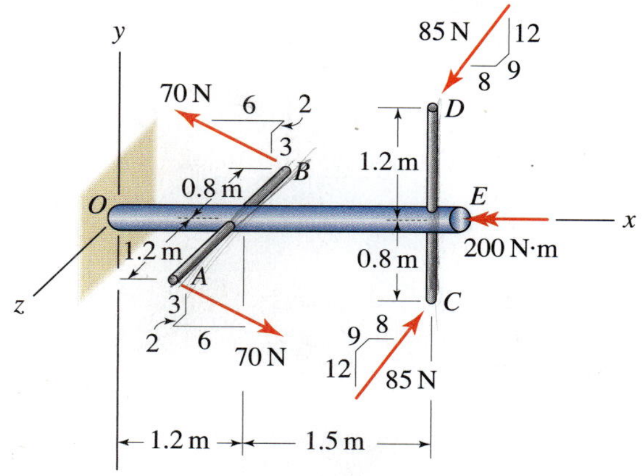

Figure P4.51–P4.53

Problem 4.52

Determine the distance between the lines of action of the 70 N forces. *Hint:* Use the vector approach to determine the moment of this couple. Then note that the magnitude of this result is equal to (70 N)d, where d is the distance between the lines of action for the two forces.

Problem 4.53

Determine the distance between the lines of action of the 85 N forces. See the hint in Prob. 4.52.

Problem 4.54

The input shaft of a speed reducer supports a 200 N·m moment, and the output shafts support 300 and 500 N·m moments.

(a) Determine the resultant moment applied by the shafts to the speed reducer.

(b) If the speed reducer is bolted to a surface that lies in the xy plane, speculate on the characteristics of the forces these bolts apply to the speed reducer. In other words, do you expect these forces to constitute couples only, or may they be more general? *Remark:* Problems such as this are discussed in detail in Chapter 5.

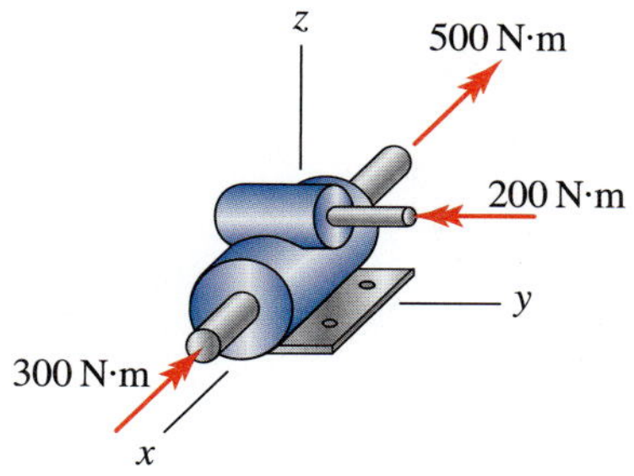

Figure P4.54

Problem 4.55

Satellites and other spacecraft perform attitude positioning using thrusters that are fired in pairs so as to produce couples. If thrusters at points A, B, C, and D each produce 3 N forces, determine the resultant moment and hence the axis through the center of mass about which an initially nonrotating satellite will begin to rotate.

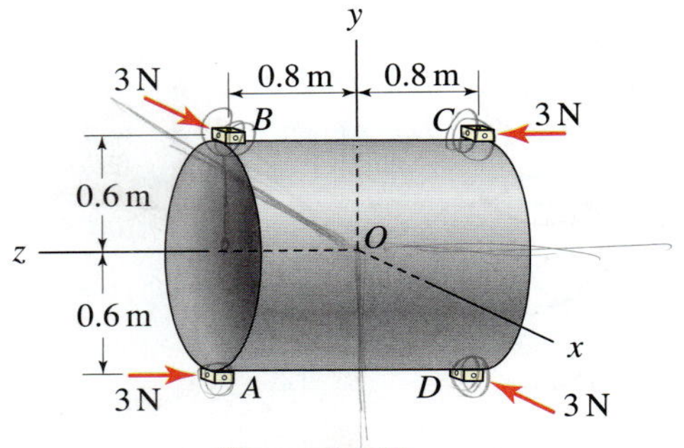

Figure P4.55

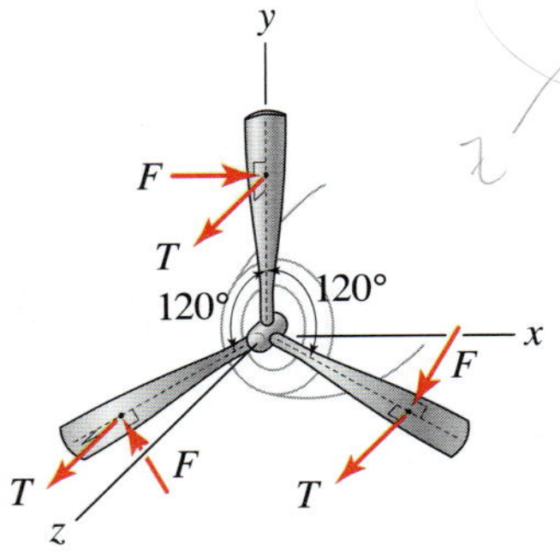

Figure P4.56

Problem 4.56

Forces F and T exerted by air on a rotating airplane propeller are shown. Forces F lie in the xy plane and are normal to the axis of each propeller blade, and thrust forces T act in the z direction. Show that the forces F can be represented as a couple or system of couples.

Problem 4.57

If the structure shown is subjected to couple forces applied at points A and B and the force applied at A is $\vec{F} = (8\,\hat{\imath} + 10\,\hat{\jmath} - 40\,\hat{k})$ lb, determine the moment of the couple about line a. Line a has direction angles $\theta_x = 72°$, $\theta_y = 36°$, and $\theta_z = 60°$.

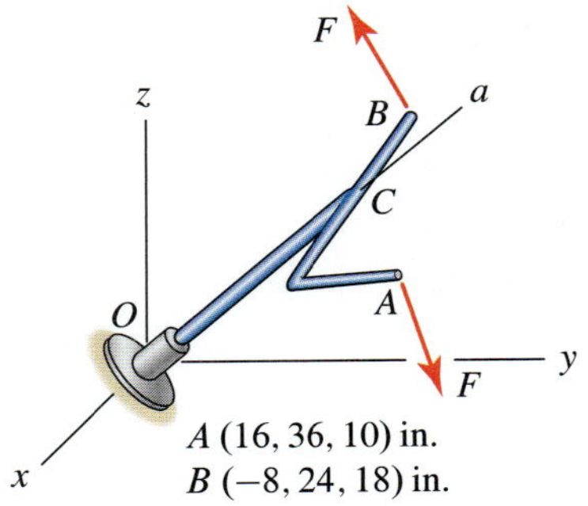

Figure P4.57 and P4.58

Problem 4.58

The structure shown has two forces of magnitude F applied. If $F = 42$ lb and the forces are to be a couple, determine the orientation they should have so that the moment of the couple about line a is as large as possible. Line a has direction angles $\theta_x = 72°$, $\theta_y = 36°$, and $\theta_z = 60°$.

4.4 Equivalent Force Systems

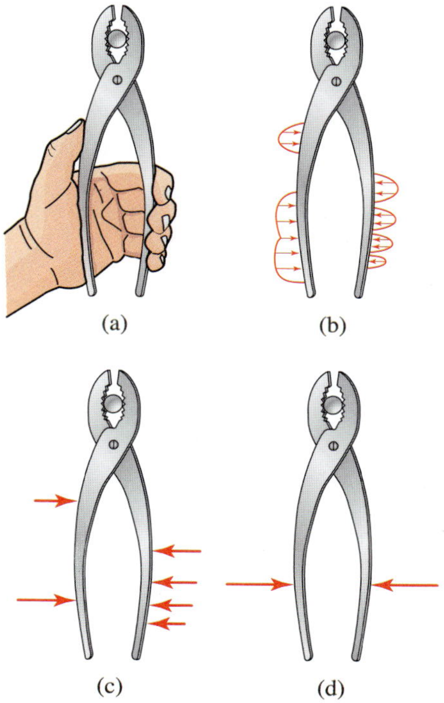

Figure 4.25
(a) A person uses a hand to grip a pair of pliers. (b)–(d) Examples of equivalent force systems.

Equivalent force systems are extremely important in mechanics and are used by engineers daily. Consider, for example, the pliers shown in Fig. 4.25(a), and imagine we wish to determine the force developed at the jaws as we squeeze the handles. Each finger and the palm of our hand apply a complex pressure distribution to the handles, as shown in Fig. 4.25(b). It would be very unfortunate if the complexities of these distributions needed to be included in all analyses. For many purposes, such as for determining the force developed at the jaws, these pressure distributions may be replaced by equivalent forces as shown in Fig. 4.25(c) and (d) with no loss of accuracy.

This section defines conditions under which two force systems are equivalent and describes techniques for replacing one force system by another. Concepts are first presented as definitions for rigid bodies, using physical arguments as justification. Implications of these concepts for deformable bodies are discussed. The section closes with a short discussion of Newton's laws, which provides a theoretical justification and interpretation of concepts.

Transmissibility of a force

Before discussing the principle of transmissibility of a force, we first define the terms *external effects* of a force applied to an object and *internal effects* of a force applied to an object. *External effects* refer to the response of the object as a whole. For example, if the object's position is fixed in space by supports, external effects include the support reactions. If the object is unsupported so that it can move in space, external effects refer to the object's displacement, velocity, and acceleration. *Internal effects* refer to the internal forces supported by a body, and if the body is deformable, the deformations that the body experiences.

The *principle of transmissibility* states that the external effects of a force applied to a rigid body are the same, regardless of the point of application of the force along its line of action. This principle is developed in Fig. 4.26 as follows. In Fig. 4.26(a), a force vector $\vec{F}$ is applied at point A on an object or

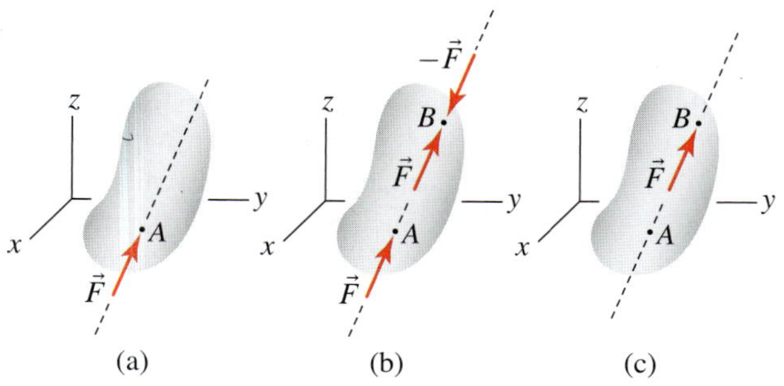

Figure 4.26. Transmissibility of force.

structure. In Fig. 4.26(b), two additional forces $\vec{F}$ and $-\vec{F}$ are applied at point B where B lies on the line of action of the force at A; since the forces at B have equal magnitude and opposite direction and are applied at the same point, they have no effect on the object. In Fig. 4.26(c) the forces $\vec{F}$ at A and $-\vec{F}$ at B have been canceled, leaving only the force $\vec{F}$ at B. We now conjecture that if the body is rigid, the external effects of $\vec{F}$ on the body are the same whether $\vec{F}$ is applied to point A or to point B.

Using definitions to follow, we will see that the principle of transmissibility of a force states conditions that give special cases of *equivalent force systems*, and that all of the force systems shown in Fig. 4.26 are equivalent. Furthermore, at the end of this section, the reason why equivalent force systems are called *equivalent* is discussed, and this provides justification for why the principle of transmissibility of a force is valid.

Rigid versus deformable objects. To be precise, the principle of transmissibility only holds for forces applied to rigid bodies. If an object is rigid, the distance between any two points, such as points A and B in Fig. 4.26, is constant regardless of where forces are applied to the object. Because the geometry of the object does not change, the position of lines of action of forces does not change, and hence the external effects on the rigid body, which are determined by summing forces and moments, are the same. Note that the principle of transmissibility makes no claims about internal forces. In fact, the internal forces depend very much on exactly where a particular force is applied along its line of action. But if the body is rigid, then regardless of the internal forces, the body does not change shape and hence the external effects of the force are unchanged.

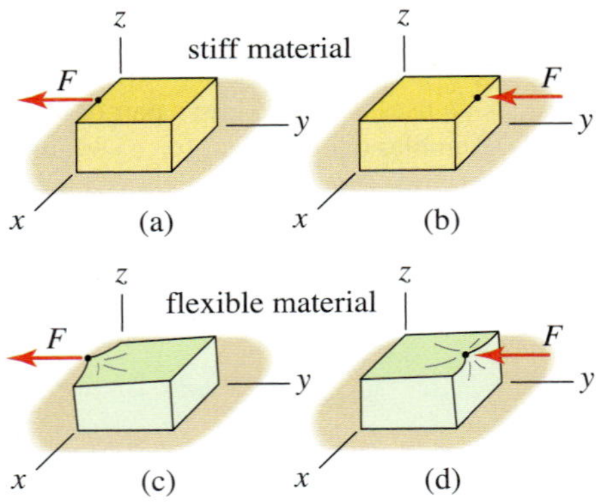

Figure 4.27
A block of material resting on a rough surface and subjected to a force. (a) and (b) The material is stiff and the difference in the block's shape for the two loadings is very small. (c) and (d) The material is flexible, and the block's shape is significantly different for the two loadings.

If a body is deformable, which is always the situation in nature, then the body will generally change shape when forces are applied. Furthermore, the deformed shape of the body depends on the distribution of internal forces within the body. The body's change of shape causes lines of action of forces to be repositioned, and hence the external effects of forces are different. Even though all bodies in nature are deformable, for many purposes objects and structures may be idealized as being rigid and the principle of transmissibility may be applied.

To illustrate further, consider a block of material resting under its own weight on a rough surface as shown in Fig. 4.27. If the material is stiff, then there is very little change of geometry for the loadings shown in Fig. 4.27(a) and (b), and for many practical purposes the object may be idealized as being rigid. The principle of transmissibility may be applied, and the response of the object, such as the force F that will cause the block to slip, is the same regardless of where F is applied along its line of action. If the material is flexible, the object may undergo significant changes of shape when subjected to the forces shown in Fig. 4.27(c) and (d). Such objects often cannot be idealized as being rigid since the deformation that occurs may have a strong effect on its response.

Equivalent force systems

The concept of an equivalent force system is developed in Fig. 4.28. In Fig. 4.28(a), a rigid object or structure has a force vector $\vec{F}$ applied at point A. In Fig. 4.28(b), two additional forces $\vec{F}$ and $-\vec{F}$ are applied at point B; since the forces at B have equal magnitude and opposite direction and are applied at the same point, they have no effect on the object. Forces $\vec{F}$ at A and $-\vec{F}$

at B are a couple and can be replaced by moment $\vec{M}_B = \vec{r} \times \vec{F}$, where $\vec{r}$ is a position vector from anywhere on the line of action of the force at B to anywhere on the line of action of the force at A (Fig. 4.28(b) shows a vector $\vec{r}$ from points B to A in particular). Thus, in Fig. 4.28(c) a new force system at point B has been developed that is equivalent to the original force system shown in Fig. 4.28(a). Further, since the moment of a couple is a free vector, $\vec{M}_B$ may be applied at any point, such as point C shown in Fig. 4.28(d).

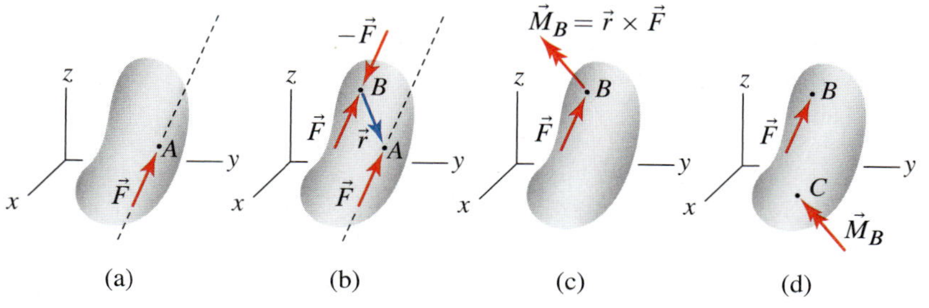

Figure 4.28. Construction of an equivalent force system at point B.

To generalize, if an object or structure has an arbitrary number of forces and/or moments applied as shown in Fig. 4.29, an *equivalent force system* at a point A consists of a resultant force $\vec{F}_R$ and a resultant moment $\vec{M}_R$ where

$$\vec{F}_R = \sum_{i=1}^{n} \vec{F}_i, \qquad \vec{M}_R = \sum_{i=1}^{n} \vec{r}_i \times \vec{F}_i + \sum_{i=1}^{m} \vec{M}_i, \tag{4.15}$$

where n is the number of forces that are applied, m is the number of moments that are applied, and $\vec{r}_i$ is a position vector from point A to anywhere on the line of action of $\vec{F}_i$. As discussed above, $\vec{M}_R$ is a free vector and may be positioned anywhere.

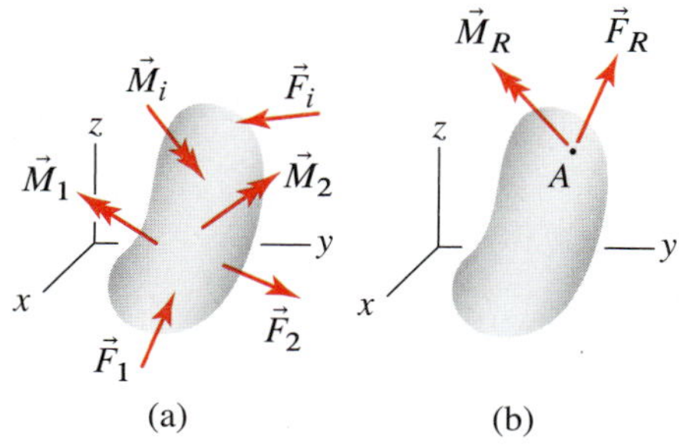

Figure 4.29 Construction of an equivalent force system at point A.

Equation (4.15) can be used as a test to determine if two force systems are equivalent. That is, *two force systems are equivalent* if

$$(\vec{F}_R)_{\text{system 1}} = (\vec{F}_R)_{\text{system 2}} \quad \text{and} \quad (\vec{M}_R)_{\text{system 1}} = (\vec{M}_R)_{\text{system 2}}, \tag{4.16}$$

where the moment summation point used for determining $(\vec{M}_R)_{\text{system 1}}$ and $(\vec{M}_R)_{\text{system 2}}$ must be the same. Using Eq. (4.15), we may state Eq. (4.16) more explicitly as *two force systems are equivalent* if

$$\Big(\sum_{i=1}^{n_1} \vec{F}_i\Big)_{\text{system 1}} = \Big(\sum_{i=1}^{n_2} \vec{F}_i\Big)_{\text{system 2}} \quad \text{and}$$
$$\Big(\sum_{i=1}^{n_1} \vec{r}_i \times \vec{F}_i + \sum_{i=1}^{m_1} \vec{M}_i\Big)_{\text{system 1}} = \Big(\sum_{i=1}^{n_2} \vec{r}_i \times \vec{F}_i + \sum_{i=1}^{m_2} \vec{M}_i\Big)_{\text{system 2}} \tag{4.17}$$

where n_1 and m_1 are the number of forces and moments in system 1, respectively, n_2 and m_2 are the number of forces and moments in system 2, respectively, and the moment summation points used for both force systems must be the same.

Some special force systems

Various special force systems, categorized as *concurrent*, *coplanar*, and *parallel*, are shown in Figs. 4.30–4.32. Observe that for each of these force systems, there exists a point (i.e., point B in Figs. 4.30–4.32) where an equivalent force system exists that consists of a *single force* only.

Remarks

- **Concurrent force system.** In these force systems, such as shown in Fig. 4.30, no resultant moment is produced by the forces about the point where the lines of action of the forces intersect. When an object is subjected to a concurrent force system, it may be idealized as a particle for purposes of equilibrium analysis, and this was studied extensively in Chapter 3.

- **Coplanar force system.** In these force systems, such as shown in Fig. 4.31, all forces lie in the same plane, and all moments are perpendicular to that plane. Use of Eq. (4.15) allows determination of an equivalent force system at point A, as shown in Fig. 4.31(b), consisting of $\vec{F}_R$ and $\vec{M}_R$. At point B, as shown in Fig. 4.31(c), the equivalent force system consists of $\vec{F}_R$ only; the location of B relative to A is determined so that $\vec{F}_R$ produces the proper moment (i.e., $M_R = F_R d \Rightarrow d = M_R/F_R$).

- **Parallel force system.** In these force systems, such as shown in Fig. 4.32, all forces are parallel, and all moments are perpendicular to the direction of the forces. Use of Eq. (4.15) allows determination of an equivalent force system at point A, as shown in Fig. 4.32(b), consisting of $\vec{F}_R$ and $\vec{M}_R$. At point B, as shown in Fig. 4.32(c), the equivalent force system consists of $\vec{F}_R$ only; the location of B relative to A is determined so that $\vec{F}_R$ produces the proper moment (i.e., $M_R = F_R d \Rightarrow d = M_R/F_R$).

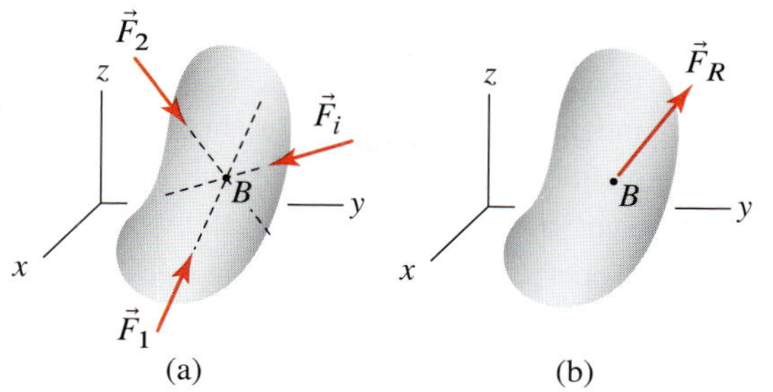

Figure 4.30. Concurrent force system: the lines of action of all forces intersect at a common point.

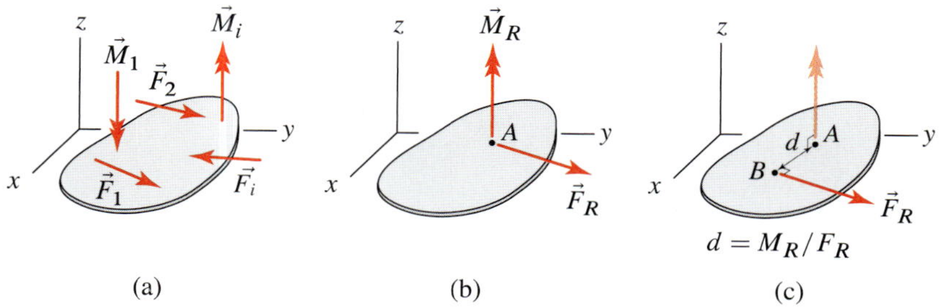

Figure 4.31. Coplanar force system: all forces lie in the same plane, and all moments are perpendicular to that plane.

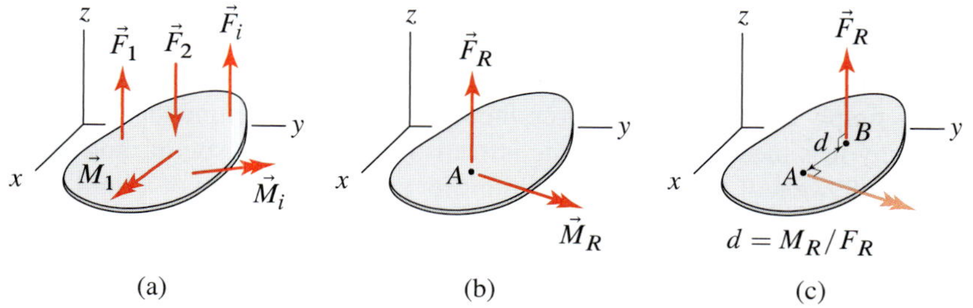

Figure 4.32. Parallel force system: all forces are parallel, and all moments are perpendicular to the direction of the forces.

Helpful Information

Positioning a force system to eliminate a moment. As discussed in connection with coplanar and parallel force systems, by suitable repositioning of the line of action of the resultant force $\vec{F}_R$, it is possible to construct an equivalent force system that has no moment, as shown in Figs. 4.31(c) and 4.32(c). However, on occasion you may find the distance d referred to in these figures to be large enough that the force is repositioned *off* the object or structure. While this is theoretically acceptable, it presents a practical problem on how such a force can be applied to the object or structure. Similar comments apply to the construction of a wrench force system, as shown in Fig. 4.34.

Wrench equivalent force systems

A feature of concurrent, coplanar, and parallel force systems is that for these systems you can always find an equivalent force system consisting of a single force only. More general force systems usually cannot be simplified to this extent, but it is useful for us to address the following question: What is the simplest force system to which any general force system can always be reduced? The answer is a force system called a *wrench*, which consists of a resultant force $\vec{F}_R$ and a resultant moment $\vec{M}_R$ that is parallel to $\vec{F}_R$, as shown in Fig. 4.33.

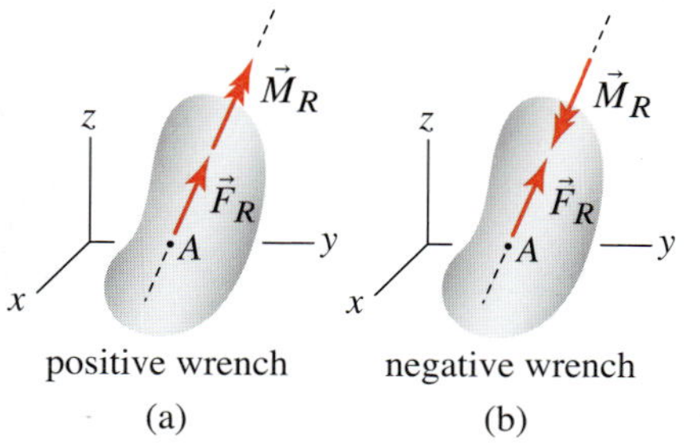

Figure 4.33
Wrench force systems. A *positive wrench* has $\vec{F}_R$ and $\vec{M}_R$ positive in the same direction. A *negative wrench* has $\vec{F}_R$ and $\vec{M}_R$ positive in opposite directions.

A wrench force system is constructed as follows. Starting with force $\vec{F}_R$ and moment $\vec{M}_R$ at point A as shown in Fig. 4.34(a), we resolve $\vec{M}_R$ into components parallel and perpendicular to $\vec{F}_R$ as shown in Fig. 4.34(b). Noting that $\vec{F}_R$ and $\vec{M}_{R\perp}$ constitute a parallel force system, $\vec{M}_{R\perp}$ may be eliminated by relocating $\vec{F}_R$ to point B where its location relative to A is determined so that $\vec{F}_R$ produces the proper moment (i.e., $M_{R\perp} = F_R d \Rightarrow d = M_{R\perp}/F_R$). Note that because $\vec{M}_{R\parallel}$ is a free vector, it can also be relocated from A to B. Example 4.15 illustrates further details of determining a wrench equivalent force system.

Helpful Information

Replacing a general force system by a *wrench*. Example 4.15 gives a two-step process for determining a wrench equivalent force system.

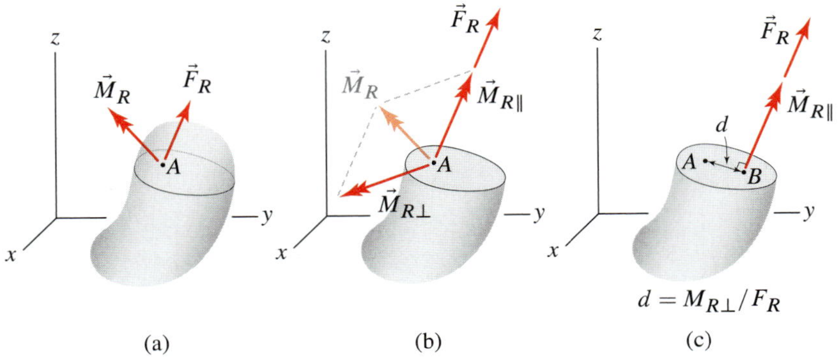

Figure 4.34. Construction of a wrench force system.

Why are equivalent force systems called *equivalent*?

If you like, you may consider the concept of an equivalent force system as being simply a definition. However, there is good reason behind the definition and why it is meaningful, and this is seen from Newton's law of motion. In Chapter 5, we will see that the conditions for static equilibrium of a rigid body are $\sum \vec{F} = \vec{0}$ and $\sum \vec{M} = \vec{0}$. If the first of these equations is not satisfied, the body will have a translational acceleration; and if the second equation is not satisfied, the body will have a rotational acceleration. Regardless of whether the body is in static equilibrium or is undergoing accelerations, the response in terms of external effects (e.g., reaction forces, accelerations, etc.) is the same for *all* systems of external forces for which $\sum \vec{F}$ and $\sum \vec{M}$ are the same, and the definition of equivalent force systems given in Eq. (4.15) is a statement of these conditions. Since the principle of transmissibility of a force is a special case of equivalent force systems, this explanation also shows why it is valid.

End of Section Summary

In this section, the concepts of equivalent force systems were described. Some of the key points are as follows:

- The *principle of transmissibility* states that the effects of a force applied to a rigid body are the same regardless of the point of application of the force along its line of action.

- Two force systems are said to be *equivalent* if the sum of forces for the two systems is the same and if the sum of moments (about any common point) for the two force systems is the same. If two force systems are equivalent, their external effects on a rigid body or structure, such as support reactions, are the same.

- Some special force systems that frequently occur were defined, including *concurrent, coplanar*, and *parallel force systems*. Coplanar and parallel force systems have the common feature that they have an equivalent force system that consists of a single force only (no moment).

- A *wrench* force system consists of a force and moment, where these have the same line of action. Any force system can be represented by an equivalent force system that is a wrench.

EXAMPLE 4.11 *Determination of an Equivalent Force System*

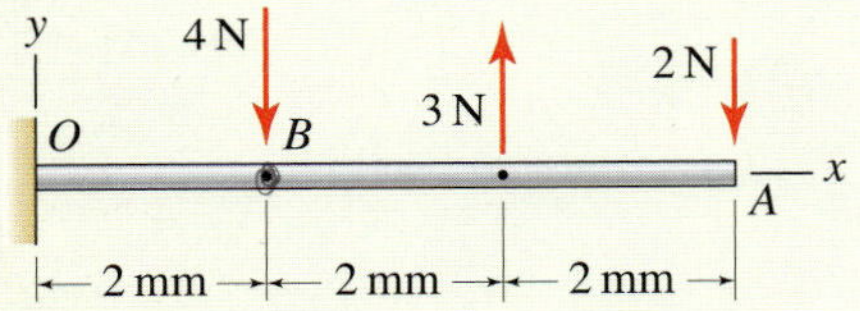

Figure 1

Determine an equivalent force system

(a) at point A,

(b) at point B,

(c) consisting of a single force only, and specify the x coordinate of the point where the force's line of action intersects the x axis.

Helpful Information

Reaction forces applied to the beam. The structure shown in Fig. 1 is supported at point O, and this support applies *reaction* forces and moments to the structure so that it is in equilibrium. However, our focus throughout this section has been on determining equivalent systems of *external* forces. Thus, whether an object is supported or unsupported is irrelevant to the task of determining equivalent systems of external forces.

SOLUTION

Road Map For all three parts of this problem, an equivalent system of external forces will be developed using Eq. (4.15) on p. 223.

Part (a)

Governing Equations & Computation We apply Eq. (4.15) with a scalar approach to obtain

$$F_{Rx} = \sum F_x = \boxed{0,} \tag{1}$$

$$F_{Ry} = \sum F_y = -4\,\text{N} + 3\,\text{N} - 2\text{N} = \boxed{-3\,\text{N},} \tag{2}$$

$$M_{RA} = \sum M_A = (4\,\text{N})(4\,\text{mm}) - (3\,\text{N})(2\,\text{mm}) = \boxed{10\,\text{N}\cdot\text{mm}.} \tag{3}$$

This force system is illustrated in Fig. 2.

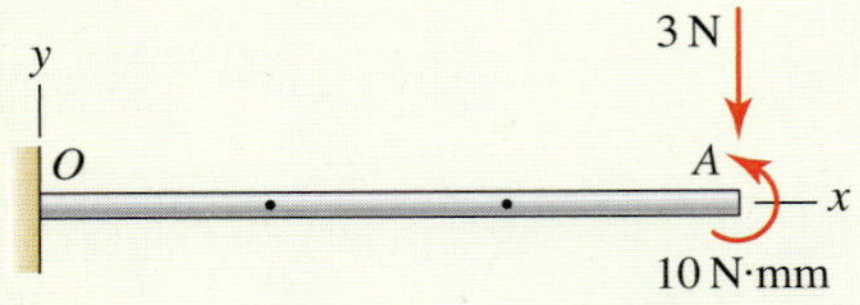

Figure 2
This force system is equivalent to that shown in Fig. 1.

Part (b)

Governing Equations & Computation For an equivalent force system at point B, the resultant forces F_{Rx} and F_{Ry} are unchanged from Eqs. (1) and (2), and referring to the forces shown in Fig. 1, the resultant moment about B is

$$M_{RB} = \sum M_B = (3\,\text{N})(2\,\text{mm}) - (2\,\text{N})(4\,\text{mm}) = \boxed{-2\,\text{N}\cdot\text{mm}.} \tag{4}$$

This force system is illustrated in Fig. 3, where M_{RB} is shown as $2\,\text{N}\cdot\text{mm}$ clockwise rather than $-2\,\text{N}\cdot\text{mm}$ counterclockwise. As an alternative to using the force system shown in Fig. 1 to write Eq. (4), we could have used the force system shown in Fig. 2 to write $M_{RB} = \sum M_B = 10\,\text{N}\cdot\text{mm} - (3\,\text{N})(4\,\text{mm}) = -2\,\text{N}\cdot\text{mm}$ which, as expected, agrees with Eq. (4).

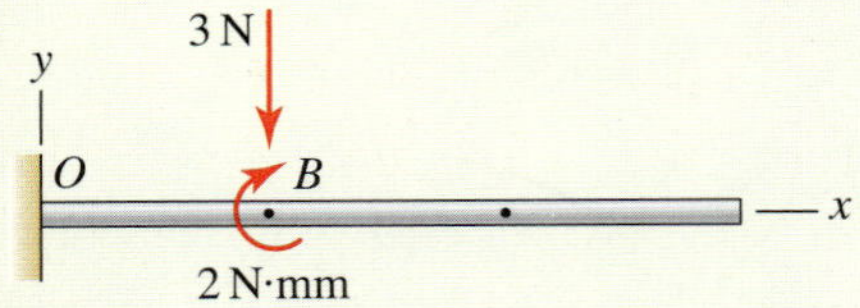

Figure 3
This force system is equivalent to those shown in Figs. 1 and 2.

Part (c)

Governing Equations & Computation A force system that consists of a single force only is shown in Fig. 4. The distance d is determined, so this force system is equivalent to the force systems shown in Figs. 1 through 3. Thus, selecting point O as a convenient location to sum moments for the force systems shown in Figs. 2 and 4 provides

$$\left(\sum M_{RO}\right)_{\text{Fig. 2}} = \left(\sum M_{RO}\right)_{\text{Fig. 4}}$$

$$-(3\,\text{N})(6\,\text{mm}) + 10\,\text{N}\cdot\text{mm} = -(3\,\text{N})d \tag{5}$$

$$\Rightarrow \quad d = \boxed{2.67\,\text{mm}.} \tag{6}$$

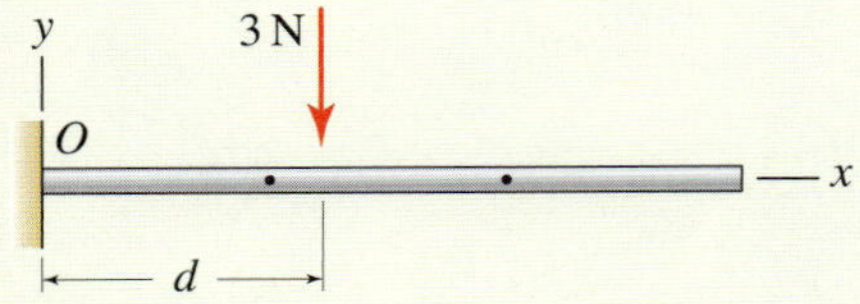

Figure 4
With the appropriate value of d, this force system is equivalent to those shown in Figs. 1–3.

Discussion & Verification The force systems shown in Figs. 1 through 3 are all equivalent, and the force system shown in Fig. 4 is also equivalent to these provided $d = 2.67\,\text{mm}$. When developing an equivalent force system that consists of a single force only, as in Part (c), sometimes you may find that the line of action of the force *does not* intersect the structure.

EXAMPLE 4.12 *Determining if Force Systems Are Equivalent*

Determine which of the force systems shown are equivalent.

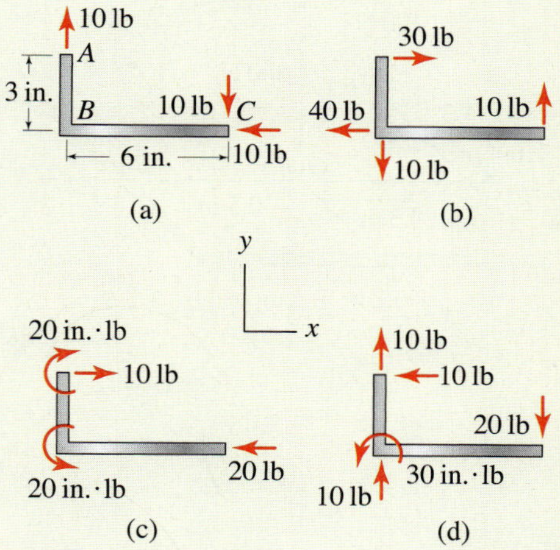

Figure 1

SOLUTION

Road Map For two force systems to be equivalent, both the resultant force and the resultant moment, taken about any convenient point, must be the same. Thus for each of the force systems shown, we will evaluate the resultant forces in the x and y directions and the resultant moment, and those force systems for which all of these are the same are equivalent.

Governing Equations & Computation Using point B as a convenient location for moment summation, we use Eq. (4.15) on p. 223 to find

Force system (a)

$$F_{Rx} = \sum F_x = -10\,\text{lb}, \tag{1}$$

$$F_{Ry} = \sum F_y = 10\,\text{lb} - 10\,\text{lb} = 0, \tag{2}$$

$$M_{RB} = \sum M_B = -(10\,\text{lb})(6\,\text{in.}) = -60\,\text{in.}\cdot\text{lb}. \tag{3}$$

Force system (b)

$$F_{Rx} = \sum F_x = -40\,\text{lb} + 30\,\text{lb} = -10\,\text{lb}, \tag{4}$$

$$F_{Ry} = \sum F_y = -10\,\text{lb} + 10\,\text{lb} = 0, \tag{5}$$

$$M_{RB} = \sum M_B = -(30\,\text{lb})(3\,\text{in.}) + (10\,\text{lb})(6\,\text{in.}) = -30\,\text{in.}\cdot\text{lb}. \tag{6}$$

Force system (c)

$$F_{Rx} = \sum F_x = 10\,\text{lb} - 20\,\text{lb} = -10\,\text{lb}, \tag{7}$$

$$F_{Ry} = \sum F_y = 0, \tag{8}$$

$$M_{RB} = \sum M_B = 20\,\text{in.}\cdot\text{lb} - 20\,\text{in.}\cdot\text{lb} - (10\,\text{lb})(3\,\text{in.}) = -30\,\text{in.}\cdot\text{lb}. \tag{9}$$

Force system (d)

$$F_{Rx} = \sum F_x = -10\,\text{lb}, \tag{10}$$

$$F_{Ry} = \sum F_y = 10\,\text{lb} + 10\,\text{lb} - 20\,\text{lb} = 0, \tag{11}$$

$$M_{RB} = \sum M_B = 30\,\text{in.}\cdot\text{lb} + (10\,\text{lb})(3\,\text{in.}) - (20\,\text{lb})(6\,\text{in.}) = -60\,\text{in.}\cdot\text{lb}. \tag{12}$$

Discussion & Verification Force systems (a) and (d) have the same resultant force and moment about point B, and hence they are equivalent to one another. Also force systems (b) and (c) have the same resultant force and moment about point B, and hence they are equivalent to one another. In summary:

Force systems (a) and (d) are equivalent, and force systems (b) and (c) are equivalent.

EXAMPLE 4.13 *Determination of an Equivalent Force System*

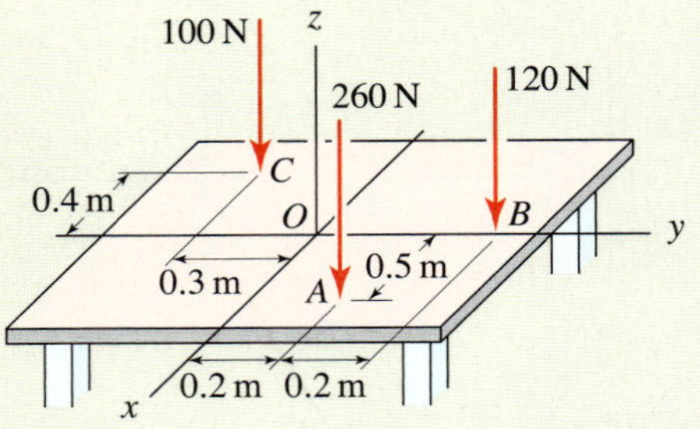

Figure 1

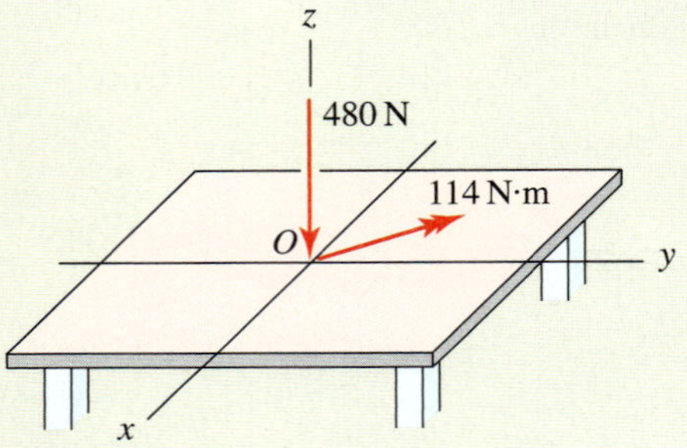

Figure 2
This force system is equivalent to that shown in Fig. 1.

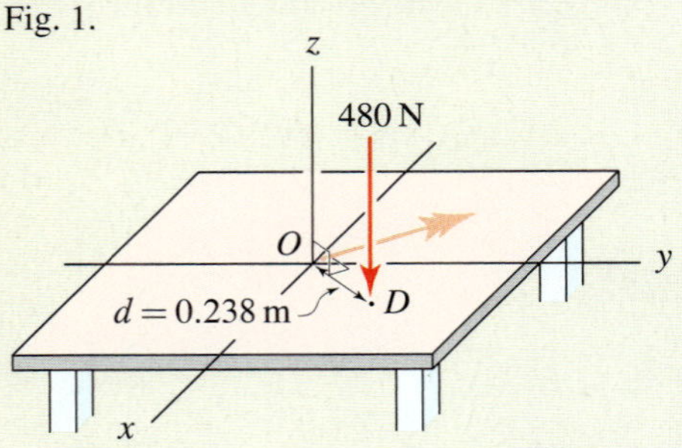

Figure 3
This force system is equivalent to those shown in Figs. 1 and 2.

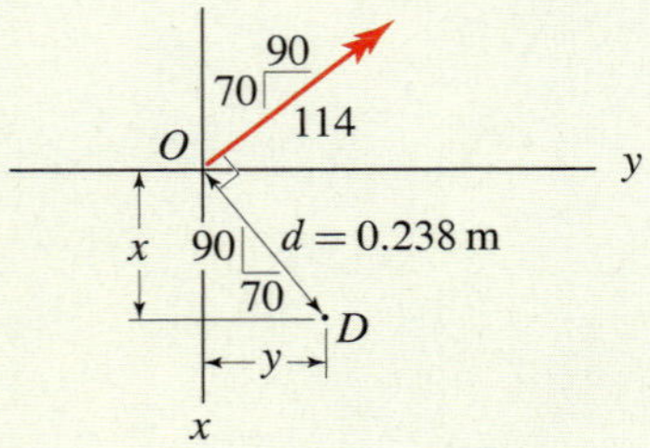

Figure 4
Use of similar triangles to locate point D.

A table supports the vertical forces shown.

(a) Determine an equivalent force system at the center of the table, point O.

(b) Determine an equivalent force system consisting of a single force, and specify the x and y coordinates of the point where the force's line of action intersects the table.

SOLUTION

Road Map Both scalar and vector approaches are effective for Part (a) of this problem, and we will use a vector approach. For Part (b), a straightforward scalar evaluation will provide the necessary location of an equivalent force system that consists of a single force only.

Part (a)

Governing Equations & Computation With the following force and position vectors

$$\vec{F}_A = -260\,\hat{k}\text{ N}, \qquad \vec{r}_{OA} = (0.5\,\hat{\imath} + 0.2\,\hat{\jmath})\text{ m}, \tag{1}$$

$$\vec{F}_B = -120\,\hat{k}\text{ N}, \qquad \vec{r}_{OB} = 0.4\,\hat{\jmath}\text{ m}, \tag{2}$$

$$\vec{F}_C = -100\,\hat{k}\text{ N}, \qquad \vec{r}_{OC} = (-0.4\,\hat{\imath} - 0.3\,\hat{\jmath})\text{ m}, \tag{3}$$

we use Eq. (4.15) on p. 223 to evaluate the resultant force and moment at point O as

$$\vec{F}_R = \vec{F}_A + \vec{F}_B + \vec{F}_C$$

$$= \boxed{-480\,\hat{k}\text{ N},} \tag{4}$$

$$\vec{M}_{RO} = \vec{r}_{OA} \times \vec{F}_A + \vec{r}_{OB} \times \vec{F}_B + \vec{r}_{OC} \times \vec{F}_C$$

$$= \boxed{(-70\,\hat{\imath} + 90\,\hat{\jmath})\text{ N·m},} \tag{5}$$

$$M_{RO} = \sqrt{(-70)^2 + (90)^2}\text{ N·m} = 114\text{ N·m}. \tag{6}$$

This force system is illustrated in Fig. 2.

Part (b)

Governing Equations & Computation As shown in Fig. 4.32 on p. 225, an equivalent force system consisting of a single force is obtained by moving $\vec{F}_R$ to a new position, point D, where D is located a distance d perpendicular to the plane containing $\vec{F}_R$ and $\vec{M}_{RO}$

$$d = \frac{M_{RO}}{F_R} = \frac{114\text{ N·m}}{480\text{ N}} = 0.238\text{ m}. \tag{7}$$

This force system is shown in Fig. 3. To determine the coordinates of D, similar triangles with the geometry shown in Fig. 4 may be used to write

$$x = (0.238\text{ m})\frac{90}{114} = \boxed{0.188\text{ m},} \tag{8}$$

$$y = (0.238\text{ m})\frac{70}{114} = \boxed{0.146\text{ m}.} \tag{9}$$

Discussion & Verification The force systems shown in Figs. 1 through 3 are all equivalent. When developing an equivalent force system that consists of a single force only, as in Part (b), sometimes you may find that the line of action of the force *does not* intersect the structure.

EXAMPLE 4.14 *Determination of an Equivalent Force System*

A casting supports the forces and moment shown where the 100 N forces are parallel to the xy plane and the moment at C has direction angles $\theta_x = 60°$, $\theta_y = 60°$, and $\theta_z = 135°$. Determine the equivalent force system acting at point O.

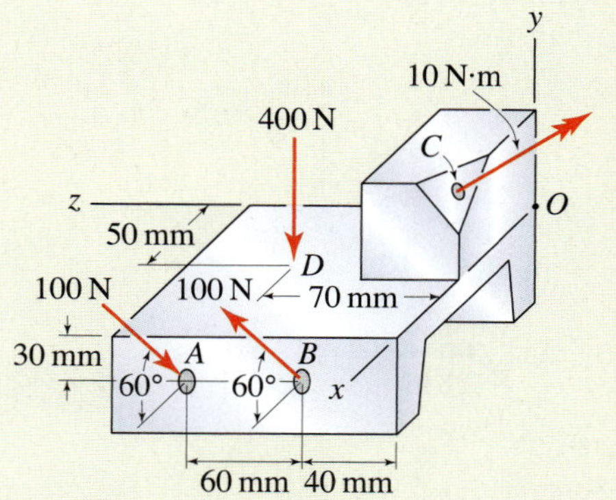

Figure 1

SOLUTION

Road Map Because of the complexity of the geometry, a vector solution is preferable to a scalar solution. Note that the two 100 N forces have equal magnitude and opposite direction, hence they are a couple.

Governing Equations & Computation We first write expressions for force, position, and moment vectors:

$$\vec{F}_B = (100\,\text{N})(\cos 60°\,\hat{\imath} + \sin 60°\,\hat{\jmath}), \qquad \vec{r}_{AB} = -60\,\hat{k}\text{ mm}, \tag{1}$$

$$\vec{F}_D = -400\,\hat{\jmath}\,\text{N}, \qquad \vec{r}_{OD} = (50\,\hat{\imath} + 70\,\hat{k})\text{ mm}, \tag{2}$$

$$\vec{M}_C = (10\,\text{N·m})(\cos 60°\,\hat{\imath} + \cos 60°\,\hat{\jmath} + \cos 135°\,\hat{k}). \tag{3}$$

Taking care to convert millimeter dimensions to meter dimensions, we use Eq. (4.15) on p. 223 to obtain the resultant force and moment at point O as

$$\vec{F}_R = \vec{F}_D = \boxed{-400\,\hat{\jmath}\,\text{N},} \tag{4}$$

$$\vec{M}_{RO} = \vec{r}_{AB} \times \vec{F}_B + \vec{r}_{OD} \times \vec{F}_D + \vec{M}_C$$

$$= \boxed{(38.2\,\hat{\imath} + 2\,\hat{\jmath} - 27.1\,\hat{k})\,\text{N·m},} \tag{5}$$

$$M_{RO} = \sqrt{(38.2)^2 + (2)^2 + (-27.1)^2}\,\text{N·m} = 46.9\,\text{N·m}. \tag{6}$$

In writing Eqs. (4) and (5), we have noted that the two forces at A and B are a couple. Hence they produce no net force and are not included in the expression for $\vec{F}_R$ in Eq. (4). The equivalent force system at point O is illustrated in Fig. 2.

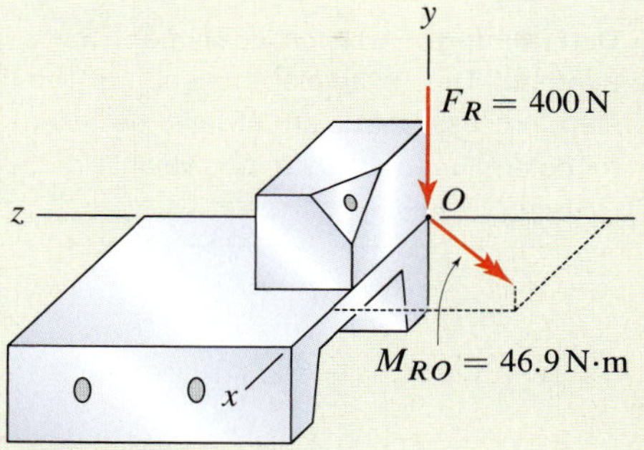

Figure 2

This force system is equivalent to that shown in Fig. 1.

Discussion & Verification The force systems shown in Figs. 1 and 2 are equivalent. As discussed above, we took advantage of the properties of a couple when writing the expressions for $\vec{F}_R$ and $\vec{M}_{RO}$ in Eqs. (4) and (5). If you did not recognize that these forces are a couple, then you would need to include the two 100 N forces in the expression for $\vec{F}_R$, and in the expression for $\vec{M}_{RO}$ you would replace $\vec{r}_{AB} \times \vec{F}_B$ by $\vec{r}_{OA} \times \vec{F}_A + \vec{r}_{OB} \times \vec{F}_B$. Nonetheless, you would have obtained the same results for $\vec{F}_R$ and $\vec{M}_{RO}$ in Eqs. (4) and (5).

EXAMPLE 4.15 *Determination of a Wrench Equivalent Force System*

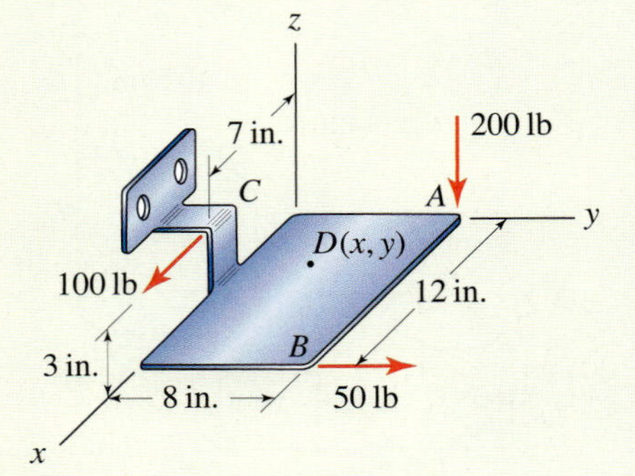

Figure 1

Helpful Information

Determining a *wrench* equivalent force system. This example presents a two-step process that can always be used to determine a wrench equivalent force system.

Replace the three forces by a wrench force system, specifying the force and moment of the wrench and the x and y coordinates of point D where the wrench's line of action intersects the xy plane.

SOLUTION

Road Map The construction of a wrench equivalent force system can be accomplished using a two-step procedure, as illustrated here.

Governing Equations & Computation

Step 1: Determine an equivalent force system at point D. By using the following force and position vectors

$$\vec{F}_A = -200\,\hat{k}\text{ lb}, \qquad \vec{r}_{DA} = -x\,\hat{\imath} + (8\text{ in.} - y)\,\hat{\jmath}, \tag{1}$$

$$\vec{F}_B = 50\,\hat{\jmath}\text{ lb}, \qquad \vec{r}_{DB} = (12\text{ in.} - x)\,\hat{\imath} + (8\text{ in.} - y)\,\hat{\jmath}, \tag{2}$$

$$\vec{F}_C = 100\,\hat{\imath}\text{ lb}, \qquad \vec{r}_{DC} = (7\text{ in.} - x)\,\hat{\imath} - y\,\hat{\jmath} + 3\,\hat{k}\text{ in.}, \tag{3}$$

the resultant force and moment at point D are

$$\vec{F}_R = \vec{F}_A + \vec{F}_B + \vec{F}_C = (100\,\hat{\imath} + 50\,\hat{\jmath} - 200\,\hat{k})\text{ lb}, \tag{4}$$

$$F_R = \sqrt{(100)^2 + (50)^2 + (-200)^2}\text{ lb} = 229\text{ lb}, \tag{5}$$

$$\begin{aligned}\vec{M}_{RD} &= \vec{r}_{DA} \times \vec{F}_A + \vec{r}_{DB} \times \vec{F}_B + \vec{r}_{DC} \times \vec{F}_C \\ &= [-1600\text{ in.}\cdot\text{lb} + (200\text{ lb})y]\,\hat{\imath} + [300\text{ in.}\cdot\text{lb} - (200\text{ lb})x]\,\hat{\jmath} \\ &\quad + [600\text{ in.}\cdot\text{lb} - (50\text{ lb})x + (100\text{ lb})y]\,\hat{k}.\end{aligned} \tag{6}$$

If we specify values for x and y, Eqs. (4) and (6) provide the resultant force and moment at that location. However, these are not likely to have the same direction and hence will not be a wrench force system. In the next step of this solution, we determine the coordinates of point D so that $\vec{M}_{RD}$ is parallel to $\vec{F}_R$.

Step 2: Make the equivalent force system a wrench. The requirement that $\vec{F}_R$ and $\vec{M}_{RD}$ be parallel is stated by $\vec{F}_R/F_R = \vec{M}_{RD}/M_{RD}$. Each of the x, y, and z components of this equation can be written separately.

$$x\text{ component:}\quad \frac{100\text{ lb}}{229\text{ lb}} = \frac{-1600\text{ in.}\cdot\text{lb} + (200\text{ lb})y}{M_{RD}}, \tag{7}$$

$$y\text{ component:}\quad \frac{50\text{ lb}}{229\text{ lb}} = \frac{300\text{ in.}\cdot\text{lb} - (200\text{ lb})x}{M_{RD}}, \tag{8}$$

$$z\text{ component:}\quad \frac{-200\text{ lb}}{229\text{ lb}} = \frac{600\text{ in.}\cdot\text{lb} - (50\text{ lb})x + (100\text{ lb})y}{M_{RD}}. \tag{9}$$

Solving Eqs. (7)–(9) for unknowns x, y, and M_{RD} provides

$$\boxed{x = 2.76\text{ in.}, \quad y = 5.48\text{ in.}, \quad M_{RD} = -1160\text{ in.}\cdot\text{lb}.} \tag{10}$$

and, from Eq. (5), the magnitude of the force in the wrench force system is

$$\boxed{F_R = 229\text{ lb}.} \tag{11}$$

The negative value for M_{RD} indicates $\vec{M}_{RD}$ is in the direction opposite $\vec{F}_R$, and hence this is a *negative wrench* as defined in Fig. 4.33 on p. 225. This wrench force system is shown in Fig. 2.

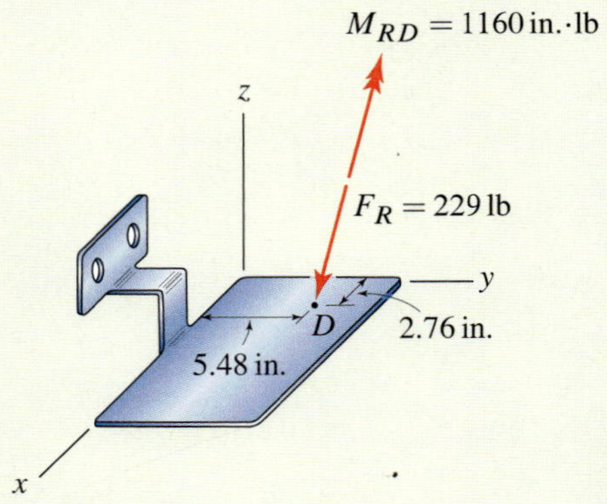

Figure 2
A wrench equivalent force system.

Discussion & Verification The force systems shown in Figs. 1 and 2 are equivalent. When determining a wrench force system, we generally identify the location of the wrench by finding *one* point on its line of action (point D in this example). Often, we will find where the wrench's line of action intersects a plane, such as the xy, yz, or zx plane. This way one of the coordinates of a point on the wrench's line of action is known (in this example the z coordinate of point D is known to be zero). A common mistake is to let all three coordinates of a point on the wrench's line of action be unknowns; then there are four unknowns but still only three equations available to determine them and hence a unique solution is not possible.

PROBLEMS

Problems 4.59 and 4.60

Determine which of the force systems shown, if any, are equivalent.

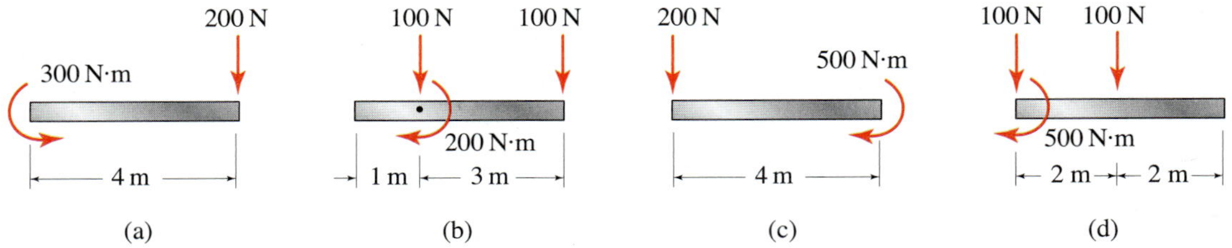

Figure P4.59

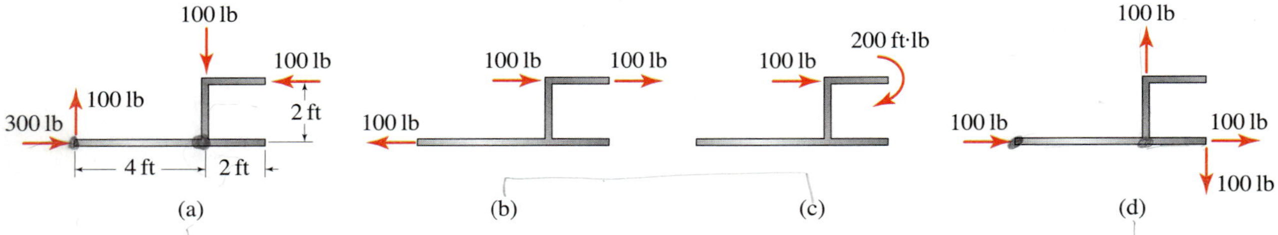

Figure P4.60

Problem 4.61

Determine values for forces F and P and moment M, if possible, so that the force systems shown in Fig. P4.61(b)–(d) are equivalent to the force system shown in Fig. P4.61(a).

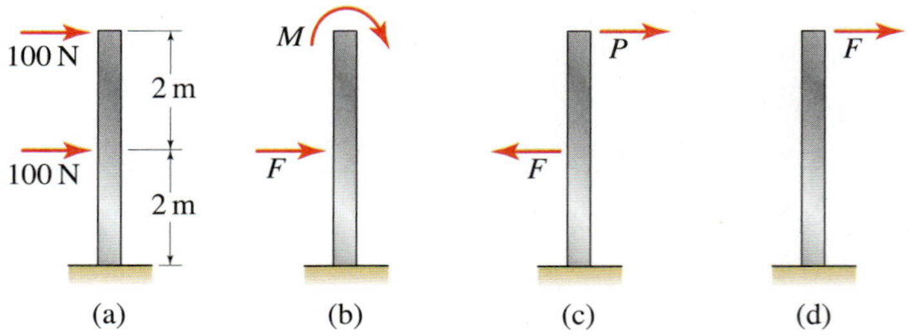

Figure P4.61

Problem 4.62

Determine values for forces F and P, if possible, so that the force systems shown in Fig. P4.62(b)–(d) are equivalent to the force system shown in Fig. P4.62(a).

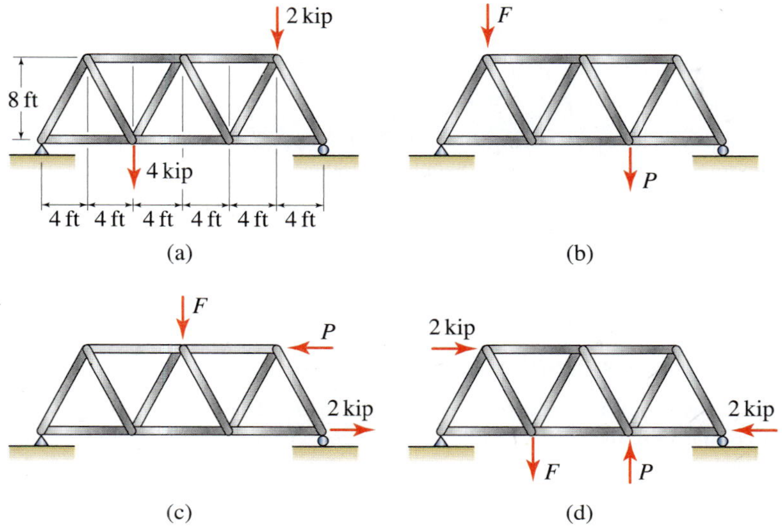

Figure P4.62

Problem 4.63

The floor of an airplane cargo bay is shown in a horizontal position.

(a) Determine an equivalent force system at point O.

(b) Determine an equivalent force system consisting of a single force, and specify the x and z coordinates of the point where the force's line of action intersects the floor.

(c) Keeping points A–D at the locations shown, suggest a repositioning of the forces so that the location of the force system described in Part (b) is closer to point O.

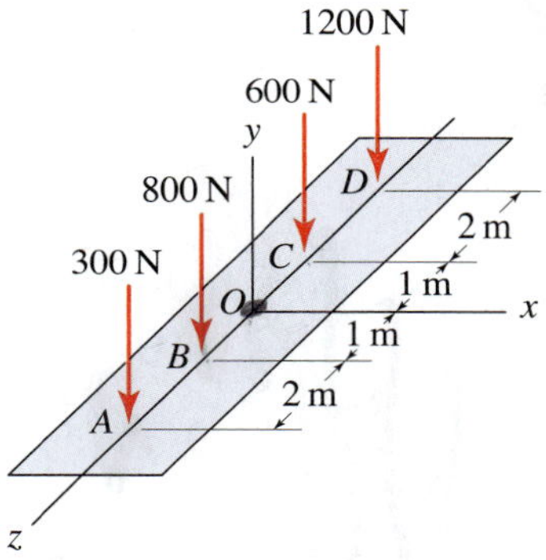

Figure P4.63

Problem 4.64

In a vehicle collision reconstruction analysis, an engineer estimates the tire forces shown. Determine an equivalent force system at point O.

Problem 4.65

For Prob. 4.64, determine an equivalent force system consisting of a single force, specifying the x coordinate of where the force's line of action intersects the x axis.

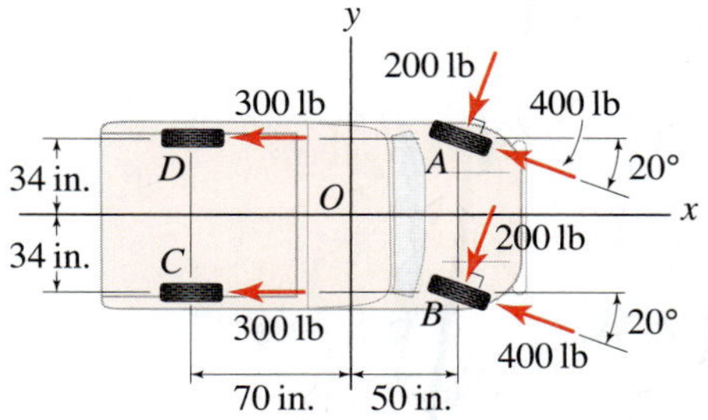

Figure P4.64 and P4.65

Problem 4.66

If $\alpha = 30°$, $W = 1$ kN, and $Q = 2$ kN in Fig. P4.12 on p. 195, determine an equivalent force system consisting of a single force and specify the distance from point C where the force's line of action intersects member CBD.

Problem 4.67

If $F = 200$ N and $P = 300$ N in Fig. P4.13 on p. 195, determine an equivalent force system consisting of a single force and specify the x coordinate of the point where the force's line of action intersects the x axis.

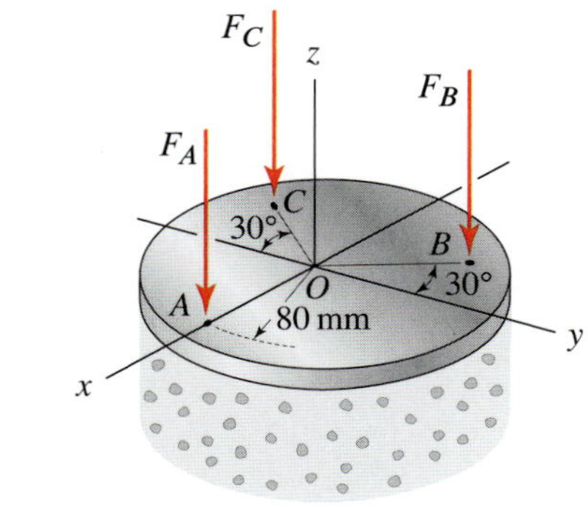

Figure P4.68

Problem 4.68

A thick circular plate is used in a machine called a *gyratory compactor* to determine the mechanical properties of hot asphalt concrete. The plate is instrumented with load cells at points A, B, and C where each load cell is located at the same 80 mm radial distance from point O. If the load cells measure forces $F_A = 500\,\text{N}$, $F_B = 600\,\text{N}$, and $F_C = 700\,\text{N}$, determine

(a) an equivalent force system at point O,

(b) an equivalent force system consisting of a single force, specifying the x and y coordinates of the point where the force's line of action intersects the xy plane.

Problem 4.69

A boat trailer is subjected to the forces shown where the forces at points A–E are vertical and the forces at points F and G lie in the yz plane. Determine

(a) an equivalent force system at point O,

(b) an equivalent force system consisting of a single force, specifying the x and y coordinates of the point where the force's line of action intersects the xy plane.

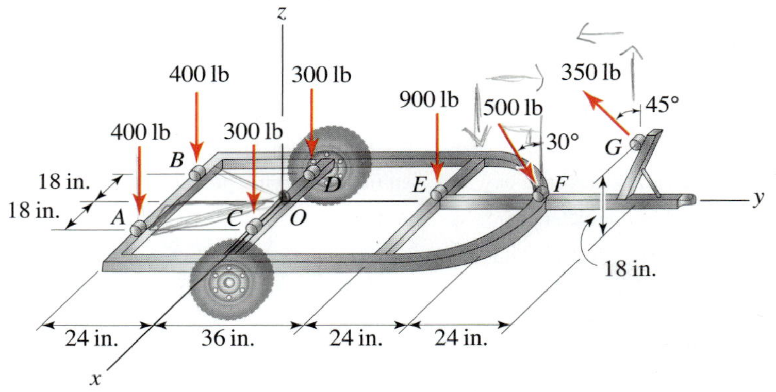

Figure P4.69

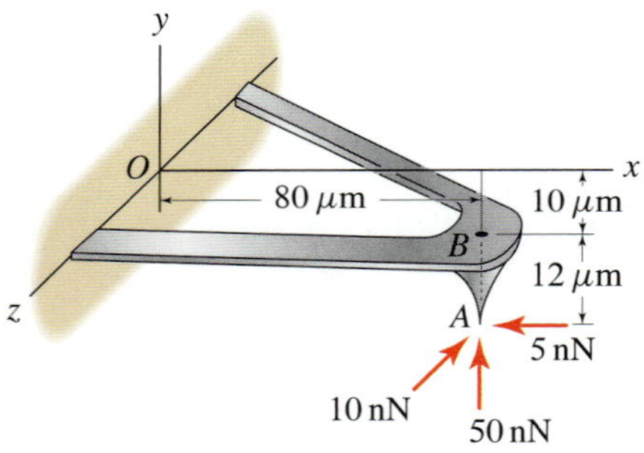

Figure P4.70 and P4.71

Problem 4.70

The tip of an atomic force microscope (AFM) is subjected to the forces shown. Use a vector approach to determine

(a) an equivalent force system at point B,

(b) an equivalent force system at point O.

Problem 4.71

Repeat Prob. 4.70, using a scalar approach.

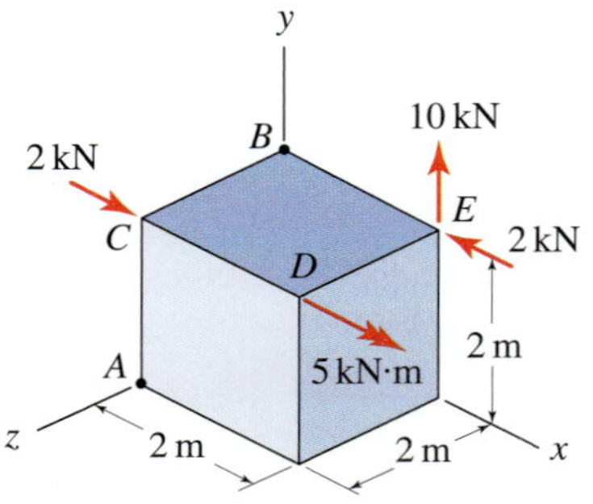

Figure P4.72 and P4.73

Problem 4.72

Determine an equivalent force system at point A, using

(a) a vector approach,

(b) a scalar approach.

Problem 4.73

Repeat Prob. 4.72 to find an equivalent force system at point B.

Problems 4.74 and 4.75

Using inspection, if possible, determine a wrench equivalent force system and specify the x and y coordinates of the point where the wrench's line of action intersects the xy plane. Express your answers in terms of parameters such as F, P, and r.

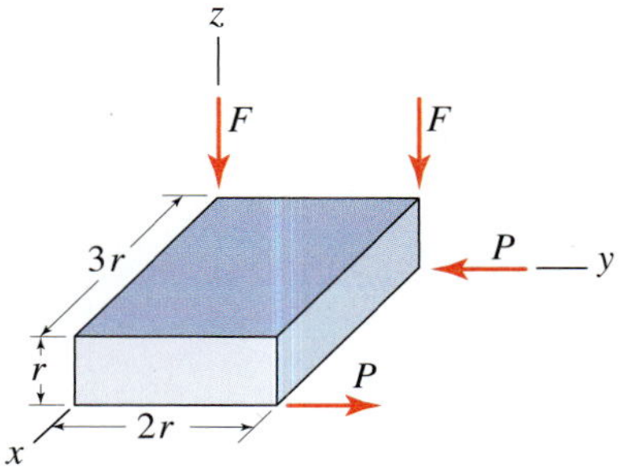

Figure P4.74

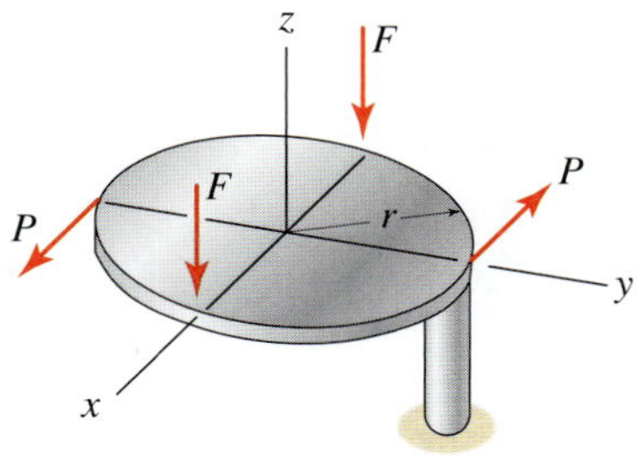

Figure P4.75

Problem 4.76

The object shown is subjected to a wrench having $\vec{F}_R = (2\,\hat{\imath} + 3\,\hat{\jmath} - 4\,\hat{k})$ N and $\vec{M}_R = (4\,\hat{\imath} + 6\,\hat{\jmath} - 8\,\hat{k})$ N·m and whose line of action intersects the xy plane at $x = 2$ m and $y = 1$ m. Determine an equivalent force system at point A, stating this in vector form.

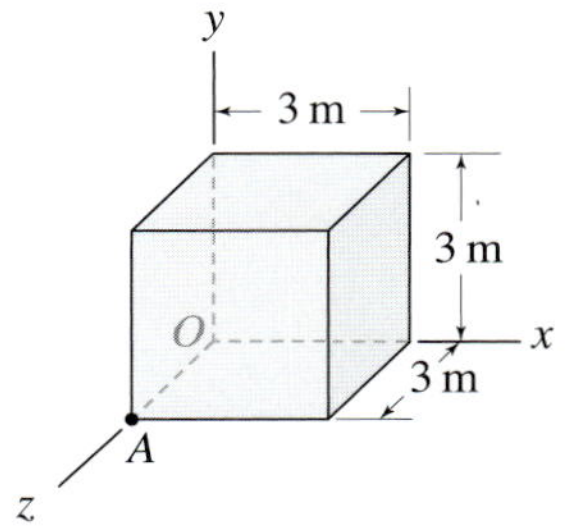

Figure P4.76

Problem 4.77

Determine a wrench equivalent force system and specify the x and y coordinates of the point where the wrench's line of action intersects the xy plane.

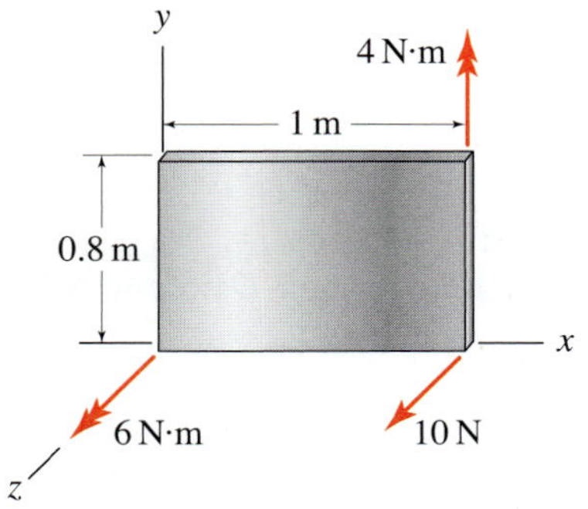

Figure P4.77

Problem 4.78

Determine a wrench equivalent force system and specify the x and y coordinates of the point where the wrench's line of action intersects the xy plane. Express your answers in terms of parameters such as F, a, and b.

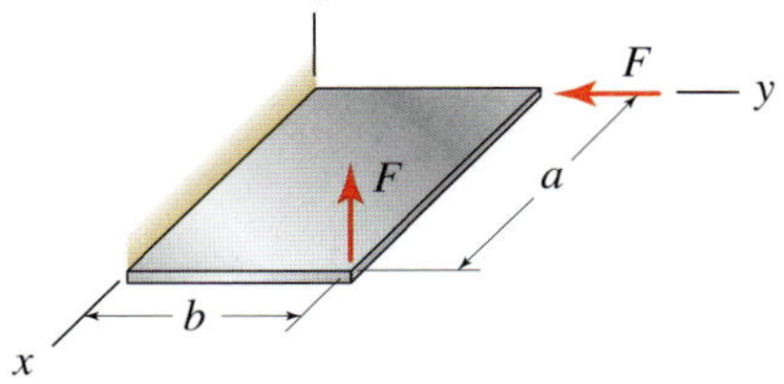

Figure P4.78

4.5 Chapter Review

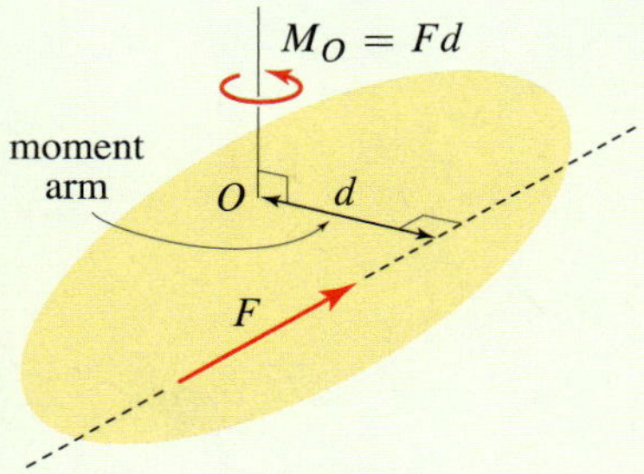

Figure 4.35
Scalar definition of the moment of a force.

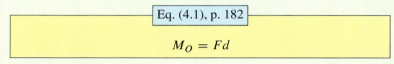

Important definitions, concepts, and equations of this chapter are summarized. For equations and/or concepts that are not clear, you should refer to the section numbers cited for additional details.

Moment of a force — scalar approach. As shown in Fig. 4.35, the *moment of a force* $\vec{F}$ about a point O is a vector (represented by the twisting action shown), and the magnitude of this moment is M_O, which is given by

Eq. (4.1), p. 182

$$M_O = Fd$$

where

F is the magnitude of the force;
d is the perpendicular distance from point O to the line of action of $\vec{F}$ and is called the *moment arm*; and
M_O has units of *force* times *length*.

The direction of the moment is not provided by Eq. (4.1), but is understood to be as follows. The line of action of the moment is parallel to the axis through point O that is perpendicular to the plane containing $\vec{F}$ and the moment arm. The direction of the moment along the line of action is given by the direction of the thumb of your right hand when your fingers curl in the twisting direction of the moment.

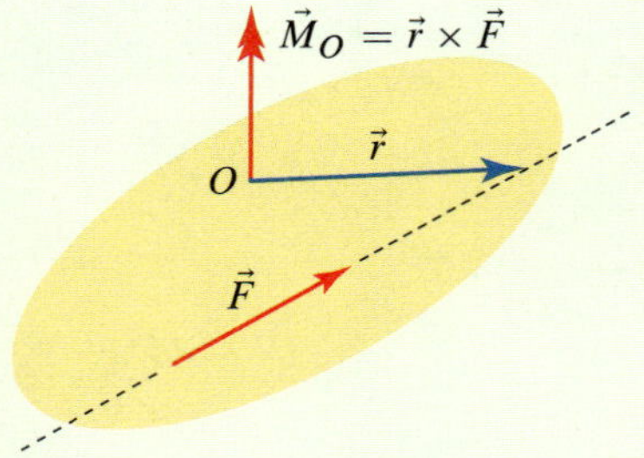

Figure 4.36
Vector definition of the moment of a force.

Moment of a force — vector approach. As shown in Fig. 4.36, the *moment of a force* $\vec{F}$ about a point O is denoted by $\vec{M}_O$ and is defined as

Eq. (4.2), p. 183

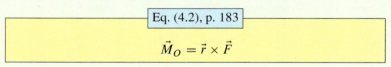

$$\vec{M}_O = \vec{r} \times \vec{F}$$

where

$\vec{F}$ is the force vector;
$\vec{r}$ is a position vector from point O to *any* point on the line of action of $\vec{F}$.

Varignon's theorem. *Varignon's theorem*, also known as the *principle of moments*, is a restatement of the distributive property of the cross product. The principle states that the moment of a force is equal to the sum of the moments of the vector components of the force. Thus, if $\vec{F}$ has vector components $\vec{F}_1$, $\vec{F}_2$, and so on, then the moment of $\vec{F}$ about a point A is given by

Eq. (4.10), p. 185

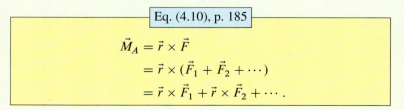

$$\begin{aligned}\vec{M}_A &= \vec{r} \times \vec{F}\\ &= \vec{r} \times (\vec{F}_1 + \vec{F}_2 + \cdots)\\ &= \vec{r} \times \vec{F}_1 + \vec{r} \times \vec{F}_2 + \cdots .\end{aligned}$$

The principle of moments is most commonly used for scalar evaluations of the moment of a force. Often, the vector components will be orthogonal, but the principle is also valid for nonorthogonal vector components.

Moment of a force about a line. The *moment of a force about a line* is defined to be the *component of the moment that is in the direction of the line.* The moment of a force about a line is discussed in Section 4.2 and can be evaluated by using vector and scalar approaches, as follows.

To determine the moment M_a of a force F about a line (or direction) a as shown in Fig. 4.37:

Vector approach:

1. Select a point P at a convenient location on line a. Determine the moment of $\vec{F}$ about P, using $\vec{M}_P = \vec{r} \times \vec{F}$, where $\vec{r}$ is a position vector from P to any point on the line of action of $\vec{F}$.*

2. $M_a = \vec{M}_P \cdot \hat{u}$, where $\hat{u}$ is a unit vector in the direction of a. To express this moment as a vector quantity, evaluate $\vec{M}_a = M_a\,\hat{u}$.

Note: Steps 1 and 2 may be combined to yield M_a directly by using the scalar triple product, $M_a = (\vec{r} \times \vec{F}) \cdot \hat{u}$.

Scalar approach:

1. Resolve F into components $F_\perp$ and $F_\parallel$ that are perpendicular and parallel, respectively, to a plane containing line a.

2. $M_a = F_\perp d$, where d is the moment arm (shortest distance) between line a and the line of action of F. (**Note:** $F_\parallel$ produces no moment about a, so you may skip its evaluation altogether.)

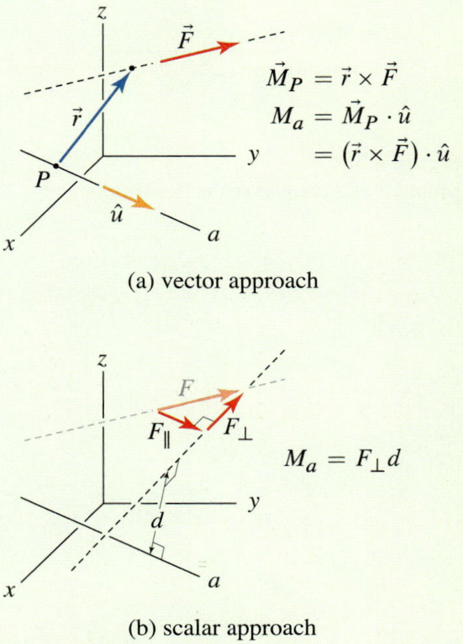

Figure 4.37
Vector and scalar approaches for determining the moment of a force about a line.

Couple. A *couple* is defined to be a system of two forces of equal magnitude and opposite direction and whose lines of action are separated by a distance. The *moment of a couple* (sometimes also called a *couple moment*) is the moment produced by the couple. A couple that is applied to a body produces a moment, but does not apply any net force to the body. The moment of a couple can be evaluated using both vector and scalar approaches as follows.

Moment of a couple — vector approach. Consider a couple consisting of two parallel forces $\vec{F}$ and $-\vec{F}$ as shown in Fig. 4.38. The moment $\vec{M}$ of this couple is

Eq. (4.11), p. 209

$$\vec{M} = \vec{r}_{AB} \times \vec{F}$$
$$= \vec{r}_{BA} \times (-\vec{F})$$

where

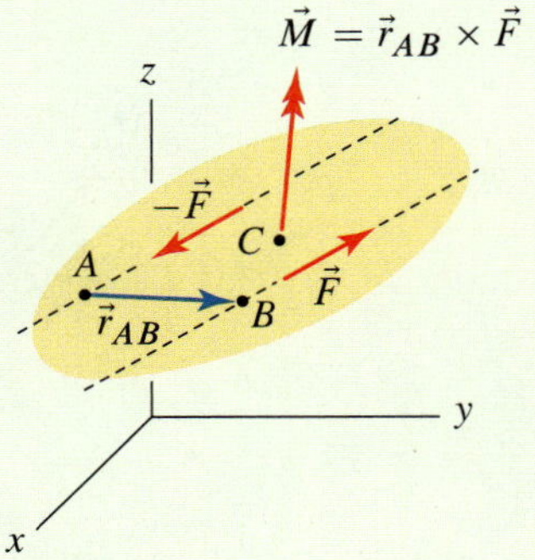

Figure 4.38
Moment of a couple: vector description.

* This procedure suggests using the cross product to determine $\vec{M}_P$. As an alternative, especially for problems with simple geometry, you could use a scalar approach to determine the vector expression for $\vec{M}_P$, to be followed by taking the dot product as described in Step 2.

$\vec{r}_{AB}$ and $\vec{r}_{BA}$ are position vectors,
A is *any* point on the line of action of $-\vec{F}$,
B is *any* point on the line of action of $\vec{F}$.

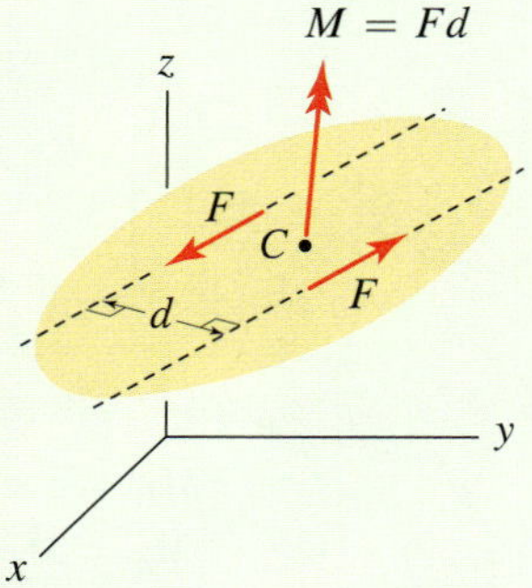

Figure 4.39
Moment of a couple: scalar description.

Moment of a couple—scalar approach. Consider a couple consisting of two parallel forces having the same magnitude F as shown in Fig. 4.39. The magnitude of the moment of this couple is

Eq. (4.12), p. 209

$$M = Fd$$

where

d is the perpendicular (shortest) distance between the forces' lines of action,
the direction of the moment is perpendicular to the plane containing the forces.

Two couples are said to be *equivalent* if the moment vectors they produce are identical (both the magnitude and direction of the moments must be the same). If an object or structure has multiple couples applied, a *resultant couple moment* may be determined by summing the individual couple moment vectors. The moment of a couple is a *free vector*, meaning the moment may be positioned anywhere on an object or structure. Further explanation of this subtle feature is given throughout Section 4.3.

Transmissibility of a force. The *principle of transmissibility* of a force, described in Section 4.4, states that the external effects of a force applied to a rigid body are the same, regardless of the point of application of the force along its line of action.

Equivalent force systems. If an object or structure has an arbitrary number of forces and/or moments applied as shown in Fig. 4.40, an *equivalent force system* at a point A consists of a resultant force $\vec{F}_R$ and a resultant moment $\vec{M}_R$ where

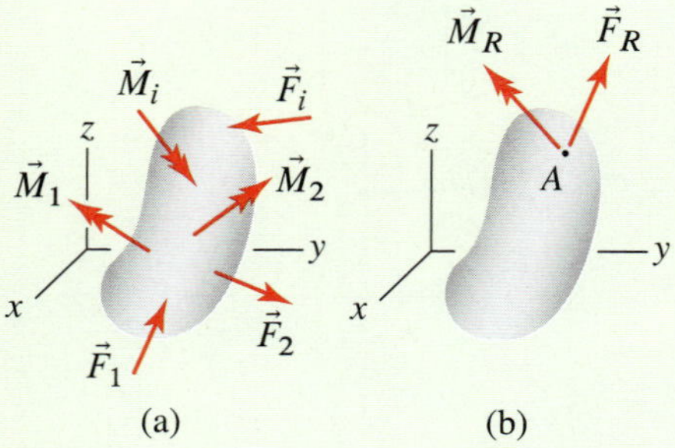

Figure 4.40
Construction of an equivalent force system at point A.

Eq. (4.15), p. 223

$$\vec{F}_R = \sum_{i=1}^{n} \vec{F}_i,$$

$$\vec{M}_R = \sum_{i=1}^{n} \vec{r}_i \times \vec{F}_i + \sum_{i=1}^{m} \vec{M}_i,$$

where n is the number of forces that are applied, m is the number of moments that are applied, and $\vec{r}_i$ is a position vector from point A to anywhere on the line of action of $\vec{F}_i$. Because $\vec{M}_R$ is a free vector, it may be positioned anywhere on the object or structure.

Two force systems are equivalent if they have the same resultant force and produce the same resultant moment about any common point. Thus, force system 1 and force system 2 are equivalent if

Eq. (4.16), p. 223

$$(\vec{F}_R)_{\text{system 1}} = (\vec{F}_R)_{\text{system 2}} \quad \text{and}$$
$$(\vec{M}_R)_{\text{system 1}} = (\vec{M}_R)_{\text{system 2}},$$

where the moment summation points used for determining $(\vec{M}_R)_{\text{system 1}}$ and $(\vec{M}_R)_{\text{system 2}}$ must be the same. Equation (4.16) may be stated more explicitly as *two force systems are equivalent* if

Eq. (4.17), p. 223

$$\left(\sum_{i=1}^{n_1} \vec{F}_i\right)_{\text{system 1}} = \left(\sum_{i=1}^{n_2} \vec{F}_i\right)_{\text{system 2}} \quad \text{and}$$
$$\left(\sum_{i=1}^{n_1} \vec{r}_i \times \vec{F}_i + \sum_{i=1}^{m_1} \vec{M}_i\right)_{\text{system 1}} = \left(\sum_{i=1}^{n_2} \vec{r}_i \times \vec{F}_i + \sum_{i=1}^{m_2} \vec{M}_i\right)_{\text{system 2}}$$

where n_1 and m_1 are the number of forces and moments in system 1, respectively, n_2 and m_2 are the number of forces and moments in system 2, respectively, and the moment summation point used for both force systems must be the same.

Some special force systems. Several categories of force systems, including *concurrent*, *coplanar*, *parallel*, and *wrench force systems*, are defined and studied in Section 4.4. If a particular force system is coplanar or parallel, then an equivalent force system that consists of a single resultant force with appropriate position can always be found. For more general force systems, an equivalent force system called a *wrench* can always be found.

REVIEW PROBLEMS

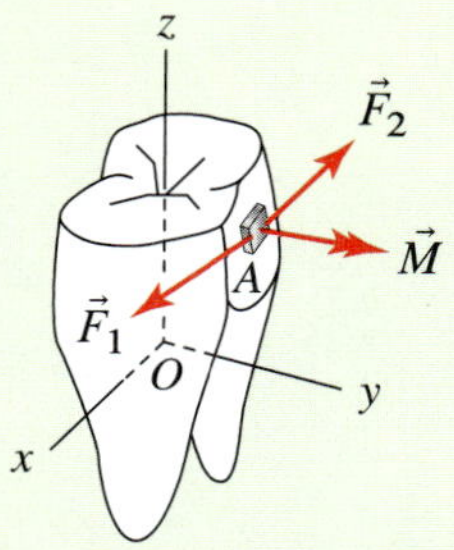

Figure P4.79

Problem 4.79

In orthodontics, teeth are repositioned by applying forces to them for prolonged periods of time. If point A has coordinates $(2, 7, 8)$ mm, $\vec{F}_1 = (0.8\,\hat{\imath} - 0.3\,\hat{\jmath} - 0.1\,\hat{k})$ N, $\vec{F}_2 = (-0.7\,\hat{\imath} - 0.1\,\hat{\jmath} + 0.05\,\hat{k})$ N, and $\vec{M} = (0.1\,\hat{\imath} + 1.4\,\hat{\jmath} - 0.3\,\hat{k})$ N·mm, determine an equivalent force system at point O.

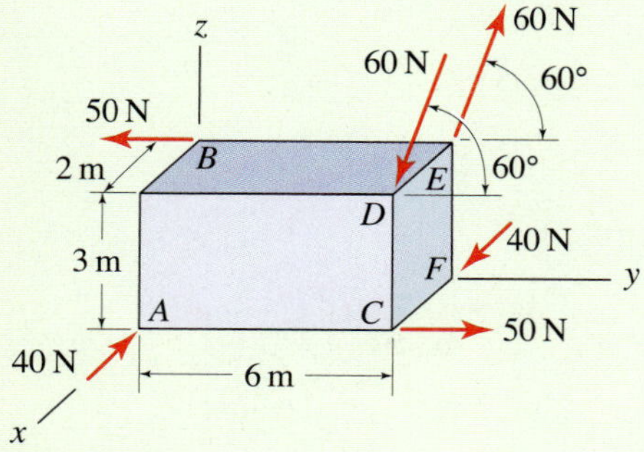

Figure P4.80

Problem 4.80

The 60 N forces lie in planes parallel to the yz plane. Determine the resultant couple moment vector for the force system shown using a

(a) vector approach,

(b) scalar approach.

Problem 4.81

Two force systems are applied to a right circular cylinder. Points A to D lie on the xz plane, and points E and F lie on the yz plane. Determine if these force systems are equivalent.

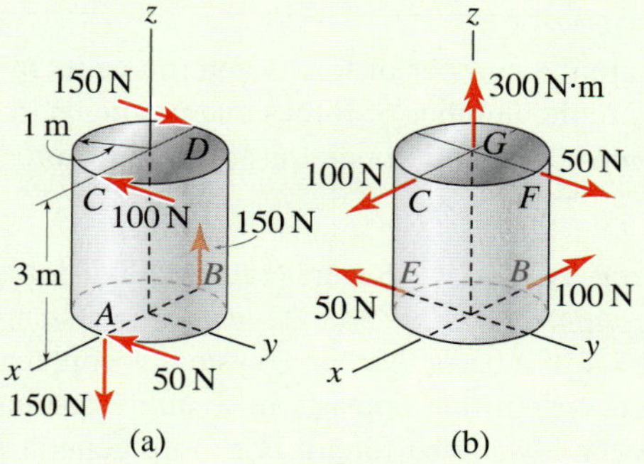

Figure P4.81

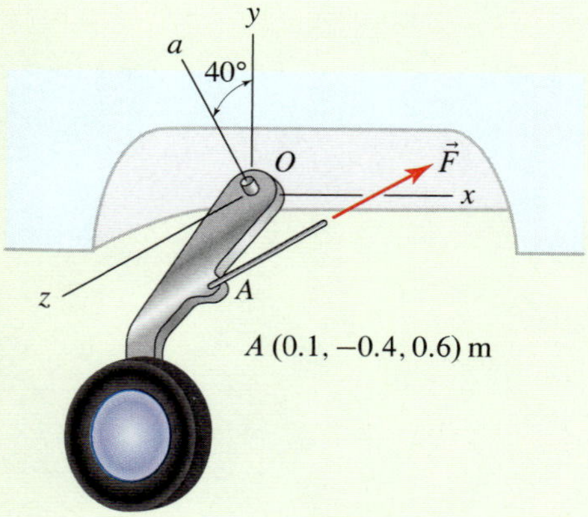

Figure P4.82

Problem 4.82

The landing gear for a fighter jet rotates about line a so that it retracts into the fuselage. Point A has the coordinates given, line a lies in the yz plane, and $\vec{F} = (300\,\hat{\imath} + 400\,\hat{\jmath} - 200\,\hat{k})$ N.

(a) Determine the moment of $\vec{F}$ about line a.

(b) Determine a new direction for line a so that the moment of $\vec{F}$ about this line is as large as possible.

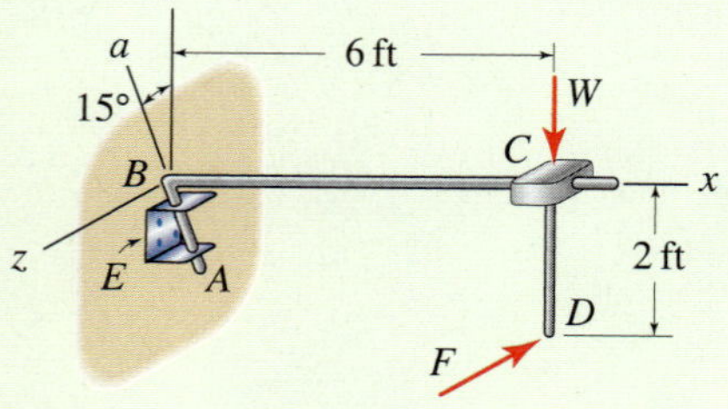

Figure P4.83

Problem 4.83

The device shown is a pointer that mounts on the front of a tractor to help its operator position the tractor relative to other rows of seeds that have been planted. Bracket E is bolted to the front of the tractor, which drives in the z direction. The bracket supports the bent boom ABC, the end C of which has a weight W and a pointer CD. The boom is allowed to rotate about line a, which lies in the yz plane, so that if the boom strikes an obstruction, the boom will rotate backward and upward to help avoid damage to it. If $W = 50$ lb, determine the force F, which acts in the $-z$ direction, that will cause the boom to begin rotating about line a.

Problem 4.84

A jack stand for supporting an automobile or truck during servicing is shown. The stand has three legs with equal radial positioning. The height h of the stand can be adjusted between 9 and 15 in. If the stand supports a 1500 lb vertical force, determine the smallest horizontal force F that will cause the moment about line AB to be zero and hence cause the stand to tip.

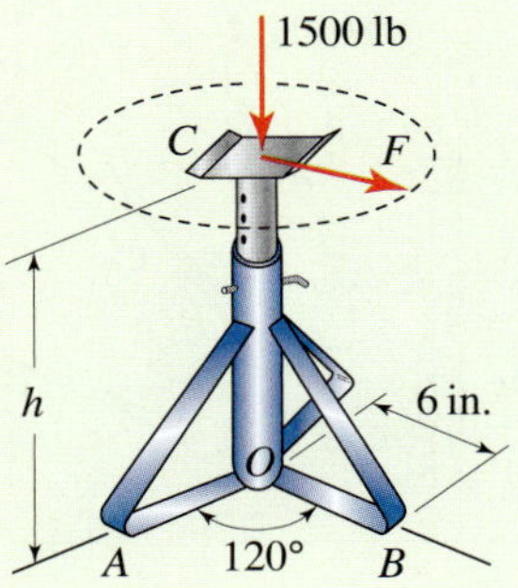

Figure P4.84

Problem 4.85

Rather than the traditional horizontal and vertical stabilizers, some aircraft such as the Bonanza 35 single-engine airplane (pictured) and the F-117 Stealth fighter feature a V tail. Points A and B are located at $A(112, 7, 0)$ mm and $B(62, -93, 0)$ mm, and direction a has direction angles $\theta_x = 150°$, $\theta_y = 60°$, and $\theta_z = 90°$. If the forces are $\vec{F}_1 = (40\,\hat{\imath} - 100\,\hat{\jmath} + 280\,\hat{k})$ N and $\vec{F}_2 = (50\,\hat{\jmath} + 120\,\hat{k})$ N, determine the resultant moment of $\vec{F}_1$ and $\vec{F}_2$ about line a.

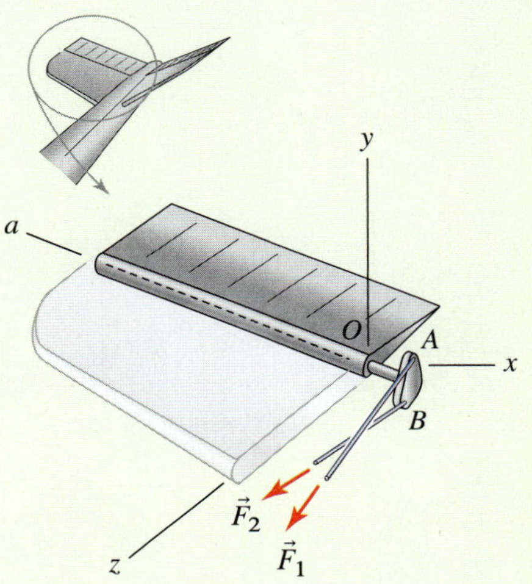

Figure P4.85

Problem 4.86

Determine values for P, Q, R, S, and M_D so that the two force systems shown are equivalent.

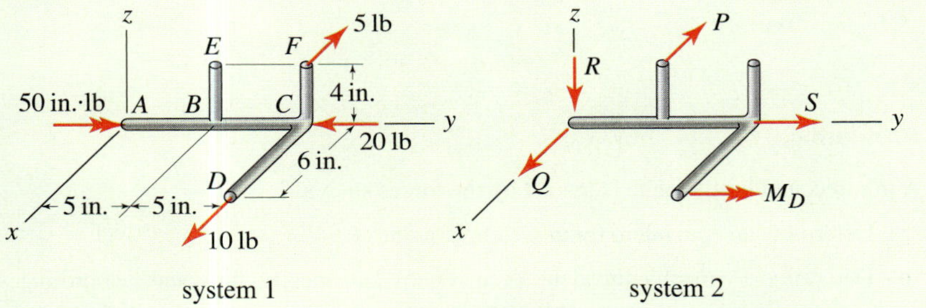

Figure P4.86

Problem 4.87

A beam is subjected to the three forces shown.

(a) Determine an equivalent force system at the midspan of the beam $x = 3$ m.

(b) Determine an equivalent force system consisting of a single force, and specify the x coordinate of the point where its line of action intersects the x axis.

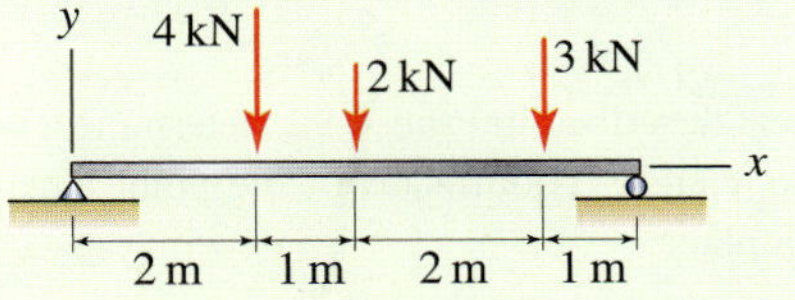

Figure P4.87

Problem 4.88

A speed control mechanism for a small gasoline engine is shown.

(a) Determine an equivalent force system at point O.

(b) Determine an equivalent force system consisting of a single force, and specify the y coordinate of the point where its line of action intersects the y axis.

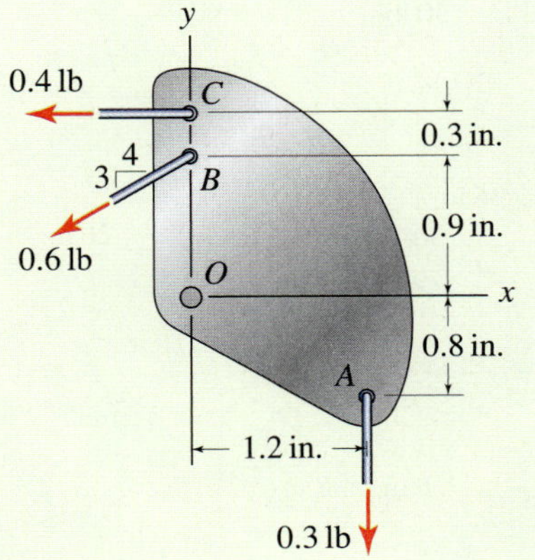

Figure P4.88

Problem 4.89

A *weldment* is a structure or component built by welding an assembly of pieces together. The weldment shown is subjected to three cable forces. Determine an equivalent force system consisting of a single force, and specify the x coordinate of the point where its line of action intersects the x axis.

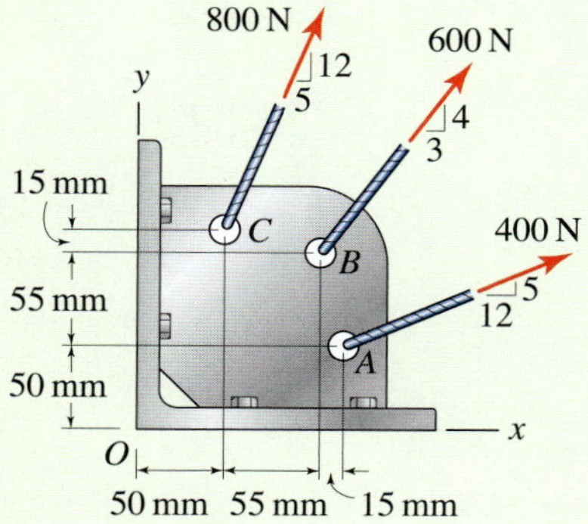

Figure P4.89

Problem 4.90

The wing of a jet supports a 900 lb force due to weight of an engine and 300 lb and 600 lb forces due to weight of fuel.

(a) Determine an equivalent force system at point A.

(b) Determine an equivalent force system consisting of a single force, and specify the x and z coordinates of the point where its line of action intersects the xz plane.

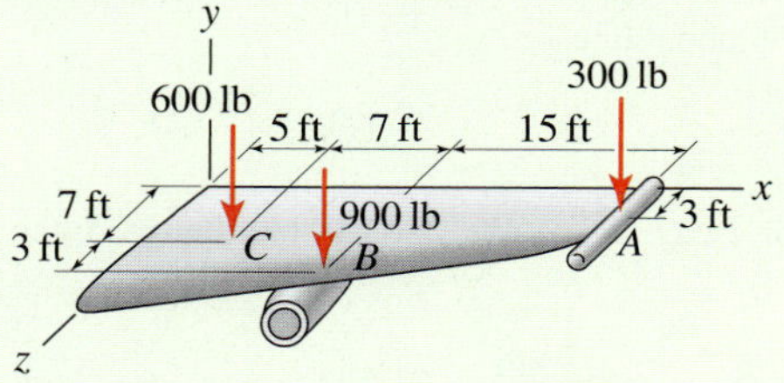

Figure P4.90

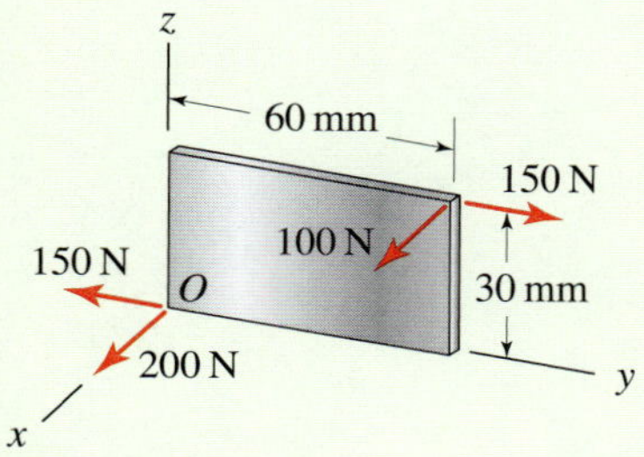

Figure P4.91

Problem 4.91

A thin rectangular flat plate is loaded by the forces shown.

(a) Determine an equivalent force system at point O.

(b) Determine a wrench equivalent force system, and specify the y and z coordinates of the point where the wrench's line of action intersects the yz plane.

Problem 4.92

A seat of a roller coaster is subjected to the forces shown during a turn. The force at A is vertical, the forces at B and C are parallel to the x and z directions, respectively, and the forces at D and E lie in planes parallel to the yz plane. Determine an equivalent force system at point O, where the seat is attached to the car.

Problem 4.93

For the roller coaster seat described in Prob. 4.92, determine a wrench equivalent force system and specify the x and z coordinates of the point where the wrench's line of action intersects the xz plane.

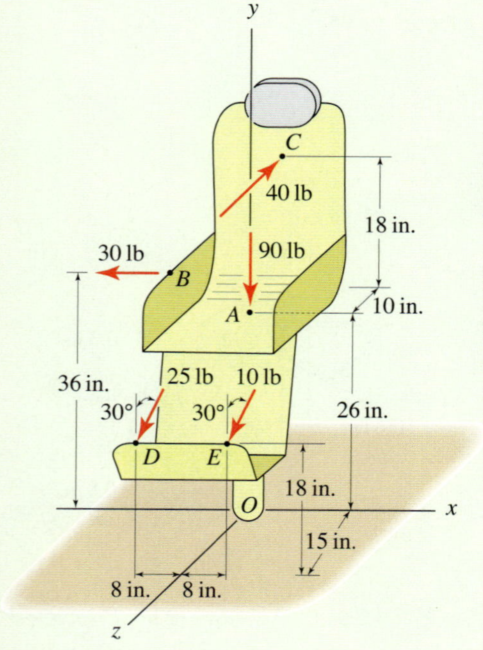

Figure P4.92 and P4.93

5 Equilibrium of Bodies

This chapter discusses equilibrium of a single rigid body. The rigid body may be a single object, or may be an assemblage of numerous members whose arrangement is such that the assemblage as a whole can be treated as a single rigid body. Because all materials are deformable, there are no true rigid bodies in nature. Nonetheless, it is rather remarkable how often a rigid body idealization can be used.

5.1 Equations of Equilibrium

We begin this chapter with a general discussion of the equations of equilibrium for bodies in three dimensions. Subsequent sections consider applications to both two-and three-dimensional problems.

Shown in Fig. 5.1 are objects whose equilibrium we will analyze. The object may be a single body of solid material as shown in Fig. 5.1(a) or may consist of an arrangement of numerous members as shown in Fig. 5.1(b). In either case, we idealize the object to be a *rigid body* so that the distance between any two points within the object remains constant under all circumstances of loading and/or motion. The object will have forces and/or moments applied to it, and will often also have *supports* that fix its position in space. Note that there are many objects whose equilibrium we are interested in that are unsupported or are only partially supported against motion. Examples include aerospace vehicles, ships, automobiles rolling on a highway, moving components in machines, and so on. Our objective in this chapter is to determine conditions under which equilibrium of a single rigid body is obtained. In the case of objects that consist of an assemblage of numerous members, we may also be interested in the forces supported by the individual members, and this topic is taken up in Chapter 6. Similarly, for objects that consist of solid

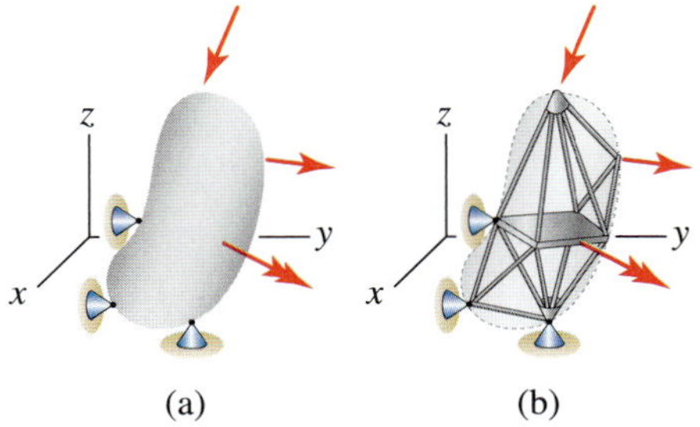

Figure 5.1
Examples of objects that may be idealized as rigid bodies.

material only, we may also be interested in the internal forces that develop within the material, and this topic is taken up in Chapter 8.

The equations governing the static equilibrium of a body are

$$\sum \vec{F} = \vec{0} \quad \text{and} \quad \sum \vec{M}_P = \vec{0} \tag{5.1}$$

where the summations include all forces and moments that are applied to the body, and the moments are evaluated about any convenient point P that we may select. In scalar form, Eq. (5.1) is

$$\begin{aligned} \sum F_x &= 0 & \qquad & \sum M_{Px} = 0 \\ \sum F_y &= 0 & \text{and} \qquad & \sum M_{Py} = 0 \\ \sum F_z &= 0 & & \sum M_{Pz} = 0. \end{aligned} \tag{5.2}$$

The force equilibrium equations $\sum \vec{F} = \vec{0}$ are identical to those used for equilibrium of particles. The *moment equilibrium equations* $\sum \vec{M}_P = \vec{0}$ are new, and the rationale for these is that if they are not satisfied, the body will undergo angular accelerations due to the twisting action of moments. Both equations are postulates, and no fundamental proof of their validity exists. Rather, we must accept these as laws that nature follows.

5.2 Equilibrium of Rigid Bodies in Two Dimensions

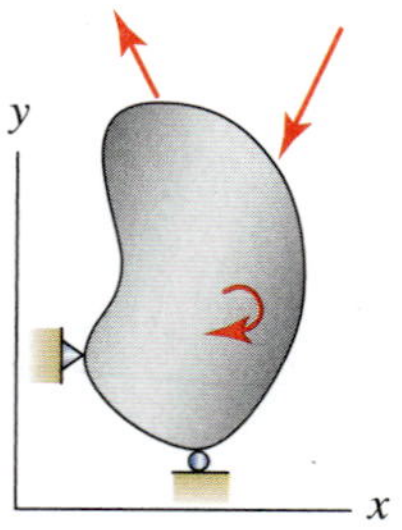

Figure 5.2
An object in two dimensions in equilibrium.

The object shown in Fig. 5.2 can be modeled using two dimensions if all forces, including the forces applied by the supports to the object, lie in the same plane and all moments have direction perpendicular to this plane. In two dimensions, with x and y being the in-plane coordinates as shown in Fig. 5.2, the equations $\sum F_z = 0$, $\sum M_{Px} = 0$, and $\sum M_{Py} = 0$ in Eq. (5.2) are always satisfied, leaving

$$\sum F_x = 0, \quad \sum F_y = 0, \quad \text{and} \quad \sum M_{Pz} = 0. \tag{5.3}$$

For brevity, in the remainder of this section, as well as for two-dimensional problems in general, we will drop the z subscript in the moment equilibrium equation. As stated above, the summations in Eq. (5.3) include *all* forces and moments applied to the body, including the *reaction forces*.

Reactions

A *reaction* is a force or moment exerted by a support on a structure. Supports and the reactions they produce take many forms as shown in Fig. 5.3. In all cases, the reactions for a particular support may be determined by considering the motion the support prevents. That is, if a support prevents translation (or rotation) in a certain direction, it can do so only by producing a reaction force

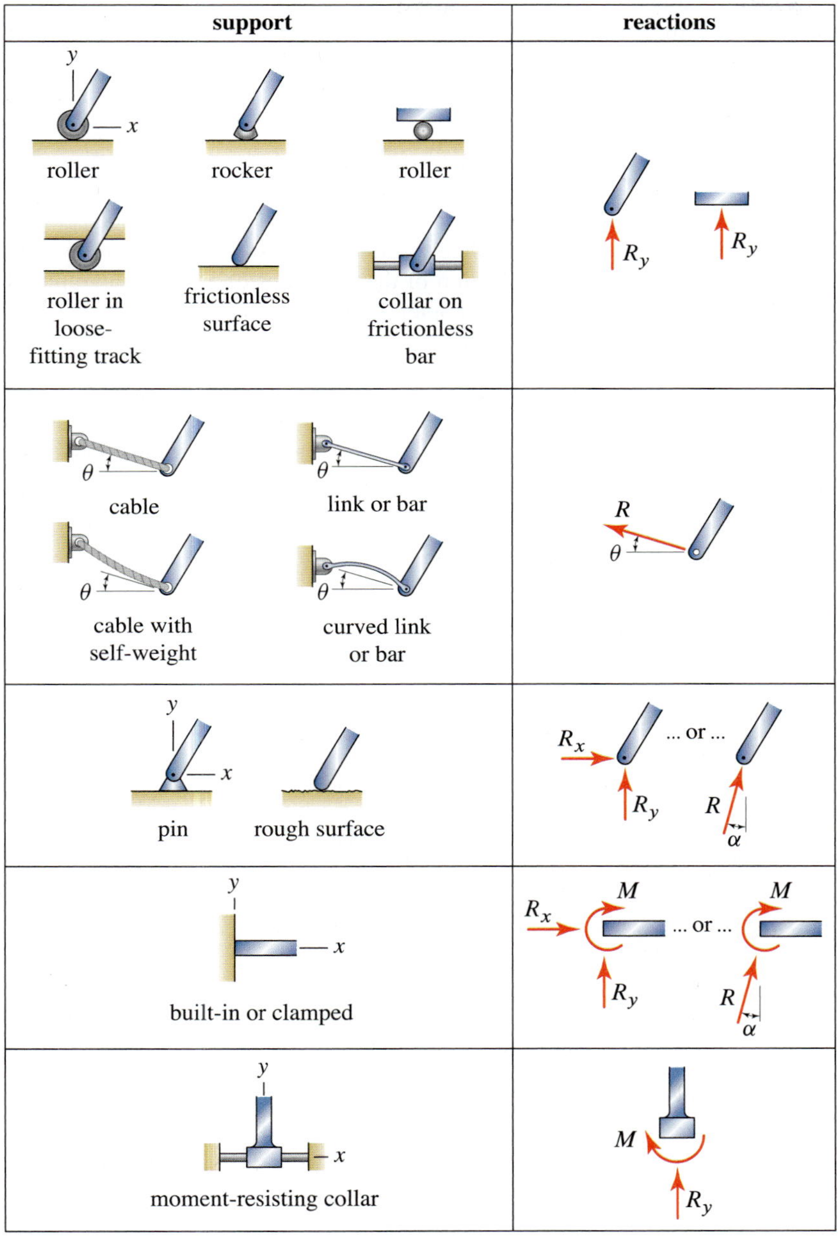

Figure 5.3. Common supports for bodies in two dimensions and their associated reaction forces. Cable, link, and bar support members are assumed weightless except where otherwise noted.

(or moment) in that direction. For example, in the case of a pin support, the pin prevents translation of the bar in both the x and y directions and thus must produce reaction forces in each of these directions. The pin does not prevent rotation of the bar, and hence there is no moment reaction. It is not necessary to memorize the reactions shown in Fig. 5.3. Rather, you should reconstruct these as needed.

Free body diagram (FBD)

A free body diagram (FBD) is a sketch of a body or structure that includes all forces and moments that are applied to the body or structure. The FBD is an essential aid for accurate application of the equations of equilibrium, and all of the remarks made on FBDs in Chapter 3 are applicable here.

Procedure for Drawing FBDs

1. Decide on the body (or portion of a body) whose equilibrium is to be analyzed.
2. Imagine this body is "cut" completely free (separated) from its environment. That is:
   - In 2D, think of a closed line that completely encircles the body.
   - In 3D, think of a closed surface that completely surrounds the body.
3. Sketch the body.
4. Sketch the forces:
   (a) Sketch the forces that are applied to the body by the environment (e.g., weight).
   (b) Wherever the cut passes through a structural member, sketch the forces that occur at that location.
   (c) Wherever the cut passes through a support (i.e., where a support is removed from the body), sketch the reaction forces and moments that occur at that location.
5. Sketch the coordinate system to be used. Add pertinent dimensions and angles to the FBD to fully define the locations and orientations of all forces.

The order in which the forces are sketched in Step 4 is irrelevant. For complicated FBDs, it may be difficult to include all of the dimensions and/or angles in Step 5. When this is the case, some of this information may be obtained from a different sketch.

In this chapter we consider equilibrium of an entire body or structure only. Hence, when we construct the FBD, the cut taken in Step 2 of the above procedure will encompass the entire object. Thus, in this chapter the forces cited in Step 4(b) do not arise. Occasionally, objects are supported by cables and/or bars, and when cutting through these supports to draw a FBD, we categorize their forces as reactions.

Once the FBD is drawn, equilibrium equations may be written and solved to obtain the solution to the problem. In this process, drawing the FBD is the most important step, since once a proper FBD is available, writing the equilibrium equations and solving these are routine. The solution of a typical problem is outlined in the following example.

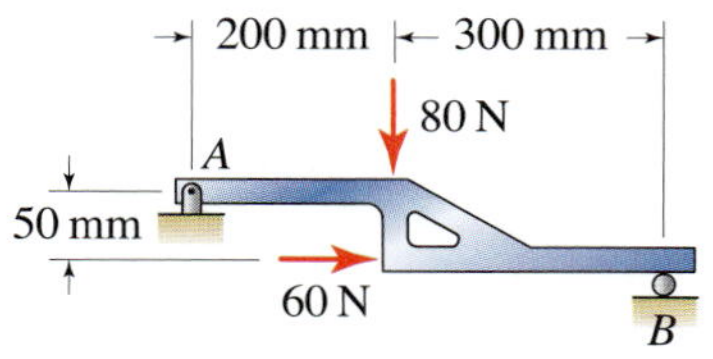

Figure 5.4
A structure with supports at points A and B.

Mini-Example. Determine the support reactions for the structure shown in Fig. 5.4.

Solution. The completed FBD is shown in Fig. 5.5 and it is constructed as follows. We first sketch the structure and then choose an xy coordinate system. At each support that is cut through (or removed from the structure) we introduce the appropriate reactions. Thus, for the pin support at A we introduce reactions A_x and A_y, and at the roller support at B we introduce reaction B_y.

Next, we use Eq. (5.3) to write the equilibrium equations

$$\sum F_x = 0: \qquad A_x + 60\text{ N} = 0, \tag{5.4}$$

$$\sum F_y = 0: \qquad A_y + B_y - 80\text{ N} = 0, \tag{5.5}$$

$$\sum M_A = 0: \qquad B_y(500\text{ mm}) - (80\text{ N})(200\text{ mm}) + (60\text{ N})(50\text{ mm}) = 0. \tag{5.6}$$

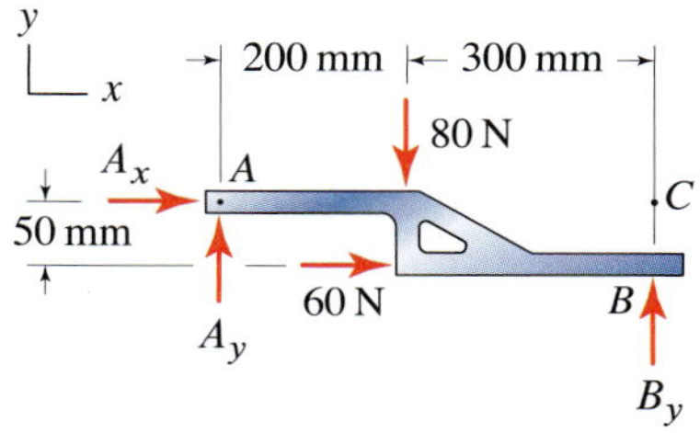

Figure 5.5
Free body diagram showing all forces applied to the object.

Solving Eqs. (5.4)–(5.6) provides

$$A_x = -60\text{ N}, \quad A_y = 54\text{ N}, \quad \text{and} \quad B_y = 26\text{ N}. \tag{5.7}$$

Helpful Information

Direction for moment summation. When writing moment equilibrium equations in two-dimensional problems, such as Eq. (5.6), we will always take *counterclockwise* to be the *positive* direction for moments.

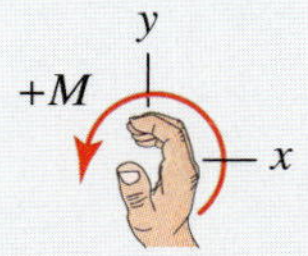

Figure 5.6

This choice is consistent with the right-hand rule for the xy coordinate system shown.

Remarks. The following remarks pertain to the foregoing example.

- **Coordinate system.** While we will most often use a coordinate system whose directions are horizontal and vertical, occasionally other choices may be more convenient and will be used.

- **Direction of reactions in FBD.** When putting reaction forces and moments in the FBD, we often do not know the actual directions these forces will have until after the equilibrium equations are solved. In Fig. 5.5 we have elected to take A_x, A_y, and B_y to be positive in their positive coordinate directions. After solving the equilibrium equations, we may find some of these are negative (such as A_x above), meaning they actually act in the directions opposite those shown in Fig. 5.5. When you encounter such a situation, you should resist the temptation to revise the direction of those reactions in the FBD, as this is likely to lead to errors since all equilibrium equations also need to be revised. Rather, simply allow those reactions to have negative values.

- **Number of unknowns.** After we draw the FBD, it is a good idea to count the number of unknowns. In Fig. 5.5, there are three unknowns, namely, A_x, A_y, and B_y; and since there are three equilibrium equations, we expect to have a determinate system of algebraic equations that can be solved to obtain a unique solution for the unknowns. If more or fewer than three unknowns are present in the FBD and you rule out any errors in your FBD, then the object you are modeling is partially fixed or is statically indeterminate, as discussed in the next section.

- **Selection of moment summation point.** In Eq. (5.6) moments were evaluated about point A. While any point can be used, the merit of point A is that the unknown reactions at A produce no moment about point A and hence do not enter the moment equilibrium equation, leaving B_y as the only unknown in that expression, which is easily solved for. When you sum moments, it is helpful to use a moment summation point that omits as many unknowns as possible.

- **Alternative equilibrium equations.** Rather than using $\sum F_y = 0$, Eq. (5.5), we could use in its place the moment equilibrium equation

$$\sum M_C = 0: \quad -A_y(500 \text{ mm}) + (80 \text{ N})(300 \text{ mm}) + (60 \text{ N})(50 \text{ mm}) = 0 \tag{5.8}$$

where point C is located at the intersection of the lines of action of A_x and B_y, as shown in Fig. 5.5. Observe that the moment summation point does not need to lie on the object. The merit of using point C for moment summation is that it will provide an equation that contains A_y as the only unknown. Solving Eqs. (5.4), (5.6), and (5.8) provides the same results given in Eq. (5.7). Additional comments on the use of alternative equilibrium equations follow.

Alternative equilibrium equations

As demonstrated in the foregoing example, it is possible to replace either or both of the $\sum F_x = 0$ and $\sum F_y = 0$ equations with moment equilibrium equations where moment summation is taken about different points. The merit of doing this is that, with appropriate selection of moment summation points, an equation system that is easier to solve can often be obtained.

While it might seem as if "trading" a $\sum F_x = 0$ and/or $\sum F_y = 0$ equation for a moment equilibrium equation violates the fundamental principle of equilibrium stated in Eq. (5.3), in reality, writing multiple moment equilibrium equations, if properly done, still ensures $\sum F_x = 0$ and $\sum F_y = 0$. To explain further, consider the three equilibrium equations $\sum F_x = 0$, $\sum F_y = 0$, and $\sum M_P = 0$. Each of these equations may be multiplied by any nonzero number, and the resulting equations may be added or subtracted. None of these manipulations change the basic fact that there are still three independent equations whose solutions are the same as those of the original equations. Replacing the $\sum F_x = 0$ (or $\sum F_y = 0$) equation with the moment equilibrium equation $\sum M_A = 0$, subject to the minor restrictions discussed below on where point A may be located, is identical to multiplying the $\sum F_x = 0$ (or $\sum F_y = 0$) equation by a suitable nonzero number and adding the result to the $\sum M_P = 0$ equation. To illustrate, in the foregoing example, the $\sum M_C = 0$ expression, Eq. (5.8), is identical to Eq. (5.6) after the product of Eq. (5.5) and 500 mm is subtracted from it.

The various alternative equilibrium equations that may be used in place of Eq. (5.3) are

$$\sum F_x = 0, \quad \sum M_A = 0, \quad \text{and} \quad \sum M_P = 0. \tag{5.9}$$

Points A and P must have different x coordinates.

$$\sum M_A = 0, \quad \sum F_y = 0, \quad \text{and} \quad \sum M_P = 0. \tag{5.10}$$

Points A and P must have different y coordinates.

$$\sum M_A = 0, \quad \sum M_B = 0, \quad \text{and} \quad \sum M_P = 0. \tag{5.11}$$

Points A, B, and P may not lie on the same line.

The restrictions on locations of moment summation points given in Eqs. (5.9) through (5.11) ensure that the three equilibrium equations that are written are independent and hence have a unique solution. In practice, however, these rules usually do not need to be explicitly consulted, as the strategy you are likely to use to write your equations to find your unknowns is almost always sufficient to ensure that the rules cited above are adhered to.

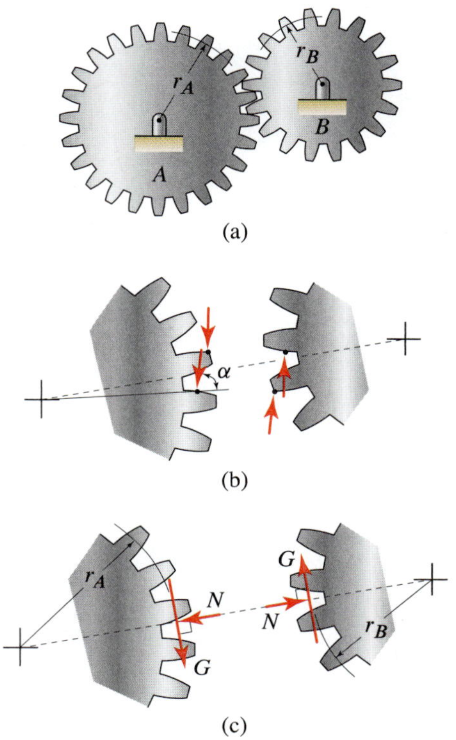

Figure 5.7
(a) Meshing spur gears with pitch radii r_A and r_B. (b) Multiple teeth on each gear are usually in contact at the same time, and this example shows two teeth on each gear in simultaneous contact. The force on a particular tooth is not perpendicular to the gear's radius ($\alpha \neq 90°$). (c) Regardless of the number of teeth in contact, all tooth forces on a gear may be represented by a single tangential force G and a normal force N, positioned on the line connecting the gears' centers and at the pitch radius of each gear. For many applications, N is neglected.

Gears

Gears are manufactured in an enormous variety of shapes and forms, and methods of analysis and design of gears are a specialized topic. In this book we consider the simplest and most common type of gears, called *spur gears*, as shown in Fig. 5.7(a). Radii r_A and r_B are the effective radii of the gears and are called the *pitch radii*. In general, two meshing gears have multiple teeth in contact at the same time. If, for example, gear A shown in Fig. 5.7(a) rotates counterclockwise and applies power to gear B, then the forces between the contacting teeth are as shown in Fig. 5.7(b). The shape of gear teeth is such that the force supported by an individual tooth is not perpendicular to the gear's radius at the point of contact (i.e., angle α shown in Fig. 5.7(b) is not 90°). All of the forces acting on the teeth of a particular gear can be vectorially summed to obtain an equivalent force system that represents all the tooth forces as shown in Fig. 5.7(c). This force system is positioned on the line connecting the gears' centers and is at the pitch radius of each gear. Note that in addition to the tangential force G, there is a normal force N that tends to push the gears apart. While N is smaller than G, it is not insignificant and its size as a fraction of G depends on the shape of the gears.* When dealing with gears in this book, we will usually neglect N. While not perfect, this is a common simplification that is useful and adequate for many purposes. Example problems at the end of this section elaborate further on this issue.

Interesting Fact

Standards for gears. The American Gear Manufacturers Association (AGMA) is an organization of gear manufacturers from around the world that develops standards for the manufacture, performance, and use of gears. When possible, an engineer will select gears from the standard shapes and sizes that are commercially available. Sometimes special purpose gears are needed, and these must be custom designed and manufactured.

* One of the primary measures of gear shape is the *pressure angle*. The pressure angle is selected by a gear designer (among several industry-standard values) so that the gear has acceptable force transmission ability, noise, life, and so on. The pressure angle also determines how large N is as a fraction of G.

Examples of correct FBDs

Figure 5.8 shows several examples of properly constructed FBDs. Comments on the construction of these FBDs follow.

Bulletin board. After sketching the bulletin board, we apply the 40 N weight through its center of gravity at point C. At point A, the support does not permit horizontal and vertical motion, and therefore there are two reactions A_x and A_y. At point B, the board cannot move perpendicular to the interface between the screw and wire hanger, and therefore there is a reaction B in the direction shown.

Lifting machine. After sketching the frame of the lifting machine, we apply the 200 lb force and the 40 lb weight through its center of gravity at point C. At point A, the gear applies a tangential force A_y to the frame and the normal force between the gear and frame is neglected. At point B the support does not permit horizontal translation and rotation, and therefore there are two reactions consisting of force B_x and moment M_B.

Folding desk. We are instructed to neglect the weights of members. Member AB prevents motion of point A in the direction along line AB, and hence there must be a reaction force F_{AB} whose line of action is line AB. Similarly, member CD produces a reaction F_{CD} whose line of action is line CD. At point E we assume the contact is smooth, in which case there is one reaction force E that is perpendicular to the interface between the desk and the seat in front of it. While we are instructed that a person can apply a 75 lb force, no information is given on the position or orientation of this force. Thus, we have chosen to give this force a position and orientation so that it maximizes the severity of the loading on the desk (i.e., it is placed at the edge of the desk, perpendicular to the desktop).

Examples of incorrect and/or incomplete FBDs

Figure 5.9 shows several examples of incorrect and/or incomplete FBDs. Comments on how these FBDs must be revised follow, but before reading these, you should study Fig. 5.9 to find as many of the needed corrections and/or additions on your own as possible.

Pickup truck

1. The 15 kN weight should be vertical.
2. The rear wheels should have a reaction force A_x in the x direction.

Hand truck

1. The reaction force at B should be directed from point B through the bearing of the wheel. We also note that if the bearing of the wheel is frictionless, then there will be no friction force at B.
2. The force F applied by the operator's hand is properly shown, but it should be noted that both its magnitude and direction are unknown.

examples of correct FBDs

Bulletin board. A bulletin board with 40 N weight and center of gravity at point *C* hangs on a wall by screws at points *A* and *B*. Draw the FBD for the bulletin board.

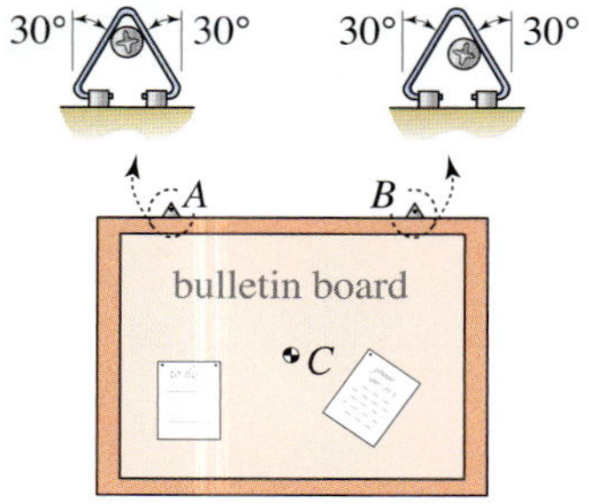

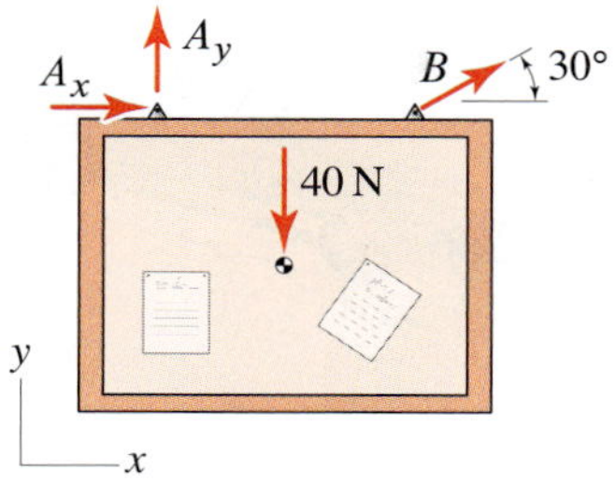

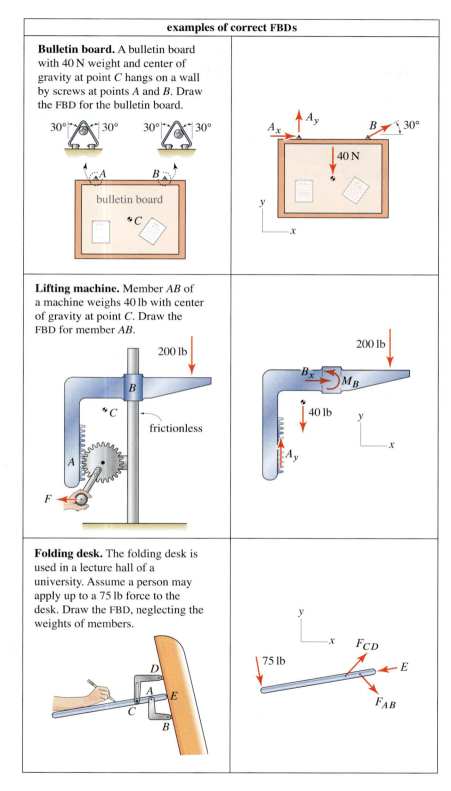

Lifting machine. Member *AB* of a machine weighs 40 lb with center of gravity at point *C*. Draw the FBD for member *AB*.

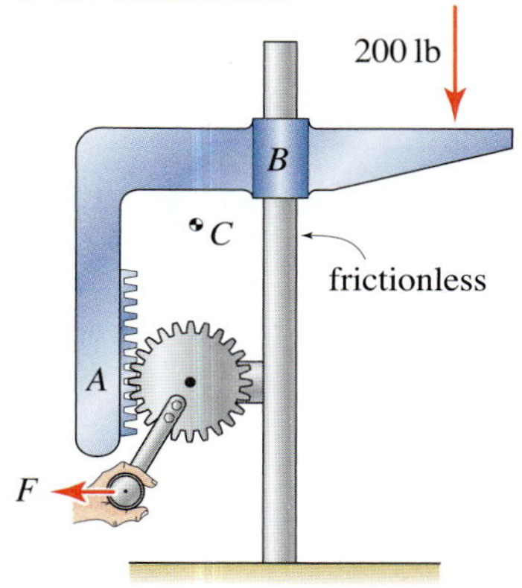

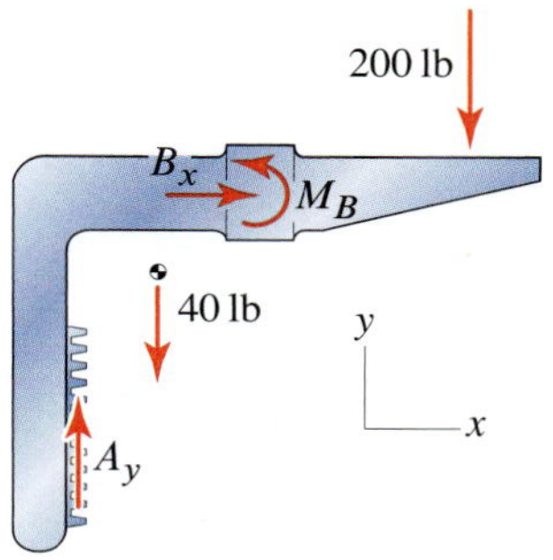

Folding desk. The folding desk is used in a lecture hall of a university. Assume a person may apply up to a 75 lb force to the desk. Draw the FBD, neglecting the weights of members.

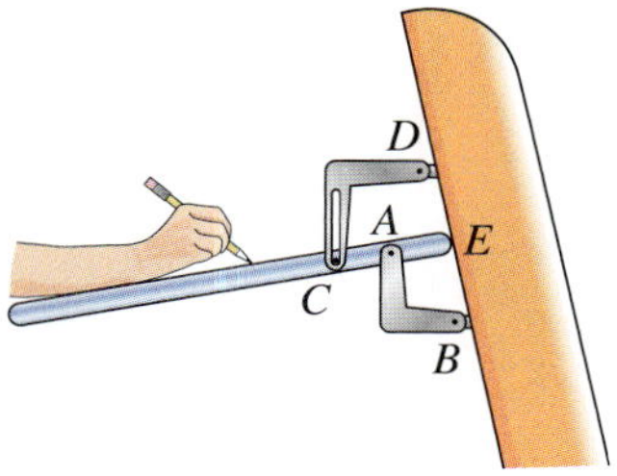

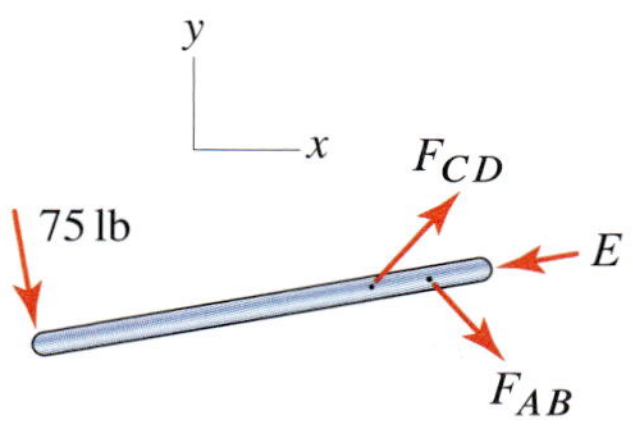

Figure 5.8. Examples of properly constructed FBDs.

examples of incorrect and/or incomplete FBDs	
Pickup truck. A rear-wheel-drive pickup truck drives up an incline with steady speed. The truck weighs 15 kN with center of gravity at point C. Draw the FBD for the truck. 	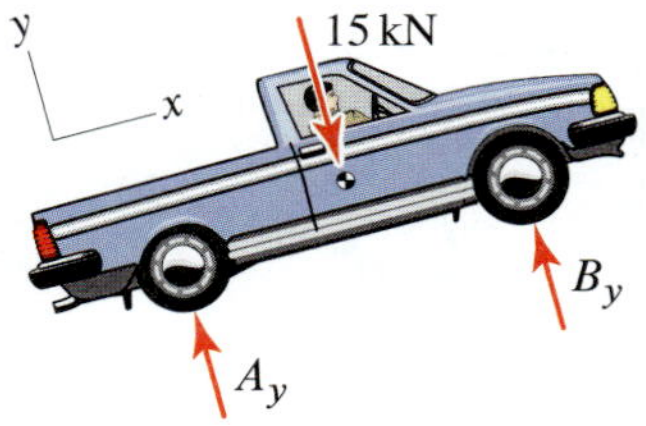
Hand truck. A hand truck with 75 N weight and center of gravity at point C is used to move a keg weighing 400 N with center of gravity at point D. Draw the FBD if the truck is on the verge of rolling over the bump at point B. 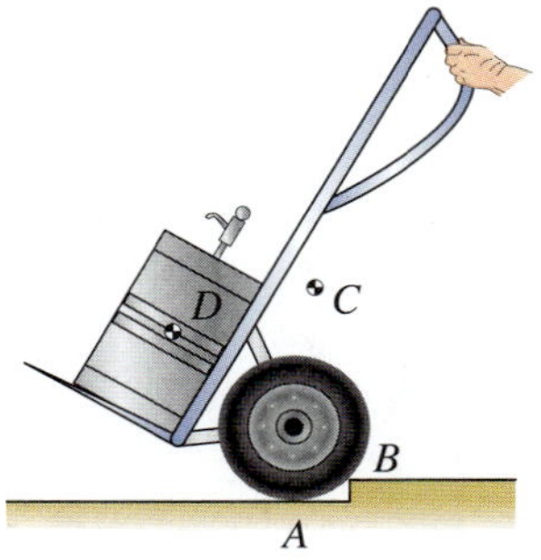	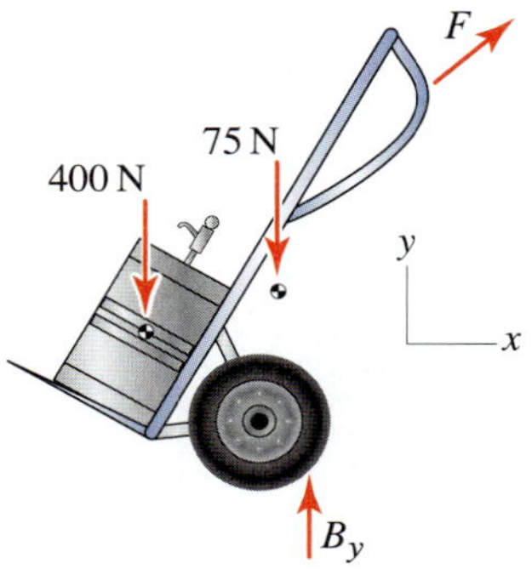
Machine control. Slotted bracket *EBD* is used to control the height of a machine at point D. Draw the FBD of member *EBD*, neglecting its weight. 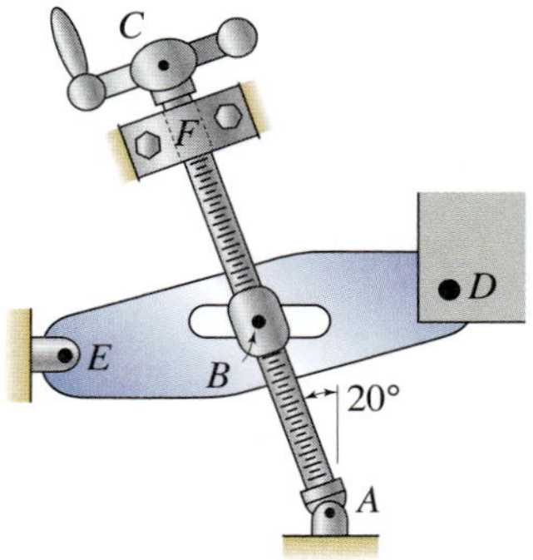	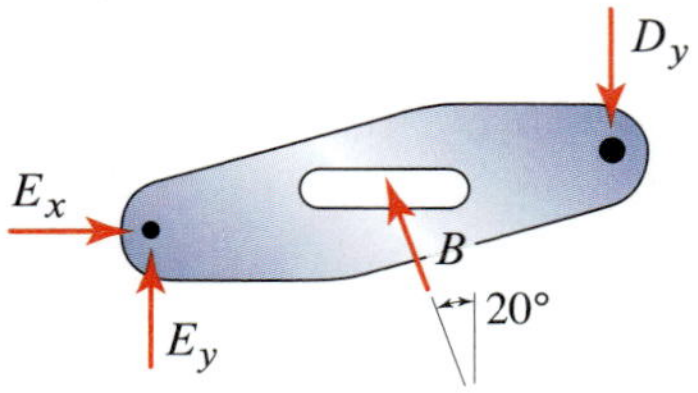

Figure 5.9. Examples of incorrect and/or incomplete FBDs.

Machine control

1. Although not really a deficiency with the FBD, a coordinate system should be selected and shown.

2. The reaction force at B should be perpendicular to the slot (i.e., vertical).

3. The pin at D should also have a horizontal force D_x.

End of Section Summary

In Sections 5.1 and 5.2, the equations governing static equilibrium of a rigid body were reviewed, and analysis procedures for bodies in two dimensions were discussed. Some of the key points are as follows:

- Both vector and scalar approaches can be used for problems in two dimensions. However, most often a scalar approach will be the most straightforward.

- The equilibrium equations for a body in two dimensions, with x and y being the in-plane directions, are $\sum F_x = 0$, $\sum F_y = 0$, and $\sum M_P = 0$ where P is the moment summation point you select.

- Alternative sets of equilibrium equations are available where the $\sum F_x = 0$ and/or $\sum F_y = 0$ equations are replaced by additional moment equilibrium equations, subject to certain restrictions on where the moment summation points may be located. In some problems, use of these alternative equilibrium equations will reduce the amount of algebra required to solve for the unknowns. It is important to remember that a body in two dimensions has only three independent equilibrium equations available.

- A *free body diagram* (FBD) is a sketch of a body and all of the forces and moments applied to it. The FBD is an essential tool to help ensure that all forces are accounted for when writing the equilibrium equations.

- Gears are common, and are often used in connection with shafts. Typically, when two gears mesh, multiple teeth from each gear are in simultaneous contact. However, for many purposes these forces may be replaced with an equivalent force system consisting of a tangential force, and possibly a normal force, as shown in Fig. 5.7.

EXAMPLE 5.1 *Curved Link Support*

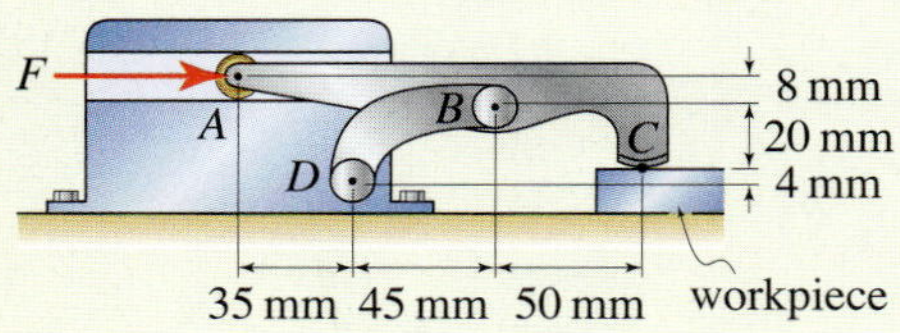

Figure 1

A device for clamping a flat workpiece in a machine tool is shown. If a 200 N clamping force is to be generated at C, and if the contact at C is smooth (no friction), determine the force F required and the reaction at B.

SOLUTION

Road Map Determining the force F required to produce a 200 N clamping force will be accomplished by analyzing the equilibrium of the clamp ABC. Member BD is viewed as a support for the clamp (as shown in Fig. 5.3), and thus it is not necessary to explicitly address its equilibrium.

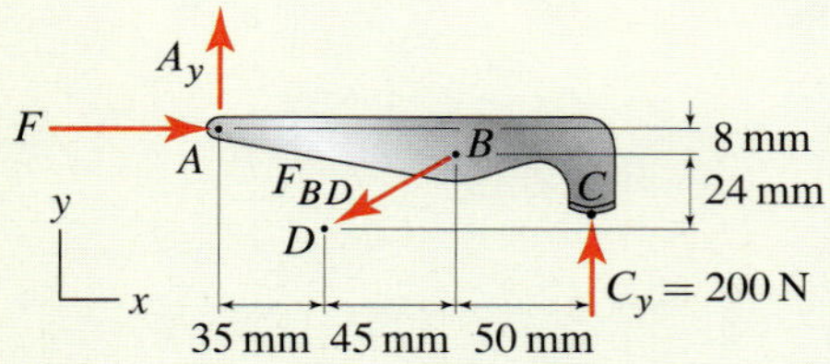

Figure 2
Free body diagram.

Modeling The completed FBD for the clamp is shown in Fig. 2 and it is constructed as follows. We first sketch clamp ABC and then choose an xy coordinate system. The roller support at A prevents translation of the clamp in the y direction, and thus there must be a y direction reaction A_y that enforces this constraint. Similarly, at the smooth support at C the only reaction is C_y, which we desire to be 200 N.

The curved link supports the clamp at point B, and the reaction forces and/or moments at point B are determined by consulting Fig. 5.3 on p. 247, or better yet by the discussion given in Fig. 3. Hence, the reaction force due to the link is F_{BD} with direction from points B to D.

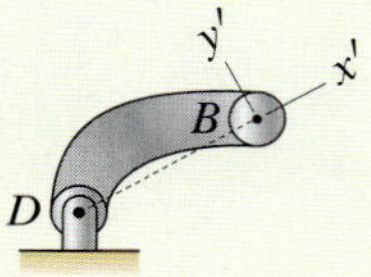

Figure 3
The curved link prevents translation of point B in the x' direction, and therefore the link must apply a reaction force to the clamp in this direction. Translation of point B in the y' direction is possible, therefore there is no reaction force in this direction. Further, rotation of the clamp about point B is possible, therefore there is no moment reaction.

Governing Equations & Computation Prior to writing equilibrium equations, we resolve the reaction at B into horizontal and vertical components so that moment arms are more easily obtained. The revised FBD is shown in Fig. 4.

When writing equilibrium equations, we usually try to write them so that each equation contains only one unknown, which may then be found immediately. When this is possible, simultaneous algebraic equations do not need to be solved. In Fig. 4, we observe that both the $\sum F_x = 0$ and $\sum F_y = 0$ equations involve two unknowns each, and thus these equations are not especially convenient to start with, although no harm is done if you do this. We next consider $\sum M = 0$ and we look for a point where the lines of action of two (or more) unknown forces intersect; in Fig. 4 point A is such a point.* Thus, writing $\sum M_A = 0$, with positive moments taken counterclockwise, we immediately find F_{BD} as follows:

$$\sum M_A = 0: \quad -F_{BD}\left(\frac{45}{51}\right)(8\,\text{mm}) - F_{BD}\left(\frac{24}{51}\right)(80\,\text{mm})$$

$$+ (200\,\text{N})(130\,\text{mm}) = 0 \quad \Rightarrow \quad \boxed{F_{BD} = 581.6\,\text{N.}} \qquad (1)$$

Now that F_{BD} is known, the equations $\sum F_x = 0$ and $\sum F_y = 0$ may be written in any order to obtain the remaining unknowns

$$\sum F_x = 0: \qquad F - F_{BD}\left(\frac{45}{51}\right) = 0 \Rightarrow \quad \boxed{F = 513.2\,\text{N,}} \qquad (2)$$

$$\sum F_y = 0: \qquad A_y - F_{BD}\left(\frac{24}{51}\right) + 200\,\text{N} = 0 \Rightarrow \quad \boxed{A_y = 73.7\,\text{N.}} \qquad (3)$$

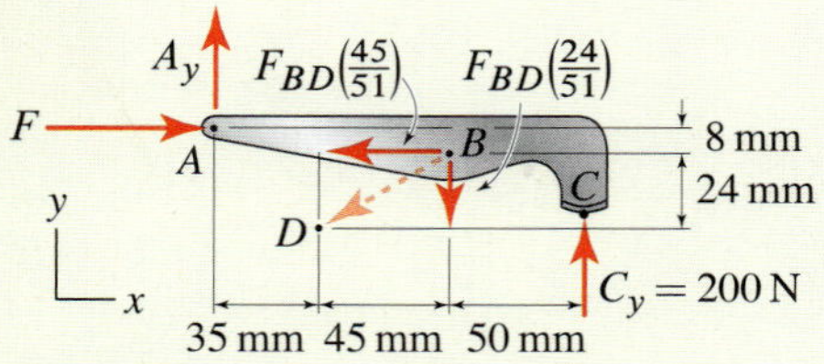

Figure 4
Revised FBD with F_{BD} resolved into horizontal and vertical components $\left[\text{where } \sqrt{(45)^2 + (24)^2} = 51\right]$ so that moment arms are more easily obtained.

* The intersection of the lines of action of A_y and F_{BD} shown in Fig. 2 is another such point, as is the intersection of the lines of action of F and F_{BD}. However, determining the locations of these intersection points requires some work, and for this reason, point A is a better choice.

Discussion & Verification

- Intuitively, we expect the solution of our equilibrium equations to show that $F > 0$, and indeed this is obtained. If we had obtained a negative value for F, we would immediately suspect an error and would correct this before continuing with our solution.
- Once we verify that our solution passes the preceding casual but essential check, we should then verify that *all* unknowns (F, F_{BD} and A_y) satisfy *all* of the equilibrium equations. If any of the equilibrium equations is not satisfied, then a math error has been made. However, this check does not guarantee that the equilibrium equations have been accurately written. Thus, it is essential that the FBD be accurately drawn and that the equilibrium equations accurately include all forces and moments from the FBD.
- An additional check on our solution that is usually easy to carry out is to use the FBD of Fig. 2 or 4 to evaluate the moment of all forces about additional points to verify that moment equilibrium is satisfied. For example, if we evaluate moments about point C in Fig. 4, a correct solution must show that $\sum M_C = 0$.

EXAMPLE 5.2 *Two-dimensional Idealization of a Three-dimensional Problem*

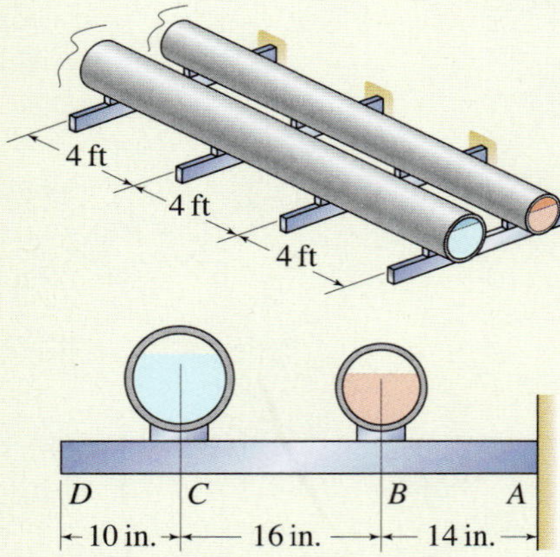

Figure 1

A large number of identical uniform cantilever beams, each with 40 in. length and 10 lb/ft weight, are used to support 4 ft sections of pipes B and C. A typical beam $ABCD$ is shown. If pipes B and C plus their contents weigh 25 lb/ft and 35 lb/ft, respectively, determine the reactions at the built-in end of the cantilever beams.

SOLUTION

Road Map We will idealize this problem to be two-dimensional and will draw an FBD of one cantilever beam, assuming it supports the weights of a 4 ft length of each pipe plus its own weight. Our goal is to determine the reactions at the built-in support.

Modeling The FBD is shown in Fig. 2 and is constructed as follows. Beam $ABCD$ is sketched first, and an xy coordinate system is chosen. The reactions at the built-in support at point A may be determined by consulting Fig. 5.3, but it easier to simply construct these by considering the motion that the support prevents. That is, the support at A prevents horizontal translation of the beam, and hence there must be a reaction A_x. The support also prevents vertical translation of the beam, and hence there must be a reaction A_y. Finally, the support prevents rotation of the beam, and hence there must be a moment reaction M_A. The beam's total weight is

$$W_E = \left(10\ \frac{\text{lb}}{\text{ft}}\right)(40\ \text{in.})\left(\frac{1\ \text{ft}}{12\ \text{in.}}\right) = 33.33\ \text{lb}, \tag{1}$$

and since the beam is *uniform*, meaning its cross-sectional dimensions and material are the same over its entire length, this weight is placed at the *center of gravity* of the beam, which is its midpoint E. The weights of 4 ft lengths of pipes B and C are

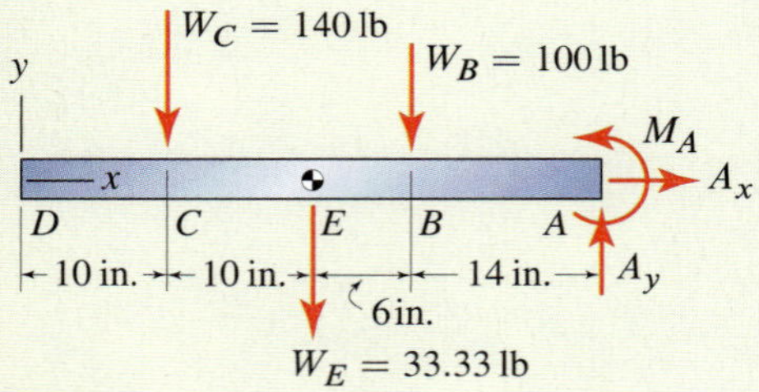

Figure 2
Free body diagram.

$$W_B = \left(25\ \frac{\text{lb}}{\text{ft}}\right)(4\ \text{ft}) = 100\ \text{lb}, \quad W_C = \left(35\ \frac{\text{lb}}{\text{ft}}\right)(4\ \text{ft}) = 140\ \text{lb}. \tag{2}$$

Governing Equations & Computation Writing equilibrium equations provides the reactions at A as

$$\sum F_x = 0: \quad A_x = 0 \tag{3}$$

$$\Rightarrow \boxed{A_x = 0,} \tag{4}$$

$$\sum F_y = 0: \quad A_y - 140\ \text{lb} - 33.33\ \text{lb} - 100\ \text{lb} = 0 \tag{5}$$

$$\Rightarrow \boxed{A_y = 273\ \text{lb},} \tag{6}$$

$$\sum M_A = 0: \quad M_A + (140\ \text{lb})(30\ \text{in.}) + (33.33\ \text{lb})(20\ \text{in.}) + (100\ \text{lb})(14\ \text{in.}) = 0 \tag{7}$$

$$\Rightarrow \boxed{M_A = -6270\ \text{in.}\cdot\text{lb} = -522\ \text{ft}\cdot\text{lb}.} \tag{8}$$

Discussion & Verification

- Because of the loading and geometry, we expect $A_y > 0$ and $M_A < 0$, which are obtained. We should also verify that A_x, A_y, and M_A satisfy all of the equilibrium equations.
- The biggest assumption made is that an individual beam is responsible for supporting the weight of a 4 ft length of each pipe only. Whether or not this assumption is valid cannot be verified with the limited data given here. However, the intent of using multiple beams to support the pipes is that each beam shares equally in supporting the weight. This assumption is reasonable if the two pipes are sufficiently flexible and/or all of the support beams are accurately aligned.

EXAMPLE 5.3 *Two-dimensional Idealization of a Three-dimensional Problem*

The rear door of a minivan is hinged at point A and is supported by two struts; one strut is between points B and C, and the second strut is immediately behind this on the opposite side of the door. If the door weighs 350 N with center of gravity at point D and it is desired that a 40 N vertical force applied by a person's hand at point E will begin closing the door, determine the force that each of the two struts must support and the reactions at the hinge.

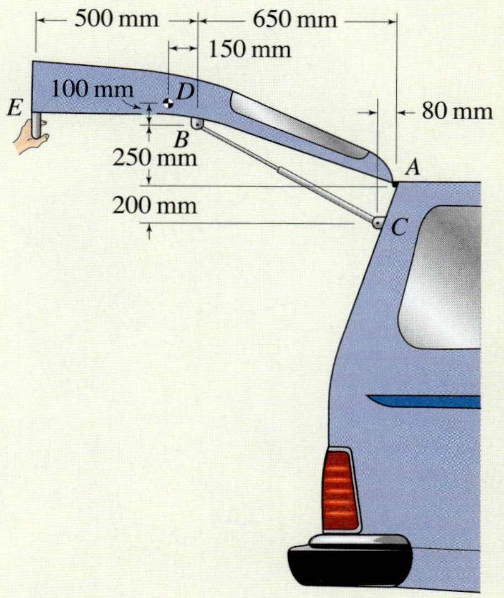

Figure 1

SOLUTION

Road Map Although the problem is really three-dimensional, a two-dimensional idealization is sufficient and will be used here. We will neglect the weights of the two struts since they are probably very small compared to the weight of the door.

Modeling The FBD is shown in Fig. 2 and is constructed as follows. The door is sketched first and an xy coordinate system is chosen. The person's hand at E applies a 40 N downward vertical force, and the 350 N weight of the door is a vertical force that acts through point D. The hinge (or pin) at A has horizontal and vertical reactions A_x and A_y. The force in *one* strut is F_{BC}, with a positive value corresponding to compression. Thus, the total force applied by the two struts is $2F_{BC}$. In Fig. 2, the horizontal and vertical components of the strut force are determined using similar triangles with the geometry shown.

Governing Equations & Computation Summing moments about point A is convenient because it will produce an equilibrium equation where F_{BC} is the only unknown:

$$\sum M_A = 0: \quad (40\text{ N})(1.150\text{ m}) + (350\text{ N})(0.800\text{ m})$$

$$- 2F_{BC}\left(\frac{450}{726.2}\right)(0.650\text{ m}) + 2F_{BC}\left(\frac{570}{726.2}\right)(0.250\text{ m}) = 0 \quad (1)$$

$$\Rightarrow \boxed{F_{BC} = 789.1\text{ N.}} \quad (2)$$

Thus, the force supported by each of the two struts is $F_{BC} = 789.1$ N.

The reactions at point A are found by writing the remaining two equilibrium equations

$$\sum F_x = 0: \quad -2F_{BC}\left(\frac{570}{726.2}\right) + A_x = 0 \quad (3)$$

$$\Rightarrow \boxed{A_x = 1239\text{ N,}} \quad (4)$$

$$\sum F_y = 0: \quad -40\text{ N} - 350\text{ N} + 2F_{BC}\left(\frac{450}{726.2}\right) + A_y = 0 \quad (5)$$

$$\Rightarrow \boxed{A_y = -588.0\text{ N.}} \quad (6)$$

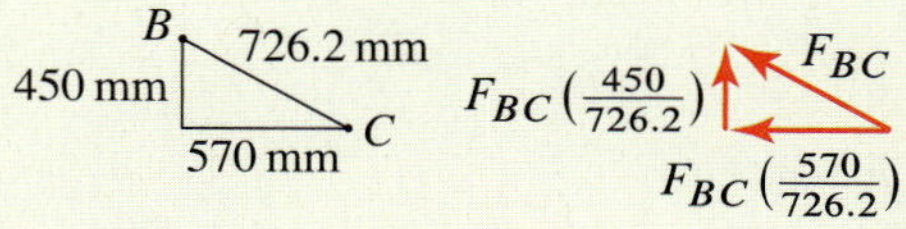

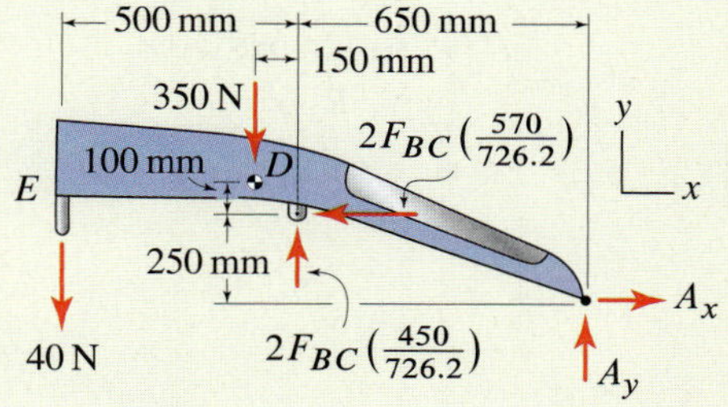

Figure 2
Free body diagram.

Discussion & Verification

- Because of the geometry and loading for this problem, we intuitively expect the struts to be in compression. Since the strut force F_{BC} was defined to be positive in compression, we expect the solution to Eq. (1) to give $F_{BC} > 0$, which it does.
- You should verify that the solutions are mathematically correct by substituting F_{BC}, A_x, and A_y into all equilibrium equations to check that each of them is satisfied. However, this check does not verify the accuracy of the equilibrium equations themselves, so it is essential that you draw accurate FBDs and check that your solution is reasonable.

EXAMPLE 5.4 *Numerous Roller Supports*

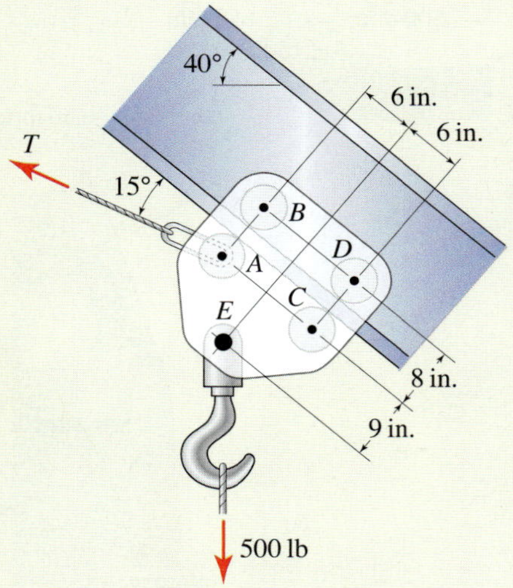

Figure 1

A trolley rolls on the flange of a fixed I beam to move a 500 lb vertical force. The trolley has a total of eight rollers; four of these are shown in Fig. 1 at points A, B, C, and D, and the remaining four are located immediately behind these points on the opposite side of the trolley. Thus, there is one pair of rollers located at A, another pair at B, and so on. The rollers at A and B are a loose fit so that only one pair will make contact with the flange of the I beam, and similarly for the rollers at C and D, as well as for those on the opposite side of the trolley. Determine the cable force T and the reactions at each of the four pairs of rollers.

SOLUTION

Road Map At the outset of this problem it is not known whether the pair of rollers at A or the pair of rollers at B will make contact. Similarly, it is not known whether the pair of rollers at C or the pair of rollers at D will make contact. To proceed, we will assume the rollers at B and D make contact (the rollers at A and C could just as well have been chosen), and after the analysis is complete, we will interpret our results to determine which rollers are actually in contact.

Modeling The FBD is shown in Fig. 2. Because the geometry is given with respect to the I beam's axis of orientation, use of an xy coordinate system with this same orientation will allow for easy determination of moment arms. In drawing the FBD, we have assumed that the upper pairs of rollers at B and D make contact, and we label their reactions as R_1 and R_2 in the FBD. After analysis is complete, we must examine the signs of R_1 and R_2; a positive value indicates the assumption was correct, while a negative value means the assumption was incorrect and that the adjacent pair of rollers actually makes contact.

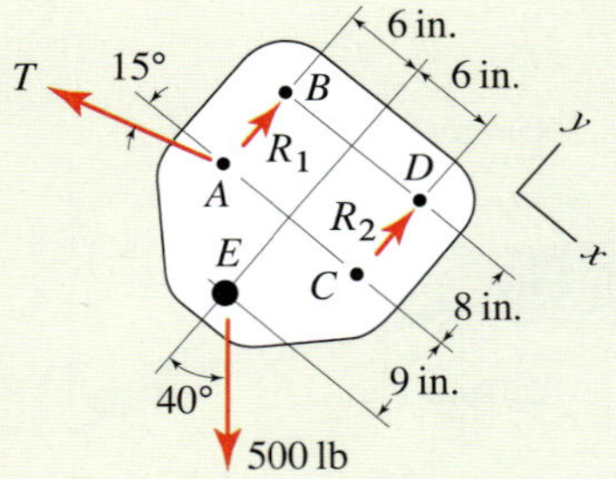

Figure 2
Free body diagram assuming contact is made at roller pairs at B and D.

Governing Equations & Computation The following sequence of equilibrium equations provides for easy determination of the unknowns:

$$\sum F_x = 0: \quad -T\cos 15° + (500\text{ lb})\sin 40° = 0 \tag{1}$$

$$\Rightarrow \boxed{T = 332.7\text{ lb},} \tag{2}$$

$$\sum M_A = 0: \quad -(500\text{ lb})\cos 40°(6\text{ in.}) + (500\text{ lb})\sin 40°(9\text{ in.}) + R_2(12\text{ in.}) = 0 \tag{3}$$

$$\Rightarrow R_2 = -49.53\text{ lb}, \tag{4}$$

$$\sum F_y = 0: \quad R_1 + R_2 - T\sin 15° - (500\text{ lb})\cos 40° = 0 \tag{5}$$

$$\Rightarrow R_1 = 518.7\text{ lb}. \tag{6}$$

Discussion & Verification Since R_1 is positive, the assumption of contact at the pair of rollers at B was correct and thus we write $A_y = 0$ and $B_y = R_1 = 518.7$ lb where positive B_y is measured in the positive y direction. Since R_2 is negative, it is the pair of rollers at C rather than the pair at D that actually makes contact; hence we write $C_y = R_2 = -49.53$ lb and $D_y = 0$ where positive C_y is measured in the positive y direction. Thus, our solutions for the cable tension and the reactions at each pair of rollers may be summarized as

$$\boxed{\begin{array}{c} A_y = 0, \quad B_y = 518.7\text{ lb}, \quad C_y = -49.53\text{ lb}, \\ D_y = 0, \quad \text{and} \quad T = 332.7\text{ lb}. \end{array}} \tag{7}$$

EXAMPLE 5.5 *Loose-fitting Gears*

A drum for mixing material rotates clockwise under the power of a geared motor at A. The drum weighs 320 lb and is supported by a bearing at point B, and the material being mixed weighs 140 lb with center of gravity at point D. If gear A is a loose fit with the gear on the drum, determine the reactions at point B and the gear tooth force required to operate the machine. Assume the machine operates at steady speed and the material being mixed maintains the same shape and position as the drum rotates.

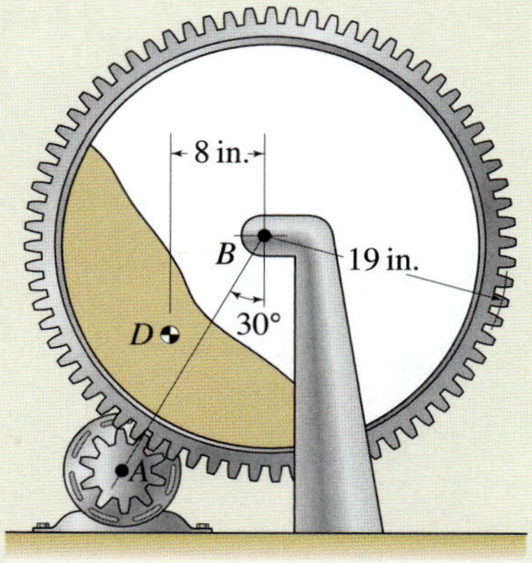

Figure 1

SOLUTION

Road Map Because the drum rotates at constant speed and the material being mixed maintains the same shape and position, this is a problem of static equilibrium. To determine the support reactions at bearing B and the gear tooth force, we will analyze the equilibrium of the drum and the material being mixed.

Modeling The FBD is shown in Fig. 2, where we simplify the gear tooth force (as discussed in Fig. 5.7) to be a single force G tangent to the gear and positioned at the pitch radius of the gear, which is 19 in. from point B. The 320 lb weight of the drum is positioned at the drum's center, point B, and the weight of the material being mixed is placed at its center of gravity, point D.

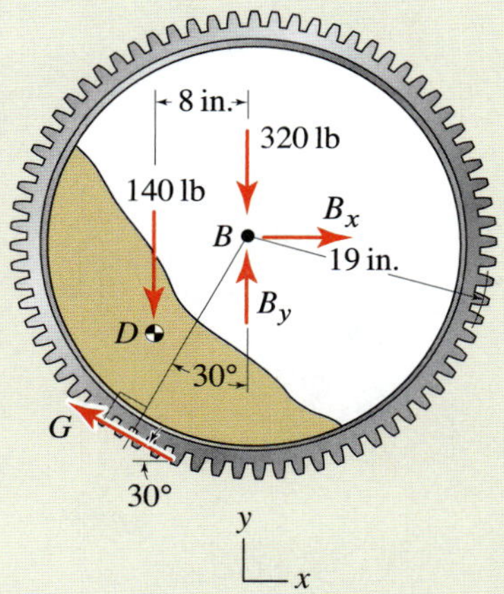

Figure 2
Free body diagram.

Governing Equations & Computation Using the FBD in Fig. 2, the equilibrium equations and solutions are as follows:

$$\sum M_B = 0: \quad -G(19\,\text{in.}) + (140\,\text{lb})(8\,\text{in.}) = 0 \tag{1}$$

$$\Rightarrow \boxed{G = 58.95\,\text{lb},} \tag{2}$$

$$\sum F_x = 0: \quad B_x - G\cos 30° = 0 \tag{3}$$

$$\Rightarrow \boxed{B_x = 51.05\,\text{lb},} \tag{4}$$

$$\sum F_y = 0: \quad B_y - 320\,\text{lb} - 140\,\text{lb} + G\sin 30° = 0 \tag{5}$$

$$\Rightarrow \boxed{B_y = 430.5\,\text{lb}.} \tag{6}$$

Discussion & Verification

- If the drum does not rotate at uniform speed, then accelerations are not zero and concepts of dynamics must be used to determine the gear tooth force and the reactions at B. If the material being mixed does not maintain the same shape and position as the drum rotates, then at a minimum the moment arm in Eq. (1) changes with time, and depending on how the shape and/or position changes with time, the problem may also be dynamic.
- The idealization of the gear tooth force consisting of only a tangential force is appropriate, provided the gears are a loose fit, because both gears rotate on bearings (or axes) that are fixed in space. You should contrast this situation with that in Example 5.6.

EXAMPLE 5.6 *Gears Actively Pressed Together*

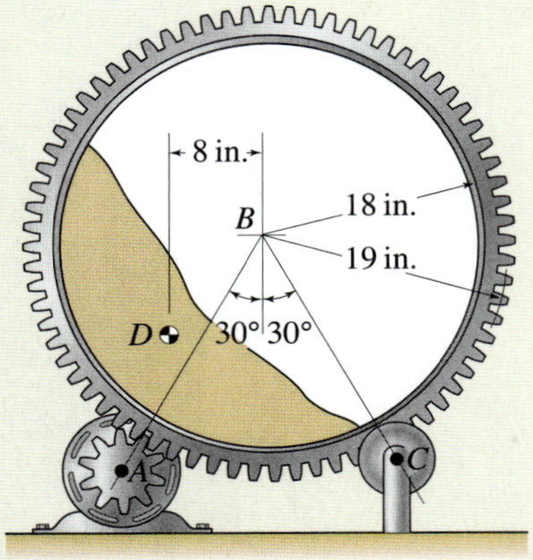

Figure 1

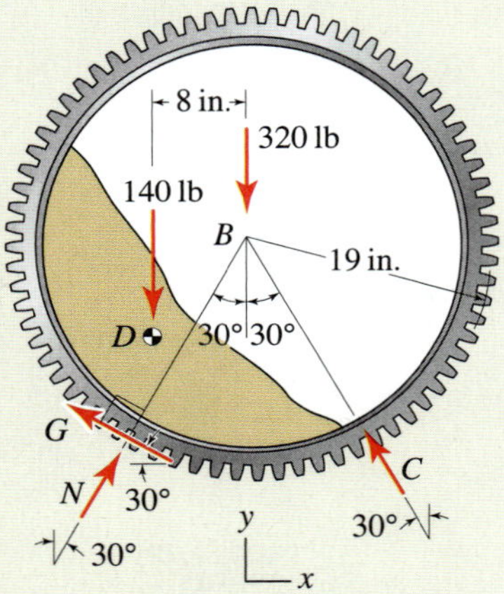

Figure 2
Free body diagram.

The mixing machine described in Example 5.5 is modified so that the drum is supported by a motor-powered gear at point A and a frictionless roller at point C. Determine the reactions at C and the gear tooth force required to operate the machine.

SOLUTION

Road Map Because the drum rotates at constant speed and the material being mixed maintains the same shape and position, this is a problem of static equilibrium. To determine the support reactions at the roller C and the gear tooth forces, we will analyze the equilibrium of the drum and the material being mixed. In contrast to Example 5.5, here the gears are actively pressed together, and thus the gear tooth forces consist of both tangential and normal forces.

Modeling The FBD is shown in Fig. 2 where the gear tooth forces consist of a tangential force G and a normal force N. The 320 lb weight of the drum is positioned at its center, point B, and the weight of the material being mixed is placed at its center of gravity, point D.

Governing Equations & Computation Using the FBD in Fig. 2, the equilibrium equations are

$$\sum M_B = 0: \quad -G(19\,\text{in.}) + (140\,\text{lb})(8\,\text{in.}) = 0 \tag{1}$$

$$\Rightarrow \boxed{G = 58.95\,\text{lb},} \tag{2}$$

$$\sum F_x = 0: \quad N\sin 30^\circ - G\cos 30^\circ - C\sin 30^\circ = 0, \tag{3}$$

$$\sum F_y = 0: \quad N\cos 30^\circ + G\sin 30^\circ + C\cos 30^\circ - 320\,\text{lb} - 140\,\text{lb} = 0. \tag{4}$$

Equation (1) was solved immediately for G, and the result is the same as that obtained in Example 5.5. With the solution for G known, Eqs. (3) and (4) can be solved simultaneously to obtain

$$\Rightarrow \boxed{N = 299.6\,\text{lb} \quad \text{and} \quad C = 197.5\,\text{lb}.} \tag{5}$$

Discussion & Verification The main difference between this example problem and Example 5.5 is the method by which the drum is supported, and this is reflected in the way the gear tooth forces are idealized. In Example 5.5 both gears rotated on fixed bearings and it was therefore possible to model the gear tooth force as consisting of a single tangential force only. In this example, only the gear at A rotates on a fixed bearing, and the gear on the drum is actively pressed into the other gear under the action of the 320 and 140 lb forces. Thus, the normal component of the gear tooth force N plays an important role in supporting the drum, and it cannot be neglected.

PROBLEMS

Problem 5.1

A freezer rests on a pair of supports at A and a pair of supports at B. Weights for various parts of the freezer are shown. Neglecting friction, determine the reactions at each pair of supports.

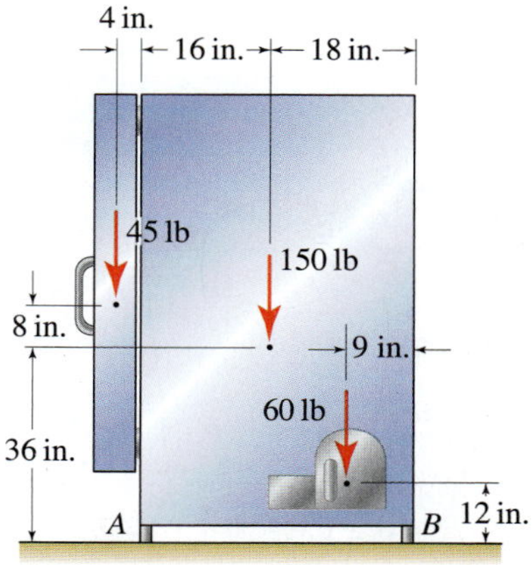

Figure P5.1

Problem 5.2

Weights of various parts of a computer are shown. Neglecting friction, determine the reactions at the supports at points A and B.

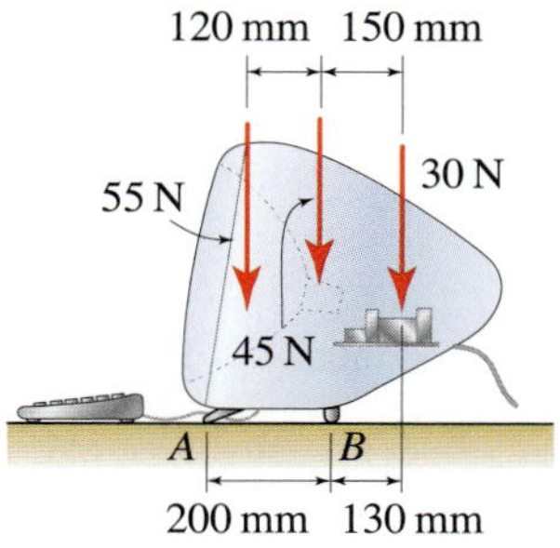

Figure P5.2

Problem 5.3

The tool shown is used in a gluing operation to press a thin laminate to a thicker substrate. If the wheels at points A and B both have 2 in. diameter and have frictionless bearings, and a 20 lb vertical force is applied to the handle of the tool, determine the forces applied to the top of the laminate and the bottom of the substrate.

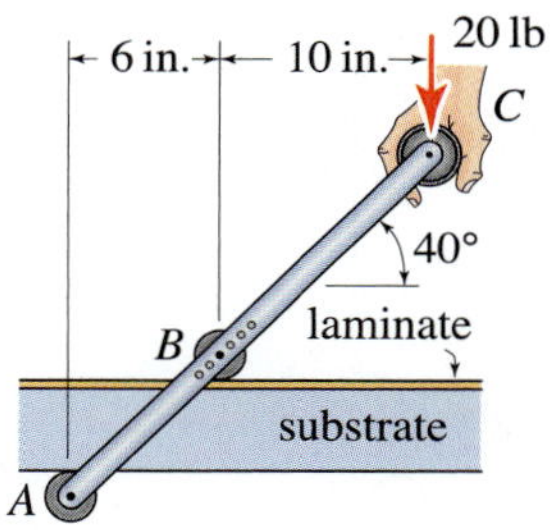

Figure P5.3

Problem 5.4

An escalator is driven by a chain connected to point A that supports a force F. The rollers at points A and B have frictionless bearings and ride in a loose-fitting track. If a person weighing 200 lb is being lifted at a steady speed and other weights may be neglected, determine the required chain force F and the reactions at the wheels at A and B. If the person is being lifted at a variable speed, will the force F be different than that calculated earlier? Explain.

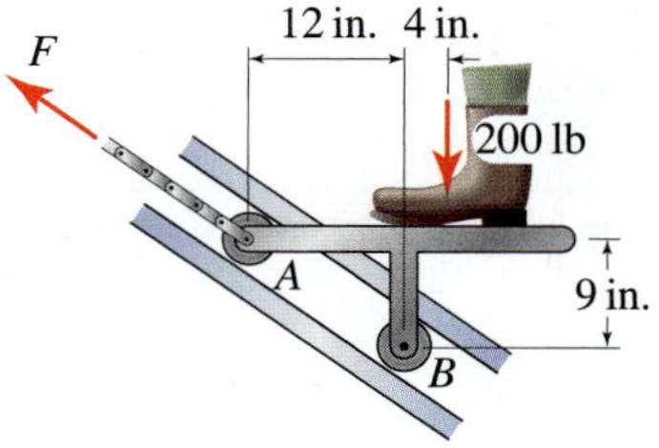

Figure P5.4

Problem 5.5

Bar $ABCD$ is supported by a pin at point C and a cable from points D to E. Determine the reactions at C and the force supported by the cable.

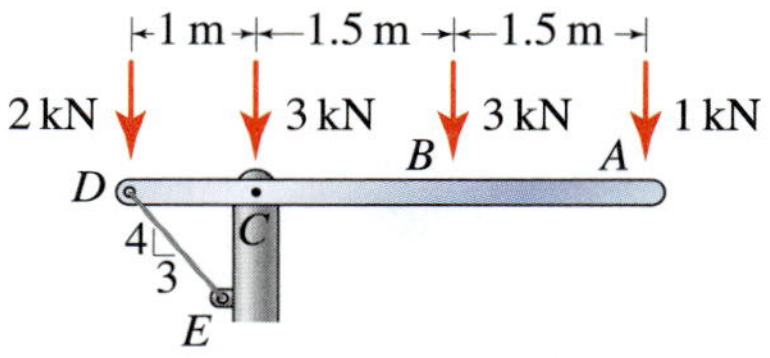

Figure P5.5

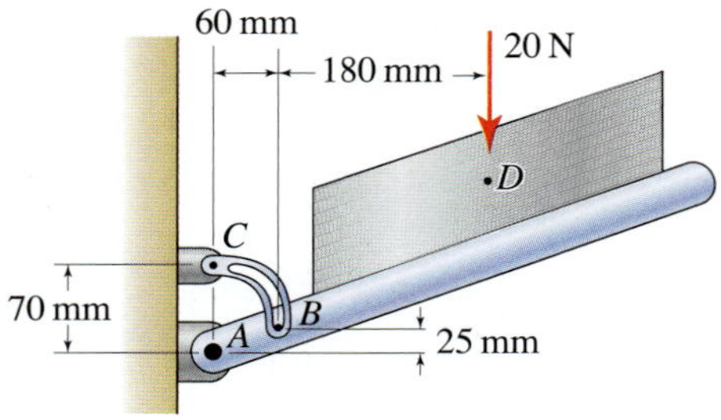

Figure P5.6

Problem 5.6

A foldable tray for the paper supply of a photocopy machine is shown. The tray is supported by a single hinge at A and two slotted links (one on each side of the tray). If the stack of paper weighs 20 N and other weights may be neglected, determine the reactions at the hinge and at point B for *one* of the links.

Problem 5.7

The boat, motor, and trailer have weights $W_b = 600$ lb, $W_m = 125$ lb, and $W_t = 300$ lb, respectively. If the distance from the front of the boat to the hitch is $d = 4$ ft, determine the vertical reaction at point A and the reaction on each of the wheels.

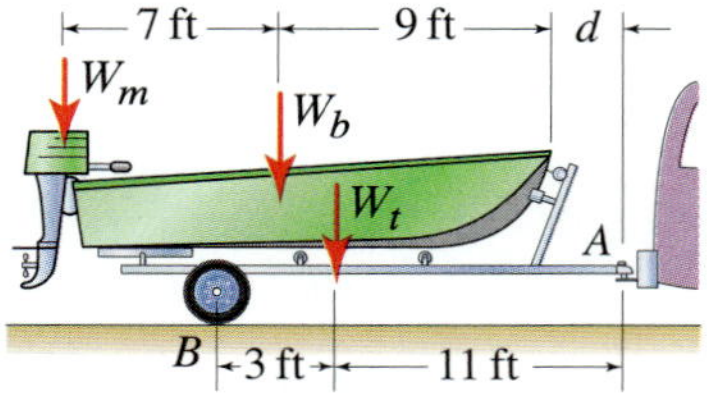

Figure P5.7 and P5.8

Problem 5.8

For the boat trailer of Prob. 5.7, determine the distance d so that the vertical reaction at point A is 100 lb.

Problem 5.9

A hand cart weighing 800 N is used for moving heavy loads in a warehouse. If each axle (pair of wheels) can support a maximum of 10 kN, determine the largest weight W that may be supported and the position d where it should be placed, assuming both axles are loaded to their capacity.

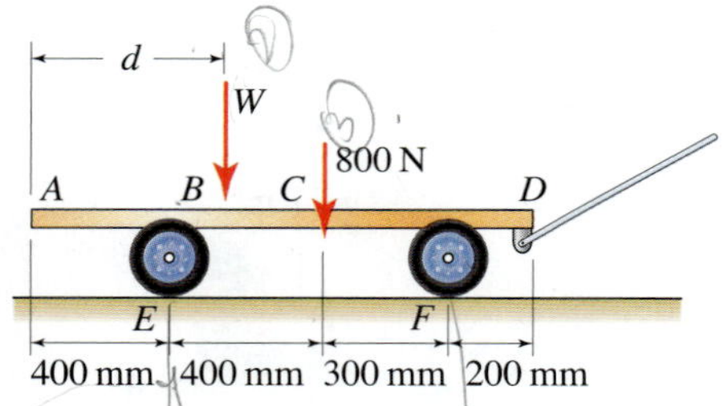

Figure P5.9

Problems 5.10 and 5.11

A variety of structures with pin, roller, and built-in supports are shown. In Figs. P5.10(c) and P5.11(c), the rollers at point D allow vertical translation and constrain horizontal translation and rotation. Determine all reactions. Express your answers in terms of parameters such as L, F, P, and/or M.

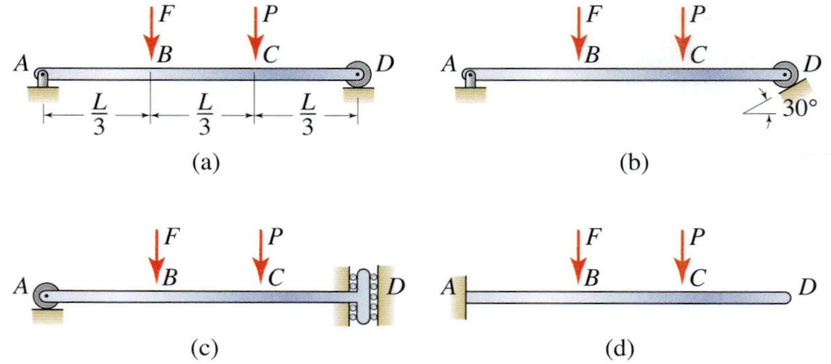

Figure P5.10

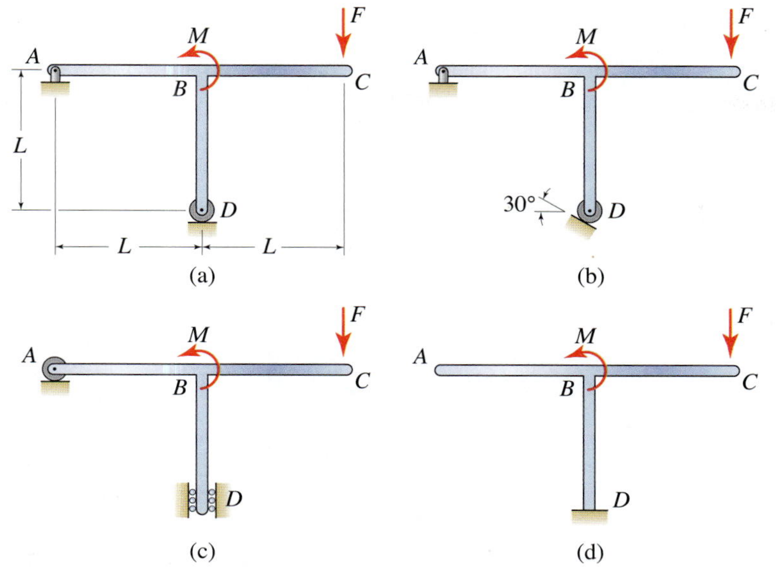

Figure P5.11

Problems 5.12 and 5.13

In the structures shown, all members have the same 2 m length. Determine all support reactions.

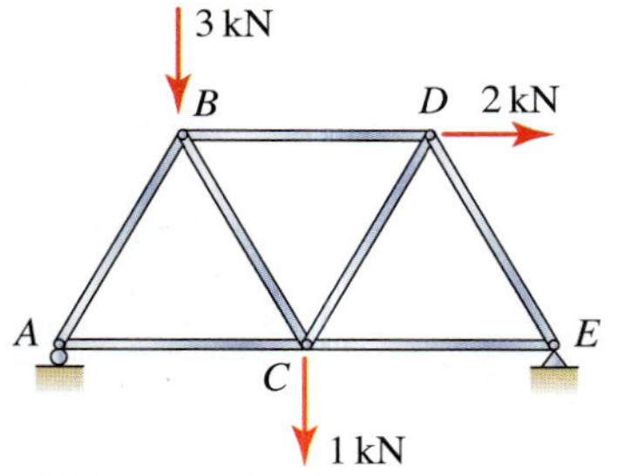

Figure P5.12

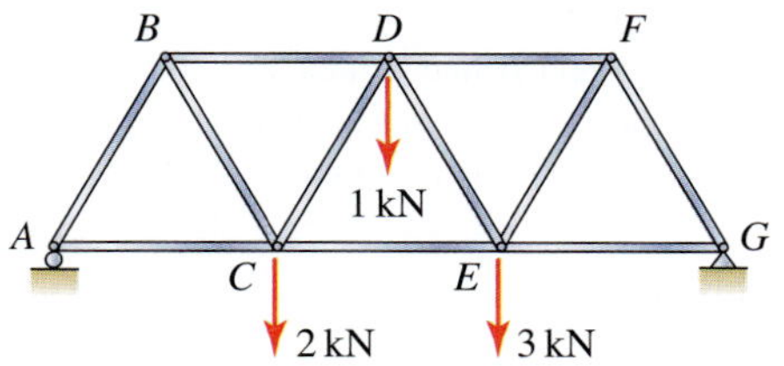

Figure P5.13

Problem 5.14

The top chord of the truss is subjected to a uniform distributed load which gives rise to the forces Q shown. If $Q = 1$ kip, determine all support reactions.

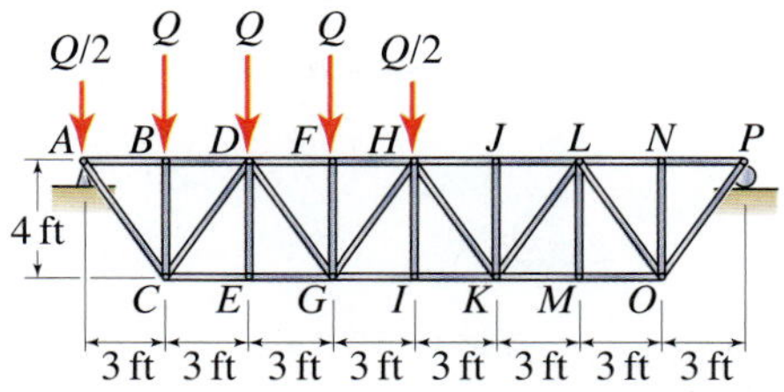

Figure P5.14

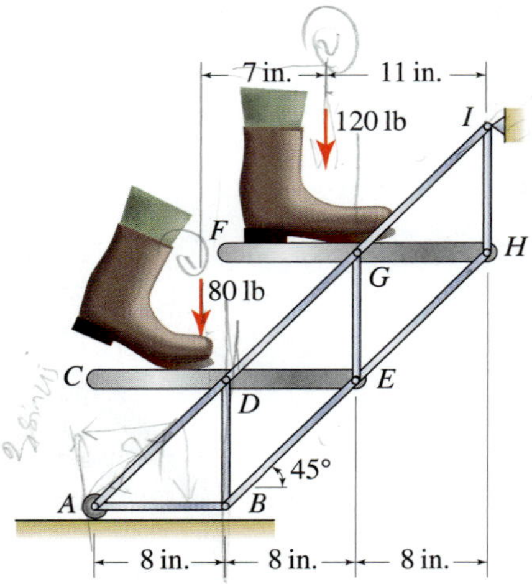

Figure P5.15

Problem 5.15

Determine the support reactions for the short flight of stairs due to the forces shown on the two steps.

Problem 5.16

The forklift has a vehicle weight of $W_V = 15{,}000$ lb, fuel weight of $W_F = 300$ lb, and operator weight of $W_O = 160$ lb. If $P = 2000$ lb, determine the reactions on each pair of wheels.

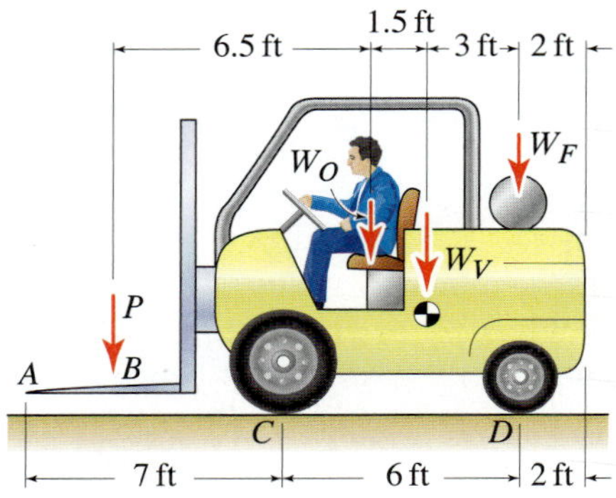

Figure P5.16

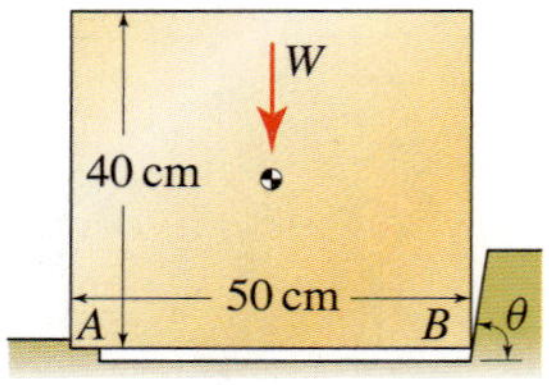

Figure P5.17

Problem 5.17

A fixture for positioning 40 cm by 50 cm size cardboard boxes with weight $W = 150$ N in a packaging company is shown. If the weight of the box and its contents acts through the center of the box, determine the largest value of θ if the magnitude of the reactions at points A and B may not exceed 800 N, at which point the box begins to crush. The notch at point A prevents horizontal and vertical translation of the box and assume the contact at point B is frictionless.

Problem 5.18

Bar ABC is supported by a roller at B and a frictionless collar at A that slides on a fixed bar DE. Determine the support reactions at points A and B. Express your answers in terms of parameters such as P, a, etc.

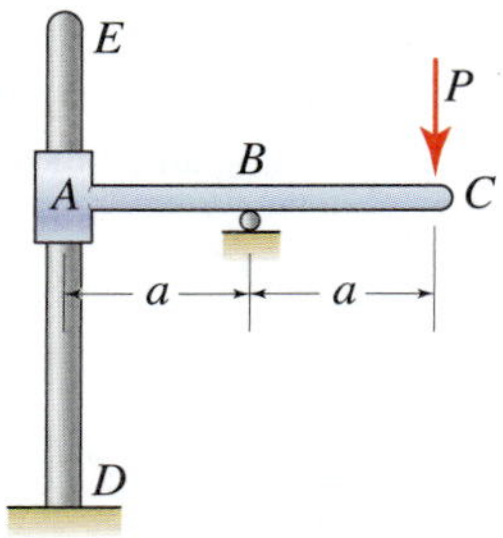

Figure P5.18

Problem 5.19

Structure $ABCDE$ is supported by a roller at D and a frictionless collar at A that slides on a fixed bar FG. Determine the support reactions at points A and D. Express your answers in terms of parameters such as P, Q, a, b, c, etc.

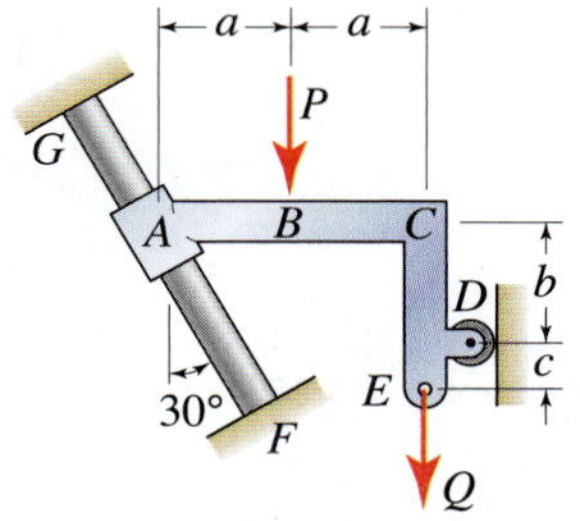

Figure P5.19

Problems 5.20 through 5.22

A motor and mounting hardware with weight $W = 65$ lb are supported by rollers A, B, C, and D. Each pair of rollers is loose-fitting so that only one roller of the pair will make contact with the fixed rail. This problem may be idealized as two-dimensional if the torque the motor applies to the pulleys is neglected. Determine which rollers make contact and the reactions at these rollers if:

Problem 5.20 $P = 0$ and $Q = 125$ lb.

Problem 5.21 $P = 200$ lb and $Q = 0$.

Problem 5.22 $P = 200$ lb and $Q = 125$ lb.

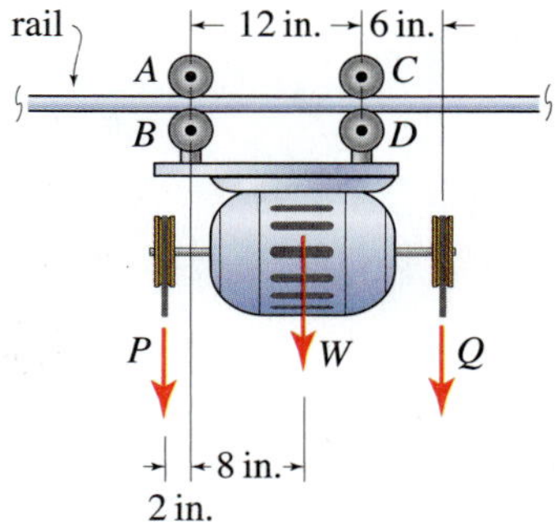

Figure P5.20–P5.22

Problem 5.23

During assembly in a factory, a compressor with weight $W = 120$ lb rests on a bench. A cover plate is to be attached using four bolts. To speed production, a special machine that simultaneously tightens all four bolts is used. Determine the largest torque M that may be simultaneously applied to each bolt before the compressor begins to tip.

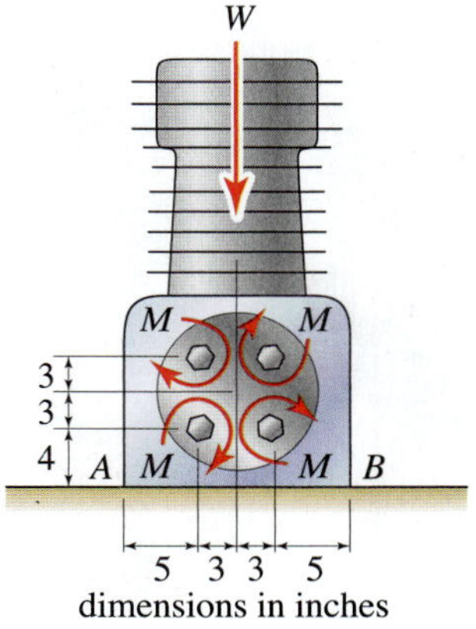

Figure P5.23

Problem 5.24

A tractor is fitted with a hole-drilling attachment. The tractor has weight $W_V = 2000$ lb, the operator has weight $W_O = 180$ lb, and the supplemental weights at the front of the tractor weigh $W_E = 300$ lb.

(a) Determine the largest thrust that may be produced by the drilling attachment at point A.

(b) Describe some simple ways the drilling thrust determined in Part (a) can be increased (e.g., addition of more weight at E, removal of weight at E, repositioning of weight from E to D, and so on).

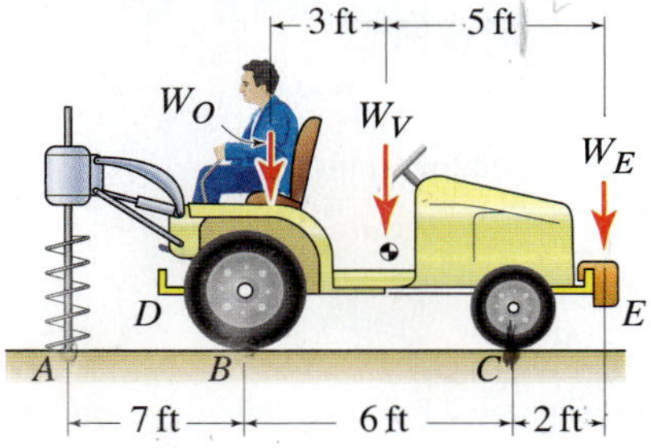

Figure P5.24

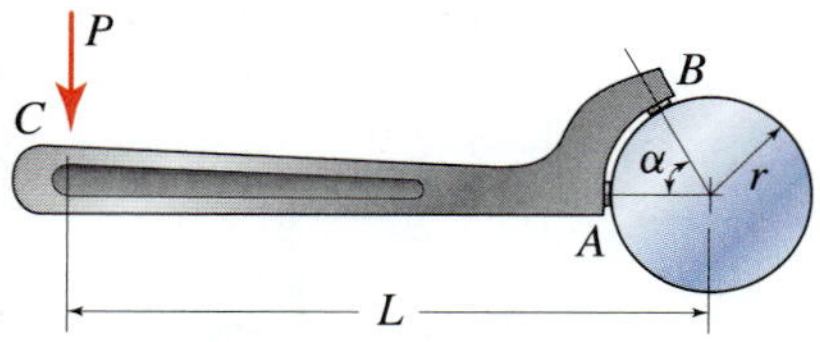

Figure P5.25 and P5.26

Problem 5.25

A spanner wrench is used to apply torque to circular shafts and other similar shapes. Such wrenches are routinely used in the setup of tools such as milling machines, lathes, and so on. The wrench makes contact with the shaft at point A, which may be assumed to be frictionless, and at B where a small pin fits into a hole in the shaft. If $P = 80$ N, $L = 120$ mm, $r = 25$ mm, and $\alpha = 120°$, determine the reactions at points A and B.

Problem 5.26

In the spanner wrench of Prob. 5.25, determine the range of values for angle α so that the pin at B will not slip out of its hole when force P is applied for

(a) $L/r = 4$.

(b) Any value of L/r.

Problem 5.27

An arbor press is used to apply force to the workpiece at point C. The length h of handle DE can be adjusted by sliding it through a hole in shaft F such that 50 mm $\leq h \leq$ 250 mm. The press has weight $W = 350$ N and simply rests on the rough surface of a table. Force P is applied perpendicular to the handle.

(a) If the handle is horizontal ($\alpha = 0°$), determine the smallest force P that will cause the press to tip about point A.

(b) For any possible position of the handle, determine the smallest force P and the corresponding value of α that will cause the press to tip about point A.

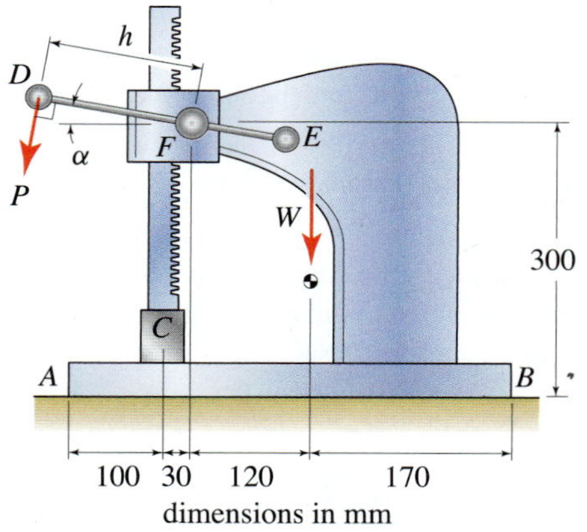

Figure P5.27

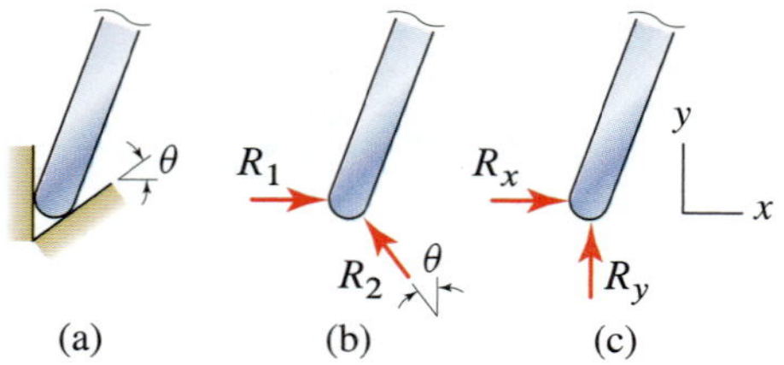

Figure P5.28

Problem 5.28

A portion of a structure is supported by a frictionless V-shaped notch as shown. Which are the proper support reactions: those shown in Fig. P5.28(b) or those in Fig. P5.28(c), or are both of these acceptable? Explain.

Note: Concept problems are about *explanations*, not computations.

Problem 5.29

An antenna used for research at a university is shown. It has the feature that it is easily raised and lowered so that it can be outfitted with different equipment. The antenna is pinned to a supporting frame at point A, and it is raised and lowered using the gear at B to which a hand crank is attached. The antenna and its attached sector gear weigh 300 and 120 lb, respectively, and the 100 lb horizontal force at point C models the effect of wind loads during a storm. If the gear at B is locked so it does not rotate, determine the gear tooth force and the reactions at A.

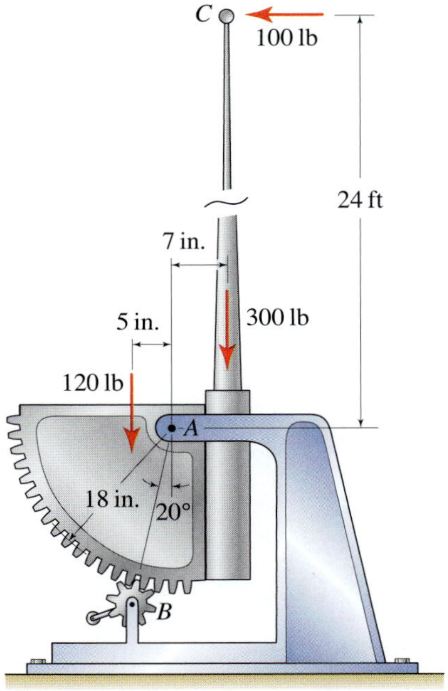

Figure P5.29

Problem 5.30

Frame $ABCD$ is rigidly attached to a gear E, which engages two parallel geared tracks F and G that are fixed in space. The gear has 4 in. pitch radius. If the gear has weight $W = 8$ lb and $P = 10$ lb, determine the tooth forces between gear E and each track. *Hint:* A normal force is present either at F or G but not both.

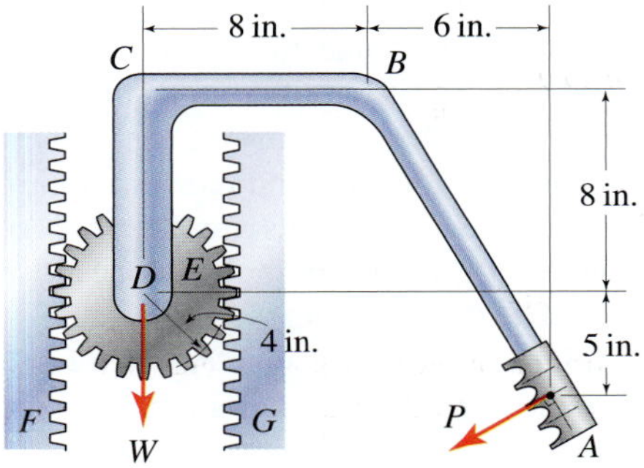

Figure P5.30

Problem 5.31

Frame BCD is rigidly attached to a gear A, which engages a geared track E that is fixed in space. The gear has 30 mm pitch radius, and portions BC and CD of the frame are perpendicular. If the gear has weight $W = 6$ N and $P = 4$ N, determine the value of Q needed so that angle $\theta = 30°$, and also determine the reactions (tooth forces) between the gear and track.

Problem 5.32

In the frame and gear of Prob. 5.31, if $W = 6$ N, $P = 4$ N, and $Q = 8$ N, determine the value of θ when the assembly is in static equilibrium and the reactions (tooth forces) between the gear and track.

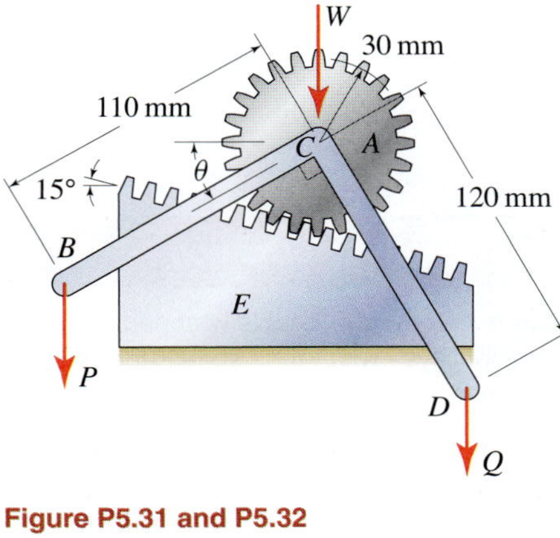

Figure P5.31 and P5.32

5.3 Equilibrium of Bodies in Two Dimensions—Additional Topics

We begin this section by more thoroughly examining the theoretical underpinnings of the equilibrium analyses performed in the previous section. We then introduce springs and the inclusion of deformable members in equilibrium problems. Finally, we present a very brief introduction to statically indeterminate problems in an example.

Why are bodies assumed to be rigid?

Throughout statics, when analyzing the equilibrium of bodies, objects, and structures, we almost always assume they are rigid (springs are a notable exception). This assumption is made out of necessity for the same reasons as those discussed in Section 3.2 in connection with cables and bars being idealized as inextensible. When we write equilibrium equations such as $\sum F_x = 0$, $\sum F_y = 0$ and $\sum M_P = 0$, the geometry of the object is needed so that forces can be resolved into components and moment arms can be determined. If the object is deformable, as real objects are, then this geometry is usually unknown and the equations of static equilibrium alone are too few to solve for the unknown reactions *and* the new geometry. Structures or objects that are stiff typically undergo little change of geometry when forces are applied, and hence they can often be idealized as rigid. For structures that are flexible, we usually have no alternative but to incorporate deformability into our analysis, and this topic is addressed in subjects that follow statics and dynamics, such as mechanics of materials.

Helpful Information

Cable Geometry. A cable wrapped around two pulleys with the same radius is common.

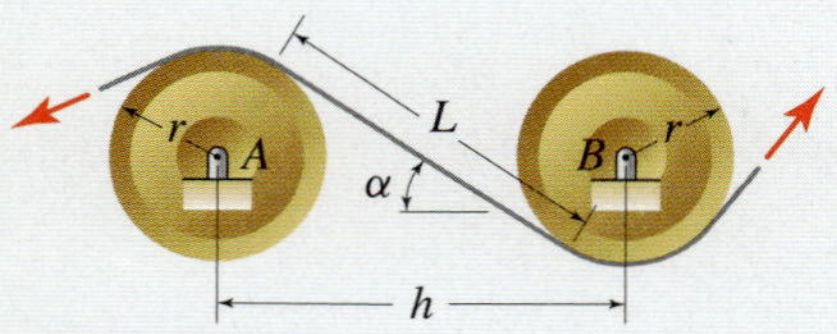

Figure 5.10

With r being the radius of the pulleys and h being the distance between the pulleys' bearings, the orientation α and the length L of the cable segment between the pulleys are

$$\alpha = \sin^{-1}\frac{2r}{h}, \quad L = h\cos\alpha. \tag{5.12}$$

Derivation of these expressions is given as an exercise in Prob. 5.33.

Treatment of cables and pulleys

Cables and pulleys are common components in structures, and there are several options for how these may be treated in an analysis. Consider the example shown in Fig. 5.11(a), where force W is specified and the support reactions are to be determined. Since the pulleys are idealized as frictionless and the cable is continuous, perfectly flexible, and weightless, all portions of the cable support the same tensile force which is equal to W. Three possible FBDs are shown in Fig. 5.11(b)–(d). In the first of these, Fig. 5.11(b), the pulleys are left on the structure, and when we take the cut to draw the FBD, the forces shown are exposed. In the next two FBDs, Fig. 5.11(c) and (d), the pulleys are removed from the structure. That is, when taking the cut to produce these FBDs, we remove the pulleys and this exposes the forces that the pulleys apply to the structure. In Fig. 5.11(c), the pulley forces are simply "shifted" to the bearings of the pulleys. Justification for this is seen in the FBDs of the pulleys themselves shown on the right-hand side of Fig. 5.11(c), where the bearing forces have been constructed so that each pulley is in equilibrium. The FBD shown in Fig. 5.11(d) is very similar to that shown above it except the bearing forces are further resolved into x and y components. For most purposes, you will likely find the FBD approaches in Fig. 5.11(b) and (c) to be the most useful.

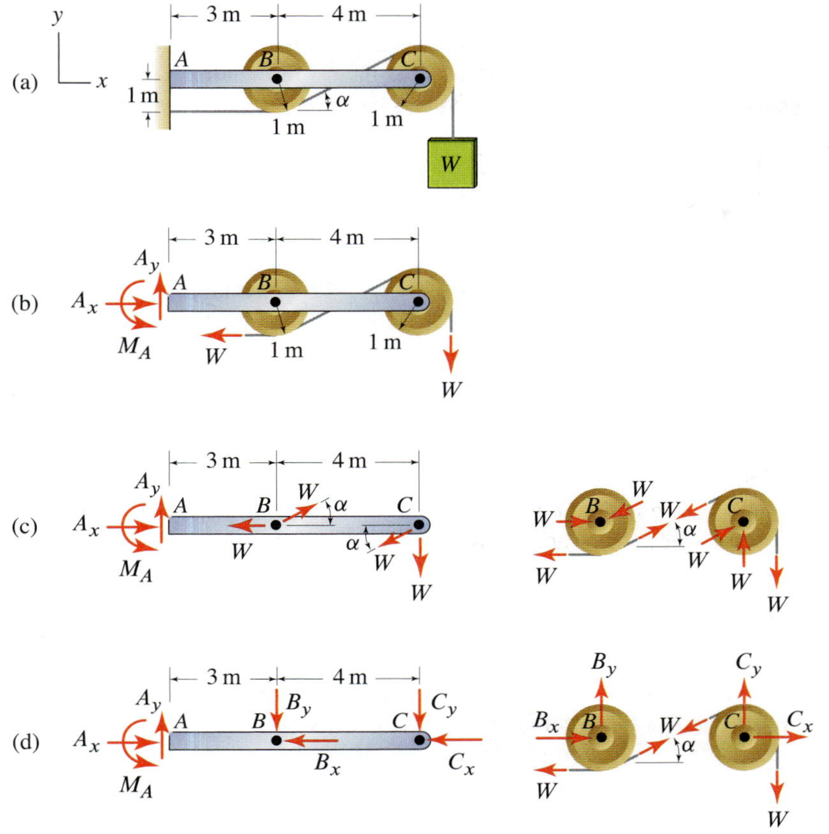

Figure 5.11. (a) Cantilever beam structure with frictionless pulleys at points B and C. (b)–(d) Free body diagrams that may be used to determine the reactions at point A. Because the pulleys have the same radius, angle α may be easily computed using Eq. (5.12).

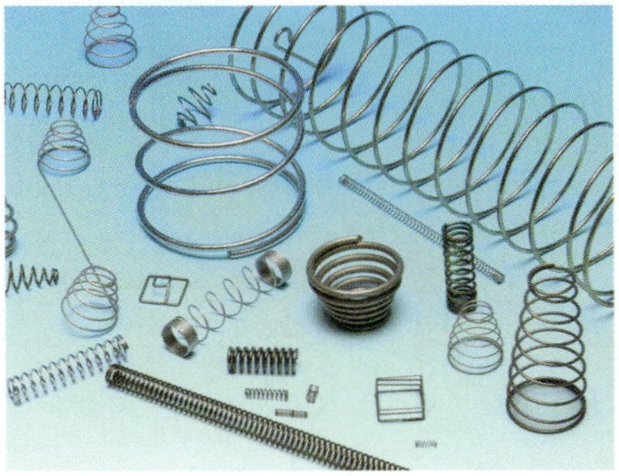

Figure 5.12
An assortment of small coil springs made of wire.

Springs

Springs are one of the few deformable members we consider in statics. Some examples of springs are shown in Fig. 5.12, and a schematic representation is shown in Fig. 5.13. All of the remarks made about springs in Section 3.2 are applicable here, and for linear elastic behavior, the spring law is

$$\begin{aligned} F_s &= k\delta \\ &= k(L - L_0) \end{aligned} \tag{5.13}$$

where

F_s is the force supported by the spring;
δ is the elongation of the spring from its unstretched or undeformed length;
k is the spring stiffness (units: force/length);
L_0 is the initial (unstretched) spring length;
L is the final spring length.

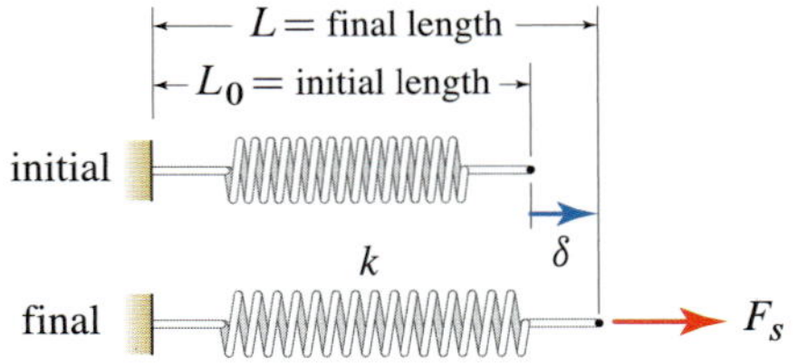

Figure 5.13
A spring produces a force F_s that is proportional to its elongation δ. Such springs are sometimes called *axial springs*.

The spring stiffness k is always positive. Because the force F_s and the elongation δ in Eq. (5.13) are directed along an axis, or line, these springs are sometimes called *axial springs* to differentiate them from torsional springs, which are discussed next. Equation (5.13) can be written using other sign conventions for the positive directions of F_s and δ, but this may require introducing a negative sign in Eq. (5.13) as discussed in Section 3.2.

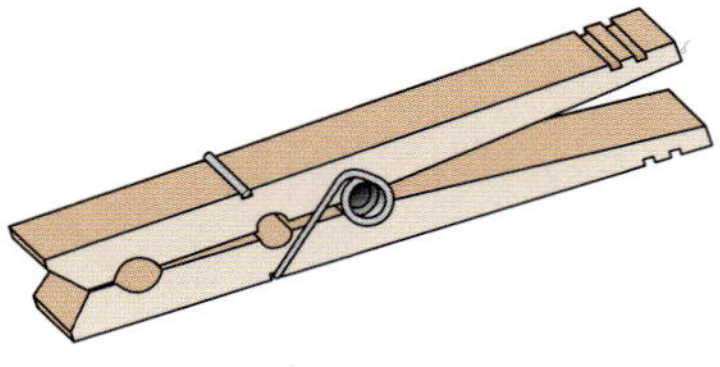

Figure 5.14
Example of a torsional spring used in a clothespin.

Torsional springs are also common, and a simple example of a torsional spring in use is the clothespin shown in Fig. 5.14. Figure 5.15 shows some small torsional springs made of wire, and you should contrast the construction of these with the axial springs shown in Fig. 5.12. A torsional spring produces a moment that is proportional to the relative rotation, or twist, of the spring. A torsional spring is shown schematically in Fig. 5.16, and for linear elastic behavior, the *spring law* is

$$\boxed{M_t = k_t \theta} \tag{5.14}$$

where

M_t is the moment produced by the spring;
θ is the rotation of the spring measured in radians from the unloaded geometry;
k_t is the spring stiffness (units: moment/radian).*

By definition, the *spring stiffness* k_t is always positive. Note that units for the stiffness of a torsional spring are different than units for the stiffness of an axial spring. Equation (5.14) assumes M_t and θ are positive in the same direction. If you choose to take M_t and θ to be positive in opposite directions, then a negative sign must be introduced in Eq. (5.14).

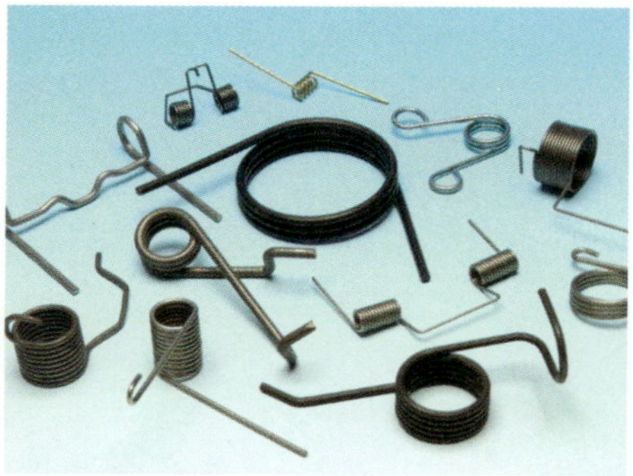

Figure 5.15
An assortment of small torsional springs made of wire.

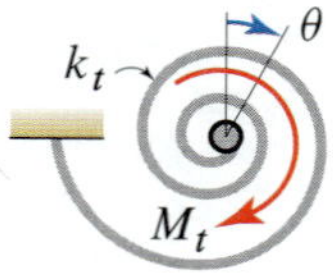

Figure 5.16
A torsional spring produces a moment M_t that is proportional to its rotation θ.

Superposition

Superposition is a concept that can be used to replace a problem having complex loading by a sum (or superposition) of problems having simpler loading. Often, each of the problems with the simpler loading is easier to analyze than the original problem. Even when the problems with the simpler loading are no easier to analyze than the original problem, superposition may still be useful. The following example illustrates superposition.

Mini-Example. The cantilever beam shown in Fig. 5.17 is loaded by two 1 kN forces. Use superposition to determine the moment reaction at point A.
Solution. The loading for this problem can be broken into the two simpler loadings shown in Fig. 5.17, load case 1 and load case 2, which we will simply call case 1 and case 2. By drawing FBDs and writing equilibrium equations, you can verify that the moment reactions at point A for each load case are $(M_A)_{\text{case 1}} = 2\text{ kN·m}$ and $(M_A)_{\text{case 2}} = 4\text{ kN·m}$. The total moment reaction is the sum of those for each load case. Thus,

$$\begin{aligned}(M_A)_{\text{total}} &= (M_A)_{\text{case 1}} + (M_A)_{\text{case 2}} \\ &= 2\text{ kN·m} + 4\text{ kN·m} \\ &= 6\text{ kN·m}.\end{aligned} \tag{5.15}$$

* The stiffness for torsional springs is sometimes given using degrees to measure rotation. However, for analytical work it is best to use radians.

The same remarks apply to all other reactions so that, for example, the vertical force reaction at point A is also the sum of the reactions for each load case.

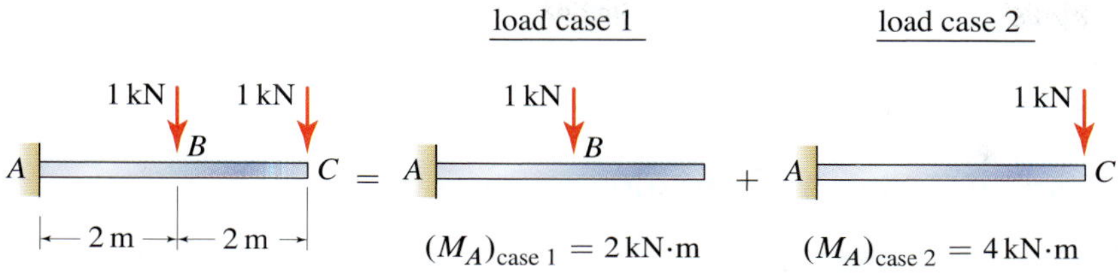

Figure 5.17. Example illustrating superposition.

To further demonstrate the utility of superposition, imagine we wish to determine the moment reaction at A for the loading shown in Fig. 5.18. Rather than reanalyze this problem from the beginning, we may use our previous superposition analysis where we simply scale the results for each of the constituent load cases. Thus, for Fig. 5.18, the total moment at point A is

$$\begin{aligned}(M_A)_{\text{total}} &= 3(M_A)_{\text{case 1}} + 5(M_A)_{\text{case 2}}\\ &= 3(2\ \text{kN}\cdot\text{m}) + 5(4\ \text{kN}\cdot\text{m})\\ &= 26\ \text{kN}\cdot\text{m}. \qquad (5.16)\end{aligned}$$

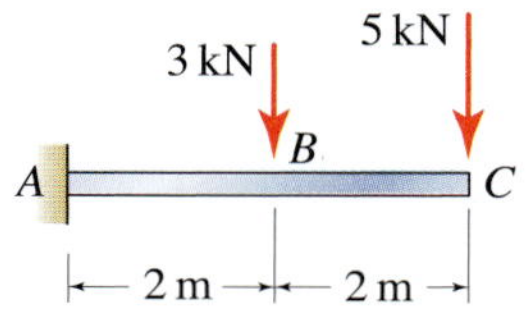

Figure 5.18
Using the results of the superposition solution for Fig. 5.17, this example is very easy to analyze.

Supports and fixity

The way an object is supported determines its *fixity* and whether it is statically determinate or indeterminate. These concepts are defined as follows:

Complete fixity. A body with *complete fixity* has supports that are sufficient in number and arrangement so that the body is completely fixed in space and will undergo no motion (either translation or rotation) in any direction under the action of *any* possible set of forces.

Partial fixity. A body with *partial fixity** has supports that will allow the body to undergo motion (translation and/or rotation) in one or more directions. Whether or not such motion occurs depends on the forces and/or moments that are applied and whether the body is initially in motion.

No fixity. A body with *no fixity* has no supports and is completely free to translate and rotate in space.

All objects fall into one of the above three categories. Rather than the word *fixity*, it is common to refer to a body's ability to undergo *rigid body motion*. A body in two dimensions can undergo three types of rigid body motion: translation in two orthogonal directions and rotation, or any combination of these. Thus, equivalent nomenclature for complete fixity, partial fixity, and no fixity is no rigid body motion capability, partial rigid body motion capability, and

* Some statics textbooks describe partial fixity as "improper supports." However such nomenclature is undesirable since it implies that all objects should be fully fixed when, of course, this is not the case.

full rigid body motion capability, respectively. For a body in two dimensions, complete fixity requires supports having a total of three or more reaction forces and/or moments with suitable arrangement. Partial fixity results when there is only one or two reaction forces, or may also result if there are three or more reaction forces and/or moments but they do not have sufficient arrangement to fully prevent motion.

Static determinacy and indeterminacy

Static determinacy and indeterminacy were introduced in Example 3.6 on p. 142. These are defined as follows:

Statically determinate body. For a *statically determinate body*, the equilibrium equations of statics are sufficient to determine all unknown forces and/or other unknowns that appear in the equilibrium equations.

Statically indeterminate body. For a *statically indeterminate body*, the equilibrium equations of statics are not sufficient to determine all unknown forces and/or other unknowns.

All objects that are in static equilibrium fall into one of the above two categories. For a statically indeterminate body, there are more unknowns than there are equilibrium equations with which to determine them, and in general none of the unknowns can be found. On occasion, however, the equilibrium equations may be sufficient so that some (but not all) of the unknowns in a statically indeterminate problem can be found.

A simple rule of thumb to help ascertain whether an object is statically determinate or indeterminate is to compare the number of unknowns to the number of equilibrium equations, and we call this *equation counting*. With n being the number of unknowns, the rule of thumb for a single body in two dimensions is as follows:

If $n < 3$	The body is statically determinate, and it can have partial fixity ($n = 1$ or 2) or no fixity ($n = 0$).	
If $n = 3$	The body is statically determinate if it has full fixity. The body is statically indeterminate if it has partial fixity.	(5.17)
If $n > 3$	The body is statically indeterminate, and it can have full fixity or partial fixity.	

Successful use of equation counting, Eq. (5.17), requires some judgment on your part, which is why it is called a *rule of thumb* rather than a *rule*. Nonetheless, it is quite useful and will be employed throughout this book with enhancements as needed. Thorough understanding of equation counting here will make these subsequent enhancements self-evident. The basis for this rule of thumb and subtleties in its application are explored in the following example.

Mini-Example. A body with a variety of support schemes is shown in Fig. 5.19. For each support scheme, specify whether the body has partial or full fixity and whether it is statically determinate or indeterminate.

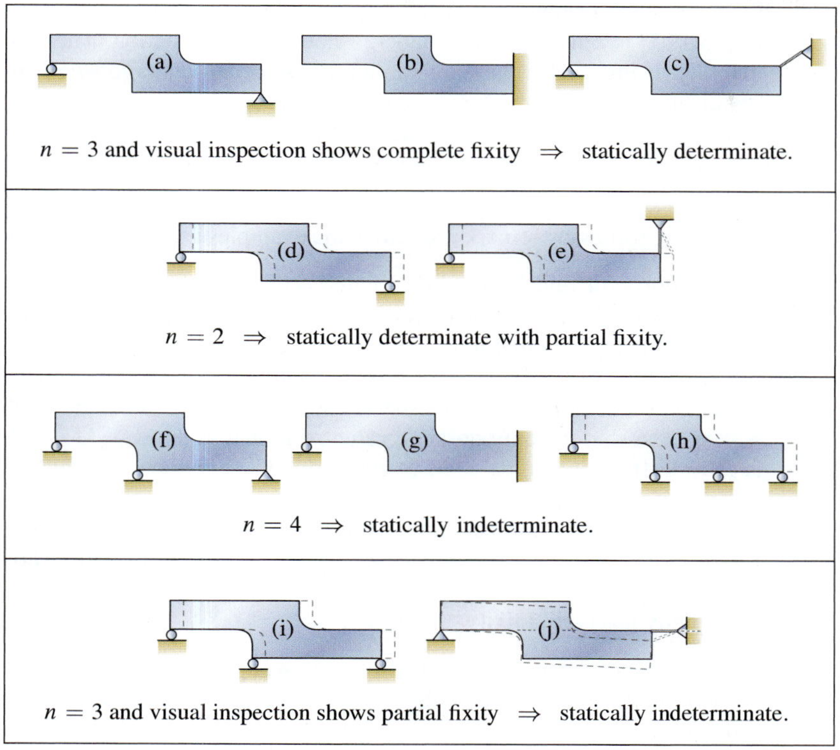
(a)
(b)
(c)
$n = 3$ and visual inspection shows complete fixity $\Rightarrow$ statically determinate.
(d)
(e)
$n = 2$ $\Rightarrow$ statically determinate with partial fixity.
(f)
(g)
(h)
$n = 4$ $\Rightarrow$ statically indeterminate.
(i)
(j)
$n = 3$ and visual inspection shows partial fixity $\Rightarrow$ statically indeterminate.

Figure 5.19. Examples of equation counting to determine if a body is statically determinate or indeterminate. For bodies that are not fully fixed, a possible displaced position is shown by the dashed outlines.

Solution. In Fig. 5.19(a), the roller has one unknown reaction force and the pin has two, for a total of $n = 3$. Inspection of the body shows it has complete fixity. That is, the supports are sufficient so that the body cannot translate horizontally, cannot translate vertically, and cannot rotate. Hence, Eq. (5.17) indicates the body is statically determinate. Similar remarks apply to Fig. 5.19(b) and (c).

The bodies shown in Fig. 5.19(d) and (e) each have two unknown reactions, and Eq. (5.17) indicates both bodies are statically determinate with partial fixity. Inspection of Fig. 5.19(d) shows this body can undergo horizontal translation while the body of Fig. 5.19(e) can undergo horizontal translation and rotation. When stating that these bodies are statically determinate, we presume the loading is such that the problems are indeed static rather than dynamic. For example, if the body of Fig. 5.19(d) has a net horizontal force applied to it, then it will undergo horizontal acceleration and Newton's law $F = ma$ must be used.

The bodies shown in Fig. 5.19(f), (g) and (h) each have four unknown reactions, and Eq. (5.17) indicates that all three bodies are statically indeterminate.

Inspection of Fig. 5.19(f) and (g) shows that both of these have complete fixity. Inspection of Fig. 5.19(h) shows that it can undergo horizontal translation, and hence it has only partial fixity.

The bodies shown in Fig. 5.19(i) and (j) each have three unknown reactions, but examination shows that each has partial fixity. The motion capability in Fig. 5.19(i) is obvious; it can undergo horizontal translation. The motion capability in Fig. 5.19(j) is more subtle; the body can undergo *small* rotation about the pin before the link's orientation changes enough to restrain further motion. Hence, Eq. (5.17) indicates that both bodies are statically indeterminate. ■

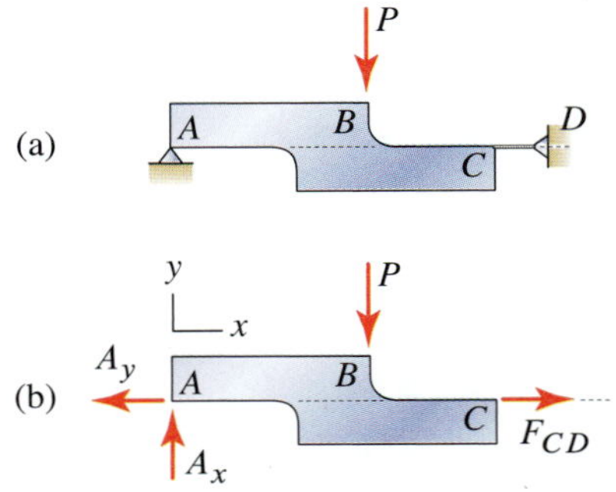

Figure 5.20
Use of a force to determine a body's fixity.

Remarks. The following remarks pertain to the examples shown in Fig. 5.19.

- The supports for Fig. 5.19(c) and (j) are very similar, and it is perhaps subtle why the latter of these has only partial fixity. Another perspective that might help show that it has partial fixity is to consider the response of the body of Fig. 5.19(j) when it is subjected to a force P, as shown in Fig. 5.20. After we draw the FBD and sum moments about point A, it is very clear that the only possible value of P that satisfies $\sum M_A = 0$ is $P = 0$. If a nonzero value of P were applied, the supports would be incapable of developing reactions that could equilibrate P and dynamic motion would occur, at least until such time that the geometry of link CD changes. Use of a force to determine fixity is helpful, but is not foolproof. A force with the wrong position or orientation may not reveal that a body has partial fixity.

- Whether or not a body has rigid body motion capability has nothing to do with the forces that are applied to it. However, if a body does have rigid body motion capability, whether or not motion occurs depends strongly on the forces and/or moments that are applied.

- Equation counting is easy, but successful use relies on your ability to examine a body to determine its fixity. Essentially, equation counting provides only a scalar characterization of a body's supports. A more complete characterization would need to include the vector aspect of the supports, namely, the positions and directions of the reactions. But such a method of analysis is considerably more complex.

Two-force and three-force members

If a body or structural member is subjected to forces at two points or three points only, as described below, then when in equilibrium the orientation of the forces supported by the body has special properties. These situations are defined as follows:

Two-force member. A body subjected to forces at two points (no moment loading and no distributed forces such as weight) is called a *two-force member*. The special feature of a two-force member is that, when in equilibrium, the two forces have the same line of action and opposite direction. Examples are shown in Fig. 5.21.

Three-force member. A body subjected to forces at three points (no moment loading and no distributed forces such as weight) is called a *three-force* *member*. The special feature of a three-force member is that, when in equilibrium, the lines of action of all three forces intersect at a common point. If the three forces are parallel (this is called a *parallel force system*), then their point of intersection can be thought of as being at infinity. Examples are shown in Fig. 5.22.

If a body is not a two-force or three-force member, then we will refer to it as either a *zero-force member*, if it is subjected to no force at all, or a *general multiforce member*, if it is subjected to forces at more than three points and/or has moment loading and/or has distributed loading. Zero-force members will be routinely encountered in truss structures, discussed in Chapter 6, and we will see they can play an important role in improving the strength of a truss.

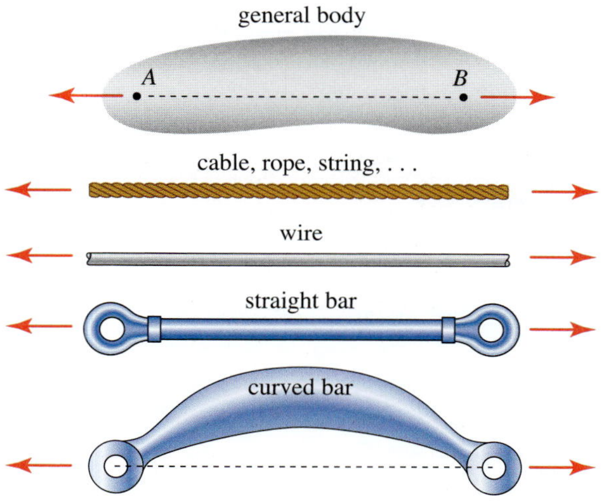

Figure 5.21. Examples of two-force members. In all cases, the two forces have equal magnitude and opposite direction and have the same line of action.

Common Pitfall

Geometry of a two-force member. A common misconception is that a two-force member is always straight. While two-force members are often straight, they can be curved or have other geometry. Chapter 8 discusses the merits of two-force members that are straight versus two-force members that are not straight (see Prob. 8.22 on p. 466).

Remarks on two-force members

- The characteristics of a two-force member can be proved using the following argument. Consider the general body shown in Fig. 5.21, and imagine the forces at A and B may not have the orientations shown. Moment equilibrium about A is satisfied only if the line of action of the force at B passes through A. Similarly, moment equilibrium at B is satisfied only if the force at A passes through B. Hence the two forces must have the same line of action, which is the line containing A and B. Then summing forces in the direction from A to B shows that the two forces must have equal magnitude and opposite direction.
- The support reactions for cable, link, and bar supports were shown in Fig. 5.3 on p. 247, and these are repeated in Fig. 5.23.* In Section 5.2, the support reactions for these members were obtained by considering the motion that the support prevents. Alternatively, the reaction forces

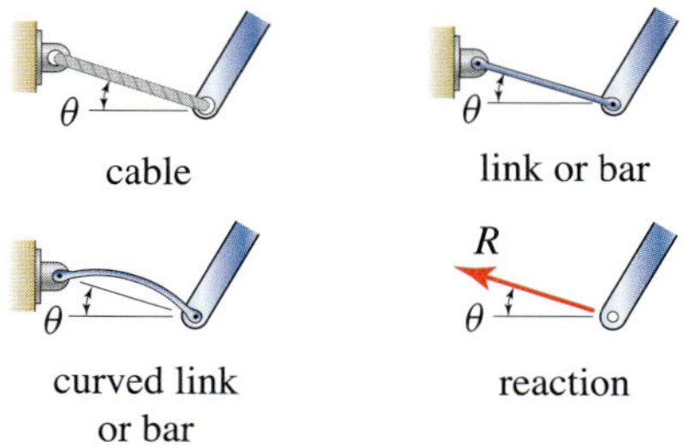

Figure 5.23
Support reactions for cable, link, and bar supports repeated from Fig. 5.3 on p. 247. Because all of these supports are two-force members, the reaction force R is directed between the endpoints of these members.

* The cable with self-weight shown in Fig. 5.3 is not a two-force member, and hence it is not shown here.

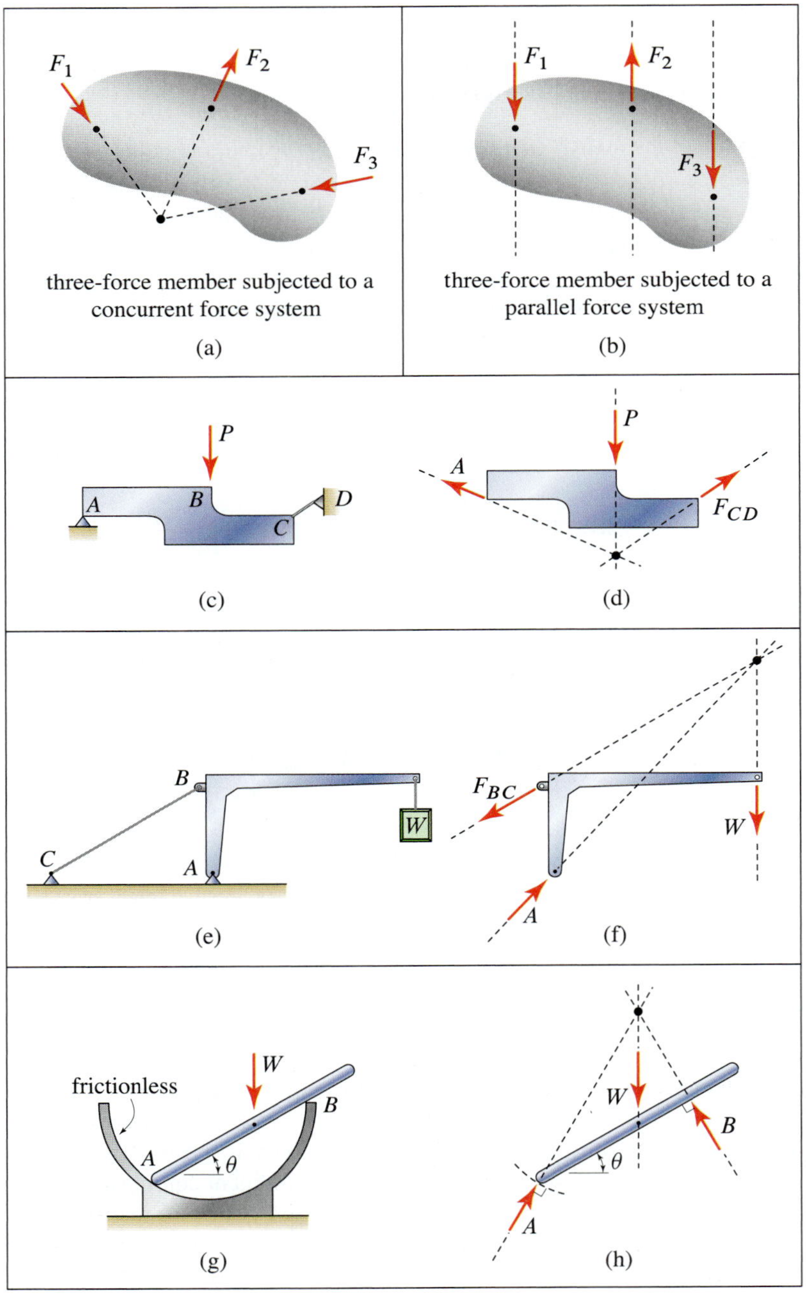

Figure 5.22. Examples of three-force members. (a) Three-force member with a concurrent force system: all forces intersect at a common point. (b) Three-force member with a parallel force system. (c)–(h) Examples of three-force members with concurrent force systems. In (g) and (h), the bar's equilibrium position has a value of θ so that all forces intersect at a common point.

for these members can be obtained by recognizing that the cable, link, and bar supports are all two-force members; hence the reaction force R is directed between the endpoints of these members.

- The ability to identify two-force members is important and can help simplify the analysis of complex problems, such as with frames and machines, discussed in Chapter 6.

Remarks on three-force members

- The characteristics of a three-force member with a concurrent force system can be proved using the following argument. Consider the body shown in Fig. 5.22(a), and imagine force F_3 may not have the orientation shown. Moment equilibrium about the intersection point of the lines of action of F_1 and F_2 is satisfied only if the line of action of F_3 passes through this point. The same argument can be made by considering the orientations of F_1 and F_2 to be different. Hence, moment equilibrium for a three-force member with a concurrent force system requires the lines of action of the three forces to intersect at a common point.

- Although the forces applied to a three-force member can be parallel, such as shown in Fig. 5.22(b), there are no especially significant remarks that can be made about this situation. Rather, analysis proceeds in the usual fashion where we write equations such as $\sum F_x = 0$, $\sum F_y = 0$, and $\sum M_P = 0$.

End of Section Summary

In this section, some of the finer points regarding static equilibrium of a rigid body were reviewed. Some of the key points are as follows:

- When equilibrium equations are written, the geometry of the structure in the equilibrium position must be used. By assuming a structure or object is rigid, the geometry before and after forces are applied is the same, and hence the solution to the equilibrium equations provides an exact solution for the problem.

- When a structure or object includes pulleys and cables, there are several options for how these may be treated when FBDs are drawn. Sometimes it is most effective if a particular pulley is left on the structure and the cut that is used to draw the FBD is taken through that pulley's cables. In other cases it may be most effective to take a cut that removes a pulley from the structure, in which case the forces that the pulley applies to the structure must be included in the FBD; this is often called *shifting* the cable forces to the bearing of a pulley. A more detailed discussion of these issues is given in connection with Fig. 5.11.

- Springs were described in detail in Section 3.2 and were reviewed again in this section.

- A *torsional spring* is a deformable member that produces a moment M_t proportional to the amount of twist θ of the spring. The *spring law* is $M_t = k_t\theta$ where k_t is called the *spring stiffness*, $k_t \geq 0$, and k_t has units of moment/radian. In writing this equation it is assumed that M_t and θ are positive in the same direction.

- *Superposition* was described. In simple terms, superposition is most often used to replace a problem having complex loading by two or more simpler problems that are easier to analyze. The solution to the original problem is obtained by superposing the solutions to the simpler problems.

- An object's *fixity* is determined by the type, number, and arrangement of its supports. All objects have either *full fixity*, *partial fixity*, or *no fixity*. Throughout statics and dynamics, we rely on visual inspection of a structure and its supports to determine its fixity. Rather than the above nomenclature, it is also common to refer to an object's ability to undergo *rigid body motion*.

- *Statically determinate* and *statically indeterminate* structures and objects were described. For a statically determinate structure or object, the equations of equilibrium are sufficient to determine all unknowns. For a statically indeterminate structure or object, the equations of equilibrium are not sufficient to determine all unknowns. *Equation counting* was described as an effective way to determine whether a structure was statically determinate or indeterminate.

- *Two-force* and *three-force* members were described. If a member is a two-force or three-force member, then the forces it supports have special properties. The ability to identify two-forces members is especially important as this will allow for simplifications that make the analysis easier.

EXAMPLE 5.7 *Springs*

A wind tunnel is used to experimentally determine the lift force L and drag force D on a scale model of an aircraft. The bracket supporting the aircraft is fitted with an axial spring with stiffness $k = 0.125$ N/mm and a torsional spring with stiffness $k_t = 50$ N·m/rad. By measuring the deflections δ and θ of these springs during a test, the forces L and D may be determined. If the geometry shown in Fig. 1 occurs when there is no airflow, and if the springs are calibrated so that $\delta = 0$ and $\theta = 0°$ when there is no airflow, determine L and D if $\delta = 2.51$ mm and $\theta = 1.06°$ are measured.

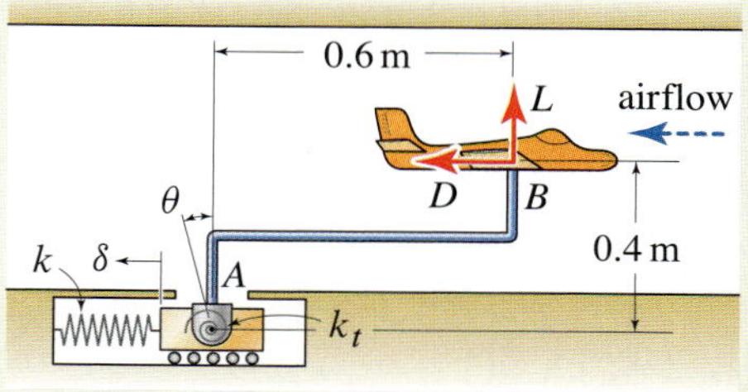

Figure 1

SOLUTION

Road Map This problem involves spring elements, and these have equations that govern their load–deformation response. Thus, the problem-solving methodology used here will be enhanced to emphasize that in addition to the need for equilibrium equations, force laws that describe the behavior of the springs are needed.

Modeling The FBD is shown in Fig. 2 where we assume that bar AB, which supports the aircraft, is slender enough that it does not develop any lift or drag forces.

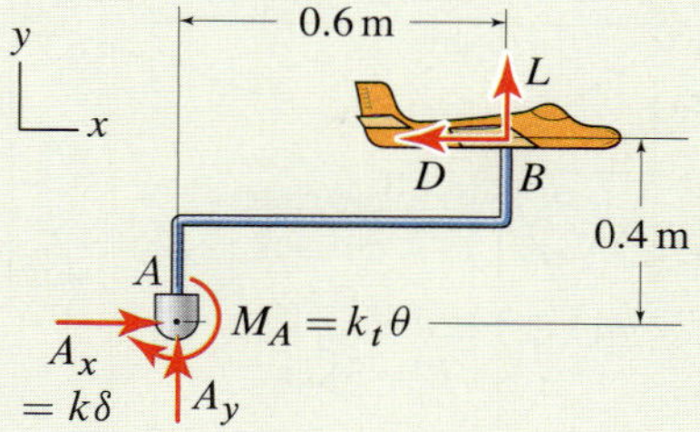

Figure 2
Free body diagram.

Governing Equations

Equilibrium Equations Summing forces in the x direction and summing moments about point A provide

$$\sum F_x = 0: \quad A_x - D = 0, \tag{1}$$

$$\sum M_A = 0: \quad -M_A + D(400 \text{ mm}) + L(600 \text{ mm}) = 0. \tag{2}$$

In writing Eq. (2), we assume the deformation of the torsional spring is small so that the geometry of the support bracket AB is essentially unchanged from its original geometry.

Force Laws The force supported by the axial spring is related to its deformation δ, and the moment supported by the torsional spring is related to its rotation θ, by

$$A_x = k\delta, \tag{3}$$

$$M_A = k_t\theta. \tag{4}$$

Computation Using Eq. (3), Eq. (1) may be solved for:

$$D = k\delta = \left(0.125\frac{\text{N}}{\text{mm}}\right)(2.51 \text{ mm}) = \boxed{0.314 \text{ N.}} \tag{5}$$

Using Eq. (4) and the solution for D just obtained, Eq. (2) may be solved for:

$$L = \frac{-D(400 \text{ mm}) + k_t\theta}{600 \text{ mm}}$$

$$= \frac{-(0.314 \text{ N})(400 \text{ mm}) + (50 \frac{\text{N}\cdot\text{m}}{\text{rad}})(\frac{10^3 \text{ mm}}{\text{m}})(\frac{\pi \text{ rad}}{180°})1.06°}{600 \text{ mm}} = \boxed{1.33 \text{ N.}} \tag{6}$$

Figure 3
A model of a Boeing 787 Dreamliner is prepared for testing in a large wind tunnel.

Discussion & Verification If the springs are not sufficiently stiff, then δ and/or θ may be substantially larger and the original geometry cannot be used when writing Eq. (2). An additional disadvantage of a soft torsional spring is that if θ is large, then the angle of attack of the aircraft also changes appreciably, which is undesirable. Since δ and θ are known in this problem, it is easy to verify that the difference between the original geometry and the deformed geometry is small. You can explore this issue greater detail in Probs. 5.53 through 5.56.

EXAMPLE 5.8 *Free Body Diagram Choices for Pulleys*

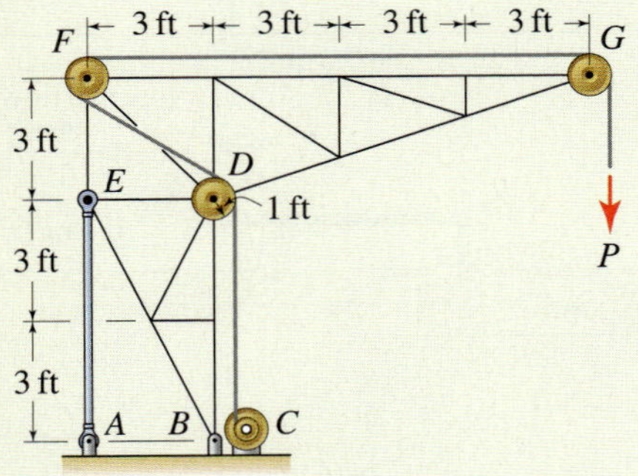

Figure 1

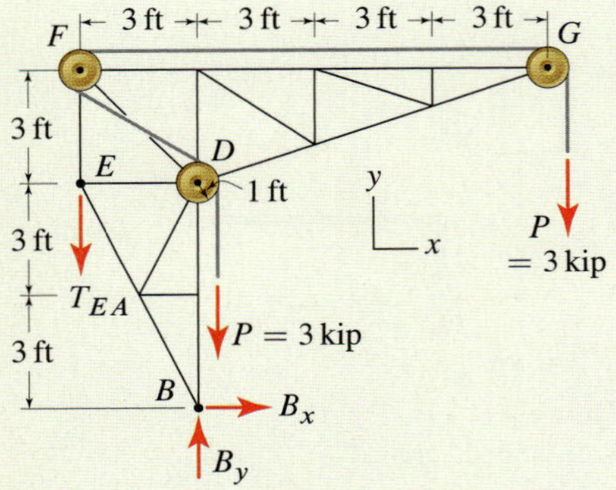

Figure 2
Free body diagram leaving all pulleys on the structure.

The stationary crane is supported by a pin at point B and a bar between points A and E. A winch at point C is used to raise and lower loads. The pulleys at points D, F, and G have 1 ft radius. Determine the support reactions due to the vertical force P if $P = 3$ kip.

SOLUTION

Road Map The structure has a single cable wrapped around pulleys at D, F, and G. Assuming these pulleys are frictionless and neglecting the weight of the cable, the force supported by the cable has the same value P throughout its length. When treating the cable and pulleys in FBDs, we may choose to leave these on the FBD, or to remove them, or to use a combination of these. Our choice of approach will be determined by the ease with which moment arms for the cable forces can be determined.

Modeling In considering the FBD options as discussed in connection with Fig. 5.11 on p. 271, we first consider leaving the cable and all the pulleys on the structure. The resulting FBD is shown in Fig. 2 where we observe that all moment arms are easily obtained, and thus it is not necessary to consider other options for drawing the FBD.

Governing Equations & Computation Using the FBD in Fig. 2, the equilibrium equations and solutions are

$$\sum F_x = 0: \quad B_x = 0 \tag{1}$$

$$\Rightarrow \quad \boxed{B_x = 0,} \tag{2}$$

$$\sum M_B = 0: \quad T_{EA}(3\text{ ft}) - (3\text{ kip})(1\text{ ft}) - (3\text{ kip})(10\text{ ft}) = 0 \tag{3}$$

$$\Rightarrow \quad \boxed{T_{EA} = 11.0\text{ kip},} \tag{4}$$

$$\sum F_y = 0: \quad -T_{EA} + B_y - 3\text{ kip} - 3\text{ kip} = 0 \tag{5}$$

$$\Rightarrow \quad \boxed{B_y = 17.0\text{ kip}.} \tag{6}$$

Discussion & Verification

- In the FBD of Fig. 2, a positive value of T_{EA} corresponds to tension in bar AE, and a positive value of B_y corresponds to vertical compression in the pin support at B. Intuitively, we expect bar AE to be in tension and the pin support at B to be in compression, and indeed our solutions show $T_{EA} > 0$ and $B_y > 0$.
- In this example, the weight of the crane was neglected, and thus, the reactions we computed are those due to supporting the force P only. The weight of the crane is probably not small. Problem 5.58 asks you to determine the reactions due to only the weight of the crane and then to use superposition with the results of this example to determine the total support reactions.

EXAMPLE 5.9 *Free Body Diagram Choices for Pulleys*

The stationary crane is supported by a pin at point B and a bar between points A and E. A winch at point C is used to raise and lower loads. The pulleys at points D, F, and G have 1 ft radius. Determine the support reactions due to the force P if $P = 3$ kip and $\theta = 20°$.

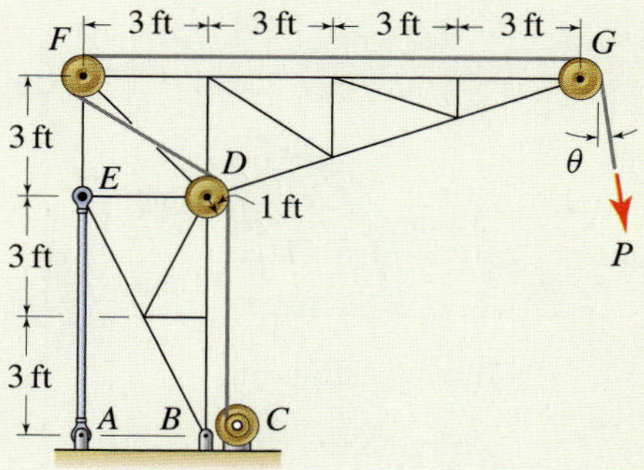

Figure 1

SOLUTION

Road Map This problem is identical to that in Example 5.8, except that the cable segment at pulley G is oriented 20° from the vertical. The structure has a single cable wrapped around pulleys at D, F, and G. Assuming these pulleys are frictionless and neglecting the weight of the cable, the force supported by the cable has the same value P throughout its length. We will begin by using an FBD where all of the pulleys are left on the structure and will consider if the moment arms needed are easy to obtain. If the moment arms are not easy to obtain, we will then consider if it is more convenient to use an FBD where some or all of the pulleys are removed from the structure.

Modeling In considering the FBD options as discussed in connection with Fig. 5.11 on p. 271, we first consider leaving the cable and all the pulleys on the structure, and the resulting FBD is shown in Fig. 2. Unfortunately, determining the moment arm for the cable force at pulley G in this FBD is tedious. Thus, we draw a new FBD, as shown in Fig. 3, where the pulley at G is removed, and we observe that all moment arms are now easily obtained. Note that the horizontal forces at pulley F and point G do not have the same line of action.

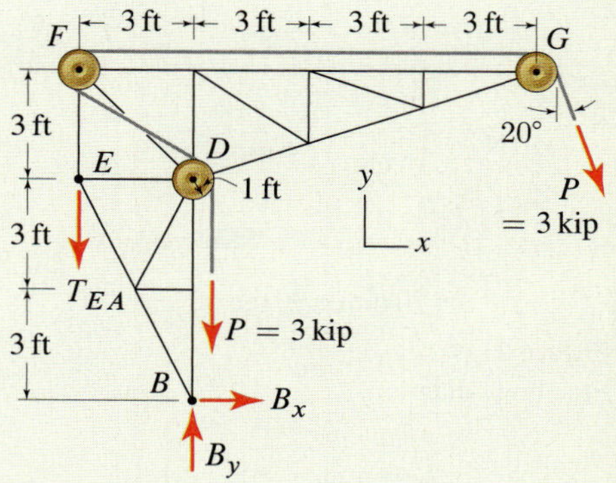

Figure 2
Free body diagram leaving all pulleys on the structure.

Governing Equations & Computation Using the FBD in Fig. 3, the equilibrium equations and solutions are

$$\sum F_x = 0: \quad B_x + 3\text{ kip} - 3\text{ kip} + (3\text{ kip})(\sin 20°) = 0 \tag{1}$$

$$\Rightarrow \quad \boxed{B_x = -1.03\text{ kip},} \tag{2}$$

$$\sum M_B = 0: \quad T_{EA}(3\text{ ft}) - \underbrace{(3\text{ kip})(1\text{ ft})}_{\text{pulley } D} - \underbrace{(3\text{ kip})(10\text{ ft})}_{\text{pulley } F} + \underbrace{(3\text{ kip})(9\text{ ft})}_{\text{pulley } G}$$

$$- \underbrace{(3\text{ kip})(\cos 20°)(9\text{ ft})}_{\text{pulley } G} - \underbrace{(3\text{ kip})(\sin 20°)(9\text{ ft})}_{\text{pulley } G} = 0 \tag{3}$$

$$\Rightarrow \quad \boxed{T_{EA} = 13.5\text{ kip},} \tag{4}$$

$$\sum F_y = 0: \quad -T_{EA} + B_y - 3\text{ kip} - (3\text{ kip})(\cos 20°) = 0 \tag{5}$$

$$\Rightarrow \quad \boxed{B_y = 19.4\text{ kip}.} \tag{6}$$

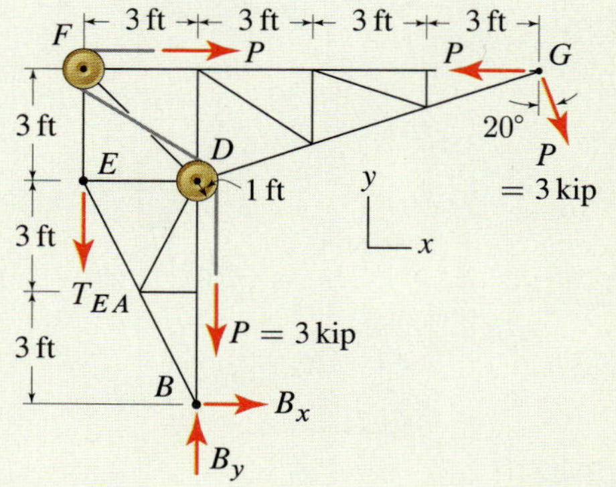

Figure 3
Free body diagram where the pulley at G has been removed. Observe that the horizontal forces at pulley F and point G do not have the same line of action.

Discussion & Verification

- The comments made in the Discussion & Verification in Example 5.8 also apply here, and indeed we find that bar AE is in tension ($T_{EA} > 0$) and the pin at B is in vertical compression ($B_y > 0$).
- Comparing our answers in Eqs. (2), (4), and (6) with those in Example 5.8 shows that the reactions in this example are slightly larger in magnitude. This can be explained by comparing the FBD shown in Fig. 2 of Example 5.8 with the FBD shown in Fig. 2 of this example, where it is seen that the moment arm for the force P at pulley G is larger in this example, hence the larger reaction forces found in this example are expected.

EXAMPLE 5.10 *Equilibrium of a Three-force Member*

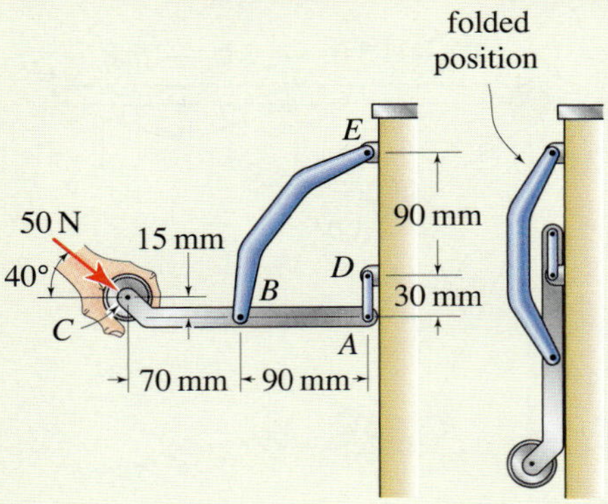

Figure 1

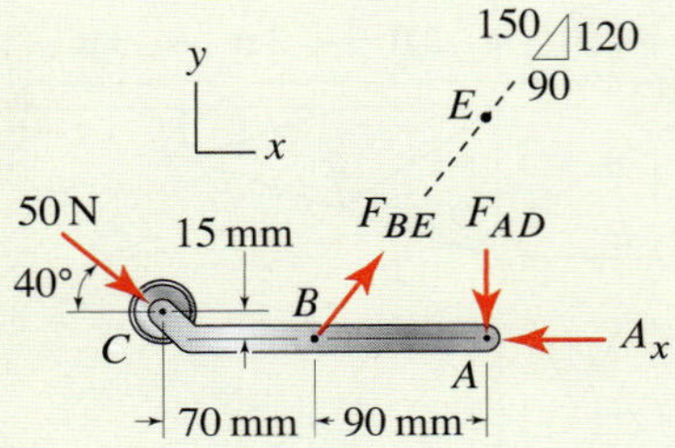

Figure 2
Free body diagram.

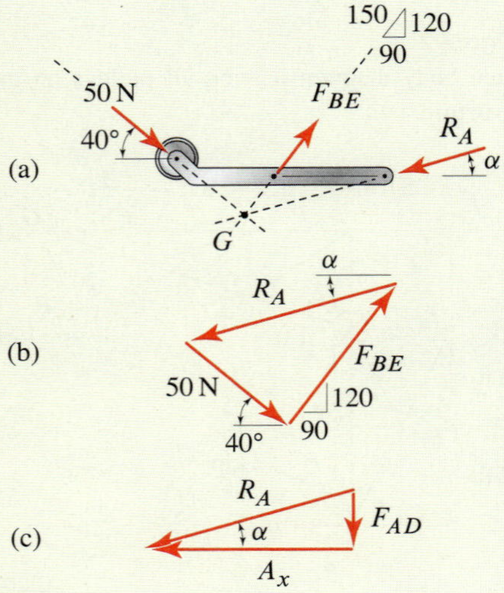

Figure 3
Alternate solution for a three-force member. (a) Free body diagram. (b) A closed force triangle enforces $\sum \vec{F} = \vec{0}$. (c) Once R_A is determined, A_x and F_{AD} may be determined.

A handle for pushing a cart is shown. The handle has the feature that it may be easily folded against the side of the cart when it is not needed. Determine the forces supporting handle ABC when the 50 N force is applied.

SOLUTION

Road Map This problem is readily solvable by the methods and approaches discussed in Section 5.2. Namely, an FBD is drawn, and equilibrium equations are written and solved to determine the support reactions. However, once the FBD is drawn, a close examination of this shows that member ABC is a three-force member, and this offers an alternative solution approach. Both solutions are carried out and contrasted.

Solution 1

Modeling Members AD and BE are both two-force members. Thus the force supported by member AD acts along the line connecting points A and D, and similarly, the force supported by member BE acts along the line connecting points B and E. Taking advantage of these features of two-force members gives the FBD shown in Fig. 2, where we have assumed that the contact between the handle and cart at point A is frictionless.

Governing Equations & Computation Using the FBD in Fig. 2, equilibrium equations can be written and immediately solved as follows:

$$\sum M_B = 0: \quad -(50\,\text{N})(\cos 40°)(15\,\text{mm}) + (50\,\text{N})(\sin 40°)(70\,\text{mm}) - F_{AD}(90\,\text{mm}) = 0 \tag{1}$$

$$\Rightarrow \boxed{F_{AD} = 18.61\,\text{N},} \tag{2}$$

$$\sum F_y = 0: \quad (-50\,\text{N})(\sin 40°) + F_{BE}\left(\frac{120}{150}\right) - F_{AD} = 0 \tag{3}$$

$$\Rightarrow \boxed{F_{BE} = 63.44\,\text{N},} \tag{4}$$

$$\sum F_x = 0: \quad (50\,\text{N})(\cos 40°) + F_{BE}\left(\frac{90}{150}\right) - A_x = 0 \tag{5}$$

$$\Rightarrow \boxed{A_x = 76.37\,\text{N}.} \tag{6}$$

Solution 2

Modeling This solution begins with all of the modeling considerations from Solution 1, resulting in the FBD shown in Fig. 2. Examination of member ABC in Fig. 2 shows it is a three-force member, because it has forces applied at three points only (points A, B, and C). To carry out a solution that exploits the properties of a three-force member, we combine the two forces A_x and F_{AD} at point A in Fig. 2 into a single force, which we will call R_A and whose magnitude is $R_A = \sqrt{A_x^2 + F_{AD}^2}$. Both the magnitude and orientation of R_A are unknown. The FBD in Fig. 2 is revised to give the FBD shown in Fig. 3(a), where we use the fact that the lines of action of the three forces applied to a three-force member must intersect at a common point, which is point G in Fig. 3(a).

Governing Equations & Computation Using the FBD in Fig. 3(a), the equilibrium equation $\sum M_G = 0$ is automatically satisfied, and the equilibrium equation $\sum \vec{F} = \vec{0}$ is satisfied using the force polygon shown in Fig. 3(b) where the three forces are added head to tail to form a closed polygon. The calculations needed to determine the unknowns F_{BE}, R_A, and α from Fig. 3(b) are tedious, so we will only summarize the procedure. The 50 N force and F_{BE} both have known lines of action, and in the FBD shown in Fig. 3(a), we must determine the location where these lines of action intersect, point G. This calculation involves applications of the laws of sines and/or cosines. Because member ABC is a three-force member, the line of action of R_A must also intersect point G, and this provides the orientation α that this force must have. To determine the value of R_A, we construct the force polygon (for a three-force member this will always be a triangle) as shown in Fig. 3(b) (this enforces $\sum F_x = 0$ and $\sum F_y = 0$). Application of the laws of sines and/or cosines provides the value of R_A. Finally, A_x and F_{AD} are determined by resolving R_A into components as shown in Fig. 3(c).

Discussion & Verification Occasionally, the second solution approach may provide a clever solution for a particular three-force member problem. However, the first solution approach is more methodical and is very robust (that is, it can always be used, regardless of whether a member is a three-force member or not). Furthermore, even for three-force member problems, the first solution approach will usually be more straightforward than the second.

EXAMPLE 5.11 *Introduction to a Statically Indeterminate Problem*

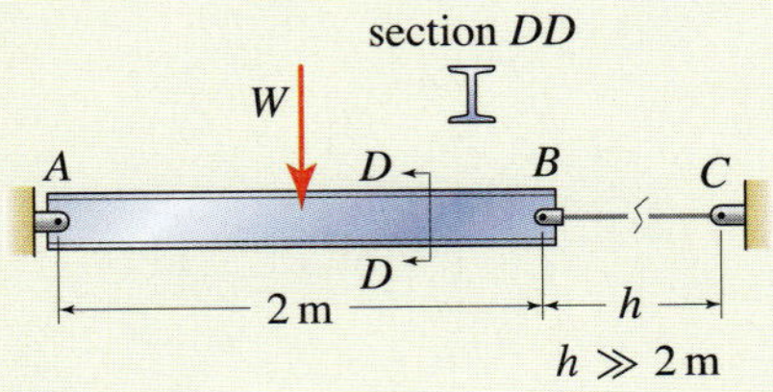

Figure 1

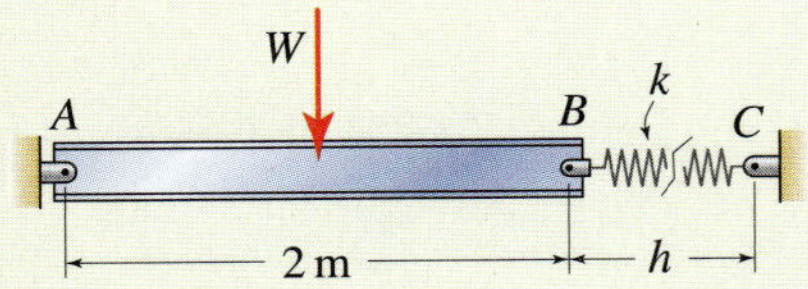

Figure 2
Cable BC is modeled with a spring having stiffness k.

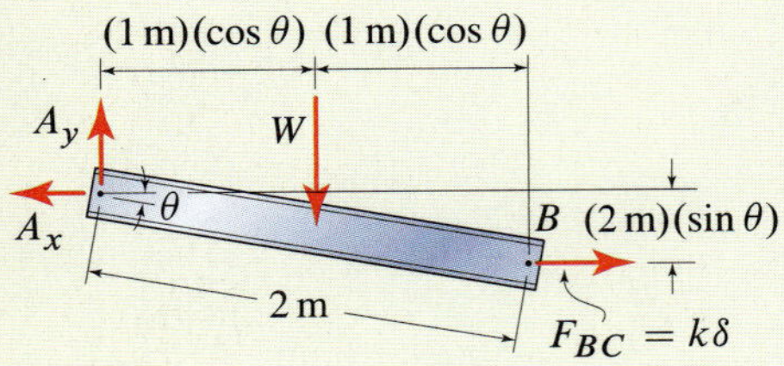

Figure 3
Deformed position of the beam where the spring's length is assumed to be large enough that the spring force is always horizontal.

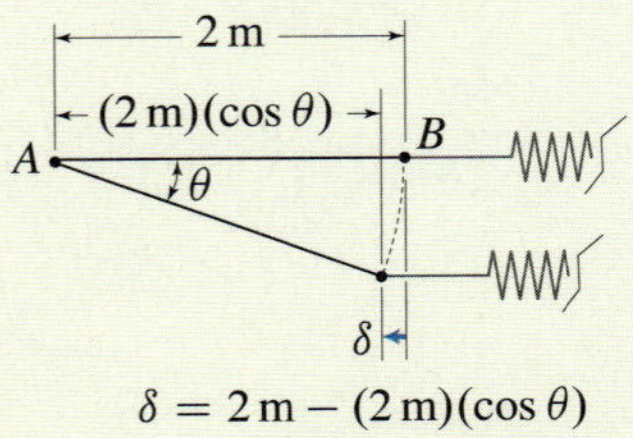

Figure 4
Computation of the spring's elongation δ.

An I beam with weight W acting at its center is supported by a long cable from points B to C. When the beam and cable are horizontal, the cable supports zero force. Determine the support reactions if $W = 1$ kN.

SOLUTION

Road Map There are two unknown reactions at pin A, and the force supported by cable BC is unknown. Hence, $n = 3$ in Eq. (5.17) on p. 274. Examination of the structure shows that it has only partial fixity; the beam can undergo small rigid body rotation about point A. If this is not clear, then drawing an FBD as suggested in Fig. 5.20 may help show that it is partially fixed. Equation (5.17) therefore indicates the beam is statically indeterminate, and it is therefore not possible to determine all the unknowns by using just the three equations of static equilibrium.

While the analysis of statically indeterminate problems is normally beyond the limits of statics, we can occasionally perform accurate analysis by using simple springs to introduce deformability into the model for the problem.

Modeling The cable in Fig. 1 is probably considerably more deformable than the I beam, and thus it is reasonable to model the beam as being rigid and the cable as a spring with stiffness k, as shown in Fig. 2. In this model, we will take $k = 100$ kN/m, although more generally it is possible to precisely specify this stiffness, given the cable's cross-sectional area, length, and material. The FBD for the beam is shown in Fig. 3 where we have assumed the spring's length is large enough that its force is always horizontal.

Governing Equations

Equilibrium Equations Summing moments about point A provides

$$\sum M_A = 0: \quad -W(1\,\text{m})(\cos\theta) + F_{BC}(2\,\text{m})(\sin\theta) = 0. \tag{1}$$

Force Laws The spring force F_{BC} is related to the spring's deformation δ by

$$F_{BC} = k\delta. \tag{2}$$

Kinematic Equations Using Fig. 4, the spring's deformation δ is related to the rotation θ of the I beam by

$$\delta = (2\,\text{m})(1 - \cos\theta). \tag{3}$$

Computation Combining Eqs. (1) through (3) gives

$$W(1\,\text{m})(\cos\theta) - k(2\,\text{m})(1 - \cos\theta)(2\,\text{m})(\sin\theta) = 0, \tag{4}$$

where $W = 1$ kN, $k = 100$ kN/m, and θ is the unknown to be determined. Equation (4) is easily solved using computer mathematics software to obtain $\boxed{\theta = 0.171\text{ rad,}}$ which is equivalent to $\boxed{9.77^\circ.}$ Alternatively, a reasonably accurate solution can be obtained by simply plotting the value of the left-hand side of Eq. (4) versus θ to determine the value of θ for which this expression is approximately zero. Once the solution for θ is known, the spring force and reactions are easily determined as follows. Using Eqs. (2) and (3), the spring force is $\boxed{F_{BC} = 2.903\text{ kN.}}$ Then writing equations for $\sum F_x = 0$ and $\sum F_y = 0$ provides $\boxed{A_x = 2.90\text{ kN and } A_y = 1\text{ kN.}}$

Discussion & Verification Use of springs to develop models for structures in the fashion demonstrated here is common and very powerful.

PROBLEMS

Problem 5.33

Derive Eq. (5.12) on p. 270.

Problems 5.34 and 5.35

Determine the reactions at point A and the force supported by the cable.

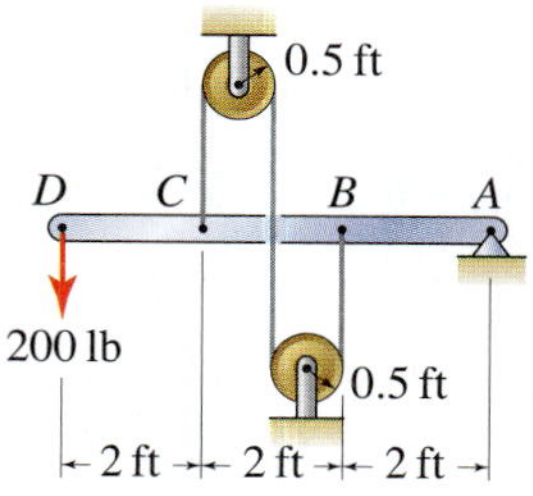

Figure P5.34

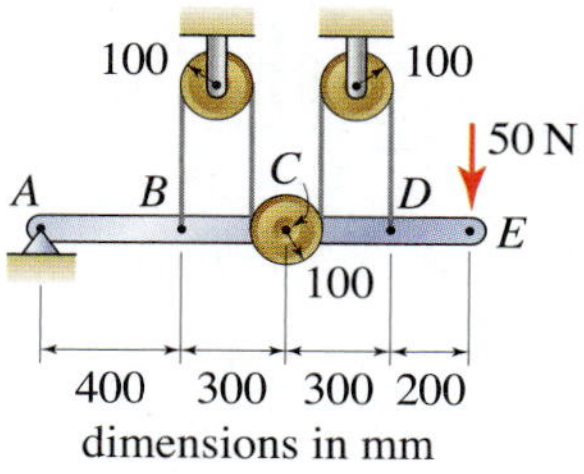

Figure P5.35

Problems 5.36 and 5.37

The crane is supported by a pin at point A and a roller at point B. A winch at point C is used to raise and lower loads. The pulleys all have 350 mm radius, and cable segment ED is horizontal. Determine the support reactions if $P = 10$ kN and

(a) $\theta = 0°$.

(b) $\theta = 30°$.

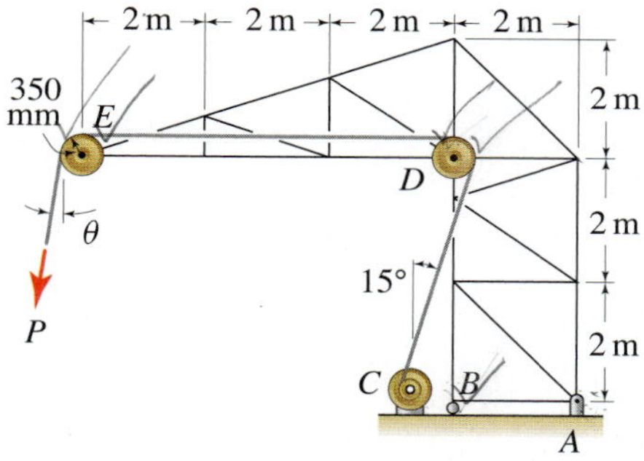

Figure P5.36

Problem 5.38

A walkway for loading and unloading ships at a wharf is shown. The elevation of the walkway is controlled by cable BCD, which is attached to a drum on a geared motor at B. If the 1 kip force is vertical and is positioned halfway between points A and C, determine the forces supported by cables BCD and DE, the reactions at A, and the force supported by bar DF.

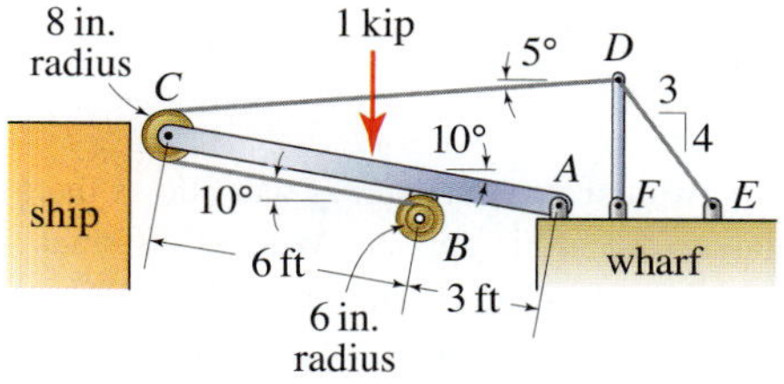

Figure P5.38

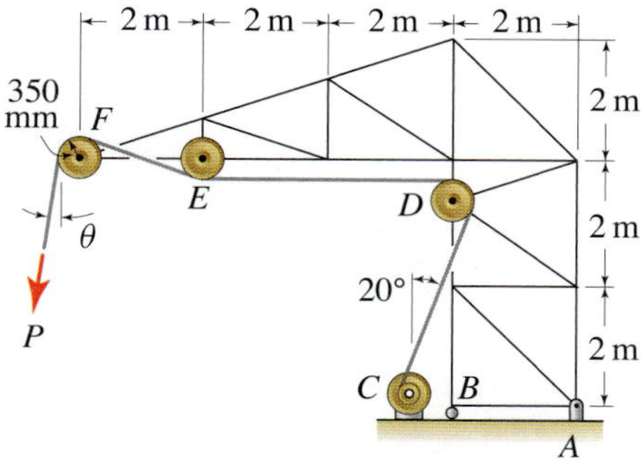

Figure P5.37

Problem 5.39

Repeat Prob. 5.38 if the 1 kip vertical force is positioned at the bearing of pulley C.

Problems 5.40 through 5.42

In the structure shown, member $ABCD$ is supported by a pin at C and a cable that wraps around pulley E, which is frictionless.

(a) Specify if member $ABCD$ has complete fixity or partial fixity and whether it is statically determinate or statically indeterminate.

(b) Draw the FBD for member $ABCD$, and determine the cable tension in terms of force F and length L. Comment on any difficulties that might arise in your analysis.

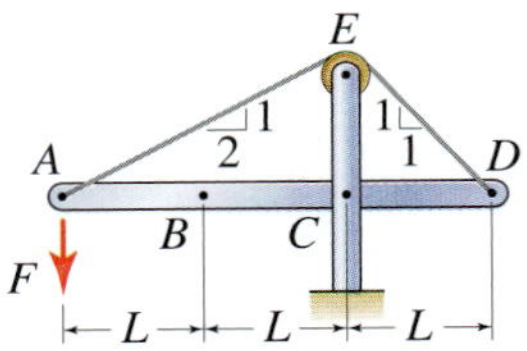

Figure P5.40

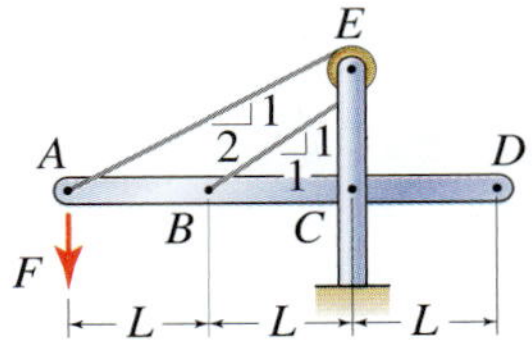

Figure P5.41

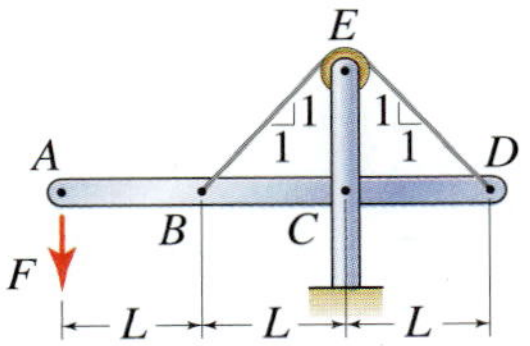

Figure P5.42

Problem 5.43

An office chair has a compressed spring that allows the chair to tilt backward when a sufficiently large force F is applied. When F is small, the stop at point A prevents the chair from tilting forward. If the spring has 10 lb/in. stiffness and 15 in. unstretched length, determine the value of F that will cause the chair to begin tilting backward.

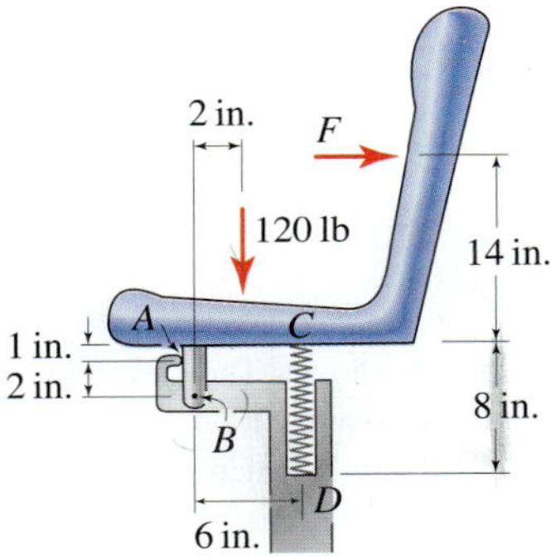

Figure P5.43

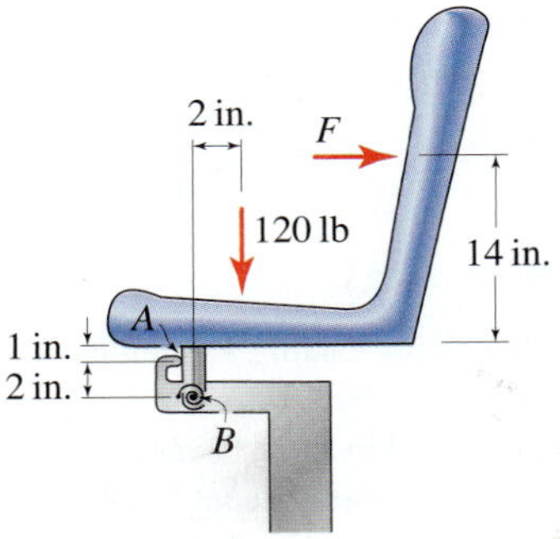

Figure P5.44

Problem 5.44

An office chair has a prewound torsional spring that allows the chair to tilt backward when a sufficiently large force F is applied. When F is small, the stop at point A prevents the chair from tilting forward. If the spring has 100 in.·lb/rad stiffness and is prewound by 3/4 of a turn, determine the value of F that will cause the chair to begin tilting backward.

Problem 5.45

A model for a 110 V electrical wall switch is shown. You may use your discretion to specify an appropriate value for force Q to operate the switch, and to decide if Q is always horizontal or is always perpendicular to line BCD. Specify appropriate values for the spring stiffness and the initial length of the spring.

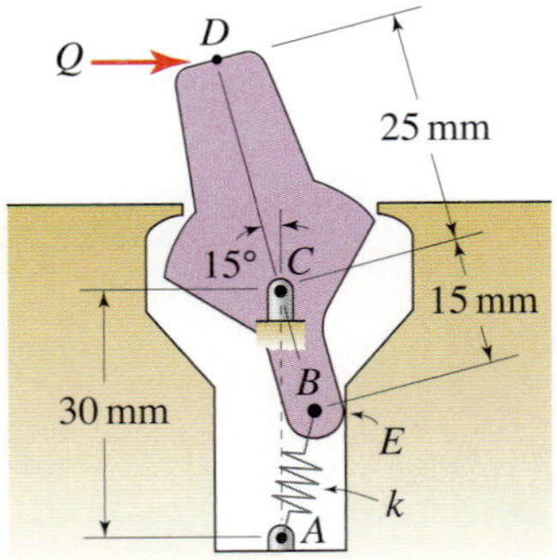

Figure P5.45

Problem 5.46

A stiff fiberglass antenna is supported by a coiled spring at point A that has torsional stiffness $k_t = 50\ \text{N}\cdot\text{m/rad}$. Force F at point B is always horizontal and models wind forces on the antenna. Determine the rotation θ of the antenna if

(a) $F = 5\,\text{N}$.

(b) $F = 50\,\text{N}$.

(c) Discuss why the answer for Part (b) is not 10 times greater than the answer for Part (a).

Hint: When appropriate in this problem, you should use the original geometry when writing equilibrium equations. When this simplification is employed, you should discuss its validity.

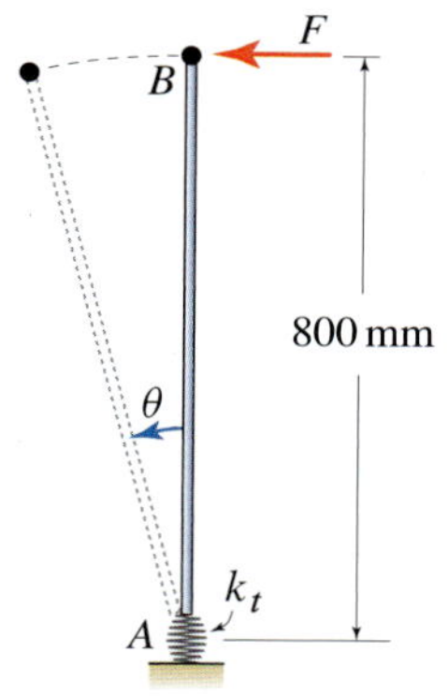

Figure P5.46

Problems 5.47 and 5.48

Draw the FBD for the structure shown. Then write the four equilibrium equations $\sum F_x = 0$, $\sum F_y = 0$, $\sum M_C = 0$, and $\sum M_D = 0$. If possible, solve these equations to determine the support reactions. Discuss the difficulties that arise.

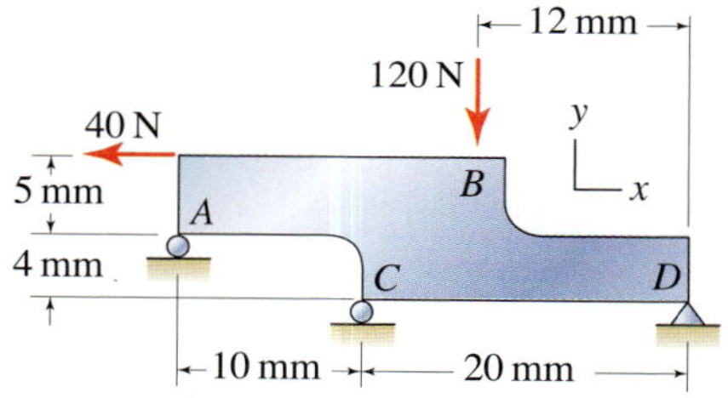

Figure P5.47

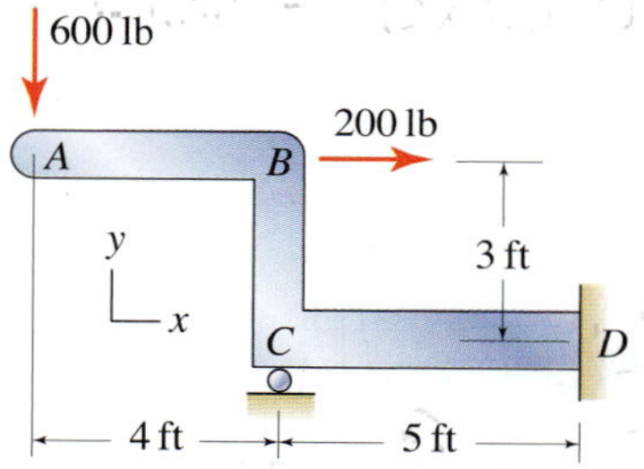

Figure P5.48

Problem 5.49

The I beam shown is statically indeterminate. Under certain circumstances, it may be appropriate to use a model where the I beam is rigid and the roller supports at points B and C are replaced by vertical springs of equal stiffness so that the support reactions may be determined. Do this and find the reactions at points A, B, and C.

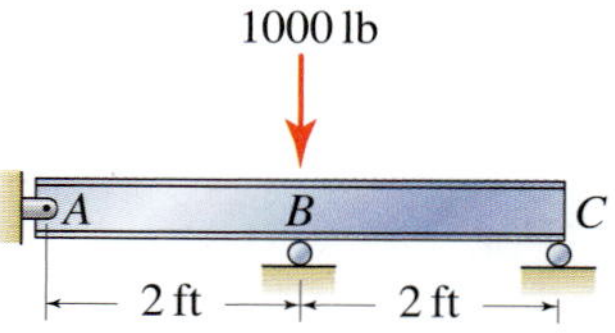

Figure P5.49

Problem 5.50

The I beam shown in Fig. P5.50(a) is statically indeterminate. Under certain circumstances, it may be appropriate to use the model in Fig. P5.50(b) where the I beam is rigid, the built-in support at point A is replaced by a pin and torsional spring with stiffness k_t, and the roller support at point C is replaced by a vertical spring with stiffness k. Use this model to determine the reactions at points A and C. Express your answers in terms of parameters such as F, L, k, k_t, etc.

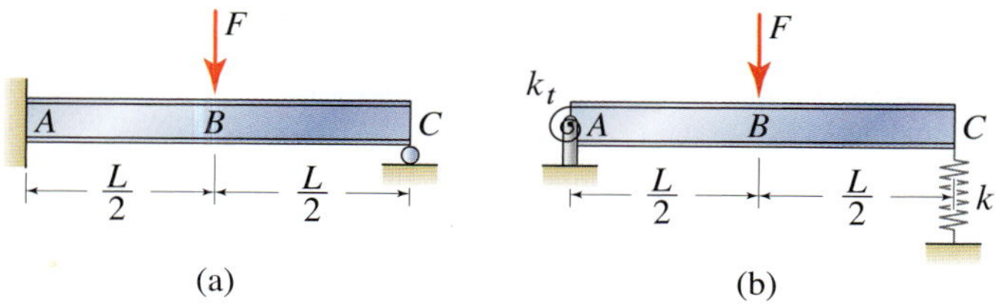

Figure P5.50

Problems 5.51 and 5.52

For each object shown, specify whether it has partial fixity or full fixity and whether it is statically determinate or statically indeterminate.

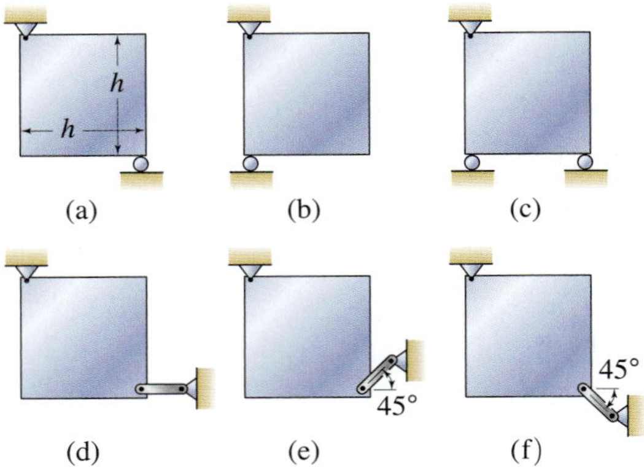

Figure P5.51

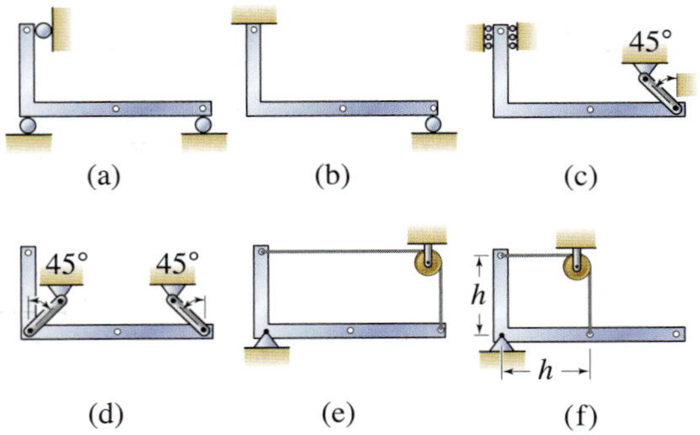

Figure P5.52

Problem 5.53

Without solving, speculate on the difficulty of each of Probs. 5.54 through 5.56.
Note: Concept problems are about *explanations*, not computations.

Problem 5.54

Repeat Example 5.7 on p. 281, using the actual geometry when writing the equilibrium equations to determine the lift and drag forces L and D. Assume L is vertical and D is horizontal. In your opinion, are the differences between your answers here and those in Example 5.7 acceptable? Describe some ways the design of the wind tunnel of Example 5.7 could be changed so that these differences are reduced (e.g., change of dimensions, spring stiffnesses, etc.).

Problem 5.55

In Example 5.7 on p. 281, if $L = 2$ N and $D = 0.3$ N, determine δ and θ, using the original geometry when writing the equilibrium equations.

Problem 5.56

In Example 5.7 on p. 281, if $L = 2$ N and $D = 0.3$ N, determine δ and θ, using the actual geometry when writing the equilibrium equations. Assume L is vertical and D is horizontal.

Problem 5.57

Can the solution to Prob. 5.22 on p. 267 be obtained by superposing the solutions to Probs. 5.20 and 5.21? Explain.
Note: Concept problems are about *explanations*, not computations.

Problem 5.58

Consider the structure from Example 5.8 on p. 282, shown again here where W is the weight of the structure with center of gravity at point H.

(a) If $W = 2$ kip, determine the support reactions due to the weight of the structure only (i.e., $P = 0$).

(b) Use superposition of the results from Part (a) and Example 5.8 to determine the total values of the support reactions when $W = 2$ kip and $P = 3$ kip.

(c) Use superposition of the results from Part (a) and Example 5.8 to determine the total values of the support reactions when $W = 1.8$ kip and $P = 4$ kip.

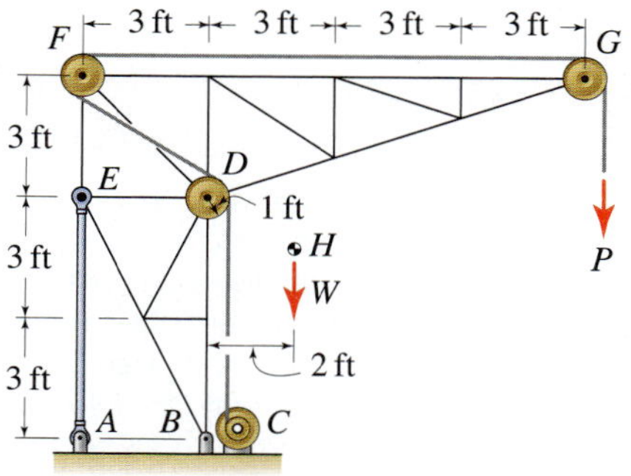

Figure P5.58

Problem 5.59

For each of the support schemes shown in Fig. 5.19 on p. 275, apply a vertical downward force P, at location B as shown in Fig. 5.20, and specify if the object is a two-force, three-force, or general multiforce member.

Problems 5.60 through 5.72

Identify each of the members cited below as a zero-force, two-force, three-force, or multiforce member.

Problem 5.60 Members ABC and BD in Example 5.1 on p. 256.

Problem 5.61 Member $ABCD$ in Example 5.2 on p. 258.

Problem 5.62 Door $ABDE$ and strut BC in Example 5.3 on p. 259.

Problem 5.63 Plate $ABCDE$ in Example 5.4 on p. 260.

Problem 5.64 Drum and contents in Example 5.5 on p. 261.

Problem 5.65 Drum and contents in Example 5.6 on p. 262.

Problem 5.66 Member ABC in Prob. 5.3 on p. 263.

Problem 5.67 Step AB in Prob. 5.4 on p. 263.

Problem 5.68 Members $ABCD$ and DE in Prob. 5.5 on p. 263.

Problem 5.69 Tray AB and link BC in Prob. 5.6 on p. 264.

Problem 5.70 Member ABC in Prob. 5.18 on p. 266.

Problem 5.71 Member $ABCDE$ in Prob. 5.19 on p. 267.

Problem 5.72 Wrench ABC in Prob. 5.25 on p. 268.

5.4 Equilibrium of Bodies in Three Dimensions

The equations governing static equilibrium of a body in three dimensions were stated in Section 5.1 and are repeated here:

$$\sum \vec{F} = \vec{0} \quad \text{and} \quad \sum \vec{M}_P = \vec{0} \tag{5.18}$$

where subscript P denotes the moment summation point you select. In scalar form, Eq. (5.18) is

$$\begin{aligned} \sum F_x &= 0 & & \sum M_{Px} = 0 \\ \sum F_y &= 0 & \text{and} \quad & \sum M_{Py} = 0 \\ \sum F_z &= 0 & & \sum M_{Pz} = 0. \end{aligned} \tag{5.19}$$

As illustrated in the examples of this section, it is possible to use different points for writing the moment summation expressions in Eq. (5.19).

Fundamentally, the analysis of rigid body equilibrium in three dimensions is the same as analysis in two dimensions. The major differences in three dimensions are that FBDs are usually more intricate, reactions are more complex, and the number of unknowns to be determined and the number of equations to be solved are greater. Comments made in Section 5.2 on alternative equilibrium equations also apply here.

Reactions

Common supports and their associated reactions are shown in Fig. 5.25. In all cases, the reaction forces and moments for a particular support may be determined by considering the motion the support prevents. For example, in the case of a bar supported by a pin, the pin prevents motion of the bar in the x, y, and z directions, and thus the pin must produce reaction forces in each of these directions. The pin also prevents rotation of the bar about the y and z axes, and hence there must also be moment reactions about these axes. The pin does not prevent rotation about the x axis; therefore there is no moment reaction about this axis. It is not necessary to memorize the reactions shown in Fig. 5.25. Rather, you should reconstruct these as needed.

More on bearings

As shown in Fig. 5.25, a bearing nominally has two moment reactions. Moment reactions in bearings are sometimes undesirable, especially when two or more bearings are used to support a rotating shaft. The two most common ways to eliminate moment reactions in bearings are by use of a special type of bearing called a *self-aligning* bearing or by use of two or more *perfectly aligned bearings*. Self-aligning bearings are described in the Helpful Information margin note on this page. Perfectly aligned bearings are conventional bearings that are required to be perfectly aligned so that it is justified to assume the moment reactions are zero.

Helpful Information

Moment directions in the scalar approach. In two-dimensional problems all moment vectors are parallel to one another because they are all perpendicular to the plane of the problem. As a consequence, the moment directions may be fully described using words like *clockwise* and *counterclockwise*. In three-dimensional problems, moment vectors generally have components in all three coordinate directions, and if a scalar approach is used, you must decide on a sign convention for each of these directions. Thus, words like *clockwise* and *counterclockwise* have little meaning, and a more robust scheme must be used, as illustrated in the examples of this section.

Helpful Information

Self-aligning bearings. Moment reactions in bearings are sometimes undesirable, and there are some ways these may be eliminated. A special type of bearing, called a *self-aligning* bearing, is common, and an example is shown below in Fig. 5.24.

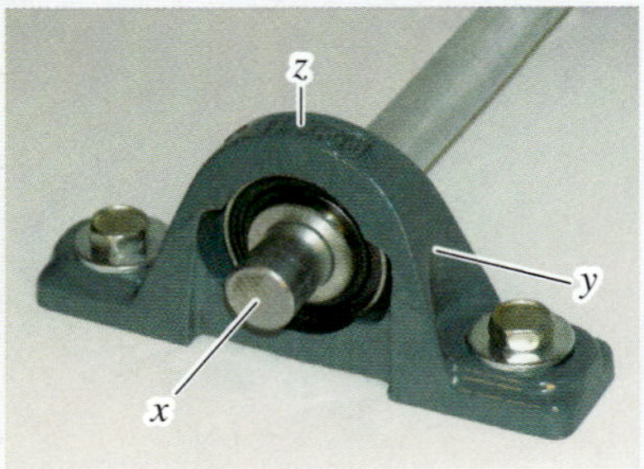

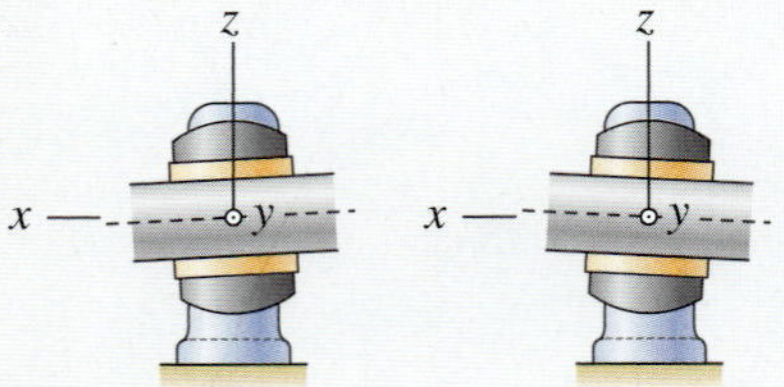

Figure 5.24

A self-aligning bearing allows the shaft to undergo small rotations about the y and z axes so that reactions M_y and M_z are zero.

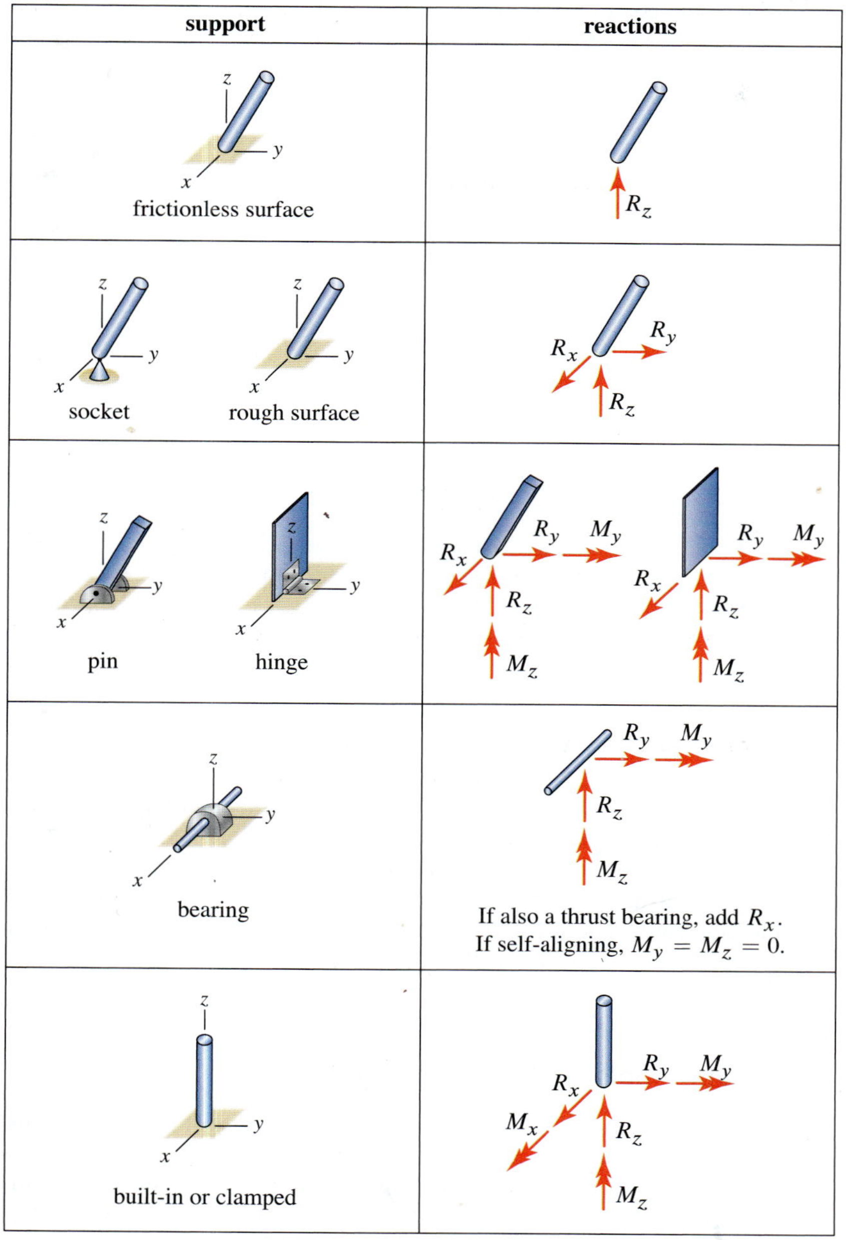

support
reactions
frictionless surface
R_z
socket
rough surface
R_x
R_y
R_z
pin
hinge
R_x
R_y
M_y
R_z
M_z
bearing
R_y
M_y
R_z
M_z
If also a thrust bearing, add R_x.
If self-aligning, $M_y = M_z = 0$.
built-in or clamped
R_x
M_x
R_y
M_y
R_z
M_z

Figure 5.25. Common supports for bodies in three dimensions and the associated reactions.

For example, consider a shaft supported by two bearings as shown in Fig. 5.26(a). Normally, bearings will have moment reactions, and the FBD for this is shown in Fig. 5.26(b). This FBD has eight unknown reactions and is statically indeterminate because the six available equilibrium equations are not sufficient to determine all of the reactions. If the bearings at A and C are *self-aligning* or are *perfectly aligned* (subject to the warning in the margin note), then the moment reactions are zero and the FBD is shown in Fig. 5.26(c) which is statically determinate.

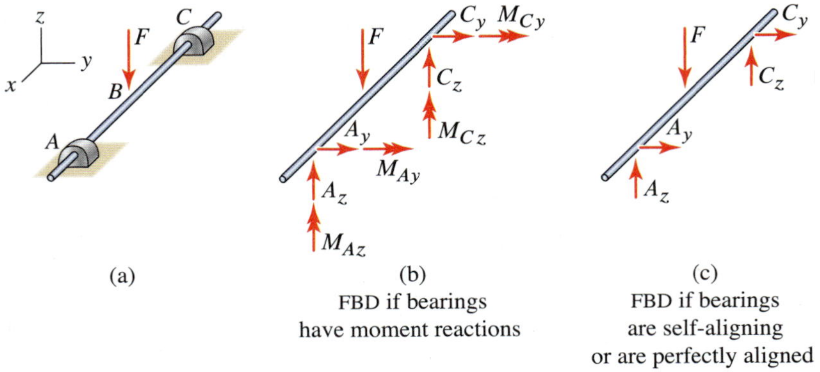

Figure 5.26. (a) A shaft supported by two bearings. (b) The FBD if the bearings have moment reactions. (c) The FBD if the bearings are *self-aligning* or are *perfectly aligned*—observe that the moment reactions are zero.

Common Pitfall

Warning: perfectly aligned bearings. You must be very cautious when assuming the moment reactions for a bearing are zero under the premise that the bearings are perfectly aligned. Despite your assumption, there are many reasons for which the moment reactions may not be zero including poor construction (i.e., the bearings are not as well aligned as you anticipated), misalignment that develops due to wear or vibration, or large deflection of the shaft or object being supported that will cause moment reactions to develop, even if the bearings are perfectly aligned.

Bearings are occasionally designed to prevent axial motion of the object they support, and such bearings are called *thrust bearings*. For example, if the bearing shown in Fig. 5.25 is a thrust bearing, the bearing will not allow translation of the shaft in the x direction, and thus the bearing will have a reaction force R_x in addition to those shown in Fig. 5.25.

The remarks made here for perfectly aligned bearings also apply to objects that are supported by two or more hinges. In particular, it is possible to assume that the moment reactions for the hinge shown in Fig. 5.25 are zero if two or more hinges are perfectly aligned.

Scalar approach or vector approach?

The scalar approach requires good visualization ability, skill is needed to correctly identify moment arms, and it is necessary to be consistent with positive and negative directions for moments. Problems with simple geometry can often be effectively solved using a scalar approach. The vector approach can be used for both simple and complex problems and does not require careful visualization, and positive and negative moment directions are automatically accounted for. While the selection of an analysis approach for a particular problem is your choice, you should still be comfortable with both approaches. Most of the example problems of this section use both approaches, and you should contrast the merits of these approaches to help you learn which is more effective for a particular problem.

Solution of algebraic equations

If a scalar approach is used, a thoughtful strategy on the order in which equilibrium equations are written, and the selection of moment summation points and axes, may provide equilibrium equations whose unknowns are uncoupled or are weakly coupled. Uncoupled or weakly coupled equations can be solved with a minimum of algebra. If a vector approach is used and an effective moment summation point can be identified, the resulting equilibrium equations may also be uncoupled or weakly coupled. For both scalar and vector approaches, more complex problems will usually result in equilibrium equations that are highly coupled and are tedious to solve by hand. For such situations, you are encouraged to use computer software or a programmable calculator to solve the equilibrium equations.

Examples of correct FBDs

Figure 5.27 shows several examples of properly constructed FBDs. Comments on the construction of these FBDs follow.

Cable–supported cantilever. After sketching the structure, we apply the 200 lb force at D, and the 50, 30, and 20 lb weights of portions OB, BC, and CD of the structure, respectively; because each of these portions is uniform, these weights are applied at the center of each portion. Because the cable passes over a frictionless ring at A, and because we are neglecting the weight of the cable, the force T supported throughout the cable is the same and is taken to be positive in tension. The support at O prevents the structure from translating in the x, y, and z directions, and hence there are reaction forces O_x, O_y, and O_z. Similarly, the support at O prevents the structure from rotating about the x, y, and z axes, and hence there are reaction moments M_{Ox}, M_{Oy}, and M_{Oz}. Observe that this FBD is statically indeterminate: there are seven unknown forces and moments and only six equilibrium equations.

Storage chest. After sketching the lid, we apply its 15 lb weight. This weight is vertical, and because the lid is uniform, it acts through the center of the lid, point E. The cord force is T_{CD}, taken to be positive in tension. Each hinge prevents translation in all three coordinate directions and prevents rotation about the y and z axes; hence there are three force reactions and two moment reactions at each hinge. Rather than use reactions A_y, A_z, M_{Ay}, and M_{Az}, we could have used reactions A_n, A_t, M_{An}, and M_{At}, where n and t are normal and tangent directions, respectively, to the lid (similar remarks apply to the reactions at B). The FBD as shown in Fig. 5.27 is statically indeterminate: there are 11 unknown forces and moments and only six equilibrium equations. If the hinges are perfectly aligned, we may then assume the four moment reactions are zero. If we further assume that A_x or B_x is zero (this assumption is not warranted unless further information is given in the problem statement), then the problem becomes statically determinate.

Aircraft landing gear. After sketching the landing gear, we apply the 200 and 300 N weights. Member AB is a two-force member, so the force it applies to the landing gear is directed along line AB, and we have selected F_{AB} to be positive in tension. The bearing at O prevents translation in all directions, which gives rise to reaction forces O_x, O_a, and O_b, where a is the direction

examples of correct FBDs

Cable-supported cantilever. The structure is built in at O and is also supported by a single cable that passes through a frictionless ring at A. Portions OB, BC, and CD of the structure are each uniform and weigh 50 lb, 30 lb, and 20 lb, respectively, and the weight supported at D is 200 lb. Draw the FBD for member $OBCD$.

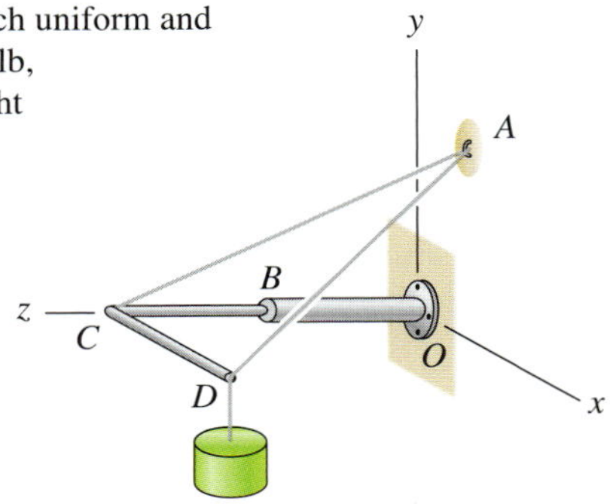

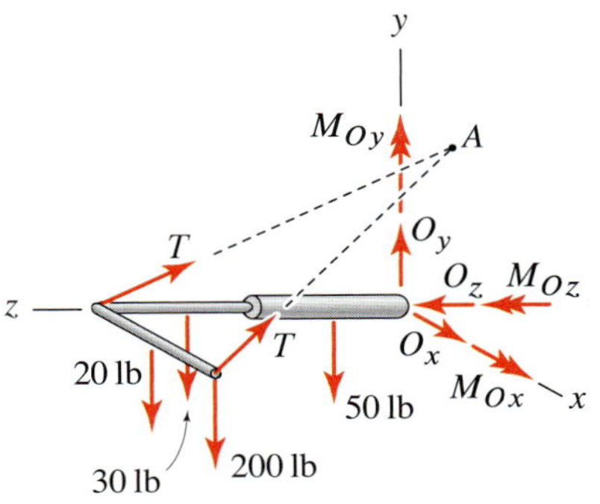

Storage chest. The lid of a storage chest is uniform with 15 lb weight. It is supported by hinges at points A and B, and a cord between points C and D. Draw the FBD for the lid.

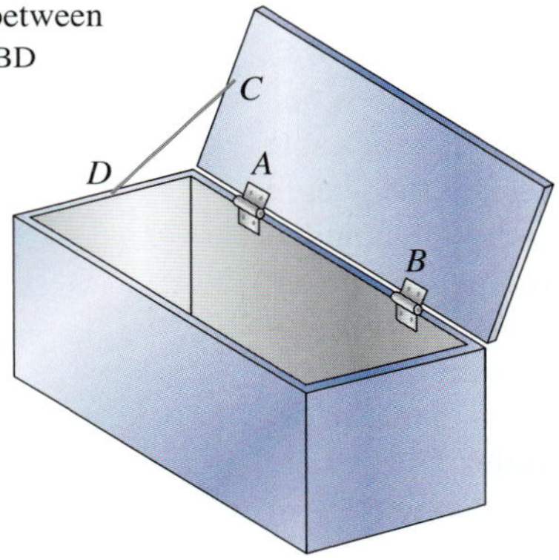

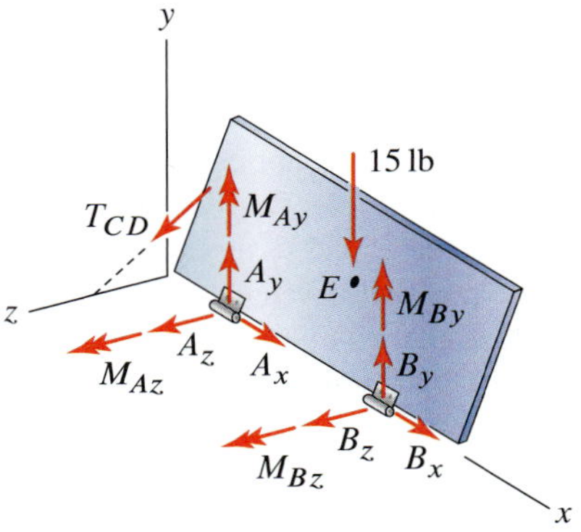

Aircraft landing gear. The landing gear of a jet pivots on a bearing at O and is actuated by two-force member AB. Member OCA has 300 N weight whose line of action passes through point C, and the tire and wheel weigh 200 N. Draw the FBD of member OCA and the wheel.

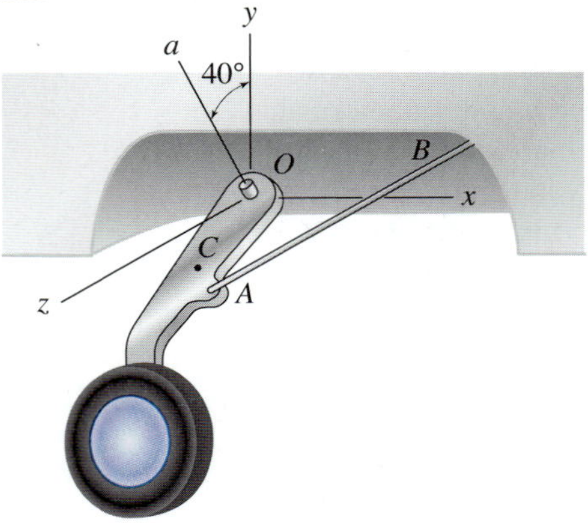

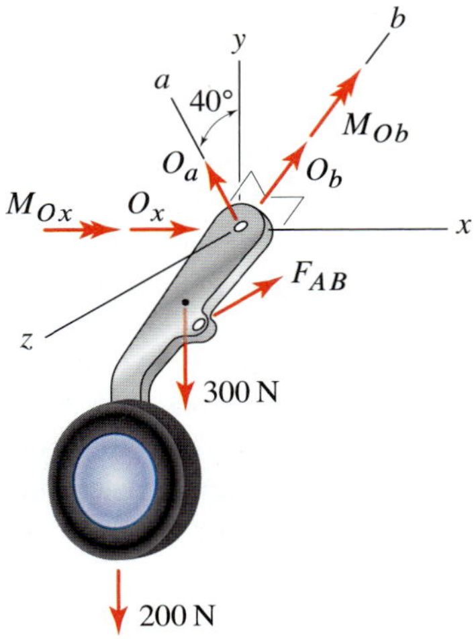

Figure 5.27. Examples of properly constructed FBDs.

about which the landing gear pivots and b is perpendicular to the x and a directions. The bearing also prevents rotation about the x and b directions, so there are reaction moments in these directions. This problem has six reactions and six equilibrium equations, so we expect it to be statically determinate.

Examples of incorrect and/or incomplete FBDs

Figure 5.28 shows several examples of incorrect and/or incomplete FBDs. Comments on how these FBDs must be revised follow, but before reading these, you should study Fig. 5.28 to find as many of the needed corrections and/or additions on your own as possible.

Beam

1. The forces shown at points B and C have the proper directions and positions where C is midway between A and B, but their magnitudes should be $(50\,\text{kg})(9.81\,\text{m/s}^2) = 490.5\,\text{N}$ at B and $(25\,\text{kg})(9.81\,\text{m/s}^2) = 245.3\,\text{N}$ at C.

2. The beam is free to translate in the x and y directions, so A_x and A_y are not reactions and should not be shown.

3. The beam is free to rotate about the z axis, so M_{Az} is not a reaction and should not be shown.

4. The support prevents the beam from rotating about the x and y axes, so moment reactions M_{Ax} and M_{Ay} must be included.

In summary, the reactions at the support consist of A_z, M_{Ax}, and M_{Ay} only. The beam has partial fixity since the number of reactions is less than the number of equilibrium equations.

Shifting fork

1. The 5 N force applied by the gear at G to the fork is properly shown, but there is also a z direction force G_z at that location. If the fit between the fork and the shaft at G is not loose, then additional reaction forces and moments are possible.

2. Because of the thrust collar at H, there is a reaction F_x.

3. The shifting fork may freely rotate on shaft AB; therefore M_{Fx} is not a reaction and should not be shown.

4. Shaft AB prevents the shifting fork from rotating about the y axis; therefore a moment reaction M_{Fy} must be included.

In summary, the reactions on the shifting fork consist of F_x, F_y, F_z, M_{Fy}, M_{Fz}, and G_z only. This problem has six reactions and six equilibrium equations, so we expect it to be statically determinate. This problem is revisited again in Example 8.3 on p. 463, and the correct FBD is shown there in Fig. 2.

examples of incorrect and/or incomplete FBDs

Beam. A uniform beam with 25 kg mass is supported by a slotted block at A that is built in. The slot is frictionless and allows the beam to slide freely. End B of the beam supports a 50 kg mass. Draw the FBD for the beam.

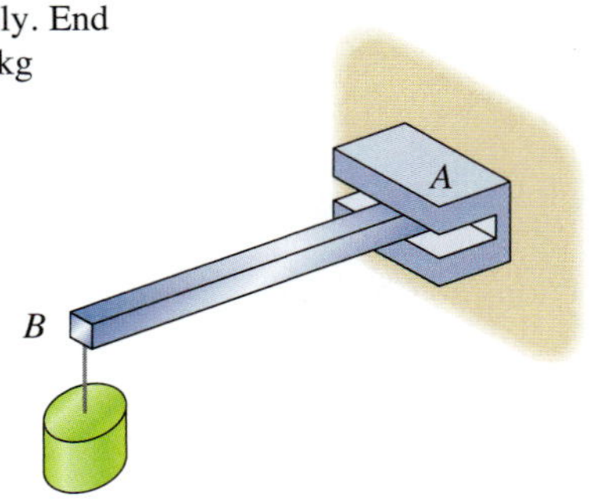

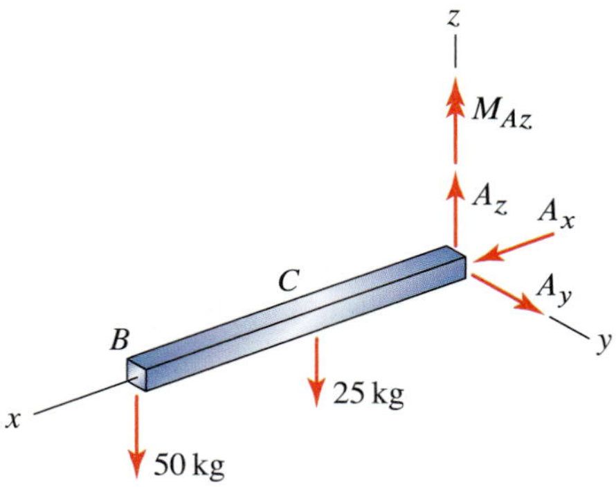

Shifting fork. Member EFG is a shifting fork used in a transmission to move the gear at G along shaft CD. It is actuated by the 10 and 20 N forces at E, and the gear at G applies a 5 N force to the fork in the direction from D to C. The fork is supported by a fixed shaft AB that has a thrust collar at H. Draw the FBD for the shifting fork, assuming the fork is a loose fit on the shaft at G.

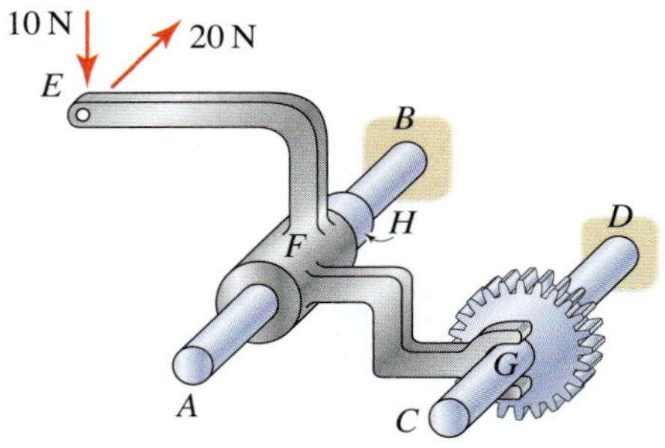

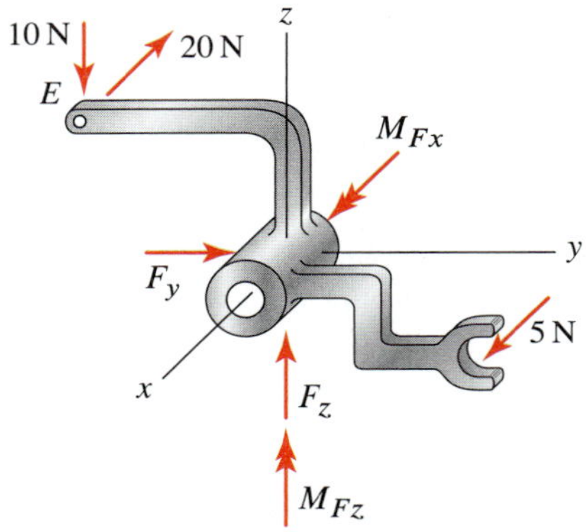

Crop planting positioner. The device shown is a pointer that mounts on the front of a tractor to help its operator position the tractor relative to other rows of seeds that have been planted. Bracket E is bolted to the front of the tractor, which drives in the z direction. The bracket supports the bent boom ABC, the end C of which has a weight W and a pointer CD. The boom is allowed to rotate about line a, which lies in the yz plane, so that if the boom strikes an obstruction, the boom will rotate backward and upward to help avoid damage to it. Draw the FBD of the boom and pointer, lumping all of the reaction forces at A and B into a single system of forces and/or moments at point B.

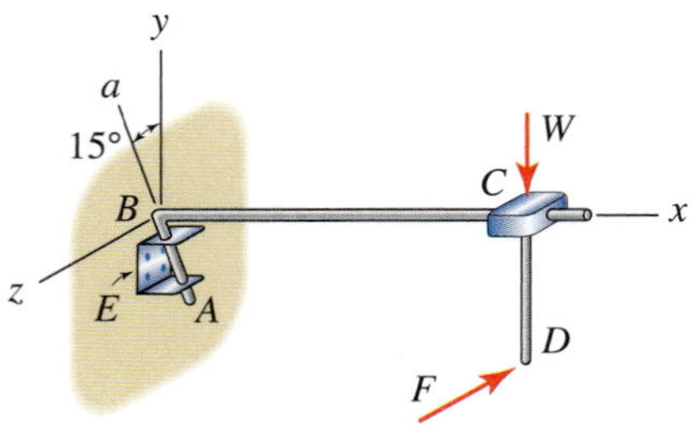

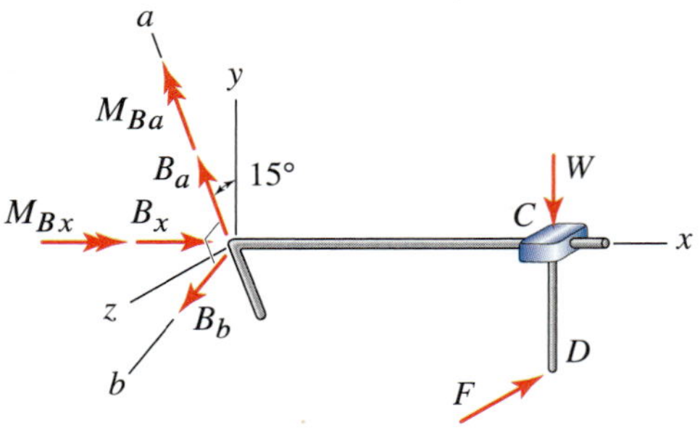

Figure 5.28. Examples of incorrect and/or incomplete FBDs.

Crop planting positioner

1. The support at E allows the boom to rotate about the a direction; therefore M_{Ba} is not a reaction and should not be shown.

2. The support at E prevents the boom from rotating about the b direction; therefore a moment reaction M_{Bb} must be included.

In summary, the reactions at B consist of B_x, B_a, B_b, M_{Bx}, and M_{Bb} only. This problem has five reactions and six equilibrium equations, so it has partial fixity. Although the problem description does not provide these details, most likely the device has a stop at E so that if $F = 0$, rotation about the a direction is prevented.

End of Section Summary

In this section, static equilibrium of a body in three dimensions was discussed. Some of the key points are as follows:

- Problems in three dimensions with simple geometry can often be effectively solved using a scalar approach. But very often a vector approach will be more tractable.

- If a scalar approach is used, a sign convention for positive moments about the three Cartesian coordinate directions must be adopted (the right-hand rule is customary), and great care must be exercised when summing moments.

- In general, a bearing has moment reactions, and often these play an important role in the equilibrium of a structure. In structures with multiple bearing supports, moment reactions at the bearings may be undesirable and may be eliminated by using special bearings called *self aligning bearings* or by using conventional bearings but requiring their axes to be *perfectly aligned*. If you design or analyze a structure and assume reaction moments at bearings are zero by virtue of perfect alignment, you must be very careful to ensure this assumption is valid.

- In general, equilibrium of a single rigid body requires the solution of a system of six algebraic equations, which can be tedious to solve by hand. Software for solving such systems of equations can be very helpful.

EXAMPLE 5.12 *Socket and Cable Supports*

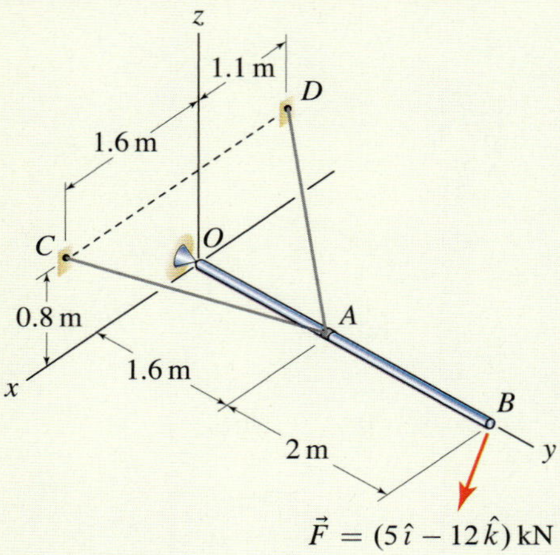

Figure 1

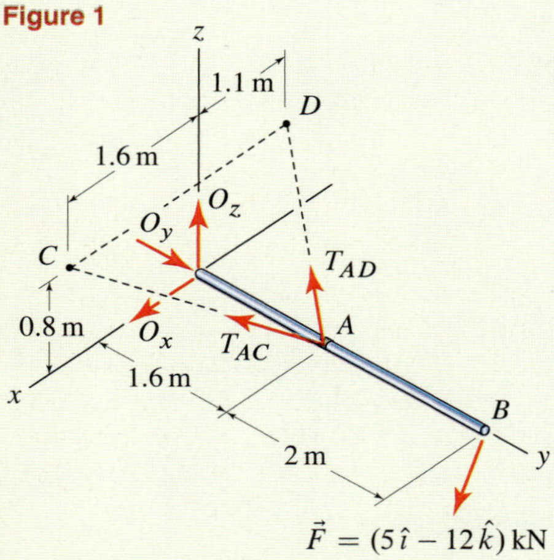

Figure 2
Free body diagram for the vector approach.

Interesting Fact

Partial fixity or full fixity? You might have noticed there are only five unknowns in this example, while there are six equilibrium equations. Generalizing the equation counting idea described in Section 5.3 to three dimensions, we see that the boom has only partial fixity. Examining the FBD in Fig. 2 reveals that none of the forces prevent rotation of the boom about the y axis. However, this partial fixity may be an artifact of our modeling idealization, and the real structure may have some minor support details that render it fully fixed.

Boom OAB is supported by a socket at point O and two cables. Determine the support reactions at O and the forces supported by the two cables.

SOLUTION

Road Map Both vector and scalar solutions are effective, and both are used.

Vector solution

Modeling The FBD is shown in Fig. 2 where T_{AC} and T_{AD} are defined to be positive in tension. The reactions at the socket support may be determined by consulting Fig. 5.25, but it is easier to simply construct these by considering the motion the support prevents. That is, the support at O prevents translation of the boom in each of the x, y, and z directions, and hence there must be reaction forces O_x, O_y, and O_z in these directions. The socket allows rotation of the boom about all three coordinate axes, and thus, there are no moment reactions.

Governing Equations & Computation The equilibrium equations to be used are $\sum \vec{F} = \vec{0}$ and $\sum \vec{M}_P = \vec{0}$. Careful consideration of which of these equations to write first and where to locate point P may help reduce the algebra required to solve for the unknowns. In this problem, $\sum \vec{M}_O = \vec{0}$ is an effective choice: unknowns O_x, O_y, and O_z pass through the moment summation point and hence do not enter the expression, leaving T_{AC} and T_{AD} as the only unknowns.* With the following vector expressions,

$$\vec{r}_{OA} = 1.6\,\hat{\jmath}\ \text{m}, \qquad \vec{T}_{AC} = T_{AC}\frac{1.6\,\hat{\imath} - 1.6\,\hat{\jmath} + 0.8\,\hat{k}}{2.4}, \tag{1}$$

$$\vec{r}_{OB} = 3.6\,\hat{\jmath}\ \text{m}, \qquad \vec{T}_{AD} = T_{AD}\frac{-1.1\,\hat{\imath} - 1.6\,\hat{\jmath} + 0.8\,\hat{k}}{2.1}, \tag{2}$$

we obtain

$$\sum \vec{M}_O = \vec{0}: \quad \vec{r}_{OA} \times \vec{T}_{AC} + \vec{r}_{OA} \times \vec{T}_{AD} + \vec{r}_{OB} \times \vec{F} = \vec{0}, \tag{3}$$

$$\Big[(1.28\,\text{m})\,\hat{\imath} + (-2.56\,\text{m})\,\hat{k}\Big]\frac{T_{AC}}{2.4} + \Big[(1.28\,\text{m})\,\hat{\imath} + (1.76\,\text{m})\,\hat{k}\Big]\frac{T_{AD}}{2.1} + \Big[(-43.2)\,\hat{\imath} + (-18.0)\,\hat{k}\Big]\,\text{kN·m} = \vec{0}. \tag{4}$$

Grouping all terms that multiply $\hat{\imath}$ in Eq. (4) provides the first of the following equations, and grouping all terms that multiply $\hat{k}$ provides the second:

$$\left(\frac{1.28\,\text{m}}{2.4}\right)T_{AC} + \left(\frac{1.28\,\text{m}}{2.1}\right)T_{AD} - 43.2\,\text{kN·m} = 0, \tag{5}$$

$$\left(\frac{-2.56\,\text{m}}{2.4}\right)T_{AC} + \left(\frac{1.76\,\text{m}}{2.1}\right)T_{AD} - 18.0\,\text{kN·m} = 0. \tag{6}$$

Solving Eqs. (5) and (6) provides the solutions

$$\Rightarrow \quad \boxed{T_{AC} = 23.0\,\text{kN} \quad \text{and} \quad T_{AD} = 50.75\,\text{kN}.} \tag{7}$$

The reactions at point O can now be determined by writing

$$\sum \vec{F} = \vec{0}: \quad O_x\,\hat{\imath} + O_y\,\hat{\jmath} + O_z\,\hat{k} + \vec{T}_{AC} + \vec{T}_{AD} + \vec{F} = \vec{0}, \tag{8}$$

which is easily solved for

$$\Rightarrow \quad \boxed{O_x = 6.25\,\text{kN}, \quad O_y = 54.0\,\text{kN}, \quad \text{and} \quad O_z = -15.0\,\text{kN}.} \tag{9}$$

*$\sum \vec{M}_A = \vec{0}$ is another effective choice: unknowns T_{AC}, T_{AD}, and O_y pass through the moment summation point, leaving O_x and O_z as the only unknowns.

Remark. A clever solution for this problem is obtained by examining the FBD in Fig. 2 and recognizing that the sum of moments about line OD, which of course must be zero for equilibrium, involves only *one* unknown,* namely, T_{AC}. To carry out this solution, we first evaluate the moment about some point on line OD, such as point O, in which case Eq. (3) is written, and then we take the dot product of this with a unit vector in the OD direction to write

$$\sum M_{OD} = 0: \quad (\vec{r}_{OA} \times \vec{T}_{AC} + \vec{r}_{OA} \times \vec{T}_{AD} + \vec{r}_{OB} \times \vec{F}) \cdot \hat{u}_{OD} = 0, \tag{10}$$

$$(\vec{r}_{OA} \times \vec{T}_{AC} + \vec{r}_{OB} \times \vec{F}) \cdot \hat{u}_{OD} = 0, \tag{11}$$

where the latter expression can be written because we know $\vec{T}_{AD}$ produces no moment about line OD (if you are uncertain about this statement, you should evaluate $(\vec{r}_{OA} \times \vec{T}_{AD}) \cdot \hat{u}_{OD}$ to verify that it is zero). Problem 5.79 explores this solution in greater detail.

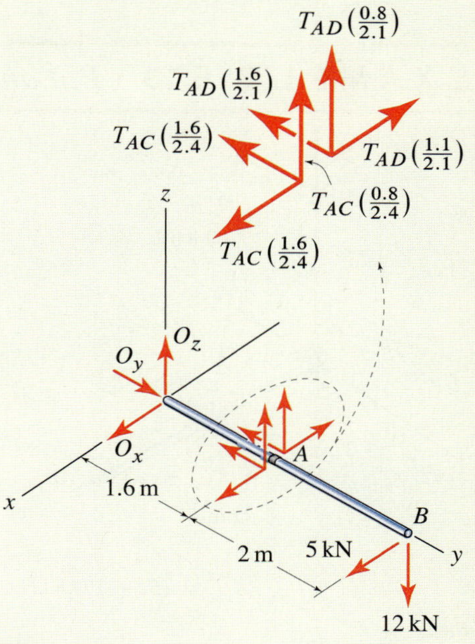

Figure 3
Free body diagram for the scalar approach.

Scalar solution

Modeling The FBD is shown in Fig. 3 where for convenience the cable forces have been resolved into x, y, and z components.

Governing Equations & Computation In this solution we will elect to sum moments about axes passing through point A. With a scalar solution it is necessary to be absolutely consistent with a sign convention for positive and negative moments, and we will generally take moments to be positive in the positive coordinate directions according to the right-hand rule. Thus, we write†

$$\sum M_{Ax} = 0: \quad -O_z(1.6\,\text{m}) - (12\,\text{kN})(2\,\text{m}) = 0 \Rightarrow \boxed{O_z = -15.0\,\text{kN},} \tag{12}$$

$$\sum M_{Az} = 0: \quad O_x(1.6\,\text{m}) - (5\,\text{kN})(2\,\text{m}) = 0 \Rightarrow \boxed{O_x = 6.25\,\text{kN},} \tag{13}$$

$$\sum F_x = 0: \quad O_x + T_{AC}\left(\frac{1.6}{2.4}\right) - T_{AD}\left(\frac{1.1}{2.1}\right) + 5\,\text{kN} = 0, \tag{14}$$

$$\sum F_z = 0: \quad O_z + T_{AC}\left(\frac{0.8}{2.4}\right) + T_{AD}\left(\frac{0.8}{2.1}\right) - 12\,\text{kN} = 0. \tag{15}$$

The first two equations were immediately solved for O_z and O_x, and the last two equations can be solved simultaneously to obtain

$$\Rightarrow \quad \boxed{T_{AC} = 23.0\,\text{kN} \quad \text{and} \quad T_{AD} = 50.75\,\text{kN}.} \tag{16}$$

Finally, the last unknown is obtained from

$$\sum F_y = 0: \quad O_y - T_{AC}\left(\frac{1.6}{2.4}\right) - T_{AD}\left(\frac{1.6}{2.1}\right) = 0 \quad \Rightarrow \quad \boxed{O_y = 54.0\,\text{kN}.} \tag{17}$$

As expected, all of these solutions agree with those obtained using a vector approach.

Discussion & Verification We expect both cables to be in tension, and indeed our solution shows $T_{AC} > 0$ and $T_{AD} > 0$. After these simple checks, we should verify that our solutions satisfy all of the equilibrium equations.

Helpful Information

More on +/− moment directions. If the convention for positive and negative moment directions in the scalar approach is not clear, then let's examine Eq. (13) more closely. Consider a z' direction that is parallel to the z axis and that passes through point A.

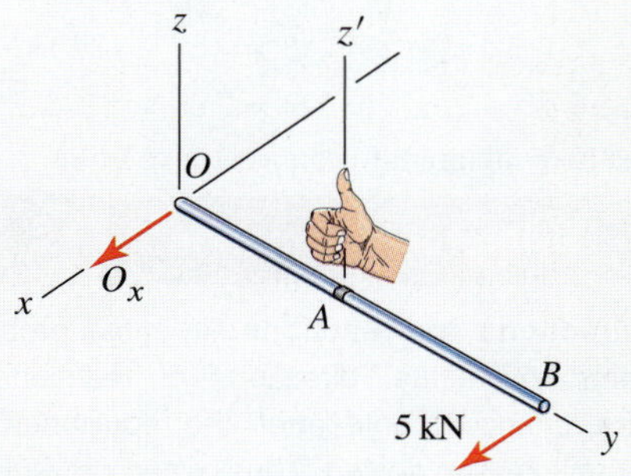

If you point the thumb of your right hand in the z' direction, then your fingers define the direction for positive moment about the z axis through point A. When writing Eq. (13), observe that the moment produced by O_x about the z' axis is in the same direction as your fingers curl, hence it is positive, while the moment produced by the 5 kN force is in the opposite direction, hence it is negative.

* Similarly, the sum of moments about line OC will involve T_{AD} as the only unknown.
† In words, $\sum M_{Ax}$ means "sum of moments about the x axis passing through point A."

EXAMPLE 5.13 *Pin and Cable Supports*

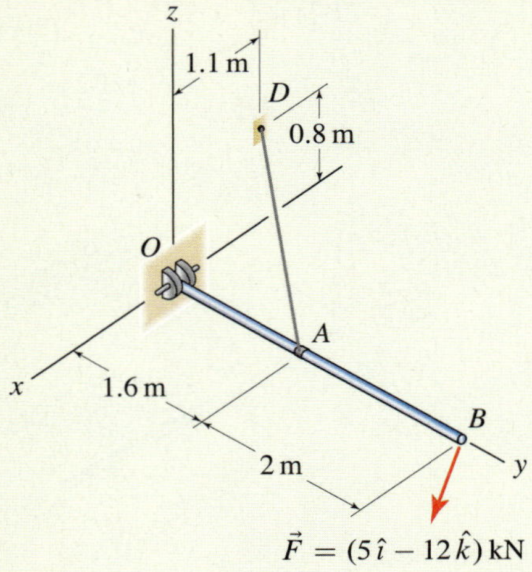

Figure 1

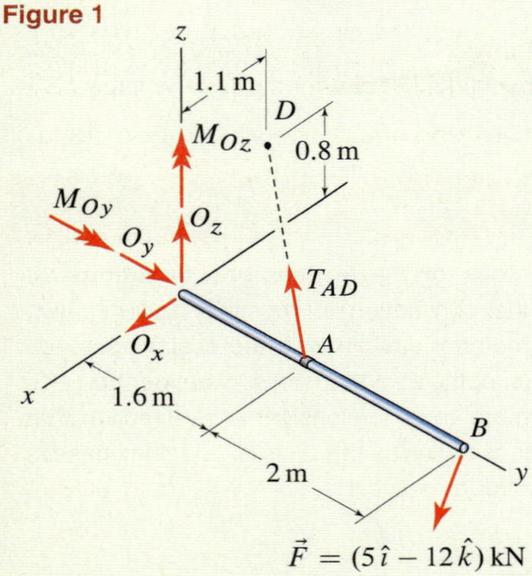

Figure 2
Free body diagram for the vector solution.

Helpful Information

Directions for reactions. In most problems the actual directions of reactions are unknown until after the equilibrium equations are solved. Thus, when drawing an FBD, we will routinely take reactions (both forces and moments) to be positive in the positive coordinate directions, such as shown at point O in Fig. 2. Doing so makes it easy to write a vector expression for the reaction, which in this example is $\vec{O} = O_x\hat{\imath} + O_y\hat{\jmath} + O_z\hat{k}$. Other conventions can be used, provided you are careful when writing vector expressions. For instance, if O_x is taken to be positive in the opposite direction shown in Fig. 2, then you write $\vec{O} = -O_x\hat{\imath} + O_y\hat{\jmath} + O_z\hat{k}$.

Boom OAB is supported by a pin at point O and a cable. Determine the support reactions at O and the force supported by the cable.

SOLUTION

Road Map This problem is similar to Example 5.12, with the difference being in the details of how the boom is supported. Both vector and scalar solutions are effective and both are illustrated.

Vector solution

Modeling The FBD is shown in Fig. 2. The support at O prevents translation of the boom in each of the x, y, and z directions, hence there must be reaction forces O_x, O_y, and O_z in these directions. The pin prevents rotation of the boom about the y and z axes, hence there must be reaction moments M_{Oy} and M_{Oz} about these axes. The cable force T_{AD} is taken to be positive in tension.

Governing Equations & Computation An effective choice for the first equilibrium equation to write is $\sum\vec{M}_O = \vec{0}$ because reaction forces O_x, O_y, and O_z do not enter this equation. However, reaction moments M_{Oy} and M_{Oz} do enter this equation, as does T_{AD}. Nonetheless, $\sum\vec{M}_O = \vec{0}$ provides a system of three scalar equations that can be solved to find the three unknowns it contains. With the following vector expressions

$$\vec{r}_{OA} = 1.6\,\hat{\jmath}\text{ m},\quad \vec{r}_{OB} = 3.6\,\hat{\jmath}\text{ m},\quad \vec{T}_{AD} = T_{AD}\frac{-1.1\,\hat{\imath} - 1.6\,\hat{\jmath} + 0.8\,\hat{k}}{2.1} \tag{1}$$

we obtain

$$\sum\vec{M}_O = \vec{0}:\quad M_{Oy}\,\hat{\jmath} + M_{Oz}\,\hat{k} + \vec{r}_{OA}\times\vec{T}_{AD} + \vec{r}_{OB}\times\vec{F} = \vec{0} \tag{2}$$

$$M_{Oy}\,\hat{\jmath} + M_{Oz}\,\hat{k} + \Big[(1.28\text{ m})\,\hat{\imath} + (1.76\text{ m})\,\hat{k}\Big]\frac{T_{AD}}{2.1} + \Big[(-43.2)\,\hat{\imath} + (-18.0)\,\hat{k}\Big]\text{ kN·m} = \vec{0}. \tag{3}$$

Grouping all terms that multiply $\hat{\imath}$ in Eq. (3) provides the first of the following equations, grouping terms that multiply $\hat{\jmath}$ provides the second, and grouping terms that multiply $\hat{k}$ provides the third:

$$\left(\frac{1.28\text{ m}}{2.1}\right)T_{AD} - 43.2\text{ kN·m} = 0, \tag{4}$$

$$M_{Oy} = 0, \tag{5}$$

$$M_{Oz} + \left(\frac{1.76\text{ m}}{2.1}\right)T_{AD} - 18.0\text{ kN·m} = 0. \tag{6}$$

Equations (4)–(6) are easily solved for:

$$\Rightarrow\quad \boxed{T_{AD} = 70.9\text{ kN},\quad M_{Oy} = 0,\quad\text{and}\quad M_{Oz} = -41.4\text{ kN·m}.} \tag{7}$$

The reactions at point O can now be determined by writing

$$\sum\vec{F} = \vec{0}:\quad O_x\,\hat{\imath} + O_y\,\hat{\jmath} + O_z\,\hat{k} + \vec{T}_{AD} + \vec{F} = \vec{0} \tag{8}$$

which is easily solved for:

$$\Rightarrow\quad \boxed{O_x = 32.1\text{ kN},\quad O_y = 54.0\text{ kN},\quad\text{and}\quad O_z = -15.0\text{ kN}.} \tag{9}$$

Scalar solution

Modeling The FBD is shown in Fig. 3 where for convenience the cable force has been resolved into x, y, and z components.

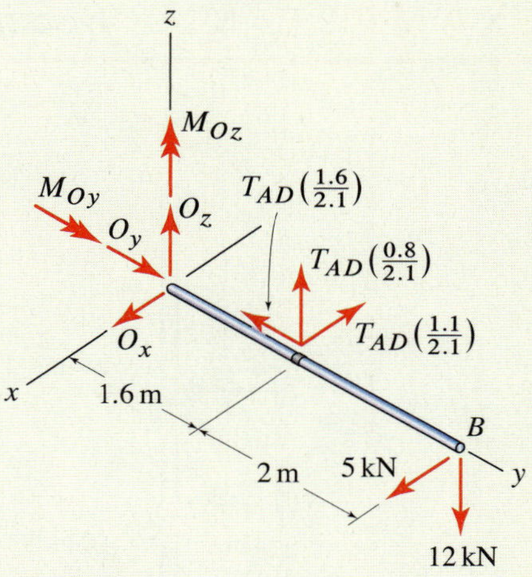

Figure 3
Free body diagram for the scalar solution.

Governing Equations & Computation A variety of different strategies for the order in which the equilibrium equations are written and solved are effective for this problem. In the following equations, we take forces and moments to be positive in the positive coordinate directions.

$$\sum M_{Ox} = 0: \quad T_{AD}\left(\frac{0.8}{2.1}\right)(1.6\,\text{m}) - (12\,\text{kN})(3.6\,\text{m}) = 0, \tag{10}$$

$$\Rightarrow \boxed{T_{AD} = 70.9\,\text{kN},} \tag{11}$$

$$\sum M_{Oy} = 0: \quad M_{Oy} = 0, \tag{12}$$

$$\Rightarrow \boxed{M_{Oy} = 0,} \tag{13}$$

$$\sum M_{Oz} = 0: \quad M_{Oz} + T_{AD}\left(\frac{1.1}{2.1}\right)(1.6\,\text{m}) - (5\,\text{kN})(3.6\,\text{m}) = 0, \tag{14}$$

$$\Rightarrow \boxed{M_{Oz} = -41.4\,\text{kN}\cdot\text{m},} \tag{15}$$

$$\sum F_x = 0: \quad O_x - T_{AD}\left(\frac{1.1}{2.1}\right) + 5\,\text{kN} = 0, \quad \Rightarrow \boxed{O_x = 32.1\,\text{kN},} \tag{16}$$

$$\sum F_y = 0: \quad O_y - T_{AD}\left(\frac{1.6}{2.1}\right) = 0, \quad \Rightarrow \boxed{O_y = 54.0\,\text{kN},} \tag{17}$$

$$\sum F_z = 0: \quad O_z + T_{AD}\left(\frac{0.8}{2.1}\right) - 12\,\text{kN} = 0, \quad \Rightarrow \boxed{O_z = -15.0\,\text{kN}.} \tag{18}$$

As expected, all of these solutions agree with those obtained using a vector solution.

Discussion & Verification We expect the cable to be in tension, and indeed our solution shows $T_{AD} > 0$. We should also verify that our solutions satisfy all of the equilibrium equations.

Common Pitfall

Summing moments. In both the vector and scalar approaches, a common error when summing moments, such as about point O in Eqs. (2), (12), and (14) of this example, is to neglect the moment reactions, the misconception being they make no contribution because their lines of action pass through the moment summation point.

EXAMPLE 5.14 *Bearing Supports*

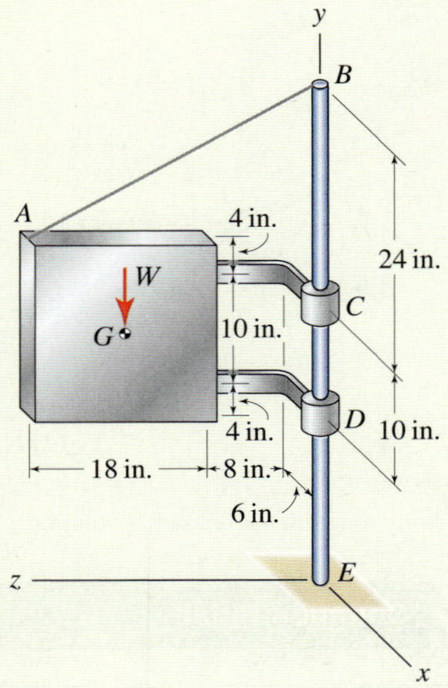

Figure 1

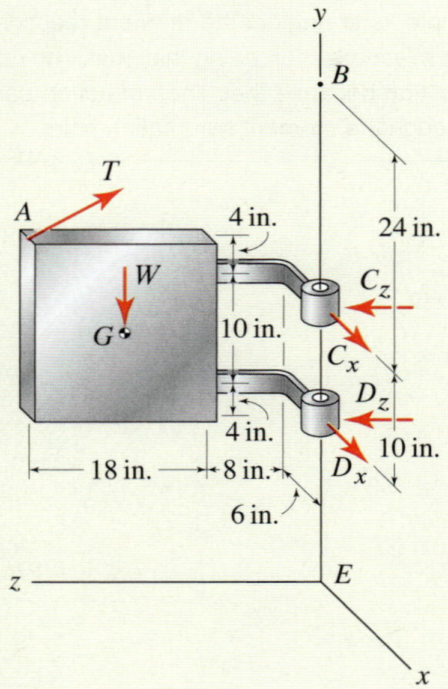

Figure 2
Free body diagram for the vector solution.

A heavy door seals a furnace used to heat-treat metal parts. The door's weight is $W = 200$ lb which acts through point G located at the center of the 18 in. by 18 in. door. Determine the force supported by cable AB and the reactions at the bearings at points C and D.

SOLUTION

Road Map Our objective is to analyze equilibrium of the furnace door. The door is supported by a cable and two bearings, and appropriate modeling idealizations regarding the bearings will be needed to obtain a statically determinate problem. Both vector and scalar solutions are effective and both are illustrated.

Vector solution

Modeling The FBD of the furnace door is shown in Fig. 2 where we have assumed both bearings are perfectly aligned (or are self-aligning) so that the bearings have no moment reactions. If this assumption is not valid, then moment reactions are present at the bearings and this problem is statically indeterminate and cannot be solved using only the equations of static equilibrium.

Governing Equations In anticipation of summing moments about point D, the required position vectors* and force vectors are

$$\vec{r}_{DG} = (-6\,\hat{\imath} + 5\,\hat{\jmath} + 17\,\hat{k})\text{ in.}, \qquad \vec{W} = -200\,\hat{\jmath}\text{ lb}, \tag{1}$$

$$\vec{r}_{DC} = 10\,\hat{\jmath}\text{ in.}, \qquad \vec{C} = C_x\,\hat{\imath} + C_z\,\hat{k}, \tag{2}$$

$$\vec{r}_{DA} = (-6\,\hat{\imath} + 14\,\hat{\jmath} + 26\,\hat{k})\text{ in.}, \tag{3}$$

$$\vec{T} = T\frac{6\,\hat{\imath} + 20\,\hat{\jmath} - 26\,\hat{k}}{\sqrt{1112}} = T(0.1799\,\hat{\imath} + 0.5998\,\hat{\jmath} - 0.7797\,\hat{k}). \tag{4}$$

Summing moments† about point D gives

$$\sum \vec{M}_D = \vec{0}: \quad \vec{r}_{DG} \times \vec{W} + \vec{r}_{DC} \times \vec{C} + \vec{r}_{DA} \times \vec{T} = \vec{0} \tag{5}$$

$$(3400\text{ in.}\cdot\text{lb})\,\hat{\imath} + (1200\text{ in.}\cdot\text{lb})\,\hat{k} + C_z(10\text{ in.})\,\hat{\imath} - C_x(10\text{ in.})\,\hat{k} + T\big[(-26.51\text{ in.})\,\hat{\imath} + (-6.119\text{ in.})\,\hat{k}\big] = \vec{0}. \tag{6}$$

Because there are no $\hat{\jmath}$ terms in Eq. (6), moment equilibrium about the y axis is automatically satisfied, regardless of the values of T and the reactions. Collecting $\hat{\imath}$ and $\hat{k}$ terms in Eq. (6) provides the following two equations, respectively

$$T(-26.51\text{ in.}) + C_z(10\text{ in.}) + 3400\text{ in.}\cdot\text{lb} = 0, \tag{7}$$

$$T(-6.119\text{ in.}) - C_x(10\text{ in.}) + 1200\text{ in.}\cdot\text{lb} = 0. \tag{8}$$

Equations (7) and (8) contain three unknowns, thus additional equilibrium equations must be written before these unknowns can be determined. Equilibrium of forces requires

$$\sum \vec{F} = \vec{0}: \quad \vec{W} + \vec{C} + \vec{D} + \vec{T} = \vec{0}. \tag{9}$$

---

* Rather than $\vec{r}_{DA}$, the vector $\vec{r}_{DB} = 34\,\hat{\jmath}$ in. could be used, and this would provide for easier evaluation of the cross product in Eq. (5).

† In Eq. (5), you may find it convenient to evaluate $\vec{r}_{DC} \times \vec{C}$ by inspection (i.e., scalar approach) rather than by formal evaluation of the cross product.

Substituting for the force vectors in Eq. (9), where $\vec{D} = D_x\,\hat{\imath} + D_z\,\hat{k}$, and collecting terms multiplying $\hat{\imath}$, $\hat{\jmath}$, and $\hat{k}$ provide

$$(0.1799)T + C_x + D_x = 0, \tag{10}$$
$$(0.5998)T - 200\text{ lb} = 0, \tag{11}$$
$$(-0.7797)T + C_z + D_z = 0. \tag{12}$$

Computation Equation (11) may immediately be solved for T. Then Eqs. (7) and (8) may be solved for C_x and C_z, followed by solution of Eqs. (10) and (12) for D_x and D_z. These solutions are

$$\boxed{\begin{aligned} &\Rightarrow \quad T = 333.4\text{ lb},\\ &C_x = -84.03\text{ lb}, \quad C_z = 544.0\text{ lb},\\ &D_x = 24.01\text{ lb}, \quad D_z = -284.0\text{ lb}.\end{aligned}} \tag{13}$$

Scalar solution

Modeling The FBD is shown in Fig. 3 where we have assumed both bearings are perfectly aligned (or are self-aligning) so that the bearings have no moment reactions. For convenience, the cable force has been resolved into x, y, and z components.

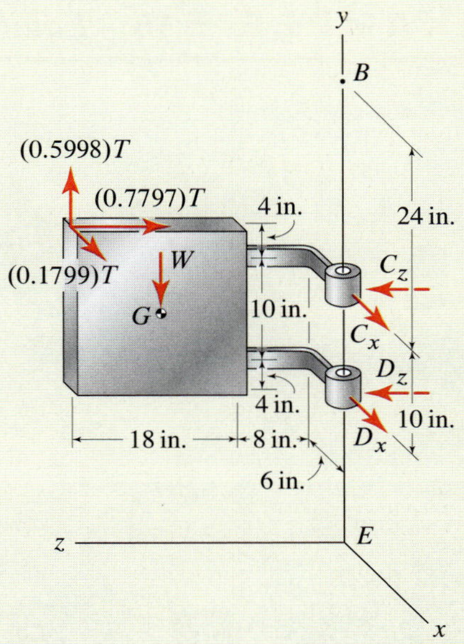

Figure 3
Free body diagram for the scalar solution.

Governing Equations & Computation The following strategy for writing and solving equilibrium equations is one of several that will allow for determination of the unknowns with minimal algebra:

$$\sum F_y = 0: \quad (0.5998)T - 200\text{ lb} = 0 \quad \Rightarrow \quad \boxed{T = 333.4\text{ lb},} \tag{14}$$

$$\sum M_{Dx} = 0: \quad -(0.7797)T(14\text{ in.}) - (0.5998)T(26\text{ in.})$$
$$+ (200\text{ lb})(17\text{ in.}) + C_z(10\text{ in.}) = 0 \tag{15}$$
$$\Rightarrow \boxed{C_z = 544.0\text{ lb},} \tag{16}$$

$$\sum M_{Dz} = 0: \quad -(0.5998)T(6\text{ in.}) - (0.1799)T(14\text{ in.})$$
$$+ (200\text{ lb})(6\text{ in.}) - C_x(10\text{ in.}) = 0 \tag{17}$$
$$\Rightarrow \boxed{C_x = -84.03\text{ lb},} \tag{18}$$

$$\sum F_x = 0: \quad (0.1799)T + C_x + D_x = 0 \quad \Rightarrow \quad \boxed{D_x = 24.01\text{ lb},} \tag{19}$$

$$\sum F_z = 0: \quad -(0.7797)T + C_z + D_z = 0 \quad \Rightarrow \quad \boxed{D_z = -284.0\text{ lb}.} \tag{20}$$

As expected, all solutions agree with those obtained using the vector approach.

Discussion & Verification We expect the cable to be in tension, and indeed our solution shows $T > 0$. We should also verify that our solutions satisfy all of the equilibrium equations.

EXAMPLE 5.15 *Equilibrium Analysis Over a Range of Motion*

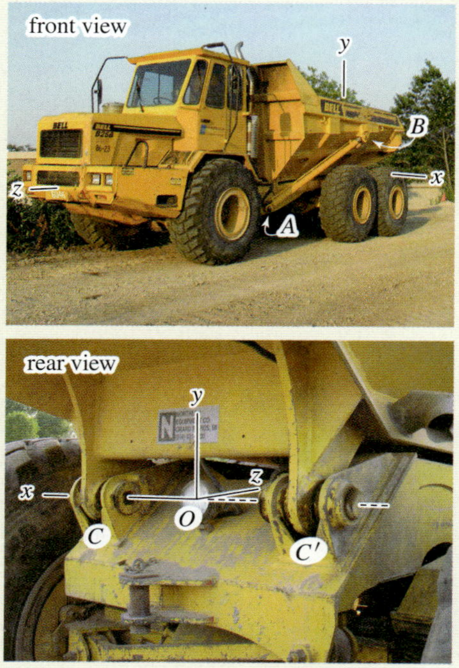

Figure 1

Helpful Information

Symmetry. For a problem to be *symmetric*, it must have symmetric geometry, loading, and supports. In other words, the geometry, loading, and supports that are present on one side of the problem's *plane of symmetry* must be a perfect mirror image of those on the other side. Taking advantage of symmetry makes a problem easier to analyze because the number of unknowns is reduced.

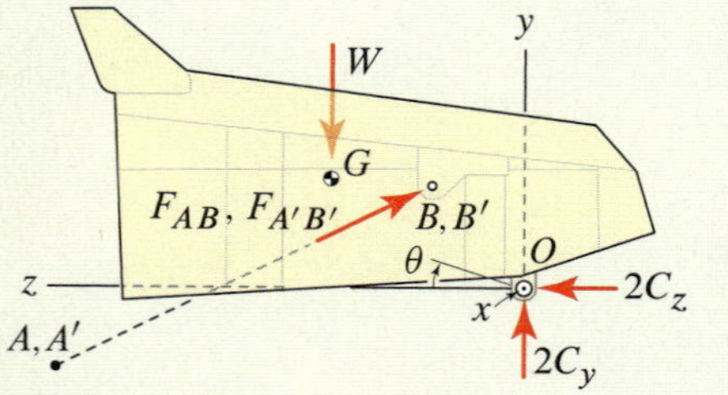

Figure 2

Free body diagram when the tilt angle is $\theta = 0°$. The x direction is perpendicular to the figure. Forces F_{AB} and $F_{A'B'}$ have components in the x direction that are not visible in this two-dimensional drawing.

The dump of a heavy-duty articulated dump truck is actuated by two symmetrically positioned hydraulic cylinders. One of the cylinders is visible in Fig. 1 between points A and B, while the cylinder on the opposite side of the truck is not visible. The dump's tilt angle is denoted by θ, where the dump is fully lowered when $\theta = 0°$ and is fully raised when $\theta = 80°$. If the dump and its contents weigh 400 kN with center of gravity at point G, determine the largest force the hydraulic cylinders must produce to raise the dump through its full range of motion. Coordinates of points A, B, and G are

$$
\begin{aligned}
&A\left[0.3\,\text{m},\ -0.5\,\text{m},\ 3.2\,\text{m}\right],\\
&B\left[1.6\,\text{m},\ (1.1\,\text{m})\sin(50° + \theta),\ (1.1\,\text{m})\cos(50° + \theta)\right],\\
&G\left[0,\ (1.8\,\text{m})\sin(30° + \theta),\ (1.8\,\text{m})\cos(30° + \theta)\right].
\end{aligned}
\tag{1}
$$

SOLUTION

Road Map We will neglect the weights of the two hydraulic cylinders, since they are likely very small compared to the weight of the dump and its contents. Thus, the hydraulic cylinders are two-force members. We will also assume the contents of the dump remain fixed in shape and position within the dump as it is being raised. With this assumption, the weight of the dump and its contents is constant, and its line of action always passes through point G, whose position is known from Eq. (1) as a function of tilt angle θ. We also assume the truck is on level ground.

Modeling Inspection of Fig. 1 shows the dump is supported by bearings at points C and C', and we assume these bearings are perfectly aligned so there are no moment reactions. Assuming the forces produced by the two hydraulic cylinders are the same, this problem is *symmetric* about the yz plane; and when drawing the FBD shown in Fig. 2, we may take the reactions at points C and C' to be the same. In Fig. 2, the hydraulic cylinder forces are taken to be positive in compression, and points A' and B' are defined to be symmetrically positioned on the opposite side of the dump from points A and B, respectively.

Governing Equations Using the following position vectors

$$
\begin{aligned}
\vec{r}_{AB} &= [1.3\,\text{m}]\,\hat{\imath} + \left[(1.1\,\text{m})\sin(50° + \theta) + 0.5\,\text{m}\right]\hat{\jmath}\\
&\quad + \left[(1.1\,\text{m})\cos(50° + \theta) - 3.2\,\text{m}\right]\hat{k},
\end{aligned}
\tag{2}
$$

$$
\begin{aligned}
r_{AB} &= \Big\{[1.3\,\text{m}]^2 + \left[(1.1\,\text{m})\sin(50° + \theta) + 0.5\,\text{m}\right]^2\\
&\quad + \left[(1.1\,\text{m})\cos(50° + \theta) - 3.2\,\text{m}\right]^2\Big\}^{1/2},
\end{aligned}
\tag{3}
$$

$$
\begin{aligned}
\vec{r}_{A'B'} &= [-1.3\,\text{m}]\,\hat{\imath} + \left[(1.1\,\text{m})\sin(50° + \theta) + 0.5\,\text{m}\right]\hat{\jmath}\\
&\quad + \left[(1.1\,\text{m})\cos(50° + \theta) - 3.2\,\text{m}\right]\hat{k},
\end{aligned}
\tag{4}
$$

$$
r_{A'B'} = r_{AB},
\tag{5}
$$

vector expressions for the hydraulic cylinder forces and the 400 kN weight can be written as

$$
\vec{F}_{AB} = F_{AB}\frac{\vec{r}_{AB}}{r_{AB}}, \qquad \vec{F}_{A'B'} = F_{A'B'}\frac{\vec{r}_{A'B'}}{r_{A'B'}}, \qquad \vec{W} = -400\,\text{kN}\,\hat{\jmath}.
\tag{6}
$$

Summation of moments about point O is given by

$$\sum \vec{M}_O = \vec{0}: \quad \vec{r}_{OA} \times \vec{F}_{AB} + \vec{r}_{OA'} \times \vec{F}_{A'B'} + \vec{r}_{OG} \times \vec{W} = \vec{0}, \tag{7}$$

where the position vectors are

$$\vec{r}_{OA} = (0.3\,\text{m})\,\hat{\imath} - (0.5\,\text{m})\,\hat{\jmath} + (3.2\,\text{m})\,\hat{k}, \tag{8}$$

$$\vec{r}_{OA'} = (-0.3\,\text{m})\,\hat{\imath} - (0.5\,\text{m})\,\hat{\jmath} + (3.2\,\text{m})\,\hat{k}, \tag{9}$$

$$\vec{r}_{OG} = \left[(1.8\,\text{m})\sin(30^\circ + \theta)\right]\hat{\jmath} + \left[(1.8\,\text{m})\cos(30^\circ + \theta)\right]\hat{k}. \tag{10}$$

Despite the perhaps formidable appearance of the vector expressions in this problem, the cross products in Eq. (7) are easy to evaluate. Using $F_{AB} = F_{A'B'}$ and $r_{AB} = r_{A'B'}$, Eq. (7) results in only an $\hat{\imath}$ term whose coefficient must be zero for equilibrium, thus

$$\frac{2F_{AB}}{r_{AB}}\left[(-0.55)\cos(50^\circ + \theta) - (3.52)\sin(50^\circ + \theta)\right] + (720\,\frac{\text{kN}}{\text{m}})\cos(30^\circ + \theta) = 0. \tag{11}$$

Computation Solving Eq. (11) for F_{AB} provides

$$F_{AB} = \frac{r_{AB}}{2}\,\frac{(720\,\frac{\text{kN}}{\text{m}})\cos(30^\circ + \theta)}{(0.55)\cos(50^\circ + \theta) + (3.52)\sin(50^\circ + \theta)}, \tag{12}$$

where r_{AB} is given by Eq. (3). To complete this problem, we plot F_{AB} from Eq. (12) for values of θ over the range 0° to 80° to obtain the results shown in Fig. 3. Thus, we see the largest force the hydraulic cylinders must produce is about $\boxed{320\text{ kN},}$ and this occurs when the dump first starts to open, when $\theta = 0^\circ$. A more accurate value of F_{AB} is obtained by evaluating Eq. (12) with $\theta = 0^\circ$, which produces $\boxed{F_{AB} = 318\text{ kN}.}$

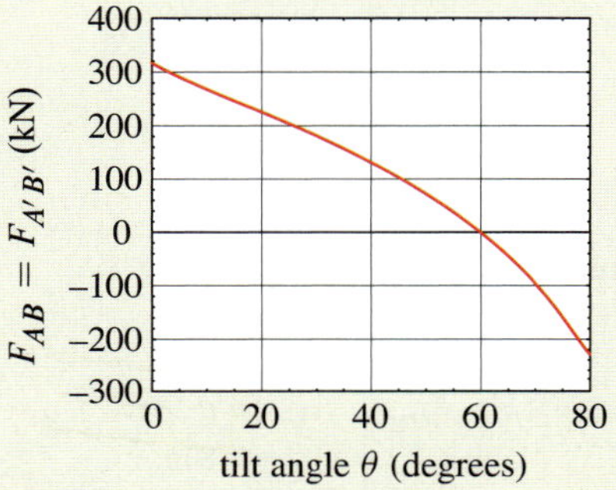

Figure 3
Force in hydraulic cylinders AB and $A'B'$ as a function of tilt angle.

Discussion & Verification

- This solution assumes the contents of the dump remain fixed in shape and position within the dump as it is being raised. For purposes of determining the largest hydraulic cylinder force needed, this assumption is useful. A more accurate model would allow the weight of the dump's contents to change with tilt angle θ. Exactly how the weight varies with θ is a difficult problem, but a simple yet useful model might let this weight vary in some fashion (perhaps linear) from full value when $\theta = 0^\circ$, to zero when $\theta = 80^\circ$. Additionally, we may also consider allowing the center of gravity for the dump's contents to change as a function of θ.
- If desired, the reactions at bearings C and C' as function of θ can be determined by writing
$$\sum \vec{F} = \vec{0}: \quad \vec{F}_{AB} + \vec{F}_{A'B'} + \vec{W} + 2\vec{C} = \vec{0}, \tag{13}$$
where $\vec{C} = C_y\,\hat{\jmath} + C_z\,\hat{k}$.
- Although not illustrated here, a scalar solution for this problem is also effective.
- Problem 5.74 gives some additional suggestions for verifying the accuracy of the solution to this example.

PROBLEMS

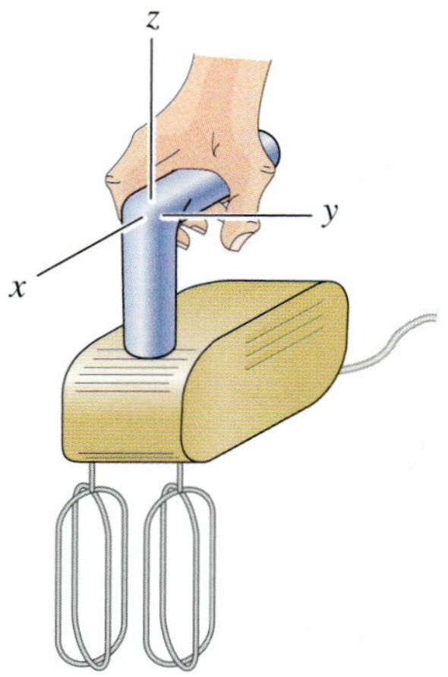

Figure P5.73

Problem 5.73

A handheld mixer for blending cooking ingredients is shown. To minimize operator fatigue, it is desirable for the reaction forces on the operator's hand to be as small as possible. With this goal in mind, should the beaters rotate in the same direction or opposite directions? Assume each beater produces a moment about the z axis, and if needed, assume reasonable values for dimensions. Explain your reasoning.

Note: Concept problems are about *explanations*, not computations.

Problem 5.74

When you evaluate the solution of a problem, it is always necessary to verify the accuracy of the solution, and when possible, performing simple checks can help with verification. In Example 5.15 on p. 306, consider the specific position $\theta = 0°$, analyze the problem afresh to determine the hydraulic cylinder force required to begin opening the dump, and verify that the results of Example 5.15 are in agreement. As another check, Fig. 3 in Example 5.15 shows the hydraulic cylinder force is zero for a particular tilt angle. Explain why this occurs and, if possible, perform a simple analysis that confirms the value of θ at which this force is zero.

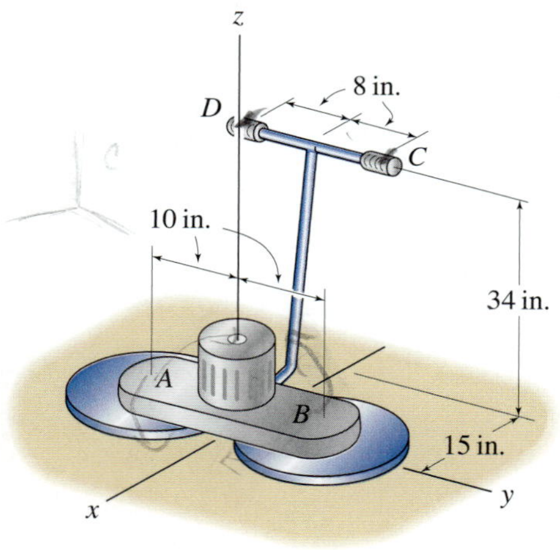

Figure P5.75

Problem 5.75

A machine for sanding wood floors is shown. The machine weighs 80 lb with center of gravity along the z axis. At each sanding drum a moment of 60 in.·lb is applied to the machine in the direction opposite the rotation of the drum. Assume the operator's hands, positioned at points C and D, can apply forces in the positive or negative x direction. Determine the forces on the operator's hands if

(a) Both sanding drums rotate about the positive z direction.

(b) The sanding drums at A and B rotate about the positive and negative z directions, respectively.

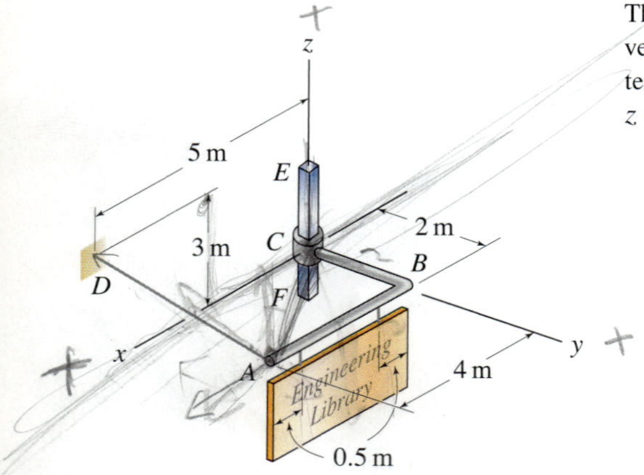

Figure P5.76

Problem 5.76

Bar EF has a square cross section and is fixed in space. The structure ABC has negligible weight and has a collar at C that has a square hole that slides freely on bar EF. The structure ABC supports a uniform rectangular sign with weight 1 kN (the two vertical edges of the sign align with points A and B). Determine the magnitude of the tension in cable AD and all of the reaction components at C referred to the x, y, and z directions provided.

Problem 5.77

Bar $ABCDE$ is supported by cable BF, a ball and socket at A, and a self-aligning bearing at E. Determine the tension in cable BF and the reactions at points A and E.

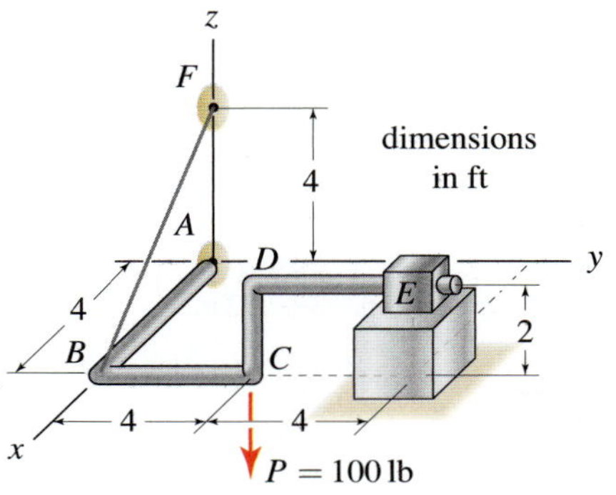

Figure P5.77

Problem 5.78

Vertical bar ED has circular cross section and is built in at E. Member ABC is a single member that lies in a horizontal plane, with portion BC parallel to the z axis and with cable CD attached to point C. The collar at A can freely slide in the y direction and can freely rotate about the y axis.

(a) Does the structure ABC have complete fixity or partial fixity, and is it statically determinate or statically indeterminate? Explain.

(b) When point B is subjected to a downward vertical force of 18 lb, determine the force supported by the cable and all support reactions at A.

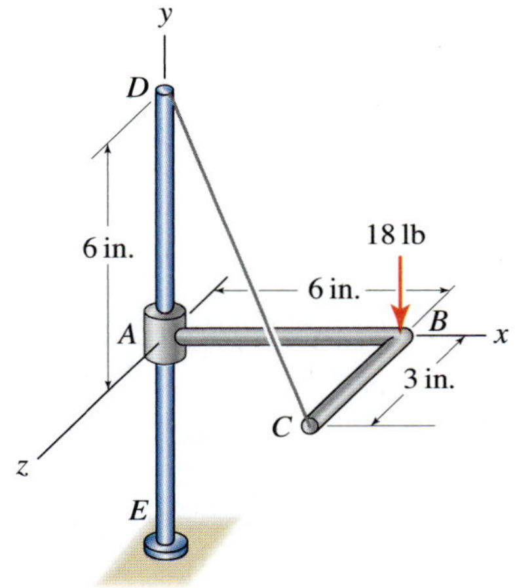

Figure P5.78

Problem 5.79

Follow the suggestion made in Eqs. (10) and (11) of Example 5.12 on p. 301 to find the tension in cable AC by summing moments about line OD.

Problem 5.80

In Prob. 5.79, find the tension in cable AD by summing moments about line OC.

Problem 5.81

Member $AGDB$ is supported by a cable DE, a self-aligning bearing at A, and a self-aligning thrust bearing at B.

(a) Draw the FBD for $AGDB$, labeling all forces and moments.

(b) Rate the solution strategies listed below for ease of obtaining the magnitude of the tension T_{DE} in cable DE. Rate the best choice as number 1, second-best choice number 2, and so on. If a solution strategy does not work, then label it with zero.

Rating	Solution strategy
____	Write $\sum \vec{F} = \vec{0}$ and $\sum \vec{M}_B = \vec{0}$, solve for T_{DE}.
____	Write $\vec{r}_{AG} \times \vec{W} + \vec{r}_{AD} \times \vec{T}_{DE} = \vec{0}$, solve for T_{DE}.
____	Write $\sum \vec{F} = \vec{0}$ and $\sum \vec{M}_O = \vec{0}$, solve for T_{DE}.
____	Write $(\vec{r}_{AG} \times \vec{W} + \vec{r}_{AD} \times \vec{T}_{DE}) \cdot \vec{r}_{OC} = 0$, solve for T_{DE}.
____	Write $\sum \vec{F} = \vec{0}$ and $\sum \vec{M}_A = \vec{0}$, solve for T_{DE}.

Note: Concept problems are about *explanations*, not computations.

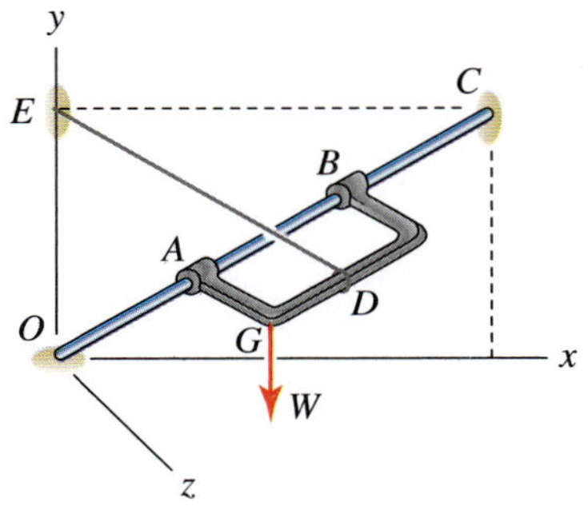

Figure P5.81

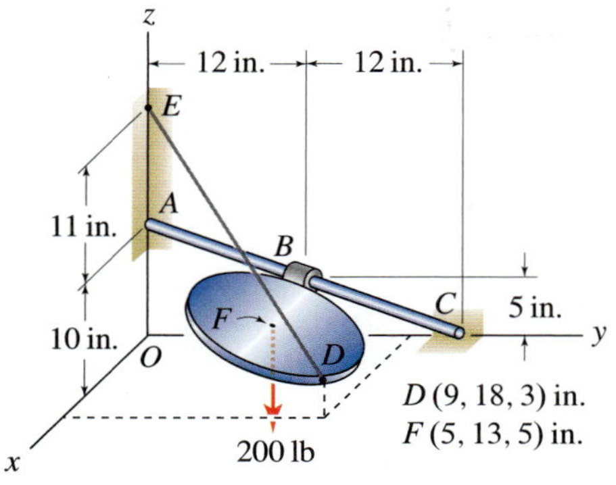

Figure P5.82

Problem 5.82

A circular plate with weight $\vec{W} = -200\,\hat{k}$ lb acting at its center, point F, is supported by cord DE and a thrust bearing at B. Shaft AC is fixed and frictionless.

(a) Draw the FBD for the plate, labeling all forces and moments.

(b) Rate the solution strategies listed below for ease of obtaining the magnitude of the tension T_{DE} in cable DE. Rate the best choice as number 1, second-best choice number 2, and so on. If a solution strategy does not work, then label it with zero.

Rating	Solution strategy
____	Write $\sum \vec{F} = \vec{0}$ and $\sum \vec{M}_B = \vec{0}$, solve for T_{DE}.
____	Write $\vec{r}_{BF} \times \vec{W} + \vec{r}_{BD} \times \vec{T}_{DE} = \vec{0}$, solve for T_{DE}.
____	Write $\sum \vec{F} = \vec{0}$ and $\sum \vec{M}_O = \vec{0}$, solve for T_{DE}.
____	Write $(\vec{r}_{BF} \times \vec{W} + \vec{r}_{BD} \times \vec{T}_{DE}) \cdot \vec{r}_{AC} = 0$, solve for T_{DE}.
____	Write $\sum \vec{F} = \vec{0}$ and $\sum \vec{M}_D = \vec{0}$, solve for T_{DE}.

Note: Concept problems are about *explanations*, not computations.

Problem 5.83

Determine the cable tension for the circular plate of Prob. 5.82.

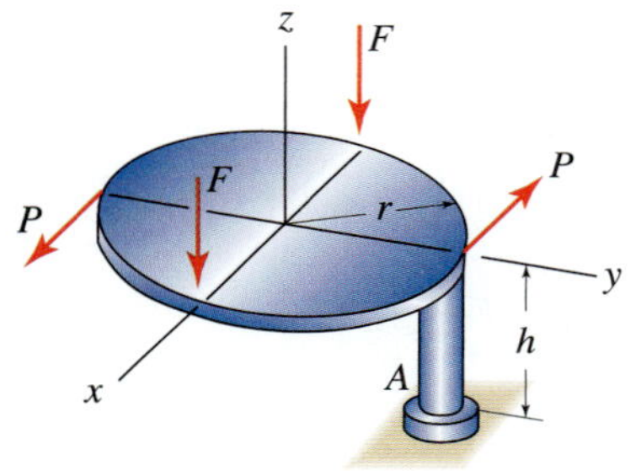

Figure P5.84

Problem 5.84

A circular plate of radius r is welded to a post with length h that is built in at point A. Determine the reactions at point A. Express your answers in terms of parameters such as r, h, F, and P.

Problem 5.85

Object $ABCDEF$ is a sliding door that is supported by a frictionless bearing at A and a wheel at F that rests on a frictionless horizontal surface. The object has weight $W = 800$ N, which acts at the midpoint of the rectangular region $BCDE$. Determine all support reactions.

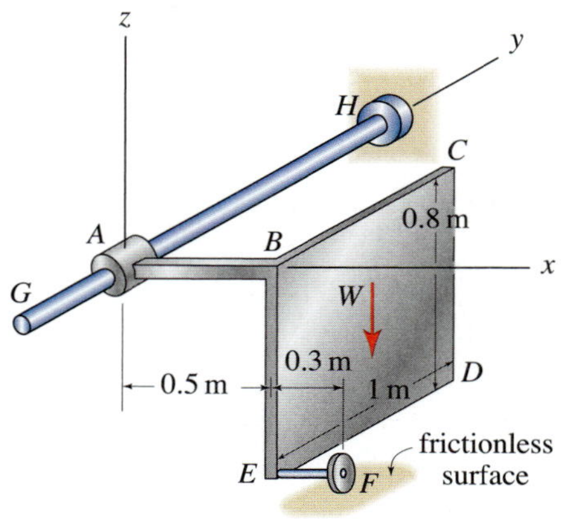

Figure P5.85

Problem 5.86

The control surface of an aircraft is supported by a thrust bearing at point C and is actuated by a bar connected to point A. The 1 kN force acts in the negative y direction, and the line connecting points A and B is parallel to the y axis. Determine the value of force F needed for equilibrium and all support reactions.

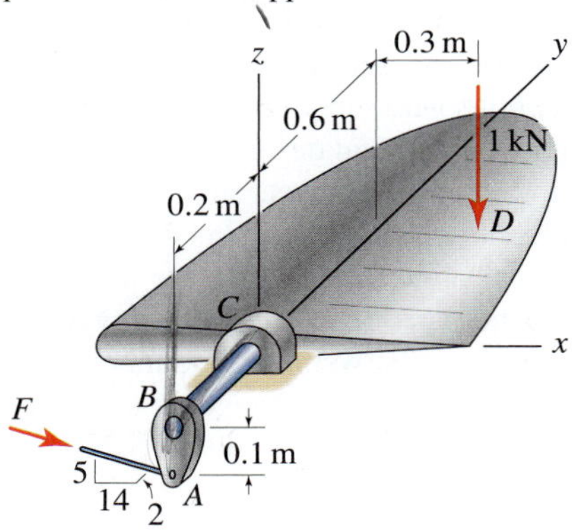

Figure P5.86

Problem 5.87

An L-shaped bar is supported by a bearing at A and a smooth horizontal surface at B. Determine the reactions at A and B.

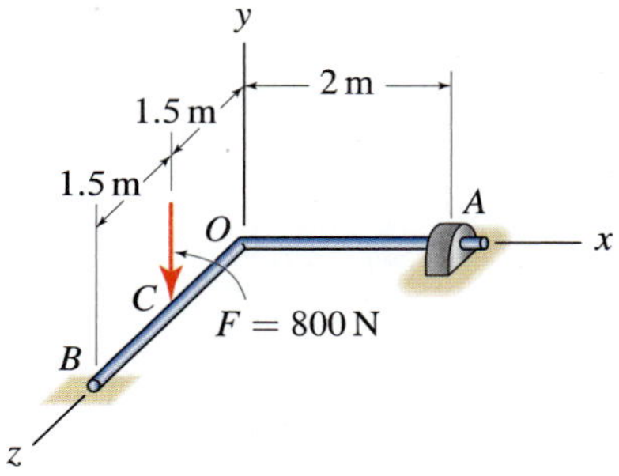

Figure P5.87

Problem 5.88

Structure $ABCD$ is supported by a collar at D that can rotate and slide along bar EF which is fixed and is frictionless. Structure $ABCD$ makes contact with smooth surfaces at A and C where the normal direction $\vec{n}$ to the surface at A lies in a plane that is parallel to the xy plane. Force P is parallel to the y axis. If $P = 10$ kN, determine the reactions at A, C, and D.

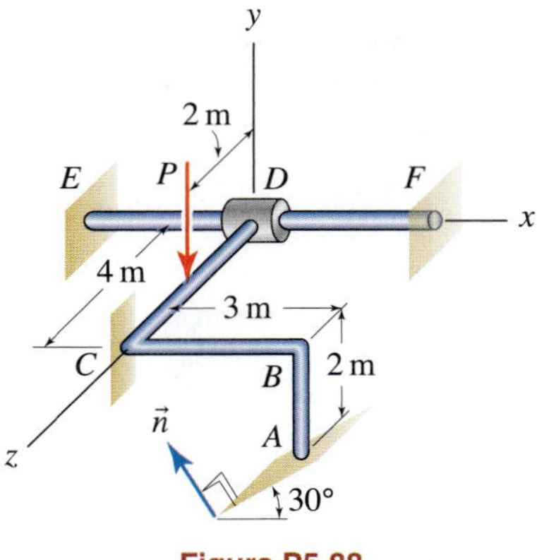

Figure P5.88

5.5 Engineering Design

Engineering design is a subject in its own right, and you will study this subject in detail as you advance in your education. The objective of *engineering design* is to develop the description and specifications for how a structure, machine, device, procedure, and so on are to be produced so that needs and requirements that have been identified are met. The objective of *engineering design theory* is to establish structured procedures for design that help you develop the most effective and optimal design possible in a timely manner. Good references are available on this subject, such as Ullman (2003), Dominick et al. (2001), and Middendorf and Engelmann (1998).*

Design is an iterative process. Whether the design is of a simple paper clip or a fuel pump for a Space Shuttle engine, a process is certain to be followed. This process may vary between individual engineers and may depend on the specific nature of what is being designed, the magnitude of the project, and other factors; but the process shown in Fig. 5.29 is commonly employed. While each of the elements in Fig. 5.29 is present in almost all design activities, there may be reordering of tasks and/or additional components needed. For example, in civil construction a project will often begin with a request for proposals (RFP) being issued by a government unit where the specifications for a project are stated. Companies that respond to the RFP will conduct preliminary design work and perform a thorough cost analysis, and only after their proposal has been accepted will detailed design work begin. In *concurrent design*, the process shown in Fig. 5.29 is interjected with input from interested parties, such as other professionals, customers, and so on, so that the final design is more certain to satisfy the needs of all affected parties. In concurrent design, tasks such as developing the manufacturing systems needed to fabricate the design may take place while the design in still being developed.

In the following paragraphs, we discuss some of the items of Fig. 5.29 in greater detail.

- **Problem identification.** Most designs begin with the identification of a need that is not met. When such a need has been identified, it is necessary to thoroughly review the state of the art to confirm the need, why it exists, why this need has not been recognized and/or addressed by others, and so on. Very simply, you must establish that there is indeed an opportunity for you to make a contribution before investing substantial time and resources in design.

- **Problem evaluation.** Here you must identify *all* of the needs your design must satisfy, not just those that might be new or novel. The needs consist of all the requirements, specifications, goals, and so on that your design must satisfy. Some needs may be achievable only at the expense of other needs (e.g., strength and light weight), so value decisions may be required to develop needs that are realistic and achievable. In the course of doing this, you must review applicable standards, codes, patents, industry practices, and so on. Depending on what you

* D. G. Ullman, *The Mechanical Design Process*, 3rd ed., McGraw-Hill, New York, 2003. P. G. Dominick, J. T. Demel, W. M. Lawbaugh, R. J. Freuler, G. L. Kinzel and E. Fromm, *Tools and Tactics of Design*, John Wiley & Sons, New York, 2001.W. H. Middendorf and R. H. Engelmann, *Design of Devices and Systems*, 3rd ed., Marcel Dekker, New York, 1998.

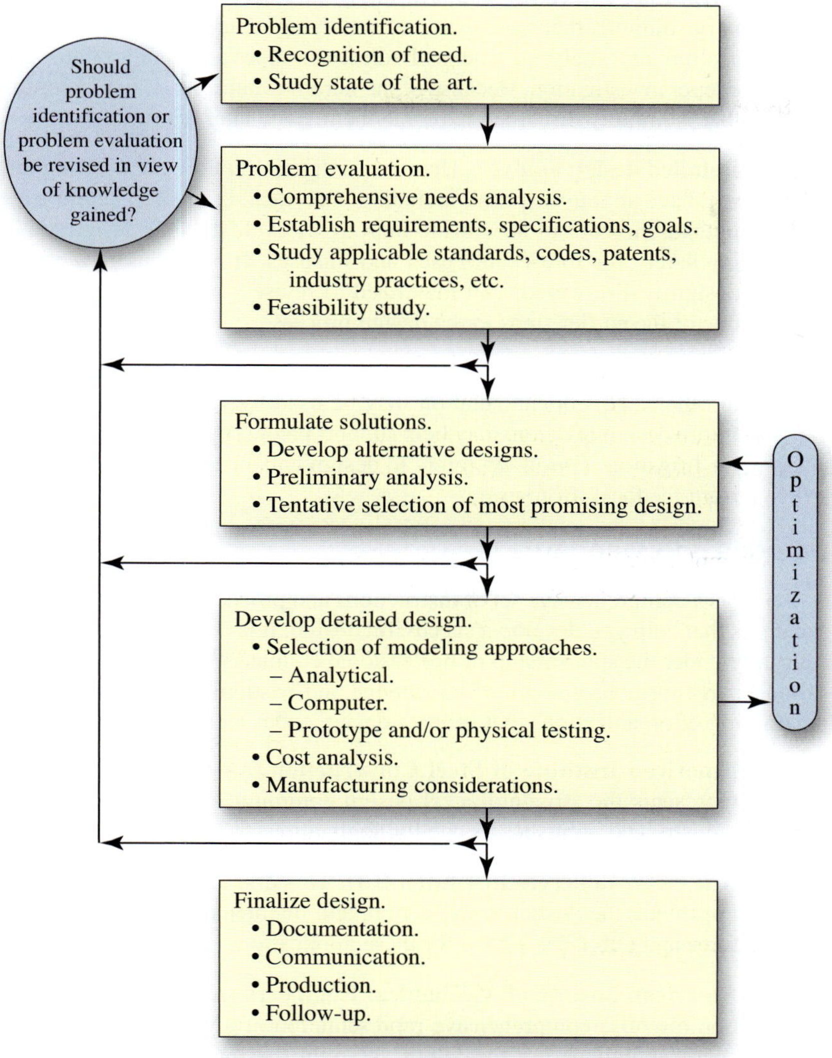

Figure 5.29. A process for engineering design.

are designing, you may be required by law to conform to certain standards. Other standards may be voluntary, but acceptance of your design by customers may depend on compliance with these. Additional comments on standards and codes follow later in this section. Once all of your needs and requirements have been established, you may be able to conduct a feasibility study to determine if your ideas are possible. In a feasibility study, you might consider if it is physically possible to satisfy your needs and objectives, although this may be difficult to fully assess until more careful design work is done. Other issues commonly considered in a feasibility study include a company's manufacturing capability, compatibility with existing product line, marketing, and economics.

- **Formulate solutions.** You will likely want to develop several preliminary solutions that are capable of satisfying your needs. Based on your preliminary analyses, you will select the most promising of these for deeper investigation. Because of the iterative nature of design, a solution that is initially discarded here may be revived later for further study.

- **Detailed design analysis.** Here you will use all of the engineering tools you have at your disposal, including analytical, computer, and/or experimental methods of analysis to thoroughly evaluate your design, including safety for foreseeable uses and misuses. If a serious flaw in your design is discovered, you may need to revise your design or return to one of the earlier steps in the design process.

- **Finalize design.** Your design must be documented and communicated to others. This documentation may be used by you or others for some future design revision, may help support patent rights, or may be needed for litigation. *Follow-up* refers to design enhancements that may result from feedback from users.

Codes and standards

A *code* is a comprehensive set of instructions and procedures for specific applications that help you develop a successful design. It is developed by engineers usually under the sponsorship of professional engineering societies, and it embodies an enormous amount of knowledge and wisdom gained over years. A sampling of organizations that sponsor design codes follows:

- **American Institute of Steel Construction (AISC).** This organization represents the structural steel design community and the construction industry and publishes a code for applications to framed steel structures.

- **American Concrete Institute (ACI).** This organization is dedicated to improving the design, construction, and maintenance of concrete structures and publishes a code for these applications.

- **American Society of Mechanical Engineers (ASME).** This organization provides comprehensive representation of a wide range of the mechanical engineering specialties and publishes numerous codes, one of which governs the design of boilers, pressure vessels, and piping, including applications to nuclear reactor components.

Use of these design codes is often voluntary (codes for nuclear components are mandatory), but if they are suitable for your design work, then they should be followed.

A *standard* is a minimum performance measure that a design should meet, but typically it will not tell you how to develop the design. In addition to the organizations stated earlier, the following organizations develop and publish numerous standards:

- **Society of Automotive Engineers (SAE).** This organization represents a wide range of interests pertaining to transportation and develops standards that apply to land, sea, air, and space vehicles.

- **American Institute of Aeronautics and Astronautics (AIAA).** This organization is the principal society representing aerospace engineers and scientists, and it publishes numerous standards governing performance of aircraft and spacecraft.

- **American National Standards Institute (ANSI).** This organization develops safety standards based on consensus of affected parties, including manufacturers and users. Use of ANSI standards is voluntary, but if applicable standards are available, these should be used.
- **Consumer Product Safety Commission (CPSC).** This is a U.S. federal agency that sets voluntary and mandatory safety standards for consumer products. They have the power to issue product recalls and can ban hazardous products.
- **Occupational Safety and Health Administration (OSHA).** This is a U.S. federal agency that encourages employers and employees to cooperate to reduce workplace hazards. They develop voluntary and mandatory workplace safety and health standards and enforce these.

The U.S. federal agencies cited above set standards for the United States only. The other organizations cited have participation and impact throughout other parts of North America. Most other countries, especially in other parts of the world, have similar organizations of their own. There are efforts to develop uniform standards that are applicable worldwide, and a notable organization that promotes this is the *International Organization for Standardization* (ISO). This organization has membership drawn from the standards organizations of approximately 150 countries.

Sometimes the difference between codes and standards is not distinct. You must be familiar with all codes and standards that could affect your design. Your design must conform to mandatory standards, and if voluntary standards are available, these should be followed if at all possible. If it is not possible to follow voluntary standards or if standards do not exist, then suitable performance and safety measures must be established. Whether or not you use a code and/or standards to develop a design, the responsibility for the safety and performance of your design is yours.

Design problems

For the design problems in this section and throughout this book, imagine you are employed as an intern working under the supervision of an engineer who asks you to conduct a design study, or to add details to a design that is started. You will be presented with a problem that is suitable to your level of knowledge along with some pertinent data. Sometimes, the information provided may not be complete, and you may not be instructed on everything that needs to be done. As shown in Fig. 5.29, this is not an attempt to be artificially vague, but rather is a reflection of how design and modeling of real life problems are carried out. It is in this spirit that the design problems throughout this book are presented, and you may need to make reasonable assumptions or seek out additional information on your own. Furthermore, our work in this book will focus on design based on first principles. That is, we will use the laws of physics plus good judgment to establish a design. We generally will not survey and apply standards that may be available, because this is not central to the theme of this book. Your work should culminate in a short written technical report that is appropriate for an engineer to read, where you present your design, state assumptions made, and so on. Appendix A of this book gives a brief discussion of technical writing that may be helpful.

EXAMPLE 5.16 *Engineering Design*

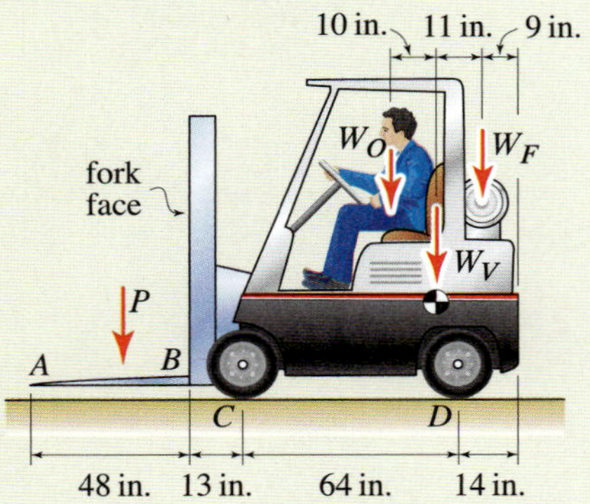

Figure 1

The forklift has a vehicle weight of $W_V = 12{,}500$ lb, fuel weight W_F, which is 300 lb when fully fueled, and operator weight W_O. The drive wheels at C are supported by a fixed axle, and the wheels at D steer the vehicle. Determine the *capacity rating* for this forklift, assuming it is operated at slow speed on a smooth, level surface. The capacity rating is the maximum load this forklift is designed to lift.

SOLUTION

Road Map The following needs must be satisfied by the capacity rating we determine:

- The capacity rating should be as large as possible while still being safe.
- The forklift should not tip forward while supporting a load, regardless of whether the operator is on or off the vehicle.
- The rear tires should support enough force so they are capable of steering the forklift. This requirement applies when the forklift is being driven, and thus the operator may be assumed to be seated on the forklift.

All our analyses will take the fuel weight to be zero, as the forklift is likely to see occasional use when its fuel supply is low, and this leads to a more conservative capacity rating. While somewhat arbitrary, but reasonable, we will assume the steering requirement is satisfied if the forces supported by the wheels at D are at least 10% of the forces supported by the wheels at C.* When we analyze this requirement, an operator's weight may be included. To obtain a conservative capacity rating, we must use the minimum possible operator weight, and this is difficult to determine with any precision. Obviously, this weight is small compared to the forklift's weight, and it is conservative to take it as zero. If this is the case, then the no tipping requirement will automatically be satisfied if the steering requirement is met.

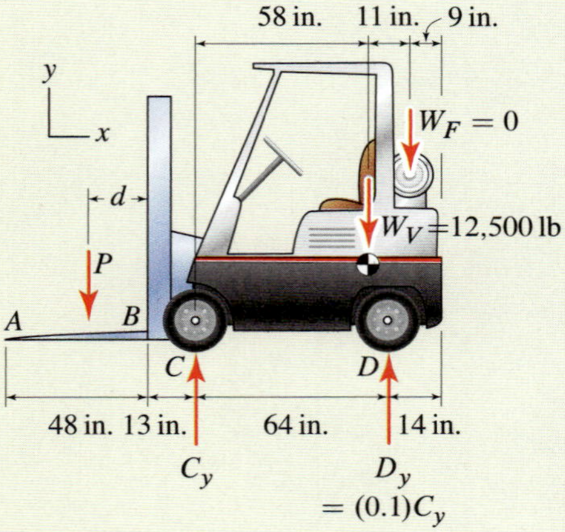

Figure 2
Free body diagram.

Modeling The FBD is shown in Fig. 2 where we have imposed the constraint $D_y = (0.1)C_y$ to enforce the need for steering ability, and the fuel and operator weights are zero. Clearly, as distance d from the load to the fork face increases (d is called the *load center*), the value of P that can be lifted decreases. The most conservative idea is to place P at the worst possible position, which is at $d = 48$ in. However, a quick investigation of industry practice shows that for purposes of specifying capacity rating, manufacturers customarily position P at the midpoint of the forks, which is $d = 24$ in.. In the following calculation we will consider both values of d, although the "official" capacity rating will be based on $d = 24$ in.

Governing Equations Using the FBD in Fig. 2, the equilibrium equations are

$$\sum M_C = 0: \quad P(d + 13\text{ in.}) - (12{,}500\text{ lb})(58\text{ in.}) + (0.1)C_y(64\text{ in.}) = 0, \qquad (1)$$

$$\sum F_y = 0: \quad C_y + (0.1)C_y - P - 12{,}500\text{ lb} = 0. \qquad (2)$$

Computation Solving Eqs. (1) and (2) for P provides

$$P = 12{,}500\text{ lb}\,\frac{58\text{ in.} - \left(\frac{0.1}{1.1}\right)64\text{ in.}}{d + 13\text{ in.} + \left(\frac{0.1}{1.1}\right)64\text{ in.}}. \qquad (3)$$

* Some research of industry practices may yield a better criterion for ensuring steerability.

Discussion & Verification When $d = 24$ in., we obtain $P = 15{,}230$ lb from Eq. (3), which we will round down slightly to obtain the capacity rating for the forklift as

$$\begin{array}{c}\textbf{Capacity rating} = 15{,}000\,\text{lb} \\ \text{(load positioned 24 in. from the fork face).}\end{array} \tag{4}$$

Remarks

- The capacity rating in Eq. (4) will be used for advertising and marketing the forklift to customers. It will also be prominently displayed on the forklift, along with other important information, so that the operator is sure to see it and understand its meaning.
- If the forklift is used with $d = 48$ in., the load that can safely be lifted, as given by Eq. (3), reduces to 9762 lb. This number will be rounded down and will be displayed on the machine so that the operator knows that the maximum load that can safely be lifted is about 9000 lb when the load is positioned at 48 in. from the fork face.
- To complete our work, we will write a short report that details our assumptions calculations, and conclusions.

DESIGN PROBLEMS

General Instructions. In all problems, write a brief technical report following the guidelines of Appendix A, where you summarize all pertinent information in a well-organized fashion. It should be written using proper, simple English that is easy to read by another engineer. Where appropriate, sketches along with critical dimensions should be included. Discuss the objectives and constraints considered in your design, the process used to arrive at your final design, safety issues if appropriate, and so on. The main discussion should be typed, and figures, if needed, can be computer-drawn or neatly hand-drawn. Include a neat copy of all supporting calculations in an appendix that you can refer to in the main discussion of your report. A length of a few pages, plus appendix, should be sufficient.

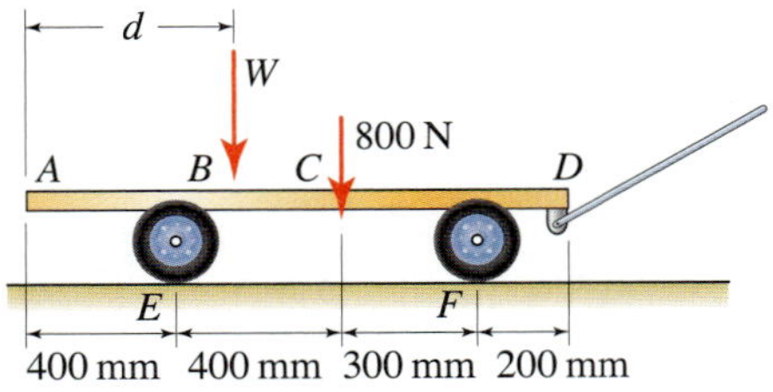

Figure DP5.1

Design Problem 5.1

A hand cart for moving heavy loads in a warehouse is shown. If each axle (pair of wheels) can support a maximum of 10 kN and if the wheels are not allowed to lift off the pavement, determine the largest weight W that may be supported for any position $0 \leq d \leq 1.3$ m.

Design Problem 5.2

Specify the weight W, width w, and depth d for the base of a particular model of a fluorescent desk lamp. For this lamp, the shade and lightbulb assembly C weighs 1.8 lb, and the movable arm AB, including the attachments at A and B, weighs 1.5 lb with center of gravity approximately midway between points A and B. For the weight of the base W that you select, the dimensions d and w should be such that the lamp is as tip-resistant as possible.

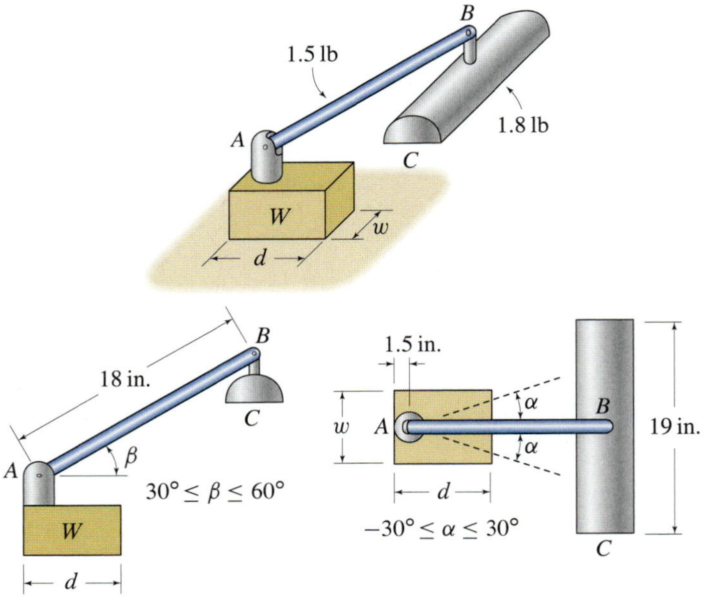

Figure DP5.2

Design Problem 5.3

The electric fan shown has the following specifications:

- The entire unit weighs $W = 15$ lb with the center of gravity shown.
- The fan produces a maximum thrust of $T = 6$ lb.
- The height h can be adjusted by the user between 24 and 48 in.
- The fan can be rotated to any horizontal position desired by the user.

Specify the dimension b for a base having three equally spaced legs. In your work, assume that the weight of the legs is already included in the 15 lb weight of the unit. Your design should consider a reasonable degree of safety against overturning. Also make at least two suggestions for how the stability of the fan can be improved without increasing the weight W or increasing the dimension b.

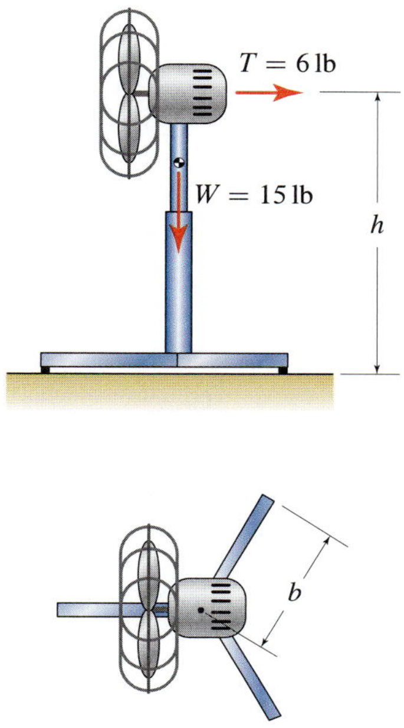

Figure DP5.3

Design Problem 5.4

The chute of a concrete truck for delivering wet concrete to a construction site is shown. The length of the chute may be changed by adding or removing segments BC and CD. Chute segments AB, BC, and CD each weigh 50 lb, and the maximum length of the chute is 144 in. The chute has semicircular shape with 8 in. inside radius, and the hydraulic cylinder GH is used to raise and lower the chute such that $0° \leq \theta \leq 50°$. Specify the force capacity of the hydraulic cylinder GH.

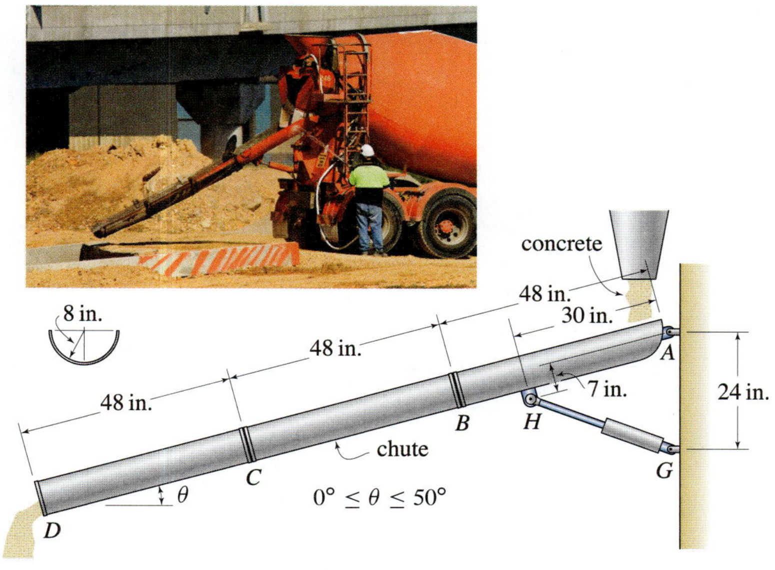

Figure DP5.4

Design Problem 5.5

A telescopic boom aerial lift is shown. The lift is designed to support one worker, plus miscellaneous tools and supplies that are likely to be used, and is to be operated on a hard level surface. Portion BC of the boom is retractable and supports a work platform whose floor is always horizontal. Turret D can rotate 360° on base E. The design of the lift is essentially complete, and weights of the various components are shown. The masses corresponding to these weights are

Boom AB:	$m_{AB} = 450$ kg,
Boom BC plus platform at C:	$m_{BC} = 500$ kg,
Turret D:	$m_D = 1800$ kg,
Base E:	$m_E = 1400$ kg.

In addition to these are the mass of the worker, tools, and supplies m_C and the mass of a counterweight m_F whose center of gravity is to be placed below point A. You are asked to do the following:

- Specify the lifting mass rating m_C.
- Specify the mass of the counterweight m_F.
- Specify the force capacity of the hydraulic cylinder between points G and H.

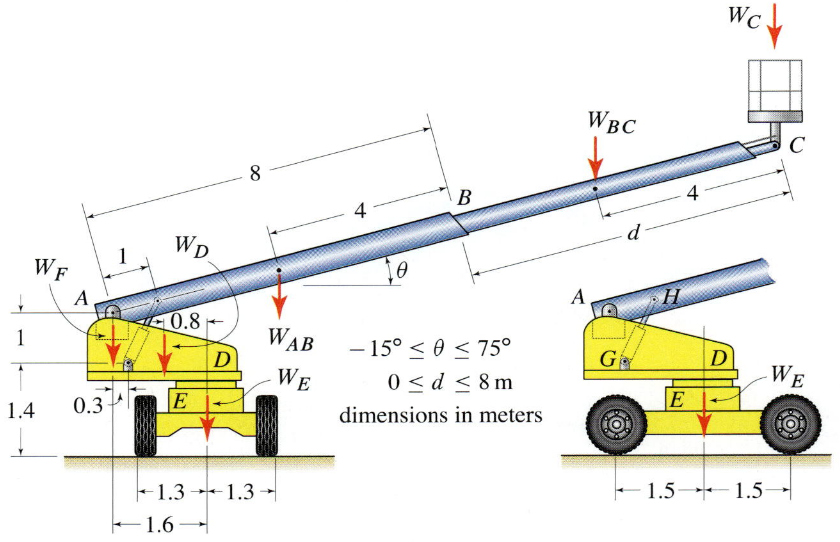

Figure DP5.5

5.6 Chapter Review

Important definitions, concepts, and equations of this chapter are summarized. For equations and/or concepts that are not clear, you should refer to the original equation and page numbers cited for additional details.

Equations of equilibrium. In three dimensions, the equations governing the static equilibrium of a body are

Eq. (5.1), p. 246

$$\sum \vec{F} = \vec{0} \quad \text{and} \quad \sum \vec{M}_P = \vec{0}$$

where the summations include all forces and moments that are applied to the body. In scalar form, Eq. (5.1) is

Eq. (5.2), p. 246

$$\begin{aligned} \sum F_x &= 0 & \qquad & \sum M_{Px} = 0 \\ \sum F_y &= 0 & \text{and} \qquad & \sum M_{Py} = 0 \\ \sum F_z &= 0 & & \sum M_{Pz} = 0. \end{aligned}$$

In two dimensions, with x and y being the in-plane coordinates, the equations $\sum F_z = 0$, $\sum M_{Px} = 0$, and $\sum M_{Py} = 0$ in Eq. (5.2) are always satisfied, and the remaining equilibrium equations are (with subscript z dropped from the moment equation)

Eq. (5.3), p. 246

$$\sum F_x = 0, \quad \sum F_y = 0, \quad \text{and} \quad \sum M_P = 0.$$

If desired, the $\sum F_x = 0$ (and/or $\sum F_y = 0$) equation in Eq. (5.3) may be replaced by an additional moment equilibrium equation provided a suitable moment summation point is selected, as discussed in Section 5.2. Similar remarks apply to alternative equilibrium equations in three dimensions.

Springs. A linear elastic spring, shown schematically in Fig. 5.30, produces an axial force F_s that is proportional to its change of length δ according to

Eq. (5.13), p. 271

$$\begin{aligned} F_s &= k\delta \\ &= k(L - L_0) \end{aligned}$$

where k is the spring stiffness (units: force/length), δ is the elongation of the spring from its unstretched length, L_0 is the initial (unstretched) spring length, and L is the final spring length. Because the force F_s and the elongation δ in Eq. (5.13) are directed along an axis, or line, these springs are sometimes called *axial springs* to differentiate them from torsional springs. Springs were discussed extensively in Chapter 3, and key points were repeated in Section 5.3.

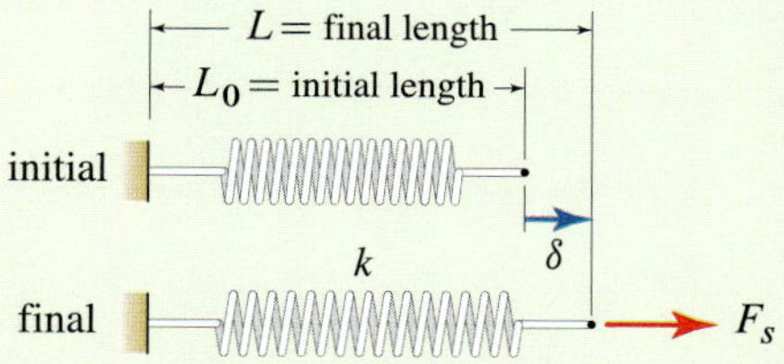

Figure 5.30
A spring produces a force F_s that is proportional to its elongation δ. Such springs are sometimes called *axial springs*.

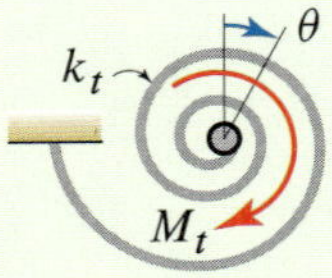

Figure 5.31
A torsional spring produces a moment M_t that is proportional to its rotation θ.

Torsional springs. A linear elastic *torsional spring*, shown schematically in Fig. 5.31, produces a moment M_t that is proportional to the relative rotation, or twist, θ according to

Eq. (5.14), p. 272

$$M_t = k_t \theta$$

where k_t is the *spring stiffness* (units: moment/radian).

Supports and fixity. *Fixity* refers to an object's ability to move in space as a rigid body. All objects fall into one of the following three categories:

Complete fixity. A body with *complete fixity* has supports that are sufficient in number and arrangement so that the body is completely fixed in space and will undergo no motion (either translation or rotation) in any direction under the action of *any* possible set of forces.

Partial fixity. A body with *partial fixity* has supports that will allow the body to undergo motion (translation and/or rotation) in one or more directions. Whether or not such motion occurs depends on the forces and/or moments that are applied and whether the body is initially in motion.

No fixity. A body with *no fixity* has no supports and is completely free to move in space under the action of forces that are applied to it.

Static determinacy and indeterminacy. An object is either *statically determinate* or *statically indeterminate*, as follows:

Statically determinate body. For a *statically determinate body*, the equilibrium equations of statics are sufficient to determine all unknown forces and/or other unknowns that appear in the equilibrium equations.

Statically indeterminate body. For a *statically indeterminate body*, the equilibrium equations of statics are not sufficient to determine all unknown forces and/or other unknowns.

A simple rule of thumb to help ascertain whether an object is statically determinate or indeterminate is to compare the number of unknowns to the number of equilibrium equations, and we call this *equation counting*. With n being the number of unknowns, the rule of thumb for a single body in two dimensions is

Eq. (5.17), p. 274

If $n < 3$	The body is statically determinate, and it can have partial fixity ($n = 1$ or 2) or no fixity ($n = 0$).
If $n = 3$	The body is statically determinate if it has full fixity. The body is statically indeterminate if it has partial fixity.
If $n > 3$	The body is statically indeterminate, and it can have full fixity or partial fixity.

Successful use of equation counting requires good judgment on your part, which is why it is called a *rule of thumb* rather than a *rule*.

Two-force and three-force members. If a body or structural member is subjected to forces at two points or three points only, as described below, then when in equilibrium the orientations of the forces supported by the body has special properties. These situations are defined as follows:

Two-force member. A body subjected to forces at two points (no moment loading and no distributed forces such as weight) is called a *two-force member*. The special feature of a two-force member is that, when in equilibrium, the two forces have the same line of action and opposite directions.

Three-force member. A body subjected to forces at three points (no moment loading and no distributed forces such as weight) is called a *three-force member*. The special feature of a three-force member is that, when in equilibrium, the lines of action of all three forces intersect at a common point. If the three forces are parallel (this is called a *parallel force system*), then their point of intersection can be thought of as being at infinity.

If a body is not a two-force or three-force member, then we refer to it as either a *zero-force member*, if it is subjected to no force at all (these are often encountered in trusses, discussed in Chapter 6), or a *general multiforce member*, if it is subjected to forces at more than three points and/or has moment loading and/or has distributed loading.

Engineering design. Engineering design was discussed in Section 5.5, and a process for developing a design was described (see Fig. 5.29 on p. 313). *Codes* and *standards* were described, and an overview of some professional and government organizations that sponsor codes and standards and/or have regulatory power over performance and safety were reviewed.

REVIEW PROBLEMS

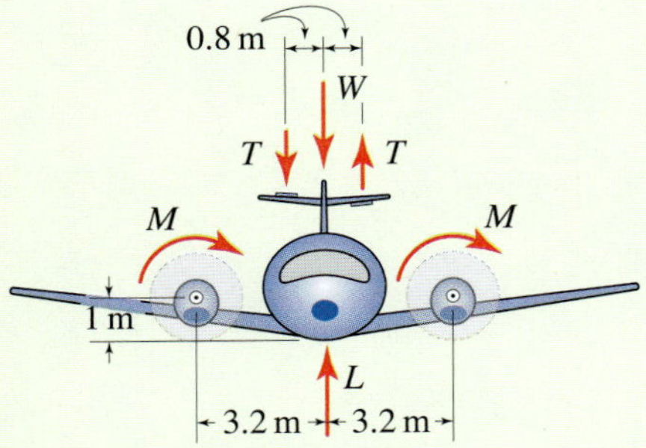

Figure P5.89 and P5.90

Problem 5.89

The propellers of the twin engine airplane shown rotate in the same direction, and each propeller exerts a moment $M = 1.3$ kN·m on the wings of the plane. To equilibrate this moment, trim tabs on the vertical stabilizer are used to produce trim forces T. Determine the value of T, assuming the trim forces are vertical.

Problem 5.90

In Prob. 5.89, in what direction do the propellers rotate? Specify clockwise or counterclockwise with respect to the view shown in Fig. P5.89, and explain your reasoning. **Note:** Concept problems are about *explanations*, not computations.

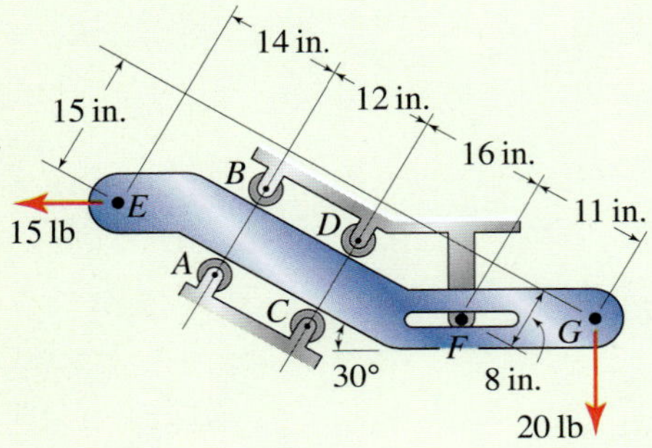

Figure P5.91

Problem 5.91

A bracket is supported by a loose-fitting pair of rollers at points A and B, and another loose-fitting pair at C and D, and a frictionless pin at F. The forces at E and G are horizontal and vertical, respectively. Determine the reactions at the pin and each of the four rollers.

Problem 5.92

A semicircular geared bracket is subjected to a vertical 80 N force at point C. The bracket is supported by frictionless pins at A and B and a gear at D. The pins and gear are fixed to plate E, and the gear at D is not allowed to rotate. Determine the tangential force supported by the gear at D and the reactions at pins A and B.

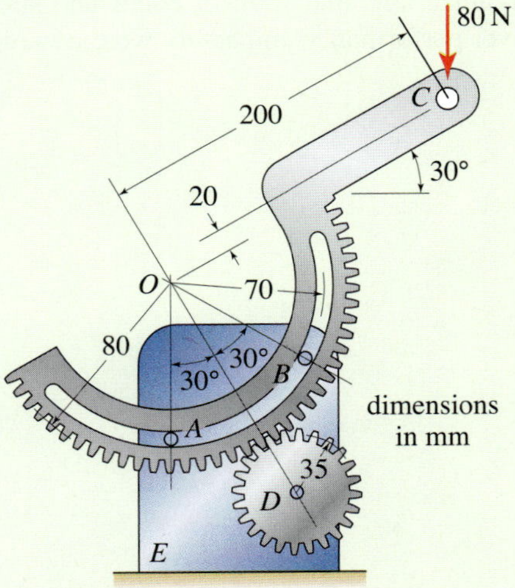

Figure P5.92

Problem 5.93

A frame supports three frictionless pulleys that guide a tape that runs at constant speed. Determine the reactions at support A if $\alpha = \beta = 0°$.

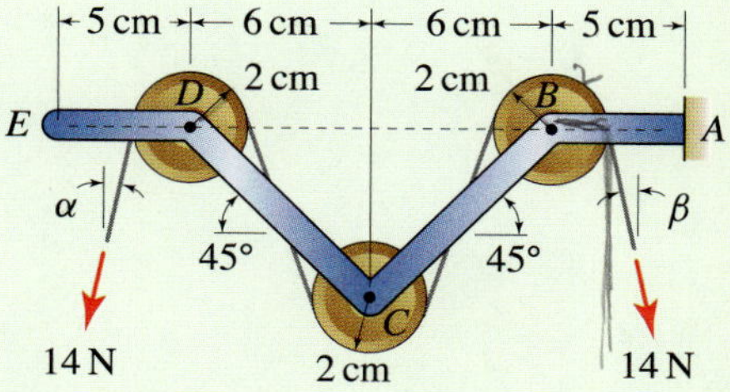

Figure P5.93–P5.95

Problem 5.94

Repeat Prob. 5.93 if $\alpha = 30°$ and $\beta = 0°$.

Problem 5.95

Repeat Prob. 5.93 if $\alpha = 30°$ and $\beta = 45°$.

Problem 5.96

The trigger of a high-pressure washer gun is shown. The torsional spring at point A has stiffness $k_t = 1100$ N·mm/rad and is prewound by 6° when it is installed (i.e., when the trigger makes contact at point G). The axial spring DE has stiffness $k_a = 0.1$ N/mm and has 40 mm unstretched length. The trigger operates the washer on low pressure when $0° < \theta \le 15°$, and when $15° < \theta \le 25°$, the trigger operates the washer on high pressure. Assume force F remains horizontal with the same line of action for all trigger positions.

(a) Determine the force F that causes the trigger to begin movement.

(b) Determine the force F that causes the trigger to first make contact with the plunger at C ($\theta = 15°$).

(c) Determine the force F required to fully pull the trigger ($\theta = 25°$). Assume the plunger at C contacts the back of the trigger at a right angle when $\theta = 25°$.

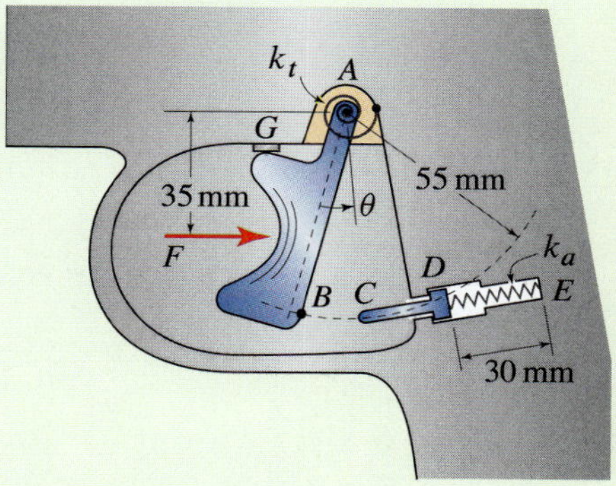

Figure P5.96

Problem 5.97

Member $ABCD$ has negligible weight.

(a) If member $ABCD$ is to be a two-force member, which (if any) of F_B, F_D, and M_C must be zero?

(b) If member $ABCD$ is to be a three-force member, which (if any) of F_B, F_D, and M_C must be zero?

(c) If $F_D = 0$, $M_C = 0$, and $F_B > 0$, draw the FBD for member $ABCD$ and sketch the force polygon corresponding to $\sum \vec{F} = \vec{0}$.

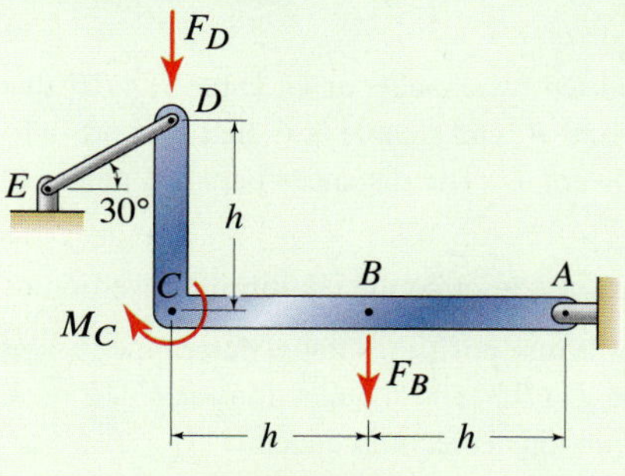

Figure P5.97

Problems 5.98 and 5.99

Draw the FBD for each object shown, and specify whether it has partial fixity or full fixity and whether it is statically determinate or statically indeterminate. Assume cables, if present, are in tension.

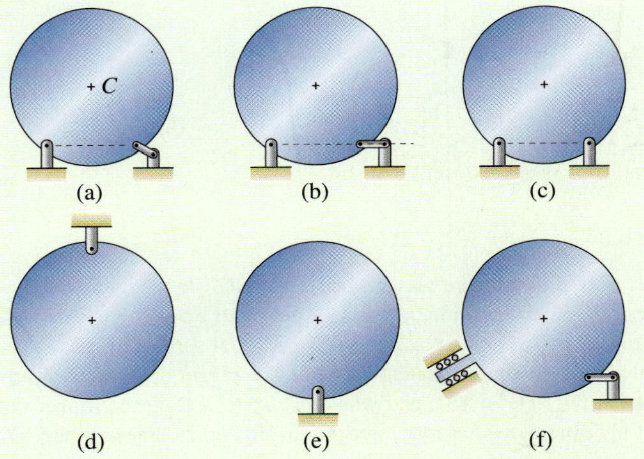

Figure P5.98

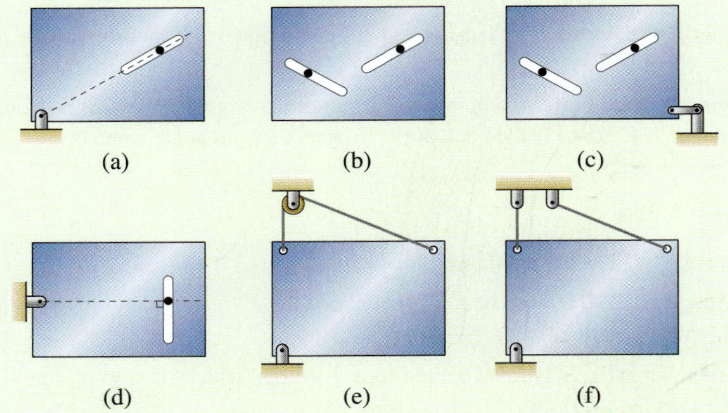

Figure P5.99

Problem 5.100

Boom $OABC$ is supported by a socket at O, cable $EABF$ that passes through small frictionless loops at A and B, and a cable at C that supports a force T_1 and whose line of action is directed toward D. The distances between points O and A, A and B, and B and C are equal.

(a) If $T_1 = 0$, qualitatively describe the equilibrium position of the boom.

(b) For the static equilibrium position shown, determine the value of T_1, the force T_2 supported by cable $EABF$, and the reactions at O. *Hint:* A numerical solution of the equilibrium equations is recommended.

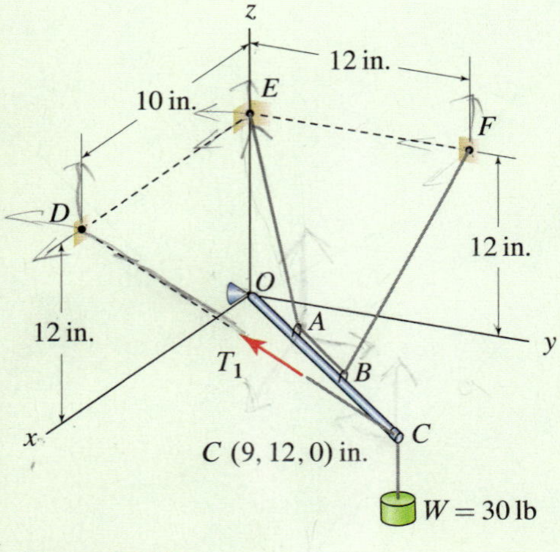

Figure P5.100

Problem 5.101

A machine for sawing concrete is shown. It is supported by a cutting disk at point C and two wheels at points A and B (the wheel at B is not shown). The wheels at A and B are separated by a 0.8 m distance along the x axis. Determine the dimension d where the cutting disk should be located so that the force supported by wheel A is 20% of the force supported by wheel B.

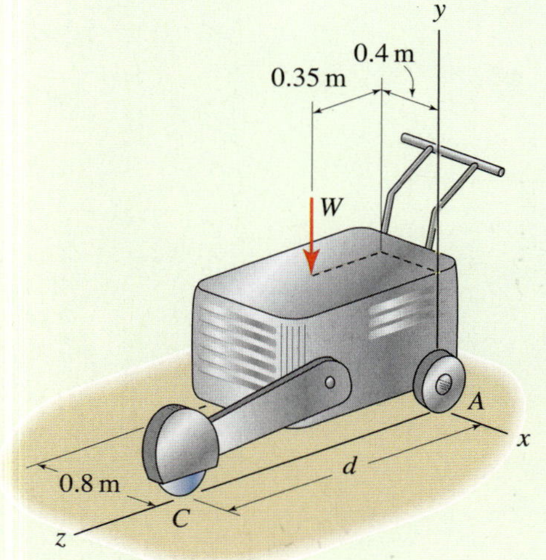

Figure P5.101

Problem 5.102

A structure consists of a thin flat plate and two short bars with bearing supports at A and B, where the bearing at B is self-aligning. The plate is loaded at its center by a 100 lb vertical force and by a 40 lb force in the x direction at one of the corners.

(a) Does the plate have complete fixity or partial fixity, and is it statically determinate or statically indeterminate? Explain.

(b) Determine all reactions at A and B.

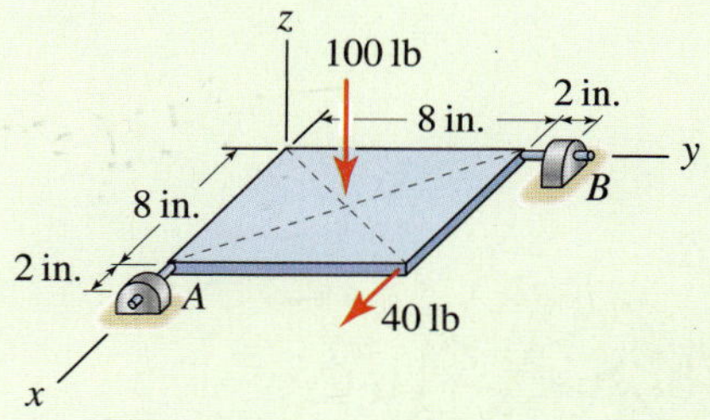

Figure P5.102

Problem 5.103

Bar $ABCD$ is supported by a cable AED, which passes over a frictionless pulley at point E, and a collar B that slides without friction on a vertical shaft with square cross section. If the tip A is subjected to a 5 kN vertical force, determine the tension in the cable and all support reactions at collar B.

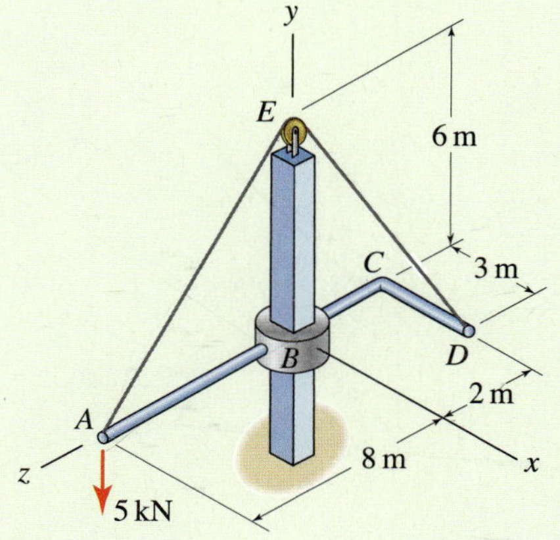

Figure P5.103

6 Structural Analysis and Machines

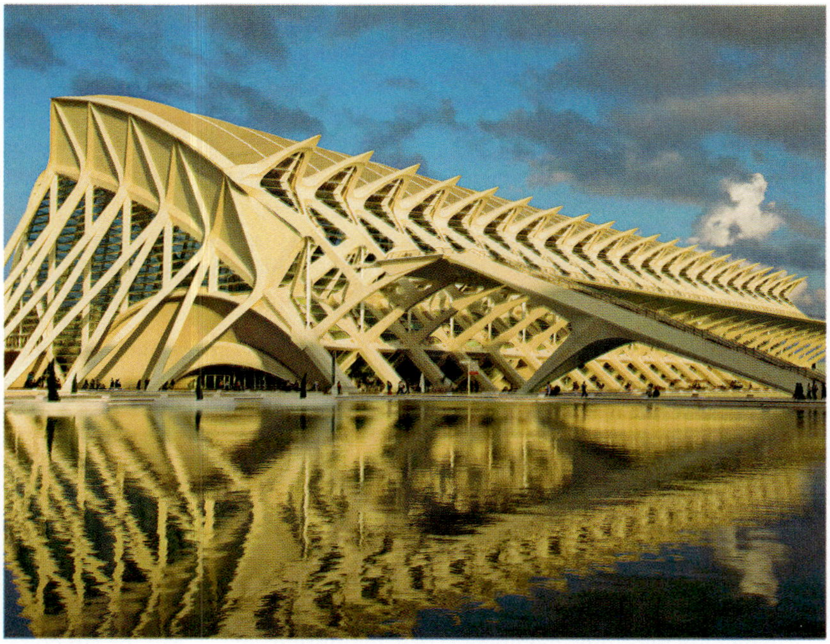

Structures and machines often consist of an assemblage of numerous members. In this chapter, we analyze the equilibrium of these. One of the goals is determination of the forces supported by the individual members of a structure or machine. Design considerations, especially for trusses, are also discussed.

Before any detailed discussion in this chapter, it is useful to reflect on the meaning of *structure* and *machine*. We use the word ***structure*** to describe an arrangement of material and/or individual members that, as a whole, is intended to support forces that are applied to it. The word ***machine*** describes an arrangement of material and/or individual members where the goal is usually transmission of motion and/or force. These definitions are broad and have considerable overlap, and frequently a particular device could appropriately be called either a structure or a machine. Although there are frequent exceptions, a structure will often use a stationary arrangement of members (that is, the individual members will have little or no motion relative to one another), while a machine will have members with significant relative motion.

6.1 Truss Structures and the Method of Joints

A ***truss*** is a structure that consists of two-force members only, where members are organized so that the assemblage as a whole behaves as a single object.

(a) (b)

Figure 6.1. Examples of truss structures. (a) The roof of this highway maintenance building under construction uses 124 identical wooden roof trusses with 16 in. spacing. (b) Mores Creek bridge over Lucky Peak Reservoir near Boise, Idaho.

Some examples of truss structures are shown in Fig. 6.1. Observe that trusses use very little material, yet they are very strong.

Throughout this section and the next, we discuss truss structures in two dimensions, and these are called *plane trusses*. For a two-dimensional structure to be a plane truss, it must have the following characteristics:

- All members must be connected to one another by frictionless pins, and the locations of these pins are called *joints*.
- Each member may have no more than two joints.
- Forces may be applied at joints only.
- The weight of individual members must be negligible.

If all of the above characteristics are satisfied, then it is guaranteed that all members of the structure are two-force members. The important consequence of having only two-force members is that equilibrium analysis of the structure reduces to equilibrium analysis of a system of particles where the number of particles equals the number of joints in the truss. Some common types of truss structures are shown in Fig. 6.2.

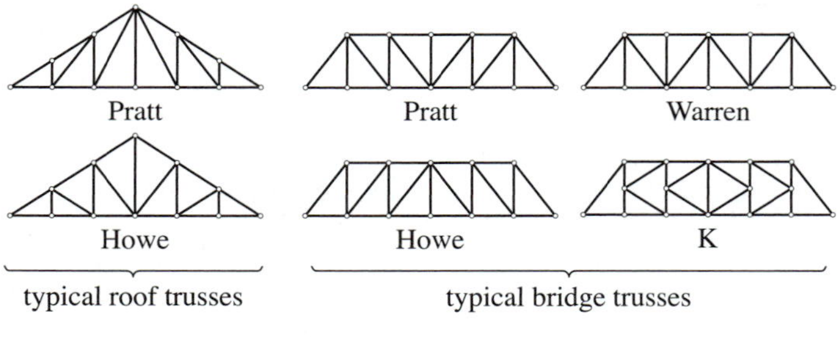

Figure 6.2. Some common types of truss structures.

Figure 6.3
While most truss structures are composed entirely of straight members, this truss uses a curved upper chord.

Recall from Chapter 5 that a member does not need to be straight to be a two-force member. While trusses are most often constructed using straight members, they may contain members with curved or other complex shapes, and Fig. 6.3 shows such an example.

When may a structure be idealized as a truss?

Relatively few real life structures fully conform to the definition of a truss, the main reason usually being that connections between members are not pins. Figure 6.4 shows some examples of typical connections in structures. These

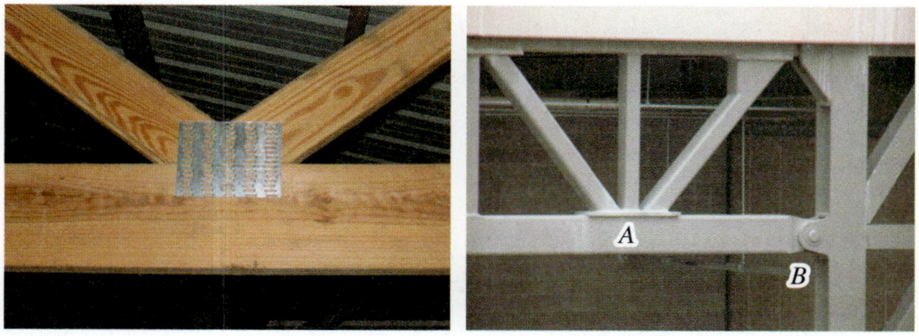

(a) (b)

Figure 6.4. Typical connections between members of a structure. (a) In this wooden roof truss, a rectangular nail plate is used to connect members. (b) In this steel truss, connection A is welded while connection B is pinned.

connections are capable of supporting moments, and thus the members that emanate from such connections usually are not two-force members. For precise analysis, structures with connections such as these should be modeled as *frame structures* which are discussed in Section 6.4. If a structure is sufficiently stiff, then the moments at connections such as those shown in Fig. 6.4 may be small, and it may be reasonable to assume they are zero, in which case the structure could be modeled as a truss. To summarize, if a particular structure would qualify as a truss except for the connection details, then in engineering practice these structures are often modeled as trusses anyway and the methods of analysis discussed in this section and the next are used. This idealization is far from perfect, but is widely used, and with good judgment and generous factors of safety, it is very successful.

Another common departure from the definition of a truss is that structures often have forces applied at locations other than joints. In Section 6.2, a method is discussed where such loads are replaced by equivalent loads positioned at joints so that truss analysis can be performed.

Method of joints

In general, the objectives in analysis of a truss are determination of the support reactions for the truss and determination of the forces supported by the individual members of the truss. In the *method of joints*, a truss is analyzed by treating each joint as a particle. If the entire truss is in equilibrium, then every joint within the truss is also in equilibrium (indeed, *all* material within the truss is in equilibrium). Analysis proceeds by drawing FBDs of the joints throughout the truss, writing equilibrium equations for each of these, and solving the equilibrium equations for the unknowns. Prior to this, it may be necessary or desirable to determine the reactions for the truss as a whole, and the methods of Chapter 5 can be used for this. Analysis by the method of joints is illustrated in the following example.

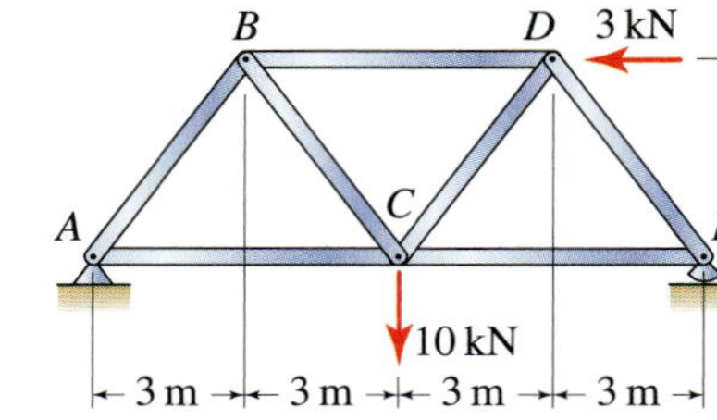

Figure 6.5
A plane truss.

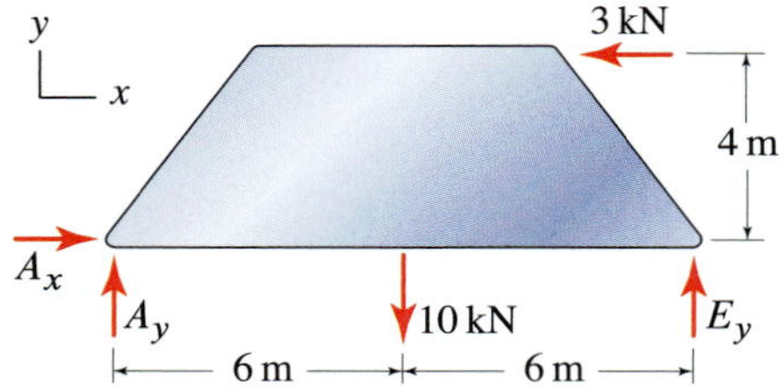

Figure 6.6
Free body diagram to determine the support reactions.

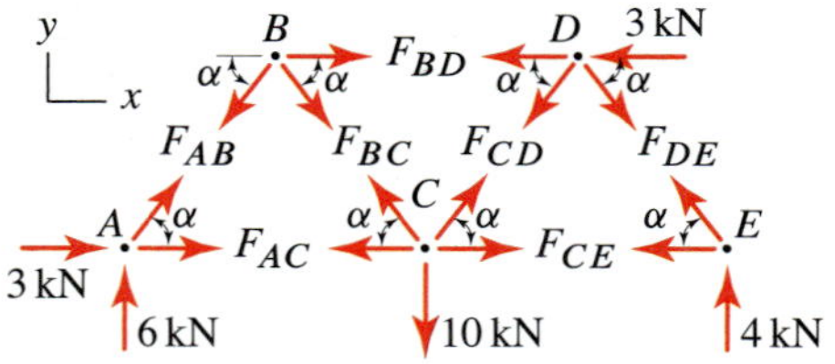

Figure 6.7
Free body diagrams of joints (pins) in the truss. All member forces are defined to be positive in tension. For example, a positive value of F_{AB} means that member AB is in tension.

■ **Mini-Example.** Use the method of joints to determine the force supported by each member of the truss shown in Fig. 6.5.

Solution. We begin by obtaining the support reactions for the truss by drawing the FBD of the structure as a whole as shown in Fig. 6.6. Writing equilibrium equations provides

$$\sum M_A = 0: \quad -(10\,\text{kN})(6\,\text{m}) + E_y(12\,\text{m}) + (3\,\text{kN})(4\,\text{m}) = 0 \quad \Rightarrow \quad E_y = 4\,\text{kN}, \tag{6.1}$$

$$\sum F_y = 0: \quad A_y + E_y - 10\,\text{kN} = 0 \quad \Rightarrow \quad A_y = 6\,\text{kN}, \tag{6.2}$$

$$\sum F_x = 0: \quad A_x - 3\,\text{kN} = 0 \quad \Rightarrow \quad A_x = 3\,\text{kN}. \tag{6.3}$$

Next, we draw FBDs of each joint (pin) as shown in Fig. 6.7. Since each joint has a concurrent force system, each joint is treated as a particle in equilibrium, and hence each joint permits *two* equilibrium equations to be written. Among the many joints that the truss has, a good strategy is to always (if possible) select a joint that has no more than two unknowns, since then the unknowns may be immediately solved for. Among the five possible joints shown in Fig. 6.7, joints A and E are the only two having only two unknowns. Noting that $\alpha = \tan^{-1}(4/3) = 53.1°$, we select joint A and write*

Joint A:

$$\sum F_y = 0: \quad 6\,\text{kN} + F_{AB}\sin\alpha = 0 \quad \Rightarrow \quad F_{AB} = -7.5\,\text{kN}, \tag{6.4}$$

$$\sum F_x = 0: \quad 3\,\text{kN} + F_{AB}\cos\alpha + F_{AC} = 0 \quad \Rightarrow \quad F_{AC} = 1.5\,\text{kN}. \tag{6.5}$$

We may now repeat this procedure for joint E, but we will continue with joint B instead, as now it has only two unknowns. Using the FBD for joint B gives

Joint B:

$$\sum F_y = 0: \quad -F_{AB}\sin\alpha - F_{BC}\sin\alpha = 0 \quad \Rightarrow \quad F_{BC} = 7.5\,\text{kN}, \tag{6.6}$$

$$\sum F_x = 0: \quad -F_{AB}\cos\alpha + F_{BC}\cos\alpha + F_{BD} = 0 \quad \Rightarrow \quad F_{BD} = -9.0\,\text{kN}. \tag{6.7}$$

Joints C, D, and E remain to be analyzed, and of these, joints D and E both have two unknowns. Furthermore, examination of the FBDs for joints D and E shows that E will entail slightly less algebra to solve for its unknowns, and thus we elect to write equilibrium equations for it.

Joint E:

$$\sum F_y = 0: \quad 4\,\text{kN} + F_{DE}\sin\alpha = 0 \quad \Rightarrow \quad F_{DE} = -5\,\text{kN}, \tag{6.8}$$

$$\sum F_x = 0: \quad -F_{CE} - F_{DE}\cos\alpha = 0 \quad \Rightarrow \quad F_{CE} = 3\,\text{kN}. \tag{6.9}$$

There is now only one unknown remaining, namely, F_{CD}, and either joint C or D may be used to determine it. We select D because its FBD has one less force than the FBD for joint C. Thus

Joint D:

$$\sum F_y = 0: \quad -F_{CD}\sin\alpha - F_{DE}\sin\alpha = 0 \quad \Rightarrow \quad F_{CD} = 5\,\text{kN}. \tag{6.10}$$

* Because of the nice geometry of this truss, we could avoid computing α and use $\cos\alpha = 3/5$ and $\sin\alpha = 4/5$.

Remarks

- This solution began with finding the support reactions for the entire structure. Depending on the geometry of the structure and how it is supported, finding the reactions at the outset may not be needed. In this example, it was necessary because otherwise all of the FBDs shown in Fig. 6.7 have three or more unknowns. Example 6.3 considers a structure where the support reactions are not needed prior to analysis by the method of joints.

- In the FBD shown in Fig. 6.6, the truss is represented as a solid object to emphasize that for the analysis of support reactions, the arrangement of members within the structure is irrelevant *provided* it is sufficient to support the forces and reactions that are applied to it. When this assumption is not true, this phase of the solution must be revised, as discussed later.

- In the FBDs of Fig. 6.7, all of the member forces are defined so that a positive value corresponds to tension. In the solutions obtained here, members AC, BC, CE, and CD support positive forces, hence they are in tension; and members AB, BD, and DE support negative forces, hence they are in compression.

- To obtain the solutions for all unknowns, only one equilibrium equation for joint D was used, and neither of the two equilibrium equations for joint C was used. To help check the accuracy of your solution, you should write these three equilibrium equations and verify that they are satisfied. If they are not, then an error has been made.

- In view of the preceding comment, you might wonder why all of the unknowns could be determined without using *all* of the available equilibrium equations. Recall from Section 5.3 that having more equilibrium equations than unknowns usually indicates a mechanism. However, the truss of Fig. 6.5 is clearly not a mechanism: it is fully fixed and is statically determinate. Indeed, because the support reactions were found at the outset, the three equilibrium equations in question were exhausted in writing Eqs. (6.1)–(6.3), and in fact the number of unknowns and the number of available equilibrium equations are the same. For this reason, the three "extra" equilibrium equations for joints D and E were not needed to find the unknowns.

Helpful Information

Method of joints. The method of joints is really nothing more than the method of analysis discussed in Chapter 3, namely, equilibrium of a system of particles.

Interesting Fact

Computer analysis. Although not discussed in this book, the method of joints can be automated for solution using a computer. With knowledge of statics, mechanics of materials, and a basic understanding of matrices, developing such software is surprisingly straightforward. Computer programs for truss analysis see widespread use.

Zero-force members

A truss member (or any member) that supports no force is called a *zero-force member*. Trusses often contain many zero-force members. While it might seem that such members play no role in strengthening a truss, in fact they may be very important, as discussed in Section 6.2. Here we discuss a method to identify zero-force members by inspection. Consider the situation shown in Fig. 6.8; joint A connects three truss members, two of the members are collinear (AB and AD), and no external force is applied to joint A. With the xy coordinate system shown, summing forces in the y direction provides

$$\sum F_y = 0: \quad F_{AC} \sin\alpha = 0 \tag{6.11}$$

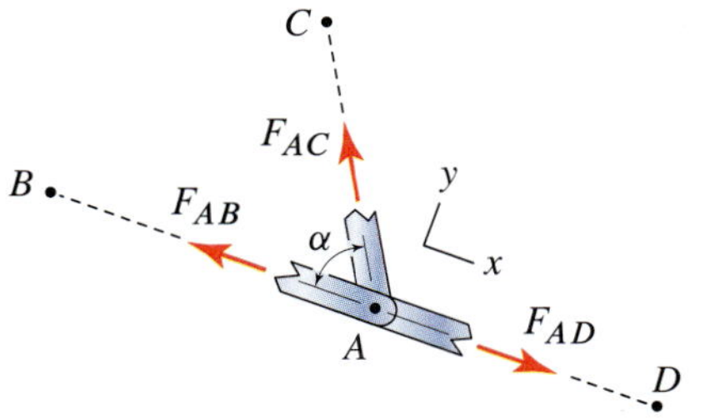

Figure 6.8
Geometry of members in a truss allowing a zero-force member to be recognized by inspection.

Unless $\alpha = 0°$ or $180°$, in which case member AC is collinear with the other two members, $F_{AC} = 0$ and member AC is a *zero-force member*. Note that if joint A has an external force applied to it, then if that force has any component in the y direction shown in Fig. 6.8, then $F_{AC} \neq 0$. Similarly, if joint A has more than three members connected to it, then in general $F_{AC} \neq 0$.

Identification of zero-force members in a truss. Figure 6.8 defines a pattern that allows zero-force members in a truss to be easily identified. To use this, we examine the joints in a truss, and if a particular joint

- Has three members connected to it,
- Two of these members are collinear, and
- The joint has no external force applied to it,

then the noncollinear member is a zero-force member.

Rather than memorize this list, it is easier to simply understand the rationale behind a zero-force member; namely, Fig. 6.8 and Eq. (6.11). Furthermore, understanding these features allows you to recognize other situations in which a zero-force member occurs. For example, if member AD in Fig. 6.8 is not present, then summing forces in the y direction shows that member AC is zero-force, and with this result, summing forces in the x direction shows that member AB is also a zero-force member.

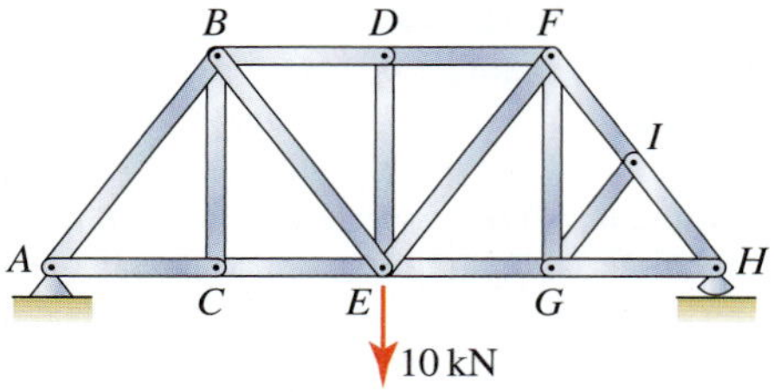

Figure 6.9
Example of a truss containing several zero-force members.

Mini-Example. Identify the zero-force members in the truss shown in Fig. 6.9.

Solution. By inspection of each joint, we identify the following zero-force members:

Examination of joint C shows that member BC is zero-force.
Examination of joint D shows that member DE is zero-force.
Examination of joint I shows that member GI is zero-force.
Because member GI is zero-force, examination of joint G then shows that member FG is zero-force.

Remark. For a member such as DE, the presence of the 10 kN force applied to joint E is a common source of confusion, as intuition may suggest (wrongly) that member DE must participate in supporting the 10 kN force. Note that our conclusion that member DE is a zero-force member is based entirely on the conditions at joint D.

Before closing, we note that most of our discussion focused on *sufficient* conditions for a member to be zero-force. That is, if an equilibrium equation for a particular joint can be written in the form of Eq. (6.11), and $\alpha \neq 0°$ or $180°$, then this is a sufficient condition to conclude that a member is zero-force, and furthermore it is possible to easily identify such members by inspection. However, it is not a *necessary* condition. That is, a truss may have other zero-force members. For example, in the truss shown in Fig. 6.9, if other forces were applied at joints B and/or F and they had proper magnitudes and directions, then it is possible that by chance other members, in addition to the four members already cited, could be zero-force members. However, it is typically not possible to identify such members by inspection; rather if any such members are present, they are seen to be zero-force only after the equilibrium equations have been solved. This situation arises in Example 6.3.

Typical truss members

The vast majority of trusses are constructed entirely of straight, slender members. The reason is that straight members are very efficient at supporting axial forces. Because of this, straight members can usually be *slender*, meaning the length of a member is substantially greater than the size of its cross section. While a truss may contain members that are not straight, such as the curved upper chord shown in Fig. 6.3, such members are considerably less efficient at supporting forces because they also experience *bending*, and therefore substantially more material is required so that the member is strong enough. When curved members are used in a truss, it is almost always for nonstructural reasons such as aesthetics, ease of manufacture, and so on. Efficiency of straight members versus curved members for supporting loads is discussed in detail in Chapter 8 where internal forces and bending are treated (see Prob. 8.22 on p. 466). An additional, important factor controlling strength of straight truss members is the possibility of buckling due to compressive forces. Buckling is briefly discussed in Section 6.2.

> **Helpful Information**
>
> **An experiment: straight versus non-straight truss members.** Take a metal or plastic clothes hanger and cut it at the two locations shown. Then, use your hands to apply forces to the non-straight portion. Clearly it is very flexible and will undergo significant *bending*. In contrast, use your hands to apply tensile forces to the straight portion and observe that no visible deformation is produced, even if you apply as much force as possible. The straight member is very efficient at supporting the forces applied to it, whereas the non-straight member is not as efficient because of bending.

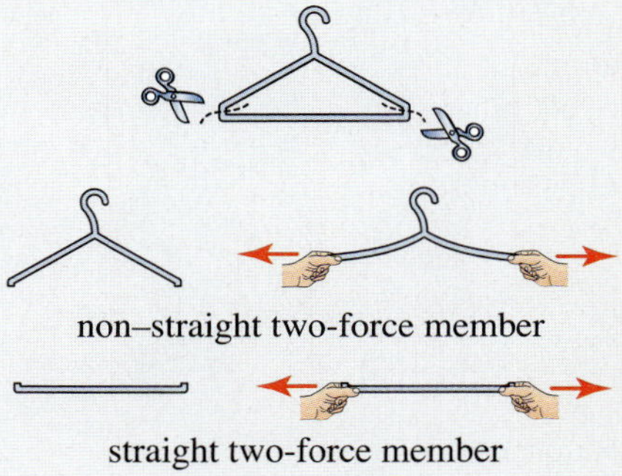

End of Section Summary

In this section, *truss structures* are defined and the *method of joints* is developed for determining the forces supported by individual members of a truss. Some of the key points are as follows:

- A *truss* is a structure that consists of two-force members only, where members are organized so that the assemblage as a whole behaves as a single object. For all members of a structure to be two-force members, the structure must have the following characteristics:

  - All members must be connected to one another by frictionless pins, and the locations of these pins are called *joints*.
  - Each member may have no more than two joints.
  - Forces may be applied at joints only.
  - The weight of individual members must be negligible.

  Real structures usually do not satisfy all of these requirements. Nonetheless, many real structures are idealized as trusses.

- In the *method of joints*, the force supported by each member of a truss is determined by drawing FBDs for each joint and then requiring that all joints be in equilibrium by writing the equations $\sum F_x = 0$ and $\sum F_y = 0$ for each joint.

- A truss member (or any member) that supports no force is called a *zero-force member*. Trusses often contain many zero-force members, and the role played by these in strengthening a truss is discussed in Section 6.2.

EXAMPLE 6.1 Truss Analysis by the Method of Joints

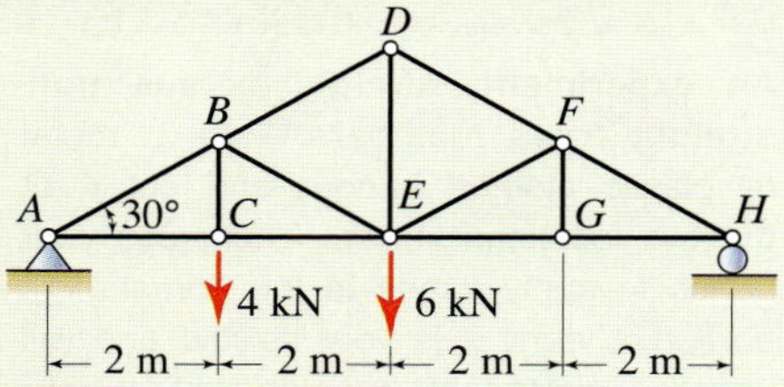

Figure 1

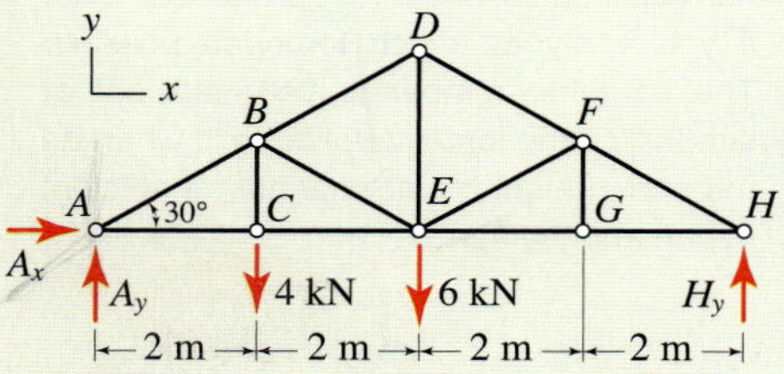

Figure 2
Free body diagram for determining the support reactions.

The structure shown consists of 13 members where the hollow circles indicate joints. Determine the force supported by each member of the structure.

SOLUTION

Road Map In this structure, we will neglect the weights of individual members under the assumption they are small compared to the 4 and 6 kN forces. Because forces are applied at joints only and all members are connected to one another by frictionless pins, all members of the structure are two-force members and hence the structure is a truss. To analyze this truss, we begin by determining the support reactions, followed by use of the method of joints to determine the force supported by each member.

Modeling To determine the reactions for the truss as a whole, we draw the FBD for the entire structure as shown in Fig. 2.

Governing Equations & Computation Using the FBD shown in Fig. 2, we write and solve the following equilibrium equations:

$$\sum M_A = 0: \quad -(4\,\text{kN})(2\,\text{m}) - (6\,\text{kN})(4\,\text{m}) + H_y(8\,\text{m}) = 0 \;\Rightarrow\; H_y = 4\,\text{kN}, \quad (1)$$

$$\sum F_y = 0: \quad A_y + H_y - 4\,\text{kN} - 6\,\text{kN} = 0 \;\Rightarrow\; A_y = 6\,\text{kN}, \quad (2)$$

$$\sum F_x = 0: \quad A_x = 0 \;\Rightarrow\; A_x = 0. \quad (3)$$

Modeling Now that the support reactions are known, we proceed with the method of joints. Examination of joint G shows that member FG is zero-force, and with this knowledge, examination of joint F then shows that member EF is zero-force. Recognizing zero-force members at the outset of an analysis may save you some work, but in the following analysis we will use this knowledge as a partial check of our solution. To use the method of joints, we draw FBDs of all joints in the truss, as shown in Fig. 3. In these FBDs, all member forces are taken to be positive in tension. For example, if we find that F_{AB} is positive, then member AB is in tension, whereas if we find that F_{AB} is negative, then member AB is in compression.

Governing Equations & Computation We begin writing equilibrium equations at joint A, because there are only two unknowns there (joint H is an equally good choice). Following joint A, we proceed to joint C, as there are then only two unknowns there, and so on.

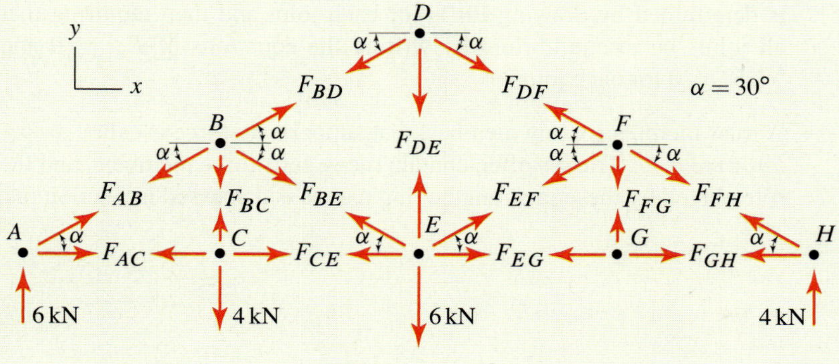

Figure 3. Free body diagrams of joints where $\alpha = 30°$.

Joint A:

$$\sum F_y = 0: \quad F_{AB} \sin 30^\circ + 6\,\text{kN} = 0 \quad \Rightarrow \quad \boxed{F_{AB} = -12\,\text{kN},} \tag{4}$$

$$\sum F_x = 0: \quad F_{AC} + F_{AB} \cos 30^\circ = 0 \quad \Rightarrow \quad \boxed{F_{AC} = 10.39\,\text{kN}.} \tag{5}$$

Joint C:

$$\sum F_x = 0: \quad -F_{AC} + F_{CE} = 0 \quad \Rightarrow \quad \boxed{F_{CE} = 10.39\,\text{kN},} \tag{6}$$

$$\sum F_y = 0: \quad F_{BC} - 4\,\text{kN} = 0 \quad \Rightarrow \quad \boxed{F_{BC} = 4\,\text{kN}.} \tag{7}$$

Joint B:

$$\sum F_x = 0: \quad -F_{AB} \cos 30^\circ + F_{BD} \cos 30^\circ + F_{BE} \cos 30^\circ = 0, \tag{8}$$

$$\sum F_y = 0: \quad -F_{AB} \sin 30^\circ + F_{BD} \sin 30^\circ - F_{BE} \sin 30^\circ - F_{BC} = 0, \tag{9}$$

$$\Rightarrow \quad \boxed{F_{BD} = -8\,\text{kN} \quad \text{and} \quad F_{BE} = -4\,\text{kN}.} \tag{10}$$

Joint D:

$$\sum F_x = 0: \quad -F_{BD} \cos 30^\circ + F_{DF} \cos 30^\circ = 0 \Rightarrow \boxed{F_{DF} = -8\,\text{kN},} \tag{11}$$

$$\sum F_y = 0: \quad -F_{BD} \sin 30^\circ - F_{DE} - F_{DF} \sin 30^\circ = 0 \Rightarrow \boxed{F_{DE} = 8\,\text{kN}.} \tag{12}$$

Joint H:

$$\sum F_y = 0: \quad F_{FH} \sin 30^\circ + 4\,\text{kN} = 0 \quad \Rightarrow \quad \boxed{F_{FH} = -8\,\text{kN},} \tag{13}$$

$$\sum F_x = 0: \quad -F_{FH} \cos 30^\circ - F_{GH} = 0 \quad \Rightarrow \quad \boxed{F_{GH} = 6.928\,\text{kN}.} \tag{14}$$

Joint G:

$$\sum F_y = 0: \quad F_{FG} = 0 \quad \Rightarrow \quad \boxed{F_{FG} = 0,} \tag{15}$$

$$\sum F_x = 0: \quad -F_{EG} + F_{GH} = 0 \quad \Rightarrow \quad \boxed{F_{EG} = 6.928\,\text{kN}.} \tag{16}$$

Joint F:

$$\sum F_x = 0: \quad -F_{DF} \cos 30^\circ - F_{EF} \cos 30^\circ + F_{FH} \cos 30^\circ = 0 \tag{17}$$

$$\Rightarrow \quad \boxed{F_{EF} = 0.} \tag{18}$$

Discussion & Verification

- Notice that we were able to determine all of the member forces without using one of the equilibrium equations for joint F and the two equations for joint E. As a check on our solution, you should write these three equations and verify that the member forces we determined satisfy all of them.
- In Eqs. (15) and (18), two members were found to have zero force, and these are the same members that were identified by inspection at the outset of our analysis as being zero-force members. This is a useful partial check of solution accuracy.
- This solution neglected the weight of the truss. An approximate way to include this weight is illustrated in Example 6.2.

EXAMPLE 6.2 *A Truss with Curved Members and Weights of Members*

The structure shown consists of five pin-connected members (members AC and BD do not intersect). In addition to the 4 kip force shown, the individual members of the structure have the weights listed in Table 1. Idealize the structure to be a truss and determine the forces supported by each memeber.

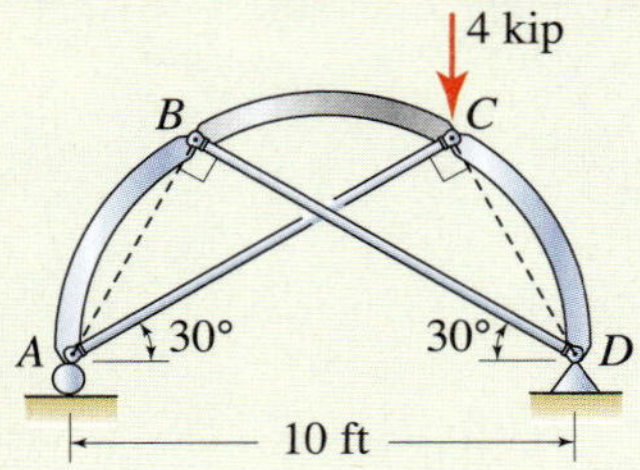

Figure 1

Table 1
Weights of individual members.

Member	Weight
AB	$W_{AB} = 400\,\text{lb}$
AC	$W_{AC} = 100\,\text{lb}$
BC	$W_{BC} = 200\,\text{lb}$
BD	$W_{BD} = 100\,\text{lb}$
CD	$W_{CD} = 400\,\text{lb}$

SOLUTION

Road Map Because the weights listed in Table 1 are distributed forces, the members of this structure *are not* two-force members and hence this structure *is not* a truss. Nonetheless, because these weights are small compared to the 4 kip force applied at Joint C, it is possible to approximate this structure as a truss, and to do so under these circumstances is common.

Modeling The weights (distributed forces) listed in Table 1 will be replaced with forces that are applied to joints by using the following procedure. For each member, that member's weight is proportioned equally between the two joints it connects to. Thus, for member AB, one-half of its weight will be applied to joint A and one-half will be applied to joint B. Similarly, for member AC, one-half of its weight will be applied to joint A and one-half will be applied to joint C. Doing this for all members results in the following forces at joints

$$P_A = (W_{AB} + W_{AC})/2 = 250\,\text{lb}, \tag{1}$$
$$P_B = (W_{AB} + W_{BC} + W_{BD})/2 = 350\,\text{lb}, \tag{2}$$
$$P_C = (W_{AC} + W_{BC} + W_{CD})/2 = 350\,\text{lb}, \tag{3}$$
$$P_D = (W_{BD} + W_{CD})/2 = 250\,\text{lb}. \tag{4}$$

Treatment of distributed forces in this fashion is sometimes called *load lumping* because a distributed force is being replaced by concentrated forces.*

To determine the reactions for the truss as a whole, we draw the FBD shown in Fig. 2. Then, to use the method of joints to determine the force supported by each member, we draw the FBDs shown in Fig. 4 where member forces are taken to be positive in tension.

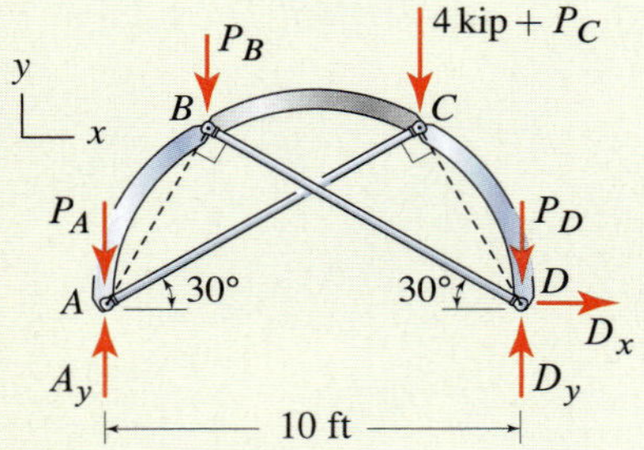

Figure 2
Free body diagram for determining the support reactions.

Governing Equations & Computation To find the reactions for the truss as a whole, we use the FBD shown in Fig. 2 and write the following equilibrium equations (with the aid of Fig. 3 to determine moment arms):

$$\sum M_A = 0: \quad -P_B(2.5\,\text{ft}) - (4000\,\text{lb} + P_C)(7.5\,\text{ft}) + (D_y - P_D)(10\,\text{ft}) = 0 \quad \Rightarrow \quad D_y = 3600\,\text{lb}, \tag{5}$$

$$\sum F_y = 0: \quad A_y + D_y - P_A - P_B - 4000\,\text{lb} - P_C - P_D = 0 \quad \Rightarrow \quad A_y = 1600\,\text{lb}, \tag{6}$$

$$\sum F_x = 0: \quad A_x = 0 \quad \Rightarrow \quad A_x = 0. \tag{7}$$

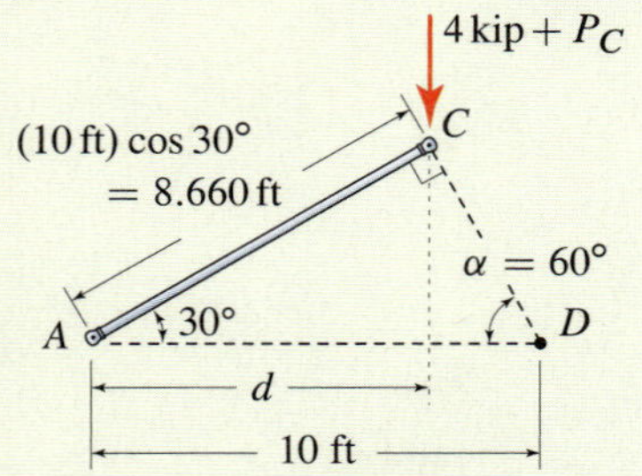

$d = (8.660\,\text{ft})\cos 30° = 7.5\,\text{ft}$

Figure 3
Geometry of member AC and joints A, C, and D to determine moment arms for writing equilibrium equations.

* Problem 7.85 shows that the load lumping procedure outlined here provides an equivalent force system for straight members with uniform weight distributions. For members that are not straight (as in this example) or if the weight distribution is not uniform, this load lumping procedure is an approximation, although it is commonly used and provides good results provided the shape and weight distribution of members does not deviate greatly from being straight and uniform, and the weights of members are small compared to other forces the structure supports.

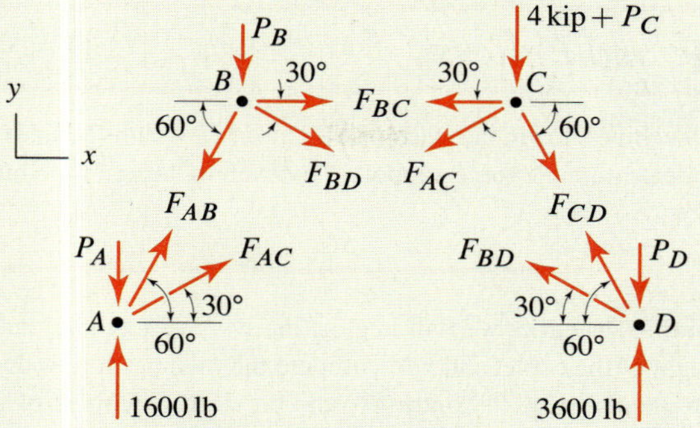

Figure 4. Free body diagrams of joints.

To apply the method of joints, we use the FBDs of all the joints in the truss, as shown in Fig. 4, where member forces are taken to be positive in tension. We then write and solve the equilibrium equations for the joints as follows:

Joint *A*:

$$\sum F_x = 0: \qquad F_{AB}\cos 60^\circ + F_{AC}\cos 30^\circ = 0, \tag{8}$$

$$\sum F_y = 0: \qquad F_{AB}\sin 60^\circ + F_{AC}\sin 30^\circ + 1600\,\text{lb} - P_A = 0, \tag{9}$$

$$\Rightarrow \boxed{F_{AB} = -2338\,\text{lb} \quad \text{and} \quad F_{AC} = 1350\,\text{lb}.} \tag{10}$$

Joint *B*:

$$\sum F_y = 0: \qquad -F_{AB}\sin 60^\circ - F_{BD}\sin 30^\circ - P_B = 0 \Rightarrow \boxed{F_{BD} = 3350\,\text{lb},} \tag{11}$$

$$\sum F_x = 0: \quad -F_{AB}\cos 60^\circ + F_{BD}\cos 30^\circ + F_{BC} = 0 \Rightarrow \boxed{F_{BC} = -4070\,\text{lb}.} \tag{12}$$

Joint *C*:

$$\sum F_y = 0: \qquad -F_{AC}\sin 30^\circ - F_{CD}\sin 60^\circ - 4000\,\text{lb} - P_C = 0 \tag{13}$$

$$\Rightarrow \boxed{F_{CD} = -5802\,\text{lb}.} \tag{14}$$

Discussion & Verification We were able to determine all of the member forces without using one of the equilibrium equations for joint C or the two equilibrium equations for joint D. As a check on our solution, you should write these three equations and verify that the member forces we determined satisfy all of them.

EXAMPLE 6.3 *A Truss with a Cable and Pulleys*

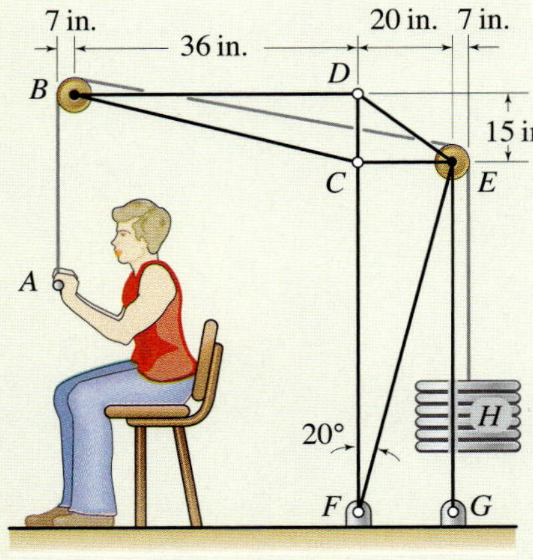

Figure 1

In the exercise machine shown, the stack of weights at H weighs 50 lb. If cable segment AB is vertical, determine the force supported by each member of the machine.

SOLUTION

Road Map In this structure, we will neglect the weights of individual members, including the weight of the cable, and will compute the member forces due to the person lifting the 50 lb weight only. In contrast to the previous examples of this section, in this problem the geometry of the structure is such that the support reactions are not needed before we can proceed with the method of joints. Further comments on how to recognize such situations are given at the end of this problem.

Modeling To use the method of joints, we draw FBDs of all the joints in the truss, as shown in Fig. 2. In these FBDs, all member forces are taken to be positive in tension. Also, we have elected to leave the pulleys on the joints. Alternatively, we could have removed the pulleys by shifting the cable forces to the bearings, as discussed in connection with Fig. 5.11 on p. 271. Indeed, if either cable segment AB or EH were not vertical, this would be the preferred approach because the moment arms would be easier to obtain. Finally, we will assume the pulleys at B and E are frictionless, and we will neglect the weight of the cable itself so that the tensile force supported by the cable is the same throughout its entire length and is equal to the 50 lb weight at H. Since the pulleys at B and E have the same radius, the orientation of cable segment BE is given by $\alpha = \tan^{-1}(15\,\text{in.}/56\,\text{in.}) = 15.00°$.

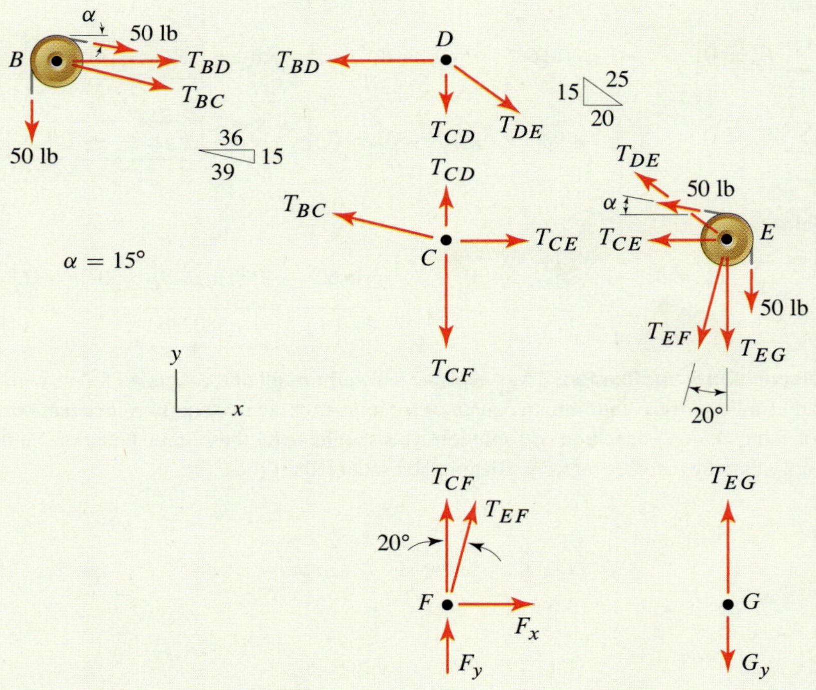

Figure 2. Free body diagrams of joints.

Governing Equations & Computation We begin writing equilibrium equations at joint B, because there are only two unknowns there. Furthermore, examination of the FBD for joint B shows that the $\sum F_y = 0$ equation will contain only one unknown, whereas the $\sum F_x = 0$ will contain two, thus we will write the $\sum F_y = 0$ expression first. Then we will proceed to joint D, as there are only two unknowns there (once T_{BD} is found), and so on.

Joint B:

$$\sum F_y = 0: \quad -50\,\text{lb} - (50\,\text{lb})\sin\alpha - T_{BC}\left(\frac{15}{39}\right) = 0, \tag{1}$$

$$\Rightarrow \quad \boxed{T_{BC} = -163.6\,\text{lb},} \tag{2}$$

$$\sum F_x = 0: \quad (50\,\text{lb})\cos\alpha + T_{BC}\left(\frac{36}{39}\right) + T_{BD} = 0, \tag{3}$$

$$\Rightarrow \quad \boxed{T_{BD} = 102.8\,\text{lb}.} \tag{4}$$

Joint D:

$$\sum F_x = 0: \quad -T_{BD} + T_{DE}\left(\frac{20}{25}\right) = 0, \quad \Rightarrow \quad \boxed{T_{DE} = 128.4\,\text{lb},} \tag{5}$$

$$\sum F_y = 0: \quad -T_{CD} - T_{DE}\left(\frac{15}{25}\right) = 0, \quad \Rightarrow \quad \boxed{T_{CD} = -77.06\,\text{lb}.} \tag{6}$$

Joint C:

$$\sum F_x = 0: \quad -T_{BC}\left(\frac{36}{39}\right) + T_{CE} = 0, \quad \Rightarrow \quad \boxed{T_{CE} = -151.0\,\text{lb},} \tag{7}$$

$$\sum F_y = 0: \quad T_{BC}\left(\frac{15}{39}\right) + T_{CD} - T_{CF} = 0, \quad \Rightarrow \quad \boxed{T_{CF} = -140.0\,\text{lb}.} \tag{8}$$

Joint E:

$$\sum F_x = 0: \quad -T_{CE} - T_{DE}\left(\frac{20}{25}\right) - (50\,\text{lb})\cos\alpha - T_{EF}\sin 20^\circ = 0, \tag{9}$$

$$\Rightarrow \quad \boxed{T_{EF} = 0,} \tag{10}$$

$$\sum F_y = 0: \quad T_{DE}\left(\frac{15}{25}\right) + (50\,\text{lb})\sin\alpha - 50\,\text{lb} - T_{EF}\cos 20^\circ - T_{EG} = 0, \tag{11}$$

$$\Rightarrow \quad \boxed{T_{EG} = 40.00\,\text{lb}.} \tag{12}$$

Figure 3
An exercise machine similar to that shown in Fig. 1 in use.

Discussion & Verification

- The forces supported by all members were determined without using the equilibrium equations for joints F and G. If desired, the equilibrium equations for these joints may also be written to obtain the support reactions, which are found to be $F_x = 0$, $F_y = 140.0\,\text{lb}$, and $G_y = 40.0\,\text{lb}$.
- After Eq. (9) is solved, member EF is found to be a zero-force member. Because neither of joints E or F fits the pattern shown in Fig. 6.8 on p. 333, it is not possible to use inspection to determine that this member is zero-force.
- This problem was convenient to solve without first determining the support reactions because joint B had only two unknowns, and thus it was possible to begin the analysis at this location. In contrast, examination of the FBDs for Examples 6.1 and 6.2 shows that if the reactions are not obtained first, then the FBDs for all joints in those examples contain three or more unknowns.

PROBLEMS

Problems 6.1 and 6.2

Determine the force supported by each member of the truss if $P = 1000$ lb.

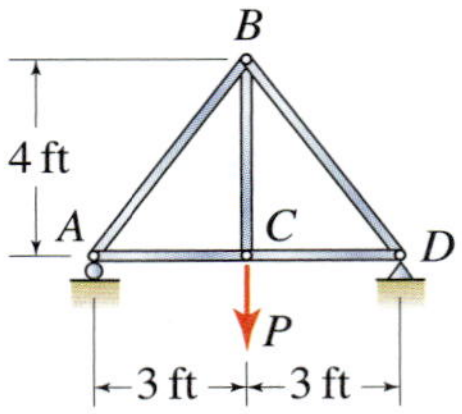

Figure P6.1

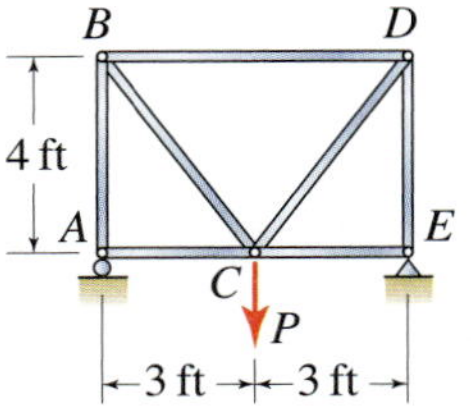

Figure P6.2

Problems 6.3 and 6.4

For the truss shown in Fig. P6.1, each member can support a maximum tensile force of 4000 lb and a maximum compressive force of 2000 lb.

Problem 6.3 If $P = 1000$ lb, determine the factor of safety for the truss.

Problem 6.4 Determine the largest positive value of P that can be applied.

Problems 6.5 and 6.6

For the truss shown in Fig. P6.2, each member can support a maximum tensile force of 6000 lb and a maximum compressive force of 4000 lb.

Problem 6.5 If $P = 1000$ lb, determine the factor of safety for the truss.

Problem 6.6 Determine the largest positive value of P that can be applied.

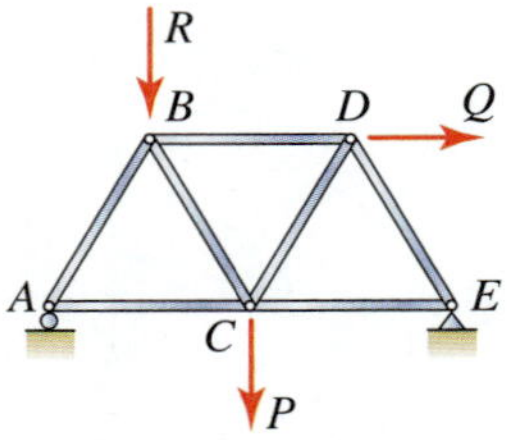

Figure P6.7

Problem 6.7

All members of the truss have the same length. Determine the force supported by each member if $P = 1$ kN, $Q = 2$ kN, and $R = 3$ kN.

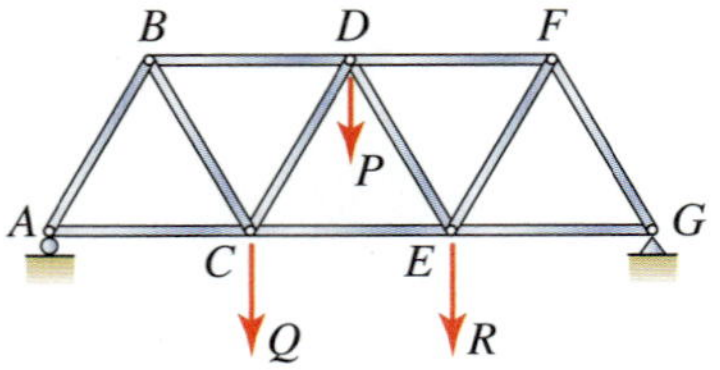

Figure P6.8

Problem 6.8

All members of the truss have the same length. Determine the force supported by members CE and DF if $P = 1$ kN, $Q = 2$ kN, and $R = 3$ kN.

Problems 6.9 and 6.10

Determine the force supported by each member of the truss if $P = 4$ kN and $Q = 1$ kN.

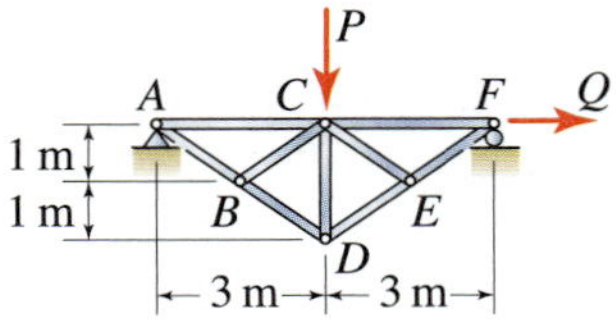

Figure P6.9

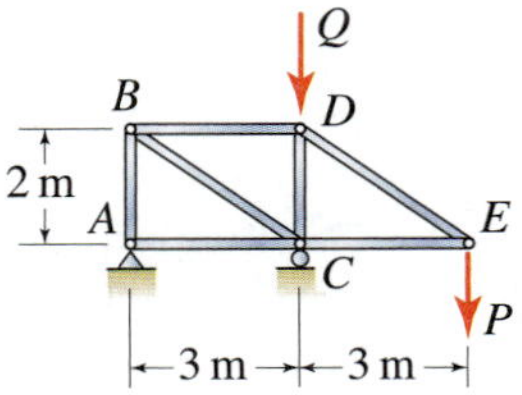

Figure P6.10

Problems 6.11 and 6.12

For the truss shown in Fig. P6.9, each member can support a maximum tensile force of 15 kN and a maximum compressive force of 10 kN.

Problem 6.11 If $P = 4\,\text{kN}$ and $Q = 1\,\text{kN}$, determine the factor of safety for the truss.

Problem 6.12 If $Q = P/4$, determine the largest positive value of P that can be applied.

Problems 6.13 and 6.14

For the truss shown in Fig. P6.10, each member can support a maximum tensile force of 20 kN and a maximum compressive force of 12 kN.

Problem 6.13 If $P = 4\,\text{kN}$ and $Q = 1\,\text{kN}$, determine the factor of safety for the truss.

Problem 6.14 If $Q = P/4$, determine the largest positive value of P that can be applied.

Problem 6.15

Determine the force supported by each member of the truss. Express your answers in terms of P and Q.

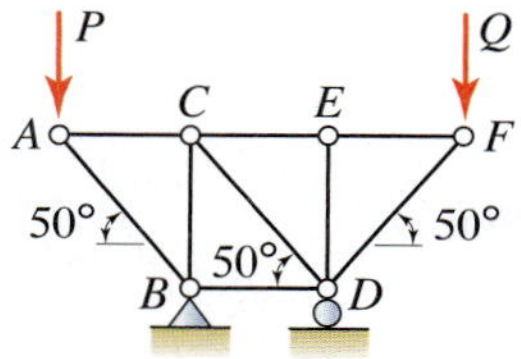

Figure P6.15 and P6.16

Problem 6.16

Each member can support a maximum tensile force of 15 kip and a maximum compressive force of 10 kip. If $0 \le P \le 4\,\text{kip}$ and $0 \le Q \le 4\,\text{kip}$, determine the lowest factor of safety for the truss. *Hint:* The answer is the lowest factor of safety among the three load cases $P = 4\,\text{kip}$ and $Q = 0$, $P = 0$ and $Q = 4\,\text{kip}$, and $P = Q = 4\,\text{kip}$.

Problem 6.17

Determine the force supported by each member of the truss. Express your answers in terms of P and Q.

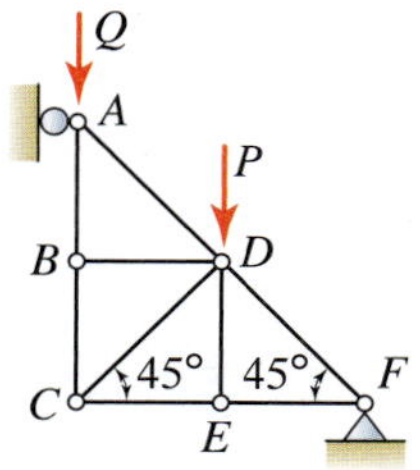

Figure P6.17 and P6.18

Problem 6.18

Each member can support a maximum tensile force of 3 kN and a maximum compressive force of 2 kN. If $0 \le P \le 400\,\text{N}$ and $0 \le Q \le 400\,\text{N}$, determine the lowest factor of safety for the truss. *Hint:* The answer is the lowest factor of safety among the three load cases $P = 400\,\text{N}$ and $Q = 0$, $P = 0$ and $Q = 400\,\text{N}$, and $P = Q = 400\,\text{N}$.

Problems 6.19 and 6.20

The structure consists of seven pin-connected members. Determine the force supported by all members. Express your answers in terms of parameters such as F and L.

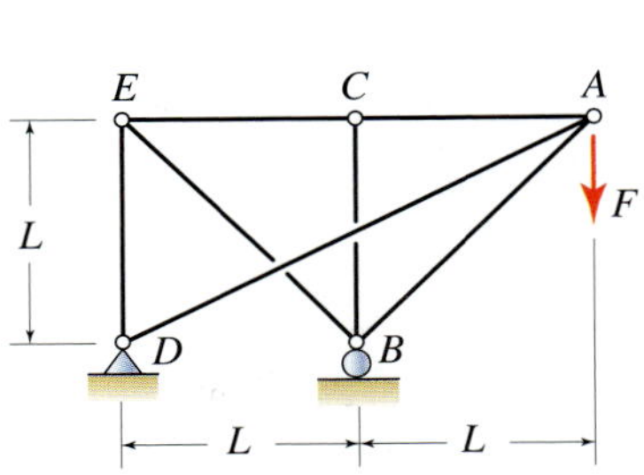

Figure P6.19

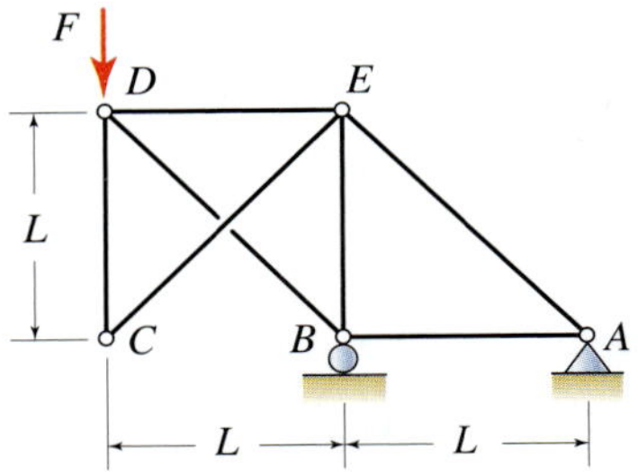

Figure P6.20

Problems 6.21 and 6.22

The structure consists of pin-connected members. Determine the force supported by all members.

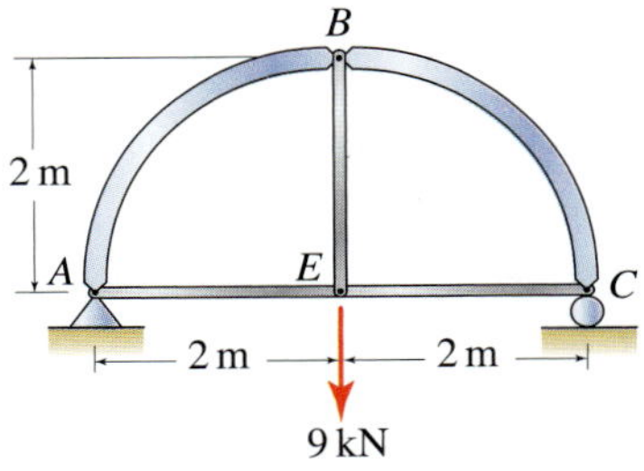

Figure P6.21

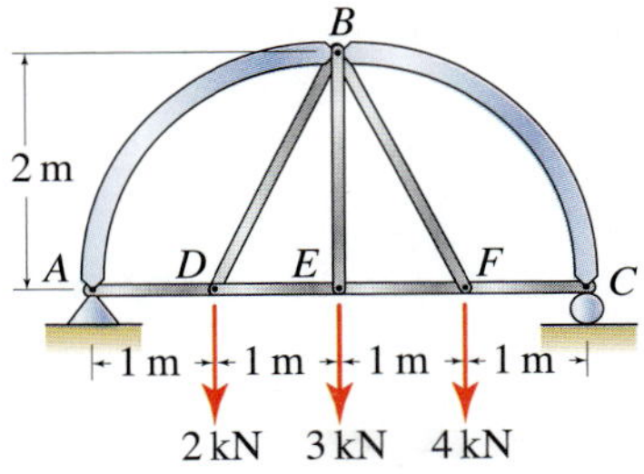

Figure P6.22

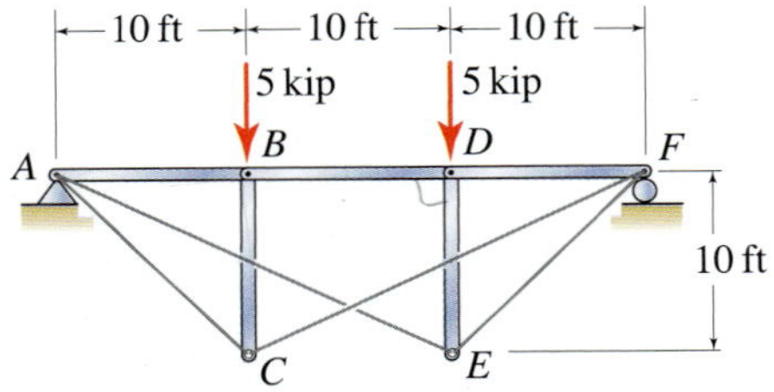

Figure P6.23

Problem 6.23

A *Bollman* truss is shown. Determine the force supported by each of the five bars and four cables.

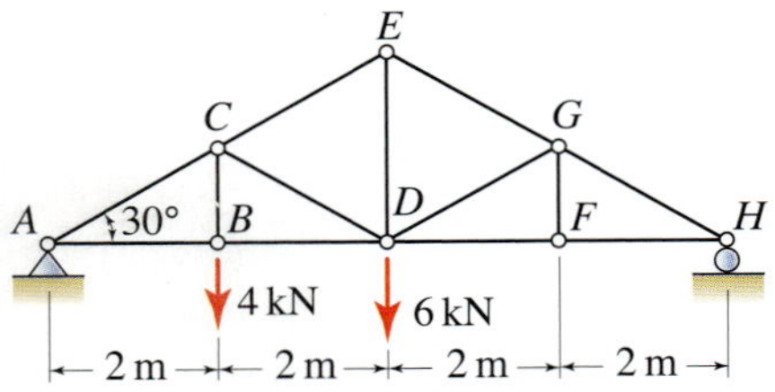

Figure P6.24

Problem 6.24

(a) By inspection, identify the zero-force members in the truss.

(b) Determine the force supported by all members of the truss.

Problem 6.25

(a) By inspection, identify the zero-force members in the truss.

(b) Determine the force supported by member FG and all of the members to the left of it.

(c) Determine the force supported by all of the members to the right of member FG.

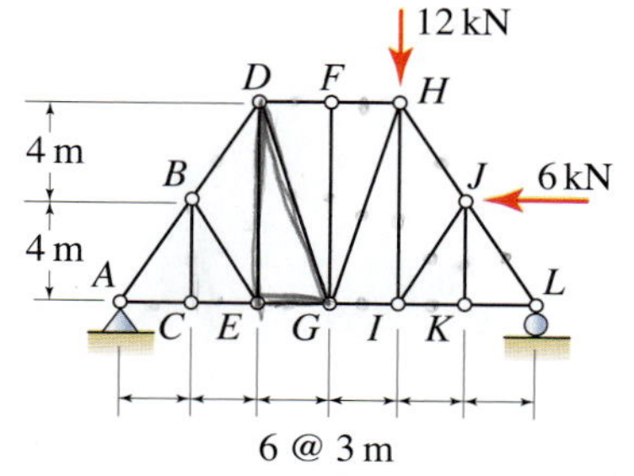

Figure P6.25

Problem 6.26

(a) By inspection, identify the zero-force members in the truss.

(b) Determine the force supported by member CH and all of the members to the left of it.

(c) Determine the force supported by all of the members to the right of member CH.

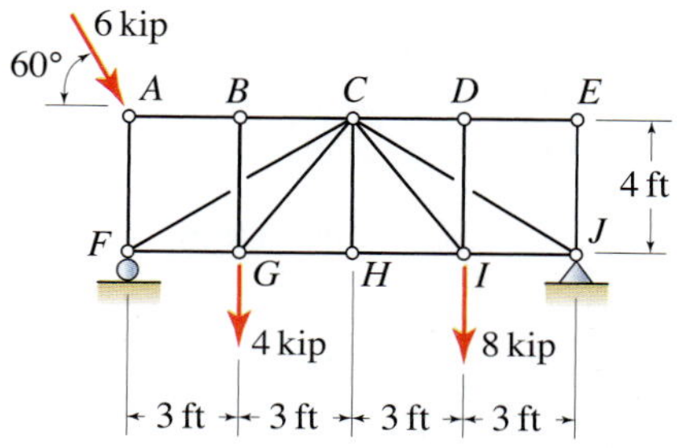

Figure P6.26

Problem 6.27

The truss has frictionless pulleys at points B, J, and K. Cable segment BL is vertical and $W = 1$ kip.

(a) By inspection, identify the zero-force members in the truss.

(b) Determine the force supported by member CE.

(c) Determine the force supported by member GI.

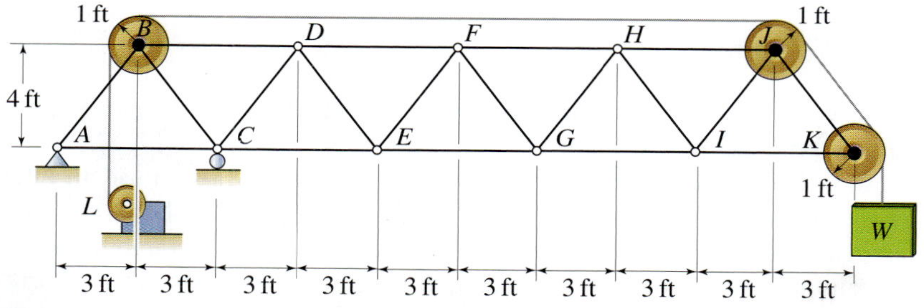

Figure P6.27

6.2 Truss Structures and the Method of Sections

The *method of sections* is an effective and popular approach for determining the forces supported by individual members of a truss, especially when forces in only a portion of the members must be found. In this method, we select a member whose force we want to determine. Then a cut is taken that passes through this member, subdividing the truss into two parts. If the entire truss is in equilibrium, then both parts of the truss are also in equilibrium. Analysis proceeds by drawing an FBD for one of the parts of the truss, writing equilibrium equations, and solving these for the unknowns. Prior to this, it may be necessary or desirable to determine the reactions for the truss as a whole. Analysis by the method of sections is illustrated in the following example.

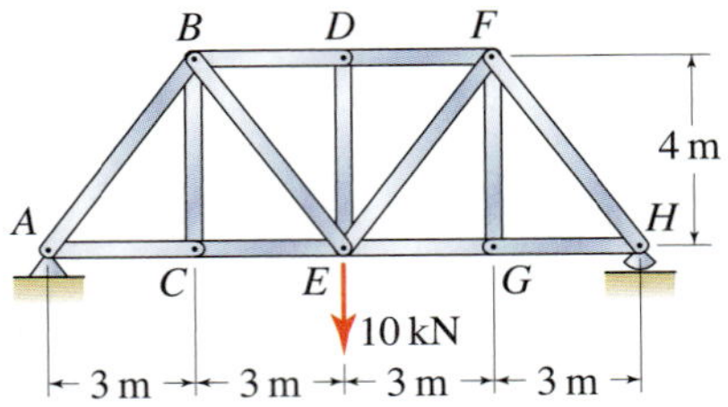

Figure 6.10
A plane truss.

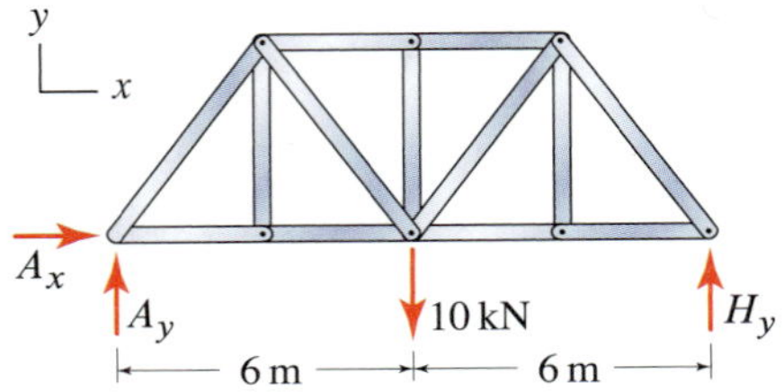

Figure 6.11
Free body diagram to determine the support reactions.

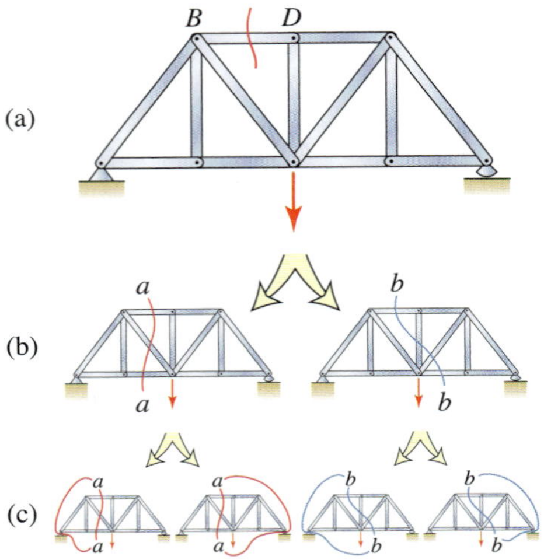

Figure 6.12
Possible cuts that will allow the force in member BD to be determined by the method of sections. (a) The cut starts by passing through member BD. (b) Two possible paths (among many) for continuing the cut started in (a). (c) Four possible paths for completing the cut so that an FBD can be drawn.

Mini-Example. Use the method of sections to determine the force supported by member BD of the truss shown in Fig. 6.10.

Solution. We begin by obtaining the support reactions for the truss by drawing the FBD of the structure as a whole, as shown in Fig. 6.11. Writing equilibrium equations provides

$$\sum M_A = 0: \quad -(10\,\text{kN})(6\,\text{m}) + H_y(12\,\text{m}) = 0 \;\Rightarrow\; H_y = 5\,\text{kN}, \tag{6.12}$$

$$\sum F_y = 0: \quad A_y + H_y - 10\,\text{kN} = 0 \;\Rightarrow\; A_y = 5\,\text{kN}, \tag{6.13}$$

$$\sum F_x = 0: \quad A_x = 0 \;\Rightarrow\; A_x = 0. \tag{6.14}$$

Because the geometry and loading for this problem are symmetric, we could have determined these reactions by inspection.

In order to find the force in member BD, it is necessary to use an FBD where the cut passes *through* member BD, and Fig. 6.12(a) shows the start of such a cut. In Fig. 6.12(b), two possible paths *aa* and *bb* (among many) for continuing the cut are shown. Figure 6.12(c) shows completed cuts where in each case, the cut is a closed line that fully encompasses the left-hand or right-hand portion of the structure.

Once a cut has been taken and an FBD of a portion of the structure has been drawn, there are three equilibrium equations that may be written. Thus, a good strategy is to select a cut (when possible) so that no more than three members are cut since then the unknown forces for these members may be immediately solved for by using the three equilibrium equations that are available. Comparing the two cuts shown in Fig. 6.12(b), observe that cut *aa* passes through three members and hence its FBD will have three unknowns, while cut *bb* passes through four members and hence its FBD will have four unknowns.* Thus, cut *aa* is the better choice.

Using cut *aa*, we draw the FBD for the left-hand and right-hand portions of the structure as shown in Fig. 6.13, where member forces F_{BD}, F_{BE}, and F_{CE} are positive in tension. Normally, we would draw only one of these FBDs, and we usually select the side of the truss that has fewer forces and/or more straightforward geometry. Using the FBD for the left-hand portion of the structure, we notice that point E is a convenient location for summing moments,

* If you recognized that member DE is a zero-force member, then cut *bb* has only three unknowns and is therefore as good a choice as cut *aa*.

because two of the unknown forces pass through this point, leaving only F_{BD}. Thus, taking positive moment counterclockwise, we write

$$\sum M_E = 0: \ -(5\,\text{kN})(6\,\text{m}) - F_{BD}(4\,\text{m}) = 0 \ \Rightarrow \ F_{BD} = -7.5\,\text{kN}. \quad (6.15)$$

Although this problem does not ask for the forces in members BE and CE, we will nonetheless determine these by writing and solving the following equilibrium equations

$$\sum M_B = 0: \quad -(5\,\text{kN})(3\,\text{m}) + F_{CE}(4\,\text{m}) = 0 \ \Rightarrow \ F_{CE} = 3.75\,\text{kN}, \quad (6.16)$$

$$\sum F_x = 0: \quad F_{CE} + F_{BE}\left(\frac{3}{5}\right) + F_{BD} = 0 \ \Rightarrow \ F_{BE} = 6.25\,\text{kN}. \quad (6.17)$$

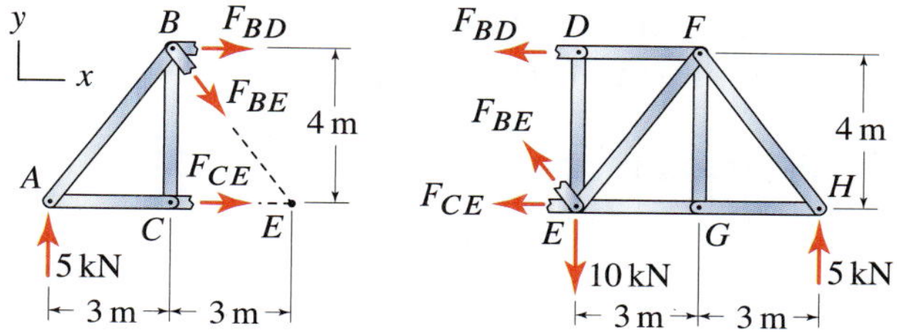

Figure 6.13. Free body diagrams.

Remarks

- Compared to the method of joints, the method of sections is often easier to use when the forces in only a portion of a structure's members are desired. If the method of joints were used for this example, FBDs would need to be drawn and equilibrium equations written for joints A, C, and B (in that order) before F_{BD} could be determined.
- When a cut is taken, the truss is subdivided into two (or more) parts. Regardless of which part you choose to draw an FBD for, there are only three equilibrium equations that may be used to determine the unknown forces. For example, if the FBD shown in Fig. 6.13 for the right-hand portion of the structure is used, the $\sum M_E = 0$, $\sum M_B = 0$, and $\sum F_x = 0$ expressions that are written are identical to Eqs. (6.15)–(6.17). If different moment summation points are used, then the resulting equilibrium equations will differ from Eqs. (6.15) and (6.16), but nonetheless they will be a linear combination of these so that there are only three independent equations available for determining the unknown forces.
- To further elaborate on the consequence of the last remark, imagine that cut bb shown in Fig. 6.12 was used. The FBD for this cut has four unknown member forces. As a strategy to determine the four unknowns, you might consider writing three equilibrium equations for the left-hand

portion of the truss and then writing an additional equilibrium equation for the right-hand portion. However, in view of the foregoing remark, this strategy will fail because of the four equilibrium equations, only three are independent and thus it is not possible to determine all four unknowns in this fashion.

- If the problem had asked for the forces in members BD and DE, then the method of sections with cut bb shown in Fig. 6.12 (or a similar cut) would be needed, because it passes through both of these members. However, as explained in the preceding remarks, the method of sections with cut bb will not be sufficient to determine these forces. If the truss is statically determinate (discussed later in this section), then multiple applications of the method of sections and/or the method of joints will be needed. In this example, we could first use the method of sections with cut aa to find F_{BD} followed by the method of sections with cut bb to find the remaining unknowns. Alternatively, first we could use the method of joints for joint D to find F_{DE}. Then we could use the method of sections with cut bb; because F_{DE} is known, the three remaining unknowns can then be determined.

Treatment of forces that are not at joints

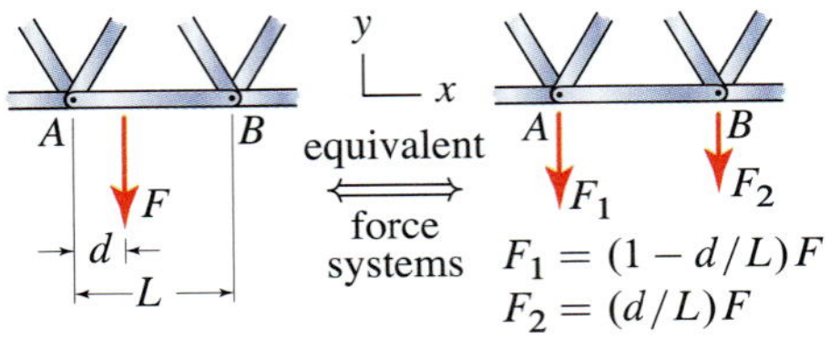

Figure 6.14
For many purposes, a force that is not located at a joint may be replaced by an equivalent force system where forces are positioned at joints.

In real life structures, forces often occur at locations other than joints. For many purposes such forces may be replaced by equivalent force systems, so that truss analysis may be used. Consider, for example, the vertical force F shown in Fig. 6.14. To see that F_1 and F_2 have the values reported in Fig. 6.14, we use the usual procedure to construct an equivalent force system, given by Eq. (4.16):

$$\left(\sum F_y\right)_{\text{system 1}} = \left(\sum F_y\right)_{\text{system 2}}$$

$$F = F_1 + F_2, \tag{6.18}$$

and

$$\left(\sum M_A\right)_{\text{system 1}} = \left(\sum M_A\right)_{\text{system 2}}$$

$$-Fd = -F_2 L. \tag{6.19}$$

Solving Eqs. (6.18) and (6.19) provides

$$F_1 = (1 - \frac{d}{L})F \quad \text{and} \quad F_2 = \frac{d}{L}F. \tag{6.20}$$

You may wish to verify that Eq. (6.20) provides the proper results for F_1 and F_2 for the cases $d = 0$ and $d = L$.

Although we say the two force systems shown in Fig. 6.14 are *equivalent*, exactly how equivalent they really are depends on what we are interested in. In this case, when member AB is subjected to F (with $0 < d < L$), it experiences *bending*, while when it is subjected to F_1 and F_2, it does not. Nonetheless, the two force systems are equivalent in the sense that effects that are *external* to member AB, such as reactions and the forces supported by all other members, are the same. The issue of when two "equivalent" force systems really are equivalent is sometimes subtle, and you may not be able to fully understand these subtleties until you study internal forces in Chapter 8.

Nonetheless, the discussion of why equivalent force systems are called equivalent in Section 4.4 always applies and is the source for resolving any questions of interpretation.

Static determinacy and indeterminacy

In a *statically determinate* truss, the equations of equilibrium are sufficient to determine the forces supported by all members of the truss and the support reactions. In a *statically indeterminate* truss, the equations of equilibrium are not sufficient to determine all of these. Compared to a statically determinate truss, a statically indeterminate truss has extra members and/or supports. As a result, the forces supported by each member and possibly the support reactions also depend on the material the members are made of and the size of their cross sections. Methods of analysis for statically indeterminate trusses are covered in more advanced subjects.

A simple rule of thumb, called *equation counting*, can be used to help determine whether a truss is statically determinate or indeterminate. The rule is developed by comparing the number of unknowns for a truss to the number of equilibrium equations. Consider a plane truss having

$$\begin{aligned} m &= \text{number of members,} \\ r &= \text{number of support reactions,} \\ j &= \text{number of joints.} \end{aligned} \tag{6.21}$$

Each member has one unknown force and each support reaction has one unknown force, so that $m + r$ is the total number of unknowns. For a plane truss, each joint has two equilibrium equations so that $2j$ is the total number of equations. Thus, the rule of thumb is as follows:

If $m + r < 2j$	The truss is a mechanism and/or has partial fixity.	
If $m + r = 2j$	The truss is statically determinate if it has full fixity. The truss is statically indeterminate if it has partial fixity.	(6.22)
If $m + r > 2j$	The truss is statically indeterminate, and it can have full fixity or partial fixity.	

- If $m + r < 2j$, not all of the equilibrium equations can be satisfied, implying that some of them may not be satisfied, in which case motion will occur. If the truss has only partial fixity, then the truss as a whole may undergo a rigid body motion. If the truss has full fixity, then it is a mechanism, meaning it may collapse. The presence of a mechanism in a truss is disastrous, unless the truss is intended to be part of a machine. Trusses are almost always supported so that they are fully fixed. Thus, the situation $m + r < 2j$ is rarely permitted.

- If $m + r = 2j$, there are as many unknowns as equilibrium equations. Barring the possibility of having members and/or supports with insufficient arrangement, all unknowns can be determined and all equilibrium equations can be satisfied.

- If $m+r > 2j$, not all of the unknowns can be determined, and additional equations, in the form of models that characterize the deformability of all members of the truss, must be introduced.

Several examples of equation counting are shown in Fig. 6.15.

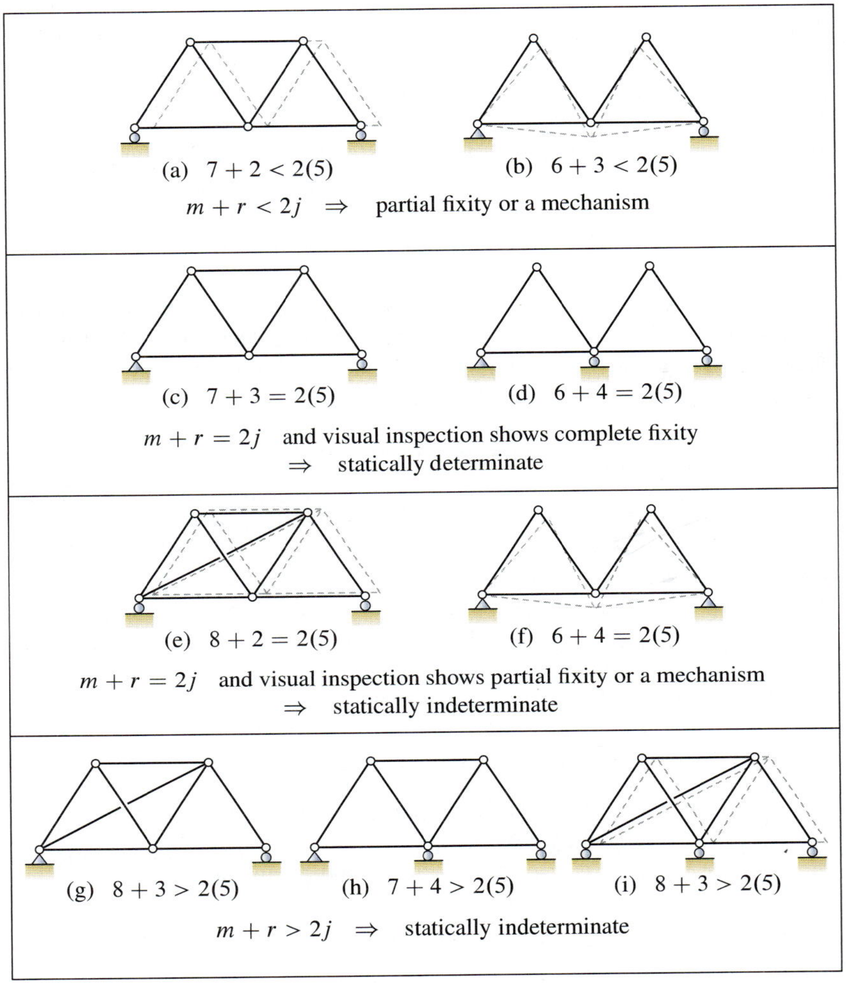
(a) $7+2<2(5)$
(b) $6+3<2(5)$
$m+r<2j$ ⇒ partial fixity or a mechanism
(c) $7+3=2(5)$
(d) $6+4=2(5)$
$m+r=2j$ and visual inspection shows complete fixity
⇒ statically determinate
(e) $8+2=2(5)$
(f) $6+4=2(5)$
$m+r=2j$ and visual inspection shows partial fixity or a mechanism
⇒ statically indeterminate
(g) $8+3>2(5)$
(h) $7+4>2(5)$
(i) $8+3>2(5)$
$m+r>2j$ ⇒ statically indeterminate

Figure 6.15. Examples of equation counting to determine if a truss is statically determinate or statically indeterminate. For trusses that are not fully fixed or are mechanisms, a possible displaced position is shown by the dashed outlines.

Design considerations

Simple, compound, and complex trusses

In a *simple truss*,* members are arranged beginning with three members joined at their ends to one another in the shape of a triangle, as shown in Fig. 6.16, and adding to this two new noncollinear members for each new joint that is added. All of the trusses shown in Fig. 6.2 on p. 330 are simple trusses. A *compound truss* is formed by interconnecting two or more simple trusses to form a stable structure, where stability is provided by a sufficient number and arrangement of members that connect the simple trusses. A *complex truss* is formed by interconnecting two or more simple trusses, along with a sufficient number and arrangement of supports to make the structure stable. These three truss categories are contrasted in Fig. 6.17.

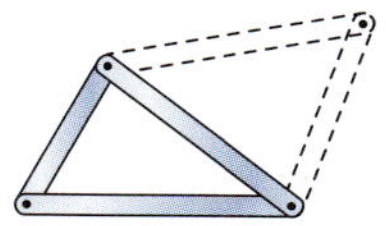

Figure 6.16
In a *simple truss* the arrangement of members begins with three members that form a triangle and adding to this two new members (shown by dashed lines) for each new joint that is added.

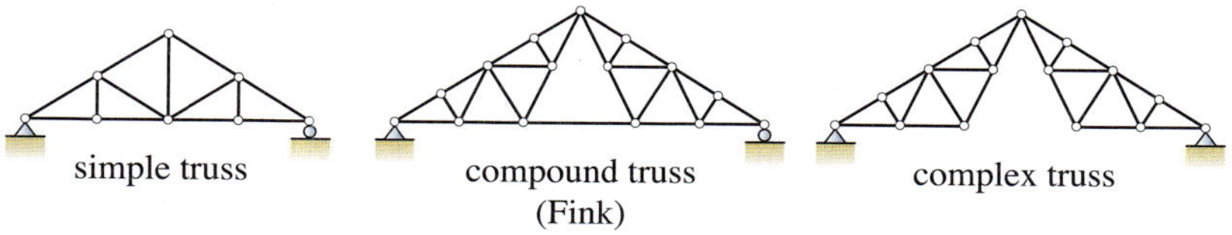

Figure 6.17. Examples of simple, compound, and complex trusses. The compound truss shown is called a Fink truss.

A truss does not need to be a simple truss to be a good design. For example, the compound and complex trusses shown in Fig. 6.17 and the Baltimore truss shown in Fig. 6.18 are not simple trusses, but they are nonetheless popular and effective designs. However, one of the main features of a simple truss is guaranteed stability: if the pattern of construction described in Fig. 6.16 is used, the resulting truss will have no mechanisms; and if it is fully supported, it will be stable. The worst blunder you can make in designing a truss is to produce a structure that is unstable. At a minimum, you will be very embarrassed, and at worst, if the structure is built, people may be injured or killed. Note that as the arrangement of members in a truss becomes more complex, it becomes more difficult to ensure there are no mechanisms. Use of equation counting in Eq. (6.22) is helpful, but is not foolproof.

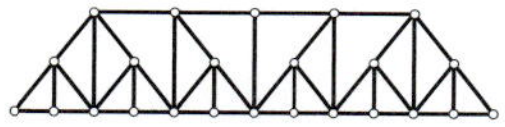

Figure 6.18
The Baltimore truss is not a simple truss, but it is a popular and effective design.

Buckling of truss members

A typical truss uses remarkably little material, yet provides impressive strength. Because truss members are usually slender, they are susceptible to *buckling* under compressive forces. To understand this phenomenon better, perform the experiment shown in Fig. 6.19, using a wood or metal yardstick (or meterstick), ideally one that is very straight. Under low compressive force, the yardstick remains straight. But under sufficiently high compressive force, you should observe the yardstick suddenly buckles. When it buckles, it begins to bend and loses strength. In fact, the loss of strength is very dramatic and for this reason, buckling failures in structures are often catastrophic. Buckling is a fascinating subject and is crucially important. The following example explores buckling

* Additional details and more advanced methods of analysis are contained in T. Au and P. Christiano, *Fundamentals of Structural Analysis*, Prentice-Hall, Englewood Cliffs, NJ, 1993 and E. C. Rossow, *Analysis and Behavior of Structures*, Prentice-Hall, Upper Saddle River, NJ, 1996.

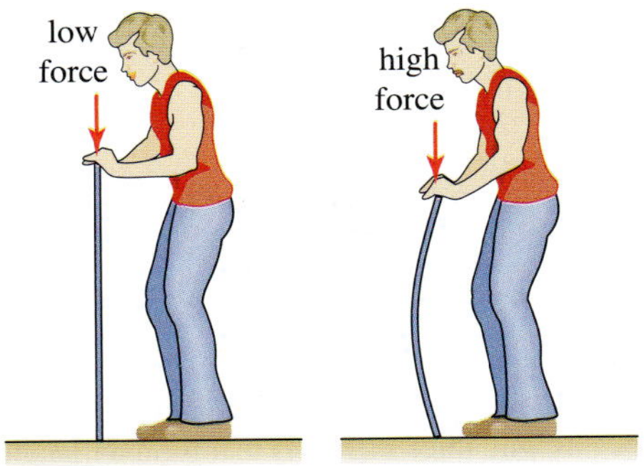

Figure 6.19
An experiment showing how a yardstick (or meterstick) buckles under high compressive force.

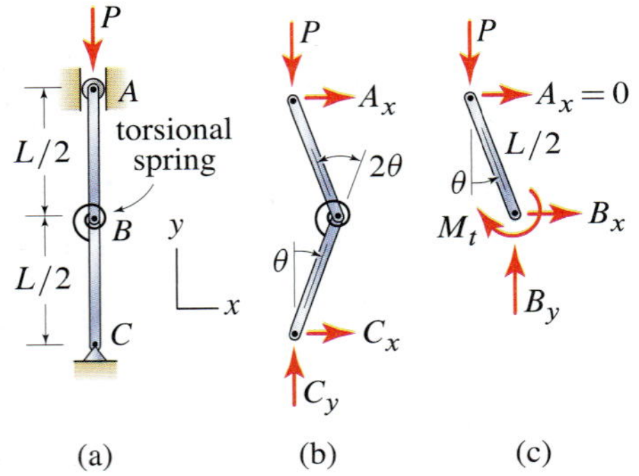

Figure 6.20
A simple model to study buckling.

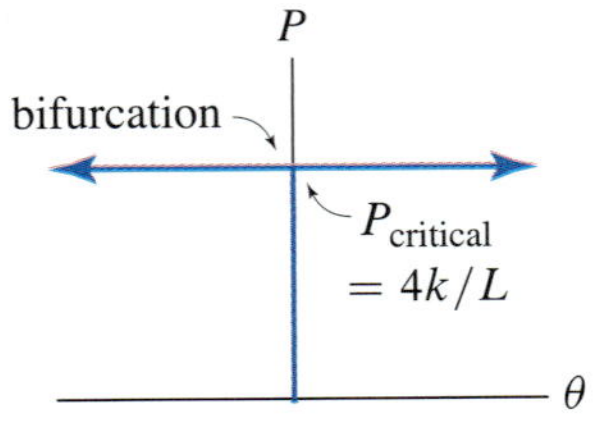

Figure 6.21
Load-deformation results for buckling of the model shown in Fig. 6.20.

for a simple problem, but its results are characteristic of response for more complicated structures such as columns and truss members.

Mini-Example. A simple model for buckling analysis of an initially straight truss member is shown in Fig. 6.20(a) where bars AB and BC are rigid, and the ability of the structure to deform is modeled by the torsional spring. Determine the value of P for which the structure buckles.

Solution. Although this is a crude model for a straight truss member, it is easy to analyze and its results are revealing. The FBD for the buckled structure is shown in Fig. 6.20(b). Note that because of the torsional spring, members AB and BC are not two-force members, although the structure ABC as a whole is. For this reason (or by writing $\sum M_A = 0$ and $\sum M_C = 0$), we determine $A_x = C_x = 0$, and $\sum F_y = 0$ gives $C_y = P$. Next, we remove* member AB from the structure, giving the FBD shown in Fig. 6.20(c). Then

$$\sum M_B = 0: \quad -M_t + P\frac{L}{2}\sin\theta = 0. \tag{6.23}$$

Noting that the torsional spring is twisted by 2θ, the spring law, Eq. (5.14) on p. 272, gives $M_t = k(2\theta)$. Assuming small angles with θ measured in radians, $\sin\theta \approx \theta$, and Eq. (6.23) becomes

$$\left(2k - P\frac{L}{2}\right)\theta = 0. \tag{6.24}$$

Equation (6.24) has two solutions:

$$\text{Solution 1:} \qquad \theta = 0 \qquad \text{trivial solution,} \tag{6.25}$$

$$\text{Solution 2:} \qquad 2k - P\frac{L}{2} = 0 \qquad \text{buckling solution.} \tag{6.26}$$

Solution 1 is called the *trivial solution* because it says that structure ABC remains straight. Solution 2 applies when

$$P_{\text{critical}} = 4k/L, \tag{6.27}$$

where the subscript *critical* is added to emphasize that this value of P causes the structure to buckle. When $P = P_{\text{critical}}$, according to Eq. (6.24), θ can have arbitrary value. Observe that if the spring becomes stiffer, P_{critical} increases whereas if the length of the structure increases, P_{critical} decreases.

This solution and its ramifications are summarized in Fig. 6.21. Let force P slowly increase from zero value. As long as $P < P_{\text{critical}} = 4k/L$, equilibrium [Eq. (6.24)] requires $\theta = 0$. When $P = P_{\text{critical}} = 4k/L$, equilibrium is satisfied regardless of the value of θ, which can have arbitrary positive or negative values. Thus, as P reaches P_{critical}, the structure undergoes an abrupt change from $\theta = 0$ to $\theta \neq 0$. Further, there is no particular preference for θ to have a positive or negative value, and this branching of the response is often called a *bifurcation*. Any attempt to apply a force greater than P_{critical} will cause a dynamic event in which the member rapidly collapses. ■

The buckling analysis of a truss member is more complicated than the foregoing example, and this is a topic studied in mechanics of materials. The result

* Section 6.4 discusses procedures for taking a structure such as this apart for purposes of determining the forces supported by multiforce members.

for the buckling load is $P_{\text{critical}} = c/L^2$ where c is a constant that depends on the member's material and cross-sectional geometry and L is the length of the member. The most important observation is the strong dependence of the buckling load on length: if the length of a member is doubled, its buckling load decreases by a factor of 4! For this reason, long compression members are generally undesirable in trusses.

Use of zero-force members

Consider the example analyzed at the start of this section, repeated again in Fig. 6.22. By inspection, members BC, DE, and FG are zero-force members. You are probably wondering why zero-force members are present in the truss, and if they could be eliminated without reducing the strength of the truss.

To answer these questions requires knowledge of which members are in tension and which are in compression, and these are designated in Fig. 6.22 using + and − symbols, respectively. In Fig. 6.22, members BD and DF both have 3 m length, and as such, the compressive load at which they will buckle is proportional to $1/(3\,\text{m})^2$. If member DE were not present, then joint D would not be present, member BF would have 6 m length, and its buckling load would be *4 times* lower than that for members BD and DF, presuming the material and cross-sectional geometry of the members are the same. Thus, even though member DE is a zero-force member, its effect on increasing strength is remarkable. Members AC and CE are in tension, and buckling is not an issue for them. Thus, it may be possible to eliminate member BC without reducing the strength of the truss. Similarly, it may be possible to eliminate member FG. It is important to note that there may be reasons for which members BC and FG are important. For instance, the loading in this example is very simple, and in real life applications, joints C and G may be subjected to loads, either continuously or periodically, in which case members BC and/or FG would not be zero-force members.

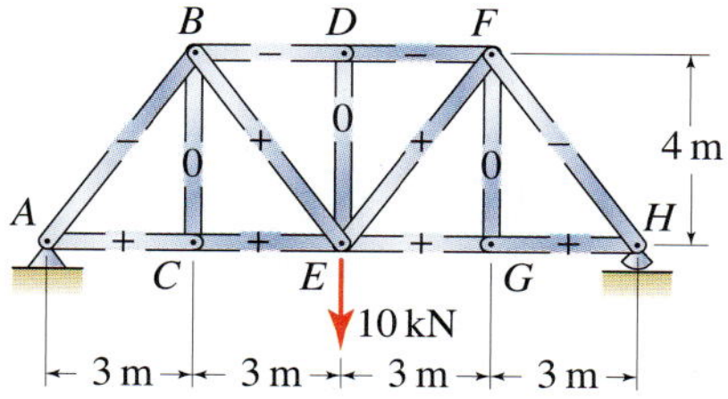

Figure 6.22
A truss subjected to a 10 kN force. Members that are in tension and compression are identified by + and − symbols, respectively, and zero-force members are identified by 0.

> **Concept Alert**
>
> **Buckling of truss members.** Because of buckling, long compression members are undesirable in trusses. Zero-force members are effective for reducing the length of compression members, and this can substantially improve the strength of a truss.

Statically determinate versus statically indeterminate trusses

Both statically determinate and statically indeterminate trusses are popular, and both have advantages and disadvantages. Statically determinate trusses are usually straightforward to design and analyze using the methods discussed in this chapter. Statically indeterminate trusses are usually more difficult to design and analyze. One of the primary features of statically indeterminate trusses is the potential for greater safety because failure of an individual structural member or support does not necessarily mean the entire structure will fail. Figure 6.23 discusses an example of failure in a statically determinate structure.

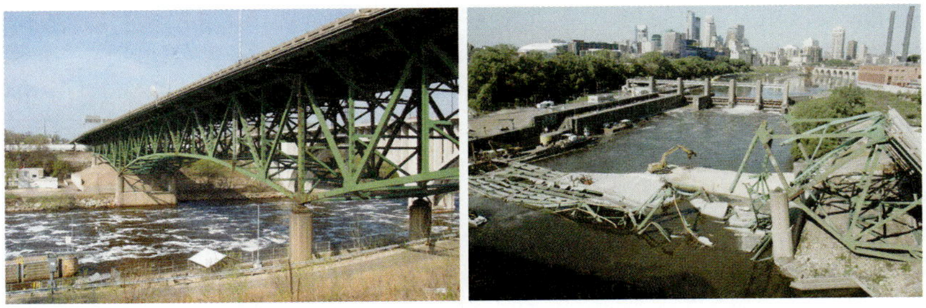

Figure 6.23. Failure of the Interstate 35W bridge in Minneapolis, Minnesota. This bridge opened to traffic in 1967 and collapsed without warning during the evening rush hour on August 1, 2007, killing 13 people and injuring over 100 more. The National Transportation Safety Board's (NTSB's) investigation determined that some of the gusset plates were undersized and that failure of some of these initiated collapse of the bridge. One of the criticisms of this bridge's design is that it was statically determinate, and hence failure of one member or connection is sufficient to cause collapse of the entire structure.

End of Section Summary

In this section, the *method of sections* is developed for determining the forces supported by individual members of a truss, and several characteristics affecting the design and performance of trusses are discussed. Some of the key points are as follows:

- In the *method of sections*, a cut is passed through the truss, and the FBD that results is required to be in equilibrium by writing the equations $\sum F_x = 0$, $\sum F_y = 0$, and $\sum M = 0$.
- In a *statically determinate* truss, the equations of equilibrium are sufficient to determine the forces supported by all members of the truss and the support reactions. In a *statically indeterminate* truss, there are more unknowns than the number of equilibrium equations, and hence, not all of the unknowns (perhaps none of them) can be determined. *Equation counting* is an effective way to determine whether a truss is statically determinate or indeterminate.
- A *simple truss* has members arranged in a triangular pattern, as shown in Fig. 6.16. *Compound* and *complex* trusses consist of two or more simple trusses connected to one another. Simple, compound, and complex trusses are popular, but other designs for trusses are also common and can perform very well. One of the main features of simple, compound, and complex trusses is that they are always stable, presuming that no members or connections fail.
- Because truss members are usually straight and slender, compression members are susceptible to *buckling*. Furthermore, the force at which buckling occurs in a straight member decreases very rapidly as a member becomes longer (the buckling load is proportional to $1/L^2$, where L is the member's length). Thus, it is undesirable to have long compression members in a truss. Zero-force members can be very effective in reducing the length of compression members and thus improving the overall strength of a truss.

EXAMPLE 6.4 *Truss Analysis by the Method of Sections*

The top chord of the steel truss is subjected to a combined dead load and live load of 6 kN/m. Determine the force supported by member FH. All angles are 60°.

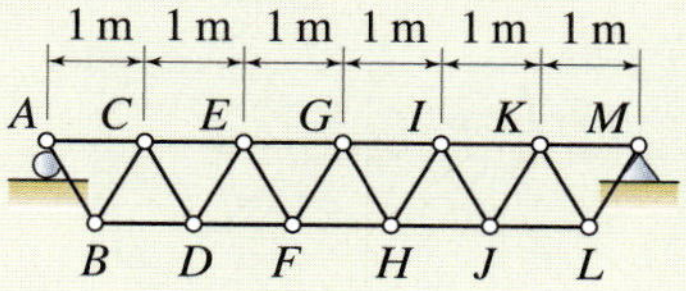

Figure 1

SOLUTION

Road Map Because of the 6 kN/m distributed load, the members of the top chord of this structure are not two-force members, and hence this structure is not a truss. Nonetheless, the structure will be approximated as a truss by lumping the distributed load into forces at joints. After this, support reactions will be determined followed by use of the method of sections. Before proceeding, we determine if the truss is statically determinate or indeterminate. Examination of Fig. 1 shows that the numbers of members, support reactions, and joints are $m = 23$, $r = 3$, and $j = 13$, respectively. Application of Eq. (6.22) on p. 349 shows $m + r = 2j$, and since the truss is fully fixed, it is statically determinate.

Modeling Member AC is subjected to a vertical force of (6 kN/m)(1 m) = 6 kN. Of this, half will be applied to joint A and half to joint C. Similarly, member CE is subjected to a 6 kN force, and half will be applied to joint C and half to joint E, and so on for all members of the top chord of the truss. The final forces are shown in Fig. 2. To determine the support reactions, the FBD of the whole structure is drawn in Fig. 2.

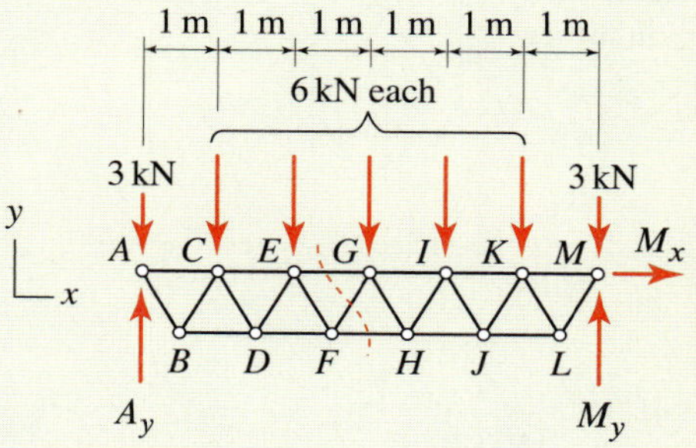

Figure 2

Free body diagram for determination of the support reactions.

Governing Equations & Computation In anticipation of using the method of sections, we will need the reactions for only one side of the truss, and we will use the left-hand side (for this problem, the right-hand side is an equally good choice). Using the FBD of Fig. 2, the only equilibrium equation needed is

$$\sum M_M = 0: \quad -A_y(6\,\mathrm{m}) + (3\,\mathrm{kN})(6\,\mathrm{m}) + (6\,\mathrm{kN})(5\,\mathrm{m}) + (6\,\mathrm{kN})(4\,\mathrm{m}) + (6\,\mathrm{kN})(3\,\mathrm{m}) + (6\,\mathrm{kN})(2\,\mathrm{m}) + (6\,\mathrm{kN})(1\,\mathrm{m}) = 0, \tag{1}$$

$$\Rightarrow A_y = 18\,\mathrm{kN}. \tag{2}$$

Modeling To use the method of sections to determine the force supported by member FH, we must take a cut that passes through this member. As usual, once the cut through FH is started, it must be closed. In Fig. 2, we elect to cut through the three members shown (EG, FG, and FH), and then we complete the cut so that it encompasses the left-hand portion of the structure. The resulting FBD is shown in Fig. 3 where member forces are taken to be positive in tension. The unknown force F_{FH} can quickly be determined by applying moment equilibrium about point G:

$$\sum M_G = 0: \quad -(18\,\mathrm{kN})(3\,\mathrm{m}) + (3\,\mathrm{kN})(3\,\mathrm{m}) + (6\,\mathrm{kN})(2\,\mathrm{m}) + (6\,\mathrm{kN})(1\,\mathrm{m}) + F_{FH}(0.866\,\mathrm{m}) = 0, \tag{3}$$

$$\Rightarrow \boxed{F_{FH} = 31.18\,\mathrm{kN}.} \tag{4}$$

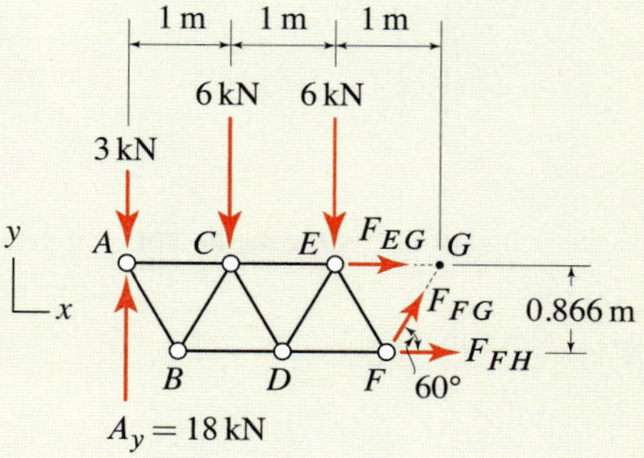

Figure 3

Free body diagram using the method of sections.

Although not asked for, forces supported by members EG and FG are easily found:

$$\sum F_y = 0: \quad 18\,\mathrm{kN} - 3\,\mathrm{kN} - 6\,\mathrm{kN} - 6\,\mathrm{kN} + F_{FG}\sin 60° = 0 \tag{5}$$

$$\Rightarrow F_{FG} = -3.464\,\mathrm{kN}, \tag{6}$$

$$\sum F_x = 0: \quad F_{EG} + F_{FG}\cos 60° + F_{FH} = 0 \tag{7}$$

$$\Rightarrow F_{EG} = -29.44\,\mathrm{kN}. \tag{8}$$

Discussion & Verification You should also try to estimate the effort required to use the method of joints to solve this problem. Doing so will help you judge which of these methods, or if a combination of them, is most efficient for a particular problem.

EXAMPLE 6.5 *Truss Analysis by the Method of Sections*

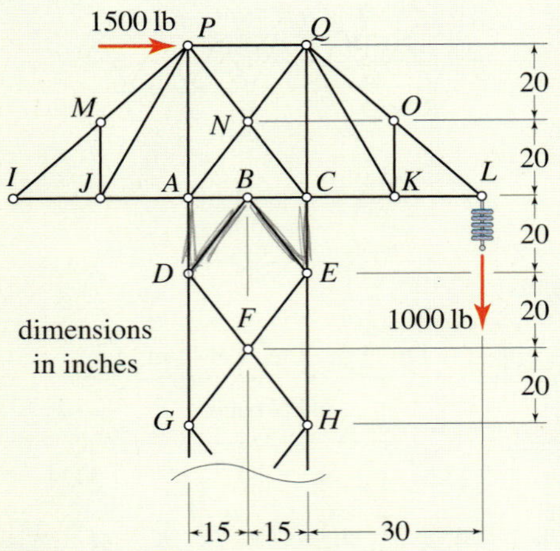

Figure 1

A transmission tower for supporting electric wires is shown. During construction,* the tower will, at certain times, support the load of only one wire (the 1000 lb vertical force at L), and may also be subjected to the force of a storm (crudely modeled by the 1500 lb horizontal force at P). To help determine if the tower has a sufficient factor of safety for this loading scenario, determine the forces supported by members AD, BD, BE, and CE.

SOLUTION

Road Map A quick inspection of the members of interest in Fig. 1 indicates that because of the number of unknowns involved with any candidate cut (some possible cuts are shown in Fig. 2), use of the method of sections will not be as straightforward as in Example 6.4. Although we will use the method of sections in this example, more generally you should also consider the method of joints, or a combination of these, and select the approach that is most efficient.†

Before proceeding, by inspection we observe that there are eight zero-force members present in Fig. 1: JM, JP, KO, KQ, IJ, AJ, IM, and MP. Given the complexity of this truss, we may also suspect that it is statically indeterminate. While we do not have details of the entire structure and thus cannot make a definitive determination, if we imagine that joints G and H are supported by a pin and roller, respectively, then the truss has the number of members $m = 31$, number of support reactions $r = 3$, and number of joints $j = 17$, and the truss is fully fixed. Application of Eq. (6.22) on p. 349 shows $m + r = 2j$; and since the truss is fully fixed, it is statically determinate, and we should be able to determine the desired member forces.

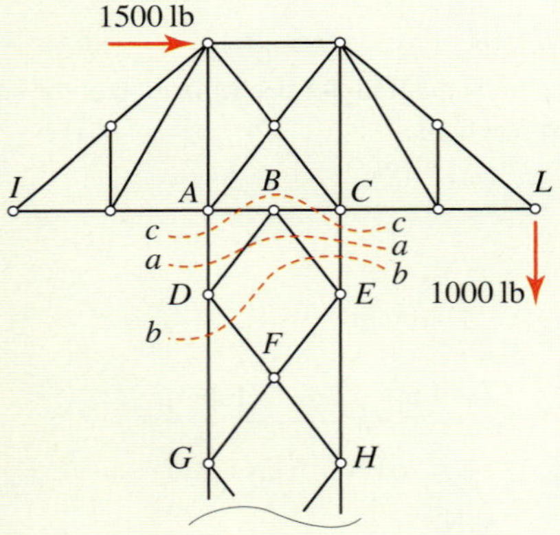

Figure 2
Various cuts for use by the method of sections.

Modeling Cut aa shown in Fig. 2 passes through all of the members we are interested in. But, because the resulting FBD will contain four unknowns, additional FBDs will be needed so that additional equilibrium equations can be written. Cut bb could be used, and while this also results in an FBD with four unknowns, two of these are the same as those from cut aa. Thus, this solution strategy will work: draw two FBDs, one for cut aa and one for cut bb, and write equilibrium expressions to obtain six equations with six unknowns.

While the foregoing strategy is straightforward, some additional thought gives a clever solution using cut cc, which results in the FBD shown in Fig. 3: while it has four unknowns, it is nonetheless possible to determine two of these. Then cut aa, with the FBD shown in Fig. 4, can be used to obtain the remaining unknowns.

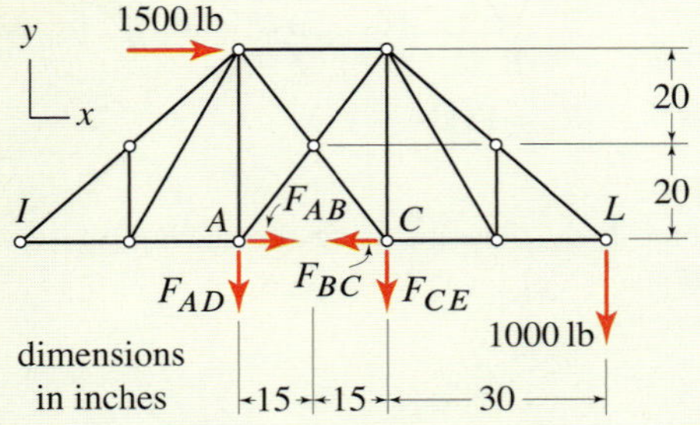

Figure 3
Free body diagram for cut cc.

Governing Equations & Computation Using cut cc, the FBD is shown in Fig. 3. We cannot determine F_{AB} and F_{BC}, but we can determine the remaining unknowns as follows:

$$\sum M_A = 0: \; -(1500\,\text{lb})(40\,\text{in.}) - (1000\,\text{lb})(60\,\text{in.}) - F_{CE}(30\,\text{in.}) = 0 \quad (1)$$

$$\Rightarrow \boxed{F_{CE} = -4000\,\text{lb},} \quad (2)$$

$$\sum F_y = 0: \quad -F_{AD} - F_{CE} - 1000\,\text{lb} = 0 \quad (3)$$

$$\Rightarrow \boxed{F_{AD} = 3000\,\text{lb}.} \quad (4)$$

* Safety during construction is a difficult and sometimes overlooked aspect of the overall design of structures.

† After examining Fig. 1, hopefully you determined that for this problem, the method of sections is likely more efficient than the method of joints.

Then consider the FBD for cut *aa*, shown in Fig. 4. Since there are now only two unknowns, only two equilibrium equations are needed, and we elect to use the following:

$$\sum F_x = 0: \quad 1500\,\text{lb} - F_{BD}\left(\frac{15}{25}\right) + F_{BE}\left(\frac{15}{25}\right) = 0, \quad (5)$$

$$\sum F_y = 0: \quad -F_{AD} - F_{BD}\left(\frac{20}{25}\right) - F_{BE}\left(\frac{20}{25}\right) - F_{CE} - 1000\,\text{lb} = 0. \quad (6)$$

Solving Eqs. (5) and (6) provides

$$\Rightarrow \quad \boxed{F_{BD} = 1250\,\text{lb} \quad \text{and} \quad F_{BE} = -1250\,\text{lb}.} \quad (7)$$

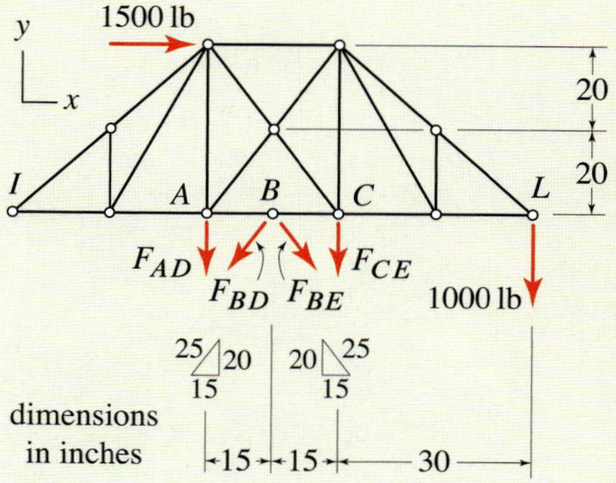

Figure 4
Free body diagram for cut *aa*.

Discussion & Verification As an alternative to the solution followed here, use of cuts *aa* and *bb* is also a good solution strategy, although it entails more algebra. Nonetheless, you may wish to use this strategy, and you should obtain the same solution. If you are using a computer to solve the equilibrium equations, then the extra algebra is irrelevant.

PROBLEMS

Problems 6.28 and 6.29

All members of the truss have the same length. Determine the force supported by members BD, CD, and CE.

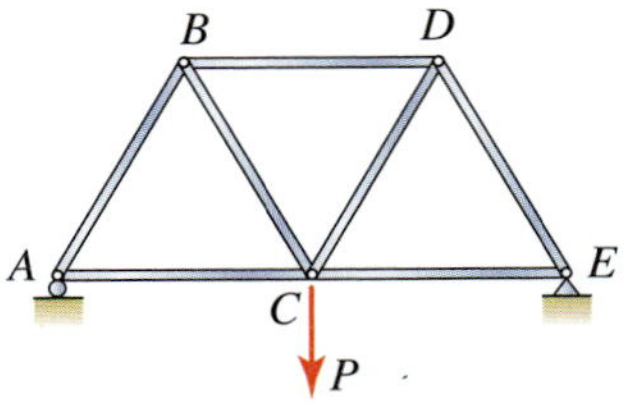

Figure P6.28

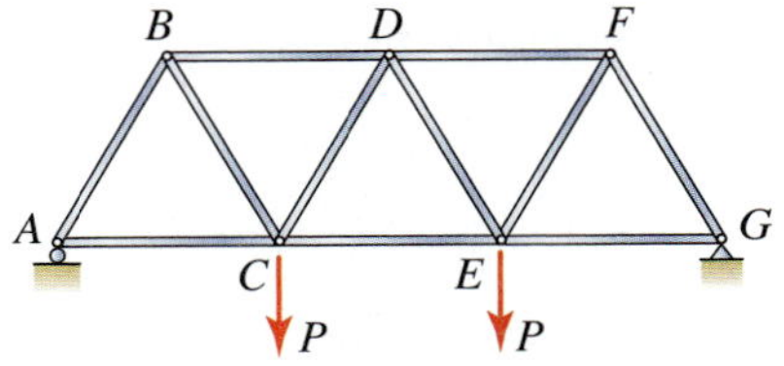

Figure P6.29

Problem 6.30

(a) By inspection, identify the zero-force members in the truss.

(b) Of the zero-force members identified in Part (a), which could possibly be eliminated without reducing the strength of the truss? Explain.

(c) Find the force supported by member GH.

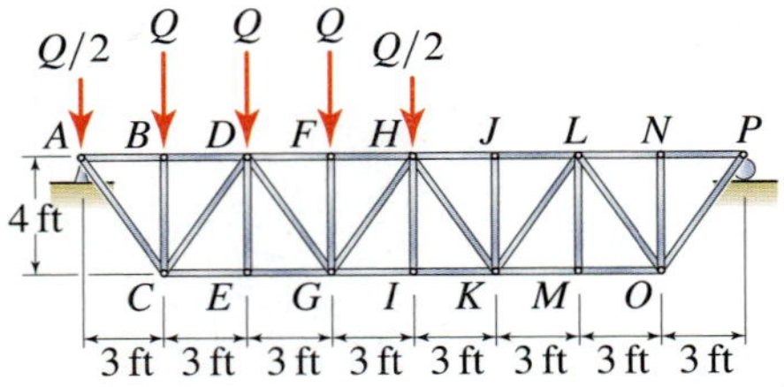

Figure P6.30

Problem 6.31

(a) By inspection, identify the zero-force members in the truss.

(b) Of the zero-force members identified in Part (a), which could possibly be eliminated without reducing the strength of the truss? Explain.

(c) Find the force supported by member FH.

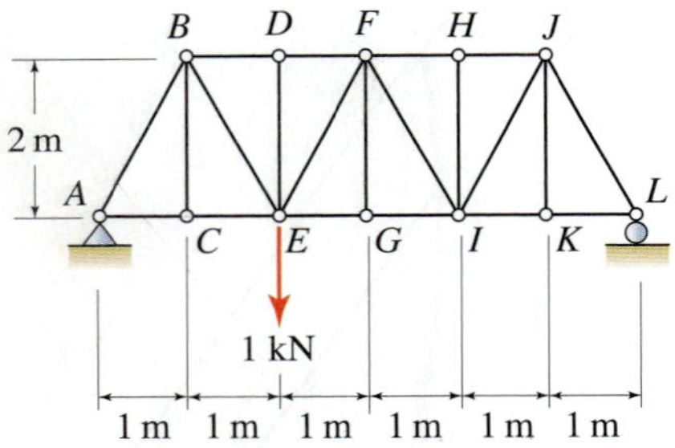

Figure P6.31

Problem 6.32

(a) By inspection, identify the zero-force members in the truss.

(b) Of the zero-force members identified in Part (a), which could possibly be eliminated without reducing the strength of the truss? Explain.

(c) Find the force supported by member FG.

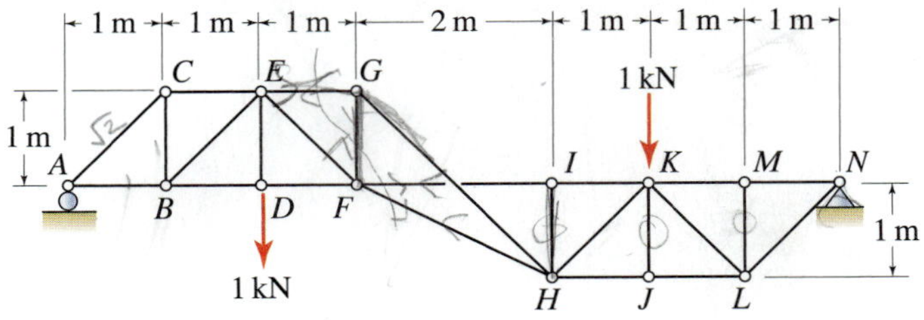

Figure P6.32

Problem 6.33

The structure has a frictionless pulley at G.

(a) By inspection, identify the zero-force members in the truss.

(b) Of the zero-force members identified in Part (a), which could possibly be eliminated without reducing the strength of the truss? Explain.

(c) Find the force supported by member GH.

(d) Find the force supported by member JK.

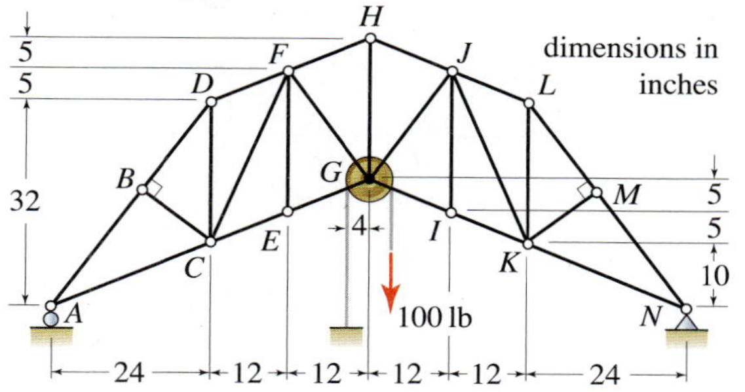

Figure P6.33

Problem 6.34

The electric power transmission tower supports two wires that apply 4 kip vertical forces, and the 8 kip horizontal force crudely models wind loading during a storm.

(a) Determine if the truss is statically determinate or indeterminate.

(b) Determine the force supported by the four members that emanate from joint D.

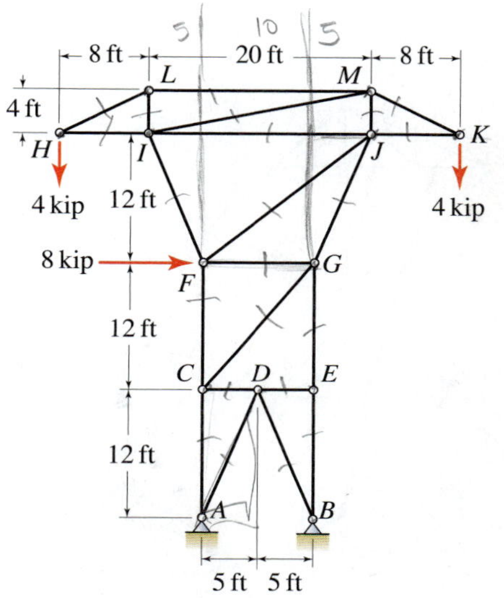

Figure P6.34

Problem 6.35

(a) Determine if the truss is statically determinate or indeterminate.

(b) Determine the force supported by the five members that emanate from joint G.

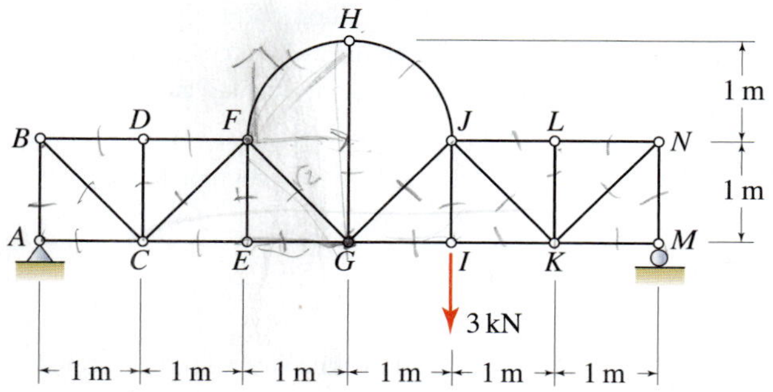

Figure P6.35

Problem 6.36

The boom of a tower crane is shown. The pulleys at A, B, and Q are frictionless, and $W = 10\,\text{kN}$.

(a) Determine the force supported by cable JT.

(b) Determine the force supported by the five members that emanate from joint J.

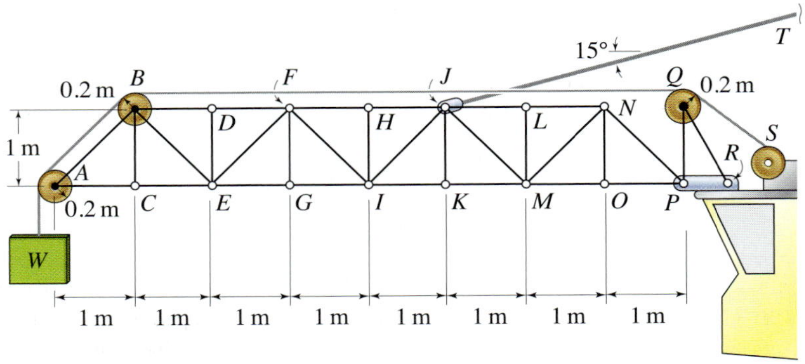

Figure P6.36

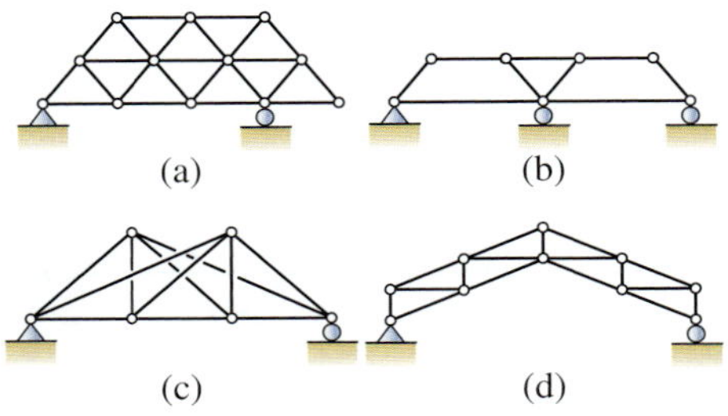

Figure P6.37

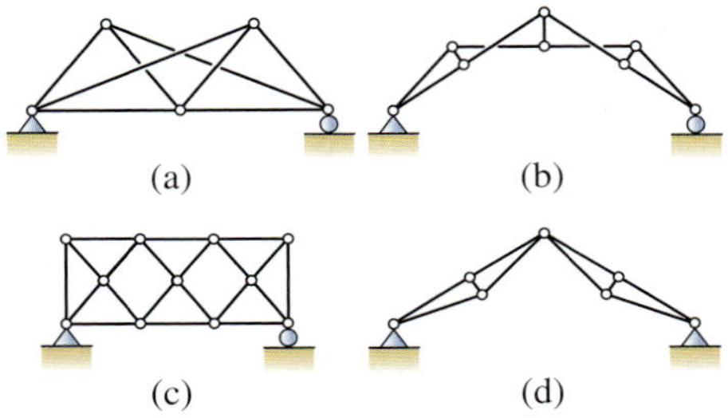

Figure P6.38

Problems 6.37 and 6.38

Determine if each truss is statically determinate, statically indeterminate, or a mechanism.

Problem 6.39

A *gambrel* room-in-attic truss is shown.

(a) Use equation counting, Eq. (6.22) on p. 349, to show this truss is a mechanism.

(b) This truss type is popular for many applications. Discuss why this truss design does not collapse in view of the fact that it is a mechanism according to truss theory, as shown in Part (a).

(c) If the roller support at J is replaced by a pin support, equation counting indicates that the number of equations and the number of unknowns is the same. With a pin support at J, is the structure no longer a mechanism according to truss theory? Explain.

Note: Concept problems are about *explanations*, not computations.

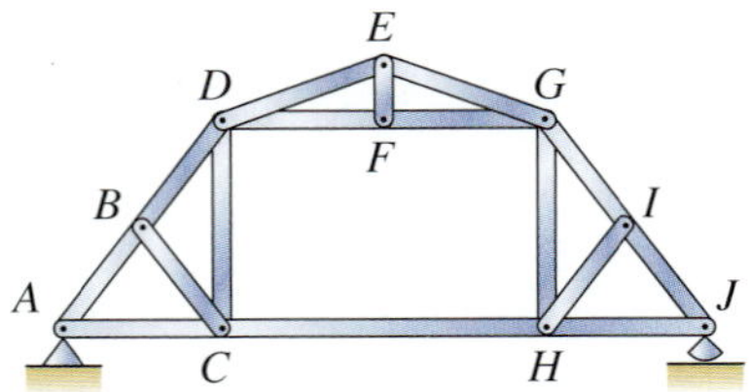

Figure P6.39

Problem 6.40

The two truss designs differ only in their depths.

(a) Determine the forces supported by all members of both trusses, and compare values for corresponding members in each truss.

(b) Offer reasons why truss (b) is the better design.

(c) Offer reasons why truss (a) is the better design.

(d) Overall, which of the two trusses do you believe is stronger, assuming all members are of the same material and have the same cross-sectional shape?

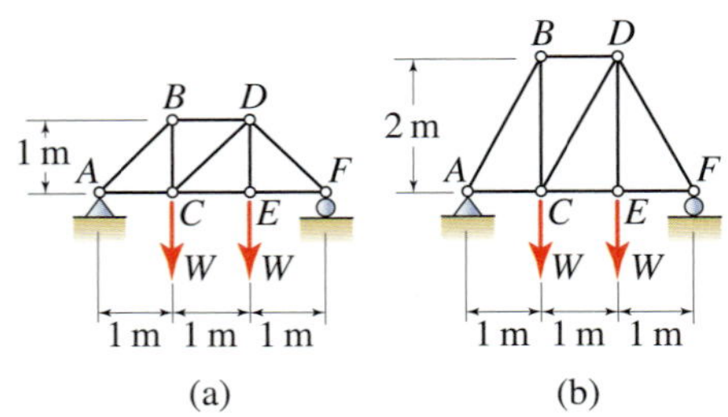

Figure P6.40

DESIGN PROBLEMS

General Instructions. In all problems, write a brief technical report, following the guidelines of Appendix A, where you summarize all pertinent information in a well-organized fashion. It should be written using proper, simple English that is easy to read by another engineer. Where appropriate, sketches along with critical dimensions should be included. Discuss the objectives and constraints considered in your design, the process used to arrive at your final design, safety issues if appropriate, and so on. The main discussion should be typed, and figures, if needed, can be computer-drawn or neatly hand-drawn. Include a neat copy of all supporting calculations in an appendix that you can refer to in the main discussion of your report. A length of a few pages, plus appendix, should be sufficient.

Design Problem 6.1

Design a footbridge for crossing a small stream. The bridge is intended for residential use only. The bridge consists of two identical trusses, spaced 3 ft apart. For ease of fabrication, each truss is to be constructed of one size of welded steel pipe from Table 3.3 on p. 164, with all members having the same 3 ft length. The 100 lb forces represent the dead loads. You are to specify the maximum safe live load for the bridge and the diameter of pipe to be used.

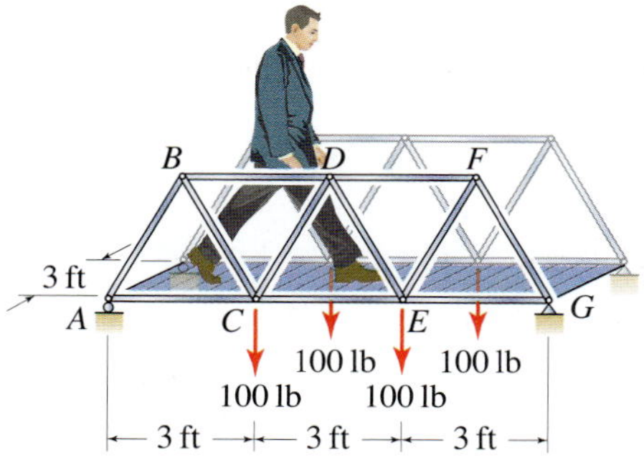

Figure DP6.1

Design Problem 6.2

Design a truss to be made of welded steel pipe to support the loads that are given. The truss is to be supported at joints A and E and should have joints at points B, C, and D (joints at other locations are also permissible). For ease of fabrication, the truss should be constructed from only one size of pipe from Table 3.3 on p. 164. You may consider the weight of the truss as already being approximately accounted for through the loads that are given, and no additional factors of safety beyond those already incorporated in Table 3.3 are needed. While it is not necessary to determine the truss design with minimum weight, it is desirable to use a small-diameter pipe so that the truss is reasonably economical.

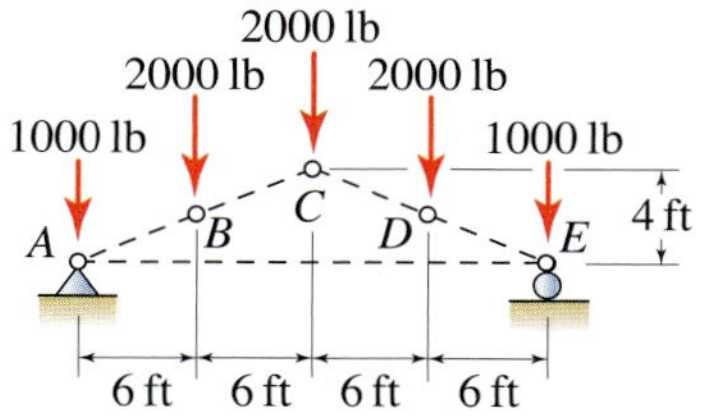

Figure DP6.2

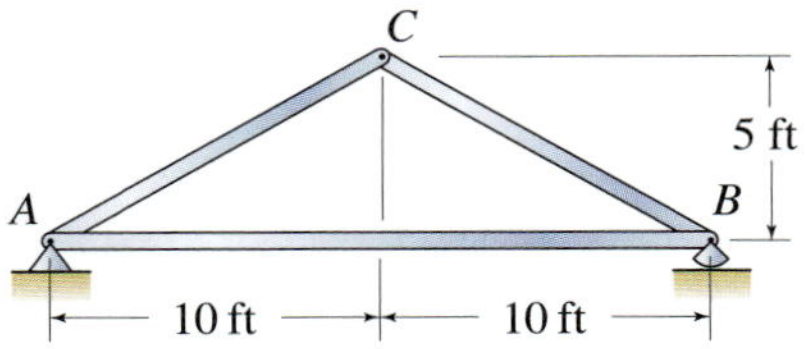

Figure DP6.3

Table 6.1
Allowable loads for dimensioned lumber where L is the length of the member.

Dimensioned Lumber		
	Allowable load	
	2 × 4 in.	2 × 6 in.
Tension:	8,000 lb	14,000 lb
Compression:		
$0 < L \le 2$ ft	8,000 lb	14,000 lb
2 ft $< L \le 4$ ft	2,000 lb	3,000 lb
4 ft $< L \le 6$ ft	1,000 lb	1,500 lb
6 ft $< L \le 8$ ft	0	1,000 lb

Design Problem 6.3

Design a simply supported wooden roof truss for residential construction. The truss must have the span and height shown, and it will support the following loads:

- The bottom chord of the truss is unloaded.
- The top chord of the truss is subjected to a dead load of 40 lb/ft of chord length and a maximum live load of 80 lb/ft of chord length (all loads are vertical).
- All other loads can be neglected.

The members of the truss are to be wood with 2 × 4 in. and/or 2 × 6 in. cross section (nominal dimensions). For ease of fabrication, the top chord should be entirely one dimension of member, and similarly for the bottom chord. The wood members have the allowable strengths given in Table 6.1. You are to specify the member size for both the top and bottom chords, and specify the size and placement of additional truss members between the top and bottom chords that you determine are needed. Your design does not need to be fully optimized in terms of minimizing the amount of wood used, but your design should be reasonably economical (e.g., a truss constructed entirely of 2 × 6 in. wood is probably not economical). If you determine that it is not possible to achieve the strength needed using 2 × 4 in. and/or 2 × 6 in. members, then you should specify the number of identical trusses that should be attached to one another so that the specified loads can be supported.

6.3 Trusses in Three Dimensions

A *space truss* is a three-dimensional structure that consists of two-force members only, where members are organized so that the assemblage as a whole behaves as a single object. Some examples of space trusses are shown in Fig. 6.24.

(a) (b) (c) (d)

Figure 6.24. Examples of space trusses. (a) The base of the Eiffel Tower in Paris, France, showing an intricate arrangement of individual truss members that form larger structural members, which serve as truss members for the entire structure. (b) The Expo Center, constructed in Vancouver, Canada, for the 1986 World Exposition, uses a spherical truss structure. (c) A view from the base of a tower that supports electric wires. (d) A large crane in a lowered position in anticipation of an approaching storm.

For a three-dimensional structure to be a space truss, it must have the following characteristics:

- All members must be connected to one another by frictionless ball-and-socket joints.
- Each member may have no more than two joints.
- Forces may be applied at joints only.
- The weight of individual members must be negligible.

If all of the those characteristics are satisfied, then it is guaranteed that all members of the structure are two-force members. Note that while trusses are most often constructed using straight members, they may contain members with curved or other complex shapes, provided they are two-force members.

> **Interesting Fact**
>
> **Computer modeling.** Because of the complexity of most space trusses, computer methods of analysis such as the *finite element method* are usually used. These computer programs automate the handling of data, generation of equilibrium equations, and solving for unknowns. However, you are responsible for modeling, determination of loads, interpretation of results, and—*hugely important*—assessing the accuracy of results. The methods of analysis discussed for trusses in this book give you many of the tools you need to interpret the results of computer models and to help ensure their accuracy.

For practical applications, the definition of a space truss is restrictive, because in real life, members are usually attached to one another using connections that are not ball-and-socket joints. Similar to the discussion in Fig. 6.4 on p. 331, the connections in real structures are usually capable of supporting moments, and thus the members that emanate from such connections usually are not two-force members. For precise analysis, structures with moment-resisting connections should be modeled as *frame structures*, which are discussed in Section 6.4. Nonetheless, if a particular structure would qualify as a truss except for the connection details, in engineering practice these structures are often modeled as trusses anyway.

Fundamentally, the methods for equilibrium analysis of space trusses are the same as those for plane trusses. That is, both the method of joints and method of sections can be used. The major differences for space trusses are that FBDs are usually more intricate, reactions are more complex, and the number of unknowns to be determined and the number of equations to be solved are greater.

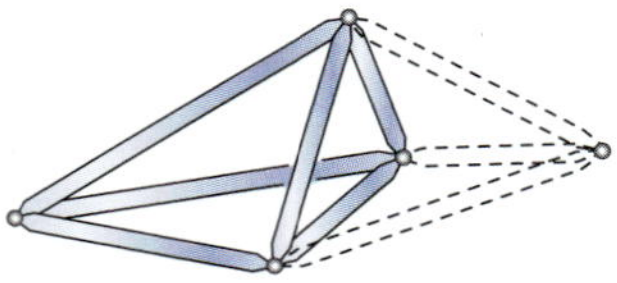

Figure 6.25
In a *simple space truss* the arrangement of members begins with six members that form a tetrahedron, and adding to this three new members (shown by dashed lines) for each new joint that is added.

Stability of space trusses and design considerations

A significant difference between space trusses and plane trusses is the difficulty in assessing *stability*. If a space truss is unstable, it is often difficult, even for experienced engineers, to identify this defect based on inspection only. An effective approach to help avoid instability is to design space trusses starting with the tetrahedral arrangement of members shown in Fig. 6.25; such structures are called *simple space trusses*.

While there are advanced analytical methods to help determine the stability of a truss, the simple rule of thumb called *equation counting*, which was introduced in Section 6.2 for plane trusses, is very useful with the modifications described here. To help determine whether a truss is statically determinate or statically indeterminate, and whether it is stable or unstable, we compare the number of unknown forces to the number of equilibrium equations. Consider a space truss having

$$\begin{aligned} m &= \text{number of members}, \\ r &= \text{number of support reactions}, \qquad (6.28) \\ j &= \text{number of joints}. \end{aligned}$$

Each member has one unknown force, and each support reaction has one unknown force, so that $m + r$ is the total number of unknowns. For a space truss, each joint has three equilibrium equations so that $3j$ is the total number of equations. Thus, the rule of thumb* is as follows:

If $m + r < 3j$	The truss is a mechanism and/or has partial fixity.	
If $m + r = 3j$	The truss is statically determinate if it has full fixity. The truss is statically indeterminate if it has partial fixity.	(6.29)
If $m + r > 3j$	The truss is statically indeterminate, and it can have full fixity or partial fixity.	

Use of this rule of thumb requires that we inspect a truss structure to determine its fixity. A truss (or any object) in three dimensions has the potential for six types of rigid body motion: translation in each of the x, y, and z directions and rotation about each of the x, y, and z axes, or a combination of these. Thus, a structure that is fully supported against rigid body motion (i.e., fully fixed) requires a minimum of six support reactions that are properly arranged. If a structure in three dimensions has less than six support reactions, then it will have only partial fixity. Several examples are shown in Fig. 6.26.

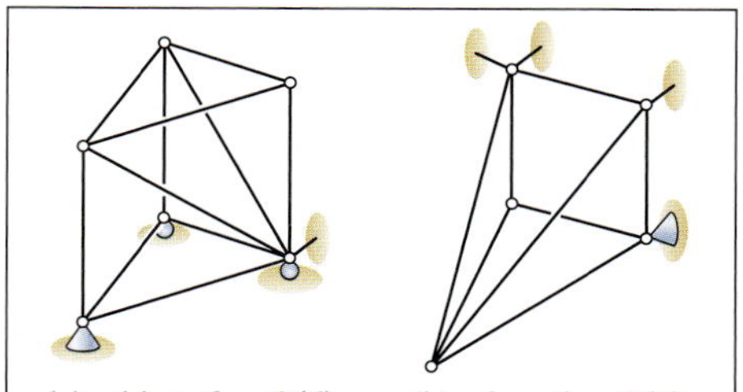

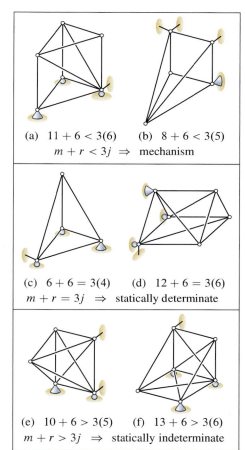

(a) $11 + 6 < 3(6)$ (b) $8 + 6 < 3(5)$
$m + r < 3j \Rightarrow$ mechanism

(c) $6 + 6 = 3(4)$ (d) $12 + 6 = 3(6)$
$m + r = 3j \Rightarrow$ statically determinate

(e) $10 + 6 > 3(5)$ (f) $13 + 6 > 3(6)$
$m + r > 3j \Rightarrow$ statically indeterminate

Figure 6.26
Examples of equation counting to determine if a truss is statically determinate or indeterminate and whether it is stable. All trusses shown here have six support reactions that provide full fixity.

End of Section Summary

In this section, methods of analysis for space trusses were described. Some of the key points are as follows:

- The methods for equilibrium analysis of space trusses are the same as those for plane trusses. That is, both the *method of joints* and *method of sections* can be used.
- A *simple space truss* has members arranged in a tetrahedral pattern, as shown in Fig. 6.25. One of the main features of simple space trusses is that they are always stable.
- A rule of thumb called *equation counting*, when used with good judgment, can be effective for determining the stability and static determinacy of a space truss.

* The comments following Eq. (6.22) on p. 349 also apply here (with $2j$ replaced by $3j$).

EXAMPLE 6.6 *Space Truss Analysis by the Method of Joints*

A boom for a crane is shown in a horizontal position. If $W = 1$ kN, determine the force supported by member DG.

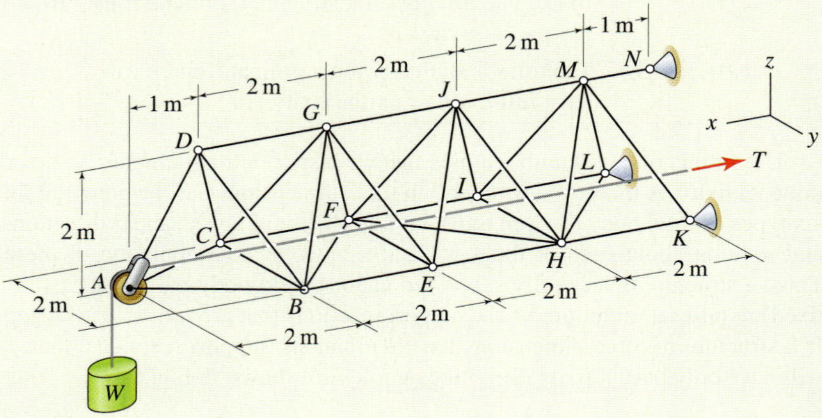

Figure 1.

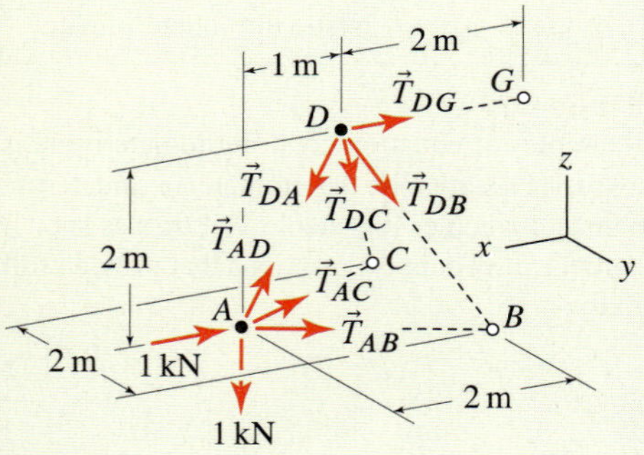

Figure 2
Free body diagrams of joints A and D.

SOLUTION

Road Map Before beginning our analysis, we determine if the truss is statically determinate or indeterminate. Examination of Fig. 1 shows that the number of members is $m = 33$, number of support reactions is $r = 9$, number of joints is $j = 14$, and the truss is fully fixed. Application of Eq. (6.29) on p. 365 provides $m + r = 3j$, and since the truss is fully fixed, it is statically determinate.

To use the method of joints to determine the force supported by member DG, we examine joint D in Fig. 1 to see that there are four members that emanate from it, and thus its FBD will involve four unknowns, while only three equilibrium equations are available. Examination of joint A shows that its FBD will have three unknowns, and therefore this is an ideal joint to begin with.

Modeling The FBDs for joints A and D are shown in Fig. 2, where member forces are positive in tension, and the weight of the members is neglected. Assuming the pulley is frictionless and the cable is weightless, the force supported throughout the entire cable is the same, with value $T = 1$ kN. Either we can leave the pulley attached to the truss,* or as shown in Fig. 2, we can remove the pulley by shifting the pulley forces to the bearing of the pulley.

Governing Equations & Computation Referring to the FBD for joint A shown in Fig. 2, we observe the equilibrium equation $\sum F_z = 0$ will have only one unknown, namely, T_{AD}. The vector expression for the force member AD exerts on joint A is

$$\vec{T}_{AD} = T_{AD}\left(\frac{-\hat{\imath} + 2\hat{k}}{\sqrt{5}}\right). \tag{1}$$

* If we elect to leave the pulley on the truss, then we will need to assume a value for the radius of the pulley. The radius we select, however, is arbitrary, since all member forces, reactions, etc., will be the same as those found in this solution regardless of the pulley's radius. If you are uncertain that these comments are true, or how to proceed if the pulley is left on the truss, then you should review Fig. 5.11 on p. 271.

Using a scalar approach to sum forces in the z direction provides
Joint A:

$$\sum F_z = 0: \quad -1\,\text{kN} + T_{AD}\Big(\frac{2}{\sqrt{5}}\Big) = 0 \quad \Rightarrow \quad T_{AD} = 1.118\,\text{kN}. \tag{2}$$

We omit finding the remaining unknowns for joint A since they are not needed for equilibrium of joint D.

Referring to the FBD for joint D shown in Fig. 2, we see that it will probably be easier to write the equilibrium equations using a vector approach. Vector expressions for the forces the emanate from joint D are

$$\vec{T}_{DA} = -\vec{T}_{AD} = (1.118\,\text{kN})\Big(\frac{\hat{\imath} - 2\hat{k}}{\sqrt{5}}\Big), \tag{3}$$

$$\vec{T}_{DB} = T_{DB}\Big(\frac{-\hat{\imath} + \hat{\jmath} - 2\hat{k}}{\sqrt{6}}\Big), \tag{4}$$

$$\vec{T}_{DC} = T_{DC}\Big(\frac{-\hat{\imath} - \hat{\jmath} - 2\hat{k}}{\sqrt{6}}\Big), \tag{5}$$

$$\vec{T}_{DG} = T_{DG}(-\hat{\imath}). \tag{6}$$

The equilibrium equations for joint D are
Joint D:

$$\sum \vec{F} = \vec{0}: \quad \vec{T}_{DA} + \vec{T}_{DB} + \vec{T}_{DC} + \vec{T}_{DG} = \vec{0}. \tag{7}$$

Substituting Eqs. (3) through (6) into the above along with $T_{AD} = 1.118\,\text{kN}$ and grouping terms in the x, y, and z directions provide the following three equations

$$T_{DB}\Big(\frac{-1}{\sqrt{6}}\Big) + T_{DC}\Big(\frac{-1}{\sqrt{6}}\Big) + T_{DG}(-1) = (-1.118\,\text{kN})\Big(\frac{1}{\sqrt{5}}\Big), \tag{8}$$

$$T_{DB}\Big(\frac{1}{\sqrt{6}}\Big) + T_{DC}\Big(\frac{-1}{\sqrt{6}}\Big) = 0, \tag{9}$$

$$T_{DB}\Big(\frac{-2}{\sqrt{6}}\Big) + T_{DC}\Big(\frac{-2}{\sqrt{6}}\Big) = (1.118\,\text{kN})\Big(\frac{2}{\sqrt{5}}\Big). \tag{10}$$

Solving these equations provides

$$\Rightarrow \quad T_{DB} = T_{DC} = -0.6124\,\text{kN} \quad \text{and} \quad \boxed{T_{DG} = 1.000\,\text{kN}.} \tag{11}$$

Discussion & Verification Because the loading for this truss is simple, based on inspection we expect members AD and DG to be in tension, and members DB and DC to be in compression, and indeed our solution shows this. Further, because of the symmetry of the problem, we expect the force supported by members DB and DC to be the same, and our solution also shows this. Other checks include substituting the solutions into the original equilibrium equations to verify that all of these are satisfied. However, this does not provide a check that the equilibrium equations are correct. Therefore, it is essential that FBDs be accurately drawn and equilibrium equations be correctly written.

EXAMPLE 6.7 *Space Truss Analysis by the Method of Sections*

For the structure in Example 6.6 with $W = 1$ kN, determine the force supported by all members that emanate from the supports at points K, L, and N.

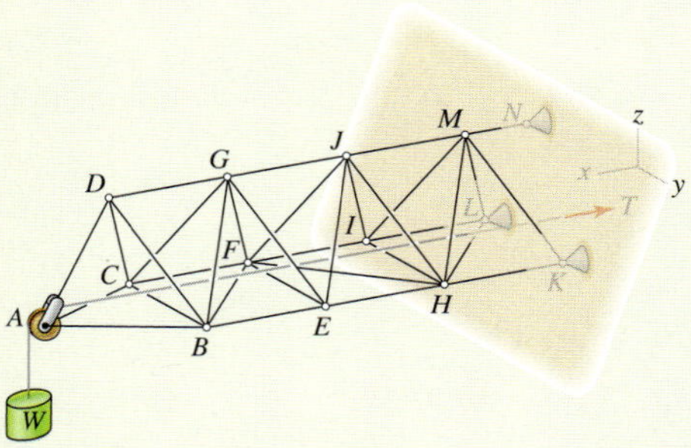

Figure 1
To use the method of sections, a cut is taken that passes through the structure, intersecting the members of interest.

SOLUTION

Road Map We will use the method of sections with a cut that passes through the members of interest, as shown in Fig. 1. Observe that the cut passes through six members. Since six equilibrium equations are available, we expect to be able to determine all of the unknown member forces using the FBD that results from this cut.

Modeling Using the cut shown in Fig. 1, the FBD for the left-hand portion of the structure is shown in Fig. 2, where member forces are positive in tension, and the weight of the members is neglected. Assuming the pulley is frictionless and the cable is weightless, the force supported throughout the entire cable is the same, with value $T = 1$ kN. We can either leave the pulley attached to the truss or, as shown in Fig. 2, remove the pulley by shifting the pulley forces to the bearing of the pulley. In this problem, removing the pulley from the truss will be slightly easier since we will be able to combine the two 1 kN forces at point A in Fig. 2 into a single force vector and thus will be able to evaluate its moment using one cross product.

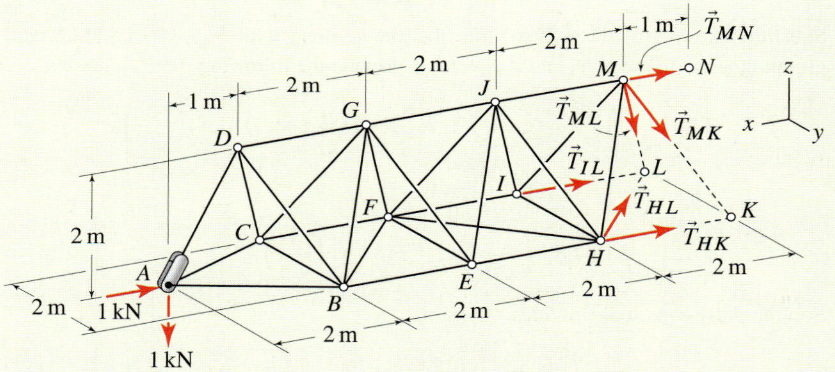

Figure 2. Free body diagram.

Governing Equations & Computation Before we write equilibrium equations, vector expressions for truss member forces and the cable forces at A, which we will call $\vec{F}$, are needed:

$$\vec{T}_{HK} = T_{HK}(-\hat{\imath}), \qquad \vec{T}_{HL} = T_{HL}\Big(\frac{-2\hat{\imath} - 2\hat{\jmath}}{\sqrt{8}}\Big), \tag{1}$$

$$\vec{T}_{IL} = T_{IL}(-\hat{\imath}), \qquad \vec{T}_{MK} = T_{MK}\Big(\frac{-\hat{\imath} + \hat{\jmath} - 2\hat{k}}{\sqrt{6}}\Big), \tag{2}$$

$$\vec{T}_{ML} = T_{ML}\Big(\frac{-\hat{\imath} - \hat{\jmath} - 2\hat{k}}{\sqrt{6}}\Big), \qquad \vec{T}_{MN} = T_{MN}(-\hat{\imath}), \tag{3}$$

$$\vec{F} = (1\text{ kN})(-\hat{\imath} - \hat{k}). \tag{4}$$

Equilibrium of forces will provide three scalar equations with six unknowns, while equilibrium of moments about point M will provide three scalar equations with three unknowns; thus, we begin by summing moments. Using the following position vectors

$$\vec{r}_{MA} = (7\hat{\imath} - 2\hat{k})\,\text{m}, \qquad \vec{r}_{MH} = (\hat{\imath} + \hat{\jmath} - 2\hat{k})\,\text{m}, \qquad \vec{r}_{MI} = (\hat{\imath} - \hat{\jmath} - 2\hat{k})\,\text{m}, \quad (5)$$

we sum moments about point M:

$$\sum \vec{M}_M = \vec{0}: \quad \vec{r}_{MA} \times \vec{F} + \vec{r}_{MH} \times (\vec{T}_{HK} + \vec{T}_{HL}) + \vec{r}_{MI} \times \vec{T}_{IL} = \vec{0}. \quad (6)$$

Carrying out the cross products in Eq. (6), canceling the meter unit, and grouping terms in the x, y, and z directions give the following equations:

$$\left(\frac{-4}{\sqrt{8}}\right)T_{HL} = 0, \quad (7)$$

$$2T_{HK} + \left(\frac{4}{\sqrt{8}}\right)T_{HL} + 2T_{IL} = -9\,\text{kN}, \quad (8)$$

$$T_{HK} - T_{IL} = 0. \quad (9)$$

Equations (7)–(9) are easily solved to obtain

$$\Rightarrow \quad \boxed{T_{HL} = 0 \quad \text{and} \quad T_{HK} = T_{IL} = -2.250\,\text{kN}.} \quad (10)$$

Equilibrium of forces provides

$$\sum \vec{F} = \vec{0}: \quad \vec{F} + \vec{T}_{HK} + \vec{T}_{HL} + \vec{T}_{IL} + \vec{T}_{MK} + \vec{T}_{ML} + \vec{T}_{MN} = \vec{0}. \quad (11)$$

By using the solutions given in Eq. (10) and grouping terms in the x, y, and z directions, Eq. (11) gives the following equations:

$$\left(\frac{-1}{\sqrt{6}}\right)T_{MK} + \left(\frac{-1}{\sqrt{6}}\right)T_{ML} - T_{MN} = -3.500\,\text{kN}, \quad (12)$$

$$\left(\frac{1}{\sqrt{6}}\right)T_{MK} + \left(\frac{-1}{\sqrt{6}}\right)T_{ML} = 0, \quad (13)$$

$$\left(\frac{-2}{\sqrt{6}}\right)T_{MK} + \left(\frac{-2}{\sqrt{6}}\right)T_{ML} = 1\,\text{kN}. \quad (14)$$

Equations (12)–(14) are easily solved to obtain

$$\Rightarrow \quad \boxed{T_{MN} = 4.000\,\text{kN} \quad \text{and} \quad T_{MK} = T_{ML} = -0.6124\,\text{kN}.} \quad (15)$$

Discussion & Verification

- Because of the complexity of most space trusses, verification of the solution can be challenging. In this problem, based on inspection, we expect member MN to be in tension and members HK and IL to be in compression, and indeed our solution shows this. Further, because of the symmetry of the problem, we expect the forces supported by members HK and IL to be the same and the forces supported by members MK and ML to be the same, and our solution also shows this.
- A simple partial check of our solution's quantitative accuracy can be obtained by using a scalar approach to evaluate moment equilibrium about line KL in the FBD of Fig. 2. Taking moments to be positive in the y direction, we obtain

$$\sum M_{KL} = 0: \; (1\,\text{kN})(8\,\text{m}) - T_{MN}(2\,\text{m}) = 0 \Rightarrow T_{MN} = 4.000\,\text{kN}, \quad (16)$$

  which agrees with the result found earlier. You should study Fig. 2 to see if other scalar equilibrium equations can be written to help check the remaining solutions.

PROBLEMS

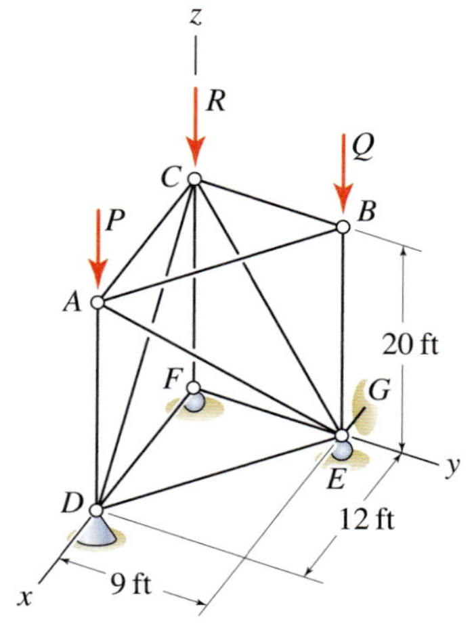

Figure P6.41

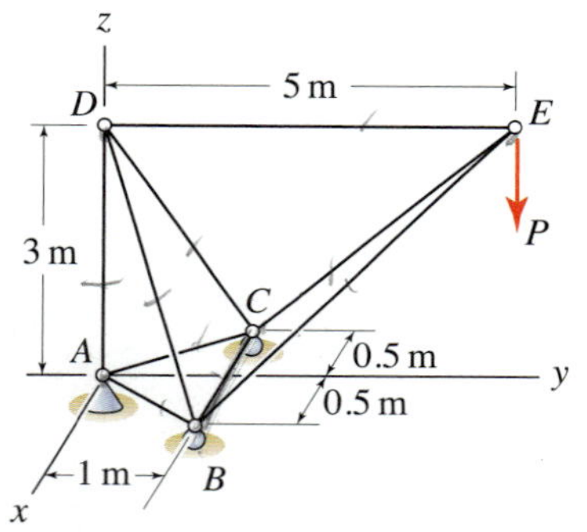

Figure P6.42

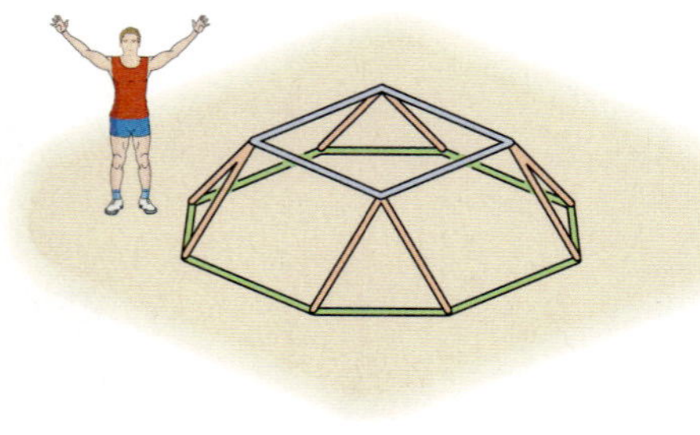

Figure P6.43

Problem 6.41

The truss shown has a socket support at point D, rollers at points E and F, and a link from E to G that prevents motion in the x direction.

(a) Does the truss have partial fixity or full fixity?

(b) Determine if the truss is statically determinate or statically indeterminate.

(c) Determine the force supported by all members of the truss if $P = 1$ kip, $Q = 2$ kip, and $R = 3$ kip.

Problem 6.42

The truss shown has a socket support at point A and rollers at points B and C. Points A, B, and C lie in the xy plane, and point E lies in the yz plane.

(a) Does the truss have partial fixity or full fixity?

(b) Determine if the truss is statically determinate or statically indeterminate.

(c) Determine the force supported by all members of the truss if $P = 1$ kN. Force P is vertical.

Problem 6.43

Imagine you have been retained by an attorney to serve as an expert witness for possible litigation regarding an accident that occurred on the playground structure shown. You have been given only preliminary information on this structure, including the following:

- It was designed as a truss structure.
- It had been in service for several years, without incident.
- It collapsed while an adult was climbing on it, causing serious injury to the person.
- It was thought to be in good condition at the time of the accident.
- Detailed information on the design, fabrication, installation, and so forth has not yet been made available to you.

Until you have fully studied all of the information available, it is not possible for you to render an informed opinion on this structure and/or the accident that occurred. Nonetheless, the attorney retaining you is anxious to informally hear your initial impressions. With the information provided, what would you tell the attorney? In answering this, consider responses such as the structure may have been defective in its design, it may have been defective in its fabrication or installation, the adult probably should not have been on the structure and thus he or she is probably responsible for its failure, or I really don't know until I have further information.

Problems 6.44 through 6.48

The truss shown has a socket support at point L, a roller at point K, and a cylindrical roller at point A that prevents motion in the y and z directions.

Problem 6.44 Determine if the truss is statically determinate or statically indeterminate.

Problem 6.45 Use the method of joints to determine the force supported by member DG if $P = 40\,\text{kN}$ and $Q = 0\,\text{kN}$.

Problem 6.46 Use the method of joints to determine the force supported by member MJ if $P = Q = 10\,\text{kN}$.

Problem 6.47 Use the method of sections to determine the force supported by member JG if $P = 40\,\text{kN}$ and $Q = 0\,\text{kN}$.

Problem 6.48 Use the method of sections to determine the force supported by member JG if $P = Q = 10\,\text{kN}$.

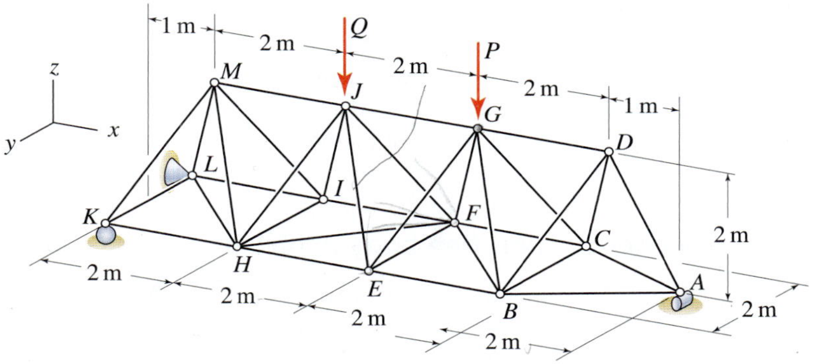

Figure P6.44–P6.48

6.4 Frames and Machines

Analysis of frames and machines is discussed in this section, and these are defined as follows:

Frame. A *frame* is a structure that contains one or more three-force and/or multiforce members. Typically, a frame is fully fixed in space and uses a stationary arrangement of members with the goal of supporting forces that are applied to it.

Machine. A *machine* is an arrangement of members where typically the members can have significant motion relative to one another. The usual goal of a machine is transmission of motion and/or force. While a machine can be composed entirely of two-force members, in which case it may be analyzed as a truss, in this section we are interested in machines that contain one or more multiforce members.

Figure 6.27
The mountain bike frame shown is an example of a frame structure where multiforce members are used and the connections between members are often moment-resisting.

These definitions are broad and have considerable overlap. Nonetheless, when multiforce members are involved, the methods of static analysis for frames and machines are the same. Throughout statics, we assume that if motion is possible, it is slow enough that inertia can be neglected. If this is not the case, then $\sum \vec{F} \neq \vec{0}$, $\sum \vec{M} \neq \vec{0}$, and methods of analysis from dynamics must be used.

Figure 6.28
This microelectromechanical system (MEMS) machine was fabricated at Sandia National Laboratory. The machine features a mirror whose tilt angle can be adjusted by a motor (not shown) that drives gears and a rack and pinion. The largest gear is about 150 μm in diameter (a human hair is about 100 μm in diameter). The methods discussed in this section can be used to analyze such systems. A large challenge is controlling contact and friction forces in MEMS.

Analysis procedure and free body diagrams (FBDs)

The main objective in structural analysis of frames and machines is determination of the forces supported by all members. Depending on the arrangement and loading of members, a particular problem may involve a combination of particles in equilibrium (discussed in Chapter 3), two-force members in equilibrium (truss analysis, discussed in Sections 6.1 and 6.2), and multiforce members in equilibrium (discussed in Chapter 5). Thus, there are no new technical tools needed to analyze frames and machines. Rather, the challenge is synthesis of the many skills you have already developed.

Determination of the forces supported by the individual members of a frame or machine requires that each member be separated from its neighboring members, with an FBD being drawn for each. The forces of action and reaction between the various members must be shown in the FBDs so that Newton's third law is satisfied. Using these FBDs, equilibrium equations are then written for each member, and presuming the frame or machine is statically determinate, these equations are solved to determine the forces supported by all members. Because frames and machines usually involve numerous members and because many of these members will typically be three-force members and/or multiforce members,* numerous FBDs are usually needed and sometimes these are complex. You will likely find the FBDs for frames and machines to be among the most challenging you will encounter in statics, and all of the skills you have cultivated for drawing FBDs will be needed.

To draw the FBDs needed for the analysis of a frame or machine, the procedure for drawing the FBD for a single body, discussed in Section 5.2,

* If only two-force members were involved, then the structure or machine would be a truss, and the methods of Chapter 3 and Sections 6.1 and 6.2 would be sufficient.

is repeatedly applied to each of the members of the frame or machine. This procedure is as follows:

Procedure for Drawing FBDs

1. Disassemble the frame or machine into individual members.

For each member:

2. Imagine this member is "cut" completely free (separated) from its environment. That is:
    - In 2D, think of a closed line that completely encircles the member.
    - In 3D, think of a closed surface that completely surrounds the member.
3. Sketch the member.
4. Sketch the forces.
    (a) Sketch the forces that are applied to the member by the environment (e.g., weight).
    (b) Wherever the cut passes through a structural member, sketch the forces that occur at that location.
    (c) Wherever the cut passes through a support (i.e., where a support is removed from the member), sketch the reaction forces and moments that occur at that location.
5. Sketch the coordinate system to be used. Add pertinent dimensions and angles to the FBD to fully define the locations and orientations of all forces.

The order in which the forces are sketched in Step 4 is irrelevant. For complicated FBDs, it may be difficult to include all of the dimensions and/or angles in Step 5. When this is the case, some of this information may be obtained from a different sketch.

Examples of correct FBDs

Figure 6.29 shows several examples of properly constructed FBDs for frames and machines. Comments on the construction of these FBDs follow.

Traffic signal pole. After sketching each member, we apply external forces, which consist of the weights W_C and W_H. If you want to consider the weights of other members (e.g., the weight of member ABF may be relatively large), then their weights can also be included in the FBDs. Next, we apply the appropriate reactions for the built-in support at point A. Then we search for two-force members and identify cables DE and GF, and we assign their forces as T_{FG} and T_{DE}, respectively. Finally, at locations where members are joined to one another, we include the appropriate forces, ensuring that forces of action and reaction between interacting bodies are equal in magnitude and opposite in direction (i.e., Newton's third law is satisfied).

Helpful Information

Two-force members. When drawing FBDs for frames and machines, it is helpful to first identify two-force members and label their unknowns. Doing this provides for less complicated FBDs, fewer equilibrium equations that need to be written, and fewer unknowns to be determined.

Common Pitfall

Statics versus dynamics. Consider the structure shown, where member AB is pinned to member BC and force F is known.

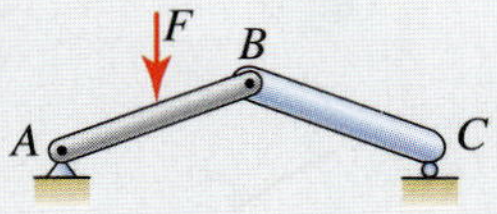

The following FBD

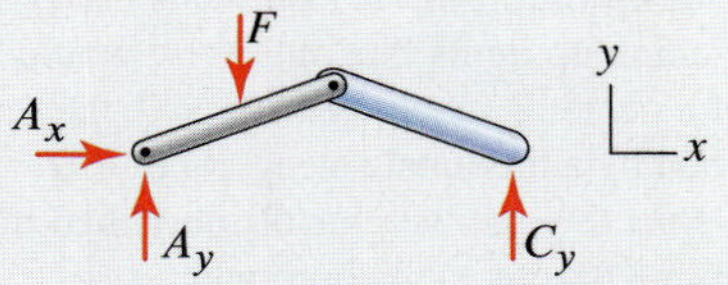

is properly drawn. However, a common pitfall for problems such as this is to write equilibrium equations such as $\sum M_A = 0$ to determine the reaction C_y. The error in writing equilibrium equations such as this is that this is *not* a problem of static equilibrium. Clearly, in this structure with this loading, members AB and BC will undergo accelerations, and hence $\sum M_A \neq 0$, and concepts of dynamics are needed to determine reactions, velocities, accelerations, etc.

examples of correct FBDs

Traffic signal pole. A pole supporting traffic and pedestrian signals is shown. Neglecting the weights of individual members, except for the two signals, draw FBDs of each component.

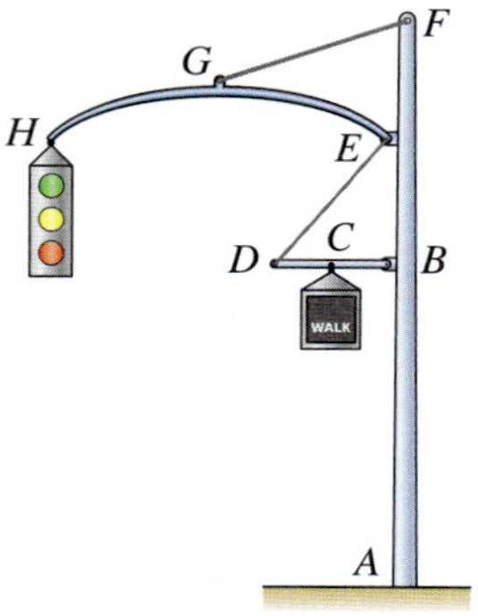

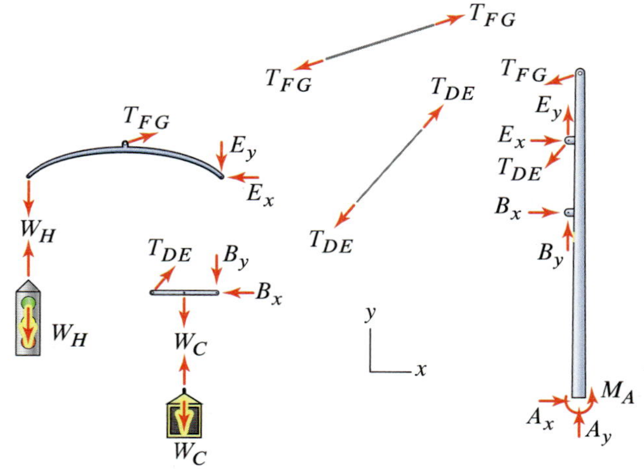

Aircraft service vehicle. A vehicle for servicing aircraft is shown. Draw FBDs so that the forces supported by members of the lift mechanism due to the weights of containers *J* and *K* may be determined.

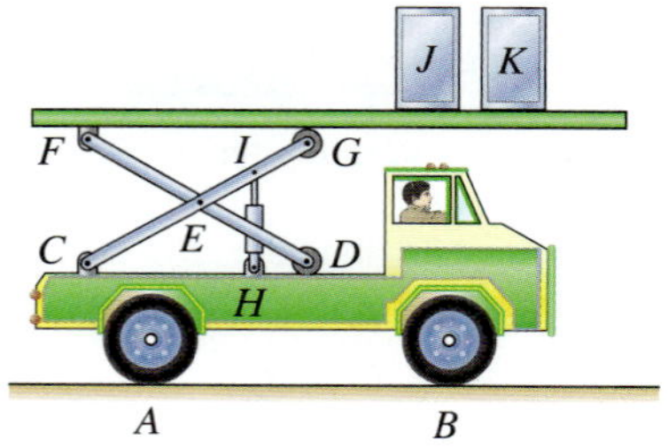

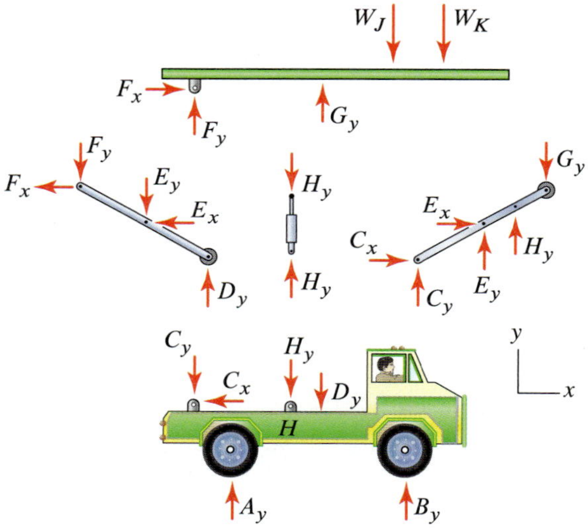

Pickup tool. The tool shown is used for picking up items that otherwise would be out of a person's reach. When $P = 0$, the tool is held open by a torsional spring at point *D*. Neglecting the weights of individual members, draw FBDs of each component.

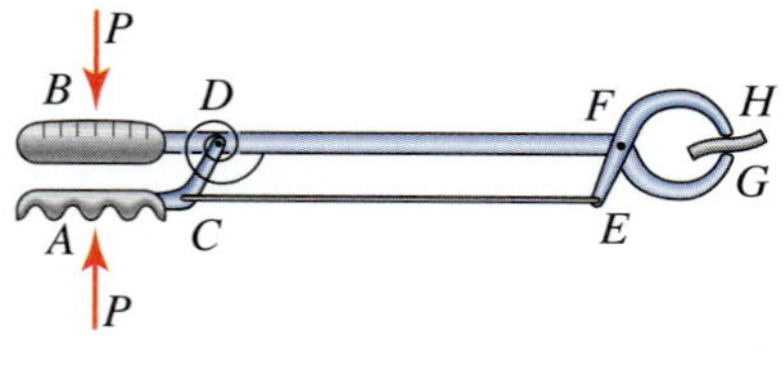

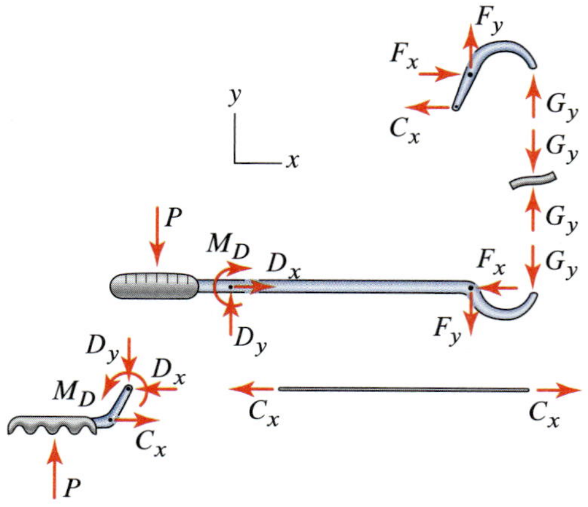

Figure 6.29. Examples of properly constructed FBDs.

Aircraft service vehicle. We are instructed to consider the weights of the crates at J and K only, although clearly the weights of several other components are large. When sketching each member's shape, we elect to leave the wheels on the vehicle. Assuming the vehicle is parked on a level surface, the support reactions consist of A_y and B_y. The only two-force member is the hydraulic cylinder HI and the force it supports is labeled H_y. Finally, at locations where members are joined to one another, we include the appropriate forces, ensuring that forces of action and reaction between interacting bodies are equal in magnitude and opposite in direction.

Pickup tool. We are instructed to neglect the weights of individual members. Rod CE and the item GH being gripped are the only two-force members, and the forces they support are labeled C_x and G_y, respectively. At pin D, in addition to the usual two forces D_x and D_y, we have a moment M_D due to the torsional spring.

Examples of incorrect and/or incomplete FBDs

Figure 6.30 shows several examples of incorrect and/or incomplete FBDs for frames and machines. Comments on how these FBDs must be revised follow, but before reading these, you should study Fig. 6.30 to find as many of the needed corrections and/or additions on your own as possible.

Frame with pulley

1. The forces on the pulley must be revised in one of the following ways:
   - Either the force W applied to the bearing of the pulley must be placed on the right-hand rim of the pulley, or
   - The force W oriented at 30° on the left-hand rim of the pulley must be shifted to the bearing.
2. The connection forces at pin B must also include vertical forces B_y.
3. The support reactions at A must also include a moment reaction M_A.

Scraper

1. The connection at E allows no rotation about the z axis (direction out of the plane of the figure); therefore a moment M_E must be applied to the scoop and tractor such that they are in opposite directions.
2. The front tire exerts a horizontal force on the ground due to the axle torque M_D, and therefore a horizontal force B_x is needed between the tire and ground.

Ratchet pruner

1. Although not really a deficiency with the FBDs, a coordinate system should be selected and shown.
2. With the information given, the positions of the two 10 lb forces on the handles are uncertain. Either we must do more detailed work to precisely locate these (this may be difficult), or we must make a reasonable decision for these positions. In these FBDs, the latter approach is used, and while the locations of the 10 lb forces are approximate, they nonetheless will probably be very acceptable for purposes of analysis

examples of incorrect and/or incomplete FBDs

Frame with pulley. The pulley at C is frictionless and points B–D are frictionless pins. Neglecting the weights of individual members, draw FBDs of each component.

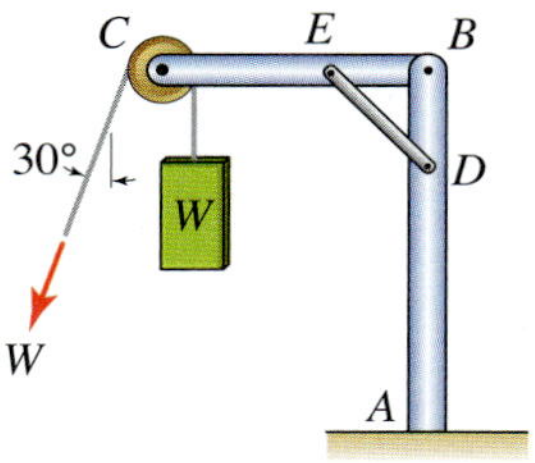

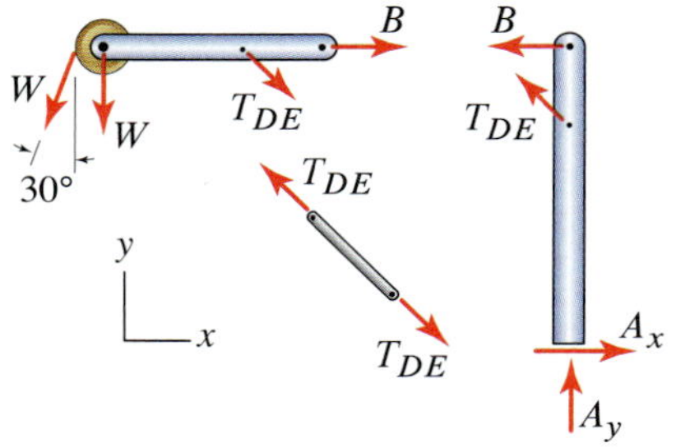

Scraper. A scraper for moving earth is shown. The center of gravity for the scoop and its contents is at point G, and the center of gravity for the tractor is at point H. The rear wheels at C roll freely while the front wheels at D are powered by the tractor's engine. Draw FBDs of the wheels, scoop, and tractor.

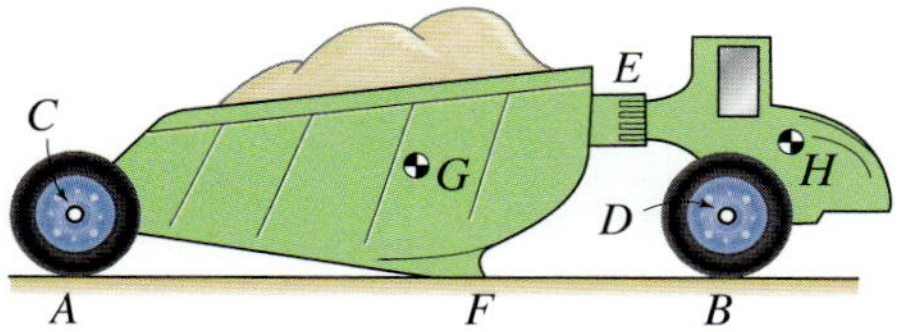

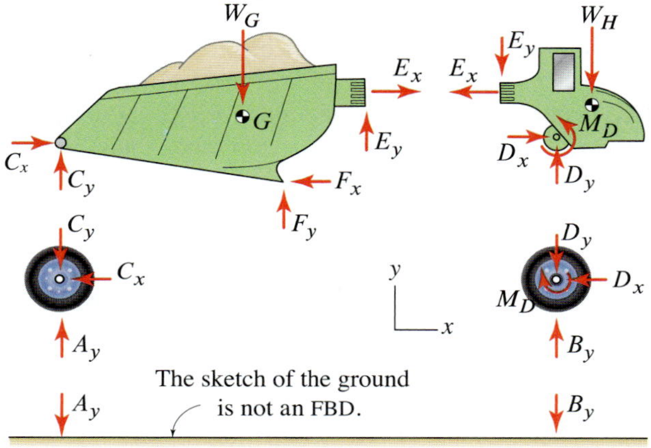

Ratchet pruner. A ratchet pruner for gardening is shown. All connections are pins. If 10 lb forces are applied to the handles of the pruner, draw FBDs of each component.

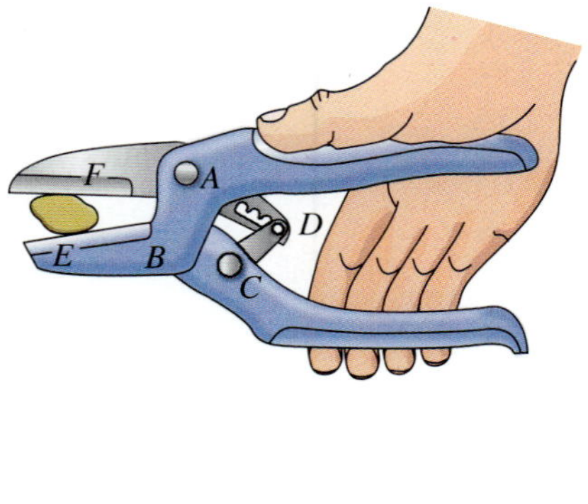

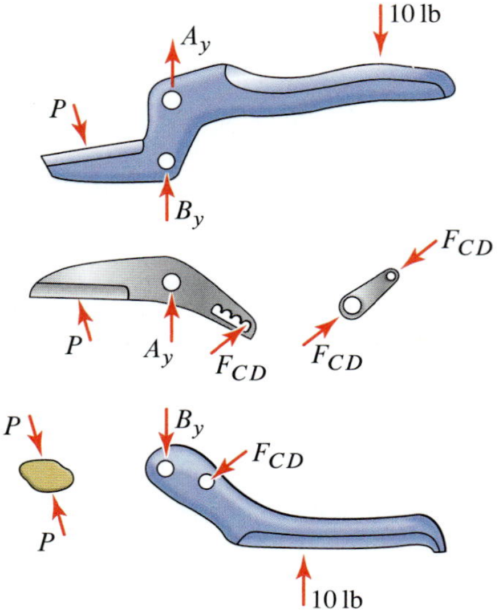

Figure 6.30. Examples of incorrect and/or incomplete FBDs.

and design. The main deficiency shown by the 10 lb forces in these FBDs is that they have different lines of action; if this is the case, then the pruner will tend to rotate due to the moment they produce. Hence, the FBDs must be revised so that the two 10 lb forces have the same line of action.

3. The connection forces at pin A must also include horizontal forces A_x.

4. The connection forces at pin B must also include horizontal forces B_x.

End of Section Summary

In this section, methods of analysis for frames and machines are discussed. Some of the key points are as follows:

- A *frame* is a structure that contains one or more three-force and/or multiforce members. Typically, a frame is fully fixed in space and uses a stationary arrangement of members with the goal of supporting forces that are applied to it.

- A *machine* is an arrangement of members where typically the members can have significant motion relative to one another. The usual goal of a machine is transmission of motion and/or force. Throughout statics, when motion is possible, we assume it is slow enough that inertia can be neglected. When this is not the case, methods of analysis from dynamics must be used.

- When drawing FBDs for frames and machines, it is helpful to first identify two-force members and label their unknowns. Doing this provides for less complicated FBDs, fewer equilibrium equations that need to be written, and fewer unknowns to be determined.

EXAMPLE 6.8 *Analysis of a Bridge*

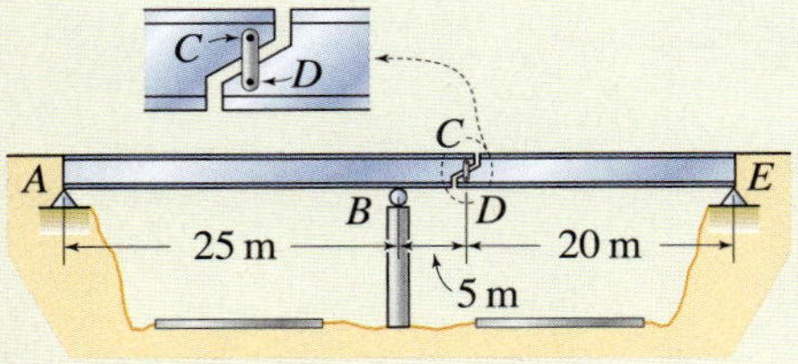

Figure 1

The design shown is common in multispan highway bridges. If the bridge supports a uniform vertical load of 8 kN/m, determine the support reactions and the force supported by the link CD.

SOLUTION

Road Map This bridge has a beam from A to C and a beam from D to E. Because the two beams, when connected by link CD, do not behave as a single body, multiple FBDs will be required.

Modeling The FBDs are shown in Fig. 2 and are constructed as follows. We start by sketching beams AC and DE and link CD, which we assume is vertical. We examine the structure for two-force members and find that link CD is the only two-force member present (disregarding the vertical support column at B). We label the force supported by link CD in the FBDs as T_{CD}, where positive corresponds to tension. Next, we add the support reactions to the FBDs. Finally, beam AC has 30 m length, and therefore the force it supports due to the distributed load is $(8\text{ kN/m})(30\text{ m}) = 240\text{ kN}$, which we place at the center of beam AC. Similarly, beam DE has 20 m length, and therefore the force it supports is $(8\text{ kN/m})(20\text{ m}) = 160\text{ kN}$, which we place at the center of beam DE.

> **Helpful Information**
>
> **Two-force members.** When we draw FBDs, identifying and treating two-force members first will simplify the analysis because all the equilibrium equations for the two-force members are satisfied. For example, the FBD for link CD shown in Fig. 2 satisfies $\sum F_x = 0$, $\sum F_y = 0$, and $\sum M = 0$, leaving only the equations for the remaining FBDs to be written and solved.

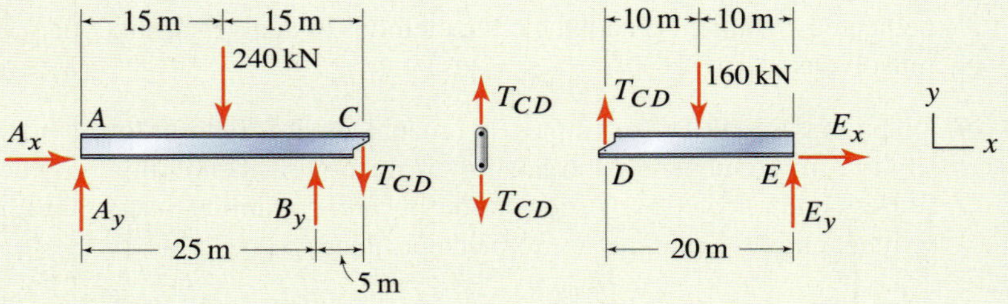

Figure 2. Free body diagrams.

Governing Equations & Computation Examining Fig. 2 shows that beam AC has four unknowns, while beam DE has only three. Thus, we will begin writing equilibrium equations for beam DE first, followed by equations for beam AC.

Member *DE*:

$$\sum F_x = 0: \quad E_x = 0 \quad \Rightarrow \quad \boxed{E_x = 0,} \tag{1}$$

$$\sum M_E = 0: \quad -T_{CD}(20\text{ m}) + (160\text{ kN})(10\text{ m}) = 0 \quad \Rightarrow \quad \boxed{T_{CD} = 80.0\text{ kN},} \tag{2}$$

$$\sum F_y = 0: \quad T_{CD} - 160\text{ kN} + E_y = 0 \quad \Rightarrow \quad \boxed{E_y = 80.0\text{ kN}.} \tag{3}$$

Member *AC*:

$$\sum F_x = 0: \quad A_x = 0 \quad \Rightarrow \quad \boxed{A_x = 0,} \tag{4}$$

$$\sum M_A = 0: \quad -(240\text{ kN})(15\text{ m}) + B_y(25\text{ m}) - T_{CD}(30\text{ m}) = 0 \tag{5}$$

$$\Rightarrow \quad \boxed{B_y = 240\text{ kN},} \tag{6}$$

$$\sum F_y = 0: \quad A_y - 240\text{ kN} + B_y - T_{CD} = 0 \quad \Rightarrow \quad \boxed{A_y = 80.0\text{ kN}.} \tag{7}$$

Discussion & Verification

- **Does the solution appear to be reasonable?** To the extent possible, you should verify that your solution is reasonable. For example, beam DE is simply supported, and with a 160 kN load at its midspan, the solutions $T_{CD} = 80.0\,\text{kN}$ and $E_y = 80.0\,\text{kN}$ given by Eqs. (1) and (2) are clearly correct.
- **Verification of the solution.** You should verify that the solution is mathematically correct by substituting forces into all equilibrium equations to check that each of them is satisfied. However, this check does not verify the accuracy of the equilibrium equations themselves, so it is essential that you draw accurate FBDs and check that your solution is reasonable.

EXAMPLE 6.9 *Analysis of a Basketball Hoop*

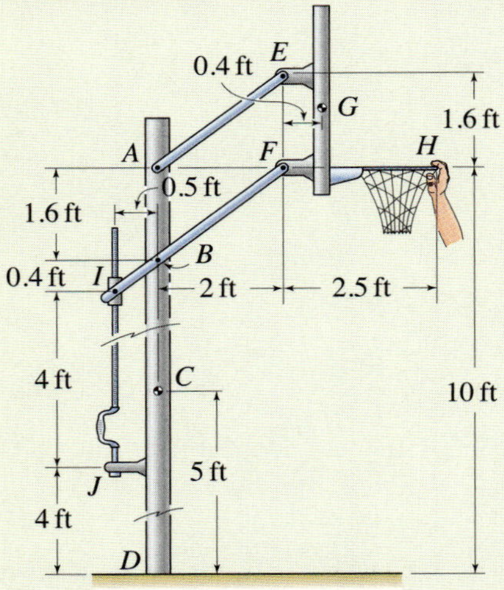

Figure 1

Helpful Information

Hanging on the rim. If the person hanging on the rim is in static equilibrium, then the force applied to the rim is his or her weight. If the person is in motion (e.g., swinging or jumping), then because of inertia, the force applied is usually different than his or her weight, with magnitude and direction that are time-dependent.

A basketball hoop whose rim height is adjustable is shown. The supporting post $ABCD$ weighs 90 lb with center of gravity at point C, and the backboard-hoop assembly weighs 50 lb with center of gravity at point G. The height of the rim is adjustable by means of the screw and hand crank IJ, where the screw is vertical. If a person with 180 lb weight hangs on the rim, determine the support reactions at D and the forces supported by all members.

SOLUTION

Road Map To determine the support reactions at the base of the structure, we first draw an FBD of the entire structure, using a cut that passes through point D only, and we then write and solve the equilibrium equations for this FBD. To determine the forces supported by the individual members, we will draw FBDs of each member followed by writing and solving equilibrium equations for these.

Modeling To determine the support reactions for the structure, we draw an FBD, using a cut that passes through the structure at point D only, as shown in Fig. 2. The 90 and 50 lb weights are vertical forces placed at their respective centers of gravity, and we assume the weights of the other components are negligible. We assume the person hanging on the rim is in static equilibrium and thus applies a 180 lb vertical force at point H.

Governing Equations & Computation Using the FBD in Fig. 2, we write and solve the following equilibrium equations:

$$\sum F_x = 0: \qquad D_x = 0 \quad \Rightarrow \quad \boxed{D_x = 0,} \tag{1}$$

$$\sum F_y = 0: \qquad D_y - 90\,\text{lb} - 50\,\text{lb} - 180\,\text{lb} = 0 \quad \Rightarrow \quad \boxed{D_y = 320\,\text{lb},} \tag{2}$$

$$\sum M_D = 0: \qquad M_D - (50\,\text{lb})(2.4\,\text{ft}) - (180\,\text{lb})(4.5\,\text{ft}) = 0, \tag{3}$$

$$\Rightarrow \quad \boxed{M_D = 930\,\text{ft}\cdot\text{lb}.} \tag{4}$$

Modeling To determine the forces supported by the individual members, we draw FBDs for each member as shown in Fig. 3. These FBDs are constructed by first sketching the members. We then identify any two-force members and label their forces first. In Fig. 3, members AE and IJ are the only two-force members, and we label their forces as T_{AE} and T_{IJ}, respectively. After the FBDs for the two-force members are completed, we add to the remaining FBDs the forces from the two-force members such that Newton's third law is satisfied. That is, we apply force T_{AE} to the post $ABCD$, and to the backboard-hoop, making sure they are applied in directions opposite to the forces on member AE. Similarly, we apply force T_{IJ} to the post $ABCD$ and to member IBF. The FBDs are completed by adding the support reactions at D and adding horizontal and vertical forces at pins where structural members are attached.

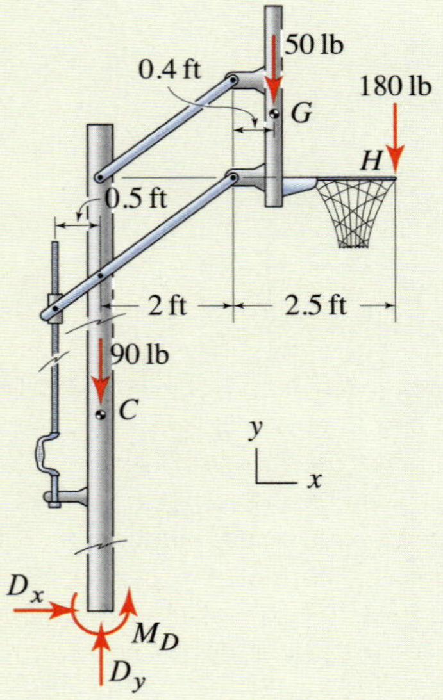

Figure 2
Free body diagram of the entire structure for determining the support reactions.

Governing Equations & Computation By treating two-force members as described here, members AE and IJ automatically satisfy $\sum F_x = 0$, $\sum F_y = 0$, and $\sum M = 0$. Of the three members remaining, the backboard-hoop assembly has the fewest unknowns (three), and thus we begin writing and solving equilibrium equations for it. Using the geometry shown in Fig. 1, the horizontal component of the force supported by member AE is $T_{AE}(2\,\text{ft})/\sqrt{(2\,\text{ft})^2 + (1.6\,\text{ft})^2} = T_{AE}(2/2.561)$, and the vertical component is $T_{AE}(1.6/2.561)$. Thus, equilibrium equations can be written as

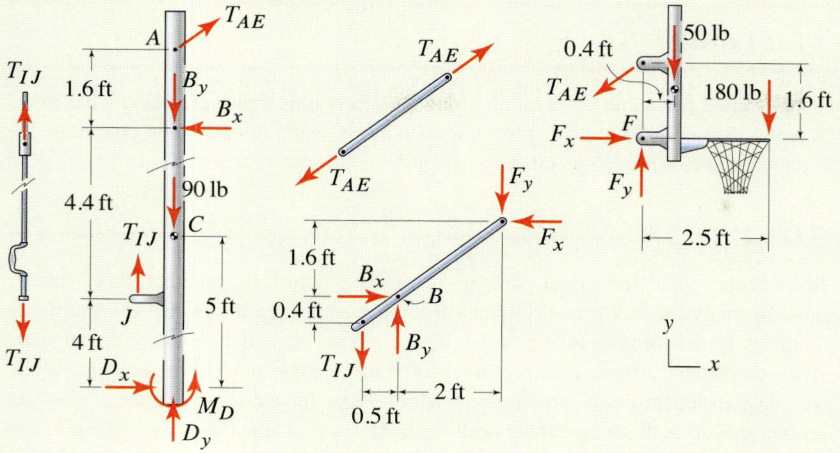

Figure 3. Free body diagrams for determining the forces supported by each member.

Member *EFH*:

$$\sum M_F = 0: \quad T_{AE}\left(\frac{2}{2.561}\right)(1.6\,\text{ft}) - (50\,\text{lb})(0.4\,\text{ft}) - (180\,\text{lb})(2.5\,\text{ft}) = 0, \tag{5}$$

$$\Rightarrow \boxed{T_{AE} = 376.2\,\text{lb},} \tag{6}$$

$$\sum F_x = 0: \quad F_x - T_{AE}\left(\frac{2}{2.561}\right) = 0 \;\Rightarrow\; \boxed{F_x = 293.8\,\text{lb},} \tag{7}$$

$$\sum F_y = 0: \quad F_y - T_{AE}\left(\frac{1.6}{2.561}\right) - 50\,\text{lb} - 180\,\text{lb} = 0 \;\Rightarrow\; \boxed{F_y = 465.0\,\text{lb}.} \tag{8}$$

The FBDs remaining for members $ABCD$ and IBF now have three unknowns each, and either could be considered next. We will use member IBF, and we write

Member *IBF*:

$$\sum M_B = 0: \quad T_{IJ}(0.5\,\text{ft}) + F_x(1.6\,\text{ft}) - F_y(2\,\text{ft}) = 0, \tag{9}$$

$$\Rightarrow \boxed{T_{IJ} = 920.0\,\text{lb},} \tag{10}$$

$$\sum F_x = 0: \quad B_x - F_x = 0 \;\Rightarrow\; \boxed{B_x = 293.8\,\text{lb},} \tag{11}$$

$$\sum F_y = 0: \quad -T_{IJ} + B_y - F_y = 0 \;\Rightarrow\; \boxed{B_y = 1385\,\text{lb}.} \tag{12}$$

With Eq. (12), all of the unknowns have been determined, and it is not necessary to write the equilibrium equations for the post $ABCD$.

Discussion & Verification

- You should check that the solutions appear to be reasonable. In this problem the loading is simple enough that we expect members AE and IJ to be in tension, and indeed our solution shows this. By inspection, the reactions at the base of the post, point D, also have the proper signs.
- In this solution we found the support reactions first. Doing this used three equilibrium equations, and thus it was possible to find all of the remaining unknowns without writing the equilibrium equations for the post $ABCD$. To help verify the solution's accuracy, you could write and solve these equations and you should find that D_x, D_y, and M_D agree with the results in Eqs. (1), (2), and (4).

Helpful Information

Strategy for finding reactions. Rather than begin our solution by finding the support reactions, we could have proceeded directly to the FBDs in Fig. 3, and this eventually would have produced the same reactions. This was the procedure used in Example 6.8 because in that problem it was not possible to determine the support reactions at the outset.

EXAMPLE 6.10 *Analysis of a Ladder*

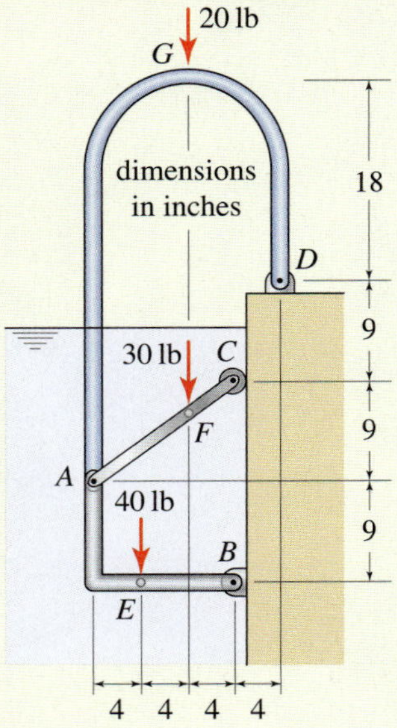

Figure 1

A ladder used by swimmers to climb in and out of a pool is shown. The ladder has rungs at points E and F. Neglecting forces applied by the water to the ladder, determine the forces supported by each member.

SOLUTION

Road Map Since the ladder contains at least one multiforce member, this structure must be analyzed as a frame. In fact, all three members of the ladder are multiforce members (three-force members, to be precise). What is challenging here is that point A is complicated—three members are joined by a single pin at this location. While we could disassemble the structure and draw FBDs for the three members only, for problems such as this where three or more members are supported by a single pin, it is especially helpful to draw a separate FBD for the pin. Hence, four FBDs will be used in our solution.

Modeling We disassemble the structure into its three members plus the pin at A, which leads to the FBDs shown in Fig. 2. These FBDs are constructed as follows. Beginning with member AGD, point D is pin-supported so it has the usual two orthogonal reactions D_x and D_y. In addition, the member is supported by a pin at A; the pin applies forces A_{1x} and A_{1y} to the member, and the member applies these same forces to the pin, but in opposite directions in accord with Newton's third law. Next, member AFC is supported by a roller at C, so there is one reaction that is perpendicular to the wall of the pool. In addition, the member is supported by a pin at A: the pin applies forces A_{2x} and A_{2y} to the member, and the member applies these same forces to the pin, but in opposite directions. The FBD for member AEB is similarly drawn.

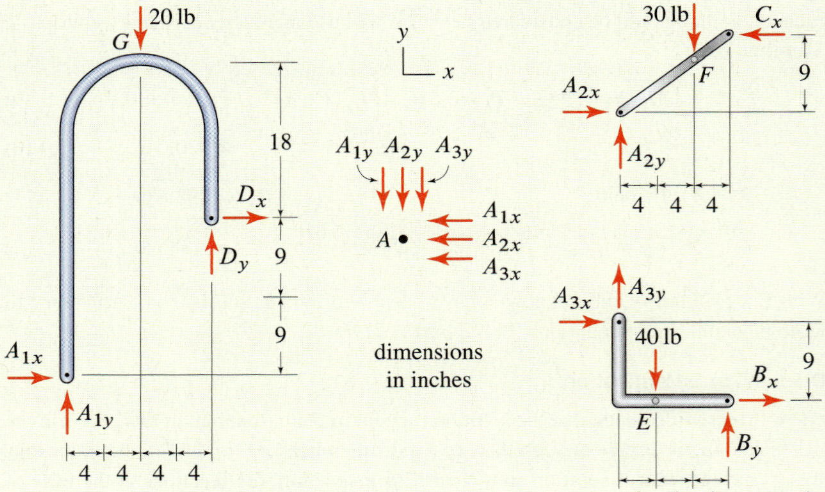

Figure 2. Free body diagrams.

Governing Equations & Computation Examination of the FBDs shows that member AFC has only three unknown forces, and hence we choose this to begin writing equilibrium equations for it. Thus,

Helpful Information

Three or more members supported by a pin. When we draw FBDs of frames and machines where three or more members are supported by a single pin, such as the pin at point A in this example, it is helpful to also draw an FBD of the pin.

Member *AFC*:

$$\sum M_A = 0: \quad -(30\,\text{lb})(8\,\text{in.}) + C_x(9\,\text{in.}) = 0 \quad \Rightarrow \quad \boxed{C_x = 26.67\,\text{lb},} \tag{1}$$

$$\sum F_y = 0: \quad A_{2y} - 30\,\text{lb} = 0 \quad \Rightarrow \quad \boxed{A_{2y} = 30.00\,\text{lb},} \tag{2}$$

$$\sum F_x = 0: \quad A_{2x} - C_x = 0 \quad \Rightarrow \quad \boxed{A_{2x} = 26.67\,\text{lb}.} \tag{3}$$

Examination of the remaining FBDs shows they each contain four unknowns, so there is no preferred order for treating them since a system of coupled algebraic equations will need to be solved. Thus, for the remaining three FBDs, the equilibrium equations are as follows:

Member *AGD*:

$$\sum M_A = 0: \quad -(20\,\text{lb})(8\,\text{in.}) - D_x(18\,\text{in.}) + D_y(16\,\text{in.}) = 0, \tag{4}$$

$$\sum F_x = 0: \quad A_{1x} + D_x = 0, \tag{5}$$

$$\sum F_y = 0: \quad A_{1y} + D_y - 20\,\text{lb} = 0. \tag{6}$$

Member *AEB*:

$$\sum M_A = 0: \quad B_x(9\,\text{in.}) + B_y(12\,\text{in.}) - (40\,\text{lb})(4\,\text{in.}) = 0, \tag{7}$$

$$\sum F_x = 0: \quad A_{3x} + B_x = 0, \tag{8}$$

$$\sum F_y = 0: \quad A_{3y} + B_y - 40\,\text{lb} = 0. \tag{9}$$

Pin A:

$$\sum F_x = 0: \quad -A_{1x} - A_{2x} - A_{3x} = 0, \tag{10}$$

$$\sum F_y = 0: \quad -A_{1y} - A_{2y} - A_{3y} = 0. \tag{11}$$

Equations (4)–(11) are a system of eight equations with eight unknowns. While these may be solved manually, the use of a computer is more effective, and provides

$$\boxed{\begin{array}{llll} B_x = -19.6\,\text{lb} & B_y = 28.0\,\text{lb} & D_x = 46.2\,\text{lb} & D_y = 62.0\,\text{lb} \\ A_{1x} = -46.2\,\text{lb} & A_{1y} = -42.0\,\text{lb} & A_{3x} = 19.6\,\text{lb} & A_{3y} = 12.0\,\text{lb} \end{array}} \tag{12}$$

Discussion & Verification

- **Verification of solution.** You should verify that the solution is mathematically correct by substituting forces into all equilibrium equations to check that each of them is satisfied. However, this check does not verify the accuracy of the equilibrium equations themselves, so it is essential that you draw accurate FBDs and check that your solution is reasonable.
- **Alternate FBDs.** The use of a separate FBD for the pin at A was helpful in this problem. However, other FBDs, provided they are properly drawn, can also be used. For example, if pin A is left on member AGD so that the remaining two members directly apply forces to AGD, then the FBDs are as shown in Fig. 3. You should sketch for yourself the FBDs that result if pin A is left on member AFC, or member AEB.
- **The bigger picture.** The solution outlined here is the first step (a major first step) in fully designing and/or analyzing this pool ladder. In mechanics of materials, a subject that follows statics, you will use the forces determined here to investigate if the ladder is strong enough, if its deformations are acceptable, and so on.

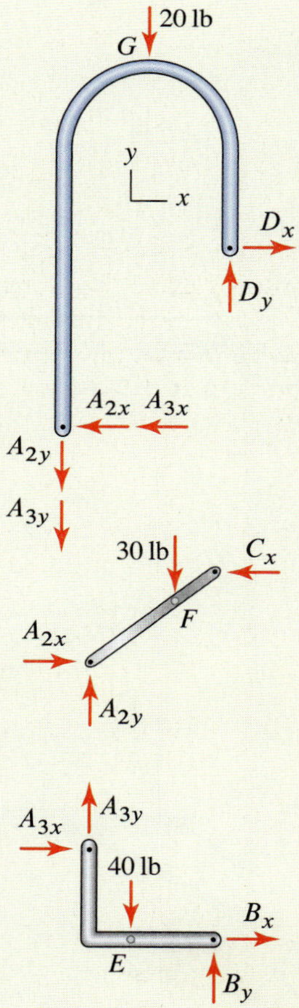

Figure 3
Alternate FBDs where pin A and its forces are left on member AGD.

PROBLEMS

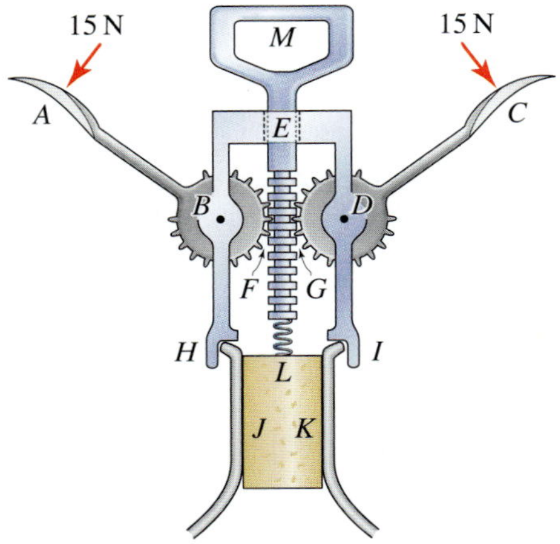

Figure P6.49

Problem 6.49

A cork puller for removing the cork from a bottle is shown. Draw FBDs for all of the members of the cork puller and the cork (i.e., members AB, CD, LM, $HBEDI$, and JKL), labeling all unknowns.

Problem 6.50

The keyboard mechanism for a piano consists of a large number of parts, but for purposes of modeling the transmission of force and motion from the key to the piano wire, the model shown can be used. Draw FBDs for all of the members (i.e., members ABC, DCE, EF, and GFH), labeling all unknowns.

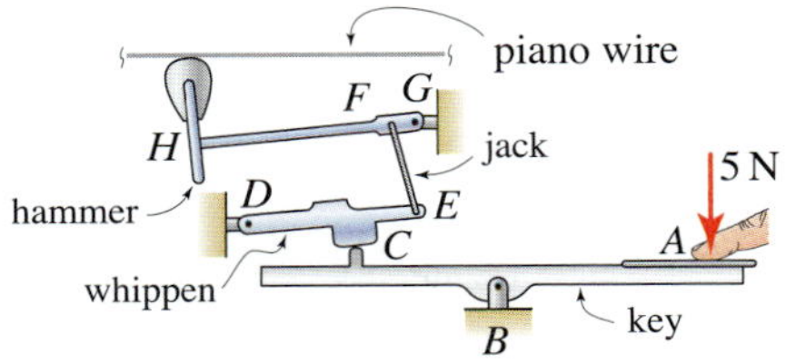

Figure P6.50

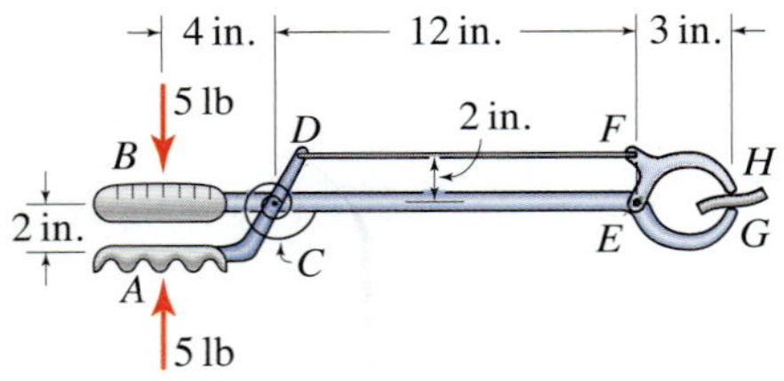

Figure P6.51

Problem 6.51

The tool shown is used for picking up items that otherwise would be out of a person's reach. In the position shown, the torsional spring at C supports a moment of 10 in.·lb which tends to open the jaws of the tool. Neglect the weights of the individual members and neglect friction at points H and G. Determine the forces supported by all members of the tool and the force applied by the tool to the item being gripped at H and G.

Problem 6.52

Compare the designs for the pickup tools shown in Fig. P6.51 and in Fig. 6.29 on p. 374. If the dimensions of the members in these tools are approximately the same, speculate on which of these tools will be capable of applying the greater force to the item being gripped at H and G before one of the members of the tool fails.
Note: Concept problems are about *explanations*, not computations.

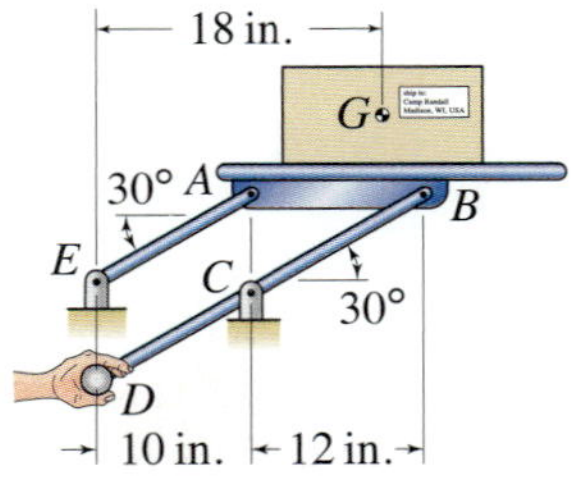

Figure P6.53 and P6.54

Problems 6.53 and 6.54

The elevation and tilt angle of table AB are controlled by an operator's hand at point D. The table supports a box weighing 10 lb with center of gravity at point G. Determine the force the operator must apply to keep the table in equilibrium, and determine the force supported by all members of the machine.

Problem 6.53 The force applied by the operator is vertical.

Problem 6.54 The force applied by the operator is horizontal.

Problems 6.55 and 6.56

The elevation and tilt angle of table AB are controlled by an operator's hand at point D. The table supports a box weighing 40 N with center of gravity at point G. Determine the force the operator must apply to keep the table in equilibrium, and determine the force supported by all members of the machine.

Problem 6.55 The force applied by the operator is vertical.

Problem 6.56 The force applied by the operator is horizontal.

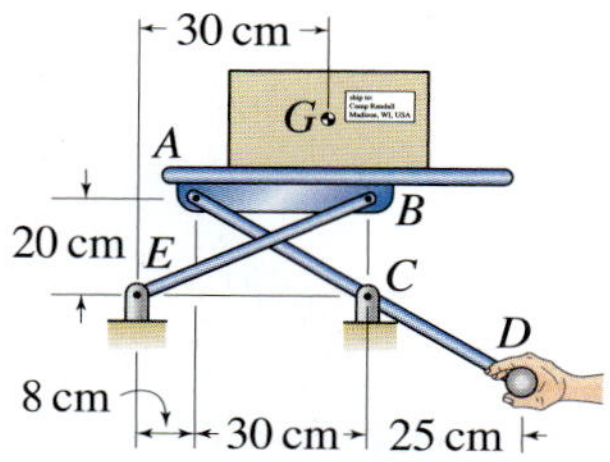

Figure P6.55 and P6.56

Problem 6.57

A hydraulically powered lift supports the two boxes shown. Each box weighs 200 lb with center of gravity in the middle of each box. Neglect the weight of individual members.

(a) Draw four FBDs, one each for members AEF, ABC, ED, and BE, labeling all forces.

(b) Determine the force in the hydraulic actuator BE required to keep the lift in equilibrium.

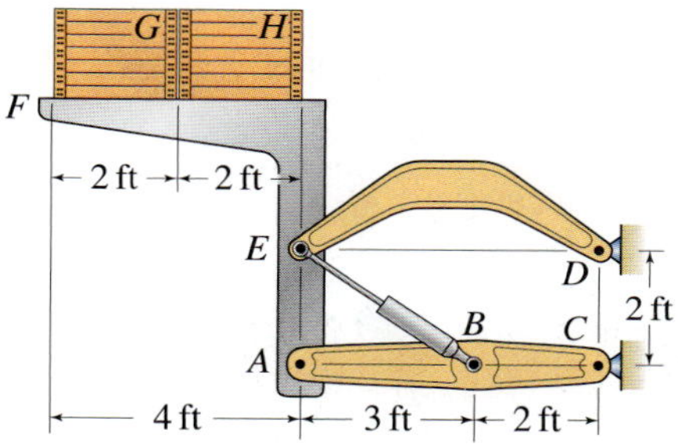

Figure P6.57

Problem 6.58

The shovel of an end loader has pins at points A, B, C, and D. The scoop supports a downward vertical load of 2000 lb, which is not shown in the figure, at point E.

(a) Draw four FBDs, one each for parts AB and AC, hydraulic cylinder AD, shovel CDE, and member BD, labeling all forces.

(b) Determine the force the hydraulic cylinder AD must generate to keep the shovel in equilibrium.

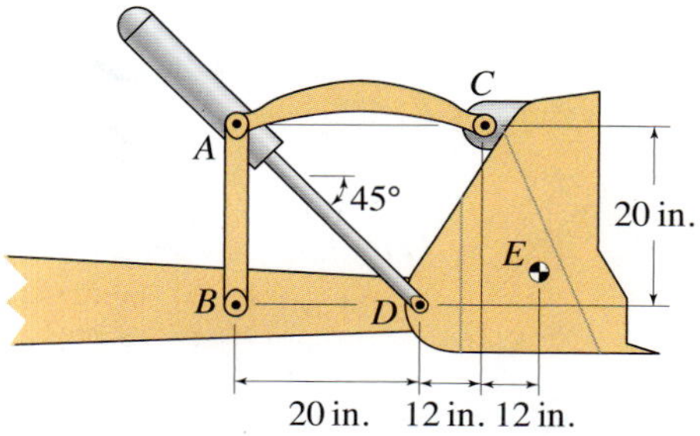

Figure P6.58

Problem 6.59

A backhoe is shown. Hydraulic cylinders GF and BC are horizontal and vertical, respectively. Neglect the weights of members.

(a) Draw six FBDs, one each for members DEF and FCA, shovel AB, and hydraulic cylinders BC, CE, and FG.

(b) If each of the three hydraulic cylinders is capable of producing a 1000 lb force (in both tension and compression), determine the largest weight W that can be supported in the position shown.

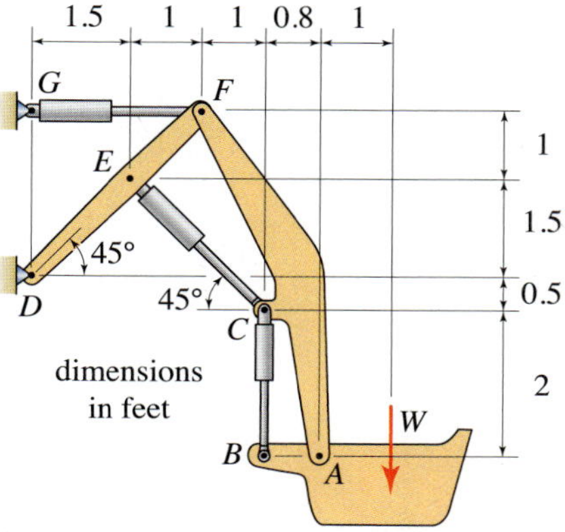

Figure P6.59

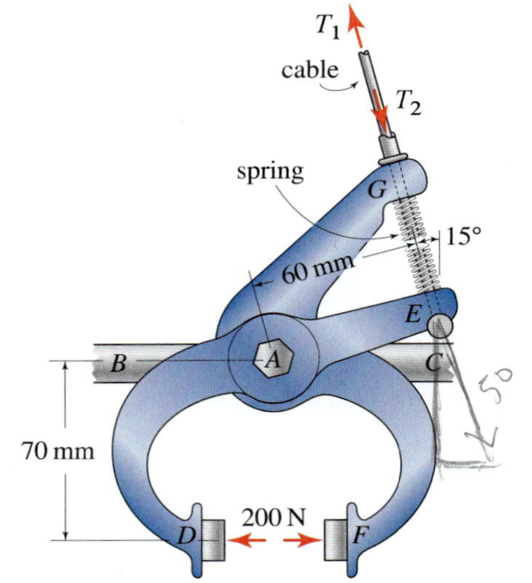

Figure P6.60

Problem 6.60

The hand brake for a bicycle is shown. Portions DE and FG are free to rotate on bolt A which is screwed into the frame BC of the bicycle. The brake is actuated by a shielded cable where T_1 is applied to point E and T_2 is applied to point G. A spring having 50 N compressive force is placed between points E and G so that the brake stays open when it is not being used. Assume the change in the spring's force is negligible when the brake is actuated to produce the 200 N forces at points D and F.

(a) Draw three FBDs, one each for DE and FG and bolt A, labeling all forces.

(b) Determine the necessary cable forces T_1 and T_2.

(c) Determine the forces exerted by DE and FG on bolt A.

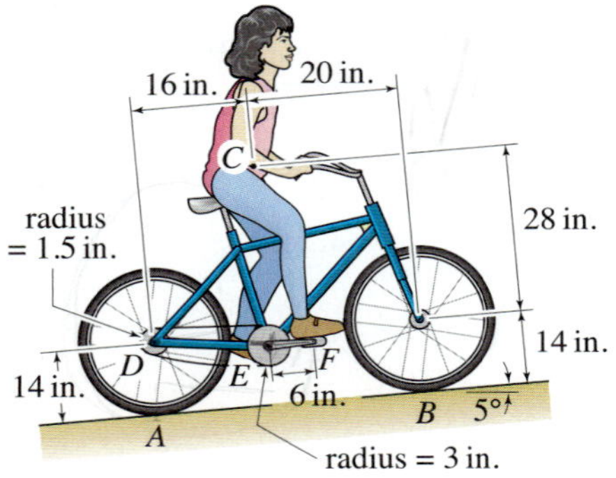

Figure P6.61

Problem 6.61

A bicycle is pedaled up a gentle incline. The rider and bicycle weigh 120 lb with center of gravity at point C. For the position shown, determine the force the rider must apply to the pedal at F for the bicycle to move at constant speed. Assume the rider applies a force to the pedals using her right foot only, the force applied by the rider to pedal F is perpendicular to the crank EF, and the lower portion of the chain between sprockets D and E is slack.

Problem 6.62

The linkage shown is used on a garbage truck to lift a 2000 lb dumpster. Points A–G are pins, and member ABC is horizontal. This linkage has the feature that as the dumpster is lifted and rotated, the front edge A of the dumpster is simultaneously lowered, placing it at a more convenient height for emptying.

(a) Draw five FBDs, one each for members ABC, CD, and EDG, dumpster H, and hydraulic cylinder CF, labeling all forces.

(b) For the position shown, when the dumpster just fully lifts off the ground, determine the force in hydraulic cylinder CF.

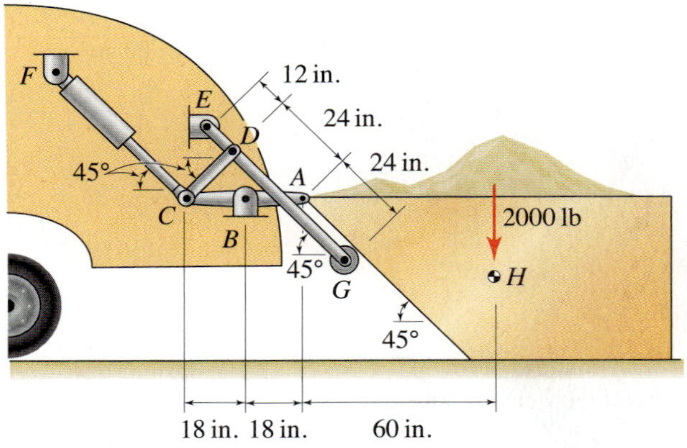

Figure P6.62

Problem 6.63

An end loader for a small garden tractor is shown. All connections are pinned. The only significant weight is W.

(a) Draw six FBDs, one each for members BC, DG, and AEF, plate CDE, hydraulic cylinder EG, and shovel AB, labeling all forces.

(b) If the hydraulic cylinder EG is capable of producing a 3000 lb force, determine the largest weight W that may be lifted in the position shown.

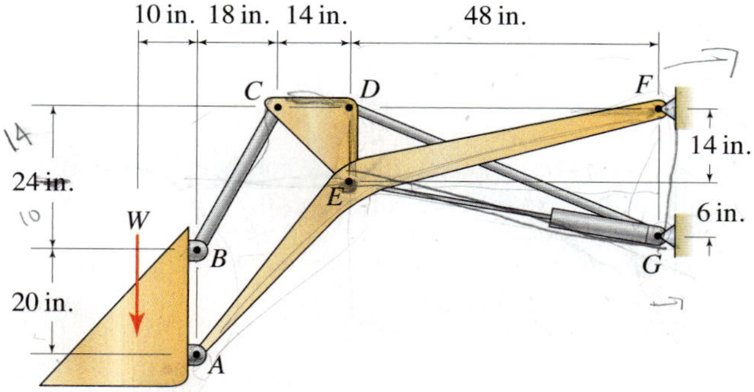

Figure P6.63

Problem 6.64

A prosthetic arm and hand assembly is shown. Points B and H are fixed to the arm. The hand is actuated by a pneumatic cylinder BF that opens and closes the hand. The spring FH helps keep the hand in alignment with the arm. If the person holds a bag of groceries that weighs 30 N and grips the bag at A and D with a 10 N force, determine the forces supported by the pneumatic cylinder and the spring. Assume that while holding the bag of groceries, the arm and hand have the geometry shown where the pneumatic cylinder is horizontal.

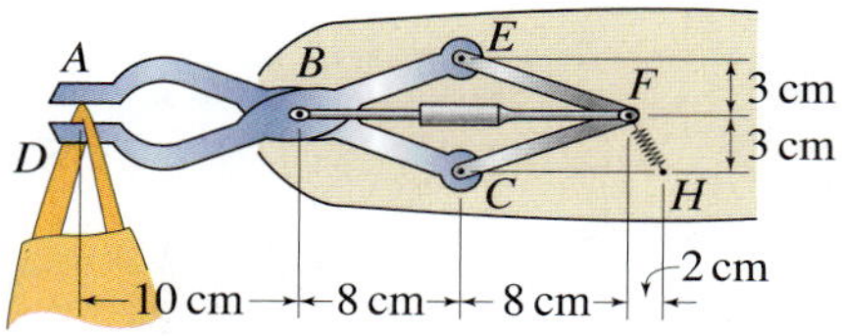

Figure P6.64

Problem 6.65

Three common designs for a bridge to span a multilane highway are shown. Design 1 is the same as that considered in Example 6.8 on p. 378 where link CD is used to connect the two spans of the bridge. Design 2 uses two simply supported I beams. Design 3 uses a single I beam from points A to E with a roller support at midspan. Comment on the pros and cons of each of these designs, considering issues such as the following:

- Which are straightforward to analyze for determining the support reactions and which may be difficult to analyze?
- Which may be easier to construct?
- Which may be more sensitive to unexpected support motion, such as if the support at B settles with time?
- Which may be safer, such as if the support at B were accidentally struck by a vehicle passing under the bridge?

Note: Concept problems are about *explanations*, not computations.

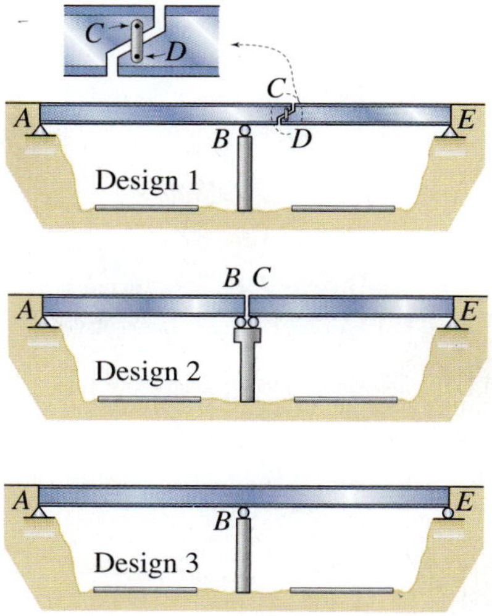

Figure P6.65

6.5 Chapter Review

Important definitions, concepts, and equations of this chapter are summarized. For equations and/or concepts that are not clear, you should refer to the original equation and page numbers cited for additional details.

Trusses. A *truss* is a structure that consists of two-force members only, where members are organized so that the assemblage as a whole behaves as a single object. For a two-dimensional structure, for all members to be two-force members, the structure must have the following characteristics:

- All members must be connected to one another by frictionless pins, and the locations of these pins are called *joints*.
- Each member may have no more than two joints.
- Forces may be applied at joints only.
- The weight of individual members must be negligible.

A structure having the above characteristics is called a *plane truss*.

Method of joints. In the *method of joints*, the force supported by each member of a truss is determined by drawing FBDs for each joint, and then requiring that each joint be in equilibrium by writing the equations $\sum F_x = 0$ and $\sum F_y = 0$ for each joint.

Method of sections. In the *method of sections*, the force supported by members of a truss is determined by taking a cut through the structure, and the FBD that results is required to be in equilibrium by writing the equations $\sum F_x = 0$, $\sum F_y = 0$, and $\sum M = 0$.

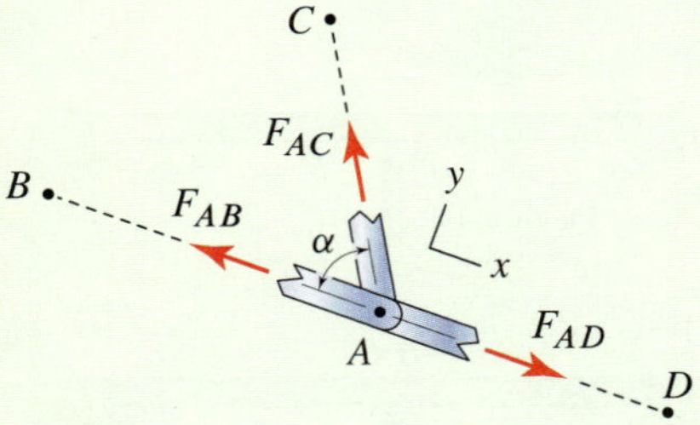

Figure 6.31
Geometry of members in a truss allowing a zero-force member to be recognized by inspection.

Zero-force members. A truss member (or any member) that supports no force is called a *zero-force member*. Zero-force members in a truss can be identified by finding joints that match the pattern shown in Fig. 6.31. If

- A particular joint has three members connected to it,
- Two of these members are collinear, and
- The joint has no external force applied to it,

then the noncollinear member at that joint is a zero-force member.

Zero-force members can be very effective in improving the overall strength of a truss by reducing the length of compression members and thus improving their resistance to buckling.

Statically determinate and indeterminate trusses. In a *statically determinate* truss, the equations of equilibrium are sufficient to determine the forces supported by all members of the truss and the support reactions. In a *statically indeterminate* truss, the equations of equilibrium are not sufficient to determine all of these. A simple rule of thumb to help determine whether a truss is statically determinate or indeterminate is to compare the number of unknowns to the number of equilibrium equations, and we call this *equation counting*. For a plane truss having

Eq. (6.21), p. 349

m = number of members,
r = number of support reactions,
j = number of joints,

the rule of thumb is as follows:

Eq. (6.22), p. 349

If $m + r < 2j$	The truss is a mechanism and/or has partial fixity.
If $m + r = 2j$	The truss is statically determinate if it has full fixity. The truss is statically indeterminate if it has partial fixity.
If $m + r > 2j$	The truss is statically indeterminate, and it can have full fixity or partial fixity.

Successful use of equation counting requires good judgment on your part, which is why it is called a rule of thumb rather than a rule.

Simple, compound, and complex trusses. A *simple truss* has members arranged in a triangular pattern, as shown in Fig. 6.16 on p. 351. *Compound* and *complex* trusses consist of two or more simple trusses connected to one another. While simple trusses (and compound and complex trusses) are popular, other designs for trusses are also common and can perform very well. One of the main features of simple trusses (and compound and complex trusses) is that they are always stable.

Buckling. Because truss members are usually straight and slender, compression members are susceptible to *buckling*. Furthermore, the force at which buckling occurs in a straight member decreases very rapidly as a member becomes longer (the buckling load is proportional to $1/L^2$ where L is the length of the member). Thus, it is undesirable to have long compression members in a truss.

Space trusses. A *space truss* is a three-dimensional structure that consists of two-force members only, where members are organized so that the assemblage as a whole behaves as a single object. The comments made earlier regarding the requirements for two-force members in a plane truss also apply, except in a space truss, members must be connected to one another by ball-and-socket joints. Space trusses can be analyzed using the *method of joints* and the *method of sections*, and the *equation-counting* rule of thumb can also be used, except each joint has three equilibrium equations.

Frames and machines. Frames and machines are defined as follows, along with some tips for drawing FBDs.

Frame. A *frame* is a structure that contains one or more three-force and/or multiforce members. Typically, a frame is fully fixed in space and uses a stationary arrangement of members with the goal of supporting forces that are applied to it.

Machine. A *machine* is an arrangement of members where typically the members can have significant motion relative to one another. The usual goal of a machine is transmission of motion and/or force. Throughout statics when motion is possible, we assume it is slow enough that inertia can be neglected. When this is not the case, methods of analysis from dynamics must be used.

FBDs. When we draw FBDs for frames and machines, it is helpful to first identify two-force members and label their unknowns. Doing this provides for less complicated FBDs, fewer equilibrium equations that need to be written, and fewer unknowns to be determined.

REVIEW PROBLEMS

Problem 6.66

Member AGB is a single member with pin connections at points A, G, and B. Similarly, member BHC is a single member with pin connections at points B, H, and C. All other connections are also pins.

(a) Is this structure a truss? Explain.

(b) Determine the forces supported by all truss members.

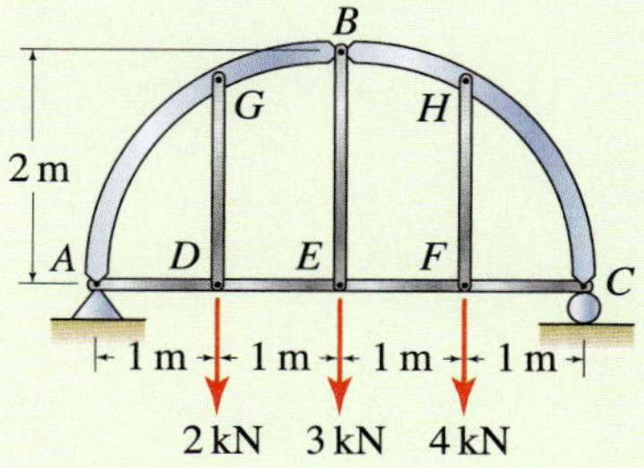

Figure P6.66

Problem 6.67

(a) Determine if the truss is statically determinate or indeterminate.

(b) By inspection, identify the zero-force members in the truss.

(c) Regardless of your answer to Part (a), determine the force supported by member QI.

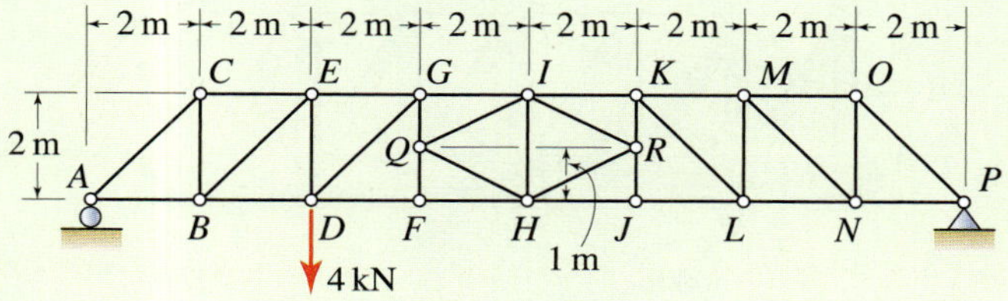

Figure P6.67

Problem 6.68

(a) Determine if the truss is statically determinate or indeterminate.

(b) By inspection, identify the zero-force members in the truss.

(c) Regardless of your answer to Part (a), determine the force supported by member DG.

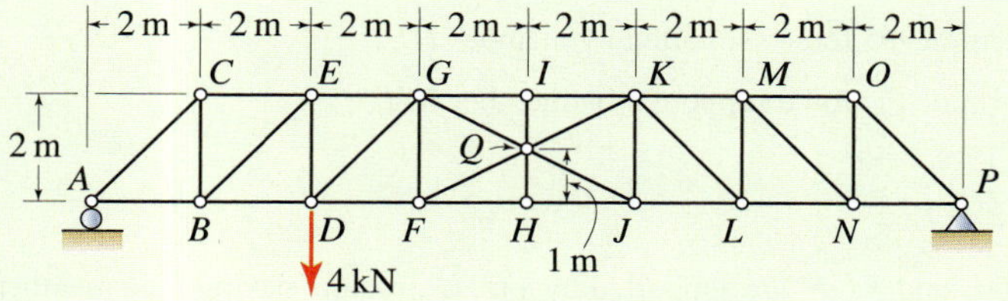

Figure P6.68

Problem 6.69

Compared to the truss shown in Fig. P6.69(a):

- The truss in (b) subdivides members AB and BC in half.
- The truss in (c) subdivides members AB, BC, AD, and DC in half.
- The truss in (d) subdivides members AB and BC into thirds.

Consider failure due to in-plane buckling only. If the truss in Fig. P6.69(a) fails when $P = 1440$ lb, specify (if possible) the values of P at which the trusses in Fig. P6.69(b), (c), and (d) will fail.

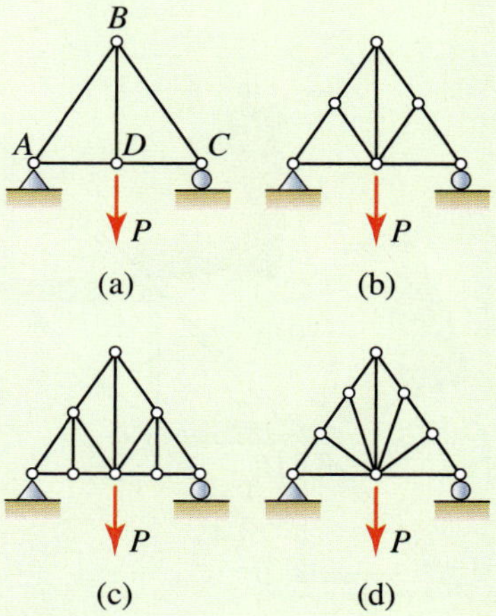

Figure P6.69

Problem 6.70

The boom hangs in the vertical xz plane and is supported by a socket at point N and cables OB and BP. The bottom flange of the boom is equipped with a roller system so that a load can be moved along its length. Even though the top and bottom flanges of the boom are continuous members, idealize the structure as a truss, and determine the force supported by the two cables and member CD when the 600 lb load is positioned at point C.

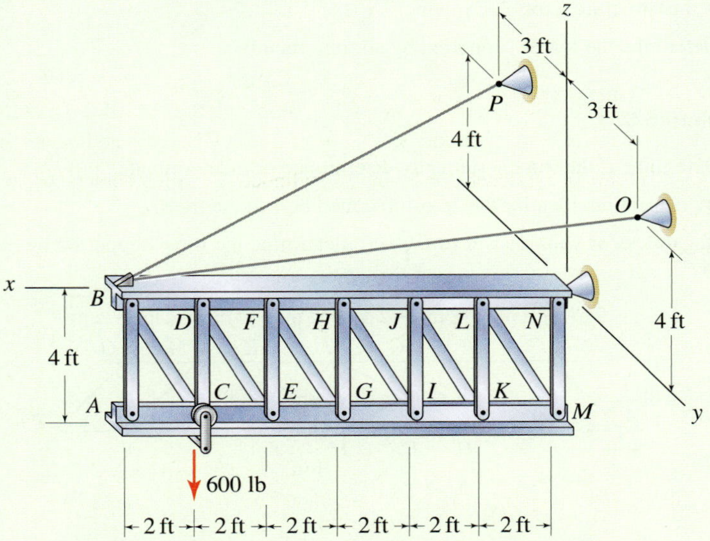

Figure P6.70

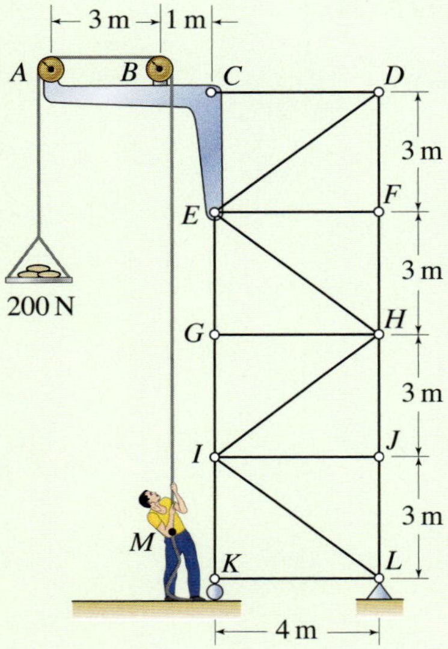

Figure P6.71

Problem 6.71

A hoist for lifting building materials is attached to a scaffold. The hoist has frictionless pulleys at points A and B, and both pulleys have 300 mm radius.

(a) By inspection, identify the zero-force members in the truss.

(b) Determine the force supported by member DE.

(c) Determine the force supported by member EH.

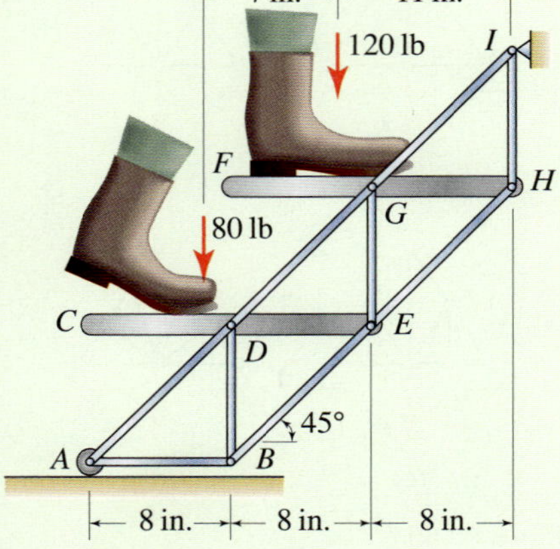

Figure P6.72

Problem 6.72

Steps CDE and FGH are supported by a truss structure having nine members. Determine the forces supported by all nine members of the truss.

Problem 6.73

A lifting platform is supported by members JKF and LH and hydraulic cylinder KL. All connections are frictionless pins.

(a) Determine the force the hydraulic cylinder must support for equilibrium.

(b) Use the method of joints to determine the force supported by member GH.

(c) Use the method of sections to determine the force supported by member BC.

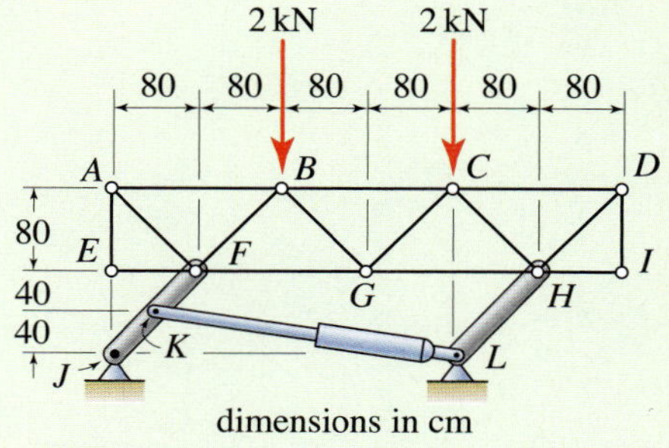

Figure P6.73

Problem 6.74

A small trailer-mounted dumper is shown. All connections are pinned and member EF is horizontal. The hydraulic cylinder simultaneously tilts the dump and opens the gate.

(a) Draw five FBDs, one each for hydraulic cylinder AE, members BDE and EF, gate FG, and the dump, labeling all forces.

(b) Determine the force supported by hydraulic cylinder AE.

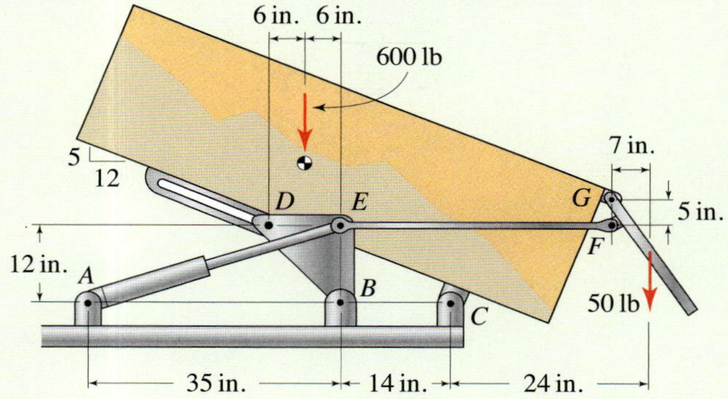

Figure P6.74

7 Centroids and Distributed Force Systems

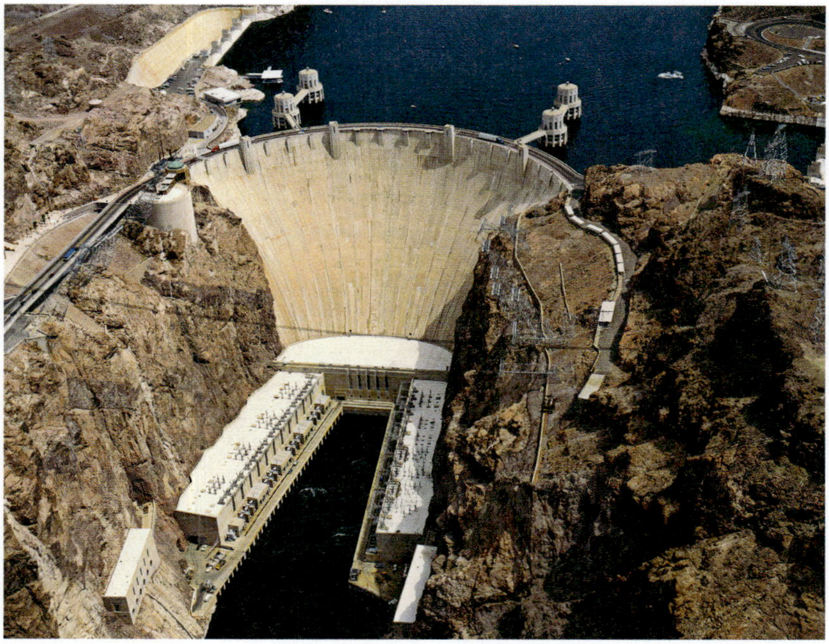

This chapter begins with discussions of centroid, center of mass, and center of gravity. The common element among these is that centroid, center of mass, and center of gravity represent *average positions* of distributions. Applications of these concepts to distributed force systems, fluid pressure loading, and gas pressure loading are then considered.

7.1 Centroid

Introduction – center of gravity

While centroids are the focus of this section, it is more intuitive to begin with a short discussion dealing with center of gravity.

Consider the chapter opening photo, shown on this page, of Hoover Dam, located on the Colorado River on the border between Arizona and Nevada. The dam is made primarily of concrete, but also includes steel piping, turbine generators, and so on. Every atom of material in the dam has a weight due to gravity, and in aggregate, these yield the total weight of the dam, which of course is quite substantial. The dam retains water in the reservoir, and this water applies pressure to the dam whose direction and intensity vary with every point on the face of the dam. In aggregate, this water pressure gives rise to a substantial force. For many analysis and design purposes,

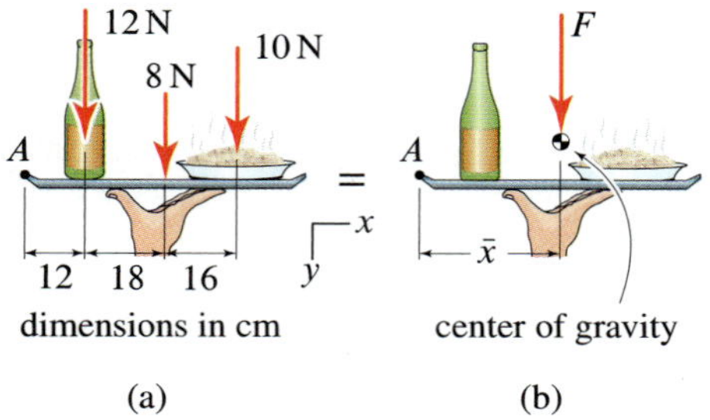

Figure 7.1
(a) A tray with wine and pasta. (b) Construction of an equivalent force system provides the weight of the collection of objects and the location of this weight, which is called the *center of gravity*.

it is sufficient to know only the total weight of the dam and the total force exerted by the water on the dam. Indeed, each of these constitutes an equivalent force system—that is, part of an equivalent force system. The remaining information needed is the location of the line of action for the weight, called the *center of gravity*, and the location of the line of action for the fluid force, sometimes called the *center of pressure*. Among the topics discussed in this chapter are methods for determining the center of gravity and the center of pressure. Thus, the concepts of this chapter build on and enhance the concepts of equivalent force systems discussed in Chapter 4. However, the concepts of this chapter go beyond forces and equivalent force systems. For example, the centroid of an area is important in governing the strength of a beam, and while the centroid is mathematically similar to the center of gravity, it is nonetheless very different. The underlying concept of this chapter is determination of the average position of a distribution, where the distribution may be shape, mass, weight, fluid pressure, gas pressure, and so on.

We may quantify some of the ideas discussed above by considering a simple example, as shown in Fig. 7.1(a), where a waiter at a restaurant brings you wine and pasta on a tray. The center of gravity is defined to be the average position of the weight distribution of the objects the waiter carries. If the waiter is any good at his job, he will be sure that the center of gravity of the tray, wine, and pasta is positioned directly over the hand he uses to carry these. If this is not the case, then the tray will fall as he attempts to carry it. In Fig. 7.1(a), the weights of the wine, tray, and pasta are 12, 8, and 10 N, respectively, with the positions shown. We would like to replace these weights with a single force, which is to be located at the center of gravity for the collection of objects. Determination of the force and its position (the center of gravity) can be accomplished by constructing an equivalent force system, as shown in Fig. 7.1(b). Using Eq. (4.16) on p. 223 for constructing an equivalent force system provides

$$\sum (F_y)_{\text{system 1}} = \sum (F_y)_{\text{system 2}} \tag{7.1}$$

$$12\,\text{N} + 8\,\text{N} + 10\,\text{N} = F \tag{7.2}$$

$$\Rightarrow \quad F = 30\,\text{N}, \tag{7.3}$$

$$\sum (M_A)_{\text{system 1}} = \sum (M_A)_{\text{system 2}} \tag{7.4}$$

$$(12\,\text{N})(12\,\text{cm}) + (8\,\text{N})(30\,\text{cm}) + (10\,\text{N})(46\,\text{cm}) = F\bar{x} \tag{7.5}$$

$$\Rightarrow \quad \bar{x} = 28.1\,\text{cm}. \tag{7.6}$$

The force $F = 30\,\text{N}$ represents the entire weight of the collection of objects, and the center of gravity $\bar{x} = 28.1\,\text{cm}$ is the location of the line of action for this force. The important idea here is that the *center of gravity* is the *average position of the weight distribution*.

Equations (7.1)–(7.6) can be generalized to give the center of gravity for a collection of an arbitrary number of objects as

$$\bar{x} = \frac{\sum_{i=1}^{n} \tilde{x}_i W_i}{\sum_{i=1}^{n} W_i} \tag{7.7}$$

where n is the number of objects, W_i is the weight of object i, and $\tilde{x}_i$ is the location of W_i (i.e., the moment arm) measured from the origin of the coordinate system. In Eq. (7.7), we use the superposed tilde ~ to emphasize that $\tilde{x}_i$ is the position of the center of gravity of object i, and this notation will help avoid confusion with the coordinate x.

In Eq. (7.7), the products $W_i \tilde{x}_i$ are called the *first moments* of the weights W_i. The concept underlying Eq. (7.7) can be generalized so that by evaluating the first moment of any distribution, the average position of that distribution may be determined. For example, by replacing W_i in Eq. (7.7) with the mass m_i of each object, $\bar{x}$ then measures the average position of the mass for a collection of objects, and this is called the *center of mass*. Similarly, by replacing W_i in Eq. (7.7) with the volume V_i of each object, $\bar{x}$ then measures the average position for the volume of a collection of objects, and this is called the *centroid*.

Concept Alert

Centroid, etc. *Centroid*, *center of mass*, and *center of gravity* all measure the average position of distributions. Furthermore, the calculation procedures for determining these are almost identical.

- The *centroid* is the average position of a distribution of shapes.
- The *center of mass* is the average position of a distribution of mass.
- The *center of gravity* is the average position of a distribution of weight.

Centroid of an area

The *centroid* is defined to be the average position of a distribution of shapes. If the distribution consists of a single shape, the shape can be a line (straight or curved), an area, or a volume. For a distribution (or collection) of shapes, the distribution can include multiple lines, or multiple areas, or multiple volumes—but not combinations of these. For the present, we consider the area shown in Fig. 7.2. By considering the area to be a collection of composite areas A_i, where the centroid of each of these is located at $\tilde{x}_i$ and $\tilde{y}_i$, the position of the centroid for the entire area, $\bar{x}$ and $\bar{y}$, can be determined by generalizing Eq. (7.7) to obtain

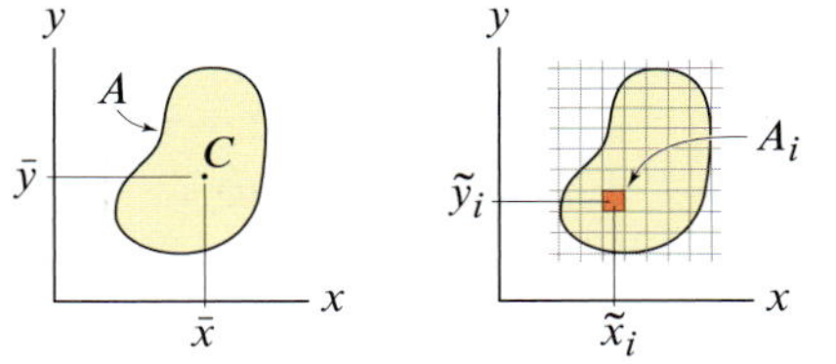

Figure 7.2
The centroid C of area A is located at $\bar{x}$ and $\bar{y}$. Area A can be considered as consisting of composite areas A_i, where the centroid of each of these is located at $\tilde{x}_i$ and $\tilde{y}_i$.

$$\bar{x} = \frac{\sum_{i=1}^{n} \tilde{x}_i A_i}{\sum_{i=1}^{n} A_i} \quad \text{and} \quad \bar{y} = \frac{\sum_{i=1}^{n} \tilde{y}_i A_i}{\sum_{i=1}^{n} A_i}, \tag{7.8}$$

where n is the number of composite shapes that constitute the entire area, A_i is the area of composite shape i, and $\tilde{x}_i$ and $\tilde{y}_i$, are the locations of the centroid of A_i measured from the origin of the coordinate system. In Eq. (7.8), the numerator $\sum \tilde{x}_i A_i$ is called the *first moment of the area about the y axis*, and $\sum \tilde{y}_i A_i$ is called the *first moment of the area about the x axis*.

By taking the limits of Eq. (7.8) as $A_i \to 0$, the summations become integrals and Eq. (7.8) becomes

$$\bar{x} = \frac{\int \tilde{x}\, dA}{\int dA} \quad \text{and} \quad \bar{y} = \frac{\int \tilde{y}\, dA}{\int dA}, \tag{7.9}$$

where $\tilde{x}$ and $\tilde{y}$, are the locations of the centroid of area element dA measured from the origin of the coordinate system. In Eq. (7.9), the numerator $\int \tilde{x}\, dA$ is called the *first moment of the area about the y axis*, and $\int \tilde{y}\, dA$ is called the *first moment of the area about the x axis*.

Centroid of a line

Consider the line shown in Fig. 7.3. By considering the line to be a collection of composite lines with lengths L_i, where the centroid of each of these is

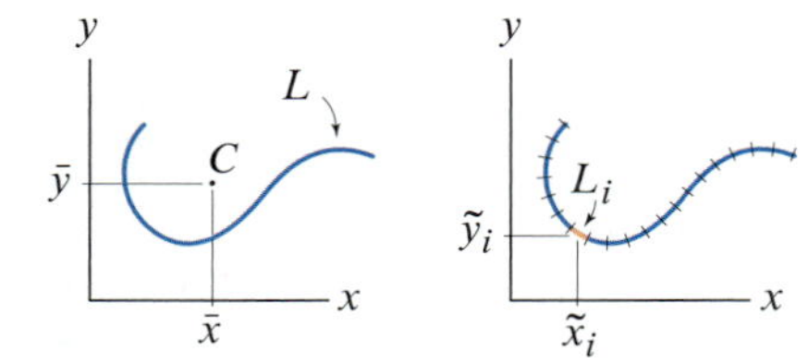

Figure 7.3
The centroid C of a line with length L is located at $\bar{x}$ and $\bar{y}$. Length L can be considered as consisting of composite lengths L_i, where the centroid of each of these is located at $\tilde{x}_i$ and $\tilde{y}_i$.

located at $\tilde{x}_i$ and $\tilde{y}_i$, the position of the centroid for the entire line, $\bar{x}$ and $\bar{y}$, is given by

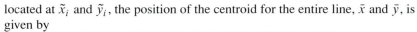

$$\bar{x} = \frac{\sum_{i=1}^{n} \tilde{x}_i L_i}{\sum_{i=1}^{n} L_i} \quad \text{and} \quad \bar{y} = \frac{\sum_{i=1}^{n} \tilde{y}_i L_i}{\sum_{i=1}^{n} L_i}, \tag{7.10}$$

where n is the number of composite lines that constitute the entire length, L_i is the length of composite line i, and $\tilde{x}_i$ and $\tilde{y}_i$ are the locations of the centroid of L_i measured from the origin of the coordinate system.

By taking the limits of Eq. (7.10) as $L_i \to 0$, this equation becomes

$$\bar{x} = \frac{\int \tilde{x}\, dL}{\int dL} \quad \text{and} \quad \bar{y} = \frac{\int \tilde{y}\, dL}{\int dL}, \tag{7.11}$$

where $\tilde{x}$ and $\tilde{y}$ are the locations of the centroid of length element dL measured from the origin of the coordinate system.

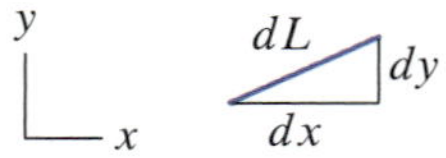

Figure 7.4
Length increment dL is related to increments dx and dy by the Pythagorean theorem.

Integration along a path. In Eq. (7.11) (and Eq. (7.18) later in this section), integrations are to be carried out along the path of a line. Such integrals are often called *line integrals* (or *path integrals*, or *curve integrals*). Rather than integrate along the path of the line, it will often be more convenient to integrate with respect to x or y, and to do this, a transformation between dL and dx or dy is needed. As shown in Fig. 7.4, length increment dL is related to increments dx and dy by the Pythagorean theorem:

$$dL = \sqrt{(dx)^2 + (dy)^2}. \tag{7.12}$$

Multiplying the right-hand side of Eq. (7.12) by dx/dx and bringing $1/dx$ within the radical give

$$dL = \sqrt{1 + \left(\frac{dy}{dx}\right)^2}\, dx. \tag{7.13}$$

Similarly, multiplying the right-hand side of Eq. (7.12) by dy/dy and bringing $1/dy$ within the radical give

$$dL = \sqrt{\left(\frac{dx}{dy}\right)^2 + 1}\, dy. \tag{7.14}$$

Rather than memorize Eqs. (7.13) and (7.14), it is perhaps easier to simply remember Eq. (7.12) along with multiplication by dx/dx or dy/dy. Example 7.5 on p. 407 illustrates the use of Eq. (7.13).

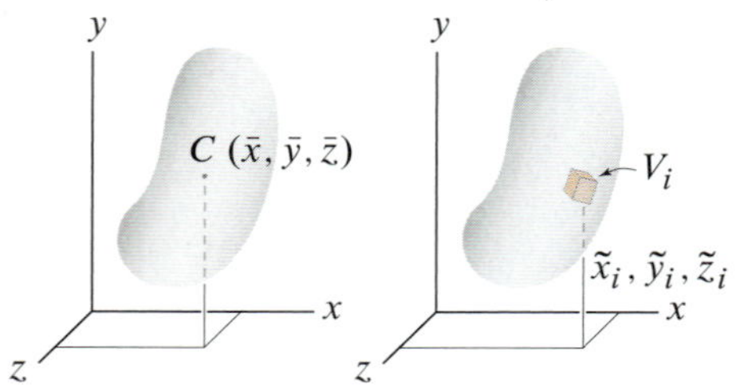

Figure 7.5
The centroid C of a volume V is located at $\bar{x}$, $\bar{y}$, and $\bar{z}$. Volume V can be considered as consisting of composite volumes V_i, where the centroid of each of these is located at $\tilde{x}_i$, $\tilde{y}_i$, and $\tilde{z}_i$.

Centroid of a volume

Consider the volume shown in Fig. 7.5. By considering the volume to be a collection of composite volumes V_i, where the centroid of each of these is located at $\tilde{x}$, $\tilde{y}$, and $\tilde{z}$, the position of the centroid for the entire volume $\bar{x}$, $\bar{y}$, and $\bar{z}$ is given by

$$\bar{x} = \frac{\sum_{i=1}^{n} \tilde{x}_i V_i}{\sum_{i=1}^{n} V_i}, \qquad \bar{y} = \frac{\sum_{i=1}^{n} \tilde{y}_i V_i}{\sum_{i=1}^{n} V_i}, \qquad \text{and} \qquad \bar{z} = \frac{\sum_{i=1}^{n} \tilde{z}_i V_i}{\sum_{i=1}^{n} V_i}, \tag{7.15}$$

where n is the number of composite volumes that constitute the entire volume, V_i is the volume of composite volume i, and $\tilde{x}_i$, $\tilde{y}_i$ and $\tilde{z}_i$ are the locations of the centroid of V_i measured from the origin of the coordinate system.

By taking the limits of Eq. (7.15) as $V_i \to 0$, this equation becomes

$$\bar{x} = \frac{\int \tilde{x}\, dV}{\int dV}, \qquad \bar{y} = \frac{\int \tilde{y}\, dV}{\int dV}, \qquad \text{and} \qquad \bar{z} = \frac{\int \tilde{z}\, dV}{\int dV}, \tag{7.16}$$

where $\tilde{x}$, $\tilde{y}$ and $\tilde{z}$ are the locations of the centroid of volume element dV measured from the origin of the coordinate system.

Helpful Information

Centroid of symmetric shapes. If a shape has an axis of symmetry, such as the y axis in the figure below, the centroid C of that shape lies on the axis of symmetry.

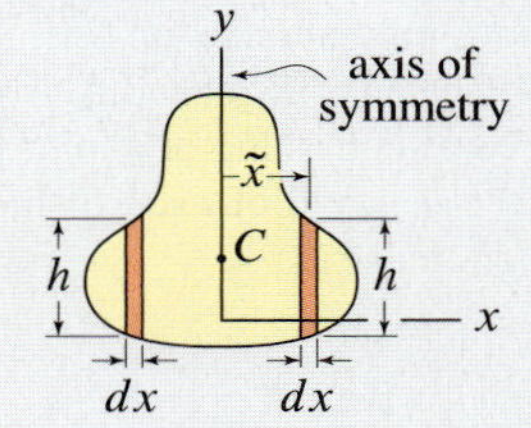

To understand why this is true, consider the two symmetrically located area elements shown. For every area element $dA = h\,dx$ to the right of the axis of symmetry, there is an equal area element to the left, with equal but negative $\tilde{x}$ value. Thus, when we evaluate $\bar{x} = \int \tilde{x}\, dA / \int dA$, the integral in the numerator is zero and hence $\bar{x} = 0$. If a shape has two axes of symmetry, then the centroid is located at the intersection of the axes of symmetry.

Unification of concepts

In this section, expressions using composite shapes and integration for the centroid of lines, areas, and volumes, in one, two, and three dimensions, have been given. The unifying idea behind all of these is that the centroid is the average position of a distribution of shape. While the number of equations presented is rather large, Eq. (7.19), discussed in the End of Section Summary on p. 400, boils all of these down into one compact expression. If you remember and understand this expression, then it is easy to extrapolate it for other applications. In addition to this, Eq. (7.12) (and/or Eqs. (7.13) and (7.14)) is also needed to determine the centroid of lines by using integration.

Which approach should I use: composite shapes or integration?

Determination of the centroid using composite shapes is usually very straightforward *provided* that the centroid position of each of the composite shapes is readily available, either by inspection or from tabulated data for common geometric shapes. To this end, the tables contained on the inside back cover of this book will be useful.

Use of integration to determine the centroid is simple in concept, but not always simple in evaluation. Nonetheless, integration can always be used. When we evaluate integrals, many choices are possible for coordinate systems (Cartesian coordinates, polar coordinates, and cylindrical coordinates are common), and area and volume integrations may require double or triple integrals. These various choices are extensively discussed in calculus. In this book, our discussion of integration for determining the centroid is more focused. In particular, we will evaluate all area and volume integrations by using single integrals. Computer mathematics software such as Mathematica and Maple can make evaluation of an integral easy, but it is still necessary for you to properly set up the integral including the limits of integration.

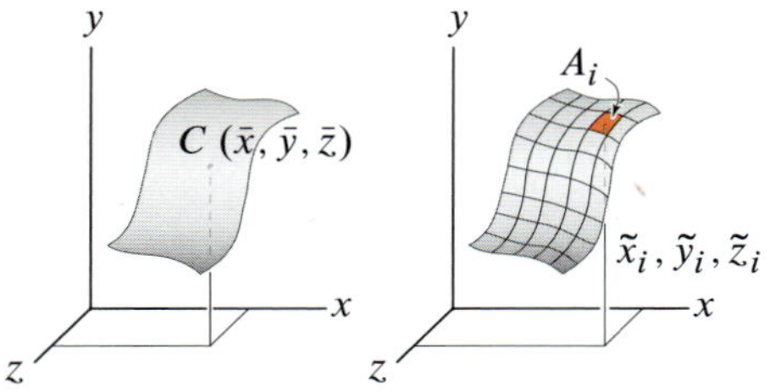

Figure 7.6
The centroid C of a surface area A in three dimensions is located at $\bar{x}$, $\bar{y}$, and $\bar{z}$. Area A can be considered as consisting of composite areas A_i, where the centroid of each of these is located at $\tilde{x}_i$, $\tilde{y}_i$, and $\tilde{z}_i$.

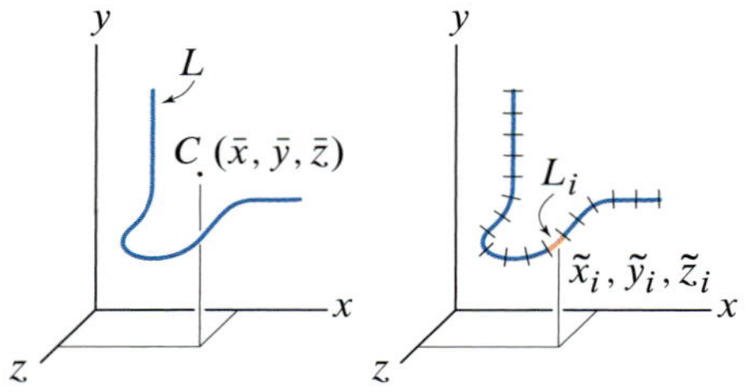

Figure 7.7
The centroid C of a line of length L in three dimensions is located at $\bar{x}$, $\bar{y}$, and $\bar{z}$. Length L can be considered as consisting of composite lengths L_i, where the centroid of each of these is located at $\tilde{x}_i$, $\tilde{y}_i$, and $\tilde{z}_i$.

Finer points: surfaces and lines in three dimensions

In this section, we have considered the centroid of lines and areas in two dimensions. However, Eqs. (7.8) through (7.11) are also applicable to lines and areas in three dimensions, provided an additional expression is written for the z location of the centroid. For example, for a surface area A in three dimensions, as shown in Fig. 7.6, the centroid positions $\bar{x}$ and $\bar{y}$ are as given in Eqs. (7.8) and (7.9), and the z position is given by

$$\bar{z} = \frac{\sum_{i=1}^{n} \tilde{z}_i A_i}{\sum_{i=1}^{n} A_i} = \frac{\int \tilde{z}\, dA}{\int dA} \tag{7.17}$$

Similarly, for a line of length L in three dimensions, as shown in Fig. 7.7, the centroid positions $\bar{x}$ and $\bar{y}$ are as given in Eqs. (7.10) and (7.11), and the z position is given by

$$\bar{z} = \frac{\sum_{i=1}^{n} \tilde{z}_i L_i}{\sum_{i=1}^{n} L_i} = \frac{\int \tilde{z}\, dL}{\int dL} \tag{7.18}$$

End of Section Summary

The *centroid* is defined to be the average position of a distribution of shapes. If the distribution consists of a single shape, the shape can be a line (straight or curved), an area, or a volume. For a distribution (or collection) of shapes, the distribution can include multiple lines, or multiple areas, or multiple volumes—but not combinations of these. The centroid can be determined by use of composite shapes or by integration.

The many formulas presented in this section can be compactly summarized. Consider the case of the centroid of an area or a distribution of areas, as shown in Fig. 7.2 on p. 397. The expressions for $\bar{x}$ in Eqs. (7.8) and (7.9) can be written together as

$$\bar{x} = \frac{\sum_{i=1}^{n} \tilde{x}_i A_i}{\sum_{i=1}^{n} A_i} = \frac{\int \tilde{x}\, dA}{\int dA}. \tag{7.19}$$

The following remarks apply to Eq. (7.19):

- The summation form is used if the centroid is to be determined using composite shapes. This approach is straightforward provided the centroid position $\tilde{x}_i$ for each of the composite shapes is readily available.
- The integral form can always be used to determine the centroid. Many of the example problems of this section illustrate this approach.
- To determine the y position of the centroid, an equation for $\bar{y}$ is written by replacing all of the x's that appear in Eq. (7.19) by y's.

- If the z position of the centroid is needed, such as for an object in three dimensions (e.g., the area in three dimensions shown in Fig. 7.6), then an equation for $\bar{z}$ is written by replacing all of the x's that appear in Eq. (7.19) by z's.

- To determine the centroid of a line or a distribution of lines (straight and/or curved), replace all of the A's that appear in Eq. (7.19) by L's.

- To determine the centroid of a volume or a distribution of volumes, replace all of the A's that appear in Eq. (7.19) by V's.

The final point to summarize is that if a shape has an axis of symmetry, then the centroid lies on the axis of symmetry. Furthermore, if a shape has two axes of symmetry, then the centroid is located at the intersection of the axes of symmetry.

EXAMPLE 7.1 *Centroid of an Area Using Composite Shapes*

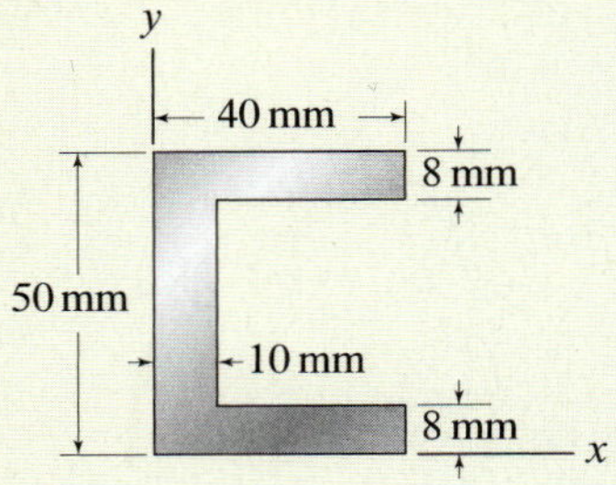

Figure 1

The cross section of an extruded aluminum channel is shown. Determine the x and y positions of the cross section's centroid.

SOLUTION

Road Map The shape of the channel's cross section is an arrangement of simple geometric shapes, namely rectangles. Thus it will be straightforward to determine the position of the centroid by using composite shapes with Eq. (7.8) on p. 397. There are several possible arrangements of composite shapes that may be used, and we will consider solutions for two of these.

Solution 1

Governing Equations & Computation In Fig. 2, the shape is subdivided into three rectangles, where A_1 is the area of rectangle 1, C_1 denotes the centroid of rectangle 1, and so on. Because the cross-sectional shape is symmetric about the line $y = 25$ mm, the centroid must lie on this line, and therefore the y position of the centroid is $\bar{y} = 25$ mm. The areas and centroid positions of each composite shape are collected in Table 1 followed by evaluation of Eq. (7.8).

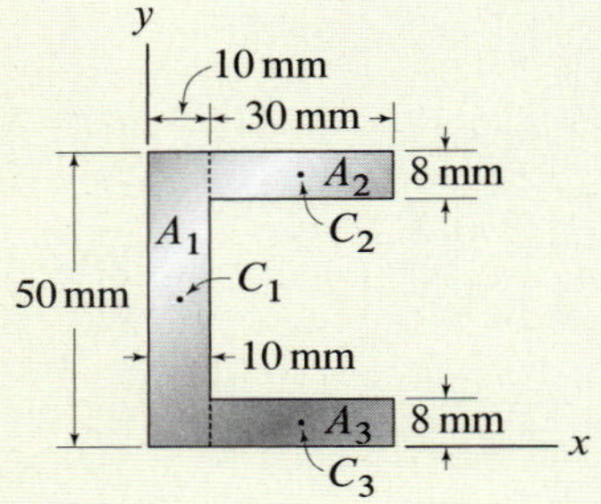

Figure 2
A selection of three composite shapes to describe the channel's cross section.

Table 1. Areas and centroid positions for composite shapes in Fig. 2.

Shape no.	A_i	$\tilde{x}_i$
1	$(10\,\text{mm})(50\,\text{mm}) = 500\,\text{mm}^2$	5 mm
2	$(30\,\text{mm})(8\,\text{mm}) = 240\,\text{mm}^2$	25 mm
3	$(30\,\text{mm})(8\,\text{mm}) = 240\,\text{mm}^2$	25 mm

$$\bar{x} = \frac{\sum_{i=1}^{3} \tilde{x}_i A_i}{\sum_{i=1}^{3} A_i} = \frac{(5\,\text{mm})(500\,\text{mm}^2) + (25\,\text{mm})(240\,\text{mm}^2) + (25\,\text{mm})(240\,\text{mm}^2)}{500\,\text{mm}^2 + 240\,\text{mm}^2 + 240\,\text{mm}^2}$$

$$= \boxed{14.8\,\text{mm.}} \qquad (1)$$

Solution 2

Governing Equations & Computation Shown in Fig. 3 is a selection of two composite shapes for the cross section of the channel, where the second of these has *negative* area. The areas and centroid positions of each composite shape are collected in Table 2 followed by evaluation of Eq. (7.8).

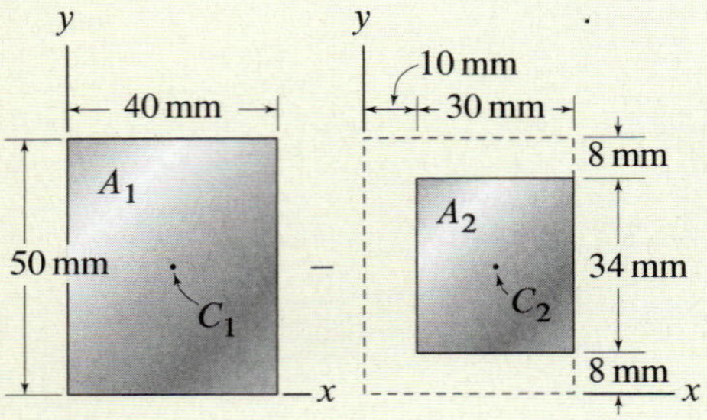

Figure 3
A selection of two composite shapes to describe the channel's cross section.

Table 2. Areas and centroid positions for composite shapes in Fig. 3.

Shape no.	A_i	$\tilde{x}_i$
1	$(40\,\text{mm})(50\,\text{mm}) = 2000\,\text{mm}^2$	20 mm
2	$-(30\,\text{mm})(34\,\text{mm}) = -1020\,\text{mm}^2$	25 mm

$$\bar{x} = \frac{\sum_{i=1}^{2} \tilde{x}_i A_i}{\sum_{i=1}^{2} A_i} = \frac{(20\,\text{mm})(2000\,\text{mm}^2) + (25\,\text{mm})(-1020\,\text{mm}^2)}{2000\,\text{mm}^2 - 1020\,\text{mm}^2} = \boxed{14.8\,\text{mm.}} \qquad (2)$$

Discussion & Verification As expected, Eqs. (1) and (2) agree. As an exercise, you should add a column for $\tilde{y}_i$ to Tables 1 and 2 and use Eq. (7.8) to show $\bar{y} = 25$ mm.

EXAMPLE 7.2 *Centroid of a Volume Using Composite Shapes*

A solid has the shape of a cylinder with a truncated cone. Determine the location of the centroid.

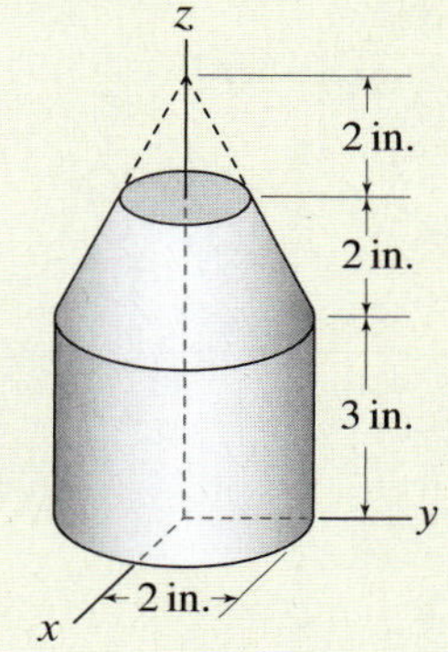

Figure 1

SOLUTION

Road Map This object consists of an arrangement of simple geometric shapes. Thus, compared to using integration, it will be easier to determine the position of the centroid by using composite shapes with Eq. (7.15) on p. 399. This object is symmetric about the xz plane, and therefore its centroid must lie in the xz plane. The object is also symmetric about the yz plane, and therefore its centroid must also lie in the yz plane. The intersection of these planes is the z axis. Therefore, we can conclude that the x and y positions of the centroid are $\bar{x} = \bar{y} = 0$.

Governing Equations & Computation In Fig. 2 the object is subdivided into three shapes. The volume and centroid position of each of the composite shapes are given in the Table of Properties of Solids on the inside back cover of this book, and these values are collected in Table 1. For example, for shape 2, the volume is $V_2 = \pi r^2 h/3$ where

Table 1. Volumes and centroid positions for composite shapes in Fig. 2.

Shape no.	V_i	$\tilde{z}_i$
1	$\pi(2\text{ in.})^2(3\text{ in.}) = 37.70\text{ in.}^3$	1.5 in.
2	$\frac{\pi}{3}(2\text{ in.})^2(4\text{ in.}) = 16.76\text{ in.}^3$	$3\text{ in.} + \frac{1}{4}(4\text{ in.}) = 4\text{ in.}$
3	$-\frac{\pi}{3}(1\text{ in.})^2(2\text{ in.}) = -2.094\text{ in.}^3$	$5\text{ in.} + \frac{1}{4}(2\text{ in.}) = 5.5\text{ in.}$

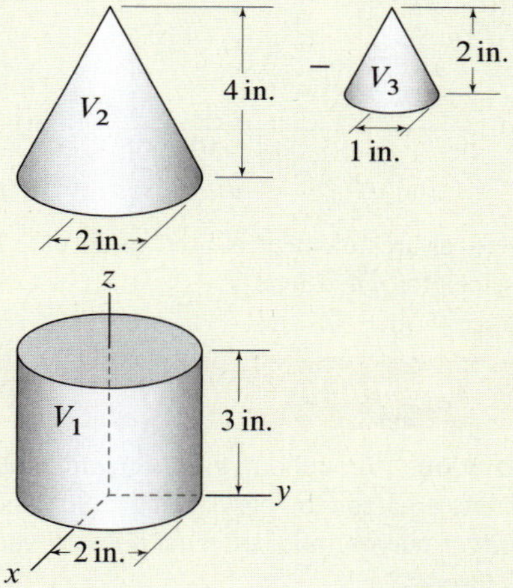

Figure 2
A selection of three composite shapes to describe the solid.

$r = 2$ in. is the radius of the cone's base and $h = 4$ in. is its height. The centroid for the cone measured from its base is $h/4$, and to this value, the distance from the origin of the coordinate system to the cone's base (i.e., 3 in.) must be added. Evaluating the expression for $\bar{z}$ in Eq. (7.15) provides

$$\bar{z} = \frac{\sum_{i=1}^{3} \tilde{z}_i V_i}{\sum_{i=1}^{3} V_i}$$

$$= \frac{(1.5\text{ in.})(37.70\text{ in.}^3) + (4\text{ in.})(16.76\text{ in.}^3) + (5.5\text{ in.})(-2.094\text{ in.}^3)}{37.70\text{ in.}^3 + 16.76\text{ in.}^3 - 2.094\text{ in.}^3}$$

$$= \boxed{2.14\text{ in.}} \qquad (1)$$

Discussion & Verification Short of resolving this problem by using a different selection of composite shapes, or using a different method (e.g., integration), there are no definitive checks of accuracy for our solution. Nonetheless, we expect $\bar{z}$ to be somewhat larger than 1.5 in., which is the centroid of volume V_1 only, and in view of this, the value found in Eq. (1) is reasonable.

Helpful Information

Solid of revolution. The shape shown in Fig. 1 is called a *solid of revolution* because it can be produced by revolving a planar shape about a straight line, called the *axis of revolution*. For example, rotating the area shown below by 360° about the z axis produces the shape shown in Fig. 1:

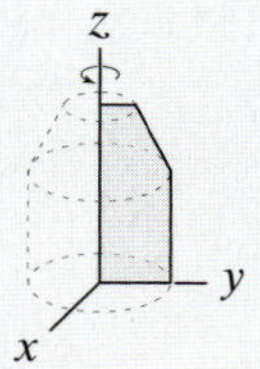

EXAMPLE 7.3 *Centroid of an Area Using Integration*

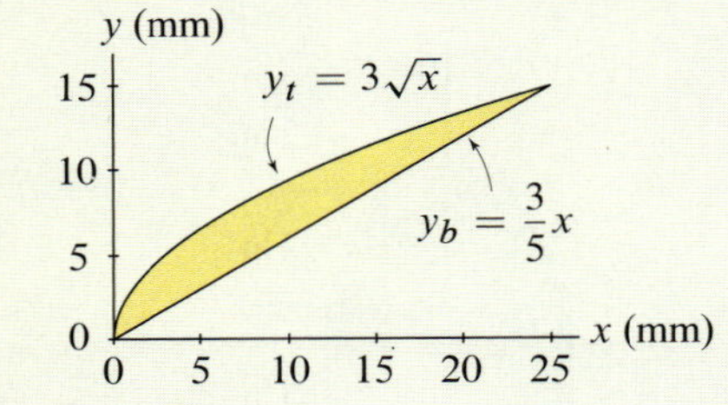

Figure 1

The cross section of a turbine blade in a pump is shown. Determine the x and y positions of the centroid.

SOLUTION

Road Map The cross section shape of the turbine blade is an area, and thus Eq. (7.9) on p. 397 will be used to determine the x and y positions of its centroid. In Eq. (7.9), three ingredients are needed: expressions for dA, $\tilde{x}$, and $\tilde{y}$. These expressions can be developed using either a vertical area element or a horizontal area element, and solutions using both of these are demonstrated.

Solution 1 – vertical area element

Governing Equations & Computation To evaluate Eq. (7.9), expressions for dA, $\tilde{x}$, and $\tilde{y}$ are needed, and these can be developed using the vertical area element shown in Fig. 2 as follows.

$$dA = (y_t - y_b)\,dx = \left(3\sqrt{x} - \tfrac{3}{5}x\right)dx, \tag{1}$$

$$\tilde{x} = x, \tag{2}$$

$$\tilde{y} = \tfrac{1}{2}(y_t + y_b) = \tfrac{1}{2}\left(3\sqrt{x} + \tfrac{3}{5}x\right). \tag{3}$$

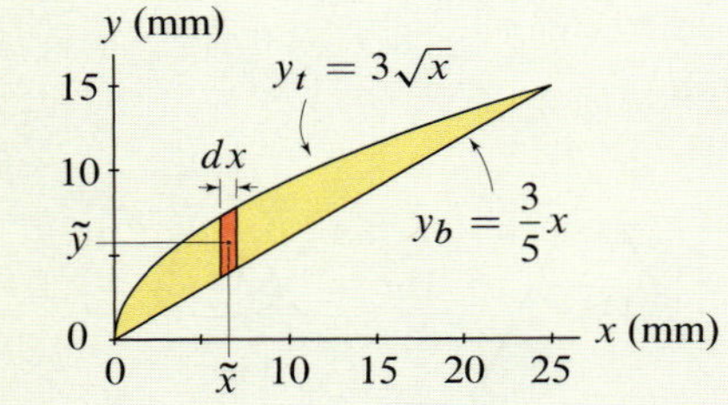

Figure 2
A vertical area element is used to develop expressions for dA, $\tilde{x}$, and $\tilde{y}$.

Substituting Eqs. (1) through (3) into Eq. (7.9), the position of the centroid is given by

$$\bar{x} = \frac{\int \tilde{x}\,dA}{\int dA} = \frac{\int_0^{25\text{ mm}} x\left(3\sqrt{x} - \frac{3}{5}x\right)dx}{\int_0^{25\text{ mm}} \left(3\sqrt{x} - \frac{3}{5}x\right)dx} = \frac{\left(\frac{6x^{5/2}}{5} - \frac{x^3}{5}\right)\Big|_0^{25\text{ mm}}}{\left(2x^{3/2} - \frac{3x^2}{10}\right)\Big|_0^{25\text{ mm}}}$$

$$= \frac{625\text{ mm}^3}{62.5\text{ mm}^2} = \boxed{10\text{ mm},} \tag{4}$$

$$\bar{y} = \frac{\int \tilde{y}\,dA}{\int dA} = \frac{\int_0^{25\text{ mm}} \frac{1}{2}\left(3\sqrt{x} + \frac{3}{5}x\right)\left(3\sqrt{x} - \frac{3}{5}x\right)dx}{\int_0^{25\text{ mm}} \left(3\sqrt{x} - \frac{3}{5}x\right)dx} = \frac{\left(\frac{9x^2}{4} - \frac{3x^3}{50}\right)\Big|_0^{25\text{ mm}}}{\left(2x^{3/2} - \frac{3x^2}{10}\right)\Big|_0^{25\text{ mm}}}$$

$$= \frac{468.75\text{ mm}^3}{62.5\text{ mm}^2} = \boxed{7.5\text{ mm}.} \tag{5}$$

Helpful Information

Notation. Throughout this problem, subscripts are used to help distinguish *top* and *bottom* curves, and *left* and *right* curves, as follows:

- Subscript t denotes "top."
- Subscript b denotes "bottom."
- Subscript r denotes "right."
- Subscript l denotes "left."

Solution 2 – horizontal area element

Governing Equations & Computation The centroid can also be determined using the horizontal area element shown in Fig. 3. Since integrations will be over y, it is necessary to express the shape of the object as functions of y. Thus, the equation for the bottom curve $y_b = (3/5)x$ is rearranged to obtain $x_r = (5/3)y$. Similarly, the equation for the top curve $y_t = 3\sqrt{x}$ is rearranged to obtain $x_l = y^2/9$. Then expressions for dA, $\tilde{x}$, and $\tilde{y}$ can be written as follows:

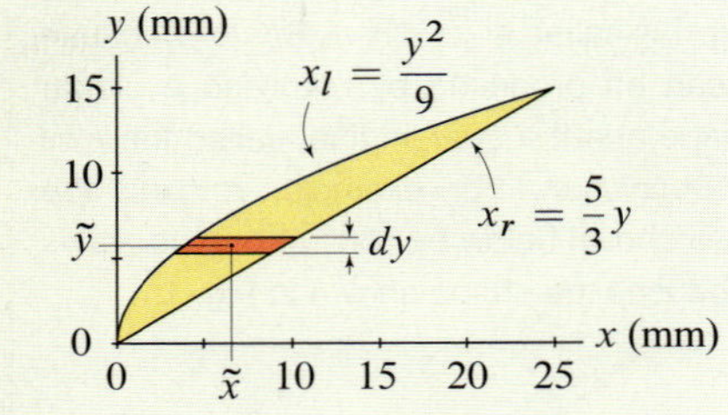

Figure 3
A horizontal area element is used to develop expressions for dA, $\tilde{x}$, and $\tilde{y}$.

$$dA = (x_r - x_l)\,dy = \left(\tfrac{5}{3}y - \frac{y^2}{9}\right)dy, \tag{6}$$

$$\tilde{x} = \tfrac{1}{2}(x_r + x_l) = \tfrac{1}{2}\left(\tfrac{5}{3}y + \frac{y^2}{9}\right), \tag{7}$$

$$\tilde{y} = y. \tag{8}$$

Substituting Eqs. (6) through (8) into Eq. (7.9), the position of the centroid is given by

$$\bar{x} = \frac{\int \tilde{x}\,dA}{\int dA} = \frac{\int_0^{15\,\text{mm}} \frac{1}{2}\left(\frac{5}{3}y + \frac{y^2}{9}\right)\left(\frac{5}{3}y - \frac{y^2}{9}\right) dy}{\int_0^{15\,\text{mm}} \left(\frac{5}{3}y - \frac{y^2}{9}\right) dy} = \frac{\left(\frac{25y^3}{54} - \frac{y^5}{810}\right)\Big|_0^{15\,\text{mm}}}{\left(\frac{5y^2}{6} - \frac{y^3}{27}\right)\Big|_0^{15\,\text{mm}}}$$

$$= \frac{625\,\text{mm}^3}{62.5\,\text{mm}^2} = \boxed{10\,\text{mm},} \tag{9}$$

$$\bar{y} = \frac{\int \tilde{y}\,dA}{\int dA} = \frac{\int_0^{15\,\text{mm}} y\left(\frac{5}{3}y - \frac{y^2}{9}\right) dy}{\int_0^{15\,\text{mm}} \left(\frac{5}{3}y - \frac{y^2}{9}\right) dy} = \frac{\left(\frac{5y^3}{9} - \frac{y^4}{36}\right)\Big|_0^{15\,\text{mm}}}{\left(\frac{5y^2}{6} - \frac{y^3}{27}\right)\Big|_0^{15\,\text{mm}}}$$

$$= \frac{468.75\,\text{mm}^3}{62.5\,\text{mm}^2} = \boxed{7.5\,\text{mm}.} \tag{10}$$

As expected, the solutions for both $\bar{x}$ and $\bar{y}$ agree with those obtained earlier in Eqs. (4) and (5).

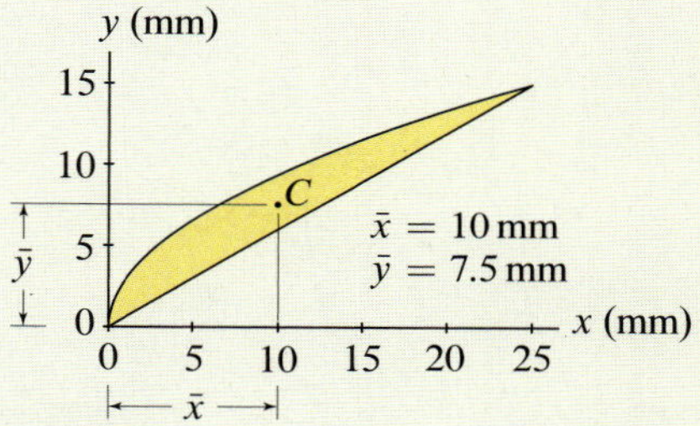

Figure 4
The centroid C of the area is located at $\bar{x}$ and $\bar{y}$.

Discussion & Verification

- The location of the centroid is shown as point C in Fig. 4, and this location appears to be reasonable.
- The solution procedure outlined here can be used for a wide variety of area centroid problems by simply using the appropriate expressions for the bounding curves y_t and y_b, or x_r and x_l. Specifically:
  - If a vertical area element is used, then expressions for the top and bottom curves y_t and y_b, respectively, as functions of x are needed, and Eqs. (1) through (3) can be used to obtain expressions for dA, $\tilde{x}$, and $\tilde{y}$.
  - If a horizontal area element is used, then expressions for the right and left curves x_r and x_l, respectively, as functions of y are needed, and Eqs. (6) through (8) can be used to obtain expressions for dA, $\tilde{x}$, and $\tilde{y}$.
- For some problems, a vertical area element may be more convenient than a horizontal area element, or vice versa. For example, for the area shown in Fig. 5, a vertical area element will be more convenient because both y_t and y_b are given by single equations. To use a horizontal area element, two different equations are needed for x_r, depending on where the area element is located.
- Software such as Mathematica and Maple can make evaluation of integrals easy, but obtaining correct results with such software requires that you provide the correct integrand and limits of integration.

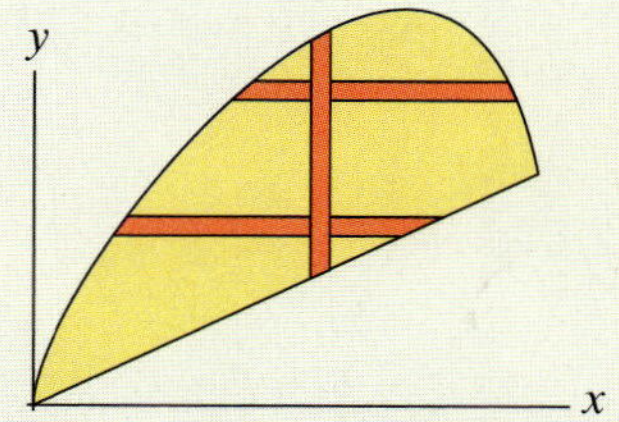

Figure 5
Example of a shape for which a vertical area element is more convenient than a horizontal area element.

EXAMPLE 7.4 *Centroid of a Line Using Integration*

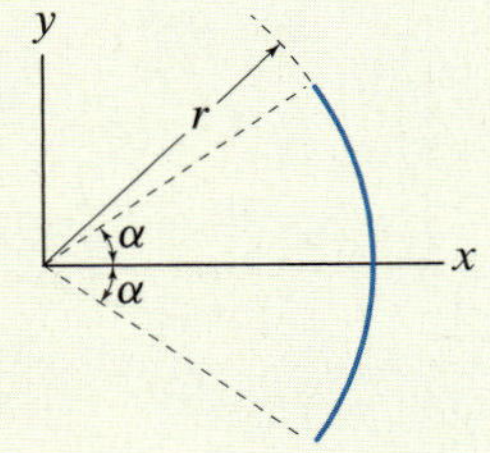

Figure 1

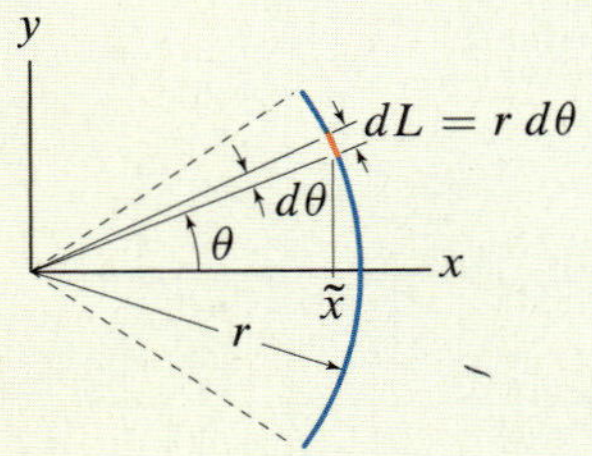

Figure 2
A line element is used to develop expressions for dL and $\tilde{x}$.

Determine the x and y positions of the centroid for the circular arc shown.

SOLUTION

Road Map Because the arc is symmetric about the x axis, the centroid must lie on the x axis. Therefore, by inspection, the y position of the centroid is $\bar{y} = 0$. To determine the x position of the centroid, we will evaluate the first expression in Eq. (7.11) on p. 398, and for developing expressions for dL and $\tilde{x}$, polar coordinates will be very convenient.

Governing Equations & Computation Expressions for dL and $\tilde{x}$ will be written using the line element shown in Fig. 2 as follows:

$$dL = r\,d\theta, \tag{1}$$

$$\tilde{x} = r\cos\theta. \tag{2}$$

Substituting Eqs. (1) and (2) into the first expression of Eq. (7.11), the x position of the centroid is given by

$$\bar{x} = \frac{\int \tilde{x}\,dL}{\int dL} = \frac{\int_{-\alpha}^{\alpha} r\cos\theta\, r\,d\theta}{\int_{-\alpha}^{\alpha} r\,d\theta} = \frac{r^2(\sin\theta)\big|_{-\alpha}^{\alpha}}{r(\theta)\big|_{-\alpha}^{\alpha}} = \frac{2r^2\sin\alpha}{2r\alpha} \tag{3}$$

$$= \boxed{\frac{r\sin\alpha}{\alpha}}. \tag{4}$$

Discussion & Verification

- For a circular arc, $r =$ constant, and the integrals in Eq. (3) are simplified, as follows. Because r is not a function of θ, the r^2 term in the numerator and the r term in the denominator of Eq. (3) may be brought outside their integrals, leaving simple functions of θ to be integrated. For a line that is not a circular arc, r is a function of θ, giving rise to generally more difficult integrals to evaluate. For such cases, the approach used in Example 7.5, where integrations are carried out with respect to x (or y), may be more convenient.
- The validity of our results can be partially checked by considering the location of $\bar{x}$ for particular values of α. When $\alpha \to 0$, the arc approaches a point, and the centroid should approach $\bar{x} = r$, which Eq. (4) does.* When $\alpha = \pi$, the arc becomes a full circle, and the centroid should become $\bar{x} = 0$, which Eq. (4) does.
- To evaluate Eq. (4) for particular values of r and α, it is necessary to measure α in radians.
- As an exercise, you can evaluate the second expression of Eq. (7.11) on p. 398 by using $\tilde{y} = r\sin\theta$ to confirm that $\bar{y} = 0$.

* Recall the small-angle approximation $\sin\alpha \approx \alpha$ when $\alpha \ll 1$ (α measured in radians).

EXAMPLE 7.5 *Centroid of a Line Using Integration*

Determine the x and y positions of the centroid for the uniform curved bar shown.

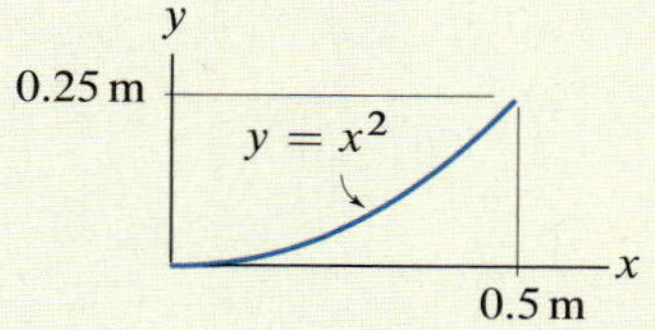

Figure 1

SOLUTION

Road Map To determine the position of the centroid, we will evaluate Eq. (7.11) on p. 398. Because the equation for the line is expressed in terms of x and y, it will be more convenient to perform integrations with respect to x or y, rather than with respect to length along the path of the line. To this end, we will use Eqs. (7.13) and (7.14) on p. 398.

Governing Equations In Eq. (7.11), $\tilde{x}$ and $\tilde{y}$ locate the centroid of length element dL, as shown in Fig. 2. An expression for dL will be written in terms of dx by using Eq. (7.13) on p. 398.* Noting that $dy/dx = 2x$,

$$dL = \sqrt{1 + \left(\frac{dy}{dx}\right)^2}\, dx = \sqrt{1 + 4x^2}\, dx, \tag{1}$$

$$\tilde{x} = x, \tag{2}$$

$$\tilde{y} = y = x^2. \tag{3}$$

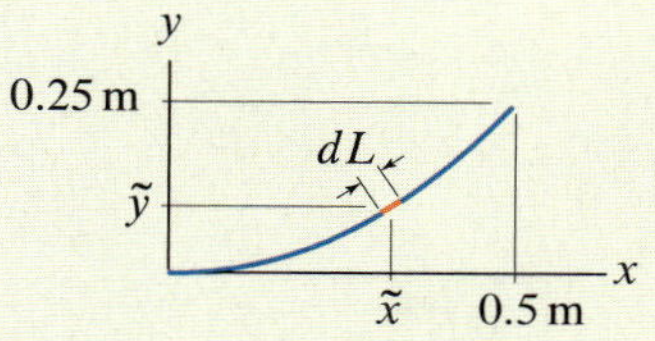

Figure 2
A line element is used to develop expressions for dL and $\tilde{x}$.

Substituting Eqs. (1) – (3) into Eq. (7.11), the x and y positions of the centroid are given by

$$\bar{x} = \frac{\int \tilde{x}\, dL}{\int dL} = \frac{\int_0^{0.5\,\mathrm{m}} x\sqrt{1 + 4x^2}\, dx}{\int_0^{0.5\,\mathrm{m}} \sqrt{1 + 4x^2}\, dx}, \tag{4}$$

$$\bar{y} = \frac{\int \tilde{y}\, dL}{\int dL} = \frac{\int_0^{0.5\,\mathrm{m}} x^2\sqrt{1 + 4x^2}\, dx}{\int_0^{0.5\,\mathrm{m}} \sqrt{1 + 4x^2}\, dx}. \tag{5}$$

Computation The integrals in Eqs. (4) and (5) are perhaps tedious to evaluate. Using software to evaluate these integrals provides

$$\bar{x} = \frac{0.1524\,\mathrm{m}^2}{0.5739\,\mathrm{m}} = \boxed{0.265\,\mathrm{m},} \tag{6}$$

$$\bar{y} = \frac{0.05252\,\mathrm{m}^2}{0.5739\,\mathrm{m}} = \boxed{0.0915\,\mathrm{m}.} \tag{7}$$

Discussion & Verification

- When using software, we must be especially cautious to check the accuracy of the answers it produces. In the intermediate results in Eqs. (6) and (7), the denominator (i.e., 0.5739 m) is the total length of the line. Note that if the line were straight, its length would be $\sqrt{(0.5)^2 + (0.25)^2}\,\mathrm{m} = 0.559\,\mathrm{m}$, and as expected, the length of the curved line is slightly greater than this. Also, if the line were straight, by inspection its centroid would be at the x and y positions 0.25 m and 0.125 m, respectively. Thus, the $\bar{x}$ and $\bar{y}$ values in Eqs. (6) and (7) are reasonable.
- As an exercise, you should repeat this example, using Eq. (7.14) on p. 398 to relate dL to dy.

* Alternatively, we could use Eq. (7.14) to write an expression for dL in terms of dy.

EXAMPLE 7.6 *Centroid of a Volume Using Integration*

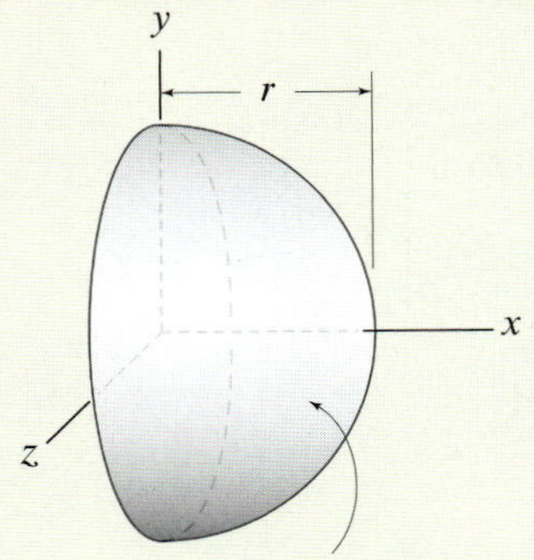

Figure 1

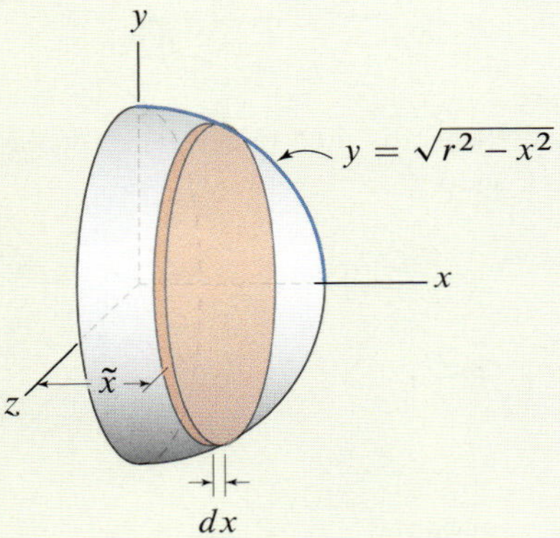

Figure 2
A thin disk volume element is used to develop expressions for dV and $\tilde{x}$. The expression for the radius y of the disk is obtained by rearranging $x^2 + y^2 + z^2 = r^2$, with $z = 0$.

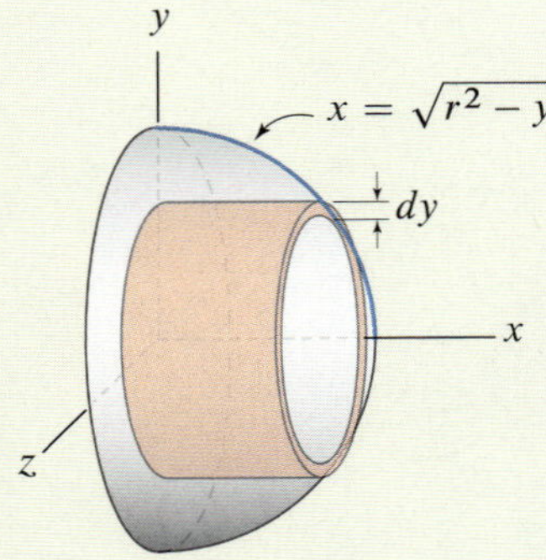

Figure 3
A thin shell volume element is used to develop expressions for dV and $\tilde{x}$. The expression for the length x of the cylinder is obtained by rearranging $x^2 + y^2 + z^2 = r^2$, with $z = 0$.

Determine the position of the centroid for the solid hemisphere of radius r shown.

SOLUTION

Road Map Because the object is symmetric about both the xy and xz planes, the centroid lies on the x axis and we can conclude that $\bar{y} = \bar{z} = 0$. To determine the x position of the centroid, we will evaluate the expression for $\bar{x}$ in Eq. (7.16) on p. 399. Expressions for dV and $\tilde{x}$ are needed, and these can be developed using either a thin disk volume element or a thin shell volume element, and both of these approaches are shown.

Solution 1 – thin disk volume element

Governing Equations & Computation Expressions for dV and $\tilde{x}$ can be written using the thin disk volume element shown in Fig. 2 as follows:

$$dV = \pi y^2\,dx = \pi(r^2 - x^2)\,dx, \tag{1}$$

$$\tilde{x} = x. \tag{2}$$

In Eq. (1), πy^2 is the area of a disk where y is the radius and dx is the thickness. The entire disk element is at the same x position, so that $\tilde{x} = x$. Substituting Eqs. (1) and (2) into the first expression of Eq. (7.16) provides the x position of the centroid as

$$\bar{x} = \frac{\int \tilde{x}\,dV}{\int dV} = \frac{\int_0^r x\pi(r^2 - x^2)\,dx}{\int_0^r \pi(r^2 - x^2)\,dx} = \frac{\pi\left(r^2\frac{x^2}{2} - \frac{x^4}{4}\right)\Big|_0^r}{\pi\left(r^2 x - \frac{x^3}{3}\right)\Big|_0^r} = \frac{\frac{\pi r^4}{4}}{\frac{2\pi r^3}{3}} = \boxed{\frac{3r}{8}}. \tag{3}$$

Solution 2 – thin shell volume element

Governing Equations & Computation Expressions for dV and $\tilde{x}$ can be written using the thin shell volume element shown in Fig. 3 as follows:

$$dV = 2\pi yx\,dy = 2\pi y\,dy\sqrt{r^2 - y^2}, \tag{4}$$

$$\tilde{x} = \frac{x}{2} = \tfrac{1}{2}\sqrt{r^2 - y^2}. \tag{5}$$

In Eq. (4), $2\pi y$ is the circumference of the shell where y is the radius, x is the length of the shell, and dy is the thickness. In Eq. (5), the x position of the centroid of the shell element $\tilde{x}$, is given by $x/2$. Substituting Eqs. (4) and (5) into the first expression of Eq. (7.16) provides the x position of the centroid as

$$\bar{x} = \frac{\int \tilde{x}\,dV}{\int dV} = \frac{\int_0^r \frac{1}{2}\sqrt{r^2 - y^2}\,2\pi y\sqrt{r^2 - y^2}\,dy}{\int_0^r 2\pi y\sqrt{r^2 - y^2}\,dy} = \boxed{\frac{3r}{8}}. \tag{6}$$

Discussion & Verification As expected, the solutions for $\bar{x}$ obtained by using the two different volume elements agree. Also, the result for the denominator of Eq. (3), namely, $2\pi r^3/3$, is the correct volume of a hemisphere (reported in the Table of Properties of Solids on the inside back cover).

PROBLEMS

General instructions. For shapes that have one or more axes or planes of symmetry, you may use inspection to determine some of the coordinates of the centroid.

Problems 7.1 through 7.6

For the area shown, use composite shapes to determine the x and y positions of the centroid.

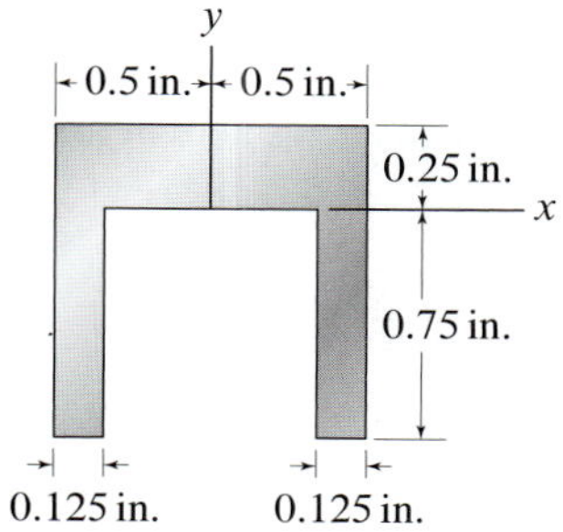

Figure P7.1

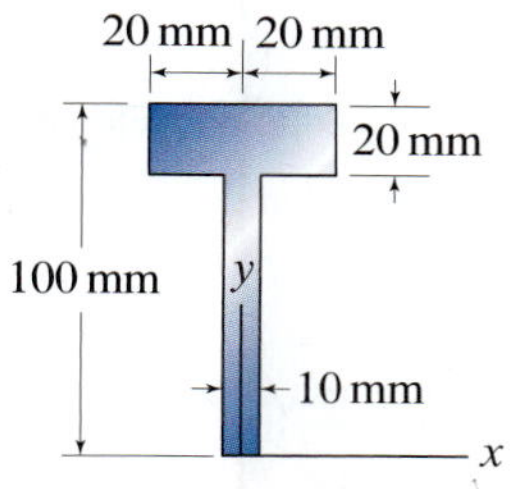

Figure P7.2

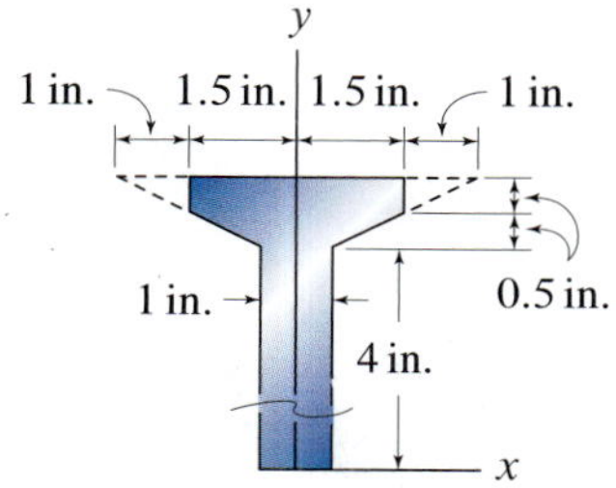

Figure P7.3

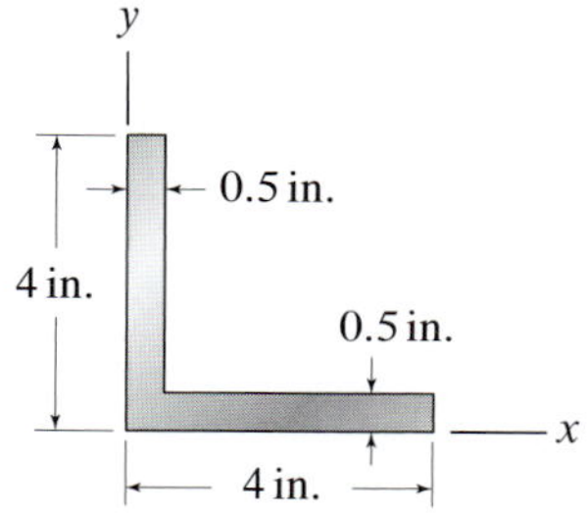

Figure P7.4

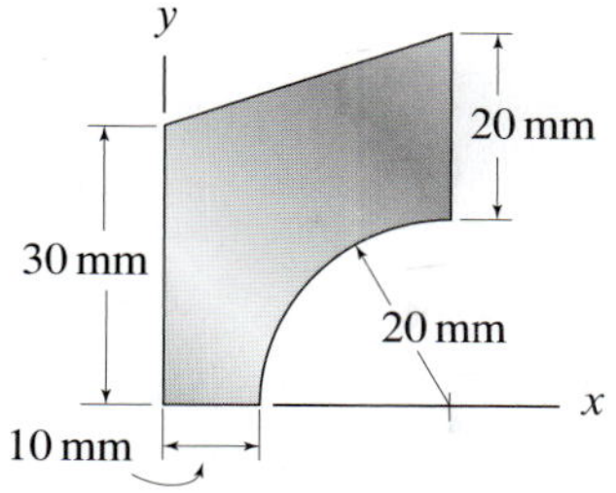

Figure P7.5

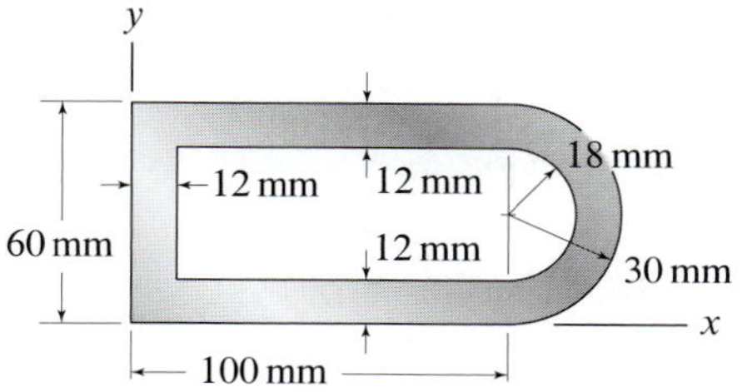

Figure P7.6

Problem 7.7

The solid shown consists of a circular cylinder and a hemisphere. Use composite shapes to determine the x, y, and z locations of the centroid.

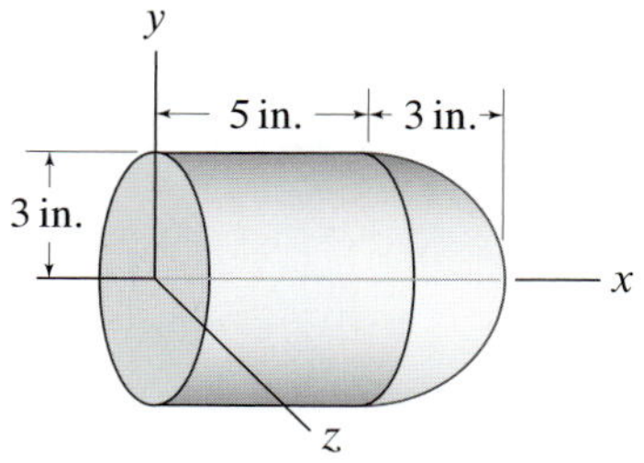

Figure P7.7

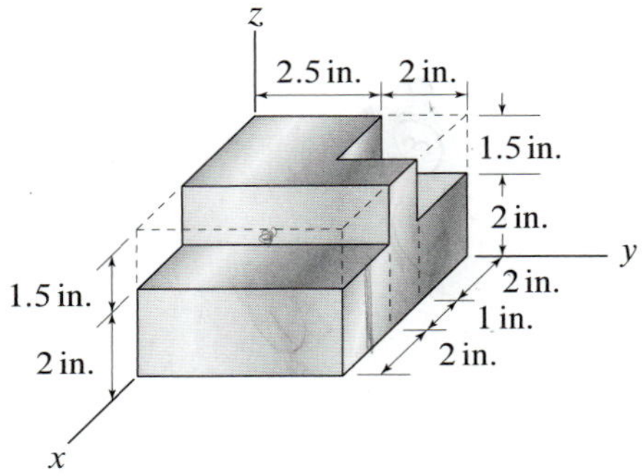

Figure P7.8

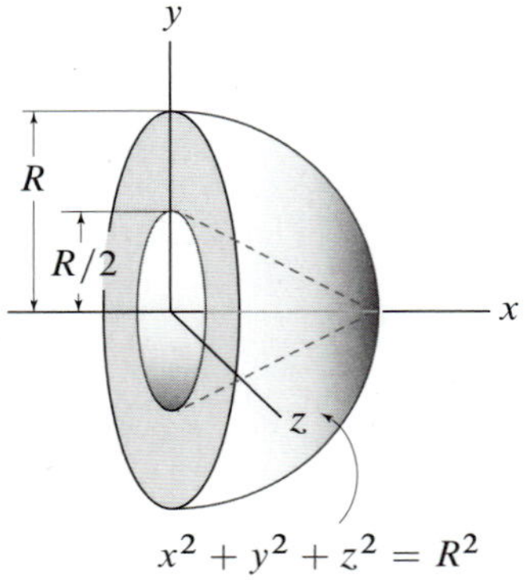

Figure P7.9

Problem 7.8

The rectangular block is a solid with rectangular cutouts. Use composite shapes to determine the x, y, and z locations of the centroid.

Problem 7.9

The solid shown consists of a hemisphere with a conical cavity. Use composite shapes to determine the x, y, and z locations of the centroid. Express your answers in terms of R.

Problems 7.10 through 7.13

For the area shown, use integration to determine the x and y positions of the centroid.

(a) Use a horizontal area element.

(b) Use a vertical area element.

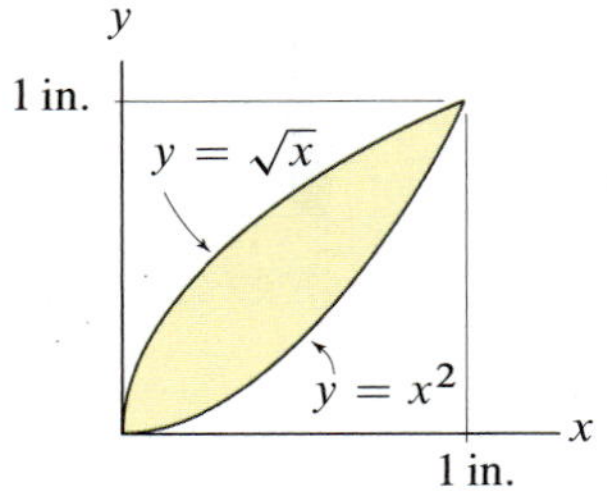

Figure P7.10

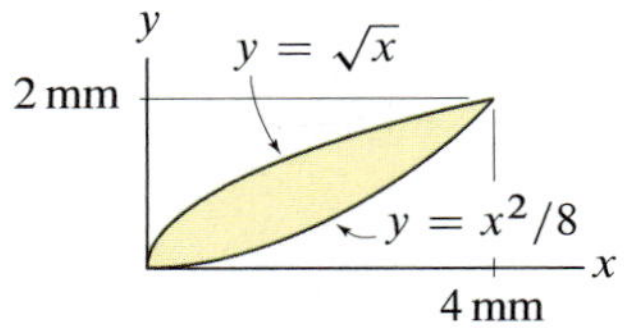

Figure P7.11

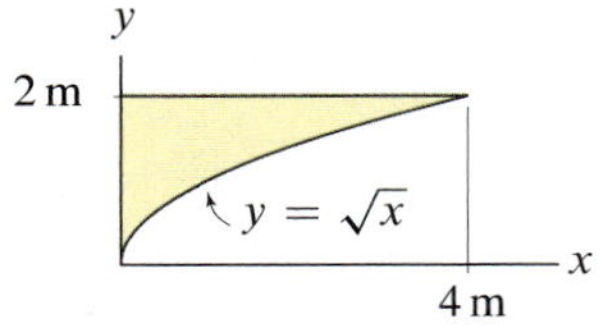

Figure P7.12

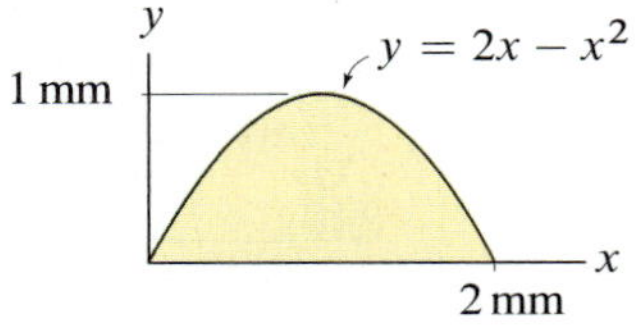

Figure P7.13

Problems 7.14 and 7.15

For the area shown, use integration to determine the x and y positions of the centroid.

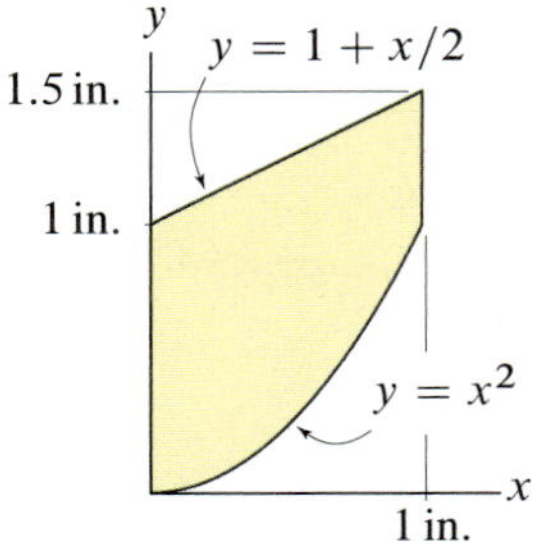

Figure P7.14

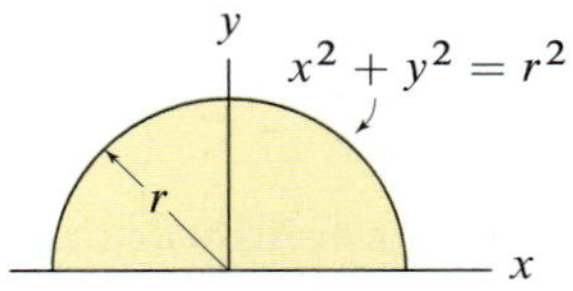

Figure P7.15

Problems 7.16 and 7.17

Determine expressions for lines y_1 and y_2, and then use integration to determine the x and y positions of the centroid.

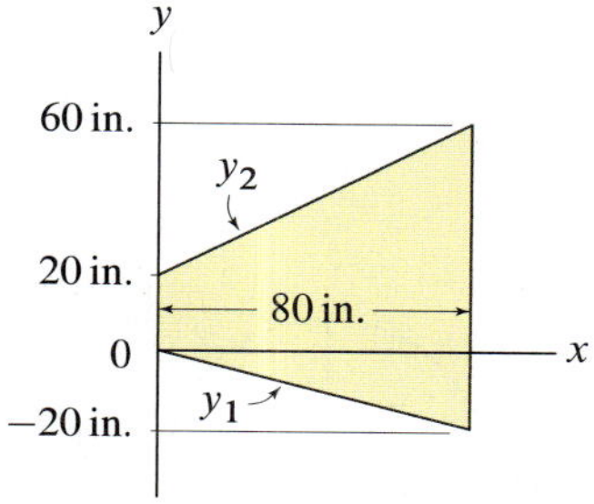

Figure P7.16

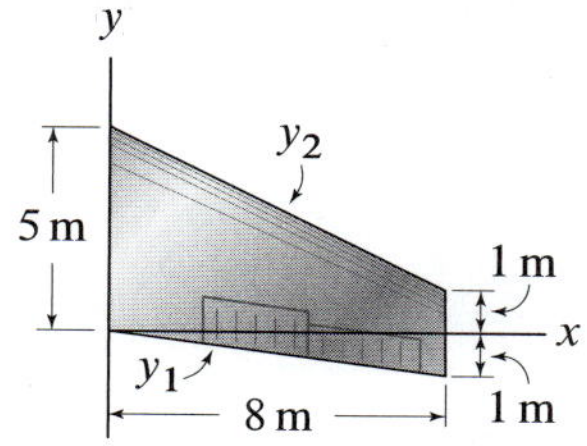

Figure P7.17

Problem 7.18

For the triangle shown, having base b and height h, use integration to show that the y position of the centroid is $\bar{y} = h/3$. *Hint:* Begin by writing an expression for the width of a horizontal area element as a function of y.

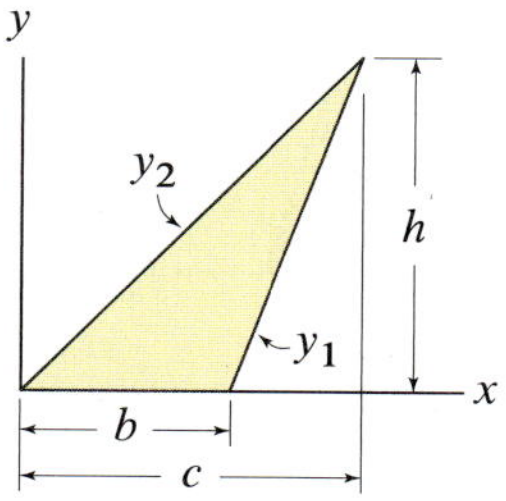

Figure P7.18

Problem 7.19

Use integration to determine the x and y positions of the centroid. Express your answers in terms of r and α.

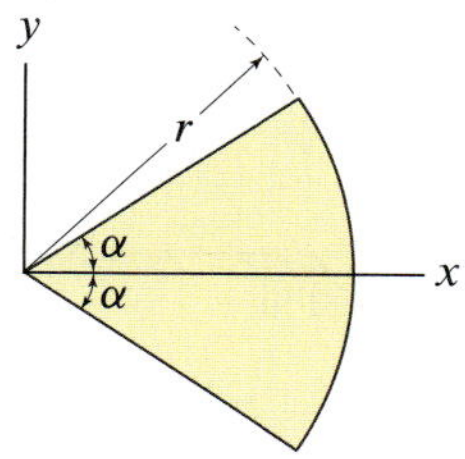

Figure P7.19

Problem 7.20

Determine constants c_1 and c_2 so that the curves intersect at $x = a$ and $y = b$. Use integration to determine the x and y positions of the centroid. Express your answers in terms of a and b.

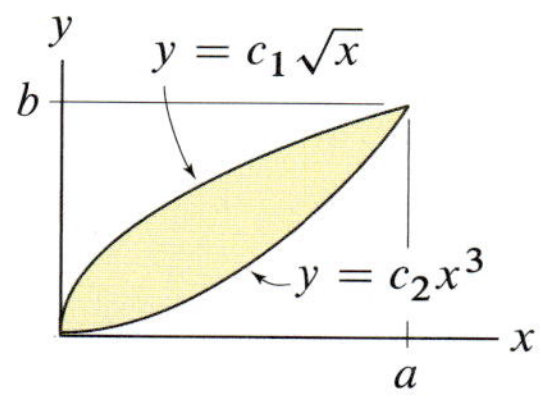

Figure P7.20

Problems 7.21 and 7.22

For the line shown:

(a) Set up the integrals for integration with respect to x, including the limits of integration, that will yield the x and y positions of the centroid.

(b) Repeat Part (a) for integrations with respect to y.

(c) Evaluate the integrals in Parts (a) and/or (b) by using computer software such as Mathematica, Maple, etc.

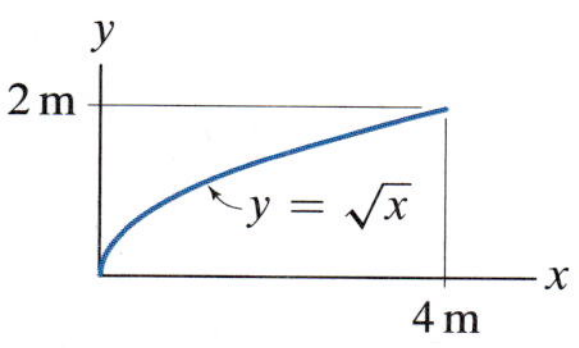

Figure P7.21

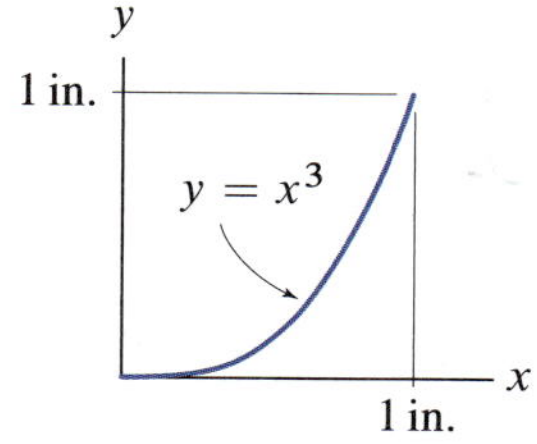

Figure P7.22

Problems 7.23 and 7.24

For the line shown:

(a) Set up the integrals for integration with respect to x, including the limits of integration, that will yield the x and y positions of the centroid.

(b) Evaluate the integrals in Part (a) by using computer software such as Mathematica, Maple, etc.

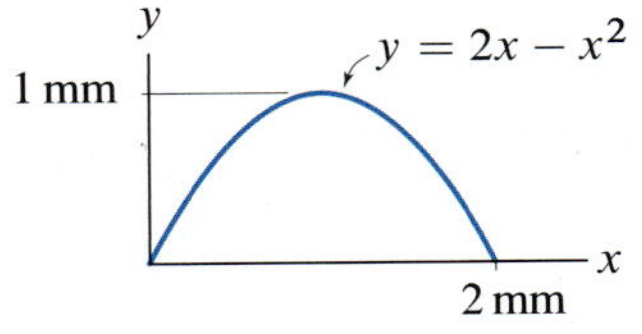

Figure P7.23

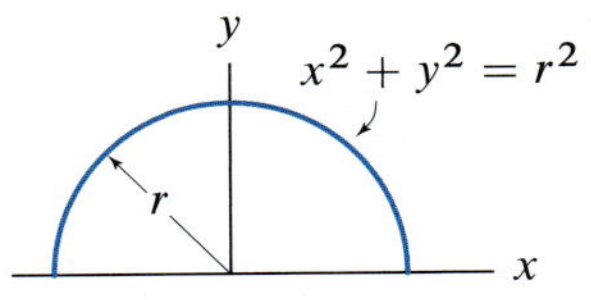

Figure P7.24

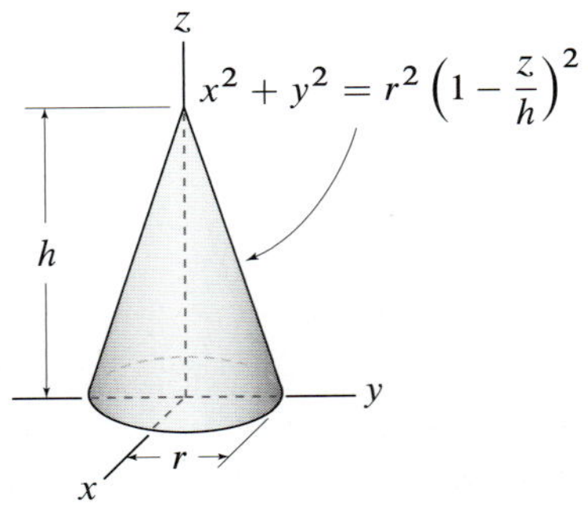

Figure P7.25

Problem 7.25

A solid cone is shown. Use integration to determine the position of the centroid. Express your answers in terms of r and h.

Problem 7.26

For the hemisphere with a conical cavity shown in Fig. P7.9 on p. 410, use integration to determine the x location of the centroid. Express your answers in terms of R.

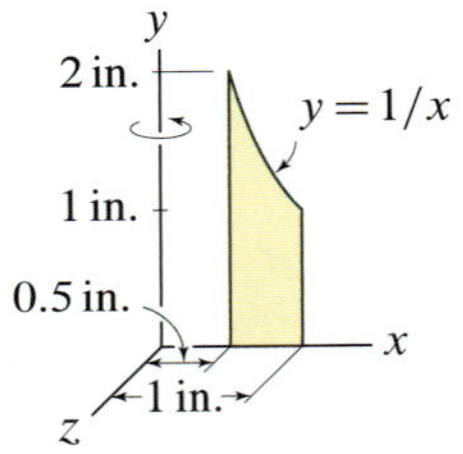

Figure P7.27

Problem 7.27

A solid of revolution is produced by revolving the shaded area shown 360° around the y axis. Use integration to determine the position of the centroid.

7.2 Center of Mass and Center of Gravity

Centroid, center of mass, and center of gravity are different, but the methods used to determine them are essentially the same, as discussed in this section. In fact, under certain circumstances, which occur often, all of these have the same values. To understand the differences between these and when their values differ or are the same, we expand on the definitions that were given in the previous section:

Centroid. The *centroid* is defined to be the average position of a shape, or a distribution of shapes. The shape, or distribution of shapes, can consist of lines (straight and/or curved), areas, or volumes, but not combinations of these. The centroid depends on geometry only and is independent of the material an object might be made of and the presence of gravity.

Center of mass. The *center of mass* is defined to be the average position of a distribution of mass. The center of mass depends on the geometry (shape) of an object and the density of the material it is made of and is independent of the presence of gravity.

Center of gravity. The *center of gravity* is defined to be the average position of a distribution of weight. The center of gravity depends on the geometry (shape) of an object, the density of the material, and the presence of gravity.

Helpful Information

Centroid, center of mass and center of gravity. Because the centroid, center of mass, and center of gravity often have the same values, some people use these words synonymously. However, these are fundamentally different—even when their values are the same—and thus we will always use proper nomenclature.

Center of mass

Using the concepts of the previous section, the center of mass for an object, or a collection of objects, in three dimensions is given by

$$\bar{x} = \frac{\sum_{i=1}^{n} \tilde{x}_i m_i}{\sum_{i=1}^{n} m_i} = \frac{\int \tilde{x}\, dm}{\int dm}, \qquad \bar{y} = \frac{\sum_{i=1}^{n} \tilde{y}_i m_i}{\sum_{i=1}^{n} m_i} = \frac{\int \tilde{y}\, dm}{\int dm},$$
$$\bar{z} = \frac{\sum_{i=1}^{n} \tilde{z}_i m_i}{\sum_{i=1}^{n} m_i} = \frac{\int \tilde{z}\, dm}{\int dm}. \qquad (7.20)$$

The above equations are also applicable to objects in two dimensions, in which case only the $\bar{x}$ and $\bar{y}$ expressions are needed, and for objects in one dimension in which case only the $\bar{x}$ equation is needed. Furthermore, Eq. (7.20) is applicable to *solids*, *surfaces*, and *wires*, and combinations of these, where these various terms are defined as follows:

A *solid* is formed when mass is distributed throughout a volume.

A *surface* is formed when mass is distributed over an area. Usually, the thickness of the surface is small compared to the other dimensions of the surface. Surfaces are often called *plates* or *shells*.

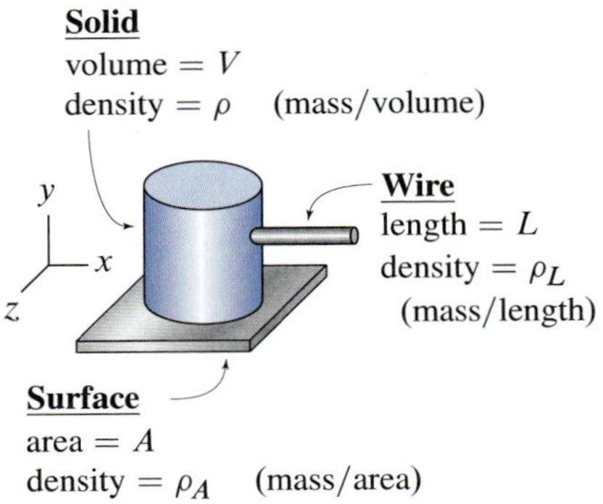

Figure 7.8
Example of an object that consists of a solid, a surface, and a wire, with some of the information needed to determine its center of mass.

A *wire* is formed when mass is distributed along a line (either straight or curved). Usually, the dimensions of the wire's cross section are small compared to the length of the wire.

For example, consider the object shown in Fig. 7.8, which consists of a solid, a surface, and a wire. Using composite shapes, the x position of the center of mass is given by the first expression in Eq. (7.20) as

$$\bar{x} = \frac{\sum_{i=1}^{n} \tilde{x}_i m_i}{\sum_{i=1}^{n} m_i} = \frac{\tilde{x}_1 \rho V + \tilde{x}_2 \rho_A A + \tilde{x}_3 \rho_L L}{\rho V + \rho_A A + \rho_L L}, \tag{7.21}$$

with similar expressions for $\bar{y}$ and $\bar{z}$. In Eq. (7.21), $\tilde{x}_1$, $\tilde{x}_2$ and $\tilde{x}_3$ are the x positions of the centers of mass for the solid, surface, and wire, respectively, for the object in Fig. 7.8. Using integration, the x position of the center of mass for the object in Fig. 7.8 is given by the first expression in Eq. (7.20) as

$$\bar{x} = \frac{\int \tilde{x}\, dm}{\int dm} = \frac{\int \tilde{x}\rho\, dV + \int \tilde{x}\rho_A\, dA + \int \tilde{x}\rho_L\, dL}{\int \rho\, dV + \int \rho_A\, dA + \int \rho_L\, dL}, \tag{7.22}$$

with similar expressions for $\bar{y}$ and $\bar{z}$.

In Eqs. (7.21) and (7.22), the densities ρ, ρ_A, and ρ_L have different definitions and these should not be confused with one another. ρ is the density of a material, according to the usual definition, with dimensions of mass/volume. In contrast, ρ_A is the density of a surface, with dimensions of mass/area. Similarly, ρ_L is the density of a wire, with dimensions of mass/length. Summaries of these definitions and relationships between them are as follows:

$$\rho = \text{density of a material} \quad \text{(dimensions: mass/volume)}, \tag{7.23}$$

$$\rho_A = \text{density of a surface} \quad \text{(dimensions: mass/area)} \quad \rho_A = \rho t, \tag{7.24}$$

$$\rho_L = \text{density of a wire} \quad \text{(dimensions: mass/length)} \quad \rho_L = \rho A, \tag{7.25}$$

where in Eq. (7.24), t is the thickness of the surface, and in Eq. (7.25), A is the cross-sectional area of the wire.

Center of gravity

As we have done throughout this book, and for the vast majority of applications in statics, we assume the special case of a *uniform gravity field*, where the acceleration due to gravity g, as given by Eq. (1.7) on p. 15, is assumed to be constant, and the direction of the attractive forces between Earth and every element of material in a body is assumed to be parallel. With this idealization, the magnitude of the weight dw of a volume dV of material is $dw = \gamma\, dV = \rho g\, dV$, where γ is the material's specific weight (dimensions: weight/volume) and ρ is the material's density (dimensions: mass/volume). Furthermore, since the direction for the weight that each volume element dV experiences is the same, it is customary to define the center of gravity to be the average position of a weight distribution in the same fashion as the center of mass is the average position of a mass distribution. Hence, for a uniform gravity field, formulas for the center of gravity are identical to Eq. (7.20) with appearances of m_i replaced by the weight of an object w_i and appearances of dm replaced by the weight element dw. Hence,

$$\bar{x} = \frac{\sum_{i=1}^{n} \tilde{x}_i w_i}{\sum_{i=1}^{n} w_i} = \frac{\int \tilde{x}\, dw}{\int dw}, \qquad \bar{y} = \frac{\sum_{i=1}^{n} \tilde{y}_i w_i}{\sum_{i=1}^{n} w_i} = \frac{\int \tilde{y}\, dw}{\int dw},$$
$$\bar{z} = \frac{\sum_{i=1}^{n} \tilde{z}_i w_i}{\sum_{i=1}^{n} w_i} = \frac{\int \tilde{z}\, dw}{\int dw}. \qquad (7.26)$$

For applications to objects in two dimensions, only the $\bar{x}$ and $\bar{y}$ expressions are needed, and for objects in one dimension only the $\bar{x}$ equation is needed. Furthermore, Eq. (7.26) is applicable to *solids*, *surfaces*, and *wires*, and combinations of these.

Substituting $w_i = gm_i$ and $dw = g\,dm$ into Eq. (7.26), and noting that g is a constant and thus can be canceled from the numerator and denominator, we see that Eqs. (7.20) and (7.26) are identical, and thus the center of mass and the center of gravity are identical when the gravity field is uniform.

In Fig. 7.9 we reconsider our earlier example of an object that consists of a solid, a surface, and a wire. Using composite shapes, the x position of the center of gravity is given by the first expression in Eq. (7.26) as

$$\bar{x} = \frac{\sum_{i=1}^{n} \tilde{x}_i w_i}{\sum_{i=1}^{n} w_i} = \frac{\tilde{x}_1 \gamma V + \tilde{x}_2 \gamma_A A + \tilde{x}_3 \gamma_L L}{\gamma V + \gamma_A A + \gamma_L L}, \qquad (7.27)$$

with similar expressions for $\bar{y}$ and $\bar{z}$. In Eq. (7.27), $\tilde{x}_1$, $\tilde{x}_2$, and $\tilde{x}_3$ are the x positions of the centers of gravity for the solid, surface, and wire, respectively. Using integration, the x position of the center of gravity for the object in Fig. 7.9 is given by the first expression in Eq. (7.26) as

$$\bar{x} = \frac{\int \tilde{x}\, dw}{\int dw} = \frac{\int \tilde{x}\gamma\, dV + \int \tilde{x}\gamma_A\, dA + \int \tilde{x}\gamma_L\, dL}{\int \gamma\, dV + \int \gamma_A\, dA + \int \gamma_L\, dL}, \qquad (7.28)$$

with similar expressions for $\bar{y}$ and $\bar{z}$.

In Eqs. (7.27) and (7.28), the specific weights γ, γ_A, and γ_L have different definitions. γ is the specific weight (or unit weight) of a material, according to the usual definition, with dimensions of weight/volume. In contrast, γ_A is the specific weight of a surface, with dimensions of weight/area. Similarly, γ_L is the specific weight of a wire, with dimensions of weight/length. Summaries of these definitions and relationships between these are as follows:

$$\gamma = \text{specific weight of a material (dimensions: weight/volume)}, \qquad (7.29)$$

Interesting Fact

Gravity gradient satellites. For some applications, such as satellite mechanics and orbital mechanics, a more accurate treatment of gravity, and hence a more appropriate definition of the center of gravity, is needed. Consider a satellite orbiting Earth.

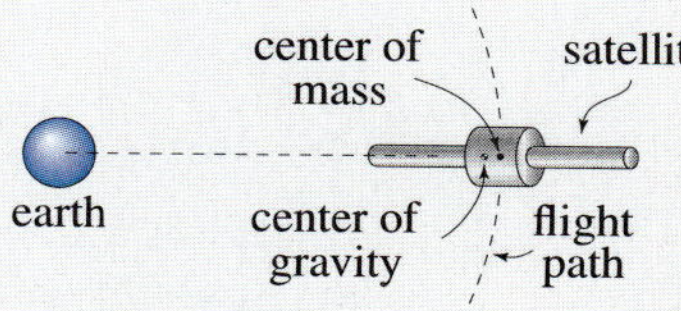

As indicated by Eq. (1.6) on p. 15, portions of the satellite's mass that are closer to Earth experience slightly higher forces of attraction (weight) than portions that are farther away. Hence, for the satellite shown here, the center of gravity is closer to Earth than the center of mass. In fact, assuming the satellite has an overall length of, say, 20 m and is in a low Earth orbit with an altitude of, say, 800 km, and depending on details of its mass distribution, the center of mass and center of gravity may differ by only about 10^{-5} m. While this difference may seem insignificant, it is often sufficient to stabilize the satellite's orientation, and such satellites are called *gravity gradient satellites*.

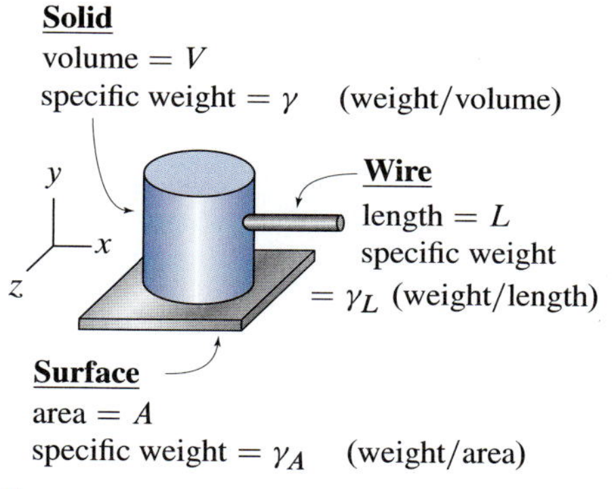

Figure 7.9
Example of an object that consists of a solid, a surface, and a wire, with some of the information needed to determine its center of gravity.

$$\gamma_A = \text{specific weight of a surface} \qquad (7.30)$$
$$\text{(dimensions: weight/area)} \quad \gamma_A = \gamma t,$$
$$\gamma_L = \text{specific weight of a wire} \qquad (7.31)$$
$$\text{(dimensions: weight/length)} \quad \gamma_L = \gamma A,$$

where in Eq. (7.30), t is the thickness of the surface, and in Eq. (7.31), A is the cross-sectional area of the wire.

End of Section Summary

The *center of mass* is defined to be the average position of a distribution of mass. The center of mass for an object can be determined using composite shapes and/or integration, and the object can be a solid, surface, or wire, or a combination of these. The center of mass is affected by an object's shape and mass distribution.

The *center of gravity* is defined to be the average position of a distribution of weight. For the vast majority of applications in statics, we assume the special case of a *uniform gravity field*, where the acceleration due to gravity g is assumed to be constant, and the direction of the attractive forces between Earth and every element of material in a body is assumed to be parallel. With this idealization, the center of gravity can be defined in similar fashion as the center of mass, and indeed, for a uniform gravity field the center of mass and center of gravity of an object are always the same.

The final point to summarize is that if an object has an axis of symmetry, which means that *both* the object's shape *and* its mass distribution are symmetric, then the center of mass lies on the axis of symmetry. Furthermore, if the object has two axes of symmetry, then the center of mass is located at the intersection of the axes of symmetry.

EXAMPLE 7.7 *Center of Mass Using Composite Shapes*

A kite consists of wooden members, paper, and string. The wooden members AB and CD have density $\rho_w = 510\,\mathrm{kg/m^3}$ and 5 mm by 5 mm square cross section. The paper $ADBC$ has density $\rho_p = 0.039\,\mathrm{kg/m^2}$. Around the perimeter of the kite (between points A, D, B, C, and A) is taut string with density $\rho_s = 2.6(10)^{-5}\,\mathrm{kg/m}$. Determine the weight of the kite and the center of mass.

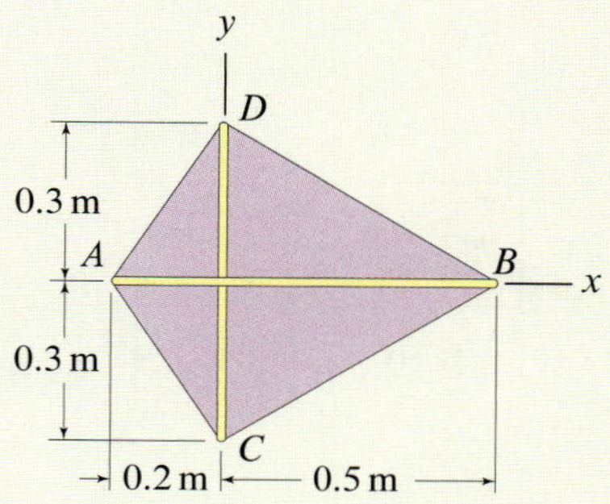

Figure 1

SOLUTION

Road Map The kite consists of an arrangement of objects having simple shapes. Thus it will be easiest to determine the center of mass by using composite shapes with Eq. (7.21) on p. 414. The kite is symmetric about the x axis, thus its center of mass must lie on the x axis, and hence $\bar{y} = 0$.

Governing Equations & Computation In Fig. 2 the kite is subdivided into eight composite shapes, or objects. The mass of each composite shape and the position of its center of mass are collected in Table 1.

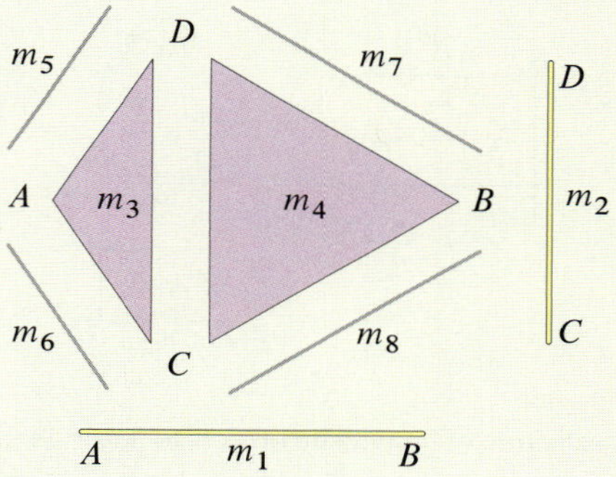

Figure 2
A selection of composite objects to describe the kite.

Table 1. Masses and center of mass positions for the composite objects in Fig. 2. Note: "w," "p," and "s" stand for wood, paper, and string, respectively.

Object no.	m_i (kg)	$\tilde{x}_i$ (m)
1 (w AB)	$510\frac{\mathrm{kg}}{\mathrm{m}^3}(0.005\,\mathrm{m})(0.005\,\mathrm{m})(0.7\,\mathrm{m}) = 0.008925$	0.15
2 (w CD)	$510\frac{\mathrm{kg}}{\mathrm{m}^3}(0.005\,\mathrm{m})(0.005\,\mathrm{m})(0.6\,\mathrm{m}) = 0.007650$	0
3 (p ADC)	$0.039\frac{\mathrm{kg}}{\mathrm{m}^2}\,\frac{1}{2}(0.6\,\mathrm{m})(0.2\,\mathrm{m}) = 0.002340$	−0.06667
4 (p BCD)	$0.039\frac{\mathrm{kg}}{\mathrm{m}^2}\,\frac{1}{2}(0.6\,\mathrm{m})(0.5\,\mathrm{m}) = 0.005850$	0.1667
5 (s AD)	$2.6(10)^{-5}\frac{\mathrm{kg}}{\mathrm{m}}\sqrt{(0.2\,\mathrm{m})^2 + (0.3\,\mathrm{m})^2} = 9.374(10)^{-6}$	−0.1
6 (s AC)	$(\ldots{}''\ldots) = 9.374(10)^{-6}$	−0.1
7 (s BD)	$2.6(10)^{-5}\frac{\mathrm{kg}}{\mathrm{m}}\sqrt{(0.5\,\mathrm{m})^2 + (0.3\,\mathrm{m})^2} = 1.516(10)^{-5}$	0.25
8 (s BC)	$(\ldots{}''\ldots) = 1.516(10)^{-5}$	0.25

Using Eq. (7.21) provides the x position of the center of mass as

$$\bar{x} = \frac{\sum_{i=1}^{n} \tilde{x}_i m_i}{\sum_{i=1}^{n} m_i} = \frac{(0.15\,\mathrm{m})(0.008925\,\mathrm{kg}) + \cdots + (0.25\,\mathrm{m})(1.516(10)^{-5}\,\mathrm{kg})}{0.008925\,\mathrm{kg} + \cdots + 1.516(10)^{-5}\,\mathrm{kg}} \tag{1}$$

$$= \frac{0.002163\,\mathrm{kg{\cdot}m}}{0.02481\,\mathrm{kg}} = \boxed{0.0872\,\mathrm{m}.} \tag{2}$$

For brevity, numbers for only the first and eighth objects are shown in Eq. (1), and you should verify the results that are given. The location of the center of mass is shown in Fig. 3. In Eq. (2), the denominator is the total mass of the kite, $m = 0.02481$ kg. Hence, the weight of the kite is

$$w = mg = (0.02481\,\mathrm{kg})(9.81\,\mathrm{m/s^2}) = \boxed{0.243\,\mathrm{N}.} \tag{3}$$

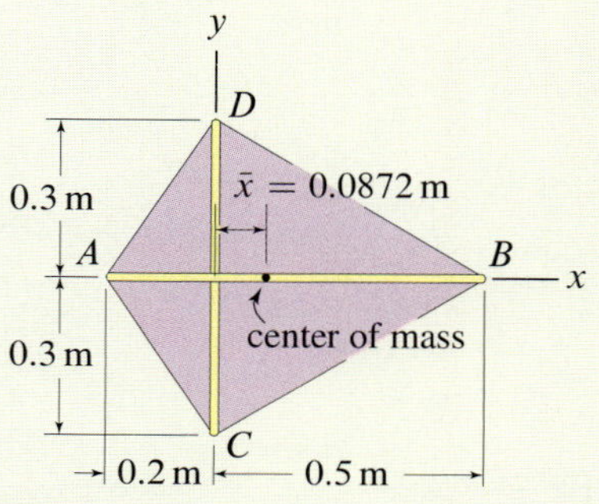

Figure 3
Location of the kite's center of mass.

Discussion & Verification We do not have a definitive check of our solution's accuracy. Nonetheless, the center of mass has a reasonable location, and the weight of the kite is reasonable (it may be helpful to express the weight in U.S. Customary units—you should find that the kite weighs slightly less than 1 ounce).

EXAMPLE 7.8 *Center of Gravity Using Composite Shapes*

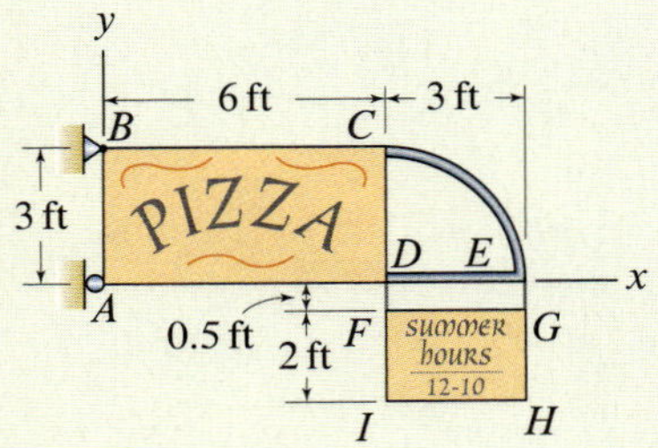

Figure 1

A sign for a restaurant is constructed of rectangular sheets of plywood $ABCD$ and $FGHI$, steel bar CED, and chains DF and EG. The plywood weighs $2\,\mathrm{lb/ft^2}$, the steel bar weighs $5\,\mathrm{lb/ft}$, and the weight of the chains is negligible.

(a) Determine the center of gravity.

(b) Determine the support reactions at A and B.

SOLUTION

Road Map Because the sign consists of an arrangement of objects having simple shapes, it will be easiest to determine the center of gravity by using composite shapes with Eq. (7.27) on p. 415. Once the center of gravity is determined, it will be straightforward to determine the support reactions.

Part (a)

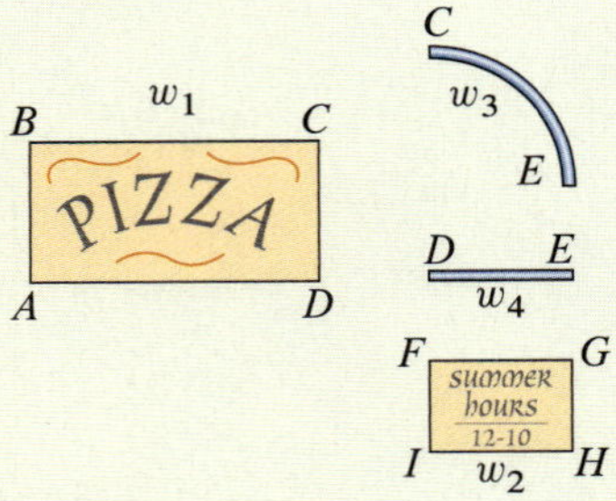

Figure 2
A selection of composite objects to describe the sign.

Governing Equations & Computation The sign is subdivided into four composite shapes, or objects, as shown in Fig. 2. The weight of each composite shape and the position of its center of gravity are collected in Table 1. For object 3 in Table 1 (the quarter circular pipe), $\pi(3\,\mathrm{ft})/2$ is the length of the pipe, and we consult the Table of Properties of Solids on the inside back cover of this book to determine $\tilde{x} = 6\,\mathrm{ft} + 2(3\,\mathrm{ft})/\pi = 7.910\,\mathrm{ft}$ and $\tilde{y} = 2(3\,\mathrm{ft})/\pi = 1.910\,\mathrm{ft}$. Using Eq. (7.27) provides the x and y positions of the center of gravity as

Table 1. Weights and center of gravity positions for composite objects. Note: "w" and "s" stand for wood and steel, respectively.

Object no.	w_i (lb)	$\tilde{x}_i$ (ft)	$\tilde{y}_i$ (ft)
1 (w $ABCD$)	$2\frac{\mathrm{lb}}{\mathrm{ft}^2}(3\,\mathrm{ft})(6\,\mathrm{ft}) = 36$	3.0	1.5
2 (w $FGHI$)	$2\frac{\mathrm{lb}}{\mathrm{ft}^2}(2\,\mathrm{ft})(3\,\mathrm{ft}) = 12$	7.5	−1.5
3 (s CE)	$5\frac{\mathrm{lb}}{\mathrm{ft}}\,\pi(3\,\mathrm{ft})/2 = 23.56$	7.910	1.910
4 (s DE)	$5\frac{\mathrm{lb}}{\mathrm{ft}}(3\,\mathrm{ft}) = 15$	7.5	0

$$\bar{x} = \frac{\sum_{i=1}^{n} \tilde{x}_i w_i}{\sum_{i=1}^{n} w_i} \tag{1}$$

$$= \frac{(3.0\,\mathrm{ft})(36\,\mathrm{lb}) + (7.5\,\mathrm{ft})(12\,\mathrm{lb}) + (7.910\,\mathrm{ft})(23.56\,\mathrm{lb}) + (7.5\,\mathrm{ft})(15\,\mathrm{lb})}{36\,\mathrm{lb} + 12\,\mathrm{lb} + 23.56\,\mathrm{lb} + 15\,\mathrm{lb}} \tag{2}$$

$$= \frac{496.9\,\mathrm{ft{\cdot}lb}}{86.56\,\mathrm{lb}} = \boxed{5.74\,\mathrm{ft},} \tag{3}$$

$$\bar{y} = \frac{\sum_{i=1}^{n} \tilde{y}_i w_i}{\sum_{i=1}^{n} w_i} \tag{4}$$

$$= \frac{(1.5\,\mathrm{ft})(36\,\mathrm{lb}) + (-1.5\,\mathrm{ft})(12\,\mathrm{lb}) + (1.910\,\mathrm{ft})(23.56\,\mathrm{lb}) + (0\,\mathrm{ft})(15\,\mathrm{lb})}{86.56\,\mathrm{lb}} \tag{5}$$

$$= \frac{81.00\,\text{ft}\cdot\text{lb}}{86.56\,\text{lb}} = \boxed{0.936\,\text{ft.}} \tag{6}$$

The denominators in Eqs. (1) through (6) are the total weight of the sign, hence

$$w = 86.6\,\text{lb} \tag{7}$$

Part (b)

Modeling The FBD for the sign is shown in Fig. 3 where the sign's 86.6 lb weight is placed at the center of gravity.

Governing Equations & Computation With the FBD in Fig. 3, equilibrium equations are written and easily solved as follows:

$$\sum M_B = 0: \quad A_x(3\,\text{ft}) - (86.6\,\text{lb})(5.74\,\text{ft}) = 0 \tag{8}$$

$$\Rightarrow \boxed{A_x = 166\,\text{lb},} \tag{9}$$

$$\sum F_x = 0: \quad A_x + B_x = 0 \tag{10}$$

$$\Rightarrow \boxed{B_x = -166\,\text{lb},} \tag{11}$$

$$\sum F_y = 0: \quad B_y - 86.6\,\text{lb} = 0 \tag{12}$$

$$\Rightarrow \boxed{B_y = 86.6\,\text{lb}.} \tag{13}$$

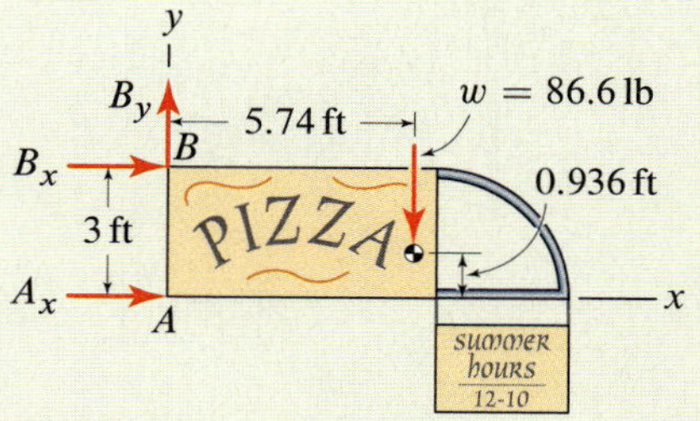

Figure 3
Free body diagram where the sign's weight is placed at the center of gravity.

Discussion & Verification

- The center of gravity for the sign, given by Eqs. (3) and (6), appears to be in a reasonable position. The reactions have proper directions and appear to have reasonable values.
- If we required the answer to only Part (b) of this example, we could use the solution as carried out above (note that the FBD shown in Fig. 3 requires determination of the center of gravity); or alternatively, it may be quicker to use the FBD shown in Fig. 4, where the center of gravity of the sign is not needed. In Fig. 4, the weight of each component object is placed at its center of gravity.

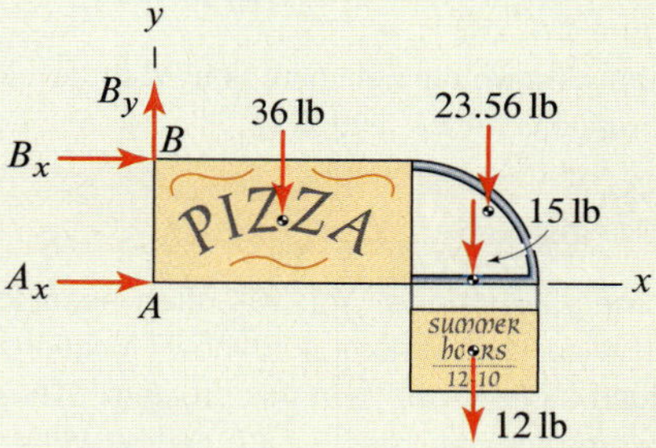

Figure 4
Free body diagram where the weight of each component object is placed at that object's center of gravity.

EXAMPLE 7.9 *Center of Gravity Using Integration*

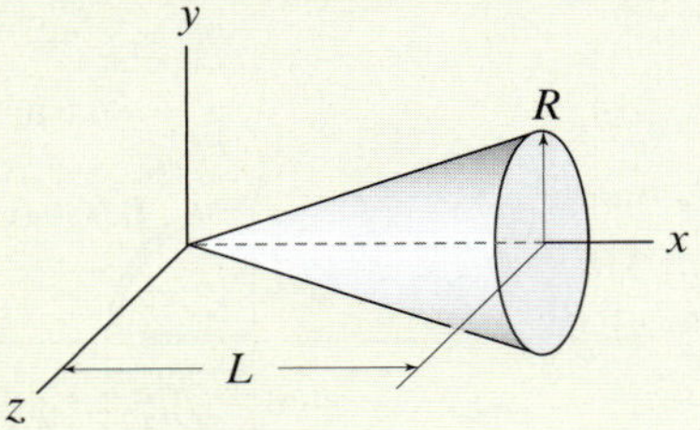

Figure 1

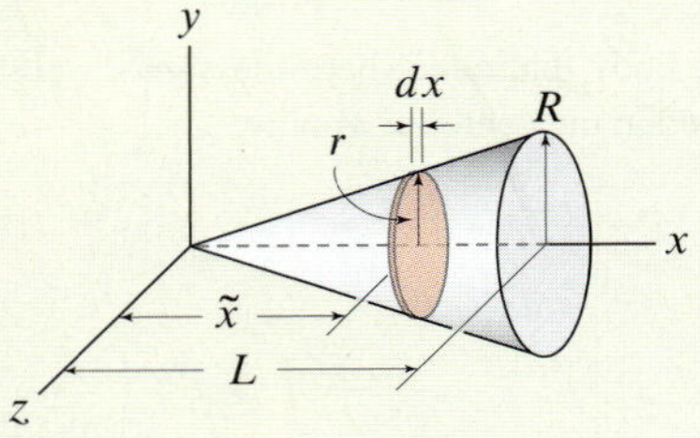

Figure 2
A thin disk volume element is used to develop expressions for dV and $\tilde{x}$.

Helpful Information

Linear functions. You will often need to develop expressions for linear functions, such as for r and γ in this problem. While expressions such as Eq. (3) can by written by inspection or by using a little trial and error, the following more rigorous procedure can always be used. Consider the expression for the specific weight γ. The problem description states that γ varies linearly, hence

$$\gamma = a + bx \tag{4}$$

where a and b are constants that are determined by noting that $\gamma = \gamma_0$ at $x = 0$ and $\gamma = \gamma_0/2$ at $x = L$. Hence,

$$\gamma = a + b(0) = \gamma_0, \tag{5}$$

$$\gamma = a + b(L) = \gamma_0/2. \tag{6}$$

Solving Eqs. (5) and (6) for a and b provides

$$a = \gamma_0 \quad \text{and} \quad b = -\frac{\gamma_0}{2L}. \tag{7}$$

Substituting a and b into Eq. (4) results in Eq. (3). Note that regardless of the approach you use to obtain a linear expression, you can always check its validity.

A solid cone is constructed of a *functionally graded* material that is designed to be hard at the tip and softer farther away. As a consequence, the specific weight of the material varies linearly from γ_0 at the tip to $\gamma_0/2$ at $x = L$. Determine the location of the center of gravity and the weight of the cone.

SOLUTION

Road Map Because the cone has specific weight that changes with position, the center of gravity will most likely not coincide with the centroid. To determine the center of gravity, we have little choice but to use integration, with the integral expressions in Eq. (7.26) on p. 415, or equivalently the volume integral expressions in Eq. (7.28). Noting that the specific weight changes with x coordinate only, the cone is symmetric about the x axis, and thus its center of gravity must lie on the x axis, and $\bar{y} = \bar{z} = 0$.

Governing Equations & Computation Using the expression $\bar{x} = \int \tilde{x}\, dw$ in Eq. (7.26) with $dw = \gamma\, dV$ gives

$$\bar{x} = \frac{\int \tilde{x}\gamma\, dV}{\int \gamma\, dV} \tag{1}$$

In Eq. (1), $\tilde{x}$ is the location of the center of gravity for the weight element $\gamma\, dV$. Because of the variable specific weight, a thin disk weight (volume) element, as shown in Fig. 2, will be more convenient than a thin shell weight (volume) element (see Prob. 7.37 for further discussion). Expressions for dV and $\tilde{x}$ are

$$dV = \pi r^2\, dx \quad \text{and} \quad \tilde{x} = x. \tag{2}$$

In Eq. (2), πr^2 is the area of a disk where, as shown in Fig. 2, r is the radius and dx is the thickness. The center of gravity of the disk element is located at $\tilde{x} = x$. Before we evaluate Eq. (1), it is necessary to express both the radius r and specific weight γ of the cone as functions of x position. These expressions are

$$r = R\frac{x}{L} \quad \text{and} \quad \gamma = \gamma_0\left(1 - \frac{x}{2L}\right). \tag{3}$$

The validity of Eqs. (3) is easily verified by evaluating each expression at $x = 0$ and $x = L$ to see that they produce the proper results. The margin note on this page provides some tips for developing expressions such as these.

Substituting Eqs. (2) and (3) into Eq. (1) provides the center of gravity

$$\bar{x} = \frac{\int \tilde{x}\gamma\, dV}{\int \gamma\, dV} = \frac{\int_0^L x\gamma_0\left(1 - \frac{x}{2L}\right)\pi\left(R\frac{x}{L}\right)^2 dx}{\int_0^L \gamma_0\left(1 - \frac{x}{2L}\right)\pi\left(R\frac{x}{L}\right)^2 dx} = \frac{\frac{3\pi L^2 R^2}{20}\gamma_0}{\frac{5\pi L R^2}{24}\gamma_0} = \boxed{\frac{18L}{25}}. \tag{8}$$

The denominator in Eq. (8) is the total weight of the cone, hence

$$\boxed{w = \frac{5\pi L R^2}{24}\gamma_0.} \tag{9}$$

Discussion & Verification The *centroid* of a solid cone, as given in the Table of Properties of Solids on the inside back cover of this book, is $3L/4$ from the tip, while the *center of gravity* in this example, from Eq. (8), is $\bar{x} = 18\,L/25 = (0.72)L$. As expected, because of the specific weight distribution in this problem, we see that the center of gravity is closer to the tip than the centroid. Of course, if the specific weight were uniform, the center of mass and the centroid would have the same location.

PROBLEMS

General instructions. For shapes that have one or more axes or planes of symmetry, you may use inspection to determine some of the coordinates of the center of mass or center of gravity.

Problem 7.28

The radio antenna shown is symmetric about its horizontal member. The horizontal member weighs 0.6 lb/ft, the vertical members weigh 0.1 lb/ft, and the circuit box weighs 8 lb with center of gravity at point A. Determine the location of the antenna's center of gravity, measured from the left-hand end.

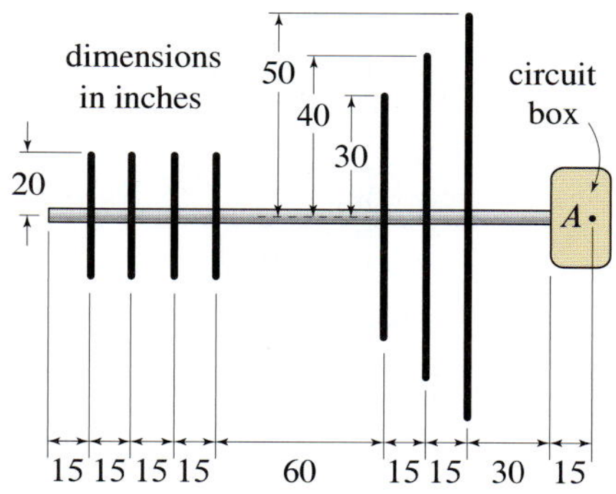

Figure P7.28

Problem 7.29

In Example 7.2 on p. 403, let the cylindrical portion be steel with specific weight $490\,\text{lb/ft}^3$ and the truncated cone portion be aluminum alloy with specific weight $170\,\text{lb/ft}^3$. Determine the center of gravity.

Problem 7.30

In Prob. 7.7 on p. 409, let the cylindrical portion be cast iron with specific weight $450\,\text{lb/ft}^3$ and the hemispherical portion be aluminum alloy with specific weight $170\,\text{lb/ft}^3$. Determine the center of gravity.

Problem 7.31

A square plate having $1.25\,\text{lb/in.}^2$ weight has a circular hole. Along the right-hand edge of the plate, a circular cross section rod having $0.75\,\text{lb/in.}$ weight is welded to it. Determine the x position of the center of gravity.

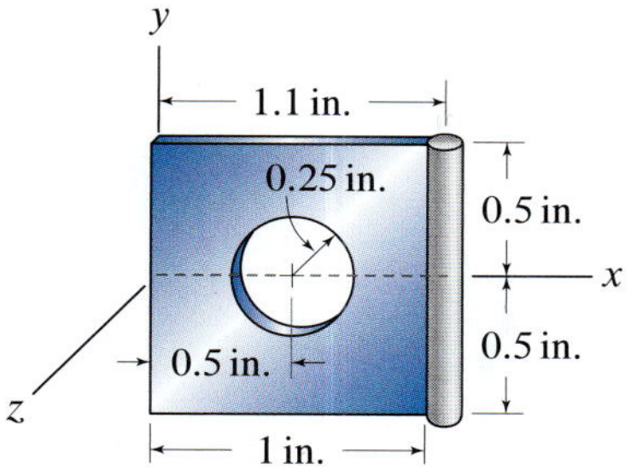

Figure P7.31

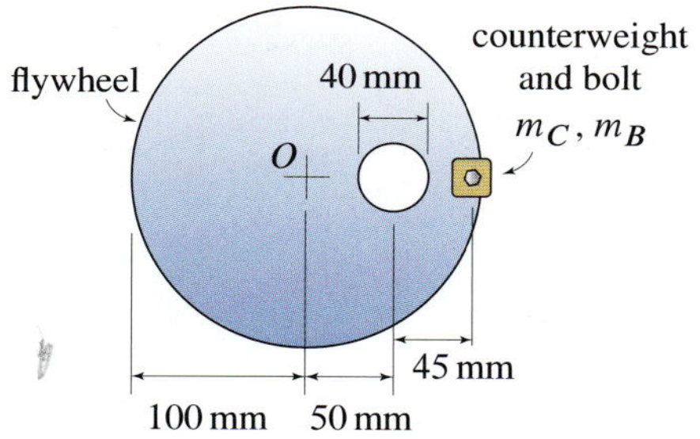

Figure P7.32

Problem 7.32

The assembly shown consists of a flywheel, a counterweight, and a bolt that attaches the counterweight to the flywheel. The flywheel is uniform with mass per area of $2\,\text{g/mm}^2$ and has a 40 mm diameter hole as shown. Determine the mass m_C of the counterweight so that the center of mass of the assembly is at point O. The counterweight is attached to the flywheel by a bolt with mass $m_B = 800\,\text{g}$ that is screwed into a 10 mm diameter hole that passes completely through the flywheel.

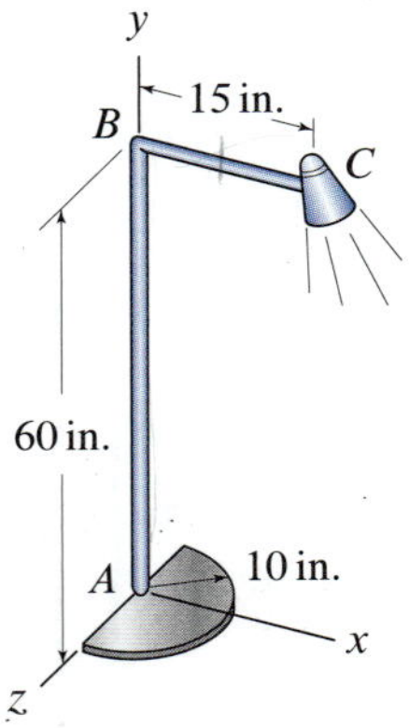

Figure P7.33

Problem 7.33

A floor lamp consists of a half-circular base with weight per area of $0.06\ \text{lb/in}^2$, tubes AB and BC, each having weight $0.05\ \text{lb/in}$, and the lamp shade at C with weight 2 lb. Tube BC is parallel to the x axis. Determine the coordinates of the center of gravity.

Problem 7.34

The assembly shown is made of a semicircular plate having weight per area of $0.125\ \text{lb/in.}^2$ and uniform rods having weight per length of $0.05\ \text{lb/in}$. Determine the coordinates of the center of gravity.

Problem 7.35

The object shown consists of a rectangular solid, a plate, and a quarter-circular wire, with masses as follows: solid: $0.0005\ \text{g/mm}^3$, plate: $0.01\ \text{g/mm}^2$, wire: $0.05\ \text{g/mm}$. Determine the coordinates of the center of mass.

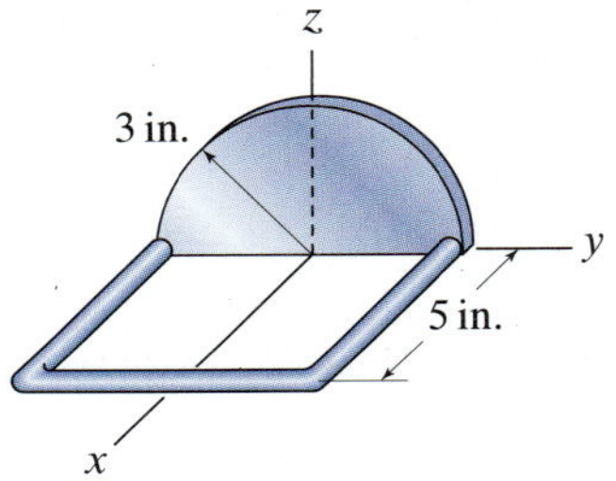

Figure P7.34

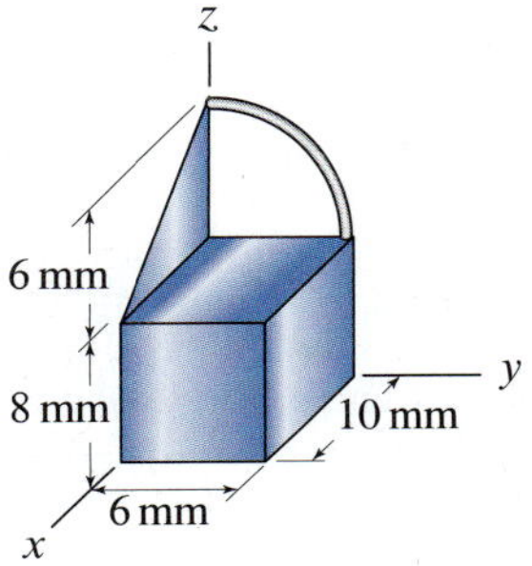

Figure P7.35

Problem 7.36

A sign frame is constructed of metal pipe having a mass per length of $4\ \text{kg/m}$. Portion CDE is semicircular. The frame is supported by a pin at B and a weightless cable CF. Determine the center of mass for the frame, the reactions at B, and the force in the cable.

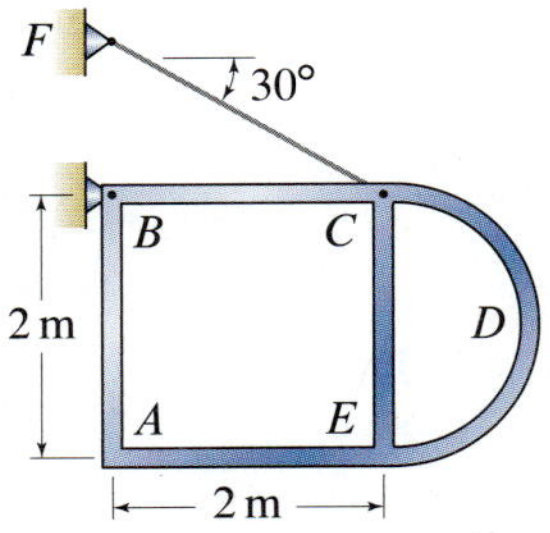

Figure P7.36

Problem 7.37

In Example 7.9 on p. 420, the center of gravity of a cone with variable specific weight is found using integration with a thin disk mass element. Discuss why, for this example, a thin disk mass element is considerably more convenient than a thin shell mass element. **Note:** Concept problems are about *explanations*, not computations.

Problems 7.38 through 7.41

For the object indicated:

(a) Fully set up the integral, including the limits of integration, that will yield the center of mass of the object.

(b) Evaluate the integral determined in Part (a), using computer software such as Mathematica, Maple, etc.

Problem 7.38 The solid hemisphere of radius r shown has density $\rho_0/2$ for $0 \le x \le r/2$ and ρ_0 for $r/2 \le x \le r$.

Problem 7.39 The hollow cone shown is constructed of a material with uniform density ρ_0 and has wall thickness that varies linearly from $2t_0$ at $x = 0$ to t_0 at $x = L$. Assume t_0 is much smaller than L and R.

Problem 7.40 A solid of revolution is formed by revolving the area shown 360° about the x axis. The material has uniform density ρ_0.

Problem 7.41 A solid of revolution is formed by revolving the area shown 360° about the y axis. The material has uniform density ρ_0.

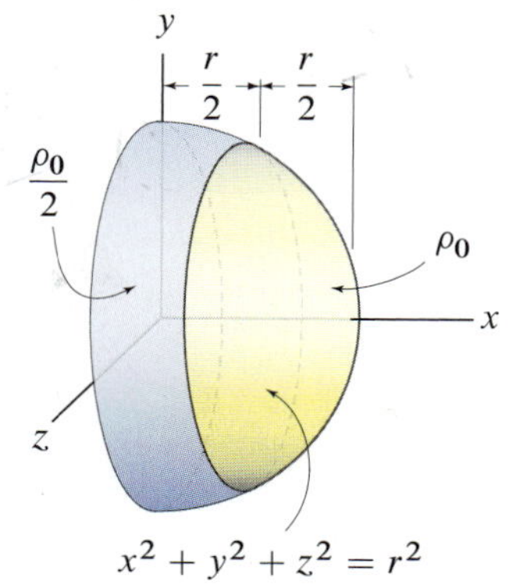

Figure P7.38

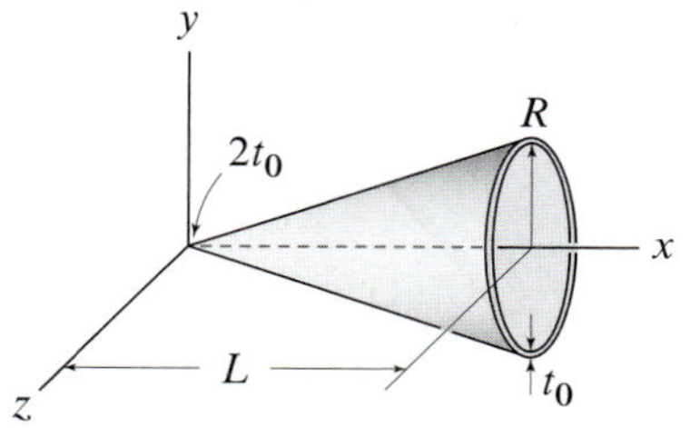

Figure P7.39

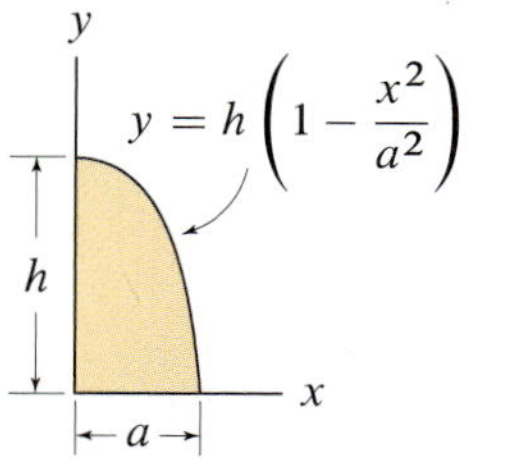

Figure P7.40 and P7.41

Problems 7.42 through 7.44

For the object indicated:

(a) Fully set up the integral, including the limits of integration, that will yield the center of gravity of the object.

(b) Evaluate the integral determined in Part (a), using computer software such as Mathematica, Maple, etc.

Problem 7.42 The solid shown has a cone-shaped cavity. The material has uniform specific weight γ_0.

Problem 7.43 A solid of revolution is formed by revolving the area shown 360° about the x axis. The material has uniform specific weight γ_0.

Problem 7.44 A solid of revolution is formed by revolving the area shown 360° about the y axis. The material has uniform specific weight γ_0.

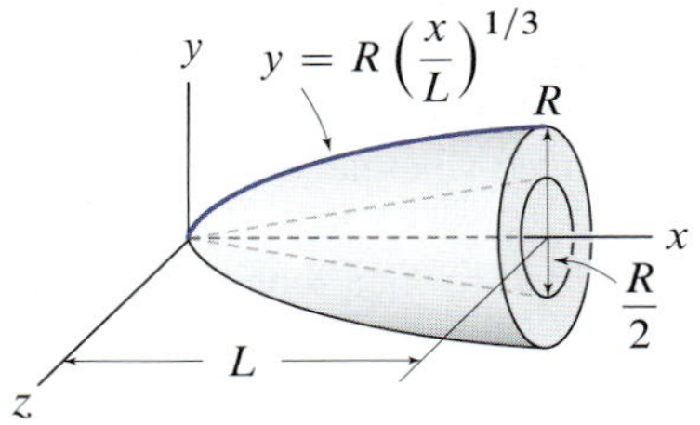

Figure P7.42

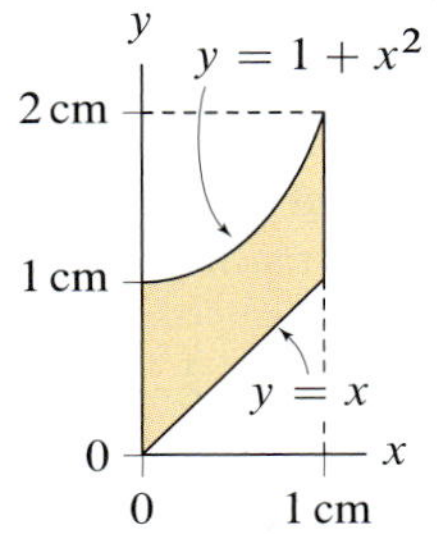

Figure P7.43 and P7.44

7.3 Theorems of Pappus and Guldinus

The two theorems discussed here are jointly attributed to Pappus of Alexandria, a mathematician who was active at the end of the 3rd century A.D. (the years of his birth, his death, and publication of his works are uncertain), and Paul Guldin, or Guldinus (original name Habakkuk Guldin) (1577–1643), who was a mathematician and astronomer. These theorems are useful and straightforward to apply for determining the area of a surface of revolution and the volume of a solid of revolution. We first present these theorems and then prove their validity.

Area of a surface of revolution

A *surface of revolution* is produced by rotating a *generating curve*, as shown in Fig. 7.10, by an angle θ (in radians) about an *axis of revolution*. The axis of revolution and the generating curve lie in the same plane, and the generating curve must not intersect the axis of revolution, although portions of the generating curve may lie on the axis of revolution. Based on physical considerations, the angle θ through which the generating curve is rotated must be positive. If $\theta = 2\pi$, a full surface of revolution is produced whereas if $0 < \theta < 2\pi$, then a partial surface of revolution is produced. Values of $\theta > 2\pi$ are normally not possible, but may be considered for unusual applications (e.g., a very thin surface wrapped upon itself multiple times).

If the generating curve has arc length L and the position of its centroid relative to the axis of revolution is $\bar{r}$, as shown in Fig. 7.10, then the area of the surface of revolution is

$$A = \theta \bar{r} L. \tag{7.32}$$

Equation (7.32) gives the area of only *one* side of the surface.

The generating curve does not need to be smooth for Eq. (7.32) to apply. For example, the generating curve shown in Fig. 7.11 consists of two straight-line segments. Equation (7.32) may be applied directly if we first determine the centroid location for the entire generating curve. An approach that is usually simpler for generating curves that consist of simple composite shapes, such as in Fig. 7.11, is to recognize that $\bar{r}L$ in Eq. (7.32) is the first moment of the generating curve's length with respect to the axis of revolution, in which case $\bar{r}L = \sum_{i=1}^{n} \tilde{r}_i L_i$, where L_i is the arc length of composite shape i (which is a line segment, either straight or curved, but usually with simple shape), $\tilde{r}_i$ is the position of the centroid of L_i measured from the axis of revolution, and n is the number of composite shapes that constitute the generating curve. Hence, Eq. (7.32) can be expressed as

$$A = \theta \sum_{i=1}^{n} \tilde{r}_i L_i. \tag{7.33}$$

For the example in Fig. 7.11, Eq. (7.33) becomes

$$A = \theta \sum_{i=1}^{2} \tilde{r}_i L_i = \theta(\tilde{r}_1 L_1 + \tilde{r}_2 L_2). \tag{7.34}$$

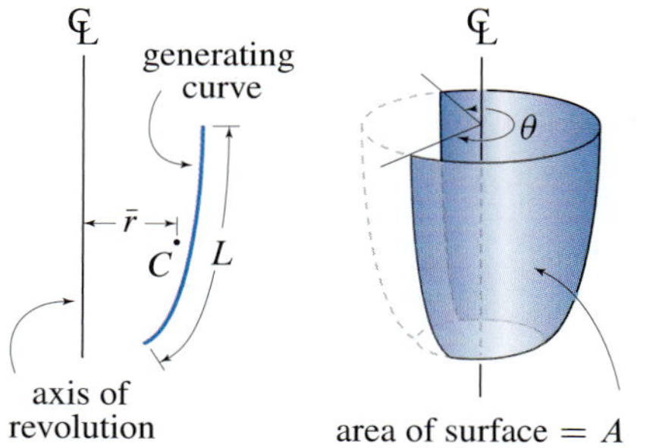

Figure 7.10
A surface of revolution is produced by rotating a generating curve by an angle θ about an axis of revolution. The generating curve has arc length L and centroid C located a distance $\bar{r}$ from an axis of revolution.

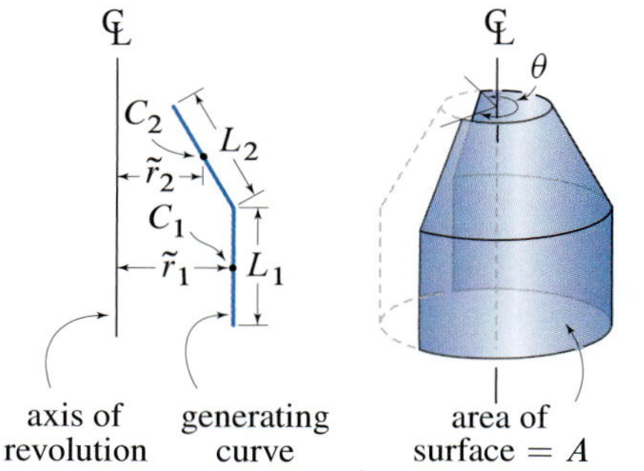

Figure 7.11
Example of a surface of revolution produced by a generating curve consisting of simple composite shapes. The two straight-line segments have lengths L_1 and L_2 and centroids located at C_1 and C_2.

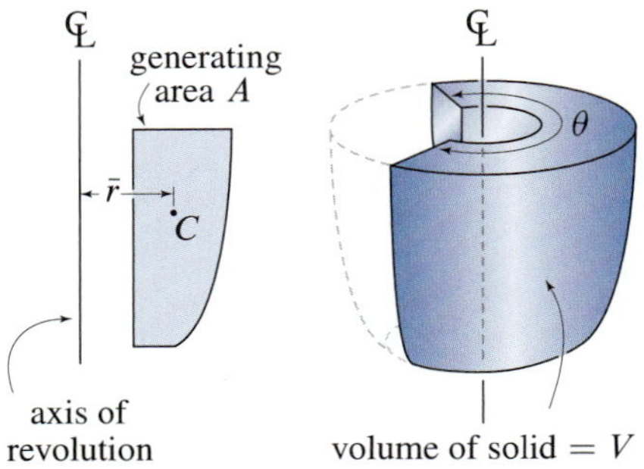

Figure 7.12
A solid of revolution is produced by rotating a generating area by an angle θ about an axis of revolution. The generating area A has its centroid C located a distance $\bar{r}$ from the axis of revolution.

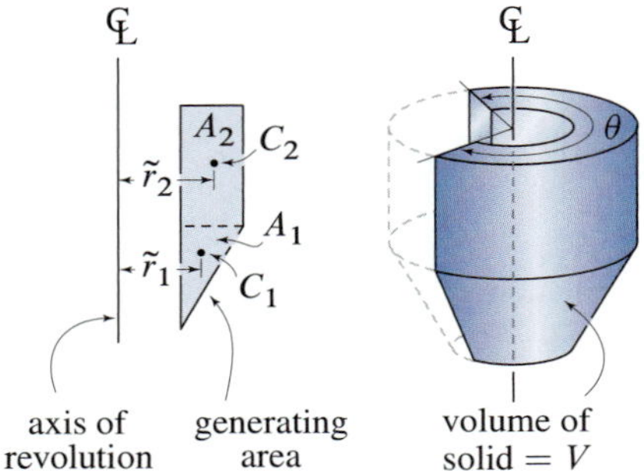

Figure 7.13
Example of a solid of revolution produced by a generating area consisting of simple composite shapes. The triangle and rectangle have areas A_1 and A_2 and centroids located at C_1 and C_2, respectively.

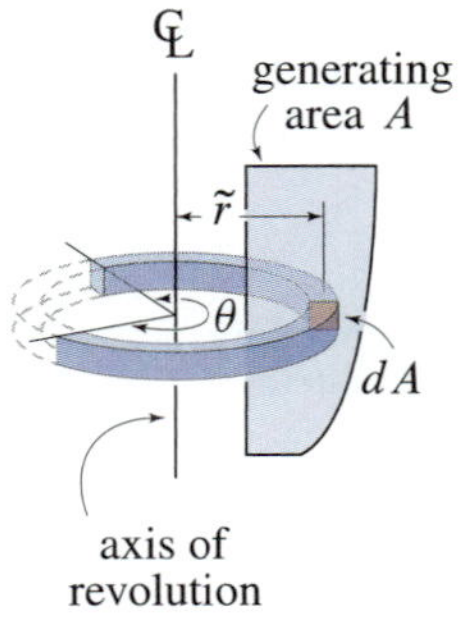

Figure 7.14
The area element dA is located a distance $\tilde{r}$ from the axis of revolution, and it gives rise to a volume $dV = \theta\tilde{r}\,dA$.

Volume of a solid of revolution

A *solid of revolution* is produced by rotating a *generating area*, as shown in Fig. 7.12, by an angle θ (in radians) about an axis of revolution. The axis of revolution and the generating area lie in the same plane, and the generating area must not intersect the axis of revolution, although portions of the generating area may lie on the axis of revolution. Based on physical considerations, the angle θ through which the generating area is rotated must be positive. If $\theta = 2\pi$, a full solid of revolution is produced whereas if $0 < \theta < 2\pi$, then a partial solid of revolution is produced. Values of $\theta > 2\pi$ are not possible.

If the centroid C of the generating area A has position $\bar{r}$ from the axis of revolution, as shown in Fig. 7.12, then the volume of the solid of revolution is

$$V = \theta\bar{r}A. \tag{7.35}$$

For applications when the generating area consists of composite shapes, such as shown in Fig. 7.13, Eq. (7.35) may be applied directly if we first determine the centroid location for the entire generating area. However, it is usually easier to replace $\bar{r}A$ in Eq. (7.35) by $\bar{r}A = \sum_{i=1}^{n} \tilde{r}_i A_i$, where A_i is the area of composite shape i, $\tilde{r}_i$ is the position of the centroid of A_i measured from the axis of revolution, and n is the number of composite shapes that constitute the generating area. Hence, Eq. (7.35) can be expressed as

$$V = \theta \sum_{i=1}^{n} \tilde{r}_i A_i. \tag{7.36}$$

For the example in Fig. 7.13, Eq. (7.36) becomes

$$V = \theta \sum_{i=1}^{2} \tilde{r}_i A_i = \theta(\tilde{r}_1 A_1 + \tilde{r}_2 A_2). \tag{7.37}$$

Proof of the Pappus–Guldinus theorems

The theorems of Pappus and Guldinus are easy to prove using calculus and the concepts of centroid defined in Section 7.1. In recognition of their contribution, we note Pappus and Guldinus developed these theorems long before calculus was invented. Because of the similarity between the proofs for a solid of revolution, and a surface of revolution, we discuss only the solid of revolution, using the generating area shown in Fig. 7.14. The area element dA is located a distance $\tilde{r}$ from the axis of revolution, and this area element gives rise to a volume $dV = \theta\tilde{r}\,dA$. The total volume of the solid of revolution is then

$$V = \int dV = \theta \int \tilde{r}\,dA. \tag{7.38}$$

Using Eq. (7.9) from Section 7.1 (p. 397), with position measured from the axis of revolution, the centroid of the generating area is $\bar{r} = \int \tilde{r}\,dA/A$, and hence Eq. (7.38) becomes

$$V = \theta\bar{r}A, \tag{7.39}$$

which is identical to Eq. (7.35).

End of Section Summary

The theorems of Pappus and Guldinus provide a straightforward means to determine the surface area and volume of objects of revolution.

A *surface of revolution* is produced by rotating a *generating curve*, as shown in Fig. 7.10, by an angle θ (in radians) about an *axis of revolution*. A *solid of revolution* is produced by rotating a *generating area*, as shown in Fig. 7.12, by an angle θ about an axis of revolution. A full surface or volume of revolution will have $\theta = 2\pi$, whereas a partial surface or volume of revolution will have $0 < \theta < 2\pi$.

EXAMPLE 7.10 *Volume and Surface Area of an Object of Revolution*

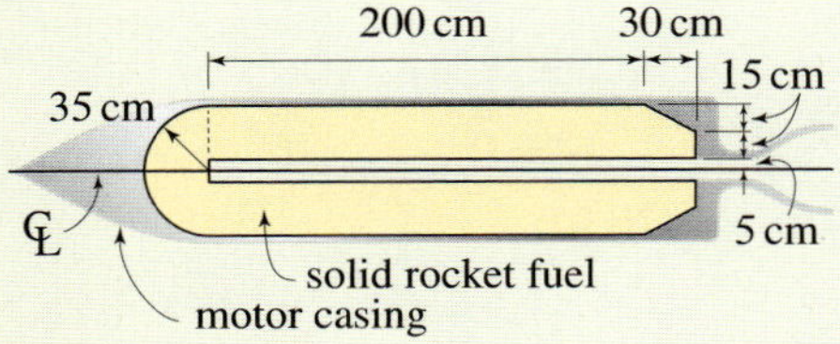

Figure 1

The cross section of a solid-fuel rocket motor is shown, where the fuel is a solid of revolution.

(a) Determine the volume of fuel.

(b) Determine the total surface area of the fuel.

SOLUTION

Road Map The fuel for the rocket is a solid of revolution, as shown in Fig. 2. We will use the theorems of Pappus and Guldinus to determine the volume of the solid and the surface area of the solid.

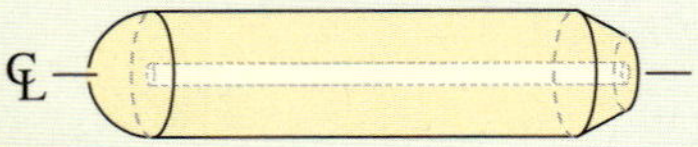

Figure 2
The fuel for the rocket is a solid of revolution.

Part (a)

Governing Equations & Computation In Fig. 3, three simple composite areas are selected to describe the generating area for a solid of revolution (i.e., the volume of the rocket's fuel). Before evaluating Eq. (7.36) on p. 426, we collect in Table 1 the area of each composite shape and the distance from the axis of revolution to the centroid of each shape.

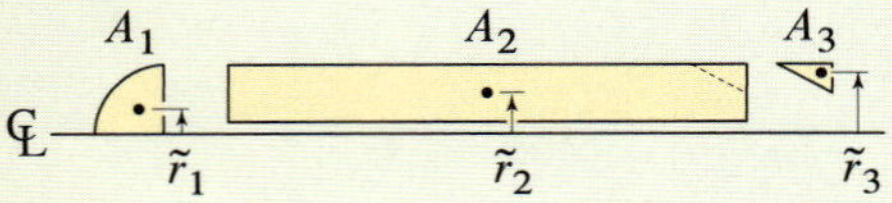

Figure 3
Three composite areas (shown with a slight horizontal separation for clarity) that describe the generating area for a solid of revolution.

Table 1. Areas and centroid positions relative to the axis of revolution for the composite shapes in Fig. 3.

Shape no.	A_i	$\tilde{r}_i$
1	$\pi(35\,\text{cm})^2/4 = 962.1\,\text{cm}^2$	$\frac{4}{3\pi}(35\,\text{cm}) = 14.85\,\text{cm}$
2	$(230\,\text{cm})(30\,\text{cm}) = 6900\,\text{cm}^2$	$20\,\text{cm}$
3	$-\frac{1}{2}(30\,\text{cm})(15\,\text{cm}) = -225\,\text{cm}^2$	$20\,\text{cm} + \frac{2}{3}(15\,\text{cm}) = 30\,\text{cm}$

Since the fuel is a full solid of revolution, $\theta = 2\pi$, and Eq. (7.36) provides

$$V = 2\pi \sum_{i=1}^{3} \tilde{r}_i A_i$$
$$= 2\pi\left[(14.85\,\text{cm})(962.1\,\text{cm}^2) + (20\,\text{cm})(6900\,\text{cm}^2) + (30\,\text{cm})(-225\,\text{cm}^2)\right]$$
$$= \boxed{9.145\times 10^5\,\text{cm}^3.} \tag{1}$$

Converting Eq. (1) from cm^3 to m^3 provides

$$V = (9.145\times 10^5\,\text{cm}^3)\left(\frac{\text{m}}{100\,\text{cm}}\right)^3 = \boxed{0.9145\,\text{m}^3.} \tag{2}$$

Part (b)

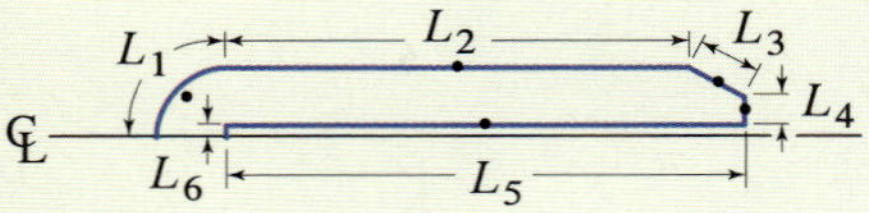

Figure 4
Six composite lines that describe the generating curve for a surface of revolution. The arc lengths of the lines are given by L_1 through L_6, and the centroid positions of the lines are indicated by the solid dots (the centroid position of L_6 is not shown).

Governing Equations & Computation In Fig. 4, six simple composite lines are identified to describe the generating curve for a surface of revolution (i.e., the surface area of the rocket's fuel). Before evaluating Eq. (7.33) on p. 425, we collect in Table 2 the arc length of each composite line and the distance from the axis of revolution to the centroid of each line.

Since the surface of the fuel is a full surface of revolution, $\theta = 2\pi$, and Eq. (7.33) provides

Table 2. Lengths and centroid positions relative to the axis of revolution for the composite shapes in Fig. 4.

Shape no.	L_i	$\tilde{r}_i$
1	$\pi(35\,\text{cm})/2 = 54.98\,\text{cm}$	$\frac{2}{\pi}(35\,\text{cm}) = 22.28\,\text{cm}$
2	$200\,\text{cm}$	$35\,\text{cm}$
3	$\sqrt{(30\,\text{cm})^2 + (15\,\text{cm})^2} = 33.54\,\text{cm}$	$20\,\text{cm} + \frac{1}{2}(15\,\text{cm}) = 27.50\,\text{cm}$
4	$15\,\text{cm}$	$5\,\text{cm} + \frac{1}{2}(15\,\text{cm}) = 12.50\,\text{cm}$
5	$230\,\text{cm}$	$5\,\text{cm}$
6	$5\,\text{cm}$	$2.5\,\text{cm}$

$$\begin{aligned} A &= 2\pi \sum_{i=1}^{6} \tilde{r}_i L_i \\ &= 2\pi\big[(22.28\,\text{cm})(54.98\,\text{cm}) + \cdots + (2.5\,\text{cm})(5\,\text{cm})\big] \\ &= \boxed{65{,}960\,\text{cm}^2.} \end{aligned} \tag{3}$$

For brevity, numbers for only the first and sixth line elements are shown in Eq. (3), and you should verify the results that are given. Converting Eq. (3) from cm^2 to m^2 provides

$$A = (65{,}960\,\text{cm}^2)\left(\frac{\text{m}}{100\,\text{cm}}\right)^2 = \boxed{6.596\,\text{m}^2.} \tag{4}$$

Discussion & Verification To help judge if our solutions are reasonable, we will compare the volume and surface area of the fuel, as given by Eqs. (1) and (3), with those for a circular cylinder with radius r and length L. Using $r = 35\,\text{cm}$ and $L = 265\,\text{cm}$, the volume of this cylinder is $V = \pi r^2 L = 1.02 \times 10^6\,\text{cm}^3$, and the surface area including the ends is $A = 2\pi r L + 2\pi r^2 = 66{,}000\,\text{cm}^2$. Observe that with the values of r and L cited, the volume of the cylinder is larger than the exact volume of the fuel, while it is unclear if the surface area of the cylinder should be larger or smaller than the exact surface area of the fuel. The volume estimate of $1.02 \times 10^6\,\text{cm}^3$ is in reasonable agreement with Eq. (1) and, as expected, is larger than Eq. (1). The surface area estimate of $66{,}000\,\text{cm}^2$ is in good agreement with Eq. (3).

PROBLEMS

General instructions. Use the theorems of Pappus and Guldinus to determine the volume and/or surface area for the following problems.

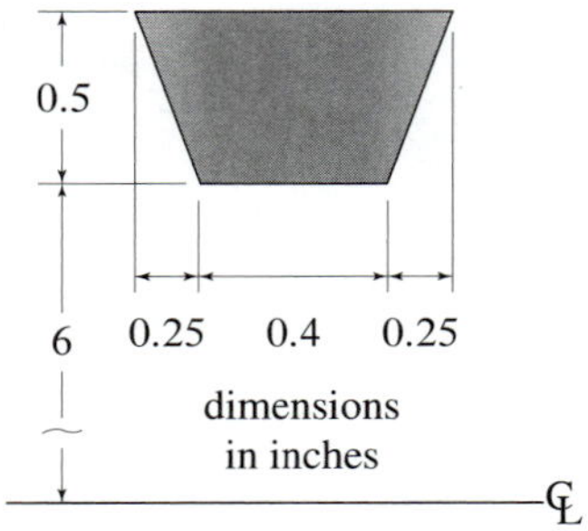

Figure P7.45

Problem 7.45

The cross section of a rubber V belt is shown. If the belt has circular shape about the axis of revolution with an inside radius of 6 in., determine the volume of material in the belt and the surface area of the belt.

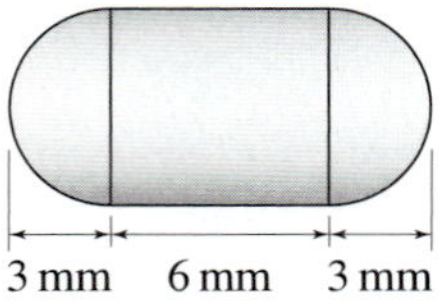

Figure P7.46

Problem 7.46

A pharmaceutical company's design for a medicine capsule consists of hemispherical ends and a cylindrical body. Determine the volume and outside surface area of the capsule.

Problem 7.47

A solid is produced by rotating a triangular area 360° about the vertical axis of revolution shown. Determine the volume and surface area of the solid.

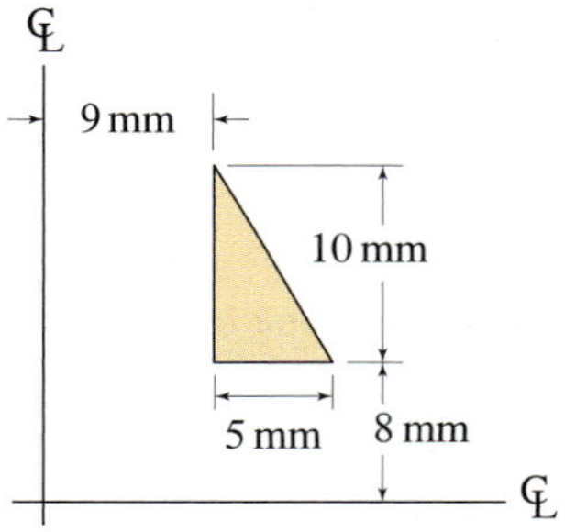

Figure P7.47 and P7.48

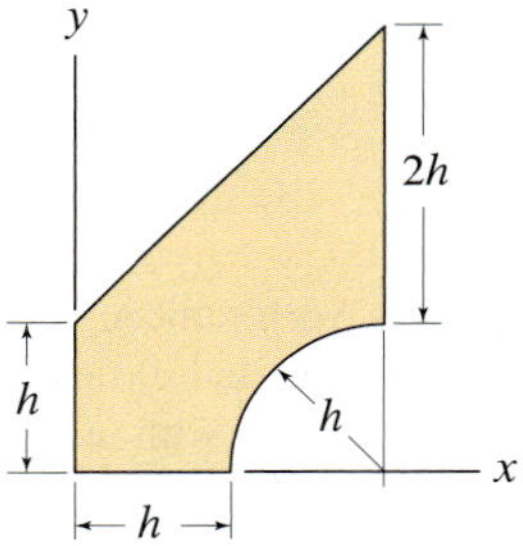

Figure P7.49 and P7.50

Problem 7.48

Repeat Prob. 7.47 if the solid is produced by rotating the triangular area about the horizontal axis shown.

Problem 7.49

A solid is generated by rotating the shaded area shown 360° about the y axis. Determine the volume and surface area of this solid in terms of the dimension h.

Problem 7.50

Repeat Prob. 7.49 if the solid is generated by rotating the shaded area about the x axis.

Problem 7.51

Determine the volume and surface area for the solid of revolution in Example 7.2 on p. 403.

Problem 7.52

A metal Sierra cup is used by campers as a multipurpose utensil for drinking, holding food, and so on. It is an object of revolution that has the generating curve shown. If the cup is made of titanium sheet that weighs $0.004\ \text{lb/in.}^2$, determine the volume of fluid the cup is capable of holding and the total weight of the cup.

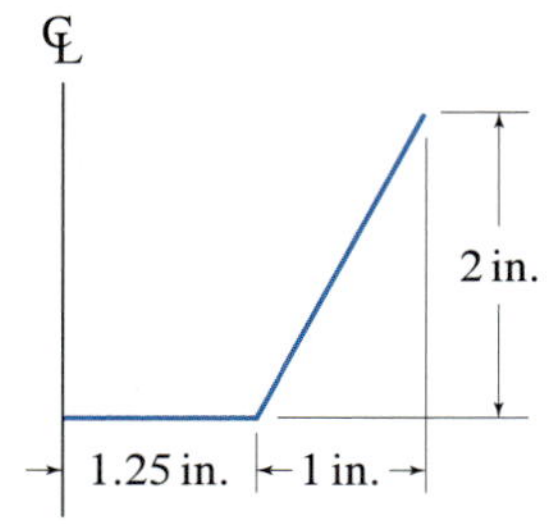

Figure P7.52

Problem 7.53

A funnel is to be made of thin sheet metal using the generating curve shown. If the funnel is to be plated with 0.005 mm thick chrome on all surfaces, determine the volume of chrome required.

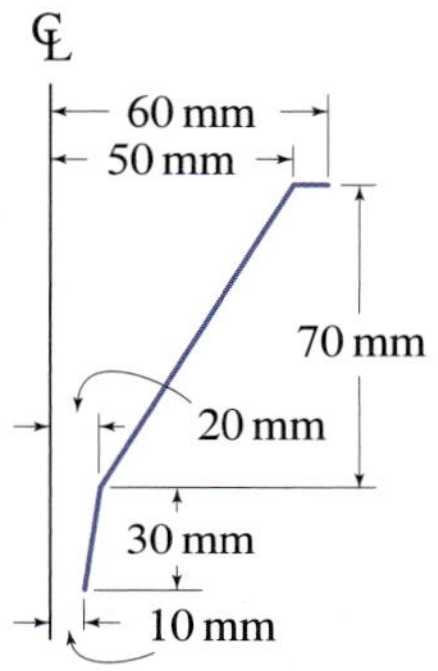

Figure P7.53

Problem 7.54

A pressure vessel is to be constructed as a solid of revolution using the generating area shown.

(a) Determine the volume of material needed to construct the pressure vessel.

(b) Determine the outside surface area.

(c) Determine the inside surface area.

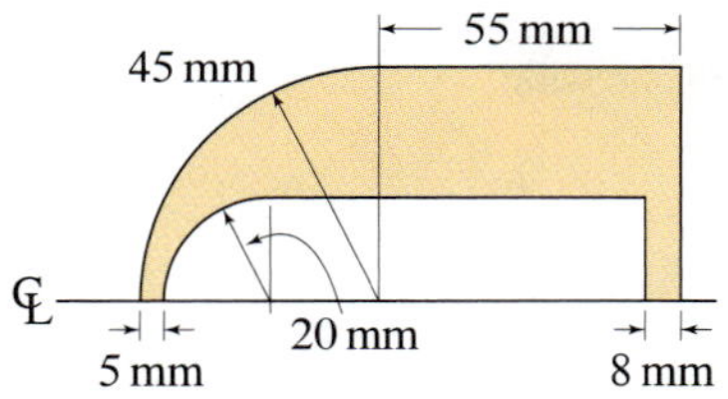

Figure P7.54

Problem 7.55

The penstock shown is retrofitted to an existing dam to deliver water to a turbine generator so that electricity may be produced. The penstock has circular cross section with 24 in. diameter throughout its length, including section BC, which is a 90° elbow. Determine the total weight of portion $ABCD$ of the penstock, including the water that fills it. The penstock is made of thin-walled steel with a weight of $0.07\ \text{lb/in.}^2$, and the water weighs $0.036\ \text{lb/in.}^3$.

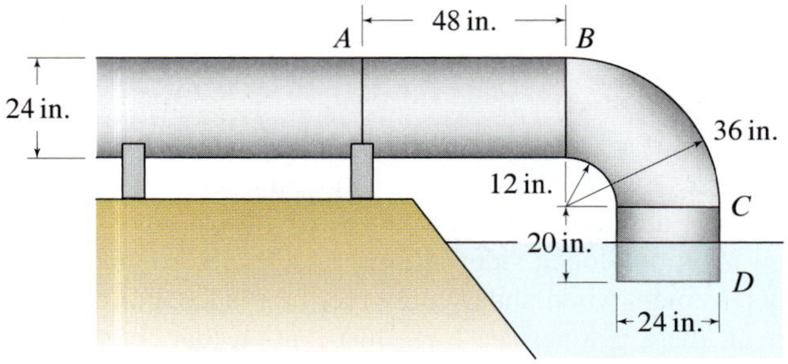

Figure P7.55

Problem 7.56

Determine the volume and surface area for the hemispherical solid with a conical cavity shown in Fig. P7.9 on p. 410.

Problem 7.57

The area of Prob. 7.12 on p. 410 is revolved 360° about the line $x = -1$ m to create a solid of revolution. Determine the volume and surface area of the solid. *Hint:* This problem is straightforward if you first solve Prob. 7.12 on p. 410 and Prob. 7.21 on p. 411.

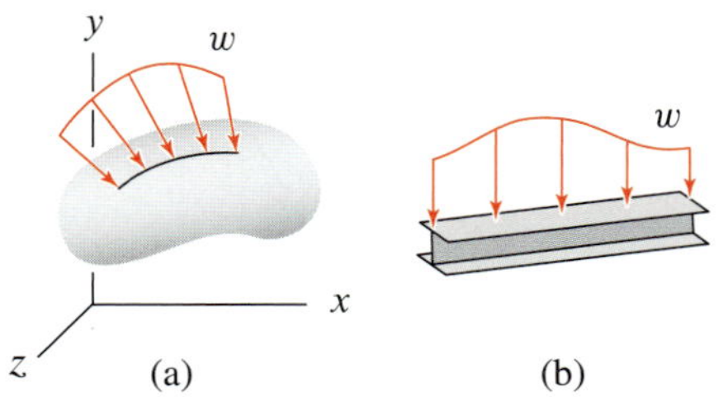

Figure 7.15
(a) A *line force* w is a distributed force that acts along a line; w has dimensions of force/length. (b) An example of a line force applied to an I beam.

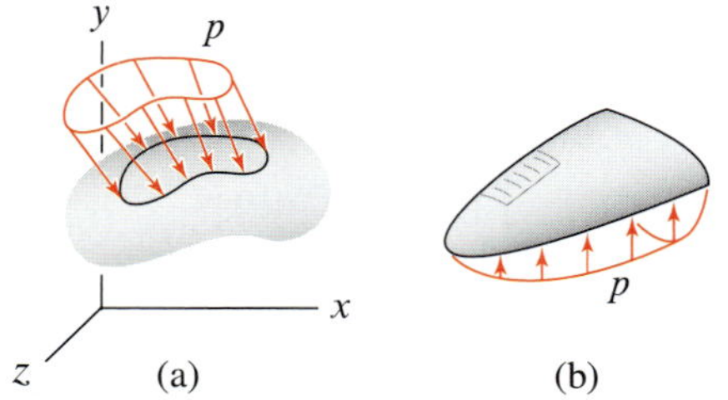

Figure 7.16
(a) A *surface force* p is a distributed force that acts over a surface; p has dimensions of force/area. (b) An example of a surface force is the air pressure applied to the bottom surface of an airplane wing.

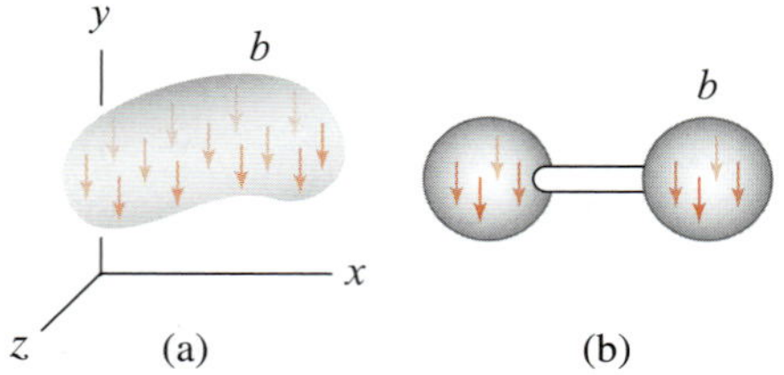

Figure 7.17
(a) A *volume force* b is a distributed force that acts throughout a volume; b has dimensions of force/volume. (b) An example of a volume force is the weight of material in a dumbbell used for weight-lifting exercises.

Helpful Information

Units for surface forces. For convenience, special units for surface forces have been defined. In the U.S. Customary system, $1\text{ psi} = 1\text{ lb/in.}^2$, and in the SI system, $1\text{ Pa} = 1\text{ N/m}^2$. These units are read as follows: "psi" means pound per square inch, and "Pa" means pascal, which is a unit named in honor of the French scientist Blaise Pascal (1623–1662).

7.4 Distributed Forces, Fluid and Gas Pressure Loading

Throughout this book, we have idealized all forces to be point forces. A *point force* is a force that is concentrated at a single point. Point forces do not exist in nature. Rather, all forces are *distributed* as discussed below. In this section, we discuss procedures for how distributed forces can be idealized, or modeled, by point forces.

Distributed forces

Distributed forces are forces that are distributed along a line, over a surface, or throughout a volume. Specific types of distributed forces are as follows:

Line force. A force distributed along a line is called a *line force*, or a *line load*. Examples of a line force w are shown in Fig. 7.15, where w has dimensions of force/length. Line forces are often used in connection with beam-type structures.

Surface force. A force distributed over a surface is called a *surface force*, or a *traction*. Examples of a surface force p are shown in Fig. 7.16, where p has dimensions of force/area. Surface forces result from fluid and gas pressure loadings, contact (with or without friction), as well as a variety of other situations.

Volume force. A force distributed throughout a volume is called a *volume force*, or a *body force*. Examples of a volume force b are shown in Fig. 7.17, where b has dimensions of force/volume. Common examples of volume forces are weight due to gravity and attractive forces within an iron object due to a magnetic field.

Line forces, surface forces, and volume forces are all vectors. Line forces and surface forces do not need to be perpendicular to the objects or structures to which they are applied, although often they will be. For example, consider standing on a gently sloped sidewalk on a cold winter day. Under one foot there is dry pavement while under your other foot is ice. For the foot touching dry pavement, there is a surface force distribution that consists of a normal direction component *and* a tangential direction component. In fact, the tangential direction component keeps you from slipping! For the foot touching ice, the surface force distribution has only a normal direction component. In nature, all forces are either surface forces or volume forces. Line forces and point forces are useful idealizations for surface forces and volume forces. In this section, we use Chapter 4 concepts on equivalent force systems and the concepts of centroid discussed earlier in this chapter to replace distributed forces by point forces.

Distributed forces applied to beams

Consider the example shown in Fig. 7.18(a) where a 30 m long cantilever beam is subjected to a distributed force. The distributed force w varies linearly from

12 kN/m at the left-hand end to 6 kN/m at the right-hand end. Using the coordinate system shown in Fig. 7.18(a), the distributed force w as a function of position x is (for tips on developing expressions for linear functions, see the Helpful Information margin note on p. 420)

$$w = 12\,\text{kN/m} - (0.2\,\text{kN/m}^2)x. \tag{7.40}$$

Note that in writing Eq. (7.40), we consider the distributed force w to be positive when acting in the $-y$ direction, and we will follow this convention throughout this book. As shown in Fig. 7.18(b), over a small length of beam dx, the distributed force w produces a force $dF = w\,dx$. Hence, the total force F produced by the distributed loading is

$$F = \int dF = \int w\,dx \tag{7.41}$$

$$= \int_0^{30\,\text{m}} \left[12\,\text{kN/m} - (0.2\,\text{kN/m}^2)x\right] dx = 270\,\text{kN}. \tag{7.42}$$

Using Eq. (7.19) on p. 400, with dA replaced by the force dF, the centroid of the distributed load, and hence the position of the line of action of the force F, is

$$\bar{x} = \frac{\int \tilde{x}\,dF}{\int dF} = \frac{\int \tilde{x}\,w\,dx}{\int w\,dx} \tag{7.43}$$

$$= \frac{\int_0^{30\,\text{m}} x\left[12\,\text{kN/m} - (0.2\,\text{kN/m}^2)x\right] dx}{270\,\text{kN}} = \frac{3600\,\text{kN·m}}{270\,\text{kN}} = 13.33\,\text{m}. \tag{7.44}$$

Using the results of Eqs. (7.42) and (7.44), the equivalent force system shown in Fig. 7.18(c) can be constructed. Then an FBD may be drawn as shown in Fig. 7.18(d), and reactions may be determined (you should verify for yourself that the reactions are $N_A = 0$, $V_A = 270\,\text{kN}$, and $M_A = -3600\,\text{kN·m}$).

The forgoing treatment demonstrates the use of integration to determine the force and centroid of a distributed load. For distributed loads that have simple shape, such as uniform and linear distributions, it may be more convenient to use composite shape concepts, as was done in earlier sections of this chapter. Returning to the cantilever beam shown in Fig. 7.18(a), either of the composite shapes shown in Fig. 7.19 may be used. In the case of Fig. 7.19(a), the "area" and centroid of the triangular distribution are $F_1 = (1/2)(\text{base})(\text{height}) = (1/2)(30\,\text{m})(6\,\text{kN/m}) = 90\,\text{kN}$ and $\tilde{x}_1 = 10\,\text{m}$. The "area" and centroid of the rectangular distribution are $F_2 = (\text{base})(\text{height}) = (30\,\text{m})(6\,\text{kN/m}) = 180\,\text{kN}$ and $\tilde{x}_2 = 15\,\text{m}$. Similar comments apply for determining the values of F_1, $\tilde{x}_1$, F_2, and $\tilde{x}_2$ for Fig. 7.19(b). For Fig. 7.19(a) and (b), you should draw FBDs and verify that the reactions are the same as those reported earlier.

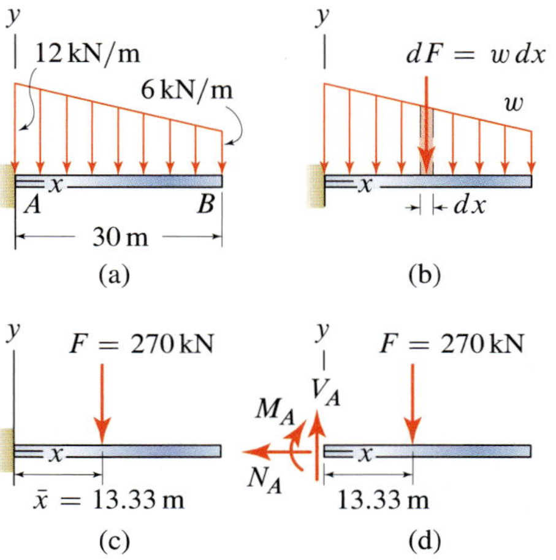

Figure 7.18
(a) A distributed force applied to a cantilever beam. (b) A distributed force w acting over a small length dx produces a force $dF = w\,dx$. (c) An equivalent force system consisting of a force F at position $\bar{x}$. (d) A FBD that can be used to determine the support reactions.

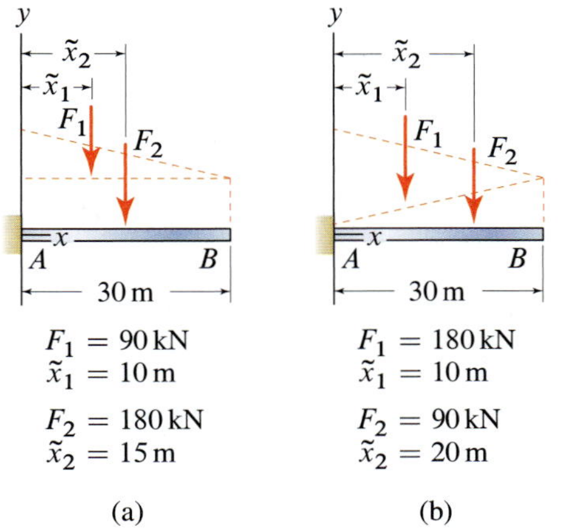

Figure 7.19
Two examples of using composite shapes to determine the forces due to the distributed load shown in Fig. 7.18(a). (a) Triangular and rectangular distributions. (b) Two triangular distributions.

Fluid and gas pressure

Pressures from fluids and gases are surface forces and are always compressive. We consider an incompressible stationary (nonflowing) fluid, as shown in Fig. 7.20. The surface of the fluid lies in the xy plane and is subjected to

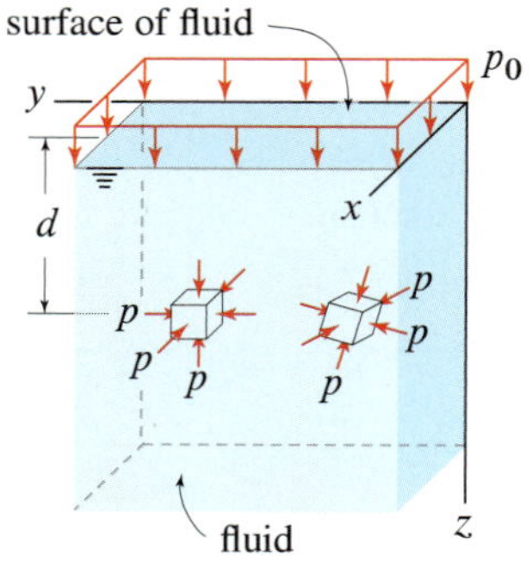

Figure 7.20
A volume of incompressible stationary fluid with density ρ. The surface of the fluid lies in the xy plane and is subjected to a pressure p_0. Infinitesimally small cubes of fluid are subjected to hydrostatic compressive pressure p.

Helpful Information

Can gas pressure be neglected in FBDs? Provided *all* surfaces of an object are subjected to the same gas pressure, for most purposes we can omit this pressure in FBDs with no error or loss of accuracy. For example, consider a glass full of your favorite beverage resting on a table, as shown in (a) below. The figure in (b) shows a two-dimensional view of the FBD including air pressure, and (c) shows the FBD omitting air pressure, where W is the weight of the glass and beverage and R is the reaction between the glass and table. Although we do not prove this, when summing forces in the vertical direction, the downward component of all forces due to the air pressure in (b) is equilibrated by the upward component of all forces due to the air pressure. Hence, the same reaction R is obtained for both FBDs, namely, $R = W$.

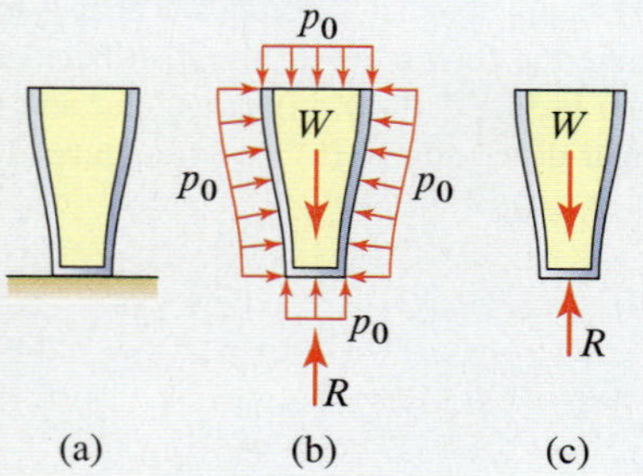

a gas pressure p_0. We consider an infinitesimally small cube of fluid located at a depth d below the surface. Regardless of the cube's orientation, such as the two different orientations shown in Fig. 7.20, all six faces of the cube are subjected to the same compressive pressure p as follows

$$\begin{aligned} p &= p_0 + \rho g d \\ &= p_0 + \gamma d, \end{aligned} \tag{7.45}$$

where

p_0 is the pressure at the surface of the fluid;
d is the depth below the fluid's surface;
g is acceleration due to gravity;
ρ is the density of the fluid;
γ is the specific weight of the fluid ($\gamma = \rho g$); and
p is the pressure at a depth d due to the fluid and gas, with dimensions of force/area.

Remarks

- The state of pressure due to gas and/or fluid loading is called *hydrostatic* because the pressure at a particular point has the same intensity in all directions.
- Observe that Eq. (7.45) is a linear function of depth d.
- Equation (7.45) gives the *absolute* pressure at a point, that is, the total pressure due to both gas and fluid loading. For many applications, only the portion of the pressure due to the fluid loading is needed, and this is often called the *gage pressure*, and Eq. (7.45) with p_0 omitted becomes

$$\begin{aligned} p &= \rho g d \\ &= \gamma d. \end{aligned} \tag{7.46}$$

- If there is no gas pressure acting on the surface of the fluid, then $p_0 = 0$ and the absolute pressure and the gage pressure are identical. However, even when the gas pressure is significant, exact analyses can often be carried out using only the gage pressure, and this is the approach normally used. The margin note on this page and Example 7.14 on p. 442 discuss this issue in greater detail.
- If there is no fluid pressure loading, then Eq. (7.45) shows that all points, regardless of position d, are subjected to the same gas pressure p_0. While this model is suitable for some applications, you should note that in contrast to fluids where the assumption of incompressibility is reasonable over a very large range of conditions, gases are highly compressible and pressures are strongly dependent on temperature.

To understand more deeply why Eq. (7.45) describes gas and fluid pressure, consider the rectangular prism of fluid shown in Fig. 7.21(a). The FBD for this prism of fluid is three-dimensional, but only a two-dimensional view is needed and this is shown in Fig. 7.21(b). Note that at point A, the pressure acting on the vertical surface is the same as the pressure acting on the horizontal surface. Identical remarks apply for points B, C, and D. Summing forces in the z direction provides

$$\sum F_z = 0: \quad p_0 a^2 + W - pa^2 = 0, \tag{7.47}$$

where W is the weight of the prism of fluid, which is equal to the product of the density of the fluid ρ, acceleration due to gravity g, and the volume of the prism $a^2 d$; thus $W = \rho g a^2 d$. In writing Eq. (7.47), the force from the gas pressure on surface AB is given by the product of pressure at that location p_0 and the area over which it acts, which is a^2. Similarly, the force on surface CD is the product of pressure at that location p and the area over which it acts, which is also a^2. Substituting $W = \rho g a^2 d$ into Eq. (7.47) and solving for p yields Eq. (7.45).

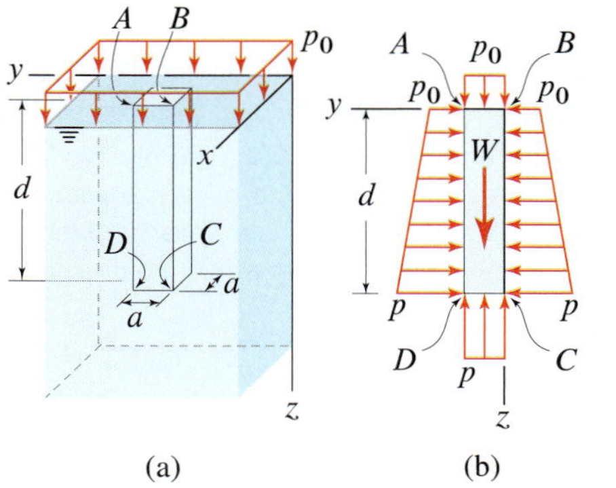

Figure 7.21
(a) A rectangular prism of fluid with height d, width a, and thickness a, whose surface is subjected to a pressure p_0. (b) A two-dimensional view of the FBD for the prism of fluid.

Forces produced by fluids

When a fluid makes contact with the surface of a body, a surface force that is compressive is exerted on the body, and this can be a significant source of loading. In the following discussion, we consider approaches for determining the forces that fluids apply to surfaces of bodies and structures.

Fluid forces on flat surfaces

In Fig. 7.22, we consider the situation of a rectangular plate submerged in an incompressible stationary fluid.

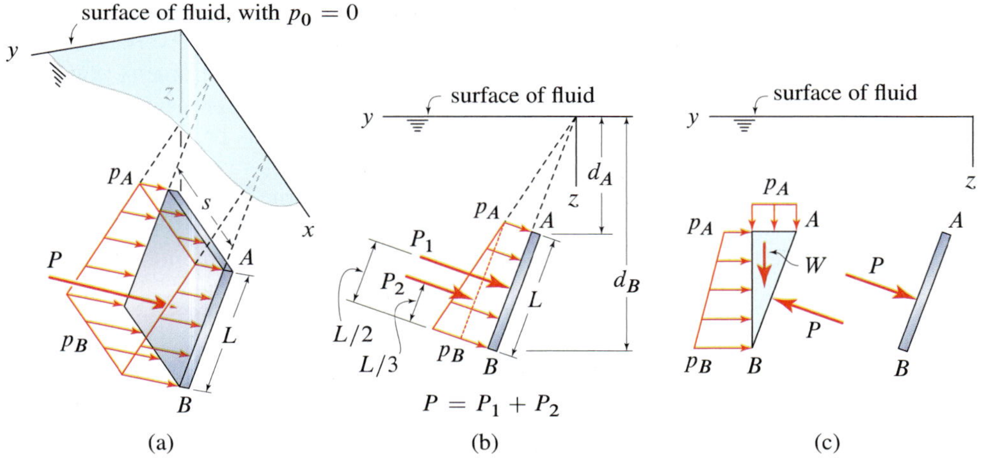

Figure 7.22. (a) Fluid pressure acting on *one* surface of a flat rectangular plate and a force P that is equivalent to the fluid pressure distribution. The other surfaces of the plate also have fluid pressures acting on them, but these pressures are not shown. (b) A view of the plate looking down the x axis showing the pressure distribution and an approach for determining the force the fluid applies to the surface of the plate. (c) An alternative approach using an FBD of a volume of fluid for determining the force the fluid applies to the surface of the plate.

The upper and lower edges of the plate, denoted by A and B, are parallel to the surface of the fluid and have depths d_A and d_B, respectively. We focus

Helpful Information

Don't confuse p and P. In Fig. 7.22, and elsewhere in this section, lowercase p represents pressure (dimensions: force/area), and uppercase P represents force. Thus, p and P are different—but they are related. Integrating pressure p over the area it acts upon yields the force P.

Concept Alert

Fluid and gas pressure. The direction of the pressure that a fluid or gas applies to the surface of a body is always perpendicular to the surface over which the pressure acts.

Common Pitfall

Errors in FBDs. A common error when drawing FBDs is to have fluid and gas pressures that are not perpendicular to the surfaces over which they act. For example, consider a gravity dam that retains water.

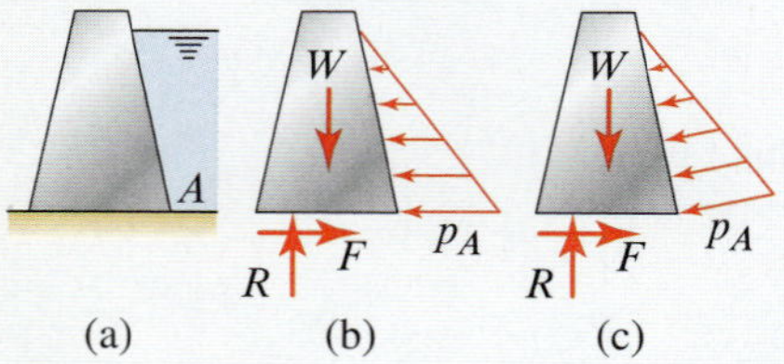

Both of the FBDs in (b) and (c) correctly have linear water pressure distributions, but the FBD in (b) is incorrect because the water pressure is not perpendicular to the surface over which it acts.

on the gage pressure, which is the pressure due to the fluid only. The pressure on the upper edge is given by Eq. (7.46) and has value $p_A = \gamma d_A$, and the pressure on the lower edge has value $p_B = \gamma d_B$. Between edges A and B the pressure distribution is linear, and, *very importantly*, the direction of the pressure is perpendicular to the surface of the plate over the entire surface of the plate.

Two approaches for determining the force P and/or P_1 and P_2 that the fluid applies to the surface of the plate are shown in Fig. 7.22(b) and (c). In Fig. 7.22(b), we use concepts of equivalent force systems from Chapter 4 and the concepts of centroid discussed earlier in this chapter to break the pressure distribution into two simple composite shapes. The first shape is a rectangular "volume" with dimensions p_A by L by s; the force corresponding to this is $P_1 = p_A L s$; and the position of the line of action of P_1 is through the centroid of a rectangular volume, namely, $L/2$ from edge B as shown. The second shape is a triangular "volume" with height $p_B - p_A$, length L, and width s; the force corresponding to this is $P_2 = (1/2)(p_B - p_A)Ls$; and the position of the line of action of P_2 is through the centroid of a triangular volume, namely, $L/3$ from edge B as shown. After P_1 and P_2 are known and their lines of action have been located, we proceed to determine similar quantities for other surfaces of the body that may be in contact with fluid. We then draw an FBD where all forces acting on the body are included, followed by writing equilibrium equations and so on.

In Fig. 7.22(c), we use a different approach wherein equilibrium concepts are used to determine the force the fluid applies to the surface of the structure. We begin by drawing an FBD of a triangular wedge of fluid; although the FBD is three-dimensional, a two-dimensional view as shown in Fig. 7.22(c) is adequate for our purposes. The wedge of fluid has weight W, the upper edge of the fluid is subjected to a uniform pressure p_A $(= \gamma d_A)$, the vertical surface is subjected to a linear pressure distribution that changes from p_A at the top to p_B $(= \gamma d_B)$ at the bottom, and rather than show the pressure distribution on surface AB, we instead show the force P that this pressure distribution produces. We then write equilibrium equations $\sum F_y = 0$ and $\sum F_z = 0$ to determine the y and z components of P, and we write $\sum M = 0$ to locate the line of action of P. Note that by determining the force P that acts on edge AB of the fluid, we have, according to Newton's third law of motion, determined the force that the fluid applies to the surface of the structure, also shown in Fig. 7.22(c).

Fluid forces on nonflat surfaces

In Fig. 7.23, we consider the situation of a curved plate with uniform width s submerged in an incompressible stationary fluid. The upper and lower edges of the plate are parallel to the surface of the fluid and have depths d_A and d_B, respectively. The pressures on the upper and lower edges have values p_A and p_B, respectively. Between edges A and B the value of the pressure has a linear distribution; for example, if edge B is twice as deep as edge A, then p_B is twice as large as p_A. Despite the linearity in the value of the pressure, because the direction of the pressure is always perpendicular to the surface of the plate, the pressure distribution is complex. To determine the force P that the fluid applies to the plate by using the approach shown in Fig. 7.23(b), it is necessary to develop expressions for the y and z components of the pressure

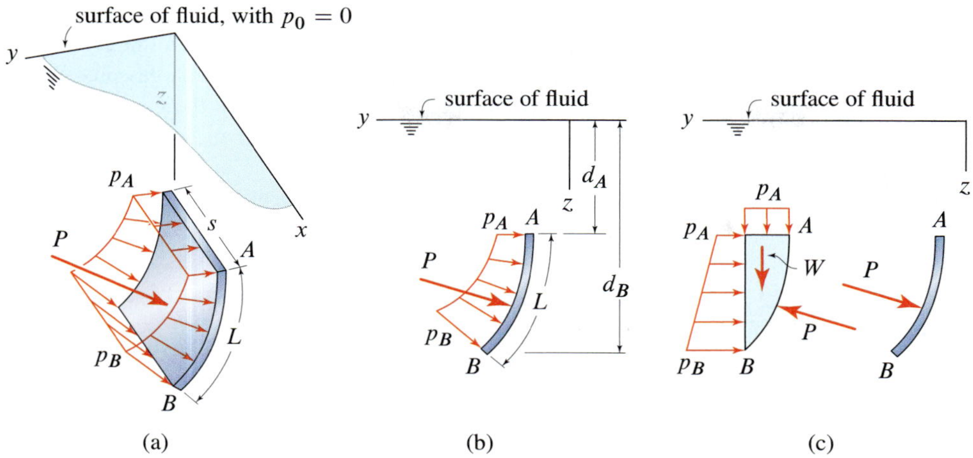

Figure 7.23. (a) Fluid pressure acting on *one* surface of a curved plate, and a force P that is equivalent to the fluid pressure distribution. The other surfaces of the plate also have fluid pressures acting on them, but these pressures are not shown. (b) A view of the plate looking down the x axis showing the pressure distribution. (c) An approach using an FBD of a volume of fluid for determining the force the fluid applies to the surface of the plate.

p_y and p_z and to integrate these over the area of the plate to obtain the force $P_y = \int p_y s\, dL$, where L is the arc length of the curved plate and s is its width, and similarly for $P_z = \int p_z s\, dL$. Once P_y and P_z are known, the force the fluid applies to the plate is $P = \sqrt{P_y^2 + P_z^2}$, and then we must proceed to determine the location of its line of action. Carrying out these calculations can be tedious.

Fortunately, for many problems the alternative procedure shown in Fig. 7.23(c) will be much easier to use. If the shape of the volume of fluid in the FBD is sufficiently simple that its volume and centroid can be readily determined, then this approach is straightforward. Observe in Fig. 7.23(b) and (c) that while the fluid pressures are always perpendicular to the surfaces they act on, the line of action of force P is not necessarily perpendicular to the surface.

Remarks

We end this discussion with a few closing remarks. We have presented two solution approaches for determining the force that a fluid applies to the surface of a structure. The examples of this section give further details on how these two solution approaches are carried out and contrast them to help you decide which approach is more convenient for a particular problem. For surfaces with more complicated shape and/or orientation than those considered here, we may have no choice but to carry out the integrations described above to determine the forces due to the fluid loading.

Forces produced by gases

Loads applied to structures from a pressurized gas are often important, and the preceding discussion of fluid pressure loading is generally applicable with the following simplifications. In a pressurized gas, the density of the gas is typically negligible* so that all surfaces of a structure over which the gas makes contact are subjected to the same pressure p_0. As with fluids, gas pressure is compressive and always acts perpendicular to the surface the gas makes contact with. Thus, Figs. 7.22 and 7.23 are applicable with $p_A = p_B = p_0$, and in Figs. 7.22(c) and 7.23(c), the weight of the volume of gas in the FBDs is $W = 0$. Example 7.14 provides further details on treatment of forces due to a pressurized gas.

End of Section Summary

In this section, distributed forces and their treatment were discussed. *Distributed forces* are forces that are distributed along a line, over a surface, or throughout a volume. A force distributed along a line is called a *line force*, or a *line load*, and this has dimensions of force/length. A force distributed over a surface is called a *surface force*, or a *traction*, and this has dimensions of force/area. A force distributed throughout a volume is called a *volume force*, or a *body force*, and this has dimensions of force/volume.

Fluid pressure is a surface force. If the fluid is incompressible and at rest, then the pressure within the fluid is given by Eq. (7.45) as $p = p_0 + \rho g d$ where p_0 is the gas pressure (constant) at the surface of the fluid. For many purposes, only the portion of the pressure due to fluid loading is important, and this pressure ($p = \rho g d$) is sometimes called the *gage pressure*. In either case, fluid pressure acts normal to the surface the fluid makes contact with.

Gas pressure is a surface force. If the gas has negligible density and is at rest, then the pressure throughout the gas has constant value p_0, and all surfaces that make contact with the gas are subjected to this pressure, which acts normal to the surfaces.

Also discussed in this section were methods for determining the forces that fluid and gas pressure distributions apply to structures.

* If the density of a gas is large enough that it cannot be neglected, then the gas is essentially a fluid and is treated as such. Further, if the gas is stationary and can be assumed to have constant density, then the methods for treatment of fluids discussed in this section can be applied. However, because gases are highly compressible, the assumption of constant density is more restrictive for gases than for fluids.

EXAMPLE 7.11 *Distributed Forces Applied to a Beam*

A bookshelf holding 11 books is supported by brackets at points A and B. The weights and thicknesses of the books are given in Table 1. Determine the support reactions.

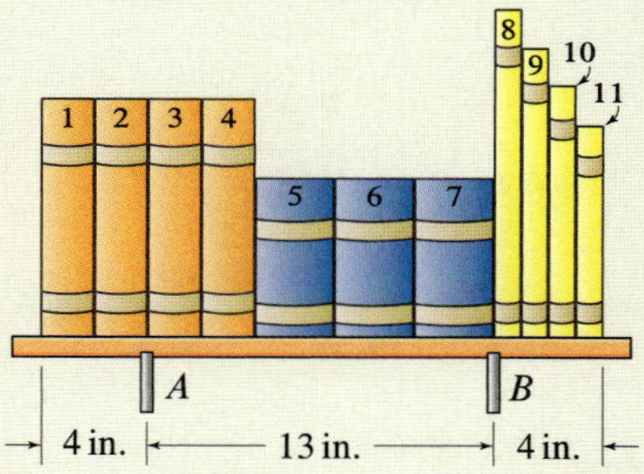

Figure 1

Table 1
Weights and thicknesses of books.

Books	Weight	Thickness
1–4	4 lb	2 in.
5–7	3 lb	3 in.
8	4 lb	1 in.
9	3 lb	1 in.
10	2 lb	1 in.
11	1 lb	1 in.

SOLUTION

Road Map We could determine the support reactions for the shelf by applying 11 individual loads according to the weights given in Table 1, but because of the large number of forces, it will be better to model the weight of the books by using distributed forces. Once this is accomplished, an FBD is drawn and equilibrium equations are written and solved.

Modeling Books 1–4 have identical weight and thickness, and thus over the portion of the shelf where these books rest, they apply a uniform distributed force of 4 lb/2 in. = 2 lb/in. Similarly, books 5–7 have identical weight and thickness, and thus over the portion of the shelf where these books rest, they apply a uniform distributed force of 3 lb/3 in. = 1 lb/in. Books 8–11 have the same thickness and show a linear variation in their weights, and thus the force they apply will be modeled using a linear distributed force that varies from 4 lb/in. at book 8 to 1 lb/in. at book 11. The model for books 8–11 is approximate, although reasonably accurate; Prob. 7.68 gives suggestions for developing a more accurate model. The distributed forces are shown in Fig. 2.

The problem statement does not provide details of the supports at A and B, and thus we will idealize the support reactions to consist of vertical forces only, and the FBD is shown in Fig. 2. At point A and/or B, there is probably also a horizontal force reaction, but the presence of such is not important to this problem.

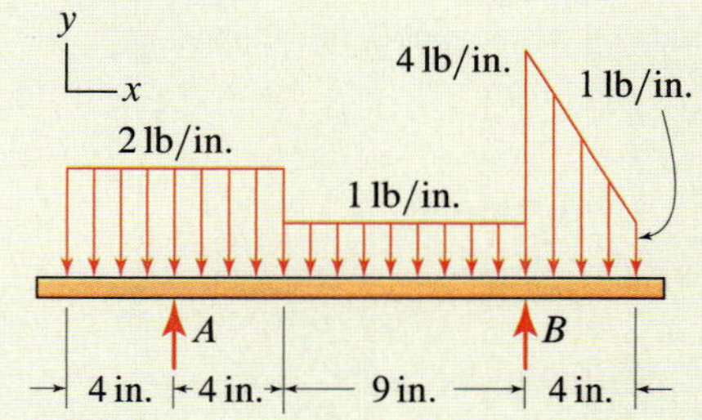

Figure 2
Free body diagram showing distributed loads.

Governing Equations & Computation Shown in Figs. 2 and 3 are FBDs; the first of these shows the distributed loads while the second has replaced these with equivalent concentrated loads as follows:

$$F_1 = (2\,\text{lb/in.})(8\,\text{in.}) = 16\,\text{lb}, \qquad F_2 = (1\,\text{lb/in.})(9\,\text{in.}) = 9\,\text{lb},$$
$$F_3 = (1\,\text{lb/in.})(4\,\text{in.}) = 4\,\text{lb}, \qquad F_4 = \tfrac{1}{2}(3\,\text{lb/in.})(4\,\text{in.}) = 6\,\text{lb}, \tag{1}$$

and the locations of these forces are shown in Fig. 3. Each of the forces given in Eq. (1) is determined by evaluating the "area" of the distributed load and placing this at the centroid of the distributed load. For example, the distributed load for books 1–4 is a rectangle with height 2 lb/in. and base 8 in.; the "area" of this rectangle is $F_1 = (2\,\text{lb/in.})(8\,\text{in.}) = 16\,\text{lb}$, and the centroid is at the interface between books 2 and 3. The distributed load for books 8–11 is broken into two simpler distributions consisting of a rectangle and a triangle, as shown by the dashed lines in Fig. 3.

The support reactions are found by writing and solving the following equilibrium equations:

$$\sum M_A = 0: \quad B(13\,\text{in.}) - F_2(8.5\,\text{in.}) - F_3(15\,\text{in.}) - F_4(14.33\,\text{in.}) = 0 \tag{2}$$

$$\Rightarrow \boxed{B = 17.1\,\text{lb},} \tag{3}$$

$$\sum F_y = 0: \quad A + B - F_1 - F_2 - F_3 - F_4 = 0 \tag{4}$$

$$\Rightarrow \boxed{A = 17.9\,\text{lb}.} \tag{5}$$

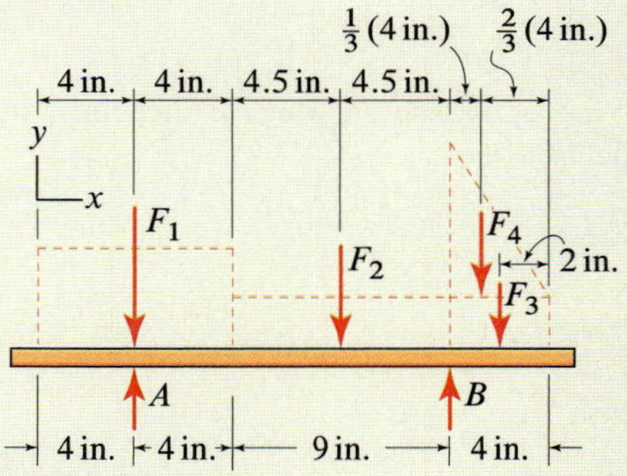

Figure 3
Free body diagram showing concentrated loads.

Discussion & Verification To help verify that F_1 through F_4 are evaluated accurately, note that their sum is equal to the total weight of the books from Table 1, which is 35 lb. Note that the sum of the reactions A and B is also 35 lb.

EXAMPLE 7.12 *Fluid Pressure Loading and FBD Choices*

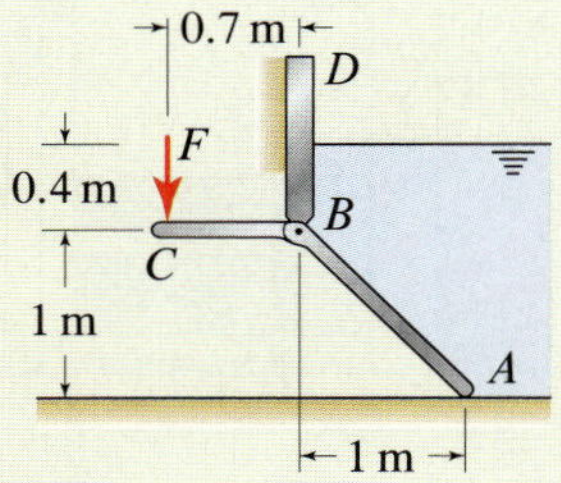

Figure 1

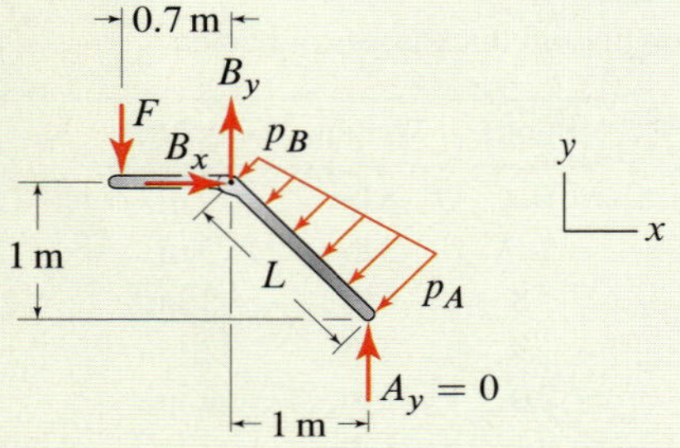

Figure 2
Free body diagram showing the pressure distribution the water applies to the gate.

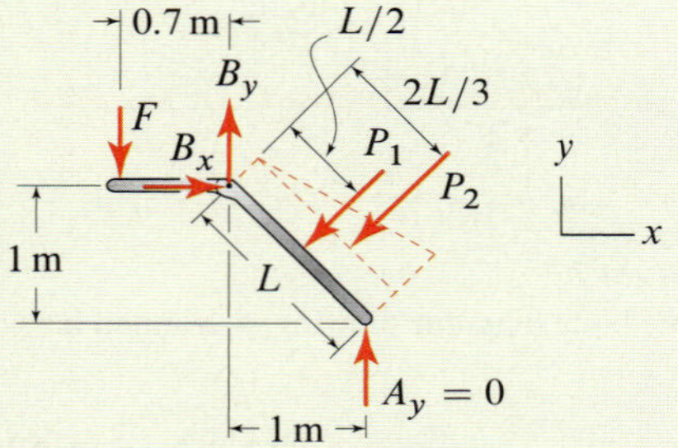

Figure 3
Free body diagram showing the forces the water applies to the gate.

Freshwater in a channel is retained by a flat rectangular gate with 0.6 m width (into the plane of the figure) that is supported by a pin at B. The vertical wall BD is fixed in position. If the weight of the gate is negligible, determine the force F required to begin opening the gate.

SOLUTION

Road Map When drawing an FBD, we can use either of the approaches shown in Fig. 7.22 on p. 435 for treating the fluid. Namely, we may draw an FBD of the gate alone, including the pressures the water applies to it, as shown in Fig. 7.22(b). Or we may draw an FBD of the gate *and* a judiciously selected volume of water, as shown in Fig. 7.22(c). Both approaches are effective for this problem, although the first approach is slightly easier and will be used here.

Modeling While the gate is a three-dimensional structure with length from A to B of $L = \sqrt{(1\,\text{m})^2 + (1\,\text{m})^2} = 1.414\,\text{m}$ and a width of 0.6 m, an FBD in two dimensions is suitable, as shown in Fig. 2. When the gate begins to open, contact at A is broken and $A_y = 0$.

Governing Equations & Computation The density of freshwater is $\rho = 10^3\,\text{kg/m}^3$, and from Eq. (7.46) on p. 434, the pressures at points A and B are

$$p_A = \rho g d_A = \left(10^3\,\frac{\text{kg}}{\text{m}^3}\right)\left(9.81\,\frac{\text{m}}{\text{s}^2}\right)(1.4\,\text{m}) = 13{,}730\,\frac{\text{N}}{\text{m}^2} = 13.73\,\frac{\text{kN}}{\text{m}^2}, \quad (1)$$

$$p_B = \rho g d_B = \left(10^3\,\frac{\text{kg}}{\text{m}^3}\right)\left(9.81\,\frac{\text{m}}{\text{s}^2}\right)(0.4\,\text{m}) = 3924\,\frac{\text{N}}{\text{m}^2} = 3.924\,\frac{\text{kN}}{\text{m}^2}. \quad (2)$$

In Fig. 3, the FBD of Fig. 2 is redrawn with the pressure distribution replaced by the forces the water applies to the structure. These forces are determined by breaking the pressure distribution into simpler composite shapes consisting of a rectangular "volume" with height p_B, length $L = 1.414\,\text{m}$, and width 0.6 m, and a triangular "volume" with height $p_A - p_B$ and the same length and width. Thus, the forces corresponding to these shapes are

$$P_1 = p_B L (0.6\,\text{m}) = 3.330\,\text{kN}, \quad P_2 = \tfrac{1}{2}(p_A - p_B) L (0.6\,\text{m}) = 4.162\,\text{kN}, \quad (3)$$

and the locations of the lines of action of these forces are shown in Fig. 3.

To determine the force F required to begin opening the gate, we sum moments about point B as follows:

$$\sum M_B = 0: \qquad F(0.7\,\text{m}) - P_1\left(\frac{L}{2}\right) - P_2\left(\frac{2L}{3}\right) = 0 \quad (4)$$

$$\Rightarrow \boxed{F = 8.97\,\text{kN}.} \quad (5)$$

If desired, the reactions B_x and B_y can be determined by writing the equilibrium equations $\sum F_x = 0$ and $\sum F_y = 0$.

Discussion & Verification The force F determined above is large, but is reasonable considering the depth of the water and the size of the gate. To contrast the solution approaches cited in the Road Map discussion, you should consider resolving this problem using an FBD that includes a volume of fluid.

EXAMPLE 7.13 *Fluid Pressure Loading and FBD Choices*

Freshwater in a channel is retained by a cylindrical gate with 0.6 m width (into the plane of the figure) that is supported by a pin at B. The vertical wall BD is fixed in position. If the weight of the gate is negligible, determine the force F required to begin opening the gate.

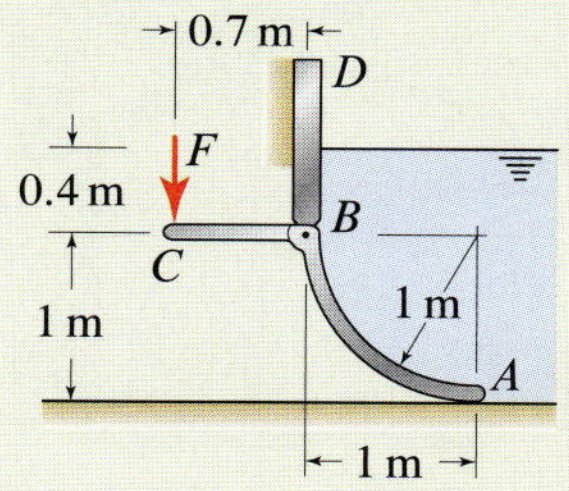

Figure 1

SOLUTION

Road Map This example differs from Example 7.12 only in the geometry of the gate. Fundamentally, we may develop an FBD using either of the approaches shown in Fig. 7.23(b) or (c) on p. 437 for treating the fluid. However, for this problem it will be considerably easier to use the latter approach where the FBD includes an appropriate volume of fluid.

Modeling An FBD of the gate and a cylindrical volume of fluid is shown in Fig. 2. When the gate begins to open, contact at A is broken and $A_y = 0$.

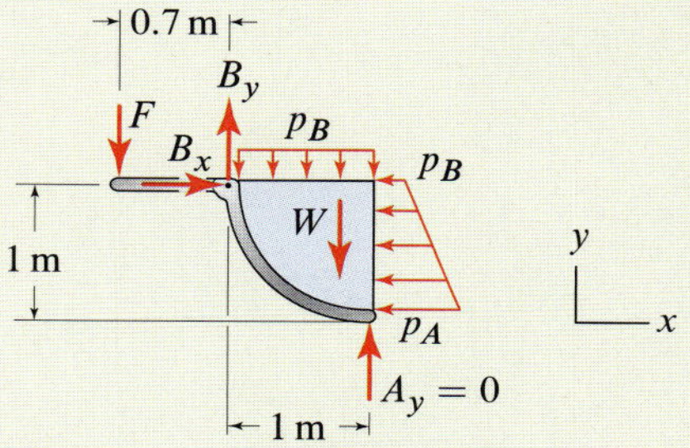

Figure 2

Free body diagram showing the gate, a cylindrical volume of fluid, and the pressure distribution acting on the volume of fluid.

Governing Equations & Computation The density of freshwater is $\rho = 10^3\ \text{kg/m}^3$, and from Eq. (7.46) on p. 434, the pressures at points A and B are

$$p_A = \rho g d_A = \left(10^3 \frac{\text{kg}}{\text{m}^3}\right)\left(9.81 \frac{\text{m}}{\text{s}^2}\right)(1.4\ \text{m}) = 13{,}730 \frac{\text{N}}{\text{m}^2} = 13.73 \frac{\text{kN}}{\text{m}^2}, \quad (1)$$

$$p_B = \rho g d_B = \left(10^3 \frac{\text{kg}}{\text{m}^3}\right)\left(9.81 \frac{\text{m}}{\text{s}^2}\right)(0.4\ \text{m}) = 3924 \frac{\text{N}}{\text{m}^2} = 3.924 \frac{\text{kN}}{\text{m}^2}. \quad (2)$$

Note that these pressures are identical to those found in Example 7.12. In Fig. 3, the FBD of Fig. 2 is redrawn with the pressure distributions replaced by the forces

$$P_1 = P_2 = p_B(1\ \text{m})(0.6\ \text{m}) = 2.354\ \text{kN}, \quad (3)$$

$$P_3 = \tfrac{1}{2}(p_A - p_B)(1\ \text{m})(0.6\ \text{m}) = 2.943\ \text{kN}, \quad (4)$$

where locations of the lines of action of these forces are shown in Fig. 3. The weight W of the cylindrical volume of water is the product of the water's density ρ, acceleration due to gravity g, and the cylindrical volume V:

$$W = \rho g V = \left(10^3 \frac{\text{kg}}{\text{m}^3}\right)\left(9.81 \frac{\text{m}}{\text{s}^2}\right)\frac{\pi(1\ \text{m})^2}{4}(0.6\ \text{m}) = 4623\ \text{N} = 4.623\ \text{kN}. \quad (5)$$

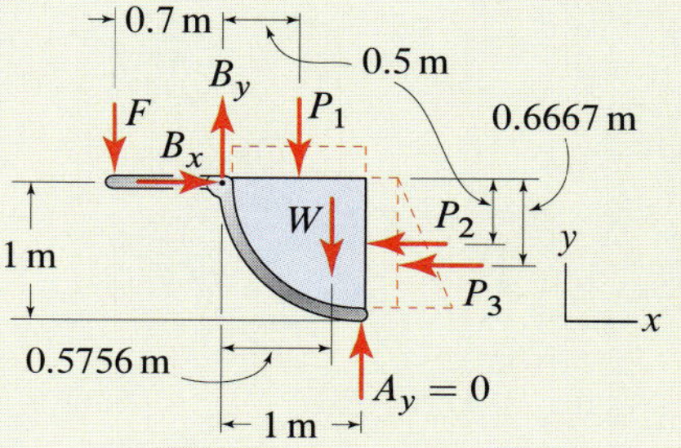

Figure 3

Free body diagram showing the forces produced by the pressure distributions.

The line of action of W is obtained by using the centroid position of a quarter-circular area from the Table of Properties of Lines and Areas on the inside back cover of this book and then evaluating $(1\ \text{m}) - (4)(1\ \text{m})/(3\pi) = 0.5756\ \text{m}$.

To determine the force F required to begin opening the gate, we sum moments about point B as follows:

$$\sum M_B = 0: \quad F(0.7\ \text{m}) - P_1(0.5\ \text{m}) - P_2(0.5\ \text{m}) - P_3(0.667\ \text{m})$$
$$-W(0.576\ \text{m}) = 0 \quad \Rightarrow \quad \boxed{F = 9.97\ \text{kN}.} \quad (6)$$

If desired, the reactions B_x and B_y can be determined by writing the equilibrium equations $\sum F_x = 0$ and $\sum F_y = 0$.

Discussion & Verification The force F determined here is slightly larger than the result for Example 7.12. The ease of carrying out the solution strategy in this example depends strongly on how easy it is to obtain the volume and centroid position of the fluid volume used in the FBD. To help you contrast this solution approach with that used in Example 7.12, the FBD of only the gate is shown in Fig. 4 where the complexity of the pressure distribution is apparent.

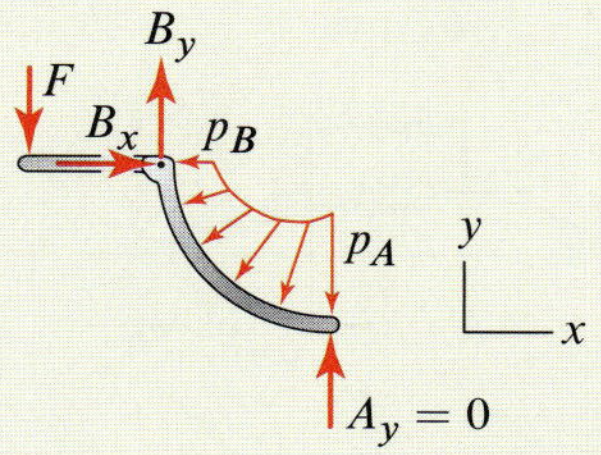

Figure 4

Free body diagram showing the pressure distribution the water applies to the gate.

EXAMPLE 7.14 *Gas Pressure Loading*

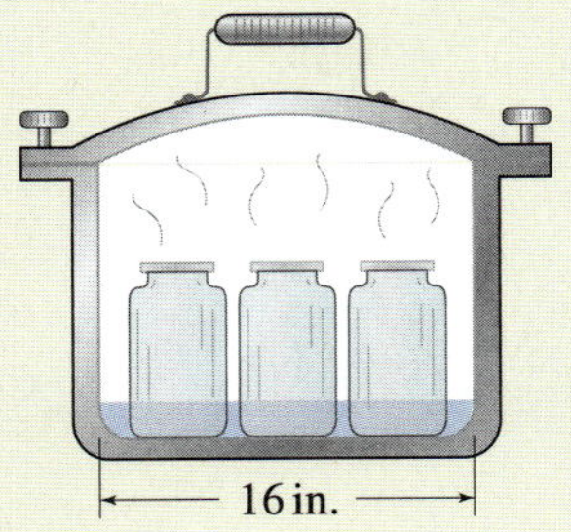

Figure 1

A *pressure cooker* is a large pot used for cooking and canning* foods under high pressure and temperature. The pressure cooker shown is cylindrical, operates at an internal pressure of 30 psi (psi = lb/in^2), and has a lid that is clamped to the base using six hand-tightened screws symmetrically located around the pot's circumference. Determine the force each screw must support due to the pressure within the cooker.

SOLUTION

Road Map To determine the force supported by the screws, we can use an FBD of the lid alone, or we can use an FBD of the lid *and* a judiciously selected volume of air. Because the surface of the lid over which the air makes contact has complex shape, the latter approach will be considerably easier. The air will be treated as a gas, so that it is assumed to have negligible weight.

Modeling We will assume that all six screws support the same force F. An FBD of the lid and a volume of air is shown in Fig. 2 where we have neglected the weight of the lid and air. To help provide clarification, the volume of air included in this FBD is shown in Fig. 3, and the uniform pressure distribution shown in the FBD acts on the flat bottom surface of this volume of air.

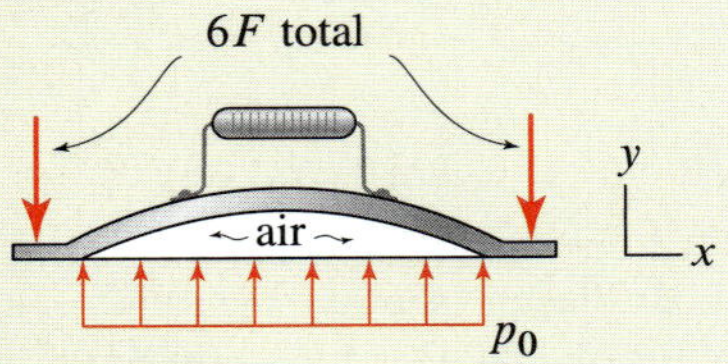

Figure 2
Free body diagram of the lid and a volume of air.

Governing Equations & Computation In Fig. 4, the FBD of Fig. 2 is redrawn with the pressure distribution replaced by the force

$$P = p_0 \pi (8\,\text{in.})^2 = 6032\,\text{lb}, \tag{1}$$

where the pressure is $p_0 = 30\,\text{lb/in.}^2$, $\pi(8\,\text{in.})^2$ is the area of the circular surface over which the pressure acts, and the line of action of P passes through the centroid of the pressure distribution.

To determine the force F supported by each of the six screws, we use the FBD of Fig. 4 to sum forces in the y direction to obtain

$$\sum F_y = 0: \quad -6F + P = 0 \quad \Rightarrow \quad \boxed{F = 1010\,\text{lb.}} \tag{2}$$

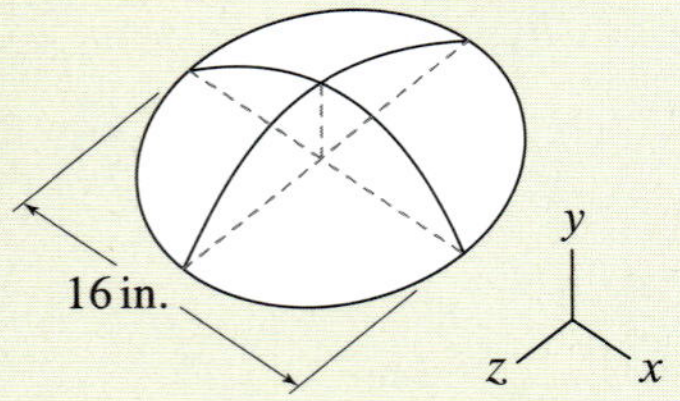

Figure 3
A sketch of the volume of air included in the FBD of Fig. 2.

Discussion & Verification

- The force P determined in Eq. (1) and the bolt force F determined in Eq. (2) are impressively large, and this underscores the need for safety when designing and operating a pressure cooker (and pressure vessels in general).
- The screw force F determined here is due to the pressurization only. In addition to this force, each of the screws supports a force due to the hand tightening, but this value is probably small compared to Eq. (2). Also, the assumption to neglect the weight of the lid is clearly acceptable in view of the results of Eq. (1).
- The pressure $p_0 = 30\,\text{lb/in.}^2$ is the *gage* pressure. In the FBDs, atmospheric pressure (approximate value at sea level is $14.7\,\text{lb/in.}^2$) is not included, because it has no net effect on equilibrium. That is, if atmospheric pressure were included, it would completely surround the objects shown in the FBDs, and when we wrote equilibrium equations such as $\sum F_y = 0$, its effects would fully cancel in these expressions.

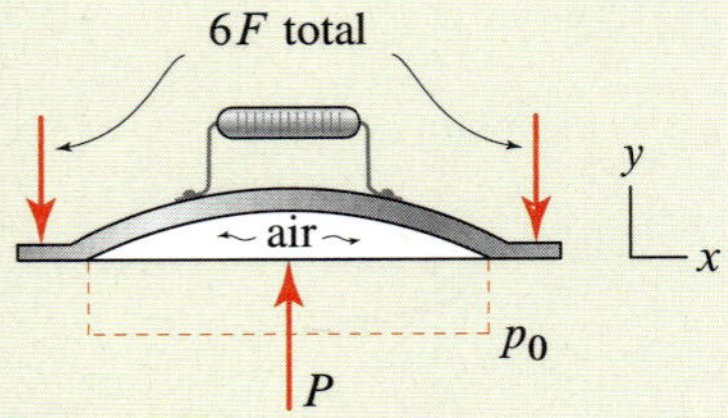

Figure 4
Free body diagram showing the force produced by the pressure distribution.

---

* *Canning* is a process whereby food in sealed jars is placed in a pressure cooker with a small amount of water that is raised to high pressure and temperature to kill harmful bacteria. After heat treatment, the jars may be stored at room temperature without spoilage of the food.

PROBLEMS

General instructions. Unless otherwise stated, in the following problems you may use integration, composite shapes, or a combination of these. The specific weight and density of water are 0.0361 lb/in.^{3} and 10^3 kg/m^3, respectively.

Problem 7.58

A shelf in a grocery store supports 100 bags of rice, each bag weighing 1 lb. Consider the arrangements shown: (a) The bags are stacked at a uniform height, (b) the bags are stacked twice as high on the right-hand side as on the left-hand side, (c) the bags are stacked twice as high on the right-hand side as on the left-hand side with a linear variation, and (d) the bags are stacked twice as high in the middle as at the two ends with linear variations. For each of the arrangements, develop an expression (or multiple expressions if needed) for the distributed force w as a function of position x.

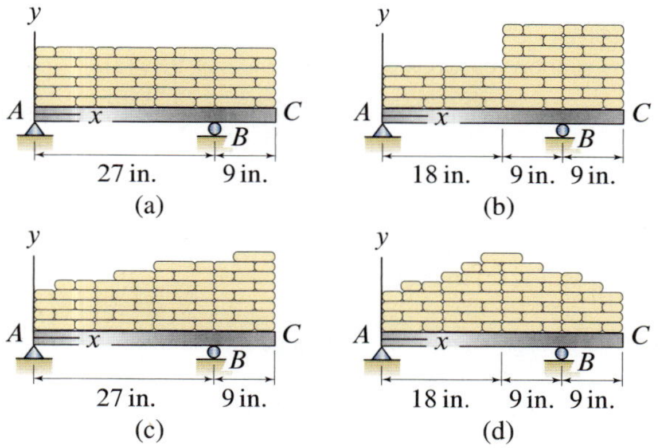

Figure P7.58

Problem 7.59

A cantilever beam supports a wall built of 1000 bricks, each brick having 5 kg mass. Consider the arrangements shown: (a) The wall has uniform height, (b) the wall is twice as high on the left-hand side as on the right-hand side, (c) the wall is twice as high on the left-hand side as on the right-hand side with a linear variation, and (d) the wall is twice as high in the middle as at the two ends with linear variations. For each of the arrangements, develop an expression (or multiple expressions if needed) for the distributed force w as a function of position x.

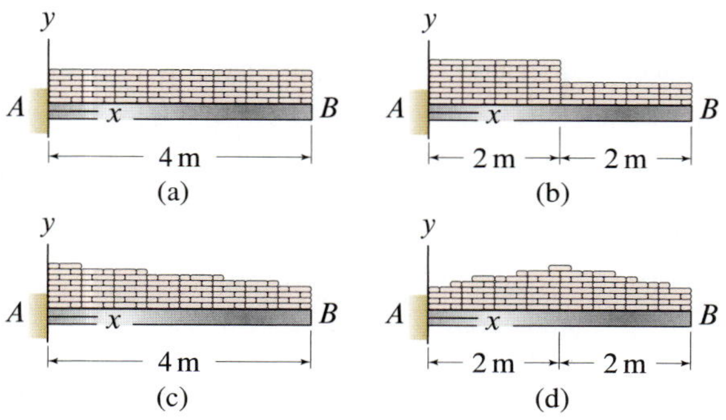

Figure P7.59

Problems 7.60 through 7.63

Determine the support reactions, using composite shapes to represent the distributed load.

Problem 7.60 Loading (a) in Prob. 7.58 on p. 443.

Problem 7.61 Loading (d) in Prob. 7.58 on p. 443.

Problem 7.62 Loading (a) in Prob. 7.59 on p. 443.

Problem 7.63 Loading (d) in Prob. 7.59 on p. 443.

Problems 7.64 through 7.67

(a) Use integration (Eqs. (7.41) and (7.43) on p. 433) with the appropriate expression(s) for the distributed load w to determine the x position of the line of action for the resultant force produced by the distributed load.

(b) Determine the support reactions using the results of Part (a).

(c) Determine the support reactions using composite shapes to represent the distributed load.

Problem 7.64 Loading (b) in Prob. 7.58 on p. 443.

Problem 7.65 Loading (c) in Prob. 7.58 on p. 443.

Problem 7.66 Loading (b) in Prob. 7.59 on p. 443.

Problem 7.67 Loading (c) in Prob. 7.59 on p. 443.

Problem 7.68

Consider modeling the weights from books 8–11 from Example 7.11 on p. 439, using the linear force distribution $w = a + bx$, as shown here. The constants a and b can be determined by requiring the two force systems shown to be equivalent. That is,

$$\int_0^{4\text{ in.}} w\,dx = \sum_{i=8}^{11} W_i \quad \text{and} \quad \int_0^{4\text{ in.}} xw\,dx = \sum_{i=8}^{11} M_{Bi},$$

where M_{Bi} is the moment about point B (the origin) of weight W_i.

(a) Using the weights and geometry of books 8–11 given in Table 1 of Example 7.11, evaluate the above expressions and solve for a and b to show that $w = \frac{35}{8}\,\frac{\text{lb}}{\text{in.}} - \left(\frac{15}{16}\,\frac{\text{lb}}{\text{in.}^2}\right)x$.

(b) Evaluate the distributed force from Part (a) at $x = 0$ and $x = 4$ in., and compare these to the values used in Example 7.11, namely, 4 lb/in. and 1 lb/in., respectively.

(c) Discuss why the distributed force from Part (a) is better than that used in Example 7.11.

(d) Using the distributed force from Part (a), determine the support reactions for the bookshelf and compare to those found in Example 7.11.

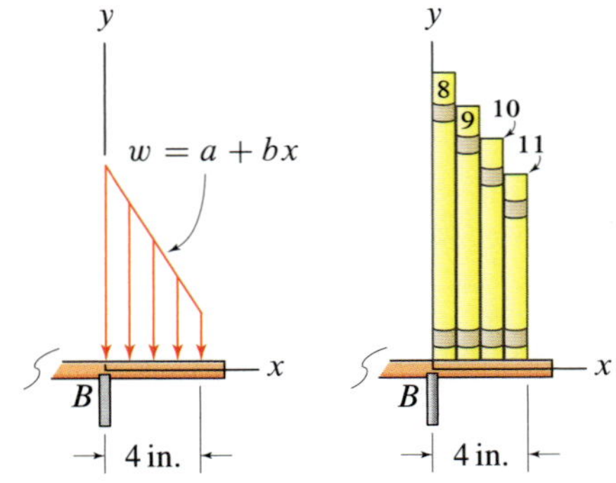

Figure P7.68

Problem 7.69

(a) For the distributed loading shown, develop an expression (or multiple expressions if needed) for the distributed force w as a function of position x.

(b) Use integration (Eqs. (7.41) and (7.43) on p. 433) with the results of Part (a) to determine the total force produced by the distributed load and the x position of its line of action.

(c) Determine the support reactions using the results of Part (b).

(d) Determine the support reactions using composite shapes for the distributed load.

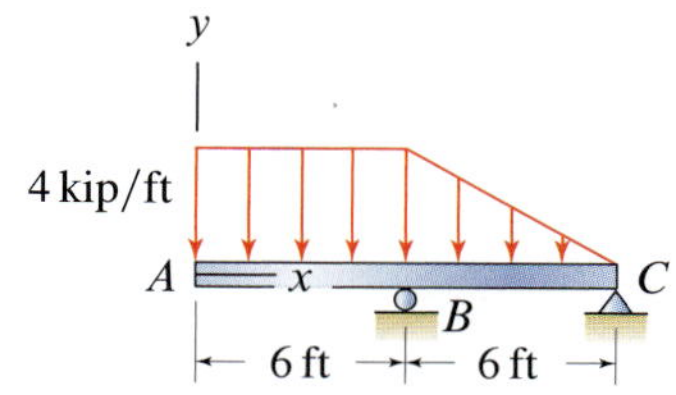

Figure P7.69

Problem 7.70

A beam is loaded by a distributed force that begins at the left-hand end as an 800 N/m uniform load with $dw/dx = 0$ and decreases to zero at the right-hand end.

(a) Determine the constants a, b, and c so that the quadratic polynomial $w = a + bx + cx^2$ describes this loading.

(b) Determine the constants d and f so that the trigonometric function $w = d\cos(fx)$ describes this loading.

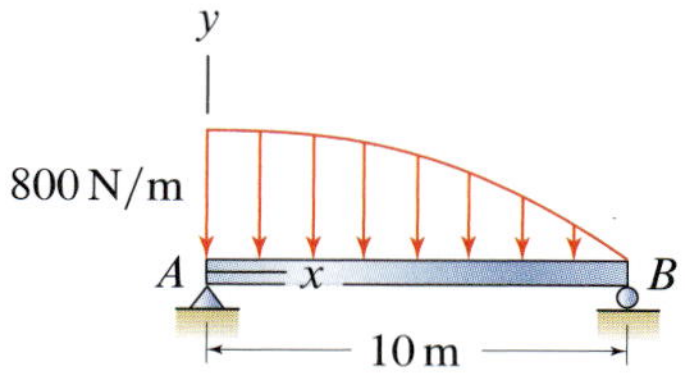

Figure P7.70 and P7.71

Problem 7.71

(a) Use integration (Eqs. (7.41) and (7.43) on p. 433) with the results of Part (a) of Prob. 7.70 to determine the total force produced by the distributed load, the x position of its line of action, and the support reactions.

(b) Use integration (Eqs. (7.41) and (7.43) on p. 433) with the results of Part (b) of Prob. 7.70 to determine the total force produced by the distributed load, the x position of its line of action, and the support reactions.

(c) Compare the results of Parts (a) and (b), and discuss if these should be the same or if differences are expected.

Problems 7.72 through 7.77

Determine the support reactions for the loading shown.

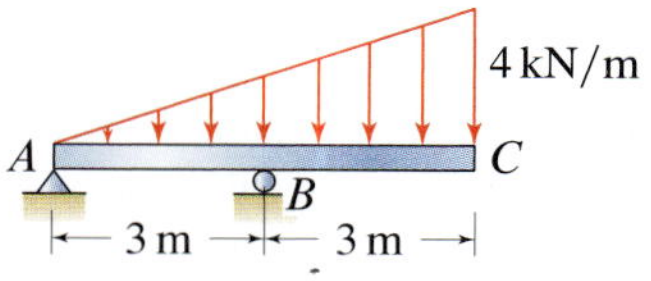

Figure P7.72

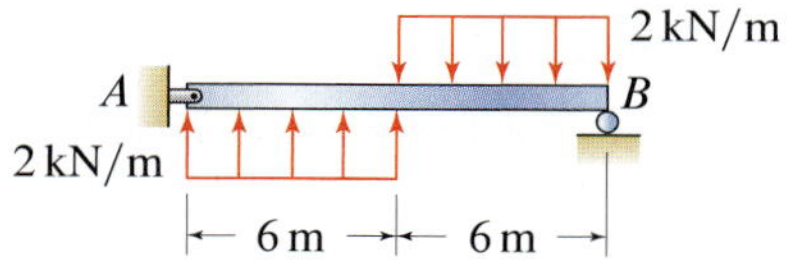

Figure P7.73

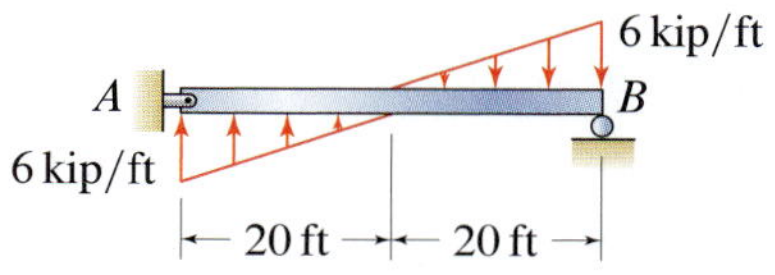

Figure P7.74

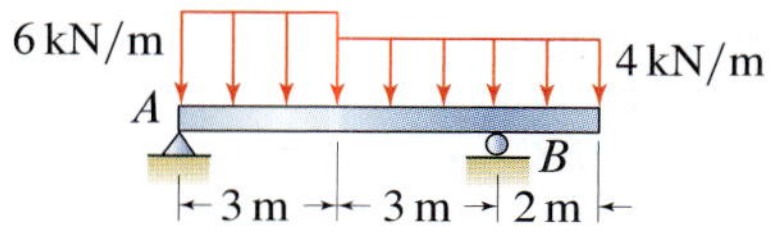

Figure P7.75

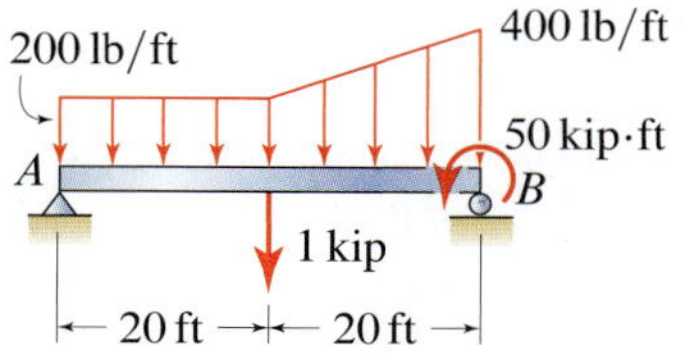

Figure P7.76

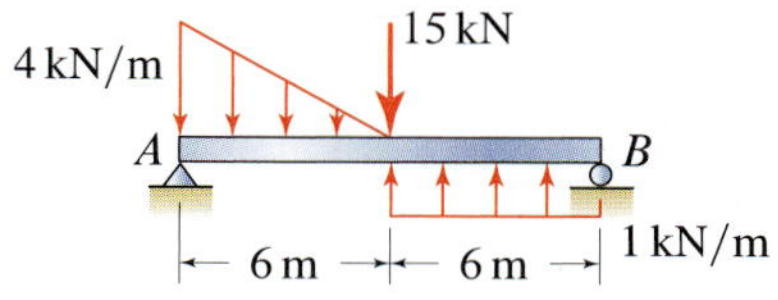

Figure P7.77

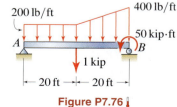

Problem 7.78

In Fig. P7.72 on p. 445, replace the pin and roller supports with a built-in support at C, and determine the support reactions.

Problem 7.79

In Fig. P7.73 on p. 445, replace the pin and roller supports with a built-in support at A, and determine the support reactions.

Problem 7.80

In Fig. P7.74 on p. 445, reposition the roller support to the midspan of the beam, and determine the support reactions.

Problem 7.81

In Fig. P7.75 on p. 445, replace the pin and roller supports with a built-in support at A, and determine the support reactions.

Problem 7.82

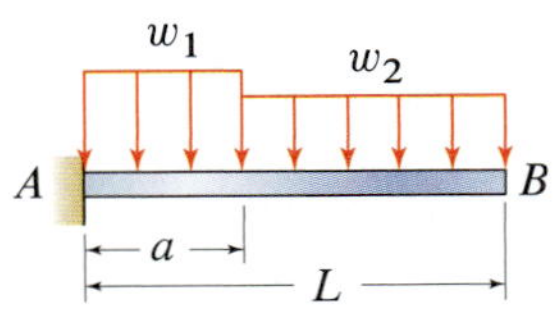

Figure P7.82

Determine the support reactions for the cantilever beam. Express your answers in terms of parameters such as w_1, w_2, a, and L.

Problem 7.83

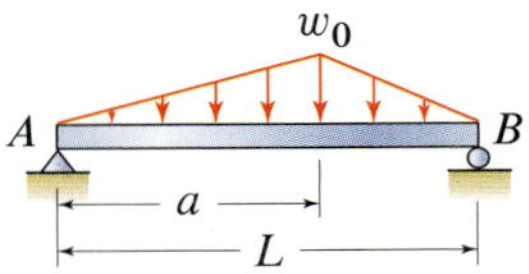

Figure P7.83

Determine the support reactions for the simply supported beam. Express your answers in terms of parameters such as w_0, a, and L.

Problem 7.84

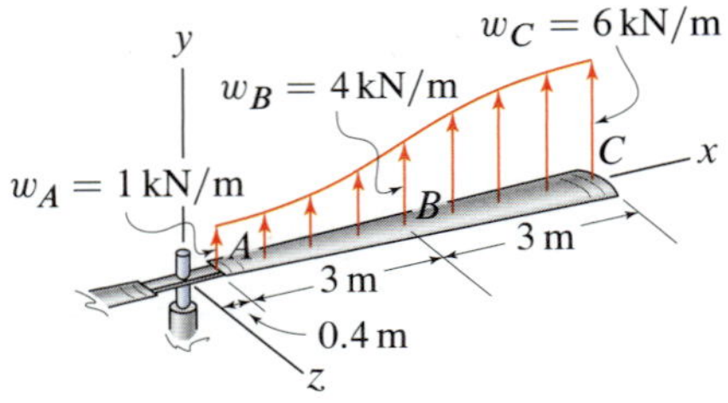

Figure P7.84

A blade of the main rotor of a hovering helicopter is subjected to the y direction distributed forces shown, where values of the distributed force are known at points A, B, and C. Determine the constants a, b, and c so that the quadratic polynomial $w = a + bx + cx^2$ describes this loading. Using this polynomial, determine the total force produced by this distribution and the x position of its line of action.

Problem 7.85

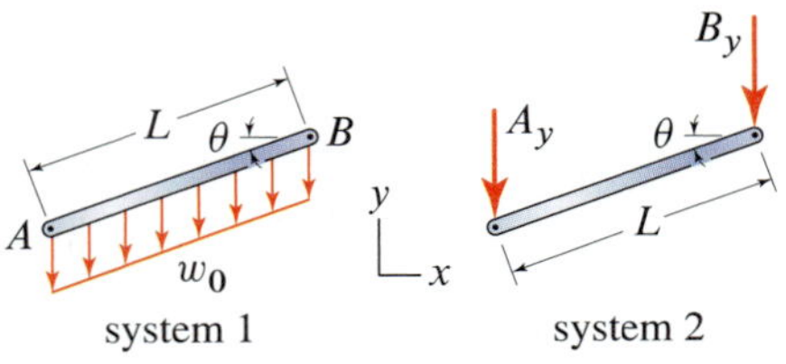

Figure P7.85

Consider a straight uniform member with length L and weight W. Two force systems for representing the weight of this member are shown. In system 1, the distributed force is uniform with value $w_0 = W/L$.

(a) Use Eq. (4.16) on p. 223 to determine A_y and B_y in terms of W so that the two force systems are equivalent.

(b) Are the results for Part (a) in agreement with the load lumping scheme described for trusses in Example 6.2 on p. 338?

(c) Without calculations, but perhaps by use of an appropriate sketch, show that the results of Part (a) *do not* apply for a member that is not straight and/or has nonuniform weight distribution. *Hint:* Redraw Fig. P7.85 using a member that is not straight or using a straight member with nonuniform weight distribution. Then argue that the results of Part (a) do not constitute an equivalent force system for this situation.

Problem 7.86

A uniform curved beam with circular shape and weight W has a built-in support at A. Determine the support reactions. Express your answers in terms of parameters such as W and R.

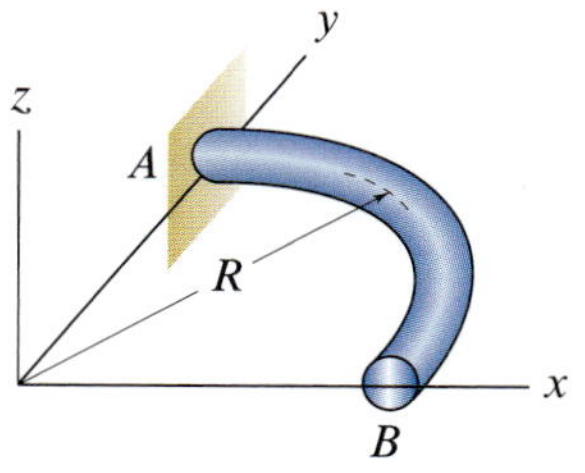

Figure P7.86

Problem 7.87

A uniform curved beam with circular shape and weight W is supported by a frictionless bearing at A and a roller at B. Determine the support reactions. Express your answers in terms of parameters such as W and R.

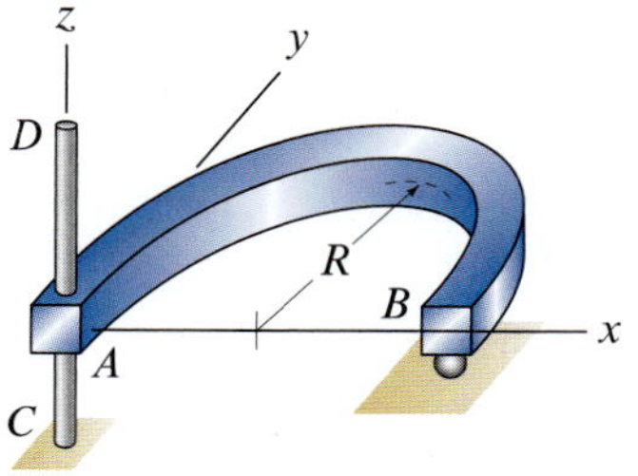

Figure P7.87

Problem 7.88

A cube of material with edge lengths d and specific weight 2γ is suspended by a cable and is submerged to a depth d in a fluid having specific weight γ. Determine the force T in the cable. Express your answer in terms of parameters such as d, γ, etc.

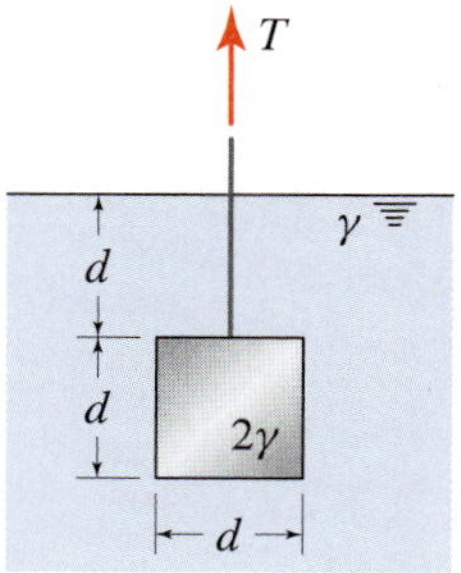

Figure P7.88

Problem 7.89

In Fig. P7.89(a), the concrete wall of a building has a small water-filled gap between it and the adjacent soil. In Fig. P7.89(b), a concrete wall is used to retain water. If the depth d of water is the same for both walls, which wall will be subjected to the larger forces due to water pressure? Explain.

Note: Concept problems are about *explanations*, not computations.

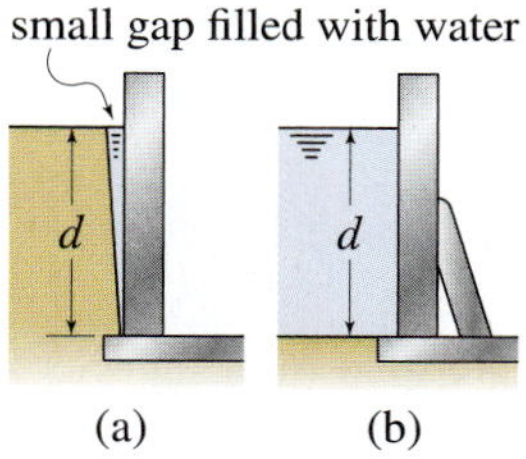

Figure P7.89

Problems 7.90 and 7.91

Water in a channel is retained by a gate with 5 in. width (into the plane of the figure). The gate is supported by a pin at B and a vertical cable at A, and the contact between the gate and the bottom of the channel at A is frictionless. The vertical wall BC is fixed in position. If the gate's 50 lb weight is uniformly distributed, determine the cable force T required to begin opening the gate.

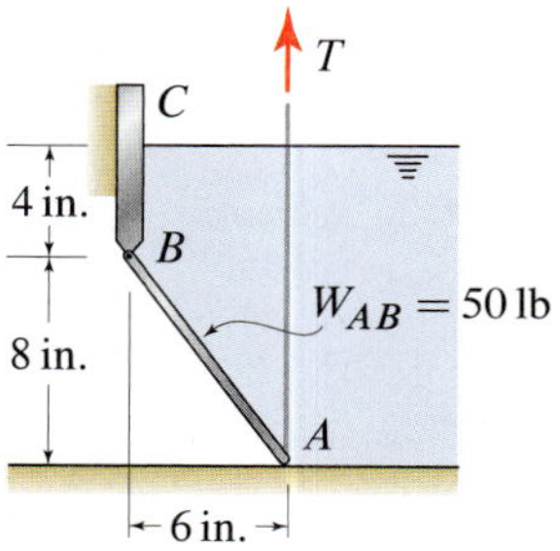

Figure P7.90

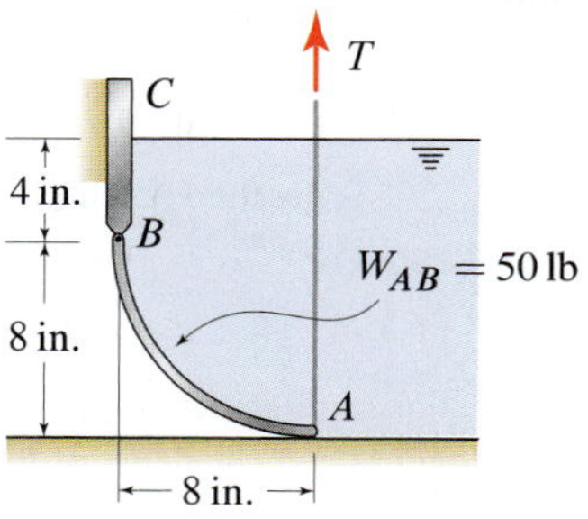

Figure P7.91

Problems 7.92 and 7.93

Water in a channel is retained by a gate with 10 cm width (into the plane of the figure). The gate is supported by a pin at B and by frictionless contact with the bottom of the channel at A. The gate is outfitted with a prewound torsional spring at B with stiffness $k_t = 75\,\text{N·m/rad}$. The vertical wall BC is fixed in position. If the gate has negligible weight, determine the amount θ_0 the spring must be prewound so that the gate will begin to open when $d = 50\,\text{cm}$.

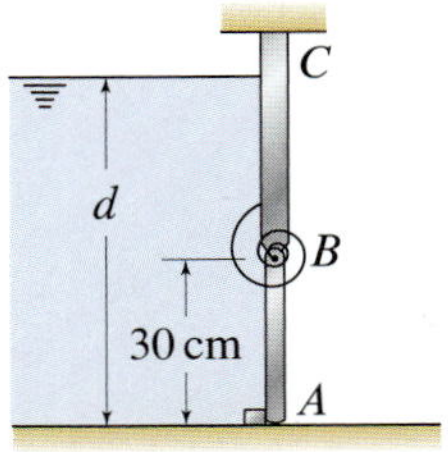

Figure P7.92

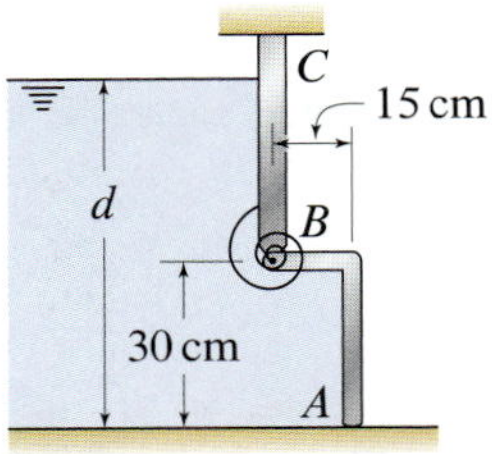

Figure P7.93

Problem 7.94

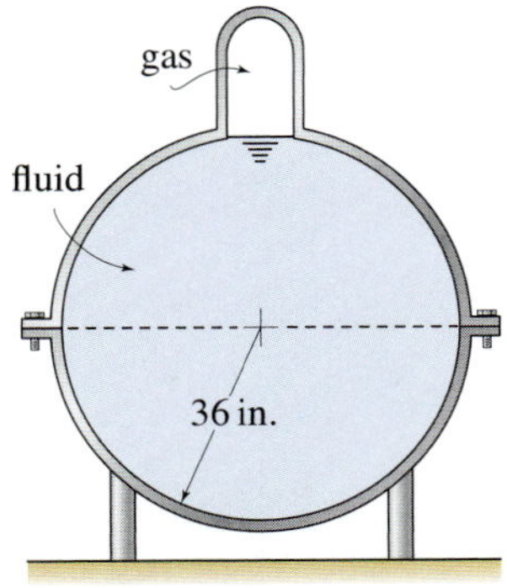

Figure P7.94

The cross section through a spherical tank is shown. The upper and lower portions of the tank are attached using 72 bolts that are uniformly spaced around the perimeter of the tank. The upper portion of the tank has a small dome that contains a gas. The fluid in the tank has specific weight of $0.06\,\text{lb/in.}^3$ and approximately spherical shape. Assuming all the bolts support the same force, determine the force each bolt supports due to the fluid and gas pressures if:

(a) The gas is not pressurized (i.e., it is at atmospheric pressure).

(b) The gas is pressurized to 5 psi.

Problem 7.95

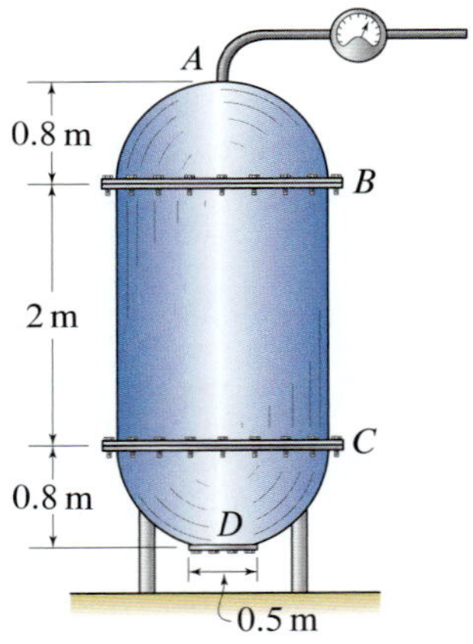

Figure P7.95 and P7.96

The tank shown has a cylindrical midsection with hemispherical ends. Each of the hemispherical ends is attached to the cylindrical midsection using 60 bolts that are uniformly spaced around the perimeter of the tank. At D, the tank has a circular access plate that is attached using 12 bolts that are uniformly spaced. The tank is fully filled with a fluid having density $900\,\text{kg/m}^3$. Assume that each of the bolts at B supports the same force, each of the bolts at C supports the same force, and each of the bolts at D supports the same force. However, the forces supported by the bolts at B, C, and D are probably different. Assume the piping that enters the tank at A is flexible and has negligible weight. Determine the force each bolt supports due to the fluid pressure if the fluid at A is at atmospheric pressure.

Problem 7.96

Repeat Prob. 7.95 if the fluid at A is at $10\,\text{kN/m}^2$ pressure.

Problem 7.97

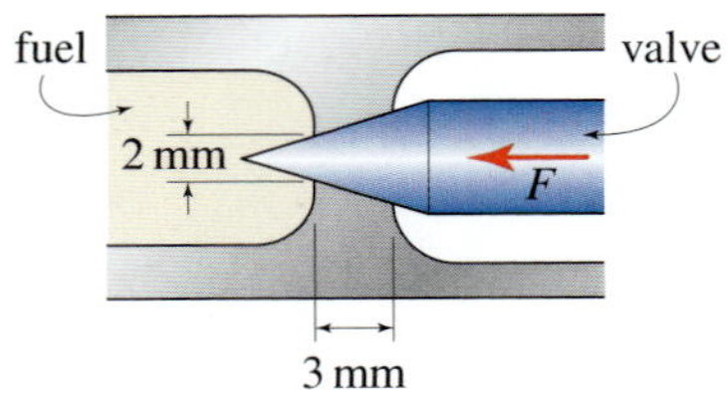

Figure P7.97

The cross section through the valve of a fuel injector for an engine is shown, where the tip of the valve has conical shape. If the fuel is at $500\,\text{kN/m}^2$ pressure, determine the force F that must be applied to keep the valve closed. *Hint:* The pressure due to weight of the fuel is negligible compared to $500\,\text{kN/m}^2$.

Problem 7.98

Grain is contained in a silo. The walls of the silo are fixed, and the door $ABCD$ can be opened to allow the grain to pour out. Door $ABCD$ is flat, with 8 in. depth (into the plane of the figure). Idealize the grain to be a fluid with $0.025\ \mathrm{lb/in.^3}$ specific weight. In the position shown, the hydraulic cylinder EG is horizontal. Neglect the weights of the individual members. Determine the force the hydraulic cylinder EG must support to keep the door in equilibrium. Report your answer, using a positive value for tension in the hydraulic cylinder and a negative value for compression.

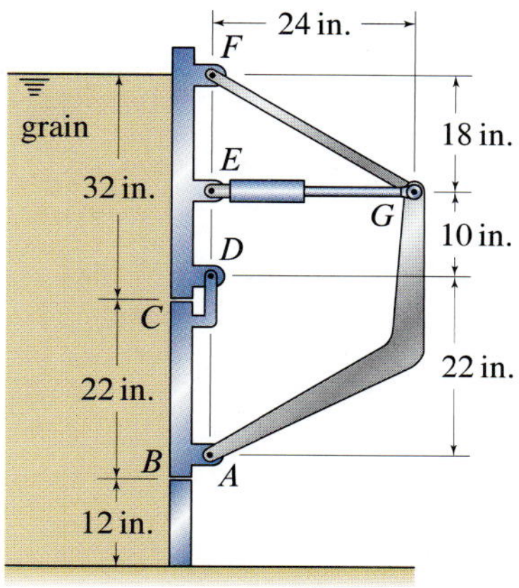

Figure P7.98

7.5 Chapter Review

Important definitions, concepts, and equations of this chapter are summarized. For equations and/or concepts that are not clear, you should refer to the original equation and page numbers cited for additional details.

Centroid. The *centroid* is defined to be the average position of a distribution of shape. The equations for determining the centroid of lines, areas, and volumes can be compactly summarized as follows. Consider the example of the centroid of an area or a distribution of areas, as shown in Fig. 7.24. The x position of the centroid of this area can be determined by using composite shapes or integration as follows

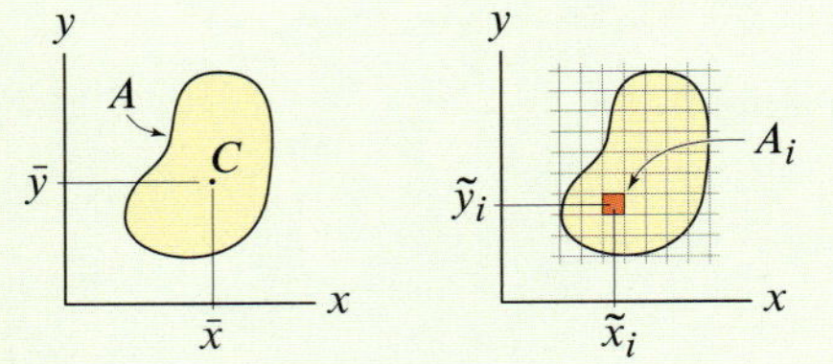

Figure 7.24
The centroid C of area A is located at $\bar{x}$ and $\bar{y}$. Area A can be considered as consisting of composite areas A_i, where the centroid of each of these is located at $\tilde{x}_i$ and $\tilde{y}_i$.

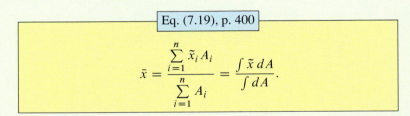

Eq. (7.19), p. 400

$$\bar{x} = \frac{\sum_{i=1}^{n} \tilde{x}_i A_i}{\sum_{i=1}^{n} A_i} = \frac{\int \tilde{x}\, dA}{\int dA}.$$

The expressions for the y position of the centroid are identical to Eq. (7.19) with all of the x's replaced by y's. Similarly, the expressions for the z position of the centroid are identical to Eq. (7.19) with all of the x's replaced by z's. To determine the centroid of a line or a distribution of lines (straight and/or curved), all of the A's that appear in Eq. (7.19) are replaced by L's. To determine the centroid of a volume or a distribution of volumes, all of the A's that appear in Eq. (7.19) are replaced by V's.

Center of Mass. The *center of mass* is defined to be the average position of a distribution of mass. The equations for determining the center of mass are identical to Eq. (7.19) with all of the A's replaced by m's; hence

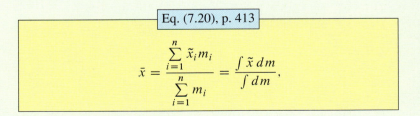

Eq. (7.20), p. 413

$$\bar{x} = \frac{\sum_{i=1}^{n} \tilde{x}_i m_i}{\sum_{i=1}^{n} m_i} = \frac{\int \tilde{x}\, dm}{\int dm},$$

with similar expressions for $\bar{y}$ and $\bar{z}$. Furthermore, the summations and integrations in Eq. (7.20) may be over a combination of volumes, surfaces, and/or wires, as shown in Eqs. (7.21) and (7.22) on p. 414.

Center of Gravity. The *center of gravity* is defined to be the average position of a distribution of weight. For objects in a uniform gravity field, the equations for determining the center of gravity are identical to Eq. (7.19) with all of the

A's replaced by w's, hence

Eq. (7.26), p. 415

$$\bar{x} = \frac{\sum_{i=1}^{n} \tilde{x}_i w_i}{\sum_{i=1}^{n} w_i} = \frac{\int \tilde{x}\, dw}{\int dw},$$

with similar expressions for $\bar{y}$ and $\bar{z}$. Furthermore, the summations and integrations in Eq. (7.26) may be over a combination of volumes, surfaces, and/or wires, as shown in Eqs. (7.27) and (7.28) on p. 415.

Theorems of Pappus and Guldinus. A *surface of revolution* is produced by rotating a *generating curve*, as shown in Fig. 7.25, by an angle θ (in radians) about an *axis of revolution*. From the theorems of Pappus and Guldinus, the area of one side of the surface of revolution is

Eqs. (7.32) and (7.33), p. 425

$$A = \theta \bar{r} L = \theta \sum_{i=1}^{n} \tilde{r}_i L_i.$$

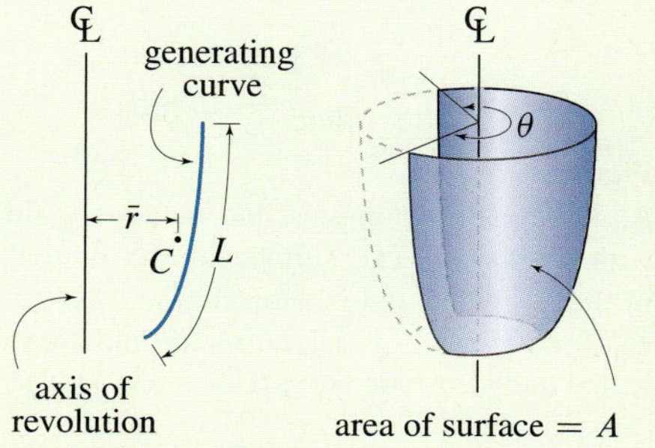

Figure 7.25
A surface of revolution is produced by rotating a generating curve by an angle θ about an axis of revolution. The generating curve has arc length L and centroid C located a distance $\bar{r}$ from an axis of revolution.

The latter expression given above, Eq. (7.33), is useful when the generating curve consists of simple composite shapes, as shown in Fig. 7.11 on p. 425.

A *solid of revolution* is produced by rotating a *generating area*, as shown in Fig. 7.26, by an angle θ about an axis of revolution. From the theorems of Pappus and Guldinus, the volume of the solid of revolution shown in Fig. 7.26 is

Eqs. (7.35) and (7.36), p. 426

$$V = \theta \bar{r} A = \theta \sum_{i=1}^{n} \tilde{r}_i A_i.$$

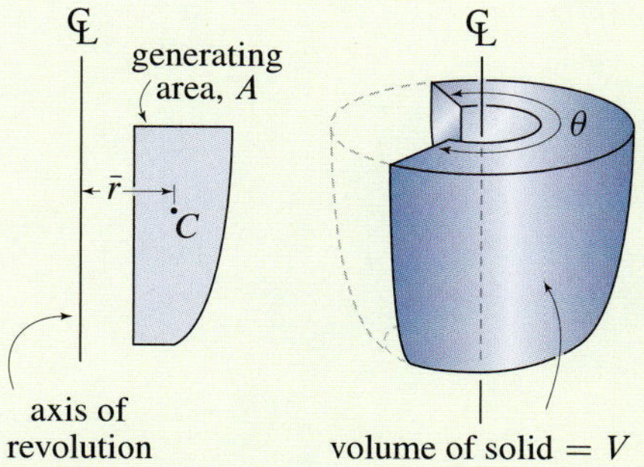

Figure 7.26
A solid of revolution is produced by rotating a generating area by an angle θ about an axis of revolution. The generating area A has its centroid C located a distance $\bar{r}$ from the axis of revolution.

The latter expression given above, Eq. (7.36), is useful when the generating area consists of simple composite shapes, as shown in Fig. 7.13 on p. 426.

Distributed Forces. *Distributed forces* are forces that are distributed along a line, over a surface, or throughout a volume. A force distributed along a line is called a *line force*, or a *line load*, and this has dimensions of force/length. A force distributed over a surface is called a *surface force*, or a *traction*, and this has dimensions of force/area. A force distributed throughout a volume is called a *volume force*, or a *body force*, and this has dimensions of force/volume. Line forces, surface forces, and volume forces, are all vectors. Line forces and surface forces do not need to be perpendicular to the objects or structures they are applied to, although often they will be.

Distributed Forces Applied to Beams. Beams are often subjected to distributed forces that are line loads. With w being the line load, with dimensions of force/length, and x being a coordinate along the length of the beam, the total force F produced by the distributed load and the position $\bar{x}$ of its line of action are

Eqs. (7.41) and (7.43), p. 433

$$F = \int dF = \int w\,dx, \qquad \bar{x} = \frac{\int \tilde{x}\,dF}{\int dF} = \frac{\int \tilde{x}\,w\,dx}{\int w\,dx}.$$

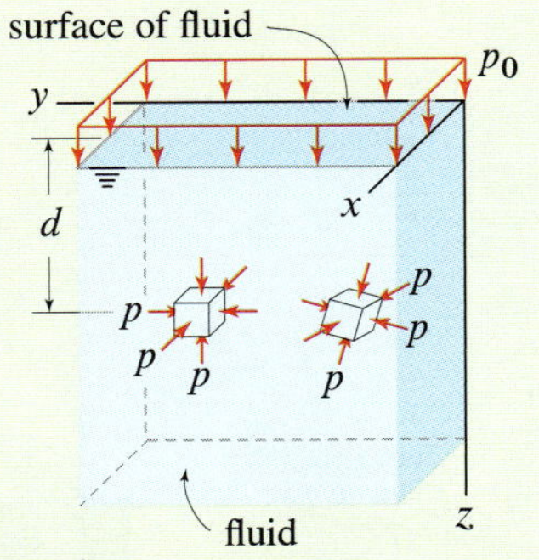

Figure 7.27

A volume of incompressible stationary fluid with density ρ. The surface of the fluid lies in the xy plane and is subjected to a pressure p_0. Infinitesimally small cubes of fluid are subjected to hydrostatic compressive pressure p.

Fluid and Gas Pressure. If a fluid is incompressible and at rest, as shown in Fig. 7.27, then the pressure within the fluid is given by

Eq. (7.45), p. 434

$$\begin{aligned} p &= p_0 + \rho g d, \\ &= p_0 + \gamma d, \end{aligned}$$

where p_0 is the gas pressure (constant) at the surface of the fluid. For many purposes, only the portion of the pressure due to fluid loading is important, and this pressure ($p = \rho g d = \gamma d$) is sometimes called the *gage pressure*.

The forces that a fluid and/or gas applies to a structure can, in principle, always be determined by integration. However, the use of composite shapes and/or a judicious FBD, as discussed in Section 7.4, will often be more straightforward. An important point is that fluid and gas pressures are always perpendicular to the structure or surface over which they act.

REVIEW PROBLEMS

General instructions. For problems involving shapes or objects having one or more axes or planes of symmetry, you may use inspection to determine some of the coordinates of the centroid, center of mass, and/or center of gravity.

Problems 7.99 through 7.101

For the area shown, use composite shapes to determine the x and y positions of the centroid.

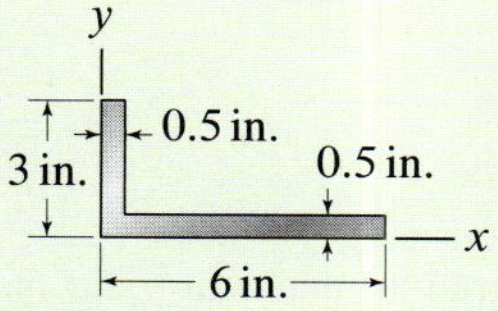

Figure P7.99

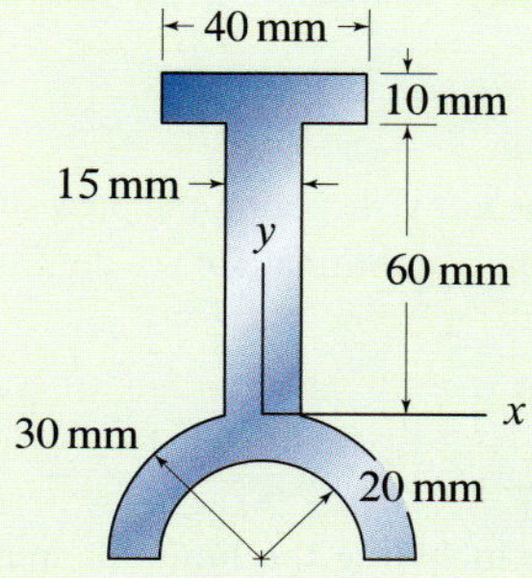

Figure P7.100

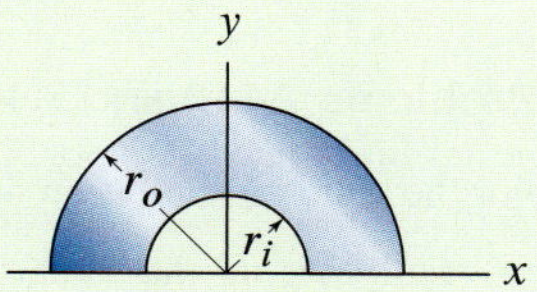

Figure P7.101

Problem 7.102

For the area shown, use integration to determine the x and y positions of the centroid.

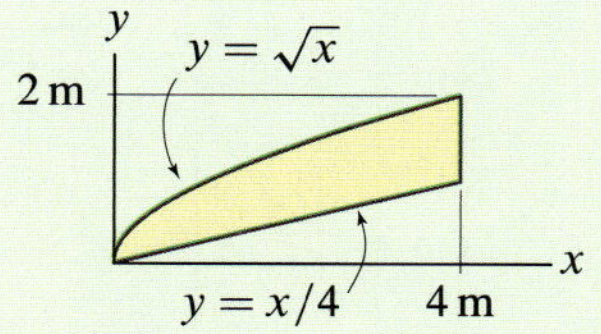

Figure P7.102

Problem 7.103

For the truncated circular cone shown, use composite shapes to determine the location of the centroid.

Problem 7.104

For the truncated circular cone shown, use integration to determine the location of the centroid.

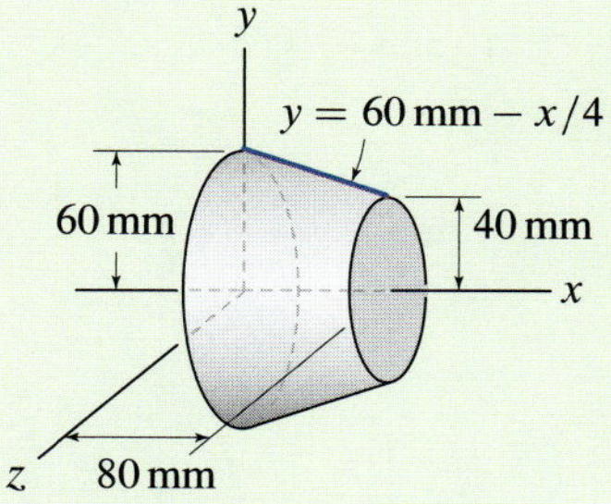

Figure P7.103 and P7.104

Problem 7.105

The truncated circular cone shown has a truncated conical hole.

(a) Fully set up the integral, including the limits of integration, that will yield the centroid of the object.

(b) Evaluate the integral determined in Part (a) using computer software such as Mathematica, Maple, etc.

Problem 7.106

The truncated circular cone shown has a truncated conical hole and is made of a material with density 0.002 g/mm^3. Let the conical hole be filled with a material with density 0.003 g/mm^3.

(a) Fully set up the integral, including the limits of integration, that will yield the center of mass of the object.

(b) Evaluate the integral determined in Part (a) using computer software such as Mathematica, Maple, etc.

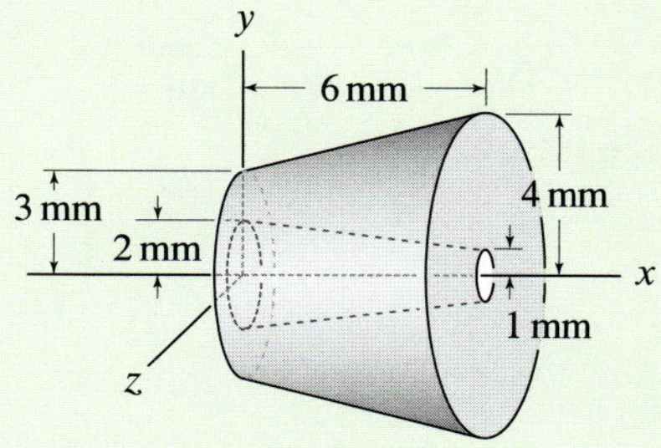

Figure P7.105 and P7.106

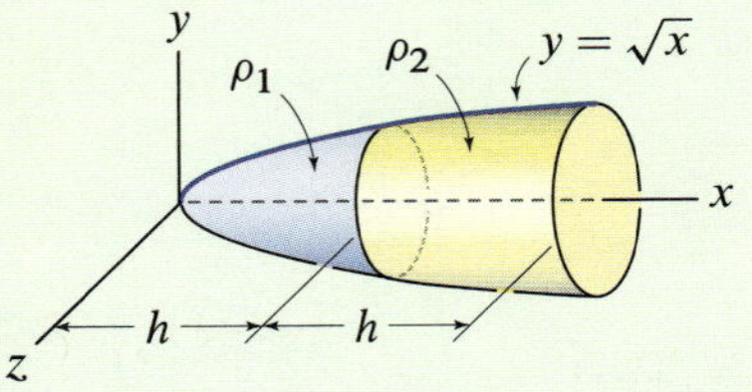

Figure P7.107

Problem 7.107

The bullet-shaped object is a solid of revolution that is composed of materials with densities ρ_1 and ρ_2. Set up the integral, including the limits of integration, that will yield the x position of the center of mass.

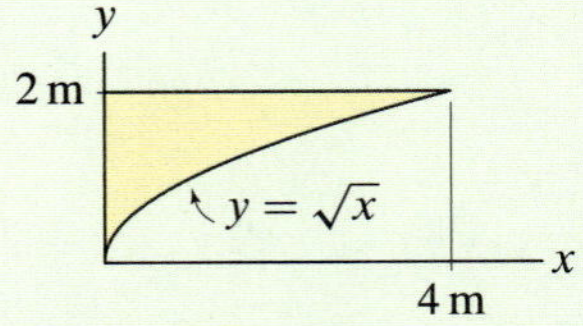

Figure P7.108 and P7.109

Problem 7.108

A solid of revolution is produced by revolving the area shown 360° about the x axis. Use integration to determine the x coordinate of the centroid.

Problem 7.109

A solid of revolution is produced by revolving the area shown 360° about the y axis. Use integration to determine the y coordinate of the centroid.

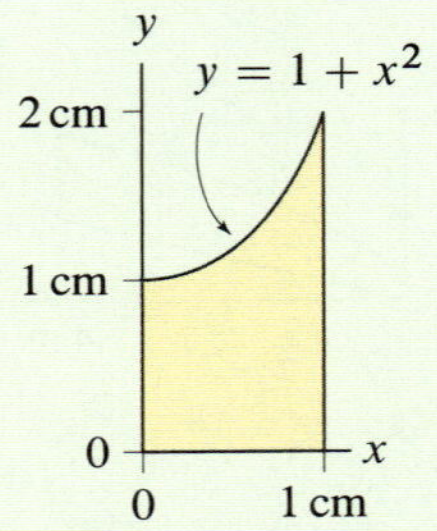

Figure P7.110 and P7.111

Problems 7.110 and 7.111

For the solid of revolution described below:

(a) Fully set up the integral, including the limits of integration, that will yield the centroid of the object.

(b) Evaluate the integral determined in Part (a) using computer software such as Mathematica, Maple, etc.

Problem 7.110 The solid of revolution is produced by revolving the area shown about the x axis.

Problem 7.111 The solid of revolution is produced by revolving the area shown about the y axis.

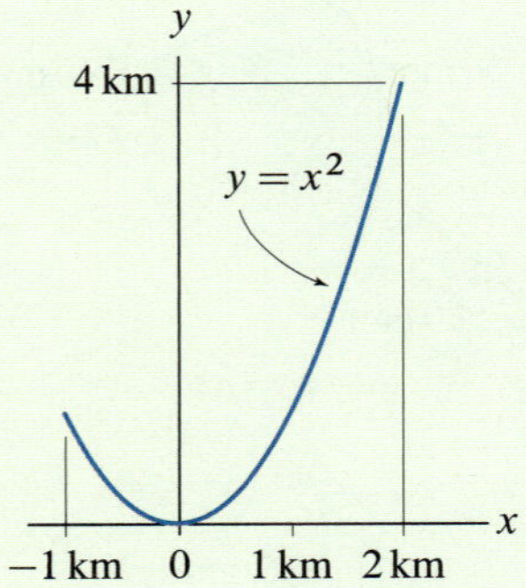

Figure P7.112

Problem 7.112

For the line shown:

(a) Set up the integrals for integration with respect to x, including the limits of integration, that will yield the x and y positions of the centroid.

(b) Evaluate the integrals in Part (a) using computer software such as Mathematica, Maple, etc.

Problem 7.113

The area of Prob. 7.15 on p. 410 is revolved 360° about the line $x = -r$ to create a solid of revolution. Determine the volume and surface area of the solid.

Problem 7.114

The area of Prob. 7.15 on p. 410 is revolved 360° about the x axis to create a solid of revolution. Determine the volume and surface area of the solid.

Problem 7.115

A solar panel has the shape of a 90° sector of a truncated right circular cone, and hence it is a surface of revolution. Use the Pappus-Guldinus theorem to determine the outside surface area of the panel.

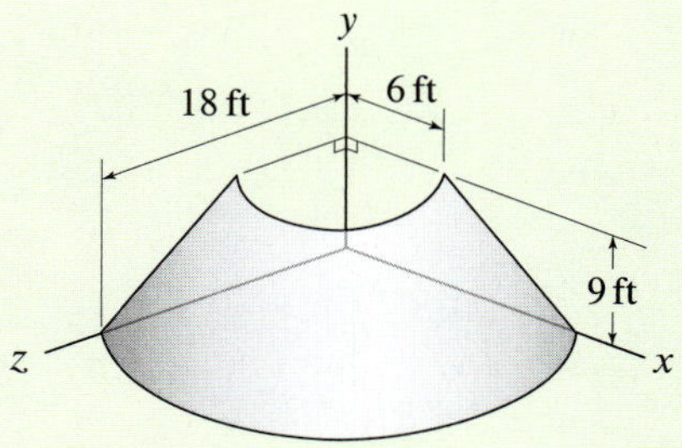

Figure P7.115

Problem 7.116

A scoop for handling animal food is shown. The scoop's shape is one-half of a truncated circular cone. Use the Pappus-Guldinus theorem to determine the volume of food the scoop will hold, assuming the food is "level." Also, disregarding the handle, determine the area of sheet metal, in cm^2, required to fabricate the scoop.

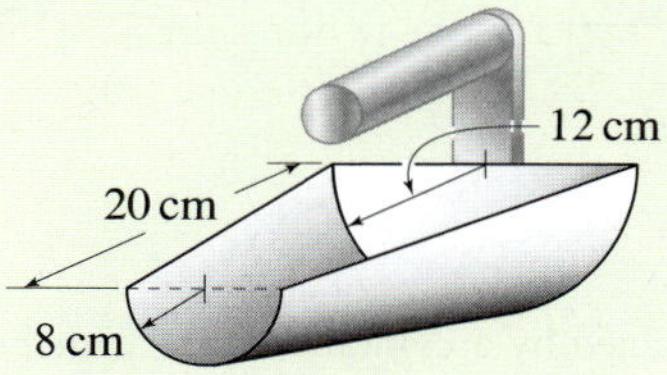

Figure P7.116

Problems 7.117 through 7.119

Determine the support reactions for the loading shown.

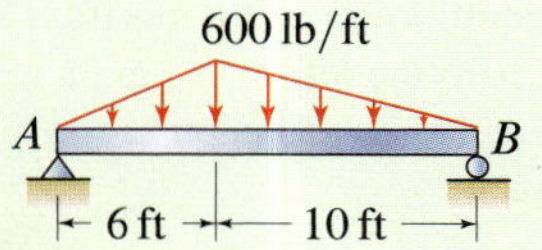

Figure P7.117

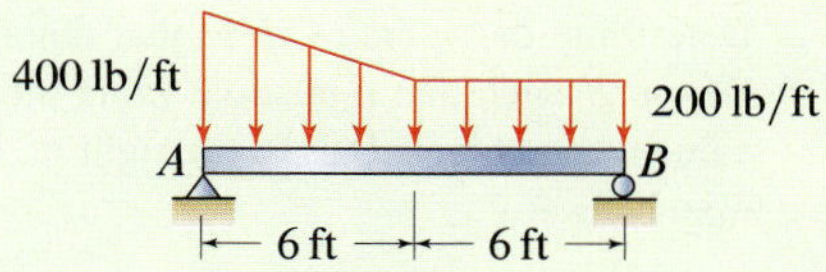

Figure P7.118

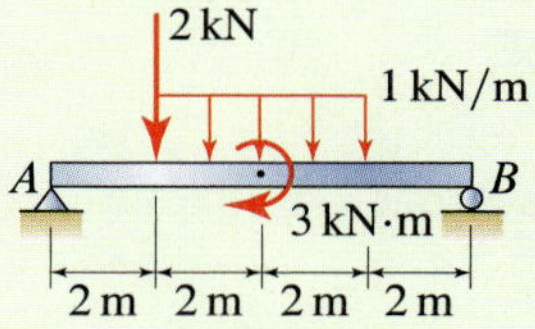

Figure P7.119

Problem 7.120

In Fig. P7.117, reposition the pin support 6 ft to the right of point A, and determine the support reactions.

Problem 7.121

In Fig. P7.118, replace the pin and roller supports with a built-in support at A, and determine the support reactions.

Problem 7.122

In Fig. P7.119, reposition the roller support 2 m to the left of point B, and determine the support reactions.

Problem 7.123

A circular plate with 21 in. radius is subjected to the pressure distribution shown. By treating the pressure distribution as a solid of revolution, use the theorems of Pappus and Guldinus to determine the total force applied to the plate.

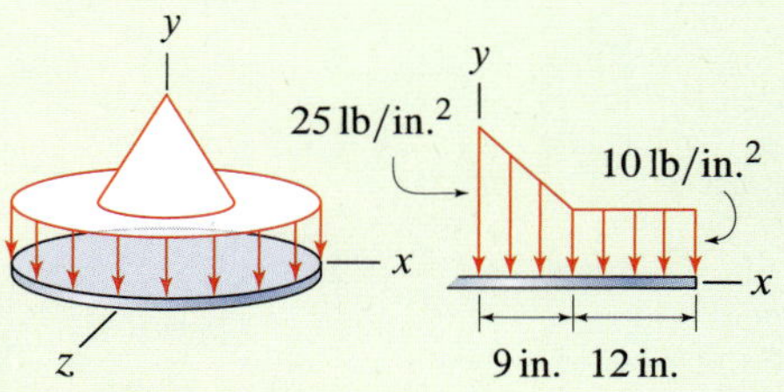

Figure P7.123

Problem 7.124

Water in a channel is retained by a gate with 0.5 ft width (into the plane of the figure). The gate is supported by a pin at A and a roller at C. The vertical wall AD is built into the bottom of the channel. If the gate has negligible weight, determine the support reactions.

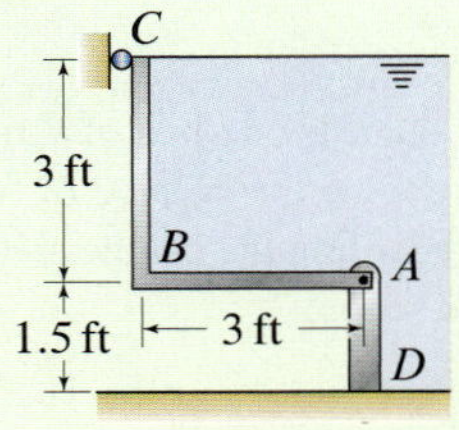

Figure P7.124

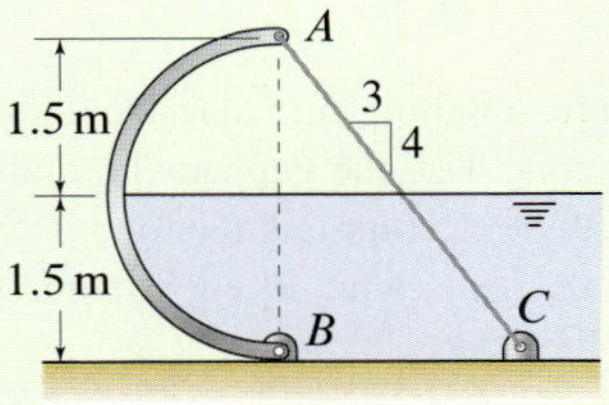

Figure P7.125

Problem 7.125

Water in a channel is retained by a cylindrical gate with 2 m width. The gate is supported by a pin at B and a cable between A and C. If the gate has negligible weight, determine the force supported by the cable and the reactions at B.

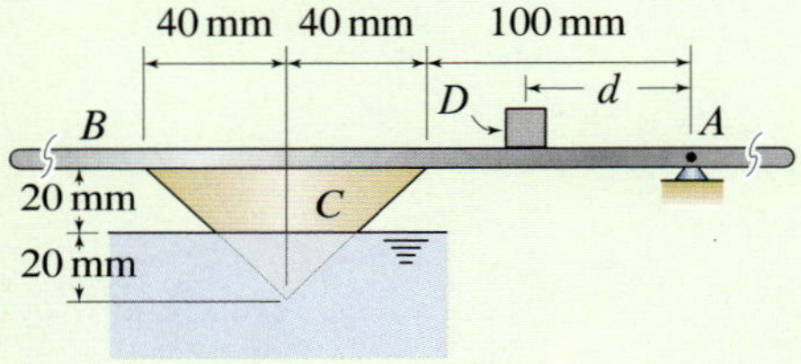

Figure P7.126

Problem 7.126

A uniform right circular cone C with 40 mm radius at its base and 0.1 N weight is attached to a beam AB with negligible weight. The cone is partially submerged in water. A block D with 0.2 N weight is placed a distance d from the support at A. Determine the value of d so that the system is in equilibrium in the position shown. Report d such that a positive value means block D is to the left of A, and a negative value means block D is to the right of A. *Hint:* The theorems of Pappus and Guldinus may be useful.

8 Internal Forces

When a structure is subjected to external forces, the various members of the structure develop internal forces within them. In fact, it's because of these internal forces that the structure is able to support the external forces that are applied to it. In earlier chapters of this book, we have routinely determined the internal forces that cables and bars support. In this chapter we discuss internal forces for more complex structural members, such as beams.

8.1 Internal Forces in Structural Members

Why are internal forces important?

Internal forces are forces and moments that develop within structural members and/or materials due to the external forces that are applied. Knowledge of the internal forces that a particular member must support is essential before the member can be designed. The design of a member includes: specification of the material(s) it is constructed of, its shape and dimensions, methods of support and/or attachment to other members, and so on. Obviously, as the internal forces that a member must support become larger, the member must be more substantial in size and/or be constructed of stronger material(s) so that it has sufficient strength.

A structural member is said to be *slender* if the dimensions of its cross section are small compared to its length. Slender members are very common, and the methods of analysis of statics and mechanics of materials are very effective for these. Many structural members are not slender, and determining internal forces and design are usually more difficult and require more advanced theory and methods of analysis. Thus, the main focus of this chapter is determination of internal forces in slender members.

Internal forces for slender members in two dimensions

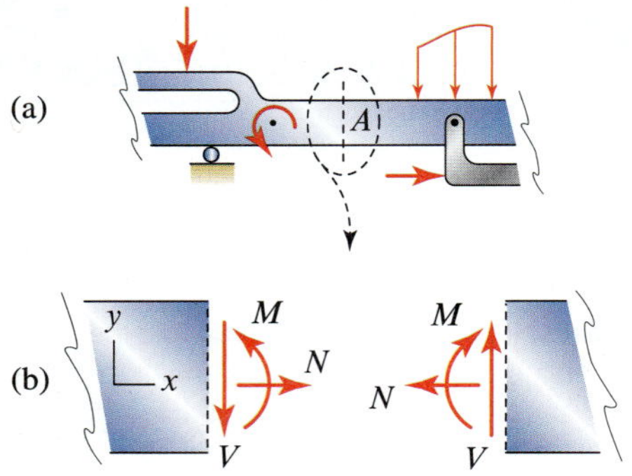

Figure 8.1
Internal forces that develop on a particular cross section of a slender member in two dimensions.

Consider the structure in two dimensions shown in Fig. 8.1(a), and imagine we are interested in the internal forces that act on cross section A. We take a cut through cross section A, thus separating the left-hand portion of the structure from the right-hand portion. The forces that develop on this cross section are called *internal forces*, as shown in Fig. 8.1(b), and these forces must be present if the material immediately to the left of the cut is to remain bonded or attached to the material immediately to the right of the cut. Another way to understand the nature of these forces is by analogy with the reaction forces for a built-in support: the material to the left of cut A may not translate in the horizontal direction relative to the material to the right of the cut, thus a force N must exist. Similarly the material to the left may not translate vertically relative to material to the right, thus a force V must exist; and material to the left may not rotate relative to material to the right, thus a moment M must exist. Observe that the internal forces in Fig. 8.1(b) satisfy Newton's third law.

Remarks

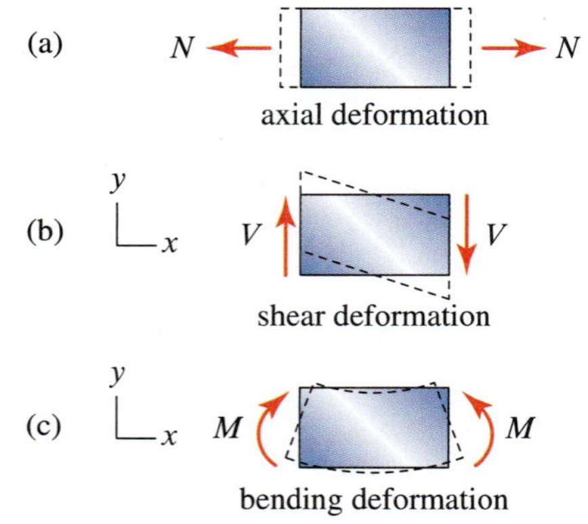

Figure 8.2
In a structure made of deformable material, internal forces N, V, and M produce the types of deformation shown by the dashed outlines.

- For several reasons, most of which are not clear until we study mechanics of materials, it is most useful to determine internal forces referenced to directions that are along and transverse to the axis of the member. Thus, we will routinely select an xy system (or tn system for members that are curved or have shape that is not straight), such as in Fig. 8.1(b), where x is in the axial direction and y is in the transverse direction. Furthermore, we will usually take the origin of the coordinate system (the $y = 0$ position) to coincide with the centroid of the cross section.
- The internal forces shown in Fig. 8.1(b) are often categorized as follows:
  - N is called the *normal force* or *axial force*. We will usually follow the sign convention shown in Figs. 8.1(b) and 8.2(a) wherein a positive value of N corresponds to tension. Normal force gives rise to the axial deformation shown in Fig. 8.2(a).
  - V is called the *shear force*. Although it's an arbitrary choice, we will usually follow the sign convention shown in Figs. 8.1(b) and 8.2(b). Shear force gives rise to the shear deformation shown in Fig. 8.2(b).
  - M is called the *bending moment*. Although it is an arbitrary choice, we will usually follow the sign convention shown in Figs. 8.1(b) and 8.2(c). Bending moment gives rise to the bending deformation shown in Fig. 8.2(c).
- Later sections of this chapter focus on structures consisting of a single straight member, and it will be straightforward (and important) to follow the sign conventions shown in Figs. 8.1 and 8.2. In this section, our main focus is determination of the absolute values of internal forces at various locations in a structure, and many of the structures we consider have multiple members and/or members that are not straight. Thus, we will

sometimes adopt different sign conventions that are more convenient (e.g., Example 8.3).

Internal forces for slender members in three dimensions

The internal forces for a slender member in three dimensions are shown in Fig. 8.3 where it is seen that on every cross section, there exist six internal forces, as follows: N is called the *normal force* or *axial force*, and it gives rise to axial deformation, V_y and V_z are called *shear forces* and these give rise to shear deformations, M_y and M_z are called *bending moments* and these give rise to bending deformations, and M_x is called a *torque* and this gives rise to twisting deformation, or torsional deformation. For convenience, the forces and moments acting on the left-hand side of the cut are taken to be positive in the positive coordinate directions, while the forces and moments on the right-hand side of the cut are positive in the negative coordinate directions, as required by Newton's third law. Other sign conventions may be used provided Newton's third law is respected.

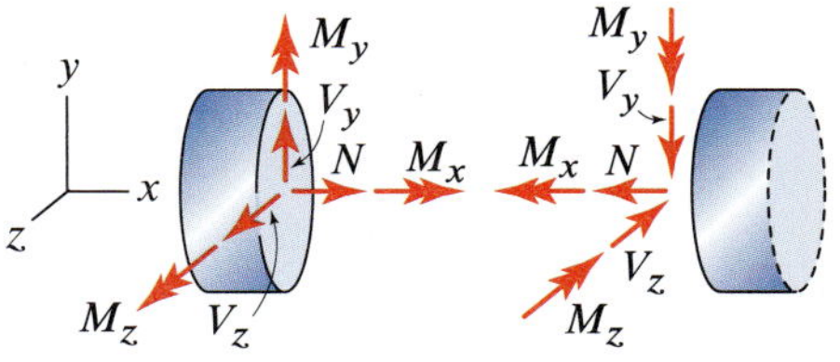

Figure 8.3
Internal forces that develop on a particular cross section of a slender member in three dimensions.

Figure 8.4. The Grand Canyon Skywalk is an observation platform on the grounds of the Hualapai Indian Nation in Arizona. It extends 70 ft over the edge of the Grand Canyon and is 4000 ft above the canyon's floor. The platform has steel framing, its deck is 3 in. thick glass, and it is designed to accommodate 120 people. Knowledge of the internal forces due to loads from the materials, spectators, and other sources is essential for the design of this structure.

Interesting Fact

Effects of internal forces on strength. All of the internal forces a member supports are important, and one of the main topics of mechanics of materials is how to design a member so that it can safely support all of these. Nonetheless, for slender members, the following generalizations are often true:

- Straight members are extremely effective at supporting axial tensile internal forces.
- Straight members are susceptible to buckling, and often they can support only low axial compressive internal forces.
- Compared to the effects of other internal forces, shear internal forces are often not of great concern. An important exception is for materials that are weak in shear, such as wood and some fiber-reinforced composite materials.
- Members are flexible in bending, and thus bending moments are of great importance.
- Members are flexible in torsion, and thus twisting moments (torques) are of great importance.

Determination of internal forces

First, we must decide where in a structure we wish to determine internal forces. Note that, in general, the internal forces on *every* cross section of *every* member of a structure are different, and usually one of our goals is to determine all of these. Sometimes it is sufficient to determine only the maximum values of the internal forces for each member. With some experience, you will often be able to identify, by inspection, the cross section or sections that experience the largest internal forces. For example, in a cantilever beam the internal forces are usually highest at the built-in support, and in a simply supported beam the internal moment is usually highest near the midspan of the beam.

Once a cross section of interest is identified, we proceed by taking a cut through that cross section and drawing an FBD of a portion of the structure, followed by writing and solving equilibrium equations. This process is then repeated for other cross sections of interest. Other approaches are possible (e.g., solution of differential equations), and some of these are discussed later in this chapter.

End of Section Summary

In this section, internal forces in slender structural members were discussed. For every cross section where the internal forces are to be determined, we select a coordinate system in which one of the coordinates (often the x coordinate) is in the axial direction of the member and the other coordinates are perpendicular to the axis of the member. Furthermore, the origin of the coordinate system is usually taken to coincide with the centroid of the cross section.

At each cross section of a member in two dimensions, there are three internal forces consisting of an axial force N, a shear force V, and a bending moment M. At each cross section of a member in three dimensions (where x is in the axial direction of the member), there are six internal forces consisting of an axial force N, two shear forces V_y and V_z, two bending moments M_y and M_z, and a torque M_x.

In this section, internal forces are determined by taking a cut through the cross section of interest, drawing an FBD of a portion of the structure, followed by writing and solving equilibrium equations.

EXAMPLE 8.1 *Internal Forces for a Two-Dimensional Problem*

A metal bottle cap opener is shown. If a vertical force of 10 N at point A is required to remove a bottle cap, determine the internal forces that develop on cross section D.

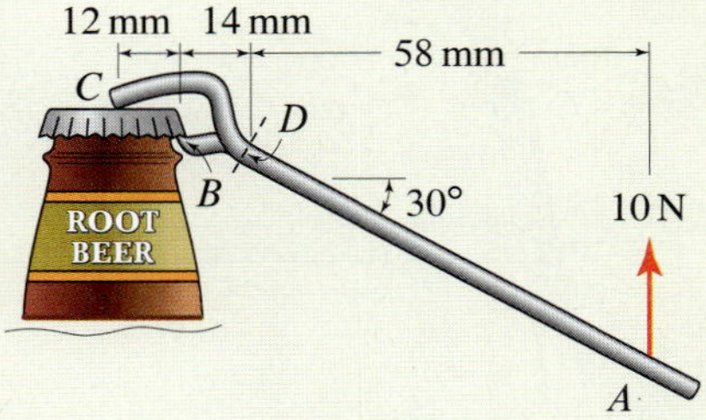

Figure 1

SOLUTION

Road Map We are especially interested in the internal forces on cross section D because, among all the locations in the bottle opener, this location will have very high internal forces. To explain further, imagine the bottle opener is made of metal that is too thin or is used to open a particularly stubborn bottle. Intuitively, we expect the bottle opener to severely bend and deform at cross section D. As engineers, our ultimate objective is to design the bottle opener (i.e., specify the material it is made of and its thickness and width) so that it is sufficiently strong for its intended use. Before this can be done, the maximum internal forces in the bottle opener must be determined.

We will neglect the weight of the bottle opener since it is clearly small compared to the other forces in the system. The forces acting on the bottle opener are obtained first. Then, to determine the internal forces on cross section D, a cut will be taken through that cross section, an FBD will be drawn, and equilibrium equations will be written and solved.

Modeling The FBD for the bottle opener is shown in Fig. 2. Although there may be horizontal force components at B and C, we will assume they are negligible compared to the vertical components B_y and C_y.

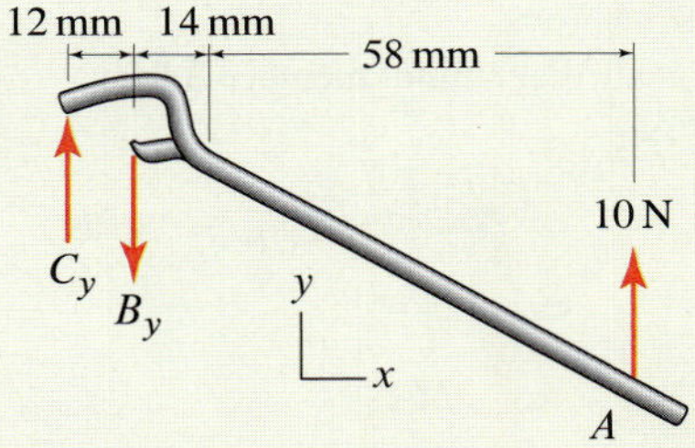

Figure 2
Free body diagram for the bottle opener.

Governing Equations & Computation Using the FBD shown in Fig. 2, we write and solve the following equilibrium equations:

$$\sum M_C = 0: \quad -B_y(12\,\text{mm}) + (10\,\text{N})(84\,\text{mm}) = 0 \quad \Rightarrow \quad B_y = 70\,\text{N}, \qquad (1)$$

$$\sum F_y = 0: \quad C_y - B_y + 10\,\text{N} = 0 \quad \Rightarrow \quad C_y = 60\,\text{N}. \qquad (2)$$

Modeling To determine the internal forces on cross section D, we take a cut through that cross section on the FBD of Fig. 2 to draw the two FBDs shown in Fig. 3. Observe that the assignment of directions for the internal forces matches those shown in Fig. 8.1(b).

Governing Equations & Computation Either of the FBDs shown in Fig. 3 may be used to determine the internal forces N_D, V_D, and M_D, and we will select the right-hand FBD, since it has simpler geometry and contains fewer forces. Using the tn coordinate system shown, where t is oriented along the axis of the member and n is in the transverse direction, we write and solve the following equations:

$$\sum F_t = 0: \quad -N_D - (10\,\text{N})(\sin 30^\circ) = 0 \quad \Rightarrow \quad \boxed{N_D = -5.00\,\text{N},} \qquad (3)$$

$$\sum F_n = 0: \quad V_D + (10\,\text{N})(\cos 30^\circ) = 0 \quad \Rightarrow \quad \boxed{V_D = -8.66\,\text{N},} \qquad (4)$$

$$\sum M_D = 0: \quad -M_D + (10\,\text{N})(58\,\text{mm}) = 0 \quad \Rightarrow \quad \boxed{M_D = 580\,\text{N·mm}.} \qquad (5)$$

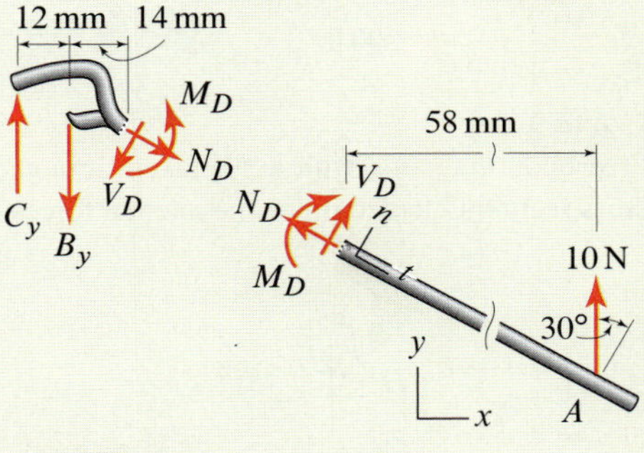

Figure 3
Free body diagrams with a cut taken through cross section D to determine the internal forces.

Discussion & Verification With some foresight, we might have anticipated using the right-hand FBD in Fig. 3, in which case the reaction forces B_y and C_y are not needed and we could have avoided writing Eqs. (1) and (2). Using the left-hand FBD in Fig. 3, you should verify that the same internal forces N_D, V_D, and M_D are obtained.

EXAMPLE 8.2 *Internal Forces for a Two-Dimensional Problem*

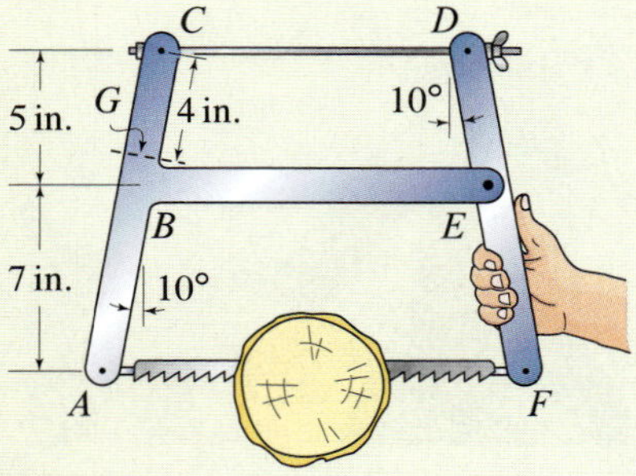

Figure 1

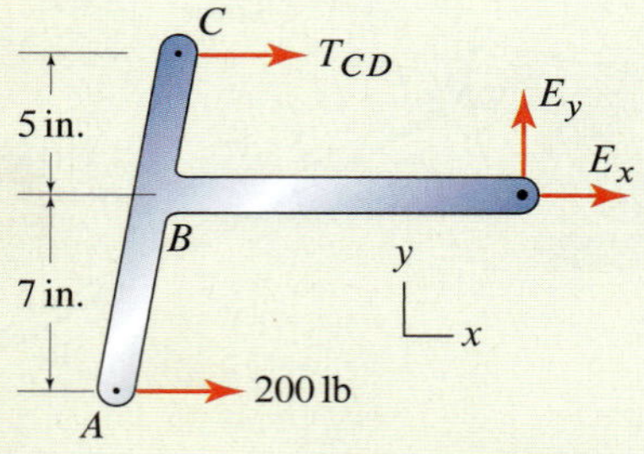

Figure 2
Free body diagram for member $ABCE$.

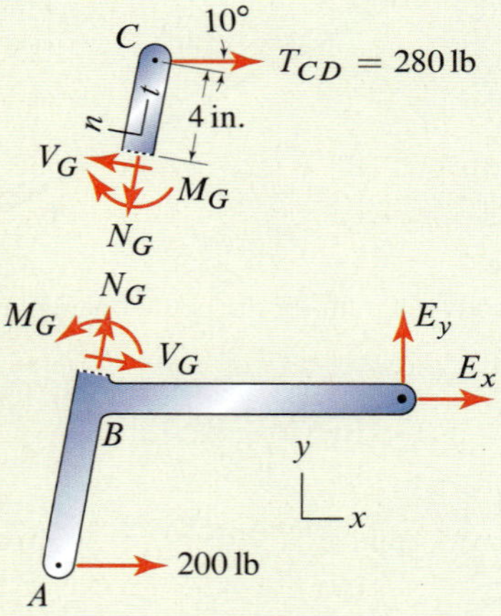

Figure 3
Free body diagrams with a cut taken through cross section G to determine the internal forces.

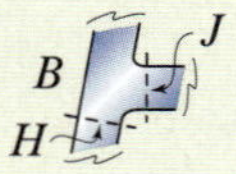

Figure 4
Additional cross sections in the vicinity of B where internal forces should be determined.

A hand saw for cutting firewood is shown. The blade of the saw is tensioned to 200 lb using the wing nut at D. Neglecting the forces due to cutting, determine the internal forces acting on cross section G.

SOLUTION

Road Map We will neglect the weights of individual members of the saw under the assumption they are small. We also neglect the forces due to cutting, although these forces may not be small, and it may be warranted to consider an analysis where these are included. The forces acting on member $ABCE$ are obtained first. To determine the internal forces on cross section G, a cut will be taken through that cross section, an FBD will be drawn, and equilibrium equations will be written and solved.

Modeling Separating member $ABCE$ from the saw gives the FBD shown in Fig. 2.

Governing Equations & Computation Using the FBD shown in Fig. 2, we write and solve the following equilibrium equations:

$$\sum M_E = 0: \quad -T_{CD}(5\text{ in.}) + (200\text{ lb})(7\text{ in.}) = 0 \quad \Rightarrow \quad T_{CD} = 280\text{ lb}, \tag{1}$$

$$\sum F_y = 0: \quad E_y = 0 \quad \Rightarrow \quad E_y = 0, \tag{2}$$

$$\sum F_x = 0: \quad T_{CD} + 200\text{ lb} + E_x = 0 \quad \Rightarrow \quad E_x = -480\text{ lb}. \tag{3}$$

Modeling To determine the internal forces on cross section G, we take a cut through that cross section on the FBD of Fig. 2 to draw the two FBDs shown in Fig. 3. Observe that the assignment of directions for the internal forces matches those shown in Fig. 8.1(b).

Governing Equations & Computation Either of the FBDs shown in Fig. 3 may be used to determine the internal forces N_G, V_G, and M_G, and we will select the upper FBD, since it has simpler geometry and contains fewer forces. Using the tn coordinate system shown, where t is oriented along the axis of the member and n is in the transverse direction, we write and solve the following equations:

$$\sum F_t = 0: \quad -N_G + T_{CD}(\sin 10^\circ) = 0 \Rightarrow \boxed{N_G = 48.6\text{ lb},} \tag{4}$$

$$\sum F_n = 0: \quad V_G - T_{CD}(\cos 10^\circ) = 0 \Rightarrow \boxed{V_G = 276\text{ lb},} \tag{5}$$

$$\sum M_G = 0: \quad -M_G - T_{CD}(\cos 10^\circ)(4\text{ in.}) = 0 \Rightarrow \boxed{M_G = -1100\text{ in.}\cdot\text{lb}.} \tag{6}$$

Discussion & Verification The internal forces on cross section at G are high. However, there are other locations in member $ABCE$, as well as member DEF, which also will likely support high internal forces, and a complete analysis requires that these also be considered. Regarding junction B, Fig. 4 identifies some additional cross sections we should also consider (see Prob. 8.2).

EXAMPLE 8.3 *Internal Forces for a Three-Dimensional Problem*

Member EFG is a shifting fork used in a transmission to move gear G along shaft CD. It is actuated by the 10 and 20 N forces at E, which act in the $-z$ and $-x$ directions, respectively, and gear G applies a 5 N force to the fork in the x direction. The fork is supported by a fixed shaft AB that has a thrust collar at H, and the fork is a loose fit on the shaft at G so that there is only a z direction reaction. Determine the internal forces acting on cross section J.

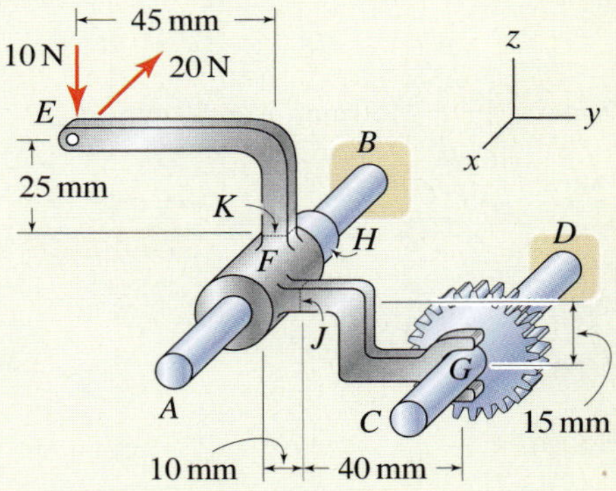

Figure 1

SOLUTION

Road Map Neglecting weight and friction, an FBD of the entire shifting fork will be drawn, followed by writing and solving equilibrium equations to determine the necessary support reactions. Observe in Fig. 1 that portion FG is cantilever-supported by the remainder of the structure, and hence the internal forces at cross section J are expected to be large. To determine the internal forces on cross section J, a cut will be taken through this cross section, an FBD will be drawn, and equilibrium equations will be written and solved.

Modeling The FBD for the shifting fork EFG is shown in Fig. 2.

Governing Equations & Computation Using the FBD shown in Fig. 2, we write and solve the following equilibrium equation for the reaction G_z:

$$\sum M_x = 0: \quad (10\,\text{N})(45\,\text{mm}) - G_z(50\,\text{mm}) = 0 \quad \Rightarrow \quad G_z = 9\,\text{N}. \tag{1}$$

By writing additional equilibrium equations, the remaining reactions can be determined (some additional dimensions in Fig. 1 may be needed). However, some foresight into the FBD that will be used to determine the internal forces on cross section J shows that the remaining reactions are not needed, and hence we will avoid determining them.

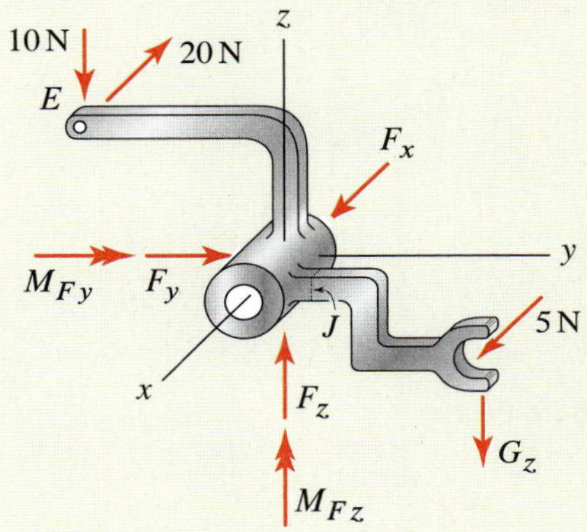

Figure 2
Free body diagram for the shifting fork.

Modeling To determine the internal forces on cross section J, we take a cut through that cross section on the FBD of Fig. 2 to draw the FBD shown in Fig. 3, where we have placed the origin of the coordinate system at the centroid of cross section J. For simplicity, we have assigned the directions for the internal forces so that all forces and moments are positive in the positive coordinate directions. Rather than the FBD of Fig. 3, we could have drawn an FBD of the left-hand portion of the shifting fork, although this will clearly have many more forces and moments and hence is not as judicious a choice as Fig. 3.

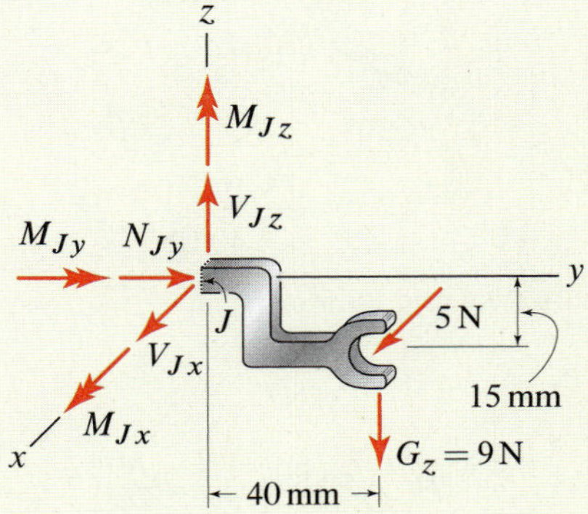

Figure 3
Free body diagrams with a cut taken through the cross section at J to determine the internal forces.

Governing Equations & Computation Using the FBD of Fig. 3, we write and solve the following equations:

$$\sum F_x = 0: \quad V_{Jx} + 5\,\text{N} = 0 \quad \Rightarrow \quad \boxed{V_{Jx} = -5\,\text{N},} \tag{2}$$

$$\sum F_y = 0: \quad N_{Jy} = 0 \quad \Rightarrow \quad \boxed{N_{Jy} = 0,} \tag{3}$$

$$\sum F_z = 0: \quad V_{Jz} - 9\,\text{N} = 0 \quad \Rightarrow \quad \boxed{V_{Jz} = 9\,\text{N},} \tag{4}$$

$$\sum M_x = 0: \quad M_{Jx} - (9\,\text{N})(40\,\text{mm}) = 0 \quad \Rightarrow \quad \boxed{M_{Jx} = 360\,\text{N}\cdot\text{mm},} \tag{5}$$

$$\sum M_y = 0: \quad M_{Jy} - (5\,\text{N})(15\,\text{mm}) = 0 \quad \Rightarrow \quad \boxed{M_{Jy} = 75\,\text{N}\cdot\text{mm},} \tag{6}$$

$$\sum M_z = 0: \quad M_{Jz} - (5\,\text{N})(40\,\text{mm}) = 0 \quad \Rightarrow \quad \boxed{M_{Jz} = 200\,\text{N}\cdot\text{mm}.} \tag{7}$$

Discussion & Verification Cross section K shown in Fig. 1 where portion EF is built into the shifting fork is also a likely location of high internal forces, and Prob. 8.23 asks you to determine these.

PROBLEMS

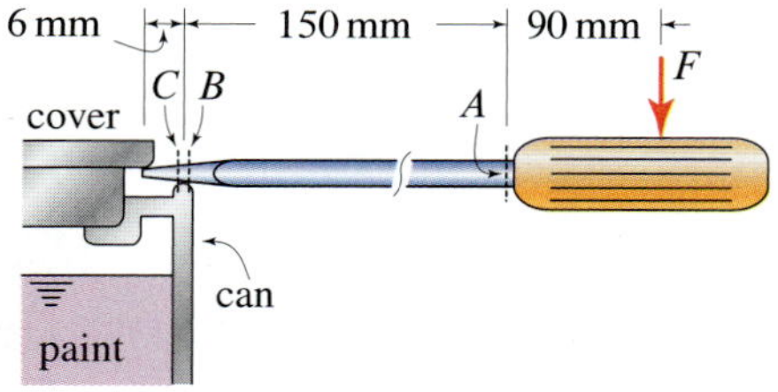

Figure P8.1

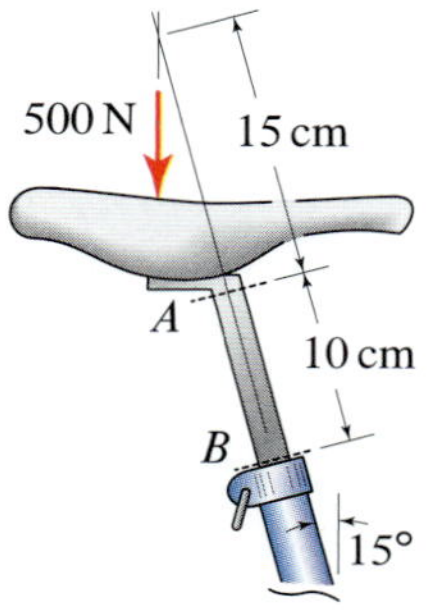

Figure P8.3

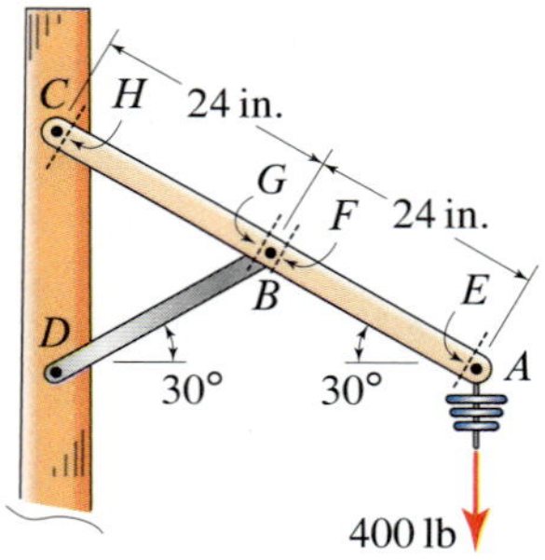

Figure P8.4 and P8.5

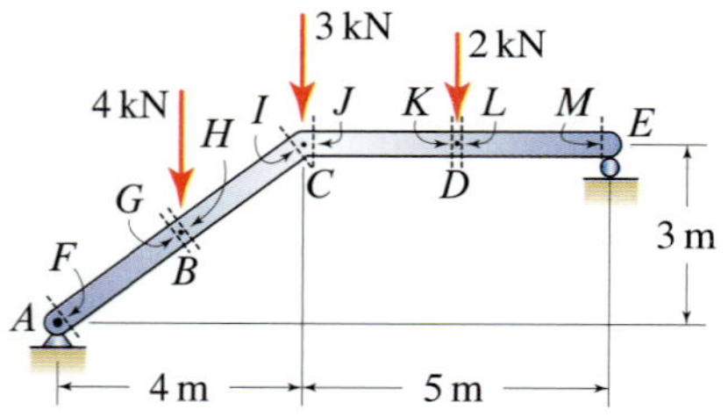

Figure P8.6 and P8.7

General instructions In the problems of this section, report the axial internal force, using positive and negative values for tension and compression, respectively, while for shear and moment internal forces, the absolute values are acceptable.

Problem 8.1

A screwdriver is used to pry open the cover of a can containing paint. If a force $F = 75\,\text{N}$ is required to open the cover, and assuming the contact forces between the screwdriver and can are vertical, determine the internal forces acting on cross sections A, B, and C of the screwdriver. Cross sections B and C are immediately to the right and left, respectively, of where the screwdriver makes contact with the can.

Problem 8.2

Determine the internal forces acting on cross sections H and J in Fig. 4 of Example 8.2 on p. 462. Cross section H is located 6 in. from point A, measured along line AB.

Problem 8.3

Determine the internal forces acting on cross sections A and B of the bicycle seat stem.

Problems 8.4 and 8.5

Member ABC and brace BD are used to suspend an electric transmission line from a utility pole. Determine the internal forces acting on:

Problem 8.4 Cross sections E and F, which are located immediately to the left of point A and to the right of point B, respectively.

Problem 8.5 Cross sections G and H, which are located immediately to the left of point B and to the right of point C, respectively.

Problems 8.6 and 8.7

The structure shown consists of a single member $ABCDE$ with a pin support at A and a roller support at E. Points B and D are at the midpoints of their respective segments. Determine the internal forces acting on:

Problem 8.6 Cross sections F, G, H, and I, which are located immediately to the right of A, the left of B, the right of B, and the left of C, respectively.

Problem 8.7 Cross sections J, K, L, and M, which are located immediately to the right of C, the left of D, the right of D, and the left of E, respectively.

Problems 8.8 and 8.9

The structure of Probs. 8.6 and 8.7 is revised to consist of two members, ABC and CDE, that are pinned to one another at C and have pin supports at A and E. Points B and D are at the midpoints of their respective members. Determine the internal forces acting on:

Problem 8.8 Cross sections F, G, H, and I, which are located immediately to the right of A, the left of B, the right of B, and the left of C, respectively.

Problem 8.9 Cross sections J, K, L, and M, which are located immediately to the right of C, the left of D, the right of D, and the left of E, respectively.

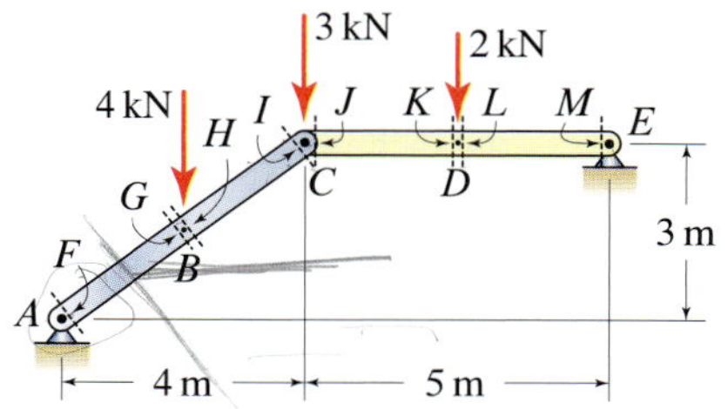

Figure P8.8 and P8.9

Problems 8.10 and 8.11

A hacksaw for cutting metal is shown. Assume contact between the frame ABC and the handle assembly occurs at points B and C only, neglect friction, and neglect the size of the pin and notch at point B. If the blade is tensioned to 100 lb, determine the internal forces acting on:

Problem 8.10 Cross sections D, E, and F, where cross section D is located immediately above point A.

Problem 8.11 Cross sections G, H, I, and J, where cross sections H and I are located immediately to the right and left of point B, respectively, and cross section J is immediately to the right of point C.

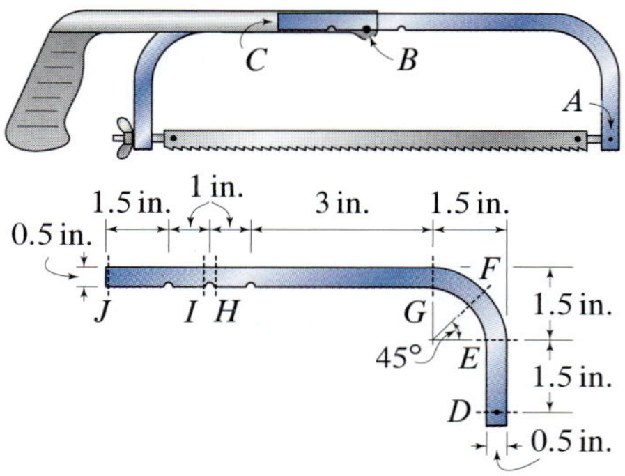

Figure P8.10 and P8.11

Problems 8.12 through 8.16

A weight $W = 2\,\text{kN}$ is supported by a cable that passes over frictionless pulleys at D and F. The cable is attached to a winch at G, and cable segment DG is vertical. Member ABC is built in at A, and members ABC, $DCEF$, and BE are attached using pins at points B, C, and E. Neglecting the weights of individual members, determine the internal forces acting on:

Problem 8.12 Cross sections H and I, located immediately to the left of point F and the right of point E, respectively.

Problem 8.13 Cross sections J and K, located immediately to the left of point E and the right of point C, respectively.

Problem 8.14 Cross sections L and O, located immediately to the left of point C and the right of point D, respectively.

Problem 8.15 Cross sections Q and P, located immediately below and above point B, respectively.

Problem 8.16 Cross section R located at the midpoint of member BE. Also determine the internal forces (i.e., the reactions) at the built-in support at A.

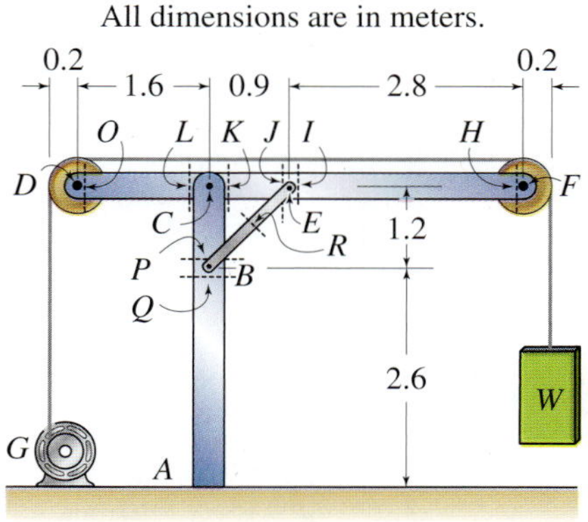

Figure P8.12–P8.16

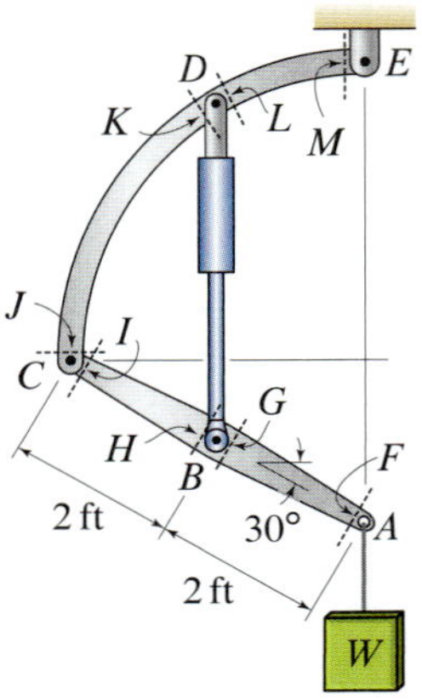

Figure P8.17–P8.20

Problems 8.17 through 8.20

A machine for lifting heavy objects on an assembly line is shown. It consists of a straight member ABC, a quarter-circular member CDE, and a hydraulic cylinder BD. In the position shown, the hydraulic cylinder is vertical. If $W = 800$ lb, and neglecting the weight of the components of the machine, determine the internal forces acting on:

Problem 8.17 Cross sections F and G, located immediately to the left of point A and to the right of point B, respectively.

Problem 8.18 Cross sections H and I, located immediately to the left of point B and to the right of point C, respectively.

Problem 8.19 Cross sections J and K, located immediately above point C and below and to the left of point D, respectively.

Problem 8.20 Cross sections L and M, located immediately to the right and above point D and to the left of point E, respectively.

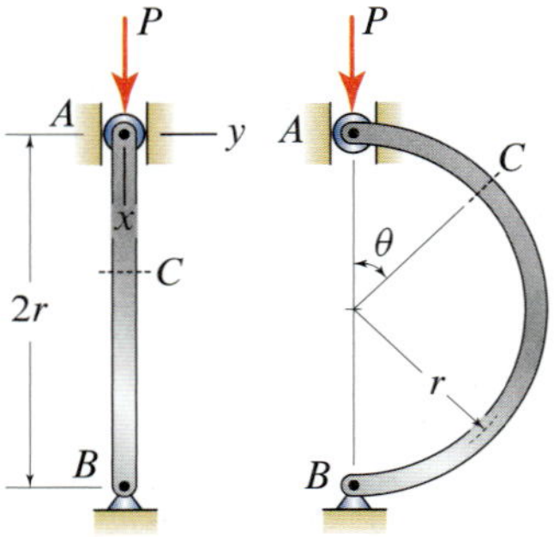

Figure P8.21 and P8.22

Problem 8.21

Two structural members are shown; one is straight and the other is semicircular. Neglect the weight of the members.

(a) For the straight member, show that the internal forces acting on cross section C are $V = 0$, $N = -P$, and $M = 0$.

(b) For the semicircular member, show that the internal forces acting on cross section C are $V = -P\cos\theta$, $N = -P\sin\theta$, and $M = -Pr\sin\theta$.

Problem 8.22

Both the straight and semicircular members shown in Fig. P8.22 are two-force members, and hence either can be used as members of a truss. However, it is most common that truss members are straight. Compare the answers to Prob. 8.21 (which are given in the problem description) to argue why straight truss members are preferable to curved truss members.

Note: Concept problems are about *explanations*, not computations.

Problems 8.23 and 8.24

Determine the internal forces acting on:

Problem 8.23 Cross section K in Fig. 1 of Example 8.3 on p. 463.

Problem 8.24 The cross section at A in Example 4.3 on p. 190.

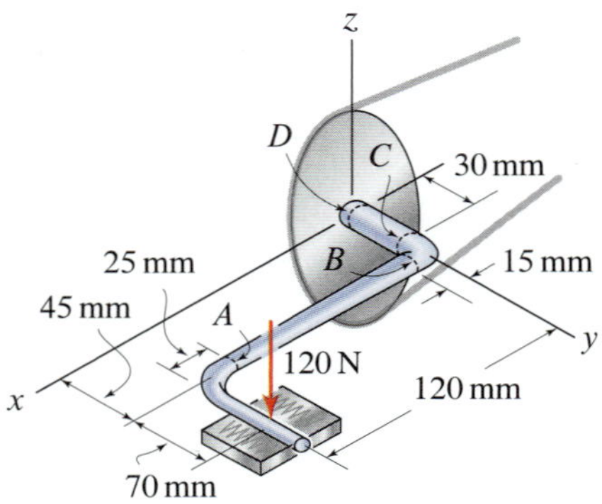

Figure P8.25–P8.28

Problems 8.25 through 8.28

One of the cranks of a child's bicycle is shown. The entire crank lies in the xy plane. Determine the internal forces acting on:

Problem 8.25 Cross section A.

Problem 8.26 Cross section B.

Problem 8.27 Cross section C.

Problem 8.28 Cross section D, which is located immediately next to the sprocket.

8.2 Internal Forces in Straight Beams

In this section and the next, we focus on determination of the internal forces in straight beams. In this section, we use equilibrium concepts to accomplish this while in Section 8.3 we use differential equations.

Determination of V and M using equilibrium

Our goal in this section is the determination of the internal forces *everywhere* throughout a straight beam, and we will accomplish this using the *equilibrium approach*. We begin by assigning an xy coordinate system, as shown in Fig. 8.5(a), where x is along the axis of the beam. Usually, the $x = 0$ position will be at one of the ends of the beam, and the $y = 0$ position is taken to coincide with the centroid of the beam's cross section. With this coordinate system, we define positive distributed force w to act in the $-y$ direction. The cross section shown in Fig. 8.5(a) is located at position x, and the directions for positive internal forces on this cross section are defined in Fig. 8.5(b).

In contrast to the previous section, where we found the internal forces at only *selected* cross sections, here we are interested in determining the internal forces *everywhere*. Hence, in this section, we will draw multiple FBDs as needed by taking cuts at arbitrary positions x, and we will use equilibrium equations to determine the internal forces N, V, and M, which in general will be functions of position x. There is considerable advantage to knowing the internal forces as functions of position. For one, it allows for easy plotting of internal forces, as discussed below. Furthermore, when you study mechanics of materials, you will see that all aspects of the behavior of a beam are governed by differential equations (some aspects of this are discussed in Section 8.3). Thus, if the shear and/or moment distribution is known, then it will be possible to determine behavior such as deflections of a deformable beam, reactions (including reactions for statically indeterminate beams), and more. All of these applications require a sign convention that must be rigorously followed, and we will use the sign convention defined in Fig. 8.5. However, other sign conventions are common, and if you consult other references such as textbooks, technical papers, or handbooks, you may see other sign conventions used.

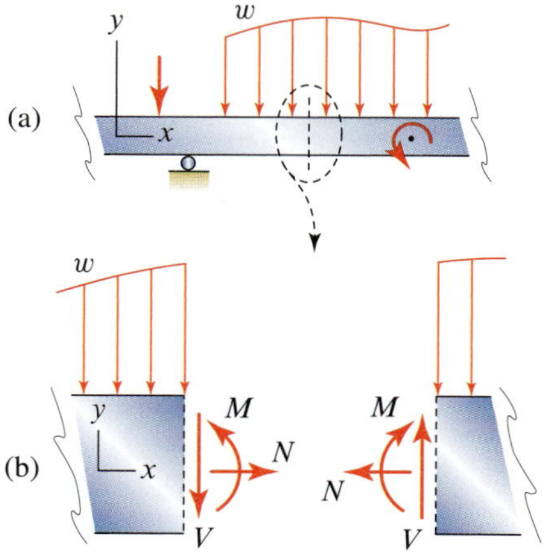

Figure 8.5
Internal forces that develop on a particular cross section of a straight beam in two dimensions.

Helpful Information

Forces in the x direction. In addition to the forces shown in Fig. 8.5, it is possible to have point forces and distributed forces that act in the x direction. If this is the case, then these forces must also be included in all FBDs that are drawn (Prob. 8.55 explores this situation). Nonetheless, beams are very often used in a horizontal position with most of the forces being due to gravity, and hence the situation depicted in Fig. 8.5, while a special case, is very common and is the focus in this chapter. Further, since distributed loads usually arise due to gravity, it is convenient to take positive distributed load w to be downward ($-y$ direction).

Shear and moment diagrams

Shear and moment diagrams are plots of the shear V and moment M as functions of position x. These diagrams help us understand how the internal forces change throughout a beam and show locations where these have large values. This knowledge helps us to design a beam that has sufficient strength, and drawing shear and moment diagrams is a routine part of the design process. For some applications, plots of the axial force are also important, although in this book our main focus is on the shear and moment.

End of Section Summary

In this section the shear and moment at *all* locations in a straight beam are determined. The equilibrium approach is used where cuts are taken at arbitrary (variable) positions as needed, FBDs are drawn, and equilibrium equations are written. Finally, the usefulness of *shear and moment diagrams* is discussed.

EXAMPLE 8.4 *Internal Forces in a Straight Beam with a Distributed Load*

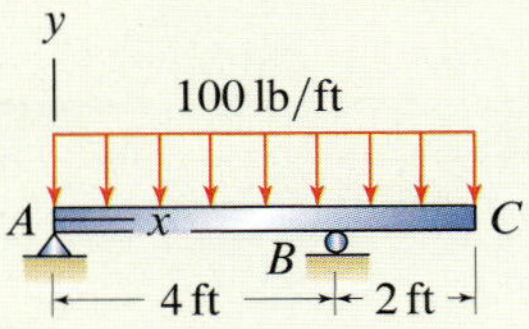

Figure 1

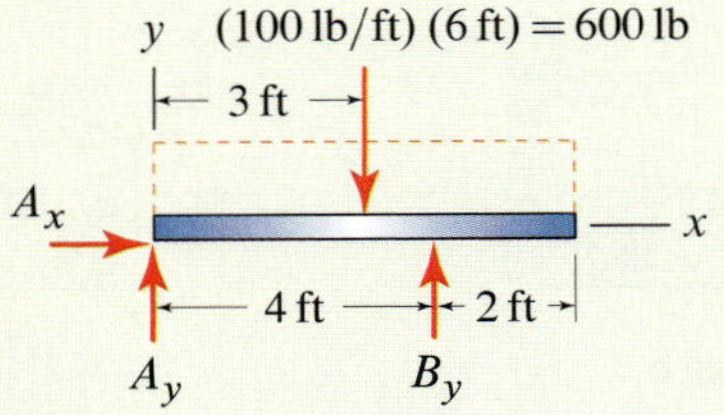

Figure 2
Free body diagram for the beam to determine the support reactions.

A simply supported beam with an overhang is subjected to a uniformly distributed load. Determine the shear and moment throughout the beam, and plot these versus position.

SOLUTION

Road Map We begin by drawing an FBD of the entire beam and writing equilibrium equations to determine the reactions at A and B. To determine the internal forces throughout the beam, we will take a cut through a cross section (subsequently called cross section D), where this cross section has general (or variable) position in the beam. We will draw FBDs as needed and apply equilibrium equations to determine the shear and moment as functions of position.

Modeling The FBD for the entire beam is shown in Fig. 2, where the distributed force has been replaced by an equivalent force consisting of a single 600 lb force placed at the centroid of the distributed force's shape (i.e., centroid of a rectangle).

Governing Equations & Computation Using the FBD shown in Fig. 2, we write and solve the following equilibrium equations to obtain the reactions.

$$\sum M_A = 0: \quad -(600\,\text{lb})(3\,\text{ft}) + B_y(4\,\text{ft}) = 0 \quad \Rightarrow \quad B_y = 450\,\text{lb}, \tag{1}$$

$$\sum F_y = 0: \quad -600\,\text{lb} + A_y + B_y = 0 \quad \Rightarrow \quad A_y = 150\,\text{lb}, \tag{2}$$

$$\sum F_x = 0: \quad A_x = 0 \quad \Rightarrow \quad A_x = 0. \tag{3}$$

> **Common Pitfall**
>
> **Equivalent forces.** The 600 lb equivalent force shown in Fig. 2 may be used *only* for determination of the support reactions A_x, A_y, and B_y. For determinations of the internal forces in Figs. 3 and 4, the original 100 lb/ft distributed loading must be used. Example 8.5 offers additional explanation.

Modeling To determine the internal forces throughout the beam, we consider cross section D, whose location is variable throughout the beam. As was done in Section 8.1, to determine the internal forces acting on cross section D, we take a cut that passes through this cross section, followed by construction of FBDs. Thus, when cross section D is located between points A and B, the two FBDs that result are shown in Fig. 3. A few important points regarding Fig. 3 are as follows:

- The length of the beam in the left-hand FBD is x, hence the force on this portion of the beam due to the distributed load is $(100\,\text{lb/ft})x$. Similarly, the length of the beam in the right-hand FBD is $6\,\text{ft} - x$, hence the force on this portion of the beam due to the distributed load is $(100\,\text{lb/ft})(6\,\text{ft} - x)$.
- The directions for the internal forces on both the left-hand and right-hand FBDs match the directions shown in Fig. 8.5(b) on p. 467.

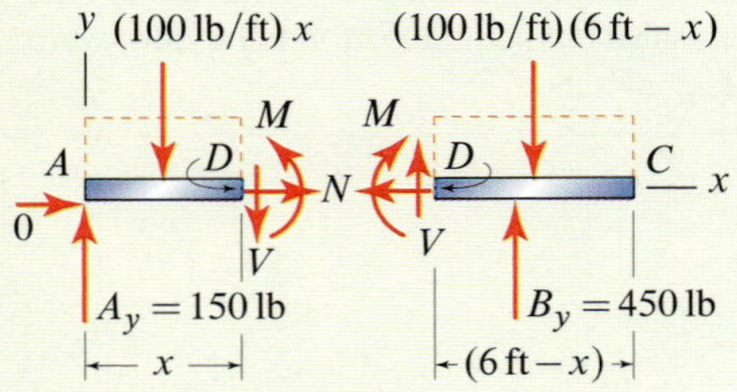

Figure 3
Free body diagrams when cross section D is between points A and B; i.e., $0 \le x \le 4\,\text{ft}$.

Governing Equations & Computation Either of the FBDs shown in Fig. 3 may be used to determine the internal forces acting on cross section D, and we will select the left-hand FBD. With some experience, we could have anticipated using this FBD, and we would have omitted drawing the FBD of the right-hand portion of the structure. We write and solve the following equations.

$$\sum F_y = 0: \quad -V - (100\,\text{lb/ft})x + 150\,\text{lb} = 0 \tag{4}$$

$$\Rightarrow \quad \boxed{V = 150\,\text{lb} - (100\,\text{lb/ft})x,} \tag{5}$$

$$\sum M_D = 0: \quad M + (100\,\text{lb/ft})x\frac{x}{2} - (150\,\text{lb})x = 0 \tag{6}$$

$$\Rightarrow \quad \boxed{M = (150\,\text{lb})x - (50\,\text{lb/ft})x^2,} \tag{7}$$

$$\sum F_x = 0: \quad N = 0 \quad \Rightarrow \quad \boxed{N = 0.} \tag{8}$$

The results obtained in Eqs. (5), (7), and (8) are valid as long as the FBD is valid, which is all cross sections whose x coordinate lies in the region $0 \le x \le 4$ ft.

Modeling When cross section D is located between points B and C, the FBDs that result are shown in Fig. 4.

Governing Equations & Computation Either of the FBDs shown in Fig. 4 may be used to determine the internal forces acting on cross section D, and we will select the right-hand FBD. We write and solve the following equations.

$$\sum F_y = 0: \quad V - (100\,\text{lb/ft})(6\,\text{ft} - x) = 0 \tag{9}$$

$$\Rightarrow \quad \boxed{V = 600\,\text{lb} - (100\,\text{lb/ft})x,} \tag{10}$$

$$\sum M_D = 0: \quad -M - (100\,\text{lb/ft})(6\,\text{ft} - x)\frac{6\,\text{ft} - x}{2} = 0 \tag{11}$$

$$\Rightarrow \quad \boxed{M = -1800\,\text{ft}\cdot\text{lb} + (600\,\text{lb})x - (50\,\text{lb/ft})x^2,} \tag{12}$$

$$\sum F_x = 0: \quad N = 0 \quad \Rightarrow \quad \boxed{N = 0.} \tag{13}$$

The results obtained in Eqs. (10), (12), and (13) are valid as long as the FBD is valid, which is for all cross sections whose x coordinate lies in the region $4\,\text{ft} \le x \le 6$ ft.

Discussion & Verification The results for shear V and moment M are plotted in Fig. 5, where Eqs. (5) and (7) are used for $0 \le x \le 4$ ft, and Eqs. (10) and (12) are used for $4\,\text{ft} \le x \le 6$ ft. These plots display some interesting features of internal forces:

- The shear at $x = 0$ is equal to the support reaction A_y.
- The change in shear at $x = 4$ ft is equal to the support reaction B_y. That is, the shear just to the right of B (200 lb) minus the shear just to the left of B (−250 lb) is equal to B_y (450 lb).
- The shear at the right-hand end is zero, because this end is unsupported *and* has no concentrated force applied.
- The moment at $x = 0$ is zero, because the support at that location has no moment reaction.
- The moment at the right-hand end is zero, because this end is unsupported *and* has no concentrated moment applied.
- The largest moment throughout the beam is −200 ft·lb, which occurs at $x = 4$ ft. In mechanics of materials, this value will play an extremely important role in the design of the beam (i.e., determination of the material and cross-sectional shape for the beam). The values of the shear force are sometimes also important, especially for materials that are weak in shear, such as wood and some composite materials.
- The largest value of the moment between A and B may be of interest. This value can be determined by evaluating Eq. (7), using trial and error with different values of x in the neighborhood of about 1.5 ft. However, a local maximum value of moment occurs where $V = 0$ (this is proved in Section 8.3).* Thus, we could solve Eq. (5) for the value of x that makes $V = 0$ and then substitute this value into Eq. (7) to obtain the moment. If you verify this for yourself, you should find that $V = 0$ at $x = 1.5$ ft and $M = 112.5$ ft·lb at this location.

* A local maximum value of the moment can also occur at an interval endpoint where $V \neq 0$, such as at $x = 4$ ft in this example.

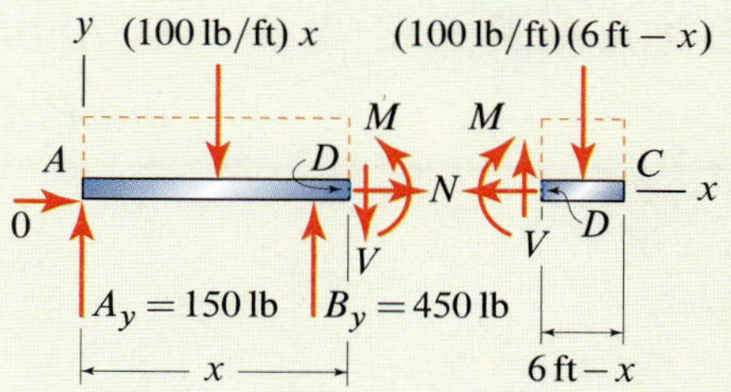

Figure 4
Free body diagrams when cross section D is between points B and C; i.e., $4\,\text{ft} \le x \le 6$ ft.

Helpful Information

V at $x = 4$ ft. The careful reader will notice that Eq. (5) is valid only for $0 \le x < 4$ ft and Eq. (10) is valid only for $4\,\text{ft} < x \le 6$ ft, because at $x = 4$ ft, V is undefined (it undergoes a discontinuity as shown in Fig. 5). Nonetheless, as a matter of practice, in this chapter we will use $\le$ signs for the regions of validity for V and M expressions with the understanding that there will often be a discontinuity in V and/or M at interval endpoints.

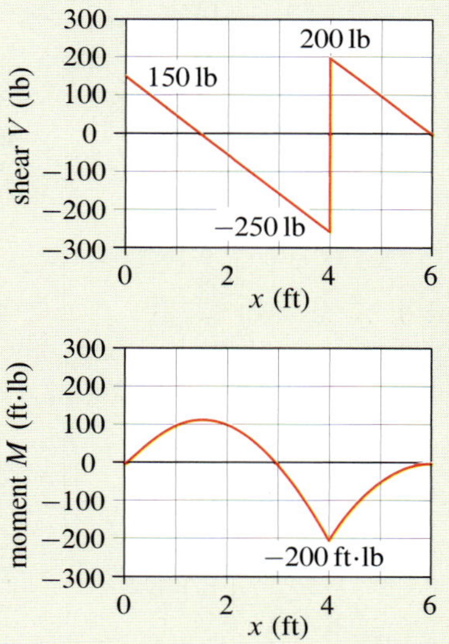

Figure 5
Shear and moment diagrams.

EXAMPLE 8.5 *A Simply Supported Beam with Distributed Load and Point Load*

Two simply supported beams with different loadings are shown.

(a) Determine the shear and moment as functions of position for the uniformly distributed loading w_0 shown in Fig. 1(a).

(b) Determine the shear and moment as functions of position for the point load P shown in Fig. 1(b).

(c) Let $P = w_0 L$. Plot the shear and moment determined in Parts (a) and (b) and comment on the differences, if any, between these.

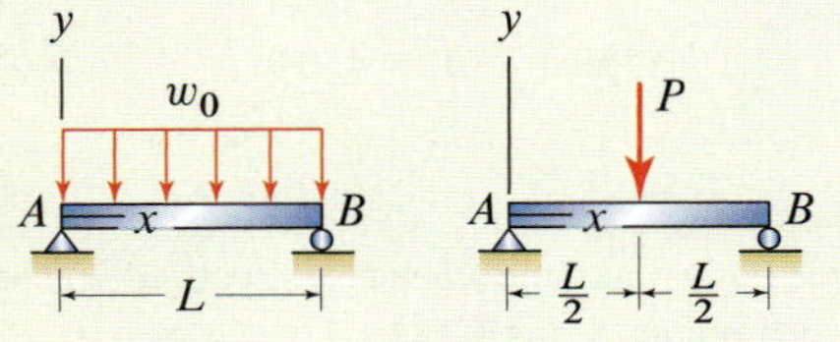

Figure 1

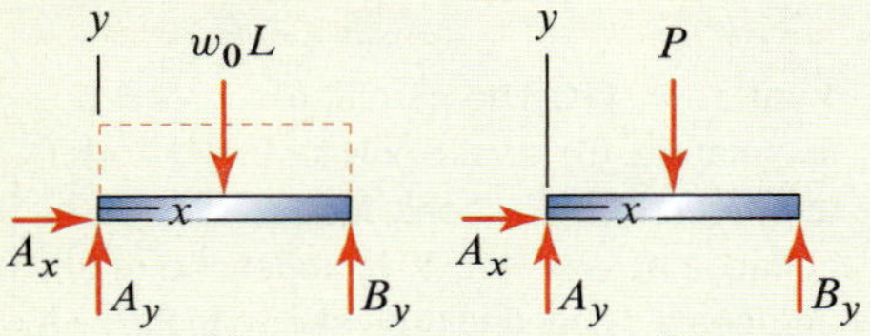

Figure 2
Free body diagrams to determine the support reactions for each beam.

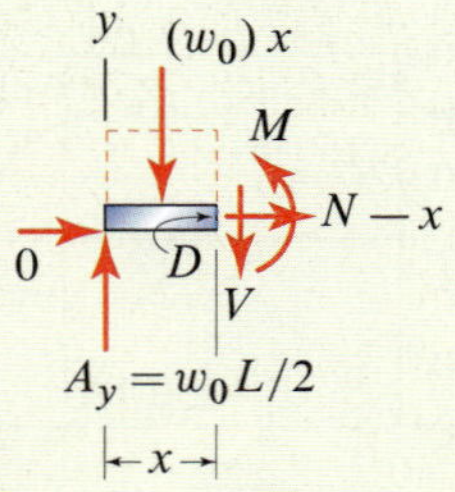

Figure 3
Free body diagram for the left-hand portion of the uniformly loaded beam of Fig. 1(a).

SOLUTION

Road Map For both Parts (a) and (b), we first determine the support reactions. We then proceed to determine the shear and moment by taking cuts through cross sections of the beam, drawing FBDs as needed, and applying equilibrium equations. Finally, for Part (c) we let $P = w_0 L$ and compare the results for the shear and moment.

Modeling For purposes of determining the support reactions for the beams in Parts (a) and (b), FBDs for each beam are drawn in Fig. 2, where in Fig. 2(a) an equivalent force of $w_0 L$ has been used in place of the original distributed loading.

Governing Equations & Computation By writing the equations $\sum M = 0$, $\sum F_y = 0$, and $\sum F_x = 0$, you should verify that the support reactions for the uniformly loaded beam in Fig. 2(a) are $A_x = 0$ and $A_y = B_y = w_0 L/2$. Similarly, you should verify that the support reactions for the beam with a point load in Fig. 2(b) are $A_x = 0$ and $A_y = B_y = P/2$. In anticipation of Part (c), note that if $P = w_0 L$, the two FBDs in Fig. 2 are identical and the support reactions are also identical.

Part (a)

Modeling For the beam with the distributed load in Fig. 1(a), we determine the internal forces by taking a cut through a cross section to draw the FBD shown in Fig. 3.

Governing Equations & Computation Using the FBD shown in Fig. 3, we write and solve the following equations.

$$\sum F_y = 0: \quad -V - w_0 x + w_0 \frac{L}{2} = 0 \quad \Rightarrow \quad \boxed{V = w_0\left(\frac{L}{2} - x\right),} \tag{1}$$

$$\sum M_D = 0: \quad M + w_0 x \frac{x}{2} - \frac{w_0 L}{2} x = 0 \quad \Rightarrow \quad \boxed{M = \frac{w_0}{2}(Lx - x^2),} \tag{2}$$

$$\sum F_x = 0: \quad N = 0 \quad \Rightarrow \quad \boxed{N = 0.} \tag{3}$$

The results obtained in Eqs. (1) through (3) are valid as long as the FBD is valid, which is for any cross section between ends A and B (i.e., $0 \le x \le L$).

Part (b)

Modeling For the beam with a point load in Fig. 1(b), we determine the internal forces by taking a cut through a cross section to draw the FBDs shown in Fig. 4.

Governing Equations & Computation Using the FBD shown in Fig. 4(a) for the region $0 \le x \le L/2$, we write and solve the following equations.

$$\sum F_y = 0: \quad -V + \frac{P}{2} = 0 \quad \Rightarrow \quad \boxed{V = \frac{P}{2},} \tag{4}$$

$$\sum M_D = 0: \qquad M - \frac{P}{2}x = 0 \quad \Rightarrow \quad \boxed{M = \frac{P}{2}x,} \tag{5}$$

$$\sum F_x = 0: \qquad N = 0 \quad \Rightarrow \quad \boxed{N = 0.} \tag{6}$$

Using the FBD shown in Fig. 4(b) for the region $L/2 \le x \le L$, we write and solve the following equations

$$\sum F_y = 0: \qquad -V - P + \frac{P}{2} = 0 \quad \Rightarrow \quad \boxed{V = -\frac{P}{2},} \tag{7}$$

$$\sum M_D = 0: \qquad M + P\left(x - \frac{L}{2}\right) - \frac{P}{2}x = 0 \quad \Rightarrow \quad \boxed{M = \frac{P}{2}(L - x),} \tag{8}$$

$$\sum F_x = 0: \qquad N = 0 \quad \Rightarrow \quad \boxed{N = 0.} \tag{9}$$

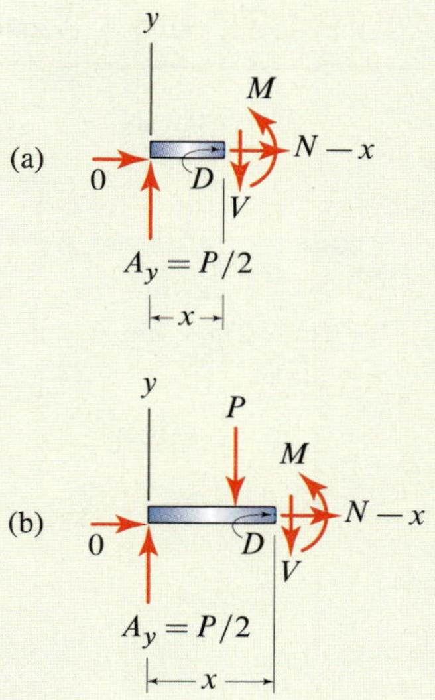

Figure 4
Free body diagrams for the beam with a point load shown in Fig. 1(b). The first FBD is valid for the region $0 \le x \le L/2$, and the second is valid for $L/2 \le x \le L$.

Part (c)

Discussion & Verification For the beam with uniformly distributed load, the shear and moment given by Eqs. (1) and (2) are plotted in Fig. 5(a), and for the beam with a point load, the shear and moment given by Eqs. (4), (5), (7), and (8) are plotted in Fig. 5(b).

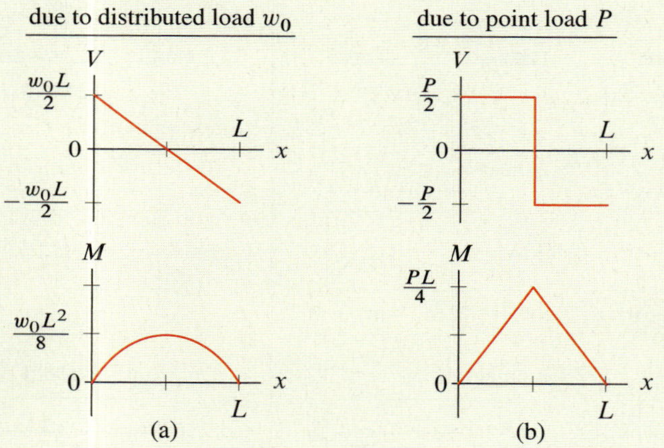

Figure 5. (a) Shear and moment diagrams for a simply supported beam with uniformly distributed load. (b) Shear and moment diagrams for a simply supported beam with a point load at midspan.

If $P = w_0 L$, then the loadings shown in Fig. 1 are equivalent according to the definition given in Chapter 4 by Eq. (4.16) on p. 223, and as mentioned earlier in this example, the FBDs in Fig. 2 are identical and the support reactions are identical. However, the internal forces shown in Fig. 5 are clearly very different. Observe that with $P = w_0 L$, all of the plots in Fig. 5 are drawn to scale, and among the many differences, the maximum value of the moment for the beam with a concentrated load is *twice as large* as that for the beam with uniform distributed loading.

Part (c) of this example demonstrates some of the subtleties and limitations of the definition of equivalent force systems given in Chapter 4. To summarize, two force systems that are equivalent, according to the definition given by Eq. (4.16) on p. 223, will produce the same *external* effects on a rigid structure (e.g., the support reactions will be the same), but *internal* effects (e.g., internal forces such as the shear and moment) are not necessarily the same.

Common Pitfall

When are equivalent force systems really *equivalent*? A common error when determining internal forces is to replace a distributed force such as w_0 shown in Fig. 1(a) by an equivalent force such as $P = w_0 L$ shown in Fig 1(b). This replacement is valid only for purposes of determining forces that are *external* to the FBD, such as support reactions. For determination of internal forces, the original loading must be used.

EXAMPLE 8.6 *Superposition*

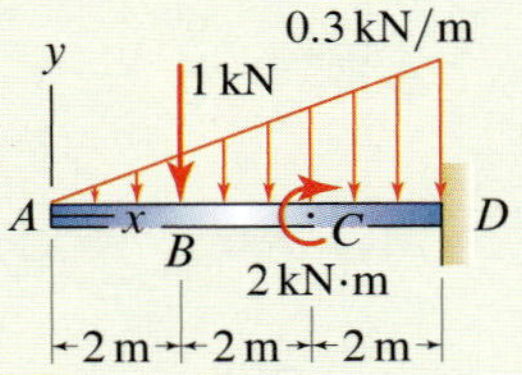

Figure 1

For the cantilever beam shown, determine the shear and moment as functions of position, and draw the shear and moment diagrams.

SOLUTION

Road Map We could follow the procedure used in earlier examples in this section, wherein we would take a cut in Fig. 1 between points A and B, draw an FBD and apply equations of equilibrium to determine the internal forces in that region, and then repeat this for cuts between B and C, and between C and D. However, we will employ an alternative solution using *superposition* where we break the original loading into simpler load cases, as shown in Fig. 2. Each of the load cases is analyzed independently, and then the total shear and moment are obtained by adding those for each of the load cases.

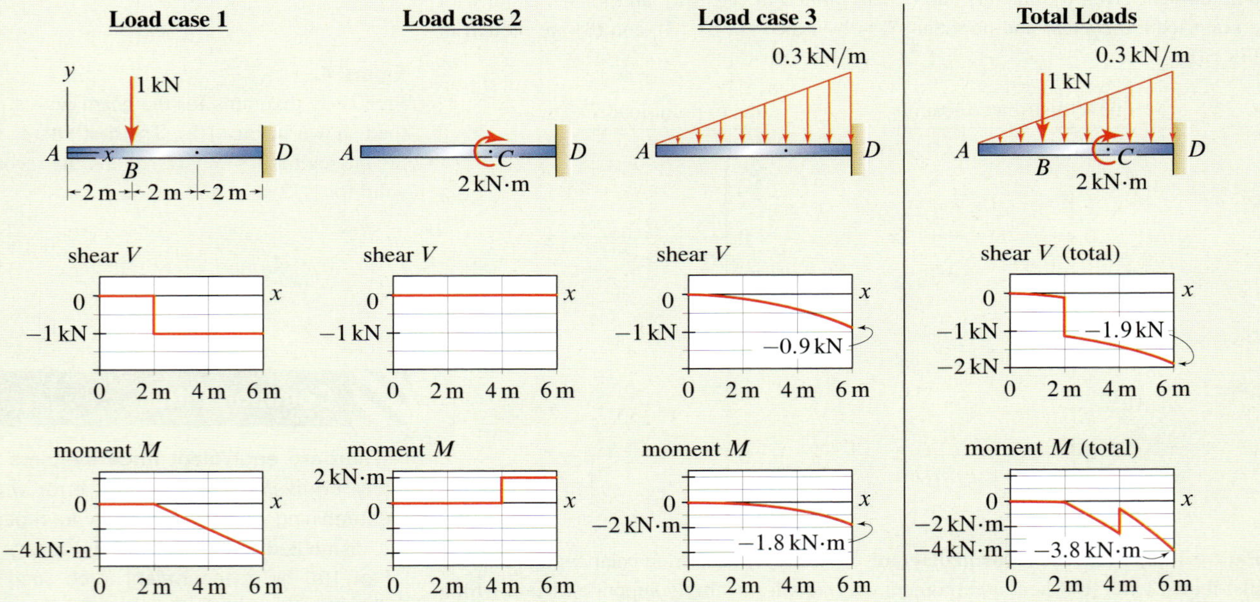

Figure 2. Selection of three load cases to be used for superposition. The shear and moment for each load case, as found in Eqs. (1)–(3) are shown. The total shear and moment, as given by Eqs. (4) and (5) are also shown.

Load case 1

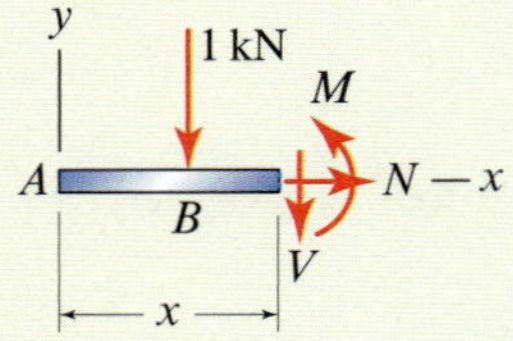

Figure 3
Free body diagram for load case 1.

Modeling For the 1 kN force applied at B, the shear and moment between A and B are clearly zero. For the region between B and D, the FBD is shown in Fig. 3.

Governing Equations & Computation Using the FBD in Fig. 3, you should write and solve the equilibrium equations $\sum F_y = 0$ and $\sum M = 0$ to obtain

$$\textbf{Load case 1:} \quad \begin{cases} V = 0, & M = 0, & \text{for } 0 \le x \le 2\,\text{m}, \\ V = -1\,\text{kN}, & M = 2\,\text{kN·m} - (1\,\text{kN})x, & \text{for } 2\,\text{m} \le x \le 6\,\text{m}. \end{cases} \tag{1}$$

Plots of V and M in Eq. (1) are shown in Fig. 2.

Load case 2

Modeling For the $2\,\text{kN·m}$ moment applied at C, the shear and moment between A and C are clearly zero. For the region between C and D, the FBD is shown in Fig. 4.

Governing Equations & Computation Using the FBD in Fig. 4, you should write and solve the equilibrium equations $\sum F_y = 0$ and $\sum M = 0$ to obtain

$$\textbf{Load case 2:} \quad \begin{cases} V = 0, \quad M = 0, & \text{for } 0 \le x \le 4\,\text{m}, \\ V = 0, \quad M = 2\,\text{kN·m}, & \text{for } 4\,\text{m} \le x \le 6\,\text{m}. \end{cases} \tag{2}$$

Plots of V and M in Eq. (2) are shown in Fig. 2.

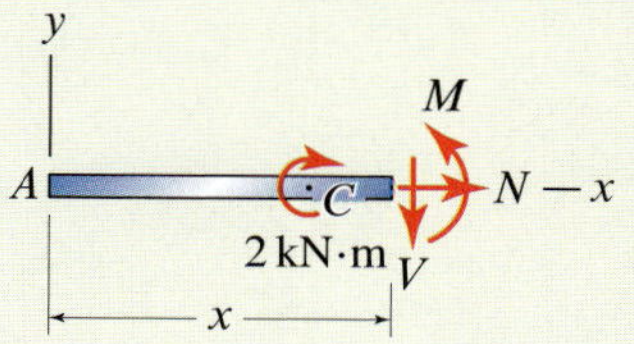

Figure 4
Free body diagram for load case 2.

Load case 3

Modeling For the linearly incresing distributed load, we take a cut between points A and D to draw the FBD shown in Fig. 5.

Governing Equations & Computation To determine the force P in the FBD of Fig. 5, first we write an expression for the distributed load as $w = (0.05\,\text{kN/m}^2)x$. This expression is easily verified by noting that $w = 0$ when $x = 0$ and $w = 0.3\,\text{kN/m}$ when $x = 6\,\text{m}$ (for tips on developing expressions for linear functions, see the Helpful Information margin note on p. 420). The force P due to the distributed load is the "area" of the triangular distributed force (i.e., 1/2 the "base" x multiplied by the "height" $(0.05\,\text{kN/m}^2)x$), hence $P = (0.025\,\text{kN/m}^2)x^2$.

Using the FBD in Fig. 5, you should write and solve the equilibrium equations $\sum F_y = 0$ and $\sum M = 0$ to obtain

$$\textbf{Load case 3:} \quad \Big\{ V = -\left(0.025\,\frac{\text{kN}}{\text{m}^2}\right)x^2, \quad M = -\frac{1}{3}\left(0.025\,\frac{\text{kN}}{\text{m}^2}\right)x^3, \quad \text{for } 0 \le x \le 6\,\text{m}. \tag{3}$$

Plots of V and M in Eq. (3) are shown in Fig. 2.

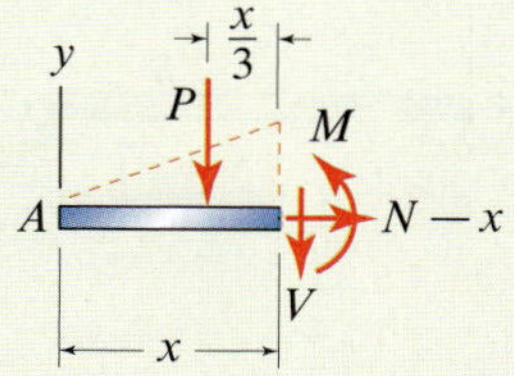

Figure 5
Free body diagram for load case 3.

Superposition for total V and M

Discussion & Verification The total shear V and moment M are obtained by adding the results given in Eqs. (1), (2), and (3), paying careful attention to use the appropriate expressions for the various regions for x. Hence,

$$\textbf{Total shear:} \quad \begin{cases} V = -\left(0.025\,\dfrac{\text{kN}}{\text{m}^2}\right)x^2, & \text{for } 0 \le x \le 2\,\text{m}, \\ V = -1\,\text{kN} - \left(0.025\,\dfrac{\text{kN}}{\text{m}^2}\right)x^2, & \text{for } 2\,\text{m} \le x \le 6\,\text{m}, \end{cases} \tag{4}$$

$$\textbf{Total moment:} \quad \begin{cases} M = -\dfrac{1}{3}\left(0.025\,\dfrac{\text{kN}}{\text{m}^2}\right)x^3, & \text{for } 0 \le x \le 2\,\text{m}, \\ M = 2\,\text{kN·m} - (1\,\text{kN})x - \dfrac{1}{3}\left(0.025\,\dfrac{\text{kN}}{\text{m}^2}\right)x^3, & \text{for } 2\,\text{m} \le x \le 4\,\text{m}, \\ M = 4\,\text{kN·m} - (1\,\text{kN})x - \dfrac{1}{3}\left(0.025\,\dfrac{\text{kN}}{\text{m}^2}\right)x^3, & \text{for } 4\,\text{m} \le x \le 6\,\text{m}. \end{cases} \tag{5}$$

The shear and moment diagrams for the total loading are obtained by summing the shear and moment diagrams for each load case, as shown in Fig. 2.

PROBLEMS

Problem 8.29

For the simply supported beam shown, let $a = 6\,\text{ft}$, $b = 4\,\text{ft}$, and $P = 1000\,\text{lb}$. Determine the shear and moment as functions of position, and draw the shear and moment diagrams.

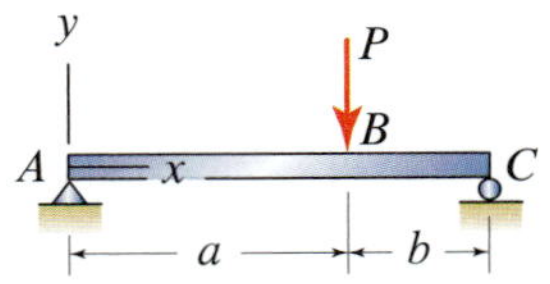

Figure P8.29–P8.31

Problem 8.30

Repeat Prob. 8.29, using $a = 3\,\text{m}$, $b = 1\,\text{m}$, and $P = 10\,\text{kN}$. Determine the shear and moment as functions of position, and draw the shear and moment diagrams.

Problem 8.31

Determine the shear and moment as functions of position. Express your answers in terms of parameters such as P, a, b, etc. Draw the shear and moment diagrams. *Hint:* The answers to this problem are given in the statement of Prob. 8.52.

Problem 8.32

A simply supported beam with two equal forces applied equidistant from the supports is called *four-point bending*. This loading arrangement is commonly used for testing beams in a laboratory.

(a) Determine the shear and moment as functions of position. Express your answers in terms of parameters such as P, L, etc. Draw the shear and moment diagrams.

(b) Comment on any interesting features the shear and moment display in the region between points B and C.

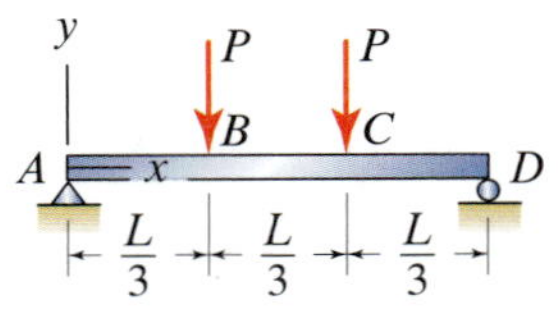

Figure P8.32

Problems 8.33 and 8.34

A simply supported beam with a 2000 ft·lb moment is shown. Determine the shear and moment as functions of position, and draw the shear and moment diagrams.

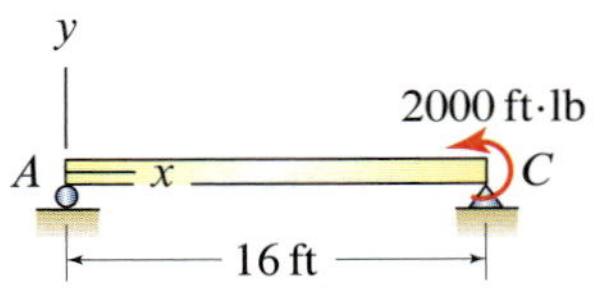

Figure P8.33

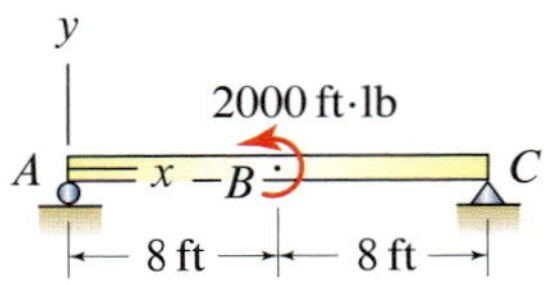

Figure P8.34

Problem 8.35

A diver stands on the end of a diving board. If the diver's mass is 70 kg, determine the shear and moment as functions of position, and draw the shear and moment diagrams.

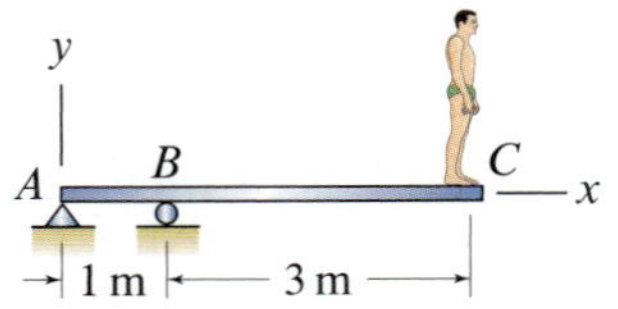

Figure P8.35

Problems 8.36 through 8.39

For the cantilever beam shown, determine the shear and moment as functions of position, and draw the shear and moment diagrams.

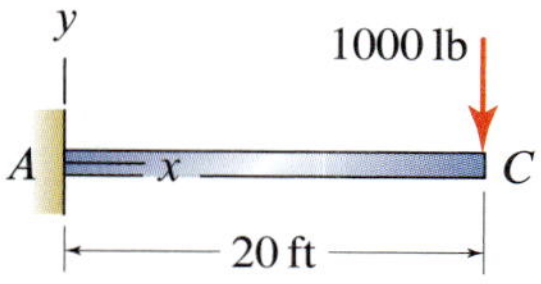

Figure P8.36

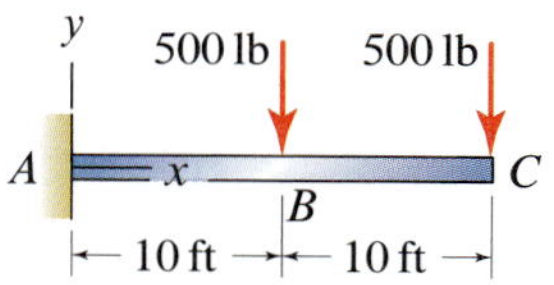

Figure P8.37

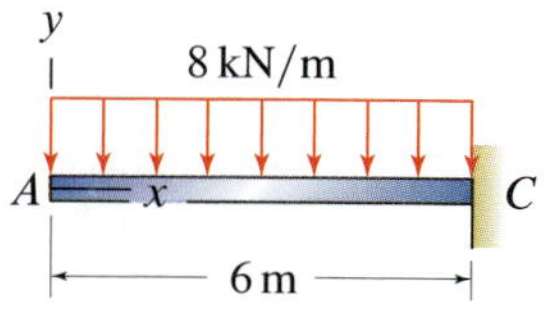

Figure P8.38

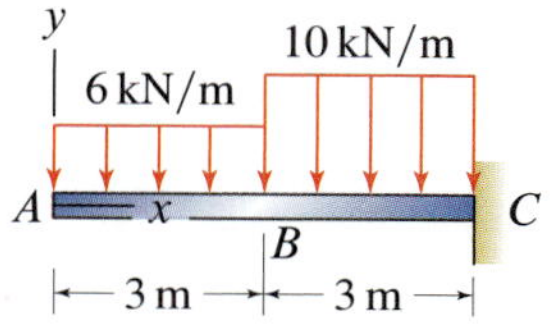

Figure P8.39

Problem 8.40

A simply supported beam has the linear distributed load shown. Determine the shear and moment as functions of position, and draw the shear and moment diagrams.

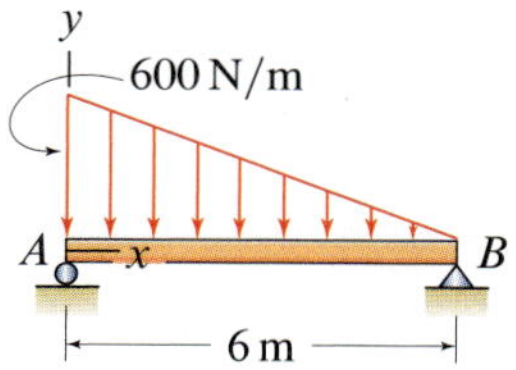

Figure P8.40

Problem 8.41

A beam with an overhang is subjected to the uniformly distributed load shown. Determine the shear and moment as functions of position, and draw the shear and moment diagrams.

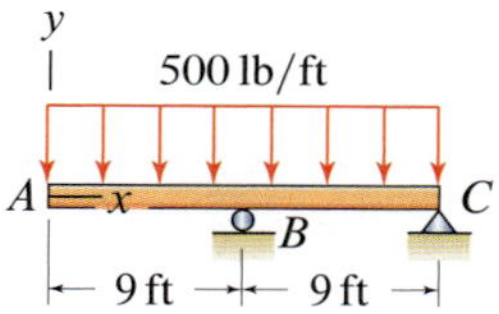

Figure P8.41

Problem 8.42

Replace the distributed load of Prob. 8.41 by a single force so that the two force systems are equivalent. Determine the shear and moment as functions of position (it should be possible to do this by inspection). Even though the loadings are equivalent, do you expect the results for this problem to be the same as those for Prob. 8.41?
Note: Concept problems are about *explanations*, not computations.

Problem 8.43

One of the beams that supports a balcony is shown. To design a beam for this purpose, it is common to use a uniformly distributed load that includes the dead loads (e.g., weight of materials) and live loads (e.g., weight of a large but reasonable number of people distributed over the balcony). If the uniformly distributed load is 2500 N/m, determine the shear and moment as functions of position, and draw the shear and moment diagrams. Idealize the supports at B and C to be a roller and pin, respectively.

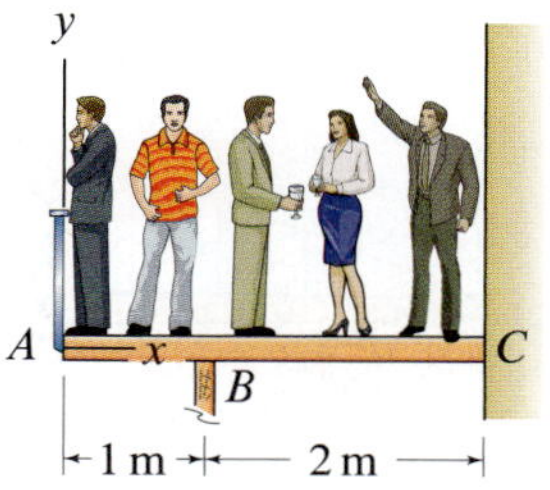

Figure P8.43

Problems 8.44 through 8.47

In Example 4.11 on p. 228, numerous equivalent force systems for a cantilever beam were developed. For each of the force systems cited below, determine the shear and moment as functions of position, and draw the shear and moment diagrams:

Problem 8.44 Figure 1 on p. 228.

Problem 8.45 Figure 2 on p. 228.

Problem 8.46 Figure 3 on p. 228.

Problem 8.47 Figure 4 on p. 228 with $d = 2.67$ mm.

Problem 8.48

Without solving Probs. 8.44 through 8.47, comment on the agreement you expect between the shear and moment distributions for each of these load cases. Are there particular points in the beam where you know the shear and moment must be the same for all of these loadings?
Note: Concept problems are about *explanations*, not computations.

Problems 8.49 and 8.50

Consider the simply supported beam shown with a uniformly distributed load and a force at midspan. Use superposition of the results from Example 8.5 on p. 470 to determine the shear and moment as functions of position, and draw the shear and moment diagrams.

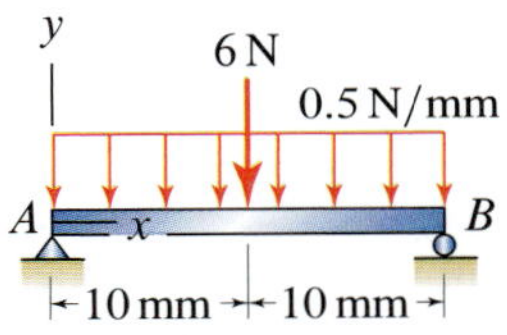

Figure P8.49

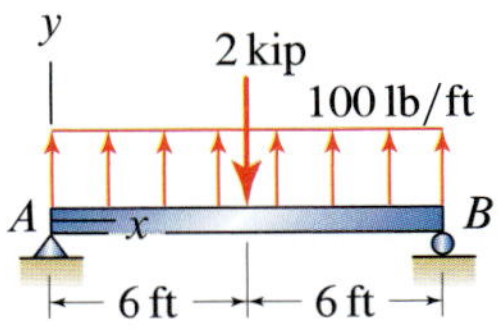

Figure P8.50

Problem 8.51

A wing of a jet is crudely modeled as a beam with the loadings shown. Use superposition of the results from Example 8.6 on p. 472 to determine the shear and moment in the wing as functions of position, and draw the shear and moment diagrams.

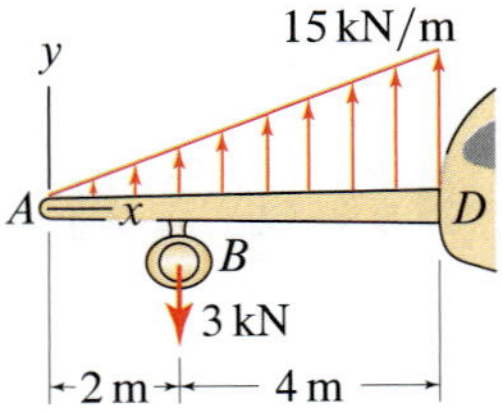

Figure P8.51

Problem 8.52

A simply supported beam is subjected to the 900 and 1200 N forces shown. Use superposition to determine the shear and moment as functions of position, and draw the shear and moment diagrams. *Hint:* The answers to Prob. 8.31 on p. 474, given below, are helpful for this problem.

$$V = \frac{Pb}{a+b}, \qquad M = \frac{Pb}{a+b}x, \qquad \text{for} \quad 0 \le x \le a,$$

$$V = -\frac{Pa}{a+b}, \qquad M = Pa\left(1 - \frac{x}{a+b}\right), \qquad \text{for} \quad a \le x \le a+b.$$

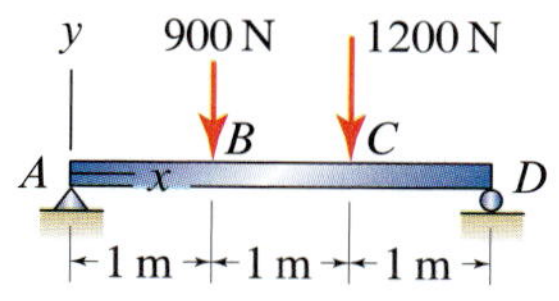

Figure P8.52

Problem 8.53

A person uses a wrench to apply a force F_A and a moment M_A to the end of a cantilever beam. The weight of the beam is represented by the uniform distributed load w_0.

(a) For $F_A \neq 0$, $M_A = 0$, and $w_0 = 0$, determine the shear and moment as functions of position. Express your answers in terms of F_A.

(b) For $F_A = 0$, $M_A \neq 0$, and $w_0 = 0$, determine the shear and moment as functions of position. Express your answers in terms of M_A.

(c) For $F_A = 0$, $M_A = 0$, and $w_0 \neq 0$, determine the shear and moment as functions of position. Express your answers in terms of w_0.

(d) If $F_A = 20$ lb, $M_A = 200$ in.·lb, $w_0 = 0.5$ lb/in., and $L = 30$ in., use superposition of the results of Parts (a) through (c) to determine the shear and moment as functions of position, and draw the shear and moment diagrams.

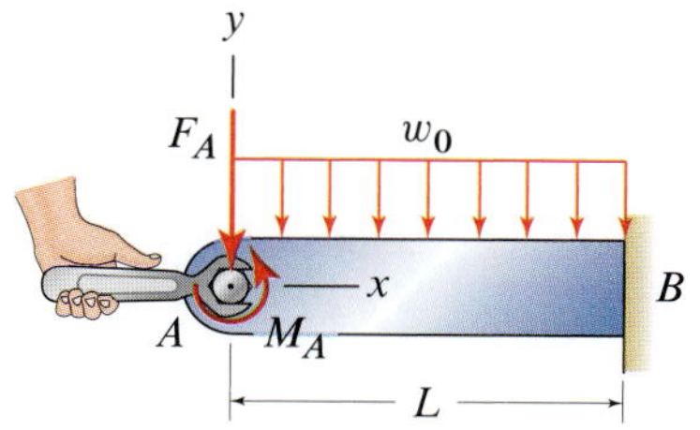

Figure P8.53

Problem 8.54

A cross section through a railroad bed is shown. The rails are supported by *ties* that are made of wood or sometimes concrete, and the ties rest on *ballast*, which is usually crushed stone. Assuming the ballast applies a uniformly distributed load to the ties, determine the shear and moment in a tie due to the 10 kip forces and the distributed load from the ballast as functions of position. Draw the shear and moment diagrams.

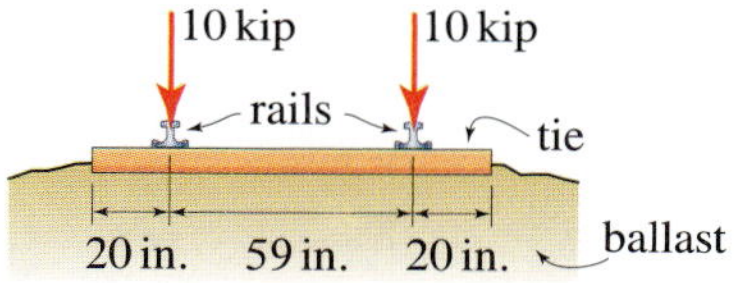

Figure P8.54

Problem 8.55

One of the beams of a staircase is to support a 200 lb/ft uniformly distributed vertical force. Determine the axial force, shear, and moment as functions of position and draw the normal force, shear, and moment diagrams.

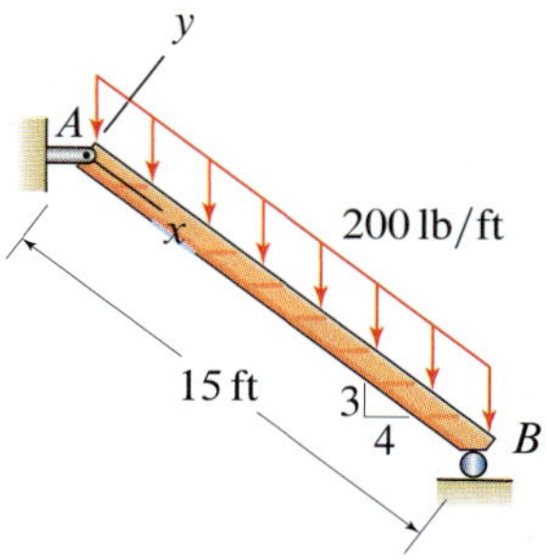

Figure P8.55

8.3 Relations Among Shear, Moment, and Distributed Force

In this section we develop differential equations that relate the distributed force, shear, and moment for a straight beam with transverse loads. These relations are useful in statics and are also useful in mechanics of materials where they are supplemented with additional differential equations that describe other aspects of the behavior of beams.

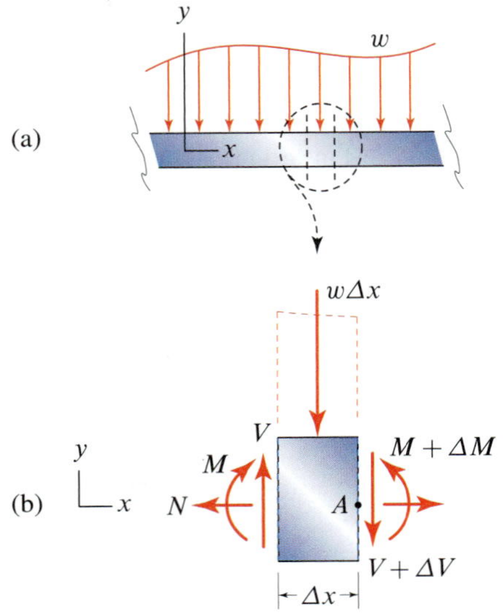

Figure 8.6
(a) A straight beam subjected to a distributed force w. (b) An FBD of a small length Δx of the beam.

Relations among V, M, and w

Consider the straight beam shown in Fig. 8.6(a); the supports for the beam and/or other forces that might be applied to it are not important for our purposes here, other than we assume the beam is in equilibrium. We take cuts through two cross sections of this beam, one at position x and the other at $x + \Delta x$, and the FBD that results is shown in Fig. 8.6(b). The internal forces on the cross section at x are the normal (axial) force N, shear force V, and moment M. The shear and moment on the cross section at $x + \Delta x$ may be different, and these are $V + \Delta V$ and $M + \Delta M$. If Δx is small, then the distributed force w in Fig. 8.6(b) is approximately uniform and the total force is therefore $w\,\Delta x$.

The small piece of beam in the FBD of Fig. 8.6(b) is in equilibrium, hence we may write

$$\sum F_y = 0: \quad V - (V + \Delta V) - w\,\Delta x = 0. \tag{8.1}$$

Rearranging the above expression and dividing by the length Δx provide

$$\frac{\Delta V}{\Delta x} = -w. \tag{8.2}$$

Taking the limit of Eq. (8.2) as $\Delta x \to 0$ provides

$$\boxed{\frac{dV}{dx} = -w.} \tag{8.3}$$

In words, Eq. (8.3) says, "The change in shear divided by the change in length of the beam at position x is equal to the negative of the distributed force at that location."

Summing moments about point A in Fig. 8.6(b) provides

$$\sum M_A = 0: \quad -M + M + \Delta M - V\,\Delta x + w\,\Delta x\,\frac{\Delta x}{2} = 0. \tag{8.4}$$

Rearranging the above expression and dividing by the length Δx provide

$$\frac{\Delta M}{\Delta x} = V - w\,\frac{\Delta x}{2}. \tag{8.5}$$

Taking the limit of Eq. (8.5) as $\Delta x \to 0$ provides

$$\boxed{\frac{dM}{dx} = V.} \tag{8.6}$$

In words, Eq. (8.6) says, "The change in moment divided by the change in length of the beam at position x is equal to the shear at that location."

> **Helpful Information**
>
> **Drawing shear and moment diagrams.** Equations (8.3) and (8.6) are useful for drawing shear and moment diagrams. In words, Eq. (8.3) says, "The slope of the shear diagram is equal to the negative of the distributed force's value," and Eq. (8.6) says, "The slope of the moment diagram is equal to the value of the shear." Example 8.9 on p. 486 illustrates the usefulness of this.

Determination of V and M using integration

Equations (8.3) and (8.6) are differential equations that can be solved to obtain the shear V and moment M, given the distributed force w. We call this the *integration approach*, and it is developed by multiplying both sides of Eq. (8.3) by dx to write

$$dV = -w\,dx. \tag{8.7}$$

Integrating both sides of the above expression provides

$$\int_{V_P}^{V} dV = -\int_{x_P}^{x} w\,dx, \tag{8.8}$$

where the limits of integration* for the right-hand integral are chosen to be x_P to x, and thus the limits of integration for the left-hand integral are V_P to V, where V_P is the shear at position x_P and V is the shear at position x. Evaluating the integral on the left-hand side of Eq. (8.8) and adding V_P to both sides provides

$$V = V_P - \int_{x_P}^{x} w\,dx. \tag{8.9}$$

Similarly, Eq. (8.6) may be written as

$$dM = V\,dx. \tag{8.10}$$

which upon integration gives

$$M = M_P + \int_{x_P}^{x} V\,dx. \tag{8.11}$$

Thus, if the distributed force w is known as a function of position, then by integration using Eq. (8.9), the shear as a function of position is obtained. Once the shear is known, then by integration using Eq. (8.11) the moment as a function of position is obtained.

Remarks

- When using Eqs. (8.9) and (8.11), we select the location of point P, whose coordinate is x_P. Point P will be a convenient location where the shear V_P and moment M_P are known or can be easily determined.
- Often, point P in Eqs. (8.9) and (8.11) will be taken to be at one of the ends of the beam, because these are locations where it will often be straightforward to determine the shear V_P and moment M_P. For example, consider the common situation of a horizontal beam of length

* Alternatively, we may use indefinite integrals, in which case Eq. (8.8) would be written as $\int dV = -\int w\,dx$. Evaluation of the integral on the left-hand side provides $V = -\int w\,dx$. Note that when the integral on the right-hand side is evaluated, a constant of integration is produced, which by comparison with Eq. (8.9) is seen to have the physical interpretation of being the shear at some location in the beam.

L with the origin of the coordinate system at the left-hand end. Taking point P at the left-hand end means $x_P = 0$, while taking point P at the right-hand end means $x_P = L$. Example 8.8 on p. 484 illustrates the use of both of these choices.

- For some problems, it may be effective to take point P to be somewhere between the ends of the beam.
- In all of the integrals written in this section, we could replace the limit of integration x by the coordinate of a point Q, which is x_Q. Then Eqs. (8.9) and (8.11) become

$$V_Q = V_P - \underbrace{\int_{x_P}^{x_Q} w\,dx}_{\substack{\text{area under } w \text{ vs.} \\ x \text{ plot between} \\ x_P \text{ and } x_Q}}, \quad \text{and} \quad M_Q = M_P + \underbrace{\int_{x_P}^{x_Q} V\,dx}_{\substack{\text{area under } V \text{ vs.} \\ x \text{ plot between} \\ x_P \text{ and } x_Q}}. \tag{8.12}$$

The integrals in the above expressions represent, respectively, the area under the w vs. x plot and the area under the V vs. x plot, between x_P and x_Q. These expressions provide a graphical method for drawing shear and moment diagrams, as illustrated in Example 8.9 on p. 486.

- Another way to interpret Eqs. (8.9) and (8.11) is that they describe the *change* of shear ΔV and the *change* of moment ΔM between positions x_P and x. Hence, Eqs. (8.9) and (8.11) can be rewritten as

$$\Delta V = -\int_{x_P}^{x} w\,dx \quad \text{and} \quad \Delta M = \int_{x_P}^{x} V\,dx. \tag{8.13}$$

As discussed in connection with Eq. (8.12), you may also elect to take $x = x_Q$, in which case ΔV and ΔM are the change in shear and moment, respectively, between points P and Q. These expressions can be useful for drawing shear and moment diagrams.

Helpful Information

Combining approaches. You may choose to use a combination of equilibrium and integration approaches to determine the shear and moment in beams. For example, consider the following beam and loading:

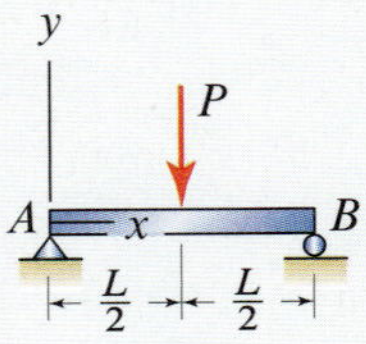

Rather than use the equilibrium approach exclusively or the integration approach exclusively (both of which are effective for this problem), you could choose to use the equilibrium approach to determine the shear, followed by use of integration to obtain the moment.

Which approach should I use?

In this chapter we have discussed two methods for determining shear and moment as functions of position:

1. *Equilibrium approach*, discussed in Section 8.2, where cuts are taken as needed, FBDs are drawn, and equilibrium equations are written and solved.
2. *Integration approach*, where the expression for the distributed force w is integrated to obtain the shear, which is then integrated to obtain the moment.

The equilibrium approach is very robust. It is straightforward for uniform distributed forces, is a bit more tedious for linear distributed forces, and becomes unwieldy when the distributed force is more complicated. The integration approach is elegant, but sometimes has subtleties. It is straightforward when there is a single function that describes a distributed force that acts over the full length of the beam *and* when the supports and/or point forces and moments are at the ends of the beam. Subtleties arise when there are multiple functions that describe the distributed force, when a beam has point forces and moments between its ends, and when a beam has supports between its ends; the examples in this section point out some of these subtleties. Very often, a combination of the equilibrium and integration approaches will be effective.

Sometimes, we will only require the shear and moment diagrams, and we will not need to know the shear and moment as functions of position. For such situations we may elect to use the equilibrium approach of Section 8.1 to determine the shear and moment at specific points within a beam, and then we use Eq. (8.3) to relate the slopes of the shear diagram to the values of the distributed force, and Eq. (8.6) to relate the slopes of the moment diagram to the values of the shear.

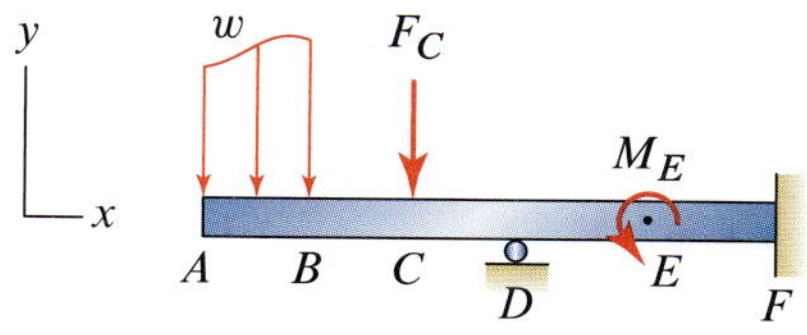

Figure 8.7
A straight beam with a variety of loadings and supports.

Tips and shortcuts for drawing shear and moment diagrams

In the remainder of this section, we discuss some of the characteristics of shear and moment distributions. Many of these characteristics you have probably noticed in the examples and problems of Section 8.2, while the expressions derived in this section provide further insights. Knowledge of these characteristics will provide you with additional tools that will be helpful for drawing shear and moment diagrams and for detecting errors.

Equations (8.9) and (8.11) on p. 479 show the following:

- In regions of a beam where the distributed force is zero, the shear is constant and the moment is linear.
- In regions of a beam where the distributed force is constant (i.e., uniform), the shear is linear and the moment is quadratic.
- In regions of a beam where the distributed force is linear, the shear is quadratic and the moment is cubic.

The following remarks pertain to Fig. 8.7:

- Point A is an unsupported end of a beam with no concentrated force and no moment applied. At A, the shear and moment are zero. This is true regardless of the presence of a distributed force w.
- At point B, a distributed force ends. The shear and moment just to the right of B are the same as those just to the left of B. The same comments apply to points where a distributed force begins.
- A concentrated force F_C acting in the negative y direction is applied at point C. The shear just to the right of C is lower than the shear just to the left of C by amount F_C. The moment just to the right of C is the same as that just to the left of C. The FBD and equilibrium equations shown in Fig. 8.8 justify the validity of these remarks.
- A roller support is positioned at point D. The shear just to the right of D is higher than the shear just to the left of D by amount D_y, where D_y is the reaction the roller applies to the beam with positive D_y acting in the positive y direction. The moment just to the right of D is the same

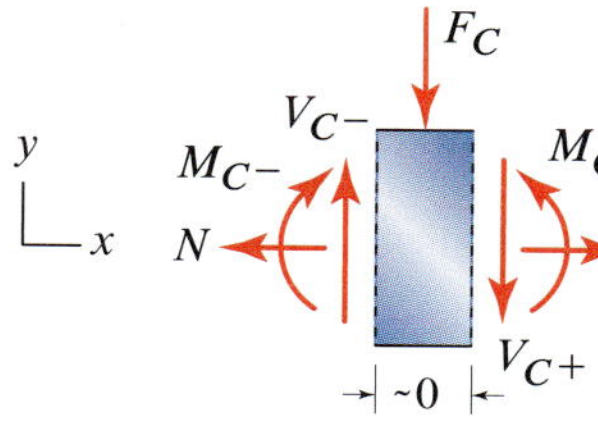

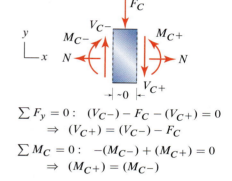

$$\sum F_y = 0: \quad (V_C-) - F_C - (V_C+) = 0$$
$$\Rightarrow \quad (V_C+) = (V_C-) - F_C$$
$$\sum M_C = 0: \quad -(M_C-) + (M_C+) = 0$$
$$\Rightarrow \quad (M_C+) = (M_C-)$$

Figure 8.8
An FBD for an infinitesimally small length of beam at point C in Fig. 8.7. F_C is the concentrated force. V_C- and M_C- are the shear and moment just to the left of C. V_C+ and M_C+ are the shear and moment just to the right of C.

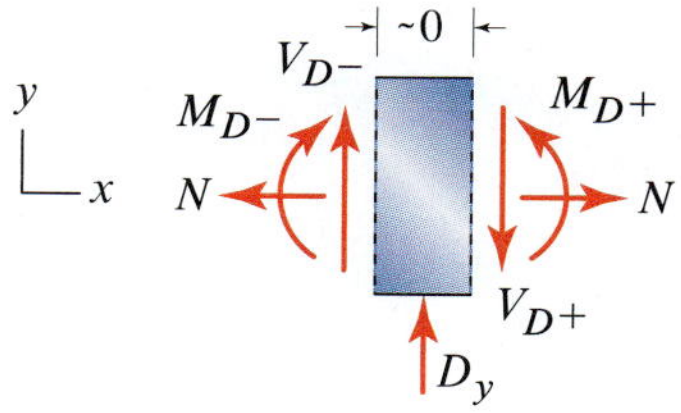

$$\sum F_y = 0: \quad (V_D-) + D_y - (V_D+) = 0$$
$$\Rightarrow \quad (V_D+) = (V_D-) + D_y$$
$$\sum M_C = 0: \quad -(M_D-) + (M_D+) = 0$$
$$\Rightarrow \quad (M_D+) = (M_D-)$$

Figure 8.9
An FBD for an infinitesimally small length of beam at point D in Fig. 8.7. D_y is the support reaction. V_D- and M_D- are the shear and moment just to the left of D. V_D+ and M_D+ are the shear and moment just to the right of D.

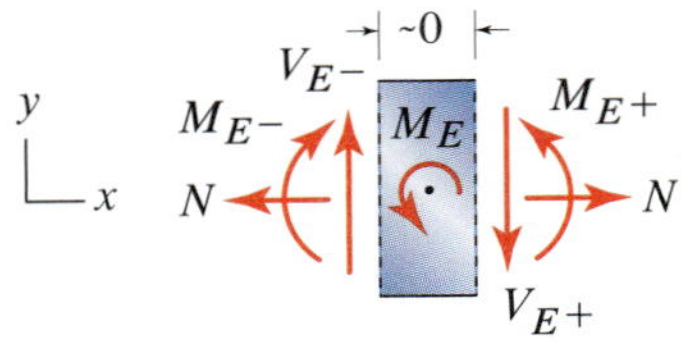

$$\sum F_y = 0: \quad (V_{E-}) - (V_{E+}) = 0$$
$$\Rightarrow \quad (V_{E+}) = (V_{E-})$$
$$\sum M_C = 0: \quad -(M_{E-}) + M_E + (M_{E+}) = 0$$
$$\Rightarrow \quad (M_{E+}) = (M_{E-}) - M_E$$

Figure 8.10
A FBD for an infinitesimally small length of beam at point E in Fig. 8.7. M_E is the moment applied at E. V_{E-} and M_{E-} are the shear and moment just to the left of E. V_{E+} and M_{E+} are the shear and moment just to the right of E.

as that just to the left of D. The FBD and equilibrium equations shown in Fig. 8.9 justify the validity of these remarks.

- A concentrated moment M_E acting counterclockwise is applied at point E. The shear just to the right of E is the same as that just to the left of E. The moment just to the right of E is lower than the moment just to the left of E by amount M_E. The FBD and equilibrium equations shown in Fig. 8.10 justify the validity of these remarks.

Design considerations

The design of a beam includes: specification of the material(s) it is constructed of, its shape and dimensions, methods of support and/or attachment to other members, and so on. When possible or convenient, beams that are commercially manufactured are used, such as the examples shown in Fig. 8.11. Some of the benefits of commercially manufactured beams are economy and rapid availability. When we use commercially manufactured beams, our task is to determine the material, size, and shape of beams that are needed. Commercially manufactured beams are available in a variety of materials, such as various grades of steel and aluminum, and a variety of shapes, such as I beam, L channel, U channel, and so on. Prefabricated reinforced concrete beams are also available. Often a beam will need to be constructed to our specifications. While cost may be higher and it may take longer to fabricate, performance may be better.

Beams are usually designed to satisfy a variety of criteria, such as these:

- *Acceptable strength*. A beam must be strong enough to support the forces applied to it without failing.
- *Acceptable deformations*. A beam should not deflect excessively.
- *Acceptable fatigue life*. A beam should be able to withstand the number of load cycles that will be applied to it over its life span.

There are often additional criteria such as cost, manufacturability, corrosion resistance, high-temperature resistance, and so on.

Figure 8.11
A steel yard showing commercially manufactured beams with an assortment of shapes.

End of Section Summary

In this section, differential equations that relate the shear, moment, and distributed force were derived for straight beams. Equation (8.3) relates the shear to the distributed force by $dV/dx = -w$, and Eq. (8.6) relates the moment to the shear by $dM/dx = V$. These relations are useful for drawing shear and moment diagrams. Also, the solutions to these differential equations were given in Eq. (8.9) where the shear is obtained by integrating the distributed force, and in Eq. (8.11) where the moment is obtained by integrating the shear.

EXAMPLE 8.7 *Integration to Determine Shear and Moment*

A simply supported beam of length L with a uniformly distributed force w_0 is shown. Use integration to determine the shear and moment as functions of position.

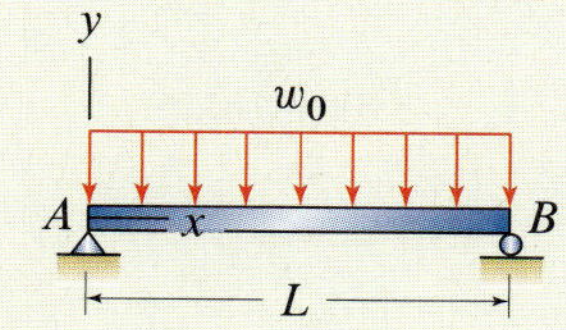

Figure 1

SOLUTION

Road Map This problem was previously solved using the equilibrium approach in Part (a) of Example 8.5 on p. 470. Here we develop the solution using integration.

To determine the shear and moment as functions of position using Eqs. (8.9) and (8.11) on p. 479, the shear V_P and moment M_P at some convenient point P must be known, and for this example, either of points A or B is such a location. We will use A, and comments are provided at the end of the solution regarding the use of B.

Modeling To determine the shear and moment at A, the reactions at that location are needed. Thus, we draw the FBD for the entire beam as shown in Fig. 2.

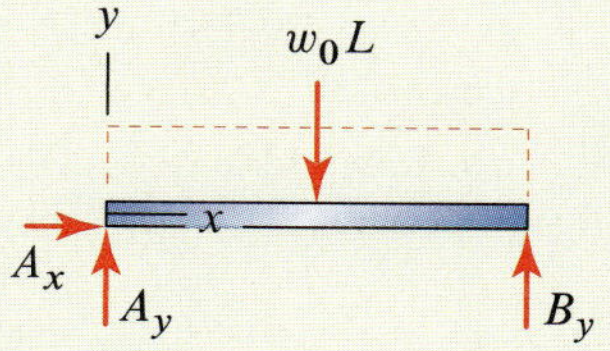

Figure 2

Free body diagram for the entire beam. By writing equilibrium equations, the support reactions are found to be $A_x = 0$ and $A_y = B_y = w_0 L/2$.

Governing Equations & Computation By writing and solving the equilibrium equations for the FBD in Fig. 2, the vertical support reaction at A is $A_y = w_0 L/2$.

Modeling To determine the shear V_A and moment M_A at point A of the beam, we take a cut just to the right of A to draw the FBD shown in Fig. 3.

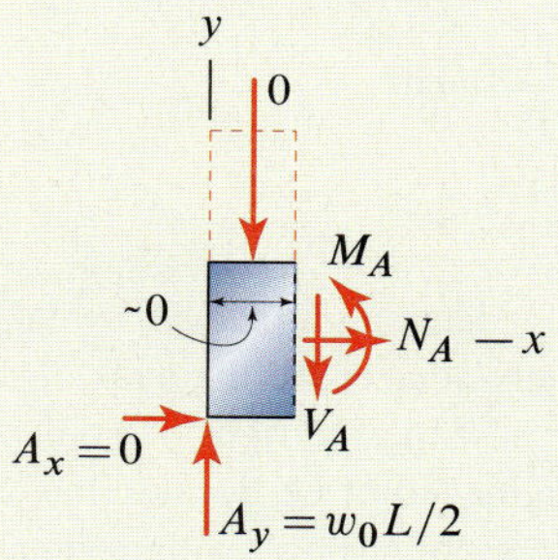

Figure 3

Free body diagram obtained by taking a cut just to the right of point A. Because the length of this portion of beam is zero, the force due to the distributed load is zero.

Governing Equations & Computation Using the FBD in Fig. 3, we write and solve the equilibrium equations,

$$\sum F_y = 0: \qquad \frac{w_0 L}{2} - V_A = 0 \quad \Rightarrow \quad V_A = \frac{w_0 L}{2}, \tag{1}$$

$$\sum M_A = 0: \qquad M_A = 0 \quad \Rightarrow \quad M_A = 0. \tag{2}$$

Noting that $V_A = w_0 L/2$ and $w = w_0$ (a constant), we see Eq. (8.9) provides

$$V = V_A - \int_{x_A=0}^{x} w_0\, dx \quad \Rightarrow \quad \boxed{V = \frac{w_0 L}{2} - (w_0)x.} \tag{3}$$

Noting that $M_A = 0$ and $V = w_0 L/2 - (w_0)x$, we see Eq. (8.11) provides

$$M = M_A + \int_{x_A=0}^{x} \left[\frac{w_0 L}{2} - (w_0)x\right] dx \quad \Rightarrow \quad \boxed{M = \frac{w_0 L}{2}x - w_0\frac{x^2}{2}.} \tag{4}$$

Discussion & Verification

- As expected, since the distributed force is uniform, the shear is linear and the moment is quadratic. Also, Eqs. (3) and (4) agree with the results of Part (a) in Example 8.5.
- Rather than take point P to be the left-hand end of the beam in Eqs. (8.9) and (8.11), we could use the right-hand end. To do this, we draw the FBD shown in Fig. 4, and we write equilibrium equations to determine $V_B = -w_o L/2$ and $M_B = 0$. In Eqs. (3) and (4), the lower limits become $x_B = L$, and V_A and M_A are replaced by V_B and M_B. The same results for V and M are obtained.
- As a partial check of our solutions, you should evaluate Eqs. (3) and (4) at $x = 0$ to verify that they yield the proper results for V_A and M_A. Another check is to determine the shear and moment at another point on the beam, such as point B in Fig. 4, to verify that the correct results are obtained there.

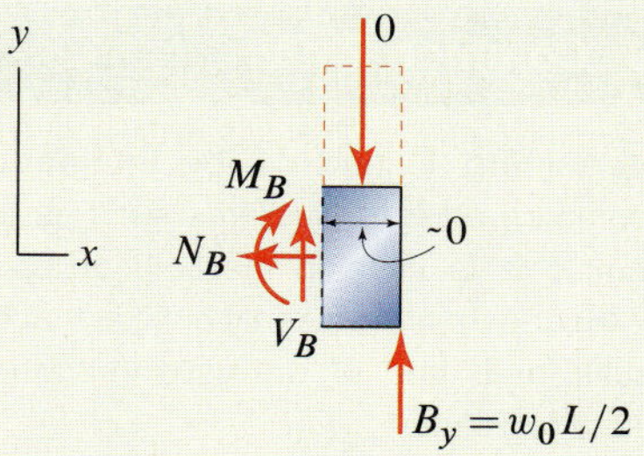

Figure 4

Free body diagram obtained by taking a cut just to the left of point B. Because the length of this portion of beam is zero, the force due to the distributed load is zero.

EXAMPLE 8.8 Integration to Determine Shear and Moment—a Subtlety

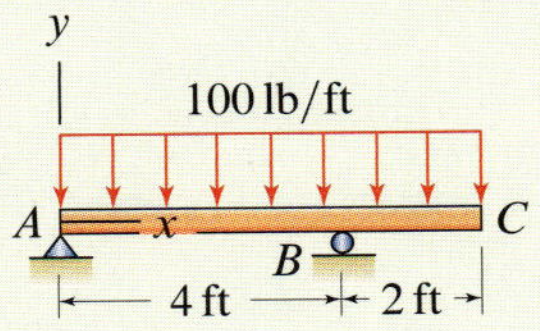

Figure 1

Reconsider the beam and loading of Example 8.4 on p. 468, where a simply supported beam with an overhang is subjected to a uniformly distributed load. Use integration to determine the shear and moment as functions of position.

SOLUTION

Road Map This problem was previously solved using the equilibrium approach in Example 8.4 on p. 468. Here we develop the solution using integration, and we point out a subtlety that arises when a support is between the ends of the beam.

In Fig.1 the distributed loading is uniform throughout the beam with value $w = 100\,\text{lb/ft}$. We will determine the shear V and moment M, using Eqs. (8.9) and (8.11) with point P taken to be at end A of the beam. The results for V and M will be valid *only* for $0 \le x \le 4\,\text{ft}$, as we know that the shear undergoes a discontinuity (i.e., jump in value) at the roller at B, and clearly this discontinuity is not contained in the results from Eq. (8.9). To determine V and M in the remainder of the beam, we will reapply Eqs. (8.9) and (8.11) with point P taken to be at end C of the beam, and the results will be valid for $4\,\text{ft} \le x \le 6\,\text{ft}$.

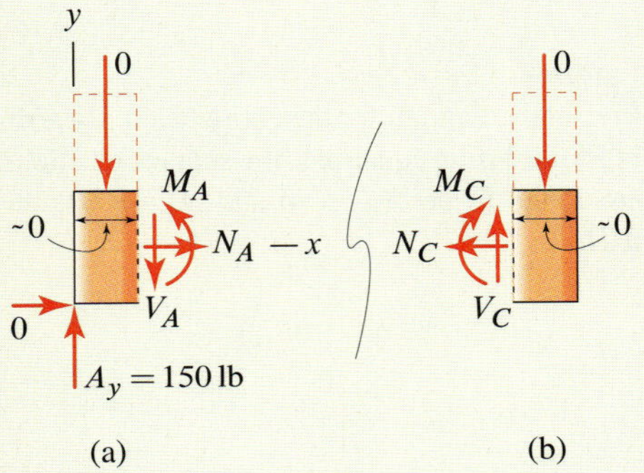

Figure 2
Free body diagrams obtained by taking cuts immediately adjacent to the ends of the beam. (a) A cut just the the right of point A. (b) A cut just to the left of C. Because the lengths of these portions of beam are infinitesimally small, the force due to the distributed load is zero.

Modeling The FBD for the entire beam was shown in Fig. 2 of Example 8.4 on p. 468, and the support reactions were determined to be $A_y = 150\,\text{lb}$ and $B_y = 450\,\text{lb}$. In this solution, we will need the shear and moment at points A and C, and to this end we draw the FBDs shown in Fig. 2.

Governing Equations & Computation Using the FBD of end A of the beam shown in Fig. 2(a), we write the equilibrium equations.

$$\sum F_y = 0: \qquad 150\,\text{lb} - V_A = 0 \quad \Rightarrow \quad V_A = 150\,\text{lb}, \tag{1}$$

$$\sum M_A = 0: \qquad M_A = 0 \quad \Rightarrow \quad M_A = 0. \tag{2}$$

Similarly, using the FBD of end C of the beam shown in Fig. 2(b), we write the equilibrium equations.

$$\sum F_y = 0: \qquad V_C = 0 \quad \Rightarrow \quad V_C = 0, \tag{3}$$

$$\sum M_C = 0: \qquad -M_C = 0 \quad \Rightarrow \quad M_C = 0. \tag{4}$$

To obtain the shear and moment for the left-hand portion of the beam, we take point P to be at end A. Noting that $V_A = 150\,\text{lb}$ and $w = 100\,\text{lb/ft}$, Eq. (8.9) provides

$$V = V_A - \int_{x_A=0}^{x} \left(100\,\frac{\text{lb}}{\text{ft}}\right) dx \quad \Rightarrow \quad \boxed{V = 150\,\text{lb} - \left(100\,\frac{\text{lb}}{\text{ft}}\right)x.} \tag{5}$$

Noting that $M_A = 0$ and V is given in Eq. (5), Eq. (8.11) provides

$$M = M_A + \int_{x_A=0}^{x} \left[150\,\text{lb} - \left(100\,\frac{\text{lb}}{\text{ft}}\right)x\right] dx \quad \Rightarrow \quad \boxed{M = (150\,\text{lb})x - \left(50\,\frac{\text{lb}}{\text{ft}}\right)x^2.} \tag{6}$$

The shear and moment determined in Eqs. (5) and (6) are valid for $0 \le x \le 4\,\text{ft}$.

Common Pitfall

Determining V and M by integration. A common error in problems such as this example is to develop the solution up to Eq. (6) and then mistakenly believe that the results for V and M are valid throughout the entire beam.

To obtain the shear and moment for the right-hand portion of the beam, we take point P to be at end C. Noting that $V_C = 0$ and $w = 100\,\text{lb/ft}$, Eq. (8.9) provides*

$$V = V_C - \int_{x_C = 6\,\text{ft}}^{x} \left(100\,\frac{\text{lb}}{\text{ft}}\right) dx \quad \Rightarrow \quad \boxed{V = 600\,\text{lb} - \left(100\,\frac{\text{lb}}{\text{ft}}\right) x.} \tag{7}$$

Noting that $M_C = 0$ and V is given in Eq. (7), Eq. (8.11) provides

$$M = M_C + \int_{x_C = 6\,\text{ft}}^{x} \left[600\,\text{lb} - \left(100\,\frac{\text{lb}}{\text{ft}}\right) x\right] dx \tag{8}$$

$$\Rightarrow \quad \boxed{M = -1800\,\text{ft·lb} + (600\,\text{lb})x - \left(50\,\frac{\text{lb}}{\text{ft}}\right) x^2.} \tag{9}$$

The shear and moment determined in Eqs. (7) and (9) are valid for $4\,\text{ft} \le x \le 6\,\text{ft}$.

Discussion & Verification

- The results for shear and moment agree with those obtained in Example 8.4. Of course, the shear and moment diagrams will also be the same, and these are shown again in Fig. 3 for purposes of the following discussion.
- Because the distributed force is uniform over the full length of the beam, the shear is linear, as expected. However, Eqs. (5) and (7) are *different* linear functions, although they share the same slope. Because $dV/dx = -w$ (i.e., the slope of the shear is equal to the negative of the distributed load), the slopes of Eqs. (5) and (7) are both $-100\,\text{lb/ft}$.
- Since the shear is linear, the moment is quadratic. However, Eqs. (6) and (9) are *different* quadratic functions.
- Observe in the moment diagram in Fig. 3 that the moment displays a local maximum at a position between 1 and 2 ft. To determine exactly where this occurs, and the value of the moment at that location, we use the expression $dM/dx = V$ as follows. At this local maximum, $dM/dx = 0$, which means that $V = 0$ at this location. Thus, we set Eq. (5) equal to zero and solve for x to obtain 1.5 ft. We then substitute $x = 1.5\,\text{ft}$ into Eq. (6) to obtain the moment 112.5 ft·lb.
- The shear V is undefined when x is exactly equal to 4 ft (point B). However, we know that the support reaction B_y at $x = 4\,\text{ft}$ is responsible for the jump that V undergoes from just to the left of B to just to the right of B. Because there is no discrete moment applied at $x = 4\,\text{ft}$, there is no discontinuity in the moment at that location. However, because the slope of M is equal to V and V has a discontinuity at $x = 4\,\text{ft}$, the slope of M changes at $x = 4\,\text{ft}$; hence, the moment curve shows a kink at that location.

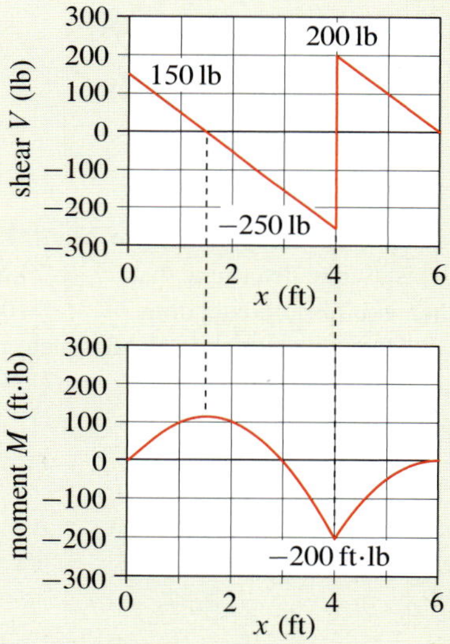

Figure 3
Shear and moment diagrams.

* You should carry out the integrations in Eqs. (7) and (8) and carefully evaluate the limits of integration, to verify the results that are reported. Incorrect evaluation of such limits of integration is a common source of error.

EXAMPLE 8.9 *Constructing Shear and Moment Diagrams*

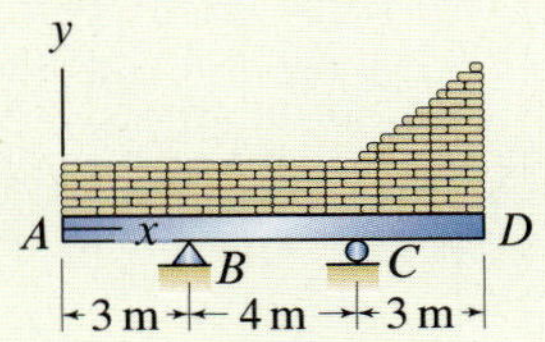

Figure 1

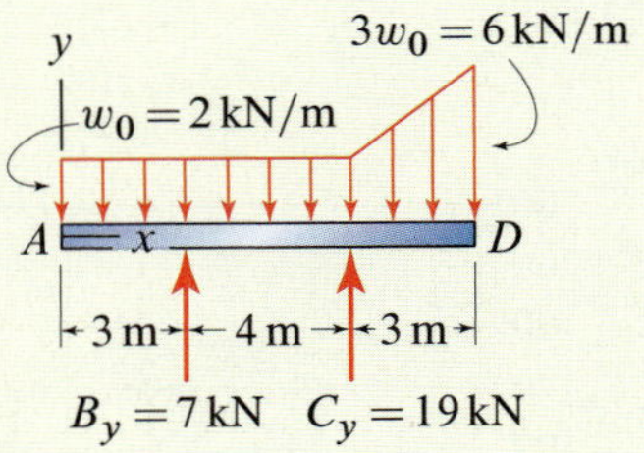

Figure 2
Free body diagram of the entire beam. Equation (1) is used to determine that $w_0 = 2\,\text{kN/m}$, and then equilibrium equations $\sum M = 0$ and $\sum F_y = 0$ are written and solved to determine that the support reactions are $B_y = 7\,\text{kN}$ and $C_y = 19\,\text{kN}$.

Beam $ABCD$ supports a wall built of concrete block. The concrete blocks have a total weight of 26 kN, and the wall is 3 times as high on the right-hand side as on the left. Draw the shear and moment diagrams.

SOLUTION

Road Map We will begin by using the description of the wall's geometry to determine the distributed force the concrete blocks apply to the beam. We will then use Eq. (8.12) on p. 480 to directly construct the shear and moment diagrams *without* explicitly determining the shear and moment as functions of position.

Modeling Given the description of the wall's geometry, we may conclude that the distributed force from the concrete block is as shown in Fig. 2, where we let the value of the distributed force at end A be w_0, and then the value at end D is $3w_0$. Since the total force from the distributed loading must equal the 26 kN weight of the concrete blocks, we may solve for w_0 by using*

$$w_0(10\,\text{m}) + \tfrac{1}{2}(3\,\text{m})(2w_0) = 26\,\text{kN} \quad \Rightarrow \quad w_0 = 2\,\text{kN/m}. \tag{1}$$

With this result, the support reactions may then be determined, and you should verify that these are $B_x = 0$, $B_y = 7\,\text{kN}$, and $C_y = 19\,\text{kN}$.

Governing Equations & Computation We will directly construct the shear and moment diagrams *without* finding V and M as functions of position. Equation (8.12) is repeated here as

$$V_Q = V_P - \underbrace{\int_{x_P}^{x_Q} w\,dx}_{\substack{\text{area under } w \text{ vs.}\\ x \text{ plot between}\\ x_P \text{ and } x_Q}} \quad \text{and} \quad M_Q = M_P + \underbrace{\int_{x_P}^{x_Q} V\,dx}_{\substack{\text{area under } V \text{ vs.}\\ x \text{ plot between}\\ x_P \text{ and } x_Q}}. \tag{2}$$

where we select points P and Q. Our strategy will be to begin by taking point P to be at the left-hand end of the beam (point A), where the shear and moment are easily found. We will then sequentially work toward the right-hand end of the beam, evaluating areas for the w vs. x diagram and constructing the shear diagram in the process. Once the shear diagram is complete, we will repeat this process to construct the moment diagram.

Shear diagram

By inspection, the shear at A is zero, because end A is unsupported and has no concentrated force applied. Hence,

$$V_A = 0, \tag{3}$$

$$V_{B-} = V_A - \left(2\,\frac{\text{kN}}{\text{m}}\right)(3\,\text{m}) = -6\,\text{kN}, \tag{4}$$

$$V_{B+} = (V_{B-}) + B_y = -6\,\text{kN} + 7\,\text{kN} = 1\,\text{kN}, \tag{5}$$

$$V_{C-} = (V_{B+}) - \left(2\,\frac{\text{kN}}{\text{m}}\right)(4\,\text{m}) = -7\,\text{kN}, \tag{6}$$

* While it is obvious that Eq. (1) is valid, we are in fact applying the first expression in Eq. (4.16) on p. 223 for construction of an equivalent force system.

$$V_{C+} = (V_{C-}) + C_y = -7\,\text{kN} + 19\,\text{kN} = 12\,\text{kN}, \tag{7}$$

$$V_D = 0. \tag{8}$$

In this example, the shear at A was determined by inspection. If you are unsure that $V_A = 0$ (or if end A had a concentrated force or a support), you draw an appropriate FBD of end A, similar to Fig. 3 on p. 483, to determine the shear at this location. Equation (2) is used to write Eqs. (4) and (6), and you should verify the areas that are computed using the w vs. x diagram shown in Fig. 2. Equations (5) and (7) are applications of the results shown in Fig. 8.9 on p. 481. Equation (8) is written by inspection, although a useful check of accuracy would be to use Eq. (2) for

$$V_D = (V_{C+}) - \left[\frac{2\,\text{kN/m} + 6\,\text{kN/m}}{2}\,(3\,\text{m})\right] = 0. \tag{9}$$

The shear diagram is shown in Fig. 3 and is constructed as follows. We first plot the values of the shear from Eqs. (3) through (8). Since w is constant from A to B and from B to C, the shear is linear in those regions; so two straight lines are drawn. As a check on our solutions, you should compute the slopes for the shear between A and B, and B and C, to verify that both are $-2\,\text{kN/m}$. Since w is linear between C and D, the shear is quadratic in that region. We use the expression $dV/dx = -w$ to determine that the slope just to the right of C is $-2\,\text{kN/m}$, and the slope at D is $-6\,\text{kN/m}$.

Moment diagram

By inspection, the moment at A is zero, because end A is unsupported and has no concentrated moment applied. Also, values of the moment will be the same on each side of the supports at B and C, so there is no need to distinguish between these. Hence,

$$M_A = 0, \tag{10}$$

$$M_B = M_A + \left[\frac{1}{2}(3\,\text{m})(-6\,\text{kN})\right] = -9\,\text{kN·m}, \tag{11}$$

$$M_C = M_B + \left[\frac{1\,\text{kN} + (-7\,\text{kN})}{2}\,(4\,\text{m})\right] = -21\,\text{kN·m}, \tag{12}$$

$$M_D = 0. \tag{13}$$

Equations (10) and (13) were written by inspection, while Eq. (2) was used to write Eqs. (11) and (12), and you should verify the areas that are computed using the V vs. x diagram shown in Fig. 3.

The moment diagram is shown in Fig. 3 and is constructed as follows. We first plot the values of the moment from Eqs. (10) through (13). Since V is linear from A to B and from B to C, the shear is quadratic in these regions, and between C and D the shear is quadratic so the moment is cubic. We use the expression $dM/dx = V$ to determine the slopes of the moment diagram, given the values of the shear, and this allows us to draw the moment diagram with the correct slopes and curvatures.

Discussion & Verification Numerous checks have been used and suggested in the course of developing this solution. Observe that the moment diagram has a local maximum at about $x \approx 3.5\,\text{m}$, and this location corresponds to $V = 0$, as expected.

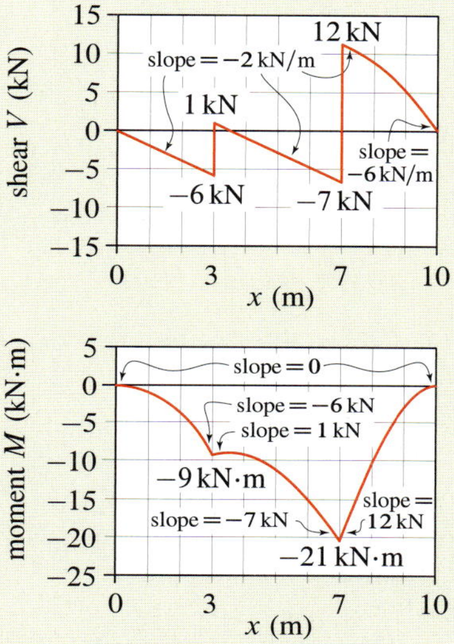

Figure 3
Shear and moment diagrams.

PROBLEMS

General instructions Use the integration approach for the following problems. Shear and moment diagrams should show the maximum values of the shear and moment and the locations where these occur.

Problems 8.56 through 8.61

Determine the shear and moment as functions of position, and draw the shear and moment diagrams.

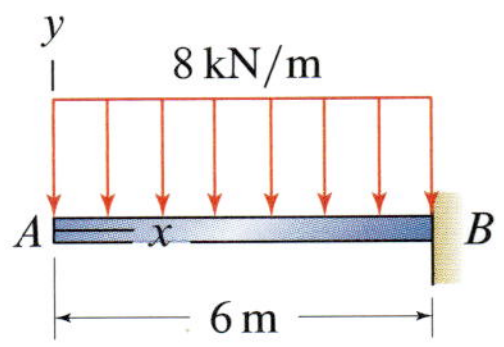

Figure P8.56

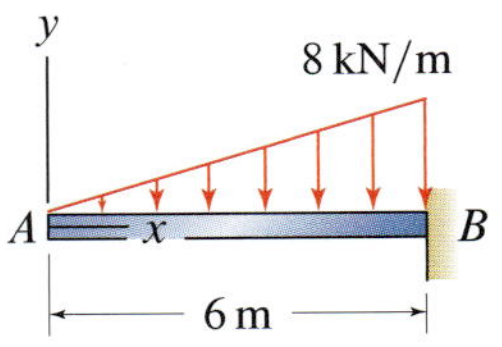

Figure P8.57

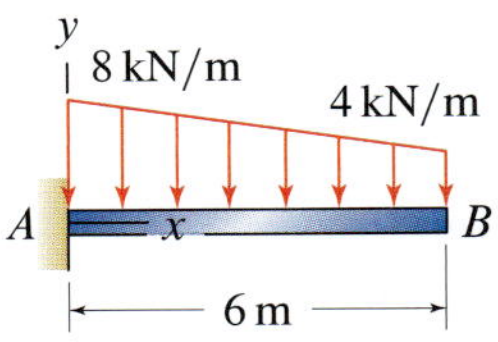

Figure P8.58

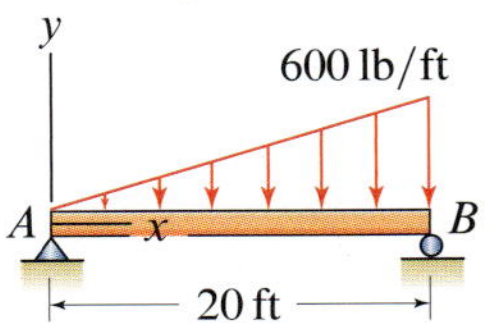

Figure P8.59

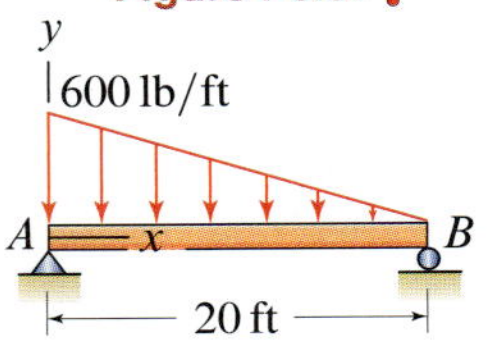

Figure P8.60

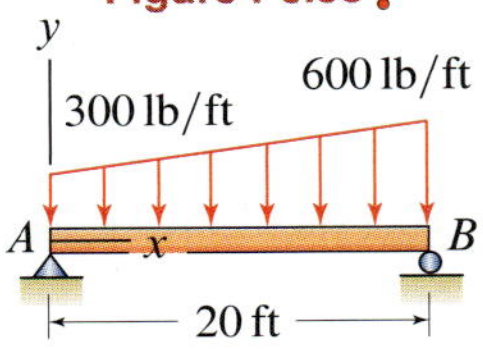

Figure P8.61

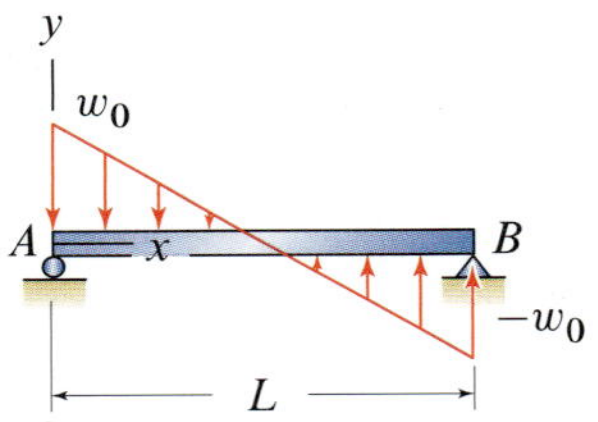

Figure P8.62

Problem 8.62

A simply supported beam has a distributed force that varies linearly from w_0 at the left-hand end to $-w_0$ at the right-hand end. Determine the shear and moment as functions of position, and draw the shear and moment diagrams.

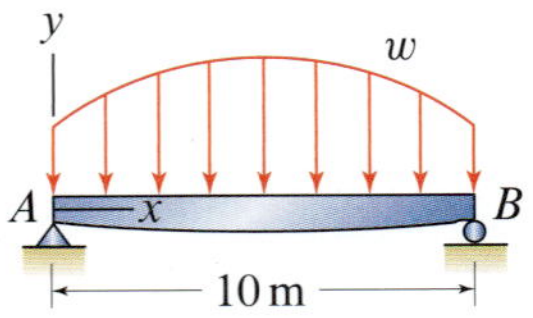

Figure P8.63

Problem 8.63

The beam is shaped so that its cross section is deeper near midspan than near the ends. As a consequence, its weight distribution is $w = (0.8\,\text{kN/m})[1 + \sin(\pi x/10\,\text{m})]$. Determine the shear and moment as functions of position due to the beam's weight distribution, and draw the shear and moment diagrams.

Problems 8.64 through 8.66

For the beam and loading shown, determine the shear and moment as functions of position, and draw the shear and moment diagrams.

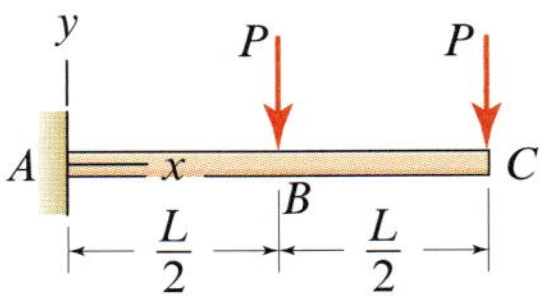

Figure P8.64

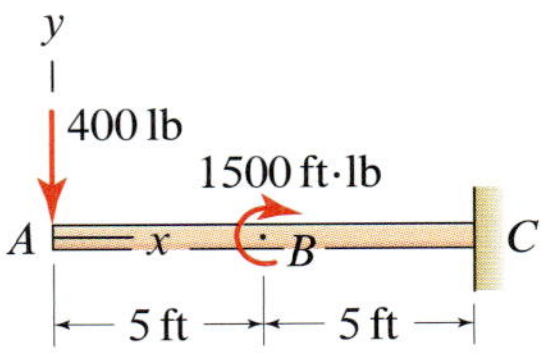

Figure P8.65

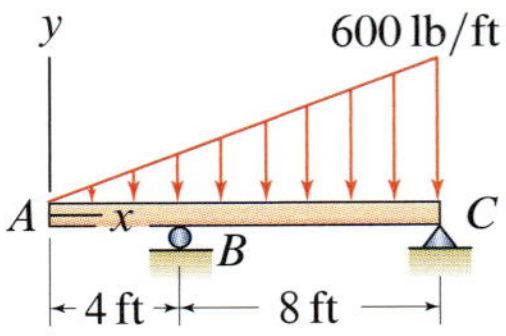

Figure P8.66

Problem 8.67

Beam *ABCD* is used to support a machine tool. The beam weighs 60 lb/ft of length, and the machine weighs 1200 lb with center of gravity at point *E*. Assuming the machine applies only vertical forces to the beam at points *B* and *C*, determine the shear and moment within beam *ABCD* as functions of position, and draw the shear and moment diagrams.

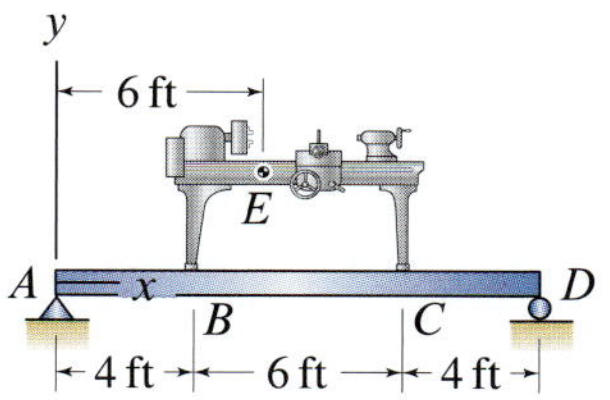

Figure P8.67

Problem 8.68

Beam *ABCD* is used to support an automobile so that it may be serviced. The beam weighs 1 kN/m of length, and the automobile weighs 10 kN with center of gravity at point *E*. Assuming the automobile's tires apply only vertical forces to the beam at points *B* and *C*, determine the shear and moment as functions of position, and draw the shear and moment diagrams.

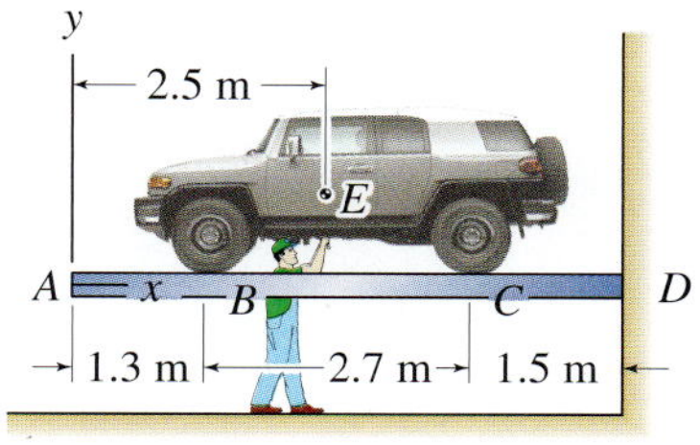

Figure P8.68

Problems 8.69 through 8.73

Draw the shear and moment diagrams for the beam and loading shown. Determination of the shear and moment as functions of position is not required.

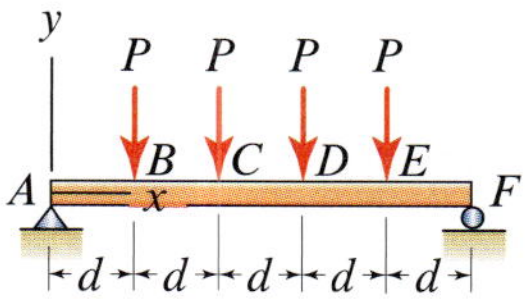

Figure P8.69

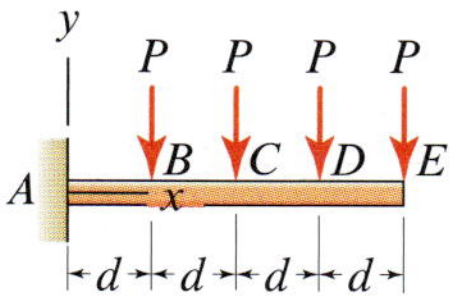

Figure P8.70

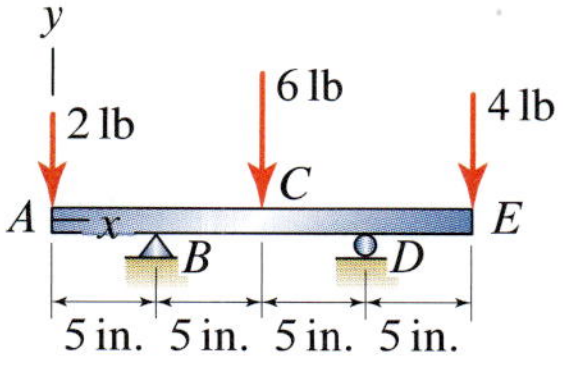

Figure P8.71

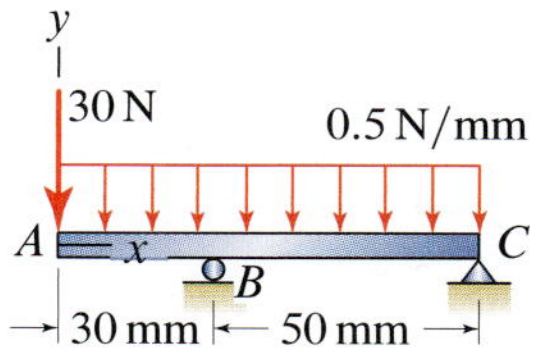

Figure P8.72

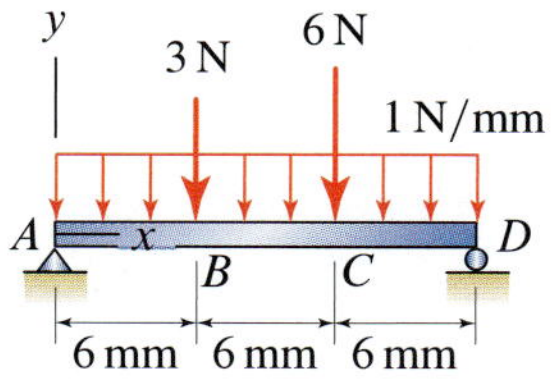

Figure P8.73

Problem 8.74

Draw the shear and moment diagrams for the bookshelf shown in Fig. 1 of Example 7.11 on p. 439. Determination of the shear and moment as functions of position is not required.

Problem 8.75

Three beams are shown, along with four possible shear diagrams and four possible moment diagrams. All forces and moments act in the directions shown. Without calculation, complete the table provided by selecting the appropriate shear and moment diagrams that correspond to each beam. Your answers may use a shear and/or moment diagram more than once. As an example, the answers for Beam 1 are provided in the table.

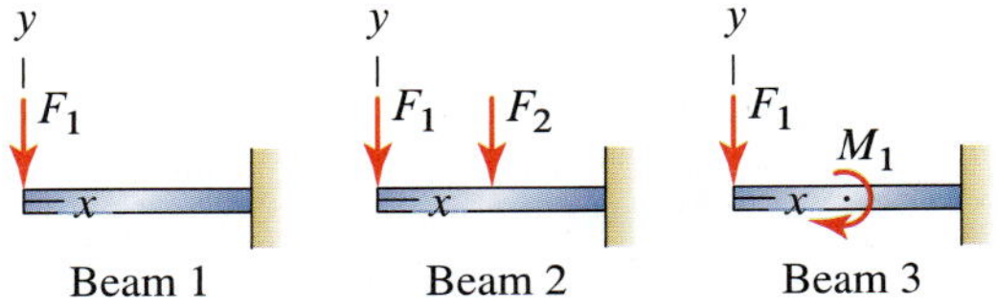

Beam	Shear diagram	Moment diagram
1	c	h
2		
3		

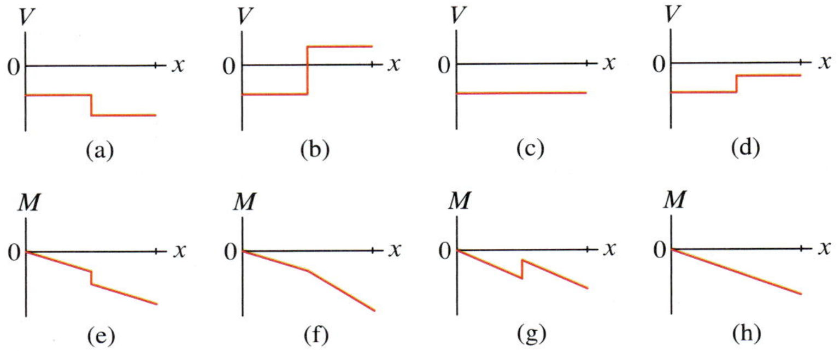

Figure P8.75

Problem 8.76

Three beams are shown, along with four possible shear diagrams and four possible moment diagrams. The distributed force acts in the direction shown. Without calculation, complete the table provided by selecting the appropriate shear and moment diagrams that correspond to each beam. Your answers may use a shear and/or moment diagram more than once.

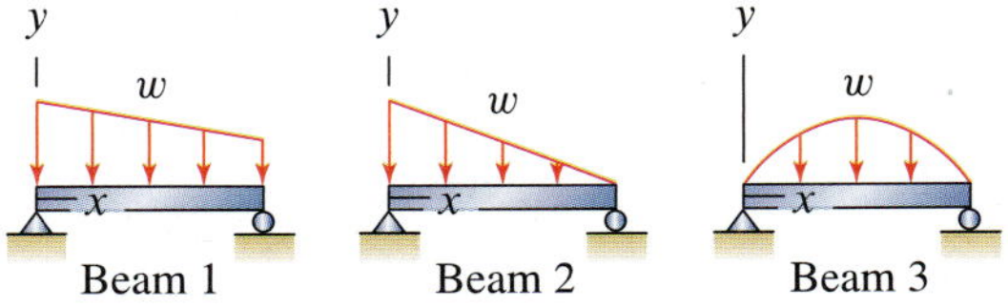

Beam	Shear diagram	Moment diagram
1		
2		
3		

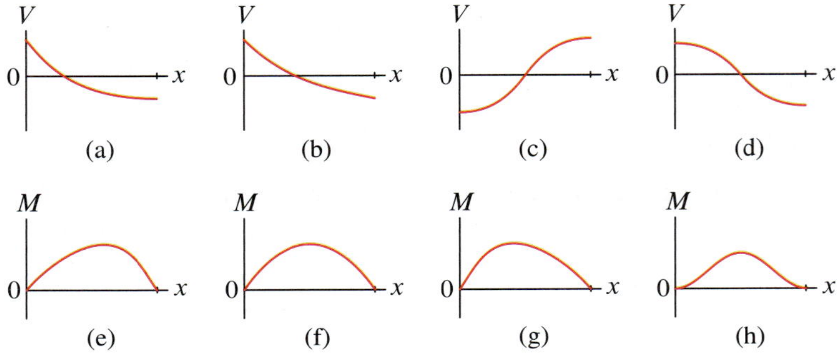

Figure P8.76

8.4 Chapter Review

Important definitions, concepts, and equations of this chapter are summarized. For equations and/or concepts that are not clear, you should refer to the original equation and page numbers cited for additional details.

Internal forces. *Internal forces* are forces and moments that develop within structural members and/or materials due to the external forces that are applied. A structural member is said to be *slender* if the dimensions of its cross section are small compared to its length. The methods of analysis discussed in this chapter are appropriate and effective for slender members. In two dimensions, the internal forces consist of two forces and one moment, as shown in Fig. 8.12, where N is called the *normal force* or *axial force*, V is called the *shear force*, and M is called the *bending moment*. In three dimensions, the internal forces consist of three forces and three moments, as shown in Fig. 8.13, where N is the *normal force* or *axial force*, V_y and V_z are *shear forces*, M_y and M_z are *bending moments*, and M_x is called the *torque*. When we find the internal forces in straight beams in two dimensions, it is necessary to follow a consistent sign convention as shown in Fig. 8.12.

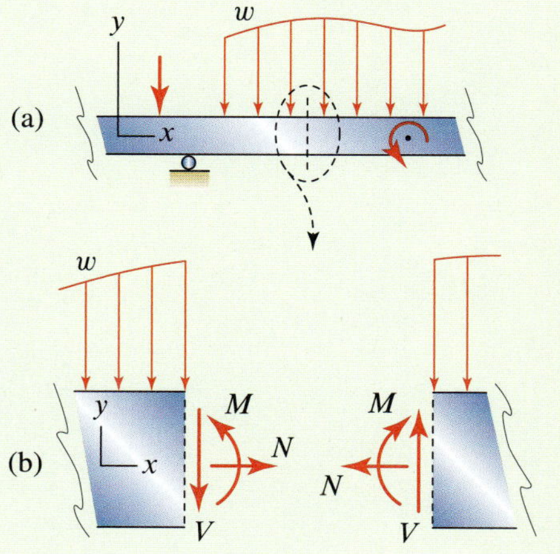

Figure 8.12
Internal forces that develop on a particular cross section of a slender member in two dimensions.

Methods for determining internal forces. Two methods for determining internal forces were discussed in this chapter: the *equilibrium approach* (used in Sections 8.1 and 8.2) and the *integration approach* (used in Section 8.3). In the equilibrium approach, cuts are taken as needed, FBDs are drawn, and equilibrium equations are written to obtain the internal forces either at specific locations in the structure or as functions of position. In the integration approach, differential equations (as summarized below) are solved to obtain the internal forces as functions of position. *Shear and moment diagrams* are plots of the shear and moment as functions of position.

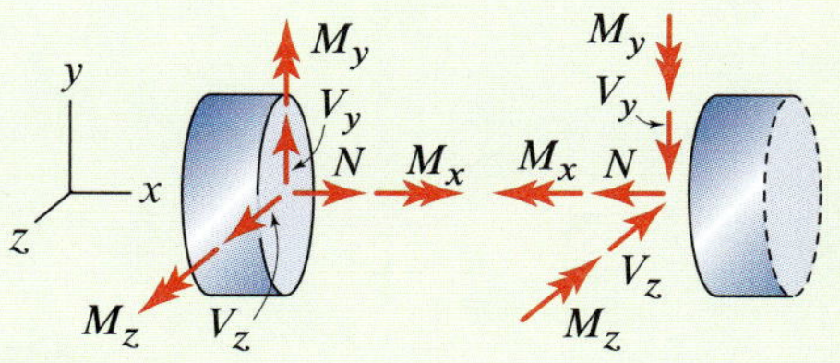

Figure 8.13
Internal forces that develop on a particular cross section of a slender member in three dimensions.

Relations among V, M, and w. By drawing an FBD of a small portion of a straight beam (as shown in Fig. 8.6 on p. 478) using the sign convention shown in Fig. 8.6, the shear, moment, and distributed force are related by

Eqs. (8.3) and (8.6), p. 478

$$\frac{dV}{dx} = -w \quad \text{and} \quad \frac{dM}{dx} = V.$$

In words, Eq. (8.3) says, "The change in shear divided by the change in length of the beam at position x is equal to the negative of the distributed force at that location," and Eq. (8.6) says, "The change in moment divided by the change in length of the beam at position x is equal to the shear at that location."

Determination of V and M using integration. Equations (8.3) and (8.6) are differential equations that can be rearranged and integrated to obtain

Eqs. (8.9) and (8.11), p. 479

$$V = V_P - \int_{x_P}^{x} w\,dx \quad \text{and} \quad M = M_P + \int_{x_P}^{x} V\,dx.$$

If the distributed force w is known as a function of position, then by integration using Eq. (8.9), the shear as a function of position is obtained. Once the shear is known, then by integration using Eq. (8.11), the moment as a function of position is obtained.

Useful forms of Eqs. (8.9) and (8.11), especially for purposes of drawing shear and moment diagrams, are

Eq. (8.12), p. 480

$$V_Q = V_P - \underbrace{\int_{x_P}^{x_Q} w\,dx}_{\substack{\text{area under } w \text{ vs.} \\ x \text{ plot between} \\ x_P \text{ and } x_Q}} \quad \text{and} \quad M_Q = M_P + \underbrace{\int_{x_P}^{x_Q} V\,dx}_{\substack{\text{area under } V \text{ vs.} \\ x \text{ plot between} \\ x_P \text{ and } x_Q}}.$$

Design considerations. Beams are usually designed to satisfy a variety of criteria, such as acceptable strength, acceptable deformations, acceptable fatigue life, low cost, manufacturability, and so on.

REVIEW PROBLEMS

Problem 8.77

A beam is supported by a roller at A and a device at C that allows vertical motion of the beam while preventing horizontal motion and rotation. Use the equilibrium approach to determine the shear and moment as functions of position and draw the shear and moment diagrams.

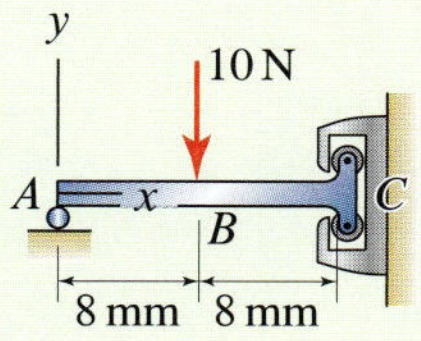

Figure P8.77

Problems 8.78 and 8.79

Members ABC and CDE are pinned to one another at C and have pin supports at A and E. Determine the internal forces acting on:

Problem 8.78 Cross sections F and G, which are located immediately to the right of A and the left of B, respectively.

Problem 8.79 Cross sections H and I, which are located immediately to the right of B and the left of C, respectively.

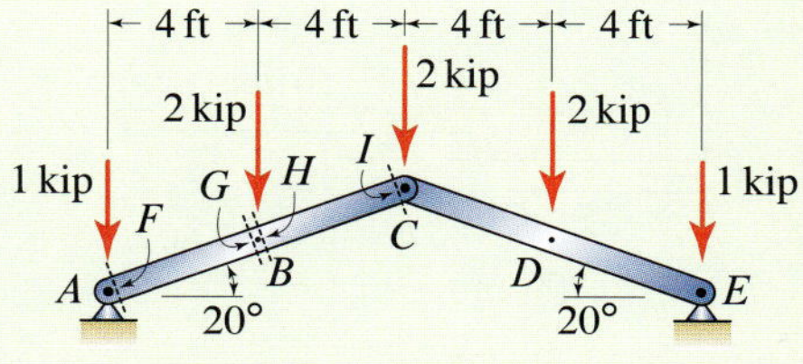

Figure P8.78 and P8.79

Problem 8.80

To provide generous leg room, a quarter-circular member AB is used to support a wall-mounted desk CD. Neglecting the weight of the members, determine the internal forces acting on cross sections E, F, and G, which are located immediately above point A, at the midpoint of member AB, and immediately to the right of point B, respectively.

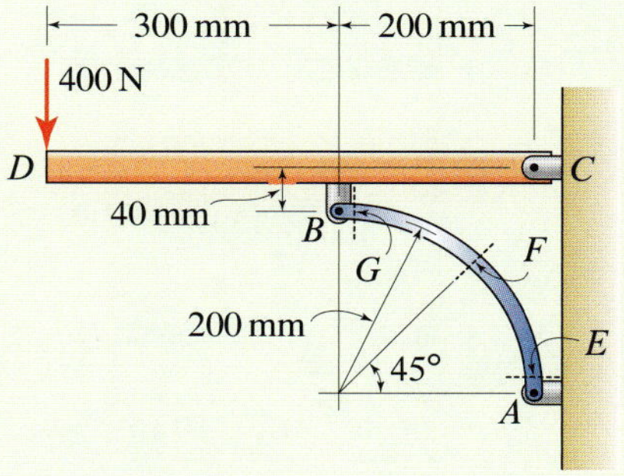

Figure P8.80

Problem 8.81

A historically important shipwreck is to be recovered, and a number of fragile wooden beams must be lifted. Specify the dimension d (in terms of length L) where the two lifting slings should be placed so that the maximum absolute value of the moment is as small as possible. Assume the beams are straight with uniform weight distribution, and the cables attached to the slings are vertical. *Hint:* The optimal value of d will give the same absolute value of the moment at the slings as at the midpoint of the beam.

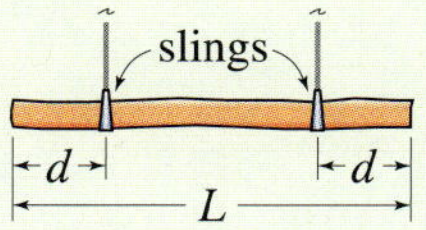

Figure P8.81

Problems 8.82 and 8.83

Use the equilibrium approach to determine the shear and moment as functions of position and draw the shear and moment diagrams.

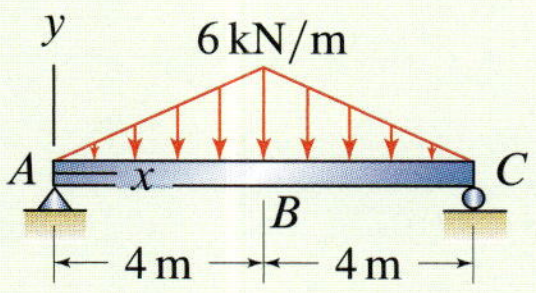

Figure P8.82

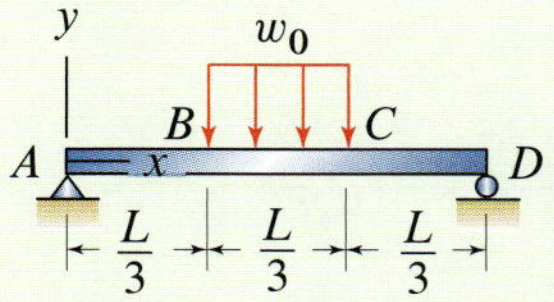

Figure P8.83

Problem 8.84

Repeat Prob. 8.82 using the integration approach.

Problem 8.85

Repeat Prob. 8.83 using the integration approach.

Problems 8.86 through 8.88

The device shown is used in a factory to support a tool that applies a 60 N vertical force at D. It has the feature that portion CD can slide to the right when the tool is needed, and can slide to the left when the tool is to be stored, such that $50\,\text{mm} \le d \le 200\,\text{mm}$. Determine the shear and moment in beam AB as functions of position and draw the shear and moment diagrams. Neglect the size of the rollers at B and C.

Problem 8.86 $d = 200\,\text{mm}$.

Problem 8.87 $d = 100\,\text{mm}$.

Problem 8.88 $d = 50\,\text{mm}$.

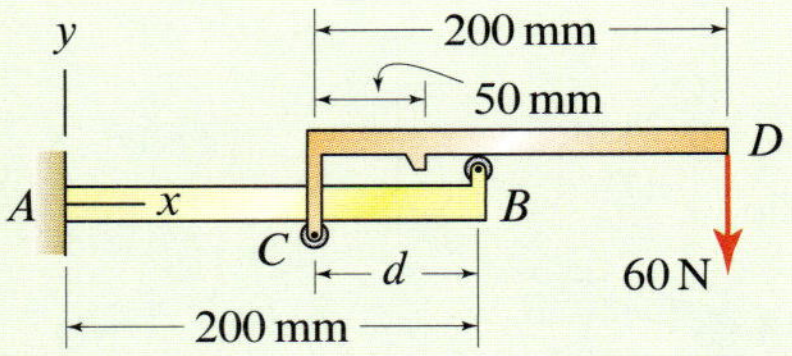

Figure P8.86–P8.88

Problem 8.89

(a) For the cantilever beam shown in Fig. P8.89(a), determine the shear and moment as functions of position. Express your answers in terms of parameters such as P, a, and b.

(b) For the cantilever beam with the three forces shown in Fig. P8.89(b), use superposition of the results of Part (a) to determine shear and moment as functions of position and draw the shear and moment diagrams.

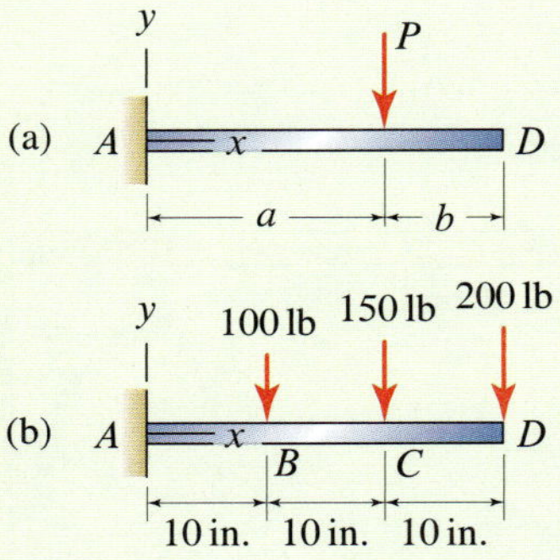

Figure P8.89

Problems 8.90 and 8.91

Draw the shear and moment diagrams for the beam and loading shown. Determination of the shear and moment as functions of position is not required.

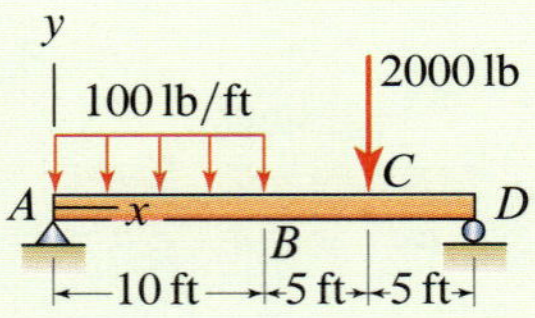

Figure P8.90

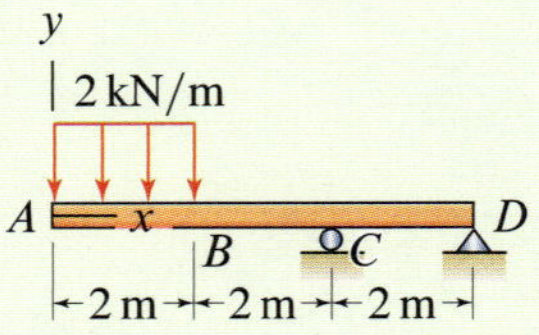

Figure P8.91

Problem 8.92

(a) Use the equilibrium approach to determine the shear as a function of position in the region $0 \le x \le 3\,\text{m}$.

(b) Use the integration approach with the results of Part (a) to determine the moment in the region $0 \le x \le 3\,\text{m}$.

(c) Values for the shear and moment just to right of B and at C are shown in the shear and moment diagrams provided. By inspection, complete these diagrams for the region $3\,\text{m} \le x \le 6\,\text{m}$. Accurately draw the shapes of the curves and label the slopes at $x = 3\,\text{m}$ and $x = 6\,\text{m}$.

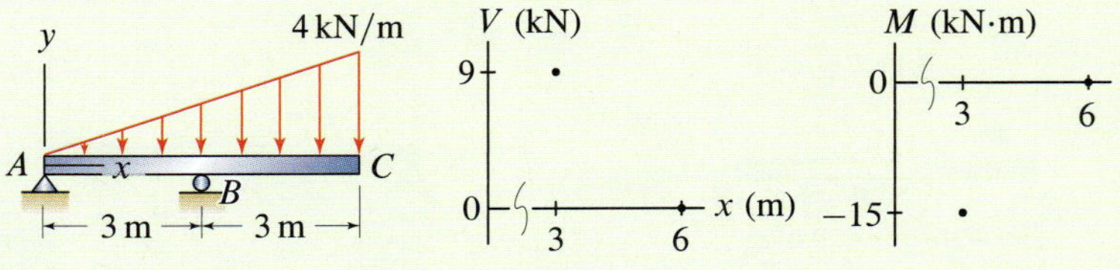

Figure P8.92

Problem 8.93

(a) Use the equilibrium approach to determine the shear as a function of position in the region 6 ft $\le x \le$ 12 ft.

(b) Use the integration approach with the results of Part (a) to determine the moment in the region 6 ft $\le x \le$ 12 ft.

(c) Values for the shear and moment at A and just to the left of B are shown in the shear and moment diagrams provided. By inspection, complete these diagrams for the region $0 \le x \le$ 6 ft. Accurately draw the shapes of the curves and label the slopes at $x = 0$ and $x =$ 6 ft.

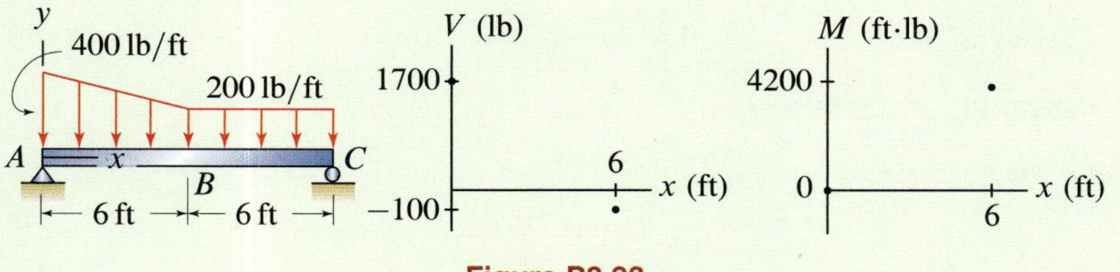

Figure P8.93

Problem 8.94

(a) Use the equilibrium approach to determine the moment as a function of position in the region $0 \le x \le$ 6 ft.

(b) Use the integration approach to determine the shear in the region 6 ft $\le x \le$ 12 ft.

(c) Values for the shear and moment just to right of B and at C are shown in the shear and moment diagrams provided. By inspection, complete these diagrams for the region 6 ft $\le x \le$ 12 ft. Accurately draw the shapes of the curves and label the slopes at $x =$ 6 ft and $x =$ 12 ft.

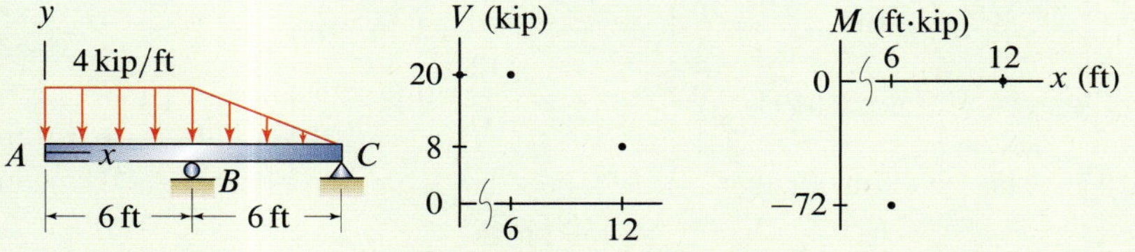

Figure P8.94

Problem 8.95

(a) Use the equilibrium approach to determine the moment as a function of position in the region $0 \le x \le 3\,\mathrm{m}$.

(b) Use the integration approach to determine the shear in the region $6\,\mathrm{m} \le x \le 9\,\mathrm{m}$.

(c) Values for the shear and moment just to the right of B and left of C are shown in the shear and moment diagrams provided. By inspection, complete these diagrams for the region $3\,\mathrm{m} \le x \le 6\,\mathrm{m}$. Accurately draw the shapes of the curves and label the slopes at $x = 3\,\mathrm{m}$ and $x = 6\,\mathrm{m}$.

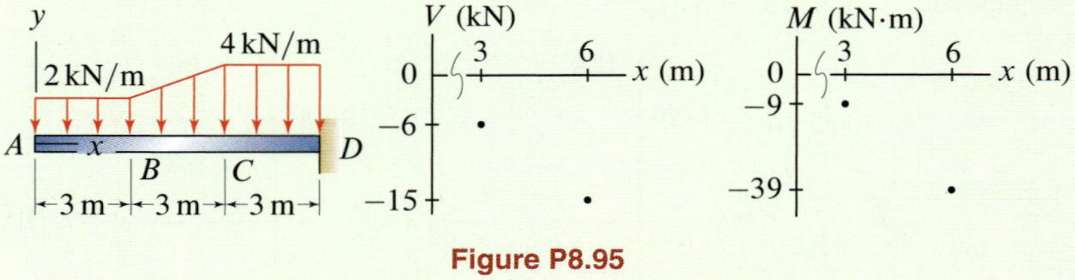

Figure P8.95

Problem 8.96

(a) Use the equilibrium approach to determine the moment as a function of position in the region $6\,\mathrm{m} \le x \le 12\,\mathrm{m}$.

(b) Use the integration approach to determine the shear in the region $0 \le x \le 6\,\mathrm{m}$.

(c) Values for the shear and moment at A, B, and C are shown in the shear and moment diagrams provided (the shear just to the left and right of B is 4 and -11 kN, respectively). By inspection, complete these diagrams. Accurately draw the shapes of the curves, and label the slopes at $x = 0$, $x = 6\,\mathrm{m}$, and $x = 12\,\mathrm{m}$.

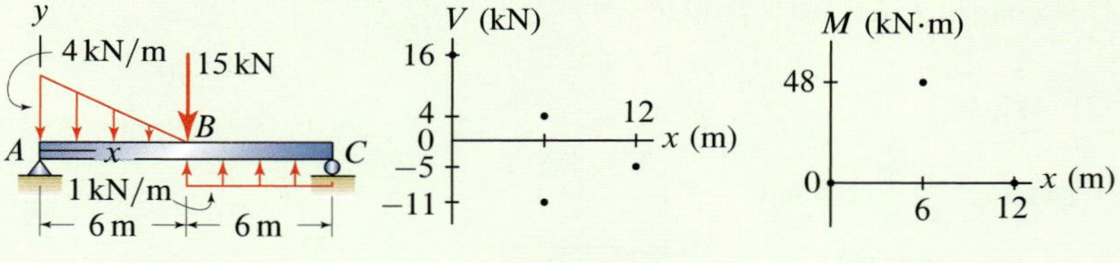

Figure P8.96

9 Friction

When two objects are in contact, friction forces generally develop between them. This chapter presents models for quantifying friction forces and discusses methods for analyzing problems with friction and sliding.

9.1 Basic Concepts

When two objects are in contact, there is often a tendency for them to slide relative to one another. When this is the case, *friction forces* that resist the sliding motion develop on the contact surfaces between the objects. A number of factors influence how large the friction forces can be, and the field of study that addresses this topic is called *tribology*, which is derived from the Greek word *tribos*, which means *rubbing*.

A brief history of tribology

Leonardo da Vinci (1452–1519) used a scientific approach in his studies of friction and recognized that friction force and normal force were proportional. In 1699, Guillaume Amontons (1663–1705), a French architect turned engineer, presented a paper to the French Academy where he reported on friction tests using various combinations of iron, copper, lead, and wood, lubricated with pork fat (suet). It is rather remarkable that he found the friction force F during sliding and the normal force N were related by $F \approx N/3$, and very

Interesting Fact

Contact surfaces. "Putting two solids together is rather like turning Switzerland upside down and standing it on Austria—the area of intimate contact will be small." These words were used by the pioneering tribologist F. P. Bowden in a BBC broadcast in 1950. Virtually all contact surfaces, despite any apparent degree of smoothness, are inherently rough at small length scales.

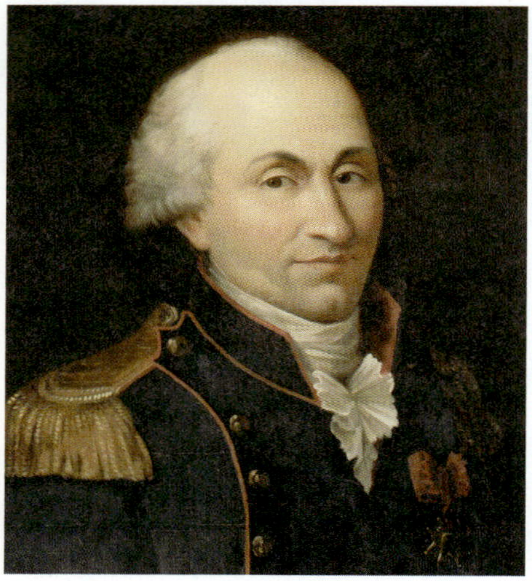

Figure 9.1
A portrait of Coulomb painted by Hippolyte Lecomte.

importantly, that the friction force was independent of the apparent area of contact. In 1781, Charles Augustin de Coulomb (1736–1806), a French physicist and engineer who also did ground-breaking work in electricity and magnetism, confirmed the findings of Amontons and distinguished between static and kinetic friction. While Coulomb's work is notable, he did not improve substantially on the findings of Amontons. Nonetheless, the friction model discussed in this chapter is usually called *Coulomb's law*, although it is occasionally referred to as *Amontons' law* or the *Amontons-Coulomb law*.*

A simple experiment

The basic features of friction between contacting bodies, and a model to quantify this phenomenon, can be developed by considering the example shown in Fig. 9.2(a), where an empty coffee cup with weight W rests on a table. Imagine using your fingers to apply a slowly increasing horizontal force P to the cup, starting from a value of zero, so as to slide the cup to the right. Between the cup and table, in addition to the normal force N, a friction force F develops, as shown in the FBD in Fig. 9.2(b). The relationship between P and the friction force F that you are likely to observe is shown in Fig. 9.2(c). In region AB of this figure, the cup undergoes no motion (this is often called *stick*), hence static equilibrium prevails and by writing $\sum F_x = 0$ we observe that $P = F$. At point B, we say that motion is *impending*, meaning that slip is about to occur. For values of P beyond point B, the interface between the cup and table is not capable of supporting friction forces that are high enough to provide equilibrium. In this regime, $P > F$, sliding occurs, the cup accelerates, and Newton's law $\sum F_x = ma_x$ must be used where the acceleration in the x direction is nonzero. If we repeat this experiment using a cup filled with enough coffee to double the weight of the original empty cup, we will find that the force at which sliding starts is approximately doubled.

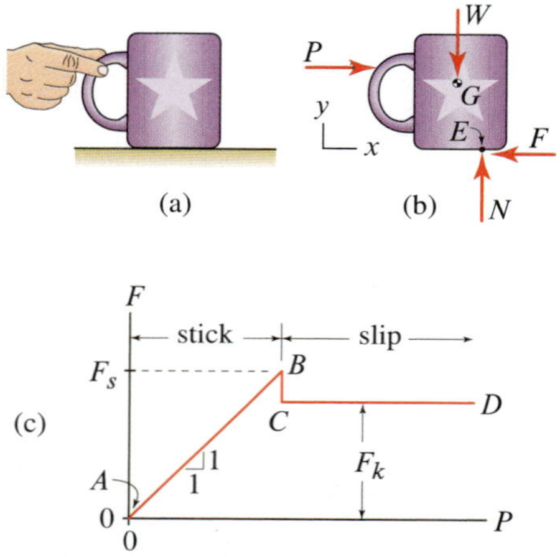

Figure 9.2
(a) An empty coffee cup with weight W and center of gravity at point G subjected to a horizontal force P. (b) Free body diagram. (c) Relationship between P and the friction force F that is likely to be observed. Notice that point E, the location where the contact forces N and F act, is generally not directly below G; writing $\sum M = 0$ will determine where this point is located.

Typically, and as shown in Fig. 9.2(c), the friction force F_s at which sliding starts is somewhat higher than the friction force F_k for sustained sliding, and these values are called the *static friction force* and the *kinetic friction force*. Figure 9.2(c) shows an instantaneous decrease of the friction force from F_s to F_k; in reality this decrease is rapid but is not instantaneous.

Figure 9.3 offers a simple theory for why $F_s > F_k$. Contact surfaces are inherently rough, and actual contact occurs at a relatively small number of prominent asperities, two of which are shown in Fig. 9.3, where the x direction is oriented along the mean plane of the interface. Before sliding, the force supported by the two asperities shown is $\vec{R}_s$ and $-\vec{R}_s$, and shortly after sliding starts, these forces change to $\vec{R}_k$ and $-\vec{R}_k$. Notice that after a small amount of sliding, the asperities of one surface have slid *up and over* those of the other surface. Hence, the x component of $\vec{R}_k$ relative to its y component is *lower* than that just before sliding. In essence, this model presumes that prior to sliding the asperities of one surface rest in the troughs of the other surface, and that during gross sliding the asperities are more likely to make contact closer to their summits. Reasons why the asperities of one surface settle into the troughs of the other surface before sliding include microvibrations and viscous deformation (i.e., time-dependent flow) of the asperities.

* Additional historical perspectives, as well as an excellent study of the field of tribology, are given by K. C. Ludema, *Friction, Wear and Lubrication*, CRC Press, Boca Raton, 1996.

Coulomb's law of friction

A model that describes the relationship between the normal force and friction force between two contacting surfaces is called *Coulomb's law*, as follows

$$|F| \le \mu_s N \quad \text{before sliding } (<) \text{ and for impending motion } (=), \tag{9.1}$$

$$|F| = \mu_k N \quad \text{after sliding begins}, \tag{9.2}$$

where

N is the normal force between the two surfaces, defined to be positive in compression, hence $N \ge 0$ always;
F is the friction force between the two surfaces, with direction that *always* opposes relative motion between the surfaces;
μ_s (Greek letter *mu*) is called the *coefficient of static friction*;
μ_k is called the *coefficient of kinetic friction*.

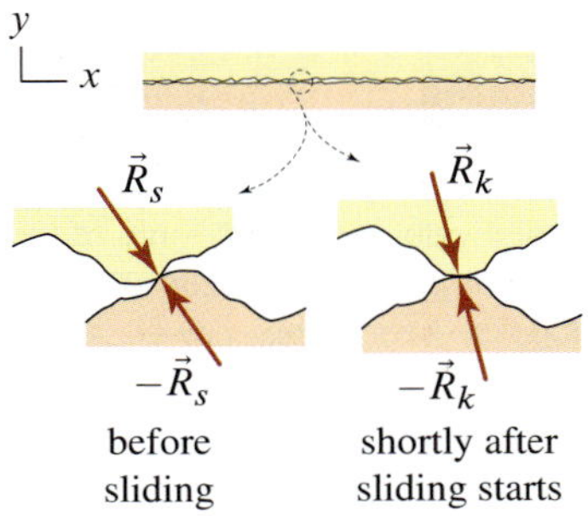

Figure 9.3
A model that explains why the static force of friction is larger than the kinetic force of friction.

Coulomb's law is independent of the apparent area of contact and the sliding velocity. Neither of these assumptions is precisely true, but for many applications they are reasonable. The absolute value of the friction force is used in Eqs. (9.1) and (9.2) for the following reason. Imagine when drawing the FBD in Fig. 9.2(b) we took F to be positive in the opposite direction shown. Presuming the cup is in equilibrium as we push on it, writing $\sum F_x = 0$ gives $F < 0$, which is perfectly fine. As P increases, F also increases in a negative sense, until its absolute value reaches $\mu_s N$ at which time motion is impending. In statics and many other subjects that follow, we draw FBDs so that by inspection or careful thought, the friction forces are drawn in proper directions to oppose sliding; when this is the case, all friction forces will have positive values and we will therefore omit the absolute value sign in Coulomb's law.

We will usually write Coulomb's law given in Eqs. (9.1) and (9.2) in the compact form

$$|F| \le \mu N, \tag{9.3}$$

where it is understood that the coefficient of static friction is used prior to sliding and when motion is impending, and the coefficient of kinetic friction is used once motion starts along with the = sign. Because of the lack of reliable values of coefficients of friction, as discussed below, we often assume $\mu_s = \mu_k$. Indeed, even very sophisticated engineering analyses frequently use this assumption. As discussed above, the absolute value sign in Eq. (9.3) will be omitted if friction forces in FBDs have the proper directions to resist sliding motion.

Concept Alert

Coulomb's law of friction. Coulomb's law of friction consists of two items:

- Equation (9.3), which states $|F| \le \mu N$, and
- The direction of friction forces *always* opposes the direction of relative motion between contacting surfaces.

The latter of these items will be accounted for when drawing FBDs, and in complicated problems this may require some careful thought.

Coefficients of friction

The friction force between two contacting surfaces is controlled by an enormous number of factors including compressive force, the combination of contacting materials, surface roughness, the processing used to create the surfaces, lubricants, the environments the surfaces are in and have previously been in, sliding history, chemistry, temperature, humidity, and many others. Consider the example of steel-on-steel contact, such as occurs between the teeth of gears in a machine. In reality, the contact is not between pure steel and pure steel.

Helpful Information

Static equilibrium. As we progress through this chapter, we should be especially mindful of the constraints underlying our work: static equilibrium requires the acceleration of a body, which is a vector, to be zero. Thus, our analyses are restricted to problems in which a body is at rest (i.e., there is no motion), or to problems where sliding is at constant speed *and* in a uniform direction with no rotation of the body.

Rather, each steel surface has a very thin layer of material containing absorbed chemicals from the manufacturing process (e.g., cutting lubricants and polishing compounds), oil that might lubricate the gears, and oxidation materials from simply being present in a natural air environment. In very large part, it is the properties of these thin surface layers that control friction. This is generally fortuitous, as the friction between atomically clean surfaces is very high, and the thin surface layers described here substantially reduce friction. In view of this, it is rather remarkable that Eq. (9.3) characterizes friction as well as it does. Nonetheless, Eq. (9.3) is *empirical*, meaning it mimics nature, but it does not describe nature. The important message in this discussion is that the coefficient of friction is not an inherent property of a material or a combination of contacting materials. The primary mission of the field of tribology is to establish how the coefficient of friction depends on various parameters, and to develop laws for friction that are more comprehensive and accurate than Coulomb's law.

Helpful Information

What value of μ should I use? Tables of general values of coefficients of friction, such as reported in Table 9.1 and in handbooks, should be viewed as rough guidelines. When a substantial investment is to be made in a design or product, when reliability must be high, or when performance depends strongly on friction, more precise knowledge of the coefficient of friction for your application should be obtained by testing. When possible, prototype devices under actual service conditions should be tested, and if this is not possible, laboratory tests with simpler geometry specimens under simulated service conditions should be performed.

Table 9.1 reports coefficients of kinetic friction (unless otherwise indicated) for some combinations of contacting materials. Notice that the range of values is large, reflecting the large number of dependencies on different variables. Values outside these ranges occur under some circumstances. Coefficients of static friction are generally equal to or larger than the coefficients of kinetic friction. In addition to the factors cited earlier, the coefficient of static friction is affected by the time of rest, and under some circumstances it may be 20–30% higher than the coefficient of kinetic friction. The coefficient of friction for alloys is generally lower than the coefficients of friction for the constituent materials (e.g., compare mild steel on self with iron on self).

Table 9.1. Coefficients of friction for various combinations of contacting materials. Unless otherwise stated, values reported are coefficients of kinetic friction for clean and unlubricated surfaces in a normal air environment at 20°C.

	Coefficient of friction[a]	
Material	On self	On mild steel
aluminum	0.8–1.2	0.5–0.6
iron	0.8–1.5	0.8–1.5
leaded bronze	–	0.2–0.4
gray cast iron	0.8–1.0	0.3–0.5
mild steel	0.7–0.9	–
rock (μ_s)	0.4–0.8	–
diamond	0.1–0.2	–
hard surface with liquid lubricant	0.02–0.05	–
hard surface with solid lubricant	0.05–0.15	–
tire rubber on pavement (μ_s)	0.4–1.4	
soft rubber on glass (μ_s)	4–10	

[a] Data taken from B. Bushan, *Introduction to Tribology,* John Wiley & Sons, New York, 2002; K. C. Ludema, *Friction, Wear and Lubrication,* CRC Press, Boca Raton, 1996; R. E. Goodman, *Introduction to Rock Mechanics*, John Wiley & Sons, New York, 1980.

Dry contact versus liquid lubrication

Coulomb's law is best suited for dry contact. When surfaces have a liquid lubricant, hydrodynamic effects occur where a portion (often sizable) of the normal force between two surfaces is supported by fluid pressure, and Coulomb's law cannot be used. Nonetheless, if a very small amount of liquid lubricant is present between two contacting surfaces, it is common to use Coulomb's law, where the effects of the lubricant are approximately accounted for by the coefficient of friction. The situation of very modest lubrication is often called *boundary lubrication*.

Angle of friction

In some engineering disciplines, such as geotechnical engineering, it is more common to characterize friction between surfaces by using the angle of friction rather than the coefficient of friction. The *angle of friction* is defined to be

$$\phi = \tan^{-1} \mu, \tag{9.4}$$

where μ is either the static or kinetic coefficient of friction, and the value of ϕ is the angle of static or kinetic friction, respectively. For example, if a particular combination of contacting materials is known to have $\mu_s = 0.35$, then from Eq. (9.4), its angle of static friction is $\phi_s = 19.3°$. The angle of friction has a straightforward physical interpretation as described in Fig. 9.4. Problem 9.6 describes another interpretation of the angle of friction, including a method for experimental determination.

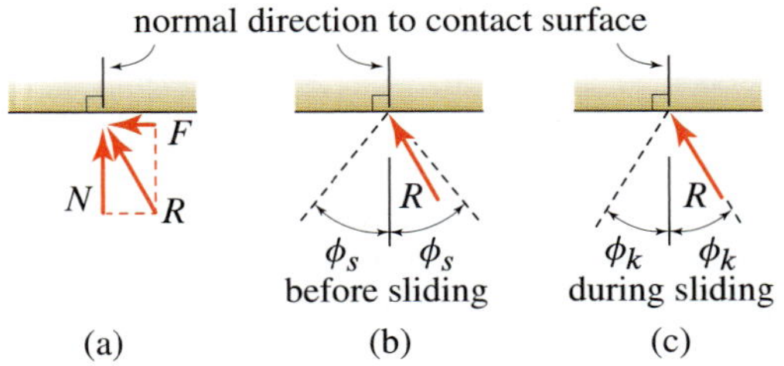

Figure 9.4. Physical interpretation of the angle of friction. (a) R is the resultant of the normal force N and friction force F acting on a surface. (b) The angle of static friction measured from the normal direction of the surface defines a cone. For an initially nonsliding interface, no sliding occurs if the resultant force lies within the cone (as shown), and impending slip occurs when the resultant force lies on the cone. (c) During sliding, the resultant force lies on the cone defined by the angle of kinetic friction.

Problems with multiple contact surfaces

In this section and the next, we consider problems having multiple contact surfaces. However, in this section we restrict our attention to the more straightforward class of problems where if one surface slides, then all surfaces must slide simultaneously. In Section 9.2, we consider more complex problems where only a subset of the surfaces slide. The category a particular problem belongs to will be straightforward to determine by considering the geometry of motion that is possible for that problem. In more precise nomenclature, the subject of geometry of motion is a called *kinematics*, and we will refer to it as such. Kinematics is discussed in detail in dynamics. For our work here, only the most elementary kinematics concepts are introduced as needed.

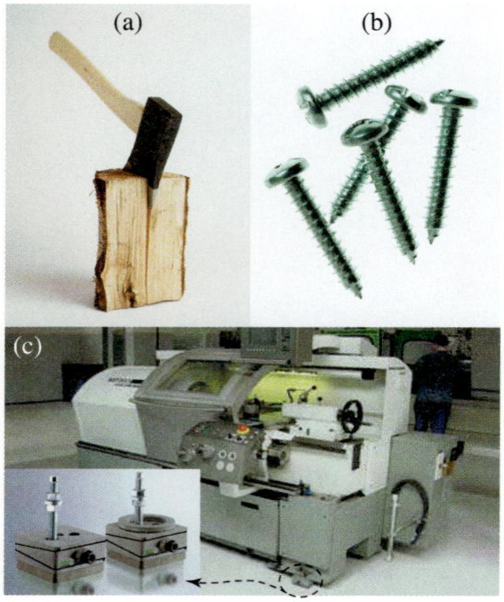

Figure 9.5
Examples of wedges. (a) An axe used for splitting firewood is a wedge. (b) A screw is a wedge wrapped in the shape of a helix. (c) A Gildemeister NEF 330 metal cutting lathe is leveled using Bilz precision leveling wedges.

Wedges

A *wedge* is a device that is useful for producing potentially large forces in the direction transverse to the faces of the wedge. Figure 9.5 shows some examples of wedges, and Examples 9.3 and 9.5 illustrate methods of analysis. A design criterion for a wedge is often that it remain in position once the insertion force is removed. For example, consider a wedge used to level one corner of a refrigerator. After the wedge is inserted, it is obviously undesirable if it slips out of position of its own accord. The performance of a wedge is dependent on its geometry and the coefficients of friction for its contact surfaces.

Coulomb's law of friction in three dimensions

Coulomb's law of friction as given by Eq. (9.3) is applicable to friction in three dimensions provided the friction force F is defined to be the resultant of the two tangential forces that generally act on a surface. This model is called *isotropic friction*, because the coefficient of friction is the same for all possible sliding directions. Consider the example shown in Fig. 9.6 where a block with the contact surface lying in the xy plane is subjected to various forces and moments such as P_1, P_2, M_1, etc. In the FBD shown in Fig. 9.6, N is the normal force, F_x and F_y are the friction forces acting in the x and y directions, and F is the resultant friction force. Coulomb's law becomes

$$F = \sqrt{(F_x)^2 + (F_y)^2} \le \mu N, \tag{9.5}$$

where the absolute value sign is not needed since F is nonnegative.

Isotropic friction may not be adequate for some problems. For example, let one or both of the surfaces in Fig. 9.6 be finished by *grinding* where x is the direction of grinding.* Such a surface likely has a lower coefficient of friction for sliding in the x direction than for sliding in the y direction, and a more sophisticated version of Coulomb's law that allows for *anisotropic friction* is needed.

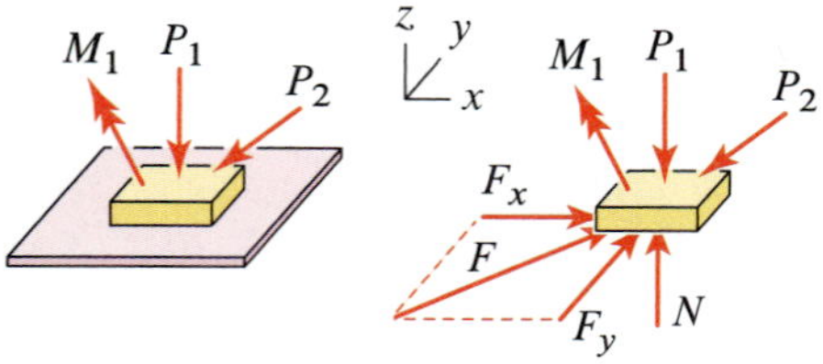

Figure 9.6
Coulomb's law of friction applied to *isotropic* friction in three dimensions.

Design considerations

The main difficulty in designs where friction is a key element is the poor predictability of frictional behavior. Coulomb's law has limited accuracy because it does not incorporate many of the phenomena that are known to influence friction, and tables of general values of coefficients of friction, such as found in handbooks, are only approximate. Even if testing is done to obtain a coefficient of friction for a particular application, over the life of a device, or even over the span of a duty cycle, the coefficient of friction may change. For example, sliding surfaces may become smoother *or* rougher due to wear, they may be affected by chemicals or wear debris, and so on. Details of how contact

* *Grinding* is a machining process used on metal and other surfaces to provide a very high-quality finish with precise tolerances.

surfaces are handled can produce unexpected changes in frictional behavior. For example, when people touch a contact surface, materials such as sweat and body oils are deposited, and this can significantly, and unexpectedly, alter friction. Designs that depend on precise frictional response are often problematic, while designs whose performance is insensitive to moderate changes in frictional behavior are preferable.

End of Section Summary

In this section, *Coulomb's law of friction* is discussed. Some of the key points are as follows:

- Coulomb's law consists of two items: the friction force F and normal force N are related by $|F| \leq \mu N$ *and* F must oppose the direction of sliding. As discussed in connection with Eqs. (9.1) and (9.2), the coefficient of static friction μ_s is used before sliding and for impending motion, and the coefficient of kinetic friction μ_k is used after sliding begins.

- We will usually draw FBDs so that the friction forces are in proper directions to resist sliding motion. When this is the case, all friction forces will have positive values and the absolute value sign in Coulomb's law will be omitted.

EXAMPLE 9.1 *Effects of Sliding Direction*

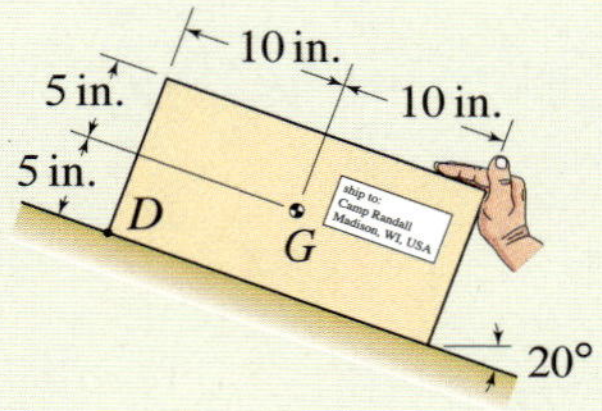

Figure 1

A steel ramp is used by a worker in a factory to move a cardboard box. If a box weighs 40 lb with center of gravity at point G, the force applied by the worker is parallel to the ramp, and the coefficient of friction between the box and ramp is $\mu_s = \mu_k = 0.3$, determine the force the worker must apply to slide the box:

(a) Up the ramp

(b) Down the ramp.

SOLUTION

Road Map We assume motion of the box is due to sliding and not tipping (Example 9.2 discusses tipping analysis and Prob. 9.5 verifies the assumption made here). We first consider the box sliding up the ramp. We draw an FBD, write equations of equilibrium, and apply Coulomb's law to obtain the solution. This procedure is then repeated for sliding down the ramp.

Part (a)

Modeling The FBD for the box is shown in Fig. 2. Very importantly, the direction of the friction force F opposes motion of the box up the ramp.

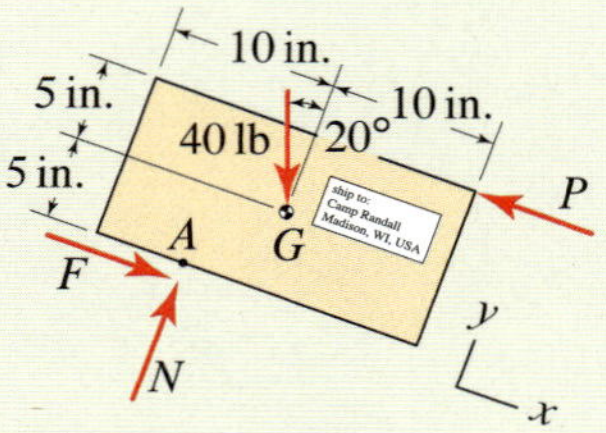

Figure 2
Free body diagram for the box sliding up the ramp. Point A is the location where the contact forces N and F act. Note that the actual forces between the box and ramp are distributed over the full contact area of the box, and N and F at point A are equivalent forces that represent these.

Governing Equations

Equilibrium Equations Using the FBD from Fig. 2, the equilibrium equations are

$$\sum F_y = 0: \qquad N - (40\,\text{lb})(\cos 20°) = 0 \quad \Rightarrow \quad N = 37.59\,\text{lb}, \tag{1}$$

$$\sum F_x = 0: \qquad F + (40\,\text{lb})(\sin 20°) - P = 0. \tag{2}$$

These two equations contain three unknowns, although we were able to solve the first of these for the normal force N.

Force Laws Assuming that sliding is occurring, Coulomb's law, Eq. (9.3) on p. 499, becomes

$$F = \mu N = (0.3)N, \tag{3}$$

where the direction of F is properly accounted for in the FBD of Fig. 2.

Computation Solving Eqs. (2) and (3) provides

$$\Rightarrow \quad \boxed{F = 11.3\,\text{lb} \quad \text{and} \quad P = 25.0\,\text{lb}.} \tag{4}$$

Part (b)

The solution procedure for sliding down the ramp is the same as that for Part (a) except that the direction of the friction force F in the FBD must be changed. The revised FBD is shown in Fig. 3. The governing equations are identical to Eqs. (1)–(3) except that F in Eq. (2) is replaced by $-F$. The solutions of these equations are

$$\Rightarrow \quad \boxed{N = 37.59\,\text{lb}, \qquad F = 11.3\,\text{lb}, \qquad \text{and} \qquad P = 2.40\,\text{lb}.} \tag{5}$$

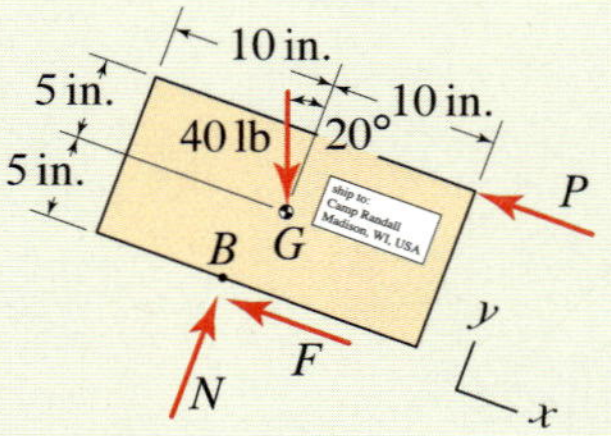

Figure 3
Free body diagram for the box sliding down the ramp. Notice that the direction of the friction force in this FBD is opposite that shown in Fig. 2 and the location B where the contact forces act is different than point A.

Discussion & Verification If $2.40\,\text{lb} \le P \le 25.0\,\text{lb}$, the box remains at rest (no sliding), and when equality prevails in this expression, motion is impending. It is possible to apply P greater than 25.0 lb, in which case the box slides up the ramp while accelerating. Similar remarks apply for sliding down the ramp.

EXAMPLE 9.2 *Sliding Motion versus Tipping Motion*

A 2 m-long concrete traffic barrier is used to provide safety for workers, pedestrians, and motorists during highway construction. If $h = 45$ cm, determine the force P that will cause motion of the barrier, and specify if the motion is due to sliding or tipping. The concrete has $0.024\,\text{N/cm}^3$ specific weight, and the interface between the barrier and pavement has static and kinetic coefficients of friction of 0.45 and 0.40, respectively.

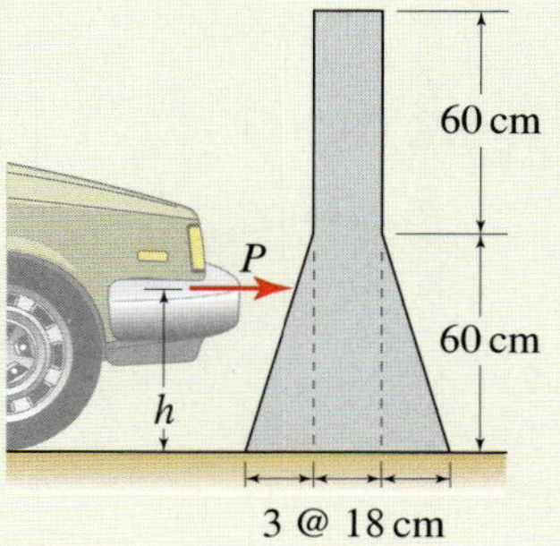

Figure 1

SOLUTION

Road Map We will assume the barrier is initially at rest when the force P is applied, and we will restrict our attention to sliding and tipping that is planar, which is reasonable if P is applied near the middle of the barrier (i.e., about 1 m from either end). We will consider motion due to sliding first, followed by motion due to tipping. For each of these, an FBD will be drawn and equations of equilibrium will be written. For the sliding analysis, Coulomb's law will be applied.

Motion due to sliding

Modeling The FBD for the barrier is shown in Fig. 2 where the direction of the friction force F opposes the direction of sliding. The weight W of the barrier is the product of specific weight and volume. You should verify that the volume of the barrier is $6.48 \times 10^5\,\text{cm}^3$. Thus, the weight is

$$W = (0.024\,\text{N/cm}^3)(6.48 \times 10^5\,\text{cm}^3) = 15.55\,\text{kN}. \tag{1}$$

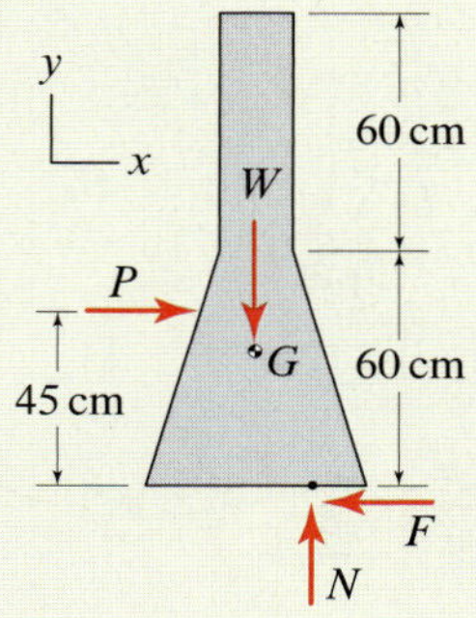

Figure 2
Free body diagram for sliding of the traffic barrier.

Governing Equations

Equilibrium Equations Using the FBD from Fig. 2, the equilibrium equations are

$$\sum F_y = 0: \qquad N - W = 0 \quad \Rightarrow \quad N = 15.55\,\text{kN}, \tag{2}$$

$$\sum F_x = 0: \qquad P - F = 0. \tag{3}$$

Force Laws Assuming that sliding is impending, Coulomb's law, Eq. (9.3), becomes

$$F = \mu N = (0.45)N, \tag{4}$$

where the coefficient of static friction is used and the direction of F is properly accounted for in the FBD of Fig. 2.

Computation Solving Eqs. (3) and (4) provides

$$\Rightarrow \qquad F = 7.00\,\text{kN} \quad \text{and} \quad P = 7.00\,\text{kN}. \tag{5}$$

Motion due to tipping

Modeling Tipping occurs when the contact forces N and F are positioned at the right-hand edge of the base, point A, as shown in the FBD in Fig. 3.

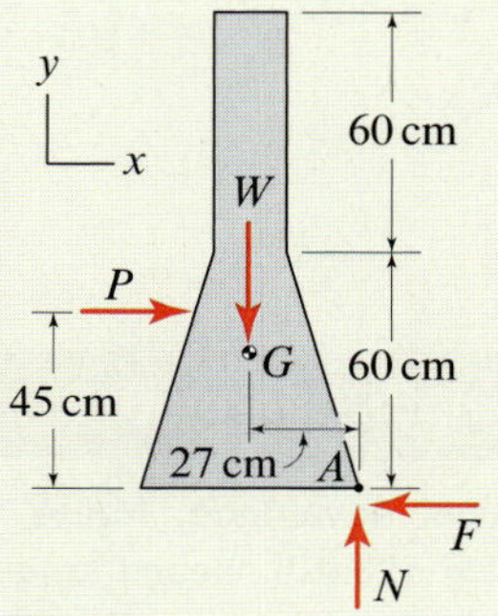

Figure 3
Free body diagram for tipping of the traffic barrier.

Governing Equations & Computation Summing moments about point A in Fig. 3 will avoid having to determine N and F. Thus,

$$\sum M_A = 0: \quad W(27\,\text{cm}) - P(45\,\text{cm}) = 0 \quad \Rightarrow \quad P = 9.33\,\text{kN}. \tag{6}$$

Discussion & Verification Motion will be impending when P reaches the smaller of Eqs. (5) and (6). Thus, when $\boxed{P = 7.00\,\text{kN, motion due to sliding is impending.}}$ If P is increased an infinitesimal amount beyond this value, the friction force F in the FBD of Fig. 2 will decrease because the coefficient of kinetic friction is less than the coefficient of static friction. Hence, if $P > 7.00$ kN, gross sliding motion occurs, with no tipping, and the barrier will accelerate.

EXAMPLE 9.3 *Wedges and Multiple Contact Surfaces*

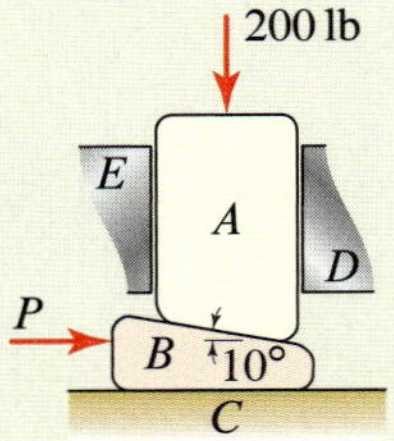

Figure 1

Two nylon blocks A and B are used to level one corner of a heavy refrigerator. The floor C and guides D and E are fixed. Block A supports a 200 lb force from the refrigerator and is a loose fit between guides D and E. The coefficient of static friction between all contact surfaces is 0.2.

(a) Determine the force P needed to raise block A.

(b) When $P = 0$, determine if the system is at rest.

SOLUTION

Road Map We will neglect the weights of the blocks, will assume the system is initially at rest, and will analyze motion due to sliding only (i.e., we assume the blocks will not tip). Because block A is a loose fit between guides D and E, it will make contact with D or E, but not both at the same time. In Part (a), block B slides to the right while block A is raised: there are three surfaces in contact and the kinematics are such that if one of these surfaces slides, then they all must slide. In Part (b), if this is a good design, there will be no sliding when $P = 0$. If sliding does occur, B will move to the left and A will be lowered, and the FBDs from Part (a) will be revised accordingly.

Part (a)

Modeling The FBD for the two blocks is shown in Fig. 2 where the directions of the friction forces oppose the directions of sliding.

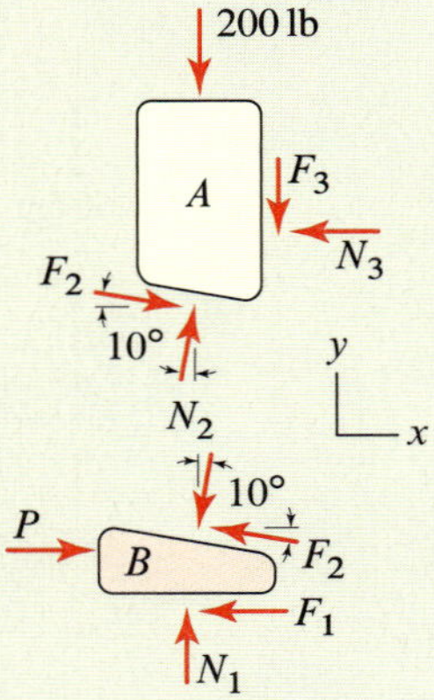

Figure 2
Free body diagrams for raising block A.

Governing Equations

Equilibrium Equations Using the FBD from Fig. 2, the equilibrium equations are

Block A:

$$\sum F_x = 0: \quad N_2(\sin 10^\circ) + F_2(\cos 10^\circ) - N_3 = 0, \tag{1}$$

$$\sum F_y = 0: \quad N_2(\cos 10^\circ) - F_2(\sin 10^\circ) - F_3 - 200\,\text{lb} = 0. \tag{2}$$

Block B:

$$\sum F_x = 0: \quad P - F_1 - N_2(\sin 10^\circ) - F_2(\cos 10^\circ) = 0, \tag{3}$$

$$\sum F_y = 0: \quad N_1 - N_2(\cos 10^\circ) + F_2(\sin 10^\circ) = 0. \tag{4}$$

Note that Eqs. (1)–(4) contain seven unknowns.

Force Laws Assuming that sliding is impending, and noting that if one surface slides then all surfaces must slide, Coulomb's law, Eq. (9.3) on p. 499, for each sliding surface provides

$$F_1 = \mu N_1 = (0.2) N_1, \quad F_2 = \mu N_2 = (0.2) N_2, \quad F_3 = \mu N_1 = (0.2) N_3, \tag{5}$$

where the coefficient of static friction is used and the directions of F_1, F_2, and F_3 have been properly accounted for in the FBDs of Fig. 2.

Computation There are now seven equations and seven unknowns. These equations are easily solved, and you should verify that the solutions are

$$\Rightarrow \quad N_1 = 217\,\text{lb}, \quad F_1 = 43.4\,\text{lb}, \tag{6}$$

$$N_2 = 228\,\text{lb}, \quad F_2 = 45.7\,\text{lb}, \tag{7}$$

$$N_3 = 84.6\,\text{lb}, \quad F_3 = 16.9\,\text{lb}, \tag{8}$$

$$\text{and} \quad \boxed{P = 128\,\text{lb}.} \tag{9}$$

> **Helpful Information**
>
> **Mechanical advantage.** Notice that the force $P = 128.0\,\text{lb}$ needed to raise block A, given by Eq. (9), is less than the 200 lb force being lifted. This feature is sometimes called *mechanical advantage*. The mechanical advantage of a wedge can be increased by decreasing the angle of the wedge (e.g., reducing the 10° angle shown in Fig. 1), or by using materials with a lower coefficient of friction.

Part (b)

Modeling We assume the system is in equilibrium with no sliding, and at the end of the solution we must verify this assumption. The FBDs are shown in Fig. 3, with the following comments. Because we assume there is no sliding, we may not use $F_1 = \mu N_1$, $F_2 = \mu N_2$, and $F_3 = \mu N_3$; rather, we are assuming $F_1 < \mu N_1$, and so on. Thus, N_1, F_1, N_2, F_2, N_3, and F_3 are six independent unknowns. While we assume there is no sliding, the directions of F_1, F_2, and F_3 are nonetheless correctly shown for block B sliding to the left and block A moving down.

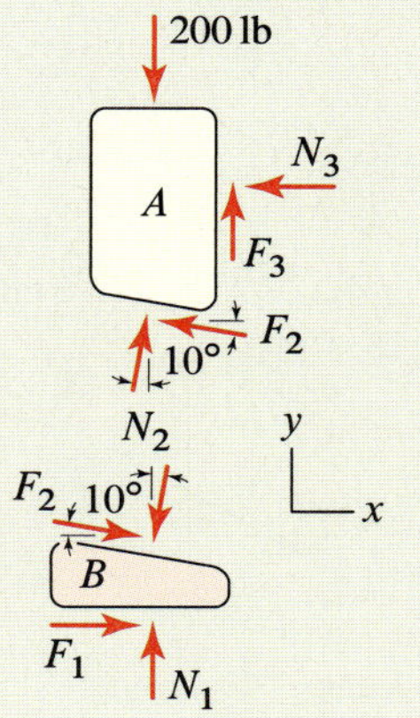

Figure 3
Free body diagrams with $P = 0$.

Governing Equations & Computation The FBDs in Fig. 3 contain six unknowns, and there are only four equilibrium equations available; hence this problem is statically indeterminate.* We proceed by assuming that block A does not contact either of the guides, which is reasonable if there is no sliding. Thus, $N_3 = F_3 = 0$. By inspection of Fig. 3, the equation $\sum F_x = 0$ for blocks A and B taken together (note that the forces N_2 and F_2 become internal to the FBD and hence they do not appear in this equilibrium equation) provides $F_1 = 0$ and writing $\sum F_y = 0$ provides $N_1 = 200\,\text{lb}$. Finally, the equilibrium equations for block A alone in Fig. 3 (with $N_3 = F_3 = 0$) are

Block A:

$$\sum F_x = 0: \quad N_2(\sin 10°) - F_2(\cos 10°) = 0, \tag{10}$$

$$\sum F_y = 0: \quad N_2(\cos 10°) + F_2(\sin 10°) - 200\,\text{lb} = 0, \tag{11}$$

$$\Rightarrow \quad N_2 = 197\,\text{lb}, \quad \text{and} \quad F_2 = 34.7\,\text{lb}. \tag{12}$$

Values for all six unknowns have been obtained (i.e., $N_1 = 200\,\text{lb}$, $F_1 = 0$, $N_2 = 197\,\text{lb}$, $F_2 = 34.7\,\text{lb}$, $N_3 = F_3 = 0$), and we must now verify that these satisfy Coulomb's law. Clearly, surface 1 does not slide, and surface 3 is not in contact. For surface 2, Coulomb's law states that sliding is impending when $F_2 = \mu N_2 = 39.4\,\text{lb}$. Since this value is larger than the result for F_2 obtained in Eq. (12), we conclude that surface 2 does not slide. Hence, the solution we obtained satisfies all equilibrium equations and Coulomb's law; therefore our initial assumptions were correct, and we may conclude that the system will be at rest when $P = 0$.

Alternate solution. As an alternative solution, we can demonstrate that a nonzero force is required to slide block B to the left. The solution procedure of Part (a) is repeated and the FBDs in Fig. 4 are used with the following comments. Force Q is defined to be positive in the $-x$ direction. At the outset of this solution, it may not be clear whether block A contacts guide D or E, and in the course of this solution you will find that contact is made at E, whose contact forces are defined to be N_4 and F_4. Observe that the directions of all friction forces oppose the directions of sliding. Problem 9.13 asks you to carry out this solution, and you should determine that $Q = 44.4\,\text{lb}$ will cause block A to be lowered. From this we may infer there is no motion when $Q = 0$, and hence when $P = 0$.

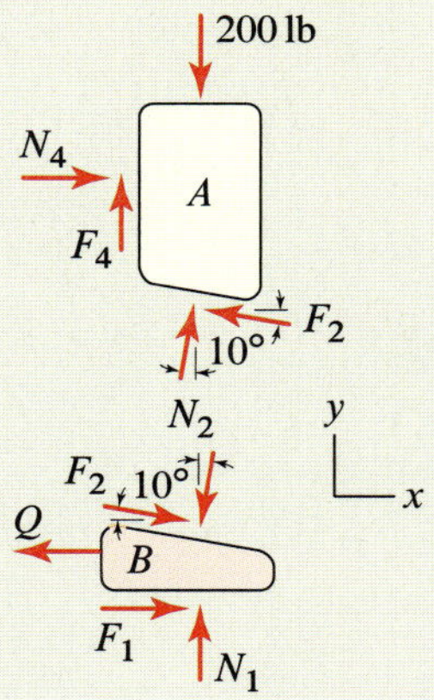

Figure 4
Free body diagrams for lowering block A.

Discussion & Verification Impending upward motion of block A occurs when $P = 128.0\,\text{lb}$, and the system is at rest when $P = 0$. Hence, this design is acceptable from the point of view that the wedge will not slip out of place when the insertion force P is removed.

* While one moment equilibrium equation could be written for each of blocks A and B, three additional unknowns in the form of the locations of the contact force systems must also be introduced; hence the problem would remain statically indeterminate.

PROBLEMS

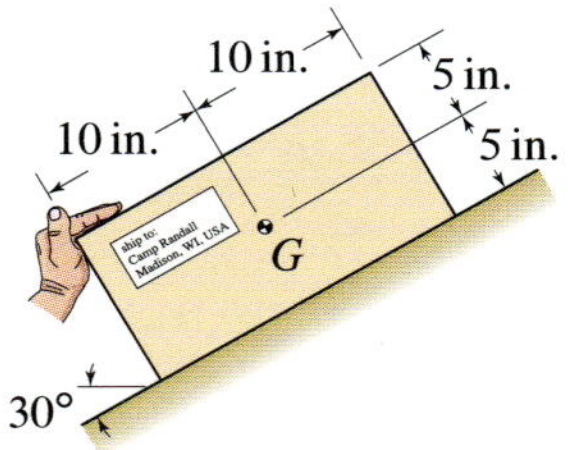

Figure P9.1 and P9.2

Problems 9.1 and 9.2

A worker applies the force described below to push a box that weighs 40 lb with center of gravity at point G. The surface between the box and ramp has coefficient of friction $\mu_s = \mu_k = 0.25$. Determine the normal force and friction force between the box and ramp, and determine if the box will slide up the ramp, down the ramp, or remain at rest.

Problem 9.1 The worker applies a 30 lb force parallel to the ramp.

Problem 9.2 The worker applies a 35 lb force that is horizontal.

Problem 9.3

The structure consists of two uniform members AB and BC, each weighing 2 kN. The members are pinned to each other at B, and the structure is supported by a pin at C and a surface at A having coefficient of static friction 1.2. Determine the largest positive value P the structure can support.

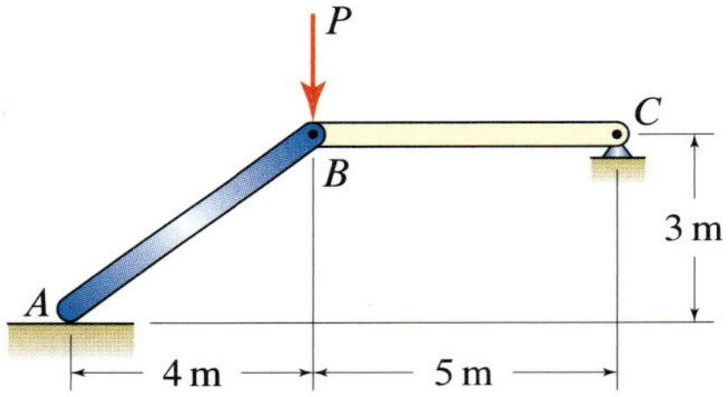

Figure P9.3

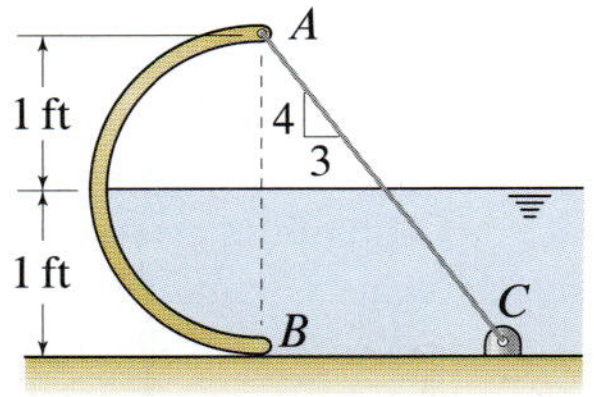

Figure P9.4

Problem 9.4

Water ($\gamma = 62.4\,\text{lb/ft}^3$) is retained by a uniform thin semicircular dam having 1 ft radius and 3 ft depth into the plane of the figure. The dam weighs 60 lb and is supported by a cable AC and by frictional contact with the bottom of the channel at B. Determine the minimum coefficient of friction so that the dam does not slip at B.

Problem 9.5

By writing $\sum M = 0$ about some convenient point, verify that the distances from the lower left-hand corner of the box (point D) to points A and B in Example 9.1 on p. 504 are 5.18 and 11.2 in., respectively. In view of these results, is motion of the box in fact due to sliding (as assumed in Example 9.1) or is it due to tipping? Explain.

Figure P9.6

Problem 9.6

The apparatus shown can be used to experimentally determine the angle of static friction, and hence the coefficient of static friction, for many combinations of contacting materials. A block of material C rests on a beam AB. Starting with $\theta = 0°$, point B is slowly lowered until block C begins to slide. Assuming block C does not tip, show that the value of θ when sliding starts is equal to the angle of friction ϕ, given by Eq. (9.4) on p. 501.

Problem 9.7

A tool chest has 800 N weight that acts through the midpoint of the chest. The chest is supported by feet at A and rollers at B. The surface has a coefficient of friction of 0.3. Determine the value of the horizontal force P necessary to cause motion of the chest to the right, and determine if the motion is sliding or tipping.

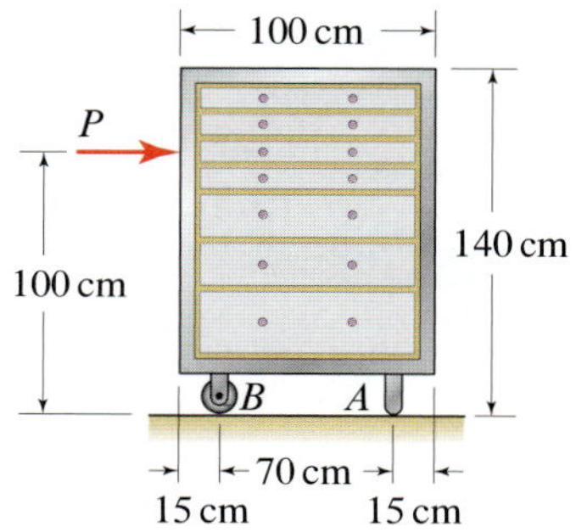

Figure P9.7

Problem 9.8

Determine the value of h in Example 9.2 on p. 505 so that sliding and tipping motion of the traffic barrier are simultaneously impending.

Problem 9.9

The machine shown is used to move boxes. Bar ABC slides horizontally in the bearing of the fixed machine housing. Points B, C, and D are pins, and point C has a frictionless roller. The flywheel E rotates clockwise under the action of moment M_E. The horizontal surface on which the box rests has coefficients of friction $\mu_s = 0.3$ and $\mu_k = 0.25$, and all other contact surfaces are frictionless. If the box weighs 900 N, determine the moment M_E that must be applied to the flywheel to initiate motion of the box, and determine if the motion is sliding or tipping.

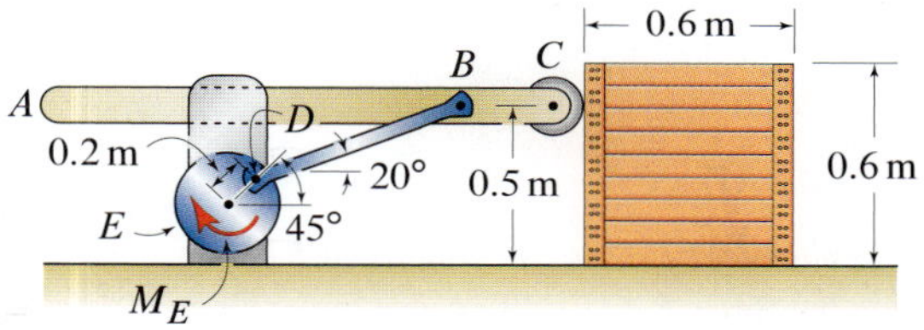

Figure P9.9

Problem 9.10

A long concrete gravity dam retains water in a reservoir. The surface between the dam and earth has coefficients of friction $\mu_s = 0.8$ and $\mu_k = 0.7$. When the reservoir is completely full (i.e., $h = 8$ m), determine if the dam is safe from both overturning (tipping) and sliding along its base. The specific weights of concrete is $\gamma_c = 25\ \text{kN/m}^3$ and the density of water is $\rho_w = 10^3\ \text{kg/m}^3$.

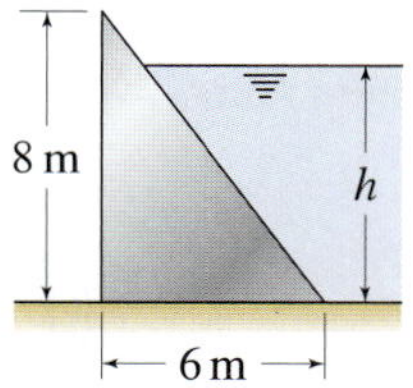

Figure P9.10

Problem 9.11

Repeat Prob. 9.10 if the reservoir is on the left-hand side of the dam.

Problem 9.12

The owner of a small concrete gravity dam is considering attaching steel plate to the face of the dam so that a greater depth of water can be retained. The surface between the dam and earth has coefficients of friction $\mu_s = 0.6$ and $\mu_k = 0.55$. Determine if the dam is safe from both overturning (tipping) and sliding along its base. The specific weights of concrete and water are $\gamma_c = 150\ \text{lb/ft}^3$ and $\gamma_w = 62.4\ \text{lb/ft}^3$, respectively. Neglect the weight of the steel plate.

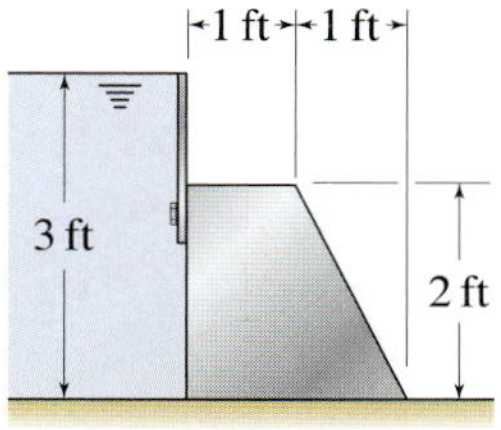

Figure P9.12

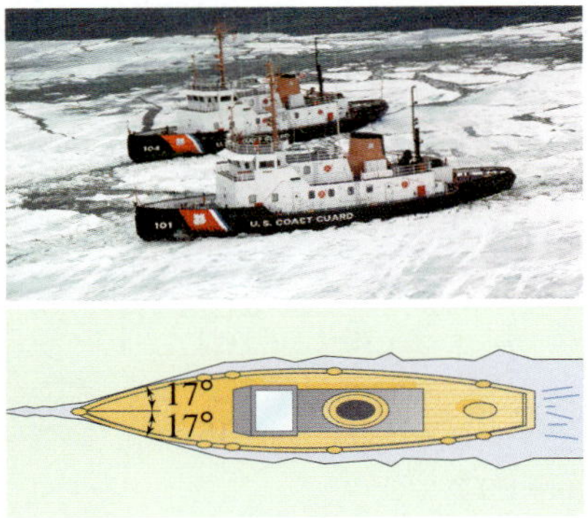

Figure P9.14

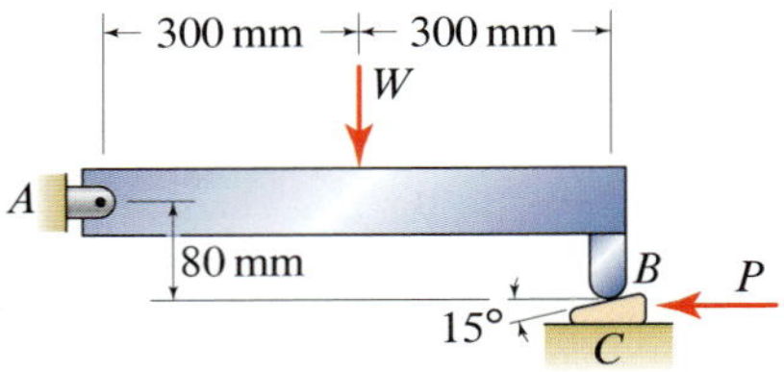

Figure P9.15

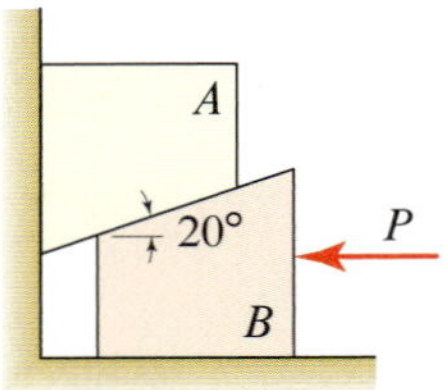

Figure P9.16 and P9.17

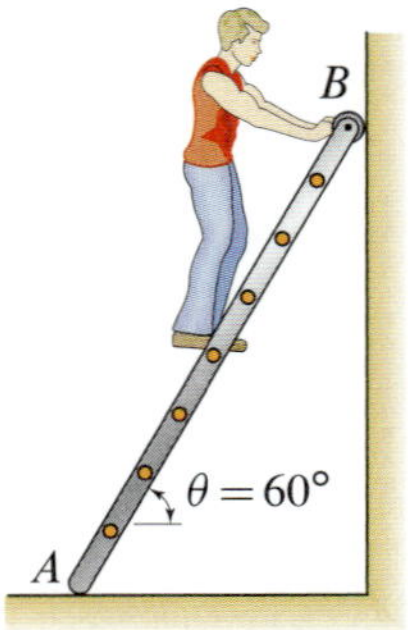

Figure P9.18 and P9.19

Problem 9.13

Carry out the alternate solution described in Example 9.3 on p. 507.

Problem 9.14

The photograph shows two U.S. Coast Guard icebreakers in an ice field, and a simple model for an ice breaking operation. If the coefficients of static and kinetic friction for contact between the ship's bow and ice are 0.08 and 0.06, respectively, and if the ship produces a thrust of 10^6 lb, determine the normal and friction forces acting on each side of the ship's bow as it moves through the ice field with constant velocity. Assume the ship makes contact with the ice only on its bow, and neglect all forces between the ship's hull and water except for the thrust.

Problem 9.15

A wedge is used to level a structure. All contact surfaces have coefficients of static and kinetic friction of 0.3 and 0.25, respectively, and $W = 500$ N. Assume the dimensions of the wedge are small. Determine the value of P to cause impending motion of the wedge:

(a) To the left.

(b) To the right.

Problems 9.16 and 9.17

Blocks A and B each have 2 kg mass. All contact surfaces have the same coefficient of friction. Determine the force P needed to cause impending motion of block B to the left if the coefficient of static friction is

Problem 9.16 0.4.

Problem 9.17 0.3.

Problem 9.18

An 8 ft long ladder has seven rungs. The rungs are spaced 1 ft apart, and the top and bottom rungs are 1 ft from their respective ends of the ladder. The top of the ladder has a roller. Neglect the weight of the ladder and assume the worker's hand applies no force to the ladder.

(a) If the worker weighs 140 lb and stands on the middle rung, determine the minimum value of the coefficient of friction so that the ladder does not slide.

(b) If the worker weighs 140 lb and stands on a different rung, does your answer to Part (a) change? Explain.

(c) If the worker weighs more than 140 lb and stands on the middle rung, does your answer to Part (a) change? Explain.

Problem 9.19

In Prob. 9.18, the roller at B is removed and the surfaces at A and B both have the same coefficient of friction. If the worker weighs 140 lb and stands on the middle rung, determine the minimum value of the coefficient of friction so that the ladder does not slide. *Hint:* The use of mathematical software is helpful, but is not required.

Problem 9.20

A table saw for cutting wood is shown. The blade rotates counterclockwise, and the operator pushes the wood into the blade using a stick to help keep his or her hand away from the blade. Despite this safety precaution, it is possible for the wood to be propelled by the blade with great force and speed into the operator, causing injury. To help prevent this accident, the saw is outfitted with an antikickback device, which weighs 0.3 lb with center of gravity at point B. Neglecting friction between the wood and saw table, determine the minimum coefficient of friction between the wood and the antikickback device so that the wood workpiece will not kick back.

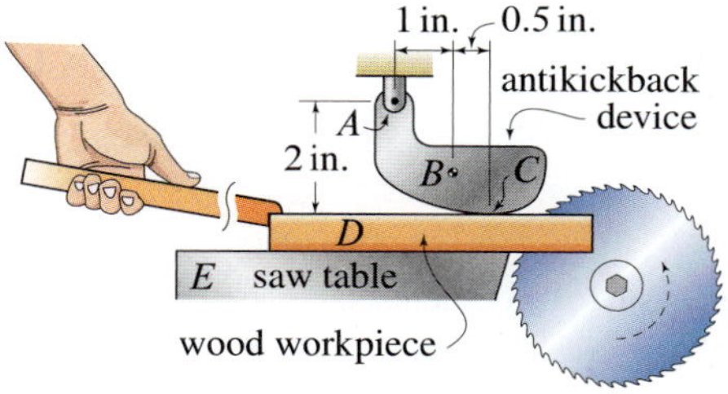

Figure P9.20

Problem 9.21

Bar ABC has square cross section and can slide in the square hole that is in collar D. Collar D is supported by fixed vertical post E that is built in at its base. The collar D can translate in the z direction and rotate about the z axis without friction; other translations and rotations are constrained. End C of the square cross section bar rests on a horizontal surface having coefficients of friction $\mu_s = 0.6$ and $\mu_k = 0.5$. If bar ABC is initially motionless, determine the positive value of P that will cause impending motion. For this value of P, also determine the reactions between bar ABC and the collar D.

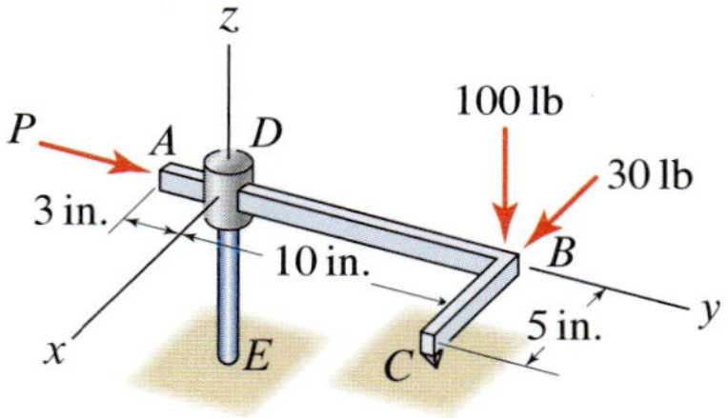

Figure P9.21

DESIGN PROBLEMS

General Instructions. In all problems, write a brief technical report following the guidelines of Appendix A, where you summarize all pertinent information in a well-organized fashion. It should be written using proper, simple English that is easy to read by another engineer. Where appropriate, sketches along with critical dimensions should be included. Discuss the objectives and constraints considered in your design, the process used to arrive at your final design, safety issues if appropriate, and so on. The main discussion should be typed, and figures, if needed, can be computer-drawn or neatly hand-drawn. Include a neat copy of all supporting calculations in an appendix that you can refer to in the main discussion of your report. A length of a few pages, plus appendix, should be sufficient.

Design Problem 9.1

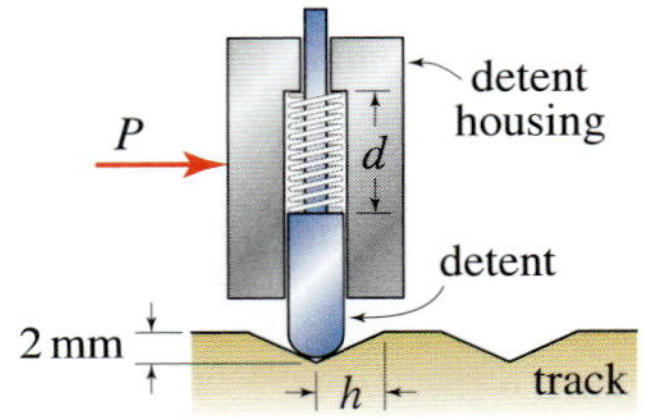

Figure DP9.1

Design a detent mechanism for precision positioning control. This mechanism consists of a spring-loaded plunger (detent) with a spherical end that is normally positioned at the bottom of the spaced notches on the track. The track is fixed in space, and the detent housing can undergo horizontal translational motion only. It is desired that the detent housing move to the next available notch when the force P is approximately 2 N. Specify the spring stiffness (in N/mm), the spring's unstretched length, and dimensions h and d where d is the length of the spring cavity when the detent is at the bottom of a notch. The coefficient of friction between the detent and the track is 0.3, and friction between the detent and the housing can be assumed to be negligible because of sufficient lubrication.

Design Problem 9.2

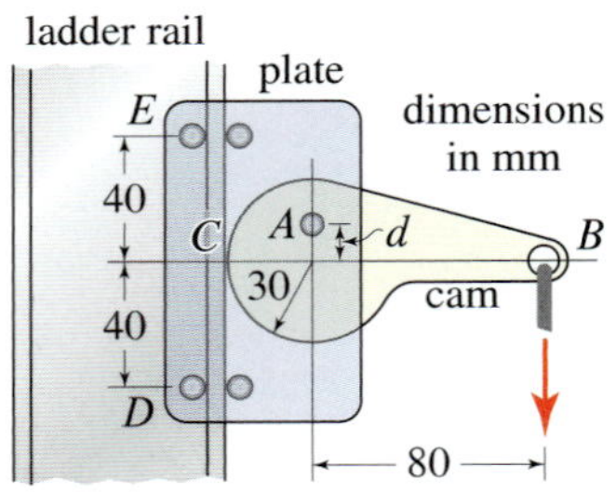

Figure DP9.2

Design a safety device for use on fixed steel ladders. The device consists of a cam, a plate, and several roller bearings attached to the plate. The device slips onto the rail of the ladder, and is attached via a short cable to a harness worn by the person using the ladder. It has the feature that if the person were to accidentally fall, the device automatically would lock in position on the ladder, and limit the distance the person will fall, whereas a small upward force on the cam will allow it to slide to a new position on the ladder. All materials are steel. The pins at E and D are equipped with roller bearings. The cam has a circular surface that contacts the flange of the ladder at C and is hinged to the plate at point A. Specify the dimension d. Note that small values of d produce large contact forces, but also lead to more rapid wear of the cam and bearings. If d is too large, it may not provide sufficient frictional resistance. Thus, you should specify a value of d with these competing factors in mind.

9.2 Problems with Multiple Contact Surfaces

This section addresses friction problems that have multiple contact surfaces where, in general, only a subset of the surfaces slide while the others do not. Such problems are challenging because we usually do not know at the outset of a solution which of the surfaces will slip. Note that it is only for surfaces that slide or have impending motion that we may write $F = \mu N$, and for surfaces that do not slide, $F < \mu N$. Recall that in Section 9.1, some problems with multiple contact surfaces were considered, but these problems were of a simpler category because they had the feature that if one surface slides, then all surfaces were required to slide. The subject of geometry of motion is called *kinematics*, and the problems considered in this section have more complicated kinematics than those in Section 9.1.

The basic concepts needed to address the problems of this section were all covered in Section 9.1. The new ingredient needed here is an understanding that only a subset of the contact surfaces will slide.

Determination of sliding directions

When you draw FBDs, it is necessary to know the directions in which sliding may occur so that friction forces can be given the proper directions. Determination of sliding directions is sometimes challenging, and simple kinematics usually help clarify this, as demonstrated in the following example.

Consider the use of a wrench to twist a pipe as shown in Fig. 9.7(a). In Fig. 9.7(b), we begin drawing FBDs; most of the forces in these FBDs are straightforward to assign, but the directions of the friction forces are, at this point, probably uncertain. As an example of how you might determine these directions, imagine that pipe C in Fig. 9.7(a) is fixed and let there be no slip at B. If handle E were to move down, then clearly jaw A also moves down and hence there is sliding at that location. The friction force F_1 the pipe applies to the jaw at A must be upward, and by Newton's third law, the friction force the jaw applies to the pipe is in the opposite direction. We now repeat this thought process by letting there be no slip at A. If handle E were to move down, then clearly jaw B moves up and hence there is sliding at that location. The friction force F_2 the pipe applies to the jaw at B must be downward, and the friction force the jaw applies to the pipe is in the opposite direction. The final FBDs are shown in Fig. 9.7(c). Once the FBDs are drawn, we proceed with the analysis by writing equilibrium equations, applying Coulomb's law, etc., to determine whether surface A or surface B will slide; Prob. 9.32 asks you to do this.

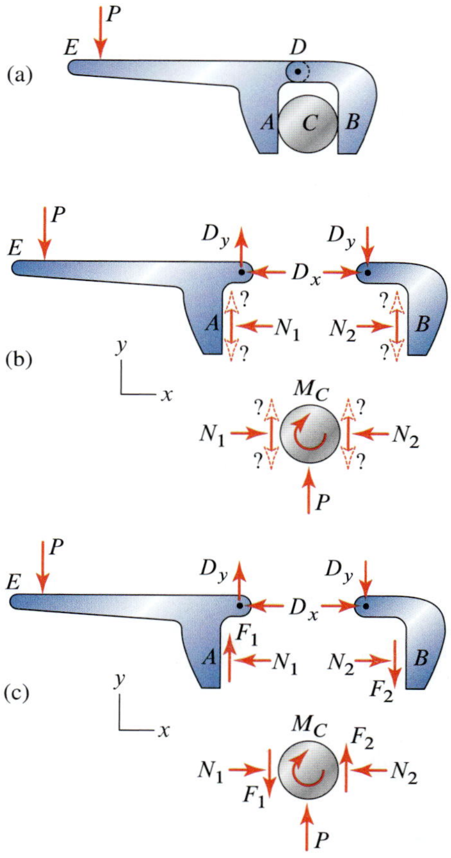

Figure 9.7
(a) A wrench is used to twist a pipe C. The wrench consists of jaw BD and handle-jaw ADE, with a pin at D. (b) Partially complete FBDs where the directions of the friction forces are uncertain, as indicated by the "?" symbols and dashed arrowheads. The force P and moment M_C on the pipe are reactions from the pipe's support that equilibrate the force P applied to handle E of the wrench. (c) Complete FBDs. Note: The FBDs can be further simplified by noticing that member BD is a two-force member.

End of Section Summary

This section considers problems with multiple contact surfaces where only a subset of the surfaces slide. Some of the key points are:

- At the outset of an analysis, it is often uncertain which surfaces will slide. Note that $F = \mu N$ may be used only for those surfaces that slide or have impending slip, while $F < \mu N$ applies for those that do not slide.

- Simple kinematics help clarify the directions for sliding so that friction forces can be given the correct directions in FBDs.

EXAMPLE 9.4 *Multiple Contact Surfaces*

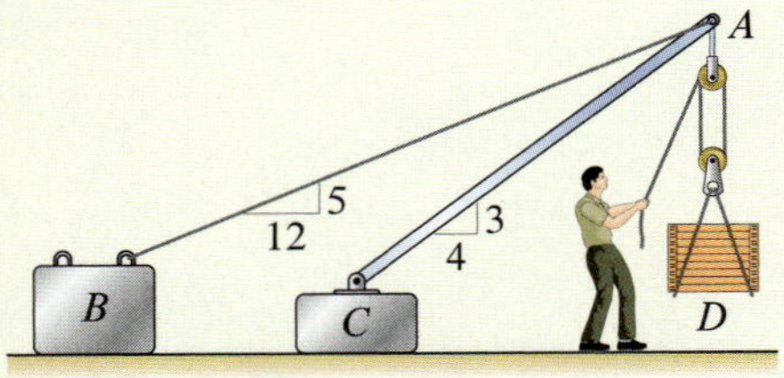

Figure 1

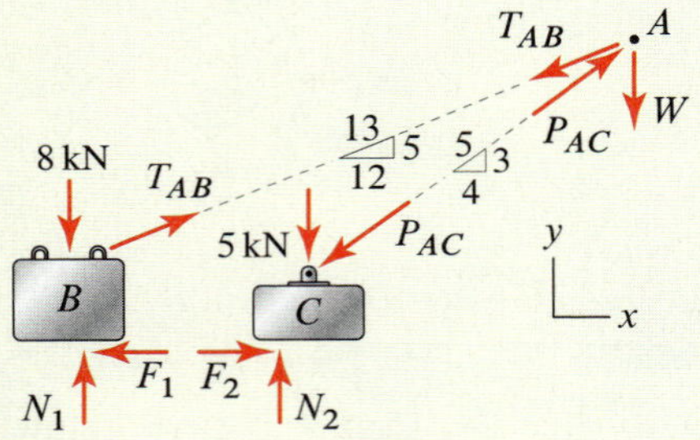

Figure 2
Free body diagrams in which the force the person applies to the pulley system is assumed to be small.

A makeshift crane consists of a cable AB, a steel bar AC, concrete blocks B and C, and a pulley system. While the crane is primitive, it is inexpensive and portable and may be especially useful in impoverished counties for lifting heavy objects on and off trucks and carts. Concrete blocks B and C weigh 8 and 5 kN, respectively, and the coefficient of friction between the blocks and the soil on which they rest is 0.5. Determine the weight W of the heaviest object D that may be lifted, and specify whether block B or C slides.

SOLUTION

Road Map We will neglect the weights of the cable, bar, and pulley system and will assume that blocks B and C do not tip. Because of the pulley system, the force applied by the person is small relative to the weight being lifted,* and we will therefore neglect the force the person applies to the cable. It is not likely that both blocks B and C will slide at the same time, and it is not clear which of these will reach its sliding threshold first. Our strategy will be to determine the forces supported by the cable and bar in terms of the weight W of object D. We will then proceed to analyze the equilibrium of block B, and then block C, to determine which of these slides, and the weight W necessary to cause this.

Modeling The FBDs for point A and blocks B and C are shown in Fig. 2 where the cable force T_{AB} is positive in tension and the bar force P_{AC} is positive in compression. Note that if block B slides, it will move to the right and if block C slides, it will move to the left, and the friction forces in these FBDs oppose these motions.

Governing Equations & Computation

Equilibrium Equations Using the FBDs in Fig. 2, equilibrium equations are written and solved as follows

$$\textbf{Point } A\textbf{:}\quad \sum F_x = 0: \quad -T_{AB}\tfrac{12}{13} + P_{AC}\tfrac{4}{5} = 0, \tag{1}$$

$$\sum F_y = 0: \quad -T_{AB}\tfrac{5}{13} + P_{AC}\tfrac{3}{5} - W = 0, \tag{2}$$

$$\Rightarrow \quad T_{AB} = \tfrac{13}{4}W \quad \text{and} \quad P_{AC} = \tfrac{15}{4}W. \tag{3}$$

$$\textbf{Block } B\textbf{:}\quad \sum F_x = 0: \quad T_{AB}\tfrac{12}{13} - F_1 = 0, \tag{4}$$

$$\sum F_y = 0: \quad -8\text{ kN} + T_{AB}\tfrac{5}{13} + N_1 = 0, \tag{5}$$

$$\Rightarrow \quad F_1 = 3W \quad \text{and} \quad N_1 = 8\text{ kN} - \tfrac{5}{4}W. \tag{6}$$

$$\textbf{Block } C\textbf{:}\quad \sum F_x = 0: \quad -P_{AC}\tfrac{4}{5} + F_2 = 0, \tag{7}$$

$$\sum F_y = 0: \quad -5\text{ kN} - P_{AC}\tfrac{3}{5} + N_2 = 0, \tag{8}$$

$$\Rightarrow \quad F_2 = 3W \quad \text{and} \quad N_2 = 5\text{ kN} + \tfrac{9}{4}W. \tag{9}$$

Note that if point A and blocks B and C are in equilibrium, then all the preceding expressions are valid, regardless of whether block B or block C slides.

Common Pitfall

Overuse of $F = \mu N$. A common error in problems with friction is to use $F = \mu N$ for *all* contact surfaces. While this is sometimes true (as in the example problems of Section 9.1), in general this is not the case. In Fig. 2 above, taking $F_1 = \mu N_1$ and $F_2 = \mu N_2$ is incorrect because it implies that both blocks have impending motion at the same time, and further, the equilibrium equations cannot be satisfied. At the end of this solution we see that when W reaches its maximum value, $F_1 = \mu N_1$ and $F_2 < \mu N_2$.

---

* Exactly what force the person applies relative to the weight being lifted depends on details of the pulley system, which are not described here.

Force Laws We now apply Coulomb's law to determine the value of W needed to cause each of the blocks to slide.

If block B slides:

$$F_1 = \mu N_1 \tag{10}$$

$$3W = (0.5)\left(8\,\text{kN} - \tfrac{5}{4}W\right) \tag{11}$$

$$\Rightarrow \quad W = 1.10\,\text{kN}. \tag{12}$$

If block C slides:

$$F_2 = \mu N_2 \tag{13}$$

$$3W = (0.5)\left(5\,\text{kN} + \tfrac{9}{4}W\right) \tag{14}$$

$$\Rightarrow \quad W = 1.33\,\text{kN}. \tag{15}$$

When W reaches the smaller of the values given by Eqs. (12) and (15), sliding is impending. Hence, the largest value of weight that may be lifted is

$$\boxed{W_{max} = 1.10\,\text{kN},} \tag{16}$$

and when this occurs, $\boxed{\text{sliding of block } B \text{ is impending.}}$

Discussion & Verification An analysis that includes the force the person applies to the cable will give more precise results than those obtained here. To carry out this analysis, details of the pulley system must be known, and the angle at which the person applies the force must be known, or more likely, a range of reasonable values should be considered.

EXAMPLE 9.5 *Multiple Contact Surfaces*

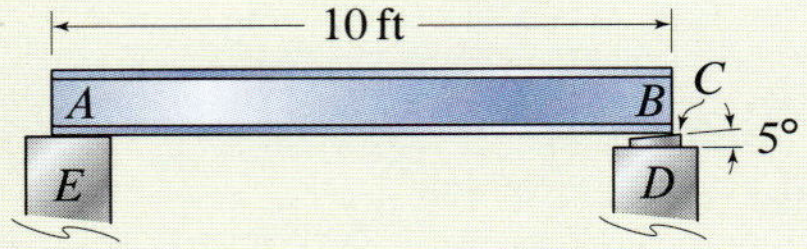

Figure 1

A 5° steel wedge C is used to level a 10 ft long uniform steel I beam AB weighing 400 lb. The ends of the beam are supported by concrete walls D and E. Determine the force needed to drive the wedge to the left, and describe the motion of the beam that results (i.e., determine if end B of the beam is lifted or if the entire beam slides to the left). The coefficients of static friction are 0.4 for steel-on-steel contact and 0.3 for steel-on-concrete contact.

SOLUTION

Road Map We will neglect the weight of the wedge. There are three potential sliding surfaces in this problem. While sliding must occur between the bottom of the wedge and the stationary wall at D, it is unclear which of the remaining two contacts slides. In the first part of our solution we determine the force needed to slide the beam and wedge together to the left. In the second part, we determine the force needed to slide only the wedge. The smaller of these values is the force that causes motion to occur.

Sliding of beam and wedge together

Modeling The FBD for the beam and wedge together is shown in Fig. 2 where the direction of the friction forces opposes sliding of the beam to the left. When writing equilibrium equations, we will assume the lines of action of F_1, F_2, and P are the same, which is reasonable if the dimensions of the wedge are small. We also assume that the support reaction at A is positioned at the end of the beam.

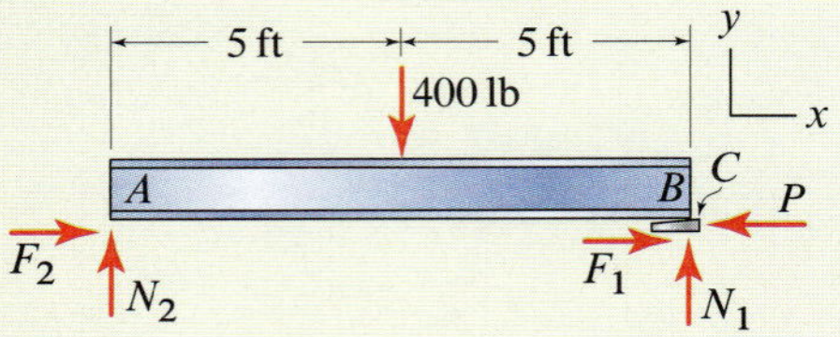

Figure 2
Free body diagram for the beam and wedge sliding together to the left.

Governing Equations

Equilibrium Equations Using the FBD in Fig. 2, the equilibrium equations are

$$\sum M_A = 0: \quad N_1(10\,\text{ft}) - (400\,\text{lb})(5\,\text{ft}) = 0 \quad \Rightarrow \quad N_1 = 200\,\text{lb}, \tag{1}$$

$$\sum F_y = 0: \quad N_1 + N_2 - 400\,\text{lb} = 0 \quad \Rightarrow \quad N_2 = 200\,\text{lb}, \tag{2}$$

$$\sum F_x = 0: \quad F_1 + F_2 - P = 0. \tag{3}$$

These three equations contain five unknowns, although we were able to solve for N_1 and N_2.

Force Laws Assuming sliding occurs at surfaces 1 and 2, Coulomb's law becomes

$$F_1 = \mu_c N_1 = (0.3)(200\,\text{lb}) = 60\,\text{lb}, \tag{4}$$

$$F_2 = \mu_c N_2 = (0.3)(200\,\text{lb}) = 60\,\text{lb}, \tag{5}$$

where μ_c is the coefficient of static friction for steel-on-concrete.

Computation Solving Eq. (3) for P provides

$$\Rightarrow \quad P = 120\,\text{lb}. \tag{6}$$

Sliding of wedge only

Modeling Here we consider the wedge sliding to the left while end B of the beam is lifted. Note that the FBD in Fig. 2 is still valid and the equilibrium equations (Eqs. (1)–(3)) are still valid, hence $N_1 = 200$ lb and $N_2 = 200$ lb. To continue, we may use an FBD of beam AB or wedge C, and we choose the latter as shown in Fig. 3.

Governing Equations

Equilibrium Equations Using the FBD shown in Fig. 3, the equilibrium equations are

$$\sum F_x = 0: \qquad F_1 + N_3(\sin 5°) + F_3(\cos 5°) - P = 0, \tag{7}$$

$$\sum F_y = 0: \qquad N_1 - N_3(\cos 5°) + F_3(\sin 5°) = 0. \tag{8}$$

Because N_1 is known from our earlier work, these two equations contain four unknowns.

Force Laws Assuming sliding occurs at surfaces 1 and 3, Coulomb's law becomes

$$F_1 = \mu_c N_1 = (0.3)(200\,\text{lb}) = 60\,\text{lb}, \tag{9}$$

$$F_3 = \mu_s N_3 = (0.4)N_3, \tag{10}$$

where μ_c and μ_s are the coefficients of static friction for steel-on-concrete and steel-on-steel, respectively, and the directions of F_1 and F_3 have been assigned in the FBD of Fig. 3 to oppose sliding motion of the wedge to the left.

Computation Solving Eqs. (7)–(10), with $N_1 = 200\,\text{lb}$, provides

$$\Rightarrow \quad N_3 = 208\,\text{lb}, \quad F_3 = 83.2\,\text{lb}, \quad \text{and} \quad P = 161\,\text{lb}. \tag{11}$$

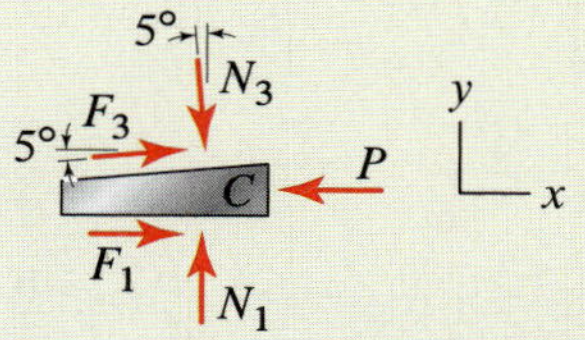

Figure 3
Free body diagram for the wedge sliding to the left.

Discussion & Verification Sliding occurs when P reaches the smaller of Eqs. (6) and (11). Thus, when $P = 120\,\text{lb}$, the beam and wedge slide together to the left. Unfortunately, this is not the desired effect. Some possible remedies that might result in end B of the beam being lifted are as follows:

- An additional weight could be temporarily added to the left-hand end of the beam while the wedge is inserted. For example, a worker could stand on end A. This will increase the normal force at end A, and hence increase the friction force there.
- A lubricant could be applied to the contact between the beam and wedge so that the coefficient of friction there is lowered.
- All the results obtained here are for static equilibrium. It may be possible to drive the wedge dynamically, such as by tapping it into position with a hammer. Analysis of this requires dynamics.

PROBLEMS

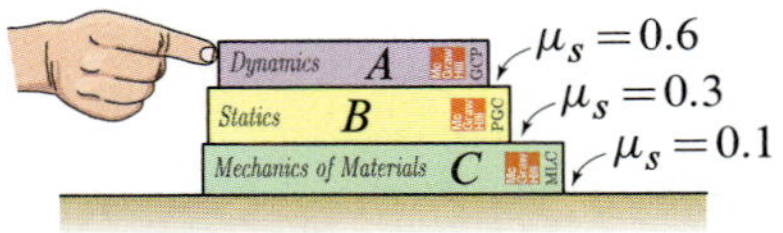

Figure P9.22

Problem 9.22

Three books rest on a table. Books A, B, and C weigh 2, 3, and 4 lb, respectively. Determine the horizontal force applied to book A that causes impending motion of any of the books to the right, and determine which books move.

Problem 9.23

Block A and spool B weigh 8 and 6 lb, respectively. The spool has a string wrapped around it to which a force P is applied. The coefficient of static friction between the spool and the contact surfaces at A and C is 0.25. Determine the value of P that causes impending motion, and determine if slip occurs at A, or C, or both locations simultaneously.

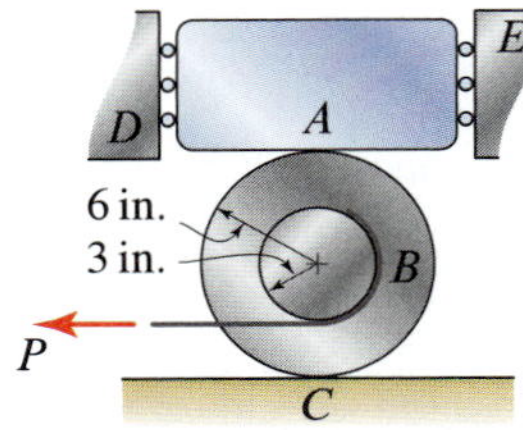

Figure P9.23

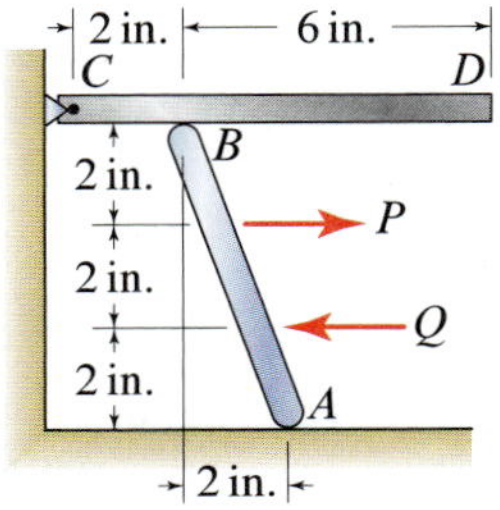

Figure P9.24–P9.26

Problems 9.24 through 9.26

Bars AB and CD are uniform and each weighs 7 lb. The coefficient of static friction at surfaces A and B is the same.

Problem 9.24 If $P = Q = 0$, determine the minimum coefficient of static friction so that the system has no motion.

Problem 9.25 If $Q = 0$ and the coefficient of static friction for all surfaces is 0.6, determine the value of P that causes impending motion of bar AB to the right.

Problem 9.26 If $P = 0$ and the coefficient of static friction for all surfaces is 0.6, determine the value of Q that causes impending motion of bar AB to the left.

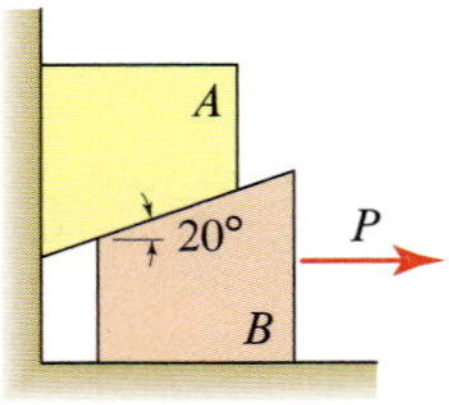

Figure P9.27 and P9.28

Problems 9.27 and 9.28

Blocks A and B each have 2 kg mass. All contact surfaces have the same coefficient of friction. Determine the force P needed to cause impending motion of block B to the right if the coefficient of static friction is

Problem 9.27 0.4.

Problem 9.28 0.3.

Problem 9.29

Blocks A and B weigh 10 and 15 N, respectively. Point C is at the midpoint of member BD, and both members AC and BD have negligible weight. The coefficient of static friction for all contact surfaces is 0.3. Determine the positive value of P that causes impending motion, and determine which of blocks A or B slides and the direction of motion.

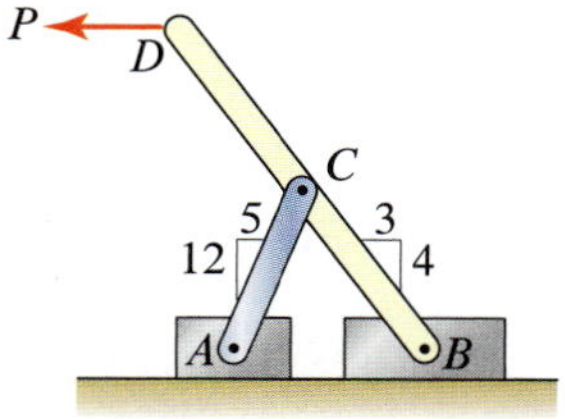

Figure P9.29

Problem 9.30

The structure consists of two uniform members AB and BC, each weighing 2 kN. The members are pinned to one another at B, and the structure is supported by surfaces at A and C having coefficients of static friction of 1.2 and 0.5, respectively. Determine the largest positive value P the structure can support.

Problem 9.31

The structure consists of two uniform members AB and BC, each weighing 2 kN and $P \geq 0$. Determine the minimum coefficient of static friction needed at surface A and the minimum coefficient of static friction needed at surface C so that neither surface slips for any value of P (this is called *self-locking*).

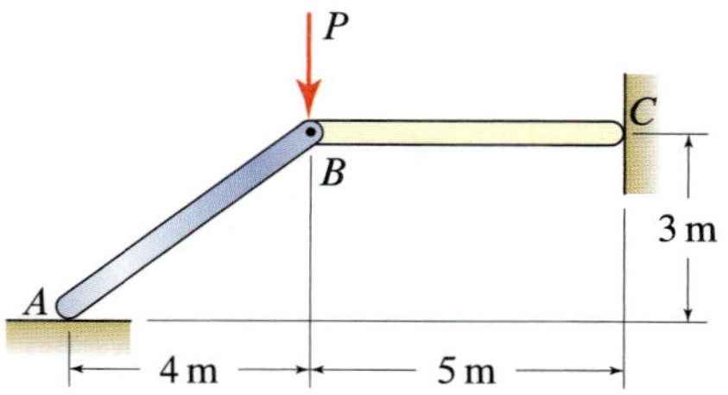

Figure P9.30 and P9.31

Problem 9.32

A wrench is used to twist a pipe C. The wrench consists of jaw BD and handle-jaw ADE, with a pin at D. The coefficient of static friction μ_s for all contact surfaces is the same. Determine the minimum value of μ_s so that there is no slip at A or B regardless of the value of force P (this is called *self-locking*).

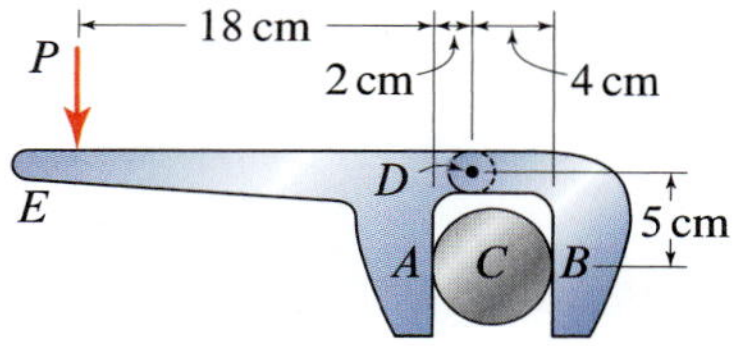

Figure P9.32

Problems 9.33 and 9.34

A truck with 1500 kg mass and center of gravity at point C is used to pull a dumpster with 1700 kg mass and center of gravity at point G. The coefficient of static friction between the tires and pavement is 1.1, and between the dumpster and pavement is 0.5. Assume the truck's engine has sufficient power and the dumpster does not tip. *Hint:* The use of mathematical software is helpful, but not required. Determine if the truck is able to pull the dumpster if:

Problem 9.33 The truck has rear-wheel drive.

Problem 9.34 The truck has four-wheel drive and end E of the cable is moved to point H.

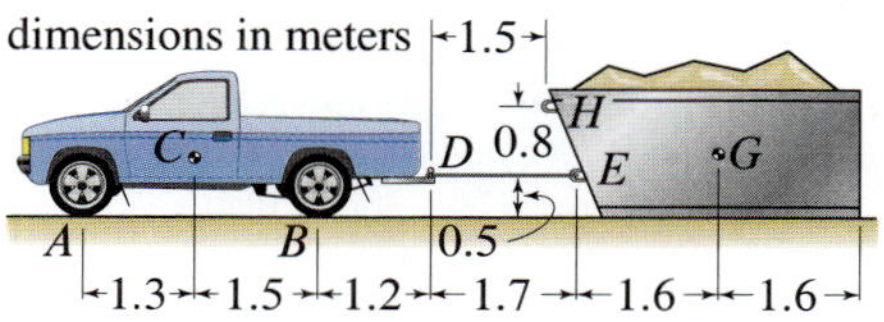

Figure P9.33 and P9.34

DESIGN PROBLEMS

General Instructions. In all problems, write a brief technical report following the guidelines of Appendix A, where you summarize all pertinent information in a well-organized fashion. It should be written using proper, simple English that is easy to read by another engineer. Where appropriate, sketches along with critical dimensions should be included. Discuss the objectives and constraints considered in your design, the process used to arrive at your final design, safety issues if appropriate, and so on. The main discussion should be typed, and figures, if needed, can be computer-drawn or neatly hand-drawn. Include a neat copy of all supporting calculations in an appendix that you can refer to in the main discussion of your report. A length of a few pages, plus appendix, should be sufficient.

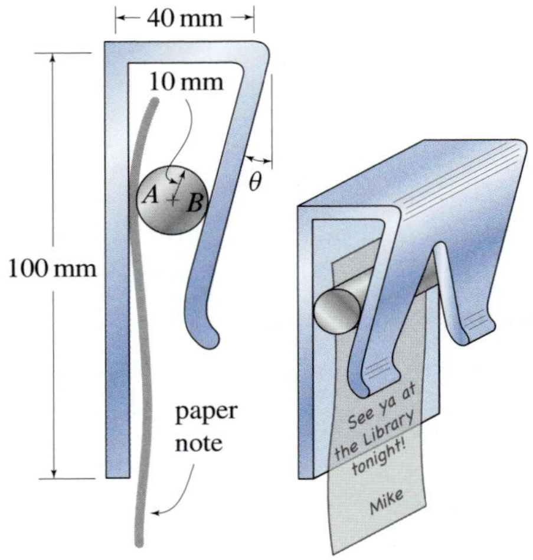

Figure DP9.3

Design Problem 9.3

Design a message holder that is to be mounted on a door or wall for holding paper notes. It is made of a 20 mm diameter circular aluminum bar and an extruded aluminum bracket. Both aluminum pieces have an anodized finish. Specify the angle θ so the holder is self-locking. That is, the note cannot be pulled out of the holder without manually lifting the circular bar. A large value of θ is probably desirable so that the circular bar may be easily lifted.

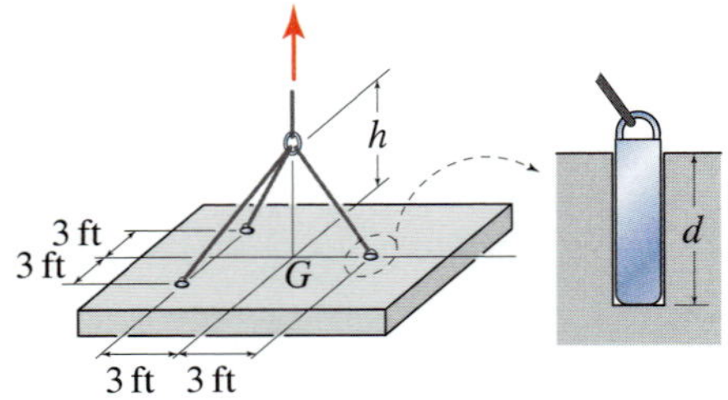

Figure DP9.4

Design Problem 9.4

A large concrete surface is to be removed by sawing it into rectangular pieces.* In each rectangular piece, three holes of depth d are drilled perpendicular to the slab's surface with the positioning shown relative to the slab's center of gravity G. In each hole a steel plug is inserted, and the slab is lifted via cables. Design this lifting scheme by specifying the depth d the holes should have and the height h so that the plugs will self-lock when the slab is lifted. Assume the concrete slab is at least 4 in. thick.

---

* A perhaps more conventional method of removal is to pulverize the slab using a tractor-mounted impact hammer. However, this method produces waste that is difficult to recycle. The scheme described here provides for waste that has useful salvage value. For example, such slabs are often used to line waterfronts for erosion control.

9.3 Belts and Cables Contacting Cylindrical Surfaces

In this section we use Coulomb's law to develop a theory for the tensile force in belts and cables sliding on cylindrical surfaces.

Figure 9.8
Engine for a Lotus Sport Exige 240R sports car. This engine has 4 cylinders and produces 243 hp. The belt wraps around 7 pulleys and is needed to operate a variety of essential engine accessories such as the power steering pump and alternator. The belt and pulleys must be designed so there is no slip.

Equilibrium analysis

Consider the situation of a flexible belt or cable in contact with a cylindrical surface, as shown in Fig. 9.9. At points A and B where the belt meets the cylindrical surface, the values of the tensile force in the belt are T_1 and T_2, respectively. We will use the convention that $T_2 \geq T_1$. Thus T_1 is the force on the *low-tension side*, and T_2 is the force on the *high-tension side*. Between points A and B the tensile force T is variable such that $T_1 \leq T \leq T_2$. In Fig. 9.9 the cylinder is fixed, and if slip occurs, the belt will move in the direction of T_2.*

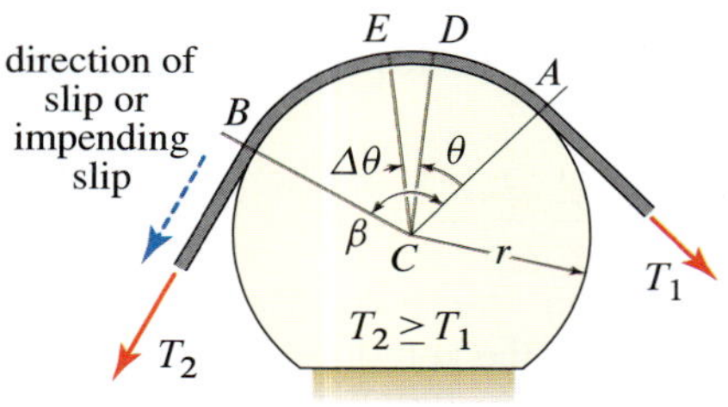

Figure 9.9
A flexible belt or cable wrapped around a fixed cylindrical surface.

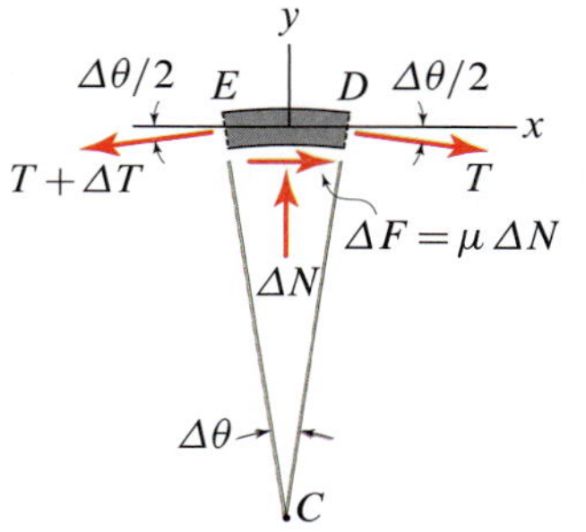

Figure 9.10
Free body diagram of a small segment of belt.

An angular coordinate θ is defined in Fig. 9.9 where $\theta = 0$ corresponds to point A on the low-tension side, and $\theta = \beta$ corresponds to point B on the high-tension side. Angle β is called the *angle of wrap*. Consider a small segment of the belt obtained by taking cuts through the belt at points D and E in Fig. 9.9. The FBD for this segment of the belt is shown in Fig. 9.10, and the equilibrium equations are

$$\sum F_x = 0: \quad -(T + \Delta T)\cos(\Delta\theta/2) + T\cos(\Delta\theta/2) + \Delta F = 0, \quad (9.6)$$

$$\sum F_y = 0: \quad -(T + \Delta T)\sin(\Delta\theta/2) - T\sin(\Delta\theta/2) + \Delta N = 0. \quad (9.7)$$

In these expressions, T is the tensile force at position θ (point D), and $T + \Delta T$ is the tensile force at position $\theta + \Delta\theta$ (point E). The normal force between the belt and cylindrical surface is ΔN, and the friction force is ΔF with direction that opposes the direction of relative slip. Assuming the belt slides on the surface or has impending slip, Coulomb's law provides $\Delta F = \mu\,\Delta N$ where

* The FBD in Fig. 9.10 and all of the expressions derived in this section are valid if Fig. 9.9 is revised so that the belt is fixed on the high-tension side and the cylinder rotates clockwise. Indeed, the FBD and all expressions are valid if both the belt and cylinder have motion provided the direction of slip or impending slip of the belt *relative* to the cylinder is in the direction of T_2.

μ is the coefficient of kinetic friction in the case of slip and μ is the coefficient of static friction in the case of impending slip. Because $\Delta\theta$ is small and is measured in radians, $\sin \Delta\theta \approx \Delta\theta$ and $\cos \Delta\theta \approx 1$. Using these expressions, dividing Eqs. (9.6) and (9.7) by $\Delta\theta$, taking the limit as $\Delta\theta \to 0$, and simplifying provide

$$\frac{dT}{d\theta} = \mu T. \tag{9.8}$$

This expression is a differential equation that describes how the tensile force T changes with position θ. It can be solved by multiplying both sides by $d\theta/T$ and integrating to write

$$\int_{T_1}^{T_2} \frac{1}{T}\, dT = \int_0^{\beta} \mu\, d\theta. \tag{9.9}$$

Assuming the coefficient of friction is uniform over the entire surface, the above integrals can be evaluated to obtain

$$\ln \frac{T_2}{T_1} = \mu\beta. \tag{9.10}$$

Taking the base e exponent ($e = 2.71828\ldots$) of both sides and solving for T_2 provide

$$\boxed{T_2 = T_1 e^{\mu\beta}, \qquad T_2 \geq T_1,} \tag{9.11}$$

where

- T_1 is the force in the low-tension side of the belt during slip or impending slip;
- T_2 is the force in the high-tension side of the belt during slip or impending slip;
- μ is the coefficient of static or kinetic friction; and
- β is the angle of wrap ($\beta \geq 0$, measured in radians).

> **Common Pitfall**
>
> **Overuse of $T_2 = T_1 e^{\mu\beta}$.** This expression relates belt tensions *only* when there is slip or impending slip. Prior to slip and impending slip, this expression cannot be used.

You should be mindful of the assumptions underlying Eq. (9.11): the surface must be cylindrical, friction is governed by Coulomb's law where the coefficient of friction is uniform, and the belt and drum are in the process of sliding relative to each other or sliding is impending. If sliding is not imminent, the most we can say is $(T_2)_{\max} > T_2 \geq T_1$ where $(T_2)_{\max} = T_1 e^{\mu\beta}$.

Problem solving Solving problems with belt friction can occasionally be challenging. To use Eq. (9.11), you must be able to identify the high-tension and low-tension sides of a belt. In other words, you must be able to determine the direction of relative slip. Example 9.6 discusses some intuitive and fundamental methods to determine the high-tension and low-tension sides of a belt. Problems with multiple contact surfaces can be challenging, and the remarks made earlier in this chapter also apply here. Depending on the kinematics of a problem, only a portion of the contacts may slip, and Eq. (9.11) may be used only for those belt-cylinder contacts that slip or have impending slip.

End of Section Summary

In this section, Coulomb's law was used to develop the relationship $T_2 = T_1 e^{\mu\beta}$ between values of the tensile forces on the high-tension side (T_2) and low-tension side (T_1) of a belt or cable in contact with a cylindrical surface under conditions of slip or impending slip. Prior to slip and impending slip, such as may occur in problems with multiple contact surfaces, this expression cannot be used.

EXAMPLE 9.6 *Belt Friction*

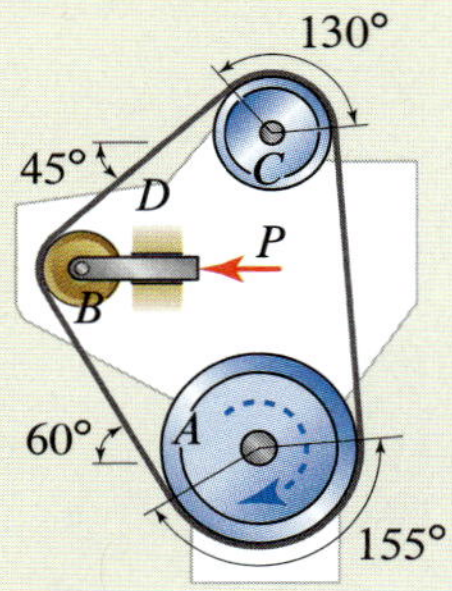

Figure 1

An accessory belt for an engine is shown. Pulley A is attached to the engine's crankshaft and rotates clockwise. The belt tensioner consists of a frictionless idler pulley at B that is mounted to a horizontal bar D that slides in a frictionless track with a horizontal force P. Pulley C operates a hydraulic pump that requires 200 N·m. The coefficients of static friction for pulleys A and C are 0.4 and 0.6, respectively, and the radii of pulleys A and C are 110 and 80 mm, respectively. Determine the minimum value of P so the belt does not slip.

SOLUTION

Road Map We will first identify the portions of the belt having uniform values of force. We then draw FBDs of selected pulleys and apply equilibrium equations to obtain some of the requirements the belt forces must satisfy. For example, the 200 N·m moment required of pulley C must be produced by the net moment of the belt forces about this pulley's bearing. Slip may occur between the belt and pulley A, or the belt and pulley C, and both of these possibilities must be considered to determine which occurs.

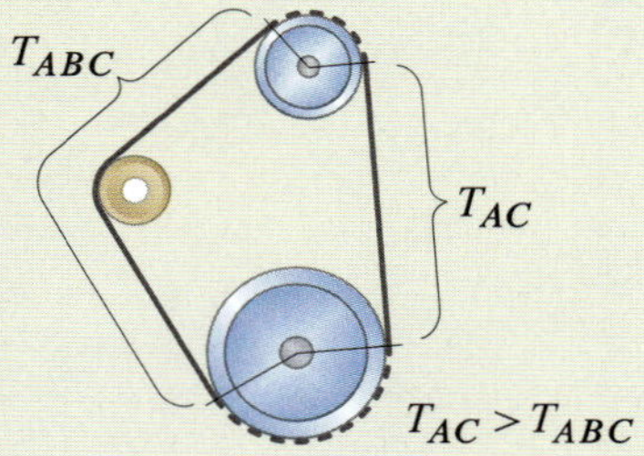

Figure 2
Regions of the belt having uniform values of force T_{ABC} and T_{AC}. The force is variable in portions of the belt shown by the dashed lines.

Modeling In Fig. 2 we identify portions of the belt that have uniform tensile force. Note that because pulley B is frictionless, the force in the entire left-hand portion of the belt from where the belt breaks contact with pulley A to where it makes contact with pulley C has the same value, T_{ABC}. The force in the right-hand portion of the belt is T_{AC}. In regions of the belt shown by dashed lines in Fig. 2, the force changes value, although the exact details of this are not needed. The FBDs for the pulleys are shown in Fig. 3.

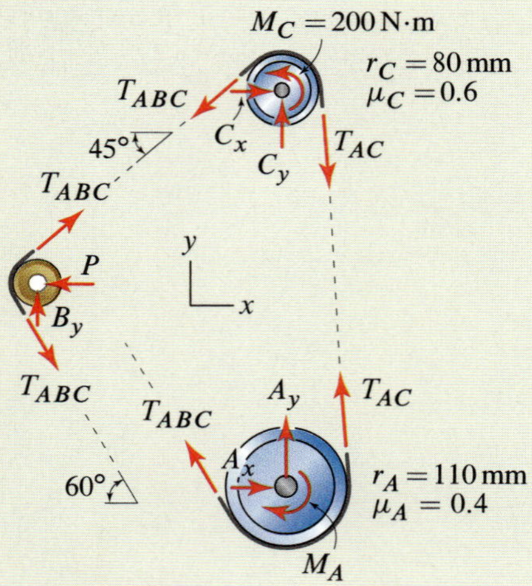

Figure 3
Free body diagrams for the pulleys.

Governing Equations

Equilibrium Equations Using the FBDs from Fig. 3, the equilibrium equations are

$$\textbf{Pulley } B\textbf{:} \quad \sum F_x = 0: \quad T_{ABC}(\cos 45°) + T_{ABC}(\cos 60°) - P = 0, \tag{1}$$

$$\Rightarrow \quad T_{ABC} = (0.8284)P, \tag{2}$$

$$\textbf{Pulley } C\textbf{:} \quad \sum M_C = 0: \quad T_{ABC}(0.08\,\text{m}) + 200\,\text{N·m} - T_{AC}(0.08\,\text{m}) = 0, \tag{3}$$

$$\Rightarrow \quad T_{AC} = T_{ABC} + 2500\,\text{N}. \tag{4}$$

Combining Eqs. (2) and (4) provides

$$T_{AC} = (0.8284)P + 2500\,\text{N}. \tag{5}$$

Equations (2) and (5) contain three unknowns T_{ABC}, T_{AC}, and P and thus cannot yet be solved.

Force Laws Before using Eq. (9.11) on p. 522 for belt friction, we must first identify the high-tension and low-tension belt forces. There are a variety of intuitive and fundamental ways to do this. You should study the problem description and Figs. 1 and 2 now to see if you can determine by inspection that $T_{AC} > T_{ABC}$.

Some explanations for why $T_{AC} > T_{ABC}$ follow:

- Pulley A is powered by moment M_A, which is supplied by the crankshaft of the engine. Hence, due to M_A, the right-hand portion of the belt (T_{AC}) is *pulled* toward the left-hand side (T_{ABC}), hence $T_{AC} > T_{ABC}$.
- You could examine Eq. (4) to determine $T_{AC} > T_{ABC}$.

- Starting with the FBD of pulley A, shown in Fig. 3 and again in Fig. 4(a), we remove the belt from the pulley to draw the two FBDs shown in Fig. 4(b) and (c). In Fig. 4(b), the friction forces, which are distributed over the contact region between the pulley and belt, must be in a direction so that the pulley is in moment equilibrium. Knowing the direction of the friction forces in Fig. 4(b), the friction forces applied to the belt, shown in Fig. 4(c), must be in the opposite direction. Consideration of moment equilibrium $\sum M_A = 0$ for the FBD in Fig. 4(c) shows that $T_{AC} > T_{ABC}$.

We now apply Eq. (9.11) for belt friction to determine the value of P needed to cause each of the pulleys to slip.

If pulley A slips: $T_{AC} = T_{ABC}\, e^{\mu_A \beta_A}$ (6)

$$(0.8284)P + 2500\,\text{N} = (0.8284)P e^{(0.4)(155°)(\pi\ \text{rad}/180°)} \quad (7)$$

$$\Rightarrow \quad P = 1547\,\text{N}. \quad (8)$$

If pulley C slips: $T_{AC} = T_{ABC}\, e^{\mu_C \beta_C}$ (9)

$$(0.8284)P + 2500\,\text{N} = (0.8284)P e^{(0.6)(130°)(\pi\ \text{rad}/180°)} \quad (10)$$

$$\Rightarrow \quad P = 1040\,\text{N}. \quad (11)$$

Comparing Eqs. (8) and (11), we observe that if P is larger than 1547 N, then neither of the pulleys will slip. When $P = 1547\,\text{N}$, slip is impending at pulley A. Hence, the smallest value of P to prevent slip is

$$\boxed{P_{min} = 1547\,\text{N},} \quad (12)$$

and when this occurs, slip is impending at pulley A.

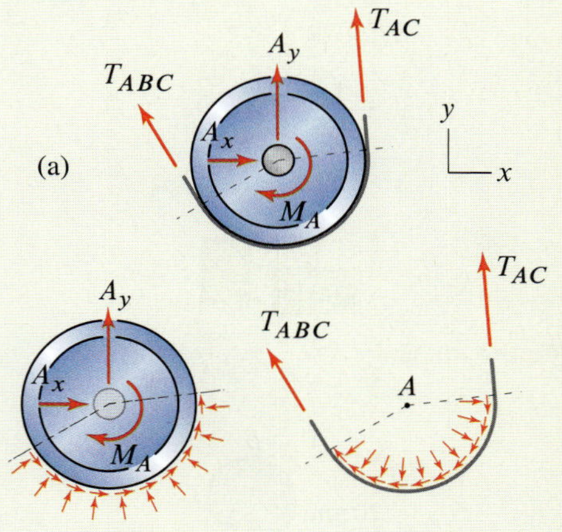

Figure 4
Free body diagrams for pulley A to help determine the high-tension and low-tension portions of the belt. (a) Free body diagram for pulley A and the belt together. (b) Free body diagram for only the pulley, showing the friction and normal forces the belt applies to the pulley. (c) Free body diagram for only the belt, showing the friction and normal forces the pulley applies to the belt.

Discussion & Verification

- If it is not clear that slip first occurs when P is the *larger* of Eqs. (8) and (11), consider the situation in which P is very large so that neither of the pulleys slips. Then as P decreases, slip will occur when the larger of Eqs. (8) and (11) is achieved.
- To provide a reasonable margin of safety against slip, we will want to use a value of P greater than 1547 N. Some of the sources of uncertainty are as follows. Pulley B may have nonnegligible friction, in which case slip at pulley A may occur at a slightly higher value of P. Probably the the greatest source of uncertainty is the possibility that the coefficients of friction might be lower at some time during the belt's life, such as might occur if the belt comes in contact with a small amount of lubricant or other fluid from the engine.
- The FBD of pulley A was not used in this solution. However, if it is necessary to determine the moment M_A the crankshaft applies to this pulley, or the reactions A_x and A_y, then this FBD would be used to write the appropriate equilibrium equations.

Common Pitfall

Measure β in radians. A common error when using the expression $T_2 = T_1\, e^{\mu\beta}$ is to forget to express the angle of wrap β in radians.

PROBLEMS

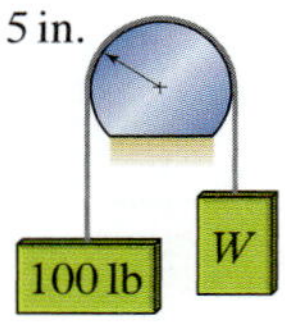

Figure P9.35

Problem 9.35

Two objects are connected by a weightless cord that is wrapped around a fixed cylindrical surface. The coefficient of static friction between the cord and surface is 0.3. One of the objects weighs 100 lb, and the other has weight W. If the system is initially at rest, determine the range of values for W for which there is no motion.

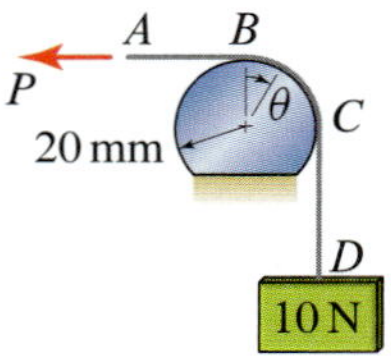

Figure P9.36 and P9.37

Problem 9.36

A 10 N weight is supported by a weightless cable where portion AB of the cable is horizontal. The coefficients of static and kinetic friction are 0.5.

(a) Determine the value of P so that downward motion of the 10 N weight is impending.

(b) If $P = 3$ N, determine the force in portion CD of the cable. Is the system in equilibrium or is there motion? Explain.

Problem 9.37

A 10 N weight is supported by a weightless cable where portion AB of the cable is horizontal. A test is conducted where it is found that $P = 4$ N will cause impending motion of the 10 N weight at D.

(a) Determine the coefficient of static friction between the cable and cylindrical surface.

(b) Determine the tension in portion BC of the cable as a function of angular position θ. Plot this function.

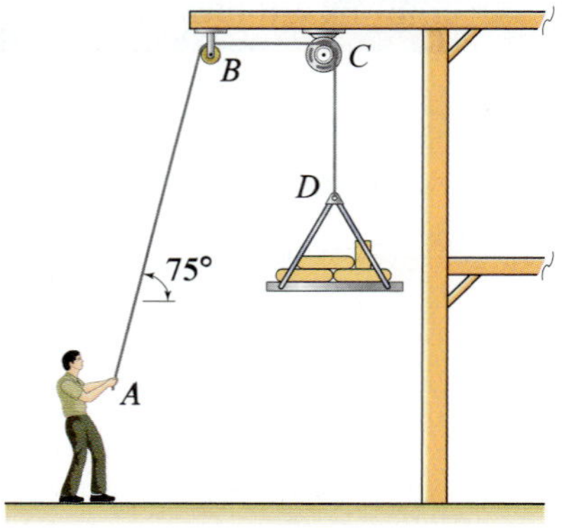

Figure P9.38 and P9.39

Problem 9.38

A hoist for lifting lightweight building materials is shown. The pulley at B is frictionless, and the pulley at C is driven by an electric motor rotating counterclockwise at a speed high enough to always cause slip between the cable and pulley C. The cable makes contact with one-fourth of the surface of pulley C, and the coefficients of static and kinetic friction are 0.25 and 0.2, respectively. If the cable has negligible weight and the building materials and platform at D have a combined weight of 100 lb, determine the force the worker must apply to the cable to

(a) Raise the building materials at a uniform speed.

(b) Lower the building materials at a uniform speed.

(c) Hold the building materials in a fixed, suspended position.

Problem 9.39

In Prob. 9.38, if the cable were given an additional full wrap around the pulley at C and if the worker can apply a force of 50 lb to the cable, determine the largest weight that may be lifted at D.

Problems 9.40 through 9.43

A brake for reducing the speed of a rotating drum is shown. The *braking moment* is defined as the resultant moment the belt produces about the drum's bearing, point A. Determine the braking moment if the coefficient of kinetic friction is 0.35 and

Problem 9.40 The drum rotates clockwise and $P = 90\,\text{N}$.

Problem 9.41 The drum rotates counterclockwise and $P = 90\,\text{N}$.

Problem 9.42 The drum rotates clockwise and $P = 20\,\text{lb}$.

Problem 9.43 The drum rotates counterclockwise and $P = 20\,\text{lb}$.

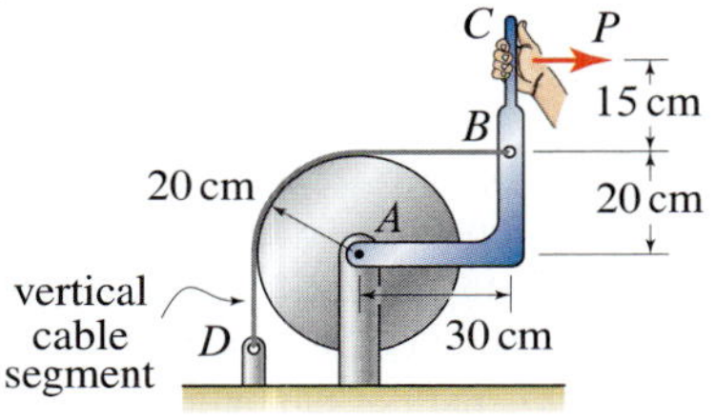

Figure P9.40 and P9.41

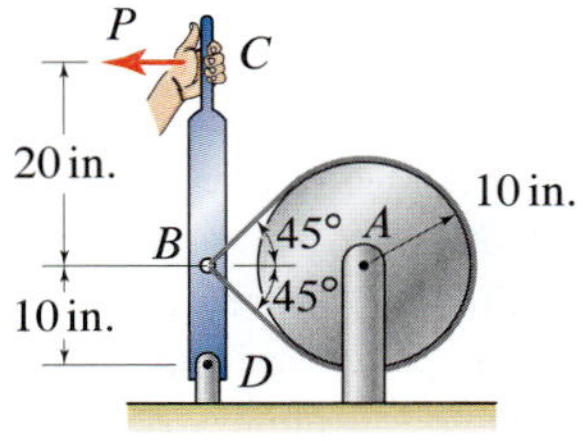

Figure P9.42 and P9.43

Problem 9.44

A spring AB with 2 N/mm stiffness is attached to a cable at B. The cable wraps around two cylindrical surfaces, and cable segments BC, DE, and GH are horizontal. All contact surfaces have coefficients of static and kinetic friction of 0.1 and 0.09, respectively. The spring is initially unstretched.

(a) Determine the value of P needed to stretch the spring by 4 mm.

(b) Once the spring is stretched by 4 mm, determine the value to which P must be reduced so that the spring begins to contract.

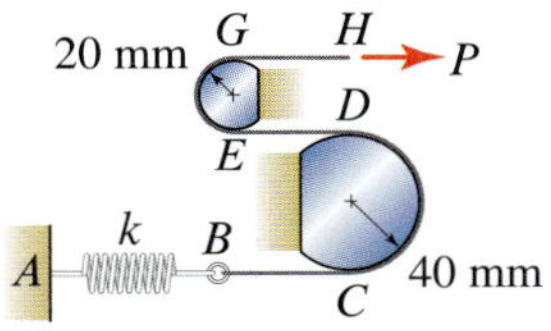

Figure P9.44

Problem 9.45

Pulley A of the treadmill is driven by a motor that can supply a moment $M_A = 200\,\text{in.}\cdot\text{lb}$. Pulley B is frictionless, and its bearing slides in a frictionless slot where force P tensions the belt. Both pulleys A and B have 1.5 in. radius, and the coefficient of static friction between pulley A and the belt is 0.3. Determine the smallest value of P so there is no slip between pulley A and the belt.

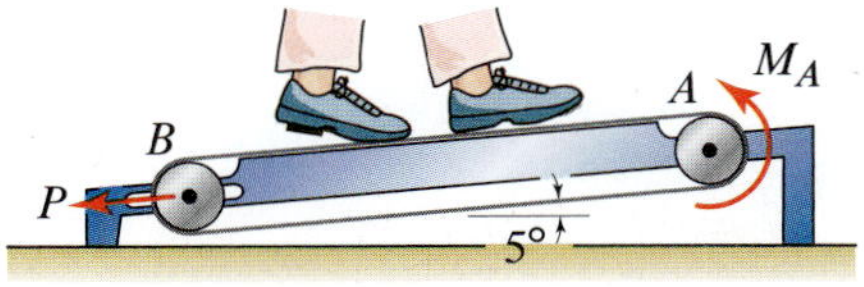

Figure P9.45

Problem 9.46

The winch is driven by an electric motor at A. The motor weighs 50 lb with center of gravity at the center of pulley A. The platform supporting the motor has negligible weight and is hinged at D. The motor turns the pulley at B via a belt between A and B. Pulley B has a spool upon which the rope BCG is wrapped. The coefficient of static friction at A is 0.3 and at B is 0.2. Assuming the motor has sufficient power, determine the largest weight W that may be lifted if there is no slip between the belt and pulleys A and B. *Hint:* The use of mathematical software is helpful, but not required.

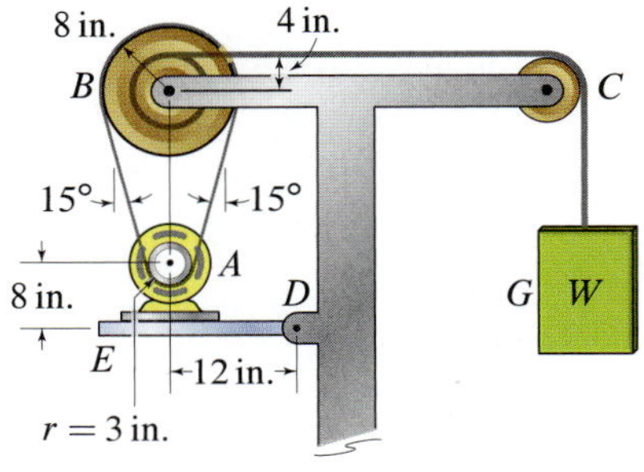

Figure P9.46

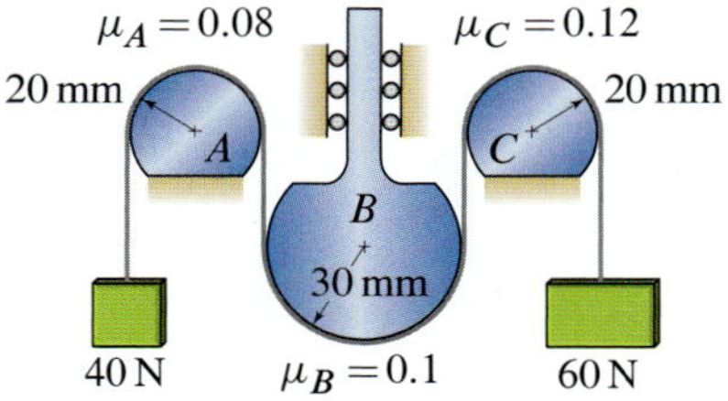

Figure P9.47 and P9.48

Problem 9.47

Surfaces A, B, and C are cylindrical with the coefficients of static friction shown. Object B can undergo vertical motion only. Determine the weight of object B that causes it to have impending downward motion, and describe which of surfaces A, B, and/or C will have impending slip.

Problem 9.48

In Prob. 9.47, determine the weight of object B that causes it to have impending upward motion, and describe which of surfaces A, B, and/or C will have impending slip.

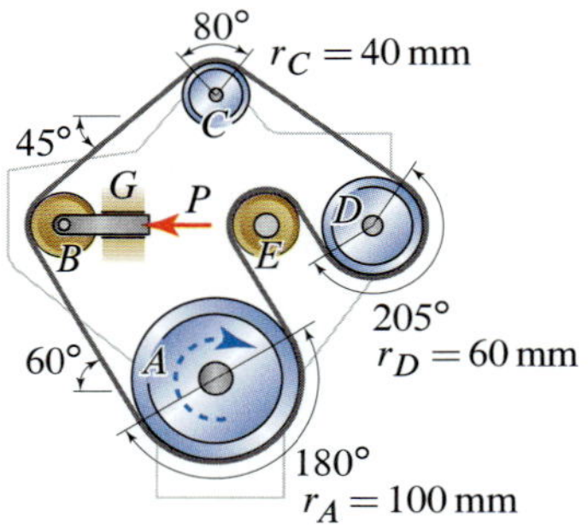

Figure P9.49

Problem 9.49

An accessory belt for an engine is shown. Pulley A is attached to the engine's crankshaft and rotates clockwise. The belt tensioner consists of a frictionless pulley at B that is mounted to a horizontal bar G that slides in a frictionless track with a horizontal force P. Pulley C operates an alternator that requires 100 N·m, and pulley D operates a hydraulic pump that requires 300 N·m. Pulley E is frictionless. The coefficients of static friction for pulleys A, C, and D are 0.5 and the radii of the pulleys are shown. Determine the minimum value of P so the belt does not slip. *Hint:* The use of mathematical software is helpful, but not required.

9.4 Chapter Review

Important definitions, concepts, and equations of this chapter are summarized. For equations and/or concepts that are not clear, you should refer to the original equation and page numbers cited for additional details.

Coulomb's law of friction. *Coulomb's law of friction* states

Eqs. (9.1) and (9.2), p. 499

$$|F| \le \mu_s N \qquad \text{before sliding } (<) \text{ and for impending motion } (=),$$
$$|F| = \mu_k N \qquad \text{after sliding begins,}$$

where:

N is the normal force between the two surfaces, defined to be positive in compression, hence $N \ge 0$ always;
F is the friction force between the two surfaces, with direction that opposes relative motion between the surfaces;
μ_s (Greek letter *mu*) is called the *coefficient of static friction*;
μ_k is called the *coefficient of kinetic friction*.

We will usually write Coulomb's law given in Eqs. (9.1) and (9.2) in the compact form

Eq. (9.3), p. 499

$$|F| \le \mu N,$$

where it is understood that the coefficient of static friction is used prior to sliding and when motion is impending, and the coefficient of kinetic friction is used once motion starts along with the = sign.

Draw FBDs so that the friction forces are in proper directions to resist sliding motion. When this is the case, all friction forces have positive values and the absolute value sign in Coulomb's law can be omitted.

Angle of friction. The *angle of friction* is defined to be

Eq. (9.4), p. 501

$$\phi = \tan^{-1} \mu,$$

where μ is either the static or kinetic coefficient of friction and the value of ϕ is the angle of static or kinetic friction, respectively.

Coulomb's law of friction in three dimensions. Coulomb's law of friction (isotropic) in three dimensions is

Eq. (9.5), p. 502

$$F = \sqrt{(F_x)^2 + (F_y)^2} \le \mu N,$$

where F_x and F_y are the friction forces in two mutually perpendicular directions along the surface and F is the resultant of these.

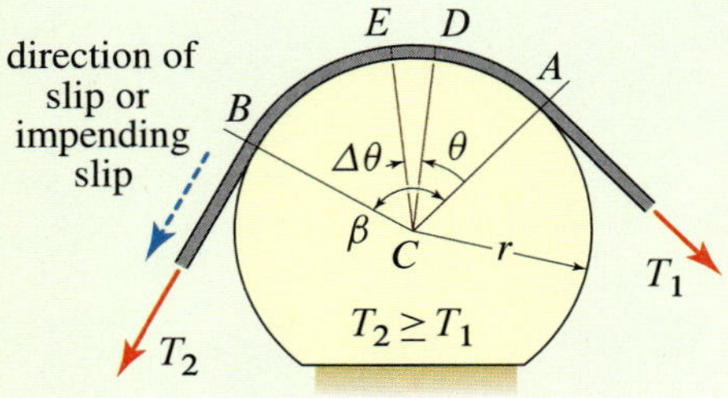

Figure 9.11
A flexible belt or cable wrapped around a fixed cylindrical surface.

Belts and cables contacting cylindrical surfaces. For slip or impending slip between a flexible belt or cable in contact with a cylindrical surface, as shown in Fig. 9.11, the forces on the *high-tension* and *low-tension* sides of the belt are related by

Eq. (9.11), p. 522

$$T_2 = T_1 e^{\mu\beta} \qquad T_2 \geq T_1$$

where

T_1 is the force in the low-tension side of the belt during slip or impending slip;
T_2 is the force in the high-tension side of the belt during slip or impending slip;
μ is the coefficient of static or kinetic friction;
β is the *angle of wrap* ($\beta \geq 0$, measured in radians).

Although Fig. 9.11 shows the cylinder as being fixed and the belt sliding or having impending slip in the direction of T_2, Eq. (9.11) is also valid if the belt is fixed on the high-tension side and the cylinder rotates; or both the belt and cylinder may have motion, in which case the direction of slip or impending slip of the belt *relative* to the cylinder is in the direction of T_2.

REVIEW PROBLEMS

Problem 9.50

A simple bracket for supporting a can of paint or other items used during construction consists of a member *ABC* and two short bars that are rigidly connected to it. The bracket is slipped onto a wooden post and is held in place by friction. Determine the minimum coefficient of static friction so that the bracket does not slip regardless of the weight supported (this is called *self-locking*).

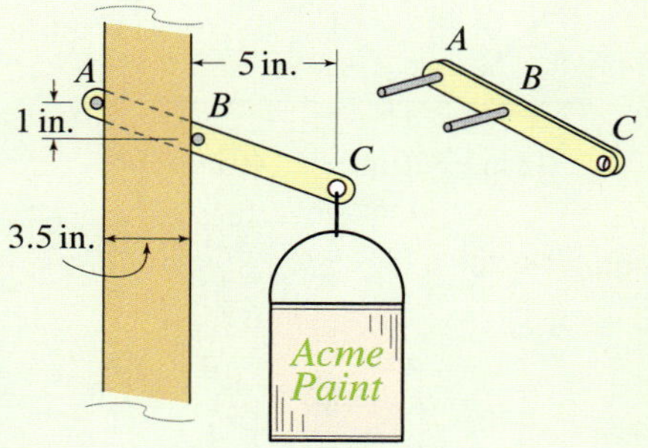

Figure P9.50

Problem 9.51

Member *CD* is used to apply a horizontal force *P* to a uniform cylinder having radius *r* and weight *W*. The contact surfaces at *A* and *B* have the same coefficient of static friction μ_s. Neglecting the weight of member *CD*, determine the force *P* necessary to cause impending motion of the cylinder. Express your answer in terms of parameters such as W, r, μ_s.

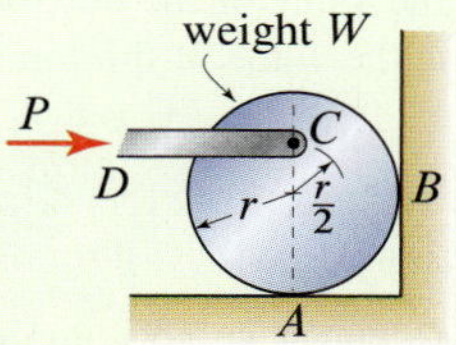

Figure P9.51

Problem 9.52

For a 3 m thick section of dam and water, determine the normal force and friction force between the dam and foundation, specify the distance from point *A* to the location of this force system, and determine the minimum value of the coefficient of friction so the dam does not slide. The density of the dam is $2400\ \mathrm{kg/m^3}$, and the density of water is $1000\ \mathrm{kg/m^3}$.

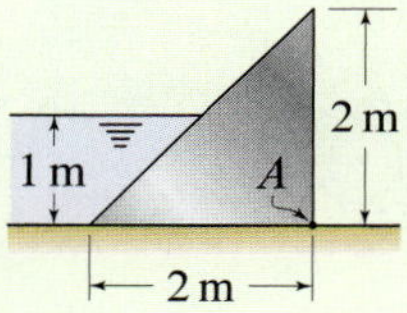

Figure P9.52

Problems 9.53 and 9.54

A roll of paper *C* with 8 N weight is supported by edges *A* and *B*. A horizontal force *P* is applied to pull a sheet of paper from the roll. Determine the value of *P* that causes motion, and determine if the motion is rolling or tipping. The coefficient of static friction between the roll of paper and edges *A* and *B* is

Problem 9.53 0.1.

Problem 9.54 0.3.

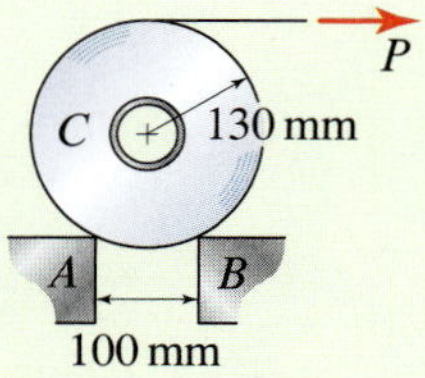

Figure P9.53 and P9.54

Problem 9.55

A car weighing 4000 lb rests on a slope with coefficient of static friction of 0.4. The car is disabled with its parking brake applied so that none of its wheels will turn or roll. A tow truck applies a *y* direction force *P* to the car. Determine the value of *P* that causes impending motion of the car, and determine the unit vector that describes the direction in which the car will begin to slide.

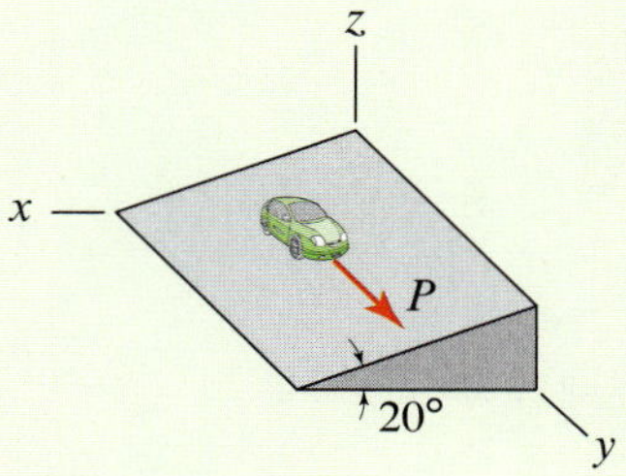

Figure P9.55

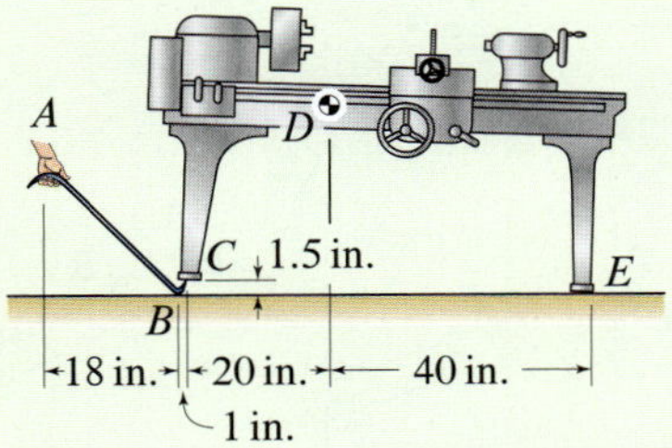

Figure P9.56

Problem 9.56

To move a lathe weighing 1200 lb with center of gravity at point D, a machinist uses a steel pry bar as shown. The legs of the lathe are iron and the floor is concrete. The coefficient of static friction for steel on iron is 0.3, and steel or iron on concrete is 0.35. If the force P applied to the pry bar at A is vertical, determine the value of P necessary to cause impending motion and describe the motion.

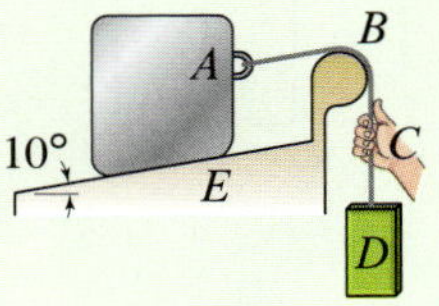

Figure P9.57

Problem 9.57

The device shown allows for easy elevation adjustments of block A on the fixed ramp E. When the user applies a sufficient upward vertical force to the cable at C, block A is lowered. When the user applies a sufficient downward vertical force to the cable at C, block A is raised. The coefficients of static and kinetic friction between block A and the ramp, and between the cable and cylindrical surface at B, are 0.1. Block A has 40 N weight, the cable has negligible weight, cable segment AB is parallel to the ramp, segment BCD is vertical, and assume block A does not tip. Determine the weight W_D of the counterweight at D so that the value of the upward force the user must apply to cause motion is the same as the value of the downward force the user must apply to cause motion. Also, determine the value of the force the user must apply to cause motion.

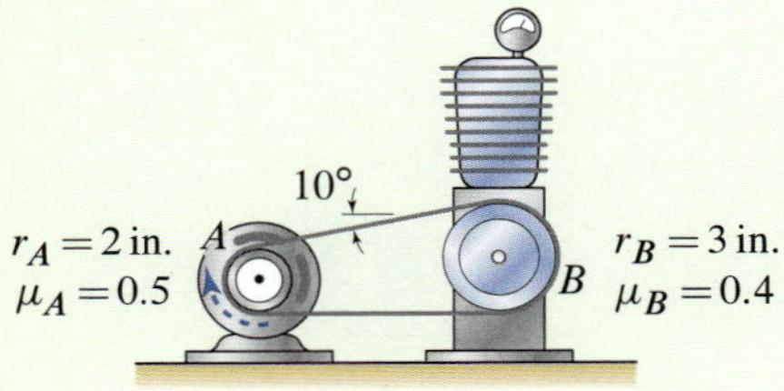

Figure P9.58

Problem 9.58

An electric motor at A is used to power an air compressor that requires a 200 in.·lb moment at pulley B to operate. When the motor is turned off, the entire belt is at a uniform tensile force T_0. When the motor is turned on, pulley A rotates clockwise and the tensile forces in the upper and lower portions of the belt change to $T_0 - \Delta T$ and $T_0 + \Delta T$, respectively. The lower portion of the belt is horizontal, and the pulley radii and coefficients of static friction are shown. Determine the initial belt tension T_0 so that neither pulley slips.

Problem 9.59

In Prob. 9.58, it was stated that when the motor is off, the tensile force in the belt is T_0, and when the motor is on, the tensile forces in the upper and lower portions of the belt change to $T_0 - \Delta T$ and $T_0 + \Delta T$, respectively. Offer justification for why the increase in force in the high-tension portion of the belt is equal to the decrease in force in the low-tension portion of the belt. *Hint:* Idealize the upper and lower portions of the belt as springs, and consider the deformations they undergo when the motor is on.

Note: Concept problems are about *explanations*, not computations.

10 *Moments of Inertia*

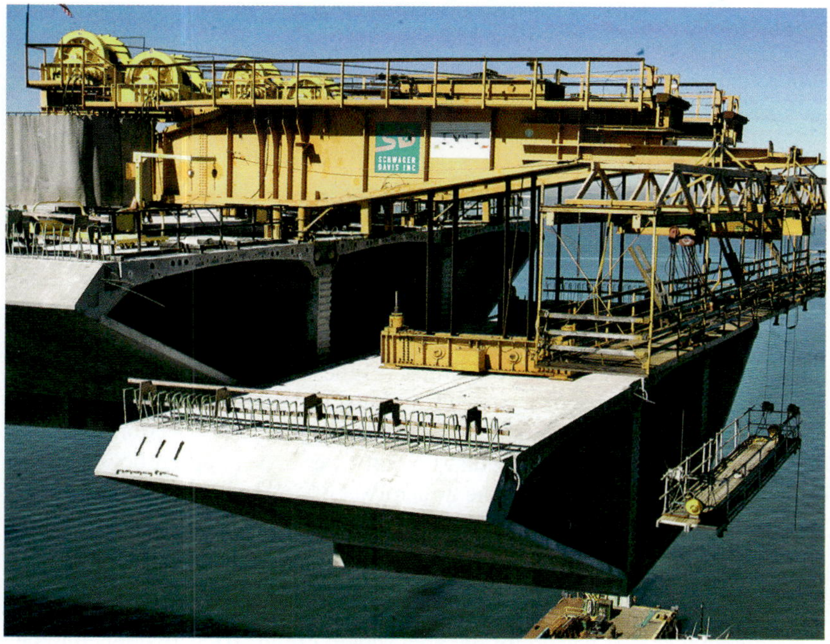

This chapter discusses area moments of inertia and mass moments of inertia. Area moments of inertia are measures of how an area is distributed about particular axes. Mass moments of inertia are measures of how mass is distributed about particular axes. Moments of inertia are used extensively in mechanics of materials, dynamics, and subjects that follow.

10.1 Area Moments of Inertia

Area moments of inertia are measures of how an area is distributed about particular axes. Before delving into the precise definitions of these, we first discuss two examples to help you understand why such a measure is useful and what this measure quantifies. The first example addresses exam scores, and the second addresses behavior of beams.

An example—test scores

Imagine you took two exams in a class, and on both exams the class average was 70% and your score was 85%. While you obviously did very well on these exams, at first glance it would appear that your performance and standing relative to your peers were identical. A study of the frequency distribution of scores for the two exams in Fig. 10.1 shows that in fact your performance on the second exam was even better than on the first because on the second exam there were fewer students who had scores higher than yours. Because the same number of students took both exams, the area under each distribution is

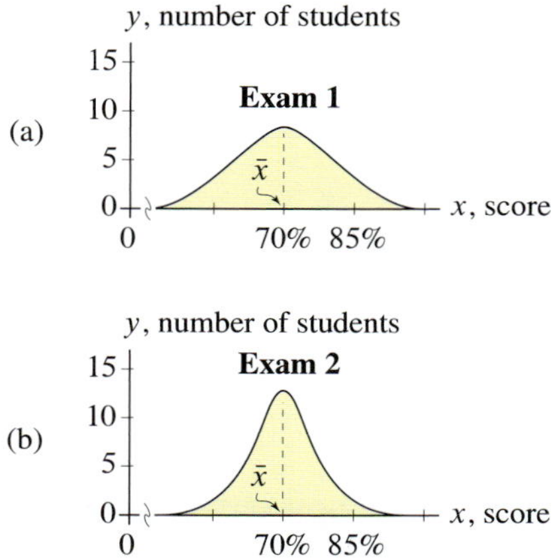

Figure 10.1
Frequency distributions for two exams. Both exams have the same average $\bar{x} = 70\%$ and the same number of students. The first exam has more students farther from the average than for the second exam, hence the *variance* and *standard deviation*, both of which are measures of a distribution, are greater for the first exam.

the same. Also, the class average is the centroid of each distribution, thus $\bar{x} = 70\%$. Common measures of distributions used in statistics are the *variance* v and *standard deviation* s, and these are related by $s = \sqrt{v}$. Because the distribution of the first exam has more students farther from the average than for the second exam, the variance and standard deviation for the first exam are greater than those for the second. Although we do not show this, the definition of the variance is identical to the area moment of inertia discussed in this section, and the standard deviation is similar to the radius of gyration.

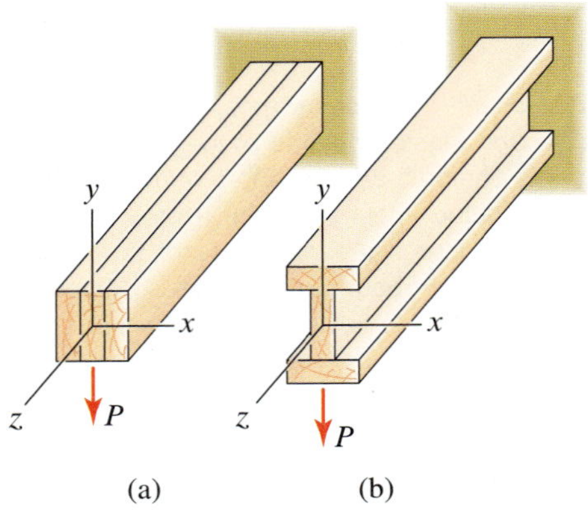

Figure 10.2
Two tip-loaded cantilever beams, each constructed by nailing or gluing three identical planks of wood together. Despite having the same length and cross-sectional area, beam (a) with the rectangular cross section will deflect substantially more than beam (b) with the I-shaped cross section.

An example—beam loading

Figure 10.2 shows two tip-loaded cantilever beams. Both beams are constructed by nailing or gluing three identical planks of wood together. Thus, both beams have the same length and cross-sectional area. When the beams are subjected to the same load P, the beam with the rectangular cross section in Fig. 10.2(a) deflects more than the beam with the I-shaped cross section (called an *I beam*) in Fig. 10.2(b). Although knowledge of mechanics of materials, along with moments of inertia from this section, is required to prove this, if each plank of wood has, for example, 2 in. by 6 in. dimensions, the rectangular cross section beam deflects almost 4 times as much as the I beam! The essential reason the I beam is so much stiffer is that more of its cross-sectional area is farther from the x axis than for the rectangular cross section. The area moment of inertia quantifies the distribution of area. Although proof of this requires mechanics of materials, the deflections of a beam are *inversely* related to the moment of inertia of its cross section.

Definition of area moments of inertia

The *area moments of inertia* for the area shown in Fig. 10.3 are defined to be

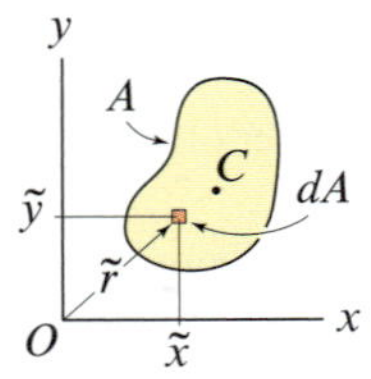

Figure 10.3
An area A with centroid C. $\tilde{x}$ and $\tilde{y}$ are the x and y positions, respectively, of the centroid of area element dA. $\tilde{r}$ is the distance from point O to area element dA.

$$I_x = \int \tilde{y}^2\, dA, \tag{10.1}$$

$$I_y = \int \tilde{x}^2\, dA, \tag{10.2}$$

$$J_O = \int \tilde{r}^2\, dA = I_x + I_y, \tag{10.3}$$

$$I_{xy} = \int \tilde{x}\tilde{y}\, dA, \tag{10.4}$$

where

$\tilde{x}$, $\tilde{y}$, and $\tilde{r}$ are defined in Fig. 10.3. Note that the definitions of these are identical to those used in Chapter 7. For example, $\tilde{x}$ is the distance (i.e., moment arm) from the y axis to the centroid of area element dA.

I_x is the *area moment of inertia about the x axis*.

I_y is the *area moment of inertia about the y axis*.

J_O is the *polar moment of inertia of the area about point O*.

I_{xy} is the *product of inertia of the area about point O*.

Concept Alert

Area Moments of inertia. Area moments of inertia are measures of how an area is distributed about particular axes. Area moments of inertia depend on the geometry of an area (size and shape) and the axes you select. Area moments of inertia are independent of forces, materials, and so on.

Remarks

- When referring to area moments of inertia, we often omit the word *area* when it is obvious from the context that we are dealing with area moments of inertia as opposed to mass moments of inertia, which are discussed later in this chapter.

- All of the moments of inertia in Eqs. (10.1)–(10.4) measure the *second moment* of the area distribution. That is, to determine I_x, I_y, and J_O in Eqs. (10.1)–(10.3), the moment arms $\tilde{y}$, $\tilde{x}$, and $\tilde{r}$, respectively, are *squared*. For I_{xy} in Eq. (10.4), the product of two different moment arms is used. Area moments of inertia are sometimes called *second moments of an area*.

- In Eqs. (10.1)–(10.4), $\tilde{x}$, $\tilde{y}$, and $\tilde{r}$ have dimensions of length, and dA has dimensions of area. Hence, all of the area moments of inertia have dimensions of $(length)^4$, such as in.^{4} or mm^4.

- Two expressions for the polar moment of inertia J_O are given in Eq. (10.3). The first is the fundamental definition, while the second is a simplification that is obtained as follows. From Fig. 10.3, it is seen that $\tilde{r}^2 = \tilde{x}^2 + \tilde{y}^2$. Substituting this expression into the definition of J_O in Eq. (10.3) and combining with Eqs. (10.1) and (10.2) show that $J_O = I_x + I_y$. Thus, if I_x and I_y are known for a particular shape, J_O may be easily determined as the sum of these rather than by evaluating the integral expression in Eq. (10.3).

- When the x and y axes pass through the centroid of a shape (point C in Fig. 10.3), we sometimes define the axes as x' and y' and we refer to the moments of inertia as *centroidal moments of inertia* with the designations $I_{x'}$, $I_{y'}$, etc. In mechanics of materials, moments of inertia about centroidal axes are especially important.

- For any area with finite size, I_x, I_y, and J_O are always positive. The product of inertia I_{xy} may be positive, zero, or negative, as discussed below.

- Evaluation of moments of inertia using composite shapes is possible, but this approach requires the *parallel axis theorem*, which is discussed in Section 10.2.

What are area moments of inertia used for?

It is useful to discuss why there are four area moments of inertia, what is different about them, and what they are used for.

Moment of inertia I_x. In Fig. 10.4, a cantilever beam is shown before and after a tip load P is applied, where P acts in the $-y$ direction. Due to P, the tip of the beam deflects in the $-y$ direction by the distance δ, *and* the cross section at the tip of the beam also rotates about the x axis. In fact, every cross section of the beam also deflects in the $-y$ direction and rotates about the x axis, although with lower values as cross sections become closer to the built-in support where there is no deflection or rotation. Because each cross section rotates about the x axis, the dispersion of the cross section's area about the x

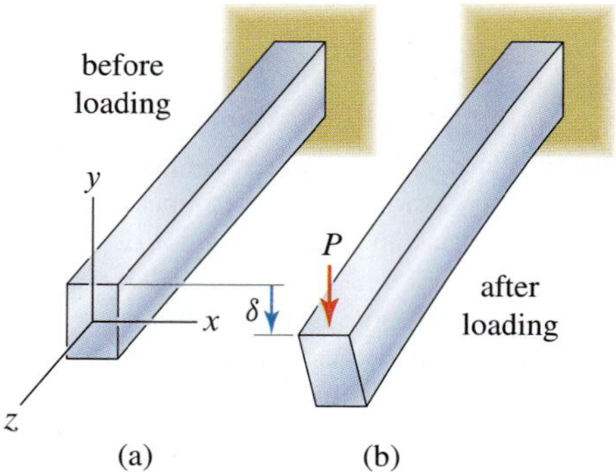

Figure 10.4
A cantilever beam with rectangular cross section before and after a tip load P is applied in the $-y$ direction.

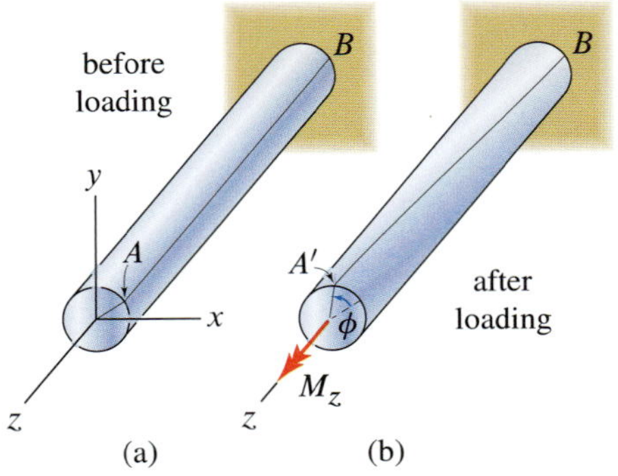

Figure 10.5
A cantilever beam (or shaft) with circular cross section before and after a tip moment M_z is applied about the z axis.

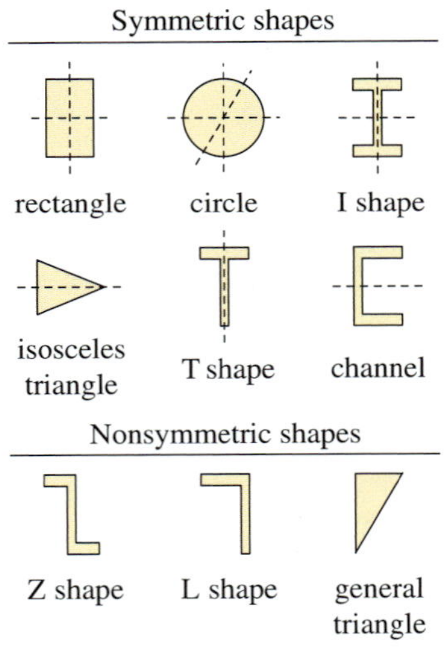

Figure 10.6
Examples of *symmetric* and *nonsymmetric* shapes. Symmetric shapes have at least one axis of symmetry (shown by dashed lines). The circular shape has an infinite number of axes of symmetry, all of which pass through the center of the circle. The product of inertia is zero for symmetric shapes when one of the coordinate axes is coincident with an axis of symmetry.

axis is important, and while you must accept this statement on faith until you study mechanics of materials, the moment of inertia about the x axis, I_x, plays an important role in how large the beam's deflections are.

Moment of inertia I_y. If Fig. 10.4 is revised so that the tip load P acts in the x direction, then deflections of the beam are in the x direction, cross sections rotate about the y axis, and the moment of inertia about the y axis, I_y, plays an important role.

Polar moment of inertia J_O. In Fig. 10.5, a cantilever beam is shown before and after a tip moment M_z acting about the z axis is applied (for this loading, it is customary to call the structure a *shaft* rather than a beam). Before the moment is applied, a line AB is drawn on the shaft. After M_z is applied, the tip of the shaft rotates about the z axis by angle ϕ (called the *angle of twist*) and point A moves to A'. In fact, every cross section of the shaft also rotates about the z axis, although with lower values as cross sections become closer to the built-in support where there is no rotation. The polar moment of inertia J_O plays an important role in determining the angle of twist ϕ.

Product of inertia I_{xy}. The product of inertia measures the asymmetry of an area with respect to the x and y axes. It can have positive, zero, or negative value, depending on the shape of an area, the selection of the x and y directions, and the location of the origin of the xy coordinate system. A *symmetric shape* has at least one axis of symmetry, a *nonsymmetric shape* has no axis of symmetry, and some examples are shown in Fig. 10.6. In engineering design, the vast majority of beams have symmetric cross section and are analyzed by taking the x and/or y direction to be an axis of symmetry for the cross section; under these circumstances $I_{xy} = 0$.

For a nonsymmetric cross section, I_{xy} is generally nonzero. Beams with nonsymmetric cross section display complicated behavior. For example, if the beam in Fig. 10.4 has the Z shape cross section shown in Fig. 10.6, then in addition to deflecting in the $-y$ direction due to load P, the tip deflects in the $-x$ direction! Such behavior is generally not desirable, hence beams are most often designed with a symmetric cross-sectional shape. The theory of beams with nonsymmetric cross sections is an advanced subject, and until you study this topic, you will not need to use the product of inertia. Hence, methods for evaluating I_{xy} are not discussed in this book.*

Radius of gyration

Rather than use area moments of inertia to quantify how an area is distributed, it is sometimes convenient to use the *radii of gyration*, which are directly related to the area moments of inertia and are defined as

$$k_x = \sqrt{\frac{I_x}{A}}, \qquad k_y = \sqrt{\frac{I_y}{A}}, \qquad k_O = \sqrt{\frac{J_O}{A}}, \tag{10.5}$$

where

* A discussion of the product of inertia and behavior of beams with nonsymmetric cross section is given in R. D. Cook and W. C. Young, *Advanced Mechanics of Materials*, 2nd ed., Prentice-Hall, Englewood Cliffs, N.J., 1998.

k_x is the *radius of gyration of the area about the x axis.*

k_y is the *radius of gyration of the area about the y axis.*

k_O is the *polar radius of gyration of the area about point O.*

A is the area of the shape.

The radii of gyration have units of *length* and have a straightforward physical interpretation, as discussed in Example 10.1.

Evaluation of moments of inertia using integration

Most of what was said in Chapter 7 with respect to evaluating the first moment of an area distribution [e.g., expressions $\int \tilde{x}\, dA$ and $\int \tilde{y}\, dA$ in Eq. (7.9) on p. 397] also applies here, with one additional restriction. In this section, we will always choose the area element dA so that *all* of the area within that element has the same moment arm $\tilde{x}$, $\tilde{y}$, or $\tilde{r}$. Figure 10.7 shows some acceptable choices of area elements for use in evaluating moments of inertia I_x and I_y, with the following comments:

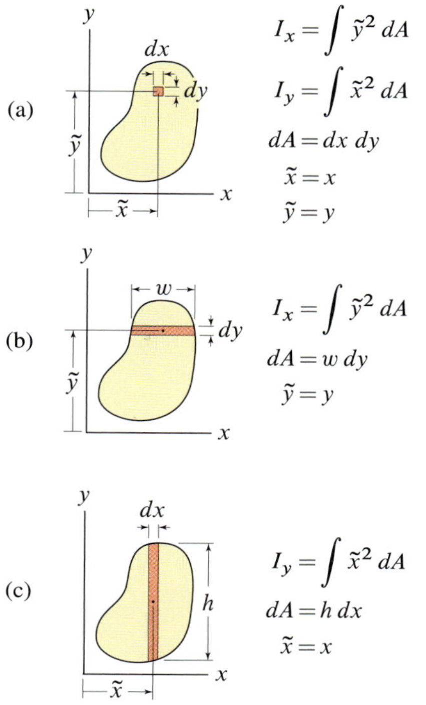

Figure 10.7
Acceptable choices for selection of area elements dA for evaluation of area moments of inertia I_x and I_y.

- The double integral approach shown in Fig. 10.7(a) has straightforward expressions for dA, $\tilde{x}$, and $\tilde{y}$. If you are comfortable with determining limits of integration and evaluating double integrals, then this approach is often very effective.

- To determine I_x using a single integration, an area element parallel to the x axis should be selected as shown in Fig. 10.7(b) so that all of the area within that element has the same moment arm $\tilde{y} = y$. The width w of the area element is generally a function of position y.

- To determine I_y using a single integration, an area element parallel to the y axis should be selected as shown in Fig. 10.7(c) so that all of the area within that element has the same moment arm $\tilde{x} = x$. The height h of the area element is generally a function of position x.

- In this section, we cannot use an area element parallel to the y axis to determine I_x because $I_x = \iint y^2\, dx\, dy \neq \int \tilde{y}^2 h\, dx$ where $\tilde{y}$ is the centroid of vertical area element $h\, dx$. In words, the moment arm squared (y^2) for area element $dx\, dy$ when integrated over y is not equal to the *average* moment arm squared ($\tilde{y}^2$) for area element $h\, dx$. Similarly, we cannot use an area element parallel to the x axis to determine I_y. In Section 10.2, the parallel axis theorem is introduced, and this restriction can be removed.

- To determine J_O using a single integration, an area element having a uniform value of radial position from point O is needed, and this requires the use of a thin circular area element as illustrated in Fig. 3 of Example 10.2. For this reason, only circular geometries may be considered if Eq. (10.3) is to be evaluated using a single integration.

Helpful Information

Choice of area element dA. In this section, to evaluate area moments of inertia I_x and I_y using a single integration, it is necessary to use an area element that is parallel to the axis about which the moment of inertia is being evaluated. Thus, I_x requires an area element parallel to the x axis, and I_y requires an area element parallel to the y axis. These restrictions can be relaxed after the parallel axis theorem is introduced in Section 10.2. To determine J_O using a single integration, an area element having a constant value of radial position from point O is needed, and this requires the use of a thin circular area element as illustrated in Fig. 3 of Example 10.2.

End of Section Summary

In this section, *area moments of inertia* I_x, I_y, J_O, and I_{xy} were defined to be the second moments of an area distribution. Some of the key points are as follows:

- Area moments of inertia are measures of how an area is distributed about particular axes. Moments of inertia have dimensions of $(length)^4$.
- The product of inertia I_{xy} is a measure of the asymmetry of an area with respect to the x and y axes. The need for I_{xy} typically arises in advanced subjects, so methods for evaluating this quantity are not discussed in this book.
- *Radii of gyration* are alternative measures of how an area is distributed. The radii of gyration are easily determined if the area moments of inertia are known, and vice versa.
- The definitions of the area moments of inertia involve integration over the area of a shape. The use of single integration to evaluate these expressions requires an area element having uniform $\tilde{x}$, $\tilde{y}$, or $\tilde{r}$ for every point within the area element. Thus, evaluation of I_x and I_y requires area elements parallel to the x and y axes, respectively, and evaluation of J_O requires a circular area element. In Section 10.2, the parallel axis theorem is introduced and this restriction can be removed.

EXAMPLE 10.1 *Area Moments of Inertia Using Integration*

A rectangular area with base b, height h, and centroid at point C is shown. Determine the area moments of inertia I_x and $I_{x'}$ and the radii of gyration k_x and $k_{x'}$.

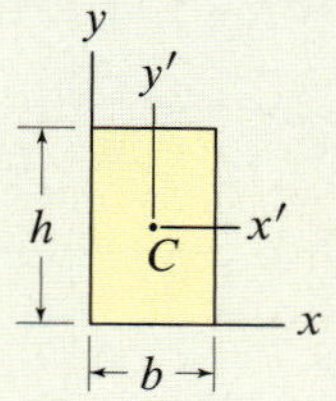

Figure 1

SOLUTION

Road Map We will use Eq. (10.1) on p. 534 to determine the moments of inertia I_x and $I_{x'}$. For I_x integration will be carried out with respect to y. For $I_{x'}$ integration will be carried out with respect to y'.

I_x

Governing Equations & Computation To determine I_x using a single integration, expressions for dA and $\tilde{y}$ are needed, and these must be developed using an area element parallel to the x axis, as shown in Fig. 2:

$$dA = b\,dy \quad \text{and} \quad \tilde{y} = y. \tag{1}$$

Substituting Eq. (1) into Eq. (10.1), the moment of inertia about the x axis is

$$I_x = \int \tilde{y}^2\,dA = \int_0^h y^2 b\,dy = \frac{by^3}{3}\bigg|_0^h = \boxed{\frac{bh^3}{3}}. \tag{2}$$

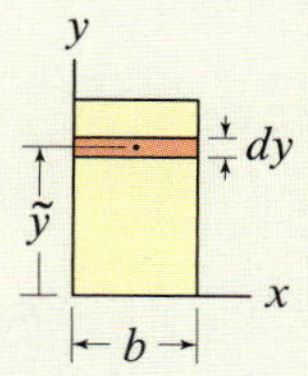

Figure 2
To determine I_x, an area element parallel to the x axis is used to develop expressions for dA and $\tilde{y}$.

$I_{x'}$

Governing Equations & Computation The expression to determine $I_{x'}$ is the same as Eq. (1) with y replaced by y' and limits of integration from $-h/2$ to $h/2$. Thus,

$$I_{x'} = \int \tilde{y}^2\,dA = \int_{-h/2}^{h/2} (y')^2 b\,dy' = \frac{b(y')^3}{3}\bigg|_{-h/2}^{h/2} = \boxed{\frac{bh^3}{12}}. \tag{3}$$

Radii of gyration

Governing Equations & Computation With the moments of inertia determined above, and noting that the cross-sectional area is $A = bh$, the radii of gyration are obtained from Eq. (10.5) on p. 536 as

$$k_x = \sqrt{\frac{I_x}{A}} = \sqrt{\frac{bh^3/3}{bh}} = \frac{h}{\sqrt{3}} = \boxed{(0.5774)h,} \tag{4}$$

$$k_{x'} = \sqrt{\frac{I_{x'}}{A}} = \sqrt{\frac{bh^3/12}{bh}} = \frac{h}{\sqrt{12}} = \boxed{(0.2887)h.} \tag{5}$$

Helpful Information

Rectangular shape. Beams often have rectangular cross-sectional shape, or shape consisting of rectangular composite shapes. Thus, the results of this example are frequently used, and most engineers commit to memory the expression $I_{x'} = bh^3/12$.

Discussion & Verification

- Comparing Eqs. (2) and (3), we see that $I_x > I_{x'}$. In general, the moment of inertia I_x about any x axis parallel to (but not coincident with) the centroidal x' axis is *larger* than $I_{x'}$. Similar remarks apply to the other moments of inertia.
- The physical interpretation of the radius of gyration is shown in Fig. 3 for the case of I_x. If the area $A = bh$ is redistributed into a thin strip, the strip must be positioned at $k_x = (0.5774)h$ from the x axis if it is to have the same moment of inertia about the x axis as the original shape shown in Fig. 1.
- To determine the moments of inertia I_y and $I_{y'}$, this solution procedure is repeated using an area element parallel to the y axis (see Prob. 10.1).

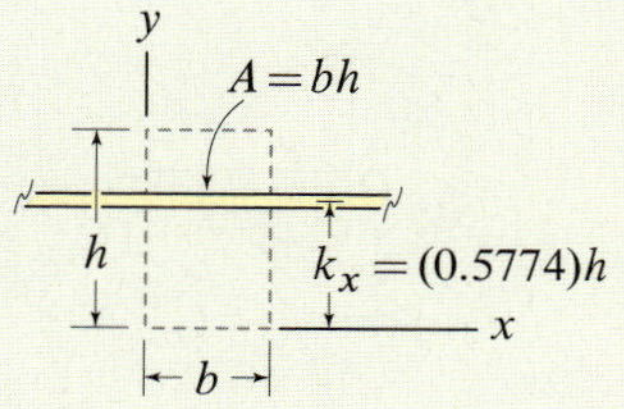

Figure 3
If the entire cross-sectional area $A = bh$ is redistributed into a thin strip parallel to the x axis, the radius of gyration k_x is the distance it must be positioned from the x axis to have the same moment of inertia as the original shape.

EXAMPLE 10.2 *Area Moments of Inertia Using Integration*

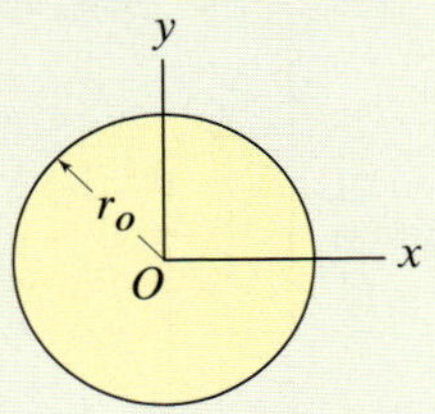

Figure 1

A circular area with outside radius r_o is shown. Determine the area moment of inertia I_y and the polar moment of inertia J_O about point O.

SOLUTION

Road Map We will use Eq. (10.2) on p. 534 to determine I_y and Eq. (10.3) to determine J_O.

I_y

Governing Equations & Computation To determine I_y using a single integration, expressions for dA and $\tilde{x}$ are needed, and these must be developed using an area element parallel to the y axis, as shown in Fig. 2:

$$dA = h\,dx \quad \text{and} \quad \tilde{x} = x. \tag{1}$$

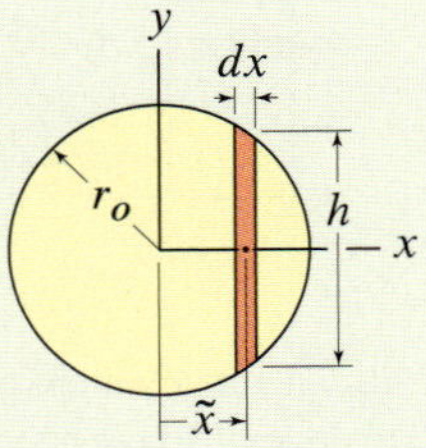

Figure 2
To determine I_y, an area element parallel to the y axis is used to develop expressions for dA and $\tilde{x}$.

The height h of the area element is related to x and radius r_o by the Pythagorean theorem $r_o^2 = x^2 + (h/2)^2$, hence

$$h = 2\sqrt{r_o^2 - x^2}. \tag{2}$$

Substituting Eqs. (1) and (2) into Eq. (10.2), and using a table of integrals or computer software to carry out the integration, the moment of inertia about the y axis is

$$\begin{aligned} I_y &= \int \tilde{x}^2\,dA = \int_{-r_o}^{r_o} x^2\, 2\sqrt{r_o^2 - x^2}\,dx \\ &= \left[\frac{-x}{2}(r_o^2 - x^2)^{3/2} + \frac{r_o^2 x}{4}\sqrt{r_o^2 - x^2} + \frac{r_o^4}{4}\sin^{-1}\left(\frac{x}{r_o}\right)\right]\Bigg|_{-r_o}^{r_o} \\ &= \boxed{\frac{\pi r_o^4}{4}}. \end{aligned} \tag{3}$$

J_O

Governing Equations & Computation To determine J_O using a single integration, an area element having a uniform value of $\tilde{r}$ is needed, and this necessitates the use of a thin circular area element as shown in Fig. 3. With r being a radial coordinate, expressions for dA and $\tilde{r}$ are obtained from Fig. 3 as

$$dA = 2\pi r\,dr \quad \text{and} \quad \tilde{r} = r. \tag{4}$$

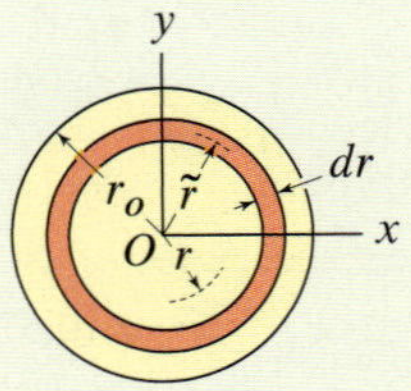

Figure 3
To determine J_O, a circular area element is used to develop expressions for dA and $\tilde{r}$.

Substituting Eq. (4) into Eq. (10.3), the polar moment of inertia about point O is

$$J_O = \int \tilde{r}^2\,dA = \int_0^{r_o} r^2\, 2\pi r\,dr = \boxed{\frac{\pi r_o^4}{2}}. \tag{5}$$

Discussion & Verification

- Because of symmetry of the circular shape in Fig. 1, it is obvious that $I_x = I_y$.
- Noting from Eq. (10.3) that $J_O = I_x + I_y$, and that $I_x = I_y$ as discussed above, we could have obtained Eq. (5) by inspection from the result in Eq. (3), or vice versa.

EXAMPLE 10.3 *Area Moments of Inertia Using Integration*

The cross section of a turbine blade in a pump is shown. Determine the area moments of inertia I_x and I_y.

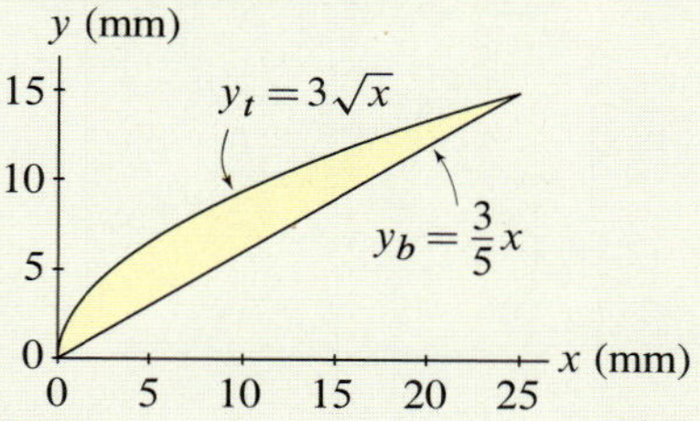

Figure 1

SOLUTION

Road Map This geometry was considered in Example 7.3 on p. 404 where the location of the centroid was determined. In this example, the location of the centroid is not needed, although knowledge of its approximate location may help with verification of the area moments of inertia, as discussed below. The expressions developed in Example 7.3 for quantities such as dA, $\tilde{x}$, and $\tilde{y}$ will be used again here.

> **Helpful Information**
>
> **Notation.** Throughout this problem, subscripts are used to distinguish *top* and *bottom* curves, and *left* and *right* curves, as follows:
>
> - Subscript t denotes "top."
> - Subscript b denotes "bottom."
> - Subscript r denotes "right."
> - Subscript l denotes "left."

I_y

Governing Equations & Computation To determine I_y using a single integration, expressions for dA and $\tilde{x}$ are needed, and these must be developed using an area element parallel to the y axis, as shown in Fig. 2:

$$dA = (y_t - y_b)\,dx = (3\sqrt{x} - \tfrac{3}{5}x)\,dx \qquad \text{and} \qquad \tilde{x} = x. \tag{1}$$

Substituting Eq. (1) into Eq. (10.2) on p. 534, the moment of inertia about the y axis is

$$I_y = \int \tilde{x}^2\,dA = \int_0^{25\text{ mm}} x^2(3\sqrt{x} - \tfrac{3}{5}x)\,dx$$
$$= \left(\frac{6x^{7/2}}{7} - \frac{3x^4}{20}\right)\Bigg|_0^{25\text{ mm}} = \boxed{8370\text{ mm}^4.} \tag{2}$$

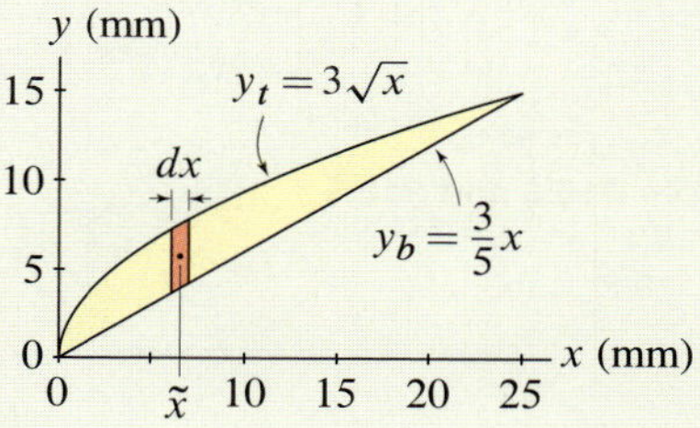

Figure 2

To determine I_y, an area element parallel to the y axis is used to develop expressions for dA and $\tilde{x}$.

I_x

Governing Equations & Computation To determine I_x using a single integration, expressions for dA and $\tilde{y}$ are needed, and these must be developed using an area element parallel to the x axis, as shown in Fig. 3:

$$dA = (x_r - x_l)\,dy = \left(\tfrac{5}{3}y - \frac{y^2}{9}\right)dy \qquad \text{and} \qquad \tilde{y} = y. \tag{3}$$

Substituting Eq. (3) into Eq. (10.1) on p. 534, the moment of inertia about the x axis is

$$I_x = \int \tilde{y}^2\,dA = \int_0^{15\text{ mm}} y^2\left(\tfrac{5}{3}y - \frac{y^2}{9}\right)dy$$
$$= \left(\frac{5y^4}{12} - \frac{y^5}{45}\right)\Bigg|_0^{15\text{ mm}} = \boxed{4220\text{ mm}^4.} \tag{4}$$

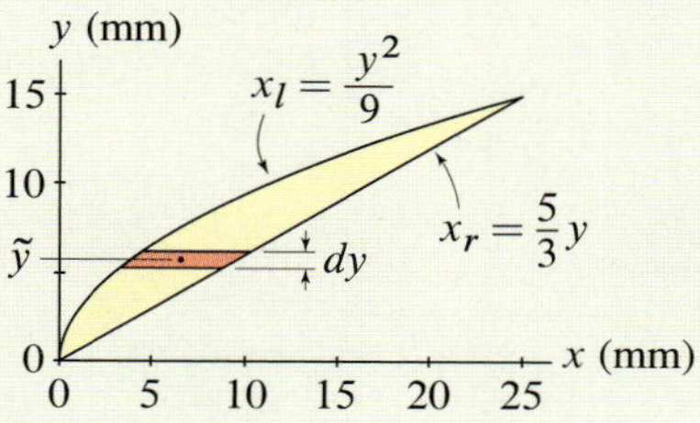

Figure 3

To determine I_x, an area element parallel to the x axis is used to develop expressions for dA and $\tilde{y}$.

Discussion & Verification

- By inspection of Fig. 1 (or by consulting the results of Example 7.3 on p. 404), the centroid of the area is farther from the y axis than from the x axis, so we expect $I_y > I_x$, and Eqs. (2) and (4) show this. Despite this observation, $I_y \neq \bar{x}^2 A$ and $I_x \neq \bar{y}^2 A$; Prob. 10.2 discusses this further.
- Knowing the moments of inertia about the centroidal x' and y' axes is also desirable. This may be accomplished by rewriting Eqs. (1)–(4) with integrations carried out with respect to x' and y', and with expressions for y_t, y_b, x_r, and x_l written in terms of x' and y'. Rather than do this, the parallel axis theorem discussed in the next section will be more convenient for evaluating $I_{x'}$ and $I_{y'}$.

PROBLEMS

Problem 10.1

For the rectangular area with base b, height h, and centroid at point C shown in Fig. 1 of Example 10.1 on p. 539, determine the area moments of inertia I_y and $I_{y'}$ and the radii of gyration k_y and $k_{y'}$.

Problem 10.2

In Example 7.3 on p. 404 the centroid $\bar{x}$, $\bar{y}$, and cross-sectional area A of a turbine blade were determined, and in Example 10.3 on p. 541 the area moments of inertia I_x and I_y were determined. Show that $I_x \neq \bar{y}^2 A$ and $I_y \neq \bar{x}^2 A$, and discuss why area moments of inertia may not be determined in this way.
Note: Concept problems are about *explanations*, not computations.

Problem 10.3

For the rectangular shape shown, determine the moments of inertia I_x and I_y.

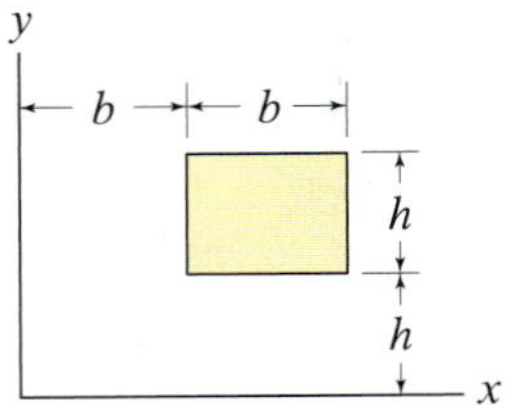

Figure P10.3

Problems 10.4 and 10.5

The cross section of a hollow tube has rectangular shape with uniform wall thickness.

Problem 10.4 Determine the moments of inertia I_x and I_y.

Problem 10.5 Determine the centroidal moments of inertia $I_{x'}$ and $I_{y'}$.

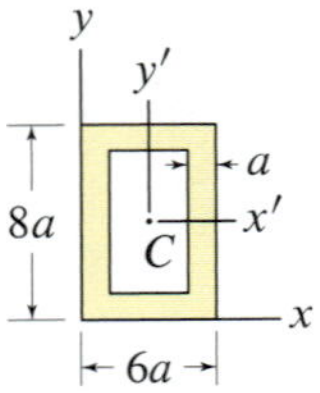

Figure P10.4 and P10.5

Problems 10.6 and 10.7

Determine I_x and I_y.

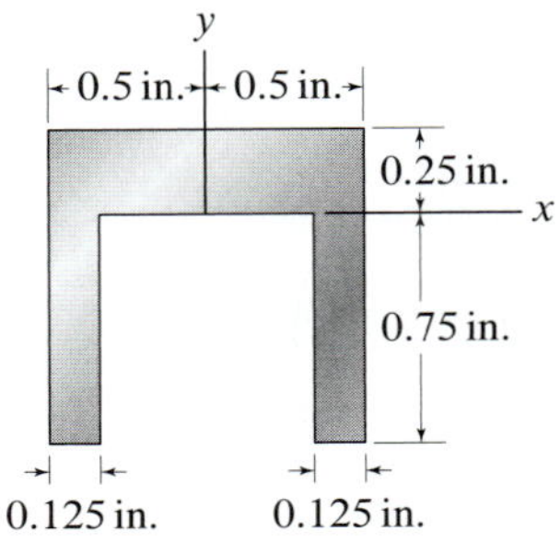

Figure P10.6

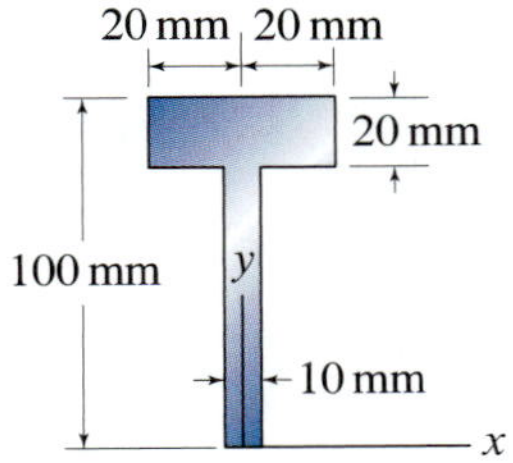

Figure P10.7

Problems 10.8 and 10.9

Determine I_x and I_y. Express your answers in terms of b and h.

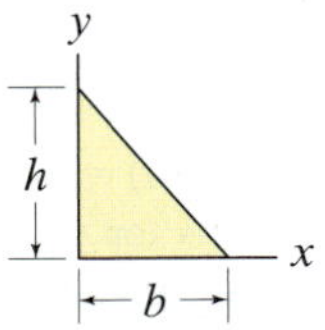

Figure P10.8

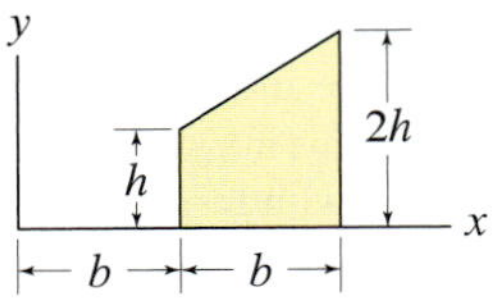

Figure P10.9

Problem 10.10

A circular area has outside radius r_o and a concentric hole with inside radius r_i. Show that the polar moment of inertia about point O is $J_O = \pi(r_o^4 - r_i^4)/2$.

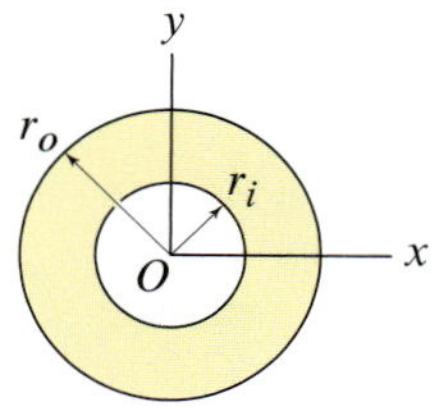

Figure P10.10

Problems 10.11 and 10.12

The area shown has outside radius r_o. Determine I_y and J_O. Comment on how your answers compare with the results of Example 10.2 on p. 540.

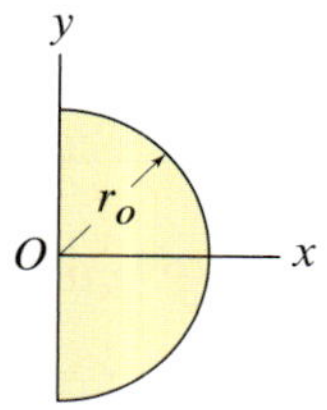

Figure P10.11

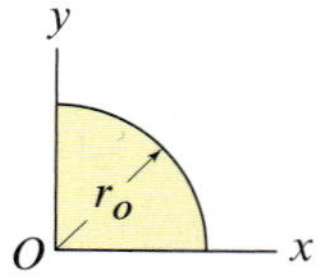

Figure P10.12

Problems 10.13 and 10.14

A semicircular area has outside radius r_o.

Problem 10.13 Determine I_x. *Hint:* The use of mathematical software is helpful, but is not required.

Problem 10.14 Determine J_A.

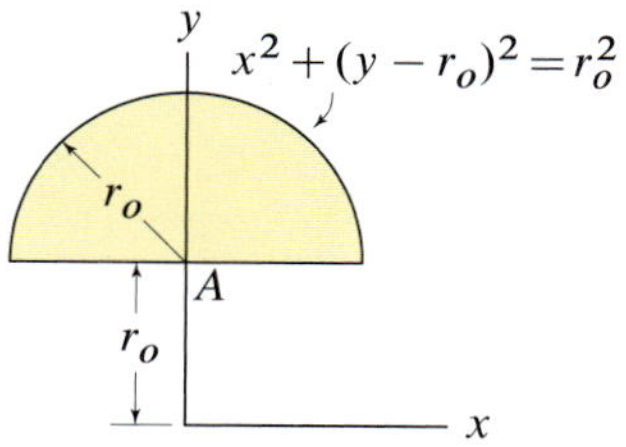

Figure P10.13 and P10.14

Problem 10.15

Consider two shafts having cross sections with the same area. The cross section for the first shaft is solid, and the cross section for the second shaft has a concentric hole whose inside radius is one-half its outside radius. Determine the ratio of the centroidal polar moments of inertia for the two shafts $J_{\text{hollow}}/J_{\text{solid}}$. *Hint:* Use the polar moment of inertia given in the statement of Prob. 10.10.

Problem 10.16

The strength of long bones, as well as slender structural members in general, is directly related to the area moments of inertia, such that if the moments of inertia increase, then strength increases. From the point of view of the ratio of strength to weight, discuss why most long bones in humans and animals, such as the human femur shown, are hollow.

Note: Concept problems are about *explanations*, not computations.

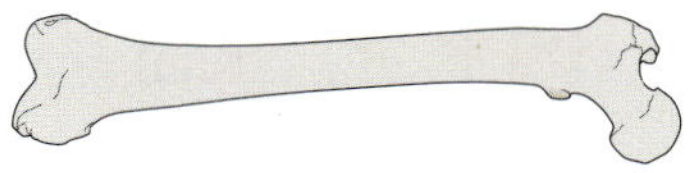

Figure P10.16

Problems 10.17 through 10.22

(a) Determine I_x.

(b) Determine I_y.

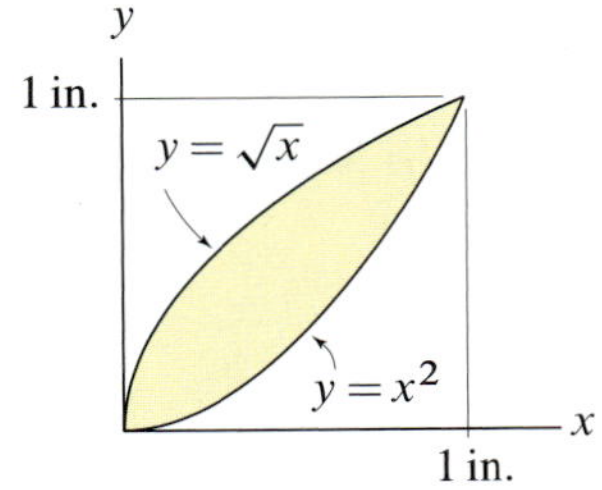

Figure P10.17

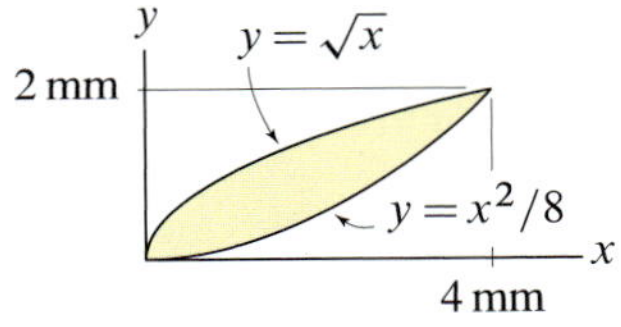

Figure P10.18

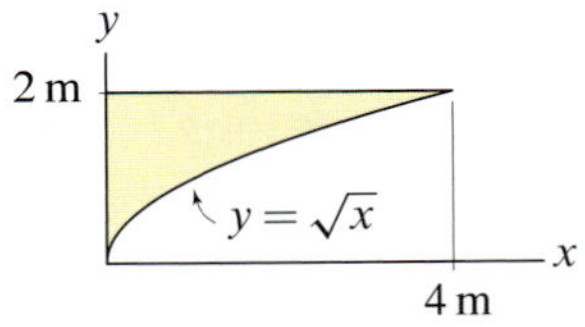

Figure P10.19

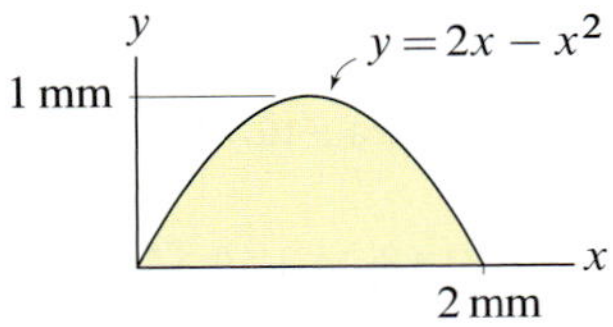

Figure P10.20

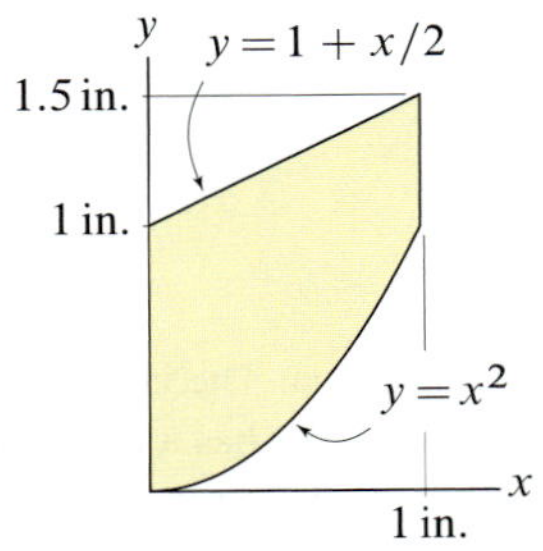

Figure P10.21

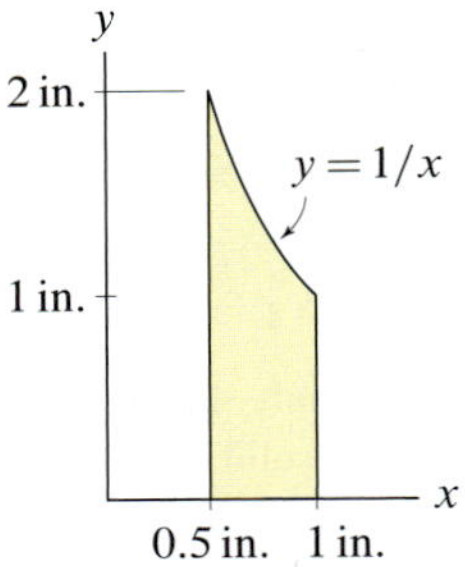

Figure P10.22

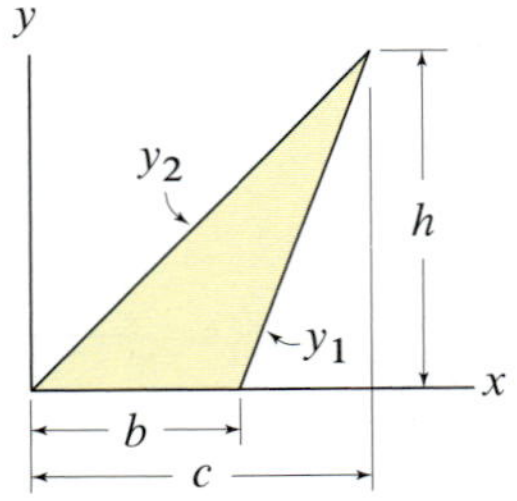

Figure P10.23

Problem 10.23

For the triangle shown, having base b and height h, show that the area moment of inertia about the x axis is $I_x = bh^3/12$.

Problems 10.24 and 10.25

Determine constants c_1 and c_2 so that the curves intersect at $x = a$ and $y = b$. Express your answers in terms of a and b. Then determine the area moment of inertia indicated.

Problem 10.24 Determine I_x.

Problem 10.25 Determine I_y.

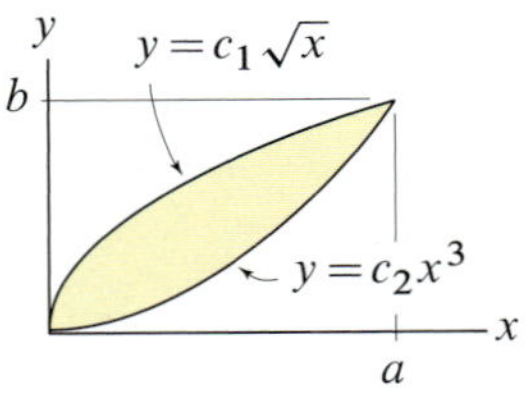

Figure P10.24 and P10.25

10.2 Parallel Axis Theorem

The parallel axis theorem relates area moments of inertia I_x, I_y, J_O, and I_{xy} to the centroidal area moments of inertia $I_{x'}$, $I_{y'}$, $J_{O'}$, and $I_{x'y'}$, where the x and x' axes are parallel and the y and y' axes are parallel. The parallel axis theorem is important and is used on a daily basis by most engineers.

Consider the area A with centroid C shown in Fig. 10.8. A centroidal $x'y'$ coordinate system is defined, with origin O' positioned at the centroid of the area. The x and y axes are parallel to the x' and y' axes, respectively, with separation distances d_x and d_y. The *parallel axis theorem* relates the moments of inertia with respect to the x and y axes to the centroidal moments of inertia as follows:

$$I_x = I_{x'} + d_x^2 A, \quad (10.6)$$

$$I_y = I_{y'} + d_y^2 A, \quad (10.7)$$

$$J_O = J_{O'} + d^2 A, \quad (10.8)$$

$$I_{xy} = I_{x'y'} + d_x d_y A. \quad (10.9)$$

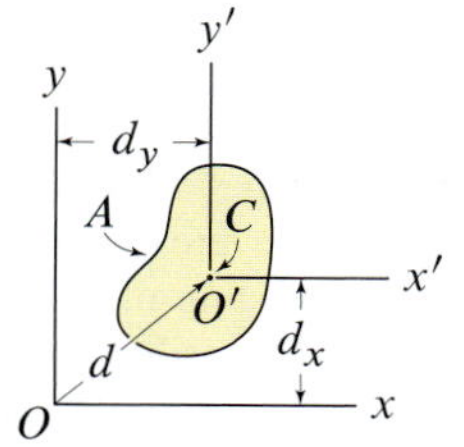

Figure 10.8
An area A with centroid at point C. The x and x' axes are parallel with separation distance d_x, and the y and y' axes are parallel with separation distance d_y.

To see how the above expressions are obtained, consider the example of determining I_x for the area shown in Fig. 10.8. Beginning with the basic definition of I_x given by Eq. (10.1) on p. 534, we use the area element shown in Fig. 10.9 to write

$$I_x = \int \tilde{y}^2 \, dA \quad (10.10)$$

$$= \int (\tilde{y}' + d_x)^2 \, dA \quad (10.11)$$

$$= \underbrace{\int (\tilde{y}')^2 \, dA}_{=I_{x'}} + 2d_x \underbrace{\int \tilde{y}' \, dA}_{=0} + d_x^2 \underbrace{\int dA}_{=A} \quad (10.12)$$

$$= I_{x'} + d_x^2 \, A. \quad (10.13)$$

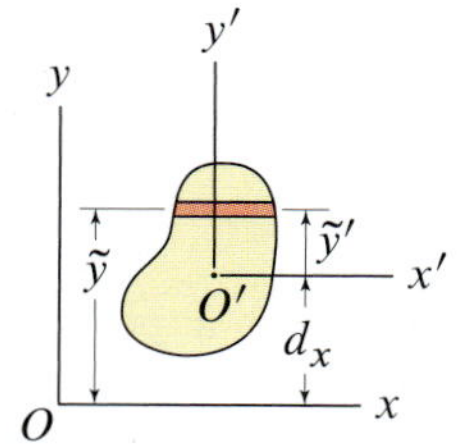

Figure 10.9
Selection of an area element for determining the moment of inertia about the x axis, I_x.

Equation (10.11) is obtained from Eq. (10.10) by noting in Fig. 10.9 that $\tilde{y} = \tilde{y}' + d_x$. Equation (10.12) is obtained by expanding $(\tilde{y}' + d_x)^2$, writing separate integrals for each term, and bringing the constant d_x outside the integrals. The first integral in Eq. (10.12) is the same as Eq. (10.1) written in terms of a centroidal coordinate, hence it yields the centroidal moment of inertia $I_{x'}$. The second integral is zero because y' is measured from the centroid of the shape, and the third integral is the area A of the shape. The proof for the other parallel axis theorems is similar to Eqs. (10.10)–(10.13).

Remarks

- It is clear from Eqs. (10.6)–(10.8) that $I_x \geq I_{x'}$, $I_y \geq I_{y'}$, and $J_O \geq J_{O'}$.
- When one is transforming the product of inertia, the signs of d_x and d_y are important. Transformation of the product of inertia arises in advanced subjects and is not explored further in this book.
- The parallel axis theorem may also be used to obtain radii of gyration, and Prob. 10.31 explores this further.

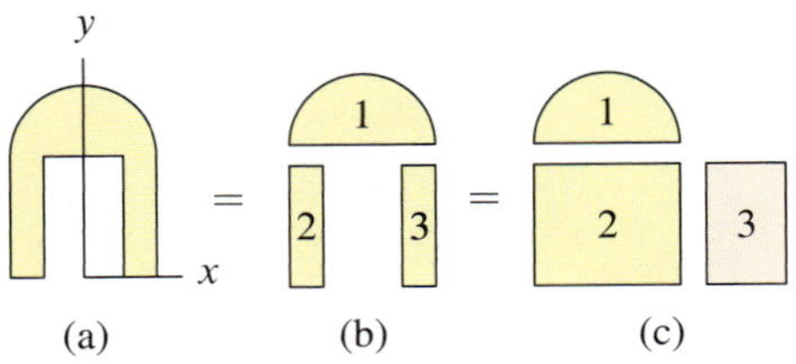

Figure 10.10
Use of composite shapes with the parallel axis theorem to determine area moments of inertia. (a) An area consisting of a semicircle and two rectangles. (b) A possible set of composite shapes. (c) Another possible set of composite shapes where shape 3 has negative area.

Common Pitfall

Parallel axis theorem. Consider the parallel axes x_1, x_2, and centroidal axis x', such as

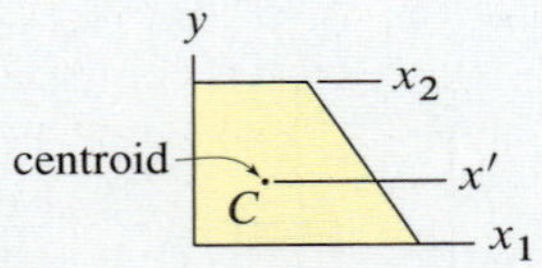

where the centroid is at point C. If I_{x_1} is known, a common error is to use the parallel axis theorem to *directly* determine I_{x_2}. In the parallel axis theorem, one of the axes is always a centroidal axis. Thus, with I_{x_1} known, the parallel axis theorem must be used *twice*, the first time to determine $I_{x'}$, and the second time to use $I_{x'}$ to determine I_{x_2}. See Prob. 10.40.

Use of parallel axis theorem in integration

The parallel axis theorem may be used to remove the restriction on the orientation and shape of the area element that was needed in Section 10.1. That is, with the parallel axis theorem, the moments of inertia I_x and I_y may be determined using a single integration with area elements that are perpendicular to the x and y axes, respectively. Example 10.4 illustrates this approach.

Use of parallel axis theorem for composite shapes

The most common use of the parallel axis theorem is for determining the moments of inertia using composite shapes. For example, consider the area shown in Fig. 10.10 where I_x is desired. The parallel axis theorem, written for composite shapes, becomes

$$I_x = \sum_{i=1}^{n} (I_{x'} + d_x^2 A)_i \qquad (10.14)$$

where n is the number of shapes, $I_{x'}$ is the moment of inertia for shape i about its centroidal x' axis, d_x is the *shift distance* for shape i (i.e., the distance between the x axis and the x' axis for shape i), and A is the area for shape i. Similar expressions may be written for I_y and J_O. To use Eq. (10.14), it is necessary to know the centroidal moment of inertia for each of the composite shapes, and this generally must be obtained by integration, as was done in Section 10.1. Fortunately, the centroidal moments of inertia for many basic shapes have been tabulated, such as in the Table of Properties of Lines and Areas on the inside back cover. Examples 10.5 and 10.6 illustrate this approach.

End of Section Summary

In this section, the *parallel axis theorem* for area moments of inertia was presented. Some of the key points are as follows:

- The parallel axis theorem relates area moments of inertia I_x, I_y, J_O, and I_{xy} to the centroidal moments of inertia $I_{x'}$, $I_{y'}$, $J_{O'}$, and $I_{x'y'}$, where the x and x' axes are parallel and the y and y' axes are parallel. A common error is to use the parallel axis theorem to relate moments of inertia between two parallel axes where neither of them is a centroidal axis.

- The parallel axis theorem has applications for determining area moments of inertia by integration, where it permits use of an area element that is perpendicular to the axis about which the moment of inertia is being determined.

- The most common application of the parallel axis theorem is for determining area moments of inertia using composite shapes. For use of composite shapes, the moment of inertia of each composite shape about its centroidal axis must be known. This may always be determined by using integration, but information for many basic shapes has been tabulated, such as in the Table of Properties of Lines and Areas on the inside back cover.

EXAMPLE 10.4 *Area Moments of Inertia Using Integration With the Parallel Axis Theorem*

Use integration with an area element perpendicular to the x axis to determine the area moment of inertia about the x axis I_x.

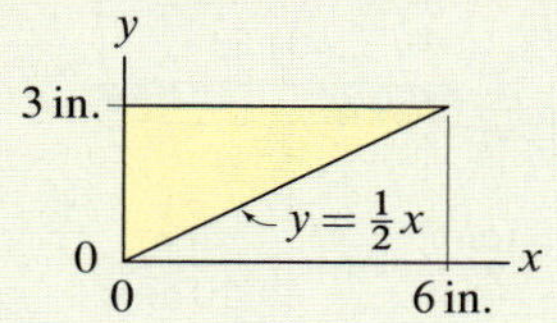

Figure 1

SOLUTION

Road Map An area element perpendicular to the x axis, in conjunction with the parallel axis theorem, may be used to determine I_x. The strategy is to write an expression for the moment of inertia of the area element *about its centroidal axis*, and to use the parallel axis theorem to obtain the moment of inertia of this area element about the x axis, followed by integration over all area elements to obtain the total moment of inertia.

Governing Equations & Computation An area element perpendicular to the x axis is shown in Fig. 2. In this approach, we evaluate the moment of inertia using

$$I_x = \int dI_x \tag{1}$$

where dI_x is the moment of inertia about the x axis for the area element shown in Fig. 2. The area element is rectangular, and the parallel axis theorem, Eq. (10.6) on p. 545, is used to write its moment of inertia about the x axis as

$$dI_x = \tfrac{1}{12}h^3\,dx + \tilde{y}^2 h\,dx. \tag{2}$$

The first term in Eq. (2) is the moment of inertia of a rectangle about its centroid, namely, $(1/12)(\text{base})(\text{height})^3$ where the base of the area element in Fig. 2 is dx and the height is h. The second term in Eq. (2) is the parallel axis shift, namely, $(\text{shift distance})^2(\text{area})$ where the shift distance is $\tilde{y}$ and the area is $h\,dx$. Letting y_t and y_b be the expressions for the top and bottom curves, respectively, that define the area in Fig. 2, we write

$$dA = (y_t - y_b)\,dx = \left(3 - \frac{x}{2}\right)dx, \tag{3}$$

$$\tilde{y} = \tfrac{1}{2}(y_t + y_b) = \tfrac{1}{2}\left(3 + \frac{x}{2}\right), \tag{4}$$

$$h = y_t - y_b = 3 - \frac{x}{2}, \tag{5}$$

Substituting Eqs. (2)–(5) into Eq. (1) provides

$$I_x = \int_0^{6\text{ in.}} \tfrac{1}{12}\left(3 - \frac{x}{2}\right)^3 dx + \int_0^{6\text{ in.}} \left[\tfrac{1}{2}\left(3 + \frac{x}{2}\right)\right]^2 \left(3 - \frac{x}{2}\right) dx$$

$$= \boxed{\frac{81}{2}\text{ in.}^4 = 40.5\text{ in.}^4} \tag{6}$$

Discussion & Verification To help judge if our answer is reasonable, we may compare our result to that for a rectangular area with x and y dimensions of 6 in. by 3 in. From the results of Example 10.1 on p. 539, or by consulting the Table of Properties of Lines and Areas on the inside back cover, the moment of inertia of a rectangular area about the axis through its base is $I_x = (1/3)(\text{base})(\text{height})^3 = (1/3)(6\text{ in.})(3\text{ in.})^3 = 54.0\text{ in.}^4$, which as expected is somewhat larger than the result in Eq. (6). Hence, our answer appears to be reasonable.

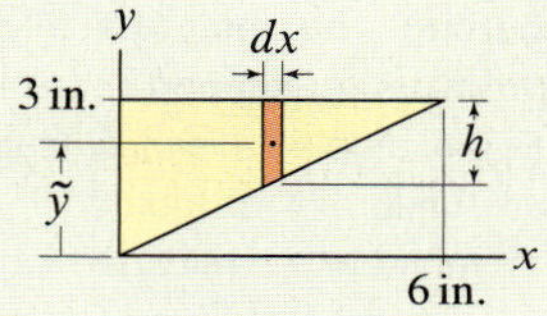

Figure 2
To determine I_x, an area element perpendicular to the x axis may be used in conjunction with the parallel axis theorem. Expressions for dA, $\tilde{y}$ and h are needed.

EXAMPLE 10.5 *Area Moments of Inertia Using Composite Shapes*

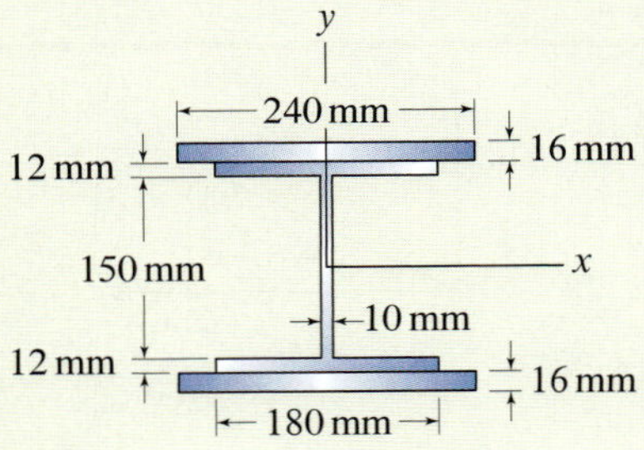

Figure 1

The cross section of an I beam with cover plates welded to its flanges is shown. The cross-sectional area is symmetric about the x and y axes. Determine the moment of inertia about the x axis of the beam's cross-sectional area with and without the cover plates.

Interesting Fact

Strengthening beams. Usually, only portions of a beam support large internal forces. For example, a simply supported beam with uniformly distributed load has maximum moment at midspan and zero moment at the supports (see Example 8.5 on p. 470). If the beam has a uniform cross section, then the cross section must be selected based on the maximum moment (and possibly the shear). Selective reinforcement using cover plates is a simple way to strengthen a beam.

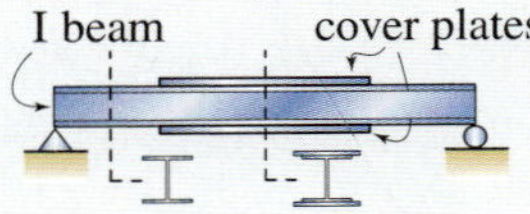

SOLUTION

Road Map The cross-sectional area consists of rectangular shapes, so the use of composite shapes for determining I_x will be convenient. We will first determine I_x for the I beam without cover plates and will then determine I_x with cover plates.

I beam without cover plates

Governing Equations & Computation Two combinations of composite shapes are shown in Fig. 2. Using Eq. (10.14) on p. 546 with the three shapes shown in Fig. 2(b), the moment of inertia about the x axis is

$$I_x = \sum_{i=1}^{3}(I_{x'} + d_x^2 A)_i \tag{1}$$

$$= \tfrac{1}{12}(10\,\mathrm{mm})(150\,\mathrm{mm})^3 + \left[\tfrac{1}{12}(180\,\mathrm{mm})(12\,\mathrm{mm})^3 + (81\,\mathrm{mm})^2(180\,\mathrm{mm})(12\,\mathrm{mm})\right]2 \tag{2}$$

$$= \boxed{3.12\times 10^7\ \mathrm{mm}^4.} \tag{3}$$

In Eq. (2), the centroidal axis of shape 1 coincides with the x axis, so the parallel axis shift for this shape is zero. Also, the contributions to I_x for shapes 2 and 3 are identical, and thus the terms within square brackets are simply doubled.

Alternate solution We may also determine I_x using the three composite shapes shown in Fig. 2(c), where the areas of shapes 2 and 3 are negative. Equation (10.14) provides

$$I_x = \tfrac{1}{12}(180\,\mathrm{mm})(174\,\mathrm{mm})^3 - \left[\tfrac{1}{12}(85\,\mathrm{mm})(150\,\mathrm{mm})^3\right]2 = \boxed{3.12\times 10^7\ \mathrm{mm}^4,} \tag{4}$$

which is identical to Eq. (3). Note that the centroidal axis of each composite shape coincides with the x axis, thus the parallel axis shift in Eq. (4) is zero for each shape.

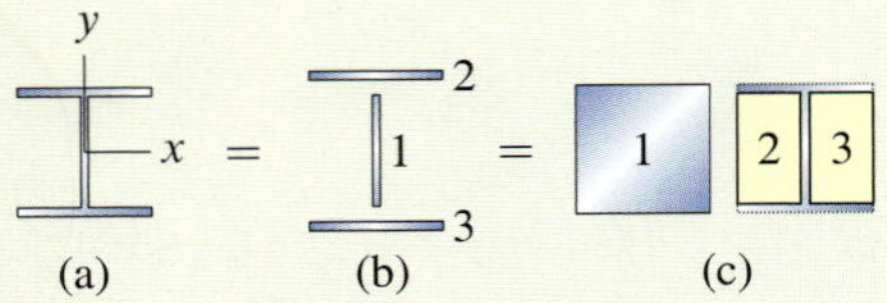

Figure 2
(a) Cross-sectional area of an I beam without cover plates. (b) and (c) Two combinations of composite shapes to describe the area.

I beam with cover plates

Governing Equations & Computation Using Eq. (10.14) with the three shapes shown in Fig. 3(b), the moment of inertia about the x axis is

$$I_x = \sum_{i=1}^{3}(I_{x'} + d_x^2 A)_i \tag{5}$$

$$= 3.121\times 10^7\ \mathrm{mm}^4 + \left[\tfrac{1}{12}(240\,\mathrm{mm})(16\,\mathrm{mm})^3 + (95\,\mathrm{mm})^2(240\,\mathrm{mm})(16\,\mathrm{mm})\right]2 \tag{6}$$

$$= \boxed{1.01\times 10^8\ \mathrm{mm}^4.} \tag{7}$$

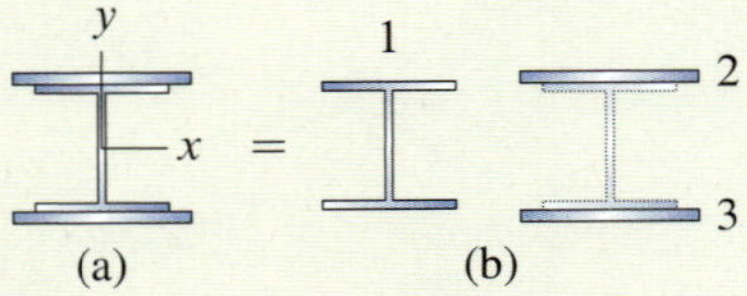

Figure 3
(a) Cross-sectional area of an I beam with cover plates. (b) A combination of composite shapes.

Discussion & Verification The addition of the cover plates increases the beam's cross sectional area by about 130%, yet I_x increases dramatically by about 220%.

EXAMPLE 10.6 *Area Moments of Inertia Using Composite Shapes*

The cross section of a bar is symmetric about the x axis. Determine d so that the origin of the coordinate system, point O, is positioned at the centroid of the area, and determine the area moment of inertia about the y axis.

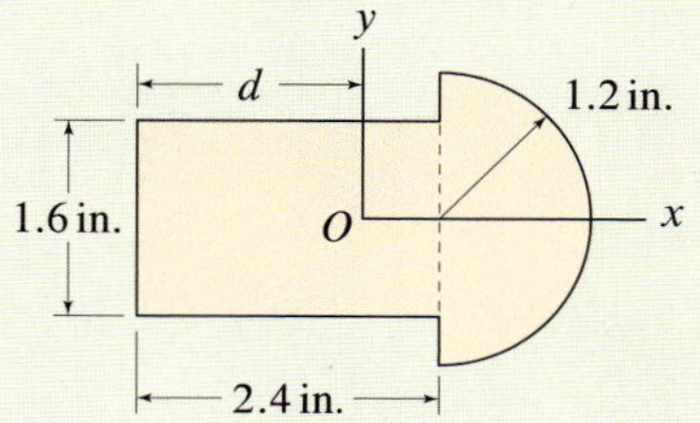

Figure 1

SOLUTION

Road Map The cross-sectional area consists of rectangular and semicircular shapes, so use of composite shapes for determining the position of the centroid and I_y will be convenient. Concepts from Section 7.1 will be used to locate the centroid, and then the parallel axis theorem will be used to determine the centroidal moment of inertia I_y.

Centroid

Governing Equations & Computation A set of composite shapes is shown in Fig. 2, where for purposes of locating the centroid, a convenient tn coordinate system is defined. The area and centroidal position of each composite shape are collected in Table 1.

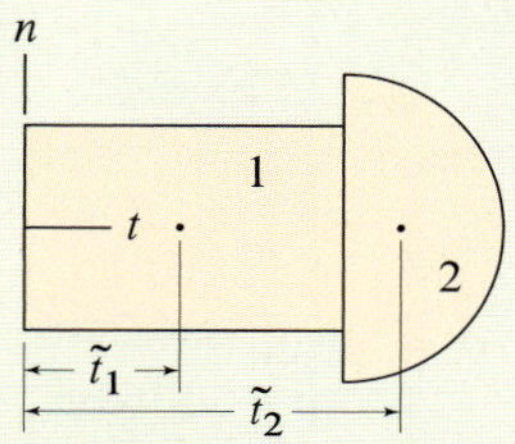

Figure 2
A combination of composite shapes and a tn coordinate system for use in determining the location of the centroid.

Table 1. Areas and centroidal positions for composite shapes in Fig. 2.

Shape no.	A_i	$\tilde{t}_i$
1	$(1.6\text{ in.})(2.4\text{ in.}) = 3.840\text{ in.}^2$	1.2 in.
2	$\frac{\pi}{2}(1.2\text{ in.})^2 = 2.262\text{ in.}^2$	$2.4\text{ in.} + \frac{4}{3\pi}(1.2\text{ in.}) = 2.909\text{ in.}$

Evaluating Eq. (7.8) on p. 397 for the location of the centroid $\bar{t}$ provides

$$\bar{t} = \frac{\sum_{i=1}^{2} \tilde{t}_i A_i}{\sum_{i=1}^{2} A_i} = \frac{(1.2\text{ in.})(3.840\text{ in.}^2) + (2.909\text{ in.})(2.262\text{ in.}^2)}{3.840\text{ in.}^2 + 2.262\text{ in.}^2} = 1.834\text{ in.} \tag{1}$$

Thus, the origin of the coordinate system in Fig. 1 should be located at $\boxed{d = 1.834\text{ in.}}$

Moment of inertia

Governing Equations & Computation To determine I_y, the same composite shapes are used, and the distance from the centroid of each composite shape to the y axis is shown in Fig. 3. From the Table of Properties of Lines and Areas on the inside back cover, the centroidal moments of inertia for each shape are

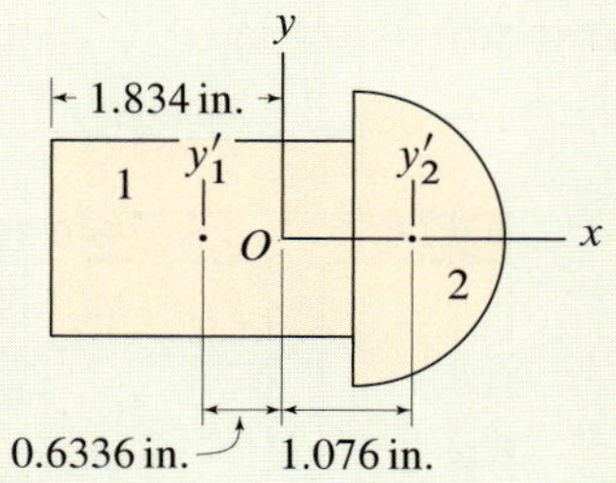

Figure 3
A combination of composite shapes to determine I_y, and distances from the centroid of each composite shape to the y axis.

$$I_{y'_1} = \tfrac{1}{12}(1.6\text{ in.})(2.4\text{ in.})^3 = 1.843\text{ in.}^4 \tag{2}$$

$$I_{y'_2} = \left(\frac{\pi}{8} - \frac{8}{9\pi}\right)(1.2\text{ in.})^4 = 0.2276\text{ in.}^4 \tag{3}$$

Using Eq. (10.14) on p. 546 with Eqs. (2) and (3), the moment of inertia about the centroidal y axis is

$$I_y = \sum_{i=1}^{2}(I_{y'} + d_y^2 A)_i = 1.843\text{ in.}^4 + (0.6336\text{ in.})^2(3.840\text{ in.}^2)$$
$$+ 0.2276\text{ in.}^4 + (1.076\text{ in.})^2(2.262\text{ in.}^2) = \boxed{6.23\text{ in.}^4} \tag{4}$$

Discussion & Verification As a rough check if our answer is reasonable, the centroidal moment of inertia of a rectangle with dimensions that approximate Fig. 1 is $(1/12)(1.6\text{ in.})(3.6\text{ in.})^3 = 6.22\text{ in.}^4$, which agrees surprisingly well with Eq. (4).

Helpful Information

Alternate strategy. Rather than evaluate I_y directly as in Eq. (4), we could evaluate I_n first (you should find $I_n = 26.75\text{ in.}^4$) and then use the parallel axis theorem to determine $I_y = I_n - d^2 A = 26.75\text{ in.}^4 - (1.834\text{ in})^2(6.102\text{ in.}^2) = 6.23\text{ in.}^4$.

PROBLEMS

Problems 10.26 and 10.27

Use integration with an area element perpendicular to the x axis to determine the area moment of inertia I_x for the shape indicated.

Problem 10.26 Figure P10.20 on p. 544.

Problem 10.27 Figure P10.21 on p. 544.

Problems 10.28 and 10.29

The beam cross sections shown are symmetric about the x and y axes. Determine the area moments of inertia I_x and I_y and the radii of gyration k_x and k_y.

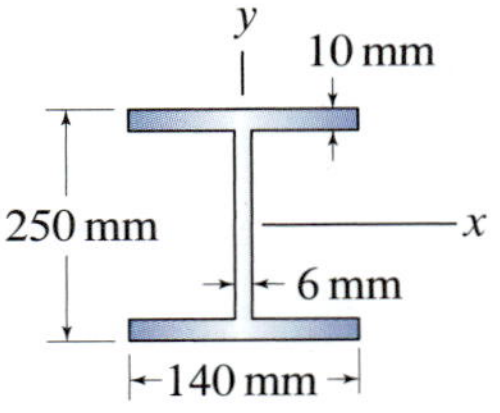

Figure P10.28

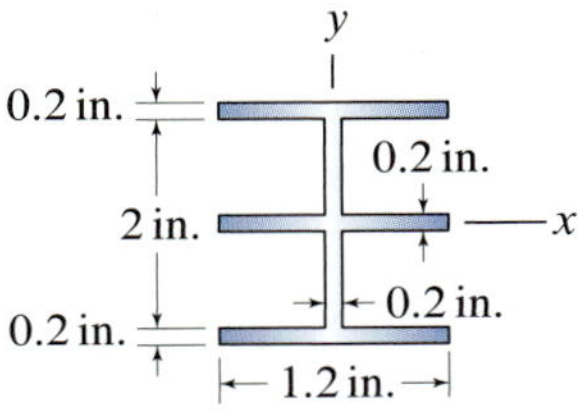

Figure P10.29

Problem 10.30

The cross-sectional dimensions for a W10 × 22 wide-flange I beam are shown. By idealizing the cross section to consist of rectangular shapes, determine the area moment of inertia I_x, and compare this to the value of $118\ \text{in.}^4$ reported in the AISC Steel Construction Manual. *Note:* The value reported by AISC accounts for the effects of small fillets where the web and flanges join.

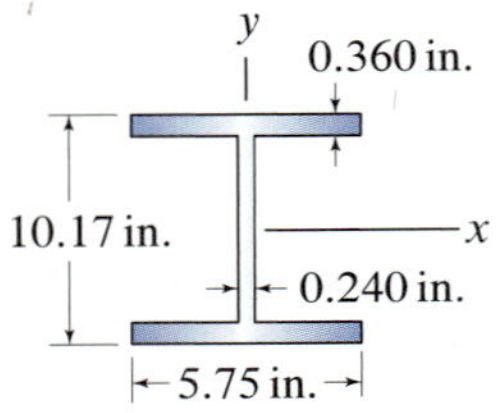

Figure P10.30

Problem 10.31

Using Eq. (10.6) on p. 545, show that the parallel axis theorem for the radius of gyration about the x axis is $k_x^2 = k_{x'}^2 + d_x^2$.

Problem 10.32

For the T shape shown, determine the radius of gyration about the x axis.

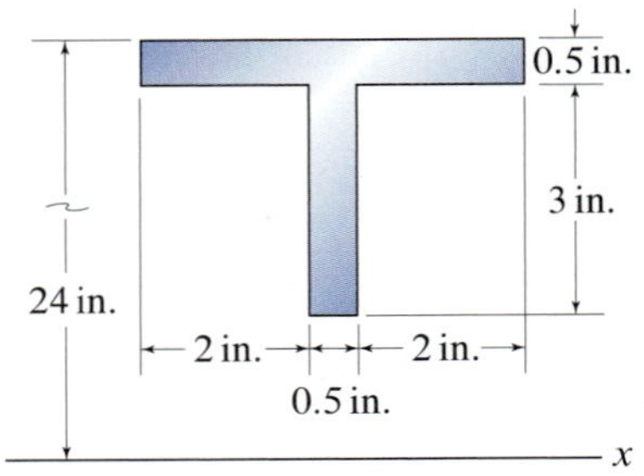

Figure P10.32

Problem 10.33

For the channel shown, determine the radii of gyration about the x and y axes.

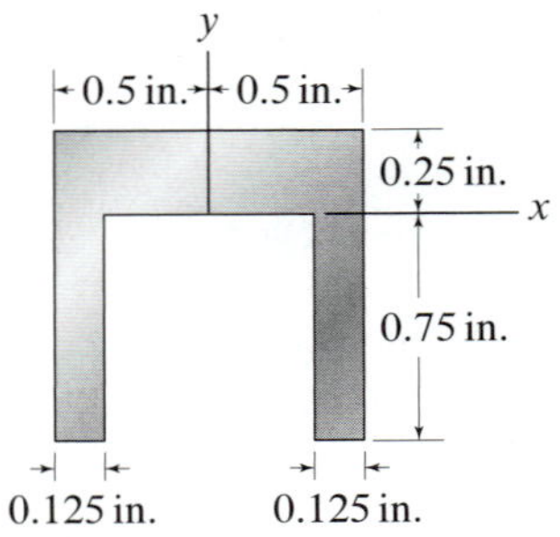

Figure P10.33

Problem 10.34

Let each of the beams shown in Fig. 10.2 on p. 534 be constructed of identical planks of wood with 2 in. by 6 in. cross-sectional dimensions. Determine the area moments of inertia for each beam's cross-sectional area about the x and y axes.

Problem 10.35

The cross sections of two beams are constructed by arranging the 2 cm by 16 cm strips of wood as shown. Determine the area moments of inertia for each beam about the horizontal and vertical axes passing through the centroid of each area.

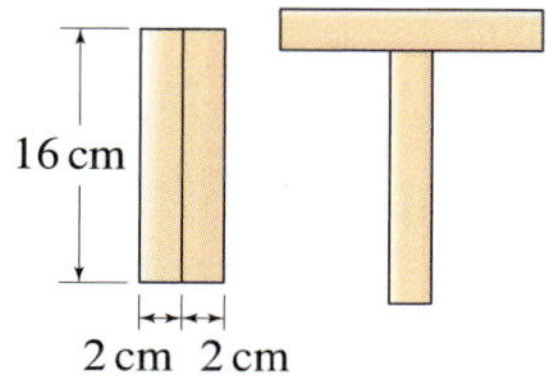

Figure P10.35

Problems 10.36 through 10.39

Determine the area moments of inertia I_x and I_y.

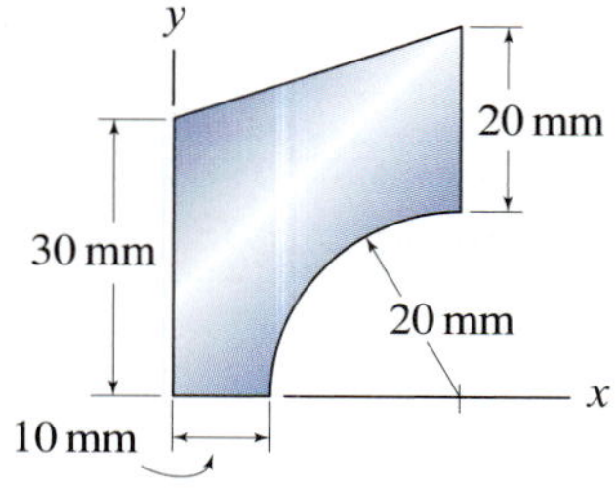

Figure P10.36

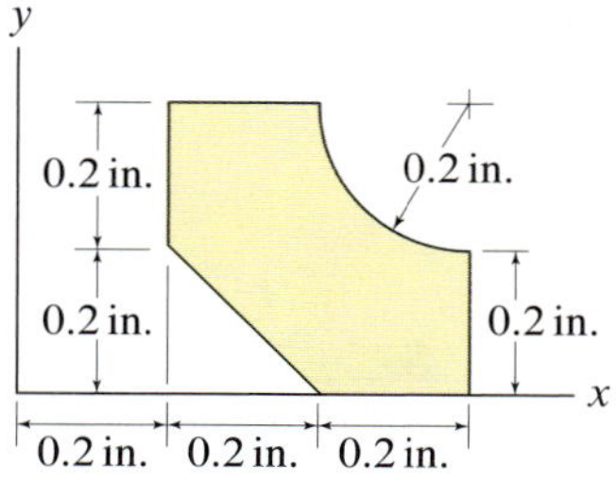

Figure P10.37

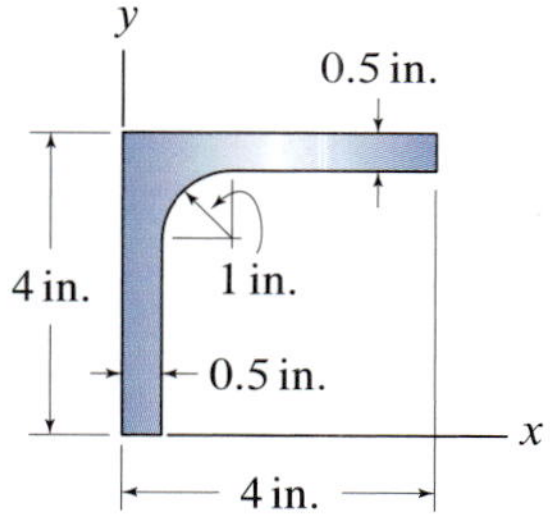

Figure P10.38

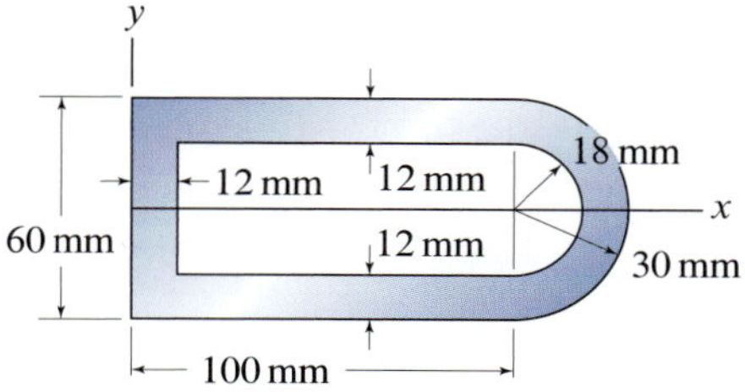

Figure P10.39

Problem 10.40

(a) Determine I_{x_1}.

(b) Use the result of Part (a) with the parallel axis theorem to determine I_{x_2}. *Hint:* See the common pitfall discussed on p. 546.

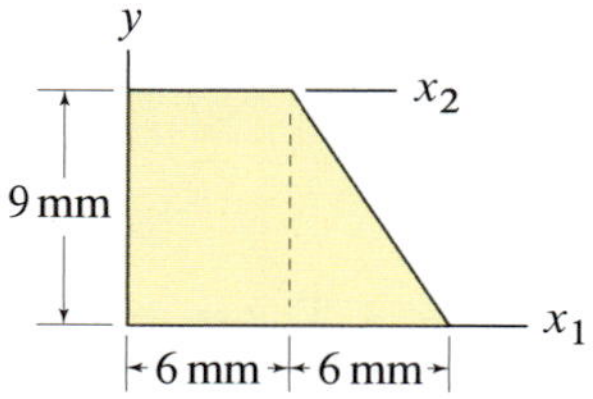

Figure P10.40

Problem 10.41

A circular hole is to be drilled through the side of a beam as shown. Describe where the hole should be positioned so that the area moment of inertia about the centroidal x' axis, at the cross section that passes through the hole, is reduced as little as possible. **Note:** Concept problems are about *explanations*, not computations.

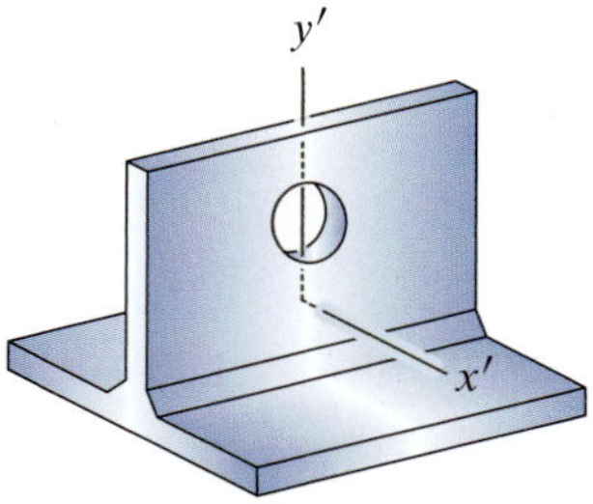

Figure P10.41

Problems 10.42 through 10.47

The cross section of the bar shown is symmetric about either the x or y axis.

(a) Determine d so that the origin of the coordinate system, point O, is positioned at the centroid of the area.

(b) Determine the area moment of inertia about the x axis.

(c) Determine the area moment of inertia about the y axis.

Hint: If you carried out some of the exercises from Section 7.1, beginning on p. 409, you may have already solved Part (a) of Probs. 10.46 and 10.47.

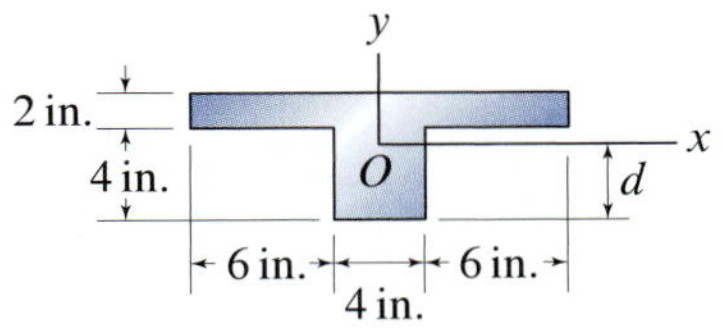

Figure P10.42

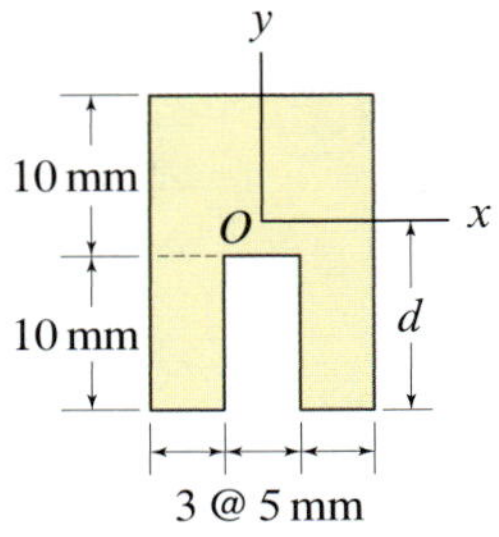

Figure P10.43

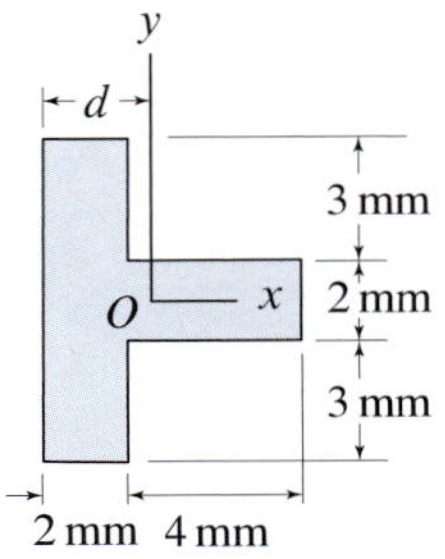

Figure P10.44

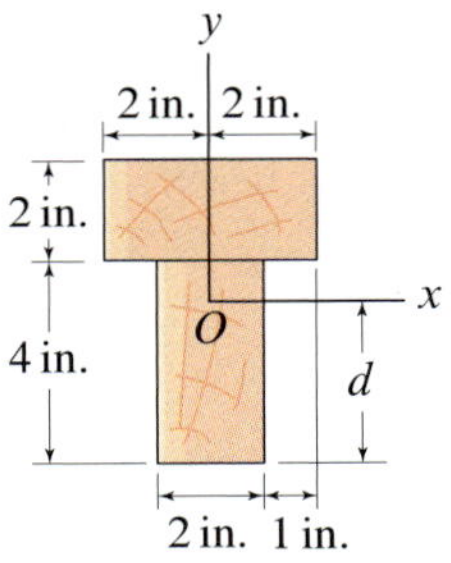

Figure P10.45

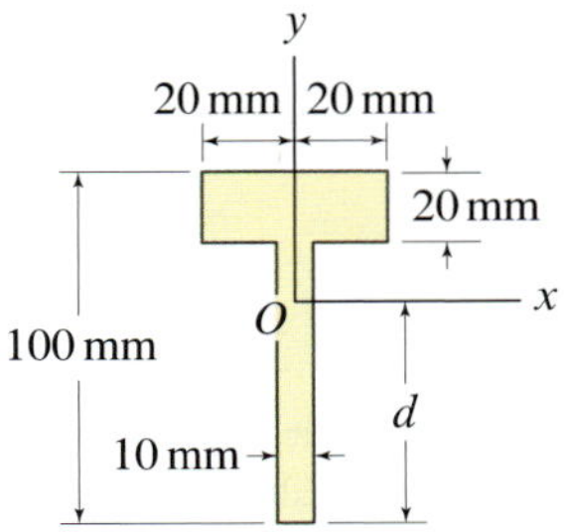

Figure P10.46

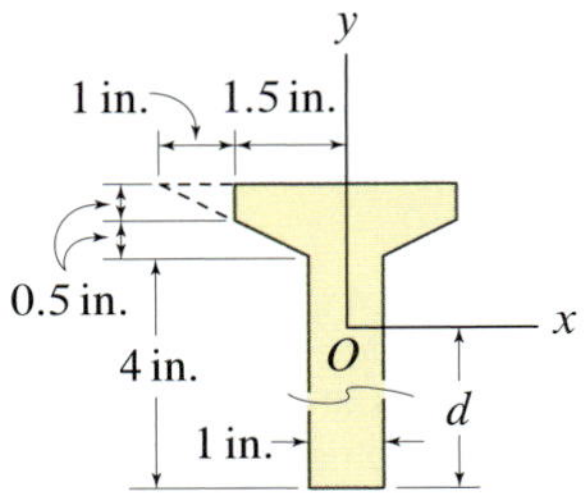

Figure P10.47

10.3 Mass Moments of Inertia

Mass moments of inertia are measures of how an object's mass is distributed about particular axes. While mass and area moments of inertia have some similarities they are different in that mass moments of inertia are inherently volume- and density-related whereas area moments of inertia are area-related. Before delving into the precise definitions of mass moments of inertia, we first discuss an example to help you understand why a measure of mass dispersion is useful and what this measure quantifies.

> **Concept Alert**
>
> **Mass moments of inertia.** Mass moments of inertia are measures of how the mass of an object is distributed about particular axes. Mass moments of inertia depend on the geometry of an object (size and shape), the density of the material(s) it is made of, and the axes you select.

An example—figure skating

Consider the figure skater shown in Fig. 10.11(a) as she spins on one skate with her arms and leg extended from her body. If friction between her skate and the ice is negligible, then she will spin at a constant rate. In Fig. 10.11(b) she draws her arms together over her head and her leg closer to her body, thus making her mass distribution more compact and thus reducing the mass moment of inertia of her body about the axis that she spins. The reduction in her mass moment of inertia between Fig. 10.11(a) and (b) causes her to spin at a higher rate; exactly how fast she spins and how this relates to the mass moment of inertia are addressed in dynamics.

(a) (b)

Figure 10.11
A figure skater spins with (a) her arms and leg extended and (b) her arms drawn together over her head and leg drawn close to her body. By making her body more compact in (b), she reduces the mass moment of inertia of her body about the axis that she spins, causing her to spin at a higher rate.

Definition of mass moments of inertia

The *mass moments of inertia* for the object shown in Fig. 10.12 are defined as

$$I_x = \int \tilde{r}_x^2\, dm = \int (\tilde{y}^2 + \tilde{z}^2)\, dm, \tag{10.15}$$

$$I_y = \int \tilde{r}_y^2\, dm = \int (\tilde{x}^2 + \tilde{z}^2)\, dm, \tag{10.16}$$

$$I_z = \int \tilde{r}_z^2\, dm = \int (\tilde{x}^2 + \tilde{y}^2)\, dm, \tag{10.17}$$

$$I_{xy} = \int \tilde{x}\tilde{y}\, dm, \tag{10.18}$$

$$I_{yz} = \int \tilde{y}\tilde{z}\, dm, \tag{10.19}$$

$$I_{xz} = \int \tilde{x}\tilde{z}\, dm, \tag{10.20}$$

where

$\tilde{r}_x$, $\tilde{r}_y$, and $\tilde{r}_z$ are defined in Fig. 10.12 as the radial distances (i.e., moment arms) from the x, y, and z axes, respectively, to the center of mass for mass element dm.

$\tilde{x}$, $\tilde{y}$, and $\tilde{z}$ are defined in Fig. 10.12 as the x, y, and z distances, respectively, to the center of mass for mass element dm.

I_x, I_y, and I_z are the *mass moments of inertia about the x, y, and z axes*, respectively.

I_{xy}, I_{yz}, and I_{xz} are the *products of inertia of the mass about the xy, yz, and xz axes*, respectively.

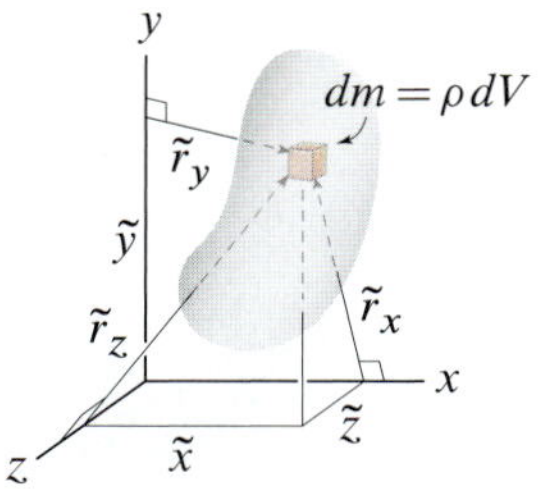

Figure 10.12
An object with mass m, density ρ, and volume V; $\tilde{r}_x$, $\tilde{r}_y$, and $\tilde{r}_z$ are radial distances from the x, y, and z axes, respectively, to the center of mass for mass element dm.

Remarks

- When referring to mass moments of inertia, we often omit the word *mass* when it is obvious from the context that we are dealing with mass moments of inertia as opposed to area moments of inertia.

- In each of Eqs. (10.15)–(10.17), two integral expressions are provided, and each is useful depending on the geometry of the object under consideration.

- The moments of inertia in Eqs. (10.15)–(10.20) measure the *second moment* of the mass distribution. That is, to determine I_x, I_y, and I_z in Eqs. (10.15)–(10.17), the moment arms $\tilde{r}_x$, $\tilde{r}_y$, and $\tilde{r}_z$ are *squared*. The second integral in each of these expressions is obtained by noting that $\tilde{r}_x^2 = \tilde{y}^2 + \tilde{z}^2$, and similarly for $\tilde{r}_y^2$ and $\tilde{r}_z^2$. For the products of inertia I_{xy}, I_{yz}, and I_{xz} in Eqs. (10.18)–(10.20), the product of two different moment arms is used.

- In Eqs. (10.15)–(10.20), $\tilde{x}$, $\tilde{y}$, and $\tilde{z}$ have dimensions of length, and dm has dimension of mass. Hence, all mass moments of inertia have dimensions of $(mass)(length)^2$, such as slug·in.^{2} or kg·m^2.

- When the x, y, and z axes pass through the center of mass of an object, we sometimes define the axes as x', y', and z', and we refer to the moments of inertia as *mass center moments of inertia* with the designations $I_{x'}$, $I_{y'}$, etc. In dynamics, moments of inertia about mass center axes are especially useful.

- For any object with positive mass and finite volume, I_x, I_y, and I_z are always positive.* The products of inertia I_{xy}, I_{yz}, and I_{xz} may be positive, zero, or negative, as discussed below.

- Evaluation of moments of inertia using composite shapes is possible using the parallel axis theorem discussed later in this section.

Helpful Information

Additional products of inertia. When you study the dynamics of bodies in three dimensions, an advanced topic, you may see that, in addition to the mass moments of inertia cited in Eqs. (10.15)–(10.20), products of inertia I_{yx}, I_{zy}, and I_{zx} are also used. However, $I_{xy} = I_{yx}$, $I_{yz} = I_{zy}$, and $I_{xz} = I_{zx}$. Despite this fact, these additional products of inertia are often used because it provides for a symmetry of notation that is convenient when writing the equations that govern the motion of a body.

What are mass moments of inertia used for?

It is useful to discuss why there are six mass moments of inertia, what is different about them, and what they are used for.

Moments of inertia I_x, I_y, and I_z. In Fig. 10.13, the International Space Station with a docked Space Shuttle is shown. Imagine that a moment M_x about the x axis is applied to the Space Station. Assuming the Space Station is rigid,† it will begin to undergo angular acceleration about the x axis. The value of the angular acceleration is directly dependent on the mass moment of inertia about the x axis I_x. Furthermore, the larger I_x is, the lower the angular acceleration will be for a given value of M_x. Similar remarks apply to moments applied about the y and z axes, and the influence that moments of inertia I_y and I_z have on angular accelerations about these axes.

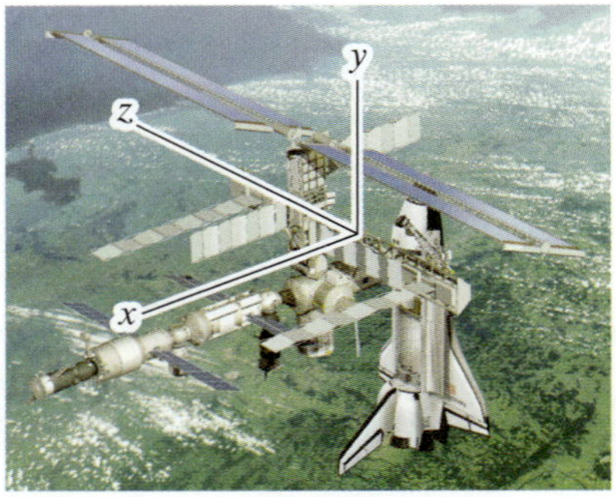

Figure 10.13
The International Space Station with a Space Shuttle docked to it.

* For the thin rod shown in the Table of Properties of Solids on the inside back cover, I_x is in fact positive, but since it is much smaller than I_y and I_z, it is usually taken to be zero.

† The International Space Station is very flexible, as are most space structures. If a moment were applied about the x axis shown in Fig. 10.13, then in addition to the rotations discussed above, the structure would vibrate. Control of vibrations in space structures is very important and receives considerable attention.

Products of inertia I_{xy}, I_{yz}, and I_{xz}. Products of inertia measure the asymmetry of a mass distribution with respect to the x, y, and z axes. Products of inertia can have positive, zero, or negative value, depending on the shape and mass distribution of an object, the selection of the x, y, and z directions, and the location of the origin of the xyz coordinate system. An object is said to be *symmetric* if it has at least one plane about which both the shape and mass distribution are symmetric. For example, if an object is symmetric about the xy plane, such as the mallet shown in Fig. 10.14, then $I_{xz} = I_{yz} = 0$ and I_{xy} may be positive, zero, or negative. If an object is symmetric about at least two of the xy, yz, and xz planes, then all of the products of inertia are zero. Thus the products of inertia are zero for a solid of revolution when one of the x, y, or z axes coincides with the axis of revolution.

Objects that have one or more nonzero products of inertia may display complicated dynamics in three-dimensional motions. For example, the Space Station is unsymmetric about the xyz axes shown Fig. 10.13, and it has nonzero products of inertia. As a consequence, if a moment M_x about the x axis is applied, the Space Station, in addition to rotating about the x axis, will rotate about the y and/or z axes. The dynamics of objects with nonzero products of inertia is an advanced subject and until you study this, you will not need to use products of inertia. Hence, methods for evaluating I_{xy}, I_{yz}, and I_{xz} are not discussed in this book.

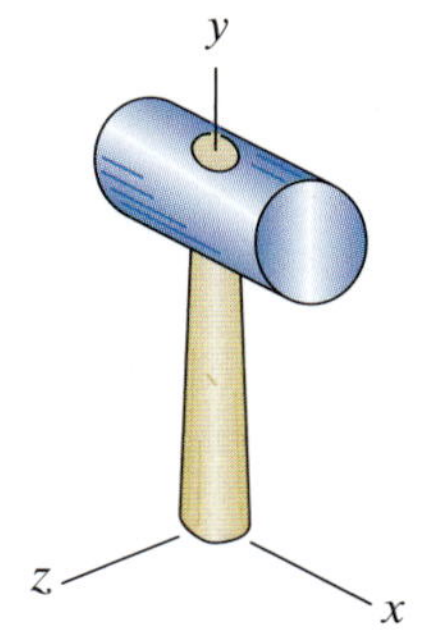

Figure 10.14
A mallet with a metal head and wood handle is shown. If the mallet's geometry and mass distribution are both symmetric about the xy plane, then the mallet is said to be a *symmetric object*. If the mallet is also symmetric about the yz plane, it may be called a *doubly symmetric object*.

Radius of gyration

Rather than use mass moments of inertia to quantify how mass is distributed, it is sometimes convenient to use the *radii of gyration*, which are directly related to the mass moments of inertia and are defined as

$$k_x = \sqrt{\frac{I_x}{m}}, \qquad k_y = \sqrt{\frac{I_y}{m}}, \qquad k_z = \sqrt{\frac{I_z}{m}}, \tag{10.21}$$

where

k_x, k_y, and k_z are the *radii of gyration of the mass about the x, y, and z axes, respectively*.

m is the mass of the object.

The radii of gyration have units of *length*, and have a straightforward physical interpretation, as discussed in Example 10.12.

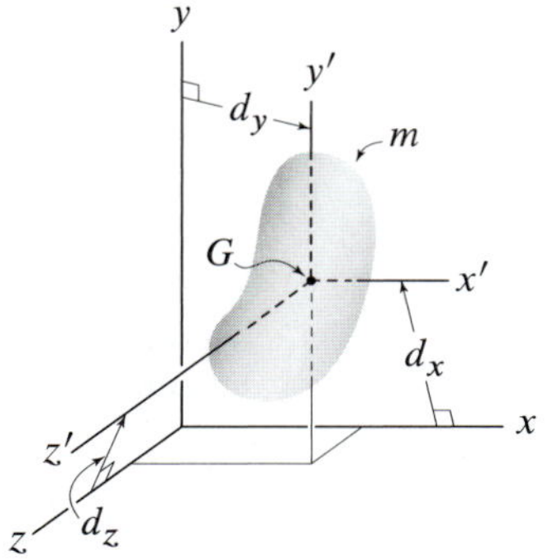

Figure 10.15
An object with mass m and center of mass at point G. The x and x' axes are parallel with separation distance d_x, the y and y' axes are parallel with separation distance d_y, and the z and z' axes are parallel with separation distance d_z.

Parallel axis theorem

The parallel axis theorem relates mass moments of inertia I_x, I_y, and I_z to the mass center moments of inertia $I_{x'}$, $I_{y'}$, and $I_{z'}$, where the x and x' axes are parallel, the y and y' axes are parallel, and the z and z' axes are parallel. The parallel axis theorem is important and is used often in dynamics and vibrations.

Consider the object with mass m and center of mass G shown in Fig. 10.15. An $x'y'z'$ coordinate system is defined, with origin at the center of mass of the object. The x, y, and z axes are parallel to the x', y', and z' axes, respectively, with separation distances d_x, d_y, and d_z, respectively. The *parallel axis theorem* relates the mass moments of inertia with respect to the x, y, and z axes to the mass center moments of inertia as follows:

$$I_x = I_{x'} + d_x^2 m, \tag{10.22}$$

$$I_y = I_{y'} + d_y^2 m, \tag{10.23}$$

$$I_z = I_{z'} + d_z^2 m. \tag{10.24}$$

The derivation of these expressions is similar to that for the area moment of inertia given by Eqs. (10.10)–(10.13) on p. 545.

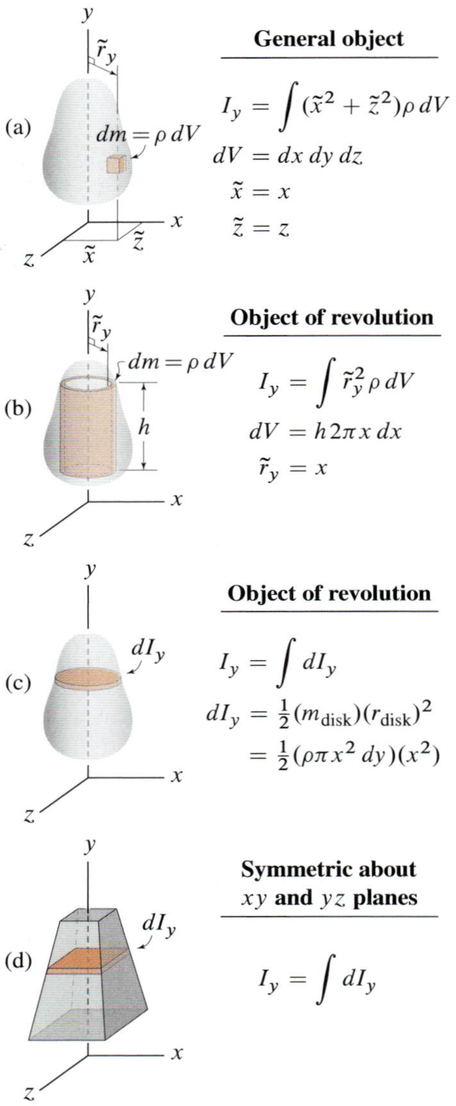

(a)
General object
$I_y = \int (\tilde{x}^2 + \tilde{z}^2)\rho\, dV$
$dV = dx\, dy\, dz$
$\tilde{x} = x$
$\tilde{z} = z$
$dm = \rho\, dV$
(b)
Object of revolution
$I_y = \int \tilde{r}_y^2 \rho\, dV$
$dV = h 2\pi x\, dx$
$\tilde{r}_y = x$
$dm = \rho\, dV$
h
(c)
Object of revolution
$I_y = \int dI_y$
$dI_y = \frac{1}{2}(m_{\text{disk}})(r_{\text{disk}})^2$
$= \frac{1}{2}(\rho\pi x^2\, dy)(x^2)$
(d)
Symmetric about xy and yz planes
$I_y = \int dI_y$

Figure 10.16
Strategies for evaluating the mass moments of inertia, using I_y as an example.

Evaluation of moments of inertia using integration

Mass moments of inertia are volume- and density-related and hence are typically more difficult to evaluate by integration than area moments of inertia. Figure 10.16 shows some strategies for determining the mass moments of inertia for an object, using I_y as an example, with comments as follows.

- **Object with general shape.** The triple integral approach shown in Fig. 10.16(a) has straightforward expressions for dV, $\tilde{x}$, $\tilde{y}$, and $\tilde{z}$. This approach is applicable to bodies with general geometry and density, and it can be used to determine I_x, I_y, and I_z. If you are comfortable with determining limits of integration and evaluating triple integrals, then this approach is robust and effective.

- **Object of revolution—thin shell volume element.** An object of revolution where the y axis is the axis of revolution is shown in Fig. 10.16(b). To determine I_y using a single integration, a thin cylindrical shell volume element can be used. The key feature of this volume element is that all material within the element has the same moment arm $\tilde{r}_y$ from the y axis. The radius of the shell is taken to be x, and the thickness is dx (these could just as well be taken as z and dz). Example 10.8 uses this approach. The object shown in Fig. 10.16(b) also has mass moments of inertia I_x and I_z, and the strategy shown in Fig. 10.16(a) or (c) must be used to determine these.

- **Object of revolution—thin disk volume element.** An object of revolution where the y axis is the axis of revolution is shown in Fig. 10.16(c). To determine I_y using a single integration, a thin circular disk volume element can be used. We consult a table of moments of inertia for common shapes, such as the Table of Properties of Solids on the inside back cover, to obtain the mass moment of inertia for a thin circular plate, which is $mr^2/2$, where m is the mass of the plate and r is its radius; this expression assumes the density of the plate is uniform. By taking m and r in this expression to be the mass and radius of the thin disk element, we obtain an expression for the mass moment of inertia for the thin disk element, which we call dI_y. Once an expression for dI_y is known, the moment of inertia for the entire object is obtained by

$$I_y = \int dI_y, \tag{10.25}$$

which requires only a single integration. This volume element may also be used to evaluate the moments of inertia I_x and I_z. Examples 10.8 and 10.9 illustrate this.

- **Symmetric object—thin plate volume element.** An object that is symmetric about both the xy and yz planes is shown in Fig. 10.16(d). A thin plate volume element is used, where the geometry of the plate depends on the object's geometry (Fig. 10.16(d) shows a rectangular plate). We consult a table of moments of inertia for common shapes, such as the Table of Properties of Solids on the inside back cover, to obtain the mass moment of inertia for the volume element dI_y and then use Eq.(10.25) to obtain the moment of inertia for the entire object. This volume element may also be used to evaluate the moments of inertia I_x and I_z. Example 10.10 illustrates this.

Evaluation of moments of inertia using composite shapes

The parallel axis theorem, written for composite shapes, becomes

$$I_x = \sum_{i=1}^{n} (I_{x'} + d_x^2 m)_i \qquad (10.26)$$

where n is the number of shapes, $I_{x'}$ is the mass moment of inertia for shape i about its mass center x' axis, d_x is the *shift distance* for shape i (i.e., the distance between the x axis and the x' axis for shape i), and m is the mass for shape i. Similar expressions may be written for I_y and I_z. To use Eq. (10.26), it is necessary to know the mass moment of inertia for each of the composite shapes about its mass center axis, and this generally must be obtained by integration or, when possible, by consulting a table of moments of inertia for common shapes, such as the Table of Properties of Solids on the inside back cover. A common error is to use the parallel axis theorem to relate moments of inertia between two parallel axes where neither of them is a mass center axis. Example 10.12 illustrates this approach.

Common Pitfall

Parallel axis theorem. Consider the parallel axes x_1 and x_2, mass center axis x', and center of mass at point G, such as shown below.

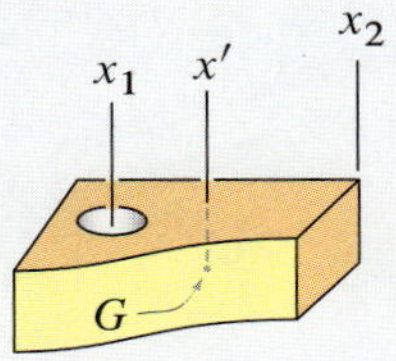

If I_{x_1} is known, a common error is to use the parallel axis theorem to *directly* determine I_{x_2}. In the parallel axis theorem, one of the axes is always a mass center axis. Thus, with I_{x_1} known, the parallel axis theorem must be used *twice*, the first time to determine $I_{x'}$ and the second time to use $I_{x'}$ to determine I_{x_2}. See Example 10.11.

End of Section Summary

In this section, *mass moments of inertia* I_x, I_y, I_z, I_{xy}, I_{yz}, and I_{xz} were defined to be the second moments of a mass distribution. Some of the key points are as follows:

- Mass moments of inertia are measures of how the mass for a particular object is distributed. Moments of inertia have dimensions of $(mass)(length)^2$.
- The products of inertia I_{xy}, I_{yz}, and I_{xz} are measures of the asymmetry of an object's mass distribution with respect to the x, y, and z axes. The need for products of inertia typically arises in advanced subjects, so methods for evaluating these quantities are not discussed in this book.
- *Radii of gyration* are alternative measures of the dispersion of mass. The radii of gyration are easily determined if the mass moments of inertia are known, and vice versa.

EXAMPLE 10.7 *Thin, Flat Plate-type Objects*

A thin rectangular plate with base b, height h, thickness t, and uniform density ρ is shown. Determine the mass moments of inertia I_x, I_y, and I_z about the center of mass for the plate, expressing each in terms of the mass m of the plate.

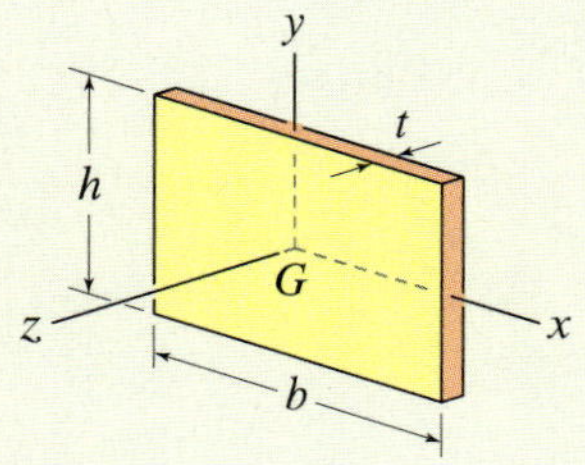

Figure 1

SOLUTION

Road Map We will use Eqs. (10.15)–(10.17) on p. 553 to determine the three moments of inertia.

I_x

Governing Equations & Computation We begin with Eq. (10.15) with $dm = \rho\, dV$

$$I_x = \int (\tilde{y}^2 + \tilde{z}^2)\rho\, dV. \tag{1}$$

If the plate is thin, then $\tilde{z}^2 \ll \tilde{y}^2$ and hence $\tilde{z}^2$ can be neglected. Using a volume element parallel to the x axis as shown in Fig. 2, we obtain

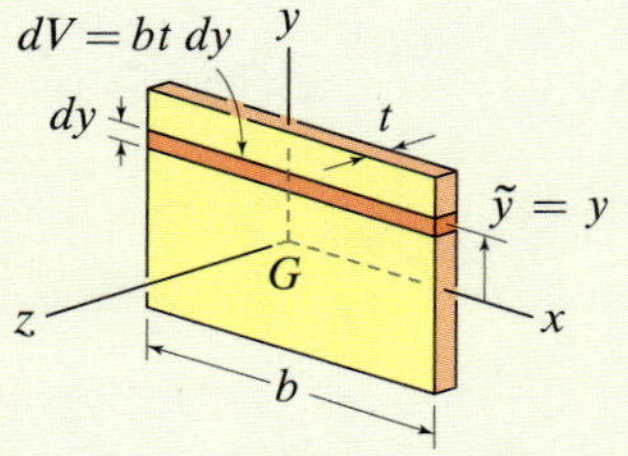

Figure 2
To determine I_x, a volume element parallel to the x axis is used to develop expressions for dV and $\tilde{y}$.

$$dV = bt\, dy \quad \text{and} \quad \tilde{y} = y. \tag{2}$$

Substituting Eq. (2) into Eq. (1), the moment of inertia about the x axis is

$$I_x = \int \tilde{y}^2 \rho\, dV = \int_{-h/2}^{h/2} y^2 \rho bt\, dy = \rho t \frac{by^3}{3}\bigg|_{-h/2}^{h/2} = \rho t \frac{bh^3}{12}. \tag{3}$$

While the above result is satisfactory, it is customary to report mass moments of inertia in terms of the total mass of an object. The mass of the thin plate is $m = \rho bht$. We express Eq. (3) in terms of mass m, using the following procedure:

$$I_x = \rho t \frac{bh^3}{12} \underbrace{\left(\frac{m}{\rho bht}\right)}_{=1} = \boxed{\frac{mh^2}{12}}. \tag{4}$$

I_y

Governing Equations & Computation We begin with Eq. (10.16)

$$I_y = \int (\tilde{x}^2 + \tilde{z}^2)\rho\, dV, \tag{5}$$

where $\tilde{z}^2$ can be neglected because $\tilde{z}^2 \ll \tilde{x}^2$. Using a volume element parallel to the y axis as shown in Fig. 3, we obtain

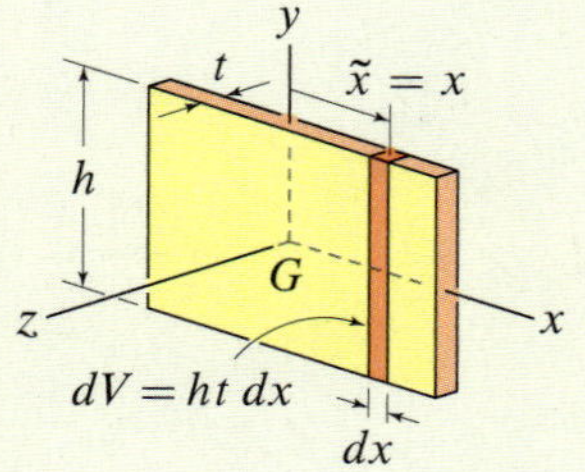

Figure 3
To determine I_y, a volume element parallel to the y axis is used to develop expressions for dV and $\tilde{x}$.

$$dV = ht\, dx \quad \text{and} \quad \tilde{x} = x. \tag{6}$$

Substituting Eq. (6) into Eq. (5), the moment of inertia about the y axis is

$$I_y = \int \tilde{x}^2 \rho\, dV = \int_{-b/2}^{b/2} x^2 \rho ht\, dx = \rho t \frac{hx^3}{3}\bigg|_{-b/2}^{b/2} = \rho t \frac{hb^3}{12}. \tag{7}$$

Expressing Eq. (7) in terms of mass m provides

$$I_y = \rho t \frac{hb^3}{12} \frac{m}{\rho bht} = \boxed{\frac{mb^2}{12}}. \tag{8}$$

I_z

Governing Equations & Computation We begin with Eq. (10.17)

$$I_z = \int (\tilde{x}^2 + \tilde{y}^2)\, \rho \, dV. \tag{9}$$

For the geometry of the plate in this problem, $\tilde{x}$ and $\tilde{y}$ are of comparable size, so neither can be neglected. This presents a difficulty if Eq. (9) is to be evaluated using a single integration. However, some ingenuity provides a straightforward solution, as follows. We break Eq. (9) into two separate integrals

$$I_z = \int \tilde{x}^2 \rho \, dV + \int \tilde{y}^2 \rho \, dV. \tag{10}$$

The first of the above integrals is identical to Eq. (7), and the second is identical to Eq. (3). Hence,

$$I_z = \frac{mb^2}{12} + \frac{mh^2}{12} = \boxed{\frac{m}{12}(b^2 + h^2).} \tag{11}$$

In essence, the first and second integrals in Eq. (10) were evaluated using different volume elements.

Discussion & Verification

- A thin plate is a common object in structures and machines. Hence, the results obtained here are usually tabulated in handbooks and texts, such as the Table of Properties of Solids on the inside back cover of this book.
- By letting h or b in Fig. 1 become small, the plate considered here becomes a thin rod. For example, if $h \to 0$, then Eqs. (4), (8), and (11) become $I_x = 0$ and $I_y = I_z = mb^2/12$. You should verify that these mass moments of inertia agree with those reported for a thin rod in the Table of Properties of Solids on the inside back cover of this book (when making this comparison, you must note that some coordinate directions and/or dimensions may be defined differently).

A Closer Look While mass moments of inertia and area moments of inertia are generally quite different, they are related for thin flat plates, such as in this example. To see this relationship, rewrite Eq. (3) as

$$I_x = \int \tilde{y}^2 \rho \, dV = \rho t \int \tilde{y}^2 \, dA. \tag{12}$$

The second integral is obtained by using $dV = t\, dA$ and factoring ρt outside of the integral because the plate has uniform density and thickness. The second integral is seen to be the definition of the area moment of inertia about the x axis (i.e., from Eq. (10.1) on p. 534, $(I_x)_{\text{area}} = \int \tilde{y}^2 \, dA$). Hence

$$I_x = \rho t (I_x)_{\text{area}} = \rho t \, \frac{bh^3}{12}, \tag{13}$$

where we have used the result from Example 10.1 on p. 539 that $(I_x)_{\text{area}} = bh^3/12$ for a rectangular area.

EXAMPLE 10.8 *Solid of Revolution—Moment of Inertia About Axis of Revolution*

Determine the mass moment of inertia about the x axis for the solid hemisphere of radius r and uniform density ρ. Express the result in terms of the mass m of the object.

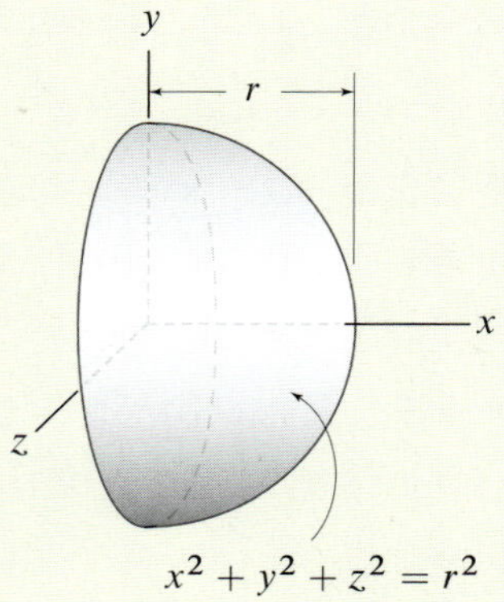

Figure 1

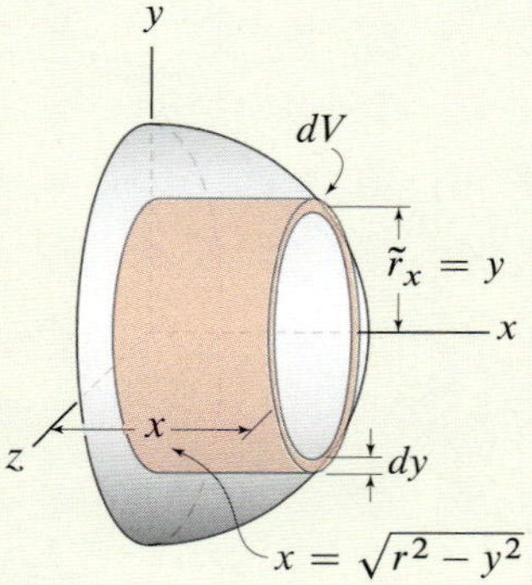

Figure 2
A thin shell volume element is used to develop expressions for dV and $\tilde{r}_x$.

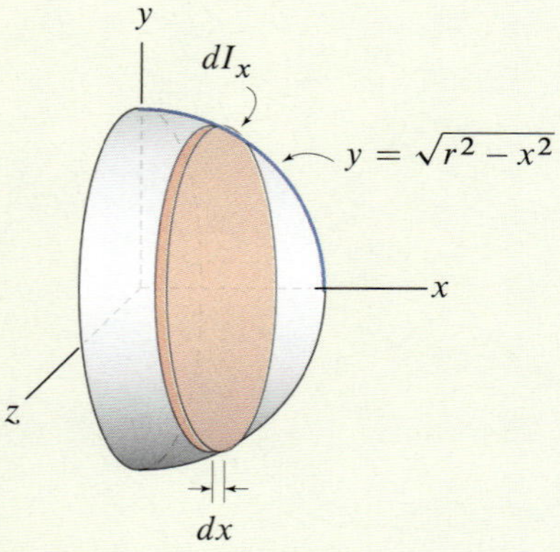

Figure 3
A thin circular disk volume element is used to develop an expression for dI_x.

SOLUTION

Road Map We will show two solutions, the first using a thin shell volume element and the second using a thin disk volume element. Note that the centroid of this object was determined in Example 7.6 on p. 408, and the expressions for dV developed in this solution are identical to those used earlier.

Solution 1—thin shell volume element

Governing Equations & Computation We begin with Eq. (10.15) with $dm = \rho\, dV$.

$$I_x = \int \tilde{r}_x^2 \rho\, dV. \tag{1}$$

The thin shell volume element shown in Fig. 2 is used to obtain expressions for dV and $\tilde{r}_x$. The merit of this volume element is that all the material in the element is at the same radial distance $\tilde{r}_x$ from the x axis. With y being the radius of the shell,

$$dV = 2\pi y x\, dy \qquad \text{and} \qquad \tilde{r}_x = y. \tag{2}$$

In Eq. (2), $2\pi y$ is the circumference of the shell, y is the radius, x is the length of the shell, and dy is the thickness. Substituting Eq. (2) into Eq. (1), with $x = \sqrt{r^2 - y^2}$, the moment of inertia about the x axis is

$$I_x = \int_0^r y^2 \rho\, 2\pi y \sqrt{r^2 - y^2}\, dy = \frac{2\pi\rho}{15}(y^2 - r^2)^{3/2}(2r^2 + 3y^2)\Big|_0^r = \frac{4\pi\rho r^5}{15}. \tag{3}$$

Using the expression for dV in Eq. (2), the mass m of a hemisphere is

$$m = \int \rho\, dV = \int_0^r \rho\, 2\pi y \sqrt{r^2 - y^2}\, dy = \frac{2\pi\rho}{3}(y^2 - r^2)^{3/2}\Big|_0^r = \frac{2\pi\rho r^3}{3}. \tag{4}$$

Finally, we express Eq. (3) in terms of mass m as follows:

$$I_x = \frac{4\pi\rho r^5}{15}\left(\frac{m}{2\pi\rho r^3/3}\right) = \boxed{\frac{2mr^2}{5}.} \tag{5}$$

Solution 2—thin disk volume element

Governing Equations & Computation We begin with Eq. (10.25) on p. 556

$$I_x = \int dI_x \tag{6}$$

where dI_x is the moment of inertia of the thin circular disk mass element shown in Fig. 3 about the x axis. From the Table of Properties of Solids, or from Fig. 10.16(c),

$$dI_x = \tfrac{1}{2}(m_{\text{disk}})(r_{\text{disk}})^2 = \tfrac{1}{2}(\rho\,\pi y^2\, dx)(y)^2 \tag{7}$$

where y is the radius of the disk. Substituting Eq. (7) into Eq. (6), with $y = \sqrt{r^2 - x^2}$, the moment of inertia about the x axis is

$$I_x = \int_0^r \tfrac{1}{2}\rho\pi (r^2 - x^2)^2\, dx = \pi\rho\left(\frac{r^4 x}{2} - \frac{r^2 x^3}{3} + \frac{x^5}{10}\right)\Big|_0^r = \frac{4\pi\rho r^5}{15}, \tag{8}$$

which is identical to Eq. (3). Hence, we obtain the same result, $\boxed{I_x = 2mr^2/5.}$

Discussion & Verification Knowing the mass moments of inertia for a cylinder and cone, it is possible to show that the result obtained here is reasonable; see Prob. 10.59.

EXAMPLE 10.9 *Solid of Revolution—Additional Moments of Inertia*

For the solid hemisphere of radius r and uniform density ρ considered in Example 10.8, shown again in Fig. 1, determine the mass moments of inertia about the y and z axes. Express the result in terms of the mass m of the object.

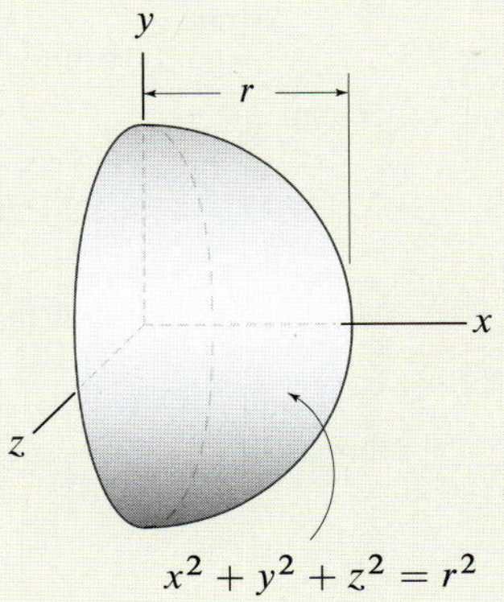

Figure 1

SOLUTION

Road Map Because of the symmetry of the object, $I_y = I_z$, and hence we will determine only one of these, arbitrarily choosing I_z. Unlike Example 10.8 where we had the choice of two volume elements, in this example we must use a thin circular disk volume element (that is, if we are to determine I_z using a single integration).

Governing Equations & Computation We begin with Eq. (10.25) on p. 556 (written in terms of z)

$$I_z = \int dI_z \tag{1}$$

where dI_z is the moment of inertia of the thin circular disk mass element shown in Fig. 2 about the z axis. Using the parallel axis theorem, Eq. (10.24) on p. 556, the moment of inertia of the thin disk element about the z axis is

$$dI_z = dI_{z'} + d_z^2\, dm \tag{2}$$

where $dI_{z'}$ is the moment of inertia of the disk element about the z' axis passing through the disk's mass center, dm is the mass of the disk, and d_z is the distance between the z and z' axes. From the Table of Properties of Solids on the back inside cover, the moment of inertia of a thin disk is

$$dI_{z'} = \tfrac{1}{4}(m_{\text{disk}})(r_{\text{disk}})^2 = \tfrac{1}{4}(\rho\pi y^2\, dx)(y)^2 \tag{3}$$

$$dm = \rho\pi y^2\, dx \tag{4}$$

$$d_z = x \tag{5}$$

where y is the radius of the disk. Using Eqs. (2)–(5) and noting that $y = \sqrt{r^2 - x^2}$, Eq. (1) becomes

$$\begin{aligned} I_z &= \int_0^r \frac{\pi\rho}{4}(r^2 - x^2)^2\, dx + \int_0^r x^2\, \pi\rho(r^2 - x^2)\, dx \\ &= \frac{\pi\rho}{4}\left(r^4 x - \frac{2r^2x^3}{3} + \frac{x^5}{5}\right)\Bigg|_0^r + \pi\rho\left(\frac{r^2x^3}{3} - \frac{x^5}{5}\right)\Bigg|_0^r \\ &= \frac{4\pi\rho r^5}{15}. \end{aligned} \tag{6}$$

Noting from Example 10.8 that the mass of the hemisphere is $m = 2\pi\rho r^3/3$, Eq. (6) expressed in terms of m is

$$I_z = \frac{4\pi\rho r^5}{15}\left(\frac{m}{2\pi\rho r^3/3}\right) = \boxed{\frac{2mr^2}{5}.} \tag{7}$$

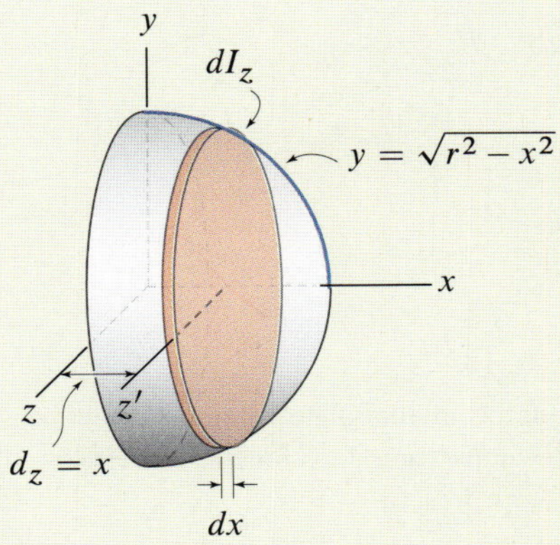

Figure 2
A thin circular disk volume element is used to develop an expression for dI_z.

Discussion & Verification Because of symmetry, $I_y = I_z$, hence $I_y = 2mr^2/5$. Comparing I_y and I_z found here with the results for I_x from Example 10.8 shows the *unexpected* result that all three of these are equal.

EXAMPLE 10.10 *Symmetric Object*

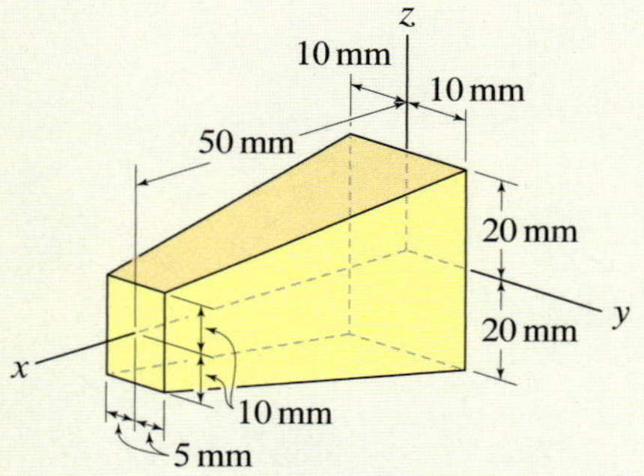

Figure 1

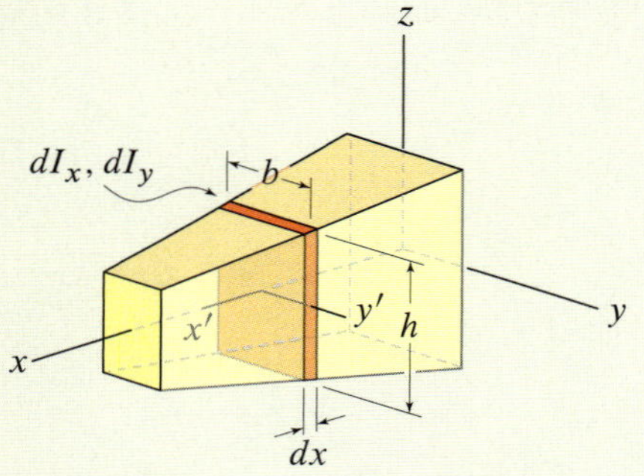

Figure 2
A thin rectangular plate volume element is used to develop expressions for dI_x and dI_y.

The tapered prism is made of aluminum and is symmetric about the xy and xz planes. Determine the mass moments of inertia about the x and y axes.

SOLUTION

Road Map We will use a thin rectangular plate volume element, with the strategy described in Fig. 10.16(d) on p. 556, to evaluate I_x and I_y.

I_x

Governing Equations & Computation We begin with Eq. (10.25) on p. 556 (written in terms of x)

$$I_x = \int dI_x \tag{1}$$

where dI_x is the moment of inertia of the thin rectangular plate mass element shown in Fig. 2 about the x axis. Note that the x axis and the x' axis passing through the mass center of the thin plate element coincide, and therefore no parallel axis shift is required. From the Table of Properties of Solids on the inside back cover,

$$dI_x = \tfrac{1}{12}(m_{\text{plate}})(b^2 + h^2). \tag{2}$$

The dimensions b and h of the thin plate element are functions of position as follows:

$$b = 20\,\text{mm} - (0.2)x \quad \text{and} \quad h = 40\,\text{mm} - (0.4)x. \tag{3}$$

The mass of the plate is the product of density ρ and the volume of the plate, hence

$$m_{\text{plate}} = \rho\, bh\, dx. \tag{4}$$

Combining Eqs. (1), (2), and (4) provides

$$I_x = \int_0^{50\,\text{mm}} \tfrac{1}{12}\rho\, bh(b^2 + h^2)\, dx. \tag{5}$$

From Table 1.4 on p. 16, the density of aluminum is $\rho = 2710\,\text{kg/m}^3 = 2.710 \times 10^{-6}\,\text{kg/mm}^3$. Substituting b and h from Eq. (3) and the value for ρ into Eq. (5) and carrying out the integration yield

$$\boxed{I_x = 7.00\,\text{kg}\cdot\text{mm}^2.} \tag{6}$$

I_y

Governing Equations & Computation We begin with Eq. (10.25) on p. 556

$$I_y = \int dI_y \tag{7}$$

where dI_y is the moment of inertia of the thin rectangular plate mass element shown in Fig. 2 about the y axis. Since the y and y' axes do not coincide, the parallel axis theorem, Eq. (10.23) on p. 556, is used to write the moment of inertia of the thin plate element about the y axis as

$$dI_y = dI_{y'} + d_y^2\, dm \tag{8}$$

where $dI_{y'}$ is the moment of inertia of the plate element about the y' axis passing through its mass center, dm is the mass of the plate element, and d_y is the distance between the y and y' axes, namely, $d_y = x$. From the Table of Properties of Solids on the inside back cover, the moment of inertia of a thin plate is

$$dI_{y'} = \tfrac{1}{12}(m_{\text{plate}})(h)^2 \tag{9}$$

where m_{plate} is given by Eq. (4). Combining Eqs. (4) and (7)–(9) provides

$$I_y = \int_0^{50\,\text{mm}} \tfrac{1}{12}\rho\, b h\, h^2\, dx + \int_0^{50\,\text{mm}} x^2\, \rho\, b h\, dx. \tag{10}$$

Substituting b and h from Eq. (3) and the value for ρ into Eq. (10) and carrying out the integration yield

$$\boxed{I_y = 41.7\,\text{kg}\cdot\text{mm}^2.} \tag{11}$$

Discussion & Verification A rough check of accuracy can be obtained by using the Table of Properties of Solids on the inside back cover to evaluate the moments of inertia for the uniform rectangular prism shown in Fig. 3, where dimensions have been chosen to approximate the shape and volume of the tapered prism. For the uniform prism shown in Fig. 3, you should verify that $I_x = 5.72\,\text{kg}\cdot\text{mm}^2$ and $I_y = 55.4\,\text{kg}\cdot\text{mm}^2$. The results for the tapered prism in Eqs. (6) and (11) are in reasonable agreement with these.

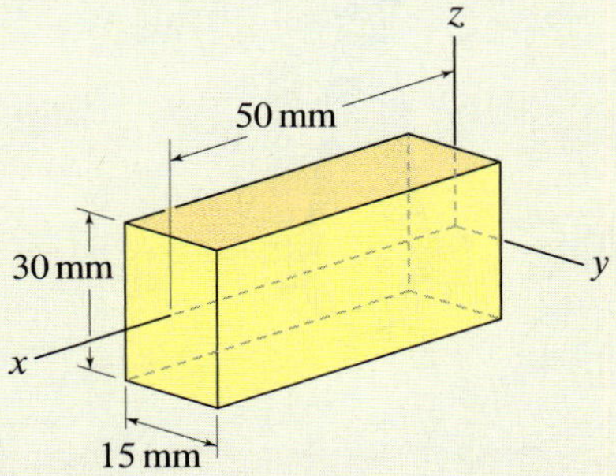

Figure 3
A uniform rectangular prism with dimensions that approximate those of the tapered prism shown in Fig. 1.

EXAMPLE 10.11 *Experimental Determination of Mass Moments of Inertia*

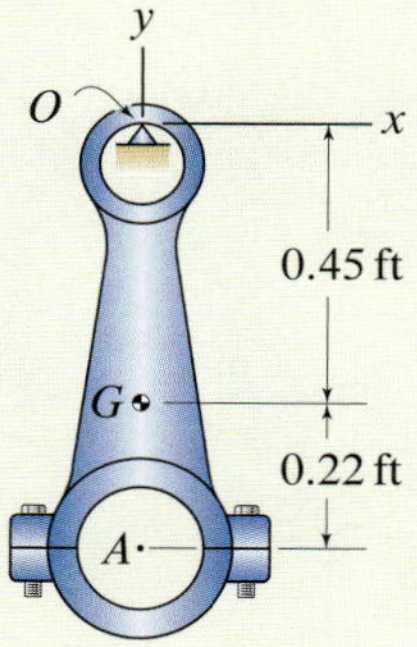

Figure 1

A connecting rod from a gasoline engine weighs 3.28 lb and has its center of gravity at point G. When the connecting rod is supported at point O and is allowed to oscillate as a pendulum, 0.816 s is required for one full cycle of motion. Determine the mass moments of inertia about points A and G.

SOLUTION

Road Map We will use Eq. (1), described in the margin note on this page, to determine I_{Oz}. We will then use the parallel axis theorem to determine the moments of inertia about points A and G. Note that, as discussed in the margin note on p. 557, to determine I_{Az} it is necessary to first know I_{Gz}.

I_{Gz}

Governing Equations & Computation Using Eq. (1), the mass moment of inertia of the connecting rod about the z axis through point O is

$$I_{Oz} = \frac{(0.816\,\text{s})^2(3.28\,\text{lb})(0.45\,\text{ft})}{4\pi^2} = 0.02489\,\text{ft}\cdot\text{lb}\cdot\text{s}^2. \tag{2}$$

To express Eq. (2) in terms of the U.S. Customary mass unit (recall from the definition in Table 1.1 on p. 9, 1 slug $\equiv$ 1 lb·s²/ft), we carry out the unit conversion

$$I_{Oz} = (0.02489\,\text{ft}\cdot\text{lb}\cdot\text{s}^2)\left(\frac{\text{slug}}{\text{lb}\cdot\text{s}^2/\text{ft}}\right) = 0.02489\,\text{slug}\cdot\text{ft}^2. \tag{3}$$

The parallel axis theorem, Eq. (10.24) on p. 556, rearranged for the moment of inertia about the z axis through the mass center G, is

$$\begin{aligned} I_{Gz} &= I_{Oz} - d_z^2\, m \\ &= 0.02489\,\text{slug}\cdot\text{ft}^2 - (0.45\,\text{ft})^2\left(\frac{3.28\,\text{lb}}{32.2\,\text{ft/s}^2}\right) \\ &= 0.02489\,\text{slug}\cdot\text{ft}^2 - 0.02063\,\text{ft}\cdot\text{lb}\cdot\text{s}^2\left(\frac{\text{slug}}{\text{lb}\cdot\text{s}^2/\text{ft}}\right) \\ &= \boxed{4.267\times10^{-3}\,\text{slug}\cdot\text{ft}^2.} \end{aligned} \tag{4}$$

I_{Az}

Governing Equations & Computation With the moment of inertia about the mass center known from Eq. (4), the parallel axis theorem can be used again to determine the moment of inertia about the z axis through point A. Thus,

$$\begin{aligned} I_{Az} &= I_{Gz} + d_z^2 m \\ &= 4.267\times10^{-3}\,\text{slug}\cdot\text{ft}^2 + (0.22\,\text{ft})^2\left(\frac{3.28\,\text{lb}}{32.2\,\text{ft/s}^2}\right) \\ &= 4.267\times10^{-3}\,\text{slug}\cdot\text{ft}^2 + 4.930\times10^{-3}\,\text{ft}\cdot\text{lb}\cdot\text{s}^2\left(\frac{\text{slug}}{\text{lb}\cdot\text{s}^2/\text{ft}}\right) \\ &= \boxed{9.20\times10^{-3}\,\text{slug}\cdot\text{ft}^2.} \end{aligned} \tag{5}$$

Discussion & Verification A rough check of accuracy can be obtained by approximating the connecting rod by a slender rod with the same mass and with length approximating that of the connecting rod. Problem 10.76 asks you to carry out this calculation.

Helpful Information

Experimental determination of mass moments of inertia. The object shown has mass m and center of gravity at point G.

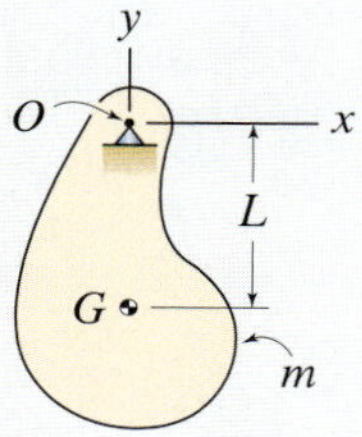

By supporting the object by a smooth pin at point O and allowing it to oscillate as a pendulum, the mass moment of inertia about the z axis through point O (i.e., perpendicular to the figure) is

$$I_{Oz} = \frac{T^2 mgL}{4\pi^2} \tag{1}$$

where T is the time required for one full cycle of motion. Thus, by locating the center of gravity for an object and carefully measuring T, m (or weight mg), and L, the mass moment of inertia is easily determined.

Remark: You will be able to derive Eq. (1) when you study dynamics. You may find this surprising, but the time T required for one full cycle of motion is independent of the amplitude of motion provided the amplitude is not too large.

EXAMPLE 10.12 *Composite shapes*

The hammer consists of a cast iron head and wood handle. The iron head has $7000\,\mathrm{kg/m^3}$ density, and its shape is a rectangular prism with a circular hole. The wood handle has $500\,\mathrm{kg/m^3}$ density, and its shape is a circular cylinder. Determine the mass moment of inertia about the z axis and the corresponding radius of gyration.

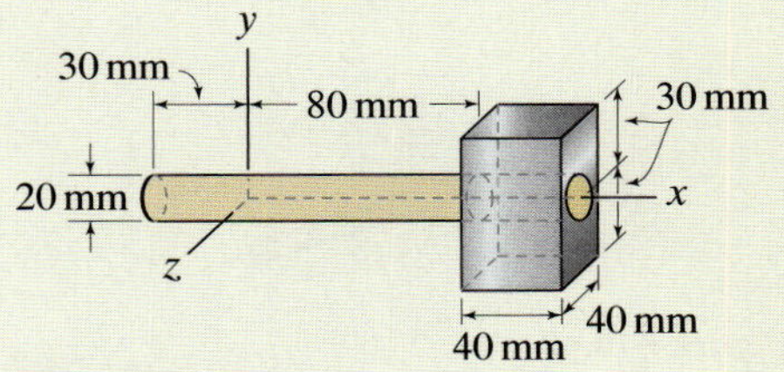

Figure 1

SOLUTION

Road Map The hammer consists of a rectangular prism and circular cylinder, so it will be convenient to determine I_z using composite shapes with the parallel axis theorem.

Governing Equations & Computation A combination of three composite shapes is shown in Fig. 2, where the mass for shape 3 is negative. The mass of each composite shape is

$$m_1 = \rho_{\mathrm{wood}} V_1 = (500\,\mathrm{kg/m^3})\pi(0.010\,\mathrm{m})^2(0.150\,\mathrm{m}) = 0.02356\,\mathrm{kg}, \tag{1}$$

$$m_2 = \rho_{\mathrm{iron}} V_2 = (7000\,\mathrm{kg/m^3})(0.040\,\mathrm{m})(0.040\,\mathrm{m})(0.060\,\mathrm{m}) = 0.6720\,\mathrm{kg}, \tag{2}$$

$$m_3 = -\rho_{\mathrm{iron}} V_3 = -(7000\,\mathrm{kg/m^3})\pi(0.010\,\mathrm{m})^2(0.040\,\mathrm{m}) = -0.08796\,\mathrm{kg}. \tag{3}$$

The parallel axis theorem for composite shapes, Eq. (10.26) on p. 557, is

$$I_z = \sum_{i=1}^{3}(I_{z'} + d_z^2\, m)_i. \tag{4}$$

For each composite shape, the Table of Properties of Solids on the inside back cover of this book is used to obtain the mass moment of inertia about the appropriate axis through the composite shape's mass center. Note that when using such a table, it is often the case that the xyz axes used in the table are different from the xyz axes used in your particular problem. Hence, you must be careful to obtain the correct expressions. Equation (4) becomes

$$\begin{aligned} I_z = {} & \frac{m_1}{12}[3(10\,\mathrm{mm})^2 + (150\,\mathrm{mm})^2] + (45\,\mathrm{mm})^2 m_1 \\ & + \frac{m_2}{12}[(40\,\mathrm{mm})^2 + (60\,\mathrm{mm})^2] + (100\,\mathrm{mm})^2 m_2 \\ & + \frac{m_3}{12}[3(10\,\mathrm{mm})^2 + (40\,\mathrm{mm})^2] + (100\,\mathrm{mm})^2 m_3 \\ & \boxed{= 6210\,\mathrm{kg{\cdot}mm^2}.} \end{aligned} \tag{5}$$

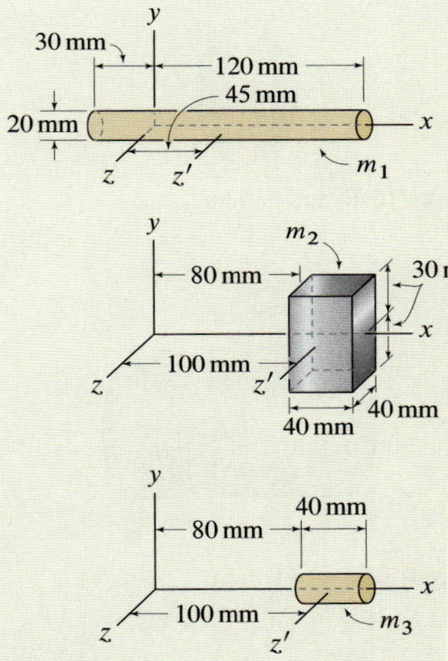

Figure 2
A combination of three composite shapes to determine I_z, and distances from the center of mass of each composite shape to the z axis. Note that $m_3 < 0$.

Noting that the total mass of the hammer is $m = m_1 + m_2 + m_3 = 0.6076\,\mathrm{kg}$, the radius of gyration is easily determined using Eq. (10.21) on p. 555 as

$$k_z = \sqrt{\frac{I_z}{m}} = \sqrt{\frac{6210\,\mathrm{kg{\cdot}mm^2}}{0.6076\,\mathrm{kg}}} = \boxed{101\,\mathrm{mm}.} \tag{6}$$

Discussion & Verification Imagine that all of the mass m of the hammer is concentrated at a point, as shown in Fig. 3. While it is physically impossible to do this, the radius of gyration k_z has the interpretation of being the distance from the z axis to the point so that the point mass has the same moment of inertia I_z as the original object. Considering that the mass of the wood handle is small compared to the iron head, and that the shift distance d_z for the wood handle in the parallel axis theorem is small compared to that for the head, we expect the radius of gyration to locate a point close to the center of mass of just the head. Thus, the value determined in Eq. (6) is reasonable.

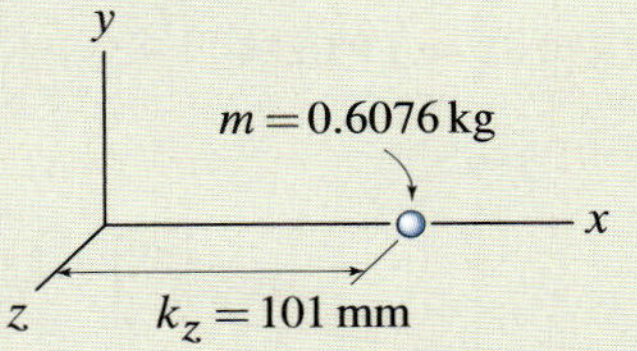

Figure 3
The radius of gyration k_z is the distance from the z axis that a point mass m should be positioned so that it has the same mass moment of inertia I_z as the original object.

PROBLEMS

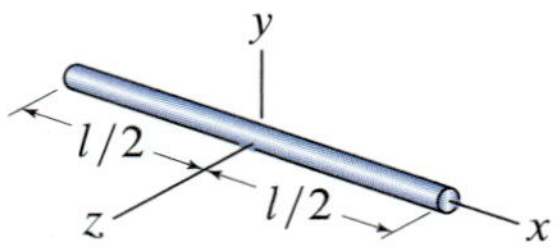

Figure P10.48

Problem 10.48

The uniform slender rod has mass m and length l. Use integration to show that the mass moment of inertia about the y axis is $I_y = ml^2/12$.

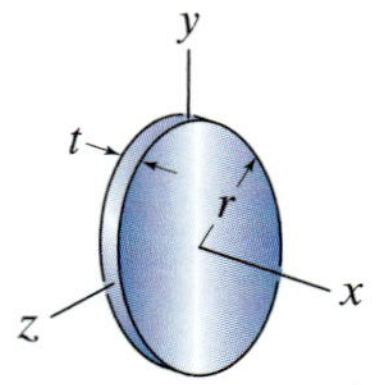

Figure P10.49 and P10.50

Problems 10.49 and 10.50

The uniform plate has thickness t, radius r, and density ρ.

Problem 10.49 Use integration with a thin cylindrical shell mass element to determine the mass moment of inertia about the x axis. Express your answer in terms of the mass m of the plate.

Problem 10.50 Assuming $t \ll r$, use integration with a mass element parallel to the y axis (as in Example 10.2 on p. 540) to determine the mass moment of inertia about the y axis. Express your answer in terms of the mass m of the plate.

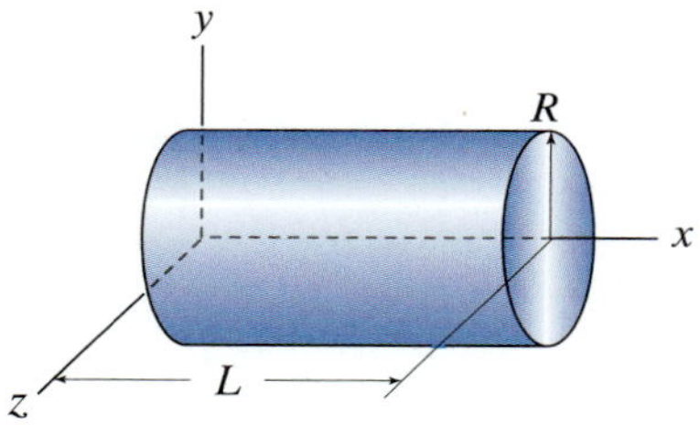

Figure P10.51 and P10.52

Problems 10.51 and 10.52

The uniform cylinder has length L, radius R, and density ρ.

Problem 10.51 Use integration with a thin cylindrical shell mass element to determine the mass moment of inertia about the x axis. Express your answer in terms of the mass m of the cylinder.

Problem 10.52 Use integration with a thin disk mass element to determine the mass moment of inertia about the y axis. Express your answer in terms of the mass m of the cylinder.

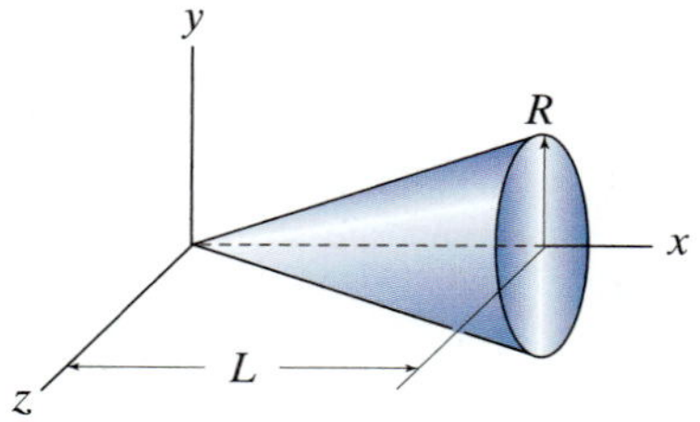

Figure P10.53 and P10.54

Problems 10.53 and 10.54

For the uniform solid cone with length L and radius R, use integration to determine the mass moment of inertia indicated, expressing your answer in terms of the mass m of the cone.

Problem 10.53 I_x.

Problem 10.54 I_z.

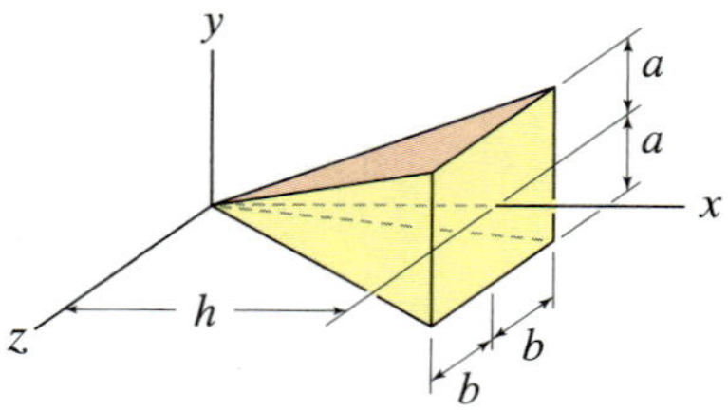

Figure P10.55 and P10.57

Problems 10.55 through 10.57

The tapered solid prism shown has density ρ and rectangular cross section. Use integration to determine the mass moment of inertia indicated, expressing your answer in terms of the mass m of the prism and parameters such as a, b, and h.

Problem 10.55 I_x.

Problem 10.56 I_y.

Problem 10.57 I_z.

Problem 10.58

The Table of Properties of Solids appearing on the inside back cover of this book shows the mass moments of inertia for a uniform sphere and hemisphere are both $I_z = 2mr^2/5$. If a sphere and hemisphere have the same radius and density, does this mean their mass moments of inertia I_z are the same? Explain.
Note: Concept problems are about *explanations*, not computations.

Problem 10.59

In Example 10.8 on p. 560, the mass moment of inertia of a hemisphere about its axis of revolution was found to be $2mr^2/5$. Show that this result is between those for a cylinder and cone, both having radius r and length r (the mass moments of inertia for these are given in the Table of Properties of Solids on the inside back cover of this book). Discuss why this result is expected.
Note: Concept problems are about *explanations*, not computations.

Problems 10.60 and 10.61

The solid hemisphere is constructed of materials with densities ρ_0 and $\rho_0/2$ as shown.

Problem 10.60

(a) Fully set up the integrals, including limits of integration, that will yield the mass moment of inertia about the x axis.

(b) Evaluate the integrals obtained in Part (a) using computer software such as Mathematica, Maple, etc.

Problem 10.61

(a) Fully set up the integrals, including limits of integration, that will yield the mass moment of inertia about the y axis.

(b) Evaluate the integrals obtained in Part (a) using computer software such as Mathematica, Maple, etc.

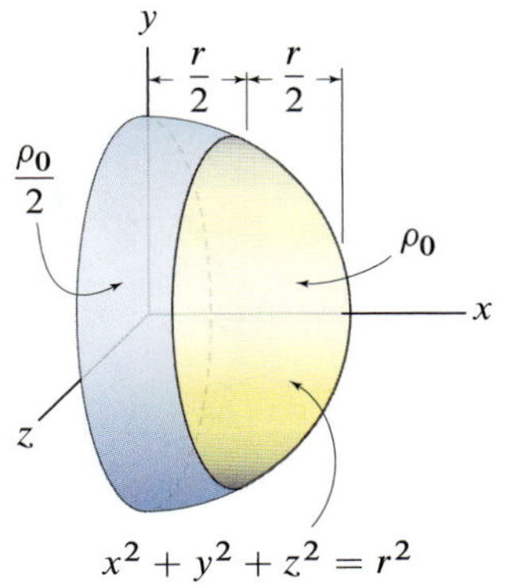

Figure P10.60 and P10.61

Problem 10.62

For the solid of revolution shown, determine the mass moment of inertia about the x axis. The material has specific weight $\gamma = 0.409\,\text{lb/in.}^3$. Report your answer using slugs and inches.

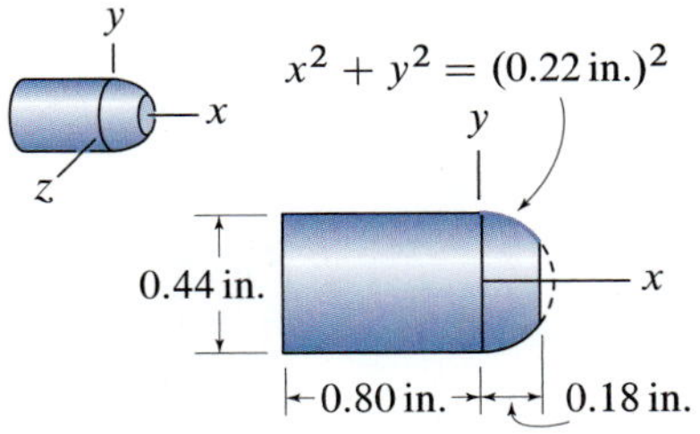

Figure P10.62

Problems 10.63 and 10.64

A solid of revolution is produced by revolving the area shown 360° around the y axis. Use integration to determine the mass moment of inertia about the axis of revolution assuming the solid has uniform density. Express your answer in terms of the mass m of the object.

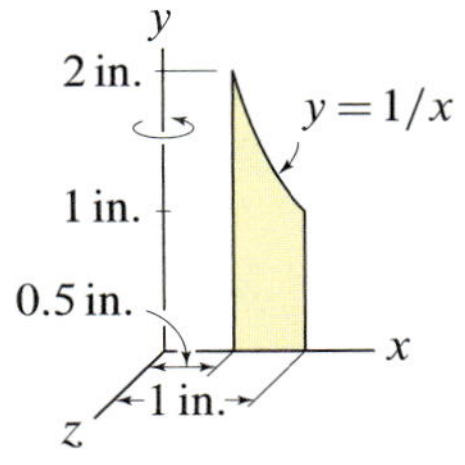

Figure P10.63

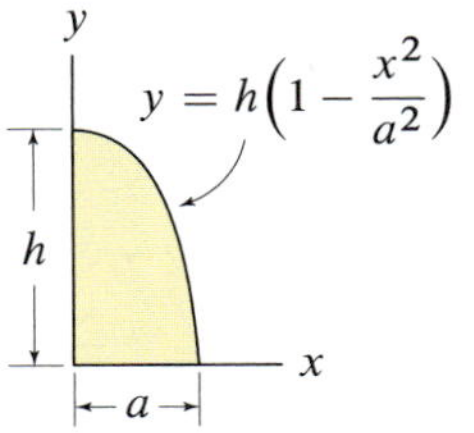

Figure P10.64

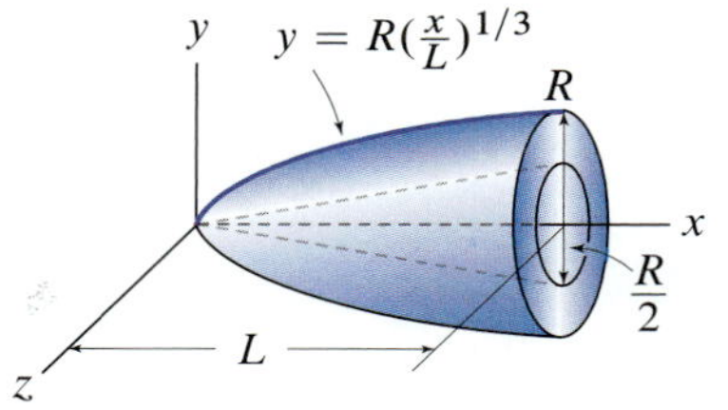

Figure P10.65 and P10.66

Problems 10.65 and 10.66

The solid shown has a cone-shaped cavity and has uniform density.

Problem 10.65

(a) Fully set up the integral, including limits of integration, that will yield the mass moment of inertia about the x axis.

(b) Evaluate the integral obtained in Part (a) using computer software such as Mathematica, Maple, etc. Express your answer in terms of the mass m of the object.

Problem 10.66

(a) Fully set up the integral, including limits of integration, that will yield the mass moment of inertia about the y axis.

(b) Evaluate the integral obtained in Part (a) using computer software such as Mathematica, Maple, etc. Express your answer in terms of the mass m of the object.

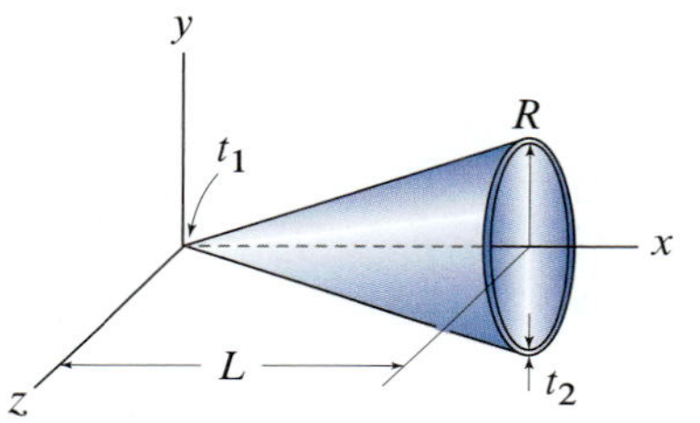

Figure P10.67 and P10.68

Problems 10.67 and 10.68

A thin-walled hollow cone has uniform density with thickness t_1 at the left-hand end and t_2 at the right-hand end.

Problem 10.67

(a) If the cone's thickness is uniform with $t_1 = t_2 = t_0$, fully set up the integral, including limits of integration, that will yield the mass moment of inertia about the x axis.

(b) Evaluate the integral obtained in Part (a) using computer software such as Mathematica, Maple, etc. Express your answer in terms of the mass m of the object.

Problem 10.68

(a) If the cone's thickness varies linearly from $t_1 = 2t_0$ at the left-hand end to $t_2 = t_0$ at the right-hand end, fully set up the integral, including limits of integration, that will yield the mass moment of inertia about the x axis.

(b) Evaluate the integral obtained in Part (a) using computer software such as Mathematica, Maple, etc. Express your answer in terms of the mass m of the object.

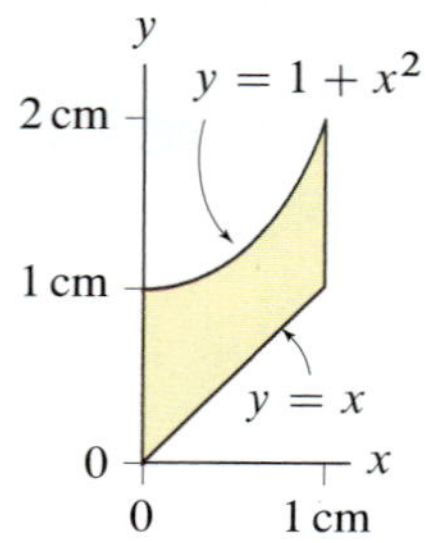

Figure P10.69 and P10.70

Problems 10.69 and 10.70

A plastic part with $1100\,\mathrm{kg/m^3}$ density is produced by revolving the area shown 360° around an axis of revolution. Determine the mass moment of inertia about the axis of revolution indicated. Express your answer in units of $\mathrm{kg{\cdot}cm^2}$.

Problem 10.69 The axis of revolution is the x axis.

Problem 10.70 The axis of revolution is the y axis.

Problems 10.71 through 10.74

The solid hemisphere shown has a cone-shaped cavity and has uniform density.

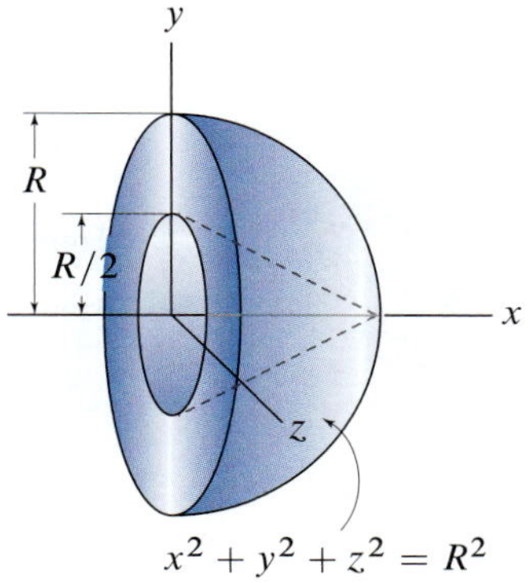

Figure P10.71–P10.74

Problem 10.71

(a) Fully set up the integral, including limits of integration, that will yield the mass moment of inertia about the x axis.

(b) Evaluate the integral obtained in Part (a) using computer software such as Mathematica, Maple, etc. Express your answer in terms of the mass m of the object.

Problem 10.72 Determine the mass moment of inertia about the x axis using composite shapes. Express your answer in terms of the mass m of the object.

Problem 10.73

(a) Fully set up the integral, including limits of integration, that will yield the mass moment of inertia about the z axis.

(b) Evaluate the integral obtained in Part (a) using computer software such as Mathematica, Maple, etc. Express your answer in terms of the mass m of the object.

Problem 10.74 Determine the mass moment of inertia about the z axis using composite shapes. Express your answer in terms of the mass m of the object.

Problem 10.75

A throwing toy is molded of uniform foam. It consists of an oblong-shaped portion, a cylindrical portion, and four rectangular fins. The density of the oblong and cylindrical shapes is $100\,\text{kg/m}^3$, and each of the fins have 1.8×10^{-3} kg mass.

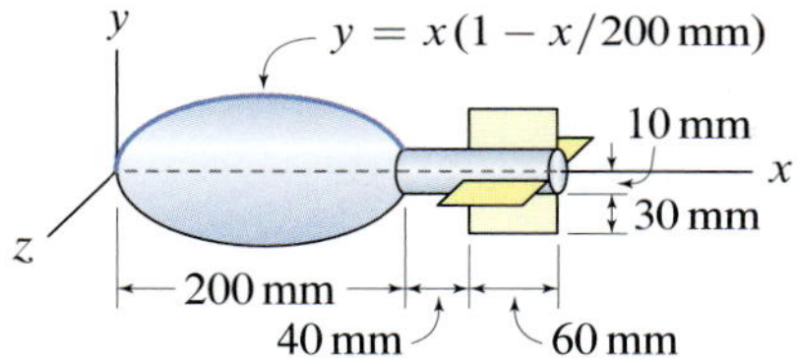

Figure P10.75

(a) Fully set up the integral, including limits of integration, that will yield the mass moment of inertia about the x axis for the oblong-shaped portion.

(b) Evaluate the integral obtained in Part (a) using computer software such as Mathematica, Maple, etc.

(c) You should find the result of Part (b) to be $79.8\,\text{kg·mm}^2$. Using this value, determine the total mass moment of inertia about the x axis for the toy.

Problem 10.76

Approximate the connecting rod in Example 10.11 on p. 564 by a uniform slender rod. Take the length of this rod to be 0.8 ft and the mass to be the same as that for the connecting rod. Determine the mass moment of inertia of this rod about its end, and compare to the value for I_{Oz} found for the connecting rod in Example 10.11.

Problems 10.77 through 10.79

A uniform rectangular prism with mass m is shown. Beginning with the appropriate mass moment of inertia given in the Table of Properties of Solids on the inside back cover of this book, use the parallel axis theorem to determine the mass moment of inertia of the prism about the axis indicated.

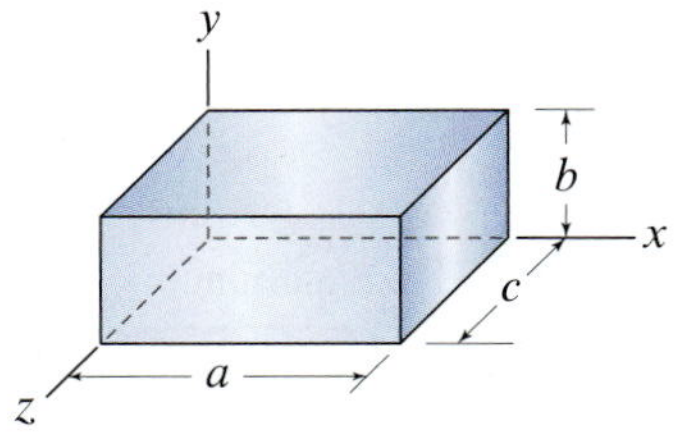

Figure P10.77–P10.79

Problem 10.77 The x axis.

Problem 10.78 The y axis.

Problem 10.79 The z axis.

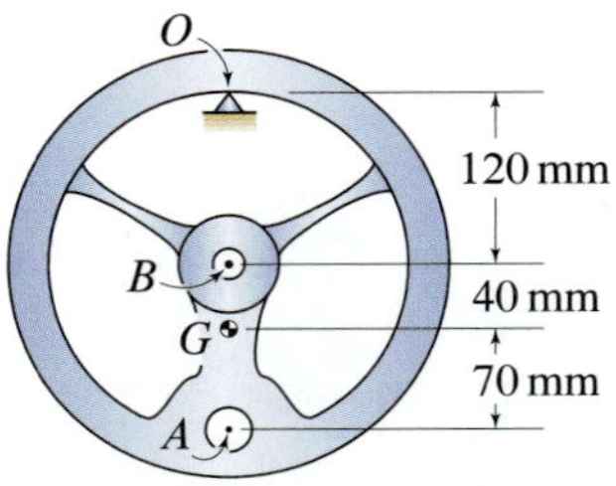

Figure P10.80

Problem 10.80

A handwheel for a machine has 0.8 kg mass and center of gravity at point G. When the handwheel is supported at point O and is allowed to oscillate as a pendulum, 1.12 s is required for one full cycle of motion. Determine the mass moment of inertia about the axis perpendicular to the figure and passing through point B. *Hint:* See the helpful information margin note on p. 564.

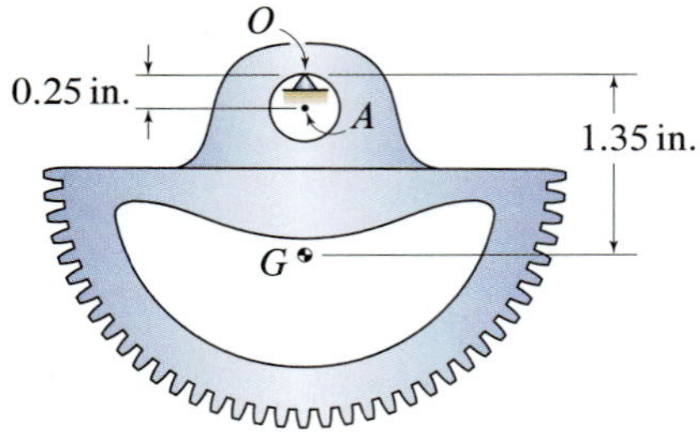

Figure P10.81

Problem 10.81

A sector gear for a machine has 0.355 lb weight and center of gravity at point G. When the sector gear is supported at point O and is allowed to oscillate as a pendulum, 0.379 s is required for one full cycle of motion. Determine the mass moment of inertia about the axis perpendicular to the figure and passing through point A. *Hint:* See the helpful information margin note on p. 564.

Problem 10.82

The plate shown has uniform thickness and is made of material with specific weight $0.7\,\mathrm{lb/in.^2}$. Determine the mass moment of inertia about the axis perpendicular to the plate and passing through point A.

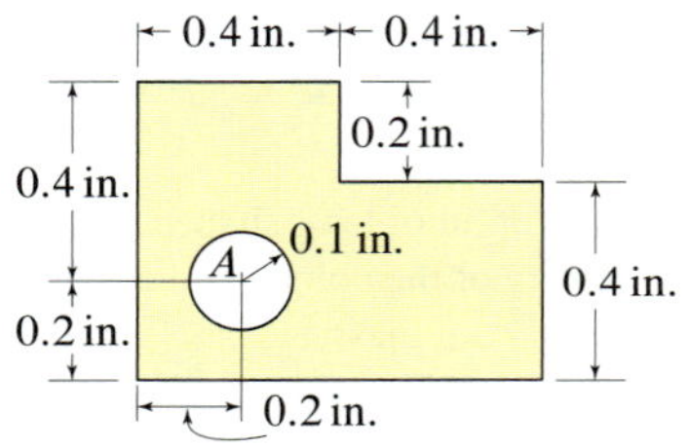

Figure P10.82

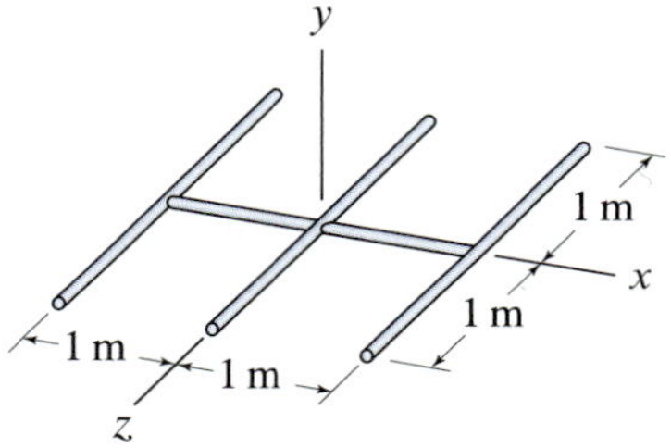

Figure P10.83

Problem 10.83

An antenna is constructed of identical small-diameter uniform rods, each having $0.25\,\mathrm{kg/m}$ mass. Determine the mass moment of inertia of the antenna about the

(a) x axis.

(b) y axis.

(c) z axis.

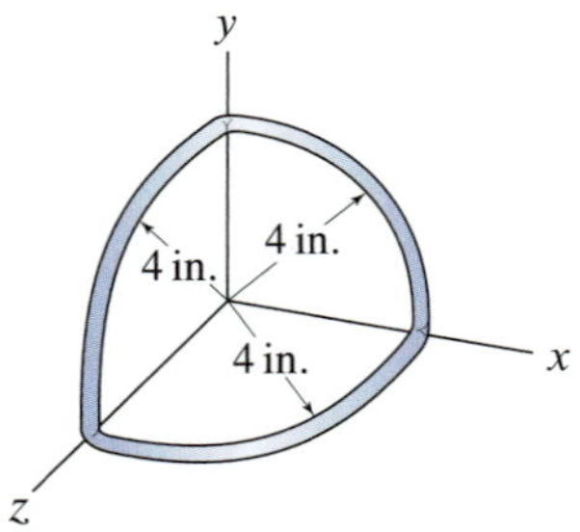

Figure P10.84

Problem 10.84

An object is constructed by welding together three small-diameter uniform identical rods, each having quarter-circular shape and 0.5 lb weight. Determine the mass moment of inertia of the object about the x axis.

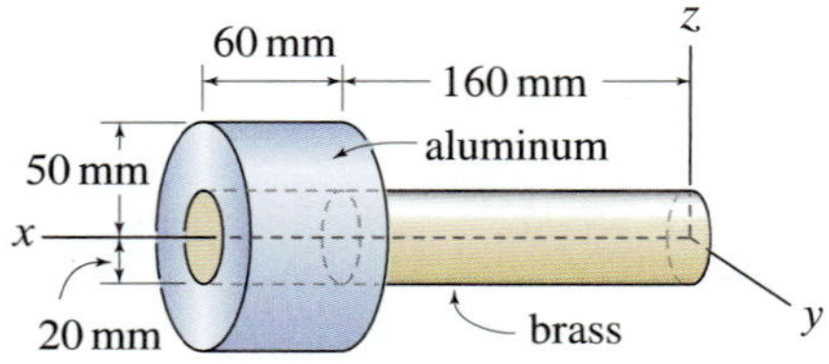

Figure P10.85 and P10.86

Problems 10.85 and 10.86

An object is constructed of a brass rod and aluminum cylinder having densities of $8500\,\mathrm{kg/m^3}$ and $2700\,\mathrm{kg/m^3}$, respectively. The brass rod fills the hole in the aluminum cylinder. Determine the mass moment of inertia of the object about the

Problem 10.85 x axis.

Problem 10.86 y axis.

Problem 10.87

The cam consists of a circular cylinder A with a circular hole to reduce its moment of inertia and a circular shaft BC about which it rotates. If the cam is made of cast iron with $7200\,\mathrm{kg/m^3}$ density, determine the mass moment of inertia about the z axis.

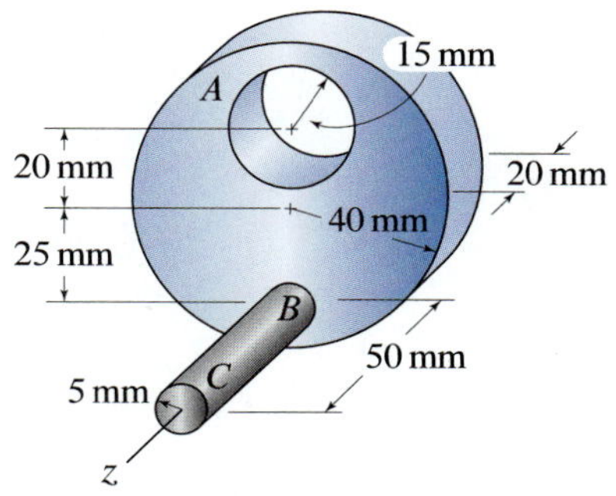

Figure P10.87

Problems 10.88 through 10.90

A bracket is constructed of a thin semicircular plate of uniform thickness having $0.05\,\mathrm{lb/in.^2}$ specific weight and a thick rectangular plate of uniform thickness having $0.25\,\mathrm{lb/in.^3}$ specific weight. The bracket is symmetric about the xz plane. Determine the mass moment of inertia of the object about the axis indicated. Report your answer using slugs and inches.

Problem 10.88 x axis.

Problem 10.89 y axis.

Problem 10.90 z axis.

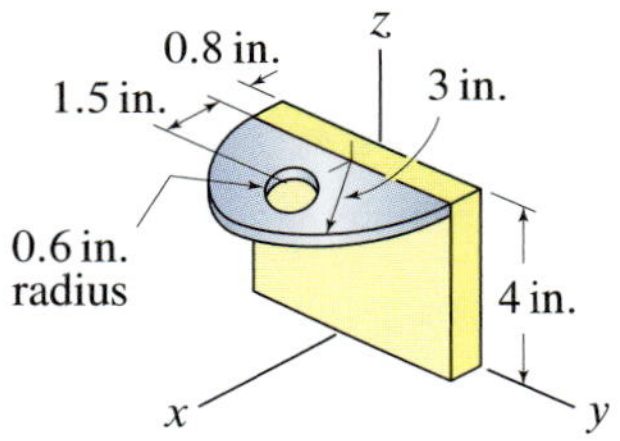

Figure P10.88–P10.90

Problems 10.91 and 10.92

An object with 6 lb weight consists of a cylinder and hemisphere of the same material. Determine the mass moment of inertia of the object about the axis indicated. Report your answer using slugs and inches.

Problem 10.91 x axis.

Problem 10.92 y axis.

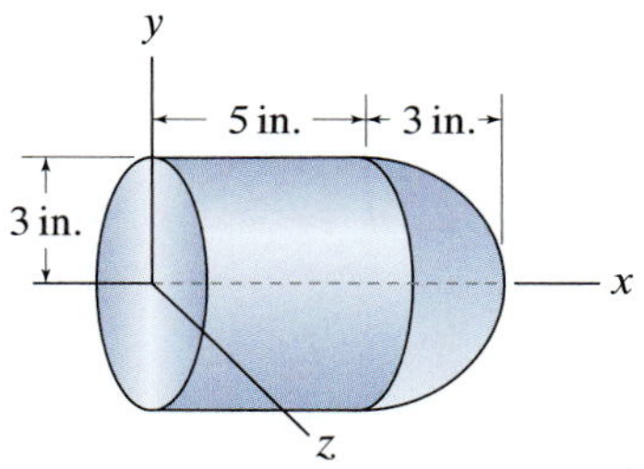

Figure P10.91 and P10.92

Problems 10.93 through 10.95

The object shown is made of uniform plastic and weighs 2.34 lb. Determine the mass moment of inertia of the object about the axis indicated. Report your answer using slugs and feet.

Problem 10.93 x axis.

Problem 10.94 y axis.

Problem 10.95 z axis.

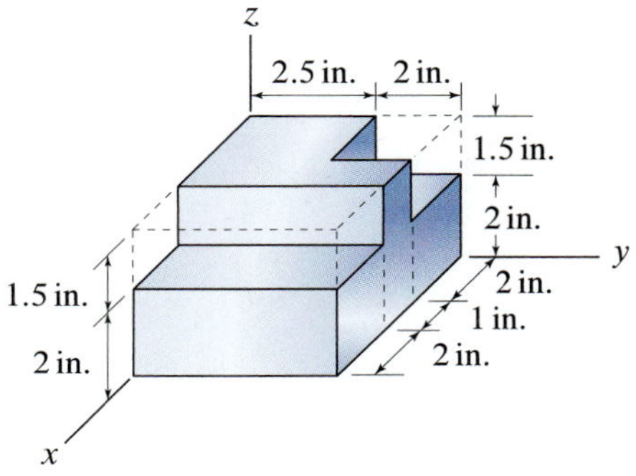

Figure P10.93–P10.95

10.4 Chapter Review

Important definitions, concepts, and equations of this chapter are summarized. For equations and/or concepts that are not clear, you should refer to the original equation and page numbers cited for additional details.

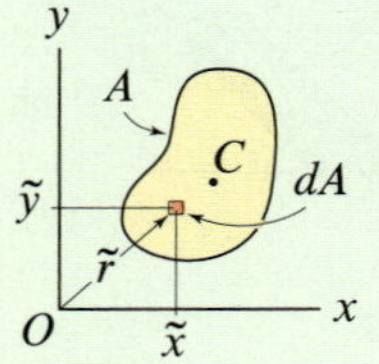

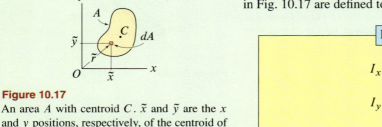

Figure 10.17
An area A with centroid C. $\tilde{x}$ and $\tilde{y}$ are the x and y positions, respectively, of the centroid of area element dA. $\tilde{r}$ is the distance from point O to area element dA.

Area moments of inertia. The *area moments of inertia* for the area shown in Fig. 10.17 are defined to be

Eqs. (10.1)–(10.4), p. 534

$$I_x = \int \tilde{y}^2\, dA,$$

$$I_y = \int \tilde{x}^2\, dA,$$

$$J_O = \int \tilde{r}^2\, dA = I_x + I_y,$$

$$I_{xy} = \int \tilde{x}\tilde{y}\, dA,$$

where

$\tilde{x}$, $\tilde{y}$, and $\tilde{r}$ are defined in Fig. 10.17. Note that the definitions of these are identical to those used in Chapter 7. For example, $\tilde{x}$ is the distance (i.e., moment arm) from the y axis to the centroid of area element dA.

I_x is the *area moment of inertia about the x axis.*

I_y is the *area moment of inertia about the y axis.*

J_O is the *polar moment of inertia of the area about point O.*

I_{xy} is the *product of inertia of the area about point O.*

Area moments of inertia have dimensions of $(length)^4$.

Area radius of gyration The *radii of gyration* of an area A are defined as

Eq. (10.5), p. 536

$$k_x = \sqrt{\frac{I_x}{A}}, \qquad k_y = \sqrt{\frac{I_y}{A}}, \qquad k_O = \sqrt{\frac{J_O}{A}},$$

where

k_x is the *radius of gyration of the area about the x axis.*

k_y is the *radius of gyration of the area about the y axis.*

k_O is the *polar radius of gyration of the area about point O.*

A is the area of the shape.

Radii of gyration have units of *length*.

Parallel axis theorem for area moments of inertia Consider the area A with centroid C shown in Fig. 10.18. A centroidal $x'y'$ coordinate system is defined, with origin O' positioned at the centroid of the area. The x and y axes are parallel to the x' and y' axes, respectively, with separation distances d_x and d_y. The *parallel axis theorem* relates the area moments of inertia with respect to the x and y axes to the centroidal area moments of inertia as follows:

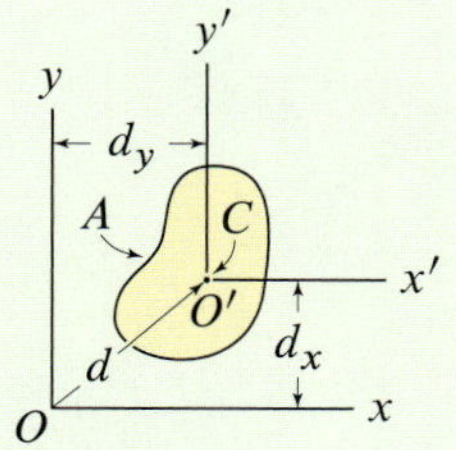

Figure 10.18
An area A with centroid at point C. The x and x' axes are parallel with separation distance d_x, and the y and y' axes are parallel with separation distance d_y.

Eqs. (10.6)–(10.9), p. 545

$$I_x = I_{x'} + d_x^2 A,$$
$$I_y = I_{y'} + d_y^2 A,$$
$$J_O = J_{O'} + d^2 A,$$
$$I_{xy} = I_{x'y'} + d_x d_y A.$$

The parallel axis theorem, written for composite shapes, is

Eq. (10.14), p. 546

$$I_x = \sum_{i=1}^{n} (I_{x'} + d_x^2 A)_i$$

where n is the number of shapes, $I_{x'}$ is the area moment of inertia for shape i about its centroidal x' axis, d_x is the *shift distance* for shape i (i.e., the distance between the x axis and the x' axis for shape i), and A is the area for shape i. Similar expressions may be written for I_y and J_O.

A common error is to use the parallel axis theorem to relate moments of inertia between two parallel axes where neither is a centroidal axis.

Mass moments of inertia The *mass moments of inertia* for the object shown in Fig. 10.19 are defined as

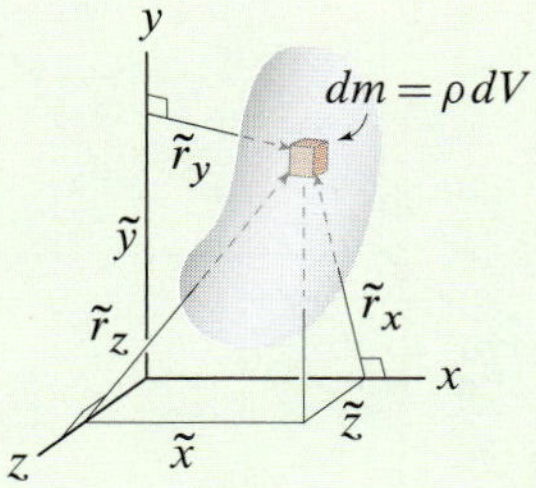

Figure 10.19
An object with mass m, density ρ, and volume V; $\tilde{r}_x$, $\tilde{r}_y$, and $\tilde{r}_z$ are radial distances from the x, y, and z axes, respectively, to the center of mass for mass element dm.

Eqs. (10.15)–(10.20), p. 553

$$I_x = \int \tilde{r}_x^2 \, dm = \int (\tilde{y}^2 + \tilde{z}^2) \, dm,$$
$$I_y = \int \tilde{r}_y^2 \, dm = \int (\tilde{x}^2 + \tilde{z}^2) \, dm,$$
$$I_z = \int \tilde{r}_z^2 \, dm = \int (\tilde{x}^2 + \tilde{y}^2) \, dm,$$
$$I_{xy} = \int \tilde{x}\tilde{y} \, dm,$$
$$I_{yz} = \int \tilde{y}\tilde{z} \, dm,$$
$$I_{xz} = \int \tilde{x}\tilde{z} \, dm,$$

where

$\tilde{r}_x$, $\tilde{r}_y$, and $\tilde{r}_z$ are defined in Fig. 10.19 as the radial distances (i.e., moment arms) from the x, y, and z axes, respectively, to the center of mass for mass element dm.

$\tilde{x}$, $\tilde{y}$, and $\tilde{z}$ are defined in Fig. 10.19 as the x, y, and z distances, respectively, to the center of mass for mass element dm.

I_x, I_y, and I_z are the *mass moments of inertia about the x, y, and z axes*, respectively.

I_{xy}, I_{yz}, and I_{xz} are the *products of inertia of the mass about the xy, yz, and xz axes*, respectively.

Mass moments of inertia have dimensions of $(mass)(length)^2$.

Mass radius of gyration The *radii of gyration* of an object with mass m are defined as

Eq. (10.21), p. 555

$$k_x = \sqrt{\frac{I_x}{m}}, \qquad k_y = \sqrt{\frac{I_y}{m}}, \qquad k_z = \sqrt{\frac{I_z}{m}},$$

where

k_x is the *radius of gyration of the mass about the x axis*.

k_y is the *radius of gyration of the mass about the y axis*.

k_z is the *polar radius of gyration of the mass about z axis*.

m is the mass of the object.

Radii of gyration have units of *length*.

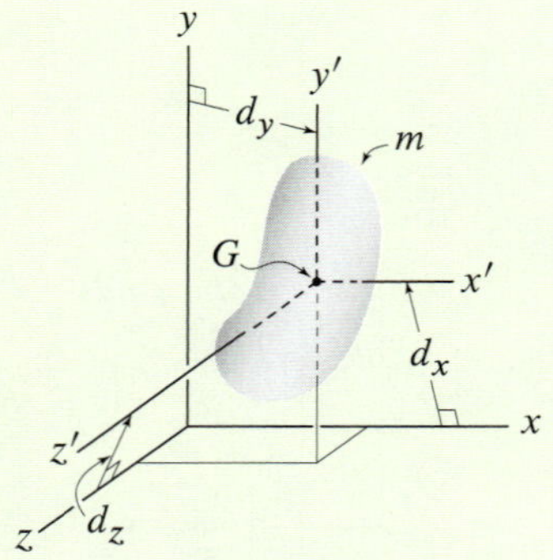

Figure 10.20
An object with mass m and center of mass at point G. The x and x' axes are parallel with separation distance d_x, the y and y' axes are parallel with separation distance d_y, and the z and z' axes are parallel with separation distance d_z.

Parallel axis theorem for mass moments of inertia Consider the object with mass m and center of mass G shown in Fig. 10.20. An $x'y'z'$ coordinate system is defined, with origin at the center of mass of the object. The x, y, and z axes are parallel to the x', y', and z' axes, respectively, with separation distances d_x, d_y, and d_z, respectively. The *parallel axis theorem* relates the mass moments of inertia with respect to the x, y, and z axes to the mass center moments of inertia as follows:

Eqs. (10.22)–(10.24), p. 556

$$I_x = I_{x'} + d_x^2\, m,$$
$$I_y = I_{y'} + d_y^2\, m,$$
$$I_z = I_{z'} + d_z^2\, m.$$

The parallel axis theorem, written for composite shapes, is

Eq. (10.26), p. 557

$$I_x = \sum_{i=1}^{n} \left(I_{x'} + d_x^2\, m\right)_i$$

where n is the number of shapes, $I_{x'}$ is the mass moment of inertia for shape i about its mass center x' axis, d_x is the *shift distance* for shape i (i.e., the distance between the x axis and the x' axis for shape i), and m is the mass for shape i. Similar expressions may be written for I_y and I_z.

A common error is to use the parallel axis theorem to relate moments of inertia between two parallel axes where neither of them is a mass center axis.

REVIEW PROBLEMS

Problem 10.96

A semicircular area has outside radius r_o and a concentric hole with inside radius r_i. Determine the polar moment of inertia of the area about point O.

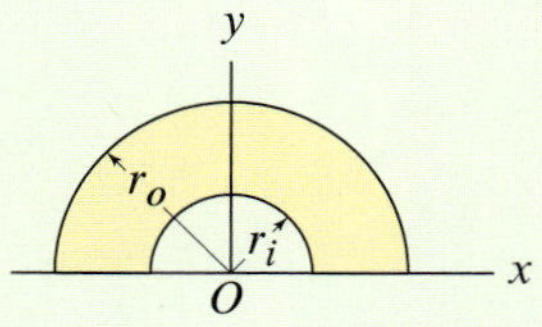

Figure P10.96

Problems 10.97 and 10.98

(a) Fully set up the integral including limits of integration that will give the area moment of inertia about the y axis.

(b) Evaluate the integrals in Part (a).

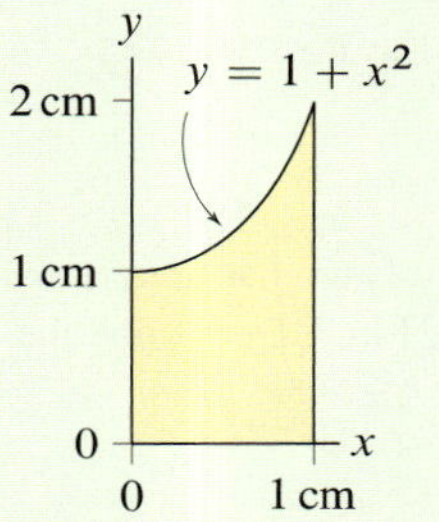

Figure P10.97

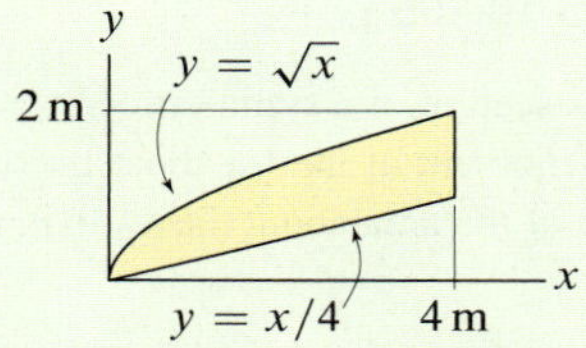

Figure P10.98

Problems 10.99 and 10.100

Determine the constants c_1, c_2, and c_3 so the curves pass through the points shown.

Problem 10.99

(a) Fully set up the integral including limits of integration that will give the area moment of inertia about the x axis.

(b) Evaluate the integral in Part (a).

Problem 10.100

(a) Fully set up the integral including limits of integration that will give the area moment of inertia about the y axis.

(b) Evaluate the integral in Part (a).

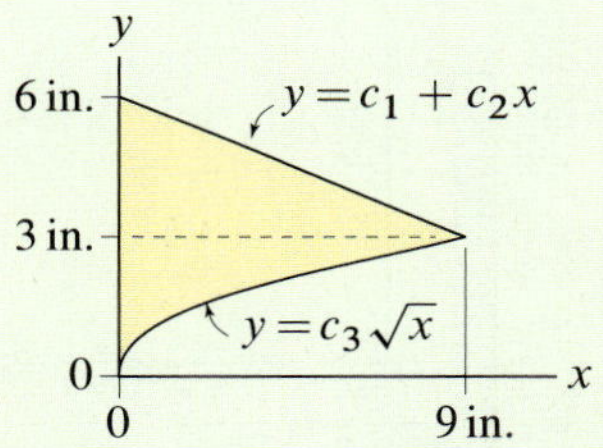

Figure P10.99 and P10.100

Problem 10.101

(a) Fully set up the integral including limits of integration that will give the area moment of inertia about the x axis.

(b) Evaluate the integral in Part (a).

Problem 10.102

(a) Fully set up the integral including limits of integration that will give the area moment of inertia about the y axis.

(b) Evaluate the integral in Part (a).

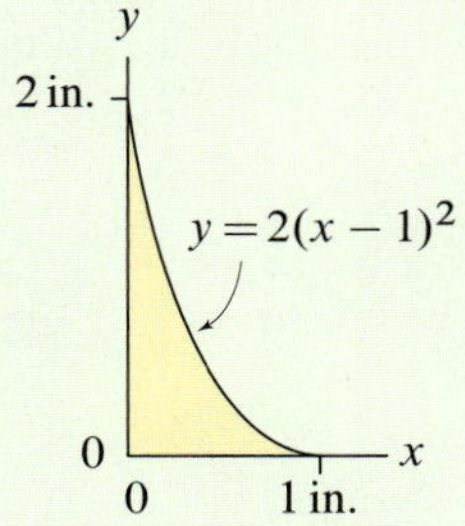

Figure P10.101 and P10.102

Problems 10.103 and 10.104

Determine the area moments of inertia I_x and I_y.

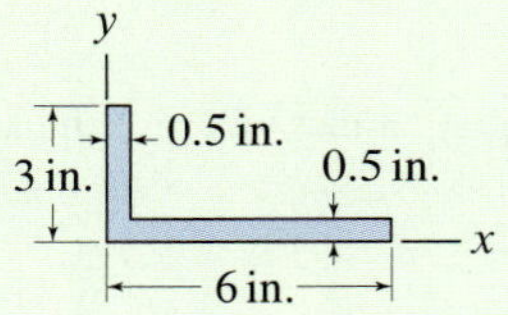

Figure P10.103

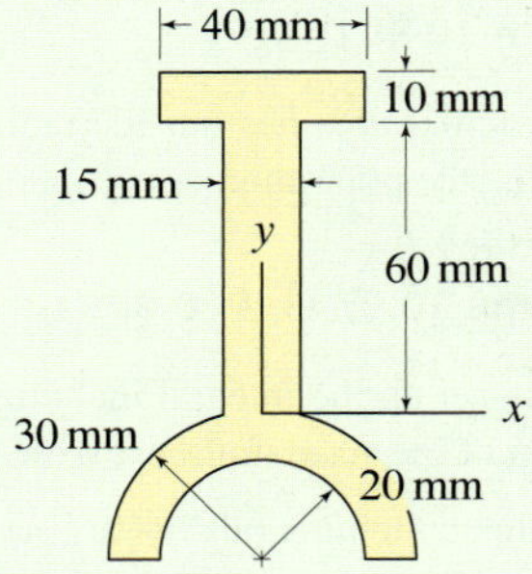

Figure P10.104

Problem 10.105

The cross section of a symmetric W8 × 15 wide-flange I beam has area $A = 4.44\ \text{in.}^2$ and area moment of inertia about the x_1 axis $I_{x_1} = 121\ \text{in.}^4$. Determine the moment of inertia of the area about the x_2 axis I_{x_2}.

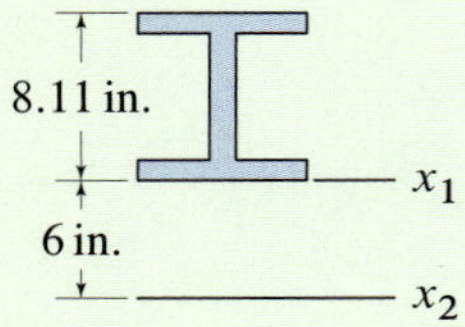

Figure P10.105

Problem 10.106

(a) Determine d so that the origin of the coordinate system, point O, is positioned at the centroid of the area.

(b) Determine the area moment of inertia about the x axis.

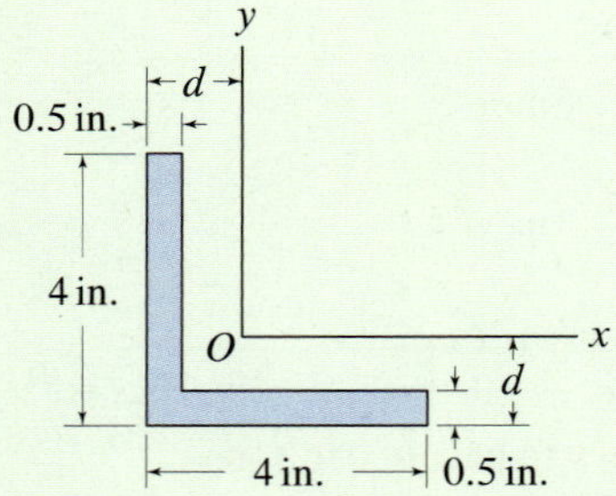

Figure P10.106

Problem 10.107

The cross section of the bar is symmetric about the x axis.

(a) Determine d so that the origin of the coordinate system, point O, is positioned at the centroid of the area.

(b) Determine the area moment of inertia about the x axis.

(c) Determine the area moment of inertia about the y axis.

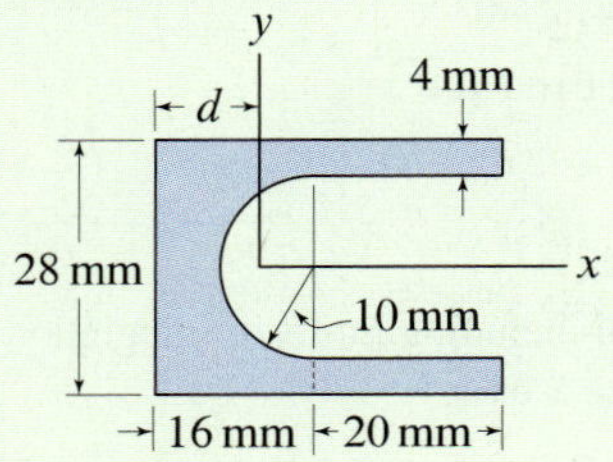

Figure P10.107

Problems 10.108 and 10.109

For the uniform solid cone with height h and radius r, use integration to determine the mass moment of inertia indicated, expressing your answer in terms of the mass m of the cone.

Problem 10.108 I_z.

Problem 10.109 I_x.

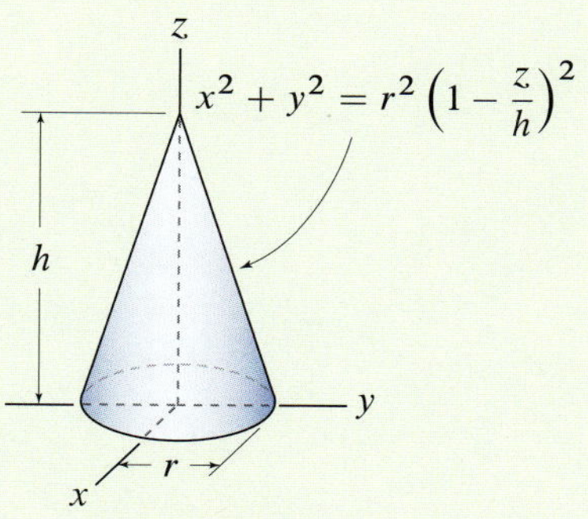

Figure P10.108 and P10.109

Problems 10.110 through 10.113

The truncated cone shown has $2000\,\mathrm{kg/m^3}$ density. Report your answers for the problems that follow using kg and mm units.

Problem 10.110 Use integration to determine the mass moment of inertia about the x axis.

Problem 10.111 Determine the mass moment of inertia about the x axis using composite shapes.

Problem 10.112 Use integration to determine the mass moment of inertia about the z axis.

Problem 10.113 Determine the mass moment of inertia about the z axis using composite shapes.

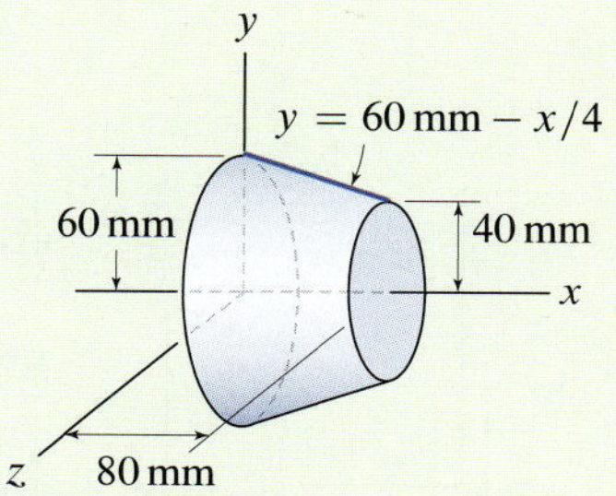

Figure P10.110–P10.113

Problems 10.114 and 10.115

The object shown has $8000\,\mathrm{kg/m^3}$ density and has a conical hole. Use integration to determine the mass moment of inertia about the

Problem 10.114 x axis.

Problem 10.115 y axis.

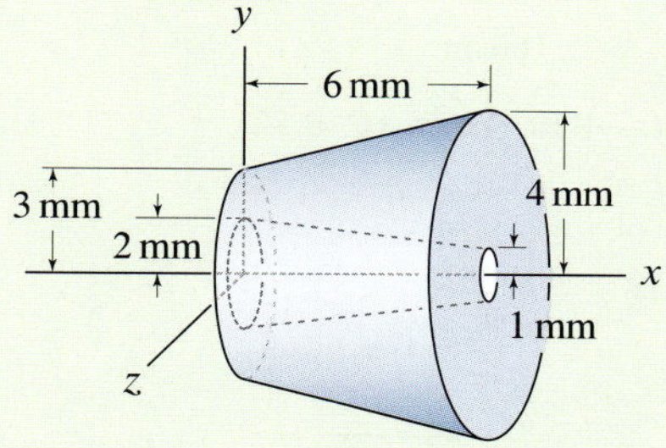

Figure P10.114 and P10.115

Problems 10.116 and 10.117

The solid of revolution consists of materials with densities ρ_1 and ρ_2.

Problem 10.116 Fully set up the integrals, including limits of integration, that will yield the mass moment of inertia about the x axis. You are not required to evaluate the integrals.

Problem 10.117 Fully set up the integrals, including limits of integration, that will yield the mass moment of inertia about the y axis. You are not required to evaluate the integrals.

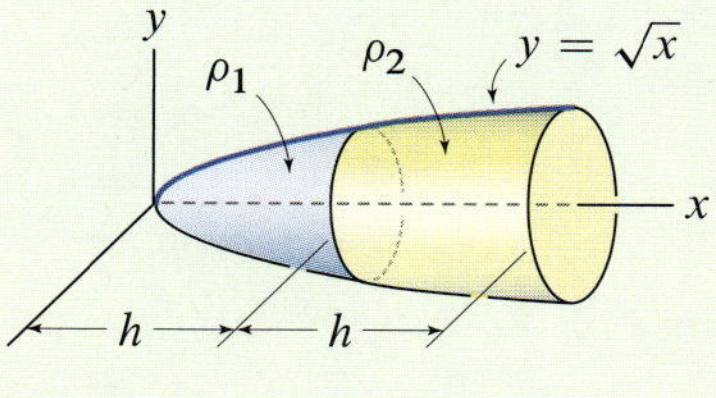

Figure P10.116 and P10.117

Problems 10.118 and 10.119

A solid of revolution is produced by revolving the area shown 360° around the y axis. Use integration to determine the mass moment of inertia about the axis of revolution assuming the solid has uniform density of $2000\,\text{kg/m}^3$.

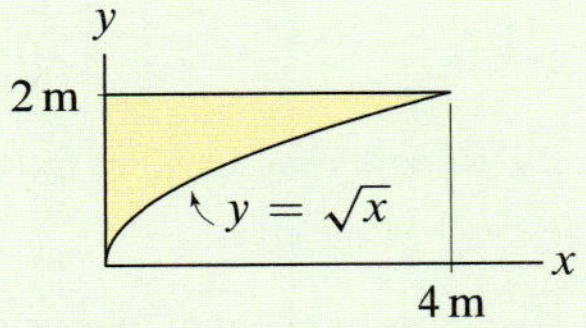

Figure P10.118

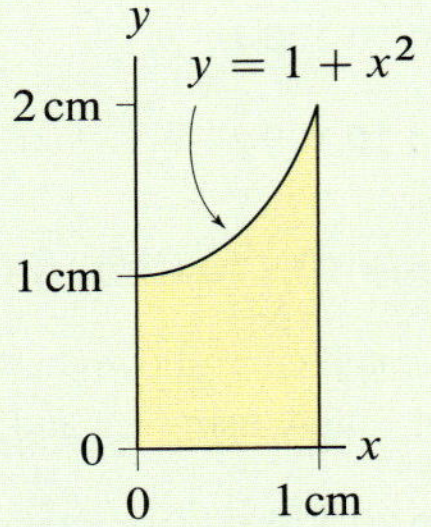

Figure P10.119

Problems 10.120 and 10.121

A beam is constructed of three identical pieces of wood, each piece having 40 mm by 160 mm by 500 mm dimensions, and 2 kg mass that is uniformly distributed. The cross section of the beam is symmetric about the x and y axes.

Problem 10.120

(a) Determine the area moment of inertia for the cross section about the x axis.

(b) Determine the mass moment of inertia for the beam about the x axis.

Problem 10.121

(a) Determine the area moment of inertia for the cross section about the y axis.

(b) Determine the mass moment of inertia for the beam about the y axis.

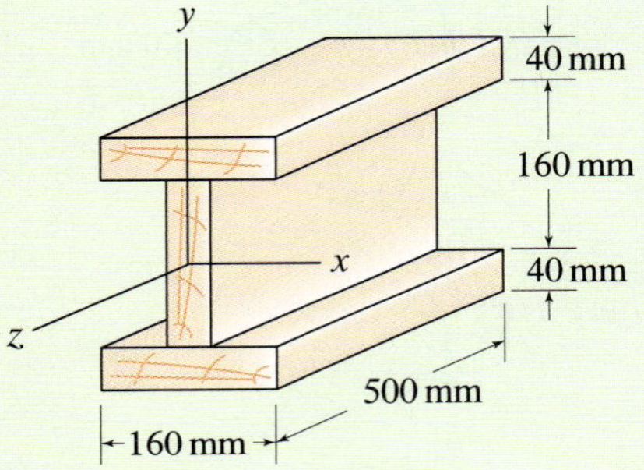

Figure P10.120 and P10.121

Problems 10.122 through 10.124

The object shown is made of thin plate with $0.02\,\text{lb/in}^2$ specific weight. Determine the mass moment of inertia about the

Problem 10.122 x axis.

Problem 10.123 y axis.

Problem 10.124 z axis.

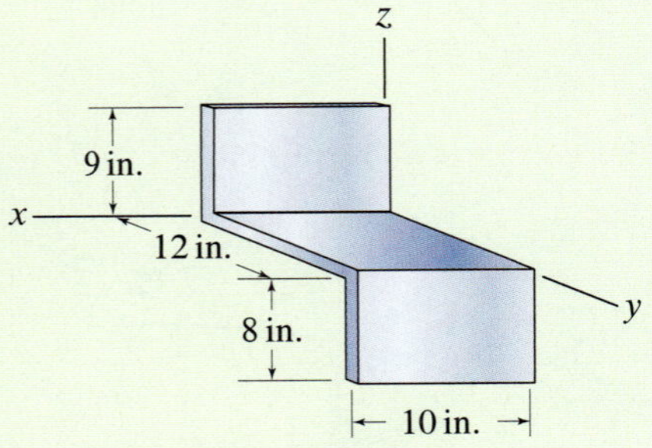

Figure P10.122–P10.124

A Technical Writing

In this appendix we use a short report as an example to discuss technical writing. The report is written by Bucky Badger, mascot of the University of Wisconsin–Madison.

A Short Guide to Technical Writing
January 1, 2009

by Bucky Badger
Dept. of Engineering Physics
University of Wisconsin–Madison

Abstract

This document is a brief overview of technical writing. It is written in a form you can use as a model for short reports that are up to a few pages in length. The abstract should briefly describe the problem you are considering and should summarize important results. For short reports, the abstract is normally only a few sentences in length.

1. Introduction

Effective written reports are necessary for communication and documentation of your analysis or design and are an essential part of your work as an engineer. Your most outstanding ideas are only as good as your ability to communicate those ideas to others. Furthermore, reports archive your ideas and may be needed in the future by you or others, to show how your design was developed, to aid in modifications, to help support patent rights, for use in litigation, and so on.

2. Technical Writing Tips

A technical report should be written using proper, simple English with correct grammar, punctuation, and spelling. It should be typewritten using a font size and line spacing that are easy to read. All pages should be numbered. The report should not be one long narrative, but rather should use sections and possibly subsections to break the material into smaller logical units. You should evaluate the intended audience and gage the technical level of your report accordingly.

The main body of your report should clearly summarize the assumptions you made, your method of analysis, and important observations. It should also discuss value decisions and economic and/or manufacturing considerations if appropriate and very clearly spell out your design recommendations and specifications. You should provide convincing justification to the reader that your findings and/or recommendations are accurate and reasonable. If equations are included in the main body of your report, they should be numbered. However, very often most of the equations, detailed calculations, and/or raw data are best placed in an appendix which should be referenced in the main body of the report.

Any articles, books, reports, published data, etc. that you use in your work should be referenced in the report using a standard format. References to Internet websites are permissible, but may be short-lived as the addresses and content of websites change frequently. Additional details on technical writing can be found in Woolever (2002). Use the spell checker in your word processor to check spelling, but be sure to thoroughly proofread your report, including the final draft, before you submit it.

3. Figures, Graphs, and Tables

Figures, graphs, and tables can convey a considerable amount of information. These should always be numbered so they can be easily referred to in your report, and they should have comprehensive captions that explain what the figure, graph, or table shows.

3.1 Figures. The time-honored Chinese proverb that "a picture is worth ten thousand words" applies here. It is considerably more effective to use a figure than to use a paragraph of text to describe what is probably self-evident in the figure. For example, imagine you are designing a wooden clothespin and must specify dimensions and the stiffness for the torsional spring. By including a sketch of the clothespin where dimensions and important parts are identified, as shown in Fig. A.1, your discussion will be considerably more straightforward and easier to comprehend.

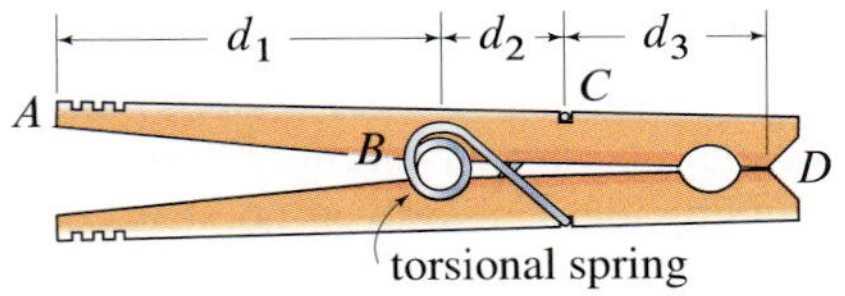

Figure A.1. Clothespin design with important parts and dimensions and identified.

3.2 Graphs. Graphs should always have labeled axes, including appropriate units. If the graph contains multiple curves, each of these should be labeled, perhaps by use of a legend.

3.3 Tables. Rows and columns of a table should always be labeled, including appropriate units.

4. Summary and Conclusions

In this final section of your report you should very clearly summarize your findings and how you reached your conclusions. As a guide, if a person were to read only this section, he or she should have a reasonable idea of your work.

References

Woolever, K. R., *Writing for the Technical Professions*, 2nd ed., Longman, 2002.

Appendix

The appendix contains detailed calculations, raw data, and other ancillary information. This is information the reader may choose not to consult, but should be included to help support your remarks and claims. It can be neatly handwritten.

End of Section Summary

This appendix used an example report to present guidelines and tips for writing short technical reports. Below is a summary of some of the most common mistakes that students make when writing technical reports.

Technical Writing Pitfalls

- Your report should not be one long narrative. Use sections and possibly subsections to break your report into logical units.
- Do not embellish your report with fictional company names, etc., or talk about yourself as a statics/engineering student, etc. Write your report as a professional document—short and to the point.
- Provide formal references for all sources of data (follow the format given in the example report).
- The vast majority of the time (almost always!), a figure that helps describe the problem you are working on is very effective. It should be in the main report.
- Very prominently and succinctly state your findings and/or design recommendations in the main body of the report. The reader should not have to search for this information.
- Do not overreference the appendix. In a short report that is only a few pages in length, you should probably reference the appendix no more than a few times. Anything that is central to your discussion in the main report (e.g., a table, graph, equation) should be present in the main report.
- Figures, graphs, and tables should always be labeled.
- Graphs should have labeled axes, including units, and a legend if the graph contains multiple curves.
- Your report should be easy to read and convincing. Ideally, a person should be able to comprehend your report with one reading.
- Proofread your entire report one last time before submitting it. Do not overrely on the spell checkers that most computer word processors have.

B Answers to Selected Problems

The answers to most even-numbered problems are provided in a downloadable PDF file at:

`www/mhhe.com/pgc`

Providing answers in this manner allows for more complex information than would otherwise be possible. In addition to final numerical or symbolic answers, selected problems have more extensive information such as free body diagrams and/or shear and moment diagrams. For example, Prob. 6.60 and the corresponding answer are shown below.

Problem 6.60

The hand brake for a bicycle is shown. Portions DE and FG are free to rotate on bolt A which is screwed into the frame BC of the bicycle. The brake is actuated by a shielded cable where T_1 is applied to point E and T_2 is applied to point G. A spring having 50 N compressive force is placed between points E and G so that the brake stays open when it is not being used. Assume the change in the spring's force is negligible when the brake is actuated to produce the 200 N forces at points D and F.

(a) Draw three FBDs, one each for DE and FG and bolt A, labeling all forces.

(b) Determine the necessary cable forces T_1 and T_2.

(c) Determine the forces exerted by DE and FG on bolt A.

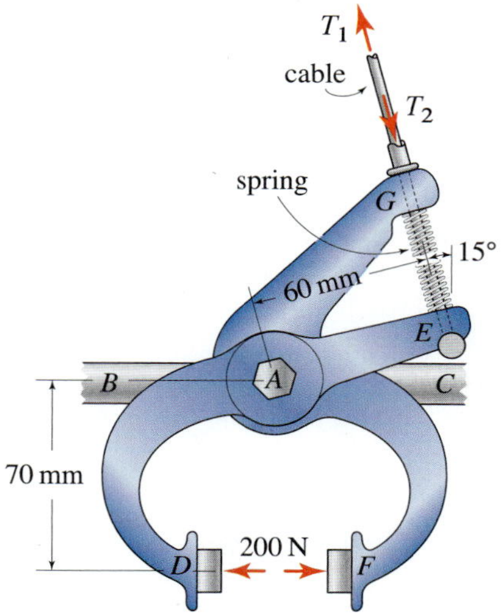

Figure P6.60

Problem 6.60 **(b)** $T_1 = 283\,\text{N}, T_2 = 283\,\text{N}$; **(c)** $A_{1x} = 260\,\text{N}, A_{1y} = -225\,\text{N}, A_{2x} = -260\,\text{N}, A_{2y} = 225\,\text{N}$.

(a)

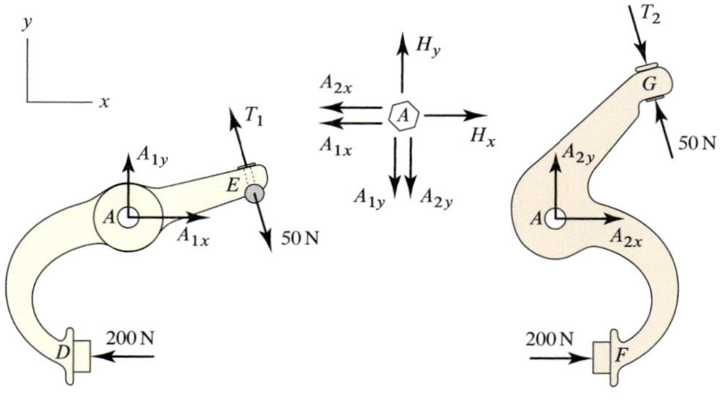

This feature not only provides more complete answers in selected circumstances, but also provides the modeling kick-start needed to get you started on some homework problems. Furthermore, the multitude of FBD answers provided give you ample opportunity to practice constructing FBDs on your own for extra problems and to enhance your ability to draw FBDs.

PHOTO

Chapter 1

Page 1, Opener: © JEAN-PHILIPPE ARLES/ Reuters/CORBIS; **p. 3, Interesting Fact:** © Godden Collection, EERC Library, University of California, Berkeley; **p. 4, Figure 1.2:** © Nicolo Orsi Battaglini/Art Resource, NY; **Figure 1.3:** © Linda Hall Library of Science, Engineering & Technology; **p. 5, Figure 1.4:** Portrait of Newton by Sr. Godfrey Kneller, 1689, © Photo by Jeremy Whitaker; **p. 18, Figure 1:** © McGraw-Hill, photo by Lucinda Dowell; **p. 20, Figure 1.8:** © M.S.C.U.A., University of Washington, Farquharson; **Figure 1.9:** © AP Photo/St. Petersburg Times, Skip O'Rourke; **p. 21, Figure 1.10:** © Anthony Correia/Getty Images; **Figure 1.11:** © Lee Lowrey/Texas A&M University; **Figure 1.12:** © AP Photo/Brian Branch-Price.

Chapter 2

Page 27, CO2: © David Sailors/CORBIS; **p. 36, Figure 1:** © McGraw-Hill, Photo by Lucinda Dowell; **p. 51, Figure 4:** © Courtesy of Stanley-Proto, Inc.; **p. 62, Figure 2.23:** © Daniel Boschung/zefa/CORBIS; **Figure 2.24:** © AP Photo/Michael Sohn; p. 66, Figure 1: NASA; **p. 75, Figure P2.26:** © Erik Isakson/ Rubberball Productions/ Getty Images, RF; **p. 98, Figure 1:** © AP Photo/Ric Francis; **p. 100, (Interesting Fact):** Department of Transportation.

Chapter 3

Page 111, CO3: © Lester Lefkowitz/Getty Images; **p. 114, Figure 3.2a:** © StockShot/ Alamy; **p. 118, Figure 3.6 (bottom):** Value RF/ © Comstock/CORBIS; **Figure 3.6 (top):** Courtesy of Performance Inc.; **p. 136, Figure 3.14:** © Taras Studio; **p. 140, Figure 1:** © Michael Plesha; **p. 143, Figure 3:** Courtesy of Missour Metals, LLC; **p. 148, Figure 3.18a and b:** © Godden Collection, EERC Library, University of California, Berkeley; **p. 161, Figure 3.20 (both):** Courtesy of Ford Motor Company.

Chapter 4

Page 181, CO4: © Ed Freeman/Getty Images; **p. 181, Figure 4.1:** © David Kimber/ The Car Photo Library; **p. 207, Figure P4.43:** © Richard Hamilton Smith/CORBIS; **p. 208, Figure 4.18a:** Courtesy of Great American Recreation Equipment, Inc.

Chapter 5

Page 245, CO5: NASA/Kennedy Space Center; **p. 271, Figure 5.12:** © Taras Studio; **p. 272, Figure 5.15:** © Taras Studio; **p. 281, Figure 3:** © Boeing, Inc.; **p. 292, Figure 5.24:** © Michael Plesha; **p. 306, Figure 1 (both):** © Michael Plesha; **p. 316, Figure 1:** © McGraw-Hill, photo by Lucinda Dowell; **p. 319, Figure DP5.4:** © William Caram/Alamy.

Chapter 6

Page 329, CO6: © Craig Roberts/Alamy; **p. 330, Figure 6.1a:** © Michael Plesha; **p. 330, Figure 6.3:** Courtesy of Continental Bridge, Inc.; **Figure 6.1b:** © David R. Frazier Photolibrary, Inc./Alamy; **p. 331, Figure 6.4 (both):** © Michael Plesha; **p. 341, Figure 3:** © Royalty-Free/Masterfile; **p. 354, Figure 6.23 (left):** © John Weeks; **Figure 6.23 (right):** © AP Photo/Jim Mone; **p. 363, Figure 6.24a and b:** © Godden Collection, EERC Library, University of California, Berkeley; **Figure 6.23c:** © Voz Noticias/CORBIS; **Figure 6.24d:** © Michael Plesha; **p. 372 Figure** 6.27: © Courtesy of Kona, USA; **Figure 6.28:** Sandia National Laboratory.

Chapter 7

Page 395, CO7: © Robert Cameron/Getty Images.

Chapter 8

Page 457, CO8: © Carl & Ann Purcell/CORBIS; **p. 459, Figure 8.4 (left):** © AP Photo/ Ross D. Franklin; **p. 459, Figure 8.4 (right):** © AP Photo/Ross D. Franklin; **p. 461, Figure 1:** © Michael Plesha; **p. 482, Figure 8.11 (top):** © age fotostock/SuperStock; **Figure 8.11 (bottom):** © Pixtal/SuperStock.

Chapter 9

Page 497, CO9: © Gene Blevins/CORBIS; **p. 498, Figure 9.1:** © Reunion De Musees Nationaux/Art Resource, NY; **p. 502, Figure 9.5a:** © Tim Ridley/Getty Images; **Figure 9.5b:** © Hans Neleman/PhotoDisc/ Getty Images; **Figure 9.5c and d:** Courtesy of Bilz Vibration Technology, Inc.; **p. 510, Figure 9.14:** United States Coast Guard; **p. 521, Figure 9.8:** © Courtesy of Lotus Engine Division.

Chapter 10

Page 533, CO10: © AP Photo/Rich Pedroncelli; **p. 553, Figure 10.11 (both):** © Michelle Wojdyla; **p. 554, Figure 10.13:** European Space Agency.

INDEX

D

F

G

H

I

J

K

R

S

U

V

W